COMPUTER ANALYSIS OF ELECTRIC POWER SYSTEM TRANSIENTS

Selected Readings

Edited by
Juan A. Martínez-Velasco
Departament d'Enginyeria Elèctrica
Universistat Politècnica de Catalunya
Barcelona - Spain

Printed in the United States of America

Editor - Juan A. Martinez-Velasco & Barbara Coburn
Typography & Layout - Jill R. Cals

Published by the Institute of Electrical and Electronics Engineers, Inc.
445 Hoes Lane, PO Box 1331, Piscataway, NJ 08855-1331.

http://www.ieee.org/eab/

CONTENTS

PREFACE

Transient phenomena in power systems are caused by switching operations, faults and other disturbances such as lightning surges. These phenomena can produce overvoltages and overcurrents, as well as abnormal waveforms and electromechanical transients. A short list of problems that the analysis of transient phenomena can solve follows:

- The coordination of power apparatus and some protective devices, such as surge arresters, can be assessed from a proper knowledge of transient overvoltages
- The rating of some protective devices, such as circuit breakers, can be determined from the overcurrents caused by faults and the transient recovery voltages produced during opening processes
- Waveforms originating during certain abnormal steady state operating conditions can lead to misoperation of HVDC control equipment.

The simulation of transient phenomena is therefore an important subject for the design of power apparatus and protective devices, as well as for the coordination between them.

Transient phenomena in power systems involve a frequency range from DC to several MHZ. A distinction is usually made between electromechanical oscillations of rotating machines, traditionally covered by transient stability studies, and electromagnetic transients. The latter type of transients can occur on a time scale that goes from microseconds to several cycles. They are a combination of traveling waves on lines, cables and buses, and of oscillations in lumped-element circuits of generators, transformers and other devices. Some electromechanical transients, such as subsynchronous resonance, for which a detailed machine models are needed, are usually included in these types of transients.

Several tools have been used over the years to analyze electromagnetic transients. In the past, miniature power system models, known as Transient Network Analyzers (TNA), were used. At present, the digital computer is the most popular tool, although TNAs are still used. The new generation of real-time digital systems are a powerful alternative in some applications for which a very high-speed simulation is required.

Many computer-based techniques have been developed to solve electromagnetic transients. They can be classified into two main groups: frequency-domain and time-domain techniques. The subject of this book is the digital simulation of electromagnetic transients in power systems using time-domain techniques. Presently, the most widely used solution method is based on the application of the trapezoidal rule and the method of characteristics, often called Bergeron's method. The aim of this book is to gather a collection of reprinted papers that highlight the present state of the art and provide an overview of the evolution of this subject.

In a book like this, it is necessary to limit the selection of papers. The book has been arranged in the following four parts:

- PART 1: SOLUTION METHODS
- PART 2: MODELING OF COMPONENTS
- PART 3: SIMULATION TOOLS
- PART 4: APPLICATIONS AND CASE STUDIES

The papers included in each part are preceded by an overview. A complementary bibliography is also included. The reader will find these additional references useful in enhancing the concepts presented in the reprinted papers.

Due to editorial constraints, the selection has been made taking into account only papers published after 1988. Just a few of the most relevant papers published before that year have been included. Their selection has been made considering their relevance to the historical development of the digital computation of electromagnetic transients in power systems.

Related to this subject, there is an appreciable number of techniques and applications. To avoid dispersion, new subjects, such as transient computation using parallel processing or real-time digital systems, have not been included.

Finally, it is worth mentioning that at the time of editing this book, the IEEE Working Group on "Modeling and Analysis of System Transients Using Digital Programs" is preparing a publication dealing with the same subject. A Task Force of this Working Group has published, or is preparing to publish, several papers providing modeling guidelines for simulation of transient phenomena. The goal of the IEEE WG is to expand these papers and collect the resulting documents in a special publication. In order to avoid redundancy with this complementary publication, none of those papers have been reprinted in this book.

Many significant books and special documents have been published that deal with the overall subject of Transient Analysis of Power Systems [1] - [13]. Several special publications and tutorial works dealing with Digital Simulation of Electromagnetic Transients are also available to readers [14] - [25].

REFERENCES

[1] R. Rudenberg, *Transient Performance of Electric Power Systems*, McGraw-Hill, New York, 1950.

[2] L.V. Bewley, *Traveling Waves on Transmission Systems*, John Wiley, New York, 1951.

[3] H.A. Peterson, *Transients in Power Systems*, John Wiley, New York, 1951.

[4] R. Rudenberg, *Electrical Shock Waves in Power Systems*, Harvard University Press, Cambridge, MA, 1968.

[5] W. Diesendorf, *Insulation Coordination in High Voltage Electric Power Systems*, Butterworth, London, 1974.

[6] J.P. Bickford, N. Mullineux and J.R. Reed, *Computation of Power System Transients*, Peter Peregrinus Ltd., London, 1976.

[7] K. Ragaller (Ed.), *Current Interruption in High-Voltage Networks*, Plenum Press, New York, 1977.

[8] K. Ragaller (Ed.), *Surges in High-Voltage Networks*, Plenum Press, New York, 1979.
[9] IEEE Seminar Report, "Power Systems Transient Recovery Voltages", 87TH0176-8-PWR.
[10] A.P. Sakis Meliopoulos, *Power System Grounding and Transients : An Introduction*, Marcel Dekker, New York, 1988.
[11] A. Greenwood, *Electrical Transients in Power Systems*, John Wiley, Second Edition, New York, 1991.
[12] A.P. Sakis Meliopoulos, "Lightning and Surge Protection", Section 27 of *Standard Handbook for Electrical Engineers*, D.G. Fink and H. W. Beaty (Ed.), 13th Edition, McGraw-Hill, New York, 1993.
[13] CIGRE WG 13.02, "Interruption of small inductive currents", December 1995.
[14] H.W. Dommel and W. Scott Meyer, "Computation of Electromagnetic Transients", *Proc. of IEEE*, vol. 62, no. 7, pp. 983-993, July 1974.
[15] "Digital Simulation of Electrical Transient Phenomena", A. Phadke (ed.), IEEE Tutorial Course, Course Text 81 EHO173-5-PWR.
[16] W. Derek Humpage, *Z-transform Electromagnetic Transient Analysis in High-Voltage Networks*, Peter Peregrinus Ltd., London, 1982.
[17] W. Derek Humpage and K. Wong, "Electromagnetic Transient Analysis in EHV Power Networks", *Proc. of IEEE*, Vol. 70, No. 4, pp. 379-402, April 1982.
[18] H.W. Dommel, *Electromagnetic Transients Program. Reference Manual (EMTP Theory Book)*, Bonneville Power Administration, Portland, 1986.
[19] EPRI Report EL-4202, "Electromagnetic Transients Program (EMTP) Primer", 1985.
[20] EPRI Report EL-4650, "Electromagnetic Transients Program (EMTP) Application Guide", 1986.
[21] EPRI Report EL-4651, "Electromagnetic Transients Program (EMTP) Workbook", 1986.
[22] EPRI Report EL-4651, "Electromagnetic Transients Program (EMTP) Workbook II", 1989.
[23] EPRI Report EL-4651, "Electromagnetic Transients Program (EMTP) Workbook III", 1989.
[24] EPRI Report EL-4651, "Electromagnetic Transients Program (EMTP) Workbook IV", 1989.
[25] CIGRE Working Group 02 (SC 33), "Guidelines for Representation of Network Elements when Calculating Transients", 1990.

PART 1

SOLUTION METHODS

PART 1
SOLUTION METHODS

INTRODUCTION

The practice of solving traveling wave problems using a digital computer was started in the early 1960s using two different techniques: Bewley's lattice diagram [A1] and Bergeron's method [A2]. These techniques were applied to solve small networks, with linear and nonlinear lumped-parameter, as well as distributed-parameter elements.

The extension to multinode networks was made by H. W. Dommel [A3], [1]. Dommel's scheme combined Bergeron's method and the trapezoidal rule into an algorithm capable of solving transients in single- and multi-phase networks with lumped and distributed parameters. This solution method was the origin of the well known Electro Magnetic Transients Program (EMTP), whose development was supported by Bonneville Power Administration (BPA).

The trapezoidal rule is used to convert the differential equations of the network components into algebraic equations involving voltages, currents and past values. These algebraic equations use a nodal approach. The resulting admittance matrix is symmetrical and remains unchanged if the integration is performed with a fixed time-step size. The solution of the transient process is then obtained using triangular factorization. This procedure can be applied to networks of arbitrary size in a very simple fashion.

Bergeron's method can be efficiently used with lossless and distortionless lines. However, parameters of actual transmission systems are frequency-dependent. The first works on frequency-dependent models were performed for telephone circuits in the 1920s [A4]. The first frequency- dependent transmission line model developed for EMTP simulations was implemented in 1973 [A5]. Much effort has been made since then, and some other frequency-dependent line models have been developed and implemented (see Part 2).

The original scheme was used to solve linear networks. However, many power components - transformers, reactors, surge arresters, circuit breakers - present a nonlinear behavior. Several modifications to the basic method were proposed to cope with nonlinear and time-varying elements [2]. These modifications were based on a current source representation, a piecewise-linear representation and the compensation method. Some of the advantages and drawbacks shown by these approaches were discussed in [2]; see also Chapter 12 in reference [18] of the Preface bibliography.

A special nonlinear power component is the synchronous generator. Three-phase synchronous generator models were implemented in mid 1970s to perform subsynchronous resonance simulations. Sophisticated interface methods based on prediction [A6] and compensation [A7] were developed to solve power network and synchronous machine equations. The BPA EMTP provided additional capabilities for simulation of rotating machinery when the Universal Machine module was implemented in the early 1980's [A8], [A9]. The two interface methods, compensation and prediction, are currently used with this module.

Built-in switch and valve models were available in the first BPA EMTP versions. However, the simulation of HVDC links raised the need for controlled valves and control systems. A major improvement was the addition of Transient Analysis of Control Systems (TACS) in 1976 [A10].

The list of selected papers for this part of the book includes works dealing with the basic solution methods implemented in the most widely used electromagnetic transients programs (EMTPs); procedures to solve numerical oscillations produced by the trapezoidal rule; initialization methods and procedures to solve the interface between power networks and control systems.

The first papers present two of Dommel's original works for digital simulation of electromagnetic transients [1], [2]. The third paper introduces the solution scheme implemented in the EMTDC [3]; it is also based on Dommel's method, but uses an innovative interface for simulation of HVDC links and machines.

The computation of electromagnetic transients with the trapezoidal rule is performed in the time-domain. Some other techniques have been developed to solve network equations using a frequency-domain solution [A11] or z-transform methods [A12]. Several alternate methodologies taking advantage of Dommel's scheme have recently been proposed. They use a hybrid frequency- and a time-domain approach [A13], or a state equation modeling [A14].

NUMERICAL OSCILLATIONS

Programs based on the trapezoidal rule are currently the most widely used for simulation of electromagnetic transients. This is due to the simplicity of this integration rule, as well as to its numerical stability. The trapezoidal rule is an A-stable method which does not produce run-off instability [A15]. However, this rule suffers from some drawbacks; it uses a fixed time-step size and can originate sustained numerical oscillations. During the last twenty years several solution to these drawbacks have been presented.

The step size determines the maximum frequency that can be simulated; therefore users have to know in advance the frequency range of the transient simulation to be performed. On the other hand, both slow and fast transients can occur at the same time in different nodes. A procedure by which two or more time steps can be used in the trapezoidal integration was presented in [A16].

In many cases, such as switching operations or transitions between segments in piecewise-linear inductances, the trapezoidal rule acts as a differentiator, and introduces sustained numerical oscillations. Several techniques have been proposed to control or reduce these numerical oscillations. One of these techniques uses additional damping to force oscillations to decay [A17]. This damping can be provided by the integration rule itself or externally, by adding fictitious resistances in parallel with inductances and in series with capacitors. This method can have an important effect on the accuracy of the solution.

Other techniques are based on the temporary modification of the solution method, only when numerical oscillations can occur, without affecting the rest of the simulation [A18], [A19].

Two papers present different approaches to solving numerical oscillations. Both of them belong to the second group; they alter the solution scheme only at discontinuities. The first paper presents the implementation of the CDA (Critical Damping Adjustment) procedure [4], previously introduced in [A19]. During a switching operation, CDA uses a backward Euler rule and two half-size integration steps. This method does not require recalculation of the admittance matrix.

The second paper presents a new scheme based on interpolation [5]. The procedure uses two-time step sizes and represents switching devices (power electronics components) by means of characteristic curves.

A modified linear interpolation to solve problems manifested not only in the network solution, but in the control system as well is presented in [A20].

INITIALIZATION

The solution of a transient phenomenon is dependent on the initial conditions with which the transient is started. Although some simulations can be performed with zero initial conditions, for instance lightning surge studies, there are many instances for which the simulation must be started from power-frequency steady-state conditions.

Capabilities to obtain the initial steady-state solution are of great importance in EMTPs. In addition, an initialization procedure can be an useful tool on its own, for instance to calculate resonant voltages due to coupling effects between parallel transmission lines.

The steady-state solution of linear networks at a single frequency is a rather simple task, and can be obtained using nodal admittance equations, as for the transient solutions. However, this task can be very complex in the presence of nonlinearities. Saturation effects in transformers and shunt reactors, rectifier loads and HVDC converter stations can produce steady-state harmonics.

The initial solution with harmonics can be obtained using some simple approaches. The simplest one is known as "brute force" approach: the simulation is started without performing any initial calculation and carried out long enough to let the transients settle down to steady-state conditions. This approach can have a reasonable accuracy, but its convergence will be very slow if the network has components with light damping. A more efficient method is to perform an approximate linear ac steady-state solution with nonlinear branches disconnected or represented by linearized models. Some EMTPs have either a "snapshot" or a "start again" feature. The state of the system is saved after a run, so later runs can be started at this point. Using a "brute force" initialization, the system is started from standstill, once it reaches the steady-state, a snapshot is taken and saved.

A significant effort has been made during the last ten years to develop efficient procedures for implementation in EMTPs and aimed at calculating ac steady-state initial conditions with the presence of nonlinear components. The techniques can be divided into two groups : frequency-domain methods and time-domain methods.

One of the first methods, known as Initialization with Harmonics (IwH), was presented in [A21].

This procedure uses an iterative solution based on the frequency-domain representation of nonlinear inductances as voltage-dependent harmonic current sources.

An improved initialization procedure, the Multiphase Harmonic Load Flow (MHLF), based on branch equations, was presented in [6], [A22]. In this method, static compensators and other nonlinear elements, under balanced or unbalanced conditions, are represented by harmonic Norton equivalent circuits. Further improvements incorporated a synchronous machine model into the initialization procedure [A23].

An improved version of the IwH method, using a Newton- Raphson harmonic steady-state calculation was presented in [7]. This method uses harmonic Norton modeling of nonlinear branches. It has been recently implemented in an EMTP. Other frequency-domain initialization methods were presented in [A24], [A25], [A26].

Several procedures have been proposed to calculate initial conditions using time-domain techniques, such as gradient and shooting methods [A27]. A paper included in this part presents a new waveform relaxation technique [7]. This method has a fast and efficient convergence in networks with nonlinear power elements and ideal diode-type devices.

Hybrid approaches to calculate initial conditions in nonlinear networks using both frequency- and time-domain techniques have also been developed [A28].

A different solution method to obtain a steady-state solution is needed when initial operating conditions are specified as power constraints. A multiphase power flow solution for EMTPs using a Newton-type method was presented in [9]. The method is based on the MHLF procedure presented in [6]. If this approach is used for EMTP initialization, sources need to be defined to drive the transient solution at those nodes for which load flow models were specified.

CONTROL SYSTEMS

The development of a section for representation of control systems in an EMTP was initially motivated by studies of HVDC links. The Transient Analysis of Control Systems (TACS) option was implemented in the BPA EMTP in 1976 [A10]. Although the main goal was the simulation of HVDC converters, it soon became obvious that TACS had many other applications, such as excitation of synchronous generators, dynamic arcs in circuit breakers, or protective relays.

Control systems are represented in TACS by block diagrams with interconnections between system elements. Control elements can be transfer functions, FORTRAN algebraic functions, logical expressions and some special devices. The solution method used by TACS is also based on the trapezoidal rule. Transfer functions in the s-domain are converted into algebraic equations in the time-domain. Components other than transfer functions can be included in a TACS section, but they are seen as nonlinear blocks and not directly added into the simultaneous solution of transfer functions.

When a nonlinear block is inside a closed-loop configuration, a true simultaneous solution is not possible. The procedure implemented in the TACS solution is simultaneous only for linear blocks (that is s-transfer functions) and sequential for nonlinear blocks. When these blocks are present, the loop is broken and the system is solved by inserting a time delay.

The resulting algebraic equations of a control system are by nature unsymmetrical. Due to this fact, the electric network and the control system were solved separately in the original TACS release. The network solution is first advanced, network variables are next passed to the control section, and then control equations are solved. Finally, the network receives control commands. The whole procedure introduces a time-step delay. This delay between the network and the control system, as well as the time delays that can exist inside the control system, are sources of different effects. Instabilities, inaccuracies and numerical oscillations produced by these delays have been reported.

Although the first release of TACS was a powerful and flexible tool, new applications have been demanding other capabilities than those implemented in the original version. One example is the new digital controls used in static compensators, HVDC converters and other FACTS devices. The execution of tasks only when needed, the simulation of conditional branching (IF-THEN-ELSE) or the manipulation of vector arrays are capabilities not available in the first TACS releases.

Much effort has been made during the last ten years to overcome main limitations and minimize problems originated by TACS. Improvements to solve internal time delays, initialization problems and some FORTRAN code limitations, were implemented and presented in [10].

Limitations in FORTRAN code capabilities were solved by developing an interface between TACS and FORTRAN subroutines. The interface presented in [11] maintains full capabilities of TACS and takes advantages of FORTRAN flexibility to represent digital controls.

Another approach to overcome these limitations was provided by MODELS. Initially known as "New TACS", MODELS program was developed to substitute the TACS program. However, it became obvious that both options provided alternate approaches, and therefore TACS was preserved. Although imbedded in the BPA EMTP and the ATP (Alternative Transients Program) version, MODELS is a general-purpose high level language that can be seen as a simulation tool on its own. This is the reason why it has been included in Part 3.

Several techniques can be used to solve simultaneously power network and control system equations and avoid problems related to the interface delay. Two procedures using compensation have been recently developed [A29], [12]. A different and simple solution using filter interposition to solve inaccuracies caused by the interface time delay was recently presented in [A30].

REFERENCES

Reprinted Papers

[1] H.W. Dommel, "Digital computer solution of electromagnetic transients in single- and multi-phase networks", *IEEE Trans. on Power Apparatus and Systems*, vol. 88, no. 2, pp. 734-741, April 1969.

[2] H.W. Dommel, "Nonlinear and time-varying elements in digital simulation of electromagnetic transients", *IEEE Trans. on Power Apparatus and Systems*, vol. 90, no. 6, pp. 2561-2567, November/December 1971.

[3] D.A. Woodford, A.M. Gole and R.Z. Menzies, "Digital simulation of dc links and ac machines", *IEEE Trans. on Power Apparatus and Systems*, vol. 102, no. 6, pp. 1616-1623, June 1983.

[4] J. Lin and J.R. Martí, "Implementation of the CDA procedure in the EMTP", *IEEE Trans. on Power Systems*, vol. 5, no. 2, pp. 394-402, May 1990.

[5] T.L.Maguire and A.M. Gole, "Digital simulation of flexible topology power electronic apparatus in power systems", *IEEE Trans. on Power Delivery*, vol 6, no. 4, pp. 1831-1840, October 1991.

[6] W. Xu, J.R. Martí and H.W. Dommel, "A multiphase harmonic load flow solution technique", *IEEE Trans. on Power Systems*, vol. 6, no. 1, pp. 174-182, February 1991.

[7] X. Lombard, J. Masheredjian, S. Lefebvre and C. Kieny, "Implementation of a new harmonic initialization method in EMTP", *IEEE Trans. on Power Delivery*, vol. 10, no. 3, pp.1343-1352, July 1995.

[8] Q.Wang and J.R. Martí, "A waveform relaxation technique for steady state initialization of circuits with nonlinear elements and ideal diodes", *IEEE Trans. on Power Delivery*, vol. 11, no. 3, pp. 1437-1443, July 1996.

[9] J.J. Allemong, R.J. Bennon and P.W. Selent, "Multiphase power flow solutions using EMTP and Newtons method", *IEEE Trans. on Power Systems*, vol. 8, no. 4, pp. 1455-1462, November 1993.

[10] R. Lasseter and J. Zhou, "TACS enhancements for the Electromagnetic Transient Program", *IEEE Trans. on Power Systems*, vol. 9, no. 2, pp. 736-742, May 1994.

[11] L.X. Bui, S. Casoria, G. Morin and J. Reeve, "EMTP TACS- FORTRAN interface development for digital controls modeling", *IEEE Trans. on Power Systems*, Vol. 7, no. 1, pp. 314-319, February 1992.

[12] S. Lefebvre and J. Mahseredjian, "Improved control systems simulation in the EMTP through compensation", *IEEE Trans. on Power Delivery*, vol. 10, no. 4, pp. 1654-1662, April 1995.

Additional References

[A1] L.O. Barthold and G.K. Carter, "Digital travelling-wave solutions. 1 - Single-phase equivalents", *AIEE Trans.*, vol. 80, pt. III, pp. 812-820, December 1961.

[A2] W. Frey and P. Althammmer, "The calculation of transients on lines by means of a digital computer", *Brown Boveri Rev.*, vol. 48, pp. 334-355, May/June 1961.

[A3] H.W. Dommel, "A method for solving transient phenomena in multi-phase systems", *Proc. of the 2nd Power Systems Computer Conference*, Stockholm, 1966.

[A4] J.R. Carson, "Wave propagation in overhead wires with ground return", *Bell Syst. Tech. J.*, vol. 5, pp. 539-554, 1926.

[A5] W. Scott Meyer and H.W. Dommel, "Numerical modelling of frequency dependent transmission-line parameters in an electromagnetic transients program", *IEEE Trans. on Power Apparatus and Systems*, vol. 93, no. 5, pp. 1401-1409, September/October 1974.

[A6] V. Brandwajn and H.W. Dommel, "A new method for interfacing generator models with an electromagnetic transients program", *Proc. of IEEE PICA*, pp. 260-265, May 1977.

[A7] G. Gross and M.C. Hall, "Synchronous machine and torsional dynamics simulation in the computation of electromagnetic transients", *IEEE Trans. on Power Apparatus and Systems*, vol. 97, no. 4, pp. 1074-1086, July/August 1978.

[A8] H.K. Lauw and W. Scott Meyer, "Universal machine modeling for the representation of rotating machinery in an electromagnetic transients program", *IEEE Trans. on Power Apparatus and Systems*, vol. 101, no. 6, pp. 1342-1352, June 1982.

[A9] H.K. Lauw, "Interfacing for universal multi-machine system modeling in an electromagnetic transients program", *IEEE Trans. on Power Apparatus and Systems*, vol. 104, no. 9, pp. 2367-2373, September 1985.

[A10] L. Dube and H.W. Dommel, "Simulation of control systems in an electromagnetic transients program with TACS", *Proc. of IEEE PICA*, pp. 266-271, 1977.

[A11] L.M. Wedepohl and S.E.T. Mohamed, "Transient analysis of multiconductor transmission lines with special reference to non-linear problems", *Proc. IEE*, vol. 117, pp. 979-988, May 1970.

[A12] W.D. Humpage, K.P. Wong, T.T. Nguyen and D. Sutanto, "z-transform electromagnetic transient analysis in power systems", *Proc. IEE*, Part C, vol. 127, pp. 370-378, November 1980.

[A13] M.D'Amore and M.S. Sarto, "A new efficient procedure for the transient analysis of dissipative power networks with nonlinear loads", *IEEE Trans. on Power Delivery*, vol. 11, no. 1, pp. 533-539, January 1996.

[A14] Y. Kang and J.D. Lavers, "Transient analysis of electric power systems : Reformulation and theoretical basis", *IEEE Trans. on Power Systems*, vol. 11, no. 2, pp. 754-760, May 1996.

[A15] S.C. Tripathy, N.D. Rao y S. Elangovan, "Comparison of stability properties of numerical integration methods for switching surges", *IEEE Trans. on Power Apparatus and Systems*, vol. 97, no. 6, pp. 2318-2326, November/December 1978.

[A16] A. Semlyen and F. de León, "Computation of electro-magnetic transients using dual or multiple time steps", *IEEE Trans. on Power Systems*, vol. 8, no. 3, pp. 1274-1281, August 1993.

[A17] F.L. Alvarado, R.H. Lasseter and J.J. Sanchez, "Testing of trapezoidal integration with damping for the solution of power transient studies", *IEEE Trans. on Power Apparatus and Systems*, vol. 102, no. 12, pp. 3783-3790, December 1983.

[A18] B. Kulicke, "Simulation program NETOMAC : Difference conductance method for continuous and discontinuous systems", *Siemens Research and Development Reports*, vol. 10, no. 5, pp. 299-302, 1981.

[A19] J.R. Martí and J. Lin, "Suppression of numerical oscillations in the EMTP", *IEEE Trans. on Power Systems*, vol. 4, no. 2, pp. 739-747, May 1989.

[A20] P. Kuffel, K. Kent and G. Irwin, "The implementation and efectiveness of linear interpolation", *Proc. of IPST'95*, pp. 499-504, Lisbon, September 3-7, 1995.

[A21] H.W. Dommel, A. Yan and S. Wei, "Harmonics from transformer saturation", *IEEE Trans. on Power Systems*, vol. 1, no. 2, pp. 209-215, April 1986.

[A22] W. Xu, J.R. Martí and H.W. Dommel, "Harmonic analysis of systems with static compensators", *IEEE Trans. on Power Systems*, vol. 6, no. 1, pp. 183-190, February 1991.

[A23] W. Xu, J.R. Martí and H.W. Dommel, "A synchronous machine model for three-phase harmonic analysis and EMTP initialization", *IEEE Trans. on Power Systems*, vol. 6, no. 4, pp. 1530-1538, November 1991.

[A24] G. Murere, S. Lefebvre, X.D. Do, "A generalized harmonic balance method for EMTP initialization", *IEEE Trans. on Power Delivery*, vol. 10, no. 3, pp. 1353-1359, July 1995.

[A25] A. Semlyen, E. Acha and J. Arrillaga, "Newton-type algorithms for the harmonic phasor analysis of non-linear power circuits in periodical steady state with special reference to magnetic non-linearities", *IEEE Trans. on Power Delivery*, vol. 3, no. 3, pp. 1090-1098, July 1988.

[A26] J. Usaola and J.G. Mayordomo, "Fast steady-state techniques for harmonic analysis", *IEEE Trans. on Power Delivery*, vol. 6, no. 4, pp. 1789-1790, October 1991.

[A27] B.K. Perkins, J.R. Martí and H.W. Dommel, "Nonlinear elements in the EMTP : Steady-state initialization", *IEEE Trans. on Power Systems*, vol. 10, no. 2, pp. 593-601, May 1995.

[A28] A. Semlyen and A. Medina, "Computation of the periodic steady state in systems with nonlinear components using a hybrid time and frequency domain methodology", *IEEE Trans. on Power Systems*, vol. 10, no. 3, pp. 1498-1504, August 1995.

[A29] A.E.A. Araujo, H.W. Dommel and J.R. Martí, "Simultaneous solution of power and control equations", *IEEE Trans. on Power Systems*, vol. 8, no. 4, pp. 1483-1489, November 1993.

[A30] X. Cao et al., "Suppresion of numerical oscillation caused by the EMTP-TACS interface using filter interposition", Paper 96 WM 094-3 PWRD, *1996 IEEE PES Winter Meeting*, January 21-25, Baltimore.

Digital Computer Solution of Electromagnetic Transients in Single- and Multiphase Networks

HERMANN W. DOMMEL, MEMBER, IEEE

Abstract—**Electromagnetic transients in arbitrary single- or multiphase networks are solved by a nodal admittance matrix method. The formulation is based on the method of characteristics for distributed parameters and the trapezoidal rule of integration for lumped parameters. Optimally ordered triangular factorization with sparsity techniques is used in the solution. Examples and programming details illustrate the practicality of the method.**

I. Introduction

THIS PAPER describes a general solution method for finding the time responses of electromagnetic transients in arbitrary single- or multiphase networks with lumped and distributed parameters. A computer program based on this method has been used at the Bonneville Power Administration (BPA) and the Munich Institute of Technology, Germany, for analyzing transients in power systems and electronic circuits [1], [2].

Paper 68 TP 657–PWR, recommended and approved by the Power System Engineering Committee of the IEEE Power Group for presentation at the IEEE Summer Power Meeting, Chicago, Ill., June 23–28, 1968. Manuscript submitted February 12, 1968; made available for printing April 10, 1968. The early stages of this work were sponsored by the German Research Association (Deutsche Forschungsgemeinschaft) while the author was with the Munich Institute of Technology.

The author is with Bonneville Power Administration, Portland, Ore.

Among the useful features of this program are the inclusion of nonlinearities, any number of switchings during the transient in accordance with specified switching criteria, start from any nonzero initial condition, and great flexibility in specifying voltage and current excitations of various waveforms.

The digital computer cannot give a continuous history of the transient phenomena, but rather a sequence of snapshot pictures at discrete intervals Δt. Such discretization causes truncation errors which can lead to numerical instability [3]. For this reason the trapezoidal rule was chosen for integrating the ordinary differential equations of lumped inductances and capacitances; it is simple, numerically stable, and accurate enough for practical purposes.

Branches with distributed parameters are assumed to be lossless; they will be called lossless lines hereafter. By neglecting the losses (which can be approximated very accurately in other ways, as will be shown) an exact solution can be obtained with the method of characteristics. This method has primarily been used in Europe, where it is known as Bergeron's method; it was first applied to hydraulic problems in 1928 and later to electrical problems (for historic notes see [5]). It is well suited for digital computers [6]–[8]. In contrast to the alternative lattice method for traveling wave phenomena [9] it offers important advantages; for example, no reflection coefficients are necessary when this method is used.

The method of characteristics and the trapezoidal rule can easily be combined into a generalized algorithm capable of solving transients in any network with distributed as well as lumped parameters. Numerically this leads to the solution of a system of linear (nodal) equations in each time step. It will be shown that lossless lines contribute only to the diagonal elements of the associated matrix; off-diagonal elements result only from lumped parameters. Thus a very fast and simple algorithm can be written when lumped parameters are excluded. However, no such restrictions are imposed. Instead, the recent impressive advances in solving linear equations by sparsity techniques and optimally ordered elimination [10] have been incorporated into an algorithm which automatically encompasses the fast solution of the restricted case and yet retains full generality.

II. Solution for Single-Phase Networks

A digital computer solution for transients is necessarily a step-by-step procedure that proceeds along the time axis with a variable or fixed step width Δt. The latter is assumed here. Starting from initial conditions at $t = 0$, the state of the system is found at $t = \Delta t, 2\Delta t, 3\Delta t, \ldots$, until the maximum time t_{max} for the particular case has been reached. While solving for the state at t, the previous states at $t - \Delta t, t - 2\Delta t, \ldots$, are known. A limited portion of this "past history" is needed in the method of characteristics, which is used for lines, and in the trapezoidal rule of integration, which is used for lumped parameters. In the first case it must date back over a time span equal to the travel time of the line; in the latter case, only to the previous step. With a record of this past history, the equations of both methods can be represented by simple equivalent impedance networks. A nodal formulation of the problem is then derived from these networks.

Lossless Line

Although the method of characteristics is applicable to lossy lines, the ordinary differential equations which it produces are not directly integrable [8]. Therefore, losses will be neglected at this stage. Consider a lossless line with inductance L' and capacitance C' per unit length. Then at a point x along the line voltage and current are related by

$$-\partial e/\partial x = L'(\partial i/\partial t) \quad (1a)$$

$$-\partial i/\partial x = C'(\partial e/\partial t). \quad (1b)$$

The general solution, first given by d'Alembert, is

$$i(x,t) = f_1(x - vt) + f_2(x + vt) \quad (2a)$$

$$e(x,t) = Z \cdot f_1(x - vt) - Z \cdot f_2(x + vt) \quad (2b)$$

with $f_1(x - vt)$ and $f_2(x + vt)$ being arbitrary functions of the variables $(x - vt)$ and $(x + vt)$. The physical interpretation of $f_1(x - vt)$ is a wave traveling at velocity v in a forward direction and of $f_2(x + vt)$ a wave traveling in a backward direction. Z in (2) is the surge impedance, v is the phase velocity

$$Z = \sqrt{L'/C'} \quad (3a)$$

$$v = 1/\sqrt{L'C'}. \quad (3b)$$

Multiplying (2a) by Z and adding it to or subtracting it from (2b) gives

$$e(x,t) + Z \cdot i(x,t) = 2Z \cdot f_1(x - vt) \quad (4)$$

$$e(x,t) - Z \cdot i(x,t) = -2Z \cdot f_2(x + vt). \quad (5)$$

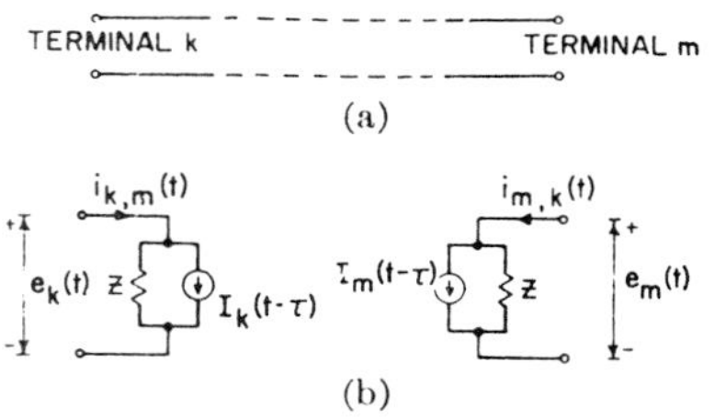

TERMINAL k TERMINAL m

(a)

$i_{k,m}(t)$ $i_{m,k}(t)$ $e_k(t)$ Z $I_k(t-\tau)$ $I_m(t-\tau)$ Z $e_m(t)$

(b)

Fig. 1. (a) Lossless line. (b) Equivalent impedance network.

Note that in (4) the expression $(e + Zi)$ is constant when $(x - vt)$ is constant and in (5) $(e - Zi)$ is constant when $(x + vt)$ is constant. The expressions $(x - vt) =$ constant and $(x + vt) =$ constant are called the characteristics of the differential equations.

The significance of (4) may be visualized in the following way: let a fictitious observer travel along the line in a forward direction at velocity v. Then $(x - vt)$ and consequently $(e + Zi)$ along the line will be constant for him. If the travel time to get from one end of the line to the other is

$$\tau = d/v = d\sqrt{L'C'} \quad (6)$$

(d is the length of line), then the expression $(e + Zi)$ encountered by the observer when he leaves node m at time $t - \tau$ must still be the same when he arrives at node k at time t, that is

$$e_m(t - \tau) + Zi_{m,k}(t - \tau) = e_k(t) + Z(-i_{k,m}(t))$$

(currents as in Fig. 1). From this equation follows the simple two-port equation for $i_{k,m}$

$$i_{k,m}(t) = (1/Z)e_k(t) + I_k(t - \tau)$$

and analogous (7a)

$$i_{m,k}(t) = (1/Z)e_m(t) + I_m(t - \tau)$$

with equivalent current sources I_k and I_m, which are known at state t from the past history at time $t - \tau$,

$$I_k(t - \tau) = -(1/Z)e_m(t - \tau) - i_{m,k}(t - \tau)$$

$$I_m(t - \tau) = -(1/Z)e_k(t - \tau) - i_{k,m}(t - \tau). \quad (7b)$$

Fig. 1 shows the corresponding equivalent impedance network, which fully describes the lossless line at its terminals. Topologically the terminals are not connected; the conditions at the other end are only seen indirectly and with a time delay τ through the equivalent current sources I.

Inductance

For the inductance L of a branch k, m (Fig. 2) we have

$$e_k - e_m = L(di_{k,m}/dt) \quad (8a)$$

which must be integrated from the known state at $t - \Delta t$ to the unknown state at t:

$$i_{k,m}(t) = i_{k,m}(t - \Delta t) + \frac{1}{L}\int_{t-\Delta t}^{t} (e_k - e_m)\,dt. \quad (8b)$$

Using the trapezoidal rule of integration yields the branch equation

$$i_{k,m}(t) = (\Delta t/2L)(e_k(t) - e_m(t)) + I_{k,m}(t - \Delta t) \quad (9a)$$

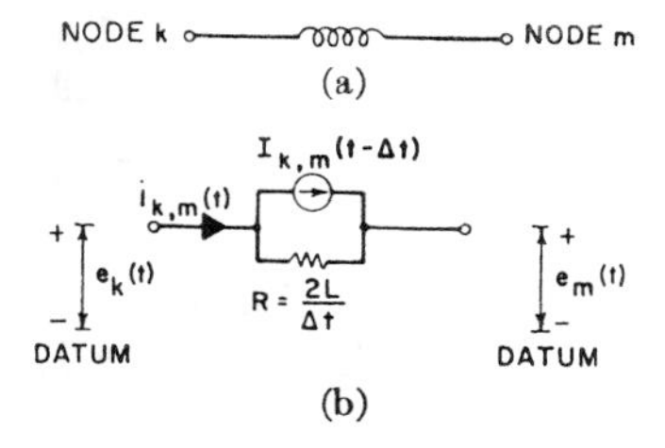

Fig. 2. (a) Inductance. (b) Equivalent impedance network.

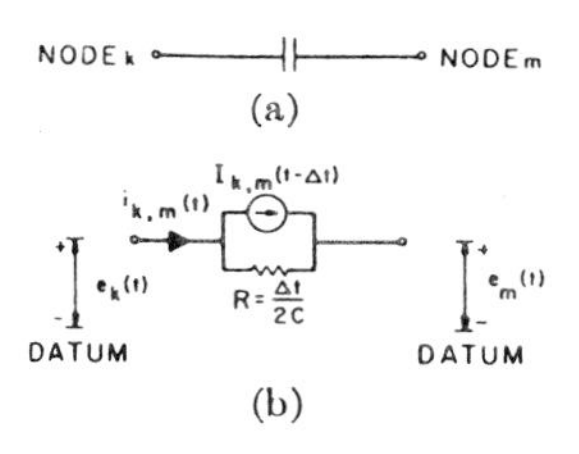

Fig. 3. (a) Capacitance. (b) Equivalent impedance network.

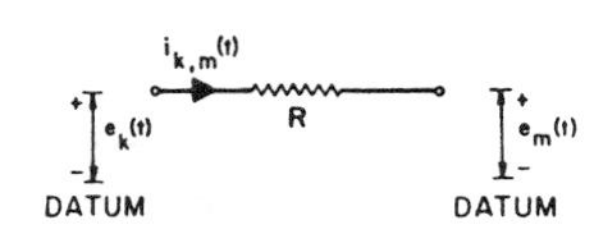

Fig. 4. Resistance.

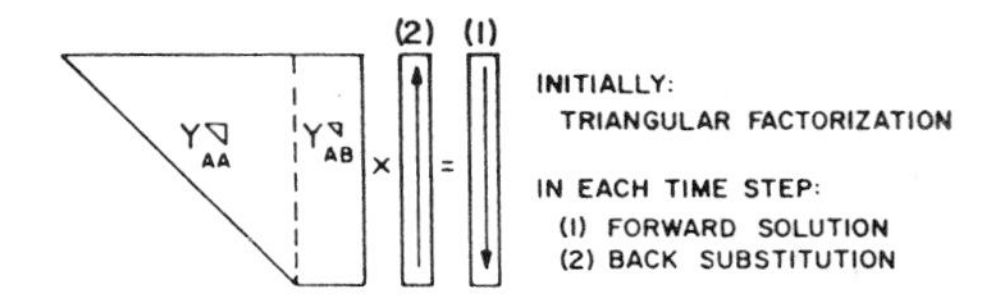

Fig. 5. Repeat solutions of linear equations.

where the equivalent current source $I_{k,m}$ is again known from the past history:

$$I_{k,m}(t-\Delta t) = i_{k,m}(t-\Delta t) + (\Delta t/2L)(e_k(t-\Delta t) - e_m(t-\Delta t)). \quad (9b)$$

The discretization with the trapezoidal rule produces a truncation error of order $(\Delta t)^3$; if Δt is sufficiently small and cut in half, then the error can be expected to decrease by the factor 1/8. Note that the trapezoidal rule for integrating (8b) is identical with replacing the differential quotient in (8a) by a central difference quotient at midpoint between $(t - \Delta t)$ and t with linear interpolation assumed for e. The equivalent impedance network corresponding to (9) is shown in Fig. 2.

Capacitance

For the capacitance C of a branch k, m (Fig. 3) the equation

$$e_k(t) - e_m(t) = \frac{1}{C}\int_{t-\Delta t}^{t} i_{k,m}(t)\,dt + e_k(t-\Delta t) - e_m(t-\Delta t)$$

can again be integrated with the trapezoidal rule, which yields

$$i_{k,m}(t) = (2C/\Delta t)(e_k(t) - e_m(t)) + I_{k,m}(t-\Delta t) \quad (10a)$$

with the equivalent current source $I_{k,m}$ known from the past history:

$$I_{k,m}(t-\Delta t) = -i_{k,m}(t-\Delta t) - (2C/\Delta t)(e_k(t-\Delta t) - e_m(t-\Delta t)). \quad (10b)$$

An equivalent impedance network is shown in Fig. 3. Its form is identical with that for the inductance. The discretization error is also the same as that for the inductance.

Resistance

For completeness we add the branch equation for the resistance (Fig. 4):

$$i_{k,m}(t) = (1/R)(e_k(t) - e_m(t)). \quad (11)$$

Nodal Equations

With all network elements replaced by equivalent impedance networks as in Figs. 1–4, it is very simple to establish the nodal equations for any arbitrary system. The procedures are well known [3] and will not be explained here. The result is a system of linear algebraic equations that describes the state of the system at time t:

$$[Y][e(t)] = [i(t)] - [I] \quad (12)$$

with

- $[Y]$ nodal conductance matrix
- $[e(t)]$ column vector of node voltages at time t
- $[i(t)]$ column vector of injected node currents at time t (specified current sources from datum to node)
- $[I]$ known column vector, which is made up of known equivalent current sources I.

Note that the real symmetric conductance matrix $[Y]$ remains unchanged as long as Δt remains unchanged. It is, therefore, preferable, though not mandatory, to work with fixed step width Δt. The formation of $[Y]$ follows the rules for forming the nodal admittance matrix in steady-state analysis.

In (12) part of the voltages will be known (specified excitations) and the others will be unknown. Let the nodes be subdivided into a subset A of nodes with unknown voltages and a subset B of nodes with known voltages. Subdividing the matrices and vectors accordingly, we get from (12)

$$\begin{bmatrix} [Y_{AA}][Y_{AB}] \\ [Y_{BA}][Y_{BB}] \end{bmatrix} \begin{bmatrix} [e_A(t)] \\ [e_B(t)] \end{bmatrix} = \begin{bmatrix} [i_A(t)] \\ [i_B(t)] \end{bmatrix} - \begin{bmatrix} [I_A] \\ [I_B] \end{bmatrix}$$

from which the unknown vector $[e_A(t)]$ is found by solving

$$[Y_{AA}][e_A(t)] = [I_{\text{total}}] - [Y_{AB}][e_B(t)] \quad (13)$$

with

$$[I_{\text{total}}] = [i_A(t)] - [I_A].$$

This amounts to the solution of a system of linear equations in each time step with a constant coefficient matrix $[Y_{AA}]$, provided Δt is not changed. The right sides in (13) must be recalculated in each time step.

Practical Computation

Equation (13) is best solved by triangular factorization of the augmented matrix $[Y_{AA}]$, $[Y_{AB}]$ once and for all before entering the time step loop. The same process is then extended to the vector $[I_{\text{total}}]$ in each time step in the so-called forward solution, followed by back substitution to get $[e_A(t)]$, as indicated in Fig. 5. Only a few elements in $[Y_{AA}]$, $[Y_{AB}]$ are nonzero; this sparsity is exploited by storing only the nonzero elements of the triangularized matrix. The savings in computer time and storage requirements can be optimized with an ordered elimination scheme [10].

Should the nodes be connected exclusively via lossless lines, with lumped parameters R, L, C only from nodes to datum, then $[Y_{AA}]$ becomes a diagonal matrix. In this case the equations could be solved separately node by node. Some programs are based on this restricted topology. However, the sparsity technique lends itself automatically to this simplification without having to restrict the generality of the network topology.

The construction of the column vector $[I_{total}]$ is mainly a bookkeeping problem. Excitations in the form of specified current sources $[i_A(t)]$ and the past history information in $-[I_A]$ are entered into $[I_{total}]$ before going into the forward solution; after $[I_{total}]$ has been built, using the still available voltages from the previous time step, specified voltage sources $[e_B(t)]$ are entered into $[e(t)]$. Excitation values may be read from cards step-by-step or calculated from standardized functions (sinusoidal curve, rectangular wave[1], etc.). The excitations may be any combination of voltage and current sources, or there may be no excitation at all (e.g., discharge of capacitor banks). After having found $[e_A(t)]$, the past history records are updated while constructing the vector $[I_{total}]$ for the next time step (see flow chart in Fig. 6). Some practical hints about recording the past history and about nonzero initial conditions may be found in Appendixes I and II.

Approximation of Series Resistance of Lines

The simplicity of the method of characteristics rests on the fact that losses are neglected. This simplicity also holds for the distortionless line, where $R'/L' = G'/C'$ (R' is the series resistance and G' the shunt conductance per unit length); the only difference is in computing I_k (and analogous I_m):

$$I_k(t-\tau) = \exp\left(-(R'/L')\tau\right)\left(-(1/Z)e_m(t-\tau) - i_{m,k}(t-\tau)\right).$$

Unfortunately, power lines are not distortionless, since G' is usually negligible (or a very complicated function of voltage if corona is to be taken into account).

The distributed series resistance with $G' = 0$ can easily be approximated by treating the line as lossless and adding lumped resistances at both ends. Such lumped resistances can be inserted in many places along the line when the total length is divided into many line sections. Interestingly, all cases tested so far showed no noticeable difference between lumped resistances inserted in few or in many places. The voltage plot in Fig. 13 was practically identical for lumped resistances inserted in 3, 65, and 300 places. In its present form, BPA's program automatically lumps $R/4$ at both ends and $R/2$ at the middle of the line (R is the total series resistance); under these assumptions the equivalent impedance network of Fig. 1 is still valid and only the values change slightly (I_m analogous to I_k):

$$Z = \sqrt{L'/C'} + \tfrac{1}{4}R$$

$$I_k(t-\tau) = ((1+h)/2)\{I_k \text{ from eq. (7b)}\} + ((1-h)/2)\{I_m \text{ from eq. (7b)}\}$$

[1] Since the trapezoidal rule is based on linear interpolation, a rectangular wave of amplitude y will always be interpreted as having a finite rate of rise $y/\Delta t$ in the first step in the presence of lumped inductances and capacitances.

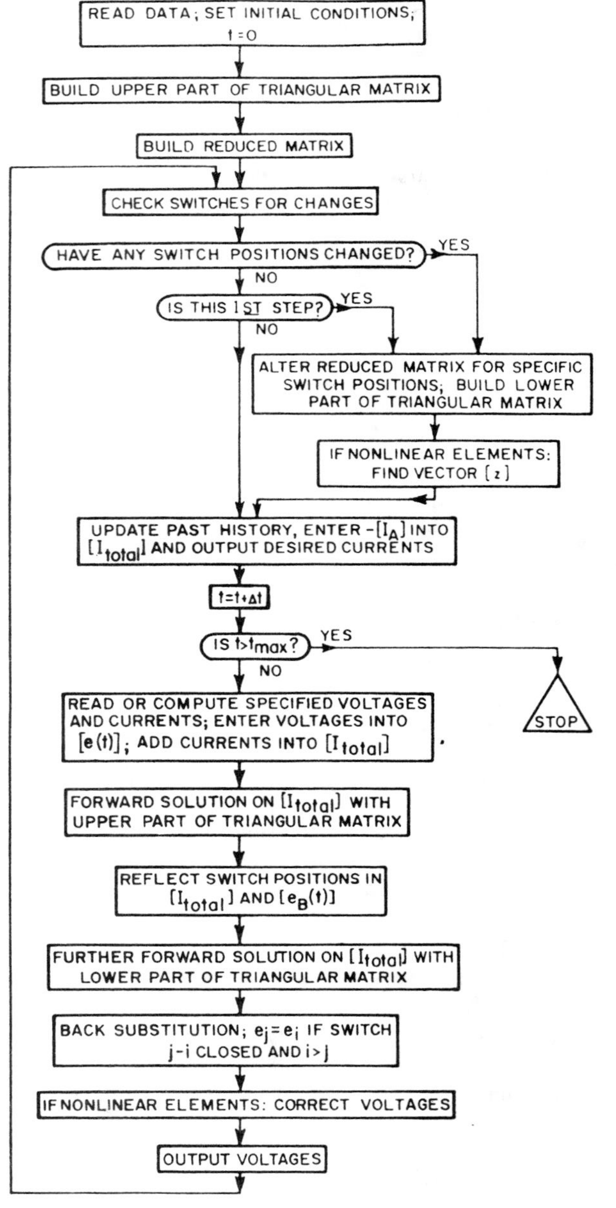

Fig. 6. Flow chart for transients program.

with

$$h = (Z - \tfrac{1}{4}R)/(Z + \tfrac{1}{4}R).$$

The real challenge for a better line representation is the frequency dependence of R' and L', which results from skin effects in the earth return [11] and in the conductors; BPA plans to explore this further (see Section IV).

Switches

The network may include any number of switches, which may change their positions in accordance with defined criteria. They are represented as ideal ($R = 0$ when closed and $R = \infty$ when open); however, any branches may be connected in series or parallel to simulate physical properties (e.g., time-varying or current-dependent resistance).

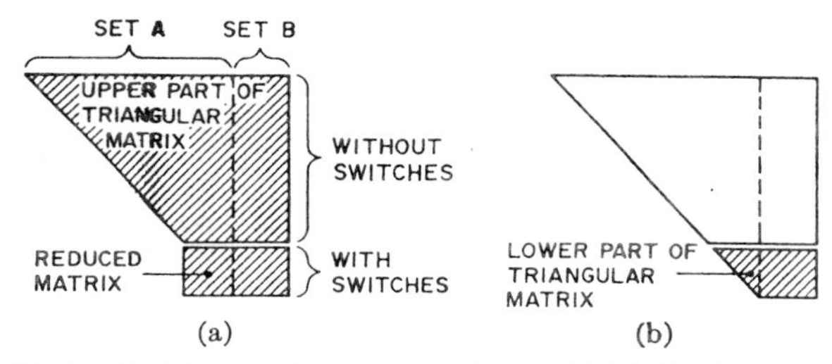

Fig. 7. Shaded areas show computation. (a) Initially. (b) After each change.

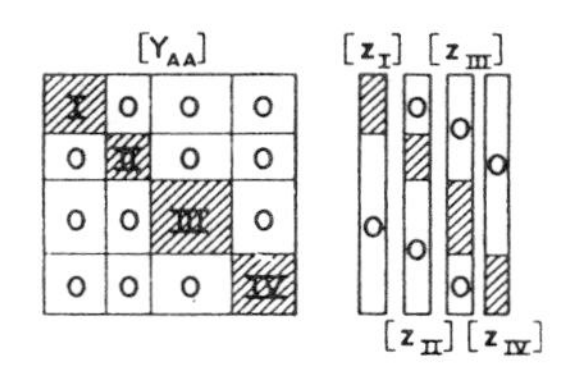

Fig. 9. Disconnected subnetworks in $[Y_{AA}]$.

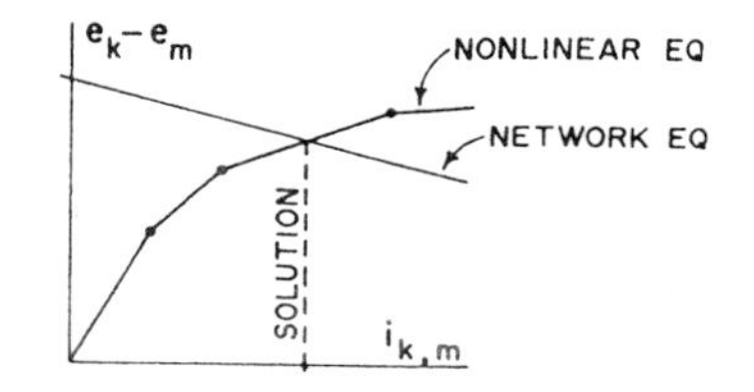

Fig. 8. Solution for nonlinear parameter.

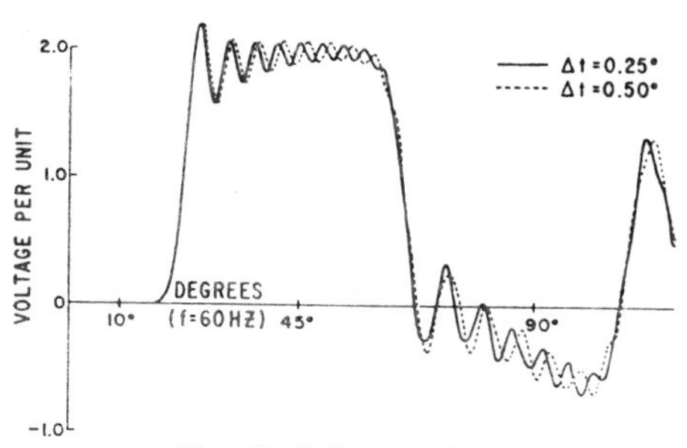

Fig. 10. Influence of Δt.

With only one switch in the network, it is best to build the matrix for the switch open and to simulate the closed position with superimposed node currents [2]. With more switches in the network, it is preferable to build $[Y_{AA}]$, $[Y_{AB}]$ anew each time a change occurs. However, it is not necessary to repeat the entire triangular factorization with each change. Nodes with switches connected are arranged at the bottom (Fig. 7). Then the triangular factorization is carried out only for nodes without switches (upper part of triangular matrix). This also yields a reduced matrix for the nodes with switches (assumed to be open). Whenever a switch position changes, this reduced matrix is first modified to reflect the actual switch positions (if closed: addition of respective rows and columns and retention of the higher numbered node in place of two nodes), then the triangular factorization is completed (lower part of triangular matrix). This scheme is included in the flow chart of Fig. 6.

Nonlinear and Time-Varying Parameters

With only one nonlinear parameter in the network, the solution can be kept essentially linear by confining the nonlinear algorithm (usually an iterative procedure) to the branch with the nonlinear parameter. To accomplish this the nonlinear parameter is not included in the matrix; its current $i_{k,m}$ is simulated with two additional node currents:

$$i_m = i_{k,m} \quad \text{and} \quad i_k = -i_{k,m}.$$

Let $[z]$ be the precalculated difference of the mth and kth columns of $[Y_{AA}]^{-1}$, which is readily obtained with a repeat solution of (13) by setting $[I_{\text{total}}] = \{0$, except $+1.0$ in mth and -1.0 in kth component$\}$ and $[e_B(t)] = 0$. Ignoring the nonlinear parameter at first, we get $[e_A^{(\text{linear})}(t)]$ from (13); the final solution follows from superimposing the two additional currents $i_k = -i_m = -i_{k,m}$:

$$[e_A(t)] = [e_A^{(\text{linear})}(t)] + [z]\cdot i_{k,m}(t). \tag{14}$$

The value $i_{k,m}$ in (14) is found by solving two simultaneous equations, the linear network equation (Thévenin equivalent)

$$e_k(t) - e_m(t) = e_k^{(\text{linear})}(t) - e_m^{(\text{linear})}(t) + (z_k - z_m)\cdot i_{k,m}(t) \tag{15}$$

and the nonlinear equation in the form of the given characteristic,

$$e_k(t) - e_m(t) = f(i_{k,m}(t)). \tag{16a}$$

BPA's program represents the nonlinear characteristic point-by-point as piecewise linear (Fig. 8), but any mathematical function could be used instead.

The nonlinear characteristic (16a) is that of a nonlinear, current-dependent resistance. If it is to represent a lightning arrester, then $i_{k,m} = 0$ until $|e_k^{(\text{linear})}(t) - e_m^{(\text{linear})}(t)|$ reaches the specified breakdown voltage of the arrester.

For a time-varying resistance, (16a) must be replaced by the simpler equation

$$e_k(t) - e_m(t) = R(t_R)\cdot i_{k,m}(t) \tag{16b}$$

where $R(t_R)$ is given as a function of the time t_R (e.g., in the form of a table). The time count t_R may be identical with the time t of the transient study, or it may start later according to a defined criterion.

The characteristic of a nonlinear inductance is usually specified as $\psi = f(i_{k,m})$. The total flux is

$$\psi(t) = \int_0^t (e_k(t) - e_m(t))\, dt + \psi(0).$$

With the trapezoidal rule of integration this becomes

$$e_k(t) - e_m(t) = (2/\Delta t) f(i_{k,m}(t)) - c(t - \Delta t), \tag{16c}$$

which simply replaces (16a). The value c must be updated with

$$c(0) = (2/\Delta t)\,\psi(0) + e_k(0) - e_m(0)$$

and then recursively

$$c(t - \Delta t) = c(t - 2\Delta t) + 2(e_k(t - \Delta t) - e_m(t - \Delta t)).$$

Generally, when a network contains more than one nonlinear parameter, the entire problem becomes nonlinear and its iterative solution quite lengthy. The algorithm remains simple, however, if the network is topologically disconnected into subnetworks, each containing only one nonlinear parameter. Disconnections give $[Y_{AA}]$ a diagonal structure with submatrices on the diagonal (Fig. 9). Note that topological disconnections are quite likely in networks containing lossless lines, since they, as well as lumped parameters from node to datum or to nodes with known voltage, do not introduce off-diagonal elements into $[Y_{AA}]$. For each subnetwork there is an independent equation (15) and each vector $[z]$ has zeros outside of that subnetwork. (Fig. 9 symbolizes nonlinear parameters I–IV in four effectively disconnected subnetworks.) Therefore each nonlinear parameter can be treated separately and exactly as above. In the superposition (14), each subnetwork will have its own $[z]$ and $i_{k,m}$. However, it is possible to compress these columns ($[z_I]$–$[z_{IV}]$ in Fig. 9) into one single column, if an address column is added to indicate the number of the subnetwork; the latter is necessary to insert the right current $i_{k,m}$ into (14). BPA's program automatically checks for violations of the disconnection rule while computing this single column together with the address column.

Accuracy

To arrive at (13), approximations have to be made only for lumped inductances and capacitances. Lossless lines and resistances are treated rigorously.

In practice, a truncation error is introduced by a lossless line whenever its travel time τ is not an integer multiple of Δt. Then some kind of interpolation becomes necessary in computing $I_k(t-\tau)$ and $I_m(t-\tau)$. One option in BPA's program uses linear interpolation, because in most practical cases the curves $e(t)$ and $i(t)$ are smooth rather than discontinuous. For cases with expected discontinuities, another option rounds the travel time τ to the nearest integer multiple of Δt. Both options raise travel times $\tau < \Delta t$ to Δt; otherwise the equivalent impedance network of Fig. 1 could not be used any more.

The trapezoidal rule of integration, used for lumped inductances and capacitances, is considered to be adequate for practical purposes, especially if the network has only a few lumped parameters. Compared with the alternative of stubline approximations [9], the results are more accurate. It is well known that the trapezoidal rule is numerically stable and has almost ideal round-off properties [12, p. 119]. When the stepwidth Δt is chosen sufficiently small to give good curve plots (points not spaced too widely), linear interpolation, on which the trapezoidal rule is based, should be a good approximation. Both requirements go hand in hand. The choice of Δt is not critical as long as the oscillations of highest frequency are still represented by an adequate number of points. Changing Δt influences primarily the phase position of the high-frequency oscillations; the amplitude remains practically unchanged (see Fig. 10 which resulted from an example similar to that of Fig. 12).

Higher accuracy could be obtained with the Richardson extrapolation [12, p. 118]. Here, the integration from $(t-\Delta t)$ to t would be carried out twice, with Δt in one step and with $\Delta t/2$ in two steps, and both results extrapolated to $\Delta t = 0$. The amount of work in each time step is thus tripled and the work for the initial triangular factorization is doubled, since two matrices, built for Δt and $\Delta t/2$, are necessary.

III. Mutual Coupling and Multiphase Networks

Lumped Parameters with Mutual Coupling

To include mutual coupling with lumped parameters the scalar quantities of a single branch are simply replaced by matrix quantities for the set of coupled branches. Consider the three coupled branches in Fig. 11 with a resistance matrix $[R]$ and an inductance matrix $[L]$. They could represent the series branches of a three-phase π-equivalent with earth return; in this case $[L]$ as well as $[R]$ would have off-diagonal elements (mutual coupling). Applying the trapezoidal rule of integration [2] yields:

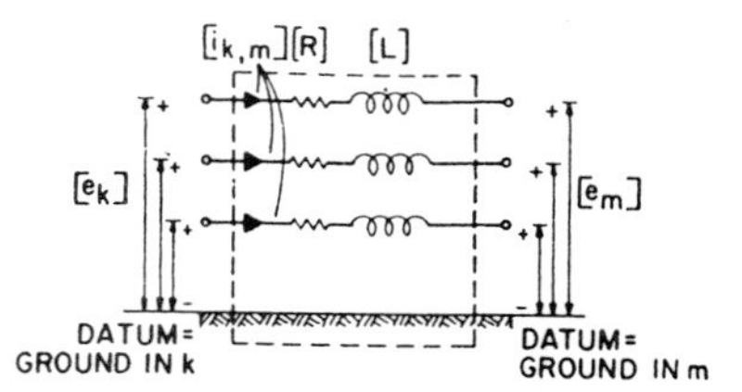

Fig. 11. Mutual coupling.

$$[i_{k,m}(t)] = [S]^{-1}([e_k(t)] - [e_m(t)]) + [I_{k,m}(t-\Delta t)] \tag{17a}$$

with $[I_{k,m}(t-\Delta t)]$ from the recursive formula

$$[I_{k,m}(t-\Delta t)] = [H]([e_k(t-\Delta t)] - [e_m(t-\Delta t)] + [S][I_{k,m}(t-2\Delta t)]) - [I_{k,m}(t-2\Delta t)]. \tag{17b}$$

All matrices in (17) are symmetric:

$$[S] = [R] + (2/\Delta t)[L]$$

$$[H] = 2([S]^{-1} - [S]^{-1}[R][S]^{-1}).$$

The only difference compared with a single branch is, that in building $[Y_{AA}]$, $[Y_{AB}]$ in (13), a matrix $[S]^{-1}$ is entered instead of a scalar value. Also in each time step a vector $[I_{k,m}]$ enters into $[I_{total}]$ instead of a scalar value.

If Fig. 11 is part of a multiphase π-equivalent representing a line section, then each set of terminals will be capacitance connected. These capacitances are actually single branches; thus no new formula is necessary. BPA's program treats them as a matrix entity $[C]$ to speed up the solution.

Lossless Multiphase Line

Equation (1) is also valid for the multiphase line if the scalars are replaced by vectors $[e]$, $[i]$ and matrices $[L']$, $[C']$. By differentiating a second time, one of the vector variables can be eliminated, which gives

$$[\partial^2 e(x,t)/\partial x^2] = [L'][C'][\partial^2 e(x,t)/\partial t^2] \tag{18a}$$

$$[\partial^2 i(x,t)/\partial x^2] = [C'][L'][\partial^2 i(x,t)/\partial t^2]. \tag{18b}$$

The solution of (18) is complicated by the presence of off-diagonal elements in the matrices, which occur because of mutual couplings between the phases. This difficulty is overcome if the phase variables are transformed into mode variables by similarity transformations that produce diagonal matrices in the modal equations [2], [13], [14]. This is the well-known eigenvalue problem. Each of the independent equations in the modal domain can then be solved with the algorithm for the single-phase line by using its modal travel time and its modal surge impedance. The transformation matrices, which give the transition to the phase domain, will generally be different for voltages and currents, e.g.,

$$[e_{phase}] = [T_e][e_{mode}] \tag{19a}$$

$$[i_{phase}] = [T_i][i_{mode}]. \tag{19b}$$

The columns in $[T_e]$, $[T_i]$ are always undetermined by a constant factor, if not normalized. A helpful relation [2], [15] is:

$$[T_i]_{unnormalized} = [C'][T_e]. \tag{19c}$$

If all diagonal elements in $[L']$ are equal to L'_{self} and all off-diagonal elements are equal to L'_{mutual} (analogous for $[C']$), then a simple transformation is possible, even if the inductances are frequency dependent [15]:

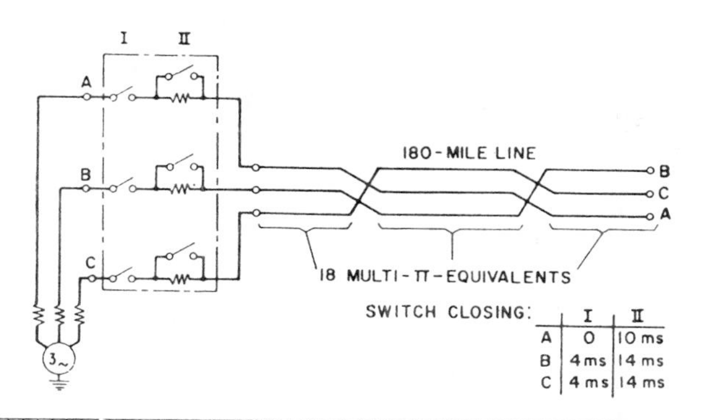

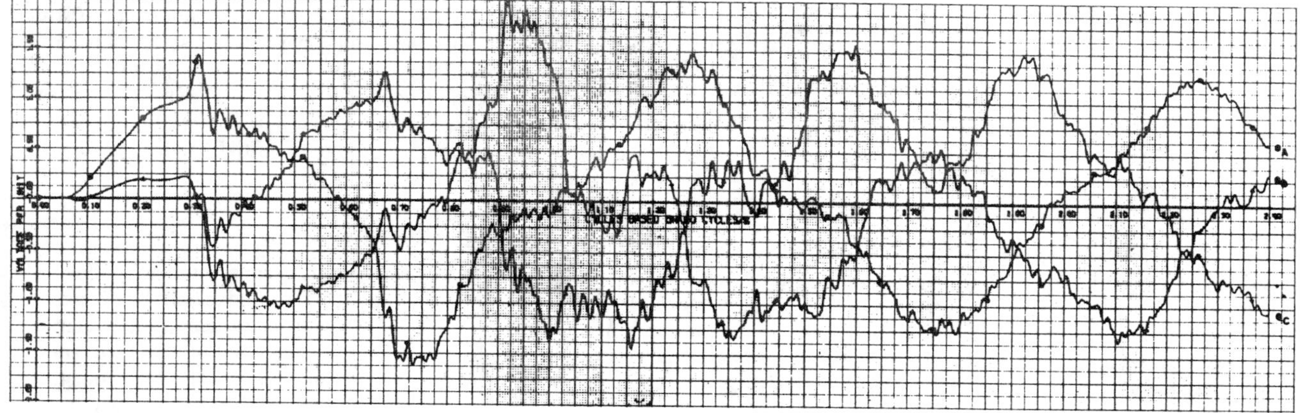

Fig. 12. Sequential closing. Network and results at the receiving end. Line energizing: 180-mile line, transposed at 60 and 120 miles. RLC for 60 Hz.

$$[T_e] = [T_i] = \begin{bmatrix} 1 & 1 & \cdots & 1 \\ 1 & 1-M & \cdots & 1 \\ \cdots & \cdots & \cdots & \cdots \\ 1 & 1 & \cdots & 1-M \end{bmatrix} \tag{20}$$

where M is the number of phases.

It can be shown [2] that the phase current vector $[i_{k,m}]$ entering the nodes at terminal k toward m can again be written as a linear vector equation

$$[i_{k,m}(t)] = [G][e_k(t)] + [I_k] \tag{21}$$

and analogous for $[i_{m,k}]$. Equation (21) is derived from a set of modal equations, subjected to the transformations (19). In building $[Y_{AA}]$, $[Y_{AB}]$ in (13), a matrix $[G]$ is entered instead of a scalar value $1/Z$. The vector $[I_k]$, which enters $[I_A]$, is calculated from the past history of the modal quantities. Since the span $(t - \tau)$ for picking up the past is different for each mode, a time argument was deliberately omitted in writing $[I_k]$. Even though the nodal equations are in phase quantities, the past history must be recorded in modal quantities.

IV. Frequency-Dependent Line Parameters

Skin effects in the earth return and conductors make the line parameters R' and L' frequency dependent [11], [14]. In multiphase lines, this affects primarily the mode associated with earth return. It is not easy to take the frequency dependence into account and at the same time maintain the generality of the program. Methods using the Fourier transform [15], [16] or the Laplace transform [17] are usually restricted to the case of a single line. Work is in progress at BPA to incorporate the frequency dependence approximately into the method of characteristics; then, instead of one value from the past history, several weighted samples will go into the computation of I_k and I_m. The weights would have to be chosen to match the frequency spectrum derived from Carson's formula [11] or from measurements on the line. In a similar approach [18], the earth return mode is passed through two RC filters before entering the node, while the others are attenuated without distortion.

V. Examples

Two simple cases are used to illustrate applications of the program. Fig. 12 shows the results for sequential closing of a three-phase, open-ended line. The curves were automatically plotted by a Calcomp plotter. For this study, the line was represented by 18 multi-π-equivalents with (coupled) lumped parameters. Fig. 13 shows the voltage at the receiving end of a single-phase line (320 miles long, $R' = 0.0376\ \Omega$/mi, $L' = 1.52$ mH/mi, $C' = 0.0143\ \mu$F/mi), that is terminated by an inductance of 0.1 H and excited with a step function $e(t) = 10$ V. The solid curve results from representing the line with 32 lumped-parameter equivalents, the dashed curve from a distributed-parameter representation.

VI. Conclusions

A generalized digital computer method for solving transient phenomena in single- or multiphase systems has been described. The method is very efficient and capable of handling very large networks. Further work is necessary to find a satisfactory way to represent frequency dependence of line parameters.

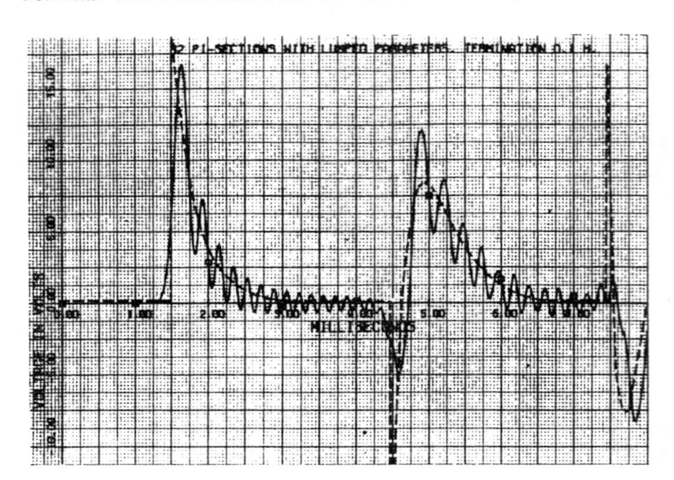

Fig. 13. Single-phase line with inductive termination.

Appendix I

Recording the Past History

The equivalent current sources I in Figs. 1–3 constitute that part of the past history, known from preceding time steps, that has to be recorded and constantly updated. They are needed in building the vector $[I_{\text{total}}]$. For each inductance and capacitance a single value $I_{k,m}(t-\Delta t)$ must be recorded, for each lossless line a double list I_k, I_m for the time steps $t-\Delta t$, $t-2\Delta t$, $\ldots$, $t-\tau$.

In updating $I_{k,m}$ for inductances and capacitances, it is faster to use recursive formulas:

$$I_{k,m}(t-\Delta t) = \pm(I_{k,m}(t-2\Delta t) + 2x)$$

(+ for inductance, − for capacitance), with $x = G(e_k(t-\Delta t) - e_m(t-\Delta t))$ and $G = \Delta t/2L$ for inductance and $G = 2C/\Delta t$ for capacitance.

These formulas are easily verified by expressing the currents in (9b) and (10b) by (9a) and (10a), respectively. To assure correct initial values in the very first time step, $I_{k,m}$ must be preset before entering the time step loop

$$I_{k,m}(\text{initial}) = i_{k,m}(0) - G(e_k(0) - e_m(0)).$$

The initial conditions $e(0)$ for voltages and $i(0)$ for currents are part of the input.

For a lossless line the values I_k, I_m must be recorded for $t-\Delta t$, $t-2\Delta t$, $\ldots$, back to $t-\tau$; they are stored in one double list, where the portion for each line has its length adjusted to its specific travel time τ. After $[e(t)]$ has been found, the double list is first shifted back one time step (entries for $t-\Delta t$ become entries for $t-2\Delta t$, etc.); then $I_k(t-\tau)$, $I_m(t-\tau)$ are computed and entered into the list. Physically, the list is not shifted; instead, the starting address is raised by 1 modulo {length of double list} [8]. The initial values for I_k, I_m must be given for $t = 0, -\Delta t, -2\Delta t, \ldots, -\tau$. The necessity to know them beyond $t = 0$ is a consequence of recording the terminal conditions only. If the conditions were also given along the line at travel time increments Δt, then the initial values at $t = 0$ would suffice.

BPA's computer program has features that help to speed up the solution. Thus a series connection of resistance, inductance, and capacitance is treated as a single branch. This reduces the number of nodes; the respective formulas can be derived by eliminating the inner nodes in the connection [2]. Likewise, single- or multiphase π-equivalents with series $[R]$ and $[L]$ matrices and with identical shunt $[C]$ matrix at both terminals are treated as one element. If the system has identical network elements (e.g., in a chain of π-equivalents), then the data are specified and stored only once.

Appendix II

Initial Conditions

BPA's computer program has two options for setting nonzero initial conditions. Voltages and currents at any point in a study can be stored and used again as initial conditions in subsequent studies that take off from that point (usually with a different Δt). They can also be computed for any sinusoidal steady-state condition with a subroutine "multiphase steady-state solution." The first option must be used if the steady-state solution is nonsinusoidal because of nonlinearities. In this case a transient study is made once and for all over a long enough time span to settle to the steady state. This gives initial conditions for all subsequent studies.

Acknowledgment

The author wants to thank his colleagues at the Bonneville Power Administration, notably Dr. A. Budner, J. W. Walker, and W. F. Tinney, for their help and for their encouragements. The idea of weighted samples to incorporate the frequency dependence of line parameters is due to Dr. A. Budner, and the subroutine to get ac steady-state initial conditions was written by J. W. Walker.

References

[1] H. Prinz and H. Dommel, "Überspannungsberechnung in Hochspannungsnetzen," presented at the Sixth Meeting for Industrial Plant Managers, sponsored by Allianz Insurance Company, Munich, Germany, 1964.

[2] H. Dommel, "A method for solving transient phenomena in multiphase system," *Proc. 2nd Power System Computation Conference 1966* (Stockholm, Sweden), Rept. 5.8.

[3] F. H. Branin, Jr., "Computer methods of network analysis," *Proc. IEEE*, vol. 55, pp. 1787–1801, November 1967.

[4] L. Bergeron, *Du Coup de Belier en Hydraulique au Coup de Foudre en Electricité*. Paris: Dunod, 1949. Transl., *Water Hammer in Hydraulics and Wave Surges in Electricity* (Translating Committee sponsored by ASME). New York: Wiley, 1961.

[5] H. Prinz, W. Zaengl, and O. Völcker, "Das Bergeron-Verfahren zur Loesung von Wanderwellen," *Bull. SEV*, vol. 16, pp. 725–739, August 1962.

[6] W. Frey and P. Althammer, "Die Berechnung elektromagnetischer Ausgleichsvorgaenge auf Leitungen mit Hilfe eines Digitalrechners," *Brown Boveri Mitt.*, vol. 48, pp. 344–355, 1961.

[7] P. L. Arlett and R. Murray-Shelley, "An improved method for the calculation of transients on transmission lines using a digital computer," *Proc. PICA Conf.*, pp. 195–211, 1965.

[8] F. H. Branin, Jr., "Transient analysis of lossless transmission lines," *Proc. IEEE*, vol. 55, pp. 2012–2013, November 1967.

[9] L. O. Barthold and G. K. Carter, "Digital traveling-wave solutions," *AIEE Trans. (Power Apparatus and Systems)*, vol. 80, pp. 812–820, December 1961.

[10] W. F. Tinney and J. W. Walker, "Direct solutions of sparse network equations by optimally ordered triangular factorization," *Proc. IEEE*, vol. 55, pp. 1801–1809, November 1967.

[11] J. R. Carson, "Wave propagation in overhead wires with ground return," *Bell Syst. Tech. J.*, vol. 5, pp. 539–554, 1926.

[12] A. Ralston, *A First Course in Numerical Analysis*. New York: McGraw-Hill, 1965.

[13] A. J. McElroy and H. M. Smith, "Propagation of switching-surge wavefronts on EHV transmission lines," *AIEE Trans. (Power Apparatus and Systems)*, vol. 81, pp. 983–998, 1962 (February 1963 sec.).

[14] D. E. Hedman, "Propagation on overhead transmission lines I—theory of modal analysis," *IEEE Trans. Power Apparatus and Systems*, vol. PAS-84, pp. 200–211, March 1965; discussion, pp. 489–492, June 1965.

[15] H. Karrenbauer, "Ausbreitung von Wanderwellen bei verschiedenen Anordnungen von Freileitungen im Hinblick auf die Form der Einschwingspannung bei Abstandskurzschluessen," doctoral dissertation, Munich, Germany, 1967.
[16] M. J. Battisson, S. J. Day, N. Mullineux, K. C. Parton, and J. R. Reed, "Calculation of switching phenomena in power systems," *Proc. IEE* (London), vol. 114, pp. 478–486, April 1967; discussion, pp. 1457–1463, October 1967.
[17] R. Uram and R. W. Miller, "Mathematical analysis and solution of transmission-line transients I—theory," *IEEE Trans. Power Apparatus and Systems*, vol. 83, pp. 1116–1137, November 1964.
[18] A. I. Dolginov, A. I. Stupel', and S. L. Levina, "Algorithm and programme for a digital computer study of electromagnetic transients occurring in power system" (in Russian), *Elektrichestvo*, no. 8, pp. 23–29, 1966; English transl. in *Elec. Technol.* (USSR), vols. 2–3, pp. 376–393, 1966.

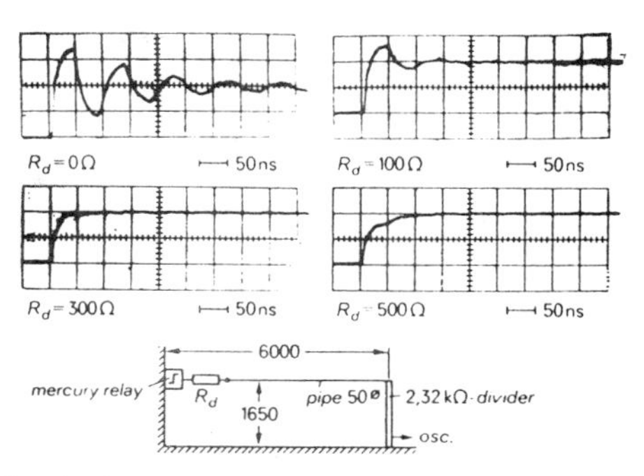

Fig. 14. Measured step response of a low-impedance voltage divider.

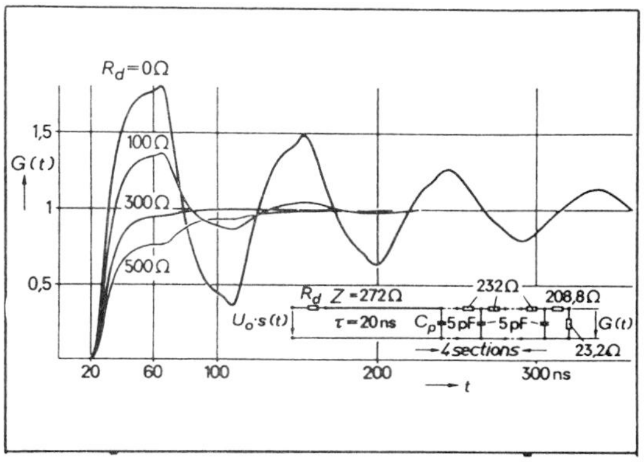

Discussion

W. Zaengl and **F. W. Heilbronner** (Hochspannungsinstitut der Technischen Hochschule München, Munich, Germany): Dr. Dommel is to be congratulated for these lucid elaborations of the treatment of electromagnetic transients. In order to demonstrate how effective this method is, we wish to append two examples of a single-phase application of the algorithm as described and the verification by experiments: 1) evaluation of the step response of an impulse voltage measuring circuit and 2) computation of the voltage breakdown in sparkgaps.

1) In high-voltage measuring techniques voltage dividers are used which cannot be constructed coaxially and are, because of voltages up to some million volts, of big dimensions. Therefore the voltage to be measured is led to the divider by metallic pipes, at the input end of which, in general, a damping resistor is connected.

For this purpose the equivalent circuit of the total measuring circuit is best represented by a lossless line (for the metallic pipe), on which traveling wave phenomena occur, and lumped parameters (for damping resistor and voltage divider). An analytic general solution to get the step response of this network is not possible.

In Fig. 14 the used measuring circuit is sketched with its dimensions. The 2.32-kΩ divider consists of stacked resistors. The output voltage, reduced by a factor of 100, is measured by an oscilloscope (Tektronix 585). Four oscillograms of the output voltage are given, resulting from various damping resistors R_d, if a voltage step generated by a mercury relay occurs at the input end of the measuring circuit.

In Fig. 15 the equivalent circuit of the test setup with its data is given and the results of the digital computation of the step response $G(t)$ with the program outlined in the paper. The surge impedance $Z = 272$ ohms and the travel time $\tau = 20$ ns result from the geometric dimensions of the pipe. The divider is represented by a multisection network of a total of five T quadripoles and an input shunt capacity $C_p = 5$ pF. In the calculation a step width Δt of $2 \cdot 10^{-9}$ seconds was used. The comparison shows a very good agreement with the experimental results of Fig. 14.

2) Whereas the solution of the foregoing problem requires no specific modification of the straightforward procedure as described in the paper, in the case of voltage breakdown, nonlinearities have to be taken into account [19]. One means of evaluating the voltage u at a time t during breakdown of a gap was given by Toepler [20]:

$$u^{(t)} = k \cdot a \cdot i^{(t)} \Big/ \int_0^t i^{(t)}\, d\xi \qquad (22)$$

i.e., the resistance of the spark is inversely proportional to the amount of charge which has flowed into the gap (a = gap spacing in cm, k = constant in the range of 10^{-4} V·s/cm, $i^{(t)}$ = current in amperes, t = time in seconds).

Fig. 15. Calculated step response of the test setup according to Fig. 14.

Fig. 16. Test setup. Front left: screened measuring cabin; front center: damped capacitive divider; front right: 80-cm rod–rod gap; center: 3-million-volt impulse generator (capacitive divider is used as load capacitance and is standing in front of the generator).

Manuscript received July 3, 1968.

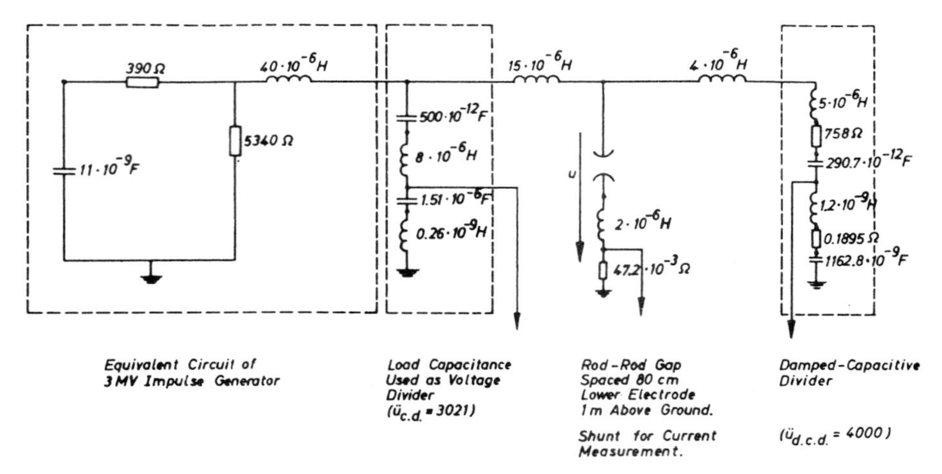

Fig. 17. Equivalent impulse circuit of Fig. 16.

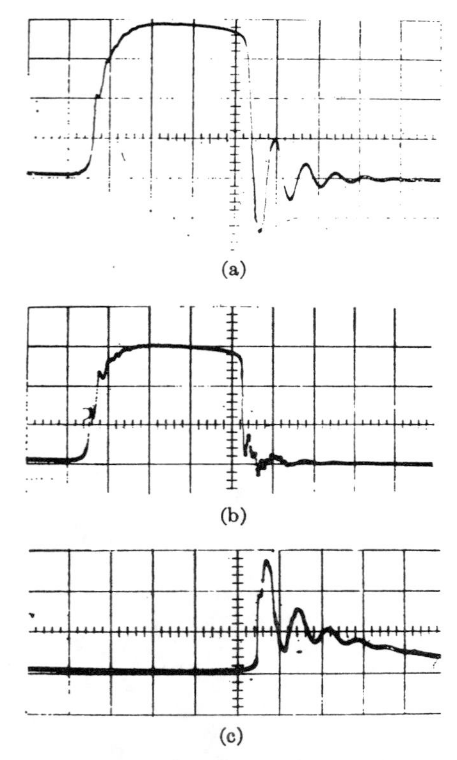

Fig. 18. Oscillograms from the voltage breakdown of a 80-cm rod–rod gap (temperature: 20°C, 716 mm Hg); horizontal deflection 10^{-6} seconds/division. (a) Capacitive divider: 138 kV per vertical division. (b) Damped capacitive divider: 183 kV per division. (c) Current shunt: 1060 amperes per vertical division.

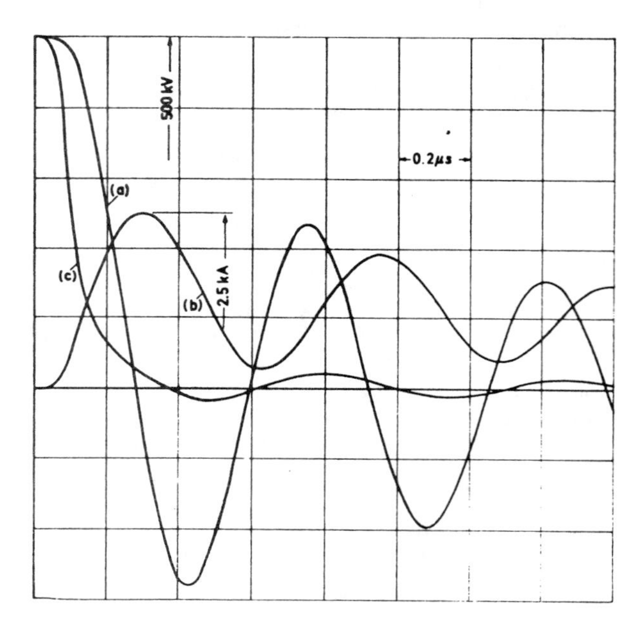

Fig. 19. Calculated voltages in different points and gap current at breakdown according to Fig. 17. (a)—voltage of capacitive divider; (b)—gap current $i^{(t)}$; (c)—gap voltage $u^{(t)}$.

Using the trapezoidal rule of integration, (22) can be rewritten as

$$u^{(t)} = \frac{k \cdot a \cdot i^{(t)}}{\mathrm{int}^{(t-\Delta t)} + \frac{1}{2}(i^{(t-\Delta t)} + i^{(t)}) \cdot \Delta t} \qquad (23)$$

where $\mathrm{int}^{(t-\Delta t)}$ is the value of the integral in the denominator of (22) up to the time $(t - \Delta t)$. This is the equivalent expression of $f(i_{k,m}(t))$ in (16a). Since the solution in connection with (15) would be of the quadratic type, it was found sufficient to linearize the problem and take the resistance of the previous time step $(t - \Delta t)$:

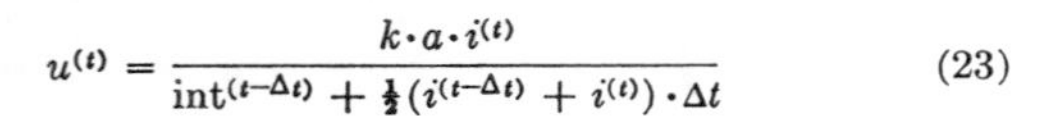

$$R^{(t-\Delta t)} = \frac{k \cdot a}{\mathrm{int}^{(t-2\Delta t)} + \frac{1}{2} \cdot (i^{(t-2\Delta t)} + i^{(t-\Delta t)}) \cdot \Delta t}. \qquad (24)$$

Thus, in terms of the paper, the voltage across a sparkgap between the nodes k and m will be

$$u^{(t)} = \frac{R^{(t-\Delta t)} \cdot (e_k(t) - e_m(t))^{(\mathrm{linear})}}{R^{(t-\Delta t)} + (z_{k,k} - z_{k,m} - z_{m,k} + z_{m,m})}. \qquad (25)$$

In order to start the process, in (24) a certain initial value of $\mathrm{int}^{(t-\Delta 2t)}$ is needed. This means in physical terms, that by some predischarges the gap must have been ionized and thus assumed some conductivity. Experience has shown that for a start the value of $R^{(t-\Delta t)}$ might be chosen a thousand times higher than the biggest resistance in the circuit.

As a demonstration, in Fig. 18 three oscillograms (Tektronix 507) of the breakdown of a 80-cm rod–rod gap in the test circuit of Figs. 16 and 17 are given. The calculated values with $k = 0.3 \cdot 10^{-4}$ V·s/cm and $\Delta t = 20 \cdot 10^{-9}$ seconds, multiplied by the corresponding divider ratios, are plotted in Fig. 19. They correspond fairly well to the oscillograms. The voltage resulting from the damped capacitive divider is within ±1 percent of u and is therefore not plotted.

Two conclusions may be drawn from a comparison of Figs. 18 and 19 and are stated without further explanation: 1) A damped capacitive divider [21] reflects the gap voltage much better than a purely capacitive divider, and 2) the common equivalent circuit of a divider may be too rough in the cases where higher harmonics occur. Then an equivalent circuit as in item 1) would be necessary.

The described application of the transient algorithm in high-voltage impulse circuits has led the discussers to various secondary problems and suggestions, of which two can be sketched here in general terms only.

1) In problems with many nodes, computer storage might be too small for building up the matrix $[Y]$. Thus the method of diacoptics is of help, especially when two major parts of the circuit are connected by a single lead which can be represented by a lumped parameter (inductance in Fig. 17).

2) If sudden changes of network parameters occur, e.g., the breakdown of a sparkgap on account of a certain overvoltage, where the resistance changes from the order of megohm to ohm in fifty to some hundred nanoseconds, it might be desirable to make the time step Δt smaller and increase it again when the rate of change is no longer of importance. Thus it is necessary to adapt the stepwidth Δt to the rate of voltage change in the network.

References

[19] F. Heilbronner and H. Kärner, "Ein Verfahren zur digitalen Berechnung des Spannungszusammenbruchs von Funkenstrecken," *ETZ-A*, vol. 89, pp. 101–108, 1968.

[20] M. Toepler, "Funkenkonstante, Zündfunken und Wanderwelle," *Arch. f. Elektrotech.*, vol. 16, pp. 305–316, 1925.

[21] W. Zaengl, "Das Messen hoher, rasch veränderlicher Stossspannungen," doctoral dissertation, Munich, 1964.

Manuscript received July 3, 1968.

D. G. Taylor and **M. R. Payne** (Central Electricity Generating Board, London, England): We have also programmed the Bergeron method for single-phase switching problems and are currently engaged in extending the treatment to multiconductor systems. Lumped L and shunt C have been represented as short lines and special "hyper-nodes" have been introduced to deal with series R and series C. Only one past history is stored which necessitates subdividing lines into sections of equal traveling time. Processed system data together with past and present values of voltage and current are stored in a structured file (in core) which is passed, using list-processing techniques, in order to advance the solution by one time step.

One advantage of subdividing lines over storing multiple past histories is that series resistance can be introduced between all sections; we have found this to be desirable in cases where the response is oscillatory and the degree of attenuation is important. The author's comments on this point would be appreciated.

A source of approximation which should be mentioned arises from the necessity for all traveling times to be integral multiples of the time increment Δt. This also applies to the method of multiple past histories since any interpolation between values is invalid. The problem is made more severe in multiconductor systems by the propagation velocities in the modes being different, in some cases by small but significant amounts. How does the author take this into account in making his initial choice of time increment, in particular for systems including asymmetrical multiconductor configurations?

In conclusion the author is to be congratulated on his adaptation of the problem for use with ordered-elimination techniques which have already made such an impact on steady-state analysis. We look forward to the author's further developments in this field, particularly with regard to the treatment of frequency dependence.

Manuscript received August 8, 1968.

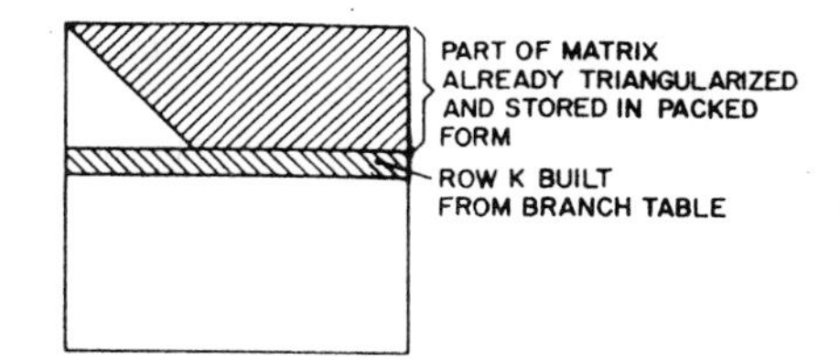

Fig. 20. Triangularization scheme.

H. W. Dommel: The author wants to thank Dr. W. Zaengl, Mr. F. W. Heilbronner, Mr. D. G. Taylor, and Mr. M. R. Payne for their valuable discussions, which illustrate the usefulness of Bergeron's method in traveling wave studies and also raise some interesting questions.

One of the main differences between the author's computer program and that of Mr. D. G. Taylor and Mr. M. R. Payne is the subdivision of the line into sections of travel time $\tau = \Delta t$ in the latter. It appears that considerable savings in computer time (but not in storage requirements) are possible when such subdivisions are avoided and multiple past histories are stored. It must be admitted, however, that lumped series resistances can be included more easily in more places with the line being subdivided, even though this can always be done with the author's program in the definition of the model at the expense of more input data. Interestingly enough, test examples showed very little or no difference at all between the insertion of lumped series resistances in few or many places (section *Approximation of Series Resistance of Lines*). Therefore, the automatic insertion at three places (terminals and midpoint) was felt to be adequate. This observation might not be true for all cases. Also, not too much significance has been placed on the approximation of distributed resistance by lumped series resistances in developing the program, since the final objective has been the approximation of the frequency dependence in the zero-sequence mode. This has not been included yet, but preliminary tests with a weighting function representation look promising.

Mr. D. G. Taylor and Mr. M. R. Payne use a stub-line (short line) representation for lumped (series and shunt) L and shunt C. It can be shown that this stub-line representation for shunt L and shunt C is equivalent with the integration of (8b), and the respective equation for C, by the trapezoidal rule over two time steps from $t - 2\Delta t$ to t (no such simple equivalence was found for series L). Since the author's method for lumped L and C is based on the trapezoidal rule of integration over one time step only from $t - \Delta t$ to t, it is more accurate than stub lines. The stub-line representation is very helpful, however, in studies involving more than one nonlinear element. As described in the section *Nonlinear and Time-Varying Parameters*, more than one nonlinear element can be handled in closed form only if they are separated by elements of finite travel time. A stub-line representation accomplishes just such a separation. As an example, a case involving a lightning arrester connected to a nonlinear inductance (transformer with saturation) can be solved by modeling the total inductance as a linear and nonlinear inductance in series, with the linear inductance placed on the side of the lightning arrester and treated as a stub line.

Mr. D. G. Taylor and Mr. M. R. Payne raise the question of errors introduced either by making all travel times an integral multiple of Δt or by using interpolation between past values. It is true that this question is even more critical in multiconductor systems with small differences in mode propagation velocities. Interpolation is indeed questionable if sudden changes occur. However, the presence of inductances and capacitances often, though not always, smoothes out sudden changes; then interpolation is a good and valid approximation. Sudden changes may also be introduced through stub-line representations and not lie in the nature of the problem. In cases where sudden changes do occur, the user has an option in which all travel times are rounded to the nearest integral multiple of Δt. As of now, the step width Δt must be chosen by the user.

In the first part of their discussion, Dr. W. Zaengl and Mr. F. W. Heilbronner show how closely computed results can agree with test results. This speaks at least as much for their good engineering judgment in selecting an equivalent model as it does for the usefulness of the computer program. Their effort to include the dynamic law of spark gaps into the program should be of interest to high-voltage engineers.

As to the specific questions raised, it is felt that the sparsity technique used (optimally ordered elimination with packed storage of nonzero elements only) is more efficient than the method of diacoptics. It was probably not made clear in the paper that the matrix $[Y]$ is never built explicitly. Rather, a branch table is used to store the information for the matrix $[Y]$. As indicated in Fig. 20 for the kth elimination step, the original row k is built from a search of the branch table (therefore, only one working row is necessary), then the elements to the left of the diagonal are eliminated with the information contained in the already available rows $1, \cdots, k-1$ of the triangularized matrix, and finally the elements $Y'_{k,k}, Y'_{k,k+1}, \cdots$ of this transformed row are added in packed form to the triangularized matrix. In a way, the method does have a built-in tearing feature similar to diacoptics in cases involving lines with distributed parameters, which disconnect the network topologically. This disconnection is more than tearing in diacoptics, since it is a true disconnection where no reconnection effect has to be introduced at a later stage of the algorithm. Thus, the use of a stub-line representation for the inductance in Fig. 17 with surge impedance $Z = L/\Delta t$ and travel time $\tau = \Delta t$, might reduce the storage requirements beyond those already achieved through sparsity. The possibility to change Δt during the computation would indeed be desirable. It is a straightforward programming task, involving changes of $[Y]$. Due to lack of time, it has not been incorporated so far.

NONLINEAR AND TIME-VARYING ELEMENTS

IN DIGITAL SIMULATION OF ELECTROMAGNETIC TRANSIENTS

Hermann W. Dommel

Bonneville Power Administration, Portland, Oregon

Abstract—Simulation programs for electromagnetic transients must provide models for lightning arresters, transformer saturation, circuit breakers and other nonlinear and time-varying effects. This paper describes techniques for keeping the solution time in networks with such elements as low as possible, by using the compensation method or network equivalents.

I. INTRODUCTION

Switching and lightning surges, ferroresonance problems and other phenomena of electromagnetic transients have traditionally been studied on transient network analyzers. Nowadays, digital computer solutions offer an alternative which is becoming more attractive for a number of reasons, such as: (1) Often, general-purpose computers are available anyhow, and (2) more generalized transients programs have been developed which do not require any programming for each individual problem. Some of these circuit analysis programs were primarily developed for electronic circuits, others for the simulation of analog computers, and some primarily for power systems.

Transient phenomena in power systems are frequently nonlinear in nature. Typical nonlinear and time-varying elements are lightning arresters, nonlinear inductances (saturation in transformers and reactors), circuit breakers, and protective gaps. Inclusion of such elements into computer programs requires modifications of linear analysis techniques. Some "practical" modification schemes will be explained in this paper. The word "practical" stands for the objective of obtaining the solution with acceptable accuracy in a minimal amount of time.

II. REVIEW OF SOLUTION METHODS FOR LINEAR NETWORKS

One program for solving transient phenomena in linear circuits[1] finds the state of the system by solving a system of linear algebraic equations,

$$[Y][e(t)] = [k] \tag{1}$$

for the vector [e(t)] of unknown node-to-datum voltages in each time step. [Y] is a constant matrix (real, symmetric). The value of [k] is known and determined from the source functions at time t and from the state of the system at previous time steps. Since many elements in [Y] are zero, sparsity techniques with optimally ordered elimination are used to triangularize [Y] once and perform a "repeat solution" in each time step.[2] Eq. (1) results from a combination of the trapezoidal rule of integration for lumped inductances and capacitances with the method of characteristics (Bergeron's method) for lines with distributed parameters. Similar programs[3] differ mainly in the permissible structure of [Y], which is strictly diagonal if lumped elements are only shunt-connected from node to datum (= ground).

The trapezoidal rule is often put aside as too simple. It is indeed less accurate than other methods for solving ordinary differential equations. However, it is numerically stable; therefore, its step-size is not restricted by the so-called smallest time constant (largest eigenvalue) barrier. This is illustrated in Fig. 1.

Paper 71 C 26-PWR - IV - A, recommended and approved by the Power System Engineering Committee of the IEEE Power Group for presentation at the 1971 PICA Conference, Boston, Mass., May 24-26, 1971. Manuscript submitted January 11, 1971, made available for printing June 9, 1971.

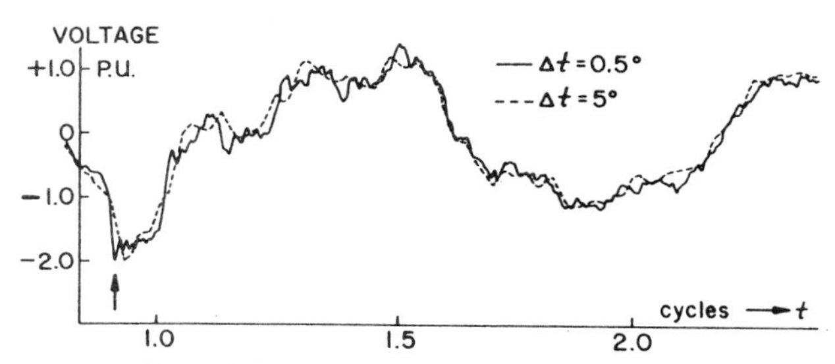

Fig. 1. Influence of step-width in trapezoidal rule.

The curves show the receiving-end voltage on one phase of an open-ended three-phase line (180 miles long) after energization. The line was represented by 18 three-phase π - circuits. One curve was computed with a step-width of 0.5° and the dotted curve with 5° (based on 60 Hz). As expected, the larger step-width is less accurate but does still give a fair and acceptable picture. It is believed that the simple trapezoidal rule of integration fulfills the objective of a fast solution with acceptable accuracy better than some supposedly more powerful methods. Fig 1 also demonstrates that the error at any one point is not a good accuracy criterion; the arrow points to a local error of the 5° - curve of 0.6 p.u., but the results are still acceptable. The trapezoidal rule is also used in the circuit analysis program MTRAC[4], which can solve circuits with square-loop magnetic cores and other nonlinear elements.

The solution of eq. (1) is formally expressed by

$$[e(t)] = [Y]^{-1}[k], \tag{2}$$

but it should be remembered that the inverse is not computed explicitly. Similar forms are obtained if the differential equations are solved by other methods (see Appendix). Therefore, the following techniques for non-linear and time-varying elements are not tied to eq. (1) but can also be used in combination with other methods.

III. TYPES OF NONLINEAR AND TIME-VARYING PARAMETERS

Time-varying Resistance and Switch.

A time-varying resistance is sometimes used to simulate circuit breakers and gaps. The ideal switch ($R = 0$ in closed position, $R = \infty$ in open position) is a special case. Attempts have been made[5] to define the resistance near current zero by differential equations, e.g., by Cassie's formula

$$R\frac{d}{dt}\left(\frac{1}{R}\right) = \frac{1}{\theta}\left(\frac{e \cdot i}{P_o} - 1\right)$$

where θ and P_o are constants. Such formulas could be used to compute R up to the last time step by numerical integration. This value could then be used as an approximation for the solution during the next time interval from $t-\Delta t$ to t. Similar formulas would be helpful for defining the voltage build-up in the current-limiting gap of modern lightning arresters, where the value $1/R(t)$ tends to decrease exponentially prior to extinction. In a detailed study of the breakdown of gaps in a multistage impulse generator[6] Toepler's law was used to find R by numerical integration.

Nonlinear Resistance.

The nonlinearity is usually expressed as a nonlinear function e=f(i). Lightning arrester blocks are generally represented this way after the breakdown of the gap. Prior to breakdown, R is infinite.

Nonlinear Inductance.

The nonlinearity is usually expressed as a nonlinear function for the total flux linkage, ψ=f(i). In BPA's transient program, this is transformed to a function e=F(i) and can then be treated the same way as a nonlinear resistance. The transformation is based on the trapezoidal rule of integration. Applying it to

$$\psi(t) = \psi(t-\Delta t) + \int_{t-\Delta t}^{t} e(u)du$$

with $\psi(t-\Delta t)$ = flux at beginning of the new-time step, gives

$$\psi(t) = \frac{\Delta t}{2} e(t) + b(t-\Delta t)$$

with the known value $b(t-\Delta t) = \psi(t-\Delta t)+\frac{\Delta t}{2} e(t-\Delta t)$. Replacing ψ in ψ=f(i) with this expression produces the desired function e=F(i). The value b can be updated recursively[1]

$$b(t-\Delta t) = b(t-2\Delta t) + \Delta t \cdot e(t-\Delta t)$$

(the factor c in ref. 1 is equivalent to $\frac{2}{\Delta t}b$). Hysteresis effects can be included[4], provided the data is available.

IV. PRINCIPLE OF NONLINEAR ANALYSIS

If the network contains branches with nonlinear and time-varying parameters, then [Y] in eq. (1) is not constant anymore, but some of its elements will be functions of current, voltage or time. In this case, the equation

$$[Y(i,e,t)][e(t)] = [k] \tag{3}$$

must be solved iteratively in each time step, e.g., by the Newton-Raphson method. In the simpler case where [Y] is only a function of time, no iteration would be necessary but [Y(t)] would still have to be triangularized anew in each time step. In either case, solution time increases considerably compared with linear circuits. Therefore, it is worthwhile to seek ways for reducing the amount of additional computations due to the nonlinear branches. This can indeed be done, either with the compensation method or by deriving network equivalents. Both methods are used simultaneously in BPA's transients program.

V. COMPENSATION METHODS

The compensation theorem in combination with triangular factorization of nodal equations can be used for transients problems to separate the nonlinear/time-varying elements from the linear part of the network.[1] It is described in more detail in ref. 7 for any nodal network formulation.

Compensation Method for a Single Branch.

Let a single branch between nodes k and m have nonlinear or time-varying parameters. The compensation theorem states that this branch can be excluded from the network and, therefore, from the matrix [Y] and, instead, be simulated by a current source i_{km}. The load convention is used here as well as in ref. 1, that is the positive direction of the current $i_{k,m}$ is defined through the branch from node k to m, whereas ref. 7 uses the source convention with the current I_{km} flowing through the network from node k to node m. The value of $i_{k,m}$ must fulfill two equations, namely the network equation of the linear system ("Thévenin" or "Helmholtz" equivalent between nodes k and m),

$$e_k(t) - e_m(t) = e_k^{(o)}(t) - e_m^{(o)}(t) - z_T \cdot i_{k,m}(t) \tag{4}$$

(superscript o indicates solution without the nonlinear branch), and the nonlinear relationship of the branch itself,

$$e_k(t) - e_m(t) = f\left(i_{k,m}, \frac{di_{k,m}}{dt}, \ldots\right). \tag{5}$$

The value of the Thévenin impedance* z_T in eq. (4) is computed only once by solving eq. (1) or (2) for [e] before entering the time step loop, with [k]= {o, except +1.0 in k-th and -1.0 in m-th component}. This produces a column vector [z] which is, in effect, the difference of the k-th and m-th columns of $[Y]^{-1}$,

$$[Y][z] = [0, \text{except} \begin{cases} +1.0 \text{ in k-th component} \\ -1.0 \text{ in m-th component} \end{cases}]$$

and then: $z_T = z_k - z_m$.

If the voltage on side m is given as a function of time (either as a voltage source or as zero if m is the node for datum = ground) then [z] is computed by setting all elements in [k] to zero except +1.0 in k-th element. Then $z_T = z_k$. The solution becomes trivial if the voltages are given on both sides; in this case, the nonlinear branch could be solved directly and its current would have no influence on the rest of the network.

The solution process in each time step is as follows:

1. Compute the node voltages $[e^{(0)}(t)]$ with a repeat solution of eq. (1), and the open circuit branch voltage $e_k^{(0)}(t) - e_m^{(0)}(t)$.
2. Solve the 2 scalar equations (4) and (5) simultaneously for $i_{k,m}$. If eq. (5) is given analytically, then the Newton-Raphson method could be used. If eq. (5) is defined point-by-point as a piece-wise linear curve (Fig. 2), then the intersection of the 2 curves must be found.
3. Find the final voltage solution by super-imposing the response to the current source $i_{k,m}$,

$$[e(t)] = [e^{(0)}(t)] - [z] \cdot i_{k,m} \tag{6}$$

Superposition is permissible because it is only used in the linear part of the network.

Step 1 is the normal solution procedure for linear circuits. Step 2 takes little extra time because it involves only 2 scalar equations. Step 3 requires N additional multiplications and additions (N = number of unknown node voltages). Therefore, the extra work of steps 2 and 3, which is a result of the nonlinearity, is rather small compared to a refactorization of [Y].

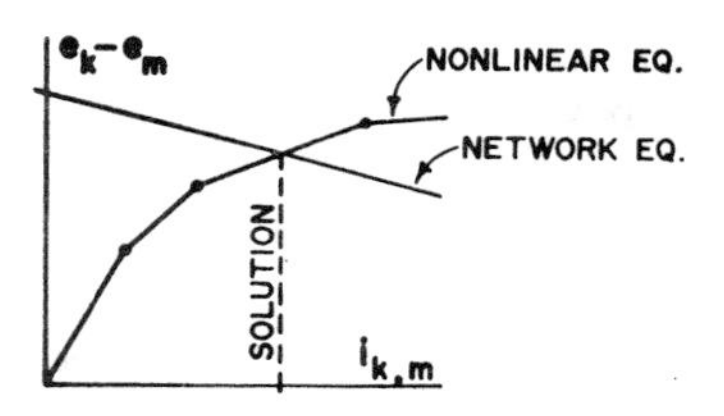

Fig. 2. Simultaneous solution of two equations.

*[z] and z_T have the opposite sign in ref. 1.

BPA's transients program uses a point-by-point representation of the nonlinearity as indicated in Fig. 2. The algorithm for finding the intersection was developed for utmost speed. The search process

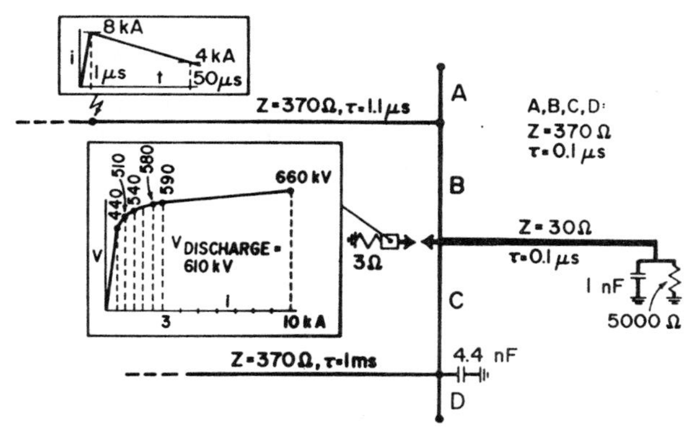

Fig. 3. Single-phase system.

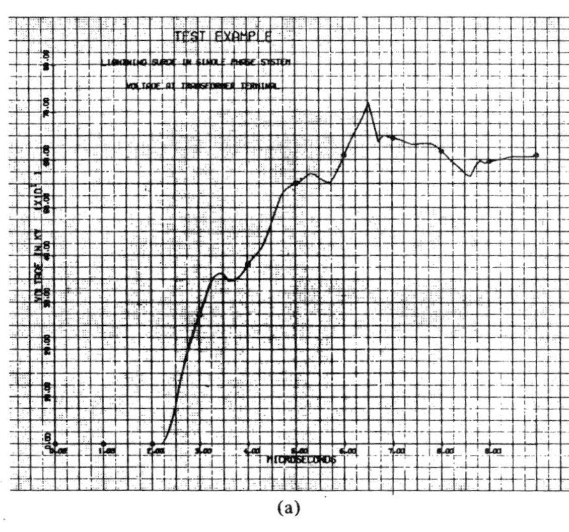

(a)

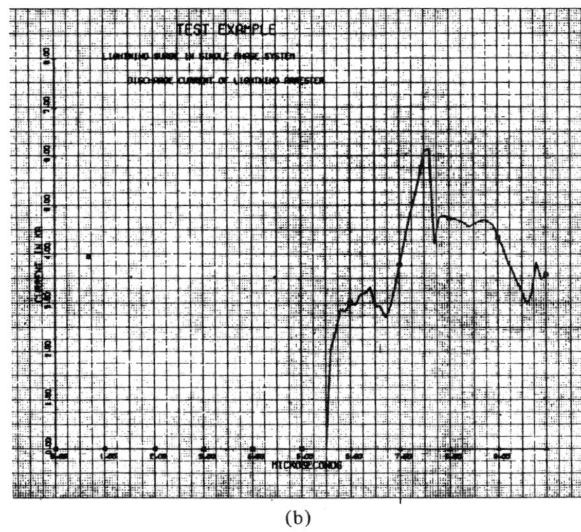

(b)

Fig. 4. (a) Voltage at transformer terminal
(b) Discharge current.

starts at the location of the intersection in the preceding time step and proceeds to the left (right) in Fig. 2 if the voltage moved down (up) as compared to the preceding time step. Note that the slope z_T remains unchanged as long as no switchings take place in the network.

The single-phase circuit in Fig. 3 was solved with the compensation method. A lightning surge, simulated as a current source with a peak value of 8 kA, strikes a line close to the substation sections A, B, C, D. The transformer is represented as R and C in parallel. It is connected to the substation via a cable (surge impedance 30 Ω, travel time 0.1 μs) and protected by a lightning arrester at the substation end of the cable.

Fig. 4a shows the voltage at the transformer terminal and Fig. 4b the discharge current of the lightning arrester. Solution time to 10 μs with a step-width of 0.1 μs was 0.6 s without plotting on a CDC 6400 computer. This example illustrates the solution technique and contains all necessary data for verification. A more realistic problem with dc lightning arresters was solved with the same technique.[8]

Step 2 of the solution process is especially simple if the branch is a time-varying resistance, where

$$i_{k,m} = \frac{e_k^{(0)}(t) - e_m^{(0)}(t)}{R(t) + z_T} .$$

An ideal switch is a special case. For open position ($R(t) = \infty$) the solution proceeds normally. If the switch is closed, then the current $i_{k,m}$ is found from

$$i_{k,m} = \frac{e_k^{(0)}(t) - e_m^{(0)}(t)}{z_T} .$$

Compensation Method for M Branches

The compensation method can also be used to simulate the effect of M (nonlinear) branches with current sources.[7] In this case, M vectors $[z^{(1)}]$, ... $[z^{(M)}]$ must be precomputed. Step 1 remains identical, but nonlinear equations must now be solved simultaneously in step 2, e.g., by Newton's method if the nonlinearities are expressed analytically. Step 3 requires M x N additional multiplications and additions now, because the response to M current sources must be superimposed,

$$[e(t)] = [e^{(0)}(t)] - [z^{(1)}] \cdot i^{(1)} \; \ldots \; - [z^{(M)}] \cdot i^{(M)} . \tag{7}$$

This technique is well suited for problems with three lightning arresters or other nonlinearities on the three phases of a three-phase bus.

It is interesting to note that lines with distributed parameters disconnect the equations for the two ends. This is not astonishing because the phenomena at one end are not immediately seen at the other end, but only τ seconds later (τ = travel time). Nonlinear elements disconnected by distributed-parameter lines can, therefore, be solved with the compensation method for a single branch.[1] In effect, the M vectors $[z^{(1)}]$, ... $[z^{(M)}]$ can be merged into one vector for storage. Unfortunately, three nonlinear elements on the three phases of one bus are not disconnected. However, three nonlinear elements each, at each end of a three-phase line, can be solved independently in two groups of three, which requires 3N mult./add. in step 3 instead of 6N mult./add.

Thévenin Impedance at Line Junctions.

Lightning arresters and nonlinear inductances (magnetizing impedance of transformers) are frequently connected from a line junction to ground (= datum node). In such cases the Thévenin impedance from the junction of distributed-parameter lines to ground is needed, which is simply

$$\frac{1}{z_T} = \frac{1}{Z_1} + \frac{1}{Z_2} \; \cdots + \frac{1}{Z_L} .$$

Z_1, ... Z_L are the surge impedances of the lines meeting at the junction. For multiphase circuits the scalars $1/Z_i$ must be replaced by inverse matrices$[Z_i]^{-1}$. The same Thévenin impedance is used in the test procedures for lightning arresters (response to steep wave fronts). A shunt inductance at the junction will alter the value slightly, depending on the step-width Δt. It contributes an equivalent resistance $2L/\Delta t$ in parallel, if the solution method of ref. 1 is used, which is seen to become infinite for very small Δt. On the other hand, a shunt capacitance contributes an equivalent resistance $\Delta t/2C$, which comes close to a short-circuit for very small Δt.

Possible Difficulties in Compensation Methods.

A low sampling rate (large step-width) tends to create more problems in nonlinear circuits than in strictly linear circuits. Any solution proceeds in discrete time steps and can, therefore, sample the nonlinear curve only in discrete points. These points will be farther apart the larger the step-width is. Since all nonlinear information between the points is lost, the solution process must fill it in with some interpolation assumption (usually linear).

Fig. 5 illustrates how a low sampling rate (wide spacing of points) can create artificial negative hysteresis or negative damping (solution proceeds from sample point 1 to 2 and back to 3). This may cause numerical amplification of oscillations in nonlinear inductances. The danger is less pronounced in nonlinear resistances since energy is always dissipated in this case. The only cure seems to be a reduction in the step-width which does, of course, increase the solution time. Such situations could be automatically detected by specifying an upper bound for permissible distance between successive ordinate points.

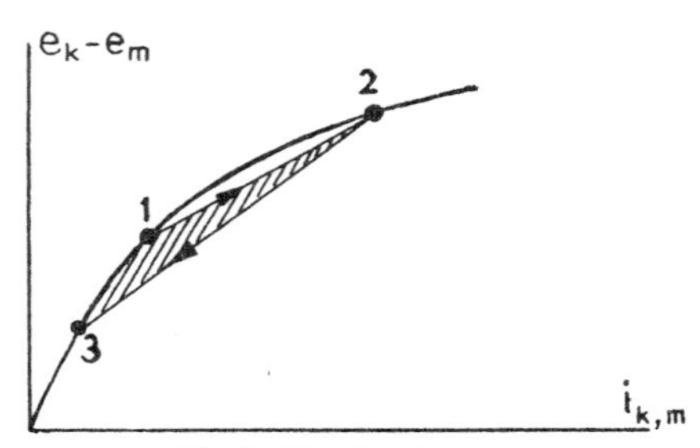

Fig. 5. Negative damping.

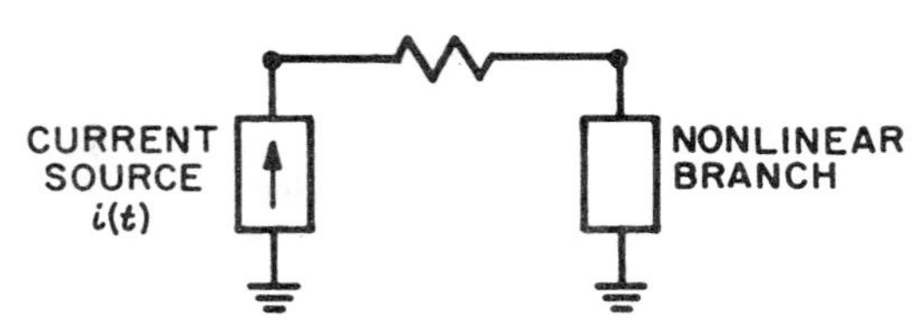

Fig. 6. Nonexistent open circuit solution

The compensation method will fail in cases where no open circuit solution exists (step 1 in the solution process). A simple example is shown in Fig. 6, where no solution exists if the nonlinear branch is removed for finding the open circuit voltage. This will lead to a matrix singularity in the solution process. Actually, a final solution exists as long as the nonlinear characteristic does not block. Such cases become solvable if the branch is split into two parallel branches, one with a linear characteristic to permit the open circuit solution, and the other with a nonlinear characteristic. Example: A nonlinear resistance $i = \tan(e)$ could be split up into two parallel branches, one with $R = 1\Omega$, and the other with the characteristic $i = \tan(e) - e$.

VI. NETWORK EQUIVALENTS

The system of nonlinear equations (3) must be solved iteratively, preferably by the Newton-Raphson method, if any elements in [Y] are functions of e or i. If they are only functions of time, then (3) can still be solved directly but [Y(t)] must be re-factorized anew in each time step. A straightforward application of the Newton-Raphson method is quite time-consuming because it requires a complete solution with triangular factorization of the entire Jacobian matrix in each iteration step. The solution time for the Newton-Raphson method, and also for re-factorization in case of time-varying parameters, is drastically reduced if the iteration and re-factorization process is confined to only those nodes which are incident to nonlinear and the time-varying parameters. This is accomplished by reducing the linear part of the network to an equivalent. Let subscripts

"1" denote the set of nodes which are incident to branches with linear parameters only, and

"2" the set of nodes incident to branches with nonlinear and time-varying branches,

and let eq. (3) be partitioned accordingly:

$$\begin{bmatrix} [Y_{11}] & [Y_{12}] \\ [Y_{21}] & [Y_{22}(i,e,t)] \end{bmatrix} \begin{bmatrix} [e_1(t)] \\ [e_2(t)] \end{bmatrix} = \begin{bmatrix} [k_1] \\ [k_2] \end{bmatrix} \tag{8}$$

Only the submatrix $[Y_{22}(i,e,t)]$ is a function of i,e, and time. Therefore, eq. (8) can be reduced to the nodes of subset 2:

$$\left([Y_{22}(i,e,t)] - [Y_{21}][Y_{11}]^{-1}[Y_{12}]\right)[e_2(t)] = [k_2] - [Y_{21}][Y_{11}]^{-1}[k_1]. \tag{9}$$

The reduction is not done in matrix form but with row-by-row elimination. First, the upper part of the matrix for nodes 1 is triangularized, using sparsity techniques and optimal ordering. Then the elimination in the lower part for nodes 2 is carried out in the columns for nodes 1 only, which produces the equivalent matrix of eq. (9). Any initial value can be used for $[Y_{22}]$ (e.g., zero). This reduction is only done once before entering the time step loop (Fig. 7a). The iterative process in each time step proceeds as follows:

(1) Use the information in the upper part of the triangularized matrix to distribute the effect of $[k_1]$ to nodes 2 in a so-called downward operation. This produces $-[Y_{21}][Y_{11}]^{-1}[k_1]$.

(2) Solve the reduced system for nodes 2, either iteratively with the Newton-Raphson method if $[Y_{22}]$ is a function of i,e, or directly if $[Y_{22}]$ is only a function of time. In the first case, the Jacobian matrix must be factorized repeatedly for the lower part, in the second case the lower part of the matrix is factorized once (Fig. 7b). The answer will be $[e_2(t)]$.

(3) Use backsubstitution in the upper part to get $[e_1(t)]$.

Compared with linear solutions, additional computer time is only required in step (2). This is kept to a minimum, since the iterative process of step 2 is confined to nodes incident to nonlinear or time-varying branches.

The simpler case where $[Y_{22}(t)]$ is a function of time requires a re-factorization of the lower part only when $[Y_{22}(t)]$ changes. This technique is used in BPA's transients program for switches.[1]

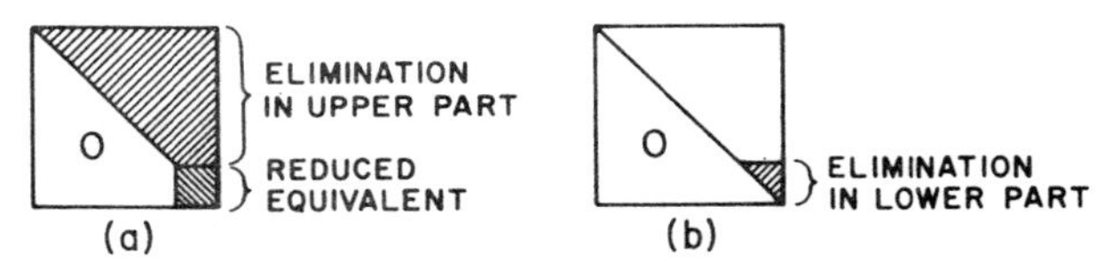

Fig. 7. Reduction Scheme
(a) Outside time step loop,
(b) In each time step.

Piecewise Linear Inductance

An inductance with infinite slope for $-\psi_s < \psi < \psi_s$ and finite slope for $|\psi| > \psi_s$ can be simulated by a linear inductance in series with an ideal switch (Fig. 8), provided the flux is always computed by integrating the voltage difference $e_k - e_m$ independent of the switch position. The switch is closed whenever $|\psi|>\psi_s$ and opened again as soon as $|\psi|<\psi_s$. A two-slope inductance (solid line in Fig. 9) is obtained by connecting a linear inductance in parallel with a "switched inductance"; adding a second "switched inductance" produces a three-slope inductance (dotted line in Fig. 9), etc. Experience indicates that two-slope or three-slope inductances are good approximations for saturation effects in transformers and reactors. Usually, the location of the "knee" is much more critical than a very detailed representation of the nonlinearity.

Two-slope inductances were used to study the ferroresonance problem of Fig. 10. After the breakers were opened to de-energize the line with another line in parallel still energized, a sustained ferroresonance occurred between the magnetizing impedance of the transformers and the coupling capacitances to the energized line. Fig. 11 shows the computed voltages at the three transformer terminals. The ferroresonance is not sustained if the transformers have delta-connected tertiary windings.

The compensation method required a much smaller step-width in solution of this case, apparently because of the described negative damping effect, which cannot occur with "switched inductances". A low sampling rate will always produce positive damping here (Fig. 12). The switch closes at point 1 (first sample point where $\psi>\psi_s$). After a solution in 2, the flux will again be less than ψ_s in point 3 which will lead to opening of the switch in the next time step (point 4). The shaded triangle indicates the artificial positive damping.

Piecewise Linear Resistance

An infinite resistance for $-e_s<e<e_s$ and a finite resistance for $|e|>e_s$ can be simulated by a linear resistance in series with an ideal switch (Fig. 13), provided a current source $i=\pm e_s/R$ (positive sign when $e>e_s$, negative sign otherwise) is connected in parallel (or a voltage source e_s in series) whenever the switch is closed. The switch is closed whenever

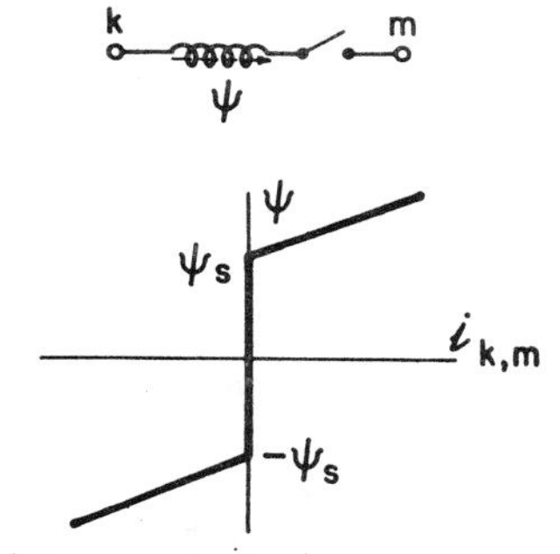

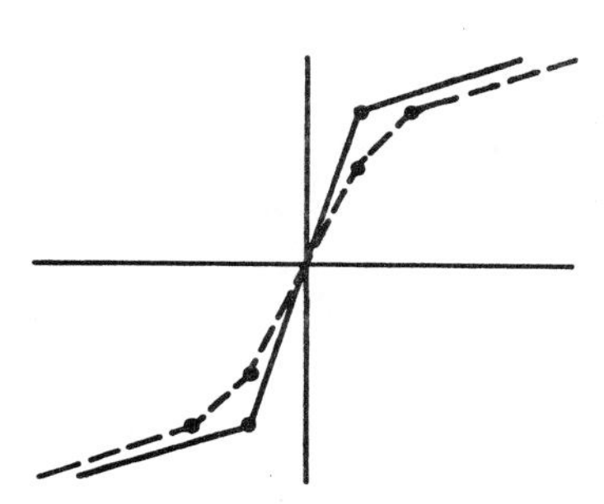

Fig. 8. "Switched inductance"

Fig. 9. Two-slope and three-slope nonlinearity.

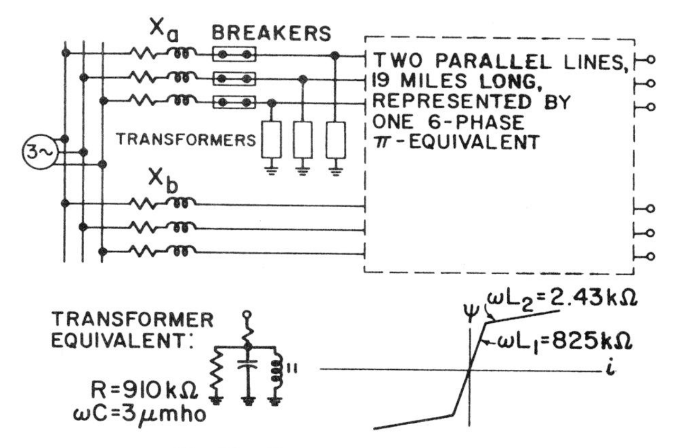

Fig. 10. Ferroresonance problem.

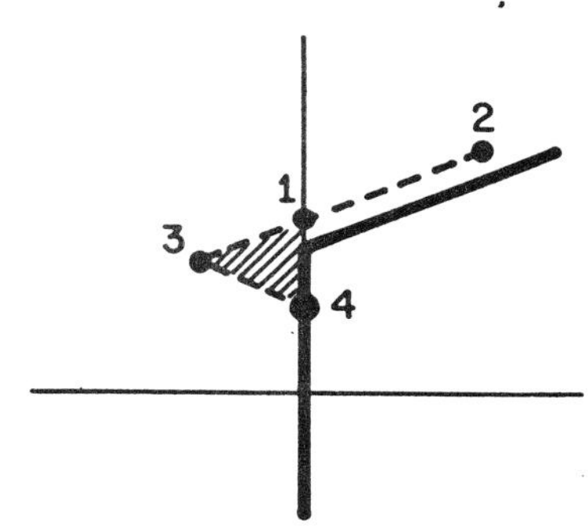

Fig. 12. Positive damping.

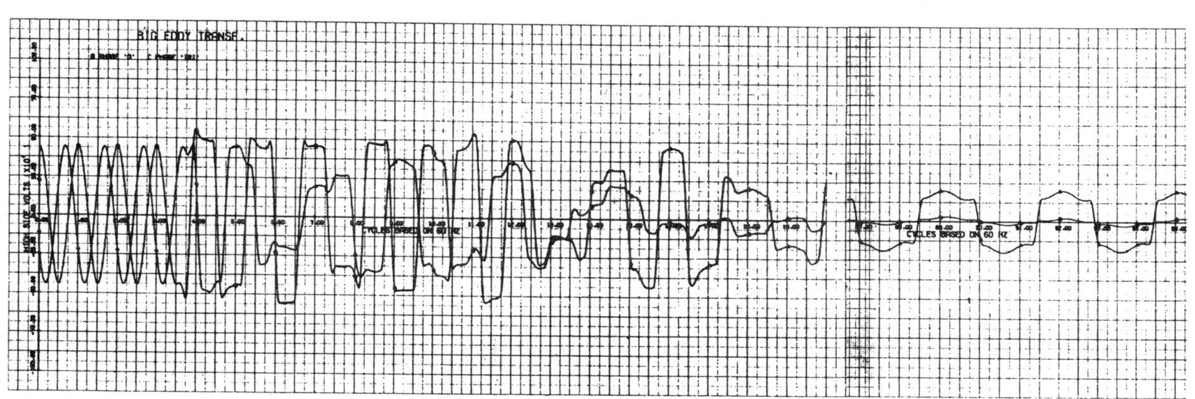

Fig. 11. Voltages at transformer terminals.

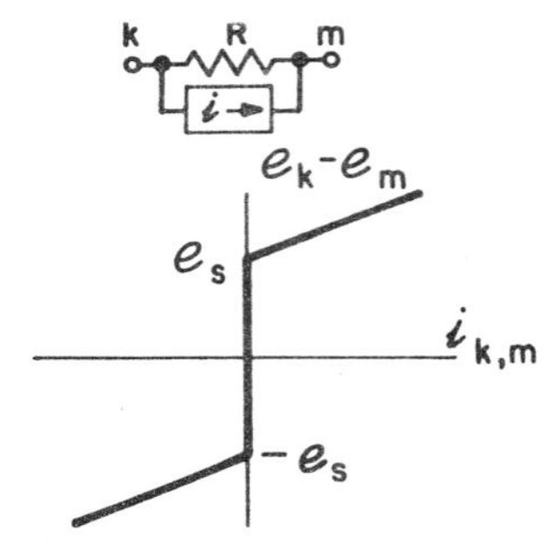

Fig. 13. "Switched resistance".

$|e|>e_s$ and opened as soon as $|e|<e_s$. Again, a two-slope resistance (solid line in Fig. 9) is obtained by connecting a linear resistance in parallel with a "switched resistance".

VII. INITIAL CONDITIONS.

Nonlinearities cause a dilemma if the transient study is to start from nonzero initial conditions, except in the case of dc initial conditions. BPA's transients program has two options. The initial conditions can either be read in or they are automatically computed in an ac steady state subroutine, using the ac source as excitation and assuming that the network is linear. A correct ac steady state solution for nonlinear networks is extremely difficult, even though there are techniques available, such as the method of harmonic balance. At this time, the user must either assume that nonlinear parameters lie within their linear region during steady state or he must use the transients program itself to run the case long enough to produce the nonsinusoidal steady state solution and have the values automatically punched on cards. These cards would then be read in to provide initial conditions for subsequent studies.

CONCLUSIONS

A straight-forward application of solution techniques for nonlinear differential equations increases the solution time for nonlinear networks considerably. The compensation method and reduction techniques make it possible to confine the time-consuming solution process for nonlinear or time-varying elements to the incident branches or nodes only. The linear and larger part of the network can then still be solved with well-known fast solution techniques. Examples illustrate the usefulness of these methods.

ACKNOWLEDGEMENT

The author is indebted to his colleagues in the System Analysis Section of BPA for their help and encouragement. The algorithm for finding the intersection (described after eq. (6) is due to A. Budner. The plotting routines and the subroutine for finding ac steady state initial conditions were developed by J.W. Walker. A.C. Legate and S. Patel made the ferroresonance study of Fig. 10 and 11. The equivalence of the fourth-order Runge-Kutta method with the fourth-order Taylor expansion of the state transition matrix in case of linear systems was pointed out to the author by K.N. Stanton, Systems Control Inc., Palo Alto, Calif.

APPENDIX

Numerical Solution of Ordinary Differential Equations.

For the sake of simplicity, let us assume an autonomous system of first order differential equations with constant coefficients,

$$[\tfrac{dx}{dt}] = [A][x].$$

The trapezoidal rule of integration gives the approximate transition function ([U] = unit matrix),

$$[x(t)] \approx \left([U] - \tfrac{\Delta t}{2}[A]\right)^{-1}\left([U] + \tfrac{\Delta t}{2}[A]\right)[x(t-\Delta t)], \tag{10}$$

which is in the form of eq. (2). An exact transition function can be obtained from the eigenvalues λ_i of [A] and the eigenvector matrix [M], assuming that all eigenvalues are distinct or that eigenvalues of multiplicity r produce exactly nullity r in the characteristic matrix ($\exp\{[\Lambda]\Delta t\}$ = diagonal matrix with elements $\exp\{\lambda_i \Delta t\}$),

$$[x(t)] = \left([M]\exp\{[\Lambda]\Delta t\}[M]^{-1}\right)[x(t-\Delta t)], \tag{11}$$

which is again in the form of eq. (2). It can also be shown that, in the case of linear systems, the well-known fourth-order Runge-Kutta method is identical with a fourth-order polynomial expansion of exp $\{[\Lambda]\Delta t\}$,

$$[x(t)] \approx \left([U] + \Delta t[A] + \tfrac{\Delta t^2}{2}[A]^2 + \tfrac{\Delta t^3}{6}[A]^3 + \tfrac{\Delta t^4}{24}[A]^4\right)\cdot[x(t-\Delta t)]. \tag{12}$$

Eq. (10) - (12) all have the same form and differ only in the numerical value of the transition matrix.

REFERENCES

[1] H.W. Dommel, "Digital computer solution of electromagnetic transients in single- and multiphase networks," *IEEE Trans. Power Apparatus and Systems*, vol. PAS-88, pp. 388-399, April 1969.
[2] W.F. Tinney and J.W. Walker, "Direct solutions of sparse network equations by optimally ordered triangular factorization," *Proc. IEEE*, vol. 55, pp. 1801-1809, November 1967.
[3] H.B. Thoren and K.L. Carlsson, "A digital computer program for calculation of switching and lightning surges on power systems," *IEEE Trans. Power Apparatus and Systems*, vol. PAS-89, pp. 212-218, February 1970.
[4] D. Nitzan, "MTRAC: Computer program for transient analysis of circuits including magnetic cores, *IEEE Trans. Magnetics*, vol. MAG-5, pp. 524-533, September 1969.
[5] T.E. Browne, Jr., "An approach to mathematical analysis of a-c arc extinction in circuit breakers," *AIEE Transactions*, vol. 77, pt. III, pp. 1508-1517, February 1959.
[6] F.W. Heilbronner, "Firing and voltage shape of multistage impulse generators," *IEEE Transactions* Paper No. 71 TP 121-PWR.
[7] W.F. Tinney, "Compensation methods for network solutions by triangular factorization," Proc. Power Industry Computer Applications Conference, Boston, Mass. May 24-26, 1971.
[8] N.G. Hingorani and S.A. Annestrand, "Insulation levels of dc filter reactors and resistors for HVDC power transmission," *IEEE Trans. Power Apparatus and Systems*, vol. PAS-89, pp. 610-618, April 1970.

Discussion

A.H. Schmidt, Jr. (United States Bureau of Reclamation, Denver, Colo. 80225): This paper describing methods of handling nonlinear and time-varying elements in a digital transient simulation is very interesting and timely. A copy of the author's transients program has been implemented here at the Bureau of Reclamation. We have found the program to be very powerful and are using it "as is" to simulate much more than just electromagnetic transients. For example, we have applied it to solving heat flow transients in buried cables, for solving mechanical vibrations in pump-turbine shafts, and are beginning to work on ways to study transients in canals.

This latter application will help us to design better canal control systems, but we must modify the program to handle special nonlinear elements. In this simulation, voltage represents the water depth or head and for a canal with sloping sides, the resistance and capacitance elements are functions of the water depth. It is hoped that the compensation method can be applied here also.

Manuscript received June 10, 1971.

The author is providing an important contribution to the solution of the transients of any nature by describing the techniques used in his transients program.

H.W. Dommel: I am indebted to Mr. Schmidt for drawing attention to the solution of transient phenomena of a nature other than electromagnetic with the transients program "as is". His ingenuity in modeling hydraulic, mechanical and heat transfer problems by using analogous electric circuits is as important as the development of the transients program itself. Making good use of an existing program requires as much creativity and engineering judgment as writing the program, if not more.

Manuscript received July 6, 1971.

The method of characteristics (Bergeron's method), which is used in the transients program for distributed parameter lines, was probably first applied to hydraulic problems[9]. Such "cross-fertilization," which was common at one time but declined somewhat because of specialization, may become important again through the use of general-purpose computer programs as shown by Mr. Schmidt's discussion.

I am not familiar enough with the nature of nonlinearities in hydraulic problems. It is hoped, however, that compensation methods or imbedding into Thevenin equivalents will help to solve nonlinearities in hydraulics also.

REFERENCE

[9] R. Löwy, Druckschwankungen in Druckrohrleitungen (Pressure variations in pressure pipe lines). Vienna: Springer, 1928, p. 84.

DIGITAL SIMULATION OF DC LINKS AND AC MACHINES

D.A. Woodford, Member, IEEE
Manitoba Hydro,
Winnipeg, Canada

A.M. Gole, Student Member, IEEE
University of Manitoba,
Winnipeg, Canada

R.W. Menzies, Senior Member, IEEE,
University of Manitoba
Winnipeg, Canada

ABSTRACT

This paper discusses the digital simulation of HVDC transmission and ac machines by an electromagnetic transients computer program. Ease in interfacing user written control system models with the network solution is presented along with discussion of means to simplify dc valve group representation in simulation. A working digital simulator has been developed enabling flexible in-house studies to be undertaken. Test results of its performance are presented.

INTRODUCTION

Since more and more transmission planning engineers have access to small, dedicated but powerful computers, electromagnetic transient simulation of power system networks by digital programs is becoming standard practice. Digital high voltage direct current (HVDC) transmission simulation is one area slow in developing, but the time is now right for its more serious consideration.

The Bonneville Power Administration's electromagnetic transients program (BPA's EMTP) is potentially a powerful tool for HVDC simulation. Because the present structure of this program intimidates the average user seeking the flexibility needed for advanced HVDC simulation, a smaller and more specialized program was developed at Manitoba Hydro. Known as MH EMTDC and based on the algorithms described by H. Dommel [1], it provides ease in interfacing user developed fortran models with an electromagnetic transients solution. MH EMTDC has the following features:

1 - The main program solves the electromagnetic transients for the network under study and calls two user written subroutines. The first subroutine interfaces the network solution of voltages, currents and branch elements for the user to process and control voltage and current sources and switch branch elements. The second subroutine allows the user to access and process any variable, solved voltage, or current for output and plotting.

2 - Disconnected subnetworks described in reference [1] are used to minimize matrix size and maximize speed of the solution.

3 - Valve group models, synchronous machines with exciters, stabilizers and governors and control circuits are available and can be assembled as subroutines by the user.

82 SM 480-2 A paper recommended and approved by the IEEE Transmission and Distribution Committee of the IEEE Power Engineering Society for presentation at the IEEE PES 1982 Summer Meeting, San Francisco, California, July 18-23, 1982. Manuscript submitted February 2, 1982; made available for printing June 4, 1982.

MH EMTDC BASIC STRUCTURE

Valve Group Subnetworks

Wherever possible, the user can take advantage of disconnected subnetworks to represent converter valve groups if the duration between time steps is small enough to assume that ac commutating bus voltages and dc current at the valve group are predictable and continuous. Then voltage sources for a valve group subnetwork are directly derived from the commutating bus ac voltages and a current source is derived from the current through the dc smoothing reactor. Solution of the valve group subnetwork produces currents for injection into the commutating busses of the ac system and a voltage source is defined on the valve side of the smoothing reactor for the dc transmission network. In Fig. 1, a simple diagram of the subnetworking process for valve groups is shown.

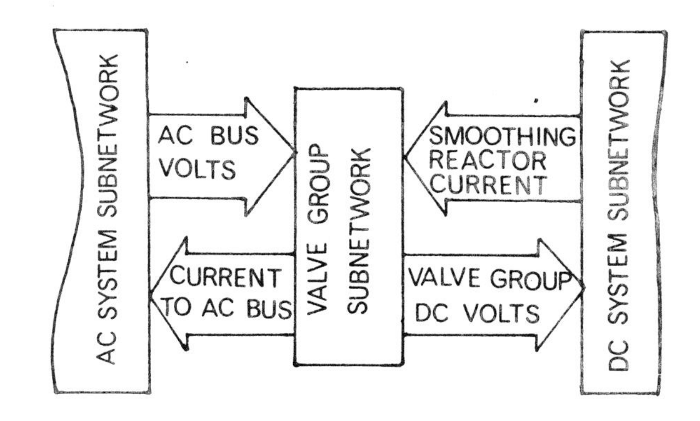

Fig. 1. Disconnected subnetwork representation for valve groups

This method of disconnection and subnetworking is useful for many dc link simulation situations but is not practical when considering ground faults within the valve group subnetwork due to its internal ground being a different reference with respect to system ground. Under these conditions, the valve group should be patched directly into the ac and dc systems.

Valve Representation

The HVDC valve is composed of one or more series strings of thyristors. Each thyristor is equipped with a resistor - capacitor damping or "snubber" circuit. One or more di/dt limiting inductors are included in series with the thyristors and their snubber circuits. It is assumed that for most simulation purposes, one equivalent thyristor, snubber circuit and di/dt limiting inductor will suffice for a valve model. The di/dt limiting inductor can usually be neglected when attempting transient time domain simulation up to about 1.5 to 2.0 kHz frequency response.

0018-9510/83/0600-1616$01.00 © 1983 IEEE

By utilizing the feature representing network branches of inductors and capacitors as resistors with an associated equivalent current source [1], a valve in a converter bridge can be reduced as depicted in Fig. 2.

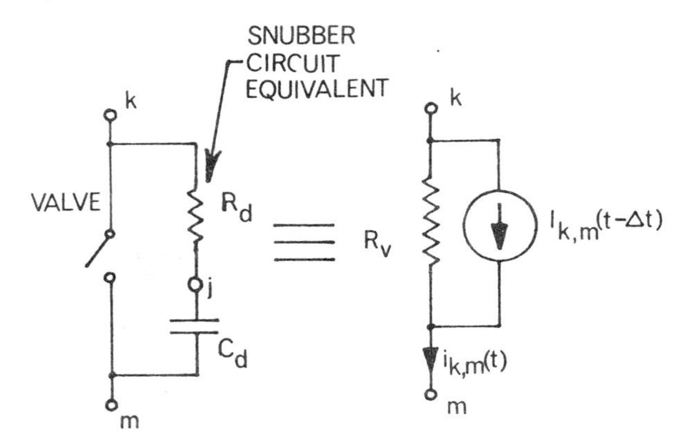

Fig. 2 Equivalencing and reduction of a converter valve

With the valve blocked (not conducting), the equivalent resistor R_V is just that derived from the snubber circuit. With the di/dt limiting inductor ignored, then from reference [1] this becomes:

$$R_V = Rd + \Delta t/2Cd \quad (1)$$

where Δt = time step duration
Rd = Snubber resistance
Cd = Snubber capacitance

With the valve de-blocked and conducting in the forward direction, the equivalent resistor R_V is changed to a low value such as one ohm:

$$R_V = 1 \quad (2)$$

The equivalent current source $I_{k,m}(t-\Delta t)$ shown in Fig. 2 between nodes k and m is determined by first defining the ratio Y as:

$$Y = \frac{\Delta t/2C_d}{R_d + \Delta t/2C_d} \quad (3)$$

From equations 10a and 10b of reference [1];

$$i_{k,m}(t) = \frac{[e_k(t) - e_m(t)] + I_{k,m}(t - \Delta t)}{R_d + \Delta t/2C_d} \quad (4)$$

then,

$$I_{k,m}(t - \Delta t) =$$

$$-Y[i_{k,m}(t - \Delta t) + 2C_d(e_j(t - \Delta t) - e_m(t - \Delta t))/\Delta t] \quad (5)$$

where,

$$e_j(t - \Delta t) = e_k(t - \Delta t) - R_d \cdot i_{k,m}(t - \Delta t) \quad (6)$$

In general, valves can be reduced to an equivalent resistor and current source as desbribed above. An extension of this treatment can be used to imbed the di/dt limiting inductor into the equivalent resistor and current source if felt necessary.

Valve Group Representation

A valve group subnetwork can be reduced to the four node equivalent circuit shown in Fig. 3. Each valve with its snubber circuit can be modelled with its equivalent resistor and current source as described above.

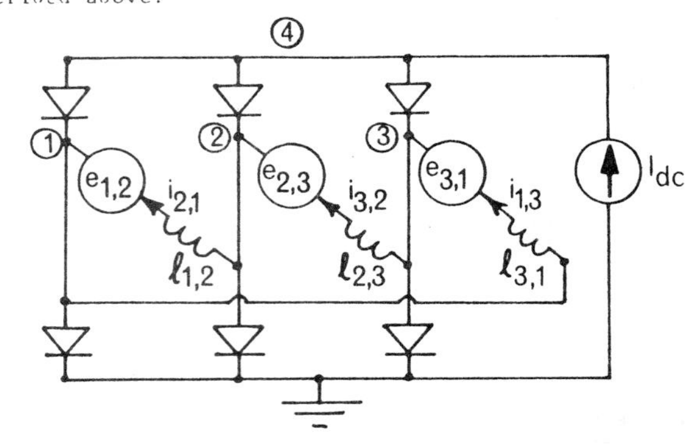

Fig. 3. Six pulse valve group subnetwork equivalent circuit

The converter transformer is represented as a delta connection of commutating reactances with voltage sources derived from the ac system commutating bus voltages. The effect of non-linear magnetizing current can be included too.

Consider just the one phase of the converter transformer represented in Fig. 3 as an inductance with its series voltage source. In Fig. 4, it is shown with two additional current sources which represent the effect of the non-linear magnetizing current. Just the phase between nodes 1 and 2 in Fig. 3 is depicted. The turns ratio n for the valve group converter transformer is;

$$n = \frac{\text{rated secondary (valve side) line volts}}{\text{rated primary (ac side) line volts}} \quad (7)$$

The subnetwork commutating voltage source $e_{1,2}$ for a star-star converter transformer at time t is;

$$e_{1,2}(t) = n.E_{A-B}(t) \quad (8)$$

where $E_{A-B}(t)$ is the predicted line-to-line ac side commutating volts for time t between phases A and B

For a star-delta converter transformer;

$$e_{1,2}(t) = \sqrt{3} . n . E_A(t) \quad (9)$$

where $E_A(t)$ is the predicted positive sequence line to neutral ac side commutating volts for time t for phase A

The commutating inductor $\ell_{1,2}$ referred to the secondary side of the converter transformer can be derived from the per unit commutating reactance, voltage and volt-amp rating of the transformer from the following expression:

$$\ell_{1,2} = \frac{3X_c \ V_s^2}{2\pi F R} \quad \text{henries} \quad (10)$$

where R = Transformer base MVA rating.
X_c = Per unit commutating reactance on the transformer base.
V_s = Rated line-to-line secondary volts in kV rms.
F = Rated frequency in hertz.

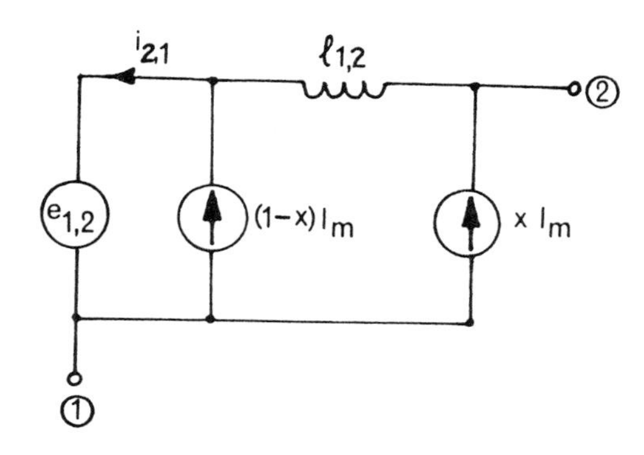

Fig. 4. Phase representation of a converter transformer

The magnetizing current for the phase can be computed as a non-linear time varying function of voltage and added as a current source to the subnetwork. The magnetizing current injection can be located anywhere between the source side and the valve side of $\ell_{1,2}$ by splitting the current source into two as shown in Fig. 4.

Let X be the desired per unit location of the injection of the magnetizing current $I_m(t)$. With X=0, $I_m(t)$ is injected on the ac side of the commutating inductor. With X=1, $I_m(t)$ is injected on the valve side. For 0<X<1, $I_m(t)$ is injected proportionately in between.

Accommodating the magnetizing component in this way provides flexibility in modelling the converter transformer as effected by its winding construction. As discussed in reference [2], the magnetizing component can then be located along $\ell_{1,2}$ to reflect the winding wound closest to the transformer core. However, with a non-linear component in each of the three phases of the one valve group subnetwork, a satisfactory solution is still possible with normal valve group and converter transformer parameters and operation despite the violation of the 'one non-linear component' disconnection rule in BPA's EMTP [1].

The secondary winding phase current $i_{2,1}(t)$ shown in Fig. 4 is transferred to the primary side of the converter transformer by the following expressions:

For a star-star transformer;

$$I_A(t) = n[i_{2,1}(t) - i_{1,3}(t)] \tag{11}$$

where $I_A(t)$ = Positive sequence current injected into the commutating bus, phase A, from the valve group sub network.

$i_{2,1}(t)$ = Secondary winding current flowing into equivalent voltage source $e_{1,2}(t)$.

$i_{1,3}(t)$ = Secondary winding current flowing into equivalent voltage source $e_{3,1}(t)$.

Similarly, for a star-delta transformer;

$$I_A(t) = \sqrt{3}.\ n.\ i_{2,1}(t) \tag{12}$$

Since $I_A(t)$ is positive sequence phase current injected each time step into phase A of the ac system commutating bus, the zero sequence component can be computed separately from the zero sequence commutating voltage and added by superposition to $I_A(t)$. Phase currents from all valve groups, synchronous machines and static compensators at the one ac bus are summed to form one element of the columnar current matrix for the ac side system solution. Then for each subnetwork, node or bus voltages are computed by linear algebraic equations which describe the state of the system at time t as per equations 12 and 13 in reference [1].

The interface between the dc system at the smoothing reactor and the valve group subnetwork is straight forward as shown in Fig. 1.

AC MACHINE MODEL

A complete ac machine model was developed by the authors from the University of Manitoba for the MH EMTDC. The attractive features of this machine model are the following:

1 - It is fairly detailed, with a complete two axis (six coil) representation, with d axis saturation included, and with additional modelling for accurate rotor side studies.

2 - **The solution can use a different timestep and integration procedure from the MH EMTDC program. In this way, it differs from other programs [3], which try to include the machine equations in a submatrix of the main system matrix.**

State Variables used for Optimal Solution

The equations are modelled in state variable form, with the machine d and q axis currents as the state variables. Referring to the d and q equivalent circuits of Fig. 5 [4], the following equations for the currents result:

$$\begin{bmatrix} Ud - \omega\Psi_q - r_a i_d \\ Uf - r_f \cdot i_f \\ - r_{kd} \cdot i_{kd} \end{bmatrix} = \begin{bmatrix} L_{md} + La & L_{md} & L_{md} \\ L_{md} & L_{md} + L_f + L_{kf} & L_{md} + L_{kf} \\ L_{md} & L_{md} + L_{kf} & L_{md} + L_{kf} + L_{kd} \end{bmatrix} \frac{d}{dt} \begin{bmatrix} i_d \\ i_f \\ i_{kd} \end{bmatrix}$$

$$= \mathcal{L}_d \frac{d}{dt} \begin{bmatrix} i_d \\ i_f \\ i_{kd} \end{bmatrix} \tag{13}$$

and

$$\begin{bmatrix} U_q + \omega\Psi_d - r_a \cdot i_q \\ -r_{kq} \cdot i_{kq} \end{bmatrix} = \begin{bmatrix} L_{mq} + La & L_{mq} \\ L_{mq} & L_{mq} + L_{kq} \end{bmatrix} \frac{d}{dt} \begin{bmatrix} i_q \\ i_{kq} \end{bmatrix}$$

$$= \mathcal{L}_q \frac{d}{dt} \begin{bmatrix} i_q \\ i_{kq} \end{bmatrix} \tag{14}$$

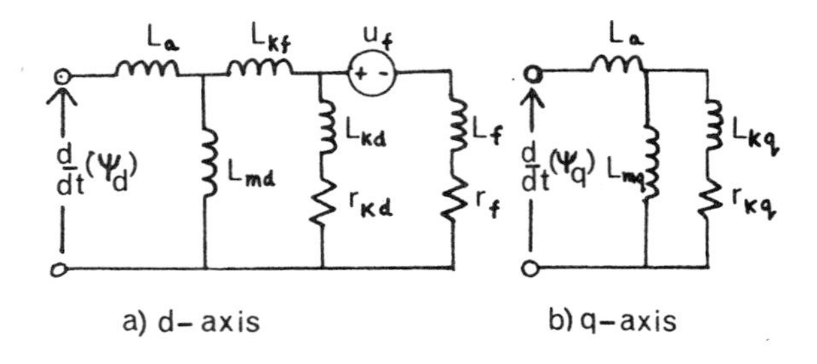

Fig. 5. AC Machine equivalent circuits [4]

Equations 13 and 14 show three coils on the d-axis and two on the q-axis, though more coils can easily be incorporated if desired. The additional inductance L_{kf} accounts for the mutual flux which links only the amortisseur and field windings, and not the stator windings. Inclusion of such flux [4, 5] has been shown necessary for an accurate representation of transient currents in the rotor circuits. Saturation is included by making the appropriate inductances (i.e. L_{md} and L_f) functions of the magnetizing current. The program determines these from the open circuit terminal characteristics of the machine.

Inverting equations 13 and 14 gives:

$$\frac{d}{dt}\begin{bmatrix} i_d \\ i_f \\ i_{kd} \end{bmatrix} = \mathcal{L}_d^{-1} \begin{bmatrix} -w.\Psi_q - r_a \cdot i_d \\ -r_f \cdot i_f \\ -r_{kd} \cdot i_{kd} \end{bmatrix} + \mathcal{L}_d^{-1} \begin{bmatrix} U_d \\ U_f \\ 0 \end{bmatrix} \qquad (15)$$

$$\begin{bmatrix} i_q \\ i_{kq} \end{bmatrix} = \mathcal{L}_q^{-1} \begin{bmatrix} \omega\Psi_d - r_a \cdot i_q \\ -r_{kq} \cdot i_{kq} \end{bmatrix} + \mathcal{L}_q^{-1} \begin{bmatrix} U_q \\ 0 \end{bmatrix} \qquad (16)$$

which are in the standard state variable form

$$\dot{\underline{X}} = A\underline{X} + B\underline{U} \qquad (17)$$

with state vector $\underline{X}$ consisting of the currents, the input vector $\underline{U}$ of applied voltages and matrices A and B.

Elements of $\mathcal{L}_d^{-1}$ are directly calculated by formula and not by inversion of matrix $\mathcal{L}d$. Additional equations for the rotor dynamics also exist, but they are in diagonal form, and are not discussed here.

In the above form, the equations are particularly easy to integrate. A block diagram of the overall scheme is shown in Fig. 6. The figure shows that the machine model makes use of phase voltages calculated by MH EMTDC to update the injected currents into MH EMTDC. Thus, computationally, it behaves like a current source. Thus, the algorithm sometimes requires a smaller time step when the machine is near open circuit conditions as the computation is liable to become unstable. Alternatively, a small capacitance or large resistance may be placed from the machine terminals to ground to prevent the machine from being totally open circuited. The physical meaning of parasitic capacitance or leakage resistance may then be applied to these newly introduced elements.

Fig. 6 also shows that merely multiplying the phase currents by an integer N simulates, from the system point of view, N identical machines operating coherently into the ac system.

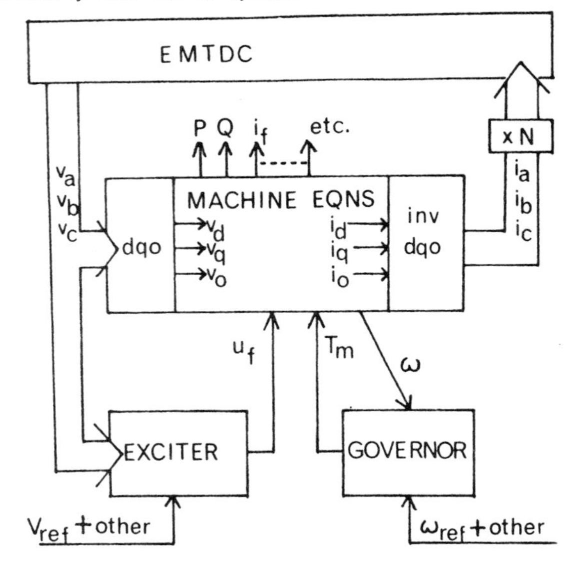

Fig. 6. Modelling scheme for the ac machine.

Other control blocks to simulate excitation and governor systems are likewise modelled, and can be selected by the main program as subroutines; the outputs of which furnish values of field voltage and mechanical torque to be used in the machine model.

Capabilities of the Model

This machine model has been found to work very satisfactorily in a number of different simulations, and can do everything that can be done on sophisticated analog simulators [6]. The model has been extensively tested for comparison with laboratory experiments on machines of small ratings, and though comparison with larger machines has not been carried out to any extent, good results are still expected to apply. The model may successfully be used in the following situations:

1 - ac side behaviour of machines including harmonic response and dynamic effects on dc links.

2 - Studies of the field circuit, in order to determine exciter stresses, self excitation, etc.

3 - Interaction between machine electrical and mechanical quantities such as subsynchronous oscillations and shaft torque impact studies.

The structure of the program, as it operates fairly independently of MH EMTDC (in terms of time step and integration procedure), allows the user to go right into the machine model and make his own changes.

As an example demonstrating terminal voltage and current, and field current responses to a disturbance, consider in Fig. 7 a dc rectifier load rejection of 190 MW on 3, 100 MW synchronous water wheel generators similar to a situation which might exist on the Manitoba Hydro system. Note the presence of harmonics in the plotted quantities which could not be detected in a phasor model often used in this type of study to observe load rejection overvoltages and self excitation.

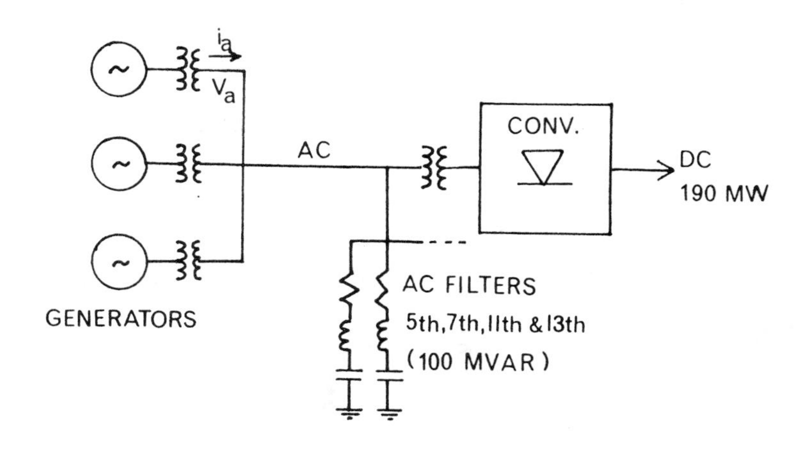

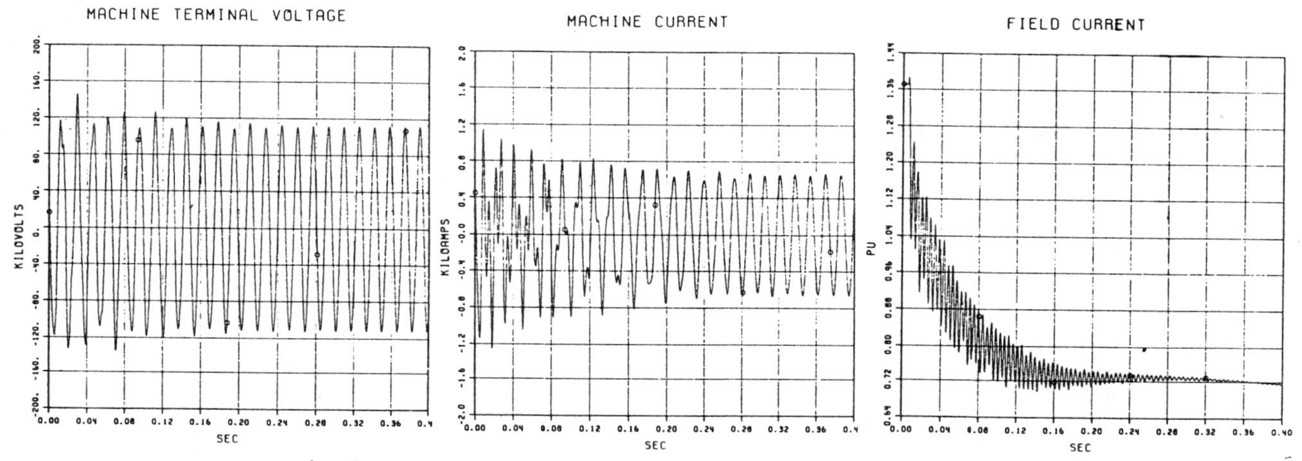

Fig. 7. dc load rejection; its effect on synchronous machines.

DC CONTROLLERS

Whether the user patches valve groups in as sub-networks or as part of a large ac and dc system network, valve firing can be accomplished by a user accessible subroutine. Although the user is free to write his own firing method, one scheme with equi-distance firing is achieved by phase locking timing pulses to the commutating bus voltages. Respective valves are turned on when a timing counter exceeds a level defined by the desired ignition angle (alpha) ordered. The alpha order is derived from user developed dc controls assembled in subroutines. Fortran functions of analog building blocks are available to help the user develop a controller. As an example, the block diagram for a basic valve group controller is shown in Fig. 8.

A simple integral plus proportional pole controller is shown in Fig. 9. Note the non-linear relationship of the alpha order output as a function of current input is "linearised" by the integral in the forward loop. Constants G and T are selected by the user for a stable response of the dc link being studied.

If necessary, the current order input to the pole controller of Fig. 9 can be derived by dividing the power order by the measured dc volts. Voltage dependant current limits can easily be added. The modelling flexibility for controls with MH EMTDC can be used to examine for example, the co-ordinated control of real and reactive power, controls for multi-terminal applications and subsynchronous dc control. It also provides a means to verify linearized design of controls.

TESTING

A good test for any dc simulator is to observe its performance in steady state and its recovery from a commutation failure caused by an ac fault at the inverter.

A single pole model in twelve pulse operation was set up on MH EMTDC as shown in Fig. 10. One phase only of the ac systems is shown although all three phases were modelled. The steady state voltage magnitude of the ac system source voltages and the initial relative phase angle of the synchronous condensor rotor were determined by separate ac and dc load-flow calculations. The system is started up in a sequence written by the user. For example:

1 - Energize ac system source voltages with synchronous condensor disconnected but with rotor spinning and locked in phase. dc valve groups blocked and bypassed.

2 - Await volts on synchronous condensor to build up, then synchronize to the system.

3 - Deblock inverter first as a rectifier, then with dc current flowing, deblock the rectifier under constant current control.

4 - When steady state is reached, the synchronous condensor rotor is freed, and if desired, dc controls can be brought on. All state variables, constants and any necessary data are stored in a file for easy restart.

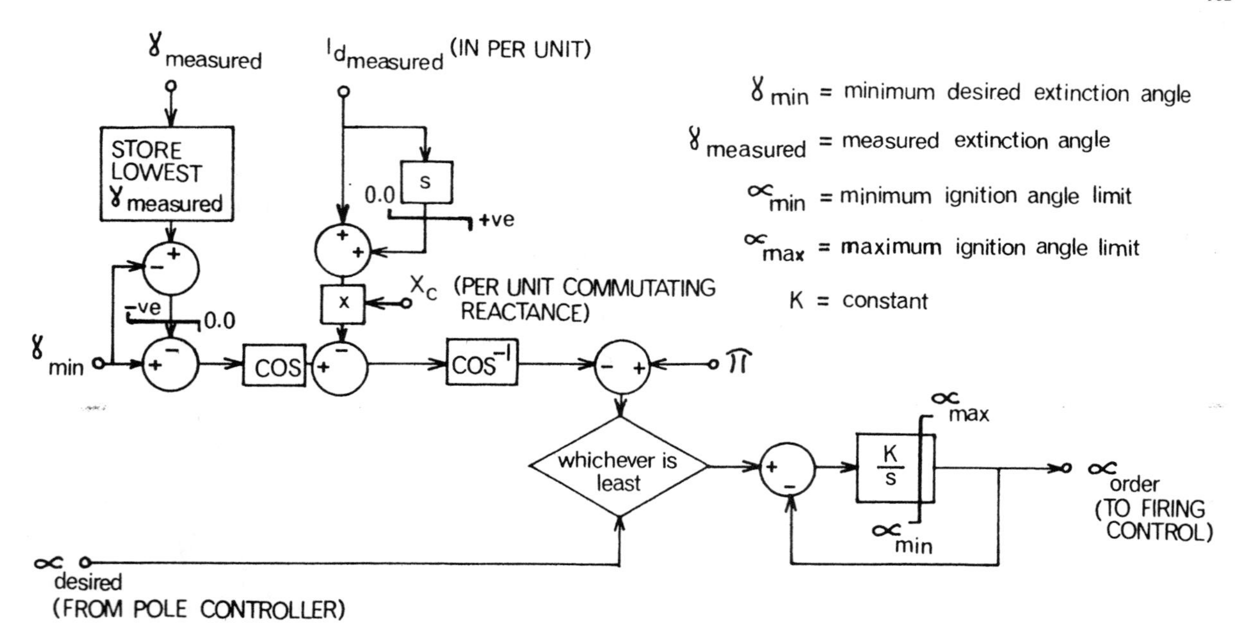

Fig. 8. Basic valve group controller for rectifiers and inverters.

Providing controls have been previously optimized, and the short circuit ratios are not too low for stable operation, steady state can usually be reached in two seconds of system time.

Test Cases

Examples to illustrate the performance of the MH EMTDC are presented. Because of interest in operation at low short circuit ratios, a value of 2.4 based on the synchronous condensor subtransient reactance X"d was selected for the inverter. For the purpose of this paper, short circuit ratio (SCR) is defined as;

$$SCR = \frac{V^2}{P_d.Z} \qquad (18)$$

where V = ac line rms commutating bus volts in kV
P_d = dc power at the converter in MW
Z = ac system short circuit impedance at the commutating bus

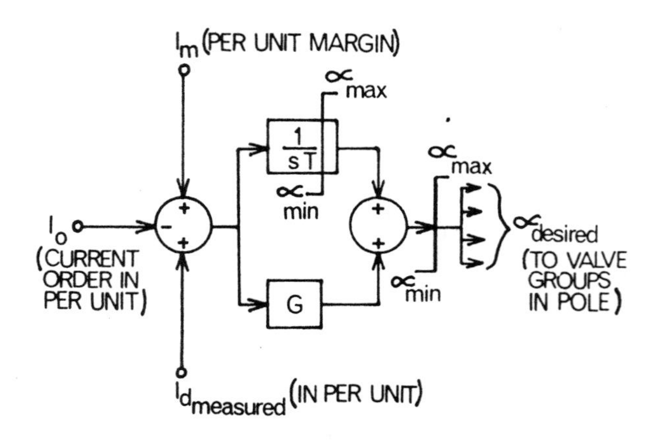

Fig. 9. Basic Pole Controller.

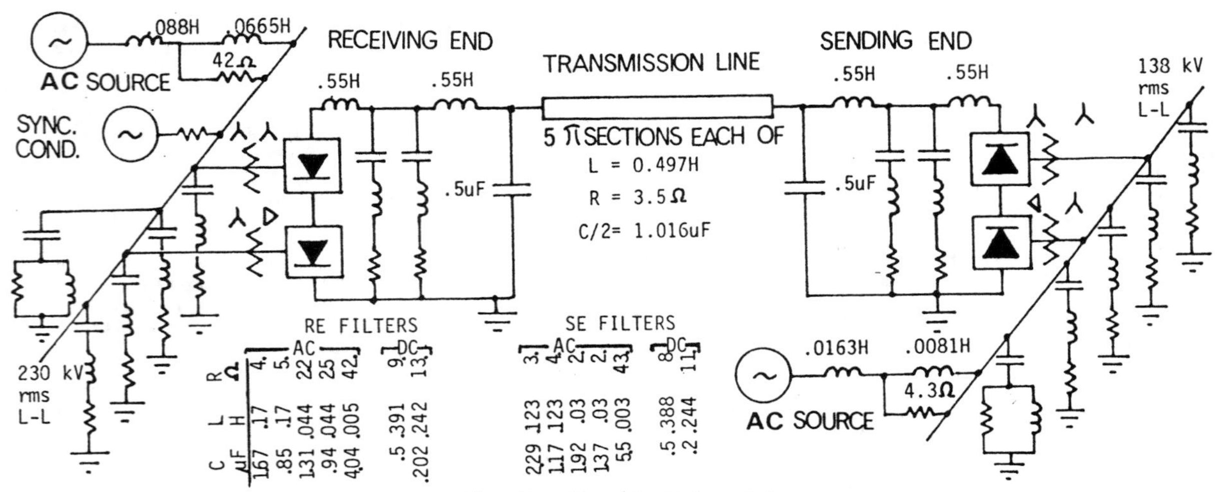

Fig. 10. Test dc link model.

In the first test case, a four cycle ac single-line-to-ground fault was applied at the inverter of the system shown in Fig. 10 with dc link operation in constant current control. Fig. 11 shows the dc current and voltage at the inverter along with the ac phase voltages at the inverter. The impact of the synchronous condensor can be observed in the modulation of the ac volts (and thus the dc volts) as a consequence of rotor swings. The peaking of volts at the inverter around the time of 0.6 seconds cause current control to transiently change from the rectifier to the inverter and back again. Also observed is the attempted commutation during the four cycles of ac fault. At least two commutation failures are evidenced by the repeated surges in dc current during the ac fault period.

A second test to demonstrate the use of EMTDC is to examine the question of whether X"d or transient reactance X'd is the most significant parameter to base the calculation of short circuit ratio as it effects inverter performance through and after an ac fault.

The synchronous condensor shown in Fig. 10 is replaced by simple positive sequence reactances to a fixed source voltage. With this reactance based firstly on X"d, and secondly on X'd, the four cycle fault above was repeated for each case and the performance of the dc link through and after the fault was observed. The dc link current at the inverter for these two cases and for the case where the synchronous condensor is modelled as in Fig. 10, are assembled together in Fig. 12. From Fig. 12, it can be seen that with X"d represented in the system replacing the synchronous condensor, the dc link performance through and immediately after the fault is closer to the case with synchronous condensor fully modelled than is where X'd is represented. (Allowance must be made for the effect of rotor swings and exciter when the synchronous condensor is modelled).

This simple study using MH EMTDC indicates that X"d rather than X'd is the better representation for use in ac system short circuit ratio calculations so far as inverter performance through ac faults is concerned.

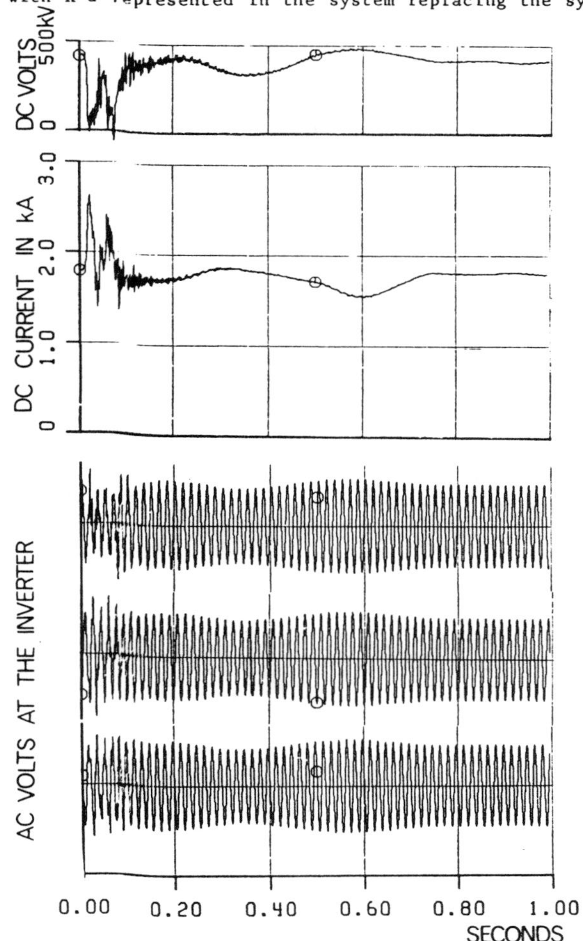

Fig. 11. 4 Cycle single-line-ground fault.

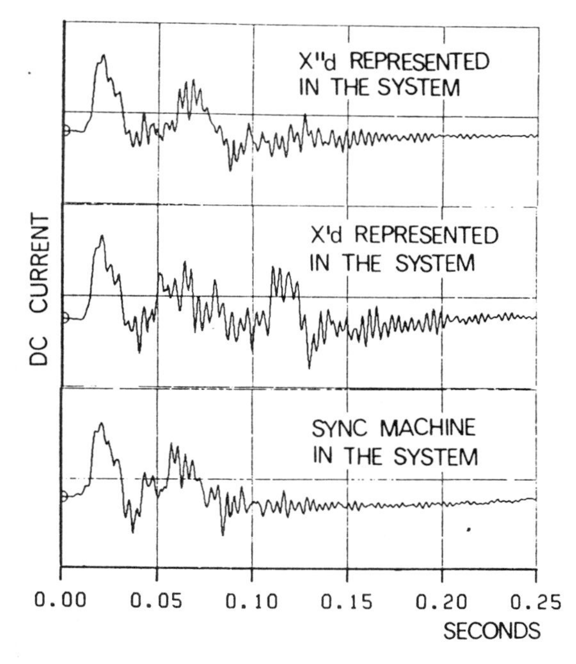

Fig. 12. Simulated Recovery of dc current after an ac inverter fault for alternative representations of X"d & Xd' and full machine model.

Comparison with an Actual dc Link

A fault similar to one fault recorded for Bipole 2 of the Nelson River DC Transmission [7] was run for comparison with this model on MH EMTDC. The actual and simulated cases are shown on similar scale in Fig. 13. It will be observed that there is more harmonic content in the simulated case compared with the actual. Since no attempt was made to provide any significant harmonic damping in the model, this case demonstrates the need for its consideration in digital simulation studies.

Taking the harmonic problem into account, it can be seen that the simulated model performance is reasonable and comparable when referenced to the performance of an actual dc link.

Computer Running Times

With the system modelled in Fig. 10 with synchronous condensor, represented fully, a one second simulation time with a twenty microsecond time step takes about twenty minutes of C.P.U. time on a PRIME 750 computer.

CONCLUSIONS

The digital dc simulator with ac machine modelling capability described here is a working computer program being used in system studies. Experience has

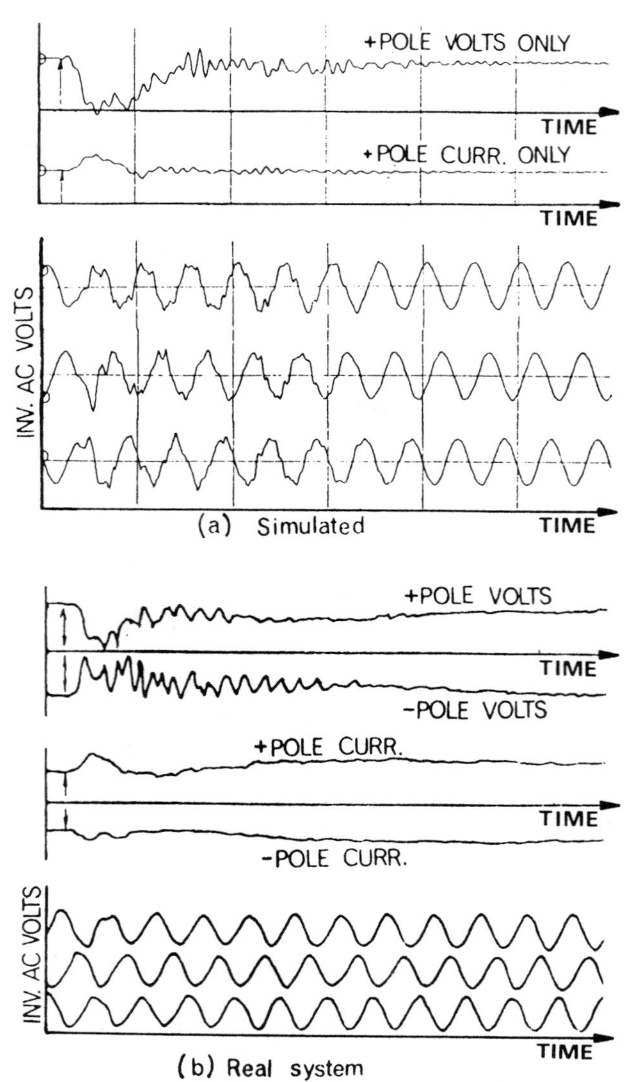

Fig. 13. Comparison of Simulator Performance with a real system for an ac fault at the inverter.

shown that a good appreciation of HVDC transmission, machines and electromagnetic transients phenomena is needed to utilize the simulator effectively. Until this appreciation is gained, the program is an effective educational tool in modern power system transient operation and analysis.

With increased use of HVDC transmission in power systems, tools such as this will enable the planner to evaluate his options in HVDC in-house without being fully dependent on equipment suppliers or at the expense of rented hardwired simulators.

ACKNOWLEDGEMENTS

This digital HVDC power system simulator program was developed with the resources of Manitoba Hydro. A grant from the Manitoba Hydro Research Committee funded contributions by the University of Manitoba for program development.

REFERENCES

[1] H.W. Dommel, "Digital Computer Solution of Electromagnetic Transients in Single - and Multiphase Networks," IEEE Transactions on Power Apparatus and Systems, vol PAS-88, No. 4, pp 388-399, April 1969.

[2] H.W. Dommel, "Transformer Models in the Simulation of Electromagnetic Transients", Proc. 5th Power Systems Computing Conference, Cambridge (England), September 1-5, 1975, Paper 3.1/4.

[3] V. Brandwajn, H.W. Dommel, "A New Method for Interfacing Generator Models with an Electromagnetic Transients Program", 1977 Power Industry Computer Applications Conference Proceedings, pp 260-265.

[4] B. Adkins and R.G. Harley, General Theory of AC Machines. Chapman and Hall, London 1975.

[5] I.M. Canay, "Causes of Discrepencies on Calculation of Rotor Quantities and Exact Equivalant Diagrams of the Synchronous Machine", IEEE Transactions on Power Apparatus and Systems, vol. PAS-88, No. 7, pp 1114-1120, July 1969.

[6] G. Jasmin, J.P. Bowles, A. Leroux, D. Mukhedkar, "Electronic Simulation of a Hydro-Generator with Static Excitation", IEEE Transactions on Power Apparatus and Systems, vol. PAS-100, No. 9, pp 4207-4215, September 1981.

[7] C.V. Thio, "AC-DC Integration Aspects and Operational Behaviour of the Nelson River HVDC System", Paper presented to CIGRE Study Committee 14 (HVDC Links), Rio De Janeiro, Brazil, August 25 and 26, 1981.

IMPLEMENTATION OF THE CDA PROCEDURE IN THE EMTP

Jiming Lin, Member, IEEE
Electric Power Research Institute
Beijing, China

José R. Martí, Member, IEEE
The University of British Columbia
Vancouver, B.C., Canada

Abstract - The CDA procedure eliminates the numerical oscillations that can occur in transients simulations that use the trapezoidal rule of integration. The CDA technique has been successfully implemented in the production code of the DCG/EPRI EMTP. This paper describes the details of this implementation for linear elements, nonlinear reactors, frequency dependent transmission lines, and synchronous machines. Simulation results involving these components are presented, showing the effectiveness of the procedure.

Keywords - Critical damping adjustment (CDA); numerical oscillations; EMTP simulations.

1 INTRODUCTION

Reference [1] presented the concept of "Critical Damping Adjustment" (CDA) to prevent numerical oscillations of the trapezoidal rule of integration at discontinuities. The CDA procedure allows the solution of the system of differential equations to proceed smoothly through discontinuities without introducing numerical damping in the solution. The advantages of this scheme over traditional techniques which introduce numerical damping in the solution were discussed in [1].

The present paper discusses the implementation of the CDA scheme for some of the major equipment models in the EMTP. These include nonlinear elements, frequency dependent transmission lines, and synchronous machines. Simulation results involving these elements are also presented.

Due to space limitations, details of the implementation of the CDA procedure for other elements in the EMTP, such as multiphase π-circuits, hysteresis loops, and the Universal Machine model, have been omitted. The required modifications for these elements are similar to the ones explained in the paper.

2 THE CRITICAL DAMPING ADJUSTMENT PROCEDURE (CDA)

To solve for the transient response of an electric circuit with the EMTP, the differential equations of each circuit component are first converted into difference equations. These difference equations involve simple algebraic relationships between voltages and currents at the time instant at which the system is solved. In a nodal formulation the network matrix equations at a given time instant t can be written as

89 SM 667-7 PWRS A paper recommended and approved by the IEEE Power System Engineering Committee of the IEEE Power Engineering Society for presentation at the IEEE/PES 1989 Summer Meeting, Long Beach, California, July 9 - 14, 1989. Manuscript submitted January 31, 1989; made available for printing June 20, 1989.

$$[G][v(t)] = [i(t)] + [h(t)], \tag{1}$$

where [G] = constant matrix of equivalent node conductances; [v(t)] = vector of node voltages; [i(t)] = vector of external source currents; and [h(t)] = vector of history terms. The history terms h(t) are known from past values of branch voltages and currents. The value of the elements of [G] and [h(t)] depends on the numerical integration (or differentiation) rule used to discretize the differential equations.

Even though the trapezoidal rule of integration has excellent characteristics in terms of accuracy and stability, it does not respond well to discontinuities. These include switching operations, discontinuities in the values of the applied sources (including at t = 0), and transitions from one segment to another in piecewise linear inductances. In these cases, trapezoidal can oscillate to the overshoot produced by the discontinuity and maintain these oscillations with little or no damping. The CDA scheme solves these problems by completely dampening out the overshoot in two $\Delta t/2$ time steps. No traces of the overshoot are seen in the results after the second $\Delta t/2$ time step, and the simulation can proceed again with trapezoidal.

The application of the CDA procedure does not interfere with the normal solution scheme in the EMTP and does not require the resetting of initial conditions or other complications.

In the EMTP solution with the CDA scheme, the system is solved normally with the trapezoidal rule at time steps t = 0, Δt, $2\Delta t$, ..., etc. until a discontinuity is scheduled to occur at time t_1^+. The network is solved normally at t_1, assuming the discontinuity has not yet occurred (solution for t_1^-). Next, the discontinuity is applied and the CDA procedure is brought in to dampen out the overshoot.

With the network modified to take into account the new condition (for example, a switch change from open to closed), the next solution point is found at $t_1+\Delta t/2$ using the backward Euler rule.

The equivalent conductances of the network elements using the backward Euler rule with a step width of $\Delta t/2$ is identical to the equivalent conductances of these elements using the trapezoidal rule with a step width of Δt. Therefore, matrix [G] in eq. (1) does not change. Only the formula to evaluate the elements in the history vector [h(t)] needs to be changed.

The network is solved a second time with the backward Euler rule using a step size $\Delta t/2$, that is, at $t_2 = (t_1+\Delta t/2)+\Delta t/2$. Afterwards, the simulation continues normally with the trapezoidal rule at $t_2+\Delta t$, $t_2+2\Delta t$, $t_2+3\Delta t$, ..., etc., until another discontinuity is encountered.

The results at $t_1+\Delta t/2$ are only mathematical quantities used by the CDA procedure but have no physical meaning. Therefore, no decisions on whether to open or close switches, etc. are made based on these results. The next decisions are made at $t_1+\Delta t$, when the overshoot has already been dampened. An exception to this rule is the case of piecewise linear inductances, for which transitions from one segment to another at $t_1+\Delta t/2$ are allowed in order to avoid excessive deviations from the characteristic curve [3].

3 LUMPED LINEAR ELEMENTS

The discretization of lumped linear elements with the trapezoidal and backward Euler rules was discussed in [1]. Figure 1 shows an inductance and its corresponding discrete-time model.

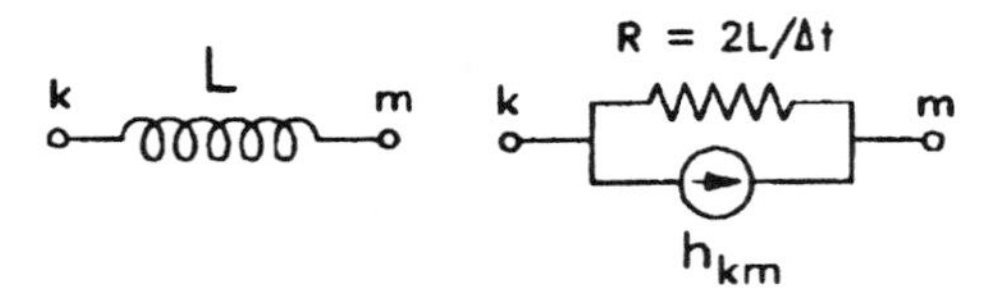

Fig. 1 Discrete-time equivalent circuit for an inductance.

The model in Fig. 1 results from applying the trapezoidal rule of integration for a step size Δt to the differential equation $v_{km}(t) = L\, di_{km}(t)/dt$. The discrete-time relationship is given by

$$i_{km}(t) = \frac{\Delta t}{2L} v_{km}(t) + h_{km}(t), \tag{2}$$

with a history term

$$h_{km}(t) = i_{km}(t-\Delta t) + \frac{\Delta t}{2L} v_{km}(t-\Delta t). \tag{3}$$

Equation (2) is inserted directly into the network nodal matrix (eq. (1)) to model the inductance branch.

Discretization of the differential equation of the inductance using the backward Euler rule for a step size $\Delta t/2$ gives exactly the same discrete-time relationship as in eq. (2). The only difference is in the history term, which is now given by

$$h'_{km}(t) = i_{km}(t-\Delta t/2). \tag{4}$$

The process of discretization of a continuous-time differential equation into a discrete-time difference equation using the trapezoidal rule for Δt and the backward Euler rule for $\Delta t/2$ is illustrated in the Appendix.

Comparing the history terms of trapezoidal (eq. (3)) and backward Euler (eq. (4)), it is seen that backward Euler uses only the history of the branch current, while trapezoidal includes the history of both the branch voltage and the branch current.

The equivalent resistance model of Fig. 1 is valid for both trapezoidal and backward Euler and therefore there is no change in the network's [G] matrix of eq. (1) due to the change of integration rule. The only change is in the history vector [h(t)].

The treatment of a capacitive branch is completely analogous to that of the inductance branch. The derivation of the corresponding relationships for general R-L-C branches is also straightforward. The procedure can also be easily extended to inductively coupled branches, which have the same relationships as the basic inductance but in matrix form.

4 NONLINEAR INDUCTANCES

A nonlinear inductance represented as straight-line segments [4] is shown in Fig. 2. The application of the CDA procedure to this element is discussed next.

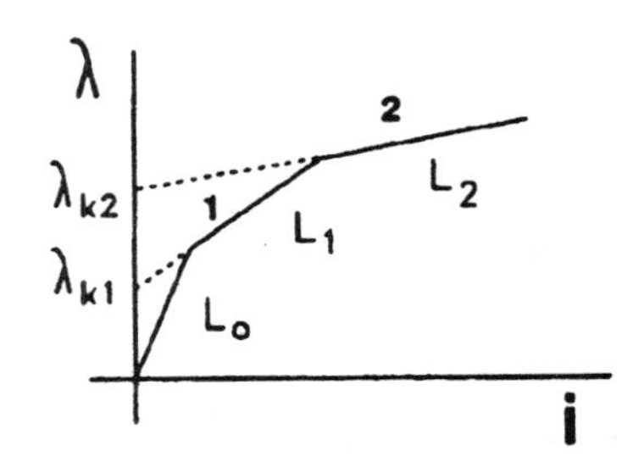

Fig. 2 Nonlinear inductance with piecewise linear characteristic.

In terms of flux linkages, the general equation for an inductance (linear or nonlinear) connected between nodes k and m is

$$v_{km}(t) = \frac{d\lambda(t)}{dt}, \tag{5}$$

where $\lambda = f(i)$.

Integrating eq. (5) with the trapezoidal rule for a full time step width Δt gives

$$\lambda(t) = \frac{\Delta t}{2} v_{km}(t) + h_\lambda(t), \tag{6}$$

where

$$h_\lambda(t) = \lambda(t-\Delta t) + \frac{\Delta t}{2} v_{km}(t-\Delta t). \tag{7}$$

Integrating eq. (5) with the backward Euler rule for $\Delta t/2$ gives the same result as in eq. (6), but with a history term given by

$$h'_\lambda(t) = \lambda(t-\Delta t/2). \tag{8}$$

The overall network solution (eq. (1)) is formulated in terms of voltages and currents. To obtain eq. (6) in terms of v and i, λ can be expressed in terms of its piecewise approximation. Assuming operation in segment 1 in Fig. 2,

$$\lambda(t) = \lambda_{k1} + L_1 i(t). \tag{9}$$

Replacing this relationship in eq. (6) gives

$$i(t) = \frac{\Delta t}{2L_1} v_{km}(t) + \frac{1}{L_1} h_\lambda(t) - \frac{\lambda_{k1}}{L_1}, \tag{10}$$

with $h_\lambda(t)$ given by eq. (7) for trapezoidal with Δt or by eq. (8) for backward Euler with $\Delta t/2$.

Equation (10) can be more simply expressed as

$$i(t) = \frac{\Delta t}{2L_1} v_{km}(t) + h(t), \tag{11}$$

with h(t) grouping the corresponding terms.

When during the normal solution with trapezoidal it is detected from the updated value of $\lambda(t)$ in eq. (9) that operation has to move from one segment to the other, for example from 1 to 2 in Fig. 2, the backward Euler rule is used for the solution in the next two $\Delta t/2$ steps with the new region parameters (L_2, λ_2).

If, after initiating the CDA procedure due to a change in operating region of the characteristic, a change to yet another region is detected after the first $\Delta t/2$ backward Euler step, three additional $\Delta t/2$ backward Euler steps are performed. The first two steps are sufficient to dampen the discontinuity created by the double switching of regions. The additional third step is performed in order to catch up with the evenly spaced (Δt) points of the normal trapezoidal solution.

TEST EXAMPLE:

Figure 3(a) shows a test system with a nonlinear reactor. Switches BL8B-BL8L close at t = 0.0194 s. Voltages at the reactor bus are shown in Figs. 3(b) and 3(c). Figure 3(b) shows the results obtained with the standard EMTP. Figure 3(c) shows the results obtained with the new EMTP using the CDA procedure. The numerical oscillations are completely eliminated without distorting the correct simulation results.

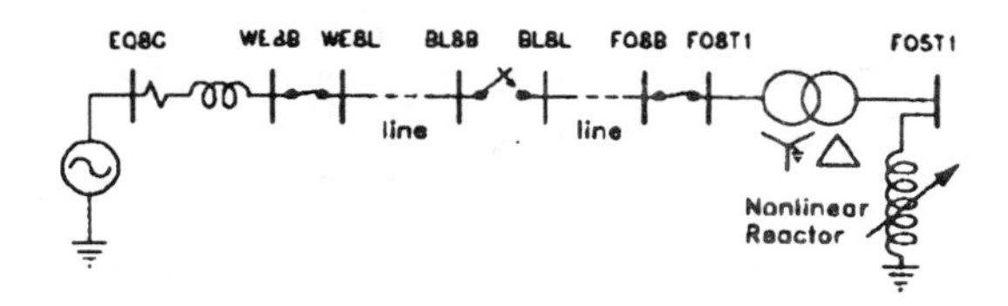

Fig. 3(a) Transient in system with nonlinear reactor.

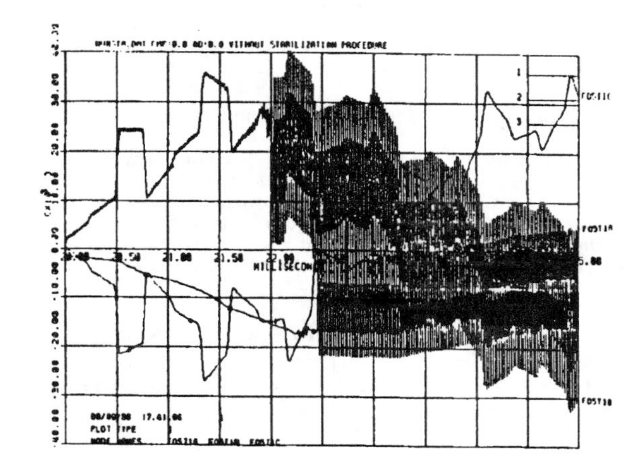

Fig. 3(b) Voltages at reactor bus. Standard EMTP.

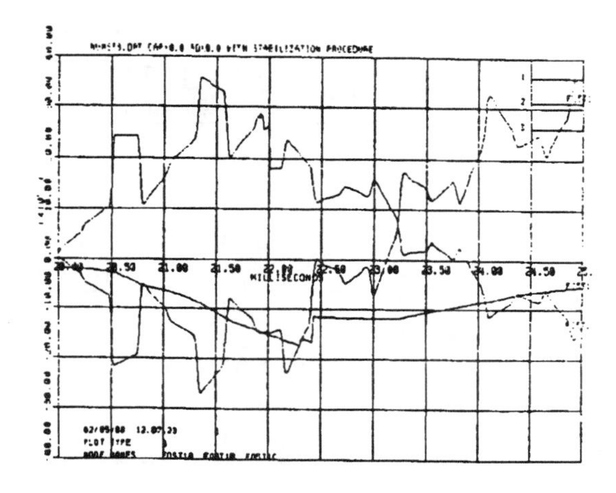

Fig. 3(c) Voltages at reactor bus. New EMTP with CDA.

5 FREQUENCY DEPENDENT TRANSMISSION LINES

A transmission line with distributed frequency dependent parameters R(ω), L(ω), G, C is shown in Fig. 4. It is assumed that the original multiphase line has been decoupled through diagonalizing matrix transformations and each mode can be analyzed separately.

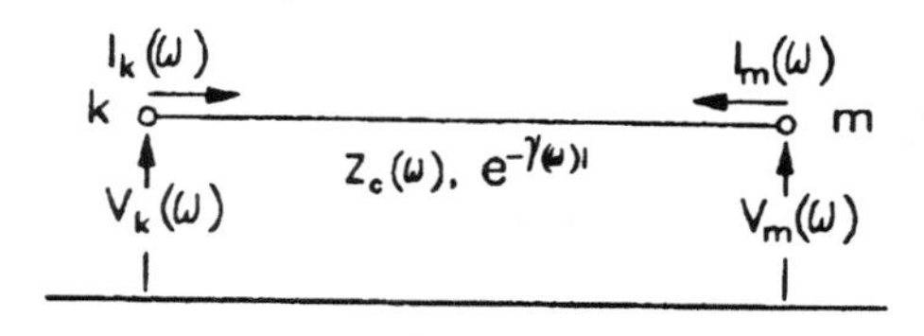

Fig. 4 Frequency dependent transmission line.

The application of the CDA procedure to the frequency dependent line model of [5] in the EMTP (fd-line model) is considered next. In the derivation of this model, voltages and currents at each line end are first related by

$$\begin{aligned} &V_m(\omega)+Z_c(\omega)(-I_m(\omega)) \\ &\quad = (V_k(\omega)+Z_c(\omega)I_k(\omega))e^{-\gamma(\omega)\ell} \\ &\quad = F_k(\omega)e^{-\gamma(\omega)\ell} \\ &\quad = F_k(\omega)A(\omega) = E_{mh}(\omega) \end{aligned} \tag{12}$$

for node m, and similarly for node k. In this equation, $Z_c(\omega) = \sqrt{[R(\omega)+j\omega L(\omega)]/[G+j\omega C]}$ is the characteristic impedance, and $A(\omega) = e^{-\gamma(\omega)\ell}$, with $\gamma(\omega) = \sqrt{[R(\omega)+j\omega L(\omega)][G+j\omega C]}$, is the propagation function for a line length ℓ.

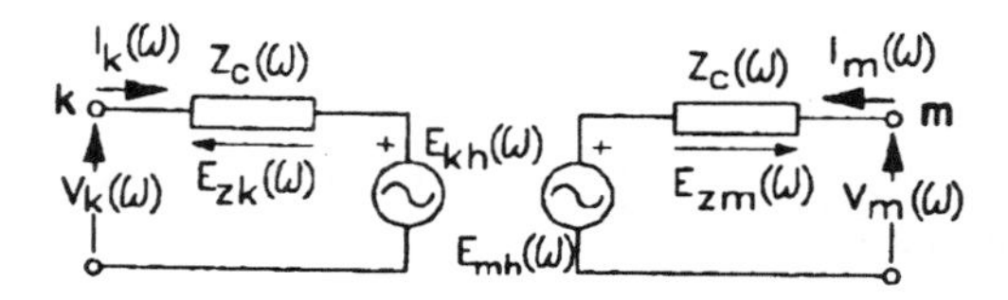

Fig. 5 Line equivalent circuit in the frequency domain.

Equation (12) for node m (and the corresponding equation for node k) give the line representation of Fig. 5.

Transferring the equivalent circuit of Fig. 5 to the time domain, and with $z_c(t) = \mathcal{F}^{-1}\{Z_c(\omega)\}$ and $a(t) = \mathcal{F}^{-1}\{e^{-\gamma(\omega)\ell}\}$, it is obtained for node m

$$v_m(t) = e_{zm}(t) + e_{mh}(t), \tag{13}$$

where

$$e_{zm}(t) = z_c(t) * i_m(t) \tag{14}$$

is the voltage drop in the equivalent impedance, and

$$e_{mh}(t) = f_k(t) * a(t) \tag{15}$$

is the equivalent history source. The symbol "*" in eqs. (14) and (15) above indicates time convolution. Similar relations are obtained for node k.

The function $f_k(t)$ in eq. (15) is formally defined in eq. (12) in the frequency domain. However, this function is more easily calculated directly from the voltage and current in the equivalent circuit at node k (Fig. 5):

$$F_k = V_k + Z_c I_k = V_k + (V_k - E_{kh}) = 2V_k - E_{kh},$$

or, transferring to the time domain,

$$f_k(t) = 2v_k(t) - e_{kh}(t). \tag{16}$$

The propagation function a(t) introduces a time delay (travelling time of the waves) on the function $f_k(t)$ (eq. (15)). As a result, the history source $e_{mh}(t)$ at time t is completely defined from past values of the quantities at node k.

Instead of performing the time domain convolutions indicated in eqs. (14) and (15), the frequency domain functions $Z_c(s)$ and A(s) ($s = j\omega$) are approximated by rational functions of s with simple negative poles and zeroes. These functions are then expanded into partial fractions. For the characteristic impedance,

$$Z_c(s) = k_0 + \sum_{j=1}^{n} \frac{k_j}{s + p_j}, \tag{17}$$

and for the propagation function,

$$A(s) = \left[\sum_{i=1}^{m} \frac{k_i}{s + p_i} \right] e^{-s\tau}, \tag{18}$$

where τ is the time delay of the line.

With $Z_c(s)$ expressed as in eq. (17), the s-plane form of eq. (14) is

$$E_{zm}(s) = k_0 I_m(s) + \sum_{j=1}^{n} E_{zmj}(s), \tag{19}$$

where each term $E_{zmj}(s)$ has the form

$$E_{zmj}(s) = \left[\frac{k_j}{s + p_j} \right] I_m(s). \tag{20}$$

These terms correspond to the first order differential equations

$$\frac{de_{zmj}(t)}{dt} + p_j e_{zmj}(t) = k_j i_m(t). \tag{21}$$

Equations (21) for each partial voltage term can now be discretized with an integration rule. Applying the trapezoidal rule with a full time step Δt gives

$$e_{zmj}(t) = \left[\frac{k_j}{2 + p_j \Delta t} \right] i_m(t) + h_{zmj}(t), \tag{22}$$

where the history term is

$$h_{zmj}(t) = \left[\frac{2 - p_j \Delta t}{2 + p_j \Delta t} \right] e_{zmj}(t - \Delta t) + \left[\frac{k_j \Delta t}{2 + p_j \Delta t} \right] i_m(t - \Delta t). \tag{23}$$

For the CDA procedure, eq. (21) is discretized with the backward Euler rule for $\Delta t/2$. This results in the same coefficient for $i_m(t)$ as in eq. (22). The only difference is in the history term, which is now given by

$$h'_{zmj}(t) = \left[\frac{2}{2 + p_j \Delta t} \right] e_{zmj}(t - \Delta t). \tag{24}$$

The partial voltage terms in eq. (22), together with k_0 in eq. (19), are added up to form the equivalent resistance in the final form of the line equivalent circuit in Fig. 6. This equivalent resistance is identical for trapezoidal with Δt and for backward Euler with $\Delta t/2$. Therefore, as in the case of lumped elements, the entries of the system [G] matrix in eq. (1) do not change during the CDA adjustment procedure.

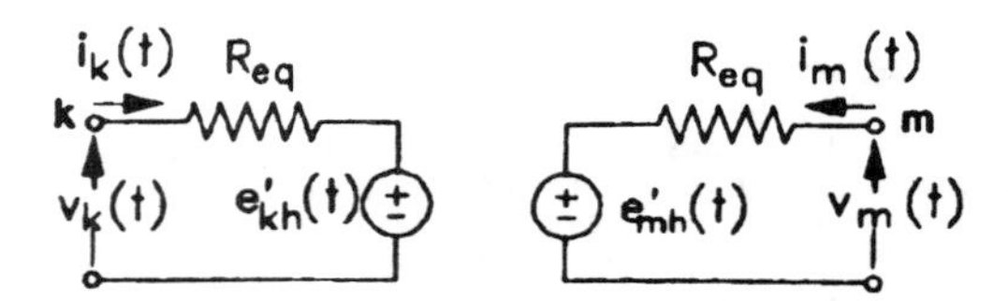

Fig. 6 Discrete-time equivalent for the fd-line model.

The above procedure to evaluate eq. (14) for $e_m(t)$ applies equally well to the evaluation of $e_{mh}(t)$ in eq. (15). In this latter case, A(s) in eq. (18) is applied to $F_k(s)$, and the resulting first order differential equations are then discretized using trapezoidal or backward Euler. Due to the time delay τ, all the terms in $e_{mh}(t)$ are history terms. The procedure can also be extended to include the frequency dependent transformation matrix model of [8].

The original implementation of the model of [5] in the EMTP used the "recursive convolution rule" to evaluate the convolutions in eqs. (14) and (15). This rule, however, is not compatible with the CDA procedure. If recursive convolution coefficients were used during the normal network solution, the equivalent resistance R_{eq} in Fig. 6 would not remain constant during the backward Euler stabilization steps. The use of the trapezoidal rule instead of the convolution rule in the present fd-line model, besides making the model compatible with the CDA procedure, results in better overall accuracy and resistance to machine truncation errors [7].

TEST EXAMPLE:

The circuit of Fig. 7(a) includes two frequency dependent line sections modelled with the JMARTI line model in the DCG/EPRI EMTP. Figures 7(b) and 7(c) show the voltages at nodes SIX-A,B,C. Switches THR-SIX open at t = 0.006 s. Figure 7(b) corresponds to the standard EMTP solution, and Fig. 7(c) corresponds to the solution with the new EMTP with the CDA procedure.

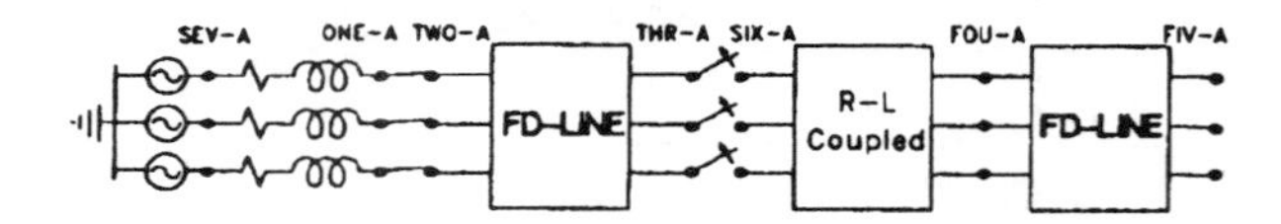

Fig. 7(a) Transients in system with frequency dependent lines.

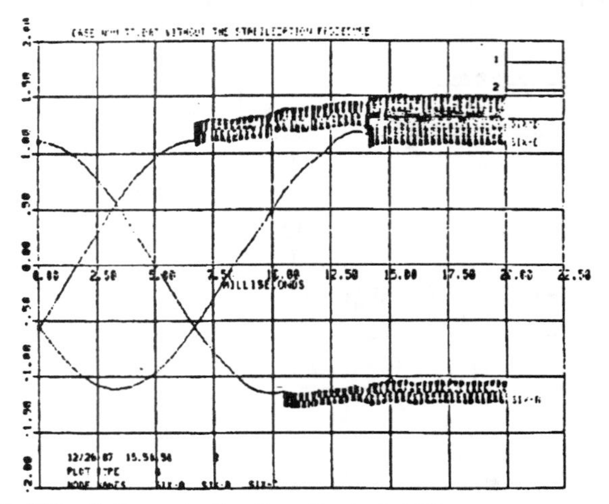

Fig. 7(b) Voltages in the line. Standard EMTP.

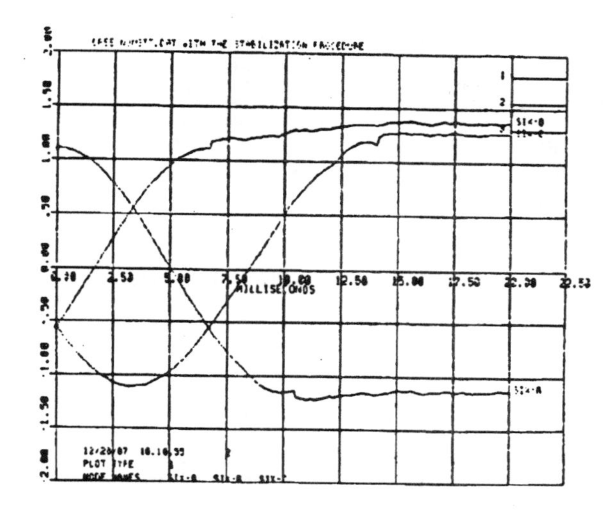

Fig. 7(c) Voltages in the line. New EMTP with CDA.

6 SYNCHRONOUS MACHINES

The three-phase synchronous machine model in the EMTP [9] involves two parts: (1) the electrical part, and (2) the mechanical part.

6.1 ELECTRICAL PART

The electrical part of the machine is modelled as seven coupled windings [6,9]. These windings are three armature windings (1,2,3), one field winding (f), two damper windings in the quadrature axis (g,Q), and one damper winding in the direct axis (D).

Transforming the equations of these seven machine windings to d,q,o variables gives

$$[v_{dqo}(t)] = -[R_{dqo}][i_{dqo}(t)] - \frac{d}{dt}[\lambda_{dqo}(t)] + \begin{bmatrix} -\omega\lambda_q(t) \\ +\omega\lambda_d(t) \\ 0 \\ 0 \\ 0 \\ 0 \\ 0 \end{bmatrix}, \tag{25}$$

with $[i_{dqo}] = [i_d, i_q, i_o, i_f, i_D, i_g, i_Q]$, and analogously for v and λ.

Consider first rows 1, 4, and 5 for the direct axis.

$$\begin{bmatrix} v_d(t) \\ v_f(t) \\ 0 \end{bmatrix} = -\begin{bmatrix} R_a & 0 & 0 \\ 0 & R_f & 0 \\ 0 & 0 & R_D \end{bmatrix}\begin{bmatrix} i_d(t) \\ i_f(t) \\ i_D(t) \end{bmatrix} - \begin{bmatrix} L_d & L_{df} & L_{dD} \\ L_{df} & L_{ff} & L_{fD} \\ L_{dD} & L_{fD} & L_{DD} \end{bmatrix}\frac{d}{dt}\begin{bmatrix} i_d(t) \\ i_f(t) \\ i_D(t) \end{bmatrix} + \begin{bmatrix} u_d(t) \\ 0 \\ 0 \end{bmatrix}, \tag{26}$$

with the speed voltage $u_d(t) = -\omega\lambda_q(t)$. In short-form notation,

$$[v(t)] = -[R][i(t)] - [L]\frac{d}{dt}[i(t)] + [u(t)]. \tag{27}$$

For the EMTP solution, the differential equations are first discretized. Discretization of eq. (27) with the trapezoidal rule for a step size Δt gives

$$[v(t)] = -\left([R] + \frac{2}{\Delta t}[L]\right)[i(t)] + [u(t)] + [h(t)], \tag{28}$$

with the history term given by

$$[h(t)] = \left(-[R] + \frac{2}{\Delta t}[L]\right)[i(t-\Delta t)] + [u(t-\Delta t)] - [v(t-\Delta t)]. \tag{29}$$

During the critical damping adjustment procedure, eq. (27) is discretized with the backward Euler rule using a step size of $\Delta t/2$. This gives the same form for eq. (28), except for the history term, which is now given by

$$[h'(t)] = \frac{2}{\Delta t}[L][i(t-\Delta t/2)]. \tag{30}$$

The derivation of the equations for the q-axis variables is analogous to the derivation for the d-axis variables, except that now both rotor voltages $v_g(t)$ and $v_Q(t)$ are zero, whereas in eq. (26) only $v_D(t)$ is zero.

For the zero sequence quantities, there is only one equation:

$$v_o(t) = -R_a i_o(t) - L_o\frac{di_o(t)}{dt}. \tag{31}$$

(The subscript of R in this equation is "a" and not "o". This latter subscript is introduced later.)

Discretizing eq. (31) with the trapezoidal rule for a step size Δt,

$$v_o(t) = -\left(R_a + \frac{2L_o}{\Delta t}\right)i_o(t) + h_o(t), \tag{32}$$

where

$$h_o(t) = \left(\frac{2L_o}{\Delta t} - R_a\right)i_o(t-\Delta t) - v_o(t-\Delta t). \tag{33}$$

Discretizing eq. (31) with the backward Euler rule for a step size $\Delta t/2$ gives the same relationship for eq. (32), but with a history term

$$h'_o(t) = \frac{2L_o}{\Delta t}i_o(t-\Delta t/2). \tag{34}$$

6.1.1 INTERFACING WITH THE EXTERNAL NETWORK

To interface eqs. (27) and (31) and the quadrature axis equation (analogous to eq. (27)) with the network connected to the machine, the following procedure is followed:

(a) Since the rotor voltages in eq. (26) are known ($v_f(t)$ = specified excitation voltage, and $v_D(t) = 0$), the two rotor currents can be eliminated. This reduces eq. (26) to a single equation:

$$v_d(t) = -R_d(t)i_d(t) + u_d(t) + h_d(t), \tag{35}$$

with $h_d(t)$ as the history term. The analogous reduction in the quadrature axis produces

$$v_q(t) = -R_q i_q(t) + u_q(t) + h_q(t). \tag{36}$$

Equation (32) is also of the same form:

$$v_o(t) = -R_o i_o(t) + h_o(t) \tag{37}$$

($R_o = R_a + 2L_o/\Delta t$).

(b) By using predicted values for the speed voltages $u_d(t)$ and $u_q(g)$ in eqs. (35) and (36), the machine becomes a simple voltage source behind R_d on the direct axis, R_q on the quadrature axis, and R_0 in the zero sequence variables. If these three independent Thevenin equivalent circuits were directly converted into the phase domain (with the d,q,o inverse transformation), the resulting phase-domain 3×3 resistance matrix would be time-dependent as well as asymmetrical. This would make the machine representation incompatible with the solution for the rest of the network.

To avoid these problems, approximations have to be made. The first approximation consists of using an average resistance $(R_d+R_q)/2$ on both d and q axes. To compensate for this, a "saliency term", $-[(R_d-R_q)/2]i_d(t)$, is added to $u_d(t)$, and $-[(R_q-R_d)/2]i_q(t)$ is added to $u_q(t)$. Predicted currents are used to evaluate these saliency terms.

These approximations will produce a simple machine model consisting of three known voltage sources behind a 3×3 equivalent resistance matrix.

(c) The complete electric network is solved together with the simple machine model of (b).

(d) From the solution of the electric network at time t, the electromagnetic torque in the machine at time t is calculated and then used to integrate the mechanical part from t-Δt to t.

Because the form of the equations remains the same, the solution procedure above is identical for the trapezoidal rule with a step width of Δt as it is for the backward Euler rule with a step width of Δt/2. Again, the only difference is in the formulas for the history terms. Also, the speed voltages and "saliency terms" must be predicted over half a time step Δt/2 for backward Euler instead of over a full step Δt for trapezoidal.

As indicated above, prediction of present value of speed voltages and saliency terms is used to simplify the machine model. In the version of the DCG/EPRI EMTP where the CDA procedure was implemented, the trapezoidal rule of integration is used for these predictions.

The application of the CDA procedure is unrelated to the prediction schemes. The only modification required in the program is to take into account the change in step width, from Δt during the normal trapezoidal solution to Δt/2 for the backward Euler CDA steps and then back to Δt to continue with the normal trapezoidal solution. During the two backward Euler steps of the CDA procedure, the trapezoidal rule is still used for the predictions, even though backward Euler is used in the discretization of the machine equations.

6.1.2 MECHANICAL PART

In the synchronous machine mechanical part, the response of the n shaft-connected rotation masses is described by the rotational form of Newton's Second Law:

$$[J]\frac{d}{dt}[\omega(t)]+[D][\omega(t)]+[K][\theta(t)] = [T_{net}(t)]+\omega_s[D_s], \tag{38}$$

with the speeds of the masses

$$[\omega(t)] = \frac{d}{dt}[\theta(t)], \tag{39}$$

and the net torque

$$[T_{net}(t)] = [T_{turbine}(t)]-[T_{gen/exc}(t)]. \tag{40}$$

where [J] = diagonal matrix of moments of inertia, [θ] = vector of angular positions of the masses, [ω] = vector of speeds, [D] = tridiagonal matrix of damping coefficients, [K] = tridiagonal matrix of stiffness coefficients, $[T_{turbine}]$ = vector of torques applied to the turbine stages, $[T_{gen/exc}]$ = vector of electromagnetic torques of generator and exciter, $[D_s]$ = vector of speed-deviation self-damping coefficient, and ω_s = synchronous velocity of the shaft-mass system.

Applying the trapezoidal rule to eqs. (38) and (39) gives

$$\left(\frac{2}{\Delta t}[J]+[D]+\frac{\Delta t}{2}[K]\right)[\omega(t)] = [T_{net}(t)]+2\omega_s[D_s]+[h(t)], \tag{41}$$

with the history term given by

$$[h(t)] = \left(\frac{2}{\Delta t}[J]-[D]-\frac{\Delta t}{2}[K]\right)[\omega(t-\Delta t)] - 2[K][\theta(t-\Delta t)]+[T_{net}(t-\Delta t)]. \tag{42}$$

Applying the backward Euler rule with step size Δt/2 to eqs. (38) and (39) gives the same result for eq. (41), with a history term

$$[h'(t)] = \frac{2}{\Delta t}[J][\omega(t-\Delta t/2)]-[K][\theta(t-\Delta t/2)]. \tag{43}$$

Again, the CDA procedure with backward Euler leads to the same discrete-time relationships as in the normal solution with trapezoidal, except for the new formulas for the history terms.

TEST EXAMPLES:

Figure 8(a) shows a test system for the synchronous machine model. Figures 8(b) and 8(c) show the voltages at the synchronous machine terminals, nodes G-A,B,C. The transient is created by first opening switches S1 at t = 0.002 s and then closing switches S2 at t = 0.04 s. The simulation in Fig. 8(b) was obtained with the standard EMTP and the simulation in Fig. 8(c) was obtained with the new EMTP with the CDA procedure.

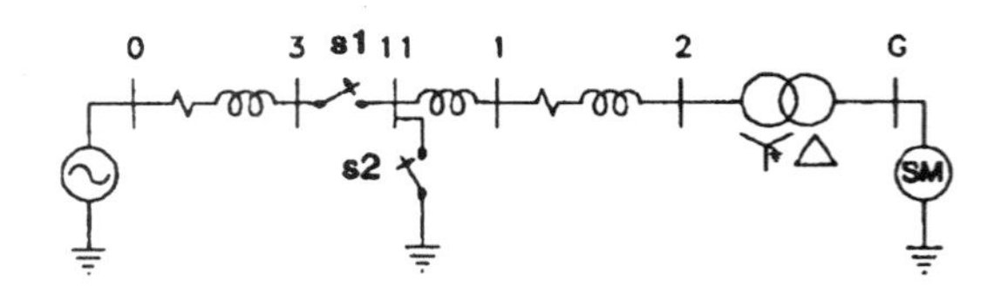

Fig. 8(a) Transients in system with synchronous machine.

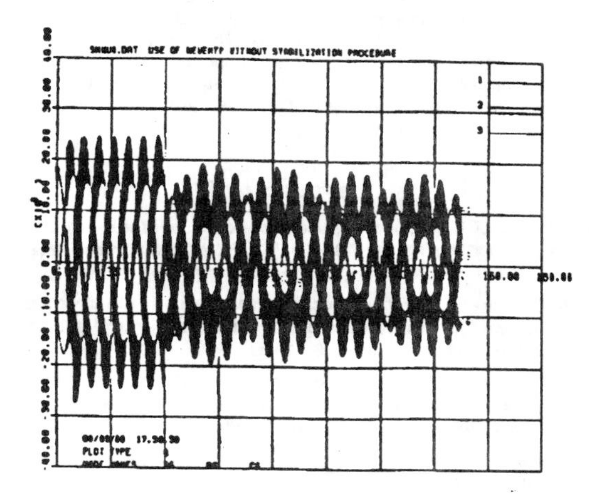

Fig. 8(b) Voltages at synchronous machine bus. Standard EMTP.

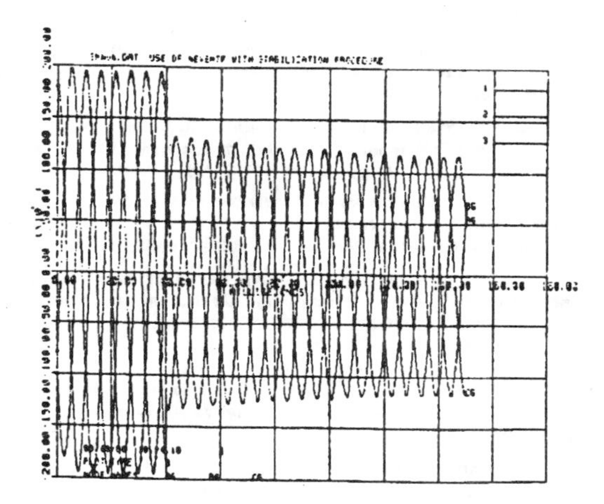

Fig. 8(c) Voltages at synchronous machine bus. New EMTP with CDA.

The principles of the Universal Machine model in the EMTP are described in references [6,10]. Even though the implementation of the CDA procedure for the Universal Machine model in the EMTP was not discussed in the paper, a test simulation has been included. Figure 9(a) shows the test circuit with an induction motor modelled with the Universal Machine model. In the simulation, switches BUS2-BUS3 open after t = 0.001 s. The results, for voltages at BUS2-A,B,C, are shown in Figs. 9(b) and 9(c). The results shown correspond to the prediction-based interface of the Universal Machine model. The simulation was also run using the compensation method interface with practically identical results.

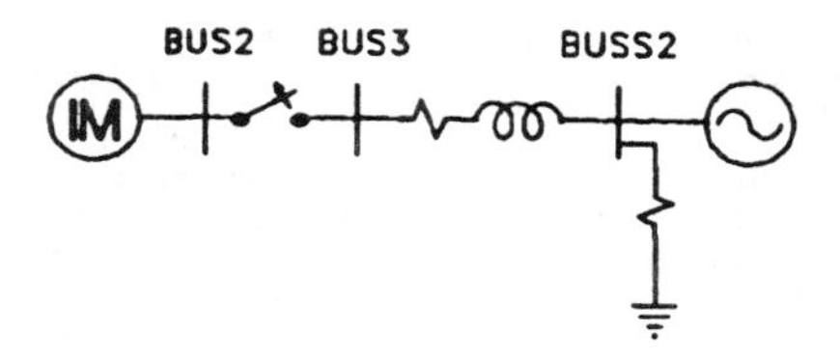

Fig. 9(a) Transients in system with induction motor (Universal Machine model).

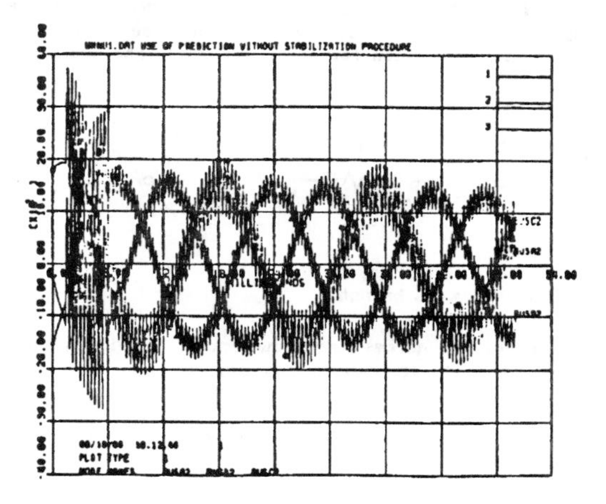

Fig. 9(b) Voltages at motor bus. Standard EMTP.

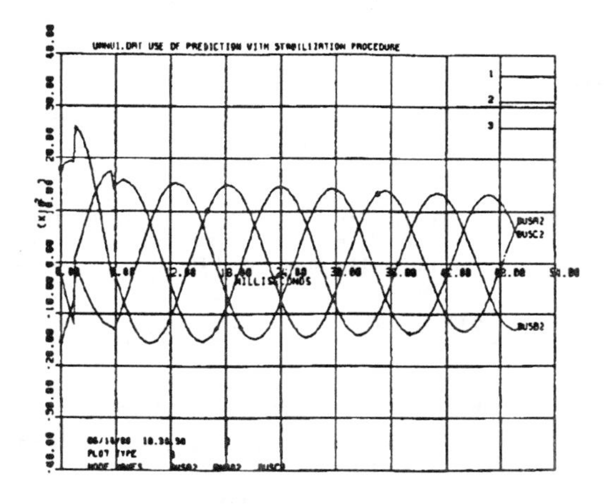

Fig. 9(c) Voltages at motor bus. New EMTP with CDA.

7 CONCLUSIONS

This paper has presented the application of the critical damping adjustment technique CDA to some of the main equipment models in the EMTP. The implementation was described for linear elements, nonlinear reactors, frequency dependent transmission lines, and for synchronous machines. As shown in the paper, even though some of the models are quite elaborate, the implementation of the CDA technique is straightforward.

The form of the discretized equations and the corresponding equivalent circuits are the same for trapezoidal with a step size of Δt and for backward Euler with a step size of $\Delta t/2$. The values of the conductances in the equivalent circuits are also the same in both cases. Only the history terms need to be changed.

Simulation results are presented showing the effectiveness of the CDA procedure in eliminating numerical oscillations in simulations involving nonlinear reactances, frequency dependent transmission lines, and synchronous machines.

APPENDIX

The relationship between the difference equations obtained with the trapezoidal rule of integration with a time step Δt and with the backward Euler rule of integration with a time step $\Delta t/2$ is illustrated next for the case of a series RL branch. The derivation for other models is analogous.

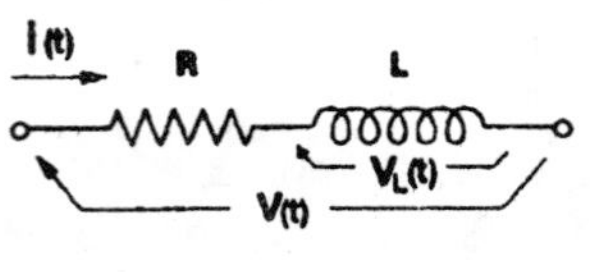

I(t) R L V_L(t) V(t)

Fig. A.1 Series RL branch.

The voltage across the inductance L in Fig. A.1 is given by $v_L = L\, di/dt$. It then follows that $v_L\, dt = L\, di$ and

$$\int_{t-\Delta t}^{t} v_L du = L[i(t) - i(t-\Delta t)]. \tag{A.1}$$

The trapezoidal rule approximation to the integral in eq. (A.1) is

$$\int_{t-\Delta t}^{t} v_L du \simeq \frac{v_L(t) + v_L(t-\Delta t)}{2} \cdot \Delta t, \tag{A.2}$$

while the backward Euler approximation is

$$\int_{t-\Delta t}^{t} v_L du \simeq v_L(t) \cdot \Delta t. \tag{A.3}$$

For the trapezoidal rule, replacing eq. (A.2) in eq. (A.1) and relating the branch voltage to the branch current, the following discrete-time relationship is obtained for the RL branch of Fig. A.1

$$v(t) = \left(R + \frac{2L}{\Delta t}\right)i(t) + \left[-v(t-\Delta t) + \left(R - \frac{2L}{\Delta t}\right)i(t-\Delta t)\right], \tag{A.4}$$

where the terms in brackets are the history terms.

For the backward Euler rule, replacing eq. (A.3) in eq. (A.1), the discrete-time relationship for the RL branch is

$$v(t) = \left(R + \frac{L}{\Delta t}\right)i(t) + \left[-\left(\frac{L}{\Delta t}\right)i(t-\Delta t)\right]. \tag{A.5}$$

If in eq. (A.5) Δt is replaced by $\Delta t/2$, the result is

$$v(t) = \left(R + \frac{2L}{\Delta t}\right)i(t) + \left[-\left(\frac{2L}{\Delta t}\right)i(t-\Delta t)\right]. \tag{A.6}$$

Comparing eqs. (A.4) and (A.6), it is seen that, except for the history terms, they are identical.

ACKNOWLEDGEMENT

The implementation of the CDA technique described in the paper was done for the DCG/EPRI version of the EMTP. The authors are grateful to DCG/EPRI and to the Canadian Electrical Association CEA for their financial support in this project and for granting permission to publish this material. Our thanks also to Dr. Hermann Dommel for his guidance during the development of the project.

REFERENCES

[1] J.R. Marti, J. Lin, "Suppression of numerical oscillations in the EMTP," 88 SM 732-0 IEEE/PES 1988 Summer Meeting, July 1988.

[2] H.W. Dommel, "Digital computer solution of electromagnetic transients in single- and multi-phase networks", IEEE Trans., Vol. PAS- 88, pp. 388-399, April 1969.

[3] H.W. Dommel, *Digital Simulation of Electrical Transient Phenomena, chapter III: Extensions of the Basic Solution Methods*, (IEEE Pub. 81EH0173-5-PWR, 1980), pp. 20-29.

[4] H.W. Dommel, "Nonlinear and time-varying element in digital simulation of electromagnetic transients," IEEE Trans., Vol. PAS-90, pp. 2561-2567, Nov/Dec 1971.

[5] J.R. Marti, "Accurate modelling of frequency-dependent transmission lines in electromagnetic transient simulations", IEEE Trans., Vol. PAS-101, pp. 147-155, January 1982.

[6] H.W. Dommel, *Electromagnetic Transients Program Reference Manual (EMTP Theory Book)*, (Bonneville Power Administration, Portland, Oregon, 1986).

[7] J.R. Marti, "Numerical integration rule and frequency-dependent line models," EMTP Newsletter, Vol. 5, No. 3, pp. 27-29, July 1985.

[8] L. Marti, "Simulation of transients in underground cables with frequency-dependent modal transformation matrices," IEEE Trans., Vol. 3, pp. 1099-1110, July 1988.

[9] V. Brandwajn, *Synchronous generator models for the analysis of electromagnetic transients*, (Ph.D. Thesis), (The University of British Columbia, Vancouver, B.C. Canada, 1977).

[10] H.K. Lauw, "Interfacing for universal multi-machine system modeling in an electromagnetic transient program," IEEE Trans., Vol. PAS-104, pp. 2367-2373, Sept. 1985.

Jiming Lin (M'88) was born in Fujian, China in 1941. He graduated in Electrical Engineering from Tsinghua University, Beijing, China in 1964.

In 1964 he joined the Electric Power Research Institute, Beijing, China. From 1964-82 he worked in the High Voltage Division in switching surges, and from 1982-86 in the Power Systems Division in power system analysis. From 1986 to 1988 he was at the University of British Columbia as a Visiting Scholar. At present he is with the Electric Power Research Institute, Beijing, China.

José R. Martí (M'71) was born in Spain in 1948. He received a M.E. degree from Rensselaer Polytechnic Institute in 1974 and a Ph.D. degree in Electrical Engineering from the University of British Columbia in 1981.

He worked for Industry from 1970 to 1972. In 1974-77 and 1981-84 he taught power system analysis at Central University of Venezuela. Since 1984 he has been with the University of British Columbia.

Discussion

Adam Semlyen (University of Toronto): I would like to commend the authors for having implemented their new approach, the Critical Damping Adjustment [1], in those parts of the EMTP where discontinuities may produce numerical oscillations. The application of the CDA to transmission lines is particularly interesting because it required abandoning the former approach of recursive convolutions in favor of a state equation formulation which is necessary for using both the trapezoidal rule and the backward Euler integration methods of the CDA.

By the use of state equations in the time domain as an equivalent of transfer functions in the frequency domain, transmission line modeling has made a full swing. Indeed, the basic frequency domain relations, in terms of a transfer function H, are of the form

$$Y = HU \tag{1}$$

and their time domain equivalent is of the form

$$y = hu + y_{hist} \tag{2}$$

which results from the discretization of the state equations

$$\dot{x} = Ax + Bu \tag{3a}$$

$$y = Cx + Du \tag{3b}$$

Equation (2) is of extreme simplicity and often translates to a Thevenin or Norton equivalent. The set {A,B,C,D} can be obtained by direct frequency domain fitting[A-D,5] of H.

The simplicity of the time domain equivalent (2) is surpassed only by the symbolic form used for convolutions:

$$y = h * u \tag{4}$$

This form can not be used directly for computations but it is convenient for derivations, as used also in the present paper. Based on such derivations, in earlier work of EMTP development full convolutions[B,F,G] and recursive convolutions[H,5] have been used. Full convolutions are computationally expensive but may be needed when a state equation approximation (or, equivalently, a rational fitting) of a given transfer function can not be obtained. Recursive convolutions are of course computationally more efficient but in general not better[D] than the state equations which are their equivalent[H].

The use of state equations does not imply restrictions to real poles even if only real arithmetic is used in the computations. The sum of a pair k of complex conjugate or real simple fractions, as in eqn.(17) of the paper, has the form

$$\frac{c''_k s + c'_k}{s^2 + \beta_k s + \gamma_k} \tag{5}$$

and can be obtained directly from the following state equations

$$\begin{bmatrix} \dot{x}_{1,k} \\ \dot{x}_{2,k} \end{bmatrix} = \begin{bmatrix} 0 & 1 \\ -\gamma_k & -\beta_k \end{bmatrix} \begin{bmatrix} x_{1,k} \\ x_{2,k} \end{bmatrix} + \begin{bmatrix} 0 \\ 1 \end{bmatrix} u \tag{6a}$$

$$y_k = [c'_k \;\; c''_k] \begin{bmatrix} x_{1,k} \\ x_{2,k} \end{bmatrix} \tag{6b}$$

by replacing the derivative by s. The resultant state equations of form (3), for all pairs k of simple fractions (5), is obtained by concatenating the corresponding matrices of equations (6): {A,B,C} = {diag(A_k), col(B_k),row(C_k)}.

The advantage of not restricting poles to the real axis[C,D] is that the same accuracy may be achieved with a lower order fitting. This may be crucial in real-time applications of transmission line modeling[I,J].

The authors' comments on the above remarks would be most appreciated.

[A] A. Semlyen, "Switching Surge Calculation with Exponential Response Functions Obtained by Direct Frequency Domain Fitting", Proceedings of the Canadian Communications and Power Conference, Montreal, November 1974, pp. 179-180.

[B] A. Morched and A. Semlyen, "Transmission Line Step Response Calculation by Least Square Frequency Domain Fitting", 1976 IEEE/PES Summer Meeting, Portland, Oregon, Paper No. A76 394-7.

[C] A. Semlyen and A. Roth, "Calculation of Exponential Propagation Step Responses - Accurately for Three Base Frequencies", IEEE Trans. on Power Apparatus and Systems, Vol. PAS-96, No. 2, March/April 1977, pp. 667-672.

[D] A. Semlyen, "Contributions to the Theory of Calculation of Electromagnetic Transients on Transmission Lines with Frequency Dependent Parameters", IEEE Trans. on Power Apparatus and Systems, Vol. PAS-100, No. 2, February 1981, pp. 848-856.

[E] A. Semlyen, "Accurate Calculation of Switching Transients in Power Systems", 1971 IEEE Winter Power Meeting, New York, Paper No. 71CP 87-PWR.

[F] W.S. Meyer and H.W. Dommel, "Numerical Modelling of Frequency-Dependent Transmission-Line Parameters in an Electromagnetic Transients Program", IEEE Trans. on Power Apparatus and Systems, Vol. PAS-99, No. 5, Sept./Oct. 1974, pp. 1401-1409.

[G] J.K. Snelson, "Propagation of Travelling Waves on Transmission Lines - Frequency Dependent Parameters", IEEE Trans. on Power Apparatus and Systems, Vol. PAS-91, No. 1, Jan/Feb. 1972, pp. 85-91.

[H] A. Semlyen and A. Dabuleanu, "Fast and Accurate Switching Transient Calculations on Transmission Lines with Ground Return Using Recursive Convolutions", IEEE Trans. on Power Apparatus and Systems, Vol. PAS-94, No. 2, March/April 1975, pp. 561-571.

[I] R.M. Mathur, Xuegong Wang, "Real-Time Digital Simulator of the Electromagnetic Transients of Power Transmission Lines", IEEE/PES 1988 Summer Meeting, Portland, Oregon, Paper No. 88 SM 584-5.

[J] Xuegong Wang, R.M. Mathur, "Real-Time Digital Simulator of the Electromagnetic Transients of Transmission Lines with Frequency Dependence", IEEE/PES 1989 Winter Meeting, New York, N.Y., Paper No. 89 WM 122-3 PWRD.

Manuscript received August 2, 1989.

J. Lin and J.R. Martí: Our thanks to Professor Semlyen for his interesting comments. Indeed, the modelling of frequency dependent transmission lines is making a full swing as the nature of frequency dependence modelling is better understood. In its present form in the DCG/EPRI EMTP, the line model of [5] has been reduced to a form that is equivalent to having a set of n linear first-order differential equations, as in the state-variable formulation of eqs. (3a) and (3b) in the Discussion. In effect, a first-order differential equation with parameters that are functions of frequency is being replaced by a set of n first-order differential equations with constant parameters. As indicated by Professor Semlyen, the link that allows this conversion is frequency-domain fitting by rational functions that can be expanded into simple first-order partial fractions. These partial fractions can correspond to either simple real poles or to pairs of complex conjugate poles.

In the work of reference [5] only simple real poles were used, mainly because of simplicity in designing the fitting algorithm. This algorithm does not need a guess of initial values of the parameters and can be easily controlled to assure convergence to a solution. Due to the smooth form of the line functions to be fitted, this simple algorithm provides very accurate results.

As Professor Semlyen points out, complex conjugate poles are equally valid alternatives to generate state equations and allow a greater degree of flexibility in the shaping of the fitting function $H(\omega)$. The availability of an extra parameter, the damping factor ζ, in addition to the corner frequency of the undampened pole, allows for a larger control of the shape of the fitting curve than what is possible with two real poles. This technique has been effectively explored by Professor Semlyen in the references mentioned in his Discussion.

The use of complex conjugate poles is particularly important when fitting transfer functions that present sharp peaks and valleys in their frequency response, such as those encountered in high-frequency transformer modelling or in network equivalents. The extension of the fitting routines of [5] to incorporate complex poles and zeroes in the fitting algorithm is currently under investigation.

Manuscript received August 31, 1989.

DIGITAL SIMULATION OF FLEXIBLE TOPOLOGY POWER ELECTRONIC APPARATUS IN POWER SYSTEMS

T.L. Maguire, Member A.M. Gole, Member

Department of Electrical and Computer Engineering
University of Manitoba
Winnipeg, Manitoba, Canada R3T 2N2

Abstract – FACTS (flexible ac transmission system) apparatuses employing many alternative configurations of power-electronic switching devices are being considered by others for increasing the power transfer capability of ac transmission lines. This paper introduces special modifications to the basic Dommel algorithm to expedite simulation of systems including arbitrary configurations of individual power-electronic switching devices. The described techniques have been implemented in a prototype transients simulation program. Two time-step sizes are used. Individual switching devices are represented according to simple characteristic curves. The modified algorithm includes iteration of a time-step when required to provide solutions on the curves. A technique is described for removing numerical oscillations of currents in capacitive loops and voltages at inductive nodes. Simulation results are presented for an example power-electronic apparatus.

Keywords: Electromagnetics Transients Simulation, Digital, Power Electronics, FACTS, Converters, Forced Commutation, Power Systems.

INTRODUCTION

This paper introduces special modifications for adapting a transients simulation program based on the trapezoidal algorithm of Dommel [1] for simulation of FACTS [2] (flexible ac transmission system) apparatuses.

The three fundamental ac transmission line conditions which could be controlled by such power-electronic apparatus are line voltage, line load angle, and line impedance. The apparatuses for accomplishing these results are respectively static compensators, fast phase shifters, and apparatuses for switching capacitors (or inductors) into series connection with the line. Dynamic control of line load angle or line impedance would allow operating the line at a higher load angle thus significantly increasing the power transfer capability of the line.

Potentially there may be many different apparatuses devised to control each of the three fundamental line conditions. In turn, the construction of each apparatus could allow variations for the configuration of individual power-electronic switching devices. Prior to installation of any given FACTS apparatus, it will be necessary to fully understand how that unique configuration of power-electronic switching devices works while embedded in the utility electrical system. In that regard, transients simulation of each configuration will be a useful tool in developing control strategies and in identifying and correcting system problems such as harmonic resonances between apparatus and the electrical system. A need therefore exists to develop techniques to expedite the transient simulation of arbitrary configurations of power-electronic devices embedded in electrical power systems.

Traditionally, the trapezoidal algorithm of Dommel [1] has been used for simulation of power systems. A primary feature of the Dommel algorithm is that a computer program can readily be prepared to automatically formulate the admittance matrix used in the solution method. Also of note is the ease with which travelling-wave transmission-line models [1] and machine models can be interfaced to the main network solution [3]. These features make the Dommel algorithm the method of choice when simulating large electrical power systems.

There are, of course, limitations to any numerical method. One limitation of this type of nodal admittance matrix analysis becomes apparent when the network to be simulated contains capacitive loops or nodes which are connected to the rest of the network only through inductive branches. In the case of capacitive loops, circumstances can arise that can lead to an erroneous numerical oscillation of two time-step duration in the numerical solution for currents in the capacitive loop. A similar numerical error can appear in the voltage solution for inductive nodes at which only inductors are incident [5].

One technique commonly used to reduce these numerical oscillations is to make slight changes to the resistive nature of the network in question in order to damp out the oscillations. However, this method of adding slight levels of resistive damping does have an effect on the accuracy of the solution. If very fast damping of the numerical oscillations is desired then damping resistances can have a significant effect on solution accuracy.

Another technique for suppressing numerical oscillations involves making a temporary modification to the solution technique when numerical oscillations arise [4]. The temporary modification includes using the results of a half-size time-step based on the backward Euler method [5]. The half-size time-step is taken following a discontinuity in the solution such as can be expected when a switch operates. The time-step length for the backward Euler method is advantageously chosen to be equal to one-half that for the trapezoidal rule algorithm because then the same conductance matrix can be used [5].

A modification to the solution technique is described in this paper which allows the continuous use of the trapezoidal rule but which also eliminates numerical oscillations. The trapezoidal algorithm is used

91 SM 414-3 PWRD A paper recommended and approved by the IEEE Transmission and Distribution Committee of the IEEE Power Engineering Society for presentation at the IEEE/PES 1991 Summer Meeting, San Diego, California, July 28 - August 1, 1991. Manuscript submitted August 29, 1990; made available for printing July 1, 1991.

exclusively for the described solution technique because it is known to be more accurate than the backward Euler method [5]. The more accurate trapezoidal algorithm is employed in the described solution technique because the oscillation suppression technique is used in every time-step rather than only at expected discontinuities.

Special care is required when handling the switching of devices such as GTO (gate-turn-off) thyristors or IGBTs (insulated gate bipolar transistors) in power system simulation packages. A GTO, for example, can be ordered to turn off when still carrying full load current which is not the case with conventional thyristors. Also, the coordinated switching of two devices that are made to switch in the same time interval (a common occurrence in several power-electronic circuit topologies) is difficult to handle when devices are represented as controlled on-off switches as is the practise in most power system simulation packages in use today. To handle such situations two basic modifications to the traditional Dommel algorithm are implemented in the prototype program.

Firstly, the program permits the use of two time-step sizes during a simulation. A specifiable large time-step size is the default. However, when it is necessary to change the characteristic for a device between the "on" and "off" characteristics then the change is carried out gradually during a sequence of small time-steps.

Secondly, the device characteristic is represented as a simple v-i characteristic curve and not as a resistance changed in one abrupt on-off step. In the case of diodes the characteristic curve is fixed. However, for GTOs and thyristors there are two characteristic curves, one being for the blocked state and the other being for the forward conducting state. The main intention here is not to represent the characteristic accurately but to improve the numerical performance of the program when devices in close proximity are switched simultaneously. Iteration of the solution within the same time-step is carried out to ensure that the solution always falls on the v-i characteristic. However the program has an automatic selection criterion that only iterates the solution when required and no iteration takes place in the period between device switchings.

The remainder of the paper begins with a discussion of the details of the representation of switching devices according to characteristic curves and the process of iteration of a time-step. Subsequent sections describe the specifics of the numerical oscillation suppression technique and the manner of operation of the prototype program. Simple examples are presented to illustrate the operation of the techniques within the program. The paper closes with the presentation of simulation results for a power delivery apparatus for application between a self-excited induction generator operable at variable speed and an isolated ac load.

REPRESENTATION OF SWITCHING DEVICES IN THE PROTOTYPE PROGRAM

In developing the prototype program a number of levels of complexity were available for consideration with respect to the representation of switching.

At the lowest level of complexity is what is referred to hereunder as a binary type switching. In binary switching a diode branch is represented by a simple resistor which is given a high resistance in the next time-step following the detection of $i \leq 0$ and is given a low resistance in the next time-step following detection of $i > 0$. For thyristors, change to a low resistance is delayed until there are both a small forward current and an "on" pulse at $(t-\Delta T)$, the beginning of the time-step. Such 'binary' switching is prone to errors when simulating arbitrary power electronic topologies. Here an alternative 'iterative' approach is used as described below.

Each switching device such as a diode, thyristor, or gate-turn-off (GTO) device is simply represented in the network by a resistance R and a series voltage source E. The values of R and E depend on the current i through the device and vary with i so as to provide a simple characteristic curve such as the solid curve illustrated in Figure 1.

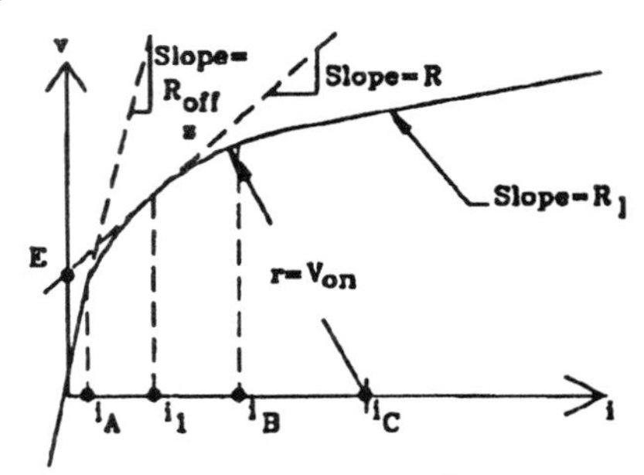

Figure 1. Typical Diode Characteristic

The characteristic illustrated in Figure 1 consists of two straight line segments which are tangent to an arc of a circle. The line segment extending into quadrant III with slope R_{off} represents the reverse-voltage blocking state of the device. The slope R_I of the line segment in quadrant I can be set to a small value to represent the loss of forward blocking capability. Conversely, slope R_I can be adjusted gradually toward slope R_{off} to represent the recovery of forward blocking capability for a thyristor.

In order to represent the characteristic properly for a given current such as i_1 in Figure 1 it is necessary to be able to obtain voltage source E and slope R as functions of the device current i. Equations for E(i) and R(i) are readily derived using basic trigonometry for characteristic curves specified in terms of R_{off}, R_I and V_{on} defined in Figure 1. The curved portion of the characteristic between i_A and i_B in Figure 1 is referred to as the switching zone because changes to E and R are required only within this zone in order to properly represent the characteristic for varying i.

The iterative aspect of the switching algorithm is described hereunder with reference to Figure 2.

It is assumed that i_1 is the solution at the beginning of a time-step in which the switching is occurring. The solution for the end of the time-step (i_S, v_S) is then computed using values for E and R evaluated with $i=i_1$. However, this gives a solution (i_S, v_S) (point X in Figure 2) which is not on the actual characteristic. (i_S, v_S) may still be an acceptable solution if the ratio $R(i_S)/R(i_1)$ is not too different from unity (indicating that the solution is not too far off the characteristic). If the solution is

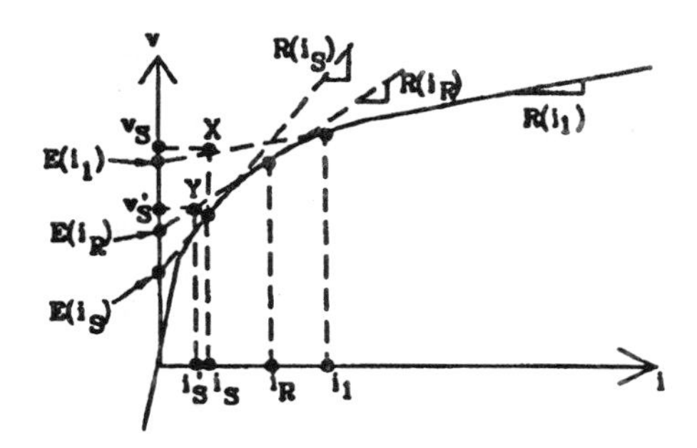

Figure 2. Explanatory Illustration of an Iteration

unacceptable the values of E and R are changed to $E(i_R)$ and $R(i_R)$ where $i_R=(i_1+i_S)/2$. The value of i_S as the candidate end of time-step solution is discarded and new solution (i_S', v_S') (point Y) is calculated by repeating the simulation time-step from i_1. It can be noted from Figure 2 that (i_S', v_S') (point Y) now falls closer to the characteristic than (i_S, v_S) (point X). If the solution is still unacceptable a further iteration with $E(i_R')$ and $R(i_R')$ where $i_R'=(i_S+i_R)/2$ is conducted to arrive at a new solution (i_S'', v_S''). Iterations are continued until the attainment of the desired convergence criterion.

Under some circumstances, i_R is not chosen midway between i_1 and i_S, but rather closer to i_1. This is to ensure that $R(i_1)$ and $R(i_R)$ are not too far apart and so that the slope R is changed gently from iteration to iteration.

This 'iterative' procedure is found to give accurate results in the modelling of device switching particularly in the coordination of the characteristics of two devices switching simultaneously. In addition, during a switching process, the timestep in the program is reduced for further accuracy. A simple circuit is illustrated in Figure 3 to aid in the further explanation of the switching algorithm used in the program.

For the circuit in Figure 3 it is assumed that a simulation has been underway for a sufficient number of large time-steps with thyristor T1 blocked so that the current i_1 through diode D1 has become steady at 10 A. In the blocked state the thyristor has a characteristic curve with $R_I = R_{off}$ in Figure 1.

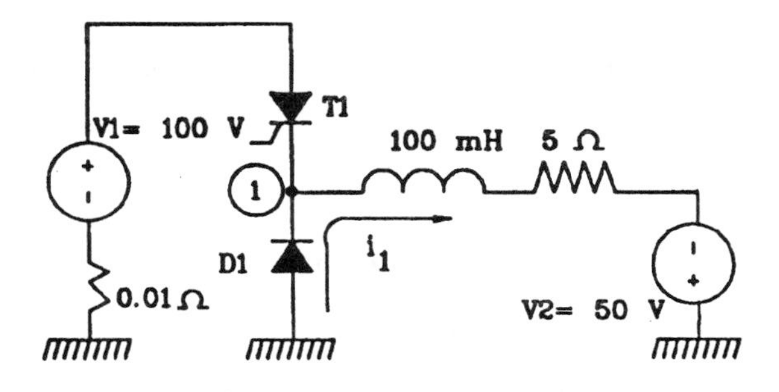

Figure 3. Circuit for Describing a Switching Operation

At the end of each large (ex. 50μS) time-step the program checks to see whether a change in the characteristic of the thyristor has been signalled by controls in a user-supplied subroutine DRIVER. When the main program receives a signal to change the thyristor characteristic to represent forward conduction, the main program can only tell that the switching was called for sometime in the large time-step. The program therefore takes action to identify the time of switching more precisely.

Upon receiving a signal to change the thyristor characteristic, the program rejects the results of a large time-step completely and resets the simulation to the condition that existed at the beginning of the rejected step. The simulation then proceeds with a scheduled number of small (ex. 1.0 μS) time-steps in order to relocate the switching event. In this manner, the commencement of the "on" switching of the thyristor characteristic can be identified within one small time-step.

Binary type switching as mentioned in the opening paragraph of this section does not work very well in this case. It is assumed that a new "on" pulse for a blocked thyristor T1 appears while current i_1 is flowing through diode D1. In that case, diode D1 will continue to be represented by a low resistance in the next time-step because i_1 is flowing at $(t-\Delta T)$. As well, thyristor T1 will be switched to a low resistance for the time-step because of the existence of forward voltage and an "on" pulse at $(t-\Delta T)$. The result of this binary device logic is that both diode D1 and thyristor T1 will be represented by low resistances for the time-step and the solution at t will include an erroneous one time-step 10000 A spike of current backward through diode D1.

Of course, in a circuit as simple as that in Figure 3, it is very easy to devise overriding logic (referred to as circuit level logic) to cause the simulation to work properly. One needs only to note that as soon as thyristor T1 is turned "on" diode D1 needs to be turned "off". This circuit level logic can be arranged so that in the very same time-step that thyristor T1 is changed to a low-resistance representation, diode D1 will be changed to a high-resistance representation (even though at $(t-\Delta T)$ the current i_1 will still be forward through the diode). This logic arranged for switching diode D1 represents a case where circuit level logic needs to be given priority over device level binary logic in order to make the simulation work properly.

As noted above, circuit level logic is very readily devised for a circuit as simple as that shown in Figure 3. On the other hand, if the circuit for simulation contains a large number of devices, then the circuit level logic required to make the simulation work properly could conceivably be very complex. It was therefore decided to provide more complex device level logic in the program so that the user of the program would not need to implement any circuit level logic.

When the iteration algorithm, discussed earlier, is employed, the erroneous result of the switching operation described above for Figure 3 is corrected without the use of circuit level logic. The erroneous current spike through diode D1 is still calculated at the end of the first solution of the time-step. However, the solution is not accepted as final because iteration of the time-step is triggered by the diode D1. This iteration is triggered due to the fact that the representation of the diode characteristic during the time-step (i.e. a low resistance R) is clearly inappropriate with a large reverse current through the diode. Continuing iteration of the time-step will occur with adjustment of diode resistance $R(i_R')$ (and source voltage $E(i_R')$ until the characteristic is represented by a large R which is appropriate for a reverse biased diode. With a large R the current through the diode D1 will be correct at the end of the time-step. The currents through devices D1 and T1 during simulation of the circuit in Figure 3 are illustrated in Figure 4.

A further enhancement used in the program involves extending the modification of device forward-resistance (R_I in Figure 1) over a sequence of small time-steps. For example, R_I can be decreased exponentially over 10 small time-steps to model the non-zero duration of the turn-on process. Likewise, the turn-off process can be extended over 20 small time-steps to model a non-zero duration of the turn-off process. This method also adds accuracy

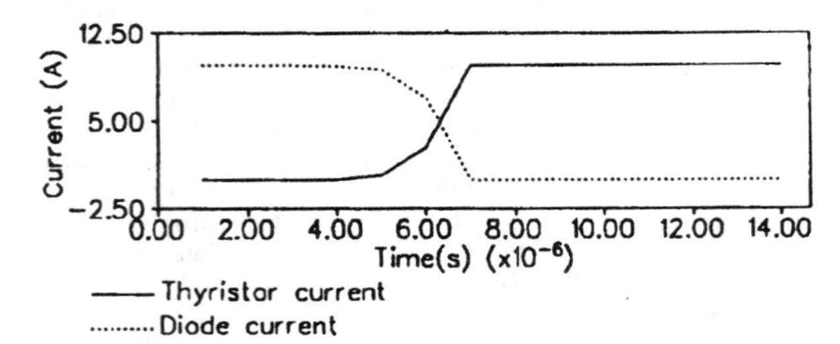

Figure 4. Device Currents During a Simple Switching Operation

to the solution as is demonstrated by the example circuit in Fig. 5. Small time-steps rather than large are always used during modifications to device forward resistance R_I in order that the modification not extend over an unacceptably long time.

Figure 5 contains an example power-electronic circuit which contains 2 GTO devices. The circuit has been simulated in order to demonstrate the effectiveness of modifying forward-resistance R_I slowly over 20 small time-steps in order to simulate the turn-off of GTO devices. Both devices are initially "off" but are switched "on" at t = 0.05 s. By t = 0.35 s a current of 500 A is established in the 0.01 H inductor. The number of volt-seconds applied to the inductor in establishing the current is 5 volt-seconds. Thus, 5 volt-seconds applied in the opposite direction is required to block the current.

When both GTO devices are turned "off" over 20 small time-steps a quasi-impulse of voltage is simulated at node 1 as illustrated in Figure 6. Integration of the impulse shows that the area under the impulse is approximately correct at 5.196 volt-seconds. This accuracy is good given the ratio of event duration (essentially 10 µS) to small time-step size (1 µS). The accuracy would not be as good with a one-step turn-off of the GTO devices. In fact, with a one-step turn-off the solution can be completely wrong. While practical GTO devices would certainly fail from the impulse in Figure 6, the example demonstrates the ability of the simulation method to handle unusual situations with accuracy.

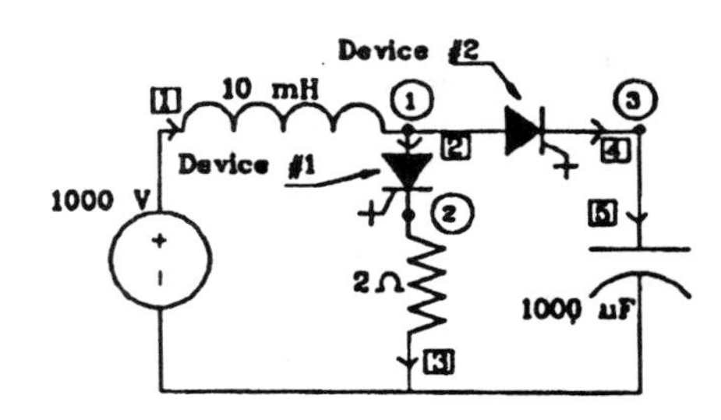

Figure 5. Circuit for Demonstrating GTO Device Turn-off Switching Operation

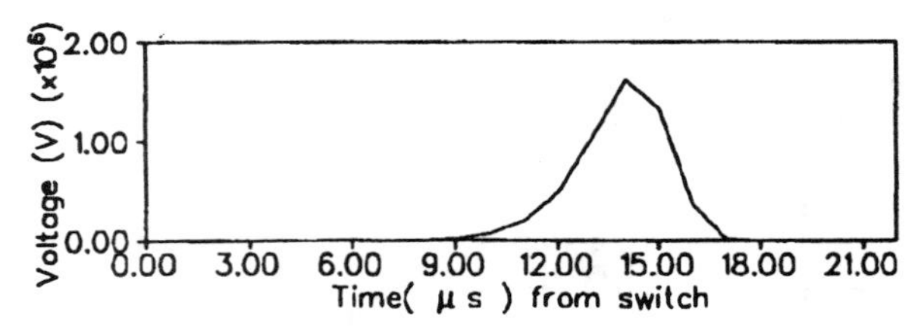

Figure 6. Node 1 Voltage Impulse for Figure 5

For many simulation cases it is adequate to have an on/off binary representation for the switching device instead of the detailed characteristic of the device. The program allows for this representation as well, and utilizing this feature often leads to a significant reduction in simulation time (typically about 40%). The user can identify which devices are required to be modeled as binary switches and which devices have to be represented in detail (such as devices in parallel or devices in series that are likely to experience simultaneous switchings).

MODIFICATION OF THE CONVENTIONAL ALGORITHM FOR SUPPRESSION OF NUMERICAL OSCILLATIONS

The prototype program is based on the Dommel algorithm [1] for nodal admittance circuit analysis.

In the Dommel [1] [6] algorithm branch equations for resistors, inductors and capacitors can all be written in the form

$$i(t) = g \cdot v(t) + I_{km}(t-\Delta T) \quad (1)$$

by the application of the trapezoidal rule of integration. This form is clearly recognizable as a conductance and parallel current source as shown in Fig. 7.

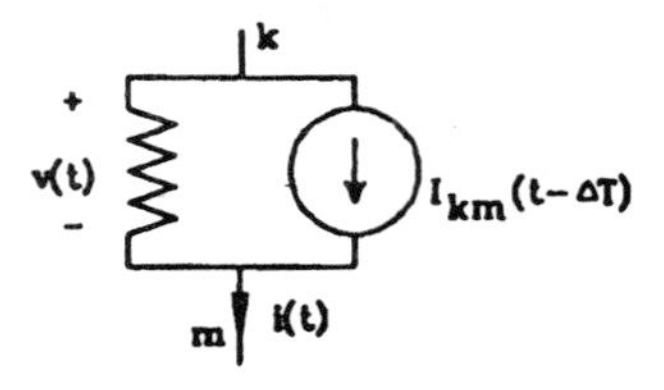

Figure 7. The Branch Equivalent in the Dommel Algorithm

Similar matrix equations can be developed for mutually coupled branches such as transformers and other elements such as transmission lines. Terms with the suffix (t-ΔT) represent (or are entirely derived from) branch voltages and currents known at the beginning of a time-step of duration ΔT. Node voltages at the end of a time-step from (t-ΔT) to t are obtained by solving for the node voltages in a network consisting of conductances g and parallel current sources $I_{km}(t-\Delta T)$. The solution for the node voltages at t involves the solution of the matrix equation

$$[G][v(t)] = [I] \qquad (2)$$

where [G] is the conductance matrix; [v(t)] is the vector of unknown voltages at t; and [I] is the vector of total current injections into each node. These injections include injections from machines or other state equation based models interfaced to the main network solution. [I] can also include terms corresponding to voltage and current sources defined for time t. Once node voltages v(t) are obtained, the application of equation (1) for each branch will yield i(t), the branch currents at the end of the time-step.

In the conventional Dommel algorithm the solution at t is complete at this point and the solution of the next time-step can begin. However, this paper will describe a modification which has been implemented in order to eliminate the two time-step numerical oscillations which are sometimes observed in the pure Dommel algorithm for capacitive loops and inductive nodes [5]. This technique also has the additional benefit of causing the iteration algorithm of the previous section to converge more quickly. The technique for suppression of numerical oscillations consists of additional action at the end of the conventional time-step described above.

The first prescribed action is to interpolate all node voltages and branch currents at (t-ΔT) with corresponding quantities at t in order to estimate branch currents [i(t-ΔT/2)] and node voltages [v(t-ΔT/2)]. All series voltage sources are also interpolated to obtain sources p(t-ΔT/2). It is also necessary to obtain the capacitor voltages at (t-ΔT/2) for branches with a series inductor and capacitor.

A time-step is next taken from (t-ΔT/2) to t using an additional [G] matrix maintained for time steps of length ΔT/2. One advantage of this technique is that current injections from machines and other state variable type models interfaced for the (t-ΔT) to t solution are the same for the (t-ΔT/2) to t solution and need not be recalculated. The limited involvement of machines and other models in the numerical oscillation suppression technique is acceptable because adequate methods exist [3] and have been used for suppression of 2 time-step numerical oscillations between the model solutions and main network solution.

Figure 8 illustrates a very simple network which can be used to demonstrate the effectiveness of the numerical oscillation suppression technique. It consists of a voltage source and inductor incident at a virtual open circuit. Application of the basic Dommel trapezoidal rule to the inductive branch with a series independent voltage source yields

$$g_L = \Delta T/_{2L} \qquad (3)$$

and

$$I_h(t-\Delta T) = g_L \cdot [p(t) + p(t-\Delta T) - v_1(t-\Delta T)] \qquad (4)$$

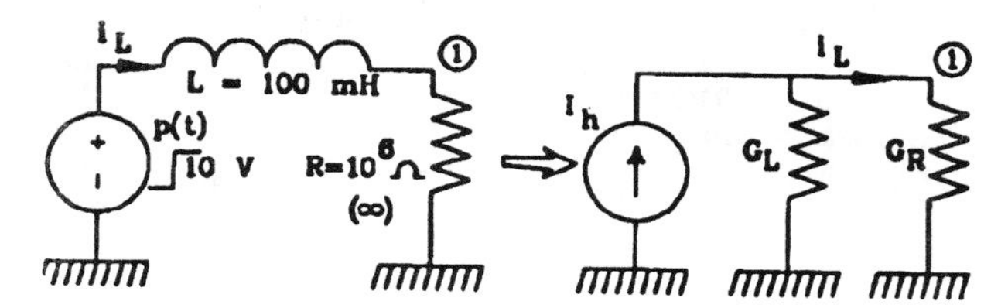

Figure 8. Circuit for Illustration of Numerical Oscillation

If initial conditions of $i_L(0)=0$ and $v_1(0)=0$ are chosen and a step voltage source p=10 volts is applied at t=0 then the conventional solution method will produce an erroneous numerical oscillation in the node 1 voltage. The node 1 voltage oscillates around the true solution of 10 volts as illustrated in Figure 9.

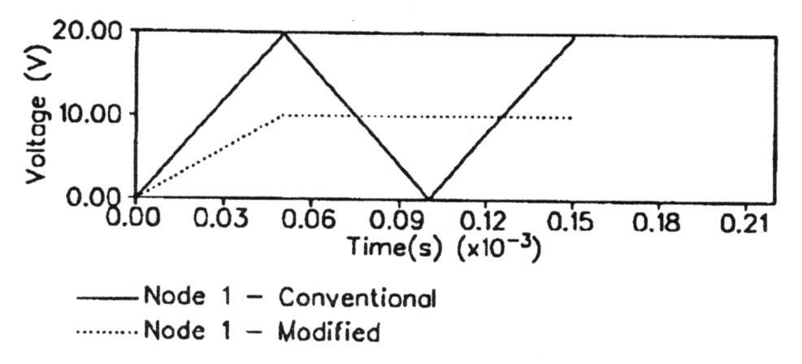

Figure 9. Removal of Inductive-Node Voltage Oscillation

On the other hand, when the numerical oscillation suppression technique is used, the erroneous numerical oscillation is removed from the solution very quickly as illustrated in the "modified" curve in Figure 9. Note that in the very next time-step the output voltage reaches essentially 10 volts.

The numerical oscillation suppression technique is similarly effective for capacitive loops.

STRUCTURE OF THE OVERALL PROGRAM

The prototype program is meant to simulate power electronic circuits in power systems and potentially large systems have to be handled. To maintain accuracy small time-steps are used to locate and carry out changes to device characteristics and iteration is used to assure that solution points fall on device characteristics. The approach of using small time-steps only during switching events and not during the entire simulation greatly reduces the computational effort. In order to further reduce the computational effort the solution for node voltages [v(t)] in equation 2 is accomplished by forward triangularization and back substitution rather than by inversion. Sparse matrix techniques have been implemented similar to those originally used in the popular EMTP program [7]. In addition, the third and most effective method of "near optimal" node numbering suggested by Tinney and Walker [7] has been implemented in order to maintain sparsity in the forward triangularization process. Row elimination is practiced in the forward triangularization [1], [7] so that when a branch conductance changes triangularization can be started at the row number associated with the smallest node number for the branch.

A further technique, not yet implemented, is to break larger networks into subsystems connected by travelling-wave transmission lines as is done in the program EMTDC [3]. In this way when one part of the network requires small time-steps for switching, the remainder of the network could take one large time-step and wait.

EXAMPLE SIMULATION

The capabilities of the program are demonstrated with the simulation of a proposed power delivery apparatus for application between a self-excited induction generator and an isolated ac load as illustrated in Figure 10 a). Such a scheme could conceivably be used to supply a remote community (such as in Northern Canada) from a local run-of-the-river type self-excited induction generator plant rated at 0.5 or 1.0 MW. The simulated apparatus was selected to have

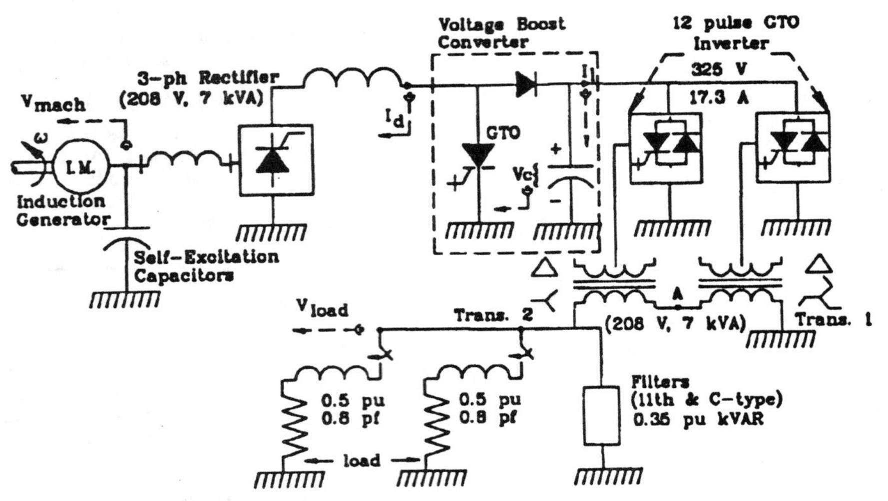

Figure 10 a). Power Circuits for Example Simulation of Power Delivery Apparatus

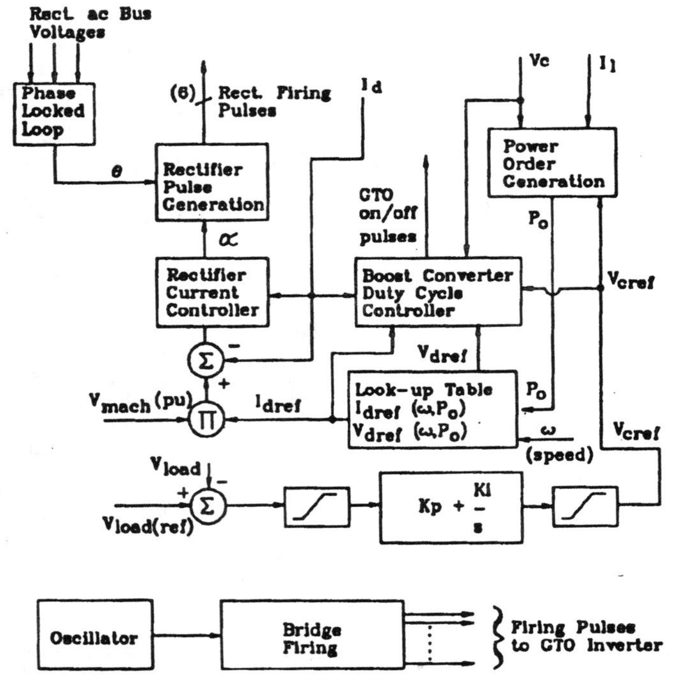

Figure 10 b). Functional Block Diagram for the Controls of the Example Apparatus

ratings which will permit a prototype apparatus (rated at 6 kW) to be implemented in the laboratory.

The example scheme is well-suited for this demonstration because it has a variety of different power system elements. The induction machine is self-excited by terminal capacitors. It has no field circuit for control of ac voltage nor does it generate a fixed 60 Hz frequency. For this reason, the generated voltage is rectified and re-invertered on the load ac side at the desired frequency (60 Hz) by means of a force-commutated voltage-type GTO inverter. Note that since the self-excited induction machine has no field-excitation circuit (unlike a synchronous machine), the machine terminal voltage can only be controlled [8] by adjusting the rectifier dc current. However the current demanded by the inverter is determined by the customer load and could be different from the rectifier current order. In fact, by selection of inverter parameters, the demanded inverter current is always less than the required rectifier current for any load less than rated. The voltage-boost converter circuit between the rectifier and inverter diverts a certain portion of the rectifier current to ground and only allows enough current to pass to the inverter as is required to keep the consumer load voltage at rated value.

The primary control objectives are to maintain the machine voltage at approximately rated value and to supply output voltage of correct frequency and magnitude at the inverter end. A functional block diagram of the controls is shown in Fig. 10 b). A fixed frequency oscillator is used to produce firing pulses for the output inverter and thus the output frequency is correctly controlled. The output ac voltage magnitude is controlled by varying the voltage V_c on the inverter dc smoothing capacitor. The reference value V_{cref} for capacitor voltage is made load dependent because the voltage drop across the inverter transformer varies with the magnitude and power factor of the load. V_{cref} is determined by a Proportional-Integral type controller acting on output voltage error. On the other hand, the machine voltage magnitude is maintained at approximately rated value by controlling the rectifier dc current and thus the reactive power drawn by the rectifier. Controlling reactive power drawn by the rectifier acts to control the machine voltage by modifying the excitation current available to the machine.

The rectifier current is controlled by a current controller as illustrated in Figure 10 b). At any given machine speed w and required power Po there is a specific combination of average dc rectifier voltage V_{dref} and current I_{dref} which maintains the self-excited machine at approximately rated voltage while delivering the required output power. It is not necessary to have machine voltage at exactly the rated value, merely within acceptable limits, because the customer is supplied from the regulated inverter output. Look-up tables of required rectifier output I_{dref} and V_{dref} as functions of w and Po have been prepared for use in the controls of the simulated apparatus. Po is the instantaneous power required to supply the load and support the output voltages and is determined from the monitored dc capacitor voltage V_c and dc current to the inverter I_L. I_{dref} from the tables is multiplied by machine voltage in p.u., V_{mach}, to produce the final rectifier current order. This makes the rectifier current order responsive to machine over-voltage or under-voltage.

The boost converter operates to pass the current of the rectifier to the dc smoothing capacitor for a controlled fraction of each period so as to maintain the desired output voltage. The voltage-boost converter controller takes in the actual dc current I_d and the actual dc capacitor voltage V_c and determines the duty cycle of the boost converter which will result in power Po being passed to the dc capacitor. The controller then sends the appropriate on/off pulses to the boost converter GTO. Limits are imposed to facilitate transient operation. The boost converter duty cycle controller also produces an estimate of the average dc voltage at the load end of the smoothing inductor as a compensation signal to assist the rectifier current controller.

Two simulation results are presented for the proposed power delivery scheme. The first result illustrates start-up of the rectifier, boost-converter, and inverter so as to control load voltage to rated value. The second result illustrates the application of full load in two steps followed by a complete load rejection as illustrated in Figure 12f. The second simulation result begins from steady operation at no load.

In the start-up simulation of the apparatus in Figure 10, the sequence of events is as follows:

a) at t=0.03 s, the machine is switched onto the self-excitation capacitors. A 0.5 p.u. voltage initial condition is assumed to hold between t=0 and t=0.03 s.
b) at t=0.04 s, the phase locked loop is freed to lock onto the machine ac voltages. The force-commutated GTO inverters are also deblocked at 0.04 s.
c) at t=0.057 s, the rectifier and boost-converter are deblocked and their associated controls are activated in order to bring the machine voltage under control. During the start-up transient the feedback control for the ac load voltage is disabled and V_{cref} is held at the minimum of its permitted range 0.954 p.u.
d) at t=0.25 s, the main start-up transient is complete and the feedback control of the ac load voltage is enabled. By 0.35 s the load voltage has risen to within 0.5% of the load reference voltage as illustrated in Figure 11 c).

The execution time for the start-up simulation was 1 minute 48 seconds per 60 Hz cycle on a Sun Sparc 1 workstation. That execution time is required for two main reasons. First of all there were 32 diodes, thyristors, and GTOs in the circuit and the boost-converter is operated at 1000 Hz. This causes a large number of switchings in the simulation. Secondly, there is a broad range of time constants involved in the circuit as is evident when switching times are contrasted to machine time constants. The overall start-up simulation therefore required 1 hour 48 minutes to complete. The large step size was 42.5 μS while the small was 1.5 μS. An option for speed-up allows some devices to be represented according to characteristic curves while others are switched in a binary (on/off) fashion. For example, representing the voltage inverters with coordinated switches reduces the execution time by approximately 40%.

Figure 11 a) illustrates the measured machine voltage during the start-up. It can be noted that the machine voltage reaches 0.75 p.u. before the rectifier deblocks at 0.057 s. Nevertheless, control of rectifier current prevents any overvoltage on the self-excited induction machine during start-up. Moreover, machine voltage settles at 1.0 p.u. (120V *l*-n) at the end of the start-up.

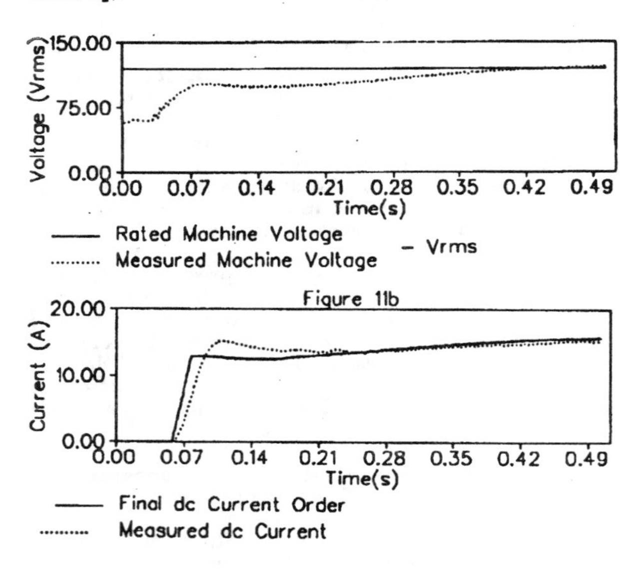

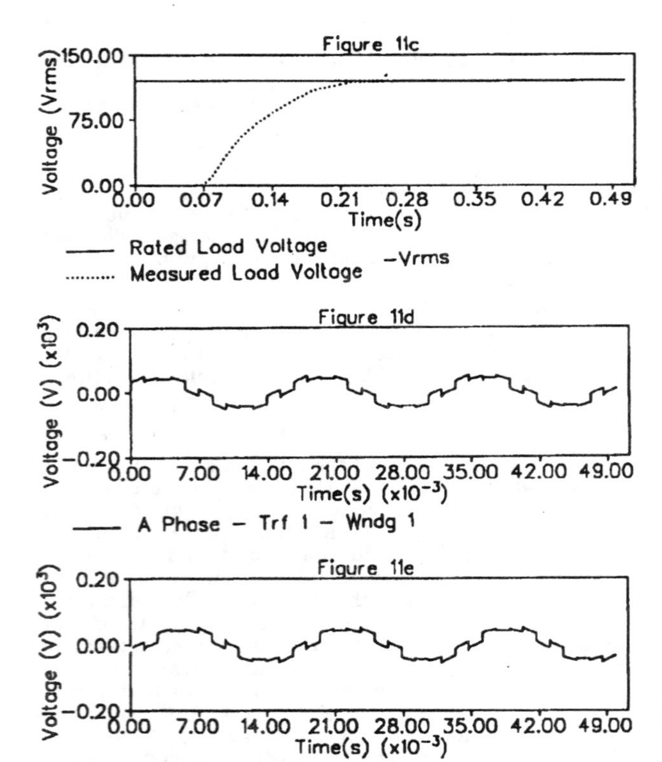

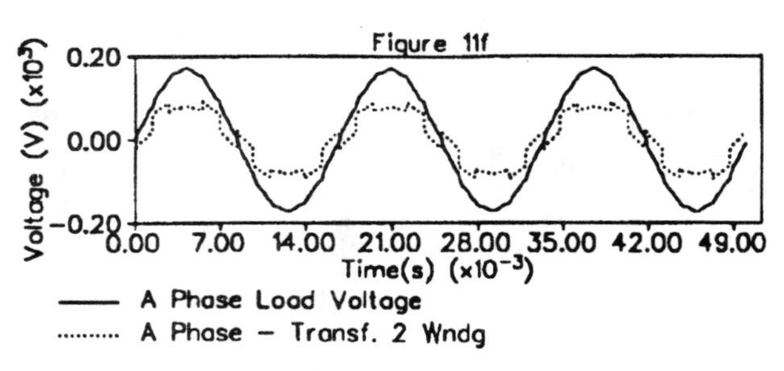

Figure 11. Simulation Results for Start-up of the Power Delivery Apparatus

Figure 11 b) illustrated the measured smoothing inductor current I_d and the final dc current order during the start-up. The final dc current order is reduced from I_{dref} (obtained from the look-up table) for most of the start-up because of the machine undervoltage illustrated in Figure 11 a). Forcing I_d to be lower than I_{dref} during start-up permits the machine voltage to build to rated value.

Figure 11 c) illustrates the build-up of load voltage during start-up. The voltage builds up with a time constant of approximately 0.1 s after the apparatus is deblocked at 0.057 s.

Figures 11 d), e) and f) illustrate the voltages on the three series-connected windings (two on trans. 1, one on trans. 2) which added together provide the A phase load voltage produced by the force-commutated GTO inverter. The waveshapes are for the last 3 cycles of start-up. Figure 11 f) also illustrates the A phase load voltage and the effect of the shunt filters. The load voltage contains harmonic voltages of 0.35% of 11th; 1.58% of 13th; 1.01% of 23rd; and 0.42% of 25th. This represents an output voltage which is of acceptable harmonic content for most applications. The winding voltages were readily simulated because the inductive nodes between the series-connected windings (such as node A in Figure 10 a) are free of numerical oscillations.

In the simulation of load changes the simulation starts with the apparatus running under no load. The sequence of events is as follows:

a) at t=0.1 s, a load of 0.5 p.u. at 0.8 p.f. is applied to the inverter.
b) at t=0.25 s, the load on the inverter is increased to 1.0 p.u. at 0.8 p.f.
c) at t=0.5 s, the load breakers open to cause a complete load rejection of the inverter load.

Figure 12 a) illustrates the ac load voltage during the load changes. A 10% dip in load voltage of approximately 2 cycle duration occurs following the increase of load to 1.0 p.u. at 0.25 seconds. There is also a longer duration 10% output overvoltage following load rejection at 0.5 seconds. The overvoltage arises out of the full-load voltage drop across the inverter transformer which disappears on load rejection and is replaced by a voltage rise due to the capacitance of the output filters. Furthermore, after complete load rejection the dc

capacitor voltage V_c cannot be reduced quickly to reduce output voltage because the only remaining outlet for energy in the dc capacitor is the inverter and ac filter losses.

Figure 12 b) illustrates measured dc current I_d as compared to the final dc current order during the load changes. I_d is significantly lower than the order for approximately 2 cycles following the increase of load to 1.0 p.u. at 0.25 seconds. This causes the 10% machine overvoltage at 0.25 seconds as illustrated in Figure 12 c).

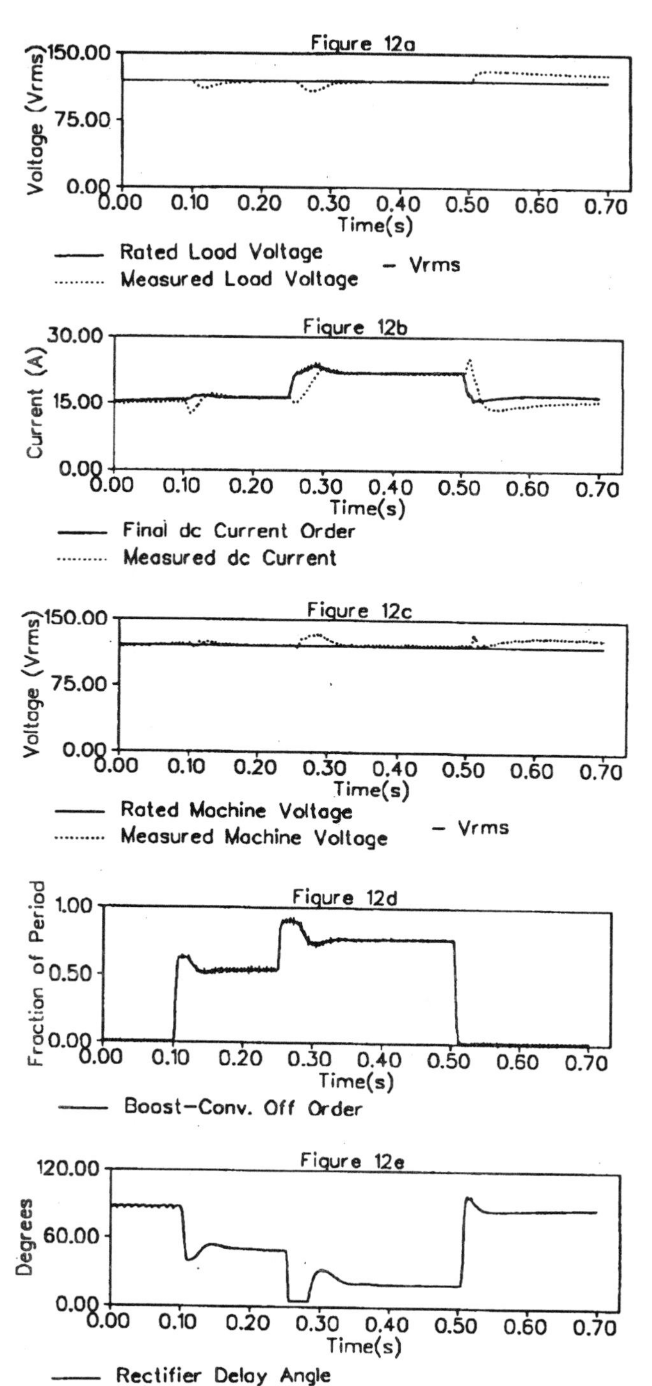

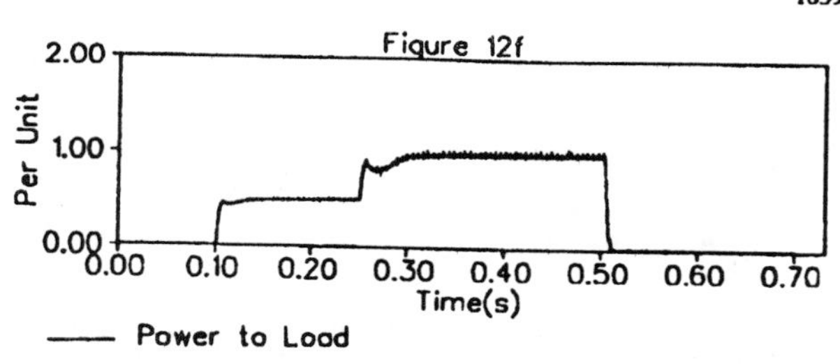

Figure 12. Simulation Results for Load Changes on the Power Delivery Apparatus

Figure 12 c) illustrates the measured machine voltage during the load changes.

Figure 12 d) illustrates the fraction off order (duty cycle) for the GTO in the boost converter during the load changes.

Figure 12 e) illustrates the rectifier delay angle during the load changes. Note that the firing angle of the rectifier goes to minimum delay of 5 degrees when the load is increased to 1 p.u. at 0.25 seconds. This angle is required in order that I_d can be increased to control the machine over-voltage which begins at that time.

CONCLUSIONS

This paper presents a method for transient simulation of power electronic apparatus embedded in larger electrical circuits. A technique has been presented for suppression of erroneous numerical oscillations which can arise in nodal admittance analysis in solutions for capacitive loops and inductive nodes. Representation of switching devices according to device characteristics has been described. In addition an iterative technique has been presented which is helpful in modelling simultaneous multiple device switchings. Recourse to a smaller timestep during switching further enhances the quality of simulation. Example simulation results are presented for a power delivery scheme which includes three power converter types which may be typical in future transmission systems.

REFERENCES

1. H.W. Dommel, "Digital Computer Solution of Electromagnetic Transients in Single and Multiphase Networks", IEEE Trans. on Power Apparatus and Systems, Vol. PAS-88, No. 4, April, pp. 388-399, (1969).

2. J. Douglas, "Switching to Silicon", IEEE Power Engineering Review, Vol. 9, No. 10, October, pp. 23-24, (1989).

3. A.M. Gole, R.W. Menzies, D.A. Woodford, and H. Turanli, "Improved Interfacing of Electrical Machine Models in Electromagnetic Transient Programs", IEEE Trans. on Power Apparatus and

Systems, Vol. PAS-103, No. 9, June, pp. 2446-2451, (1984).

4. B. Kulicke, "Simulation program NETOMAC: Difference Conductance Method for Continuous and Discontinuous Systems", Siemens Research and Development Reports, Vol. 10, No. 5, pp. 299-302, (1981).

5. H.W. Dommel (editor), *Electromagnetic Transients Program Reference Manual (EMTP THEORY BOOK)*, Portland, Oregon: Bonneville Power Administration, 1986, pp. II-1 to II-4.

6. R.W.Y. Cheung, H. Jin, B. Wu, and J.D Lavers, "A Generalized Computer-Aided Formulation for the Dynamic and Steady-State Analysis of Induction Machine Inverter Drive Systems", IEEE Trans. on Energy Conversion, Vol. 5, No. 2, June, pp. 337-343, (1990).

7. W.F. Tinney and J.W. Walker, "Direct Solutions of Sparse Network Equations by Optimally Ordered Triangular Factorization", Proc. of the IEEE, Vol. 55, No. 11, Nov., p. 1801, (1967).

8. J. Arrillaga and D.B. Watson, "Static Power Conversion from Self-Excited Induction Generators", Proc. Inst. Electr. Eng. (G.B.), Vol. 125, No. 8, pp. 743-6, Aug. (1978).

BIOGRAPHIES

T.L. Maguire (S'84,M'88) obtained the B.Sc.(E.E.) degree from the University of Manitoba in 1975 and the M.Sc. degree in Electrical Engineering from the University of Manitoba in 1986.

Mr. Maguire is currently on leave from the Manitoba HVDC Research Centre and is pursuing Ph.D. studies at the University of Manitoba. He has been employed at the Manitoba HVDC Research Centre since 1986. His research interests include electromagnetic transients simulation and power electronics.

Mr. Maguire is a Registered Professional Engineering in the Province of Manitoba and a member of the IEEE Power Engineering and Power Electronics Societies.

A.M. Gole (S'77,M'82) obtained the B. Tech. (E.E.) degree from the Indian Institute of Technology Bombay in 1978 and the Ph.D. Degree in Electrical Engineering from the University of Manitoba in 1982.

Dr. Gole is currently an Associate Professor in the Dept. of Electrical and Computer Engineering at the University of Manitoba. From Sept. 1988 to Aug. 1989 he was on sabbatical leave at IREQ in Varennes, Quebec. His research interests include HVdc transmission, power system transients simulation and power electronics.

Dr. Gole is a Registered Professional Engineer in the Province of Manitoba and a member of the IEEE Power Engineering & Power Electronics Socities.

IEEE Transactions on Power Systems, Vol.6, No.1, February 1991

A MULTIPHASE HARMONIC LOAD FLOW SOLUTION TECHNIQUE

Wenyuan Xu
Student M.IEEE

Jose R. Marti
Member, IEEE

Hermann W. Dommel
Fellow, IEEE

Department of Electrical Engineering
University of British Columbia
Vancouver, B.C.
Canada, V6T 1W5

ABSTRACT: The operation of nonlinear devices under unbalanced conditions may cause harmonic problems in power systems. A multiphase harmonic load flow solution technique for analysing such problems is described in this paper. The harmonic load flows are obtained from iterations between the Norton equivalent circuits of the nonlinear elements and the linear network solutions at harmonic frequencies. Harmonics generated by static compensators with thyristor-controlled reactors under unbalanced conditions are used to illustrate the method. The inclusion of the control characteristics of the static compensator and comparisons with field test results are described in a companion paper.

KEYWORDS: network unbalances, harmonics, harmonic load flow, multiphase solution, static compensator.

1. INTRODUCTION

The voltage and current waveforms in power systems are frequently distorted by harmonics. This distortion can cause various problems, ranging from capacitor failure to communications interference [1]. The propagation of these harmonics through the network must be assessed so that harmonic counter-measures can be properly designed. Harmonic load flow programs are an important tool for such harmonic propagation studies. The work on harmonic load flow solution techniques was pioneered by Heydt et al [1]. Their well-known HARMFLO program can be used to analyse harmonic load flows caused by rectifiers, HVDC converters, and other nonlinear devices [2, 3]. Similar techniques are described in [4, 5].

There are also many harmonic problems related to the unbalanced operation of power systems. With unbalanced conditions, the generation and propagation of harmonics are more complicated. For example, extra so-called non-characteristic harmonics may be produced. It is important to evaluate the effects of these harmonics since no filters are generally installed to alleviate them. In view of the fact that there is always some degree of unbalance and that the harmonics are more sensitive to unbalances than the fundamental frequency component, unbalanced harmonic analysis has received more attention recently. Unbalance effects have been included in the three-phase frequency scan technique proposed by Densem, Bodger and Arrillaga [5] and in the computation of non-characteristic harmonics from rectifiers [6].

90 WM 098-4 PWRS A paper recommended and approved by the IEEE Power System Engineering Committee of the IEEE Power Engineering Society for presentation at the IEEE/PES 1990 Winter Meeting, Atlanta, Georgia, February 4 - 8, 1990. Manuscript submitted August 31, 1989; made available for printing November 17, 1989.

In this paper, a multiphase harmonic load flow (MHLF) technique is described which solves the network at fundamental and harmonic frequencies in the presence of nonlinear elements and unbalances. It is based on the harmonic iteration scheme which has been used earlier to compute harmonics from HVDC converters, transformer saturation and thyristor-controlled reactors [7, 8, 9]. The harmonics caused by static compensators with thyristor-controlled reactors under balanced and unbalanced conditions are used to illustrate the method. This technique was primarily developed as an improved initialisation procedure for EMTP simulations, but it is also a useful tool by itself for multiphase harmonic load flow analysis with unbalanced conditions.

The multiphase harmonic load flow technique is simple in concept. Besides static compensators, other harmonic-producing nonlinearities with or without control specifications can be included. In order to keep the explanation of the method simple, only static compensator operation with known conduction angles is considered in this paper. The inclusion of the control characteristics of the static compensator and case study results as well as field test comparisons are presented in a companion paper [10].

2. PRINCIPLE OF HARMONIC ITERATION

To explain the principle of harmonic iteration, a static compensator with thyristor-controlled reactors (TCR) will be used as the source of harmonics. This device is essentially a reactor in series with anti-parallel thyristor valves, as shown in Figure 1. The valves conduct on alternate half-cycles of the supply frequency, for durations which depend on their firing angles α, thereby creating adjustable reactive power generation or consumption [11]. The operating range goes from no conduction at $\alpha = 180°$ to full conduction at $\alpha = 90°$. The duration of conduction is defined by the conduction angle

$$\sigma = 2(180° - \alpha).$$

If σ is less than 180°, harmonic currents are generated, as shown in Figure 1.

2.1 Harmonic Norton Equivalent Circuit of Thyristor-Controlled Reactor

To compute the harmonic currents with the harmonic iteration scheme, an equivalent linear model of the TCR shall be derived first. In general, the voltages appearing across this element will be distorted with harmonics,

$$v(t) = \sum_{h=1}^{n} |\mathbf{V}_h| cos(h\omega t + \phi_h). \quad (1)$$

With the valve fired at t_f, the current through the reactor during one half of a cycle is determined by

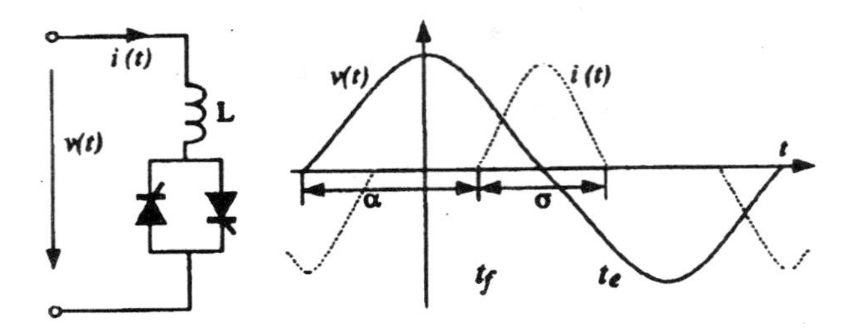

Figure 1: Thyristor-controlled reactor and waveforms.

$$L\frac{di}{dt} = v(t), \qquad i(t_f) = 0, \tag{2}$$

or

$$i(t) = \begin{cases} \sum_{h=1}^{n} |\mathbf{V}_h|(h\omega L)^{-1}[sin(h\omega t + \phi_h) - sin(h\omega t_f + \phi_h)], \\ \quad t_f \le t \le t_e; \\ 0, \quad 0 < t < t_f \text{ and } t_e < t < T/2; \end{cases} \tag{3}$$

where t_f is the instant of firing, t_e that of extinction, L the inductance of the TCR, and T the period at fundamental frequency. For the second half cycle, the current will reverse in sign (Figure 1). To obtain the current phasors as a function of the voltage phasors, Fourier analysis is required. Since the closed-form Fourier analysis of Eq. (3) is complicated due to discontinuous conduction, the distorted current is generated point-by-point from the given voltage with Eq. (3) and then analyzed with discrete Fourier analysis. This produces the harmonic content expressed by

$$i_{harmonic}(t) = \sum_{h=1}^{n} |\mathbf{I}_h| cos(h\omega t + \theta_h), \tag{4}$$

which was used as a current source model in [9] to represent the harmonic effects of the TCR. It is better to model the TCR as a Norton equivalent circuit, however. It can be shown that the equivalent inductance of a TCR for a purely sinusoidal voltage at fundamental frequency is [11]:

$$L_{eq} = \pi L(\sigma - sin\sigma)^{-1} \tag{5}$$

This equivalent inductance represents the TCR very well at fundamental frequency, and reasonably well at other frequencies. The differences between the current absorbed in $jh\omega L_{eq}$ and the actual current from Eq. (4) become the parallel current sources in the Norton equivalent circuit representation of Figure 2, with

$$Y_{h-eq} = (jh\omega L_{eq})^{-1} \tag{6}$$

$$\mathbf{I}_{h-eq} = (jh\omega L_{eq})^{-1}\mathbf{V}_h - \mathbf{I}_h \tag{7}$$

where $\mathbf{V}_h = |\mathbf{V}_h|e^{j\phi_h}$ and $\mathbf{I}_h = |\mathbf{I}_h|e^{j\theta_h}$.

For given voltages at the TCR terminal, current sources of

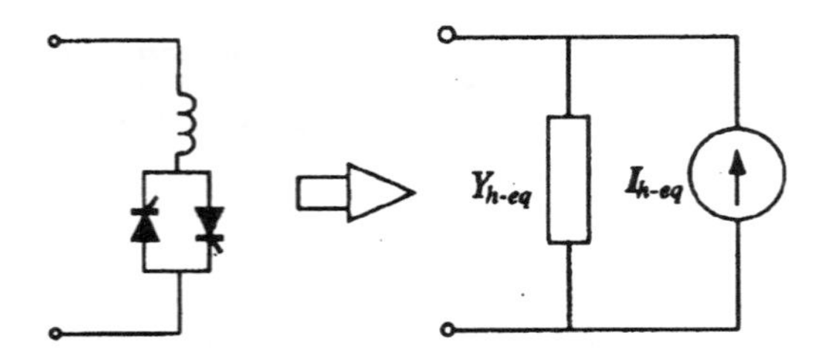

Figure 2: TCR equivalent model for multiphase harmonic load flow analysis.

the Norton equivalent circuit are easily found from Eq. (7), after $\mathbf{I}_h$ has been obtained from the Fourier analysis of Eq. (3). This relationship is symbolically expressed as

$$\mathbf{I}_{h-eq} = N(\sigma, [\mathbf{V}_1 \ \ldots \ \mathbf{V}_n]), \qquad h = 1, \ldots, n. \tag{8}$$

Note that there is no coupling among the equivalent circuits of the different harmonic frequencies. They are also independent of any network unbalances.

2.2 Harmonically-Decoupled Network Solutions

With the nonlinear TCR represented as a Norton equivalent circuit at harmonic frequencies, the node voltages of the entire network can easily be found by solving a system of multiphase node equations

$$[Y_{h-network}][\mathbf{V}_{h-network}] = [\mathbf{I}_{h-network}], \qquad h = 1, \ldots, n, \tag{9}$$

at each frequency. The nonlinear TCR effects are represented in these equations as currents in the vector $[\mathbf{I}_{h-network}]$. Once the node voltages have been obtained, improved values for the equivalent current sources can then be calculated from Eq. (8), which in turn is used to compute improved voltages. This is the process of harmonic iteration. It is continued until the changes in the equivalent current sources are sufficiently small.

This iterative process constitutes the basic idea of the multiphase harmonic load flow technique. To include the load flow constraint options of HARMFLO and similar programs, Eq. (9) at $h = 1$ must be modified into multiphase load flow equations, as described next.

3. MULTIPHASE FUNDAMENTAL FREQUENCY LOAD FLOW SOLUTIONS

The multiphase load flow solution must be able to handle unbalanced conditions. Unbalanced load flow analysis was first introduced by El-Abiad and Tarsi two decades ago [12]. Since then, much progress has been made [13, 14]. An excellent summary of the state-of-the-art of three-phase load flow analysis can be found in [13].

3.1 Modelling of Network Components

In contrast to most existing techniques, the load flow constraints for each power system component are expressed as branch equations here, instead of constraints on node quantities. Since branches can be connected in any way by the user, this provides greater flexibility.

3.1.1 Three-phase Synchronous Machines

The response of a synchronous machine is different for positive, negative, or zero sequence current injections. This must be taken into account in unbalanced load flow studies. Reference [12] has developed such a model (Figure 3(a)), with the branch equations

$$[\mathbf{I}_{km}] = [Y_s]([\mathbf{V}_k] - [\mathbf{V}_m] - [\mathbf{E}]) \tag{10}$$

where

$[\mathbf{V}_k] = [\mathbf{V}_{k-a} \ \mathbf{V}_{k-b} \ \mathbf{V}_{k-c}]^T$, voltages on side k,

$[\mathbf{V}_m] = [\mathbf{V}_{m-a} \ \mathbf{V}_{m-b} \ \mathbf{V}_{m-c}]^T$, voltages on side m,

$[\mathbf{I}_{km}] = [\mathbf{I}_{km-a} \ \mathbf{I}_{km-b} \ \mathbf{I}_{km-c}]^T$, currents from side k to side m,

$[\mathbf{E}] = [\mathbf{E}_p \ a^2\mathbf{E}_p \ a\mathbf{E}_p]^T$, internal voltages,

$[Y_s]_{mutual} = (Y_o - Y_n)/3$,

$$[Y_g]_{self} = (Y_o + 2Y_n)/3,$$
$$a = e^{-j2\pi/3}.$$

Subscripts p, n, and o indicate positive, negative, and zero sequence components, respectively. For the negative sequence reactance, $j\omega(L_d'' + L_q'')/2$ can be used. The zero sequence reactance is $j\omega L_o$. For the resistance, the armature resistance can be used, though this is not quite correct because the negative sequence resistance can be an order of magnitude larger.

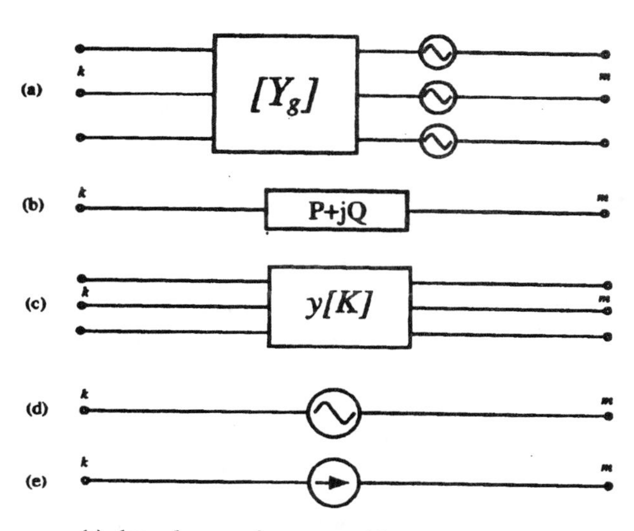

(a) three-phase synchronous machine.
(b) individual branch load. (c) three-phase static load.
(d) voltage source. (e) current source.

Figure 3: Branch models for multiphase harmonic load flow studies.

The machine internal voltage $\mathbf{E}_p$ is unknown and must be adjusted to satisfy the machine load flow constraints for terminal voltage and power output.

(a) Slack machine. The specified constraints are the magnitude and the phase angle of the positive sequence voltage at the machine terminals.

$$[T]([\mathbf{V}_k] - [\mathbf{V}_m]) = \mathbf{V}_{specified}, \tag{11}$$

where $[T] = (1/3)[1 \quad a \quad a^2]$.

Machines with these constraints correspond to the slack nodes in conventional load flow studies. Under unbalanced conditions, the negative and zero sequence voltages will be nonzero because the network sees the machine as admittances Y_n and Y_o in negative and zero sequence representations.

(b) PV machine. The specified constraints are the three-phase active power output and the magnitude of the positive sequence voltage at the machine terminals.

$$Real\{ -[\mathbf{I}_{km}]^H([\mathbf{V}_k] - [\mathbf{V}_m]) \} = P_{specified} \tag{12}$$
$$|[T]([\mathbf{V}_k] - [\mathbf{V}_m])| = V_{specified}, \tag{13}$$

where superscript ^{H} denotes conjugate transposed.

(c) PQ machine. The specified constraints are the three-phase active and the three-phase reactive power output.

$$-[\mathbf{I}_{km}]^H([\mathbf{V}_k] - [\mathbf{V}_m]) = (P + jQ)_{specified}. \tag{14}$$

The constraints for either type of machine can be generalized as

$$G([\mathbf{I}_{km}], [\mathbf{V}_k], [\mathbf{V}_m]) = F_{specified} \tag{15}$$

This equation and Eq. (10) for branch currents jointly define the three-phase machine model.

3.1.2 Multiphase Loads

Loads are usually represented as constant power consumption in single-phase (positive sequence) load flow programs. For unbalanced load flow analysis, the different response to positive, negative and zero sequence voltages and currents must be modelled, in addition to the power constraints. In view of the variety of load characteristics, four basic types of load models are proposed.

Type 1. Constant impedance load with known [Z] or [Y] branch-matrix representation.

Type 2. Load with constant active and reactive power specified as a single-phase branch (Figure 3(b)). It is defined as:

$$\mathbf{I}_{km}^H(\mathbf{V}_k - \mathbf{V}_m) = (P + jQ)_{specified}. \tag{16}$$

This load corresponds to the traditional PQ representation. However, it is defined as a branch between two nodes, rather than from node to ground as in single-phase load flow programs. This allows large flexibility in the type of connection, as for example phase-to-phase loads in delta systems.

Type 3. Static load (Figure 3(c)). In this type of load, it is assumed that the positive and negative sequence impedances are equal and that the ratio of the positive to zero sequence impedance is available. In effect, it assumes that the load impedances are balanced for the three phases. The total active and reactive power is specified, but the positive and zero sequence impedances are not explicitly known. Using symmetrical components, the branch equations for this type of load in phase quantities can be expressed as

$$[\mathbf{I}_{km}] = y[K]([\mathbf{V}_k] - [\mathbf{V}_m]) \tag{17}$$
$$[\mathbf{I}_{km}]^H([\mathbf{V}_k] - [\mathbf{V}_m]) = (P + jQ)_{specified}, \tag{18}$$

where y is an unknown admittance and $[K]$ is a known constant symmetric matrix determined from the positive to zero sequence impedance ratio. The unknown admittance must be adjusted to fulfill Eq. (18).

Type 4. Rotating machine load with unequal negative and positive sequence impedances, as in the case of induction motors. For this load, it is assumed that both the negative and zero sequence admittances are known. The positive sequence admittance is not known and is to be determined from the three-phase active and reactive power consumption. With symmetrical components, this type of load can be represented as

$$\mathbf{I}_{km-p} = Y_p(\mathbf{V}_{k-p} - \mathbf{V}_{m-p}) \tag{19}$$
$$\mathbf{I}_{km-n} = Y_n(\mathbf{V}_{k-n} - \mathbf{V}_{m-n}) \tag{20}$$
$$\mathbf{I}_{km-o} = Y_o(\mathbf{V}_{k-o} - \mathbf{V}_{m-o}) \tag{21}$$

In stead of using Y_p as an unknown, it is better to model the load as an internal voltage $\mathbf{E}_p$ behind the negative sequence admittance Y_n. $\mathbf{E}_p$ then becomes the unknown variable and Eq. (19) can be rewritten as

$$\mathbf{I}_{km-p} = Y_n(\mathbf{V}_{k-p} - \mathbf{V}_{m-p} - \mathbf{E}_p), \tag{22}$$

Transforming Eqs. (20), (21) and (22) into the phase domain and including the built-in load flow constraints, this load model can be defined as:

$$[\mathbf{I}_{km}] = [Y_g]([\mathbf{V}_k] - [\mathbf{V}_m] - [\mathbf{E}]) \tag{23}$$

$$-[\mathbf{I}_{km}]^H([\mathbf{V}_k] - [\mathbf{V}_m]) = (P + jQ)_{specified}. \tag{24}$$

where

$$[\mathbf{E}] = [\mathbf{E}_p \;\; a^2\mathbf{E}_p \;\; a\mathbf{E}_p]^T,$$
$$[Y_g]_{mutual} = (Y_o - Y_n)/3,$$
$$[Y_g]_{self} = (Y_o + 2Y_n)/3,$$

Note that with the introduction of $\mathbf{E}_p$, this type of load has the same structure as the PQ synchronous machine. It can then be simply treated as a PQ machine with negative power generation.

By specifying the load flow constraints at the branch level, the loads can be arbitrary connected between nodes or from node to ground. Loads can also be connected to the *same* node.

3.1.3 Voltage and current sources

Voltage and current sources are again represented as branches. The voltage source (Figure 3(d)) is defined as

$$\mathbf{V}_k - \mathbf{V}_m = \mathbf{E}_{specified}. \tag{25}$$

A current source between two nodes (Figure 3(e)) is defined as currents leaving two nodes,

$$\mathbf{I}_k = \mathbf{I}_{specified},$$
$$\mathbf{I}_m = -\mathbf{I}_{specified}.$$

3.1.4 Other Network Components

Overhead transmission lines, underground cables, transformers, reactors and capacitors can all be modelled as coupled π circuits. Details are well-documented in references [13] and [15]. There are no load flow constraints associated with these components.

3.2 Formulation and Solution of the Load Flow Equations

With all network components described at the branch level in the form of Norton equivalent circuits, it becomes easy to write the nodal equations for the entire network. The branch admittance matrix of each component enters the larger network admittance matrix according to well-known building rules [15], while the current sources between sides k and m enter as currents with a positive sign on side k, and with a negative sign on side m. With load flow constraints, these currents in $[\mathbf{I}_u]$ are unknown, and must be iteratively adjusted. As a result, the network equation is formed as

$$[Y][\mathbf{V}] + [\mathbf{I}_s] + [\mathbf{I}_u] = 0, \tag{26}$$

where

$[Y]$ is the network node admittance matrix constructed from the branch admittance matrices without load flow constraints,

$[\mathbf{V}]$ is the node voltage vector,

$[\mathbf{I}_s]$ is the vector of current sources leaving each node,

$[\mathbf{I}_u]$ is the vector of unknown currents (associated with load flow constraints) leaving each node.

Collecting all the related equations together, the multiphase load flow problem can be formulated as:

$$\text{network: } [f_1] = [Y][\mathbf{V}] + [\mathbf{I}_s] + [\mathbf{I}_u] = 0 \tag{27}$$
$$\text{source: } f_2 = \mathbf{V}_k - \mathbf{V}_m - \mathbf{E}_{specified} = 0 \tag{28}$$
$$\text{load-2: } f_3 = \mathbf{I}_{km}^H(\mathbf{V}_k - \mathbf{V}_m) - (P + jQ)_{specified} = 0 \tag{29}$$
$$\text{machine: } f_4 = G([\mathbf{I}_{km}], [\mathbf{V}_k], [\mathbf{V}_m]) - F_{specified} = 0 \tag{30}$$
$$\text{load-3: } f_5 = [\mathbf{I}_{km}]^H([\mathbf{V}_k] - [\mathbf{V}_m]) - (P + jQ)_{specified} = 0 \tag{31}$$
$$\text{machine: } [f_6] = [\mathbf{I}_{km}] - [Y_g]([\mathbf{V}_k] - [\mathbf{V}_m] - [\mathbf{E}]) = 0 \tag{32}$$
$$\text{load-3: } [f_7] = [\mathbf{I}_{km}] - y[K]([\mathbf{V}_k] - [\mathbf{V}_m]) = 0 \tag{33}$$

The general form of these equations can be written as

$$F([x]) = 0, \tag{34}$$

where
$$[x] = [\, \mathbf{V} \;\; \mathbf{I}_V \;\; \mathbf{I}_{L2} \;\; \mathbf{I}_M \;\; \mathbf{I}_{L3} \;\; \mathbf{E}_p \;\; y \,]^T$$
$$[F] = [\, f_1 \;\; f_2 \;\; f_3 \;\; f_4 \;\; f_5 \;\; f_6 \;\; f_7 \,]^T,$$
$$[\mathbf{I}_u] = [\, \mathbf{I}_V \;\; \mathbf{I}_{L2} \;\; \mathbf{I}_M \;\; \mathbf{I}_{L3}]^T.$$

and
$[\mathbf{I}_V]$ is the vector of currents from voltage sources,
$[\mathbf{I}_{L2}]$ is the vector of single-phase PQ load currents,
$[\mathbf{I}_M]$ is the vector of machine currents,
$[\mathbf{I}_{L3}]$ is the vector of static load currents,
$[\mathbf{E}_p]$ is the vector of machine internal voltages,
$[y]$ is the vector of static load parameter y.

Equation (34) is a set of nonlinear algebraic equations, which must be solved iteratively. Experience has shown that the Newton-Raphson method is probably the best method for conventional load flow studies [16]. It has also been chosen for the solution of Eq. (34). Rectangular coordinates are used here to separate the complex variables and equations into real form. Besides its simplicity, the rectangular representation has other advantages. For example, if there are no PV and PQ constraints, the Jacobian matrix becomes constant. The solution is then equivalent to the direct solution of the linear problem $[Y][\mathbf{V}] = [\mathbf{I}]$.

With the Newton-Raphson method, the system of linear equations

$$[J_i][\Delta x_i] = -[\Delta F(x_i)] \tag{35}$$

is solved in each iteration step, and the variables are then updated with

$$[x_{i+1}] = [x_i] + [\Delta x_i], \tag{36}$$

where
i is the iteration number,
$[J_i]$ the Jacobian matrix,
$[\Delta F(x_i)]$ the residual vector.

With Eqs. (27) to (33), Eq. (35) becomes

$$\begin{bmatrix} [\frac{\partial f_1}{\partial \mathbf{V}}] & [\frac{\partial f_1}{\partial \mathbf{I}_V}] & [\frac{\partial f_1}{\partial \mathbf{I}_{L2}}] & [\frac{\partial f_1}{\partial \mathbf{I}_M}] & [\frac{\partial f_1}{\partial \mathbf{I}_{L3}}] & 0 & 0 \\ [\frac{\partial f_2}{\partial \mathbf{V}}] & 0 & 0 & 0 & 0 & 0 & 0 \\ [\frac{\partial f_3}{\partial \mathbf{V}}] & 0 & [\frac{\partial f_3}{\partial \mathbf{I}_{L2}}] & 0 & 0 & 0 & 0 \\ [\frac{\partial f_4}{\partial \mathbf{V}}] & 0 & 0 & [\frac{\partial f_4}{\partial \mathbf{I}_M}] & 0 & 0 & 0 \\ [\frac{\partial f_5}{\partial \mathbf{V}}] & 0 & 0 & 0 & [\frac{\partial f_5}{\partial \mathbf{I}_{L3}}] & 0 & 0 \\ [\frac{\partial f_6}{\partial \mathbf{V}}] & 0 & 0 & [\frac{\partial f_6}{\partial \mathbf{I}_M}] & 0 & [\frac{\partial f_6}{\partial \mathbf{E}_p}] & 0 \\ [\frac{\partial f_7}{\partial \mathbf{V}}] & 0 & 0 & 0 & [\frac{\partial f_7}{\partial \mathbf{I}_{L3}}] & 0 & [\frac{\partial f_7}{\partial y}] \end{bmatrix} \begin{bmatrix} \Delta \mathbf{V} \\ \Delta \mathbf{I}_V \\ \Delta \mathbf{I}_{L2} \\ \Delta \mathbf{I}_M \\ \Delta \mathbf{I}_{L3} \\ \Delta \mathbf{E}_p \\ \Delta y \end{bmatrix} = - \begin{bmatrix} \Delta f_1 \\ \Delta f_2 \\ \Delta f_3 \\ \Delta f_4 \\ \Delta f_5 \\ \Delta f_6 \\ \Delta f_7 \end{bmatrix}$$

The procedure for obtaining the submatrices in this Jacobian matrix is the same as in conventional load flow techniques. Once the Jacobian matrix is obtained, Eq. (35) is solved by Gauss elimination with sparsity techniques. The largest component in the residual vector is used to test for convergence.

3.3 Initialization

Choosing an initial guess $[x_o]$ for the iterations is more complicated in the multiphase case. The traditional initialization technique, which uses 1.0 per-unit node voltage magnitudes with respective 120° phase shifts among phases a, b, and c, becomes unreliable if there are phase shifting effects through wye-delta transformer connections. Since the convergence of the Newton-Raphson method is sensitive to the initial guess, a special initialization procedure is used before entering the iteration loop.

The procedure is based on the observation that the load flow equations become linear if there are no PV or PQ constraints. To approximate the network this way, the components with PQ and PV constraints are modified as follows:

1. Machines with PV and PQ constraints and rotating machine loads are represented as admittance matrices of very small magnitude. This approximates open-circuit conditions.
2. Other loads are represented as known admittances y whose values are determined from

$$y|\mathbf{V}|^2 = (P - jQ)_{specified} \tag{37}$$

for the single-phase PQ loads, and

$$y([T][K][T]^H)|\mathbf{V}|^2 = (P - jQ)_{specified} \tag{38}$$

for three-phase static loads. The voltage magnitude $|\mathbf{V}|$ is estimated to be equal to the user-supplied rated voltage of the load.

With these approximations, the load flow solution becomes linear. Rather than writing a separate algorithm for this initialization, the normal Newton-Raphson algorithm is used with zero initial values, and the linear estimate $[x_1]$ is obtained in one iteration. This becomes the starting point for the following load flow iterations.

4. MULTIPHASE HARMONIC SOLUTIONS

To solve the network at the harmonic frequencies, it is first necessary to define how the machines and loads respond to harmonics.

4.1 Harmonic Response of Network Components

1. **Machines.** As a first approximation, it is assumed that machines do not produce harmonics. For harmonic frequencies, they can then be modelled as known admittance matrices, as suggested in reference [17]:

$$Y_{h-p} = Y_{h-n} = 1/(jhX_{1-n})$$
$$Y_{h-o} = 1/(jhX_{1-o}),$$

where h is the harmonic order, and X_{1-n} and X_{1-o} are the negative and zero sequence reactances of the machine at fundamental frequency, respectively. The internal voltage is zero for $h > 1$. A similar model is used for rotating machine loads.

For more accurate representations, it must be realized that machines act as frequency converters. For example, a negative sequence current at fundamental frequency induces second harmonics in the rotor circuits, which in turn induce third harmonic voltages in the stator. This can be taken into account with the method suggested by Semlyen, Eggleston, and Arrillaga [18].

2. **Loads.** The behaviour of loads under the combined effects of unbalanced and harmonic conditions is usually not well known. Using the load modelling techniques of reference [19], a multiphase load can be modelled as a combination of lumped R,L,C elements. If the test data needed for this representation is not available, the recommendations of reference [17] can be used. This reference suggests that the harmonic characteristics of a load can be modelled as

$$Z_{load-h} = (R_s + jX_s)//jX_p, \tag{39}$$

where

$$R_s = |V_1|^2_{rated}/P_{specified},$$
$$X_s = 0.073hR_s,$$
$$X_p = hR_s/[6.7(Q_{specified}/P_{specified}) - 0.74],$$

Using these approximations, the single-phase PQ load is replaced by Z_{load-h}, and the static load is represented as

$$Z_{h-p} = Z_{h-n} = Z_{load-h} \tag{40}$$
$$Z_{h-o} = r_h^{-1} Z_{h-p}, \tag{41}$$

where r_h is the user-supplied positive to zero sequence impedance ratio of the load at the given frequency. These sequence parameters are then transformed into phase quantities.

3. **Transmission lines.** Transmission lines are represented as exact multiphase π equivalent circuits calculated at the considered harmonic frequency [15].

4.2 Solution Technique

With the various system components represented at each harmonic frequency, the problem formulation is the same as that of Eq. (34). However, since the machines and loads are represented as constant impedances at harmonic frequencies, the problem becomes linear and iterations are not required.

5. GENERAL FORMULATION OF THE MULTIPHASE HARMONIC LOAD FLOW TECHNIQUE

Based on Sections 3 and 4, a general purpose program was written for the solution of the network with load flow constraints at fundamental and harmonic frequencies. The three solution stages, initialization, load flow, and frequency scan are all done by the Newton-Raphson solution module, with or without iterations. The harmonic iteration of Section 2 for static compensators was added by using the Norton equivalent circuits of Figure 2. The general form of the multiphase harmonic load flow (MHLF) technique can then be described with Figure 4.

The initialization is very simple: At harmonic frequencies, the equivalent currents $\mathbf{I}_{h-eq}$ representing the effects of nonlinear elements are set to zero. For the fundamental frequency load flow solution, the PV and PQ components are modified into known Y matrices. With these simplifications, the approximate node voltages without harmonic distortion ($h = 1$) are obtained in one iteration. Because the voltage harmonics are relatively small compared with the fundamental frequency components, using these node voltages as initial conditions is quite reliable.

As shown in the flow chart, the MHLF technique consists of two basic parts. The first part is the construction of harmonic

Norton equivalent circuits for the nonlinear elements (a TCR is used as an example). The second part performs the network solutions at the fundamental and harmonic frequencies. These two parts are interfaced through the process of harmonic iteration.

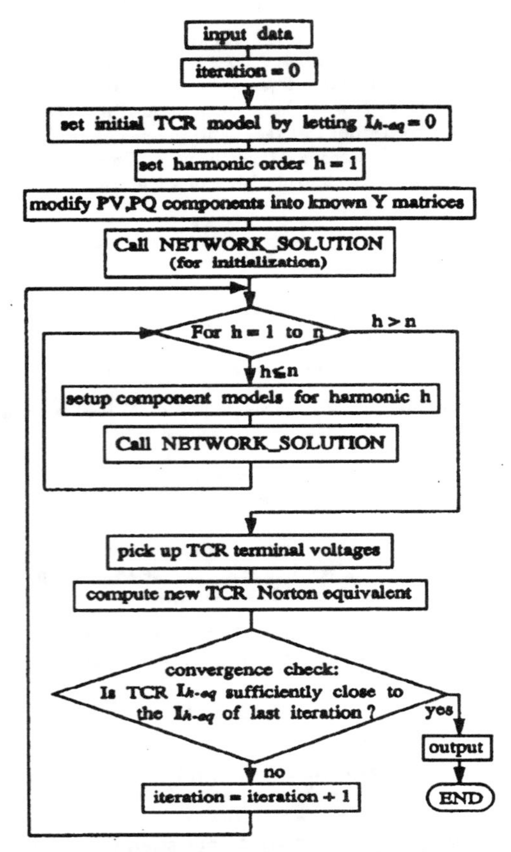

NETWORK_SOLUTION is a subroutine whose function is to form $F([x]) = 0$ equation and to solve it by Newton-Raphson method.

Figure 4: Flow chart of the MHLF technique. (A TCR is used as an example of nonlinear element.)

The main characteristics of the MHLF technique can be summarised as follows:

1. The MHLF technique is a multiphase program. It can be used either for single-phase or for three-phase harmonic load flow studies. Unbalanced operating conditions can be considered in the study.

2. Harmonics from other nonlinear elements can be analyzed with the technique. The linear equivalent circuit models for nonlinear elements not included internally in the program can be supplied by the user externally in the form of a subroutine. The MHLF program has been structured to make such interfaces easy.

3. The MHLF technique is computationally efficient. First of all, its initialization is simple and reliable. Secondly, because the Norton equivalent circuits are harmonically-decoupled, the network solutions are performed one frequency at a time. Thus the size of the Jacobian matrix is reduced considerably and the computational burden is only linearly proportional to the total number of harmonics. Thirdly, the technique has good convergence behaviour [10].

6. PRACTICAL CONSIDERATIONS FOR STATIC COMPENSATOR ANALYSIS

A practical static compensator, such as the one shown in Figure 5 [11], is more complicated than the simple TCR unit of Figure 1. First of all, there is the delta connection of the TCR's. This connection is used to filter the zero sequence harmonics (e.g. 3rd, 9th). With the multiphase representation and the Norton equivalent circuits, the modelling of delta connection is straightforward, and the circulation of the zero sequence harmonic currents is automatically taken into account. When the operating conditions are unbalanced, the non-characteristic harmonics in the delta connection are automatically obtained.

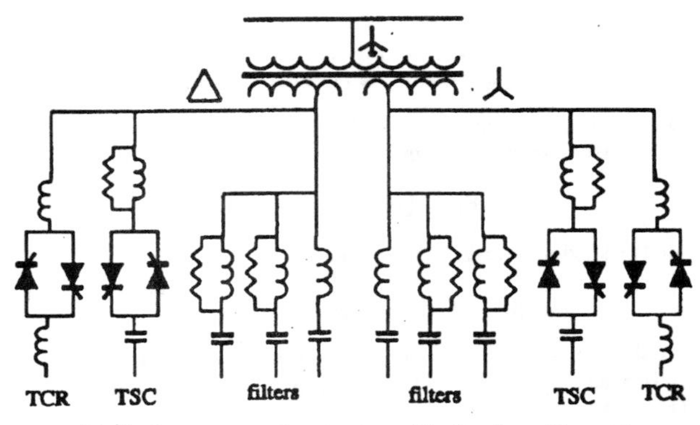

(a) Static compensator structure (single-phase diagram)

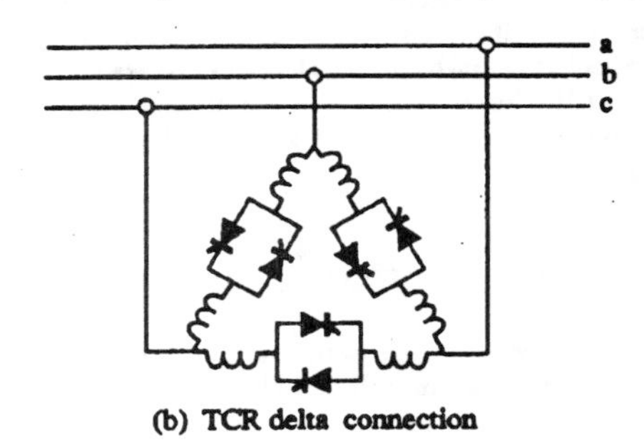

(b) TCR delta connection

Figure 5: The structure of a practical static compensator.

The second complication is the three-winding transformer connection. The phase shifting effects of the connection are critical for the cancellation of the 5th and 7th harmonics generated by the TCR's at the two secondary windings. With the multiphase modelling of transformers, the phase shift is automatically included [15]. Since the voltage ratio can also be included in the transformer model, the load flow analysis can be performed in either per-unit or physical quantities.

7. CONCLUSIONS

A multiphase harmonic load flow (MHLF) technique has been developed for the harmonic analysis of static compensators and other nonlinear devices under balanced or unbalanced conditions. This technique incorporates the harmonic iteration scheme into the multiphase framework. The MHLF technique consists of two major parts. The first part constructs harmonic Norton equivalent circuits for the nonlinear elements. The second part performs linear network solutions at fundamental and harmonic frequencies. User-supplied modules for particular nonlinear elements can also be easily interfaced with the program.

The paper describes the development of the Norton equivalent circuit for the TCR with a known conduction angle as an example of nonlinear elements modelling and the extension of this modelling to multiphase network solutions at fundamental and harmonic frequencies. The inclusion of the control characteristics of static compensators into the MHLF solution and the results of case studies are described in a companion paper [10].

8. ACKNOWLEDGEMENTS

The financial assistance of the System Engineering Division of B.C. Hydro and Power Authority is gratefully acknowledged. The authors are also indebted to TransAlta Utilities Corporation, Calgary, Alberta and to B.C. Hydro and Power Authority, Vancouver, B.C. for providing data and field test measurements.

9. REFERENCES

[1] D. Xia and G.T. Heydt, "Harmonic Power Flow Studies, Part I — Formulation and Solution, Part II — Implementation and Practical Application", *IEEE Trans. on Power Apparatus and Systems*, vol. PAS-101, pp. 1257-1270, June 1982.

[2] W. Song, G.T. Heydt and W.M. Grady, "The Integration of HVDC Subsystems into the Harmonic Power Flow Algorithm", *IEEE Trans. on Power Apparatus and Systems*, vol. PAS-103, pp. 1953-1961, Aug. 1984.

[3] W.M. Grady and G.T. Heydt, "Prediction of Power System Harmonics Due to Gaseous Discharge Lighting". *IEEE Trans. on Power Apparatus and Systems*, vol. PAS-104, pp. 554-561, March 1985.

[4] A. Semlyen, E. Acha, and J. Arrillaga, "Harmonic Norton Equivalent for the Magnetizing Branch of a Transformer", *IEE Proceedings*, Vol.134, Part C, No.2, pp. 162-169, March 1987.

[5] T.J. Densem, P.S. Bodger and J. Arrillaga, "Three Phase Transmission System Modelling for Harmonic Penetration Studies", *IEEE Trans. on Power Apparatus and Systems*, vol. PAS-103, pp. 310-317, Feb. 1984.

[6] D. Xia, Z. Shen and Q. Liao, "Solution of Non-characteristic Harmonics Caused by Multiple Factors in HVDC Transmission System", Proc. of the Third International Conference on Harmonics in Power Systems, Nashville, IN, pp. 222-228, Oct. 1988.

[7] R. Yacamini and J.C. de Oliveira "Harmonics in Multiple Convertor Systems: a Generalized Approach", *IEE Proceedings*, vol. 127, Part B, No. 2, pp. 96-106, March 1980.

[8] H.W. Dommel, A. Yan and W. Shi, "Harmonics from Transformer Saturation", *IEEE Trans. on Power Delivery*, vol. PWRD-1, pp. 209-215, April 1986.

[9] W. Xu and H.W. Dommel, "Computation of Steady-State Harmonics of Static Var Compensators", Proc. of the Third International Conference on Harmonics in Power Systems, Nashville, IN, pp. 239-245, Oct. 1988.

[10] W. Xu, J.R. Marti and H.W. Dommel, "Harmonic Analysis of Systems with Static Compensators", Paper submitted for IEEE PES Winter Meeting 1990.

[11] T.J. Miller, Ed. *Reactive Power Control in Electric Systems*, New York: John Wiley&Sons Inc., 1982.

[12] A.H. El-Abiad and D.C. Tarsi, "Load Flow Solution of Untransposed EHV Networks", *Proc. of 5th Power Industry Computer Applications Conference*, Pittsburgh, pp. 377-384, 1967.

[13] J. Arrillaga, C.P. Arnold and B.J. Harker, *Computer Modeling of Electrical Power Systems*, New Zealand: John Wiley&Sons, 1983.

[14] N.A. Wortman, D.L. Allen and L.L. Grigsby, "Techniques for the Steady State Representation of Unbalanced Power Systems", *IEEE Trans. on Power Apparatus and Systems*, vol. PAS-104, pp. 2805-2824, Oct. 1985.

[15] H.W. Dommel, *Electromagnetic Transients Program Reference Manual (EMTP Theory Book)*, Bonneville Power Administration, Aug. 1986.

[16] G.W. Stagg and A.H. El-Abiad, *Computer Methods in Power System Analysis*, New York: McGraw-Hill Book Company, 1968.

[17] CIGRE-Working Group 36-05, "Harmonics, Characteristic Parameters, Methods of Study, Estimates of Existing Values in the Network", *Electra*, no. 77, pp. 35-54, July 1981.

[18] A. Semlyen, J.F. Eggleston and J. Arrillaga, "Admittance Matrix Model of a Synchronous Machine for Harmonic Analysis", *IEEE Trans. on Power Systems* vol. PS-2, pp. 833-840, Nov. 1987.

[19] A.S. Morched and P. Kundur, "Identification and Modeling of Load Characteristics at High Frequencies", *IEEE Trans. on Power Systems*, vol. PS-2, pp. 153-160, Feb. 1987.

Wenyuan Xu (St.M'85) was born in China in 1962. He received a B.Eng. degree from Xian Jiaotong University, Xian, China in 1982 and a M.Sc. degree from the University of Saskatchewan, Saskatoon, Canada in 1985. At present, he is a Ph.D. candidate at the University of British Columbia.

Jose R. Marti (M'71) was born in Spain in 1948. He received a M.E. degree from Rensselaer Polytechnic Institute in 1974 and a Ph.D. degree in Electrical Engineering from the University of British Columbia in 1981.

From 1970 to 1972 he worked for industry. In 1974-77 and 1981-84 he taught power system analysis at Central University of Venezuela. Since 1984 he has been with the University of British Columbia.

Hermann W. Dommel was born in Germany in 1933. He received the Dipl.-Ing. and Dr.-Ing. degrees in electrical engineering from the Technical University, Munich, Germany, in 1959 and 1962 respectively. From 1959 to 1966 he was with the Technical University, Munich, and from 1966 to 1973 with Bonneville Power Administration, Portland, Oregon. Since July 1973 he has been with the University of British Columbia in Vancouver, Canada. Dr. Dommel is a Fellow of IEEE and a registered professional engineer in British Columbia, Canada.

Discussion

Adam Semlyen (University of Toronto): I would like to commend the authors for their novel approach in solving the multiphase load flow problem. Using a harmonic domain Norton equivalent for the representation of nonlinear elements has made it possible to include them in a Newton-Raphson solution of the whole system. The approach permits to solve even such difficult problems as encountered with a set of star connected nonlinear branches with isolated neutral.

In the approach of the paper, the Norton equivalent circuits are considered to be harmonically decoupled. Consequently, the load flow solution has been performed sequentially, for one harmonic at a time, a significant saving compared to a fully coupled approach. In general, however, the linearized representation of a nonlinear element in the harmonic domain[A] may couple the individual harmonics with each other. This appears clearly from equation (17a) of reference [A]

$$\mathbf{i} = \mathbf{Y}_{harm}\mathbf{v} + \mathbf{i}_N$$

which corresponds to a harmonic domain Norton equivalent with $\mathbf{i}$ and $\mathbf{v}$ being vectors of the harmonic components of currents and voltages and $\mathbf{i}_N$ the Norton current vector. The matrix $\mathbf{Y}_{harm}$ in the above equation (the Norton admittance matrix) is generally full. I wonder, therefore, what are the assumptions or procedures which permit to use the decoupled approach of the paper without loss of accuracy in the solution? Is the TCR a harmonically truly decoupled device?

In order to illustrate that the coupling between harmonics can be significant, consider the following extremely simple example. Let

$$i = \psi + \psi^3 \tag{a}$$

be the equation of a nonlinear reactor and

$$\psi = \psi_1 \cos\omega t \tag{b}$$

the input (flux linkage) of base frequency only. Direct substitution yields

$$i = (\psi_1 + \frac{3}{4}\psi_1^3)\cos\omega t + \frac{1}{4}\psi_1^3\cos 3\omega t \tag{c}$$

With $\psi_1 = \psi_{1\,base} + \Delta\psi_1$ eqn.(c) becomes

$$i = i_{base} + \Delta i \tag{d}$$

where

$$i_{base} = (\psi_1 + \frac{3}{4}\psi_1^3)_{base}\cos\omega t + \frac{1}{4}\psi_{1\,base}^3\cos 3\omega t \tag{e}$$

and

$$\Delta i = (1 + \frac{9}{4}\psi_1^2)_{base}\Delta\psi_1\cos\omega t + \frac{3}{4}\psi_{1\,base}^2\Delta\psi_1\cos\omega 3\omega t \tag{f}$$

With $\psi_{1\,base} = 1$ eqn.(f) becomes

$$\Delta i = 3.25\Delta\psi_1\cos\omega t + 0.75\Delta\psi_1\cos\omega 3\omega t \tag{g}$$

This equation shows that the same base frequency incremental input $\Delta\psi_1$ produces a third harmonic output of 23% (0.75/3.25) of the base frequency output. The resulting coupling is thus not insignificant.

[A] A. Semlyen and N. Rajakovic, "Harmonic Domain Modeling of Laminated Iron Core", IEEE Trans. on Power Delivery, Vol.4, No.1, January 1989, pp. 382-390.

Manuscript received February 5, 1990.

E. Acha (OCEPS Group, University of Durham, U.K.): I would like to congratulate the authors for this timely and most interesting paper. In particular, I would like to address the newly developed TCR model, which is simple and yet comprehensive.

TCRs are nonlinear components and their harmonic interaction with the power network is reached by iteration. This is especially true for cases in which harmonic voltage magnifications occur due to resonant conditions. One possible modelling approach is to represent each TCR branch as a voltage-dependent set of harmonic current sources, or better still, as a harmonic Norton equivalent [A]. Then, the three phase TCR model is assembled.

In this paper the authors are presenting a model, which represents each TCR branch as a voltage dependent set of harmonic current sources paralleled by a linear admittance. The linear admittance is based on fundamental frequency voltage and current considerations. This is amenable to a Norton equivalent at each harmonic frequency; as opposed to the harmonic Norton equivalent obtained in the harmonic domain. The former does not exhibit cross-couplings between harmonics, and standard factorization techniques will apply when solving the network. The latter exhibits cross-couplings between harmonics and more specialized inversion techniques will be required [B].

The authors have moved from the current source representation to the Norton equivalent representation of the TCR. Presumably, because of better convergence characteristics. Have the authors found this improvement to be significant? If so, then, the principle presented in this paper should also be applicable to other non-linear components, i.e. magnetic non-linearities and electric arcs.

The rationale behind this surmise is given below. Harmonic domain linearization lends itself to a Newton-type iterative solution, where the Jacobian matrix contains all the harmonics and the cross-couplings between harmonics. The Jacobian matrix corresponds to the admittance matrix of the harmonic Norton equivalent. During full conduction state, the TCR behaves linearly, and the admittance matrix of the harmonic Norton equivalent becomes diagonal, i.e. no cross-couplings between harmonics exist. In this condition, the Norton equivalent presented by the authors and the harmonic domain Norton equivalent will coincide. Clearly, this is not the case at other conduction states, however, a close numerical correspondance will exist if the cross-couplings between harmonics are neglected and the Jacobian matrix is evaluated during the first iteration only and then, kept constant until convergence is achieved. The Jacobian matrix is diagonally dominant and the harmonic admittance of the entire network is even more diagonally dominant. Thus, it is quite likely that the Norton equivalent representation of this paper will exhibit very good convergence characteristics even in cases of harmonic voltage magnification.

It has been found that for some non-linearities, such as iron cores, the number of iterations taken to reach convergence is the same, whether the Jacobian is evaluated at each iterative step or it remains constant after having been evaluated during the first iteration. For some other non-linearities, such as laminated iron cores, this is not the case. Nevertheless, the effects of neglecting cross-couplings between harmonics should be investigated for the case of iron cores, particularly, multi-legged transformers.

The authors must be congratulated for a most valuable paper.

[A] L.J. Bohman and R.H. Lasseter. "Harmonic Interactions in Thyristor Controlled Reactor Circuits', IEEE Transactions on Power Delivery, Vol. 4, No. 3, pp. 1919-1926, July 1989.

[B] A. Medina, J. Arrillaga and E. Acha. 'Sparsity-Oriented Hybrid Formulation of Linear Multiports and its Application to Harmonic Analysis', To be presented at the IEEE PES Winter Meeting, Atlanta, GA, Feb 4-8, 1990.

Manuscript received February 20, 1990.

Wenyuan Xu, Jose Marti and Hermann W. Dommel: We would like to thank all the discussers for their interest in the paper and for their valuable comments. We hope that the following comments will help to clarify some of the raised issues.

Professor Semlyen:

Professor Semlyen is correct in pointing out the harmonic coupling nature of nonlinear devices. The harmonic Norton equivalent curcuit for the TCR presented in the paper includes the coupling effects in the form of an iteratively adjusted harmonic current source I_{h-eq}. The system is solved for one harmonic frequency at a time. The solution is then compared with the characteristic of each nonlinear element and corrections are added to the element's harmonic current source.

Dr. Acha:

We observed significant improvement of the convergence rate by moving from a current source model to a Norton equivalent model. The improvement is due to a more accurate estimation of the fundamental frequency voltage at the TCR terminal. We completely agree with Dr. Acha's harmonic-domain linearization analysis. According to our numerous test runs, it is very likely that convergence can be further improved if the diagonal elements of the harmonic-domain Jacobian matrix are used as the equivalent admittances in the harmonic iteration scheme with nonlinear inductors has been presented in reference (I). In general, we have found that the convergence rate is inversely affected by the degrees of saturation and network harmonic voltage resonances.

Reference

[I] W. Xu, "A Multiphase Harmonic Load Flow Technique", Ph.D. Dissertation, University of British Columbia, Vancouver, B.C., Canada, February. 1990.

Manuscript received April 12, 1990.

IEEE Transactions on Power Delivery, Vol. 10, No. 3, July 1995

IMPLEMENTATION OF A NEW HARMONIC INITIALIZATION METHOD IN THE EMTP

X. Lombard[1] J. Mahseredjian[2] S. Lefebvre[2] C. Kieny[1]

[1]Direction des Etudes et Recherches
Electricité de France
Clamart, France

[2]Institut de Recherche d'Hydro-Québec (IREQ)
1800 Montée Ste-Julie, Varennes, Québec
Canada J3X 1S1

ABSTRACT : Nonlinear branches, such as saturable reactors, can generate harmonics and consequently increase the EMTP time-domain simulation time before the actual distorted steady-state is reached. This is an important initialization problem for transient analysis studies performed in steady-state operating conditions. This paper presents the implementation of a new method in the EMTP for initializing time-domain simulations. It is based on frequency domain steady-state calculations including harmonics from nonlinear branch functions.

Keywords : EMTP, Initialization, steady-state

1. INTRODUCTION

The time-domain simulation computer time of a network is closely related to its initial conditions, especially when transient analysis is performed in a distorted steady-state operating mode and the simulated system has large time-constants. A typical example [1] is the case of a severe magnetic storm that saturates power transformers. The *dc* voltage polarization of a transformer imposes an impractical EMTP [2] (Electromagnetic Transients Program) simulation time to reach steady-state.

The traditional EMTP network initialization method is provided by a fundamental frequency phasor solution with all nonlinear branches disconnected or operating on their first (passing through zero) linear segment. A major improvement is available in a more recent EMTP version through Initialization with Harmonics from nonlinear inductances [3]. This is an iterative fixed-point method based on the frequency domain representation of the nonlinear inductance as a voltage dependent harmonic current source. Besides its inability [2] to initialize for a *dc* flux component, it is shown in this paper that it may also suffer from convergence problems and cannot be applied in a more general context or for extracting ferroresonant states.

Considering that nodal analysis is applied in the EMTP formulation of frequency domain equations, other eligible methods [4]-[7] for harmonic steady-state calculations are based on a Norton equivalent representation of a nonlinear branch at each harmonic frequency.

94 SM 438-2 PWRD A paper recommended and approved by the IEEE Transmission and Distribution Committee of the IEEE Power Engineering Society for presentation at the IEEE/PES 1994 Summer Meeting, San Francisco, CA, July 24 - 28, 1994. Manuscript submitted July 21, 1993; made available for printing April 20, 1994.

This paper presents the implementation of a Newton type harmonic steady-state calculation method for initializing time-domain simulations in the EMTP. It uses a generalized and theoretically supported formulation of the harmonic Norton equivalent modelling of nonlinear branches with simplified harmonic coupling. There is no requirement for analytical formulation of nonlinear branch characteristics or precalculated knowledge of fundamental frequency behavior, as for example, in the case of a thyristor controlled reactor [6].

Practical cases with the nonlinear inductance are demonstrated.

The qualification of the Norton equivalent (NE) method as a Newton type method for solving a complete system of network equations, is also investigated.

The generalization of this paper provides a new method for the representation of *dc* components. It is applied to correctly initialize time-domain simulations where a *dc* component subsists in the reached steady-state.

Another main contribution is the ability to calculate and initialize ferroresonant states. The convergence to these states is only achieved with the presented Newton type harmonic initialization method in the EMTP.

2. CONVERGENCE PROBLEMS WITH EMTP INITIALIZATION WITH HARMONICS METHOD

Initialization with Harmonics (IwH) is an option currently available in the EMTP [2] and designed specifically for harmonics generated by the nonlinear inductance. IwH is a steady-state frequency domain solution method [3] where the nonlinear inductance is modelled as a voltage dependent harmonic current source. A preliminary step calculates the nonlinear inductance voltage phasor using its $V_{L_{RMS}} - I_{L_{RMS}}$ characteristic. This voltage phasor is used to find current harmonics from the actual nonlinear flux-current characteristic. The current harmonics are individually reinjected back into the linear network to calculate a new set of voltage phasors that are superposed to find the new flux and then the new current harmonics, until convergence. The complete procedure constitutes a fixed-point iterative method.

The circuit of Fig. 1 [8] is a nonlinear resonant circuit used to illustrate some convergence problems that may encounter the IwH method, it is also a reference circuit for the numerical section of this paper. If, for analysis purposes, all resistors are neglected, then :

$$e(t) = E\sin(\omega t) = E\cos(\omega t - 90^\circ) \quad (1)$$

$$v_L(t) = e(t) - v_C(t) \quad (2)$$

When there is no *dc* component in the circuit, the inductance current can be expressed as :

$$i_L(t) = \sum_{h=1}^{\infty} I_{L_h} \cos(h\omega t + \psi_h) \quad (3)$$

where I_{L_h} is the amplitude of harmonic component h.

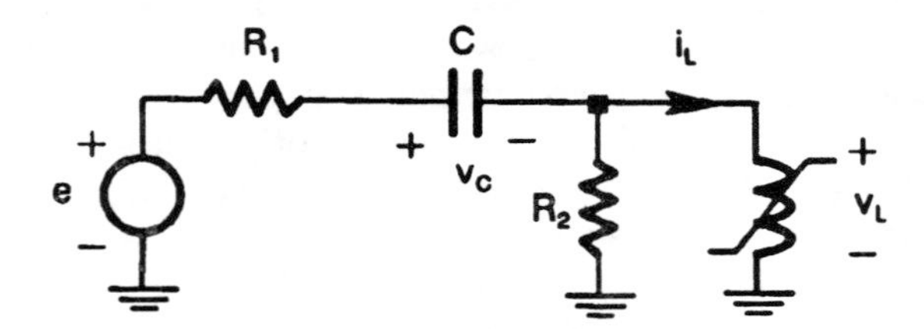

Figure 1 Nonlinear resonant circuit

To illustrate the point of this section, harmonics can be neglected, resulting in :

$$i_L(t) = I_{L_1}\cos(\omega t + \psi_1) = I_{L_1}\underline{|\,\psi_1} \quad (4)$$

$$v_L(t) = V_{L_1}\underline{|\,\psi_1 + 90^\circ} \quad (5)$$

$$v_C(t) = V_{C_1}\underline{|\,\psi_1 - 90^\circ} \quad (6)$$

Considering that $V_{C_1} = I_{L_1} / \omega C$ and combining equations (2) , (5) and (6) for two possible values of ψ_1 (180° and 0° in the right and left planes, respectively) :

$$V_{L_1} = E + \frac{I_{L_1}}{\omega C} \quad (7)$$

$$V_{L_1} = -E + \frac{I_{L_1}}{\omega C} \quad (8)$$

These two line equations are converted to RMS values and superposed on the actual $V_{L_{RMS}} - I_{L_{RMS}}$ piecewise linear inductance characteristic of Fig. 2. Two different cases are considered: $E = E_A$ and $E = E_B$ with $E_B > E_A$. The line slopes are exaggerated for illustration purposes. It can be seen that for $E = E_A$ there are three operating modes : *a* , *b* and *c*. These are in fact three different steady-state solutions for the circuit. Mode *a* is a normal mode, mode *b* is unstable and mode *c* is ferroresonant.

The IwH method currently available in the EMTP, will have no difficulty converging to the operating mode *a*. The iterative trajectory is shown by dotted lines in Fig. 2. The starting initial point is a_0 .

When the initial condition c_0 is used to find the ferroresonant state *c* , the shown iterative trajectory demonstrates the divergence (or attraction to *a*) of IwH. The EMTP IwH method is unable to find point *c* for any starting point c_0. The same conclusion is drawn for the unique operating mode *d* for $E = E_B$.

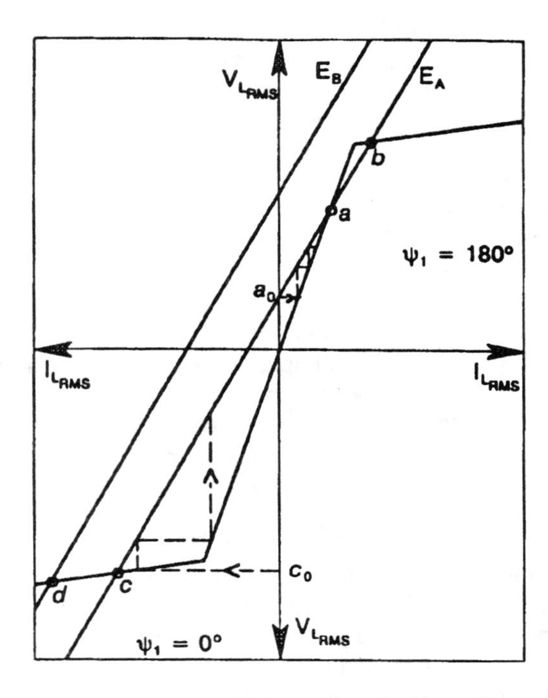

Figure 2 Operating modes for E_A and E_B Iterative trajectory of IwH

The above behavior description in the $V_{L_{RMS}} - I_{L_{RMS}}$ plane, where iterations constitute a preliminary step in IwH, can be applied in the flux-current plane for identical conclusions on the convergence properties of IwH. The IwH method being a fixed-point method is less performing, compared to a Newton type method, and guarantees convergence only if the solved function is a contraction mapping, which is not the case, for example, for operating modes *c* and *d* .

The simplified analysis of this section demonstrates the limitations of IwH. Its replacement by a Newton type general method is described next.

3. IMPLEMENTATION OF A GENERAL NEWTON TYPE METHOD

The following contributes a generalized description for a Newton type method resulting in the harmonic Norton equivalent representation. It will be referred to as the Norton equivalent (NE) method.

3. 1 The general solution method

For a given nonlinear branch b, voltage and current can be related through a general nonlinear function F_b :

$$F_b(\, v_b(t) \,,\, i_b(t) \,) = 0 \quad (9)$$

In almost all cases this function can be transformed into :

$$i_b(t) = g_b(v_b(t)) \quad (10)$$

where analytical or point-by-point knowledge of g_b is available. In a more general context, g_b or F_b can be found through direct time-domain EMTP simulation. The performance of such a treatment for initialization purposes is not simple to assess and may vary from case to case.

A general presentation of an iterative solution in the frequency domain is given in Fig. 3. The linear network is represented through separate nodal analysis equations for each harmonic h :

$$\tilde{\mathbf{Y}}_{n_h}^{(k)}\tilde{\mathbf{V}}_{n_h}^{(k+1)} = \tilde{\mathbf{I}}_{n_h}^{(k+1)} \tag{11}$$

(boldface characters are used to denote vectors and matrices). There is a total of $\mathcal{N}$ ($b = 1,\ldots,\mathcal{N}$) nonlinear branches. $\tilde{I}_{b_h}^{(k+1)}$ and $\tilde{V}_{b_h}^{(k+1)}$ are harmonic current and voltage phasors respectively and k is the iteration counter. The maximum number of harmonics actually computed is h_{max}. In a more general case a function F_b can contain a subnetwork [4] or an isolated nonlinear circuit that injects harmonics in the linear network.

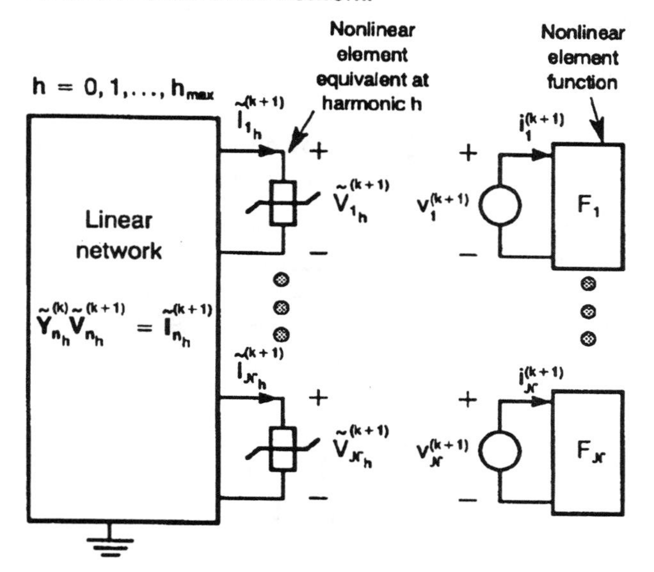

Figure 3 A general view of an iterative frequency domain steady-state solution method

The solution of the linear network equations precedes the solution for the nonlinear branches. With the formulation of equation (10), the voltage harmonics $\tilde{V}_{b_h}^{(k+1)}$ found in the network solution are used to construct $v_b^{(k+1)}$. The current harmonics are then found from g_b and used for assembling the nonlinear element harmonic equivalent in the following linear network solution, until convergence. This is called the *initialization loop*. The most simple startup is the calculation of voltage harmonics while the nonlinear branches are disconnected or replaced by fundamental frequency models. Harmonics can be present in the linear circuit from fixed value source injections.

The objective is now to derive the nonlinear branch (or more general element) equivalent circuit connected to the linear network at harmonic h. The nonlinear function g_b is linearized through truncated Taylor series expansion about the point $v_b = v_b^{(k)}$:

$$i_b^{(k+1)} = i_b^{(k)} + \left.\frac{dg_b}{dv_b}\right|_{v_b^{(k)}}\left(v_b^{(k+1)} - v_b^{(k)}\right) \tag{12}$$

The complex Fourier series of $i_b^{(k)}$, $v_b^{(k)}$ and $\left.\frac{dg_b}{dv_b}\right|_{v_b^{(k)}}$ are given by :

$$i_b^{(k)} = \sum_{h=-h_{max}}^{h_{max}} \hat{I}_{b_h}^{(k)} e^{jh\omega t}, \quad v_b^{(k)} = \sum_{h=-h_{max}}^{h_{max}} \hat{V}_{b_h}^{(k)} e^{jh\omega t}$$

and $\left.\frac{dg_b}{dv_b}\right|_{v_b^{(k)}} = \sum_{h=-h_{max}}^{h_{max}} \hat{G}_{b_h}^{(k)} e^{jh\omega t}$, respectively. When these series are substituted into (12) :

$$\sum_{h=-h_{max}}^{h_{max}} \left[\hat{I}_{b_h}^{(k+1)} - \hat{I}_{b_h}^{(k)}\right] e^{jh\omega t} \approx \left[\sum_{h=-h_{max}}^{h_{max}} \hat{G}_{b_h}^{(k)} e^{jh\omega t}\right]\left[\sum_{h=-h_{max}}^{h_{max}} \left[\hat{V}_{b_h}^{(k+1)} - \hat{V}_{b_h}^{(k)}\right] e^{jh\omega t}\right] \tag{13}$$

The hatted variables are phasors. It can be shown from the decomposition of (13) that :

$$\hat{I}_{b_h}^{(k+1)} - \hat{I}_{b_h}^{(k)} \approx \sum_{q=-h_{max}}^{h_{max}} \hat{G}_{b_{h-q}}^{(k)}\left[\hat{V}_{b_q}^{(k+1)} - \hat{V}_{b_q}^{(k)}\right] \tag{14}$$

$$h = -h_{max},\ldots,0,\ldots,h_{max}$$

A similar relation can be found from [5]. Equation (14) is expanded to become :

$$\hat{I}_{b_h}^{(k+1)} = \hat{G}_{b_0}^{(k)}\hat{V}_{b_h}^{(k+1)} + \hat{I}_{Nb_h}^{(k+1)} \tag{15}$$

where :

$$\hat{I}_{Nb_h}^{(k+1)} = \hat{I}_{b_h}^{(k)} - \hat{G}_{b_0}^{(k)}\hat{V}_{b_h}^{(k)} + \sum_{\substack{q=-h_{max}\\ q\neq h}}^{h_{max}} \hat{G}_{b_{h-q}}^{(k)}\left[\hat{V}_{b_q}^{(k+1)} - \hat{V}_{b_q}^{(k)}\right] \tag{16}$$

Since, according to Fig. 3, the linear network equations are assembled for $h = 0, 1, \ldots, h_{max}$ and considering that $\hat{I}_{b_h} = \hat{I}^{*}_{b_{-h}}$, equation (15) must be rewritten for a practical implementation :

$$\tilde{I}_{b_h}^{(k+1)} = \tilde{G}_{b_0}^{(k)}\tilde{V}_{b_h}^{(k+1)} + \tilde{I}_{Nb_h}^{(k+1)} \quad h = 0, 1, \ldots, h_{max} \tag{17}$$

where $\tilde{I}_{b_h} = 2\hat{I}_{b_h}$ for $h \neq 0$ and $\tilde{I}_{b_0} = \hat{I}_{b_0}$. The cosine series of $i_b^{(k)}$, instead of complex series, are now programmed :

$$i_b^{(k)} = \sum_{h=0}^{h_{max}} I_{b_h}^{(k)} \cos(h\omega t + \beta_{b_h}^{(k)}) \tag{18}$$

with $\tilde{I}_{b_h} = I_{b_h}\underline{|\beta_{b_h}}$ and $\beta_{b_0} \doteq 0$. The same reasoning applies to $v_b^{(k)}$ and $\left.\frac{dg_b}{dv_b}\right|_{v_b^{(k)}}$.

It is shown in Fig. 4 that equation (17) constitutes the harmonic NE circuit of branch b. Contrary to [5], there is no need to apply a sophisticated and time consuming process for calculating $\tilde{I}_{Nb_h}^{(k+1)}$ from the complicated equation (16). The complete NE circuit is determined at each iteration by simply calculating the Norton admittance $\tilde{G}_{b_0}^{(k)}$ (the mean value of $\left.\frac{dg_b}{dv_b}\right|_{v_b^{(k)}}$) and extracting $\tilde{I}_{Nb_h}^{(k+1)}$ from :

$$\tilde{I}_{Nb_h}^{(k+1)} = \tilde{I}_{b_h}^{(k)} - \tilde{G}_{b_0}^{(k)}\tilde{V}_{b_h}^{(k)} \tag{19}$$

where $\tilde{I}_{b_h}^{(k)}$ is a harmonic current component found from the excitation of g_b by $v_b^{(k)}$ and the following series expansion (18). It is important to note that the current source $\tilde{I}_{Nb_h}^{(k+1)}$ accounts for harmonic coupling [6]. Thus, this is not a harmonically decoupled solution. Moreover, equation (16) must agree with (19) at convergence. Equation (19) is not an approximation, it is the exact solution at iteration (k) found through the nonlinear element function. Note also that (19) contrary to (16) has no (k + 1) terms, thus avoiding the need for updating voltages found after each harmonic linear network solution. The solution requirements of (16) are impractical for a large network case.

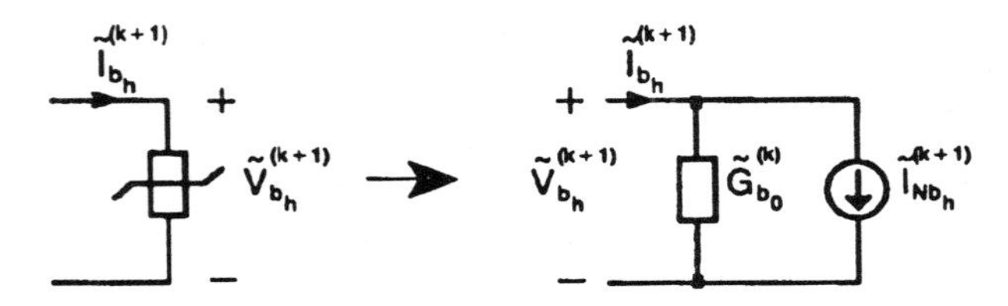

Figure 4 The harmonic Norton equivalent circuit of branch b

3.2 Example : the case of the nonlinear inductance and the presence of a dc component

This simple calculation method of the NE is now applied to the particular case of the nonlinear inductance. The nonlinear relation is now a flux−current relation :

$$i_b(t) = f_b(\phi_b(t)) \tag{20}$$

Equation (17) is applied by noticing that in steady−state conditions $\tilde{\Phi}_{b_h}^{(k+1)} = \frac{\tilde{V}_{b_h}^{(k+1)}}{jh\omega}$ and that the mean value of $\left.\frac{df_b}{d\phi_b}\right|_{\phi_b^{(k)}}$ is given by $\frac{1}{L_{b_0}^{(k)}}$:

$$\tilde{I}_{b_h}^{(k+1)} = \frac{1}{jh\omega L_{b_0}^{(k)}}\tilde{V}_{b_h}^{(k+1)} + \tilde{I}_{Nb_h}^{(k+1)} \tag{21}$$

where according to equation (17) $\tilde{G}_{b_0}^{(k)} = (jh\omega L_{b_0}^{(k)})^{-1}$. The derivative $\left.\frac{df_b}{d\phi_b}\right|_{\phi_b^{(k)}}$ is the same as $\left.\frac{di_b}{d\phi_b}\right|_{\phi_b^{(k)}}$.

Equation (21) cannot be used for $h = 0$. The particular case of a *dc* flux component due to a *dc* source presence in the linear network or *dc* current injection by nonlinear branch functions, requires a special treatment. The formulation of equation (9) which is applied here as $\mathcal{F}_b(\phi_b, i_b) = 0$, and the general Fig. 3 can easily account for this situation. The linear network solution at $h = 0$ is shown in Fig. 5 and equation (21) is rewritten in terms of ϕ_b :

$$\tilde{\Phi}_{b_0}^{(k+1)} = L_{b_0}^{(k)}\tilde{I}_{b_0}^{(k+1)} - L_{b_0}^{(k)}\tilde{I}_{Nb_0}^{(k+1)} \tag{22}$$

where $\tilde{I}_{Nb_0}^{(k+1)}$ is given by :

$$\tilde{I}_{Nb_0}^{(k+1)} = \tilde{I}_{b_0}^{(k)} - \frac{1}{L_{b_0}^{(k)}}\tilde{\Phi}_{b_0}^{(k)} \tag{23}$$

The *dc* component $\tilde{I}_{b_0}^{(k)}$ is found by inputting $\phi_b^{(k)}$ into f_b and calculating the resulting series expansion (18). The flux $\phi_b^{(k)}$ is constructed from its Fourier components $\tilde{\Phi}_{b_h}^{(k)}$. It has been observed from practical cases, that it is preferable to delay the calculation of (22) until $k = p$ where some partial convergence for *ac* components is achieved. The starting value is then given by $\tilde{\Phi}_{b_0}^{(p)} = L_{b_0}^{(p-1)}\tilde{I}_{b_0}^{(p)}$ $(p > 1)$.

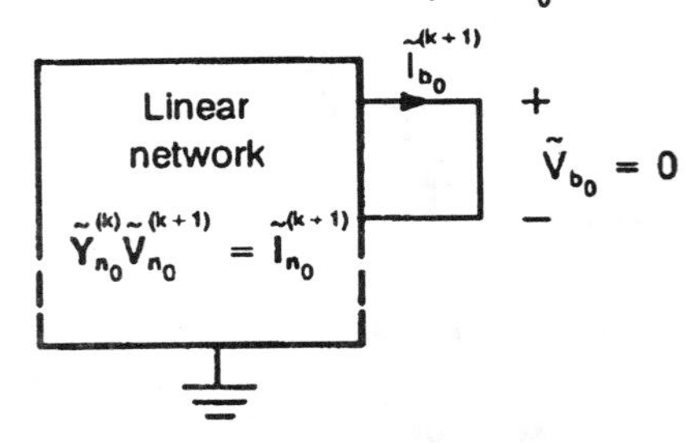

Figure 5 The linear network solution with the nonlinear inductance model for h=0

4. THE GENERAL NONLINEAR NETWORK EQUATIONS AND THE NE METHOD

This section presents an original investigation on the qualification of the NE method as a Newton type method for solving a general system of network equations.

Boldface characters are used to denote vectors and matrices. The nonlinear branches are classified first in the list of branches and identified by the subscript b. The linear branches are counted after nonlinear branches and identified by the subscript ℓ. For a general network with n nodes, $\mathcal{L}$ linear branches and $\mathcal{N}$ nonlinear branches :

$$\mathbf{A}\,\mathbf{g}(\mathbf{v}) = \mathbf{A}\,\mathbf{i}_s \tag{24}$$

where $\mathbf{A}$ is an $n \times (\mathcal{N} + \mathcal{L})$ reduced incidence matrix [9], $\mathbf{g}$ is the vector of nonlinear ($g_b \in \mathbf{g}$, $b = 1,\ldots,\mathcal{N}$) and linear ($g_\ell \in \mathbf{g}$, $\ell = 1 + \mathcal{N},\ldots,\mathcal{N} + \mathcal{L}$) branch functions, $\mathbf{v}$ is the vector of branch voltages ($v_b \in \mathbf{v}$ and $v_\ell \in \mathbf{v}$) and $\mathbf{i_s}$ is the vector of independent linear current sources. When the Newton algorithm is applied to solve equation (24), it is transformed into :

$$\mathbf{A}\left[\left.\frac{d\mathbf{g}}{d\mathbf{v}}\right|_{\mathbf{v}^{(k)}}\left[\mathbf{v}^{(k+1)} - \mathbf{v}^{(k)}\right] + \mathbf{g}(\mathbf{v}^{(k)})\right] = \mathbf{A}\mathbf{i_s} \tag{25}$$

Since equation (25) is assembled in the frequency domain, it must be independently solved for each harmonic h. The branch current vector $\mathbf{i}^{(k)} = \mathbf{g}(\mathbf{v}^{(k)})$ has harmonic components $\tilde{\mathbf{I}}_h^{(k)}$ ($\tilde{I}_{b_h}^{(k)} \in \tilde{\mathbf{I}}_h^{(k)}$). The derivation of equation (17) from (14) is applied to the decomposition and simplification of (25) :

$$\mathbf{A}\left[\tilde{\mathbf{Y}}_h^{(k)}\tilde{\mathbf{V}}_h^{(k+1)} + \tilde{\mathbf{I}}_{\mathbf{N}_h}^{(k+1)}\right] = \mathbf{A}\tilde{\mathbf{I}}_{\mathbf{s}_h} \quad h = 0, 1, \ldots, h_{max} \tag{26}$$

where $\tilde{\mathbf{Y}}_h^{(k)}$ is a diagonal matrix with : $\tilde{Y}_{bb_h}^{(k)} = \tilde{G}_{b_0}^{(k)}$ $b = 1,\ldots,\mathcal{N}$ and $\tilde{Y}_{\ell\ell_h}^{(k)} = \tilde{G}_{\ell_h}^{(k)}$ $\ell = 1 + \mathcal{N},\ldots,\mathcal{N} + \mathcal{L}$. $\tilde{\mathbf{V}}_h$ is a harmonic component of $\mathbf{v}$ and $\tilde{\mathbf{I}}_{\mathbf{s}_h}$ is a harmonic component of $\mathbf{i_s}$. For nonlinear branches $\tilde{I}_{Nb_h}^{(k+1)}$ is found from equation (19). For linear branches the NE currents are zero : $\tilde{I}_{N\ell_h} = 0$, $\ell = 1 + \mathcal{N},\ldots,\mathcal{N} + \mathcal{L}$

Equation (26) is further modified using the relation $\tilde{\mathbf{V}}_h = \mathbf{A}^t\tilde{\mathbf{V}}_{n_h}$:

$$\mathbf{A}\tilde{\mathbf{Y}}_h^{(k)}\mathbf{A}^t\,\tilde{\mathbf{V}}_{n_h}^{(k+1)} = \mathbf{A}\tilde{\mathbf{I}}_{\mathbf{s}_h} - \mathbf{A}\tilde{\mathbf{I}}_{\mathbf{N}_h}^{(k+1)} \tag{27}$$

This equation is equivalent to (11) by noticing that $\tilde{\mathbf{Y}}_{n_h}^{(k)} = \mathbf{A}\tilde{\mathbf{Y}}_h^{(k)}\mathbf{A}^t$ and $\tilde{\mathbf{I}}_{n_h}^{(k+1)} = \mathbf{A}\tilde{\mathbf{I}}_{\mathbf{s}_h} - \mathbf{A}\tilde{\mathbf{I}}_{\mathbf{N}_h}^{(k+1)}$. It appears that when the NE circuit of Fig 4 is used with the provided equations for calculating its components, then the NE method of Fig. 3 is a Newton type method for solving the nonlinear equation (24). This almost predictable conclusion has not been clearly demonstrated [4]–[6]. Some doubts were related to the formulation of the Newton equations from the nonlinear branches and the sequential solution between the linear and nonlinear parts. The NE method cannot be qualified as a true Newton–Raphson method, since the non–diagonal Jacobian matrix terms are predicted by $\tilde{I}_{Nb_h}^{(k+1)}$ from the nonlinear branch solution at iteration k, although the prediction is in agreement with the exact solution at convergence.

5. TEST CASES

The general NE method presented in this paper has been implemented in the EMTP [2] as a replacement for the previously described IwH method. A simple validation procedure is a non–initialized time–domain calculation of the steady–state. A minimal fundamental frequency initialization is also available in the standard EMTP steady–state module.

The final performance of an initialization method in the EMTP is judged upon its ability to reach the steady–state faster than a straightforward transient simulation. The computer time ratio is most dramatic for a large number of nonlinearities in large networks. Another factor is the required integration time–step.

The total simulation time must include the computer time spent in the *initialization loop*. The performance of this loop is mostly influenced by the formulation and solution of (11). Since sparse matrices are used, it is preferable to store an initial partially triangularized (without $\tilde{G}_{b_0}$) set of matrices $\tilde{Y}_{n_h}$. There is no need to retriangularize $\tilde{Y}_{n_h}$ between iterations when the changes in $\tilde{G}_{b_0}$ are negligible. The insignificant harmonic equivalents are neglected. Major parts of the existing IwH code were reusable.

5. 1 Initialization of the NE method

Unlike the fixed–point algorithm used in IwH, where it is possible for iterations to diverge regardless of the initial guess, the Newton–Kantorovich [10] theorem states that if the initial guess is close to a correct solution, then the true Newton–Raphson algorithm will always converge with a quadratic rate. Thus it may become important to provide some initial guess for the starting of the NE method. It was previously stated that all nonlinear branches are disconnected for the first iteration $k = 0$, unless a fundamental frequency model is available. For some branches a linear characteristic can be used as a simple model. The usage of the $V_{L_{RMS}} - I_{L_{RMS}}$ characteristic (as in IwH [3]) is found profitable for the case of a nonlinear inductance and increases convergence probability.

The initial guess for finding ferroresonant modes simply exploits the 180° phase difference depicted in Fig. 2. A systematic right plane to left plane approach can be implemented.

After the NE method has converged to a pre–specified tolerance, the time t is simply set to zero in all branch state variable Fourier series, to initialize and start the EMTP time–domain simulation. It is described in [3] how to account for discrepancies between steady–state and transient solutions, but it has been found from several test cases, that this fine tuning procedure can be avoided without significant impact. The time–step Δt must be correctly selected and influenced by the studied network smallest time constant.

5. 2 A ferroresonant case

The component values for the test circuit of Fig. 1 are taken from [8] where an analytical state-variable analysis is applied to solve the circuit. The purpose here is to demonstrate that the new NE method implemented in the EMTP is capable of initializing stable ferroresonant modes in addition to normal modes. The ferroresonant initialization capability is not available under the existing IwH method in the EMTP.

The time-domain simulation results for $E = 160V$ and $E = 300V$ are shown in Figures 6 and 7. In Fig. 6 the time-domain simulation for *waveforms 1* and *2* is initialized with the NE method. *Waveform 1* shows a periodic solution at fundamental frequency. This is a trivial case that can also be initialized with a single fundamental frequency solution, by replacing the nonlinear inductance by its first passing through the origin linear segment. This is an existing automatic procedure in EMTP when the NE method is bypassed.

Waveform 2 is a ferroresonant solution (see operating mode c in Fig. 2). The initialization provided by the NE method is verified by *waveform 3*. The initial flux value for this case is extracted from the NE method, and imposed manually at EMTP transient simulation startup. The precise flux initialization within the attraction domain of the second periodic solution, leads the transient solution to the desired steady-state. Arbitrary flux initialization may become a difficult task when there are several periodic solutions.

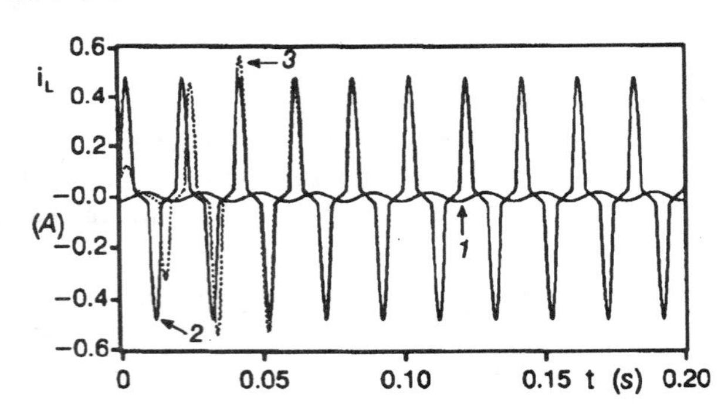

Figure 6 Two periodic solutions for E=160V, test circuit of Fig. 1

In Fig. 7 there is a single periodic state for $E = 300V$. The IwH method is once again inapplicable. The performance of the NE method initialization (*waveform 1*) is verified by a straightforward time-domain simulation (*waveform 2*). The perfect steady-state is then reached only after six periods, while the NE method forces the steady-state almost immediately. The total simulation computer time is decreased by a ratio of 3 when the NE method is used.

Besides its initialization performance, the NE method may be systematically applied to the extraction of all periodic solutions for a given nonlinear circuit. A search algorithm should be employed [11].

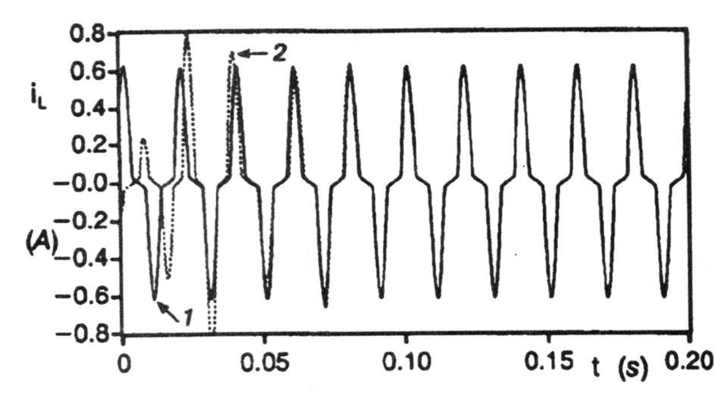

Figure 7 Periodic solution for E=300V, test circuit of Fig. 1

5. 3 A dc component case

The simple test circuit of Fig. 8 ($R_1 = 1\Omega$, $R_2 = 10\Omega$, $I_{dc} = 0.1A$ and $e = 160\sin(2\pi\ 60\ t)\ V$) is used to validate the ability of the new NE method to account for a *dc* steady-state component. The time-domain simulation results using the NE initialization are shown in Fig. 9. The steady-state is reached in the first period. In Fig.10 the EMTP simulation starts only with a 60 *Hz* initialization, the nonlinear inductance being treated as a linear segment. The waveform of Fig.10 perfectly matches the steady-state found in Fig. 9 only after ≈ 550*ms* and requires for this purpose, 15 times the computer time of Fig. 9. It is recalled that the existing IwH option is unable to initialize for a *dc* flux component in a nonlinear inductance.

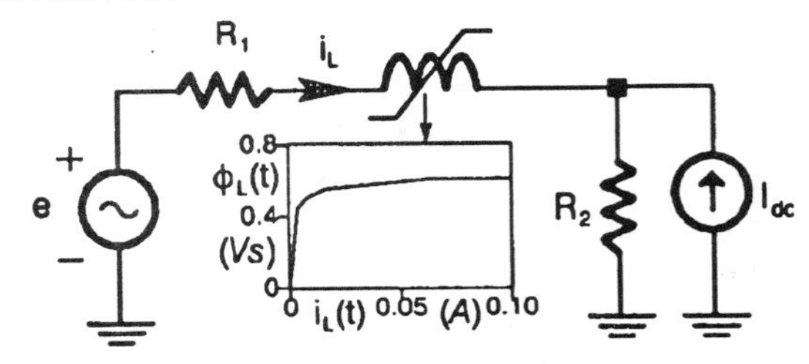

Figure 8 Injection of a dc current in a nonlinear inductance

5. 4 A large network case

A difficult case is provided by the study of magnetic storms in a large Hydro-Québec network [1] (file cscc.dat [12]). This three-phase test case contains a total of : 600 nodes, 700 branches, 45 nonlinear inductances (15 different nonlinear characteristics) for transformer magnetization, 72 *dc* current sources and 23 60 *Hz* voltage sources.

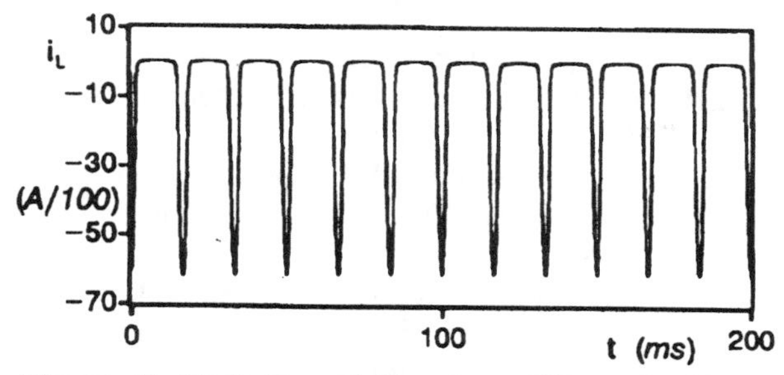

Figure 9 Periodic solution from NE method initial conditions, test circuit of Fig. 8

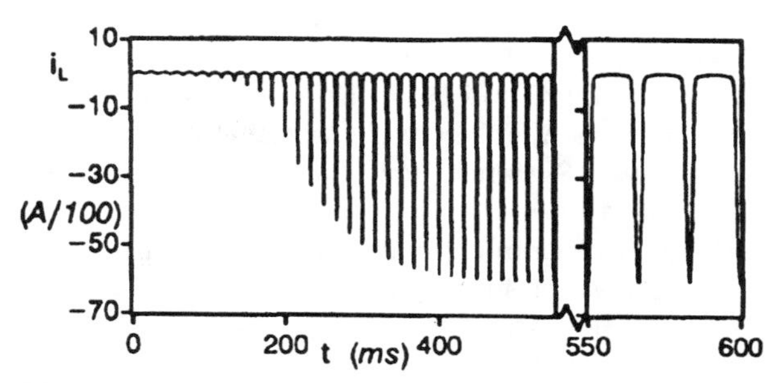

Figure 10 Simulation without harmonic initialization, test circuit of Fig. 8

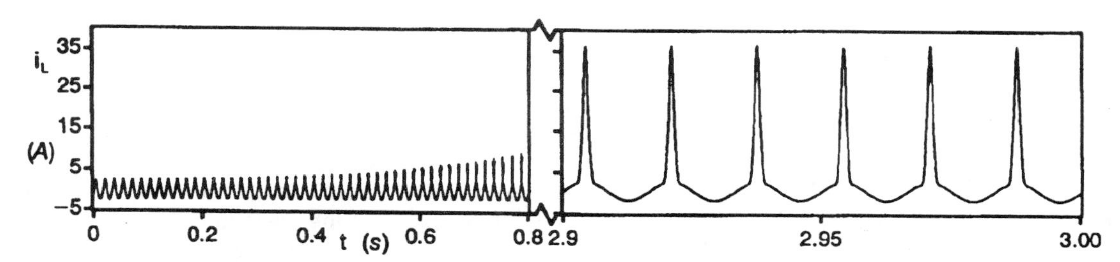

Figure 11 Simulation without harmonic initialization, a large network case

The standard EMTP simulation starts with a 60 *Hz* initialization with all nonlinear inductances operating on their first linear segment. The complete steady−state is reached only after 180 simulation periods. The current of the last to settle inductance (ref. CARt) is shown in Fig.11.

The simulation shown in Fig. 12 starts with the NE initialization and reaches the complete steady−state almost immediately. Only 4 iterations are required for convergence at a tolerance of 1.e−05. The gain in computer time is close to a ratio of 80. It implies that on a relatively fast workstation [12] the total simulation time before steady−state is reduced from 1¼ hour to a minute. This ratio is very conservative, considering the fact that the existing test case has artificially implemented deterministic steady−state accelerators [13] which are of no use to the NE method.

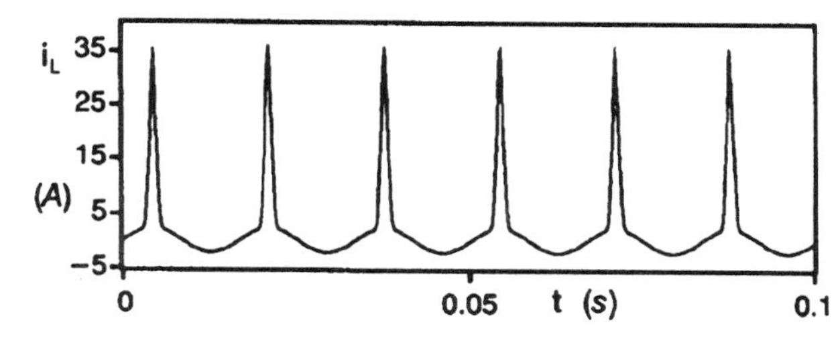

Figure 12 Simulation with harmonic initialization, a large network case

6. CONCLUSIONS

This paper has presented the implementation of a new harmonic initialization method in the EMTP : the harmonic Norton equivalent method. It is intended as a replacement for the existing IwH option, to increase numerical robustness and generality.

A new method for treating *dc* components has been presented and advantageously applied to a practical large network case with slow *dc* polarization.

The Norton equivalent method has been also applied to extract ferroresonant circuit modes.

The Norton equivalent harmonic initialization suggests a complete reprogramming of the EMTP steady−state module.

REFERENCES

[1] L. Bolduc, P. Kieffer, A. Dutil, M. Granger and Q. Bui−Van : Currents and Harmonics generated in power transformers by dc polarization. CEA Transactions on Engineering and Operation, Vol. 29, Part 1, 1990

[2] Electric Power Research Institute, EMTP Development Coordination Group, EPRI EL−6421−L : Electromagnetic Transients Program Rule Book, Version 2

[3] H.W. Dommel, A. Yan and S. Wei : Harmonics from transformer saturation. IEEE Trans. Vol. PWRD−1, No. 2, April 1986, pp. 209−215

[4] K. S. Kundert and A. Sangiovanni−Vicentelli : Simulation of nonlinear circuits in the frequency domain. IEEE Trans. on Computer−aided design, Vol. 4, No. 4, Oct. 1986, pp. 521−535

[5] A. Semlyen, E. Acha and J. Arrillaga : Newton−Type Algorithms for the harmonic phasor analysis of non−linear power circuits in periodical steady state with special reference to magnetic non−linearities.

IEEE Trans. on Power Delivery, Vol. 3, No. 3, July 1988 pp. 1090–1098

[6] W. Xu, J. Marti and H. W. Dommel : A multiphase harmonic load flow solution technique. IEEE Trans. on Power Systems, Vol. 6, No. 1, Feb. 1991, pp. 174–182

[7] W. Xu, H.W. Dommel and J. Marti : A Synchronous Machine Model for three–phase harmonic analysis and EMTP initialization. IEEE Trans. on Power Systems, Vol. 6, No. 1, Feb. 1991, pp. 174–182

[8] L. O. Chua, M. Hasler, J. Neirynck and P. Verburgh : Dynamics of a Piecewise–Linear resonant circuit. IEEE Trans. CAS, Vol. 29, No. 8, Aug. 1982, pp. 535–546

[9] G. W. Stagg and A. H. El–Abiad : Computer Methods in Power System analysis. McGraw–Hill Book Company, 1968

[10] J. M. Ortega and W. C. Rheinboldt : Iterative solution of nonlinear equations in several variables. Academic Press, Inc. 1970

[11] C. Kieny, G. Le Roy and A. Sbai : Ferroresonance study using Galerkin method with pseudo–arclength continuation method. IEEE Trans. on Power Delivery, Vol. 6, No. 4, Oct. 1991, pp.1841– 1847

[12] J. Mahseredjian : The EMTP Sun and Cray Unix versions (V2.2+), IREQ, Hydro–Québec, IREQ–93–065, March 1993

[13] L. Bolduc and P. Kieffer : Méthode améliorée pour simuler rapidement à l'aide de EMTP l'effet des courants continus établis dans les transformateurs et les réseaux. IREQ, Hydro–Québec, IREQ–4812, April 1991

BIOGRAPHIES

Xavier Lombard graduated from Ecole Nationale Supérieure d'Ingénieurs Electriciens de Grenoble in 1991, he has been with Electricité de France (Direction des Etudes et Recherches) since that date. Mr. Lombard was at IREQ Hydro–Québec in 1992 and worked on harmonic initialization within the EMTP Restructuring project.

Jean Mahseredjian (M'84) received the B.Sc.A., M.Sc.A. and Ph.D. in Electrical Engineering from Ecole Polytechnique de Montréal (Canada) in 1982, 1985 and 1990 respectively. At present he is a researcher at Institut de Recherche d'Hydro–Québec (Canada) and an associate–professor at Ecole Polytechnique de Montréal. His main research activities are modeling and numerical analysis in power systems and power electronics.

Serge Lefebvre (M'76) received the B.Sc.A. and M.Sc.A degrees in electrical engineering from Ecole Polytechnique de Montréal (Canada) in 1976 and 1977 respectively, and a Ph.D. from Purdue University (Indiana) in 1980. He is working at the Research Department of Hydro–Québec since 1981 while being an associate professor at Ecole Polytechnique de Montréal. His research interests are in power system analysis techniques, computer applications and dc systems. Dr. Lefebvre is Chairman of the IEEE working group "Dynamic performance and modeling of dc systems and power electronics".

Christophe Kieny graduated from Ecole Polytechnique (X77) and received the ingénieur diplomé degree from the Ecole Supérieure d'Electricité in 1982. He has been with Electricité de France (Direction des Etudes et Recherches) since that date. His main research activities are in the field of transformer modelling, including nonlinear and high–frequency behavior. He is currently the head of the "Numerical Models for Electrotechnics" group at EDF, where he is involved in field calculations and mathematical models for ferroresonance.

Discussion

Adam Semlyen (University of Toronto): I would like to congratulate the authors for their interesting and useful contribution regarding the calculation of the periodic steady state, with harmonics, for the purpose of initialization in the EMTP. They have used the ideas of harmonic domain analysis but have skillfully circumvented the intrinsic harmonic coupling in the resulting Norton equivalent by means of a simple approximation based on previous values. Nevertheless, the convergence of the process, while only linear, is still very good.

The new harmonic domain initialization is claimed to be superior to existing procedures that use either fixed point iterations or time domain methods for nonlinear components. While I agree with the advantages of the new, Newton-like methodology, I would like, however, to point out that some of the disadvantages of existing methods, discussed in the paper, are less severe than suggested by the authors.

First, I wish to refer to the inability of the fixed point iteration, illustrated in Figure 2, to converge to solution points *b*, *c*, or *d*. This, of course is strictly a numerical problem and it has a simple solution: invert the direction of the arrows in the solution trajectory in Figure 2 by interchanging inputs and outputs in the equations involved (they quite often are invertible). The iteration will then converge to points *b*, *c*, or *d*, but point *a* will not be reached. This convergence problem is of course unrelated to the physical stability of the operating points.

My second remark is related to the slow convergence of a *brute force* time domain simulation to a limit cycle in the case of large time constants. As discussed in the recent reference [A], there are established means to accelerate the convergence to a limit cycle (see the references in [10], and also [B]) We have found Newton's method very efficient, simple in formulation, and quite easy to implement. In addition, a time domain formulation, compared to frequency domain, is usually more natural for the (static or dynamic) nonlinear (or time varying) components of a power system. Thus, a hybrid solution (with the network equations in the frequency domain) for the calculation of the periodic steady state solution is, in my opinion, a viable alternative to a purely frequency domain approach.

Views and comments by the authors would be much appreciated.

[A] G. Murere, S. Lefebvre, and X.D. Do, "A Generalized Harmonic Balance Method for EMTP Initialization", paper no. 94 SM 439-0 PWRD, presented at the IEEE/PES Summer Power Meeting in San Francisco, California, July 1994.

[B] T.S. Parker and L.O. Chua, "Practical Numerical Algorithms for Chaotic Systems", Springer-Verlag, New York, 1989.

Manuscript received August 9, 1994.

X. Lombard, J. Mahseredjian, S. Lefebvre and C. Kieny : We thank the discusser for his comments and interest in our work.

An entire section of our paper is dedicated to the investigation on the qualification of the proposed solution method as a Newton type method. The presented method is clearly a Newton type method and has a *quadratic* rate of convergence. The *linear* rate of convergence would have resulted if a fixed-point method was used.

Two things must be kept in mind when Fig. 2 is studied. First, this is an analytic explanation of a numerical process and the load-lines do not actually exist during the numerical iterative calculations. When the iterative trajectory intersects a load-line it means that the linear network is solved for the corresponding nonlinear branch current. Second, everything should be placed in the context and limitations of the nodal analysis EMTP steady-state calculations. Inverting the iterative trajectory is not that simple. The initial guess c_0 cannot be equal to a_0 and pointing to the load-line from a voltage means solving the linear network for a nonlinear branch voltage source. The nonlinear branch is now seen as a current dependent voltage source. This will work fine for the particular case of Fig. 1, but will fail in a general case. In Fig. 1 the nonlinear inductance has a connection to ground and the corresponding voltage source can be included in nodal analysis, which is not true for a general case (see Fig. 8) of a nonlinear branch connected between two arbitrary nodes, where *modified nodal* analysis (not currently available in the EMTP) must be used. But, even then, as the discusser points out, the inverted trajectory will not locate the operating mode *a*. Implementing the direct and inverted trajectories on a trial and error basis will become time consuming in the general context of several nonlinear branches. These are the reasons for not considering the inverted iterative trajectories in the paper. The final argument against the fixed-point approach is that contrary to the Newton method, it is possible for the iteration to diverge regardless of the initial guess.

It is our opinion that the accelerated time-domain methods, where the nonlinearity is entered as a black-box, are still in a case-by-case stage and demonstrations for large networks with more than one nonlinear function are still lacking. It must be kept in mind that the steady-state prediction algorithm cannot logically use more computer time than a straightforward time-domain simulation reaching steady-state by itself.

It is obvious that the time-domain approach is the only viable choice for difficult nonlinearities where the frequency domain representation is not simple to achieve or readily available. But the method presented in this paper stressed precisely that situation: there is no need for a precalculated knowledge of frequency behavior for the assembled nonlinearities. A sentence following equation (10) indicates that the arbitrary function F_b can be found through direct time-domain simulation. In fact the demonstrated case of the nonlinear reactor proves this point. The harmonics generated by the reactor cannot be precalculated, they are actually found through the time-domain flux-current characteristic. This is exactly the same as a fully accelerated time-domain simulation: why would someone use any other lengthy accelerated-to-steady-state time-domain simulation method, if it is readily available through the simple graphical injection of the linear network's steady-state into the nonlinear time-domain function? The inclusion of other functions, such as a TCR, is also simple. If the TCR circuit is viewed as a black-box then its time-domain simulation will provide its harmonic steady-state and the corresponding Norton equivalent circuit. It is obvious that any such time-domain simulation algorithm cannot be faster than an algorithm where the TCR circuit is recognized and the distorted current is generated through the point-by-point TCR voltage function related to the firing angle. An even faster solution can be achieved if the TCR Norton equivalent inductance L_{b_0} is precalculated [6] in a preliminary analytical Fourier analysis. This inductance, contrary to the case of the nonlinear reactor, has a fixed value throughout the iterations.

Manuscript received October 26, 1994.

A Waveform Relaxation Technique for Steady State Initialization of Circuits with Nonlinear Elements and Ideal Diodes

Q. Wang (Non-Member, IEEE) J. R. Martí (Member, IEEE)

The University of British Columbia
Department of Electrical Engineering
Vancouver, B.C., Canada V6T 1Z4

***Abstract.*— This paper presents an efficient and robust technique for finding steady-state harmonic solutions in circuits with nonlinear elements and ideal diode-type devices. The problem is formulated as a two-point boundary value (TPBV) problem using the waveform relaxation technique. Ideal diodes are modelled using a Multiple Area Thevenin Equivalent technique. The overall solution is very robust and convergence is achieved within a few iterations even for very low damping and stiff systems. Examples are presented comparing the proposed technique with a previous technique and with direct "brute-force" integration.**

***Keywords*.– Steady-state solution with nonlinear elements, harmonic analysis, waveform relaxation technique.**

I. Introduction

The steady state-solution of circuits with nonlinear elements and power electronic components is an important problem, not yet fully satisfactorily resolved. Important examples are the harmonic analysis of power systems with nonlinear elements and power electronic devices and the analysis of power electronic controllers.

Harmonic balance techniques, such as those of [1, 2] are widely used for harmonic analysis of power systems. These techniques, however, are limited in their ability to model power electronic devices, and the effect of these devices is usually represented as harmonic current sources with values obtained from measurements.

In principle, a general-purpose time-domain simulation tool, like the EMTP [3] or SPICE [4] could be used to let the simulation run, starting from some initial conditions, until all transient effects have died out and the system has settled to the steady-state periodical waveform. This "brute force" approach may in practice present a number of problems, specially in cases of systems with very low damping, widely apart time constants (stiff systems), and when diode-type elements are represented as ideal devices. For example, in the common case of lightly damped power electronic circuits [5], unless artificial damping is introduces across the diode elements, and reasonable initial conditions are assumed, it may take the simulation very long times to converge, or it may fail to converge at all.

95 SM 387-1 PWRD A paper recommended and approved by the IEEE Transmission and Distribution Committee of the IEEE Power Engineering Society for presentation at the 1995 IEEE/PES Summer Meeting, July 23-27, 1995, Portland, OR. Manuscript submitted June 21, 1994; 1994; made available for printing June 7, 1995.

Several time-domain optimization techniques have been proposed for a more direct calculation of steady-state solutions in nonlinear circuits than the direct integration approach of SPICE or the EMTP. Among these techniques are the shooting method [6], the gradient method [7], and the extrapolation algorithm [8]. These techniques, however, may still present efficiency and convergence problems for certain types of steep nonlinearities and for ideal diodes.

In previously co-authored work [9], the shooting method was proposed to solve a two-point boundary value (TPBV) optimization problem, after discretization of the differential equations of the circuit with the trapezoidal rule of integration.

In the present paper, a waveform relaxation technique is proposed instead of the shooting technique for a more robust and faster convergence of the TPBV problem in stiff and very low damping systems. The problem of the representation of diode-type devices as ideal elements is effectively solved by means of a Multiple Area Thevenin Equivalent technique (MATE). The effect of the ideal diode is to join separate subareas, each with its own independent iteration procedure.

II. The Wafeform Relaxation Technique

The state of an electric network excited by inputs of period T can be described, in state-space form, by a system of n first-order (coupled) differential equations:

$$\frac{dy}{dt} = \mathbf{f}(\mathbf{y}, t), \tag{1}$$

where vector $\mathbf{y} \in R^n$ is the state of the network, and vector $\mathbf{f} \in R^n$ is periodic in t with period T, that is, $\mathbf{f}(\mathbf{y}, t) = \mathbf{f}(\mathbf{y}, t+T)$.

The aim of the steady state solution algorithm is to find the state vector $\mathbf{y}$ for which

$$\mathbf{y}(t) = \mathbf{y}(t+T). \tag{2}$$

In general, if the system characterized by the vector function $\mathbf{f}$ is not linear, solution (2) may not exist or may have a period T_1 different from T. A discussion on the existence and uniqueness of the solution to (1) and (2) can be found, for example, in [7]. In this paper, we will assume that solution (2) exists and that functions $\mathbf{f}$ in (1) are continuous both in $\mathbf{y}$ and in t. The circuit parameters are assumed time invariant. The time-dependent topological discontinuities introduced by the action of the ideal diodes are dealt with by splitting the network into subnetworks (joined by the diodes) for which the continuity conditions of (1) apply.

Periodic condition (2) defines a two-point boundary value (TPBV) problem (as opposed to the initial value problem $y(t=0)=y_0$). Several methods have been proposed for the solution of TPBV problems, among them, the collocation method, the repeated integration method, the band matrix method, the minimization method, the shooting method, and the waveform relaxation method [11].

The shooting method, proposed in our previous co-authored work [9], is a trial-and error technique where one starts with some initial guess of the variables y(0) in (2), performs the integration of the differential equations over one period T (using for instance the discrete-time solution of the EMTP), and records the result at the end of the period. A Newton correction can then be used to obtain a better guess of the new starting point. Successive integrations over one-period intervals are then performed until convergence to condition (2) is reached.

In the waveform relaxation technique proposed in this paper, the system's differential equations (1) are first replaced by finite difference equations, using for example the trapezoidal rule of integration. The values of the variables at the discrete-time points over one period T of the solution define a finite differences grid or mesh, as shown in Fig. 1. The "relaxation" concept is to adjust all the y values in the mesh *at once* (all intersection points in the mesh) towards the desired periodic condition (2). Iterations are performed, using, for example, a Newton correction technique until convergence to (2) is achieved. Visually, it is like reshaping the entire one-period of the wave at once by pulling it towards the correct steady-state form (see, for example, Fig. 5).

In the relaxation technique, optimizing iterations are performed over all the $n \times m$ variables (mesh points) defined by the finite differences mesh (Fig. 1), while, for example, in the shooting technique iterations are performed over only the n state variables at the boundary of the period (time point m in Fig. 1).

In a sense, it is surprising that the relaxation technique can be more efficient than other single-point optimizations, given the much larger number of variables to be solved for. In reality, there are several factors that make relaxation very efficient. First, since the entire waveform is adjusted at once, convergence occurs more rapidly than with single-point techniques. Second, single-point techniques, like the shooting method, require the solution of the network (EMTP-type solution) at each time point (points $i = 1, 2, \ldots m$ in Fig. 1) of the optimization interval T. The relaxation technique does not require any explicit integration solution.

In terms of overall implementation, the special banded form of the relaxation matrix permits the design of very efficient solution algorithms, both in terms of computational operations (flops count) and memory storage requirements.

III. Formulation of the Network Equations

In the time-domain solutions of the EMTP [3] and SPICE [4], the linear branches of the network (resistances, inductances, and capacitances) are converted into discrete-time equivalents consisting of a simple resistance in parallel with a history current source. Using the modified nodal analysis approach [12], the network equations at time step t can be expressed in the form

$$\begin{bmatrix} \mathbf{Y_R} & \mathbf{B} \\ \mathbf{C} & \mathbf{0} \end{bmatrix} \begin{pmatrix} \mathbf{V}(t) \\ \mathbf{Ib}(t) \end{pmatrix} = \begin{pmatrix} \mathbf{AhJh}(t) + \mathbf{AjJj}(t) \\ \mathbf{Vb}(\mathbf{Ib}(t)) \end{pmatrix}. \tag{3}$$

In this equation, $\mathbf{Y_R}$ is the matrix of nodal conductances excluding the nonlinear branches and $\mathbf{B}, \mathbf{C}, \mathbf{Ah}, \mathbf{Aj}$ are the branches incidence matrices. $\mathbf{Vb}(\mathbf{Ib}(t))$ is a functional relationship defined in equation (8). Vector $\mathbf{V}(t)$ is the node voltages and vectors $\mathbf{Ib}(t)$ and $\mathbf{Vb}(t)$ are the currents and voltages in the nonlinear branches. Current sources $\mathbf{Jh}(t)$ are the history current sources evaluated at time t, and $\mathbf{Jj}(t)$ are the external current sources at time t.

Separating in (3) the nodes with voltage sources connected, $\mathbf{V_B}$, from the other nodes, $\mathbf{V_A}$, we can write

$$\begin{bmatrix} \mathbf{Y_{RAA}} & \mathbf{B_A} \\ \mathbf{C_A} & \mathbf{0} \end{bmatrix} \begin{pmatrix} \mathbf{V_A}(t) \\ \mathbf{Ib}(t) \end{pmatrix} = \begin{pmatrix} \mathbf{Ah_A Jh}(t) + \mathbf{Aj_A Jj}(t) \\ \mathbf{Vb}(\mathbf{Ib}(t)) \end{pmatrix} - \begin{bmatrix} \mathbf{Y_{RAB}} \\ \mathbf{C_B} \end{bmatrix} \mathbf{V_B}(t). \tag{4}$$

The solution of (4) can be performed as explained in [9]. The history current sources are updated recursively at each time step of the solution by

$$\mathbf{Jh}(t) = \mathbf{UhJh}(t-\Delta t) - 2\mathbf{UhGhAh^T Vb}(t-\Delta t) \ , \tag{5}$$

where

$$\mathbf{Uh} = diag\begin{pmatrix} \mathbf{I}, \text{ when } \mathbf{Jh} \text{ is associated with an L} \\ -\mathbf{I}, \text{ when } \mathbf{Jh} \text{ is associated with a C} \end{pmatrix}, \tag{6}$$

$$\mathbf{Gh} = diag\begin{pmatrix} \frac{\Delta t}{2}L^{-1}, \text{ when } \mathbf{Jh} \text{ is associated with an L} \\ \frac{\Delta t}{2}C, \text{ when } \mathbf{Jh} \text{ is associated with a C} \end{pmatrix}. \tag{7}$$

If the nonlinear branches are represented as piecewise linear segments, the vector of branch voltages $\mathbf{Vb}(\mathbf{Ib}(t))$ can be expressed as

$$\mathbf{Vb}(\mathbf{Ib}(t)) = \mathbf{R}(t)\mathbf{Ib}(t) + \mathbf{Ek}(t) + \begin{pmatrix} \mathbf{Eh}(t) \\ 0 \end{pmatrix} \tag{8}$$

where

$$\mathbf{R}(t) = diag\left[\ldots \frac{2}{\Delta t} m_j(t) \ldots m_j(t) \ldots\right] \tag{9}$$

and

$$\mathbf{Ek}(t) = \left[\ldots \frac{2}{\Delta t} b_j(t) \ldots b_j(t) \ldots\right]. \tag{10}$$

The zero subvector in the last term of (8) corresponds to nonlinear branches with no memory (e.g., nonlinear resistances). The elements of the diagonal matrix $\mathbf{R}(t)$ are time-varying resistances which value depends on the segment of the piecewise linear characteristic where the operating point is located, $\mathbf{Ek}(t)$ are time-varying knee voltages associated with the piecewise linear segments, and $\mathbf{Eh}(t)$ are the history voltage terms in the branch description. History sources $\mathbf{Eh}(t)$ are updated as follows:

$$\mathbf{Eh}(t) = \begin{bmatrix} \mathbf{I} & \mathbf{0} \end{bmatrix} - (\mathbf{AnV}(t-\Delta t) - \mathbf{R}(t-\Delta t)\mathbf{Ib}(t-\Delta t) - \mathbf{Ek}(t-\Delta t)), \tag{11}$$

where $\mathbf{I}$ is the identity matrix, and $\mathbf{An} = \mathbf{C}$ is an incidence matrix mapping the vector of node voltages $\mathbf{V}(t)$ onto the vector of nonlinear branch voltages $\mathbf{Vb}(\mathbf{Ib}(t))$.

IV. The Finite Differences Mesh

In a step-by-step discrete-time network solver, like the EMTP, equations (2) are solved at $t = 0, \Delta t, 2\Delta t, \ldots$. With reference to Fig. 1, suppose that in a time span T (the period of the steady-state solution) there are m solution steps. These solution points, together with the value of n variables in the network, form a grid or mesh with $n \times (m-1)$ intersection points. The points $j = 1, \ldots, n$ on a vertical line at a given time point (*e.g.*, $i = 5$) are related (or "connected") to the points at the proceeding time point ($i = 4$) by equation (2). The diagram of Fig. 1 thus represents an interlocked mesh of solution points.

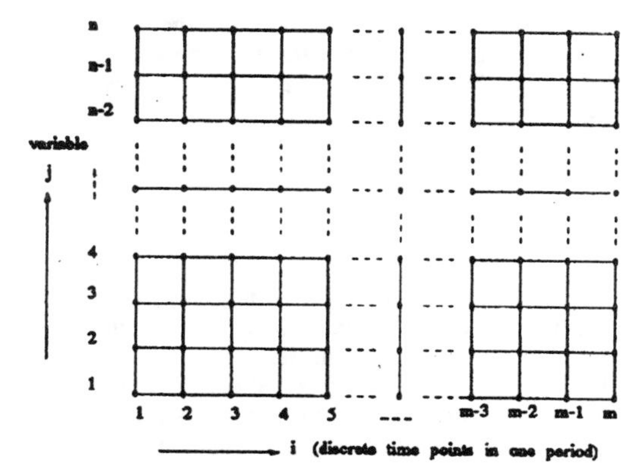

Figure 1. Finite differences mesh.

In the EMTP solution, the history sources are defined as a linear combination of voltages and currents in the corresponding branch. For example, for a capacitor C, $Jh(t) = \frac{2C}{\Delta t} v(t - \Delta t) + i(t - \Delta t)$. If a solution of period T is assumed for voltages and currents, then $Jh(t)$ will also be of period T, and, therefore, will constitute a viable choice of state variables j for the vertical axis of Fig. 1. An iteration, or "relaxation" of the values of these points towards the steady-state periodic solution established by condition (2) can then be formulated. Taking, for example, equation (5) for the history current sources, every two successive time points $i-1$ and i in the grid of Fig. 1 are related by

$$\mathbf{Jh}_i = \mathbf{UhJh}_{i-1} - 2\mathbf{UhGhAh}^T\mathbf{Vb}_{i-1} \,. \tag{12}$$

A similar relationship can be found for the history voltage sources in the nonlinear branches. Current and voltage history sources can be combined into a single vector of state variables

$$\mathbf{H}_i = \begin{bmatrix} \mathbf{Jh}_i \\ \mathbf{Eh}_i \end{bmatrix}. \tag{13}$$

The connection or relationship between $\mathbf{H}_i$ and $\mathbf{H}_{i-1}$ at the mesh points of Fig. 1 can be expressed in the general form

$$\mathbf{H}_i = \Lambda\mathbf{H}_{i-1} - \Gamma(\mathbf{H}_{i-1}) \quad i = 1, 2, \ldots, m \tag{14}$$

where

$$\Lambda = \begin{bmatrix} \mathbf{Uh} & 0 \\ 0 & 0 \end{bmatrix}, \tag{15}$$

$$\Gamma = \begin{bmatrix} 2\mathbf{UhGhAh}^T \\ [\,\mathbf{I} \;\; 0\,]\mathbf{An} \end{bmatrix} \mathbf{V}_{i-1}(\mathbf{H}_{i-1}) + \begin{pmatrix} 0 \\ [\,\mathbf{I} \;\; 0\,]\mathbf{R}_{i-1}\mathbf{I}_{i-1}\mathbf{H}_{i-1} + \mathbf{Ek}_{i-1} \end{pmatrix} \tag{16}$$

Relations (14) can be written in terms of vector-valued objective functions of the form

$$\mathbf{F}_i(\mathbf{H}_i, \mathbf{H}_{i-1}) = \mathbf{H}_i - \Lambda\mathbf{H}_{i-1} + \Gamma(\mathbf{H}_{i-1}) \quad i = 2, 3, \ldots, m \tag{17}$$

which, when (14) holds, become $\mathbf{F}_i = \mathbf{0}$.

Since each vector function in (17) has n coordinates (one for each state variable), relation (17) defines $n \times (m-1)$ objective functions, one for each intersection point in the grid. (Notice that in this context, points $i = 1$ and $i = m$ are the same point.)

Relaxation Matrix

Expanding the objective functions (17) into a Taylor series with respect to the history sources, one obtains

$$\mathbf{F}_i(\mathbf{H}_i + \Delta\mathbf{H}_i, \mathbf{H}_{i-1} + \Delta\mathbf{H}_{i-1})$$

$$= \mathbf{F}_i(\mathbf{H}_{j,i}, \mathbf{H}_{j,i-1}) + \sum_{j=1}^{n} \frac{\partial \mathbf{F}_i}{\partial \mathbf{H}_{j,i-1}} \Delta\mathbf{H}_{j,i-1}$$

$$+ \sum_{j=1}^{n} \frac{\partial \mathbf{F}_i}{\partial \mathbf{H}_{j,i}} \Delta\mathbf{H}_{j,i} + O(\Delta\mathbf{H}_{j,i-1}, \mathbf{H}_{j,i})$$

$$\approx \mathbf{F}_i(\mathbf{H}_i, \mathbf{H}_{i-1}) + \left[\frac{\partial \mathbf{F}_i}{\partial \mathbf{H}_{i-1}}\right]\Delta\mathbf{H}_{i-1} + \left[\frac{\partial \mathbf{F}_i}{\partial \mathbf{H}_i}\right]\Delta\mathbf{H}_i \tag{18}$$

With (18), an iteration procedure can be set up such that the following constraints are met at the steady-state solution point:

i) Periodicity constraint:

$$\mathbf{H}_i = \mathbf{H}_m \tag{19}$$

ii) Mesh constraints:

$$\mathbf{F}_i(\mathbf{H}_i, \mathbf{H}_{i-1}) = 0 \quad i = 2, 3, \ldots, m \tag{20}$$

With (20) in (18) and with (19), we have a system of $n \times m$ equations with $n \times m$ unknowns:

$$-\begin{bmatrix} \mathbf{I} & 0 & 0 & \cdots & 0 & -\mathbf{I} \\ \left[\frac{\partial \mathbf{F}_2}{\partial \mathbf{H}_1}\right] & \left[\frac{\partial \mathbf{F}_2}{\partial \mathbf{H}_2}\right] & 0 & \cdots & 0 & 0 \\ 0 & \left[\frac{\partial \mathbf{F}_3}{\partial \mathbf{H}_2}\right] & \left[\frac{\partial \mathbf{F}_3}{\partial \mathbf{H}_3}\right] & \cdots & 0 & 0 \\ 0 & 0 & 0 & \cdots & 0 & 0 \\ \cdots & \cdots & \cdots & \cdots & \cdots & \cdots \\ 0 & 0 & 0 & \cdots & \left[\frac{\partial \mathbf{F}_m}{\partial \mathbf{H}_{m-1}}\right] & \left[\frac{\partial \mathbf{F}_m}{\partial \mathbf{H}_m}\right] \end{bmatrix} \begin{pmatrix} \Delta\mathbf{H}_{1,1} \\ \Delta\mathbf{H}_{2,1} \\ \cdots \\ \Delta\mathbf{H}_{n,1} \\ \cdots \\ \Delta\mathbf{H}_{1,m} \\ \Delta\mathbf{H}_{2,m} \\ \cdots \\ \Delta\mathbf{H}_{n,m} \end{pmatrix} = \begin{pmatrix} \mathbf{F}_{1,1} \\ \mathbf{F}_{2,1} \\ \cdots \\ \mathbf{F}_{n,1} \\ \cdots \\ \mathbf{F}_{1,m} \\ \mathbf{F}_{2,m} \\ \cdots \\ \mathbf{F}_{n,m} \end{pmatrix} \tag{21}$$

Equation (21) defines a Newton iteration procedure:

$$\mathbf{H}_k = \mathbf{H}_{k-1} + \Delta\mathbf{H}_{k-1} \tag{22}$$

with k as the iteration count. This procedure will aim to converge to the desired solution for the history functions $\mathbf{H}_i$ in (13). The coefficients matrix in (21) is the *relaxation matrix* for the iteration procedure.

Jacobian Submatrices

The Jacobians in the main diagonal of the relaxation matrix (21) are simply the identity matrix, that is,

$$\left[\frac{\partial \mathbf{F}_i}{\partial \mathbf{H}_i}\right] = \mathbf{I}_{n \times n}. \quad (23)$$

This follows directly from (17) noticing that $\mathbf{H}_{i-1}$ is independent of $\mathbf{H}_i$.

The derivation of the expressions for the off-diagonal Jacobians $\left[\frac{\partial \mathbf{F}_i}{\partial \mathbf{H}_{i-1}}\right]$ are slightly more complicated. Their final form, however, is straightforward to evaluate. The derivation of these expressions is included in Appendix A.

V. Solution Algorithm

The form of the relaxation equation (21) can be simplified using vector notation for the state variables $\Delta \mathbf{H}_{j,i}$, using property (23) for the diagonal Jacobians, and denoting the off-diagonal Jacobians $\left[\frac{\partial \mathbf{F}_i}{\partial \mathbf{H}_{i-1}}\right]$ as $\mathbf{D}_i$:

$$-\begin{bmatrix} \mathbf{I} & \mathbf{0} & \mathbf{0} & \dots & \mathbf{0} & -\mathbf{I} \\ \mathbf{D}_2 & \mathbf{I} & \mathbf{0} & \dots & \mathbf{0} & \mathbf{0} \\ \mathbf{0} & \mathbf{D}_3 & \mathbf{I} & \dots & \mathbf{0} & \mathbf{0} \\ \mathbf{0} & \mathbf{0} & \mathbf{0} & \dots & \mathbf{0} & \mathbf{0} \\ \dots & \dots & \dots & \dots & \dots & \dots \\ \mathbf{0} & \mathbf{0} & \mathbf{0} & \dots & \mathbf{D}_m & \mathbf{I} \end{bmatrix} \begin{pmatrix} \Delta \mathbf{H}_1 \\ \Delta \mathbf{H}_2 \\ \dots \\ \Delta \mathbf{H}_m \end{pmatrix} = \begin{pmatrix} \mathbf{0} \\ \mathbf{F}_2 \\ \dots \\ \mathbf{F}_m \end{pmatrix} \quad (24)$$

Due to the special band form of (24), the solution for the incremental variables $\Delta \mathbf{H}_i$ can be found, very efficiently, using a backward-computation plus forward-substitution algorithm. For the last element $\Delta \mathbf{H}_m$,

$$\Delta \mathbf{H}_m = \left(\mathbf{I} - (-1)^{m-1} \prod_{j=2}^{m} \mathbf{D}_j\right)^{-1} \sum_{i=1}^{m-1} \left[(-1)^i \mathbf{F}_{m-i+1} \prod_{j=m-i+2}^{m} \mathbf{D}_j\right] \quad (25)$$

And since $\Delta \mathbf{H}_1 = \Delta \mathbf{H}_m$, the forward substitution for the remaining incremental variables is straightforward:

$$\Delta \mathbf{H}_{i-1} = -\mathbf{F}_{i-1} - \mathbf{D}_{i-1} \Delta \mathbf{H}_{i-2} \quad i = 2, \dots, m-1 \quad (26)$$

The above algorithm does not require unnecessary storage space for zeros in the relaxation matrix, or any operations involving those points. Additional efficiency can be gained by noticing that some of the factors in (25) are common to several of the terms and can be computed recursively.

VI. Modelling of Ideal Diodes

Ideal diodes behave like short-circuits during conduction and like open-circuits during blocking. In a general solution formulation such as (3), this behaviour poses special problems because it creates singularities in the equations. SPICE, for example, does not provide models for ideal diodes and requires that the device be represented by a given nonlinear characteristic. A device characteristic that is close to ideal (very steep nonlinearity) requires a very small solution step Δt in order to intercept the characteristic of the device. In many cases the actual characteristic of the device has little effect on the simulation results and the use of a very small Δt can be very inefficient, while the use of a larger Δt can result in nonconvergent iterations.

As recognized in the EMTP, ideal diodes are better modelled as ideal switches, which are open or closed according to the polarity of the voltage across the device.

In the work presented in this paper, ideal diodes are modelled using a Multiple Area Thevenin Equivalent technique (MATE). In this approach, the effect of ideal diodes is to join or separate independent subnetworks in the system (Fig. 9(a)). Initially, it is assumed that the diodes are not conducting and all subnetworks are disconnected. Each subnetwork is then solved separately. The voltages across the diodes are then checked to decide which diodes should conduct and which should remain open. Currents through the conducting diodes are then calculated from the combined Thevenin of all subnetworks and injected back into the subnetworks to update their internal states.

Since each subnetwork is iterated independently, the size of the integration step and the stability of the solution process are determined by the characteristics of the subnetworks and not by the characteristics of the device.

VII. Simulation Results

Circuit with Nonlinear Element

Figure 2 shows a simple case of a circuit with a nonlinear inductance. This case was originally studied in [10] to represent harmonics due to transformer saturation. This circuit was also solved in our work in [9] using the shooting algorithm. The nonlinear inductance is represented by two linear slopes L_{n1} and L_{n2}. Initially a Δt of 1.25 ms is used for the time discretization.

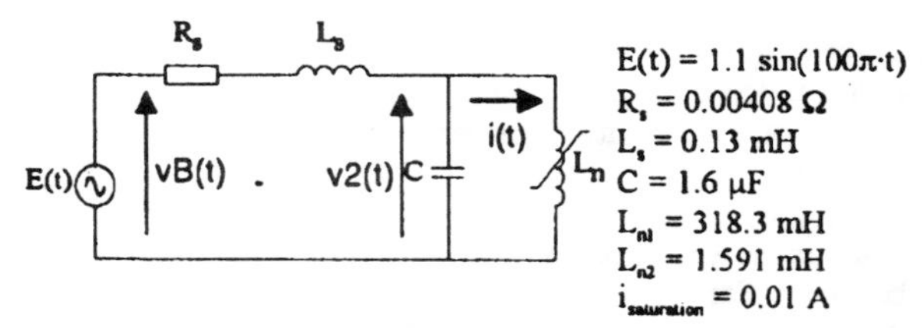

Figure 2. Circuit with a two-slope nonlinear inductance.

Figures 3, 4, and 5 show the voltage $v_2(t)$ and current $i(t)$ across the nonlinear inductance obtained with various simulation techniques for the steady-state solution of the circuit of Fig. 2. Figure 3 shows the results from direct integration ("brute-force") and was obtained by letting an EMTP simulation run until a periodic steady-state solution was reached. Figure 4 corresponds to the shooting technique of [9]. Figure 5 corresponds to the waveform relaxation technique presented in this paper. (For reference, the source voltage $v_B(t)$ is also plotted in Figs. 4, 5, and 7 for the shooting method and for the relaxation technique. For clarity in the plots, however, this voltage is not plotted in Figs. 3 and 6 for the direct integration technique.) Notice that even though all figures 3 to 7 are on a time scale, Figs. 3 and 6 correspond to a step-by-step time-domain (EMTP) simulation, while the various time cycles in Figs. 4, 5, and 7 for the shooting and relaxation techniques are the result of iterating over the same single-time-period of the signals.

Comparing the simulations of Figs. 3, 4, and 5, it can be observed that with the proposed waveform relaxation

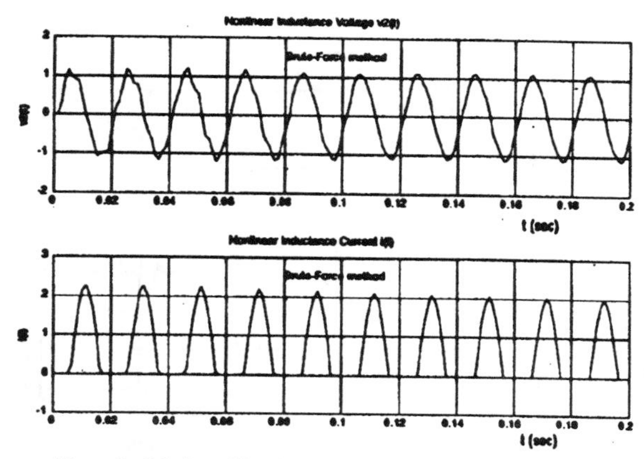

Figure 3. Solution of the circuit of Fig. 2 using direct integration.

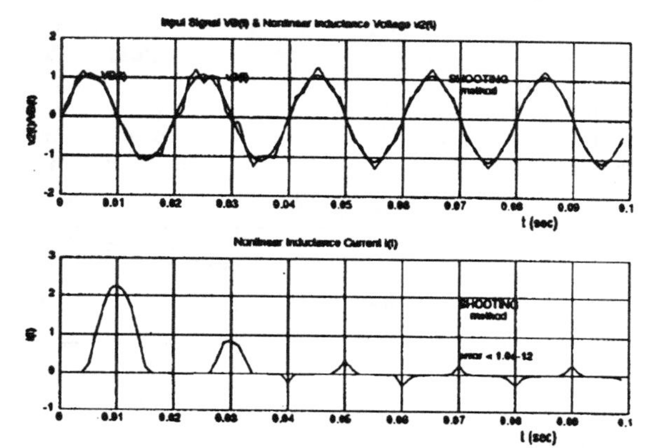

Figure 4. Solution of the circuit of Fig. 2 using the shooting method.

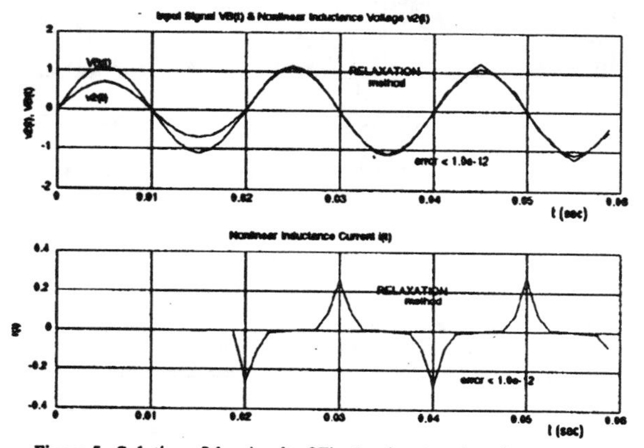

Figure 5. Solution of the circuit of Fig. 2 using the relaxation technique.

technique convergence is achieved in only two iterations, while four iterations were required with the shooting technique. The brute-force or direct integration approach was very slow to converge in this case (notice the difference in the time scales of Figs. 3 and 6 as compared to Figs. 4, 5, and 7). Table 1 shows the residual errors after each iteration for the relaxation and shooting methods. Table 1 also indicates that the total computational cost for convergence was 30,176 flops using the shooting technique, while this cost

Shooting Method		Relaxation Method	
Iteration	Residual Error	Iteration	Residual Error
0	3.3253	0	5.7667
1	1.6485	1	0.4995
2	0.3038	2	4.6168×10^{-14}
3	0.1324		
4	4.3521×10^{-14}		
flops = 30,176		*flops* = 17,219	

Table 1: Comparison of convergence speed for the circuit of Fig. 2.

was reduced to 17,219 flops with the relaxation technique.

To demonstrate the robustness of the waveform relaxation technique for low damping conditions, the resistance R_s in the circuit of Fig. 2 was reduced by a factor of ten, to $R_s = 0.000408\ \Omega$. To accurately trace the narrower reso-

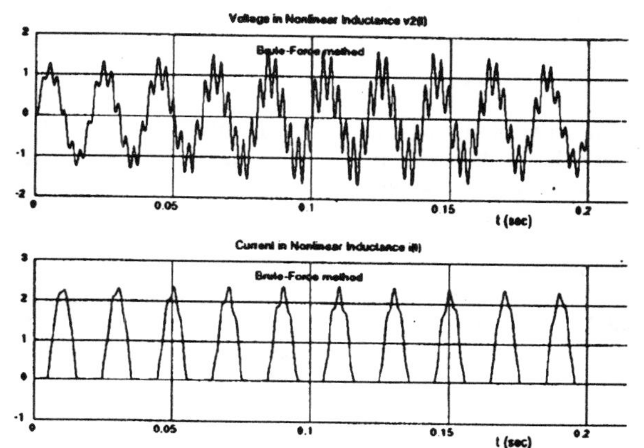

Figure 6. Solution of Example 1 with R_s=0.000408 Ω and Δt=0.125 ms using direct integration.

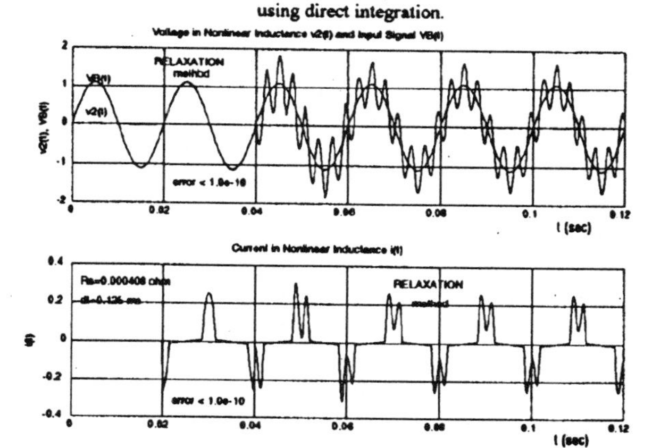

Figure 7. Solution of Example 1 with R_s = 0.000408 Ω and Δt = 0.125 ms using the relaxation technique.

nance peaks, Δt was also reduced by a factor of ten, to $\Delta t = 0.125\ ms$. Figures 6 and 7 show the results obtained for this case using brute-force integration and the relaxation technique. Note that even for these very low damping conditions (source $X_s/R_s = 123$ at 60 Hz) the relaxation technique converged very fast (in only five iterations). The shooting method failed to converge in this case.

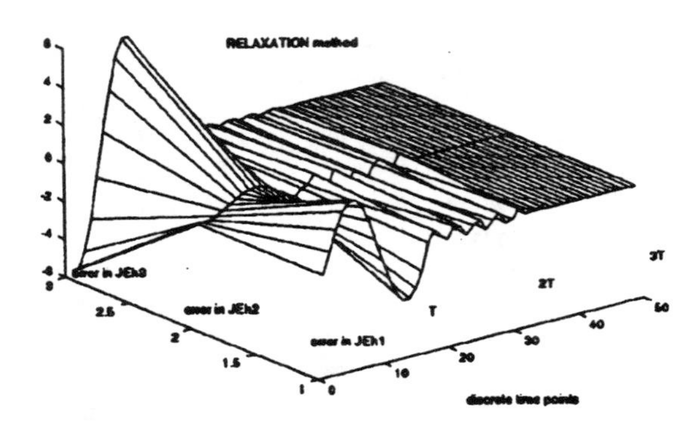

Figure 8. Mesh surface of errors in the state variables H_i at discrete time points using the relaxation technique.

Figure 8 illustrates the convergence process of the relaxation technique by means of a mesh surface of errors in the history sources H_i (21) as the process moves towards the objective of zero error.

Circuit with Ideal Diode

A rectifier circuit studied in [13] is shown in Fig. 9. Applying the Multiple Area Thevenin Equivalent technique, the circuit is split into two subnetworks, as indicated in Fig. 9(b). The diode is assumed open (non-conducting) at the initial state.

Simulations for this circuit using brute-force integration and the proposed relaxation technique are shown in Figs. 10 and 11, respectively (note the difference in the time scales

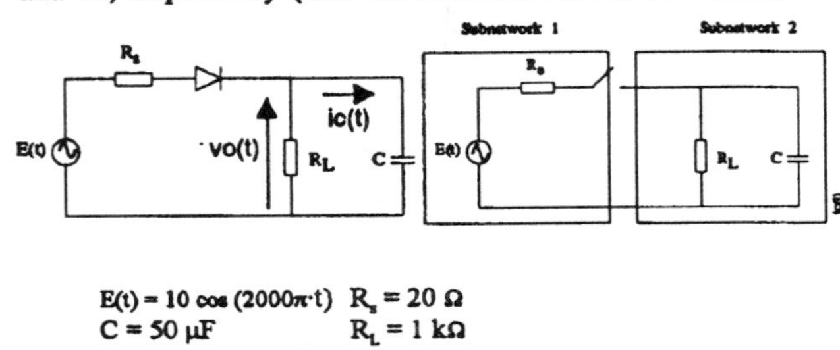

Figure 9. Rectifier circuit [13]. (a) Circuit. (b) Multi-Thevenin equivalent.

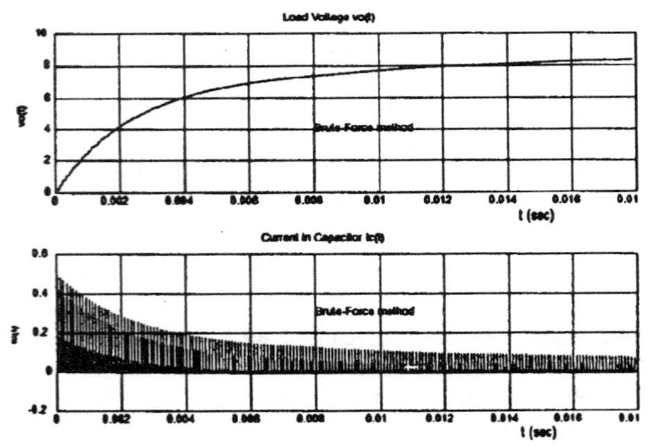

Figure 10. Solution of the circuit of Fig. 9 with C=50 μF using direct integration.

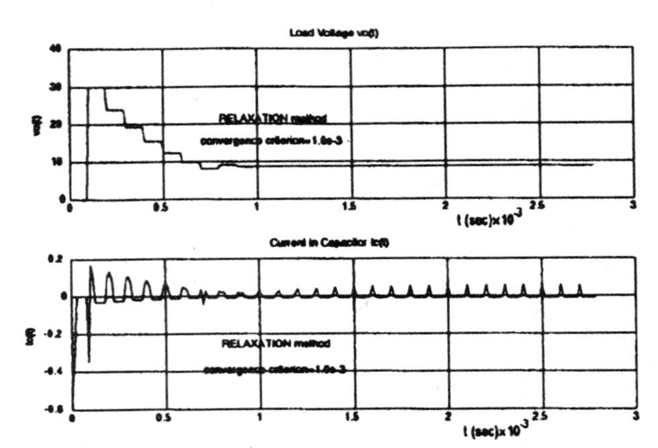

Figure 11. Solution of the circuit of Fig. 9 with C=50 μF using the relaxation technique.

of these figures). For this example, direct integration required more than 420 cycles to converge to the steady-state solution, while convergence with the relaxation technique was achieved in only 26 cycles.

Figures 12 and 13 show the results for the diode circuit of Fig. 9 after increasing the value of C by a factor of ten (note

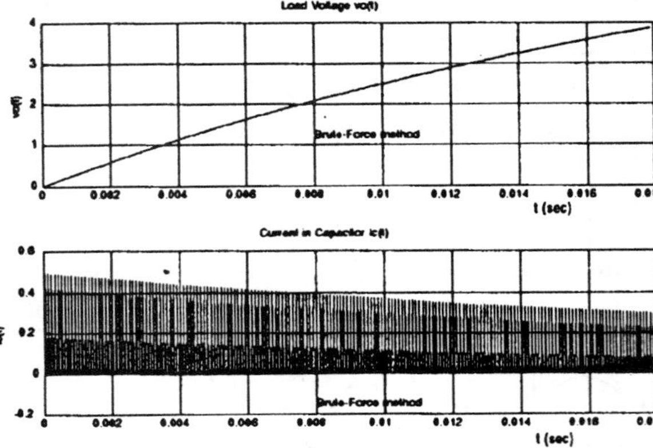

Figure 12. Solution of the circuit of Fig. 9 with C=500 μF using direct integration.

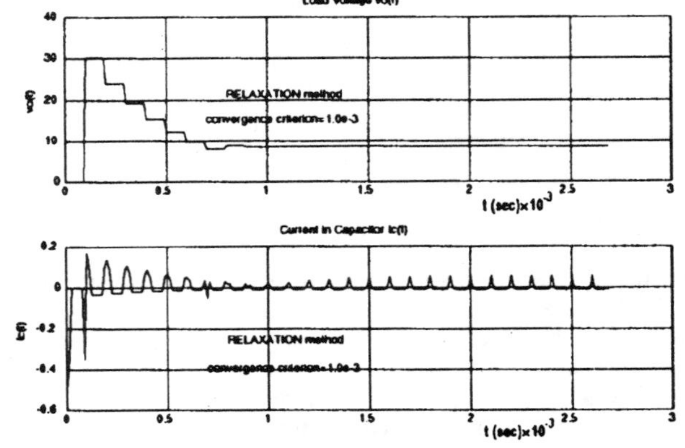

Figure 13. Solution of the circuit of Fig. 9 with C=500 μF using the relaxation technique.

the different scale factors in these figures). Direct integration now required more than 4,000 cycles to converge, while the proposed relaxation technique converged in the same number of cycles (26 cycles) as before.

Finally, it should be mentioned that the proposed technique scales well with system size. The dimensions of the Jacobians in (26) equals the dimension of the system (number of state variables); the proposed technique thus has the same computational complexity of other techniques (e.g., the shooting method), but it converges much faster.

Conclusions

An effective scheme has been developed to determine the periodic steady-state solution of circuits with nonlinear elements and ideal diodes. The scheme is based on the waveform relaxation technique in which all the points in one period of the waveform are "pulled together" towards the objective periodic state.

The procedure converges very rapidly, even for circuits with little damping, for which straight EMTP-type integration would take a very long time. By careful implementation, the scheme can be made very efficient in terms of flops count and memory storage requirements.

The technique is particularly suited for its extension to the case of ideal diodes. Using a Multiple Area Thevenin Equivalent technique, each subsystem connected by a diode is represented independently, with its own relaxation matrix. Since each subsystem is *relaxed* independently, the overall solution presents the same characteristics of robustness and rapid convergence as in no-diode systems.

References

[1] Acha, A.S.E. and Arrilaga, J., "Newton-type Algorithms for the Harmonic Analysis of Nonlinear Power Circuits in Periodical Steady State with Special Reference to Magnetic Nonlinearities," IEEE Trans. Power Delivery, Vol. PWRD-3, No. 3, pp. 1090-1098, July 1988.

[2] Xu, W., Martí, J.R., and Dommel, H.W., A Multiphase Harmonic Load Flow Solution Technique," IEEE Trans. Power Appar. and Syst., Vol. PAS-6, No. 1, pp. 174-182, February 1991.

[3] Dommel, H. W., "Digital Computer Solution of Electromagnetic Transients in Single and Multiphase Networks," IEEE Trans. Power Appar. and Syst., Vol. PAS-88, No. 2, pp. 388-396, April 1969.

[4] Nagel, L.W., *SPICE2: A Computer Program to Simulate Semiconductor Circuits*, ERL Memorandum M520, University of California, Berkeley, May 1975.

[5] Mohan, N., *Computer Exercises for Power Electronics Education*, University of Minnesota, 1990.

[6] Aprille, T.J., and Trick, T.N., "Steady-State Analysis of Nonlinear Circuits with Periodic Inputs," Proc. IEEE, Vol. 60, No. 1, pp. 108-114, January, 1972.

[7] Nakhla, M.S., and Branin, F.H., Jr., "Determining the Periodic Response of Nonlinear Systems by a Gradient Method," Int. J. Circuit Theor. Appl., Vol. 5, No. 3, pp. 255-273, 1977.

[8] Skelboe, S., "Computation of the Periodic Steady-State Response of Nonlinear Networks by Extrapolation Methods," IEEE Trans. Circ. And Syst., Vol. CAS-27, No. 3, pp. 161-175, March 1980.

[9] Perkins, B.K., Martí, J.R., and Dommel, H.W., "Nonlinear Elements in the EMTP : Steady-state Initialization," IEEE Trans. Power Systems, Paper 93 SM 508-2 PWRS, July 1993, 7 pages.

[10] Usaola, J., and Mayordomo, J.G., "Fast Steady-State Technique for Harmonic Analysis," IEEE Tran. Power Delivery, Vol. PWRD-6, No. 4, pp. 1789-1790, October 1991.

[11] Sewell Granville, *Numerical Solution of Ordinary and Partial Differential Equations*, Academic press, 1988.

[12] Ho, C.W., Ruehli, A.E., and Brennan, P.A., "The modified Nodal Approach to Network Analysis," IEEE Trans. Circ. And Syst., Vol. CAS-22, No. 6, pp. 504-509, June, 1975.

[13] Filicori, F.L. and Naldi, C.U. "An Algorithm for the periodic or Quasi-Periodic Steady-State Analysis of Non-Linear Circuits," IEEE CH 1854 - 7/83/000 - 0366, pp. 366-369.

Appendix A

The off-diagonal terms in the relaxation matrix (21) are given as follows. From (14), with (15) and (16),

$$\left[\frac{\partial \mathbf{F}_i}{\partial \mathbf{H}_{i-1}}\right] = -\Lambda\mathbf{I} + \begin{bmatrix} 2\mathbf{UhGhAh}^T \\ [\,\mathbf{I}\;\;\mathbf{0}\,]\mathbf{An} \end{bmatrix}\left[\frac{\partial \mathbf{V}_{i-1}}{\partial \mathbf{H}_{i-1}}\right] + \begin{bmatrix} \mathbf{0} \\ [\,\mathbf{I}\;\;\mathbf{0}\,]\mathbf{R}_{i-1} \end{bmatrix}\left[\frac{\partial \mathbf{Ib}_{i-1}}{\partial \mathbf{H}_{i-1}}\right]. \tag{A1}$$

Separating $\mathbf{V}$ into normal and source nodes, $\mathbf{V}_A$ and $\mathbf{V}_B$,

$$\left[\frac{\partial \mathbf{V}_{i-1}}{\partial \mathbf{H}_{i-1}}\right] = \begin{bmatrix} \frac{\partial \mathbf{V}_{Ai-1}}{\partial \mathbf{H}_{i-1}} \\ \frac{\partial \mathbf{V}_{Bi-1}}{\partial \mathbf{H}_{i-1}} \end{bmatrix} = \begin{bmatrix} \frac{\partial \mathbf{V}_{Ai-1}}{\partial \mathbf{H}_{i-1}} \\ \mathbf{0} \end{bmatrix}. \tag{A2}$$

The upper triangularization of (4) leads to

$$\begin{bmatrix} \mathbf{Y}_{RAA} & \mathbf{B}_A \\ \mathbf{0} & -\mathbf{T} \end{bmatrix}\begin{pmatrix} \mathbf{V}_A(t) \\ \mathbf{Ib}(t) \end{pmatrix} = \begin{pmatrix} \mathbf{Ah}_A\mathbf{Jh}(t) + \mathbf{Aj}_A\mathbf{Jj}(t) \\ \mathbf{Vb}(\mathbf{Ib}(t)) + \mathbf{Ah}_{A1}\mathbf{Jh}(t) + \mathbf{Aj}_{A1}\mathbf{Jj}(t) \end{pmatrix} - \begin{bmatrix} \mathbf{Y}_{RAB} \\ \mathbf{C}_B \end{bmatrix}\mathbf{V}_B(t) \tag{A3}$$

After solving for the currents in the nonlinear branches,

$$\mathbf{Ib}(t) = (\mathbf{T} + \mathbf{R}(t))^{-1}(\mathbf{Vo}(t) - \mathbf{Ek}(t)), \tag{A4}$$

where

$$\mathbf{Vo}(t) = \mathbf{C}_B\mathbf{Vb}(t) - (\mathbf{Ah}_{A1}\mathbf{Jh}(t) + \mathbf{Aj}_{A1}\mathbf{Jj}(t)) - \begin{bmatrix} \mathbf{Eh}(t) \\ \mathbf{0} \end{bmatrix}. \tag{A5}$$

The Jacobians for the nonlinear branches are then given by

$$\left[\frac{\partial \mathbf{Ib}_{i-1}}{\partial \mathbf{H}_{i-1}}\right] = (\mathbf{T} + \mathbf{R}(t))^{-1}\left[\frac{\partial \mathbf{Vo}_{i-1}}{\partial \mathbf{H}_{i-1}}\right], \tag{A6}$$

where the terms $\left[\frac{\partial \mathbf{Vo}_{i-1}}{\partial \mathbf{H}_{i-1}}\right]$ are constant. To evaluate (A6) at each time point i, only $\mathbf{R}(t)$ needs to be recalculated.

For the Jacobians $\left[\frac{\partial \mathbf{V}_{Ai-1}}{\partial \mathbf{H}_{i-1}}\right]$ in (A2), from (4),

$$\left[\frac{\partial \mathbf{V}_{Ai-1}}{\partial \mathbf{H}_{i-1}}\right] = \mathbf{Y}_{RAA}^{-1}\left(\mathbf{Ah}_A\left[\frac{\partial \mathbf{Jh}_{i-1}}{\partial \mathbf{H}_{i-1}}\right] - \mathbf{B}_A\left[\frac{\partial \mathbf{Ib}_{i-1}}{\partial \mathbf{H}_{i-1}}\right]\right). \tag{A7}$$

Since the terms $\left[\frac{\partial \mathbf{Jh}_{i-1}}{\partial \mathbf{H}_{i-1}}\right]$ are constant, the evaluation of (A7) at each time point i requires only the recalculation of $\mathbf{R}(t)$.

Biographies

Qing Wang received her B. Sc. and M. Sc. degrees in Elec. Eng. from The South China University of Technology and Beijing Polytech. University, in 1982 and 1989, respectively. She worked for the Dept. of Comp. Science and Eng. of the Beijing Institute of Information Technology from 1989 to 1991. She received a M.A.Sc. degree in Elec. Eng. at the University of British Columbia in 1994. Her research interests include computer modelling and simulation of power and power electronics circuits using SPICE and the EMTP.

José R. Martí (M'71) was born in Lérida, Spain in 1948. He received the degree of Elec. Eng. from Central University of Venezuela in 1971, the degree of M.E.E.P.E from Rensselaer Polytechnic Institute in 1974, and the Ph.D. degree from the University of British Columbia in 1981. He teaches power system analysis at the University of British Columbia, Canada, and has been involved for a number of years in the development of models and solution techniques for the transients analysis program EMTP.

MULTIPHASE POWER FLOW SOLUTIONS USING EMTP and NEWTONS METHOD

J. J. Allemong R. J. Bennon P. W. Selent

American Electric Power Service Corporation
Columbus, Ohio

ABSTRACT

This paper describes a reliable and very flexible multiphase load-flow solution process which is applicable for large transmission systems (up to 500 nodes). The process consists of an interface between the Electromagnetic Transients Program (EMTP) and a newly developed multiphase load flow algorithm that is based on the Newton-Raphson method. Subjects discussed include derivation of basic algorithm, structure of the Jacobian matrix, and convergence characteristics.

INTRODUCTION

Well known and reliable methods exist today for solving AC single-phase power flow problems. Most of these are based on the Newton-Raphson method, which has become the method of choice. Single-phase load flows always assume balanced three-phase system operation, and are ideally suited for representing large transmission networks. Studies performed usually deal with long term area planning, bulk power transfers, or outages of a major component where the unbalance effects may not be especially significant or of concern. However, as the number of high-voltage untransposed transmission lines increases, the effects of unbalance become significant and need to be properly analyzed [1]. These effects lead to the malfunction of protective relaying and the presence of significant generator negative sequence currents.

Many of these unbalance studies, requiring three-phase analysis, have been performed at American Electric Power (AEP) over the past several years. They have been, and continue to be, conducted using the three-phase and phasor solutions of the existing Electromagnetic Transients Program (EMTP). Modeling limitations exist in this method and for future studies involving system operation under unbalanced reactor or open-pole conditions, the limitations will make this procedure quite difficult. Additional limitations include poor convergence and the inability to specify load flow constraints such as, specific bus voltage regulation, PQ output for single and three phase generators and loads, etc. These limitations have also been recognized outside of AEP, by the EMTP EPRI/Development Coordination Group (EPRI/DCG). Consequently a project was initiated in which AEP, as an Associate member of the EMTP EPRI/DCG, would develop a full function three-phase load flow in EMTP.

A three phase load flow method was proposed in [3]. The intent of the present paper is to explore the viability of this method with respect to networks of practical size and complexity. The algorithm is based on Newton's method and addresses the previously stated limitations. The algorithm is capable of handling relatively large (practical) systems. A branch current method rather than a nodal method is used, resulting in greater flexibility for modeling loads and generators (e.g., delta connections). By integrating this into EMTP, full advantage is taken of EMTP's many network modeling routines (eg: transmision lines and transformers). Integration into EMTP also provides an accurate steady state network admittance matrix that does not need to be created by the load flow algorithm. Additionally, this integration provides a direct and more accurate steady-state initialization of dynamic and transient simulations. The multiphase load flow method is scheduled to be made available as part of the EPRI/DCG Version 3.0 EMTP.

DERIVATION OF BASIC ALGORITHM

Modern single-phase load flow analysis uses real and reactive power mismatches to obtain a solution. Iterations are based on the relationship between power mismatches and voltage and angle mismatches. This relationship is established through the sparse Jacobian-matrix equation:

$$\begin{bmatrix} \Delta P \\ \\ \Delta Q \end{bmatrix} = \begin{bmatrix} H & N \\ \\ J & L \end{bmatrix} \begin{bmatrix} \Delta\theta \\ \\ \dfrac{\Delta V}{V} \end{bmatrix} \tag{1}$$

Analysis in terms of polar coordinates lends itself well to problems formulated in terms of power mismatches, and three-phase load flows have been developed based on this formulation [9]. However, as will be seen, three-phase load flows require generalized models of generators and loads. It is convenient to express the equations for these models in terms of nodal currents instead of nodal powers. This formulation largely offsets the advantages of using polar coordinates.

93 WM 239-4 PWRS A paper recommended and approved by the IEEE Power System Engineering Committee of the IEEE Power Engineering Society for presentation at the IEEE/PES 1993 Winter Meeting, Columbus, OH, January 31 - February 5, 1993. Manuscript submitted August 10, 1992; made available for printing December 28, 1992.

The multiphase load flow algorithm described in this paper is based on Newton's method. Rectangular coordinates are used along with system constraints and component models based on branch quantities [3]. By using a branch representation, components can be connected in any fashion, (delta connected loads, phase-phase voltage sources, etc.), thus allowing a greater flexibility and more accurate system representation. The network components are not difficult to model using this representation, and consist of: three-phase synchronous machines, voltage sources, current sources, single-phase PQ loads, and three-phase static PQ loads. The reader is referred to [3] for a discussion of these components and their constraint equations.

Formulation of the Jacobian Matrix

The problem formulation consists of the expression of Kirchoff's Current Law and a set of constraint equations for the various non-linear elements. The effects of the linear elements (branches and shunts) are contained in the YBUS matrix while the non-linear elements contribute unknown currents. These currents are contained in a set of non-linear equations which define constraints that the solution is to satisfy. Equations 2-8 define the linear and non-linear equations used in the load flow algorithm, while Table 1 lists the constraints and controlling quantities used. The load-2 equations represent single-phase PQ loads, and the load-3 equations represent three-phase static loads.

network: $$[f_1] = [Y][V] + [I_s] + [I_u] = 0 \qquad (2)$$

source: $$f_2 = V_k - V_m - E_{specified} = 0 \qquad (3)$$

load-2: $$f_3 = I_{km}^H (V_k - V_m) - (P + jQ)_{specified} = 0 \qquad (4)$$

machine: $$f_4 = G([I_{km}], [V_k], [V_m]) - F_{specified} = 0 \qquad (5)$$

load-3: $$f_5 = [I_{km}]^H ([V_k] - [V_m]) - (P + jQ)_{specified} = 0 \qquad (6)$$

machine: $$[f_6] = -[I_{km}] - [Y_g]([V_k] - [V_m] - [E]) = 0 \qquad (7)$$

load-3: $$[f_7] = [I_{km}] - y[K]([V_k] - [V_m]) = 0 \qquad (8)$$

Where:

- [Y] is the network node admittance matrix
- [V] is the vector of node voltages
- [Is] is the vector of known current sources
- [Iu] is the vector of unknown current sources
- [E] is the vector of machine internal voltages
- [Yg] is the machine's 3x3 internal admittance matrix
- [y] is the static load parameter y
- H means complex conjugate transposed
- [K] is the symmetric ratio matrix of ZPOS to ZZERO
- k,m are the terminal ends of the branch
- P,Q are the real and reactive power, respectively

COMPONENT	CONSTRAINT	CONTROLLING QUANTITY
3-phase Loads	P, Q (3-phase)	y (equiv. admit.)
Swing Machine	|V+|, θ	E (internal voltage)
PV Machine	P 3ø , |V reg|	E (internal voltage)
PQ Machine	P,Q (3-phase)	E (internal voltage)

Table 1 - Constraints and Controlling Quantities

The elements comprising the Jacobian matrix are found by taking partial derivatives of the constraint equations with respect to node voltages, currents, machine internal voltages, and the static load parameter y. Without any restructuring or reordering, the system of equations to be solved is:

$\frac{\partial f_1}{\partial V}$ (y-bus)	$\frac{\partial f_1}{\partial I_V}$	$\frac{\partial f_1}{\partial I_L}$	$\frac{\partial f_1}{\partial I_m}$	$\frac{\partial f_1}{\partial I_{L3}}$	0	0		ΔV		Δf_1
$\frac{\partial f_2}{\partial V}$	0	0	0	0	0	0		ΔI_V		Δf_2
$\frac{\partial f_3}{\partial V}$	0	$\frac{\partial f_3}{\partial I_L}$	0	0	0	0		ΔI_L		Δf_3
$\frac{\partial f_4}{\partial V}$	0	0	$\frac{\partial f_4}{\partial I_m}$	0	0	0		ΔI_m	= -	Δf_4
$\frac{\partial f_5}{\partial V}$	0	0	0	$\frac{\partial f_5}{\partial I_{L3}}$	0	0		ΔI_{L3}		Δf_5
$\frac{\partial f_6}{\partial V}$	0	0	$\frac{\partial f_6}{\partial I_m}$	0	$\frac{\partial f_6}{\partial E}$	0		ΔE		Δf_6
$\frac{\partial f_7}{\partial V}$	0	0	0	$\frac{\partial f_7}{\partial I_{L3}}$	0	$\frac{\partial f_7}{\partial y}$		Δy		Δf_7

Figure 1 - Jacobian Matrix

Where:

- $f_{1\text{-}7}$ correspond to branch equations 2 - 8
- V is the node voltage vector
- Iv is the vector of currents from voltage sources
- I_L is the vector of currents for single-phase PQ loads
- I_m is the vector of machine currents
- I_{L3} is the vector of three-phase static load currents
- E is the machine internal voltage vector
- y is the static load admittance
- $\Delta f_{1\text{-}7}$ are the residual vectors (specified - calculated)

It is noted that the solution of (2) is conveniently carried out in rectangular coordinates, since it may be written as:

$$Y_R V_R - Y_I V_I + j(Y_R V_I + Y_I V_R) + I_{SR} + jI_{SI} + I_{UR} + jI_{UI} = 0 + j0 \quad (9)$$

Where:

Y_R, Y_I are the real and imaginary parts of the admittance matrix

V_R, V_I are the real and imaginary parts of the node voltage

I_{SR}, I_{SI} are the real and imaginary parts of the known currents

I_{UR}, I_{UI} are the real and imaginary parts of the unknown currents

The matrices Y_R and Y_I are constructed by EMTP for the linear branches. If polar coordinates were used, the solution of (2) would be much more involved.

As in the case of the full Newton method in single-phase load flows, the elements of the coefficient matrix must be re-evaluated at each iteration. Furthermore, the rectangular formulation does not lend itself to P/Θ and Q/V decoupling.[2]

Characteristics of the Jacobian Matrix

The use of constraint equations at a branch level in rectangular coordinates leads to a unique form for the coeffient matrix. In the conventional single-phase load flow, the Jacobian matrix is symmetric in structure and has the same sparsity pattern as the network topology. The three-phase load flow problem formulation, as embodied in equations (2)-(8) and the nature of the three-phase constraint equations, leads to a coefficient matrix which is unsymmetric and relatively more dense than matrices encountered in single-phase load flows. In addition, zero diagonals are present and may not be "filled" during elimination, unless these rows are moved from their natural position to the bottom of the matrix. A closer look at the constraint equations for the network (2) and voltage sources (3) reveal that the partial derivatives are constant and do not need to be re-evaluated for each iteration.

Ordering of the Equations

A characteristic of this method is that some of the equations, (voltage sources, PV and swing machines), produce zero diagonals. This affects the order in which the equations are eliminated, since the processing of rows with a zero diagonal must be delayed until the diagonal has "filled". At the same time, it is also desirable to order the processing of the rows to minimize (or reduce) the number of off-diagonal fill-ins created while also not violating the condition stated above.

If the reordering is to be determined by a conventional renumbering routine [4], the matrix structure must be made symmetric by adding zeros. Alternatively, a reordering scheme for unsymmetric matrices can be used. The authors chose to do the latter.

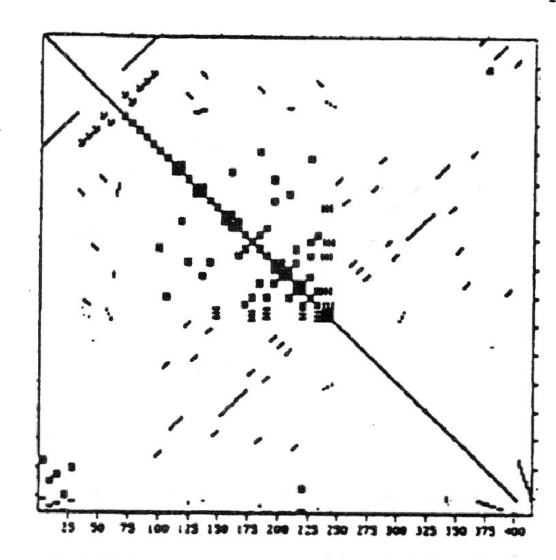

Figure 2 - LF3 Jacobian - Orig. Order Before Elimination

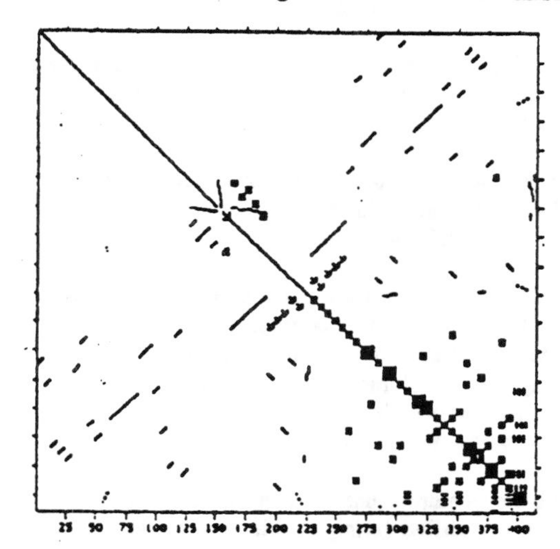

Figure 3 - Reordered LF3 Jacobian - Before Elimination

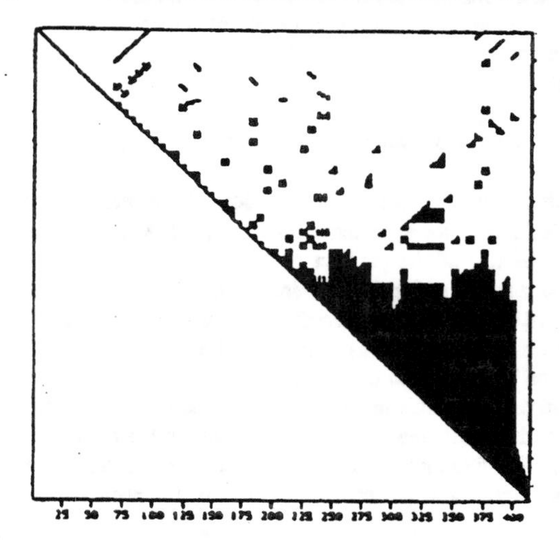

Figure 4 - Original LF3 Jacobian - After Elimination

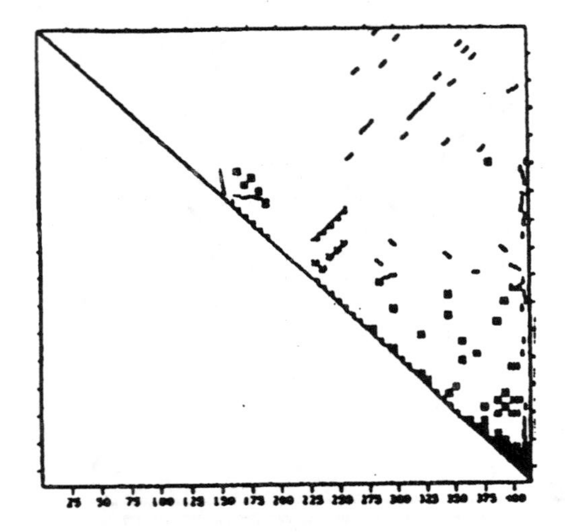

Figure 5 - Reordered LF3 Jacobian - After Elimination

The reordering is done only once, and is based on congruent permutations, that is, only diagonal elements are considered as pivots. The procedure is described below : [5]

1. For each row, i, compute the product of the number of off-diagonal entries in row i and the number of off-diagonal entries in column i.

2. Choose the row/column with the minimum product as the next one in the reordered system. Let this index be i and simulate the Gaussian elimination of the off-diagonals in column i, using the diagonal of row i as the pivot. Ties are broken by choosing the lowest row number with the minimum count.

3. Update the element counts in each row and column which are modified during the elimination. For example, suppose that element (j,i) is eliminated and that a fill element is created in row j at column k. When the existence of this fill element is recognized, the element count of row j and column k are both incremented by one.

4. Repeat steps 1-3 until all rows have been processed.

This scheme is a modest generalization of the well-known Tinney scheme 2 for symmetric matrices.[10] As a justification for the use of the product of row and column element counts as a measure for determining the ordering, the following heuristic argument is offered: When the diagonal of row i is used as a pivot, the maximum number of potential fills which can be created in other rows is equal to the number of non-zero off-diagonals in row i. Further, the number of rows in which these fills may be created is equal to the number of non-zero off-diagonals in column i. Therefore, the product of the row i count and the column i count is a measure of the number of potential fills accompanying the elimination of the non-zeros in column i.

It should be noted that in the case of a symmetric matrix, the row i and column i element counts are equal and their product is the square of this count. Thus, choosing the next row in the ordering on the basis of the square of the count is equivalent to using the count itself; from which it follows that the unsymmetric scheme, when applied to a symmetric matrix, simplifies to Tinney's scheme 2.

Based on these heuristic arguments, one suspects that this unsymmetric reordering scheme will behave, with respect to the number of fill-ins generated, similarly to the Tinney scheme 2; however more investigation of the use of this algorithm in the three-phase load flow problem may be warranted.

During development of this scheme, it was discovered that zero-diagonals could still exist during elimination. It was found that these rows corresponded to network components with voltage constraints. (i.e., voltage sources, swing machine, and the V part of PV machines). To eliminate these occurrences, the associated rows are forced to the bottom of the matrix, and are not part of the reordering scheme. Table 2 compares the number of upper triangular elements after elimination for both original and reordered Jacobian matrices. To illustrate the results further, Figures 2-5 show a Jacobian matrix, for a 124 node case before and after elimination (original order and reordered). The "black" portions of Figures 4 and 5 represent the retained upper triangle elements after elimination. As seen in Figure 4, the matrix becomes very dense after elimination on the original ordered Jacobian matrix.

	# ELEMENTS	% FULL
Original Order	21,321	24.82%
Re-ordered	3,781	4.40%

Table 2 - Comparison of Upper Triangle Sparsities

It was also discovered during the reordering, that the network admittance matrix [Y] does not change structure, rather merely shifts positions inside the Jacobian matrix. This is reasonable, since the admittance matrix is obtained in an optimal sparsity preserving order from the EMTP data assembly routines.

Initialization

Convergence of the Newton-Raphson method has proven to be very sensitive to the starting voltages. Table 3 shows the effects that the starting voltages have on the convergence of a 443 node multiphase load flow case. As seen from the table, the conventional $1.0 \angle 0$ per-unit voltage magnitude and angle is not usually applicable for the three phase load flow embedded in EMTP. This is because phase-shifting effects from wye-delta transformer connections are present in EMTP. Therefore, some special logic has been provided to calculate the initial point.

Starting Voltage Value	Convergence Behavior
1.0 $\angle 0°$	A zero diagonal occurred during iteration process
Calculated from EMTP steady state solution	Diverged after 11 iterations
Calculated from three-phase initalization routine	Converged in 5 iterations

Table 3 - Starting Voltage vs. Convergence Behavior

The initial voltages are obtained from the basic equation:

$$[Y]\,[V] = 0 \qquad (10)$$

Where:
[Y] is the network admittance matrix
[V] is the node voltage vector (unknown & known)

To solve this equation, at least one node to ground voltage must be known. (voltage sources or wye-connected swing machines). The matrices are rearranged such that all the known voltages are forced to the bottom. Equation (10) is then solved using Gaussian elimination, which is performed only on the nodes with unknown voltages. Back-substitution is then performed.

Ordinarily this method proves sufficient, however, there are some cases where this initial guess is not sufficiently close to the final solution. Reference [6] discusses various techniques to obtain a better starting point. A parameter perturbation technique that modifies the diagonal terms of nodes where single or three-phase PQ loads are represented is used. The modification involves the addition of an equivalent shunt admittance given by:

$$Y = \frac{-(P - jQ)_{specified}}{V^2} \qquad (11)$$

Where:
Y is the shunt admittance to be added
V is the node voltage
P-jQ is the specified power at the node

This leads to the following procedure for calculating the initial voltages. First, equation (10) is solved directly, resulting in a preliminary set of solved voltages. This set of voltages is then used to calculate the necessary shunt admittances given by equation (11). The appropriate diagonals in the admittance matrix are then updated. This results in a modified admittance matrix which contains approximated PQ and PV constraints. Equation (10) is then re-solved using this modified admittance matrix. This produces a set of starting voltages to begin the Newton-Raphson method. Table 4 shows a comparison of some starting voltages for the 443 node load flow case, calculated with and without the addition of the shunt admittances. As seen in Table 3, only 5 iterations were needed for convergence using this method.

Node	Solved Voltage	Starting Voltage (kV peak)	
Name	(kV peak)	w/o shunts	w/ shunts
Bus 765A	643.9	1295.0	541.7
B	640.1	1498.0	599.4
C	638.8	1170.0	571.4
Bus 345A	304.6	621.0	254.4
B	304.6	696.0	275.7
C	304.6	588.5	286.9
Bus 138A	119.9	191.3	90.4
B	119.9	246.9	103.7
C	119.9	208.2	101.2

Table 4 - Starting Voltages and Solved Voltages

Results of Three-Phase Analysis

Two multiphase load flow test cases were used to carry out testing. The first case consisted of 75 load flow nodes, all at a nominal voltage of 765-kv, with single-phase PQ loads and three-phase generators represented. The second case contained 443 load flow nodes with three-phase generators and three-phase PQ loads modeled. The voltages ranged in this case from 765-kv to 13-kv, and wye-delta transformer connections were represented.

CASE #1

This case represented AEP's 765-kv network, and was used to compare the load flow solution to actual system conditions obtained through the AEP Data Aquisition System (DAS) and State Estimator. The voltage regulation for the three PV generators consisted of the magnitude of the positive sequence voltage at two of the generator terminals. The other regulation was at a remote bus and consisted of a phase to ground magnitude. Table 5 shows magnitudes of positive sequence bus voltages from the load flow solution and actual values. Table 6 shows calculated and actual line flows eminating from a four terminal load bus. The load flow case converged in 6 iterations with a specified PQ mismatch of 130 kw.

Bus Name	LF3 |V|	Actual |V|
Rockport	101.6	101.5
Greentown	99.8	100.1
Baker	97.5	97.5
Hanging Rock	98.1	98.2
Don Marquis	97.6	97.5
N. Proctorville	98.2	98.2
Marsville	99.0	99.1
Kammer	98.2	98.2
Gavin	98.5	98.5

Table 5 - Comparison Of Positive Sequence Bus Voltages

Line Flow	LF3	Actual
HGRA - Bus 1	-338.07 - j150.68	-336.6 - j185.2
HGRA - Bus 2	324.52 + j145.18	324.5 + j146.2
HGRA - Bus 3	390.28 + j 52.33	389.0 + j100.9
HGRA - Bus 4	-374.87 - j 62.74	-376.3 - j 61.4

Table 6 - Comparison of Line Flows

As seen from the two tables, the calculated results and actual values compare quite well. There is some mismatch in the reactive power flows which can be attributed to the representation of line charging used in the case.

CASE #2

Case #2 was used to illustrate how the load flow algorithm performs under adverse conditions, such as low voltages or large phase angle differences. All loads were modeled as three phase loads and the PV generators regulated their positive sequence terminal voltages. To obtain low voltage conditions at a bus, one of three 765-kv transmission feeds was outaged, and real and reactive power loads were increased, at the bus in question as well as throughout the system.

Convergence of this case, from a flat start, occurred in 18 iterations to a specified PQ mismatch of 130 kw. Solved voltages ranged from 74% to 93%. Generator reactive power limits were enforced, resulting in generators being at their high reactive power limits and unable to maintain their scheduled voltages. Voltage unbalance, defined as the ratio of negative sequence voltage to positive sequence voltage [7], increased from 0.33% (system normal) to 1.22% (outage condition). The unbalance stated was for a particular bus, but these same types of effects were seen on all buses. As seen from the number of iterations required, the alogorithm performs quite well for large systems subjected to adverse operating conditions. The relatively high number of iterations results from the algorithm checking limits on generator reactive outputs and their associated regulated voltages. This may be reduced if the case was started from a solved set of voltages rather than a flat start.

Figure 6 is a simple diagram of the system being used in these test cases. The second test case contains significantly more nodes (i.e., 443 as compared to 75 in the first case) because of a more detailed representation, including generator step- up transformers and system step-down transformers.

Restrictions

Since the multiphase load flow algorithm interacts with EMTP, the well established EMTP rules and restrictions must be followed.[8] Two of the rules have been slightly "relaxed" for use with the multiphase load flow. They are delta connections for transformers, and the requirement for EMTP sources (type-14, etc).

The basic EMTP rule prohibiting isolated delta connection still holds true. However, connection of wye grounded three-phase synchronous generators to a delta-wye GSU can be accomplished without having to modify the corners of the delta windings by adding capacitors or resistance to ground.(see Figure 7). This type of configuration is detected for in the load flow algorithm, and applies to the steady state solution only. For transient simulations, this type of connection would have to be modified.

Also, EMTP defined sources need not exist for the load flow program provided that at least one voltage to ground is known in order to begin the solution process. This known voltage can be in the form of a SWING generator or a voltage source (EMTP or load flow). If a transient simulation is to automatically follow the load flow solution, EMTP sources would need to be defined to drive the transient solution since the load flow defined models are not present in the transient solution process.

FUTURE EFFORTS

The intent of this paper has been to show that a multiphase load flow program has been developed with a reasonable amount of flexibility and modeling capabilities. However, as more experience is gained, other topics will need to be addressed. This includes: effects of phase-shifting and tap-changing transformers, representation of system boundary conditions, extended voltage regulation (i.e. avg. phase-phase), representation of load flow component models (PQ loads, PV genera-

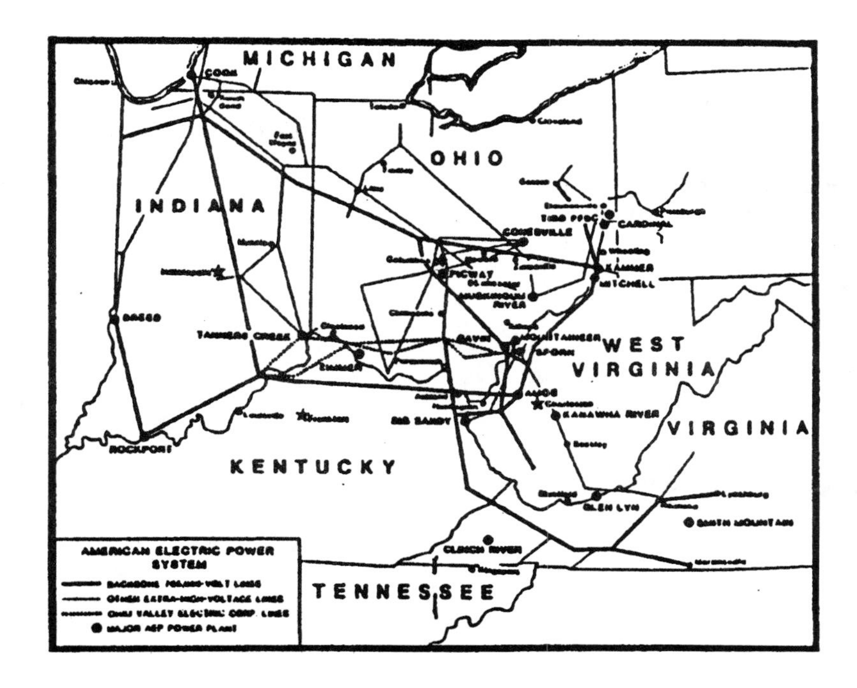

Figure 6 - System for Test Cases

tors etc.) in the transient solution. In addition, integration of this load flow and the flux-current iteration scheme [8], currently used in EMTP, needs to be investigated.

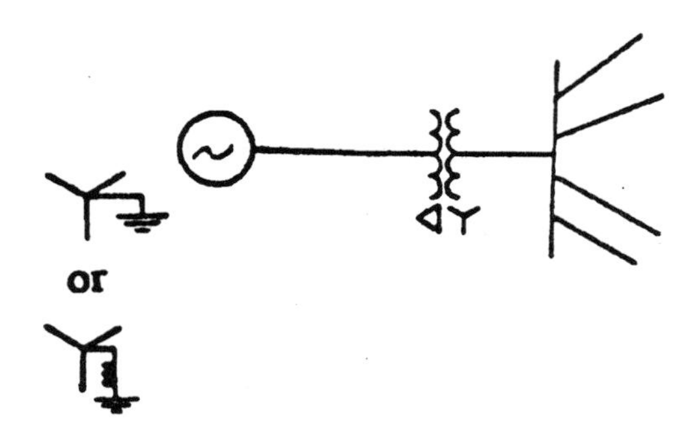

Figure 7 - Valid Load Flow Generator Connections

CONCLUSIONS

This paper has reported on a three-phase load flow process which has been incorporated into EMTP. AEP performed this work in conjunction with its membership in EPRI/DCG. For the most part, the methods employed follow those reported in [3]. In addition, careful implementation has allowed practical sized systems (several hundred nodes) to be studied. Two case studies were presented. In the first, a load flow case was formed from real-time, state estimation data and the calculated three-phase solution was seen to be very close to actual results. The second case was specifically designed to stress the algorithm. Despite low voltages, the case converged in a reasonable number of iterations, indicating that the algorithm is robust. Some discussion was also presented regarding solution initialization and modelling requirements. Future development work and user experience will undoubtedly lead to further improvements in this algorithm.

REFERENCES

[1] R.G. Wasley and M.A. Shlash, "Newton-Raphson Algorithm for 3-phase Load Flow," *IEE Proceedings*, vol. 121, No. 7, July 1974.

[2] B. Stott and O. Alsac, "Fast Decoupled Load Flow," IEEE Trans. (Power Apparatus and Systems), vol. PAS-93, no. 3, pp. 859-869, May/June 1974.

[3] W. Xu, J.R. Marti, and H.W. Dommel, "A Multiphase Harmonic Load Flow Solution Technique," IEEE Trans. (Power Systems), vol. 6, no. 2, pp. 174-182, Feb. 1991.

[4] W.F. Tinney and C.E. Hart, "Power Flow Solution by Newton's Method," IEEE Trans. (Power Apparatus and Systems), vol. PAS-86, pp. 1449-1460, Nov. 1967.

[5] H.M. Markowitz, "The Elimination Form of the Inverse and Its Application to Linear Programming," Management Science 3, 1957, pp. 255-269.

[6] H.W. Dommel, W.F. Tinney, and W.L. Powell, "Further Developments in Newton's Method for Power System Applications," Paper 70 CP 161-PWR, presented at IEEE PES Winter Meeting, New York, January 1970.

[7] W. Xu, "A Multiphase Harmonic Load Flow Solution Technique," Dissertation submitted at The University of British Columbia, February 1990.

[8] H.W. Dommel, "Electromagnetic Transients Program Reference Manual (EMTP Theory Book)," August 1986.

[9] J. Arrillaga and C.P. Arnold, "Computer Modelling of Electrical Power Systems", John Wiley & Sons, 1983.

[10] W.F. Tinney, J.W. Walker, "Direct Solutions of Sparse Network Equations by Optimally Ordered Triangular Factorization," Proc. IEEE, vol.55, No. 11, pp. 1801-1809, Nov. 1967.

TACS Enhancements for the Electromagnetic Transient Program

R.H.Lasseter, Fellow, IEEE and J. Zhou
University of Wisconsin-Madison

Abstract- Transient Analysis of Control Systems (TACS) of the Electromagnetic Transient Program (EMTP) has been enhanced in many ways. There are major changes in methods of ordering components to minimize the introduction of time step and history term errors. Initialization algorithms have been greatly enhanced to allow multiple frequency initialization. An "IF THEN ELSE" control structure and the use of user written FORTRAN routines are among the important new features available to users of TACS.
Keywords: TACS; EMTP; Control Systems; Simulation.

I. Introduction

The Electromagnetic Transient Program (EMTP) was originally developed in the late 1960's by Hermann Dommel [1]. The program was further developed at Bonneville Power Administration and has become an important industrial tool for analyzing power system transients. Currently the EMTP Development Coordination Group made up of utilities and companies from North America, Europe and Japan coordinate effort on EMTP development. The work discussed in this paper is part of this effort.

The Transient Analysis of Control Systems (TACS) was introduced to the EMTP program [2] by L. Dube in 1977. This addition allows a general representation of control systems in EMTP models. This is achieved through various system elements, such as transfer functions, algebraic functions and special devices that can be connected in an arbitrary manner to model a given control system. TACS has become important for the study of high voltage direct current transmission systems, static Var compensators, and generators when the transient responses of their respective control systems are important to the problems at hand.

This paper discusses the philosophy and implementation of enhancements to TACS. These enhancements are to be incorporated in DCG/EPRI version 3.0 of the EMTP. A different approach to basic TACS problems has been recently described by Dube and Bonfanti (3).

II. Basic TACS Solution Method

The solution methods used in previous releases of TACS assumes that the principle building block is a nth order transfer function that can be expressed as a system of first-order differential equations. The use of the trapezoidal rule of integration results in a simple algebraic equation where the derivatives of the input and output variables are eliminated. This produces a relationship for the nth order transfer function of the following general form where $u(t)$ and $x(t)$ are respectively the input and output of a given transfer function.

$$cx(t) = Kdu(t) + hist(t - \Delta t) \tag{1}$$

The coefficient K is the gain of the transfer function while c and d are calculated from the coefficients of the transfer function using recursive relationships [2]. The history term is calculated from the previous time-step solution.

A complete control system, with many elements, results in a set of equations that may be expressed in a matrix form.

$$[A_{xx}]\ [x] + [A_{xu}]\ [u] = [hist] \tag{2}$$

In this equation $[x]$ is a vector of the unknown variables or states and $[u]$ is a vector of known sources. To solve this system of equations, TACS performs a triangular factorization on $[A_{xx}]$, *and* $[A_{xu}]$. For each time step solution the $[hist]$ and source terms are formed by forward substitution followed by back substitution to obtain the system states $[x]$.

This method results in simultaneous solution for systems composed solely of transfer function blocks. For the case of a single limiter on a transfer function a simultaneous solution can also be found. This is achieved by ordering the system such that the output of the element with the limiter is the first variable found in the back substitution.

Simulation of control systems components other than transfer functions are included in TACS. These components include FORTRAN-like functions and expressions, logical functions, and special devices. In general these elements must be viewed as non-linear function blocks that are not directly included in the simultaneous solution discussed above. If these non-linear elements are included in a feedback loop, TACS breaks the loop by inserting a time delay. In most cases these time delays cause few problems. The existing TACS codes, through DCG/EPRI Version 2.0, also introduces additional time delays and in some situations incorrectly calculates the history terms. These issues can best be explained with a simple example.

0-7803-1301-1/93$03.00 © 1993 IEEE

Consider the example shown in Figure 1. This example includes a single transfer function F3 and simple gains F1 and F2. The response of this system to a step input is shown in Figure 2 under the label "simultaneous solution." The triangular form of the matrix equation for this system is shown as equation (3).

$$\begin{array}{ccccccccc} A_{11}X_1 & + & A_{12}X_2 & + & 0 & + & 0 & = & hist_{F1} \\ 0 & + & A_{22}X_2 & + & A_{23}X_3 & + & 0 & = & hist_{F2} \\ 0 & + & 0 & + & \bar{A}_{33}X_3 & + & A_{3s}U_s & = & \overline{hist_{F3}} \end{array} \quad (3)$$

The first row expresses the relationship between the input X_2 and the output X_1 for element F1. The same follows for elements F2 and F3. Due to triangulation the coefficient $\bar{A}_{33}$ is different from the original coefficient for element F3. More important is the history term, $\overline{hist_{F3}}$ which is dependent on all the states of the system, X_1, X_2, and X_3. Clearly when the history terms and the source U_s are known the output of F3 can be found. This is followed by calculating the outputs of F2 and F1. Once all states are known the history terms for the next time step are calculated. This is a very powerful method for finding simultaneous solutions for an arbitrary circuit of transfer functions.

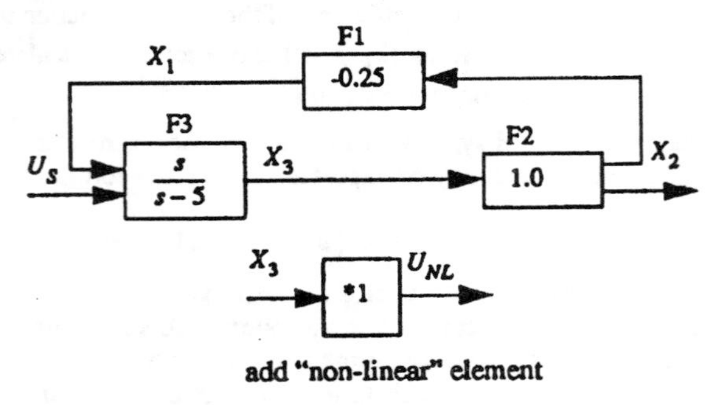

FIGURE 1. Simple Feedback Control System

More practical systems require the use of some non-linear elements. Non-linear functions are not directly included in the matrix form shown in equation 2. They are calculated as their inputs become available. Simultaneous solutions can be found for systems with non-linear elements provided these elements are not part of any feed back loop. For cases where there is feedback, which includes non-linear elements, simultaneous solutions can not be found without iteration at each time step. This can be costly in computation time and is not used by TACS. The solution method used by TACS is to decouple the feedback paths by introducing a time delay into the loop.

To understand the problems introduced by non-linear elements the circuit in Figure 1 is used. A dummy "non-linear" element is introduced which does not change the response of the circuit. This element is introduced between element F3 and F2. The current TACS treats this element as a general non-linear element. In this case the output of the "non-linear" element is assumed to be a known source, U_{NL}, not a state to be found. The resultant set of equations is shown in (4).

$$\begin{array}{ccccccccc} A_{11}X_1 & + & A_{12}X_2 & + & 0 & + & 0 & = & hist_{F1} \\ 0 & + & A_{22}X_2 & + & 0 & + & A_{23}U_{NL} & = & hist_{F2} \\ 0 & + & 0 & + & A_{33}X_3 & + & A_{3s}U_s + \bar{A}_{23}U_{NL} & = & \overline{hist_{F3}} \end{array} \quad (4)$$

The two sets of equations (3) and (4) are for the same system. The equation for F1 is not changed while F2 and F3 have different forms due to treating the output of the "non-linear" block as a source. When the equations are solved using back substitution a time delay is introduced. To find X_3 the value used for "source" U_{NL} is from the previous time step calculation. This has the effect of decoupling the loop at the input to F2 and solving the system in a sequential manner.

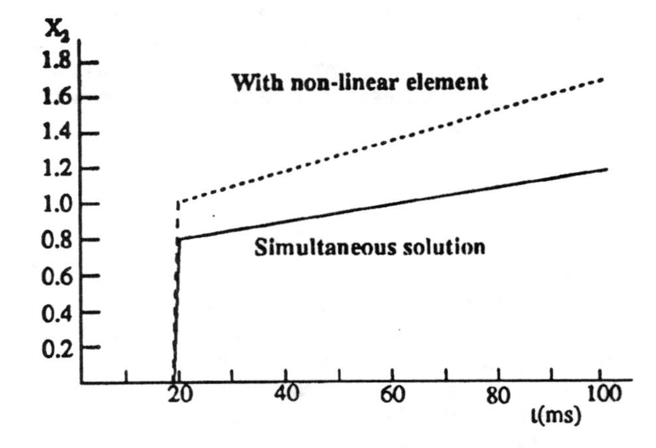

FIGURE 2. Response of System to a Step Input.

The results of including this "non-linear" element are also shown in Figure 2. This example shows a much more severe problem than a single time step delay. If the calculation was correct the added time delay would result in a solution that would oscillate about the simultaneous solution, Figure 3. In this example the feedback from F1 is lost resulting in a very different solution. This problem is due to how the history terms are calculated. In particular the history terms are calculated after all the states are found. This can become a problem when non-linear elements are introduced.

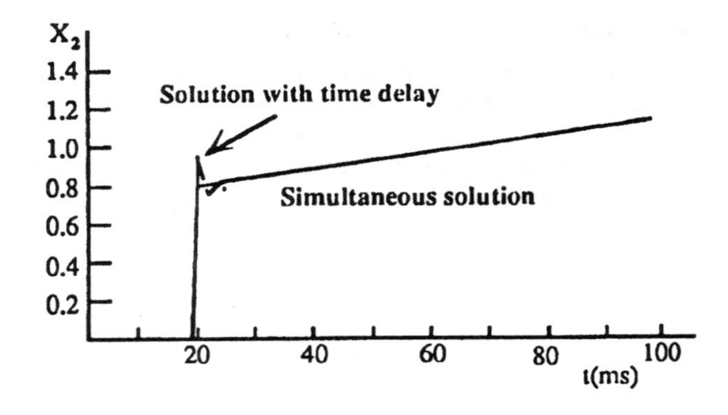

FIGURE 3. Correct Response with time delay

In this example, equation (4) indicates that X_3 is found first using the value for U_{NL} from the previous time interval. Once X_3 is known the output of the non-linear element is calculated. This new value of U_{NL} is used to find the other states followed by calculation of the new history terms. This provides an inconsistent history term for F3. Its state was found using a previous value for U_{NL}, but its history term was calculation using the updated value for U_{NL}. Problems with unnecessary time delays and history term errors have resulted in a new approach to finding TACS solutions.

III. New Solution Method for TACS

As discussed in Section II, the current TACS has problems with excessive time step delays and the calculation of history terms when non-linear elements are introduced. The objective of the current ordering method was to reduce fill-ins, saving computer storage and reduce computational time. This is still the objective in the EMTP program. The objective of the new ordering algorithm for TACS is to provide a sequential ordering of all elements that minimizes the time delays and correct errors in the history term calculation. This ordering method should also be independent of the type of element and depend only on the connections between the elements.

Sequential ordering has the problem of introducing time delays in a feedback loop when it is possible of find a simultaneous solution using the matrix methods discussed in the previous section. In this case the new algorithm must also search for sub-blocks within the sequential ordering that can be solved in a simultaneous manner. For ordering purposes these sub-blocks can then be handled as a single block, or "super block." To better explain the new algorithm the example test system shown in Figure 4 will be used as an example.

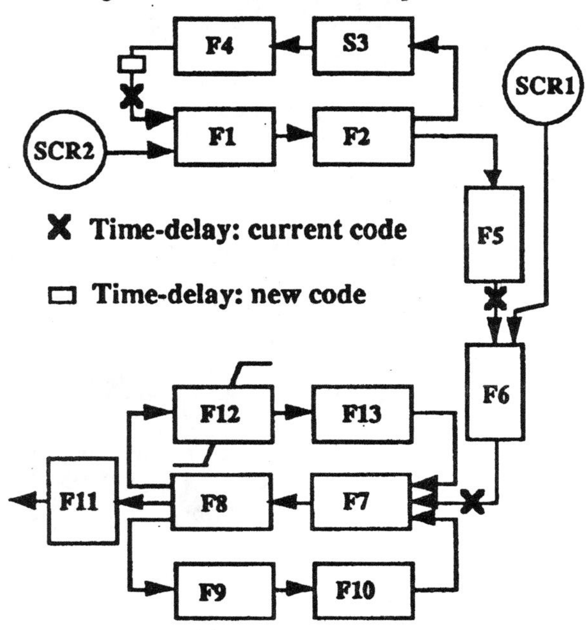

FIGURE 4. Ordering Test system.

In this system the "F" elements are transfer functions and element S3 is a non-linear element. SCR1 and SCR2 are known sources. The ordering system used in V2.0 starts with element F12 to insure correct treatment of the limiter. This would be the correct action if S3 was a linear function. In this case the existing TACS introduced three delays, one between F6 and F7, one between F5 and F6 and the other between F4 and F1. This last delay is the only one required. In a sequential only algorithm three delays are also required. One delay is placed in the F1, F2, S3, F4 loop and two delays are required in the double loop of F7, F8, F9, F10, F12, F13. This last group of elements can be solved in a simultaneous manner. The number of time delays required is now one. This delay must be in the loop containing the non-linear element.

Sequential ordering with super blocks also removes the history term problem by calculating history terms in a sequential manner. For the case of a super block, the full matrix is solved for all internal states before its history terms are calculated. This method avoids the problems of inconsistencies encountered in history terms discussed in Section II.

In the new TACS ordering scheme a matrix is used to describe the relationship between the inputs and outputs of each element in a system. A column and a row of this matrix represents how an element is connected. An element's column has information on which elements provide input signals. Each row of an element has information as to where its outputs are directed. To represent a connection between two elements an "X" will be placed at the intersection of the column representing the element receiving an input and the row representing the element which provides this signal.

The initial form of the ordering matrix for the system in Figure 4 is shown in Figure 5(a). For example, the column for F7 in Figure 5(a) indicates input from F6, F10, and F13. The row for F7 indicates that the output of element F7 is an input to F8. If an element is ordered such that the element(s) providing input precedes the element using the input(s) an "X" is placed in the upper triangle region. If this order is reversed the intersection is in the lower triangle region. For the arbitrary sequence shown in Figure 5(a) there are seven time delay errors indicated by the seven marks in the lower triangle. For example F6 precedes F5. This results in F6 using past output date from F5. The objective of any ordering scheme is to minimize the number of intersections in the low triangle. The first step in the new algorithm is to move all "source-like" elements to the top-left of the matrix. "Source-like" is defined as elements that have inputs from sources or "source-like" elements. The second step is to move all "output-like" elements to the bottom-right of the matrix. "Output-like" is defined as all elements whose output is not used by any element or used only by "output-like" elements. If a control system does not contain loops, the ordering would be complete. For this example SCR1 and SCR2 are in the input region while element F11 is in the output region. These elements are fixed in the matrix and will be neglected in the ordering of the "loop" element. In Figure 5(a) the region to be ordered is the sub-matrix between F6 and F13.

(a)

(b)

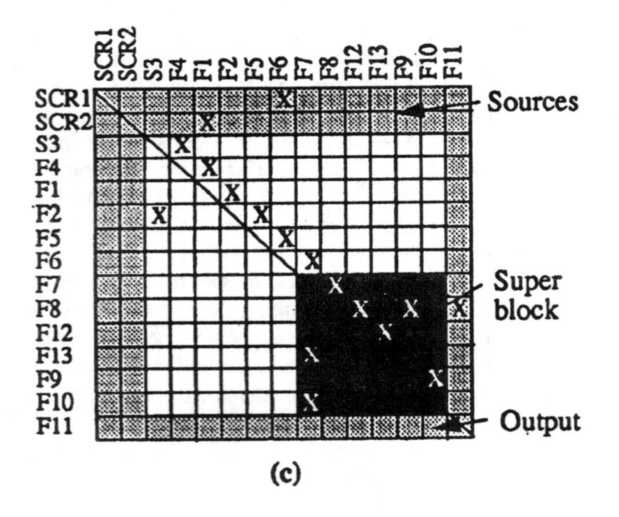

(c)

FIGURE 5. Ordering Matrix

The first ordering operation is to find a column with the maximum inputs and move this element to the top of the region. In our example this is element F7. This element is now excluded from the regions to be ordered. The next operation is to look at the outputs of the last element moved, F7 and move this element to follow F7. In our example this is F8.

Figure 5(b) shows the matrix after four forward ordering operations. The next operation is to move an element in the "output" region to a point after F13. In this case this is element F9. This continues until there are no intersections in the "output region." This implies that a loop in the control system has been ordered and there are no additional elements that are part of this loop(s).

If there are no unordered elements left, the ordering is finished. If there are elements left a backward ordering procedure is invoked. In backward ordering an element in the input region defined by the row is moved to the front of the ordered region. This continues until there are no more elements to be ordered. The maximum number of ordering steps are equal to the number of elements. The matrix after the last step is shown in Figure 5(c). Note that there are three intersections in the lower triangle region. This is the best possible solution for sequential ordering. One time step delay for each control loop.

The ordered matrix has important information. First, it provides sequential ordering with a minimum of time delays. Second, it provides information on where the time delays are placed and allows TACS to provided this information to the user. Third, the matrix allows for the identification of loops.

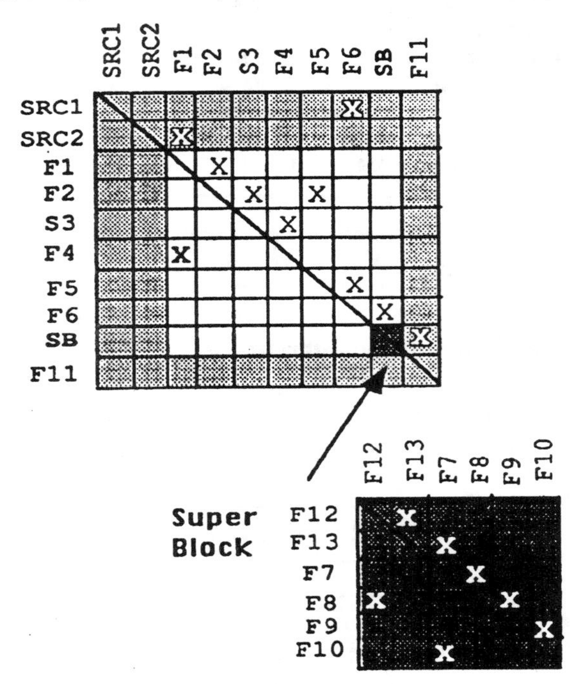

FIGURE 6. Final Form of Ordering Matrix

Elements contained in a square, with its lower left corner being an intersection point for an element in the low triangle, define a loop. For example, in Figure 5(c), F2/S3 intersection contain S3, F4, F1, and F2 comprising a control loop. The second set shown, F10/F7, contains nested loops.

Loops consisting of function blocks only with a single limiter or less can be solved simultaneously using the current methods discussed in Section II. In these cases the algorithm identifies this group of elements as a "super block" and shrinks it to a single element in the ordering matrix and reorders the system. The final ordered form is shown in Figure 6. This contains a single super block for the double loops that contain only transfer functions. The final ordering places a single time delay between elements F4 and F1, see Figure 4. Note that the super block is ordered with the limiter in the first position.

IV. Super Blocks and FORTRAN Models

The super block structure in the ordering procedure allows for the creation of other special features in TACS. In particular the enhanced TACS has two new features that are treated as "super blocks." One feature is an FORTRAN like "IF THEN ELSE" control structure in TACS. A second feature is the ability of TACS to use FORTRAN subroutines as TACS elements.

The current TACS provide many of the intrinsic functions found in FORTRAN. This allows for the creation of FORTRAN like expressions without the need to compile and link new code to the EMTP. The major shortcoming of this modeling tool is its lack of control functions to enable looping, and conditional branching. There are no GOTO, IF, DO, or CASE statements in the current TACS. The addition of an "IF THEN ELSE" structure greatly expands the class of modeling tools available to TACS

The "IF THEN ELSE" structure allows the user to create FORTRAN like models without the problems related to the use of FORTRAN code. The basic structure of this element is as follows:

```
IF (expression) THEN
  VAR1      = expression
  VAR2      = expression
etc.
ELSE
  VAR1      = expression
  VAR2      = expression
etc.
ENDIF
```

The TACS modeling data between the line starting with "IF" and the line containing "ENDIF" is considered a single super block by TACS for ordering. The "expression" is any FORTRAN-like logic or algebraic expression allowed in TACS. The variables VAR1, VAR2, etc., are names associated with the respective FORTRAN-like expressions. The variable names following the "THEN" or "ELSE" section can be unique or shared. These variables can also be internal or external to the block. The algorithm will find which variables are external and which are internal to the "IF THEN ELSE" block. In forming the super block a set of variables will be identified as input variables and another set will be identified as output variables. This block becomes as single element in the Ordering Matrix.

It is also possible to have loops within the block. In TACS the elements within the "IF THEN ELSE" block are assumed to be an independent system with inputs, outputs and internal loops that must be ordered to insure minimum number of time delays. To order this subsystem the identified input variables are handled as "sources" while the output variables are assumed to be "outputs." The remaining components are ordered using the general methods described in Section III.

The enhanced TACS can also use user written FORTRAN code as a TACS device. In this case TACS considers this user supplied device as a super block with multi-inputs and multi-outputs. The inputs and outputs of this user defined device are all TACS variables. They may connect to any other TACS component as required by the control model. Special interface routines are provided to insure correct transfer of data between the user provided FORTRAN subroutine and TACS. This user developed TACS element must also be compiled and linked to the EMTP.

V. Initialization

To reduce simulation time it is important that TACS models be initialized at a steady state operation point. This requires that all variables and related history terms are known. Theoretically TACS can not guarantee correct initialization for an arbitrary control system, but it is possible to greatly improve the current initialization system.

In the current TACS, the initialization algorithm assumes a TACS model that uses only transfer functions. It treats the dc and ac initialization separately. In each case the full system is solved simultaneously using the same methods described in Section II. Upon completing the separate calculations the initial values and history terms for the system are constructed.

For the dc steady state solution the complex frequencies, s, of the transfer functions are set to zero. The resulting matrix is used to find the dc components. For ac initialization the complex frequency, s, is set to $j\omega$. This results in an input-output matrix of complex functions. Again the necessary matrixes are set up and solved for the ac variables and history terms.

Systems modeled solely with sinusoidal sources and transfer functions can be correctly initialized in the current TACS with correction of some minor errors. Unfortunately most control systems have components that are non-linear. Non-linear elements disrupt this initialization method. The current TACS code assumes all nonlinear elements have zero output. For most TACS models this creates a situation where the initialization is not helpful. The enhanced TACS code removes current errors and provides initial values for some non-linear elements.

The principal initialization error in TACS is related to integrators, $G(s) = K/s$. Current versions of TACS can find dc steady-state values for all elements except integrators. In this case the dc output is not defined and must be provided by the user. If the user does not provide an output a default sets the output to zero. This is not always the best tactic. In cases where the integrator has an ac input this default strategy is wrong. If the input to the integrator is $\sin(\omega t)$ the expected ac output is $-\cos(\omega t)/\omega$. If a zero output is imposed by default the output then has a dc offset, $[1 - \cos(\omega t)]/\omega$. This default problem is resolved by forcing the dc component of the output to zero rather than the actual output.

The general problem of automatic initialization for control loops with non-linear elements is not theoretically solvable. There are many examples where there are no steady state solution for either an ac or dc input. Consider the system shown in Figure 7.

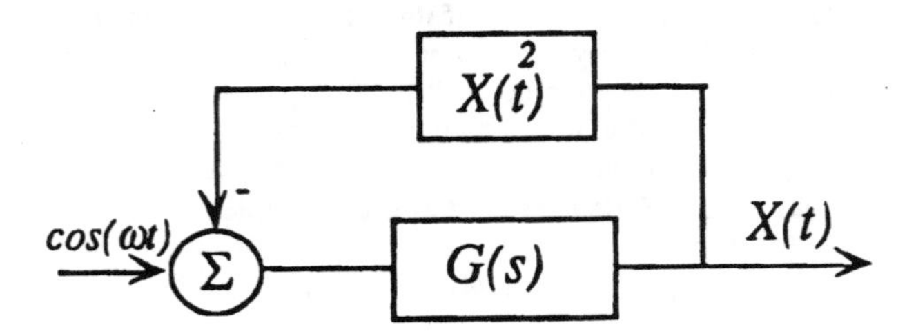

FIGURE 7. Control loop with non-linear element.

This system is a transfer function with a negative feed back being the square of the output. This system does not have a steady state solution for a general transfer function. Even when $G(s) = K$ there are no well defined ac steady solutions. The enhanced TACS takes advantage of the new solution method to provide some help but will not attempt to initialize any loops that contain non-linear elements.

The basic sequential nature of the new ordering methods coupled with the concept of super blocks allow the retention of the best features of the current methods with improvements provided for systems using non-linear elements. In the enhanced TACS any transfer function based super blocks are initialized using current methods.

Any non-linear element that has a defined ac and/or dc output for an ac or dc input can be initialized. For example a function that multiplies two inputs or squares the input will result in an ac output containing two frequencies which are different from that of the inputs. In the case of the enhanced TACS initialization is provided for all elements that have a finite number of ac/dc output components for a single ac/dc input. The outputs of any non-linear elements that have no defined ac or dc component or those which have an infinite number of ac components are set to zero.

The FORTRAN-like non-linear elements, *, -, + and **2 in TACS are initialized in the enhanced code. Some possible functions are given below;

$U_1(t) \cdot K \qquad U_1(t) + U_2(t) \qquad U_1(t) - U_2(t)$

$U_1(t) \cdot U_2(t) \qquad U_1(t)^2 \qquad \dfrac{U_1(t)}{U_{dc}}$

where $U_1(t)$ and $U_2(t)$ can be an ac or dc signal.

The algorithm is basically sequential and follows the ordering discussed earlier. Each non-linear element has both its dc and ac initial values calculated before the next element in the sequence. This allows for more that one ac frequency. For example, the operation $U_1(t) \cdot U_2(t)$, where $U_1(t)$ and $U_2(t)$ are sinusoidal functions of different frequencies has a well-defined output. The steady state solution is based on the sum and difference of the input frequencies. Since this operation has a definable steady response its initialization is included in the enhanced TACS. Consider another TACS function $\log(U_1(t))$ where $U_1(t)$ is its input. If this input is sinusoidal there is no defined ac steady state solution and TACS can not be expected to initialize this function. This algorithm through Fourier expansion also finds the initial conditions for a periodic input such as a square wave.

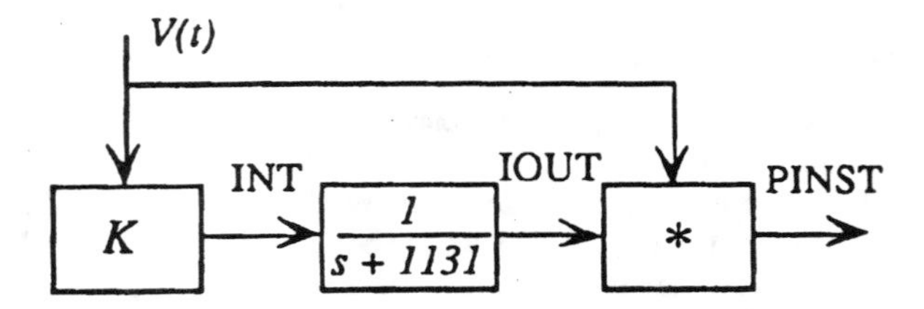

FIGURE 8. Initialization test case

These changes in TACS initialization can be best illustrated through an example, see Figure 8. In this example the input is an ac voltage. This input is scaled by a gain K and becomes the input, INT, to the transfer function. In this example INT is a cosine function with a frequency of 60 Hz and a magnitude of 1.2 Kv. The expected IOUT can be calculated.

$$IOUT(t) = V\cos(\omega t - \theta)$$

where

$$V = \frac{1.2Kv}{\sqrt{337^2 + 1131^2}} = 1.006\,volts$$

$$\theta = \operatorname{atan}\left(\frac{337}{1131}\right) = 18°$$

In the current TACS the input to the transfer function would be zero since the multiplication operation is assumed to be non-linear. This zero input would result in the transfer function being initialized to zero. This is seen in the form of the IOUT

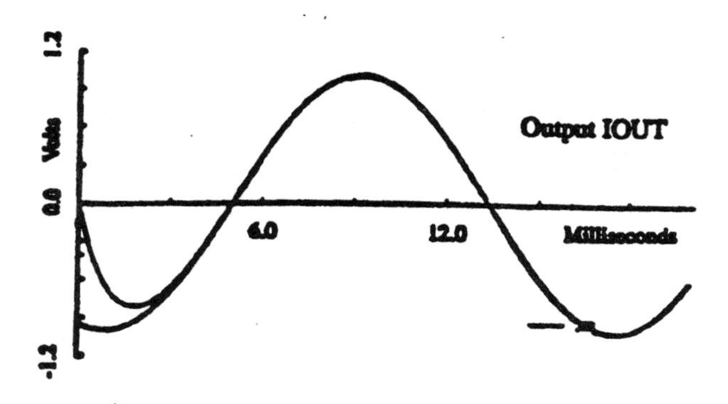

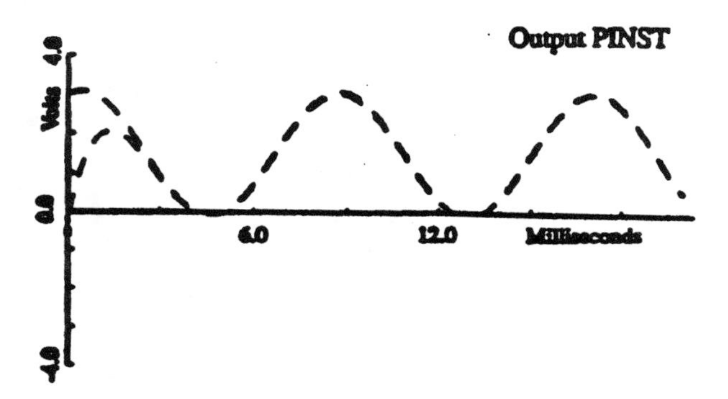

FIGURE 9. Initialization example

curve, Figure 9, which starts at zero and converges to the correct steady state in about 4 milliseconds. This convergence depends on the time constants of the transfer functions. The enhanced code has found the correct initial condition for the transfer function as shown.

The next function in the sequence multiplies two 60 Hz ac wave forms. The theoretical results consist of a dc component and an ac component at twice the input frequency. The comparison is shown in the two curves for PINST. Again the incorrectly initialized output converges to the correct value in 4 milliseconds. In this example the time for the transients to come to the correct steady state value is not large. It is also apparent that different transfer functions could result in a system that would require seconds rather than milliseconds to achieve steady state when not correctly initialized. In this case the new code would greatly reduce the time to achieve steady state. Except for such model related differences the cpu requirements of the old and new code are the same.

VI. Conclusions.

The new enhancements to TACS solve many problems that have plagued users for many years. First, the extra time delays and errors in history term calculations have been removed. Second, the location of all time delays introduced by TACS are provided in the output data for use by the user. Third, the debate over adding new devices to TACS have been squelched by the ability of TACS to use user written FORTRAN code as a TACS element. This allows libraries of special devices to be created and shared between users. Forth, the initialization has been greatly improved. Users should now find initialization useful for most TACS models.

Acknowledgment

The authors wish to acknowledge the support of this work by the EMTP Development Coordination Group. Without there support and guidance this work would not have been undertaken.

References

[1] H.W. Dommel, "Digital Computer Solution of Electromagnetic Transients in Single and Multiphase Networks," *IEEE Transactions on Power Systems*, vol. PAS-88, April 1969, pp. 388-399.

[2] L. Dube and H.W. Dommel, "Simulation of Control Systems in an Electromagnetic Transients Program with TACS," *IEEE Power Industry Computer Application Conference*, 1977, pp. 266-271.

[3] L. Dube, I. Bonfanti, "*MODELS: A New Simulation Tool in the EMTP,*" European Transactions on Electrical Power Engineering, Vol. 2, No.1, pp. 45-50, Jan./Feb. 1992

[4] L.X.Bui, S.Casoria, G.Morin and J. Reeve, "*EMTP TACS-Fortran Interface Development for digital Control Modeling,*" IEEE Trans. PAS, Vol 7, No.1, pp. 314-319, Feb. 1992.

Biographies

Robert H. Lasseter (F'92) received the Ph.D. degree in physics at the University of Pennsylvania, Philadelphia, in 1971.He was a Consultant Engineer at General Electric Company until he jointed the University of Wisconsin-Madison in 1980. His main interests are the application of power electronics to utility systems and simulation methods.

Jia-Rong Zhou is a native of Kunming, Yunnan of the People's Republic of China. He graduated from Chongqing University in Electrical Engineering in 1968. Upon graduation he worked for the Yunnan Research Institute of Scientific and Technical Information in power electronics and later as a computer programer. In 1985 he came to the University of Wisconsin as a visiting scholar. He stayed on to receive his M.S. degree from the University of Wisconsin in 1990. Currently he is with Oak Technologies Inc., Sunnyvale California.

EMTP TACS-FORTRAN INTERFACE DEVELOPMENT FOR DIGITAL CONTROLS MODELING

L. X. Bui, Senior Member, S. Casoria, Member
Institut de Recherche d'Hydro-Québec
Varennes, Canada

G. Morin, Member
Production, Transport et Distribution
Hydro-Québec, Montréal, Canada

J. Reeve, Fellow
University of Waterloo
Waterloo, Canada

Abstract

A FORTRAN interface to be used with TACS (Transient Analysis of Control Systems) in EMTP (ElectroMagnetic Transient Program) has been developed at Hydro-Québec to allow the use of FORTRAN language for simulating digital controls (for power system components such as static compensators, dc transmission and relays).

The interface has been designed without any time delay between TACS and the FORTRAN subroutines and allows users to benefit from all the flexibility of FORTRAN language to ease the task of modeling digital controls. Moreover, the property of task cyclicity in digital controls can be fully exploited to reduce the CPU time of each EMTP run. This paper presents details of the interface developed and provides results confirming its advantages for simulation of digital controls within EMTP, which was quite difficult hitherto.

Keywords: Digital Control Simulation, EMTP, TACS, HVDC Control Systems, SVC Control Systems.

INTRODUCTION

The traditional tool for simulation of electromagnetic transients in a power system is the analog simulator (TNA or dc simulator). Recently, however there has been an increasing trend towards the use of digital computer programs for this purpose. Usually designed to obtain fast solutions for a specific problem, such programs suffer from a serious drawback however: they lack the required flexibility and generality for system engineering studies, which involve the modeling of large complex systems and controls.

Over the last 15 years, the ElectroMagnetic Transient Program (EMTP) has been gaining popularity as a general-purpose program which can handle practically all types of circuit elements in a power system. To model control systems in EMTP, engineers resort to TACS [1,2], a very powerful tool for representing transfer functions, nonlinear elements, time-controlled switches, sources of any type, pseudo-FORTRAN expressions, including basic logic functions AND, OR, EQ, NE. TACS was first introduced to model synchronous-machine excitation systems and speed regulators but continuous improvements have since extended its scope to the modeling of analog controls.

In the past few years, digital controls have been used more extensively in static compensators and HVDC systems. The use of digital relays is also growing. Digital controls are basically software-oriented, which means that they are programmed essentially in a special language for control purposes, such as C language, assembler, PL/M or some in-house developed software, depending on the processor chosen for the control system concerned. These controls, using digital technology, have become more complex as designers have taken advantage of the flexibility of the software approach to build in special control functions. The use of dedicated microprocessors has permitted an increase in the number of control functions (tasks) whose execution is now limited to "only when needed". The latter feature has a considerable impact on the number of tasks performed by a single computer board. Modern controls are basically programmable real-time systems formed by different software tasks interfaced to the controlled process by D/A and A/D converters. Some faster functions are still performed by analog circuits.

At the present time, TACS solves inefficiently cyclic control functions with a time step larger than the EMTP integration time step ΔT (i.e functions corresponding to tasks executed at each interrupt request or predetermined time interval greater than ΔT), since all statements in TACS are executed at each time step. On the other hand, digital controls still use analog circuits for some high-speed functions. To increase the overall EMTP capability, interfacing digital and analog controls with a TACS-FORTRAN interface in EMTP has been investigated.

91 SM 417-6 PWRS A paper recommended and approved by the IEEE Power System Engineering Committee of the IEEE Power Engineering Society for presentation at the IEEE/PES 1991 Summer Meeting, San Diego, California July 28 - August 1, 1991. Manuscript submitted January 26, 1991; made available for printing May 17, 1991.

DIGITAL CONTROLS FOR REAL TIME SYSTEMS

Upon sensing and receiving input, a real-time system (e.g. mechanical: valve, motor etc. or electronic: transistor, thyristors) must process the input and generate responses (output) within a specified time interval. The control can be electromechanical (e.g. relays), electronic (e.g. TTL-hardwired) or microcomputer-based (e.g. hardware and software). Different hardware and software solutions exist for different applications.

For instance, in HVDC digital control systems (microcomputer-based), the time interval is typically 0.5 - 3 ms. Generally, a real-time operating system (RTOS) is used to administer processes in digital controlled systems. Faster processes requiring time intervals in the microsecond range are controlled electronically (hardwired logic).

The most important functional part of an RTOS is a task or process, which in its simplest form is a program with a single input, a straightforward calculation and a single output. However, real-time systems generally perform many processes or tasks, which can be handled in one of two ways:

1. using a dedicated CPU for every task
2. using one CPU in a multi-tasking mode.

Even if it is not time-critical, the first solution can be quite expensive. The second solution is generally adopted for this reason and also because:

1. tasks often run for brief periods between long idle durations,
2. when a task is idle, the CPU can be used for other tasks.

Therefore, the function of the RTOS is to schedule the allocation of CPU time between different tasks. Three types of scheduler are described briefly below.

In the "cyclic scheduler", all tasks are executed cyclically (e.g. a-b-c, a-b-c ...) but, since the duration of a task may vary (depending on the process conditions), the cycle time may also change, which could result in loss of synchronization with an I/O circuit.

In the "mosaic scheduler", a periodic clock (e.g. T = 5 ms) causes an interrupt, which is awaited by a special (dummy) task placed at the end (e.g. a-b-c-dummy, a-b-c-dummy ...). The only justification for the dummy task is to wait for the interrupt, before the cycle restarts, and the task duration represents wasted time.

The third type of scheduler is the "mosaic scheduler with more than one interrupt". Here, hardware interrupts may start some important tasks, usually to control faster processes. Tasks are divided into levels: hardware-interrupted tasks and tasks interrupted by the timer. Thus, at any moment, tasks in the latter level may be interrupted by a priority task in the hardware-interrupted levels. The operating system should continue to execute the interrupted task as soon as the priority task has been completed.

Finally, the software can use a high-level programming language (e.g. PL/M, C, locally developed) or one closer to the CPU language (e.g. assembler). Standardized functions (e.g. filters, integrators, limits, etc.) are built in to facilitate the design stage.

DIFFICULTIES IN SIMULATION DIGITAL CONTROL SYSTEM WITH TACS

Control systems in TACS are generally represented by a block diagram showing the interconnections among various system elements such as transfer functions, gains and algebraic and logic functions. A typical example is shown below:

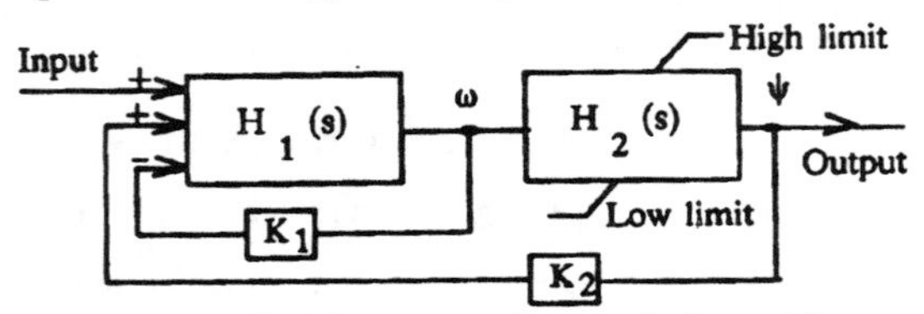

Fig. 1 : Block diagram representing a typical control system.

TACS has been designed as a toolbox of functions which users can represent in ways similar to analog computers, differential analyzers, algebraic and logical processors combined. One powerful feature of TACS is its capability to accept FORTRAN expressions (in EMTP terms, these are pseudo-FORTRAN expressions), which greatly facilitates the simulation of logic. The reason for the term "pseudo" is that these expressions are not compiled as real FORTRAN expressions but must be decoded as such in TACS. This also explains why the pseudo-FORTRAN expressions are limited to a small number of essential functions such as logic operators (OR, AND, EQ, SIN, COS, etc.). An example of a FORTRAN pseudo-expression is given below:

```
98IF3 = .NOT.A+B .LT. 3 .AND. (ISTEP .GE. 100)
```

From the user's viewpoint, this is quite adequate for simulating analog controls, including nonlinearities [3].

With the proliferation of digital controls in HVDC, static compensators and protection systems in the last few years, attempts have been made to extend the use of TACS to simulate them. However, the very flexibility offered by the software-oriented controller becomes an inconvenience in simulation with TACS. Features that are easily incorporated in digital controls, such as consideration of the past history of a variable, adaptive control, some aspects of artificial intelligence, etc., are difficult in the conventional analog approach.

Another important aspect of digital controls is the ability to use interrupt requests or different cycle times to execute different tasks as necessary (equivalent to a larger time step). Usually with EMTP and TACS, the time step is constant for the duration of the simulation. For instance, in HVDC simulations, the time step in EMTP is usually equal or less than 50 µs, but for simulating most controls, the universal use of such a fixed time step prolongs the CPU time. However, if the TACS-FORTRAN interface is available, the "execute only when needed" feature of digital controls can be fully exploited. For example, in HVDC simulation, the computation time for tasks based on firing angles and zero crossings, which can be executed at 30 electrical degree (1388 µs) intervals for 60 Hz systems in steady state, is significantly reduced.

The benefits of eventually being able to code digital controls directly into EMTP through real FORTRAN expressions may be summarized as follows:

- availability of all software functions such as:
  - DO loops
  - multi-level nesting IF
  - FORTRAN built-in functions
  - full flexibility of FORTRAN language
- use of vector array type of variable
- reduced CPU time, since tasks are executed only when needed
- use of modular structure using FORTRAN subroutines that allow:
  - repetition of internal variables (forbidden in TACS)
  - variable names up to 32 characters (as compared to 6 in TACS)
- exchange of developed models between users.

Major difficulties of digital-control simulation using TACS could be pointed out [4]. For example:

- Represention of vector arrays:

```
IF(ZERO_CROSSING) RING_COUNTER(12)=1
```

- Difficulty of simulating IF-THEN-ELSE multi-nesting logic:

```
DO 1 I=1,12
   IF(ZERO_CROSSING) THEN
   RING_COUNTER(12)=1
   ELSE RING_COUNTER(12)=0
END
```

- Execution of tasks only when needed:

As mentioned above, tasks based on zero crossings need to be computed only at 30-degree intervals in the steady state. This is impossible in TACS however since the computation must be done at every time step because there is no possibility of skipping any TACS executable statement using IF-THEN or nested IF-THEN.

In software, the executable statement is:

```
IF (ZERO_CROSSING) CALL PERIOD_TIME_CALC
```

In TACS, the whole PERIOD_TIME_CALC task will be computed at every time step, which is very CPU time consuming, since there are many such tasks in a digital control system. This is a major problem in simulating large systems, where users must save CPU time to obtain a reasonable total run time.

Since TACS has already been developed, it is prudent to retain it in order to represent any analog functions and to provide communication of variables into the main EMTP code. Modification of TACS structure to accept FORTRAN subroutines internally is not simple in our opinion and may require an important rewriting of the actual code. To solves the problem mentionned above, two alternatives are: 1- to provide a TACS-FORTRAN interface in EMTP or 2- write another control simulation package from scratch with much higher capabilities to simulate both analog and digital control systems. While the second alternative seems more elegant in the long term, the time and financial support required for the effort involved make the first considerably more attractive.

TACS - FORTRAN INTERFACE

The power system and the control system are solved in EMTP and TACS respectively. Although state variables are transferred between two programs, numerical solutions are done separately, causing a time step delay between EMTP and TACS (this delay is not related to the TACS-FORTRAN interface). In some special situations, TACS may also cause its own time delays with non-linear functions at certain positions in the control system. Usually, with a good understanding of the control system to be modelled and a judicious choice of time step, one can avoid these situations and the delays will has no effect on the accuracy of the results.

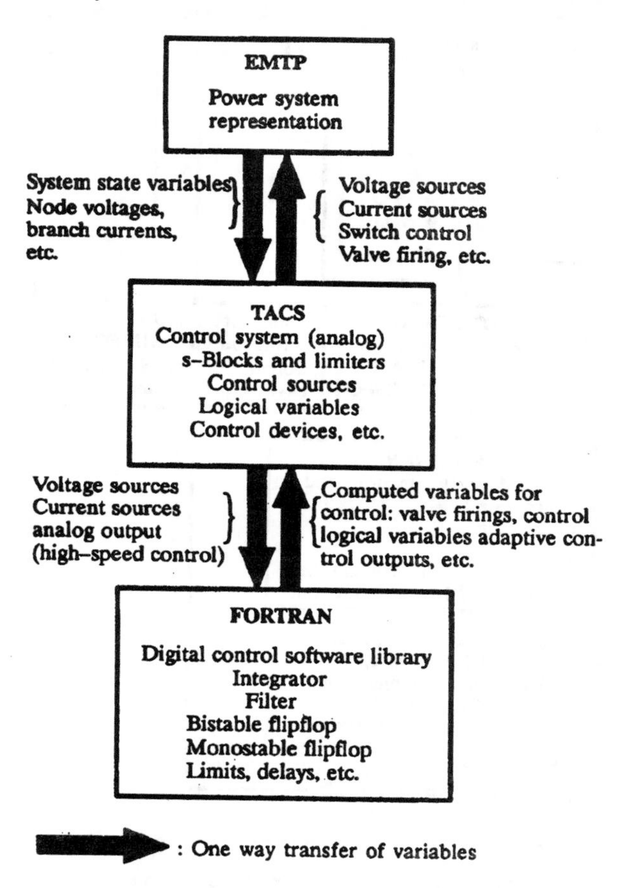

Fig. 2: EMTP, TACS and TACS-FORTRAN interface in function blocks.

In TACS, each variable is stored in an array with a pointer indicating its location so that, every time a new value is computed, it is stored in the right place for the variable in question. The solution of the set of linear equations in TACS is executed through the time step loop. In this loop, all transfer functions in the s-domain are converted into algebraic difference equations in the time domain by implicit integration of the trapezoidal rule. Before entering this time step loop, the new values of each state variable must be computed and they then serve as input to the matrix of linear algebraic equations.

Since TACS is quite suitable for simulating the analog control and needed to model high-speed functions, efforts were made to counter its weaknesses by developing the FORTRAN interface to simulate digital control in such a way that hybrid analog-digital control can be modelled in EMTP.

Previously, digital control models were interfaced with EMTP and applied to the simulation of a 4-terminal HVDC system [5]. The methodology for this paper has refined the approach of [5] in the context of contemporary digital controls and permits flexible access to TACS variables and re-entry into TACS within the same time step.

The function of the interface then is to communicate variables in two ways between TACS and FORTRAN subroutines. This is achieved by using the same location of these variables created in the TACS array of variables named XTACS through a pointer.

First, TACS computes all state variables. Then, the values of the latter are transferred to FORTRAN through the interface. The digital control is solved in FORTRAN. The new values of the control variables computed from FORTRAN are sent back to TACS through the interface. In this way, all TACS functions such as devices (frequency meter, point by point non-linearities etc.) can still be used in hybrid with FORTRAN subroutines.

Now TACS solves for the remaining logics and enters the time step loop to solve for the output of all transfer functions. The output at this step will be sent to EMTP (firing pulses, etc.) so that actions on the power system can be taken at the next time step. In order not to create any time delay between FORTRAN and TACS and vice versa, all variables between TACS and FORTRAN must be tranferred back to TACS before the next time step. Any non-digital functions, such as transducers or analog control functions are processed by TACS before the sequence of interfacing TACS and FORTRAN. The order of execution between TACS and FORTRAN is illustrated in Fig. 3.

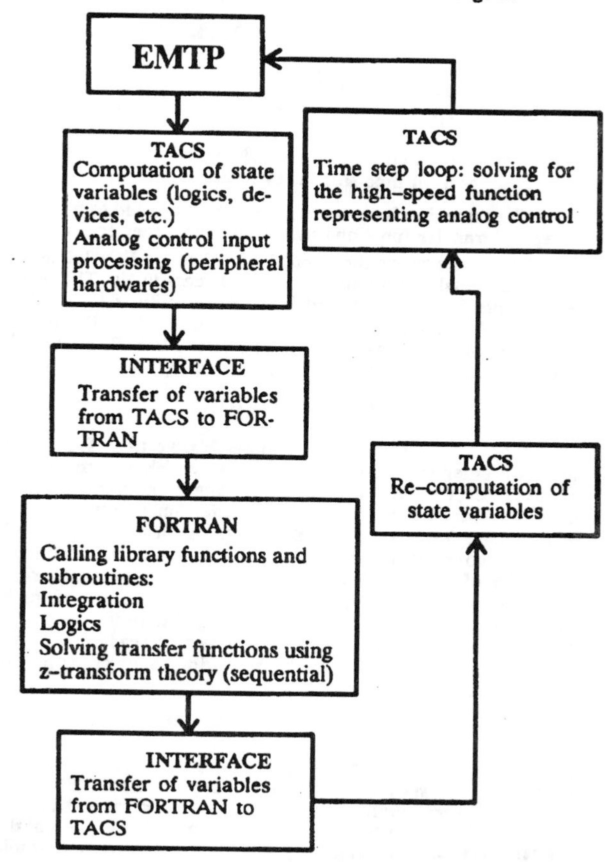

Fig. 3: TACS, interface and FORTRAN in execution order.

Once the interface has successfully transferred the variables to FORTRAN subroutines, the digital control is computed with its own subroutines solving transfer functions using the theory of z-transform converted into difference equations. For instance, a digital first order filter in cascade with a zero order hold will be treated in the following way:

$$H(s) = \frac{1 - e^{sT_o}}{s} \cdot \frac{1}{1 + sT}$$

Using z-transform, H(s) becomes:

$$H(z) = \frac{out(z)}{in(z)} = \frac{(1-e^{-T_o/T})\, z^{-1}}{1-e^{-T_o/T}\, z^{-1}}$$

Where T= time constant, T_o=sampling time

Then:

$$out(z) = (1-e^{-T_o/T})\, z^{-1}\, in(z) + e^{-T_o/T}\, z^{-1}\, out(z)$$

$$out(z) = z^{-1}\, in(z) - e^{-T_o/T} z^{-1} in(z) + e^{-T_o/T} z^{-1} out(z)$$

With a sampling time much smaller than the time constant T (T_o << T) and using McLaurin's series:

$$e^{-T_o/T} \approx 1 - T_o/T$$

$$out(z) = (1 - \frac{T_o}{T})\, z^{-1} out(z) + z^{-1} in(z) - (1\ \frac{T_o}{T})\, z^{-1} in(z)$$

$$out(z) = z^{-1} out(z) + \frac{T_o}{T}\, z^{-1} (in(z) - out(z))$$

With a time step equal to T_o :

$$out(k) = out(k-1) + \frac{T_o}{T} (in(k-1) - out(k-1))$$

The coding of such a filter in FORTRAN is then very simple:

$$out = out + \frac{T_o}{T} (in - out)$$

new value — present time step; old values — previous time step

The same development can easily be extended to treat other type of transfer functions such as second-order or lead-lag.

The structure of the control modelled by FORTRAN reflects the actual control and is prepared accordingly. The protocol for interfacing TACS and FORTRAN, using standard F77 FORTRAN, is intended to be generally applicable. The implementation of the interface on the EPRI/DCG EMTP version 1 is done in accordance with the flowchart in Fig. 4.

INITIALIZATION

Since FORTRAN is much more flexible than TACS, the user can enter any initial value for any variable in the digital control or even write a separate subroutine to compute the initial conditions for all variables for any transfer function (using backward calculation at time step zero from the desired system values). Moreover, in the case of HVDC thyristor bridge control, the order of each thyristor to be fired can be initialized right from time step zero to quickly reach steady-state conditions at the first 60 Hz cycle using the numbering system for the thyristors. In general, initialization can be done much easier than with TACS, since any variables can be overwritten at any time during the simulation.

CASE STUDY

A case study was set up in EMTP in order to compare the simulation of a six pulse rectifier feeding into a back EMF [6]. Two cases were simulated with the power system and control shown in Fig. 5. The control, for the purpose of validation of the digital interface, is a current regulator superimposed on a simple equal-firing-angle pulse generator. After steady state initialization, a dc short-circuit is applied at the back EMF node at 100

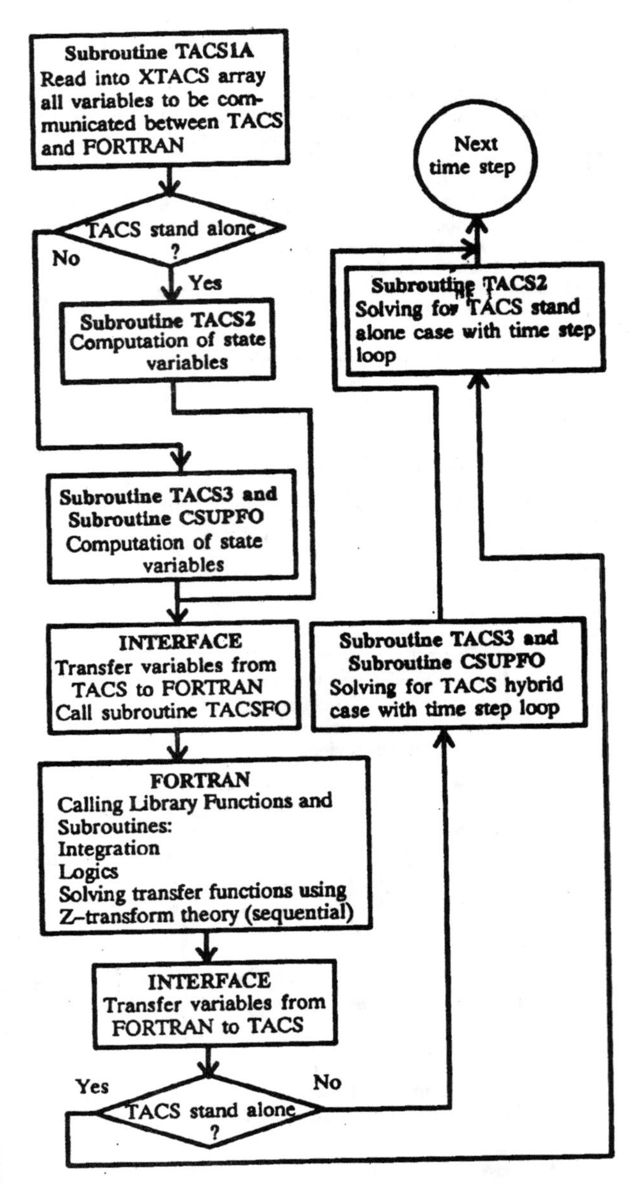

Fig. 4: Flow chart showing the implementation of FORTRAN interface into TACS.

ms. In the first case, the control was done entirely with TACS whereas, in the second case, it was entirely with FORTRAN using the technique of digital-control (TACS was used only to model two high-speed transfer functions and to transfer the values of state variables from the EMTP power system to the digital control through the interface). As the results show in Fig.6, the system node voltages, currents and the control firing angle (alpha) are identical for both cases, but the CPU time consumed for the second case is 10% lower than the first one (109 seconds as compared to 120 seconds for the first case, both cases were run on a SPARC1 SUN work-station). This rather is a modest saving in CPU time since the control modeled was very much simplified and does not fully exploit the advantage of the property of "execution only when needed". Since the saving on computing time depends on the specific control function and the

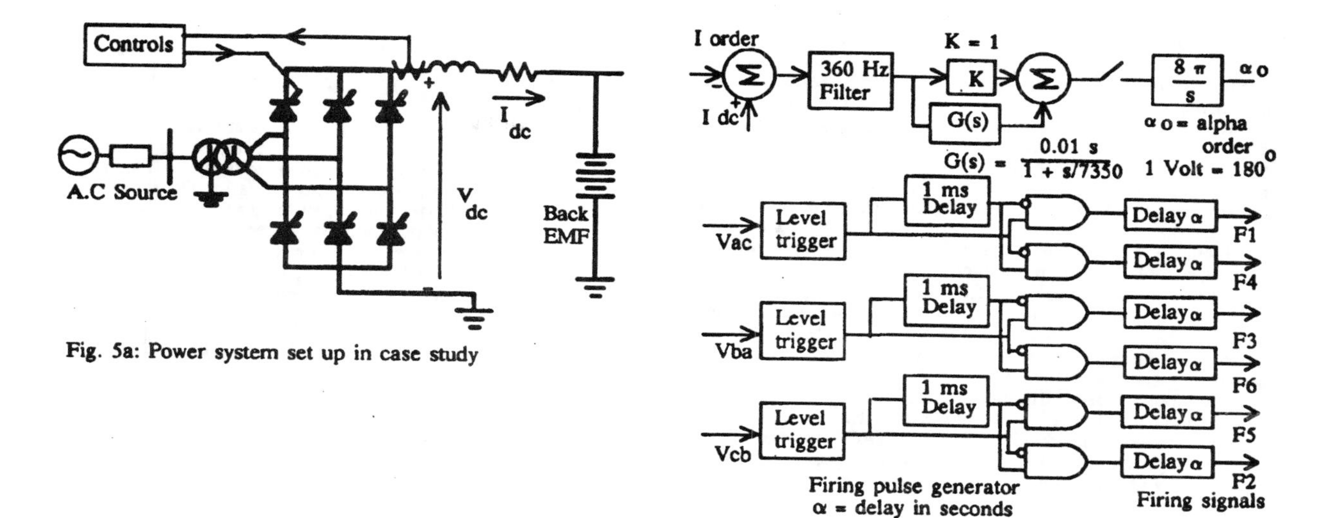

Fig. 5a: Power system set up in case study

Fig. 5b: Control system modeled for the set up above

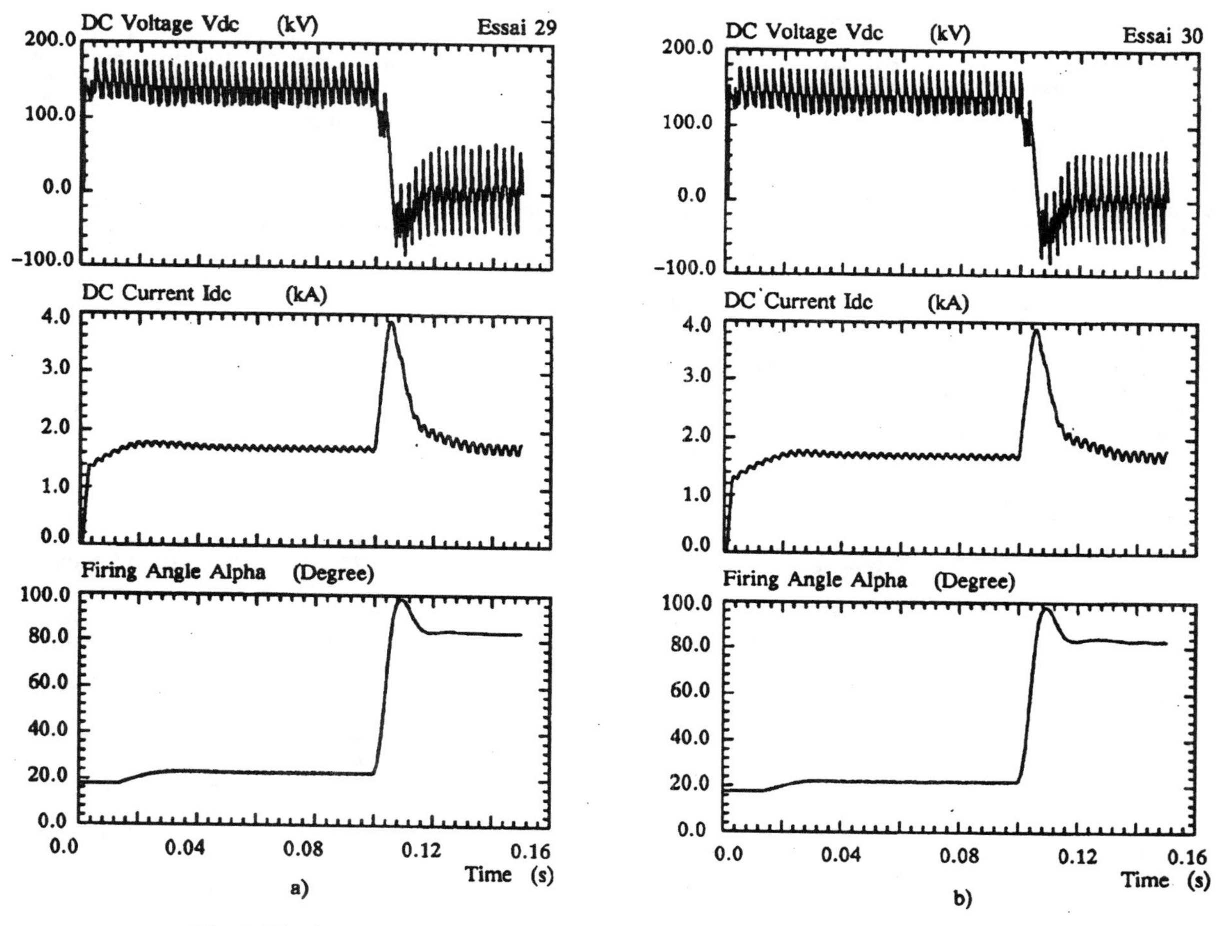

Fig. 6: Identical results showing a dc to ground fault obtained from:
a) entirely TACS simulation
b) using TACS-FORTRAN interface

transient activity under consideration, this case study does not reflect the real CPU time saving which could be realized on the simulation of a real digital control using the interface rather than using TACS. However, a typical study for an actual project involves a great many control functions, most of which are inactive at any one solution time step. It has been found that the effect of cumulative time saving is considerable and it is expected that exploiting fully the "execution only when needed" property on other tasks of the real digital control system such as voltage regulator, tap changer control ect., which are slow control loops, will add up to a much larger CPU time saving on running a EMTP case with the interface as compared to the simulation of such control using TACS. On the other hand, the difficulties shown previously in simulation of real digital control using TACS has demonstrated that it is not certain that such simulation could be realized entirely by TACS in practice since the complexity and flexibility of such software-based control has rendered the task very difficult if not impossible. Consequently, it is difficult to provide a comparison of actual digital control system between using the interface and TACS structure. The case study presented in this paper has the purpose of showing the accuracy of digital control simulation using the interface as compared to TACS and also giving a perspective saving of CPU time in a real digital control simulation case.

The method has been applied to an actual DC transmission scheme incorporating extensive digital control. In fact, this development has been used at Hydro-Québec to model the first stage (comprising two terminals) of the multi-terminal HVDC link between the La Grande generating complex and New England. The advantages demonstrated by the simple test system in this paper will be reinforced by the results for a large system including simulation of AC and/or DC faults and recovery of the above HVDC link in a future paper.

CONCLUSION

This paper has demonstrated some of the difficulties encountered in representing digital controls using TACS, including representation of some basic logic functions as compared to FORTRAN statements and the inefficiency (equations solved every time step) of TACS to model cyclic tasks and logic functions.

To overcome these limitations, a FORTRAN interface coupling TACS to FORTRAN subroutines has been developed and tested on a model of a HVDC digital control. To maintain full TACS capabilities, the interface was implemented in such a way that TACS output can be transferred directly in FORTRAN and vice-versa without adding time delays to the solution. Moreover, all functions (s-blocks, devices, etc.) available in TACS can be used in hybrid with FORTRAN subroutines. Another advantage of the interface is that, thanks to the structure of the subroutines and the standardized FORTRAN language, models developed and tested can be exchanged among users.

ACKNOWLEDGMENTS

This work was inspired by the TALKIE interface developed at the University of Waterloo. The authors would like to thank S. Lene-Smith and J. Wikston graduate students of J. Reeve, for judicious assistance. They are also grateful to Dr. P. C. S. Krishnayya of IREQ for his help and encouragement during this work. Lastly, a word of thanks should go to Lesley Kelly-Régnier of Hydro-Québec for thoughful editing of the text.

BIOGRAPHIES

L.X.Bui (S.M, 85) - graduated with a B.A.Sc. from Ecole Polytechnique de Montréal, in 1975 and received a M.Eng. degree from McGill University, Montréal in 1978. From 1975-1978, he was with the Planning Department of Hydro-Quebec. Since 1979, he has been at IREQ. His current interests are in HVDC transmission and SVC systems.

G. Morin (M, 85) - received his B.Sc.A in Electrical Engineering from University of Sherbrooke in 1978 and his M.Sc.A from Ecole Polytechnique in 1983. He is with Hydro-Québec since 1978 working in the Power System Operation Department. His main interests are harmonics, dc systems and power system transients. He is author of publications ralated to harmonic overvoltages during power system restoration. He is member of IEEE Power System Restoration Task Force, IEEE Switching Surge Group, IEEE Working Group 15.15.02 and Chairman of the Frequency Dependent Equivalents Task Force.

S. Casoria (M, 78) - was born in Cairo, Egypt in 1953. He received the B.Ing. and M.Sc.A degrees in Electrical Engineering from Ecole Polytechnique de Montréal in 1977 and 1981 respectively. Since 1981, he is with Institut de Recherche d'Hydro-Québec (IREQ) in the group Études de Réseaux. From 1988 to 1989, he was a trainee at ASEA Brown Boveri (ABB) plant in Ludvika, Sweden. During this period, he developed software subsystems implemented in the control and protection system for the Québec-New England HVDC multi-terminal transmission system. His area of professional interests include HVDC transmission and control systems.

J. Reeve (F, 81) - received the B.Sc., M.Sc., Ph.D and D.Sc. degrees from the University of Manchester, England. After employment with the English Electric Company, Stafford, involved in the development of protective relays, and 6 years as a faculty member at the University of Manchester Institute of Science and Technology, he has been with the University of Waterloo, Ontario, Canada since 1967, currently a Professor in the Department of Electrical and Computer Engineering. His research interests for 30 years have been centered mainly on aspects of dc power transmission. He is a Past Chairman of the IEEE DC Transmission Subcomittee and is currently a member of several IEEE and CIGRE Working Groups and task forces concerned with dc transmission. Dr. Reeve is President of John Reeve Consultants Ltd. He recently completed a one year assignment at IREQ concerned with the simulation of the Hydro Québec-New England multi-terminal dc system.

REFERENCES

[1]. L. Dubé, H. W. Dommel, "Simulation of Control System in an Electromagnetic Transient Program with TACS," IEEE Transactions on Power Industry and Computer Applications, 1977.

[2]. EMTP Rule Book, EPRI / DCG Version 1.0.

[3]. D. Goldsworthy, J.J. Vithayathil, "EMTP Model of an HVDC Transmission System," Proceedings of the IEEE Montech '86 Conference on HVDC Power Transmission, September 29 - October 1, 1986, pp. 39-46.

[4]. L. X. Bui, S. Casoria, G. Morin, "Modeling of Digital Controls with EMTP," CEA Meeting, March 25-29, 1989, Montréal, Canada.

[5]. J. Reeve and S.P. Chen, "Versatile Interactive Digital Simulator Based on EMTP for AC/DC Power System Transient Studies," IEEE Transactions on Power Apparatus and Systems, Vol. 103, No. 12, December 1984, pp. 3625-3633.

[6]. K. G. Fehrle, R. H. Lasseter, "Simulation of Control Systems and Application to HVDC Converters," IEEE Tutorial Course 81 EHO173-5-PWR on Digital Simulation of Electrical Transient Phenomena, 1981.

IEEE Transactions on Power Delivery, Vol. 9, No. 3, July 1994

IMPROVED CONTROL SYSTEMS SIMULATION IN THE EMTP THROUGH COMPENSATION

S. Lefebvre (Member) and J. Mahseredjian (Member)
Hydro–Québec, IREQ
Direction Technologie de réseaux
Varennes, Québec, Canada

ABSTRACT : The control systems, devices and phenomena modelled in TACS, and the electric network modeled in EMTP are solved separately with one–time–step error at the interface. This provides an efficient time–step solution, but there can be numerical stability and accuracy problems associated with the one–time–step error. This paper shows a technique which can eliminate the time delay, without having to use a simultaneous EMTP and TACS solution.

Keywords : EMTP, compensation, power electronics

1. INTRODUCTION

TACS is a section of the EMTP which allows the representation of control systems in block–diagram form. It has been used to model the control of converters, FACTS and static VAR devices, excitation systems of synchronous generators, dynamic load characteristics, etc. [1–2]

It is well known that there is a one step time delay between TACS and EMTP variables. There are many cases where such delay do not cause any trouble: fast transients in the network – slow transients in the control system. However, many examples can be created where the time delay results either in numerical instability or wrong results. An example of numerical instability is the simulation of the arc behavior in a circuit breaker (resistance simulation through a current injection) , while wrong results will be obtained in some power electronic circuits due to the delay of the trigger orders from TACS.

The information used by EMTP to advance to time t is based on the TACS solution T_{out} for time $t - \Delta T$, where ΔT is the step size. Based on the network solution E_{in} at time t, TACS computes its outputs. This is shown in Figure 1.

The concept of separate EMTP and TACS simulation is the result of preserving solution efficiency, and as much as possible modularity. The TACS solution is simultaneous for linear blocks, and sequential for nonlinear blocks or functions. The TACS equations are sparse but unsymmetrical. There are no iterations during the solution, rather the equations are ordered in such a way as to minimize any time delays that need to be introduced in loops involving nonlinear blocks. When there is freedom, the equations are then ordered to minimize fill–ins during the factorizations. Row Gaussian Elimination is used to solve the equations in the time step loop.

94 WM 084-4 PWRD A paper recommended and approved by the IEEE Transmission and Distribution Committee of the IEEE Power Engineering Society for presentation at the IEEE/PES 1994 Winter Meeting, New York, New York, January 30 - February 3, 1994. Manuscript submitted July 21, 1993; made available for printing November 29, 1993.

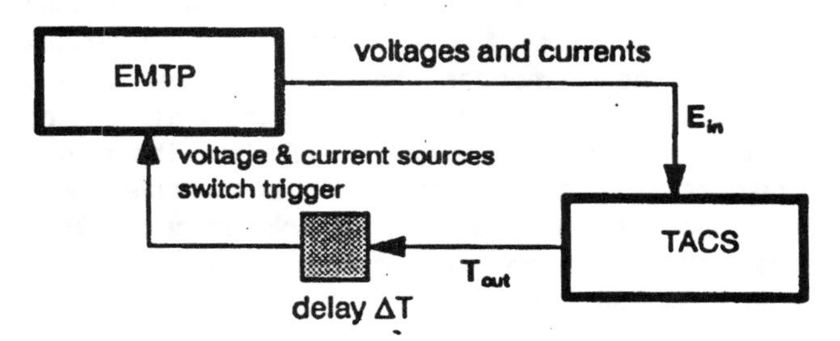

Figure 1 Solution sequence

TACS outputs to the EMTP may be in fact one or more time steps late because there are internal time delays in the TACS solution. The internal TACS time delays can often be minimized or compensated with proper control modelling. They may affect the accuracy or stability of the TACS solution, but have less impact on the EMTP–TACS interface stability. The one–time–step error at this interface can have a strong impact on the network solution. Selecting a smaller step size may sometimes be sufficient, but no simple general rule can be given to manage time delay errors though. There are even cases where irrespective of the step size, the solution is numerically unstable. This paper then does not concentrate on the internal time delays in TACS, but rather on the EMTP–TACS interface one–time–step error.

The purpose of this paper is to demonstrate the application of the compensation method to eliminate the one–time–step error, without having to have a simultaneous EMTP–TACS solution. This is an extension of the compensation technique used in EMTP for the solution of true nonlinear elements. [3]

2. BASIC ALGORITHMS

EMTP solution

The EMTP is a nodal analysis program based on the fixed time–step trapezoidal integration method. Considering only linear elements, discretized network equations are given by $\mathbf{Y_n V_n} = \mathbf{I_n} - \mathbf{I_h}$,where $\mathbf{Y_n}$ is the symmetrical nodal admittance matrix, $\mathbf{V_n}$ is the node to ground voltage vector while $\mathbf{I_n}$ and $\mathbf{I_h}$ represents respectively node current injections including current sources and 'past history' terms. For convenience in notation, assume that the impact of known voltage sources is merged with the history terms. The symmetric admittance matrix is triangularized, and the time step solution is obtained, without iteration, with forward–backward substitution. When the computation of the voltages is completed, then all history terms are updated for usage at the next time step.

Before the triangularization, it is important to order the equations in a way that will minimize the fill–ins.

There are relatively few nonlinear elements in the network, and the EMTP uses an efficient scheme for their solution that does not penalize the large linear network solution. The first manner to handle nonlinear elements is to use pseudo–nonlinearities, i.e. assume one time–step delay. This is prone to numerical difficulties. The second manner is to consider true nonlinear elements whose solution is superimposed on the solution of the linear network. This is known as the compensation method, and requires iterations within the nonlinear elements.

TACS solution

While the non–control equipment is modeled by EMTP devices, TACS is used to model the controls on these devices, or controlled voltage–current sources. A typical example is valves in power electronic circuits which can be directly modeled in the EMTP as switch–components and incorporated into the $\mathbf{Y_n}$ matrix. The switches close after the anode voltage becomes larger than the cathode voltage, as soon as a firing signal is received from TACS. The switches open as soon as the current goes through zero from a positive value. Re–triangularization is performed whenever there is a topological change in the network.

As for the network, the control systems are a mixture of linear and nonlinear elements, however the proportion of nonlinear elements in TACS is typically much higher in TACS than in EMTP. In order to preserve solution speed, the TACS solution is non iterative but introduces internal TACS time delays to take into account nonlinear components of the control circuit. This is a pseudo–nonlinear representation, but not all nonlinear components require individual time delays though. For example series nonlinearities in a loop do not have cumulative internal time delays, but only one, while some topologies can be solved without time delays[4]. If everything were linear, it would be trivial to obtain a simultaneous solution, but this is not the case in general. Because of the sequencing on nonlinear components, and because the TACS matrices are unsymmetrical, a time–step delay exists between EMTP and TACS, (Figure 1) such that independent solutions can be carried out. The pseudo–nonlinear non iterative TACS solution has been traditionally preferred over a TACS simultaneous nonlinear formulation.

The TACS matrices are unsymmetrical, triangularization yields two distinct upper and lower matrices. Because there are typically several nonlinearities in a control system, the TACS solution is not as simple as triangularization separate of the time–step loop, and a forward–backward substitution in the time–step loop. As in the EMTP, it is necessary to distinguish linear components (explicitly declared transfer functions) from the nonlinearities. Nonlinearities in TACS context, include supplemental pseudo–FORTRAN expressions which may express a linear relationship between variables.

Before entering the time–step loop, all TACS transfer functions in the s–domain are converted into algebraic difference equations in the time domain through the trapezoidal rule of integration. These equations can be written as $\mathbf{A}\,\mathbf{X} = \mathbf{b}$, where $\mathbf{A}$ is nxn and non symmetric, this is shown in Figure 2 (history terms merged on the RHS). Basically, after a simultaneous solution of the linear components is obtained, then, all supplemental devices are sequentially taken into account in a manner that reduces the internal time delays, without iterations.

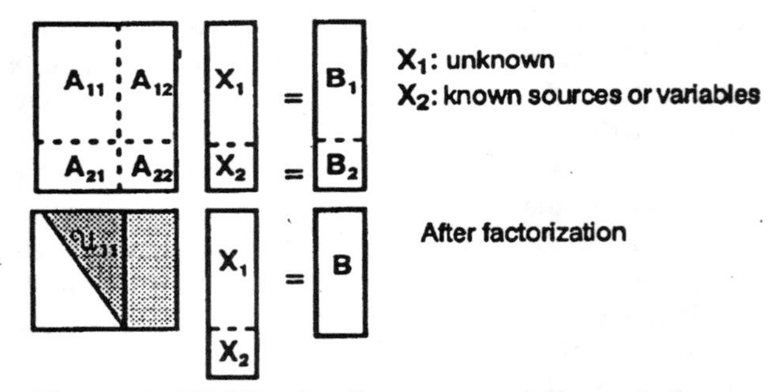

Figure 2 TACS simultaneous solution of linear components

Triangularization of the matrix **A** is performed once, then at each time step the unknown variables are found by a forward substitution on **B**, followed by back–substitution to obtain the state variables **X**. The back–substitution process is interrupted, as needed, to update the supplemental variables whenever possible. That is, when TACS does the back–substitution to solve for the output of linear components, then TACS checks the link list for outputs driving supplemental variables or devices. If any, then TACS updates the relevant supplemental variables or devices.

Ordering techniques completely differ from that of EMTP. Loops containing only function blocks are not sensitive to the ordering. In fact, if there is at most one limiter present, the internal TACS solution is simultaneous. In most cases however, the correct ordering of all TACS variables, except the sources, is crucial for reducing the internal TACS time delays and for getting as accurate results as possible without iterations. It is not simply a matter of reducing the fill–ins. Ordering is thus used to minimize the number of internal TACS time delays (prioritized over fill–ins in the triangularization), to keep the number of operations in the time–step loop as low as possible and, together with sparsity storage, to reduce the size of the memory for storing the triangularized matrices.

3. SIMULTANEOUS EMTP–TACS SOLUTION

The strategy has always been to solve EMTP and TACS separately, and with one–time–step delay, as introduced in Figure 1. This delay can create numerical instabilities, as documented in the Applications section. The internal TACS time delays may also create numerical difficulties, but the focus is not on those.

Solving the models represented in TACS simultaneously with the EMTP network is more complicated for models of power system components because the TACS matrices are unsymmetrical and there is a TACS sequential solution of nonlinear components which is closely interleaved with the solution of the linear elements themselves.

Different techniques can eliminate the one–time–step error.

<u>Predictor–corrector</u> [3]

In solving EMTP for time t, use predicted TACS values for time t. Then use the EMTP solution to correct the TACS output at time t. Repeat until convergence. In principle this is easy to implement, but the convergence characteristics are very dependent on the accuracy of the prediction. Furthermore, the entire EMTP network has to be solved repeatedly at one time step.

<u>EMTP and TACS equivalents</u>

Assume for simplicity (this restriction is for the sake of clarity only), that TACS needs E_{in} from the EMTP, and provides the current I_{out}. In the equivalent–approach, the TACS equations are reduced through partial factorization to a factorized set of internal TACS variables **X**, and an external set of variables involving the EMTP voltages E_{in} (X_2 in Figure 2) and the TACS–generated currents I_{out}. Similarly the EMTP equations are triangularized only in their upper part. This is shown in Figure 3, after factorization (history terms merged in B_1, B_2 and B_3), with $G = [G_{11}\ G_{12}]$.

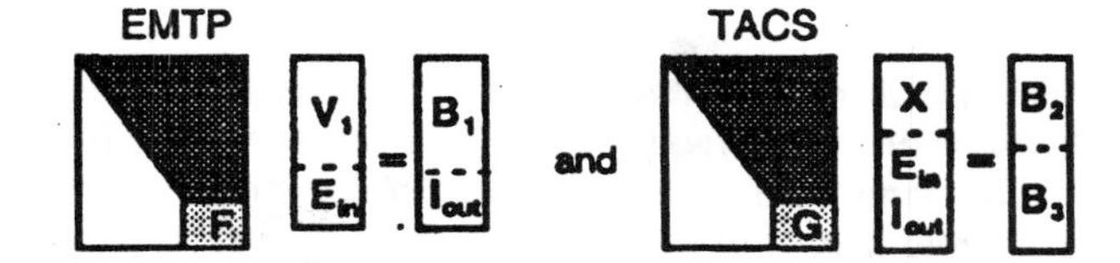

Figure 3 EMTP–TACS internal–external formulation

The matrices **F** and **G** represent exact equivalent models involving only the external variables. To obtain a simultaneous EMTP–TACS solution, first solve for the interface variables through the reduced set of equations involving external variables :

$$\begin{bmatrix} F & -\mathcal{I} \\ G_{11} & G_{12} \end{bmatrix} \begin{bmatrix} E_{in} \\ I_{out} \end{bmatrix} = \begin{bmatrix} 0 \\ B_3 \end{bmatrix} \tag{1}$$

where $\mathcal{I}$ is the identity matrix. Next, perform substitution to get the remaining internal EMTP and TACS variables. Then history terms are updated for the next time step. In general, the external variables will not be solved without iterations. Since most control systems are nonlinear, implementation is not trivial since the **G** matrix is in general a nonlinear function of E_{in} and I_{out}, and may not be explicitly available. Furthermore, it is not desirable to impact the TACS ordering linked to the nonlinear TACS components sequencing.

<u>Compensation</u>

A practical numerical solution technique is obtained with the compensation method, and with minimal modifications to the existing EMTP and TACS algorithms. Assume for simplicity, that all TACS interfaces with EMTP are modelled as equivalent resistance matrices with parallel current sources. Such models do not fit directly into the nodal EMTP equations, rather they are solved by compensation as they are treated as true nonlinearities to eliminate the one–time–step error. Although the representation seem to limit the type of interface with the EMTP, it covers all the existing TACS signals going to EMTP (current source, voltage source, trigger signal) and take into account nonlinearities of the control system. This technique is efficient and general, and does not limit the number of interfaces. This is an extension of the existing EMTP capability of solving true nonlinear elements. It is fully described in the next section.

4. EMTP–TACS INTERFACE BASED ON COMPENSATION

<u>Description of compensation technique</u>

In the TACS compensation–based methods, all interface variables can be essentially simulated as current injections I, which are super–imposed on the linear network after a solution without TACS, or with $T_{out}(t-\Delta T)$ has first been found. Figure 4 illustrates the decomposition, where there are P–true nonlinearities in

EMTP between buses k and m (solved by the EMTP compensation) and the Q–TACS interfaces, between busses i and j (solved by the TACS compensation). The TACS interfaces are assumed to be of the form V as the input and I as the output. The nonlinearities can be branches or shunt devices in the network, i.e. they do not have to be connected to ground.

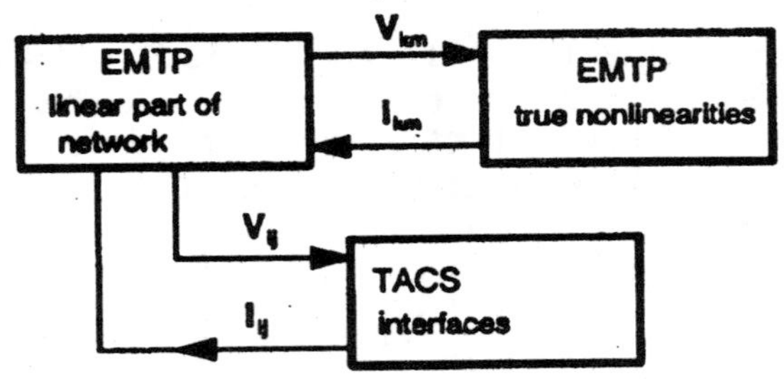

Figure 4 EMTP–TACS formulation with compensation

The EMTP true nonlinearities and the TACS interfaces are solved separately. To use the superposition principle correctly, it is necessary that the set of buses k–m and i–j be in different sub–networks. In the context of EMTP, this means that these sub–networks must be separated by at least one distributed–parameter line because, then, coupling occurs only through the history terms.

In applying compensation, once the linear network is solved, Thévenin equivalents are used to model the network at the buses where nonlinearities are connected. The equivalent EMTP network is modelled by :

$$\begin{bmatrix} V_{km} \\ V_{ij} \end{bmatrix} = \begin{bmatrix} E_{km} \\ E_{ij} \end{bmatrix} - \begin{bmatrix} Z_{km} & 0 \\ 0 & Z_{ij} \end{bmatrix} \begin{bmatrix} I_{km} \\ I_{ij} \end{bmatrix} \quad (2)$$

where E_{km} and E_{ij} are the vectors of Thévenin voltages at buses k–m and i–j; Z_{kj} and Z_{ij} are the Thévenin impedance matrices; while V_{km} and V_{ij} are the updated bus voltages. The Thévenin voltages are simply the bus voltages, without the effect of the currents I_{km} and I_{ij}. The Thévenin impedance matrices need to be recomputed each time a topological change occurs. There are standard EMTP techniques for computing the Thévenin equivalents.

The PxP matrix Z_{km} may have some topological restrictions depending on the type of nonlinear components: some type of nonlinearities are solved separately rather than iteratively with all other nonlinearities. There are no restrictions on the QxQ matrix Z_{ij}, i.e. it is a general coupled matrix. Because of the decoupling, attention focuses only on the TACS interfaces.

The Q–TACS interfaces can be modelled with coupled QxQ equations. These equations can be static or dynamic:

$$\text{static}: \quad I_{ij} = T_{out} = B(V_{ij})V_{ij} \quad (3)$$

$$\text{dynamic}: \quad I_{ij} = T_{out} = T_{out}(V_{ij}, \frac{dV_{ij}}{dt}) \quad (4)$$

where T_{out} refers to the nomenclature of Figure 1. It is assumed that all TACS dynamic interfaces can be discretized through the trapezoidal integration rule, and transformed into a static interface involving an history term :

$$\text{dynamic}: \quad I_{ij} = T_{out} = B(V_{ij})V_{ij} + I_h \quad (5)$$

Once discretized and cast under the form (5), dynamic and static TACS interfaces can be handled in the same manner. The simplest solution scheme is the Gauss iteration which consists in solving until convergence :

$$I_{ij}^{k+1} = B(V_{ij}^{k})V_{ij}^{k} + I_h \quad (6)$$

$$V_{ij}^{k+1} = E_{ij} - Z_{ij} I_{ij}^{k+1} \quad (7)$$

There are cases though where the convergence characteristics of the Gauss iterations may not be good. Since the matrices B and Z_{ij} are typically of small dimensions, the amount of computations is not an issue, the following Gauss solution scheme has been adopted as an alternative :

$$V_{ij}^{k+1} = [\, \mathcal{I} + Z_{ij} \; B(V_{ij}^{k}) \,]^{-1} [\, E_{ij} - Z_{ij} I_h \,] \quad (8)$$

$$I_{ij} = B(V_{ij}^{iterated})V_{ij}^{iterated} + I_h \quad (9)$$

This follows directly from (6)–(7), when the iteration count on the voltage is taken as k+1 in (6), except on B where it stays at k. The convergence properties of the Gauss scheme (8) are very good. When the B matrix is constant, one step is sufficient to converge (8). When B is a nonlinear matrix, iterations are required. A Newton implementation would also be possible, but this would require either the user to provide the differential of the nonlinear elements, or the program to estimate them in some manner. Since the control system generating B may be quite complex, the Gauss scheme (8) has been preferred. The scheme (6–7) is an alternative.

Modeling of interaction from TACS to EMTP

Interaction from TACS to EMTP occurs through: TACS defined EMTP slave voltage and current sources (type 60), TACS controlled switches (types 11, 12 and 13), TACS defined EMTP source scaling factor (type 17), and TACS variables used in the simulation of machines (types SM and UM).These are the existing interfaces, as taken from the EMTP rule book[1]. All interactions with the machines are not modified at this time by the compensation–based method. The type 17 source is also excluded. There are no restrictions however for including these devices; it is rather a matter of convenience at this time. All other interfaces are taken into account as follows.

Slave current source to EMTP : When the current output I_{ij} can be cast into the form (3) or (5), the application of compensation is immediate through (6–7) or (8–9). Otherwise compensation is possible by applying (7) based on the TACS computed current I_{ij}.

Slave voltage source to EMTP: This is the dual problem. Since the voltage V_{ij} is set by TACS, the slave voltage source can be replaced by a current I_{ij} given by

$$I_{ij} = Z_{ij}^{-1} \left(E_{ij} - V_{ij} \right) \qquad (10)$$

Compensation is performed by applying (7) based on these TACS computed current I_{ij}. If there are both slave voltage and current sources, which are coupled, and that need to be represented, then an hybrid formulation is feasible. A typical example is as follows :

$$\begin{bmatrix} I_{ij1} \\ V_{ij2} \end{bmatrix} = \begin{bmatrix} B_{11} & F_{12} \\ F_{21} & R_{22} \end{bmatrix} \begin{bmatrix} V_{ij1} \\ I_{ij2} \end{bmatrix} + \begin{bmatrix} I_{h1} \\ V_{h2} \end{bmatrix} \qquad (11)$$

where history terms for currents and voltages of the two sub–sets are shown. This is converted into a current injection model when it is assumed that the R_{22} matrix is non singular :

$$\begin{bmatrix} I_{ij1} \\ I_{ij2} \end{bmatrix} \begin{bmatrix} H_{11} & H_{12} \\ H_{21} & H_{22} \end{bmatrix} \begin{bmatrix} V_{ij1} \\ V_{ij2} \end{bmatrix} + \begin{bmatrix} J_{h1} \\ W_{h2} \end{bmatrix} \qquad (12)$$

$$H_{11} = B_{11} - F_{12} R_{22}^{-1} F_{21} \qquad H_{12} = F_{12} R_{22}^{-1}$$

$$H_{21} = - R_{22}^{-1} F_{21} \qquad H_{22} = R_{22}^{-1}$$

$$J_{h1} = I_{h1} - F_{12} R_{22}^{-1} V_{h2} \qquad W_{h2} = - R_{22}^{-1} V_{h2}$$

Trigger signals: these are used to control the operation of EMTP switches. The easiest way in which to apply compensation for trigger signals is to model the switch in TACS itself, and to interface with the EMTP with the calculated switch current I_{ij}. This yields a simple switch model (ideal valve) with a near zero resistance when in conduction, and a high resistance when blocked. In fact, when simulating converter bridges and power electronic circuits, it is convenient [5] to write separate circuit equations, for example with hybrid analysis instead on nodal analysis, and to interface the circuit with current sources in the EMTP.

Additional interfaces: compensation allows more generality such as branch current injections which are easily handled, as well as components which read voltages at a set of buses but inject current in another set of buses in the same sub–network or not. This last feature is useful to model dependent sources.

Modeling of interaction from EMTP to TACS

Interaction from EMTP to TACS occurs through EMTP defined TACS sources: network node voltages (type 90), switch currents (type 91), machine variables (type 92), and switch status (type 93). It has been shown above how the various interfaces are modelled. Machine variables have been excluded from the compensation. Network node voltages are implicit in the formulation of compensation. Typically currents injected into EMTP from TACS will depend on bus voltages which are updated through compensation. Switch currents and switch status are readily available when the switches are directly modeled in TACS. Then there is no problem in updating these values at the current time step. A library of switching equipments can be made available in TACS, then it is a simple matter of calling the device of choice.

5. IMPLEMENTATION

The algorithm is in fact more complicated than what is shown in the previous section because some TACS variables will generally depend on the inputs from EMTP, say V_{ij} , and we cannot wait for the next time step before updating them if the one–time–step error must be cancelled. This implies we generally need to iterate the TACS solution, at each time instant, both for the linear and nonlinear components, until convergence at the interface buses.

The flow chart for the revised TACS solution, with compensation for the currents injected into EMTP, is shown in Figure 5. In the dashed box, the final step corresponds to the standard TACS solution. Before entering this, there is a repeat solution on TACS, for this time step, whenever compensation is needed, and until the voltages at the interface buses have converged. Computation of Thévenin voltages and impedance matrix, as well as voltage updating, are performed in a manner very similar to what already exist in the EMTP for the compensation of true nonlinearities. Typically, less than 5 iterations are required for convergence. The starting guess for V_{ij} can be taken as the value of V_{ij} at the last time step, or the Thévenin voltage at this step, but the former appears to be a better choice.

The repeat solution on TACS is for adjusting the TACS sources (specifically node voltages, type 90) to the value of compensation, and for making all TACS variables consistent with the new node voltages. It is also possible to iterate TACS to reduce the internal TACS delays, but this is not the first purpose.

In Figure 5, all TACS is imposed a repeat solution. As indicated in Figure 3, this is not truly necessary, only part of the TACS equations (matrix **G** and all flagged nonlinear components) could be updated at this step. In the repeat solution there is no update of the history terms since the time step is not advanced.

In the implementation, the TACS signals I_{ij} which need to be compensated may be in many cases computed with standard TACS statements, using any combination of transfer functions, supplemental devices and pseudo

FORTRAN. It is also possible, for more complex cases, to load the user routine describing I_{ij}, and the associated history terms. The compensation is performed by specifying appropriate arguments for the interface. An example in the next section will illustrate this. The algorithm has therefore been designed and implemented so that minimal low-level intervention from the user is required.

In the EMTP, the signals I_{ij} which are TACS compensated, are marked as a special true nonlinearity. When these signals do not exist, then the TACS solution is standard, i.e. without compensation nor iterations at the same time step. This preserves solution efficiency.

6. APPLICATIONS

It is possible to use the new facilities described in this paper as a solver of user-specified components which interface with the network modelled in EMTP.

The example in Figure 6 shows a numerical instability which is cured by the compensation algorithm in TACS, but which cannot be eliminated by simply reducing the time step. With compensation, the answer is always exact, irrespective of the time step of the simulation. From 0 to ΔT, the system starts-up, and at ΔT the response is exact with or without compensation (note that the curve without compensation is scaled down by 10). Without compensation, the numerical instability is clear and growing in amplitude.

The circuit is a simple resistive circuit, part of it is modelled in EMTP, and the rest in TACS, as indicated in the Figure. In TACS, the resistance is modelled as a current injected into EMTP. The reason for the numerical instability is clear when we consider the equivalent simulation equations with the one-time-step error at the interface:

$$\begin{bmatrix} V(t) \\ I(t) \end{bmatrix} = \begin{bmatrix} 0 & -R_1 \\ 0 & -R_1/R_2 \end{bmatrix} \begin{bmatrix} V(t-\Delta T) \\ I(t-\Delta T) \end{bmatrix} + \begin{bmatrix} 1 \\ 1/R_2 \end{bmatrix} u(t) \qquad (13)$$

The discrete time system (13) has a pole which is outside the unit circle, and of magnitude R_1/R_2, independent of the step size ΔT selected. With compensation, the solution exactly matches that obtained by simulating the entire circuit in the EMTP itself.

The next example simulates two R-L circuits in TACS. There is a branch between buses 2-3, and a shunt between buses 3-0 which are modelled in TACS, with compensation. A large resistance is connected in EMTP between buses 2-3. The switch in the circuit is closed at 0.005sec. Figure 7 shows the voltages at buses 2 and 3. Results exactly match those obtained by simulating every component in the EMTP. Of course, it would be a simple matter to simulate the circuit in EMTP directly, but the example demonstrates the flexibility in adding any component in the EMTP, for specific study purposes.

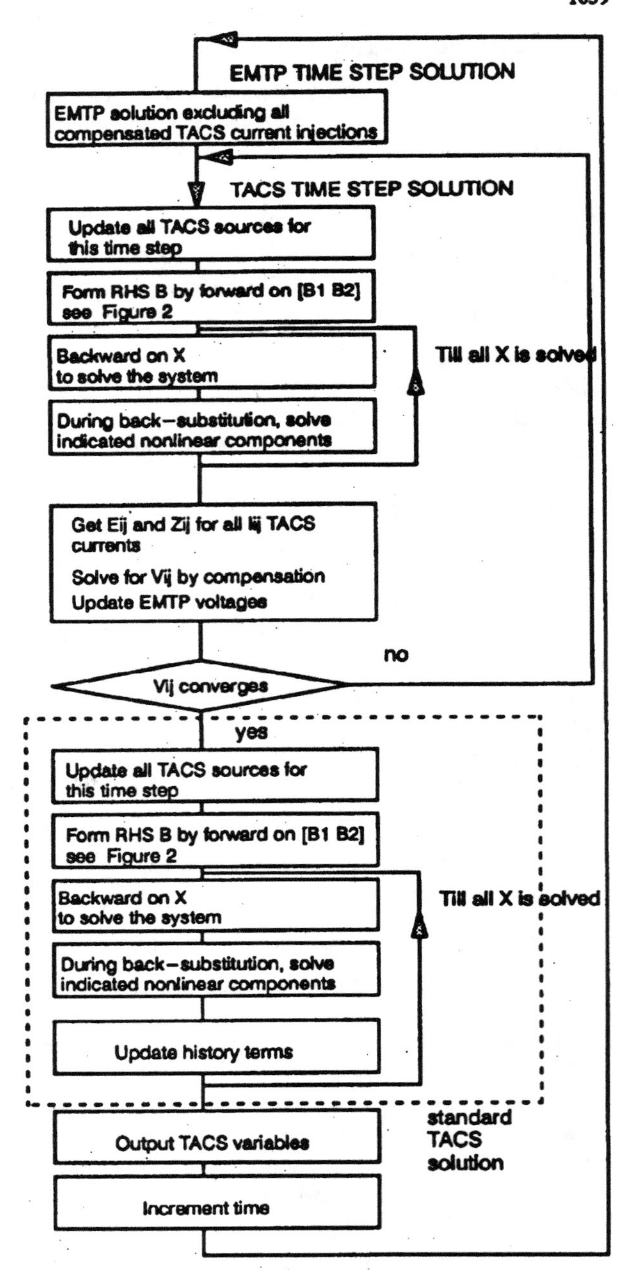

Figure 5 Flow chart of compensated TACS solution

Next consider a case where a nonlinear three-phase load is simulated in TACS, through compensation. The load is a combination of constant real power and constant impedance load. The simulation result is compared to the standard EMTP simulation with constant impedance load. Figure 8 shows the rms voltages with both types of loads. Naturally this is a low frequency load model, but the impact on the system behavior cannot be neglected. Better load models in the EMTP have the benefit of potentially improving the accuracy of simulations.

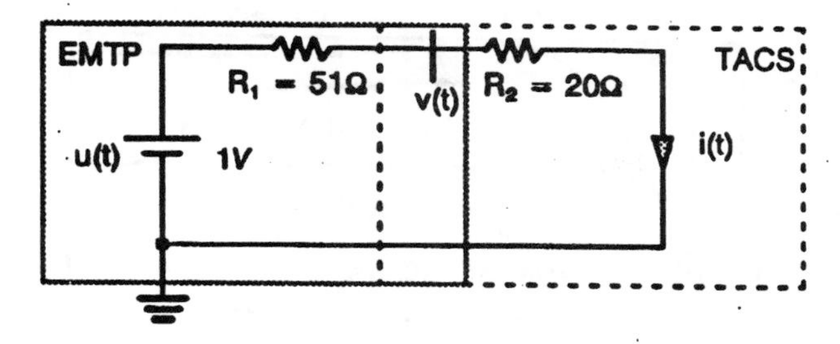

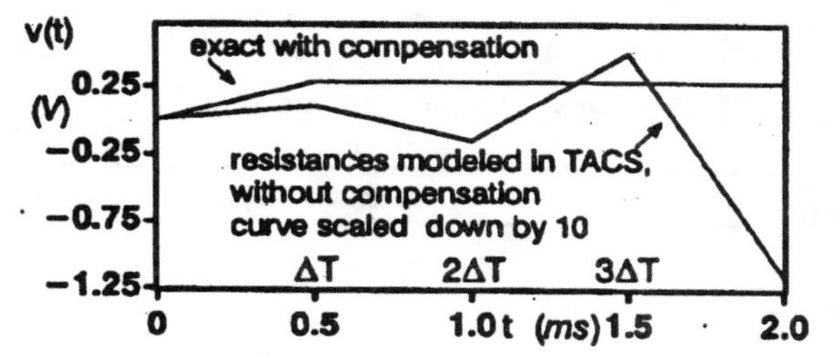

Figure 6 Circuit unstable without compensation

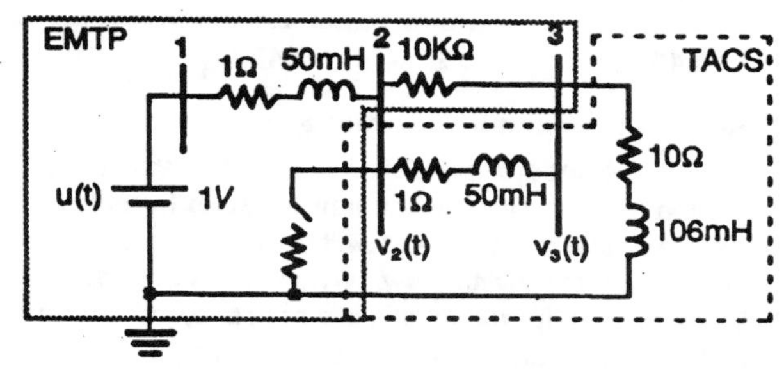

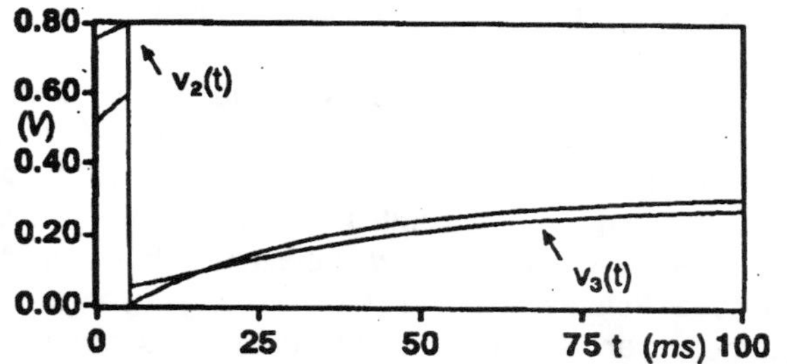

Figure 7 Two branches, with dynamics, simulated in TACS

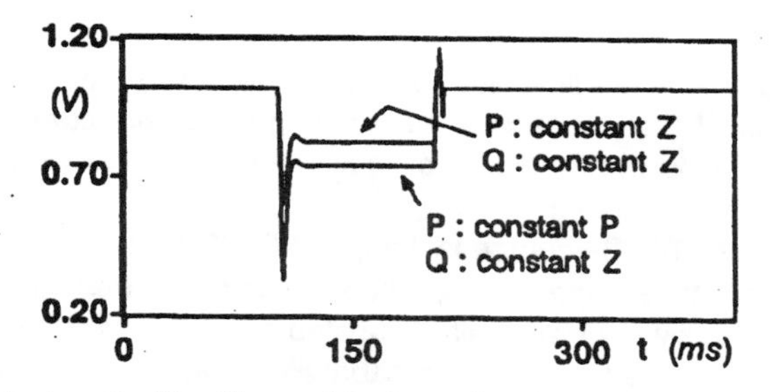

Figure 8 Nonlinear 3-phase load

The next example illustrates the elimination of time delays in valve firing. In Figure 9, a thyristor valve is simulated with the trigger pulse coming from TACS. The trigger pulse is intentionally long enough to overlap with the negative commutation voltage period. When the valve is simulated in EMTP, we get the curves marked EMTP, where the valve current starts one time-step after the enabling signal, and where the valve current goes negative for one time-step before extinction. The enabling signal indicates the conduction period: positive trigger pulse and positive current through the valve. In the standard EMTP simulation, the current is delayed not only at ignition, but also at extinction (a small negative current is allowed to flow due to the one-time-step error and the emtp-switch logic). The curves marked TACS illustrate the results when the valve in modeled through compensation in TACS. Current ignition and extinction is simultaneous with the firing pulse. Note that the current is not delayed with respect to the enabling signal.

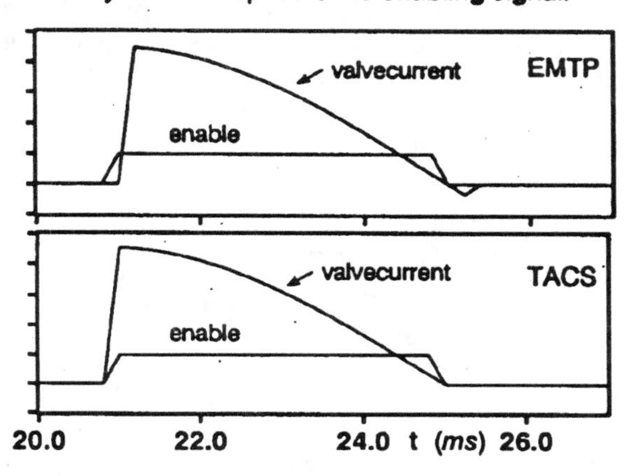

Figure 9 Thyristor valve without delay

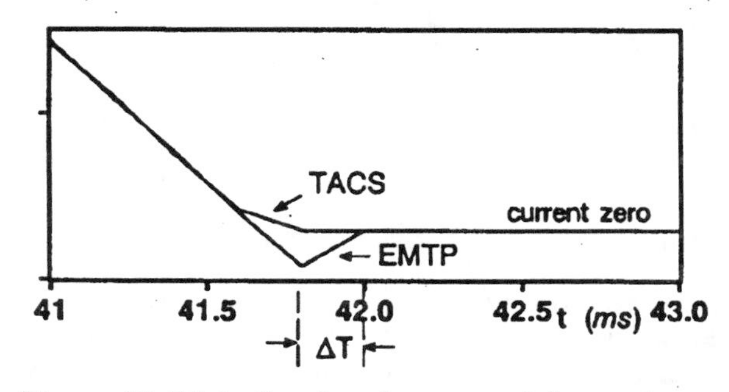

Figure 10 Distortion in valve current, large step size (current extinction of Figure 9 is zoomed in)

In Figure 9, negative current was not obtained with TACS-compensation because the valve is opened as soon as its current, at the same time step, becomes negative. This feature is the result of being able to iterate on TACS at the same time step. Of course, since the simulation is necessarily time-discretized, and since the current is forced to go to zero at the exact simulation steps, some distortion may be introduced in the valve currents. Figure 10 shows the current extinction when a larger step size is used. The exact instant of current extinction is between two integration steps. It is possible to linearly interpolate to determine this point. It would imply displacing the time mesh of the simulation in both EMTP and TACS[6]. With TACS-compensation, this is not required, but there are iterations within TACS.

The last example illustrates a case where the one-time-step error of the standard TACS-EMTP has a strong impact on the numerical accuracy, and the ease of use of the program. The circuit shown in Figure 11[7] is difficult for digital simulation because it is subjected to inductor current and capacitor voltage discontinuities. The switch sw is controlled by an external signal with period T and duty cycle d. The R-C circuit across the switch is a snubber, which is not used (because not required) with the TACS-compensation algorithm. In the continuous mode of operation, at the instant the switch sw closes, the diode D is forced to switch off, preventing a sudden capacitor voltage change. Depending on the values of R, L and C, the diode current may switch off before the switch sw closes. In practice the switch and the diode must be perfectly synchronized to get accurate results

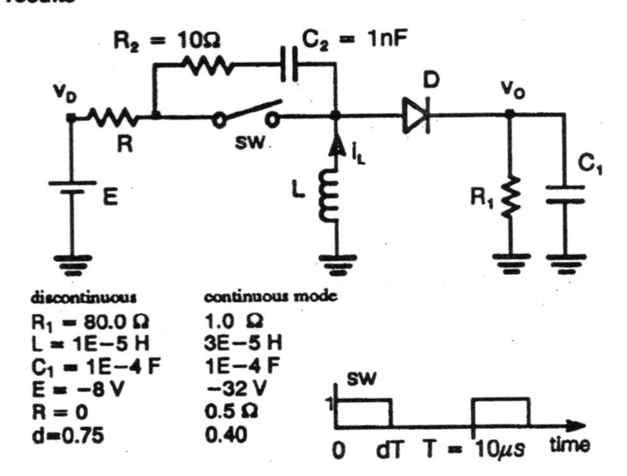

Figure 11 Buck-boost dc-dc converter

Figure 12 shows the standard EMTP-TACS simulation results are dependent on the time-step. In those simulations, for the continuous mode of operation, the switch is a type-13 TACS-controlled switch, while the diode is a type-11 ideal diode model. Diode ceases operation a time-step after the reclosing of the switch sw. Due to the one-time-step error, the capacitance partially discharges, which results in lower voltages v_0, dependent on the step size. Even more dramatic results are obtained when R=0, e.g. with the discontinuous mode data, since then the capacitance voltage instantaneously equals the source voltage. The same simulation is run with TACS-compensation (no snubber used), with the diode and the switch both modelled inside of TACS, and the result, which can be shown to be the exact solution, is insensitive to the time-step.

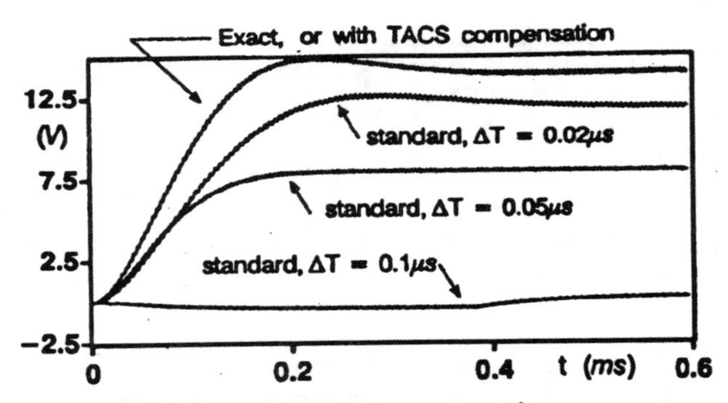

Figure 12 Voltage v_o, continuous mode

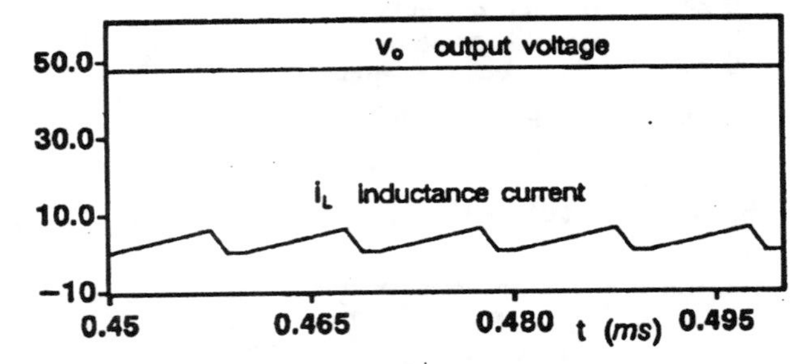

Figure 13 Discontinuous mode

In the standard simulation, the only way to get a solution less sensitive to the time-step is to replace the diode model with a type-13 switch controlled in opposite phase with the switch sw. When the converter is in discontinuous operation, this is a necessity, otherwise the diode never enters its on-state with the type-11 model. Additional logic is even required to avoid negative diode current (current must be chopped at zero). This type of modeling requires good understanding of the circuit behavior. With the TACS-compensation model, there is no distinction necessary between different modes of operation. Figure 13 illustrates results obtained in this case. It can be demonstrated that these results are exact. The steady-state results illustrate the voltage v_0 obtained and the current through the inductance. This current is typical of the discontinuous mode of operation.

7. CONCLUSIONS

This paper has presented and discussed TACS-EMTP limitations related to the one-time-step error at the interface. A new and general algorithm for alleviating the modeling difficulties has been developed and implemented.

With TACS-compensation, and TACS iterations at the same time-step, it has been illustrated that cases previously numerically unstable are stabilized. Furthermore the modeling capabilities, and ease of use, have been extended.

A new method to account for circuits of varying topologies has resulted, including the consideration of valve current chopping due to time-discretization.

The paper has demonstrated that it is possible to use TACS as a model builder and solver. The model solution is interfaced, without error, with the network modelled in EMTP. The model must generate a current injection in the EMTP, and it has been shown that many classes of model fit this. Compensated models, such as for the dc–dc converter circuit, can be constructed rather simply. As shown in the Appendix, this is done in the EMTP–TACS data file and no programming and re–compilation is required for this specially used program version.

In the implementation, all of TACS is iterated at each time step when TACS compensation is required. The computational burden can be reduced by using a scheme based on eqn(1) and Figure 3. In many cases, since the interface is less prone to errors, the integration time step can be increased thus offsetting the additional requirements of the compensation and iterations.

REFERENCES

[1] EMTP rule book, EPRI report EL–4541s–CCMP, 2 volumes.

[2] EMTP Work book IV, EPRI report EL–4651.

[3] H. W. Dommel, EMTP theory book, 2nd edition, Microtran Power System Analysis Corporation, May 1992.

[4] M. Ren–ming, 'The challenge of better EMTP TACS variable ordering,' EMTP Newsletter, Vol 4, No 4, pp. 1–6, Aug. 1984.

[5] J. Mahseredjian, S. Lefebvre., D. Mukhedkar, 'Power converter simulation module connected to the EMTP,' IEEE/PES 1990 Summer Meeting, Minneapolis, Paper 90 SM 454–9 PWRS.

[6] B. Kulicke, 'Simulation program NETOMAC: difference conductance method for continuous and discontinuous systems,' Siemens Research and Development Reports, Vol 10, No 5, 1981.

[7] D. Bedrosian, J. Vlach, 'Time–domain analysis of networks with internally controlled switches,' IEEE Trans. on Circuits and Systems, Vol 39, No 3, March 1992.

BIOGRAPHIES

Serge Lefebvre (M'76) received the BScA and MScA degrees in electrical engineering from École Polytechnique de Montréal in 1976 and 1977 respectively, and a Ph.D. from Purdue University (Indiana) in 1980. He is working at the Research Department of Hydro–Québec since 1981 while being an associate professor at École Polytechnique de Montréal. His research interest are in power system analysis techniques, computer applications, and dc systems. Dr. Lefebvre is Chairman of the IEEE working group "Dynamic performance and modeling of dc systems and power electronics".

Jean Mahseredjian (M'84) received the B.Sc.A., M.Sc.A. and Ph.D. in Electrical Engineering from Ecole Polytechnique de Montréal (Canada) in 1982, 1985 and 1990 respectively. At present he is a researcher at Institut de Recherche d'Hydro–Québec (Canada) and an associate–professor at Ecole Polytechnique de Montréal. His research interests are modeling and numerical analysis in power systems and power electronics.

APPENDIX

The EMTP–TACS data file for the dc–dc converter circuit of Figure 11 is listed below. The switch sw and the diode D are modelled as current injections CNL001 and CNL002 in the EMTP. These are generated by TACS, and the EMTP solution is compensated by a call to the TACS_INTERFACE routine, which is provided. It is simply necessary to specify the number of arguments, and the arguments in the order in which they are declared in the EMTP.

```
BEGIN NEW DATA CASE
 0.10E-6 80.0E-5
  10000    010
TACS HYBRID
C  control of the emtp switch
11VCONT   0.75
24RAMP    1.0         10.0E-6
88SIG     = VCONT .GE. RAMP
77SIG     1.0
11N00002  2.00
00SIGF    +SIG
C  if SIG > 0, close the switch T1
IF ( SIGF .GT. 0.0  ) THEN
 BNLsw1 = 100000.
ELSE
 BNLsw1 =  0.00001
ENDIF
C next find conduction pattern of diode
00CNLdio  +CNL002
IF ( CNLdio .GE. 0.0  ) THEN
 BNLOU1 =  100000.
ELSE
 BNLOU1 = .00001
ENDIF
00BNLswt  +BNLsw1
00BNLdio  +BNLOU1
C
SUBROUTINE  TACS_INTERFACE
 +N00002 +BNLswt +BNLdio   These are the arguments for the
  CNL001  CNL002            compensation algorithm
BLANK end of TACS
C NETWORK WITH TACS COMPENSATION
 VDc    PRIMRc            1.E-3
 SNUBc SWc                1.E10
           SWc           .03000
 VOc                            100.
 VOc                   80.0
C flag to indicate device with TACS-compensation
C Bus--->Bus--->Bus--->Bus--->               codex
C switch
92PRIMRcSWc                              7878.
C diode
92SWc  VOc                             7878.
BLANK CARD ENDING BRANCH DATA.
BLANK CARD ENDING SWITCHES
C Source
11VDc    -32.0
C
BLANK CARD ENDING SOURCE CARDS.
 1
BLANK
BLANK
BLANK
```

PART 2

MODELING OF COMPONENTS

PART 2
MODELING OF COMPONENTS

INTRODUCTION

An accurate simulation of every transient phenomenon requires a representation of network components valid for a frequency range that varies from DC to several MHZ. An acceptable representation of each component throughout this frequency range is very difficult, and for most components is not practically possible.

Modeling of power components, taking into account the frequency-dependence of parameters, can be practically done by developing mathematical models which are accurate enough for a specific range of frequencies. Each range of frequencies usually corresponds to some particular transient phenomena. One of the most accepted classifications of frequency ranges is that proposed by the CIGRE WG 33-02. According to the document published by this WG, frequency ranges can be classified as four groups with some overlapping

- low-frequency oscillations, from 0.1 Hz to 3 kHz
- slow-front surges, from 50/60 Hz to 20 kHz
- fast-front surges, from 10 kHz to 3 MHZ
- very-fast-front surges, from 100 kHz to 50 MHZ

Papers selected for this part of the book present modeling works for some of the most important network components—overhead lines, insulated cables, transformers, arresters, network equivalents—taking into account their frequency- dependent behavior. An additional bibliography includes references related to other components, such as rotating machines and circuit breakers.

OVERHEAD LINES

Two types of time-domain models have been developed for overhead lines and insulated cables:

a) jumped-parameter models, they represent transmission systems by lumped elements whose values are calculated at a single frequency

b) Distributed-parameter models, for which two categories can be distinguished, constant parameter and frequency- dependent parameter models.

The first type of model is adequate for steady-state calculations, although these can also be used for transient simulations in the neighborhood of the frequency at which parameters were evaluated. The most accurate models for transient calculations are those which take into account the distributed nature of parameters and consider their frequency-dependence.

A significant number of papers analyze the frequency-dependence behavior of overhead lines and insulated cables for digital simulation. Although some efficient models are presently implemented in the most widely used EMTPs, new efforts are being now devoted to the development of more efficient models raised by the necessity of fast simulations in real-time digital systems.

The first papers presenting frequency-dependent line models for digital simulation were published during the late 1960's and early 1970's [A1], [A2]. Most models were aimed at solving transmission-line equations using a time-domain solution. Those models were based on the modal theory: multiphase line equations are decoupled through modal transformation matrices, so that each mode can be separately studied as a single-phase line. For unbalanced and untransposed lines, transformation matrices are frequency dependent. However, it is possible to obtain a good accuracy using constant transformation matrices [A3], [A4].

Several approaches using modal theory have been proposed: weighting functions [A5], recursive convolutions [A6] - [A9], state-space formulation [A10]. The first paper of this section details a transmission line model using a modified recursive convolution and assuming frequency-independent transformation matrices [1]. This model is currently available in several EMTPs.

The transient solution of this model is based on the rational function approximation of the propagation and characteristic admittance functions. The order of the rational functions will depend on the line geometry, the frequency range and the desired accuracy. A high level of accuracy can only be obtained with a large number of real poles. This can slow down the simulation of large networks. Low-order fittings have been proposed as a compromise between the solution accuracy and the model simplicity [A11], [A12].

The validity and limitations of constant transformation matrices, as well as guidelines on how to choose them are discussed in [A13].

A new method of using frequency-dependent transformation matrices with a Newton-Raphson iteration technique has been recently proposed [A14]. A different solution method based on the superposition principle and the Hartley transform was presented in [A15].

Some recent works have shown that the solution of line equations can be efficiently performed using a phase-domain formulation, instead of modal-domain [A16], [A17], [A18]. The second paper in this section uses an approach based on phase-domain equations, instead of the modal domain [2]. In addition, time-domain convolution is replaced by an Auto-Regressive Moving Average (ARMA) representation. The approach is also used to solve insulated cable equations.

All the previous papers considered transmission line representations taking into account only conductor geometry. Some other parts of a transmission line, such as the towers, have an important influence on its performance in lightning studies. The concept of nonuniform transmission lines includes the effect of towers and grounding resistances. In lightning studies, towers are represented by a surge impedance with an associated travel time.

Due to the increasing interest in the simulation of lightning surges using electromagnetic transients programs, two papers related to this topic have been included here. The first presents a transmission line tower model [3]. The model was validated by a comparison of simulation results with field test measurements. The next paper proposes a nonuniform transmission line model based on a finite-difference algorithm [4]. The approach has also been validated by comparing simulation results to field measurements presented in other references. Additional literature related to tower modeling and nonuniform lines can be found in [A19] - [A23].

Corona is a source of attenuation and distortion of surges and overvoltages in overhead lines. An important effort has been made during the last 20 years to understand this effect and its representation in transient studies [A24] - [A29]. Many interesting papers dealing with corona representation in digital simulations have also been published [A30] - [A36]. The paper selected for this book presents a wide-band corona circuit model for application in digital programs [5]. The model takes into account the frequency- dependence of line parameters and the corona nonlinearity.

INSULATED CABLES

The formulation of insulated cable equations and their solutions are similar to those used with overhead lines. However, the large variety of cable designs makes it very difficult to develop a single model that represents every type of cable.

Two papers dealing with cable modeling are included in this section. The first one presents a general formulation of impedances and admittances of single-core coaxial and pipe-type cables [6]. This work was the origin of the well known *CABLE CONSTANTS* routine. This capability can be used to evaluate matrices and equivalent pi-circuits of cables and to obtain steady-state initialization at a single frequency, but it cannot be used to perform transient calculations. The validity of this formulation is restricted, as mentioned above, to transient calculations in the neighborhood of the frequency at which parameters are evaluated.

The derivation of cable impedances and admittances was the subject of some previous works. See [A37] for coaxial cables and [A38] for pipe-type cables.

The second paper presents a mathematical model taking into account the frequency-dependence of cable parameters [7]. The solution of cable equations is performed in the modal domain and assumes frequency-dependence of modal transformation matrices. The model is valid for transient simulations over a wide frequency range.

Additional works related to cable modeling and some case studies are presented in [A39] - [A42]. The solution of cable equations, taking into account frequency-dependence of parameters and carrying out calculations in the phase-domain, has been presented in some recent works [A18], [2].

POWER TRANSFORMERS

An accurate representation of a power transformer over a wide frequency range is very difficult, despite its relatively simple design. A significant effort on transformer modeling has been made during the last twenty years. Much of this effort has been devoted to the development of models used in digital simulations. A summary of some transformer modeling approaches for use in EMTPs follows :

1) The representation of single- and three-phase n-winding transformers is made in the form of a branch impedance or admittance matrix [8]; this approach cannot include nonlinear effects of iron cores. They are incorporated by connecting nonlinear inductances at winding terminals. Many built-in models currently available in several EMTPs use this type of representation; their derivation is made from nameplate data.

2) Detailed models incorporating core nonlinearities can be derived by using the principle of duality from a topology based magnetic model. This approach can create models accurate enough for low-frequency and slow-front transients.

3) Previous approaches do not consider frequency-dependent parameters; they do not represent a transformer at high frequencies, although they can be improved if lumped capacitances are connected across transformer terminals. Models taking into account frequency- dependent parameters can be divided into two groups : models with a detailed description of internal windings [A43] and terminal models, based on the fitting of the elements of a circuit that represent the transformer as seen from its terminals [A44] - [A47].

Two alternative transformer models can be used whether surge transfer has to be computed or not. Representations for both situations were proposed in the document written by the CIGRE WG 33.02, see [25] in Preface.

Three papers presenting models of three-phase transformers, valid for simulation of low-frequency and slow-front transients have been selected. The first one presents the derivation of branch impedances and admittances of three- phase transformers as coupled branches [8]. The next two papers propose models derived by using the principle of duality [9], [10].

A different approach to obtain the equivalent circuit of a three -phase five-legged transformer, valid also for low- frequency and slow-front transients, was proposed in [A48]. A hybrid model based on core topology, and consisting of electric and magnetic circuits was recently presented [A49].

The next two papers deal with high-frequency models for three-phase transformers. Reference [11] presents a transformer model which is represented by means of an equivalent network that matches the frequency response of the transformer at its terminals. Reference [12] presents a simplified model based on the classical T-form model; this model is extended to high frequencies by adding winding capacitances and representing short-circuit branches by RL frequency-dependent equivalent networks.

Transformer parameters are both nonlinear and frequency dependent. Major causes of iron core nonlinearities are saturation and hysteresis [A50] - [A55]. One of the main causes of frequency-dependence are eddy currents [A56] - [A59]. The last two reprinted papers related to transformer modeling present two important contributions in this field. Reference [13] proposes a new procedure for hysteresis representation; the paper includes an interesting review of the work performed in this area. Reference [14] describes the representation of eddy current effects, developed for the transformer model presented in [11].

Complete transformer models including saturation, hysteresis, and eddy-current losses have been proposed in [A60] - [A63].

Detailed models are needed to obtain internal transient voltage distribution. These models are reasonable accurate for insulation design, and generally consist of large networks. However, they make system models unnecessary large when the concern is the response at the transformer terminals. Some efforts have been devoted to obtain reduced transformer models, using either linear or nonlinear techniques [A64] - [A67].

A method for simulation of internal faults in power transformers using capabilities available in many EMTPs was presented and validated in [A68].

The performance of different transformer models, most of them currently implemented in many EMTPs, for the simulation of fast switching transients was analyzed in [A69].

SURGE ARRESTERS

Two basic types of surge arresters are now in use : gapped silicon-carbide arresters and gapless metal-oxide surge arresters (MOSA). Although many of the arresters are the older type gapped silicon-carbide, the majority of the new installed arresters are the gapless metal-oxide type. For this reason, only papers dealing with MOSA modeling have been selected for this book.

MOSAs have a frequency-dependent nonlinear characteristic: the voltage across the arrester is a function of both the rate of rise and the magnitude of the current conducted by the arrester.

Modeling of MOSAs was the subject of a paper written by the IEEE WG on Surge Arrester Modeling [15]. The paper recommends the representation of a MOSA by an appropriate nonlinear V-I characteristic, without including any frequency-dependence, for temporary overvoltages and switching surge transients. For lightning studies, a more sophisticated model is proposed; it can give satisfactory results for discharge currents with wide-range of time-to-crest values. The paper includes a procedure for choosing parameters of the arrester model from test data.

A different model that represents the frequency-dependence behavior by means of a nonlinear inductance in series with a nonlinear resistance was proposed in [16]. A calculation algorithm to derive parameters of the arrester model from test data is also presented.

Modeling guidelines of gapped silicon-carbide surge arresters for digital simulations of slow-front transients were presented in [A70]. The proposed model is based on current-limiting arrester design. Recommendations to adapt the model for lightning studies were also included. Additional literature related to modeling of surge arresters can be found in [A71] - [A75].

Metal-oxide varistors (MOV) models suitable for digital simulation of series compensated lines has been the subject of some recent works [A76], [A77].

NETWORK EQUIVALENTS

The simulation of transient phenomena in power systems very often requires a detailed modeling of just a small part of the system. Network equivalents can be used to represent those parts of the system for which detailed modeling is not needed. The goal is to reduce the complexity and the computation time, while the simulation accuracy is preserved.

Several procedures have been proposed since early 1970s to obtain network equivalents [A78] - [A86]. The paper selected for this book describes a method to obtain multi-port frequency-dependent equivalents [17]. Based on a previous work, it extends the concept developed in [A79] to multi-port equivalents over a wide range of frequencies.

ROTATING MACHINES

The need for detailed synchronous generator models in electromagnetic transients programs was motivated by some serious subsynchronous resonance (SSR) incidents in the early 1970's. Utilities were concerned about some problems involving interactions between synchronous generators and power systems.

One model is the simulation of torsional interactions between the mechanical turbine-generator system and the power system needs of a very detailed representation of the generator and the power system. Several dynamic three-phase synchronous generator models were developed and implemented in the BPA EMTP at mid 1970's [A87], [A88]. All those models were based on the Park's transformation for solving the electrical equations. They incorporated a detailed representation of mechanical and electrical parts, used a sophisticated solution method to solve machine-power system interface, and included an interface to control systems.

Although the development and implementation were raised by SSR problems, those models could be used for other studies, such as loss of synchronism, load rejection or transmission line reclosure.

Magnetic saturation effects were not included in the early stages. A simple and efficient representation of magnetic saturation was added to one model in the late 1970s [A89].

Interest in the analysis and simulation of renewable energy sources motivated the demand for other machine models. A very powerful and flexible module, known as Universal Machine (UM), was implemented in the BPA EMTP in 1980 [A90]. The UM module allowed the representation of up to twelve different machine models and expanded the applications of the program, i.e. the simulation of adjustable speed drives. The first UM release had several limitations that were solved in subsequent versions [A91]. An improved induction motor model for EMTP simulations was presented in [A92].

At the same time, new and improved techniques to solve machine-power system interface were developed and implemented in other EMTPs [A92].

Models currently implemented in all EMTPs are adequate for simulation of low frequency transients. They are sufficiently accurate to analyze the interaction between the machine and the power system, as well as torsional oscillations in the mechanical part. However, these models are not adequate for simulation of fast-front transients. Some switching motor operations can originate steep-front surges and cause large turn-to-turn winding stresses. Lightning surges transferred through transformers are also a source of high stresses and dielectric failures.

Recent works have proposed computer models for analyzing machine behavior in fast-front transients and predicting distribution of interturn voltages caused by steep-fronted surges [A94], [A95]. Some of these models have been represented and simulated using EMTP capabilities [A96].

Techniques to develop machine models based on their frequency response have been proposed [A47], [A97]. Most studies performed with an EMTP are dealing with large three-phase synchronous and induction machines. Works dealing with small and special machines have been recently presented [A98], [A99].

CIRCUIT BREAKERS

An accurate circuit breaker model needs representation for both opening and closing operations. A circuit breaker opens its contacts when a tripping signal is sent to it. The separation of the contacts causes the generation of an electric arc. The phenomena by which the arc is actually extinguished is very complicated. Although a large number of arc models have been proposed, there is no general acceptance for any of them.

Several approaches can be used in the development, testing and operation of circuit breakers [A100]. The most suitable approach for reproducing the arc interruption phenomenon in a transients program is the use of so called black-box models. The aim of a black-box arc model is to describe the interaction of an arc and a electrical circuit during an interruption process. They consider the arc as a two-pole, and determine the transfer function using a chosen mathematical form and fitting free parameters to measured voltage and current traces. Rather than internal processes, it is the electrical behavior of the arc which is of importance. Several levels of complexity are possible:

1) The breaker is represented as an ideal switch that opens at first zero current crossing, after the tripping signal is given. This model can obtain the voltage across the breaker; it is to be compared with a pre-specified transient recovery voltage (TRV) to withstand capability for the breaker. This model cannot reproduce any interaction between the arc and the system.

2) The arc is represented as a time-varying resistance, whose variation is determined ahead of time based on the breaker characteristic. This model can represent the effect of the arc on the system, but requires advanced knowledge of the effect of the system on the arc.

3) The most advanced models represent the breaker as a dynamically varying resistance or conductance. They can represent both the effect of the arc on the system and the effect of the

system on the arc. No precomputed TRV curves are required. These models are generally developed to determine initial arc quenching, that is to study the thermal period only, although some can also be used to determine arc reignition due to insufficient voltage withstand capability of the dielectric between breaker contacts. Their most important application cases are short line fault interruption and switching of small inductive currents.

Many models for circuit breakers, represented as a dynamic resistance, have been proposed. A survey on black-box modeling of gas (air, SF6) circuit breakers is presented in [A100]. The EMTP implementation of three dynamic arc models was presented in [A101]. Those models were adequate for gas and oil circuit breakers. All the models are useful to represent a circuit breaker during the thermal period, a physical model for representation of SF6 breakers during thermal and dielectric periods is detailed in [A102]. A model for vacuum circuit breakers is presented in [A103].

Similar models have been proposed to represent a circuit breaker in closing operations. The simplest model assumes that the breaker contacts can close on any part of the cycle. In fact, there is a closing time during which an arc may strike across the contacts as the breaker closes. A breakdown will occur if the voltage across the gap exceeds its dielectric strength. The prestrike phenomenon can be represented by an ideal switch or an arc conductance. The statistical nature of a closing operation makes this representation difficult. Not much attention has been paid to this phenomenon, and no accurate representation has been yet developed.

OTHER COMPONENTS

Capabilities currently available in most EMTPs make possible user-developed models of those components for which a built-in model has not been implemented. In fact, this is the case for some of the component models discussed above:

- A transformer model for low-frequency transients based on the principle of duality has not been implemented in any EMTP; the capabilities needed to developed such a model have been discussed in some papers [10].
- There is no built-in model for circuit breakers in most EMTPs, but its representation can be made using branches and control features, available in all EMTPs.
- Although there is a built-in surge arrester model implemented in all EMTPs, it is not adequate for lightning studies; users have to improve this model taking advantages of other capabilities.

The list of components for which a built-in model is not available in some EMTPs might include fuses [A104], [A105], instrument transformers [A106] - [A109], protective relays [A110] - [A113]. The implementation of models for instrument transformers and some types of relays in one EMTP is presented in [A114].

REFERENCES

Reprinted Papers

[1] J.R. Martí, "Accurate modelling of frequency-dependent transmission lines in electromagnetic transient simulations", *IEEE Trans. on Power Apparatus and Systems*, vol. 101, no. 1, pp. 147-155, January 1982.

[2] T. Noda, N. Nagaoka and A. Ametani, "Phase domain modeling of frequency-dependent transmission lines by means of an ARMA model", *IEEE Trans. on Power Delivery*, vol. 11, no. 1, pp. 401-411, January 1996.

[3] T. Yamada et al., "Experimental evaluation of a UHV tower model for lightning surge analysis", *IEEE Trans. on Power Delivery*, vol. 10, no. 1, pp. 393-402, January 1995.

[4] M.T. Correia de Barros and M.E. Almeida, "Computation of electromagnetic transients on nonuniform transmission lines", *IEEE Trans. on Power Delivery*, Vol. 11, no. 2, pp. 1082- 1091, April 1996.

[5] J.R. Martí, F.Castellanos and N. Santiago, "Wide-band corona circuit model for transient simulations", *IEEE Trans. on Power Delivery*, vol. 4, no. 2, pp. 1441-1449, April 1995.

[6] A. Ametani, "A general formulation of impedance and admittance of cables", *IEEE Trans. on Power Apparatus and Systems*, vol. 99, no. 3, pp. 902-910, May/June 1980.

[7] L. Martí, "Simulation of transients in underground cables with frequency-dependent modal transformation matrices", *IEEE Trans. on Power Delivery*, vol. 3, no. 3, pp. 1099-1110, July 1988.

[8] V. Brandwajn, H.W. Dommel and I.I. Dommel, "Matrix representation of three-phase n-winding transformers for steady-state and transient studies", *IEEE Trans. on Power Apparatus and Systems*, vol. 101, no. 6, pp. 1369-1378, June 1982.

[9] C.M. Arturi, "Transient simulation and analysis of a three-phase five-limb step-up transformer following an out-of-phase synchronization", *IEEE Trans. on Power Delivery*, vol. 6, no. 1, pp. 196-207, January 1991.

[10] A. Narang and R. H. Brierley, "Topology based magnetic model for steady-state and transient studies for three-phase core type transformers", *IEEE Trans. on Power Systems*, vol. 9, no. 3, pp. 1337-1349, August 1994.

[11] A. Morched, L. Martí and J. Ottevangers, "A high frequency transformer model for the EMTP", *IEEE Trans. on Power Delivery*, vol. 8, no. 3, pp. 1615-1626, July 1993.

[12] S. Chimklai and J.R. Martí, "Simplified three-phase transformer model for electromagnetic transient studies", *IEEE Trans. on Power Delivery*, vol. 10, no. 3, pp. 1316-1324, July 1995.

[13] F. de León and A. Semlyen, "Simple representation of dynamic hysteresis losses in power transformers", *IEEE Trans. on Power Delivery*, vol. 10, no. 1, pp. 315-321, January 1995.

[14] E.J. Tarasiewicz, A.S. Morched, A. Narang and E.P. Dick, "Frequency dependent eddy current models for nonlinear iron cores", *IEEE Trans. on Power Systems*, vol. 8, no. 2, pp. 588-597, May 1993.

[15] IEEE Working Group on Surge Arrester Modeling, "Modeling of metal oxide surge arresters", *IEEE Trans. on Power Delivery*, vol. 7, no. 1, pp. 302-309, January 1992.

[16] I. Kim et al., "Study of ZnO arrester model for steep front wave", *IEEE Trans. on Power Delivery*, vol. 11, no. 2, pp. 834-841, April 1996.

[17] A.S. Morched, J.H. Ottevangers and L. Martí, "Multi-port frequency dependent network equivalents for the EMTP", *IEEE Trans. on Power Delivery*, vol. 8, no. 3, pp. 1402-1412, July 1993.

Additional References

[A1] A. Budner, "Introduction of frequency-dependent line parameters into an electromagnetic transients program", *IEEE Trans. on Power Apparatus and Systems*, vol. 89, no. 1, pp. 88-97, January 1970.

[A2] J.K. Snelson, "Propagation of travelling waves on transmission lines - Frequency dependent parameters", *IEEE Trans. on Power Apparatus and Systems*, vol. 91, no. 1, pp. 85-91, January/February 1972.

[A3] P.C. Magnusson, "Travelling waves on multi-conductor open wire lines. A numerical survey of the effects of frequency dependence of modal composition", *IEEE Trans. on Power Apparatus and Systems*, vol. 92, no. 3, pp. 999-1008, May/June 1973.

[A4] R.G. Wesley and S. Selvavinayagamoorthy, "Approximate frequency-response values for transmission line transient analysis", *Proc. IEE*, vol. 121, no. 4, pp. 281-286, April 1974.

[A5] W. Scott Meyer and H.W. Dommel, "Numerical modelling of frequency dependent transmission-line parameters in an electromagnetic transients program", *IEEE Trans. on Power Apparatus and Systems*, vol. 93, no. 5, pp. 1401-1409, September/October 1974.

[A6] A. Semlyen and A. Dabuleanu, "Fast and accurate switching transient calculations on transmission lines with ground return using recursive convolutions", *IEEE Trans. on Power Apparatus and Systems*, vol. 94, no. 2, pp. 561-571, March/April 1975.

[A7] A. Ametani, "A highly efficient method for calculating transmission line transients", *IEEE Trans. on Power Apparatus and Systems*, vol. 95, no. 5, pp. 1545-1551, September/October 1976.

[A8] A. Semlyen, "Contributions to the theory of calculation of electromagnetic transients on transmission lines with frequency dependent parameters", *IEEE Trans. on Power Apparatus and Systems*, vol. 100, no. 2, pp. 848-856, February 1981.

[A9] A. Semlyen and R.H. Brierley, "Stability analysis and stabilizing procedure for a frequency dependent transmission line model", *IEEE Trans. on Power Apparatus and Systems*, vol. 103, no. 12, pp. 3579-3586, December 1984.

[A10] J.F. Hauer, "State-space modeling of transmission line dynamics via nonlinear optimization", *IEEE Trans. on Power Apparatus and Systems*, vol. 100, no. 12, pp. 4918-4924, December 1981.

[A11] L. Martí, "Low-order approximation of transmission line parameters for frequency-dependent models", *IEEE Trans. on Power Apparatus and Systems*, vol. 102, no. 11, pp. 3582-3589, November 1983.

[A12] A. Oguz Soysal and A. Semlyen, "Reduced order transmission line modeling for improved efficiency in the calculation of electromagnetic transients", *IEEE Trans. on Power Systems*, vol. 9, no. 3, pp. 1494-1498, August 1994.

[A13] J.R. Martí, H.W. Dommel, L. Martí and V. Brandwajn, "Approximate transformation matrices for unbalanced transmission lines", *Proc. of the 9th Power Systems Computer Conference*, Cascais (Portugal), 1987.

[A14] L.M. Wedepohl, H.V. Nguyen and G.D. Irwin, "Frequency- dependent transformation matrices for untransposed transmission lines using Newton-Raphson method", Paper 95 SM 602-3 PWRS, *1995 IEEE PES Summer Meeting*, July 23-27, 1995, Portland.

[A15] R. Mahmutcehajic, S. Babic, R. Gacanovic and S. Carsimamovic, "Digital simulation of electromagnetic wave propagation in a multiconductor transmission system using the superposition principle and Hartley transform", *IEEE Trans. on Power Delivery*, vol. 8, no. 3, pp. 1377-1385, July 1993.

[A16] A. Oguz Soysal and A. Semlyen, "State equation approximation of transfer matrices and its application to the phase domain calculation of electromagnetic transients", *IEEE Trans. on Power Systems*, vol. 9, no. 1, pp. 420-428, February 1994.

[A17] G. Angelidis and A. Semlyen, "Direct phase-domain calculation of transmission line transients using two-sided recursions", *IEEE Trans. on Power Delivery*, vol. 10, no. 2, pp. 941-949, April 1995.

[A18] B. Gustavsen, J. Sletbak and T. Henriksen, "Calculation of electromagnetic transients in transmission cables and lines taking frequency dependent effects accurately into account", *IEEE Trans. on Power Delivery*, vol. 10, no. 2, pp. 1076-1084, April 1995.

[A19] W.A. Chisholm, Y.L. Chow and K.D. Srivastava, "Lightning surge response of transmission towers", *IEEE Trans. on Power Apparatus and Systems*, vol. 102, no. 9, pp. 3232- 3242, September 1983.

[A20] W.A. Chisholm and Y.L. Chow, "Travel time of transmission towers", *IEEE Trans. on Power Apparatus and Systems*, vol. 104, no. 10, pp. 2922-2928, October 1985.

[A21] M. Ishii et al., "Multistory transmission tower model for lightning surge analysis", *IEEE Trans. on Power Delivery*, vol. 6, no. 3, pp. 1327-1335, July 1991.

[A22] S. Cristina and M. D'Amore, "Effect of reflection waves on polyphase non-uniform power line carrier channels : A new propagation matrix model", *IEEE Trans. on Power Apparatus and Systems*, vol. 100, no. 4, pp. 1685-1693, April 1981.

[A23] C. Menemenlis and Z.T. Chun, "Wave propagation on nonuniform lines", *IEEE Trans. on Power Apparatus and Systems*, vol. 101, no. 4, pp. 833-839, April 1982.

[A24] M.Mihailescu-Suliciu and I. Suliciu, "A rate type constitutive equation for the description of the corona effect", *IEEE Trans. on Power Apparatus and Systems*, vol. 100, no. 8, pp. 3681-3685, August 1981.

[A25] A. Inoue, "Propagation analysis of overvoltage surges with corona based upon charge versus voltage curve", *IEEE Trans. on Power Apparatus and Systems*, vol. 104, no. 3, pp. 655-662, March 1985.

[A26] X.-R. Li, O.P. Malik and Z.-D. Zhao, "A practical mathematical model of corona for calculation of transients on transmission lines", *IEEE Trans. on Power Delivery*, vol. 4, no. 2, pp. 1145-1152, April 1989.

[A27] M.A. Al-Tai et al., "The simulation of surge corona on transmission lines", *IEEE Trans. on Power Delivery*, vol. 4, no. 2, pp. 1360-1368, April 1989.

[A28] X.-R. Li, O.P. Malik and Z.-D. Zhao, "Computation of transmission lines transients including corona effects", *IEEE Trans. on Power Delivery*, vol. 4, no. 3, pp. 1816-1821, July 1989.

[A29] C. de Jesus and M.T. Correia de Barros, "Modelling of corona dynamics for surge propagation studies", *IEEE Trans. on Power Delivery*, vol. 9, no. 3, pp. 1564-1569, July 1994.

[A30] K.C. Lee, "Non-linear corona models in an electromagnetic transients program (EMTP)", *IEEE Trans. on Power Apparatus and Systems*, vol. 102, no. 9, pp. 2936-2942, September 1983.

[A31] W.-G. Huang and A. Semlyen, "Computation of electro- magnetic transients on three-phase transmission lines with corona and frequency dependent parameters", *IEEE Trans. on Power Delivery*, vol. 2, no. 3, pp. 887-898, July 1987.

[A32] H.M. Barros, S. Carneiro Jr. and R.M. Azevedo, "An efficient recursive scheme for the simulation of overvoltages on multiphase systems under corona", *IEEE Trans. on Power Delivery*, vol. 4, no. 2, pp. 1441-1449, April 1989.

[A33] P. Sarma Maruvada, D.H. Nguyen and H. Hamadani-Zadeh, "Studies on modeling corona attenuation of dynamic overvoltages", *IEEE Trans. on Power Delivery*, vol. 4, no. 2, pp. 1441-1449, April 1989.

[A34] S. Carneiro Jr. and J.R. Martí, "Evaluation of corona and line models in electromagnetic transient simulations", *IEEE Trans. on Power Delivery*, vol. 6, no. 1, pp. 334-342, January 1991.

[A35] S. Carneiro Jr., H.W. Dommel, J.R. Martí and H.M. Barros, "An efficient procedure for the implementation of corona models in electromagnetic transients programs", *IEEE Trans. on Power Delivery*, vol. 9, no. 2, pp. 849-855, April 1994.

[A36] M. Poloujadoff, J.F. Guillier and M. Rioual, "Damping model of traveling waves by corona effect along extra high voltage three phase lines", *IEEE Trans. on Power Delivery*, vol. 10, no. 4, pp. 1851-1861, October 1995.

[A37] L.A. Wedepohl and D.J. Wilcox, "Transient analysis of underground power-transmission systems", *Proc. IEE*, vol. 120, no. 2, pp. 253-260, February 1973.

[A38] G.W. Brown and R.G. Rocamora, "Surge propagation in three-phase pipe type cables. Part I - Unsaturated pipe", *IEEE Trans. on Power Apparatus and Systems*, vol. 95, no. 1, pp. 88-95, January/February 1976.

[A39] A. Ametani, "Wave propagation characteristics of cables", *IEEE Trans. on Power Apparatus and Systems*, vol. 99, no. 2, pp. 499-505, April 1980.

[A40] N. Nagaoka and A. Ametani, "Transient calculations on crossbonded cables", *IEEE Trans. on Power Apparatus and Systems*, vol. 102, no. 4, pp. 779-787, April 1983.

[A41] L. Bohmann et al., "Impedance of a double submarine cable circuit using different types of cables within a single circuit", *IEEE Trans. on Power Delivery*, vol. 8, no. 4, pp. 1668-1674, October 1993.

[A42] B. Gustavsen and J. Sletbak, "Transient sheath overvoltages in armoured power cables", *IEEE Trans. on Power Delivery*, vol. 11, no. 3, pp. 1594-1600, July 1996.

[A43] R.C. Degeneff, "A general method for determining resonances in transformer windings", *IEEE Trans. on Power Apparatus and Systems*, vol. 96, no. 2, March/April 1977.

[A44] R.C. Degeneff, "A method for constructing terminal models for single-phase n-winding transformers", Paper No. A 78 539-9, *1978 IEEE PES Summer Meeting*, July 16-21, Los Angeles.

[A45] P.T.M. Vaessen, "Transformer model for high frequencies", *IEEE Trans. on Power Delivery*, vol. 3, no. 4, pp. 1761-1768, October 1988.

[A46] V. Woivre, J.P. Artaud, A Ahmad and N. Burais, "Transient overvoltage study and model for shell-type power transformers", *IEEE Trans. on Power Delivery*, vol. 8, no. 1, pp. 212-222, January 1993.

[A47] A. Oguz Soysal, "A method for wide frequency range modeling of power transformers and rotating machines", *IEEE Trans. on Power Delivery*, vol. 8, no. 4, pp. 1802-1810, October 1993.

[A48] D.L. Stuehm, B.A. Mork and D.D. Mairs, "Five-legged core transformer equivalent circuit", *IEEE Trans. on Power Delivery*, vol. 4, no. 3, pp. 1786-1793, July 1989.

[A49] X. Chen, "A three-phase multi-legged transformer model in ATP using the directly-formed inverse inductance matrix", *IEEE Trans. on Power Delivery*, vol. 11, no. 3, pp. 1554-1562, July 1996.

[A50] W.L.A. Neves and H.W. Dommel, "On modelling iron core nonlinearities", *IEEE Trans. on Power Systems*, vol. 8, no. 2, pp. 417-425, May 1993.

[A51] W.L.A. Neves and H.W. Dommel, "Saturation curves of delta-connected transformers from measurements", *IEEE Trans. on Power Delivery*, vol. 10, no. 3, pp. 1432-1437, July 1995.

[A52] M. Vakilian and R.C. Degeneff, "A method for modelling nonlinear core characteristics of transformer during transients", *IEEE Trans. on Power Delivery*, vol. 9, no. 4, pp. 1916-1925, October 1994.

[A53] S.N. Talukdar and J.R. Bailey, "Hysteresis models for system studies", *IEEE Trans. on Power Apparatus and Systems*, vol. 95, no. 4, pp. 1429-1434, July/August 1976.

[A54] E.P. Dick and W.Watson, "Transformer models for transient studies based on field measurement", *IEEE Trans. on Power Apparatus and Systems*, vol. 100, no. 1, pp. 401-419, January 1981.

[A55] J.G. Frame, N. Mohan and T.H. Liu, "Hysteresis modeling in an electro-magnetic transients program", *IEEE Trans. on Power Apparatus and Systems*, vol. 101, no. 9, pp. 3403-3412, September 1982.

[A56] J. Avila-Rosales and F.L. Alvarado, "Nonlinear frequency dependent transformer model for electromagnetic transient studies in power systems", *IEEE Trans. on Power Apparatus and Systems*, vol. 101, no. 11, pp. 4281-4288, November 1982.

[A57] J. Avila-Rosales and A. Semlyen, "Iron core modelling for electrical transients", *IEEE Trans. on Power Apparatus and Systems*, vol. 104, no. 11, pp. 3189-3194, November 1985.

[A58] F. de León and A. Semlyen, "Time domain modeling of eddy current effects for transformer transients", *IEEE Trans. on Power Delivery*, vol. 8, no. 1, pp. 271-280, January 1993.

[A59] F. de León and A. Semlyen, "Detailed modeling of eddy current effects for transformer transients", *IEEE Trans. on Power Delivery*, vol. 9, no. 2, pp. 1143-1150, April 1994.

[A60] D.N. Ewart, "Digital computer simulation model of a steel-core transformer", *IEEE Trans. on Power Delivery*, vol. 1, no. 3, pp. 174-183, July 1986.

[A61] H. Mohseni, "Multi-winding multi-phase transformer model with saturable core", *IEEE Trans. on Power Delivery*, vol. 6, no. 1, pp. 166-173, January 1991.

[A62] D. Dolinar, J. Pihler and B. Grcar, "Dynamic model of a three-phase power transformer", *IEEE Trans. on Power Delivery*, vol. 8, no. 4, pp. 1811-1819, October 1993.

[A63] F. de León and A. Semlyen, "Complete transformer model for electromagnetic transients", *IEEE Trans. on Power Delivery*, vol. 9, no. 1, pp. 231-239, January 1994.

[A64] R.C. Degeneff, M.R. Gutierrez and P.J. McKenny, "A method for constructing reduced order transformer models for system studies from detailed lumped parameter models", *IEEE Trans. on Power Delivery*, vol. 7, no. 2, pp. 649-655, April 1992.

[A65] M. Gutierrez, R.C. Degeneff, P.J. McKennny and J.M. Schneider, "Linear, lumped parameter transformer model reduction technique", *IEEE Trans. on Power Delivery*, vol. 10, no. 2, pp. 853-861, April 1995.

[A66] R.C. Degeneff, M.R. Gutierrez and M. Vakilian, "Nonlinear, lumped parameter transformer model reduction technique", *IEEE Trans. on Power Delivery*, vol. 10, no. 2, pp. 862-868, April 1995.

[A67] R.J. Galarza, J.H. Chow and R.C. Degeneff, "Transformer model reduction using time and frequency domain sensitivity techniques", *IEEE Trans. on Power Delivery*, vol. 10, no. 2, pp. 1052-1059, April 1995.

[A68] P. Bastard, P. Bertrand and M. Meunier, "A transformer model for winding fault studies", *IEEE Trans. on Power Delivery*, vol. 9, no. 2, pp. 690-699, April 1994.

[A69] B.C. Papadias et al., "Three phase transformer modelling for fast electromagnetic transient studies", *IEEE Trans. on Power Delivery*, vol. 9, no. 2, pp. 1151-1159, April 1994.

[A70] IEEE Working Group of Surge Protective Devices Committee, "Modeling of current-limiting surge arresters", *IEEE Trans. on Power Apparatus and Systems*, vol. 100, no. 8, pp. 4033-4040, August 1981.

[A71] D.P. Carroll et al., "A dynamic surge arrester model for use in power system transient studies", *IEEE Trans. on Power Apparatus and Systems*, vol. 91, no. 3, pp. 1057-1066, May/June 1972.

[A72] S. Tominaga et al., "Protective performance of metal oxide surge arrester based on the dynamic v-i characteristics", *IEEE Trans. on Power Apparatus and Systems*, vol. 98, no. 6, pp. 1860-1871, November/December 1979.

[A73] M.V. Lat, "Analytical method for performance prediction of metal oxide surge arresters", *IEEE Trans. on Power Apparatus and Systems*, vol. 104, no. 10, pp. 2665-2674, October 1985.

[A74] C. Dang, T.M. Parnell and P.J. Price, "The response of metal oxide surge arresters to steep fronted current impulses", *IEEE Trans. on Power Delivery*, vol. 1, no. 1, pp. 157-163, January 1986.

[A75] W. Schmidt et al., "Behaviour of MO-surge-arrester blocks to fast transients", *IEEE Trans. on Power Delivery*, vol. 4, no. 1, pp. 292-300, January 1989.

[A76] D. L. Goldsworthy, "A linearized model for MOV- protected series capacitors", *IEEE Trans. on Power Systems*, vol. 2, no. 4, pp. 953-958, November 1987.

[A77] J.R. Lucas and P.G. McLaren, "A computationally efficient MOV model for series compensation studies", *IEEE Trans. on Power Delivery*, vol. 6, no. 4, pp. 1491-1497, October 1991.

[A78] A. Clerici and L. Marzio, "Coordinated use of TNA and digital computer for switching-surge studies : Transient equivalent of a complex network", *IEEE Trans. on Power Apparatus and Systems*, vol. 89, no. 6, pp. 1717-1726, November/December 1970.

[A79] A.S. Morched and V. Brandwajn, "Transmission network equivalents for electromagnetic transients studies", *IEEE Trans. on Power Apparatus and Systems*, vol. 102, no. 9, pp. 2984- 2994, September 1983.

[A80] V.Q. Do and M.M. Gavrilovic, "An iterative pole removal method for synthesis of power system equivalent networks", *IEEE Trans. on Power Apparatus and Systems*, vol. 103, no. 8, pp. 2065-2070, August 1984.

[A81] V.Q. Do and M.M. Gavrilovic, "A synthesis method for one-port and multi-port equivalent networks for analysis of power system transients", *IEEE Trans. on Power Systems*, vol. 1, no. 2, pp. 103-113, April 1986.

[A82] M. Kizilcay, "Low-order network equivalents for electromagnetic transients studies", *European Transactions on Electrical Power Engineering*, vol. 3, no. 2, pp. 123-129, March/April 1993.

[A83] A. Semlyen and M.R. Iravani, "Frequency domain modeling of external systems in an electro-magnetic transients program", *IEEE Trans. on Power Systems*, vol. 8, no. 2, pp. 527-533, May 1993.

[A84] A. Abur and H. Singh, "Time domain of external systems for electromagnetic transients programs", *IEEE Trans. on Power Systems*, vol. 8, no. 2, pp. 671-679, May 1993.

[A85] H. Singh and A. Abur, "Multiport equivalencing of external systems for simulation of switching transients", *IEEE Trans. on Power Delivery*, vol. 10, no. 1, pp. 374-382, January 1995.

[A86] J. Hong and J. Park, "A time-domain approach to transmission network equivalents via Prony analysis for electromagnetic transients analysis", *IEEE Trans. on Power Systems*, vol. 10, no. 4, pp. 1789-1797, November 1995.

[A87] V. Brandwajn and H.W. Dommel, "A new method for interfacing generator models with an electromagnetic transients program", *Proc. of IEEE PICA*, pp. 260-265, 1977.

[A88] G. Gross and M.C. Hall, "Synchronous machine and torsional dynamics simulation in the computation of electromagnetic transients", *IEEE Trans. on Power Apparatus and Systems*, vol. 97, no. 4, pp. 1074-1086, July/August 1978.

[A89] V. Brandwajn, "Representation of magnetic saturation in the synchronous machine model in an electromagnetic transients program", *IEEE Trans. on Power Apparatus and Systems*, vol. 99, no. 5, pp. 1996-2002, September/October 1980.

[A90] H.K. Lauw and W. Scott Meyer, "Universal machine modeling for the representation of rotating machinery in an electromagnetic transients program", *IEEE Trans. on Power Apparatus and Systems*, vol. 101, no. 6, pp. 1342-1352, June 1982.

[A91] H.K. Lauw, "Interfacing for universal multi-machine system modeling in an electromagnetic transients program", *IEEE Trans. on Power Apparatus and Systems*, vol. 104, no. 9, pp. 2367-2373, September 1985.

[A92] G.J. Rogers and D. Shirmohammadi, "Induction machine modelling for electromagnetic transient program", *IEEE Trans. on Energy Conversion*, vol. 2, no. 4, pp. 622-628, December 1987.

[A93] A.M. Gole, R.W. Menzies, H.M. Turanli and D.A. Woodford, "Improved interfacing of electrical machine models to electromagnetic transients programs", *IEEE Trans. on Power Apparatus and Systems*, vol. 103, no. 9, pp. 2446-2451, September 1984.

[A94] J.L. Guardado and K.J. Cornick, "A computer model for calculating steep-fronted surge distribution in machine windings", *IEEE Trans. on Energy Conversion*, vol. 4, no. 1, 95-101, March 1989.

[A95] A. Narang, B.K. Gupta, E.P. Dick, D.K. Sharma, "Measurement and analysis of surge distribution in motor stator windings", *IEEE Trans. on Energy Conversion*, vol. 4, no. 1, 126-134, March 1989.

[A96] E.P. Dick, R.W. Cheung and J.W. Porter, "Generator models for overvoltages simulations", *IEEE Trans. on Power Delivery*, vol. 6, no. 2, 728-735, April 1991.

[A97] N.J. Bacalao, P. de Arizon and R.O. Sánchez, "A model for the synchronous machine using frequency response measurements", *IEEE Trans. on Power Systems*, vol. 10, no. 1, pp. 457-464, February 1995.

[A98] A. Domijan and Y. Yin, "Single-phase induction machine simulation using the Electromagnetic Transients Program : Theory and test cases", *IEEE Trans. on Energy Conversion,* vol. 9, no. 3, pp. 535-542, September 1994.

[A99] H. Knudsen, "Extended Park's transformation for 2x3-phase synchronous machine and converter phasor model with representation of harmonics", *IEEE Trans. on Energy Conversion*, vol. 10, no. 1, pp. 126-132, March 1995.

[A100] CIGRE Working Group 13.01, "Applications of black box modelling to circuit breakers", *Electra*, no. 149, pp. 40-71, August 1993.

[A101] V. Phaniraj and A.G. Phadke, "Modelling of circuit breakers in the electromagnetic transients program", *IEEE Trans. on Power Systems*, vol. 3, no. 2, pp. 799-805, May 1988.

[A102] L. van der Sluis, W.R. Rutgers and C.G.A. Koreman, "A physical arc model for the simulation of current zero behaviour of high- voltage circuit breakers", *IEEE Trans. on Power Delivery*, vol. 7, no. 2, pp. 1016-1022, April 1992.

[A103] J. Kosmac and P. Zunko, "A statistical vacuum circuit breaker model for simulation of transients overvoltages", *IEEE Trans. on Power Delivery*, vol. 10, no. 1, pp. 294-300, January 1995.

[A104] A. Petit, G. St-Jean and G. Fecteau, "Empirical model of a current-limiting fuse using EMTP", *IEEE Trans. on Power Delivery*, vol. 4, no. 1, pp. 335-341, January 1989.

[A105] L. Kojovic and S. Hassler, "Application of current limiting fuses in distribution systems for improved power quality and protection", Paper 96 WM 070-3 PWRD, *1996 IEEE PES Winter Meeting*, January 21-25, Baltimore.

[A106] J. Lucas et al., "Improved simulation models for current and voltage transformers in relay studies", *IEEE Trans. on Power Delivery*, vol. 7, no. 1, pp. 152-159, January 1992.

[A107] M. Kezunovic et al., "Digital models of coupling capacitor voltage transformers for protective relay transient studies", *IEEE Trans. on Power Delivery*, vol. 7, no. 4, pp. 1927-1935, October 1992.

[A108] M. Kezunovic et al., "Experimental evaluations of EMTP-based current transformer models for protective relay transient study", *IEEE Trans. on Power Delivery*, vol. 9, no. 1, pp. 405-413, January 1994.

[A109] M. Kezunovic et al., "A new method for the CCVT performance analysis using field measurements, signal processing and EMTP modeling", *IEEE Trans. on Power Delivery*, vol. 9, no. 4, pp. 1907-1915, October 1994.

[A110] A. Domijan and M.V. Emami, "State space relay modelling and simulation using the Electromagnetic Transients Program and its Transient Analysis of Control Systems capability", *IEEE Trans. on Energy Conversion*, vol. 5, no. 4, pp. 697-702, December 1990.

[A111] R.E. Wilson and J.M. Nordstrom, "EMTP transient modeling of a distance relay and a comparison with EMTP laboratory testing", *IEEE Trans. on Power Delivery*, vol. 8, no. 3, pp. 984-992, July 1993.

[A112] M.T. Glinkowski and J. Esztergalyos, "Transient modeling of electromechanical relays. Part I : Armature type overcurrent relays", *IEEE Trans. on Power Delivery*, vol. 11, no. 2, pp. 763-770, April 1996.

[A113] M.T. Glinkowski and J. Esztergalyos, "Transient modeling of electromechanical relays. Part II : Plunger Type 50 relays", *IEEE Trans. on Power Delivery*, vol. 11, no. 2, pp. 771-782, April 1996.

[A114] A.K.S. Chaudhary, K.S. Tam and A.G. Phadke, "Protection system representation in the electromagnetic transients program", *IEEE Trans. on Power Delivery*, vol. 9, no. 2, pp. 700-711, April 1994.

ACCURATE MODELLING OF FREQUENCY-DEPENDENT TRANSMISSION LINES IN ELECTROMAGNETIC TRANSIENT SIMULATIONS

J.R. Marti, Member IEEE

University of British Columbia
Department of Electrical Engineering
Vancouver, B.C. V6T 1W5

ABSTRACT

The parameters of transmission lines with ground return are highly dependent on the frequency. Accurate modelling of this frequency dependence over the entire frequency range of the signals is of essential importance for the correct simulation of electromagnetic transient conditions. Closed mathematical solutions of the frequency-dependent line equations in the time domain are very difficult. Numerical approximation techniques are thus required for practical solutions. The oscillatory nature of the problem, however, makes ordinary numerical techniques very susceptible to instability and to accuracy errors. The methods presented in this paper are aimed to overcome these numerical difficulties.

I. INTRODUCTION

It has long been recognized that one of the most important aspects in the modelling of transmission lines for electromagnetic transient studies is to account for the frequency dependence of the parameters and for the distributed nature of the losses. Models which assume constant parameters (e.g. at 60 Hz) cannot adequately simulate the response of the line over the wide range of frequencies that are present in the signals during transient conditions. In most cases the constant-parameter representation produces a magnification of the higher harmonics of the signals and, as a consequence, a general distortion of the wave shapes and exaggerated magnitude peaks.

The magnification of the higher harmonics in constant-parameter representations can readily be seen from figs. 13 and 14 (described in more detail in Section VIII). These figures show the frequency response of the zero sequence mode of a typical 100-mi, 500 kV 3-phase transmission line under short-circuit and open-circuit conditions. Curves (I) correspond to the "exact" response calculated analytically from frequency-dependent parameters obtained from Carson's equations [1]. Curves (II) represent the response with constant, 60 Hz parameters.

Much effort has been devoted over the last ten years to the development of frequency-dependent line models for digital computer transient simulations. Some of the most important contributions are listed in references [2] to [8].

In theory, many alternatives are possible for the formulation of the solution to the exact line equations. In practice, however, as it is illustrated in figs. 11 and 12, the nature of a transmission line is such that its response as a function of frequency is highly oscillatory. As a consequence, the numerical problems that can be encountered in the process of solution are highly dependent on the particular approach.

The routines described in this paper avoid a series of numerical difficulties encountered in previous formulations. These routines are accurate, general, and have no stability problems. In the tests performed, over a wide range of line lengths (5 to 500 miles) for the zero and positive sequence modes, the same routines could accurately model the different line lengths and modes over the entire frequency range, from 0 Hz (d.c. conditions) to, for instance, 10^6 Hz. This is achieved without user intervention, that is, the user of these routines does not have to make value judgements to force a better fit at certain frequencies, line lengths, or modes. In transient simulations, the frequency-dependent representation of transmission lines required only 10-30% more computer time than the constant-parameter simulation.

II. TIME DOMAIN TRANSIENT SOLUTIONS

Even though the modelling of transmission lines is much easier when the solution is formulated in the frequency domain, for the study of a complete system with switching operations, non-linear elements, and other phenomena, step by step time domain solutions are much more flexible and general than frequency domain formulations.

Probably the best known example of time domain transient solutions is the Electromagnetic Transients Program (EMTP) first developed at Bonneville Power Administration (B.P.A.) from Dommel's basic work [9]. The widespread use of this program has proven its value and flexibility for the study of a large class of electromagnetic transient conditions.

The new frequency-dependent line model described in this paper has been tested in the University of British Columbia Version of the EMTP.

In the EMTP, multiphase lines are first decoupled through modal transformation matrices, so that each mode can be studied separately as a single-phase circuit. Frequency-independent transformation matrices are assumed in these decompositions. This procedure is exact in the case of balanced line configurations and still very accurate for transposed lines. In the more general case of unbalanced, untransposed lines, however, the modal transformation matrices are frequency dependent. Nevertheless, as concluded by Magnusson [10] and Wasley [11], it seems that is still possible in this case to obtain a reasonably good approximation under the assumption of constant transformation matrices.

Frequency-independent transformation matrices have been assumed in the present work.

III. SIMPLIFIED LINE MODEL

In Dommel's basic work it is assumed that the line has constant parameters and no losses. Under these simplifying assumptions the line equations are written directly in the time domain. (To account for the losses Dommel splits the total line resistance into three lumped parts, located at the middle and at the ends of the line). From d'Alembert's solution of the simplified wave equations and Bergeron's concept of the constant relationship between voltage and current waves travelling along the line, Dommel arrives at the equivalent circuit shown in fig. 1 for the line as seen from node k. An analogous model is obtained for node m. In this model R_C is the line characteristic impedance and $I_{kh}(t)$ is a current source whose value at time step t is evaluated from the known history values of the current and voltage at node m τ units of time earlier (τ is the travelling time).

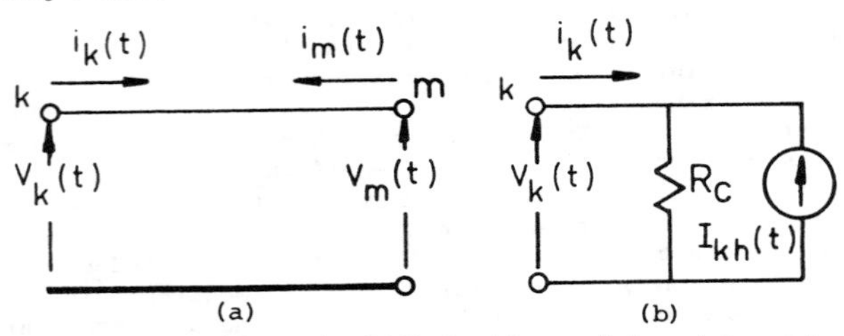

Fig. 1: Dommel's simplified line model. (a): Line mode. (b): Equivalent circuit at node k.

IV. FREQUENCY-DEPENDENT LINE MODEL: HISTORICAL REVIEW

When the frequency dependence of the parameters and the distributed nature of the losses are taken into account, it becomes very difficult, if not impossible in a practical way, to write the solution of the line equations directly in the time domain. This solution, however, can easily be obtained in the frequency domain, and is given by the well-known relations (e.g. Woodruff [12])

$$V_k(\omega) = \cosh[\gamma(\omega)\ell]V_m(\omega) - Z_c(\omega)\sinh[\gamma(\omega)\ell]I_m(\omega) \quad (1)$$

and

$$I_k(\omega) = \frac{1}{Z_c(\omega)}\sinh[\gamma(\omega)\ell]V_m(\omega) - \cosh[\gamma(\omega)\ell]I_m(\omega), \quad (2)$$

where

$$Z_c(\omega) = \sqrt{Z'(\omega)\cdot Y'(\omega)} = \text{characteristic impedance}, \quad (3)$$

$$\gamma(\omega) = \sqrt{\frac{Z'(\omega)}{Y'(\omega)}} = \text{propagation constant}, \quad (4)$$

$$Z'(\omega) = R'(\omega) + j\omega L'(\omega),\ Y'(\omega) = G'(\omega) + j\omega C'(\omega),$$

R' = series resistance, L' = series inductance,
G' = shunt conductance, C' = shunt capacitance
(primed quantities are in per unit length).

One of the first frequency-dependent line models for time-domain transient solutions was proposed by Budner [2], who used the concept of weighting functions in an admittance line model. The weighting functions in this model are, however, highly oscillatory and difficult to evaluate with accuracy.

In an effort to improve Budner's weighting-functions method, Snelson [3] introduced a change of variables to relate currents and voltages in the time domain in a way which is analogous to Bergeron's interpretation of the simplified wave equations. The new variables are defined as follows:

forward travelling functions:

$$f_k(t) = v_k(t) + R_1 i_k(t), \quad (5)$$

$$f_m(t) = v_m(t) + R_1 i_m(t), \quad (6)$$

and backward travelling functions:

$$b_k(t) = v_k(t) - R_1 i_k(t), \quad (7)$$

$$b_m(t) = v_m(t) - R_1 i_m(t), \quad (8)$$

where R_1 is a real constant defined as $R_1 = \lim_{\omega\to\infty} Z_c(\omega)$.

Equations 5 to 8 are then transformed into the frequency domain and compared with the line solution as given by eqns. 1 and 2. This idea was further developed by Meyer and Dommel [4] and resulted in the weighting functions $a_1(t)$ and $a_2(t)$ shown in fig. 2, and the equivalent line representation shown in fig. 3 for node k. In this circuit the backward travelling

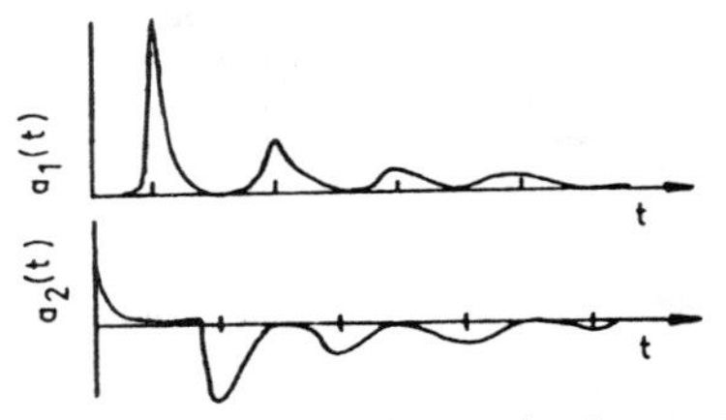

Fig. 2: Weighting functions in Meyer and Dommel's formulation.

function $b_k(t)$ is obtained from the "weighted" past history of the currents and voltages at both ends of the line and is given by the convolution integral

$$b_k(t) = \int_0^\infty \{f_m(t-u)a_1(u) + f_k(t-u)a_2(u)\}du\ . \quad (9)$$

An analogous equivalent circuit and convolution integral are obtained for node m.

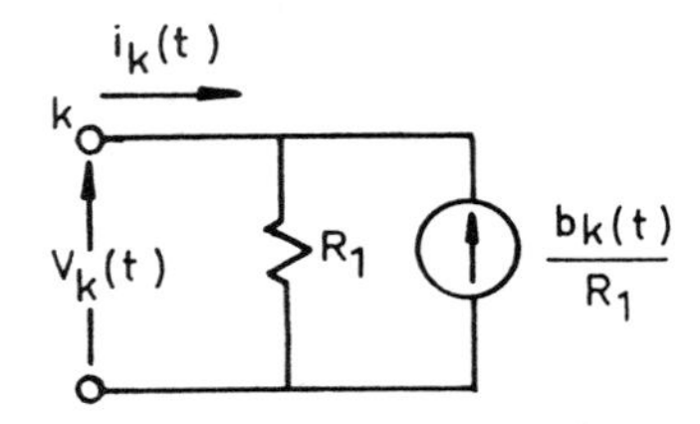

Fig. 3: Meyer and Dommel's frequency-dependent line model at node k.

Meyer and Dommel's formulation of the weighting function technique represented a considerable improvement over other weighting function methods, and has given reliable results in many cases of transient studies performed at B.P.A. This technique, however, still presents some numerical disadvantages. One of these disadvantages is the relatively time consuming process required to evaluate integral 9 at each time step of the solution. In the case study presented in [4], the running time per step for the case with frequency dependence was about three times longer than the time with no frequency dependence. Another disadvantage is the difficulty in evaluating the contribution of the tail portions of $a_1(t)$ and $a_2(t)$ to the convolution integral of eqn. 9. The successive peaks in these functions tend to become flatter and wider for increasing values along the t-axis.

Some of the main problems encountered with this method have been accuracy problems at low frequencies, including the normal 60 Hz steady state. These problems seem to be related to the evaluation of the tail portions of the weighting functions. Also, an error analysis seemed to indicate that the function $a_2(t)$ is more difficult to evaluate with sufficient accuracy than the function $a_1(t)$.

As suggested by Meyer and Dommel, the meaning of the weighting functions $a_1(t)$ and $a_2(t)$ can be visualized physically from the model shown in fig. 4. In this model the line is excited with a voltage impulse $\delta(t)$ and is terminated at both ends by the resistance R_1 of eqns. 5 to 8. Under these conditions $a_1(t)$ is directly related to the voltage at node m and $a_2(t)$ to the voltage at node k. From this model, it can be seen that the successive peaks in these functions (fig. 2) are produced by successive reflections at both ends of the line.

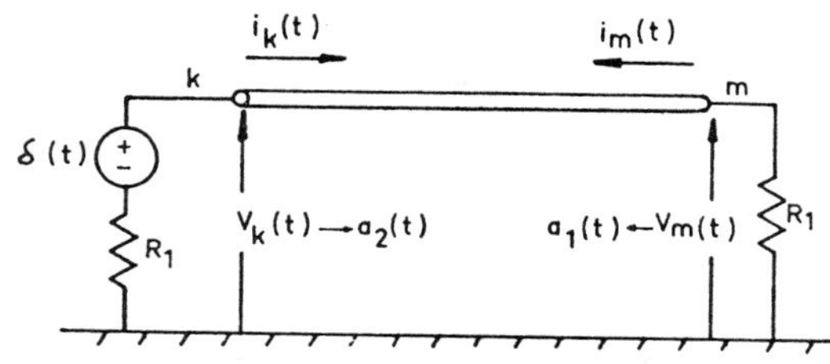

Fig. 4: Physical interpretation of Meyer and Dommel's weighting functions.

V. FREQUENCY-DEPENDENT LINE MODEL: NEW FORMULATION

The development of this model can be best explained from the physical interpretation of the concept of the weighting functions developed by Meyer and Dommel.

From the system shown in fig. 4 it can be seen that if the resistance R_1 is replaced by an equivalent network whose frequency response is the same as the characteristic impedance of the line, $Z_c(\omega)$, there will be no reflections at either end of the line. If such an equivalent network can be found, the new $a_1(t)$ weighting function will have only the first spike and the function $a_2(t)$ will become zero. This is shown in fig. 5. The form of the new weighting functions is shown in fig. 6. With this new model the problem of the tail portions and of the accurate determination of $a_2(t)$ are thus eliminated.

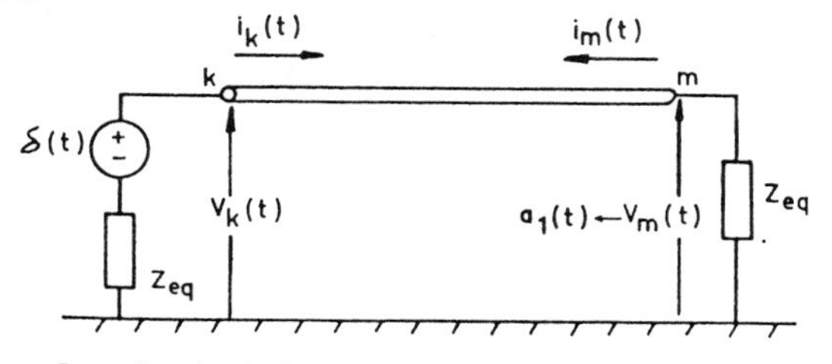

Fig. 5: Physical interpretation of the function $a_1(t)$ in the new formulation.

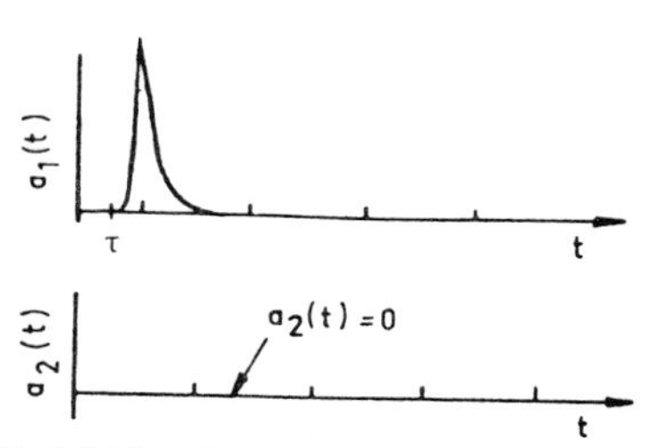

Fig. 6: Weighting functions $a_1(t)$ and $a_2(t)$ in the new formulation.

VI. MATHEMATICAL DEVELOPMENT OF THE NEW MODEL

In order to replace R_1 by Z_{eq} for the generation of the new weighting functions, the forward and backward travelling functions (eqns. 5 to 8) can be defined in the frequency domain as

$$F_k(\omega) = V_k(\omega) + Z_{eq}(\omega) I_k(\omega) \tag{10}$$

$$F_m(\omega) = V_m(\omega) + Z_{eq}(\omega) I_m(\omega) \tag{11}$$

and

$$B_k(\omega) = V_k(\omega) - Z_{eq}(\omega) I_k(\omega) \tag{12}$$

$$B_m(\omega) = V_m(\omega) - Z_{eq}(\omega) I_m(\omega) , \tag{13}$$

where $Z_{eq}(\omega)$ = impedance of linear network approximating $Z_c(\omega)$.

Comparing eqns. 10 to 13 with the general line solution in the frequency domain (eqns. 1 and 2), it follows that

$$B_k(\omega) = A_1(\omega) F_m(\omega) \tag{14}$$

and

$$B_m(\omega) = A_1(\omega) F_k(\omega) , \tag{15}$$

where

$$A_1(\omega) = e^{-\gamma(\omega)\ell} = \frac{1}{\cosh[\gamma(\omega)\ell] + \sinh[\gamma(\omega)\ell]} . \tag{16}$$

The time domain form of $A_1(\omega)$ is the function $a_1(t)$ shown in fig. 6. The time domain form of eqns. 14 and 15 is given by the convolution integrals

$$b_k(t) = \int_\tau^\infty f_m(t-u) a_1(u) du \tag{17}$$

and

$$b_m(t) = \int_\tau^\infty f_k(t-u) a_1(u) du . \tag{18}$$

The lower limit of these integrals is τ because, as it can be seen from fig. 6, $a_1(t)=0$ for $t<\tau$. (The time delay τ represents the travelling time of the fastest frequency component of the injected impulse.)

From eqns. 17 and 18 it can be seen that the values of b_k and b_m at time step t are completely defined from the past history values of the functions f_m and f_k, as long as the integration step Δt of the network solution is smaller than τ. With b_k and b_m known, the time domain form of eqns. 12 and 13 leads directly to the desired equivalent circuits at the line ends. That is, let

$$b_k(t) = E_{kh} \text{ (from history)} \tag{19}$$

and

$$b_m(t) = E_{mh} \text{ (from history)} , \tag{20}$$

then from eqns. 12 and 13,

$$v_k(t) = e_k(t) + E_{kh} \tag{21}$$

and

$$v_m(t) = e_m(t) + E_{mh} , \tag{22}$$

where $e_k(t)$ and $e_m(t)$ are the voltages across the network Z_{eq}. After converting to a modal representation, eqns. 21 and 22 give at each time step t the equivalent line models shown in fig. 7.

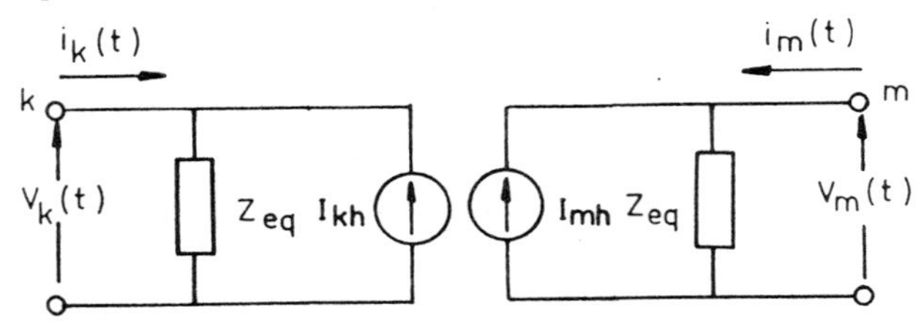

Fig. 7: New frequency-dependent line models at nodes k and m.

Synthesis of the Characteristic Impedance

The network Z_{eq} representing the line characteristic impedance $Z_c(\omega)$ is simulated by a series of Resistance-Capacitance (R-C) parallel blocks (Foster I network realization), as shown in fig. 8.

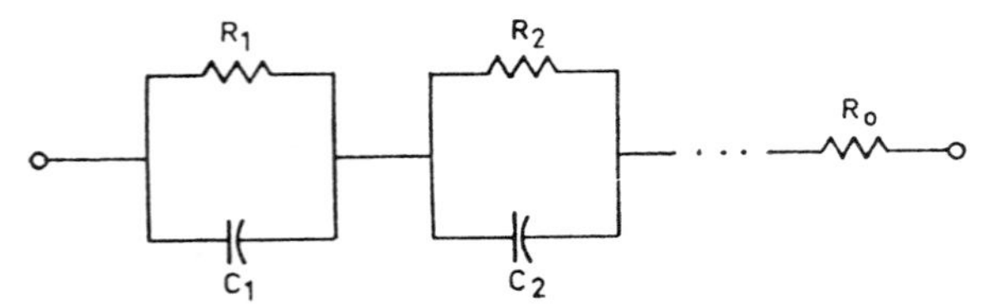

Fig. 8: Foster I realization of $Z_c(\omega)$.

The number of R-C building blocks is determined automatically by the approximating routine and depends on the particular line and mode being simulated. For the reference line studied in this paper the number of R-C blocks turned out to be 8 for the zero sequence and 9 for the positive sequence.

For the synthesis of Z_{eq} the tabular function $Z_c(\omega)$ evaluated from eqn. 3 (with the frequency dependent parameters obtained, for instance, from Carson's equations) is first approximated in the complex plane ($s=\sigma+j\omega$) by a rational function of the form

$$Z_{eq}(s) = \frac{N(s)}{D(s)} = H\,\frac{(s+z_1)(s+z_2)\ldots(s+z_n)}{(s+p_1)(s+p_2)\ldots(s+p_n)} \quad (23)$$

The break points z_i and p_i of this function are real, positive, and simple. The values of the parameters of the R-C equivalent network are obtained by expanding eqn. 23 into a series of simple fractions:

$$Z_{eq}(s) = k_o + \frac{k_1}{s+p_1} + \frac{k_2}{s+p_2} + \ldots + \frac{k_n}{s+p_n}, \quad (24)$$

from where, in fig. 8,

$$R_o = k_o$$
$$R_i = k_i/p_i$$
$$C_i = 1/k_i$$

for $i = 1, 2, \ldots, n$.

The idea of approximating the line characteristic impedance by R-C combinations (even though by a much simpler model) has been suggested by Groschupf [13], who came to the conclusion that this approximation was not necessary, and that the frequency dependence of the characteristic impedance can usually be neglected. However, it has been found in the present work that, even though the accurate simulation of $Z_c(\omega)$ is not as essential for open-ended lines, it is actually very significant in the simulation of short circuits.

Weighting Function and Past History Convolution

In order to obtain the current sources in the equivalent circuits of fig. 7, it is necessary to evaluate the weighted history functions given by eqns. 17 and 18. As noted by Semlyen ([5], [7], and [8]) the process of evaluating convolution integrals having the form of eqns. 17 and 18 can greatly be accelerated if the corresponding weighting (or "transfer") functions can be expressed as a sum of exponential terms. That is, if in general the convolution integral at time step t has the the form

$$s(t) = \int_T^\infty f(t-u)\,k e^{-\alpha(u-T)}\,du, \quad (25)$$

then s(t) can be directly obtained from the known value $s(t-\Delta t)$ at the previous time step and the known history of f at T and $(T+\Delta t)$ units of time earlier:

$$s(t) = m s(t-\Delta t) + p f(t-T) + q f(t-T-\Delta t), \quad (26)$$

where m, p, and q are constants depending on k, α, the integration step Δt, and the numerical interpolation technique. (This property is also applied by Meyer and Dommel [4] to evaluate integral 9 in the tail portion of the weighting functions $a_1(t)$ and $a_2(t)$).

Recursive evaluation of the convolution integrals has also been adopted in this work. However, as explained in Section VII, the numerical process to approximate $a_1(t)$ as a sum of exponentials is different from the one proposed by Semlyen and avoids the numerical difficulties mentioned in reference [9].

From fig. 6 it can be seen that $a_1(t)$ can be expressed as

$$a_1(t) = p(t-\tau), \quad (27)$$

where p(t) has the same form as $a_1(t)$, but is displaced τ units of time towards the origin. From the shifting property of the Fourier Transform, the corresponding frequency domain form of eqn. 27 is

$$A_1(\omega) = P(\omega)e^{-j\omega\tau}. \quad (28)$$

The function P(s) corresponding to $P(\omega)$ in the complex plane is approximated by a rational function of the form

$$P_a(s) = \frac{N(s)}{D(s)} = H\,\frac{(s+z_1)(s+z_2)\ldots(s+z_n)}{(s+p_1)(s+p_2)\ldots(s+p_m)}, \quad (29)$$

where, since $A_1(\omega)$ corresponds to the response of a passive physical system and tends to zero when $\omega\to\infty$, the number of zeroes must be smaller than the number of poles, and the real part of the poles must lie in the left-hand side of the complex plane.

After a partial-fraction expansion of eqn. 29 and subsequent transformation into the time domain, the function approximating $a_1(t)$ becomes

$$a_{1a}(t) = [k_1 e^{-p_1(t-\tau)} + k_2 e^{-p_2(t-\tau)} + \ldots + k_m e^{-p_m(t-\tau)}]\,\upsilon(t-\tau), \quad (30)$$

from which the past history integrals (eqns. 17 and 18) can be evaluated recursively.

The number of exponentials in the approximation depends on the particular line and mode. Table 1 shows the number of exponentials used to simulate the reference line for the zero and positive sequence modes and for different lengths.

5 mi		30 mi		100 mi		500 mi	
zero	pos.	zero	pos.	zero	pos.	zero	pos.
14	12	15	14	13	15	12	13

Table 1: Number of exponentials for the simulation of $a_1(t)$ for reference line.

VII. NUMERICAL TECHNIQUES

As indicated earlier, in order to simulate the characteristic impedance by an R-C equivalent network and to allow recursive evaluations of the past history convolution integrals, the frequency domain functions $Z_c(\omega)$ and $A_1(\omega)$ are approximated by rational functions.

The problem of finding a rational function to simulate the response of a network is studied in network synthesis theory. There are different numerical techniques to approximate a tabular function of frequency by means of a rational fraction of polynomials (e.g. Karni [14]).

However, most of the traditional techniques (for example, Butterworth's, Chebyshev's, Lagrange's) have mainly been applied to particular classes of problems, such as ideal filter responses. Of more recent development are more general numerical techniques, such as least-square optimizations and optimum search (e.g. gradient) algorithms.

Despite their merit for rational approximations of specific functions, programs using these routines require a series of control parameters and adjustments that depend on the particular function approximated. One of the main reasons for this is that the degree of the approximating polynomials is established beforehand and then the rational function is "forced" to fit the given curve. Specification of polynomials of larger or smaller degrees than actually required for the given function often results in numerical instability and accuracy problems. These problems are mentioned by Semlyen [8], who applies a least squares technique to simulate the system response function.

In the modelling of frequency-dependent transmission lines the form of the functions to be approximated depends on the particular line, its length, and the particular mode. An approximating function "tailor-cut" for a specific case will not generally represent the best solution for other cases.

The technique employed in this work avoids the above-mentioned problems by allowing the approximating function to "freely" adapt itself to the form of the function being approximated. This technique is based on an adaptation of the simple concept of asymptotic fitting of the magnitude function, first introduced by Bode [15]. During the process of approximation, the poles and zeros of the rational approximating function are successively allocated, as needed, while following the approximated function from zero frequency to the highest frequency at which the magnitude of the approximated function becomes practically zero or constant. The entire frequency range is thus considered and a uniformly accurate approximation is obtained. Since the poles and zeros are allocated when needed, the degree of the approximating polynomials is not pre-established, but determined automatically by the routine.

Another problem mentioned by Semlyen is the occurrence of ripples or local peaks in the approximating function. This problem is avoided here by allowing only real poles and zeros.

Some Analytical Considerations

Phase Functions:

The rational functions (23) and (29) determined by the method of asymptotic approximation have no zeros in the right-hand side of the complex plane. Under these conditions, it is shown in Fourier Transform Theory (e.g. Papoulis [16]) that the phase function is uniquely determined from the magnitude function and that the rational function belongs to the class of minimum-phase-shift functions. The agreement between the phases of $P(\omega)$ and $Z_c(\omega)$, and the phases of the corresponding rational approximations obtained in the present work shows the correctness of the minimum-phase-shift approximations.

Causality Condition:

The rational approximations $P_a(s)$ in eqn. 29 and $Z_{eq}(s)$ in eqn. 23 tend to a constant for $s=j\omega$ when $\omega \to \infty$, and have no poles in the right-hand side of the complex plane. These conditions are enough (e.g. Popoulis [16]) to assure that the corresponding time domain functions are causal (function=0 for t<0) and thus correspond to the response of a physical passive system.

VIII. NUMERICAL RESULTS

Simulation of the Characteristic Impedance and Weighting Function

Figs. 9 and 10 show the comparison between the magnitudes of $Z_c(\omega)$ and $A_1(\omega)$ for the zero and positive sequence modes of the reference 100-mi line, and the corresponding rational approximations obtained using the techniques described in this paper. The agreement shown in these figures was also found for other line lengths (5 to 500 mi) and for the corresponding phase functions.

Tests

The accuracy of the new frequency-dependent line model has been tested by analytical and field test comparisons.

Analytical Comparisons:

i) Frequency Domain Tests:

The accuracy of the new model can easily be tested in the frequency domain by connecting a single frequency voltage source at the sending end of the line, with the receiving end open or short circuited. The response of the line under these conditions can be calculated analytically from the functions $A_1(\omega)$ and $Z_c(\omega)$, and the results can be compared with those obtained using the corresponding rational approximations. These comparisons can be made over the entire frequency range.

From eqns. 1 and 2, the short-circuit current at the sending end of the line is given by

$$I_k = \frac{E_s}{Z_c} \frac{1 + A_1^2}{1 - A_1^2}, \tag{31}$$

where E_s is the applied voltage source. Similarly, the open-circuit voltage at the receiving end is given by

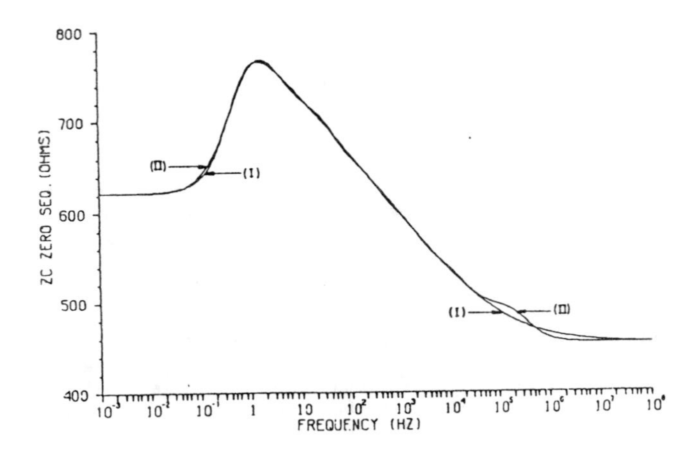

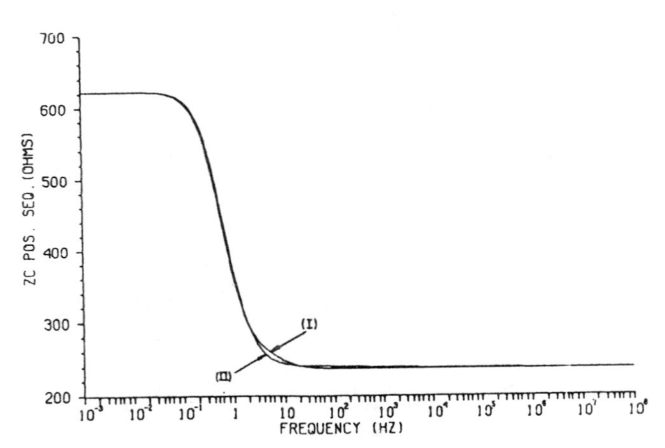

Fig. 9: Simulation of the characteristic impedance. Curves (I): Exact parameters. Curves (II): New model parameters.

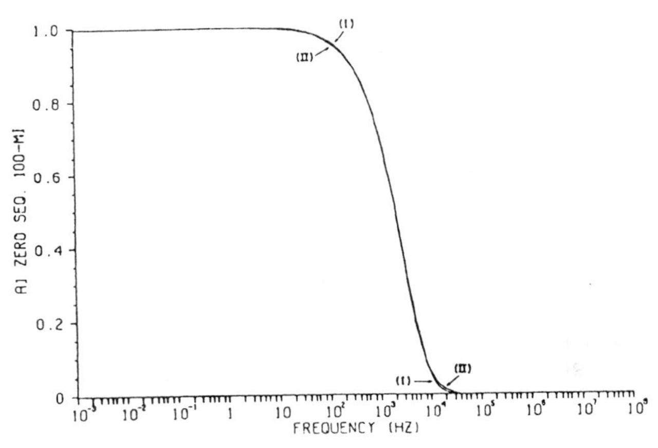

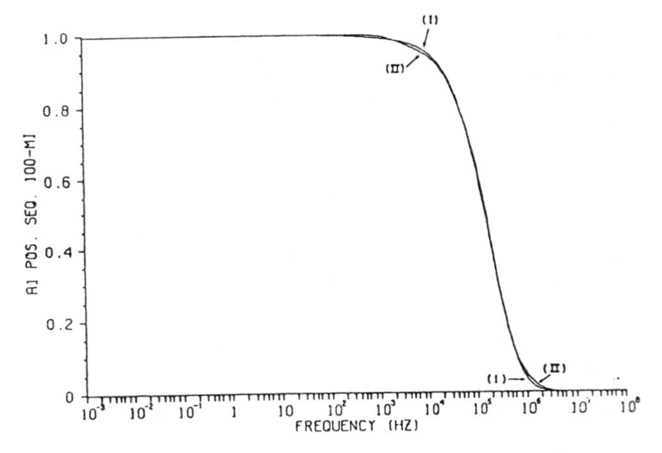

Fig. 10: Simulation of the weighting function. Curves (I): Exact parameters. Curves (II): New model parameters.

$$V_m = E_s \frac{2A_1}{1 + A_1^2} . \tag{32}$$

It is interesting to note from this last equation that the open circuit voltage is independent of the characteristic impedance. This explains why some frequency dependence models that neglect the frequency dependence of the characteristic impedance can give acceptable results if they are only tested for open-circuit conditions. On the other hand, as can be seen from eqn. 31, the correct modelling of Z_c is very important for short-circuit conditions.

The results of these comparisons are shown in figs. 11 and 12 for the zero sequence mode and a length of 100 miles. These comparisons were also made for other line lengths (from 5 to 500 miles), as well as for the positive sequence mode, with similarly good agreements. The same agreement was also found for the corresponding phase angles.

Figs. 13 and 14 show the comparison between the responses obtained using exact parameters and those assuming constant 60 Hz parameters. The limitations of the constant-parameter model for the simulation of the lower and higher frequencies is clearly illustrated in these figures.

The magnification of the higher harmonics by the constant-parameter model can also clearly be seen in the transient simulations shown in figs. 15 and 16. These figures compare the simulations using the new line model and the constant-parameter model for two cases of open-circuited line energizations. In fig. 15 the zero sequence mode of the 100-mi reference line is energized with a sinusoidal, 60 Hz, voltage source, with the peak voltage applied at t=0. In fig. 16 the line mode is energized with a unit voltage step.

ii) Time Domain Tests:

The validity and accuracy of the new line model in time domain simulations can also be assessed from single frequency open and short circuit conditions. For this purpose, the line represented by its frequency-dependent transient model was energized by a single frequency sinusoidal voltage source. Starting from the correct a.c. initial conditions (so that no disturbances exist) transient simulations using the EMTP were run. Under the indicated conditions, the time domain solutions must be perfectly sinusoidal waves with magnitude and phase as given by eqns. 31 and 32. These tests were performed for the different line lengths and modes and for frequencies along the entire frequency range. The results had the correct sinusoidal waveforms and were in complete agreement with the magnitude and phase values previously obtained in the frequency tests.

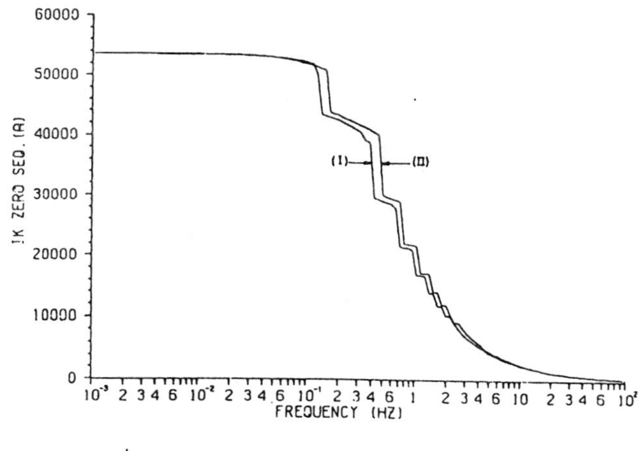

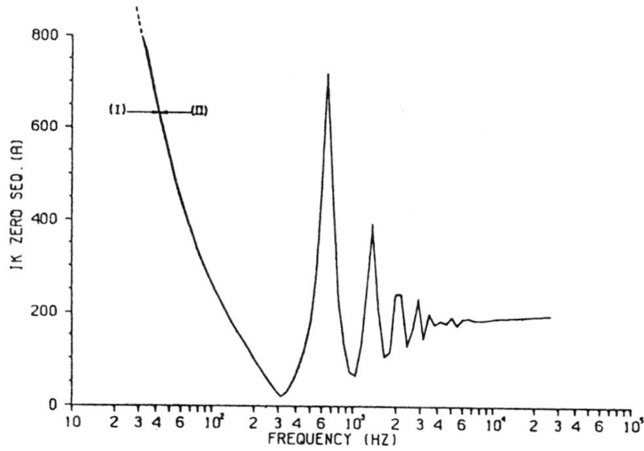

Fig. 11: Short-circuit frequency response. (Source voltage = 100 kV, I_k in amperes) Curves (I): Exact parameters. Curves (II): New model parameters.

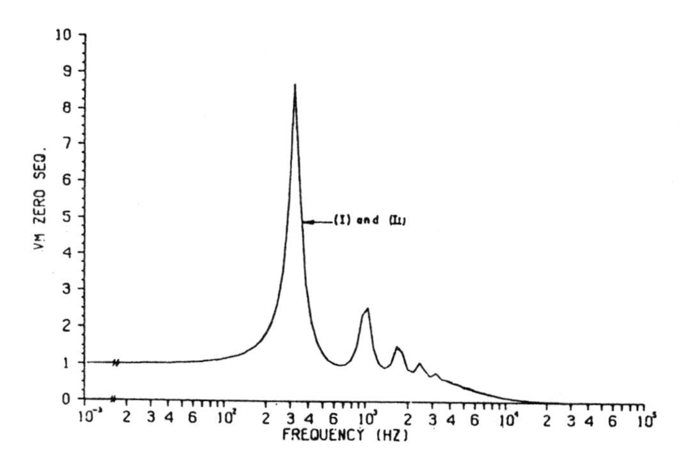

Fig. 12: Open-circuit frequency response. (Source voltage = 1.0). Curve (I): Exact parameters Curve (II): New model parameters.

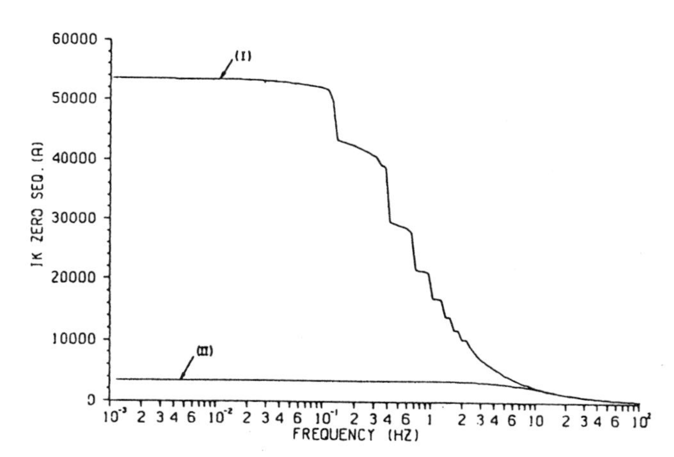

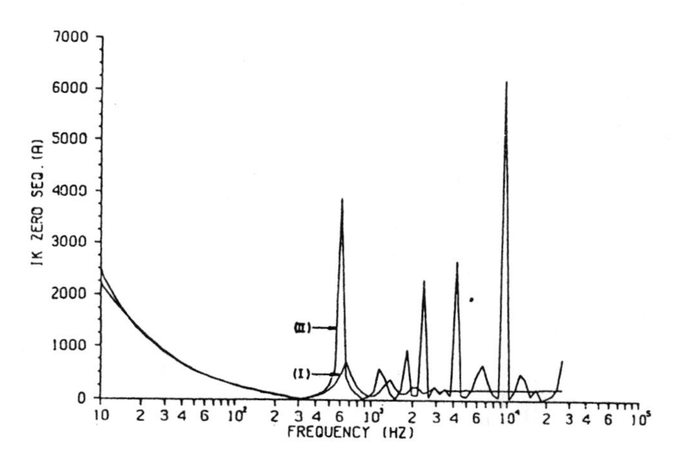

Fig. 13: Short-circuit frequency response. (Source voltage = 100 kV, I_k in amperes) Curves(I): Exact Parameters. Curves (II): Constant, 60 Hz parameters.

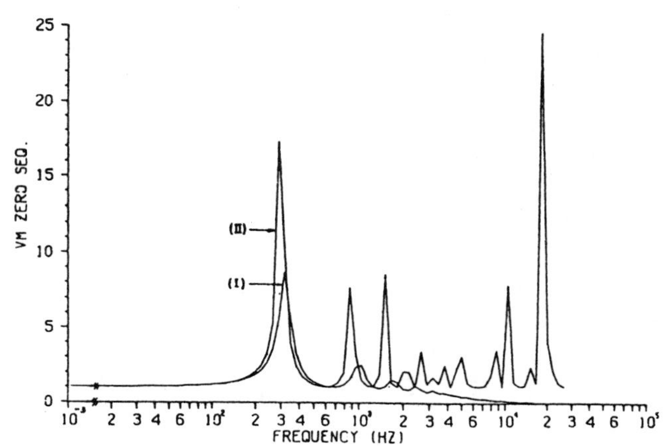

Fig. 14: Open-circuit frequency response. (Source voltage = 1.0). Curve (I): Exact parameters. Curve (II): constant, 60 Hz parameters.

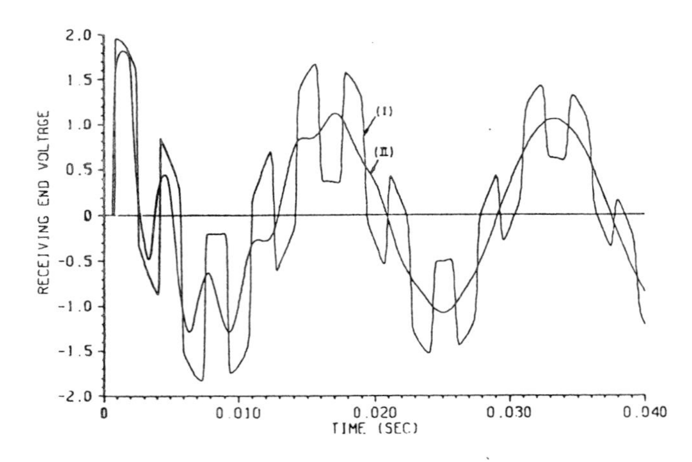

Fig. 15: Sinusoidal energization of open-circuited line (peak voltage at t=0). Curve (I) Constant, 60 Hz parameters. Curve(II):New model parameters.

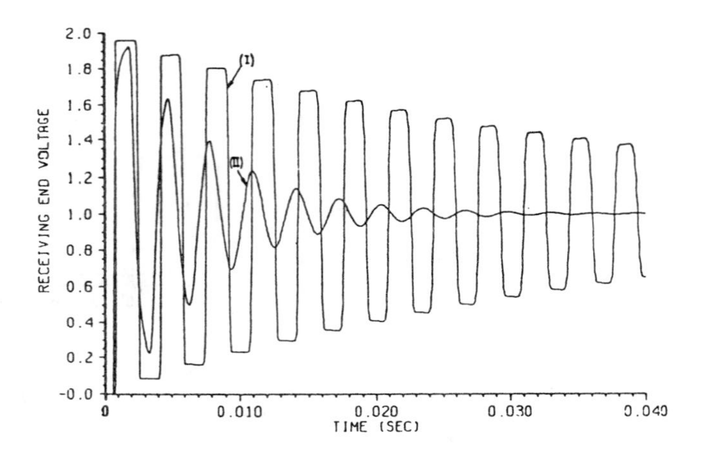

Fig. 16: Step function energization of open-circuited line. Curve (I): Constant, 60 Hz parameters. Curve (I): New model parameters.

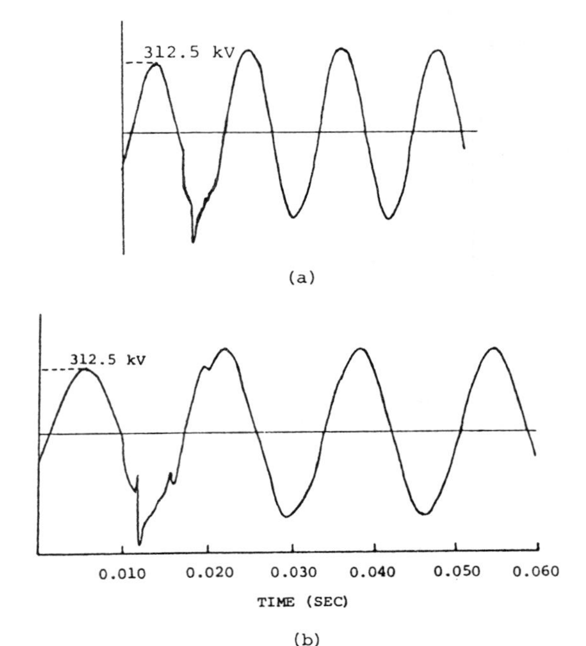

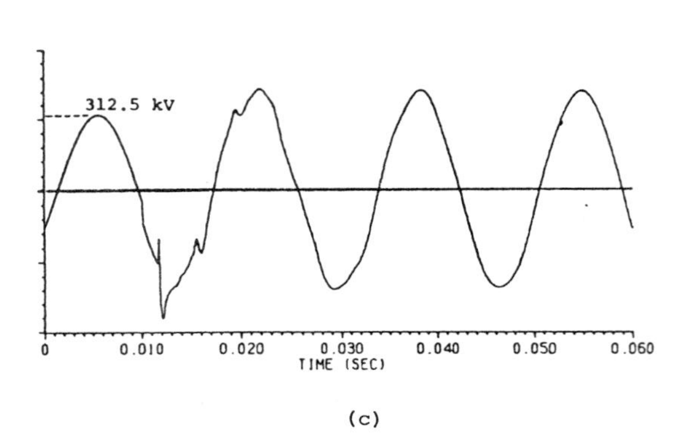

Fig. 17: Field test simulation. (a): BPA field test oscillograph. (b) BPA simulation. (c) New model simulation.

Comparison with Field Test:

The new line model was used to simulate the BPA field test described in reference [4]. This test simulates a single line to ground short circuit on an open-ended 222 km, 500 kV, 3-phase transmission line. The short circuit was applied to phase-c. The field test oscillograph for the voltage at phase-b at the end of the line is shown in fig. 17(a). To compare with BPA's digital simulation in ref. [4], the same integration step Δt=50 µsec was used, and the zero sequence mode of the line was represented by the new model described in this paper. The result of this simulation is shown in fig. 17(c). This result compares well with the field test and with BPA's simulation (fig. 17(b)). In BPA's simulation the average time per step, as compared with the solution with constant parameters, was 3.13 times longer. In the simulation with the new model this time was only 1.19 times longer.

IX. CONCLUSIONS

A new, fast, and reliable approach has been developed for the accurate modelling of transmission lines over the entire frequency range. The routines for obtaining the parameters of the model do not present the numerical difficulties encountered with previous formulations. These routines are easy to use because they do not require value judgements on the part of the user. Further work is needed in connection with the representation of unbalanced, untransposed lines with frequency-dependent modal transformation matrices.

X. ACKNOWLEDGEMENTS

The author would like to express his gratitude to Dr. H.W. Dommel, whose clear and practical thinking are always the best encouragement; to the University of British Columbia Computer Centre for its convenient and easy to use facilities; to Central University of Venezuela for their financial support during the author's leave of absence at U.B.C.; and to the Bonneville Power Administration for their constant cooperation, and for allowing the reproduction of the field test result used in this paper.

XI. REFERENCES

[1] The University of British Columbia, "Line parameters Program." Vancouver, B.C.

[2] A. Budner, "Introduction of Frequency-Dependent Line Parameters into an Electromagnetic Transients Program." IEEE Trans. Power Apparatus and Systems, vol. PAS-89, pp. 88-97, Jan. 1970.

[3] J.K. Snelson, "Propagation of Travelling Waves on Transmission Lines- -Frequency Dependent Parameters." IEEE Trans. Power Apparatus and Systems, vol. PAS-91, pp. 85-91, Jan/Feb. 1972.

[4] W.S. Meyer and H.W. Dommel, "Numerical Modelling of Frequency-Dependent Transmission-Line Parameters in an Electromagnetic Transients Program." IEEE Trans. Power Apparatus and Systems, vol.PAS-93, pp. 1401-1409, Sept/Oct. 1974.

[5] A. Semlyen and A. Dabuleanu, "Fast and Accurate Switching Transient Calculations on Transmission Lines with Ground Return Using Recursive Convolutions." IEEE Trans. Power Apparatus and Systems, vol. PAS-94, pp. 561-571, March/April 1975.

[6] A. Ametani, "A Highly Efficient Method for Calculating Transmission Line Transients." IEEE Trans. Power Apparatus and Systems, Vol. PAS-95, pp. 1545-1551, Sept/Oct. 1976.

[7] A. Semlyen and R.A. Roth, "Calculation of Exponential Step Responses - Accurately for three Base Frequencies." IEEE Trans. Power Apparatus and Systems, vol. PAS-96, pp. 667-672, March/April 1977.

[8] A. Semlyen, "Contributions to the Theory of Calculation of Electromagnetic Transients on Transmission Lines with Frequency Dependent Parameters." IEEE PES Summer Meeting, Vancouver, B.C. July 1979.

[9] H.W. Dommel, "Digital Computer Solution of Electromagnetic Transients in Single-and Multiphase Networks." IEEE Trans. Power Apparatus and Systems, vol. PAS-88, pp. 388-399, April 1969.

[10] P.C. Magnusson, "Travelling Waves on Multi-conductor Open-Wire Lines-A Numerical Survey of the Effects of Frequency Dependence of Modal Composition." IEEE Trans. Power Apparatus and Systems, vol. PAS-92, pp. 999-1008, May/June 1973.

[11] R.G. Wasley and S. Selvavinayagamoorthy, "Approximate Frequency-Response Values for Transmission-Line Transient Analysis," Proc. IEE, vol. 121, no. 4, pp. 281-286, April 1974.

[12] L.F. Woodruff, "Principles of Electric Power Transmission." 2nd Edition. New York:Wiley, 1938, pp. 105-106.

[13] E. Groschupf, "Simulation transienter Vorgänge auf Leitungssystemen der Hochspannungs-Gleichstrom-und-Drehstrom-Übertragung", Dr. -Ing, genehmigte Dissertation, Feb. 23, 1976.

[14] S. Karni, "Network Theory: Analysis and Synthesis." Boston: Allyn and Bacon, 1966, pp. 343-390.

[15] H.W. Bode, "Network Analysis and Feedback Amplifier Design." New York: Van Nostrand, 1945.

[16] A. Papoulis, "The Fourier Integral and its Applications." New York: McGraw-Hill, pp. 204-217, 1962.

José R. Martí (M'71) was born in Spain on June 15, 1948. He received the degree of Electrical Engineer from Central University of Venezuela, Caracas, Venezuela, in 1971, and the degree of M.E. in Electric Power Engineering from Rensselaer Polytechnic Institute, N.Y., in 1974. He is presently a Ph.D. candidate at the University of British Columbia, Canada.

From 1970 to 1971 he worked for Exxon in Venezuela in coordination of protective relays. From 1971 to 1972 he worked for a Consulting Engineering firm in Caracas, Venezuela, in relaying and substation design projects. In 1974 he joined the Central University of Venezuela as a professor in Power System Analysis. He is presently at U.B.C. on a leave of absence from Central University of Venezuela.

Phase Domain Modeling of Frequency-Dependent Transmission Lines by Means of an ARMA Model

T.Noda, Student Member, IEEE N.Nagaoka, Associate Member, IEEE A.Ametani, Fellow, IEEE
Department of Electrical Engineering, Doshisha University
Tanabe-cho, Tsuzuki-gun, Kyoto-pref. 610-03, Japan

***Abstract* - This paper presents a method for time-domain transient calculation in which frequency-dependent transmission lines and cables are modeled in the phase domain rather than in the modal domain. This avoids convolution due to the modal transformation, and possible numerical instability due to mode crossing. In the new approach, time domain convolutions are replaced by an ARMA (Auto-Regressive Moving Average) model that minimizes computation and is EMTP-compatible. A fast and stable method to produce the ARMA model is developed, and results are shown to agree well with both rigorous frequency-domain simulations and also field tests.**

I. INTRODUCTION

It has become important to model the frequency dependence of transmission lines and cables [1-4] accurately in order to minimize the cost of construction. Also, because electric power systems include many nonlinearities, time-domain modeling is a practical necessity.

The Electro-Magnetic Transients Program (EMTP) developed by Bonneville Power Administration (BPA) is the most widely used time-domain transient analysis program. Modal theory [5] is used in EMTP for the distributed representation of transmission lines. For this reason, weighting functions [6], exponential recursive convolution [7,10], linear recursive convolution [8], Z-transforms [9], and modified recursive convolution [11,12] all have been proposed. In those methods that relate the phase and modal domains by constant transformation matrices, the frequency dependence of the matrices is ignored. But it should not be ignored for untransposed lines or cables. In these cases, transformation matrices depend heavily on frequency. Theoretically, the frequency dependence of the matrices can be introduced into a time domain simulation by convolution [13]. However, for an n-phase transmission line, such an approach requires $2n(n-1)$ convolutions for the modal transformation at each time step. Another n convolutions are required for the modal propagation responses, and another n for the characteristic admittances. All together, $2n(n-1)+2n$ convolutions make this approach a burden on both computer time and memory. Furthermore, mode crossing (at some frequency, two or more eigenvalues become equal) enormously complicates practical implementation. On the other hand, H.Nakanishi and one of the authors proposed a transient analysis method in the phase domain rather than in the modal domain so as to minimize the numerical effort and also avoid possible numerical instability due to mode crossing [14]. Also, a direct phase-domain calculation method based on two-sided recursion was recently proposed [15].

The present paper follows the above phase domain approach, but proposes a more efficient and sophisticated method based on an ARMA model. In this approach, time domain convolutions are replaced by the ARMA model, and the computation time is greatly reduced. Because of the nature of the Z-operator, an ARMA model can economically express a phase domain response that includes discontinuities due to modal propagation [16]. A fast and stable optimization method to identify the ARMA model also is developed, by applying Householder's transformation. Because this procedure is linearized, the previous time-consuming, sometimes-unstable, nonlinear optimizations no longer are required [17,18]. Moreover, two important problems have been resolved: 1) the determination of appropriate order of an ARMA model has been solved practically using the theory of Akaike's Information Criterion (AIC) [19]; and 2) the stability of an ARMA model has been assured using Jury's method [20]. Finally, the equivalent circuit derived from the new method is compatible with programs such as EMTP, which are based on nodal admittance representation.

To confirm accuracy and practical use of the new method, transients on a 500-kV untransposed horizontal overhead line, a 500-kV untransposed vertical overhead line, and a 275-kV Pipe-type Oil-Filled (POF) cable are calculated and compared with field test results [21-23] and rigorous frequency-domain solutions [24].

II. PHASE DOMAIN MODELING

A. Phase Domain Formulation

A transmission line is represented as a multiphase distributed-parameter line shown in Fig.1 in a transient simulation. Consider a transmission line consisting of n conductors with length l. In the frequency domain, voltages and currents at distance x from the sending end are expressed in the following column vectors, $V(x,\omega) = (V_1, V_2, \cdots, V_n)^T$, $I(x,\omega) = (I_1, I_2, \cdots, I_n)^T$, where V_i, I_i : voltage and current on the i-th conductor. Let $Z(\omega)$ be the line series impedance matrix and $Y(\omega)$ the line shunt admittance matrix per unit length, evaluated in the frequency domain [1-4]. The electromagnetic wave can be described by a pair of differential equations (Telegrapher's Equation) solved as :

$$V(x,\omega) = e^{-\Gamma(\omega)x} V_f(\omega) + e^{\Gamma(\omega)x} V_b(\omega) \qquad (1)$$

$$I(x,\omega) = Y_0(\omega)\{e^{-\Gamma(\omega)x} V_f(\omega) - e^{\Gamma(\omega)x} V_b(\omega)\}, \qquad (2)$$

where $V_f(\omega)$: vector of forward traveling wave voltages, $V_b(\omega)$: vector of backward traveling wave voltages, $Y_0(\omega)$: characteristic admittance matrix, $\Gamma(\omega)$: propagation constant matrix.

95 WM 245-1 PWRD A paper recommended and approved by the IEEE Transmission and Distribution Committee of the IEEE Power Engineering Society for presentation at the 1995 IEEE/PES Winter Meeting, January 29, to February 2, 1995, New York, NY. Manuscript submitted July 25, 1994; made available for printing January 11, 1995.

Let a vector of the sending end voltages be $V_1(\omega)$ and the receiving end $V_2(\omega)$. Algebraic manipulation of (1) and (2) leads to (see Appendix A) :

$$I_1(\omega) = Y_0(\omega)V_1(\omega) - Y_0(\omega)e^{-j\omega\tau}H(\omega)\{V_2(\omega) + Z_0(\omega)I_2(\omega)\} \quad (3)$$

$$I_2(\omega) = Y_0(\omega)V_2(\omega) - Y_0(\omega)e^{-j\omega\tau}H(\omega)\{V_1(\omega) + Z_0(\omega)I_1(\omega)\}, \quad (4)$$

where $H(\omega)=e^{j\omega\tau}e^{-\Gamma(\omega)l}$: wave deformation matrix in the phase domain, τ : minimum traveling time, $Z_0(\omega)=Y_0^{-1}(\omega)$: characteristic impedance matrix. If the above equations are transformed into the time domain, they require 8 convolutions.

To reduce the computation time, the following relation (see Appendix B) :

$$Y_0(\omega)H(\omega) = H^T(\omega)Y_0(\omega) \quad (T: \text{transpose}) \quad (5)$$

is applied to (3) and (4), and we obtain

$$I_1(\omega) = Y_0(\omega)V_1(\omega) - e^{-j\omega\tau}H^T(\omega)\{Y_0(\omega)V_2(\omega) + I_2(\omega)\} \quad (6)$$

$$I_2(\omega) = Y_0(\omega)V_2(\omega) - e^{-j\omega\tau}H^T(\omega)\{Y_0(\omega)V_1(\omega) + I_1(\omega)\}. \quad (7)$$

Transforming (6) and (7) into the time domain (lower case letters are used to denote their time domain counterparts),

$$i_1(t) = \underline{y_0(t)*v_1(t)} - i_{p1}(t) \quad (8)$$

$$i_{p1}(t) = h^T(t)*\{\underline{\underline{y_0(t)*v_2(t-\tau)}} + i_2(t-\tau)\} \quad (9)$$

$$i_2(t) = \underline{\underline{y_0(t)*v_2(t)}} - i_{p2}(t) \quad (10)$$

$$i_{p2}(t) = h^T(t)*\{\underline{y_0(t)*v_1(t-\tau)} + i_1(t-\tau)\}, \quad (11)$$

where $*$: symbol to indicate matrix-vector convolution, $h(t)=F^{-1}\{H(\omega)\}$ (inverse Fourier transform). The above equations are compatible with Bergeron's expression [25]. As the terms underlined with - and = are the same, the number of convolutions is reduced to 4. Consequently, the computation time is greatly reduced using (8) to (11). The set of equations is called "Basic Equations of Phase Domain Method" in this paper.

B. Equivalent Circuit for a Time Domain Simulation

The "Basic Equations of Phase Domain Method" indicate that the equivalent circuit for a time domain simulation is expressed as an n-terminal admittance paired with an n-terminal current source (Fig.2a). It can be seen that the first term of (8) and (10) corresponds to the n-terminal admittance, and the second term to the n-terminal current source corresponding to past history obtained from (9) and (11). When including the equivalent circuit into a nodal admittance representation, how to model the frequency-dependent characteristic admittance $Y_0(\omega)$ is important. To model $Y_0(\omega)$ with ARMA models, the convolution operation $y_0(t)*v(t)$ is easily decomposed as :

$$y_0(t)*v(t) = y_{00}v(t) + y_{01}(t)*v(t-\Delta t). \quad (12)$$

Matrix y_{00} is constant, and $y_{01}(t)*v(t-\Delta t)$ is a matrix-vector convolution without instantaneous responses. From (12), the "Basic Equations of Phase Domain Method" lead to :

$$i_1(t) = y_{00}v_1(t) - i'_{p1}(t) \quad (13)$$

$$i'_{p1}(t) = i_{p1}(t) - y_{01}(t)*v_1(t-\Delta t) \quad (14)$$

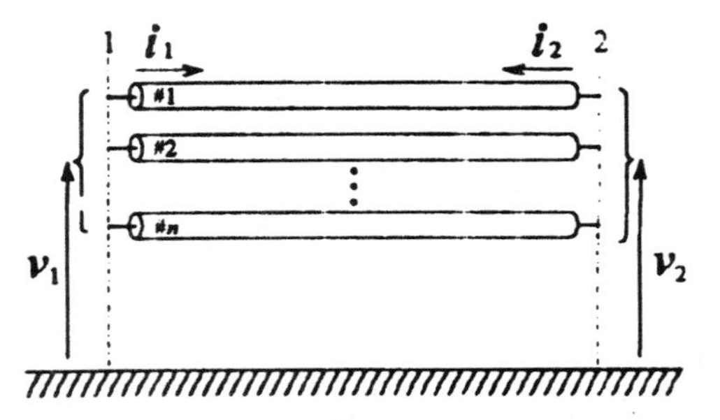

Fig.1 A multiphase distributed-parameter line

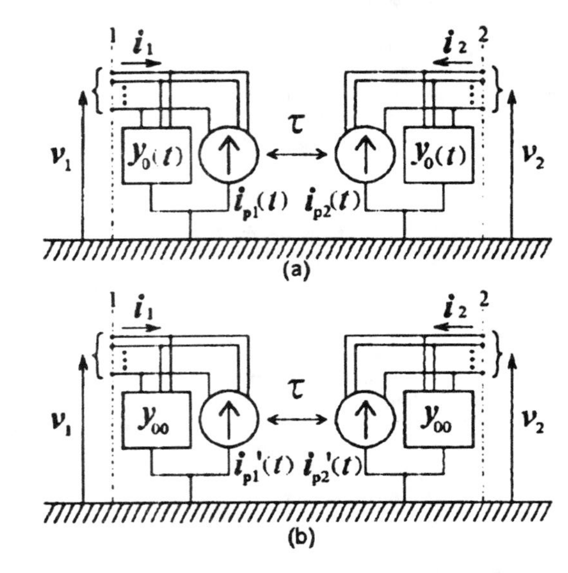

Fig.2 Equivalent circuits for a time domain simulation by phase domain modeling

$$i_2(t) = y_{00}v_2(t) - i'_{p2}(t) \quad (15)$$

$$i'_{p2}(t) = i_{p2}(t) - y_{01}(t)*v_2(t-\Delta t). \quad (16)$$

This set of equations corresponds to the equivalent circuit illustrated in Fig.2b. It can easily be introduced into programs based on nodal admittance representation, such as EMTP. First, the constant matrix y_{00} is added directly to the node conductance matrix of the whole system before the transient calculation. During the simulation, the past history current source vectors $i'_{p1}(t)$ and $i'_{p2}(t)$ are then added to the current source vector of the system at each time step.

III. CONVOLUTION BY AN ARMA MODEL

A. ARMA model

An ARMA model essentially represents a discrete-time system. Input $x(t)$ and output $y(t)$ sampled at the calculation interval Δt are denoted as $x(n)=x(t)|_{t=n\Delta t}$, $y(n)=y(t)|_{t=n\Delta t}$ (n=0,1,2,···). Using Z-transform theory, an ARMA model is defined by the following rational function of z^{-1} in the z-domain :

$$G(z)=\frac{a_0+a_1z^{-1}+\cdots+a_Nz^{-N}}{1+b_1z^{-1}+\cdots+b_Nz^{-N}}, \tag{17}$$

where a_n, b_n : coefficients of the ARMA model, N : order of the ARMA model. Because the operator z^{-n} denotes a delay of n samples, (17) is transformed into the time domain as :

$$y(n)=a_0x(n)+a_1x(n-1)+\cdots+a_Nx(n-N) \\ -b_1y(n-1)-\cdots-b_Ny(n-N). \tag{18}$$

The above equation is a time domain representation of an ARMA model, and is equivalent to applying the recursive convolution [7-12]. Using this equation, output $y(n)$ can be calculated by only $2N+1$ multiplications and $2N$ additions.

B. Phase Domain Matrix-Vector Convolution

A transient calculation based on the "Basic Equations of Phase Domain Method" requires matrix-vector convolutions in the following form :

$$y(t)=g(t)*x(t), \tag{19}$$

where $g(t)$: transfer function matrix, $x(t)$: input vector, $y(t)$: output vector. The transfer function matrix $g(t)$ corresponds to the wave deformation matrix $h(t)$ or characteristic admittance matrix $y_0(t)$ in (8) to (11). Each element of $g(t)$ is replaced by an ARMA model using an optimization method described in the following section. Once elements of $G(z)=Z\{g(t)\}$ ($Z\{\}$ denotes Z-transform) are replaced by an ARMA model, the relation between the z-domain input vector $X(z)$ and the output vector $Y(z)$ is expressed as $Y(z)=G(z)X(z)$. Therefore, the i-th element of the output vector $Y(z)$ is expressed as :

$$Y_i(z)=\sum_{j=1}^{n}G_{ij}(z)X_j(z). \tag{20}$$

Substituting (17) into (20), the inverse Z-transformation gives

$$y_i(n)=\sum_{j=1}^{n}\{\ a^{ij}{}_0x_j(n) \\ +a^{ij}{}_1x_j(n-1)+\cdots+a^{ij}{}_{N_{ij}}x_j(n-N_{ij}) \\ -b^{ij}{}_1u_{ij}(n-1)+\cdots+b^{ij}{}_{N_{ij}}u_{ij}(n-N_{ij})\ \}, \tag{21}$$

where $a^{ij}{}_n$, $b^{ij}{}_n$: coefficients of (i, j) element, N_{ij} : order of (i, j) element, $u_{ij}(n)=Z^{-1}\{G_{ij}(z)X_j(z)\}$.

IV. OPTIMIZATION TECHNIQUES

A. Linearized Least-Squares Method [16,17]

This section describes a highly efficient method called "Linearized Least-Squares (LS) Method", which approximates each element of the wave deformation matrix $H(\omega)$ and of the characteristic admittance matrix $Y_0(\omega)$ in the phase domain by an ARMA model.

Let us represent each element of $H(\omega)$ or $Y_0(\omega)$ as a set of data (ω_k, G_k) $(k=1,2,\cdots,K)$ describing a transfer function defined at a discrete angular frequency. The relation $z=\exp(j\omega\Delta t)$ replaces the transfer function (ω_k, G_k) with a z-domain transfer function (z_k, G_k). When approximating (z_k, G_k) by an ARMA model using a conventional least-squares method, the error function for the k-th datum is $err_k=G(z_k)-G_k$. If we substitute (17) into err_k, we would obtain an error function that is nonlinear due to its rational polynomial form. In this case, we would need to apply a nonlinear optimization method to identify the parameters. An initial value and a differential coefficient would be required, because the nonlinear optimization method searches for an optimum value using an iterative calculation. Moreover, assurance of convergence to an optimum value would be highly dependent on the choice of initial value.

On the other hand, the proposed Linearized LS Method chooses the following error function :

$$err'_k=(a_0+a_1z_k^{-1}+\cdots+a_Nz_k^{-N}) \\ -(1+b_1z_k^{-1}+\cdots+b_Nz_k^{-N})G_k \tag{22}$$

This error function includes all the parameters as linear parameters. This is why this method is called "Linearized LS Method", and a time consuming and sometimes unstable iterative calculation is no longer required.

To obtain the least-squares solution of (22), the conventional normal equation method is well-known. But the accuracy will get worse as the order N increases. Therefore, Householder's transformation is applied to improve the accuracy. As a result, the linearization of the error function and the application of Householder's transformation make the calculation stable, fast and accurate.

B. Samples in the z-domain

Angular frequency ω_k should be sampled logarithmically to cover a wide frequency range with a small number of samples K. The nature of the transfer function $G(z)$ implies that $G(z^*)=G^*(z)$ ("*" denotes complex conjugate) [17]. This condition in turn determines the maximum frequency in the following equation :

$$z_K=e^{j\omega_{max}\Delta t}\Rightarrow f_{max}=1/2\Delta t. \tag{23}$$

This equation corresponds to the well-known sampling theorem, and determines the maximum frequency of the data set.

C. Relative Error Evaluation

A relative type of error evaluation is selected, in order to improve the accuracy of the cut-off band [17]. This evaluation can easily be introduced into the least-squares calculation with weighting values. The weighting value for the k-th datum (ω_k, G_k) is given in the following equation (see Appendix C).

$$w_k=1/|D(z_k)G_k|^2,\ D(z)=1+b_1z^{-1}+\cdots+b_Nz^{-N} \tag{24}$$

In a practical calculation, the parameter identification must be executed after the weighting values are calculated, although a calculation of $D(z_k)$ included in w_k requires the parameters $b_1,\cdots,b_N$ which are not calculated at this time. But the parameters $b_1,\cdots,b_N$ for w_k can be evaluated roughly without using relative error evaluation, as a high accuracy for w_k is not required. Thus, the relative error evaluation method can be applied.

D. Order and Stability of an ARMA Model

In [26], it is pointed out that the order of an ARMA model cannot be determined without trial and error. The proposed method basically increases the order one by one until a best fit is found. A simple error evaluation uses standard deviation (SD) :

$$SD(N)=\sqrt{\frac{1}{K}\sum_{k=1}^{K}|G(z_k,N)-G_k|^2}. \tag{25}$$

where, for a permitted constant error ε_A and maximum order N_{max}, an order N meeting the condition $SD(N)<\varepsilon_A$ can be determined. But the existence of a model meeting this condition in the range of $N\le N_{max}$ cannot be guaranteed. In this case,

Akaike's Information Criterion (AIC) [19] provides a criterion for determining the order. The values provided in the data set (ω_k, G_k) include approximation errors and numerical calculation errors. When the errors are small, the SD will decrease monotonically as the order increases, and achieve the condition SD(N)<ε_A at a comparatively small order. When the errors are large, the model will approximate not only the true electromagnetic characteristics but also the effect of the introduced error, thus requiring too large an order. In this case, an order N which minimizes AIC(N) will give an appropriate order by estimating the error characteristics included in the data set. AIC(N) is given in the following equation.

$$\mathrm{AIC}(N) = K\ln\left\{\sum_{k=1}^{K}\left|G(z_k, N) - G_k\right|^2\right\} + 4N + 2 \qquad (26)$$

Because a model for a transient calculation has to be stable, a stability check is also required. According to (20), an output element of the matrix-vector convolution is the sum of n discrete system responses. The sum, i.e. the output, should be stable when each of its elements is stable. In the proposed method, the stability of each ARMA element is evaluated using Jury's method [20] (see Appendix D), which is especially suitable for a discrete system, as it only requires simple algebraic computations.

As a result, a model meeting the following conditions is adopted: 1) the model has a minimum order which meets the condition SD(N)<ε_A, 2) the model has an order which minimizes AIC(N) when SD(N)<ε_A cannot be achieved, 3) the model is evaluated to be stable by Jury's method.

E. Modeling of Traveling Time Differences

Because an element of a wave deformation matrix in the phase domain $H_{ij}(\omega)$ consists of n modal components, its time domain response has discontinuities every $\Delta\tau_k = \tau_k - \tau$ (τ is the traveling time of the fastest mode). But it is difficult to approximate this discontinuous response with a conventional ARMA model expressed in (17), as its impulse response with a denominator of order N is a sum of N complex exponentials. Therefore, a small modification is made in the expression of the ARMA model. The following condition determines whether the k-th mode component is dominant or not in $H_{ij}(\omega)$ (see Appendix E) :

$$\left|a_{ik}(a^{-1})_{kj}e^{-\gamma_k l}\right| > \varepsilon_M. \qquad (27)$$

where a_{ik} : (i, k) element of transformation matrix A, $(a^{-1})_{kj}$: (k, j) element of inverse transformation matrix A^{-1}, γ_k : modal propagation constant of the k-th mode, ε_M : a small constant. The above a_{ik}, $(a^{-1})_{kj}$, and γ_k are obtained directly when evaluating the propagation response matrix $e^{-\Gamma(\omega)}$. When the above condition is met, the impulse response of $H_{ij}(\omega)$ has a discontinuity at $t=\Delta\tau_k$. In this case, the ARMA model can be made to represent the discontinuity using the one sample delay nature of the Z-operator, by using the delay operator z^{-q_k} (q_k=integer($\Delta\tau_k/\Delta t$)) which corresponds to the traveling time difference applied to the corresponding numerator term. For example, supposing that $H_{ij}(\omega)$ has a discontinuity due to the k-th mode, the ARMA model with order N=5 is expressed as

$$H_{ij}(\omega) = \frac{N_0(z) + z^{-q_k}N_k(z)}{1 + b_1z^{-1} + b_2z^{-2} + b_3z^{-3} + b_4z^{-4} + b_5z^{-5}}, \qquad (28)$$

where q_k=integer($\Delta t_k/\Delta t$), $N_0(z)=a_0+a_1z^{-1}+a_2z^{-2}$, $N_k(z)=a_{qk}+a_{qk+1}z^{-1}+a_{qk+2}z^{-2}$. This representation is extended to include each mode. Note that when both the k_1-th mode and the k_2-th mode are involved dominantly in $H_{ij}(\omega)$ and the traveling time differences $\Delta\tau_{k_1}$ and $\Delta\tau_{k_2}$ are almost the same, the two modes can be represented with one q.

V. CALCULATED RESULTS

A. An Untransposed Horizontal Overhead Line

Fig.3a shows a 500-kV untransposed horizontal overhead line consisting of 3 phase wires and 2 ground wires. The voltages on the ground wires are not considered, reducing the order of the impedance and admittance matrices from 5 by 5 to 3 by 3 for the EMTP Cable Constants routine. Fig.4 shows the fitting of one element of the wave deformation matrix $H_{11}(\omega)$ using the optimization method described above. The ARMA model of this element uses an order N=9, and is accurate in both amplitude and phase. The symmetrical arrangement of the conductors reduces from 9 to 5 the number of different elements to be fitted. Table I compares the order of the elements with the order of the fitting obtained using a modal domain method [12]. The proposed optimization method uses lower orders to reproduce the given characteristics. To confirm the computation efficiency, the execution time of both the optimization and the transient calculation are also compared in Table II with those of the modal domain method. Both simulation results were obtained using the ATP version of EMTP. Fig.3b illustrates an actual test circuit where the receiving end voltages were measured [21]. The calculated results agree well with the field test results and with an exact frequency-domain simulation [24] as shown in Fig.5.

B. An Untransposed Vertical Overhead Line

The frequency dependence of a vertical overhead line is considerable. Fig.6a shows a 500-kV untransposed vertical overhead line consisting of 6 phase wires and 2 ground wires. The order reduction is also applied to neglect the ground wires. The number of fitting is reduced from 36 to 18 by the symmetrical arrangement. Fig.6b shows an actual test circuit in which the sending and receiving end voltages were measured [21]. The calculated results agree well with the field test results and the frequency-domain simulation as shown in Fig.7.

C. A Pipe-Type Oil-Filled Cable

Fig.8a shows a 275-kV pipe-type oil-filled cable. Because of its thin sheaths, the frequency dependence of the cable is also considerable. As the sheaths and the enclosing pipe touch each other, the order reduction can be applied. Fig.8b shows the test circuit in which the sending and receiving end voltages were measured [23]. The calculated results agree well with the field test results and the exact solution as shown in Fig.9.

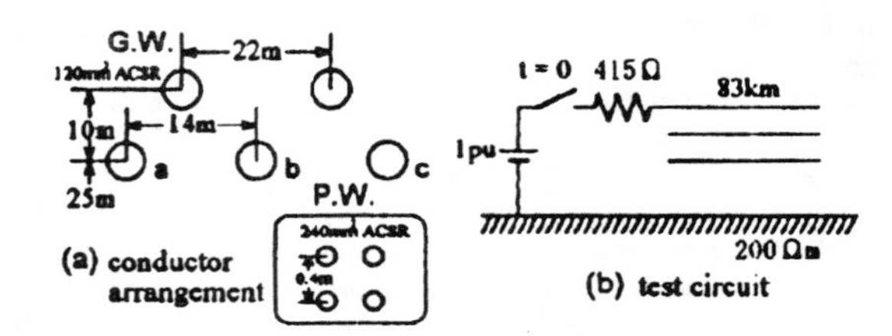

Fig.3 An untransposed horizontal overhead line

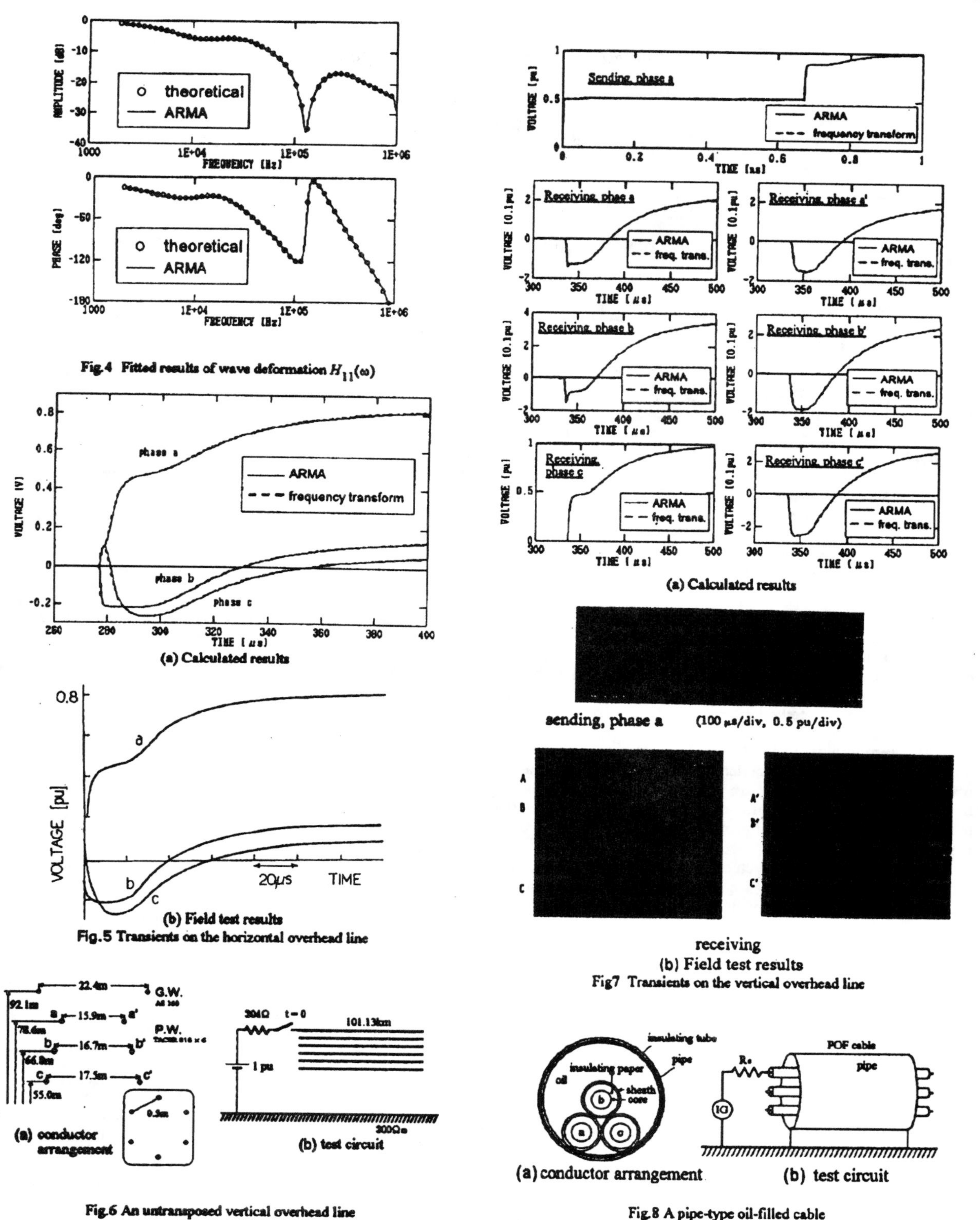

Fig.4 Fitted results of wave deformation $H_{11}(\omega)$

(a) Calculated results

(b) Field test results

Fig.5 Transients on the horizontal overhead line

(a) conductor arrangement

(b) test circuit

Fig.6 An untransposed vertical overhead line

(a) Calculated results

sending, phase a (100 μs/div, 0.5 pu/div)

receiving

(b) Field test results

Fig7 Transients on the vertical overhead line

(a) conductor arrangement

(b) test circuit

Fig.8 A pipe-type oil-filled cable

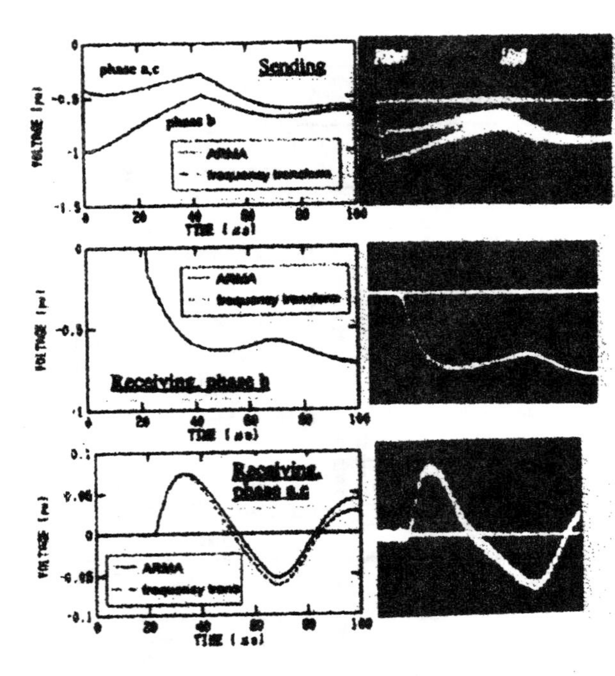

(a) Calculated results (b) Field test results

Fig.9 Transients on the pipe-type oil-filled cable

VI. CONCLUSIONS

1) This paper presents a method for representing frequency-dependent transmission lines in the phase domain rather than in the modal domain, for use in time domain transient calculations. This approach avoids convolution due to the modal transformation, and possible numerical instability due to mode crossing.

2) The approach is made efficient by replacing time domain convolution with ARMA model representation. Using the Z-operator, an ARMA model can economically express a phase domain response which has discontinuities due to modal propagation. A fast and stable method to identify the ARMA model is also developed.

3) Simulation results of actual transmission lines using the proposed method agree very well with exact frequency-domain simulations and field test measurements.

REFERENCES

[1] J.R.Carson, "Wave propagation in overhead wires with ground return," *Bell Syst. Tech., J.*, Vol. 5, pp.539-554, 1926.

[2] F.Pollaczek, "Über das Feld einer unendlich langen wechsel stromdurchflossen Einfachleitung," *E.N.T.*, Band 3 (Heft 9), pp.339-360, 1926.

[3] S.A.Schelkunoff, "The electromagnetic theory of coaxial transmission line and cylindrical shields," *Bell Syst. Tech. J.*, Vol. 13, pp.532-579, 1934.

[4] A.Ametani, "A general formulation of impedance and admittance of cables," *IEEE Trans.*, Power Apparatus and Systems, Vol. PAS-99 (3), pp.902-910, 1980.

[5] L.M.Wedepohl and S.E.T.Mohamed, "Multiconductor transmission lines : Theory of natural modes and Fourier integrals applied to transient analysis," *Proc. IEE*, Vol. 116, pp.1553-1563, 1969.

[6] W.S.Meyer and H.W.Dommel, "Numerical modeling of frequency-dependent transmission parameters in an electromagnetic transient program," *IEEE Trans.*, Power Apparatus and Systems, Vol. PAS-93, pp.1401-1409, 1974.

[7] A.Semlyen and A. Dabuleau, "Fast and accurate switching transient calculations on transmission lines with ground return using recursive convolutions," *IEEE Trans.*, Power Apparatus and Systems, Vol. PAS-94(2), pp.561-571, 1975.

[8] A.Ametani, "A highly efficient method for calculating transmission line transients," *IEEE Trans.*, Power Apparatus and Systems, Vol. PAS-95 (5), pp.1545-1549, 1976.

[9] W.D.Humpage, K.P.Wong, and T.T.Nguyen, "Z-transform electromagnetic transient analysis in power systems," *IEE Proc.*, Vol. 127, Pt. C, No.6, pp.370-378, 1980.

[10] A.Semlyen, "Contributions to the theory of calculation of electromagnetic transients on transmission line with frequency dependent parameters," *IEEE Trans.*, Power Apparatus and Systems, Vol. PAS-100 (2), pp.848-856, 1981.

[11] J.F.Hauer, "State-space modeling of transmission line dynamics via nonlinear optimization," *IEEE Trans.*, Power Apparatus and Systems, Vol. PAS-100(12), pp.4918-4925, 1981.

[12] J.R.Marti, "Accurate modelling of frequency-dependent transmission lines in electromagnetic transient simulations," *IEEE Trans.*, Power Apparatus and Systems, Vol. PAS-101 (1), pp.147-155, 1982.

[13] A.Ametani, "Refraction coefficient method for switching-surge calculations on untransposed transmission lines (Accurate and approximate inclusion of frequency dependency)," *IEEE PES Summer Meeting*, C 73-444-7, 1973.

[14] H.Nakanishi, and A.Ametani, "Transient calculation of a transmission line using superposition law," *IEE Proc.*, Vol. 133, Pt. C, No. 5, pp.263-269, 1986.

[15] G.Angelidis and A.Semlyen, "Direct Phase-Domain Calculation of Transmission Line Transients Using Two-Sided Recursions," *IEEE PES Summer Meeting*, 94 SM 465-5 PWRD, 1994.

[16] T.Noda and N.Nagaoka, "A time domain surge calculation method with frequency-dependent modal transformation matrices," *IEE Japan. Elec. Power Syst. Tech. Record*, PE-93-153, 1993.

[17] T.Noda and N.Nagaoka, "Development of ARMA models for a transient calculation using linearized least-squares method," *Trans. IEE Japan*, Vol. 114-B, No.4, 1994.

[18] T.Noda, T.Sawada, K.Fujii, and N.Nagaoka, "ARMA models for transient calculations and their identification (identification of coefficients and order)," *IEE Japan, Annual Meeting Record*, Paper No. 1373, 1994.

[19] J.S.Lim and A.V.Oppenheim, *Advanced Topics in Signal Processing*, Prentice-Hall, 1988.

[20] E.I.Jury, *Theory and Application of the z-Transform Method*, Wieley, New York, 1964.

[21] A.Ametani, T.Ono, and A.Honga, "Surge propagation on Japanese 500kV untransposed transmission line," *Proc. IEE*, Vol. 121, No.2, 1974.

[22] A.Ametani, E.Osaki, and Y.Honaga, "Surge characteristics on an untransposed vertical line," *Trans. IEE Japan*, Vol. 103-B, pp.117-124, 1983.

[23] N.Nagaoka, M.Yamamoto, and A.Ametani, "Surge propagation characteristics of a POF cable," *Trans. IEE Japan*, Vol. 105-B, pp.645-652, 1985.

[24] N.Nagaoka and A.Ametani, "A development of a generalized frequency-domain transient program--FTP," *IEEE Trans.*, Power Delivery, Vol. PWRD-

TABLE I ORDER COMPARISON

Proposed Phase Domain Method			Modal Domain Method [12]		
PHASE	$\exp(-\Gamma l)$	Y_0	MODE	$\exp(-\gamma l)$	Z_0
(1,1), (3,3)	9	4	1	12	22
(1,2), (3,2)	12	4	2	16	8
(1,3), (3,1)	11	4	3	21	27
(2,1), (2,3)	10	*			
(2,2)	10	4			

* these elements are identical to (1,2) and (3,2), because of the symmetry of Y_0

TABLE II COMPUTATION TIME COMPARISON

	Proposed Phase Domain Method	Modal Domain Method [12]
optimization	203.2 sec	374.2 sec
transient calculation	38.4 sec	32.6 sec

(IBM compatible computer 486SLC-50 + 387SX)

3(4), pp.1996-2004, 1988.
[25] H.W.Dommel, "Digital computer solution of electromagnetic transients in single- and multi-phase networks," *IEEE Trans.*, Power Apparatus and Systems, Vol. PAS-88 (4), pp.388-398, 1969.
[26] T.Yahagi, *Theory of Digital Signal Processing*, vol. 1-3, Corona Publishing, 1986.

ACKNOWLEDGMENTS

The authors are grateful to N.Mori for his assistance, and to L.Dubé, W.S.Meyer, and T.H.Liu for their valuable discussions.

APPENDICES

A. The Derivation of (3) *and* (4)

Eliminating $V_b(\omega)$ from (1) and (2), we obtain

$$Y_0(\omega)V(x,\omega)+I(x,\omega)=2Y_0(\omega)e^{-\Gamma(\omega)x}V_f(\omega). \qquad \text{(A-1)}$$

At the sending end (1, x=0) and at the receiving end (2, x=l), the following two equations are obtained from (A-1).

$$x=0: Y_0(\omega)V_1(\omega)+I_1(\omega)=2Y_0(\omega)V_f(\omega) \qquad \text{(A-2)}$$

$$x=l: Y_0(\omega)V_2(\omega)+I_2(\omega)=2Y_0(\omega)e^{-\Gamma(\omega)l}V_f(\omega) \qquad \text{(A-3)}$$

Equation (A-2) can be modified as

$$2V_f(\omega)=V_1(\omega)+Z_0(\omega)I_1(\omega). \qquad \text{(A-4)}$$

Substituting (A-4) into (A-3), we obtain (3). In the same manner, we obtain (4) by eliminating $V_f(\omega)$ from (1) and (2).

B. The Proof of $Y_0(\omega)H(\omega)=H^T(\omega)Y_0(\omega)$

Let A and B be a voltage and current modal transformation matrix respectively, and a symbol diag(a_k) denote a diagonal matrix with the k-th diagonal element a_k.

$$Y_0(\omega)e^{-\Gamma(\omega)l}=\{B\,\text{diag}(y_{0k})\,A^{-1}\}\{A\,\text{diag}(e^{-\gamma_k l})\,A^{-1}\}$$
$$=B\,\text{diag}(y_{0k})\,\text{diag}(e^{-\gamma_k l})\,A^{-1}=B\,\text{diag}(e^{-\gamma_k l})\,\text{diag}(y_{0k})\,A^{-1}$$
$$=\{B\,\text{diag}(e^{-\gamma_k l})\,B^{-1}\}\{B\,\text{diag}(y_{0k})\,A^{-1}\}$$

Using the relation B=$(A^{-1})^T$, B^{-1}=A^T,

$$=\{(A^{-1})^T\,\text{diag}(e^{-\gamma_k l})\,A^T\}\{B\,\text{diag}(y_{0k})\,A^{-1}\}$$
$$=\{A\,\text{diag}(e^{-\gamma_k l})\,A^{-1}\}^T\{B\,\text{diag}(y_{0k})\,A^{-1}\}$$
$$=\{e^{-\Gamma(\omega)l}\}^T Y_0(\omega).$$

Therefore, $Y_0(\omega)H(\omega)=H^T(\omega)Y_0(\omega)$.

C. Weighting Value for the Relative Error Evaluation

Let $N(z)$ and $D(z)$ be the numerator and denominator polynomial of a transfer function $G(z)$. The error function for the k-th datum of the Linearized LS Method is expressed as err'_k= $N(z_k)$–$D(z_k)G_k$. For relative error evaluation, the following equation is used for the weighting value of the k-th datum w_k.

$$\left|\frac{G(z_k)-G_k}{G_k}\right|=\sqrt{w_k}\left|N(z_k)-D(z_k)G_k\right| \qquad \text{(A-5)}$$

From the above equation, a weighting value for the relative error evaluation is $w_k=1/|D(z_k)G_k|$.

D. Stability Evaluation by Jury's Method

Let $D(z)=b_0+b_1z^{-1}+\cdots+b_Nz^{-N}$ be the denominator polynomial of an ARMA model. The following table called "Jury's table" denotes this method.

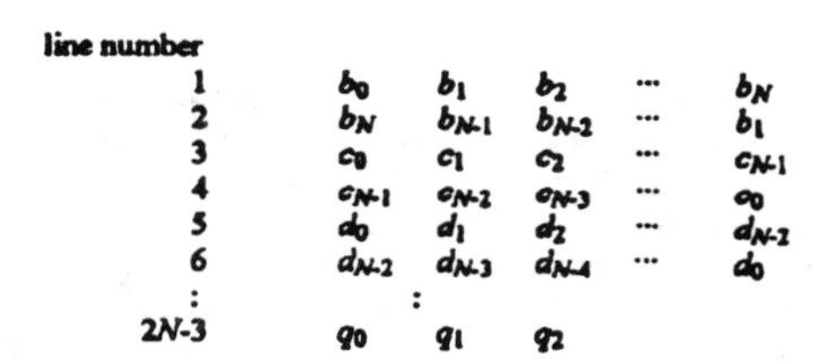

line number					
1	b_0	b_1	b_2	$\cdots$	b_N
2	b_N	b_{N-1}	b_{N-2}	$\cdots$	b_1
3	c_0	c_1	c_2	$\cdots$	c_{N-1}
4	c_{N-1}	c_{N-2}	c_{N-3}	$\cdots$	c_0
5	d_0	d_1	d_2	$\cdots$	d_{N-2}
6	d_{N-2}	d_{N-3}	d_{N-4}	$\cdots$	d_0
:		:			
2N-3	q_0	q_1	q_2		

where $c_k=b_0b_k-b_{N-k}b_N$, $d_k=c_0c_k-c_{N-k-1}c_{N-1}$, $\cdots$

Jury proved that the conditions $D(1)>0$, $(-1)^ND(-1)>0$, $b_0>|b_N|$, $c_0>|c_N|$, $d_0>|d_N|$, $\cdots$, and $q_0>|q_N|$ correspond to a stable ARMA model.

E. Evaluation of Modal Components

Phase domain wave deformation is expressed as $V_d(\omega)$=$H(\omega)$ $V_s(\omega)$, where $V_d(\omega)$: deformed voltage vector at distance l, $V_s(\omega)$: applied voltage at the sending end. Thus, an element of $H(\omega)$ can be expressed by the following equation (e_j : the j-th unit vector).

$$H_{ij}(\omega)=\{V_d(\omega)\}_{i\text{-th}},\quad \text{when } V_s(\omega)=e_j \qquad \text{(A-6)}$$

On the other hand, as the applied modal voltage is given as $V^M_s(\omega)$=$A^{-1}(\omega)V_s(\omega)$=$A^{-1}(\omega)e_j$, where $A(\omega)$ is the modal transformation matrix, the deformed modal voltage is expressed by the following modal propagation.

$$\{V_d^M(\omega)\}_{k\text{-th}}=\{A^{-1}(\omega)e_j\}_{k\text{-th}}e^{-\gamma_k l}=(a^{-1})_{kj}e^{-\gamma_k l} \qquad \text{(A-7)}$$

Transforming the above equation into the phase domain, the phase domain deformed voltage is expressed as

$$\{V_d(\omega)\}_{i\text{-th}}=\{A(\omega)V_d^M(\omega)\}_{i\text{-th}}=\sum_{k=1}^{n}a_{ik}(a^{-1})_{kj}e^{-\gamma_k l}. \qquad \text{(A-8)}$$

Substituting (A-8) into (A-6), we obtain

$$H_{ij}(\omega)=\sum_{k=1}^{n}a_{ik}(a^{-1})_{kj}e^{-\gamma_k l}. \qquad \text{(A-9)}$$

Consequently, it is possible to evaluate whether the k-th mode component is dominant or not in $H_{ij}(\omega)$ with (27).

BIOGRAPHIES

Taku Noda was born in Osaka, Japan, on July 4, 1969. He received the B.Sc. and M.Sc. degrees from Doshisha University, Kyoto, Japan in 1992 and 1994. Presently, he is a Ph.D. student at Doshisha University. Mr. Noda is a Member of IEE of Japan.

Naoto Nagaoka was born in Nagoya, Japan, on October 21, 1957. He received the B.Sc., M.Sc. and Ph.D. degrees from Doshisha University, Kyoto, Japan in 1980, 1982 and 1993. Presently, he is an Associate Professor at Doshisha University. Dr. Nagaoka is a Member of IEE of Japan.

Akihiro Ametani (M'71,SM'84,F'92) was born in Nagasaki, Japan, on February 14, 1944. He received the B.Sc. and M.Sc. degrees from Doshisha University, Kyoto, Japan, in 1966 and 1968, and the Ph.D. degree from University of Manchester, England in 1973. He was employed by Doshisha University from 1968 to 1971, the University of Manchester (UMIST) from 1971 to 1974, and also Bonneville Power Administration for summers from 1976 to 1981. He is currently a Professor at Doshisha University. His teaching and research responsibilities involve electromagnetic theory, transients, power systems and computer analysis. Dr. Ametani is a Fellow of IEE and a member of CIGRE and IEE of Japan and is a Chartered Engineer in the United Kingdom.

Discussion

Adam Semlyen (University of Toronto): I would like to congratulate the authors for their interesting contribution to the computation of electromagnetic transients. The most fundamentally interesting feature of the described methodology is the fact that the procedure, while fully reflecting the frequency dependence of the components involved and of the system, is entirely in the time domain. In principle, the discrete-time modeling adopted is the most natural representation of any linear subsystem for time domain simulation. Indeed, in this approach, just as in reference [15] of the paper, no frequency domain model has first to be built and then adequately approximated for time domain calculations.

The paper comes in close succession to our study [15] on the same topic. The computational methodology of the two papers is essentially the same but is designated by different names: *Two-Sided Recursions* (TSR) in [15] and *Auto-Regressive Moving Average* (ARMA) in this paper. The fact that these studies have been performed independently and almost simultaneously, with many strong and essential similarities both in the fundamental ideas and in some details, suggests that the new methodology (TSR or ARMA) is not simply a minor variation of existing methods but a well structured, rationally built procedure developed in response to widely perceived needs in the calculation of electromagnetic transients.

The main similarities in the two methods, in addition to the basic idea of using a small number of time domain samples in both input and output, are the following:

(a) The input-output relations are written in matrix form in the phase domain rather than for individual modes.

(b) The coefficients of the recursive process are identified directly in the frequency domain over a wide range of data points using robust linear algebra procedures for least squares problems.

(c) The requirement of testing for the stability of the procedures has been recognized, addressed, and satisfactorily solved.

There are of course also differences between the two methods. One, perhaps, pertains to the strong reliance in this paper on the z-transform itself which is not used in [15]. It is also noteworthy that the paper presents an application to the calculation of cable transients where the usefulness of the new method is most evident. The authors' comments on these issues would be highly appreciated.

Manuscript received February 22, 1995.

HUYEN V. NGUYEN, HERMANN W. DOMMEL and JOSE R. MARTI (The University of British Columbia, Vancouver, B.C., Canada) The authors should be congratulated for developing a sophisticated method which takes into account the complete frequency-dependent nature of untransposed transmission lines for time-domain programs such as the EMTP. The discussers would appreciate it if the authors could comment on the following issues:

1) Realizing the reduction in the number of convolution operations at each time step, the discussers have also been working in modelling the untransposed transmission lines directly in the phase-domain in a Ph. D. project since early of 1994. For modal parameters calculations, a diagonalization technique proposed in [A] has been used. In addition, the fitting is performed directly in the phase-domain using a modified version of the rational approximation method developed in [B]. Our experience has shown that the off-diagonal elements of the wave deformation matrix $\mathbf{H}(\omega)$ are not easily fitted. The difficulty increases for lines with strong asymmetry. We wonder if the authors have encountered the same experience. If possible, could the authors show the comparisons of the fitted and the exact functions of one of the off-diagonal elements (for example element 56) of their vertical line (Fig. 6), in both frequency- and time-domains. The graph of the elements of the wave deformation matrix in the time-domain might show the discontinuities caused by different modes as pointed out in section E of the paper. Incidentally, what is the typical value of ε_M in this section? Is it an arbitrary constant or is there a criterion for selecting it? Does it vary for different line geometries? For a typical element of the $\mathbf{H}(\omega)$ matrix of the vertical line in Fig. 6, how many q's are needed to represent accurately the travelling time differences caused by all the modes?

2) Have the authors applied their modelling technique for lines with strong asymmetry? An example is a six phase line where the left three phases are vertical while the right ones are horizontal. Another case which is very common in practice is when two or more different voltage lines share the same right-of-way and may be located one underneath the other.

3) Lastly, did the authors assume zero conductance value in their line shunt admittance matrix $\mathbf{Y}(\omega)$? If not, what was the typical value used for it?

[A] L.M. Wedepohl, H.V. Nguyen and G.D. Irwin, "Frequency-Dependent Transformation Matrices for Untransposed Transmission Lines using Newton-Raphson Method," Submitted for the 1995 IEEE Summer Power Meeting, Portland, Oregon.

[B] J.R. Marti , "Accurate Modelling of Frequency-Dependent Transmission Lines in Electromagnetic Transient Simulations," IEEE Trans. on Power Apparatus and Systems, Vol. PAS-101, pp. 147-157, 1982

Manuscript received February 28, 1995.

Thor Henriksen and **Bjørn Gustavsen** (Norwegian Electric Power Research Institute, Trondheim, Norway) : The authors are to be commended for presenting an efficient method for time domain calculation of transients on transmission lines and cables which takes into account the frequency dependency of the transfer function and the characteristic admittance. We have the following remarks and questions :

The paper deals with a general problem often encountered in transient calculations : How to include in time domain calculations an admittance or transfer function whose characteristics are known in the frequency domain? Numerical convolution can in principle always be applied, but the evaluation of the convolution integral can be very time consuming.

The above mentioned problem can be avoided by approximating the actual function in the frequency domain by an analytical function which gives the convolution integral a recursive formulation in the time domain. The authors have selected the ARMA function (17) which gives a two sided recursion formula (18). The latter seems to be identical to (4) in [15]. An alternative approach is to use a rational function (with the complex frequency s as variable). The coefficients of the polynomials can be considered as unknowns, which are found in a similar way as the coefficients in (22) in the paper. Another alternative is to use poles and zeroes as parameters.

The various approaches give differences in accuracy and computational requirements. In our opinion, differences in computational requirements are in most cases of minor importance.

Stability can be regarded as a part of the accuracy problem since the error due to the approximation in the frequency domain should be measured in the time domain. This error depends in general also on the actual transients. Have the authors compared their approach to the two alternative approaches regarding accuracy (and stability)? If so, have they also made the comparison for other applications than the ones presented in the paper?

Another point of general interest is the frequencies used when determining the parameters of the approximating function. The authors claim that these frequencies should be logarithmically spaced, but how many points are needed? Is it sufficient to use a relatively small number, or is it in general advisable to use as high number as possible? We expect that problems may arise if some elements of the transfer function contain contributions from modes having widely different velocities. (This will be the case when doing calculations on crossbonded cables as the sheath voltages cannot be neglected). Such elements will oscillate very fast in the frequency domain, and we therefore wonder what will happen if these oscillations are not resolved at high frequencies due to the use of logarithmically spaced samples.

The authors present an approach to this problem, given by (28). However, is seems to us that the numerical examples in the paper do not involve modes having widely different propagation velocities. Have the authors tried the ARMA model on such cases?

This problem could have been solved by separating the contribution from the different modes in the frequency domain, although it requires the eigenvectors to be calculated as smooth functions of frequency. The authours claim that mode crossing "enormously complicates practical implementation". However, it is according to our experience possible to overcome that problem by taking the direction of the eigenvectors into account.

Suppose that $H_{ij}(\omega)$ in (28) is decomposed in the frequency domain and that the ARMA model (28) is established separately for each mode. The coefficients b_n in eq. (28) would then be different for each mode. It is therefore suprising to observe that it is seems sufficient to use a low order as 5 in (28). Does this imply that the coefficients are forced to be the same for all modes? If so, could one expect to improve the accuracy by performing the above mentioned decomposition in the frequency domain?

The authors check the resulting ARMA model for stability by applying Jury's method. But what is done if Jury's method predicts instability?

In order to obtain a linearized Least Squares Method the authors optimize the fitting function using (22) instead of the original function G(z) (17). How does this affect the accuracy of the resulting fit?

The authors' comments would be appreciated very much.

Manuscript received February 28, 1995.

T.Noda, N.Nagaoka, and A.Ametani: The authors would like to thank the discussers for their interesting and useful comments.

<u>In reply to Prof. A.Semlyen :</u>

Because the proposed ARMA model methodology is based on the Z-transform theory, we were able to use two methods, which have been developed and fully investigated in the field of signal processing. One is the theory of AIC (Akaike's Information Criterion) for the order determination. and the other is the Jury's method for the stability evaluation. These methods have been approved to be useful and actually used in practice. Therefore, we were able to rely upon these methods and were also able to save time to develop reliable corresponding methods. Moreover, the agreements between calculation results and field tests shown in the paper approve that these methods are useful. Especially, tables I and II explain that the proposed order determination algorithm described in section IV-D is efficient by the combination use of SD (Standard Deviation), AIC, and the Jury's method.

The traveling time differences due to modal propagation should be modeled adequately, because the phase-domain propagation response of a transmission line consists of modal components. Especially, in case of a cable, it is almost impossible to model the traveling time differences by a sum of exponentials, namely, by a rational function of s in the frequency domain, because the traveling time difference between coaxial and sheath modes is considerable. To solve this problem, we proposed the modification of an ARMA model, in section IV-E, to express the phase-domain propagation response with an ARMA model, using that the operator z^{-1} is a delay operator. It should be noted that the modification is not a problem of the proposed linearized procedure of the least-squares parameter estimation. Therefore, it can be said that the usefulness of the proposed method is evident by the accuracy of cable transients shown in Fig. 7, as pointed out by the discusser.

<u>In reply to Mr. H.V.Nguyen, Prof. H.W.Dommel, and Prof. J.R.Marti :</u>

1) The proposed linearized least-squares (LS) method performs the fitting directly in the z-domain rather than in the Laplace s-domain. Therefore, operator z^{-n}, which denotes a delay of n samples, facilitates the approximation of the off-diagonal elements of the wave deformation matrix $H(\omega)$, which include discontinuities due to modal propagation as described in section IV-E. On the other hand, using a rational function of s, it is difficult to approximate an element of the off-diagonal elements, because a rational function of s essentially expresses a continuous response in the time domain. Moreover, the off-diagonal element is not always a minimum-phase function, although the rational fitting method developed in ref. [B] assumes the minimum-phase condition. On the other hand, the proposed linearized LS method does not assume the condition. Thus, we have no difficulty for fitting of the off-diagonal elements.

Fig. A shows the fitted results of an off-diagonal element $H_{56}(\omega)$ of the untransposed vertical overhead line described in section V-B. The figure shows (a) order determination procedure, namely, AIC and SD - order characteristic, (b) the fitted results in the frequency domain, and (c) step response in the time domain. Not only in the frequency domain but also in the time domain, the proposed linearized LS method shows high accuracy. Additionally, Table A shows the determined coefficients.

The value $|a_{ik}(a^{-1})_{kj} e^{-\tau_k^{ij}}|$ of eq. (27) indicates the ratio of the k-th mode in the (i, j) phase component. When $|a_{ik}(a^{-1})_{kj} e^{-\tau_k^{ij}}|$ is small, the corresponding discontinuity due to the k-th mode is small, and an independent q_k term in (28) is not necessary for an approximation of the small discontinuity. Therefore, ε_M determines the maximum error due to the modeling of traveling time differences. When a user gives the fitting error constant ε_A of the linearized LS process described in section IV-D, the modal error constant ε_M should be selected to a little greater value than ε_A, because the discontinuity which is determined not to have an independent q_k term by (27) is also approximately taken into account in the linearized LS process. ε_A = 1% and ε_M = 3% were used to perform the fitting of the three cases described in section V. Considering the above, ε_M depends only on the fitting error constant ε_A provided by a user.

In case of the untransposed vertical overhead line in section V-B, only one $q_k(-48)$ is used as indicated in Table A. Usually, in case of an overhead line, only one q_k is required, because the propagation velocity of the earth return mode is much smaller than the other modes, namely, aerial modes, and the velocities of the aerial modes are almost the same. On the other hand, in case of a cable, more q_k terms should be required.

2) Calculation examples in the text concern existing lines and cables in Japan, and we have not applied the proposed method to lines with strong asymmetry which seem not to be existing. It can be found that the asymmetry and a multi-circuit line do not make the fitting difficult, because of the reasons described in the above.

3) We assumed zero conductance value. Because the proposed method uses a least-squares fitting instead of a local asymptotic fitting, the zero conductance value does not affect the fitting procedure.

In reply to Dr.T.Henriksen and Dr. B.Gustavsen :

As explained in the discussion by Mr. Nguyen et al., it is very difficult to approximate the off-diagonal elements of the wave deformation matrix $H(\omega)$ with a rational function of s. Therefore, it is difficult to compare the accuracy of the proposed ARMA method with the rational function methods. But the time-domain accuracy of the proposed method can be confirmed by a comparison with results using the rigorous frequency domain simulation in Fig 5(a), Fig 7(a), and Fig. 9(a).

Fig. 8 shows a pipe-type oil-filled cable which involves widely different propagation velocities. Also, Fig. B(b) shows the fitted results of $H_{12}(\omega)$ of an underground single-core coaxial cable illustrated in Fig. B(a). The element $H_{12}(\omega)$ of the cable is a typical example which has widely different velocities. As pointed out by the discussers, the element oscillates very fast in the frequency domain as shown in Fig B(b). But the fitting by the proposed linearized LS method with a small number of samples is accurate enough both in the frequency domain and in the time domain as is clear in Fig. B(c). Our experience indicates that the number of samples is enough to be 10–20 points per decade. The reason, for the small number of samples, is that the

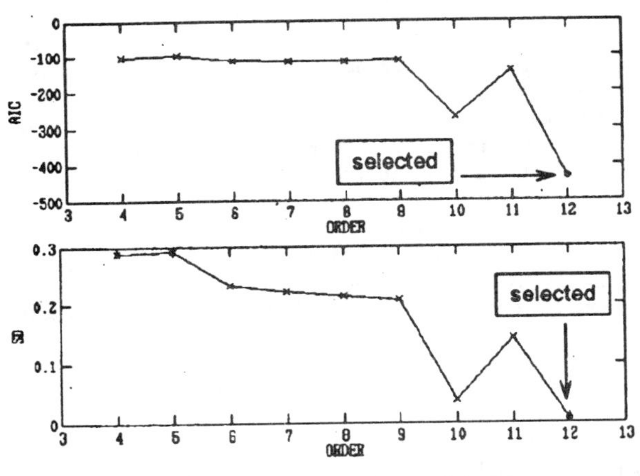

(a) AIC and SD - order characteristic

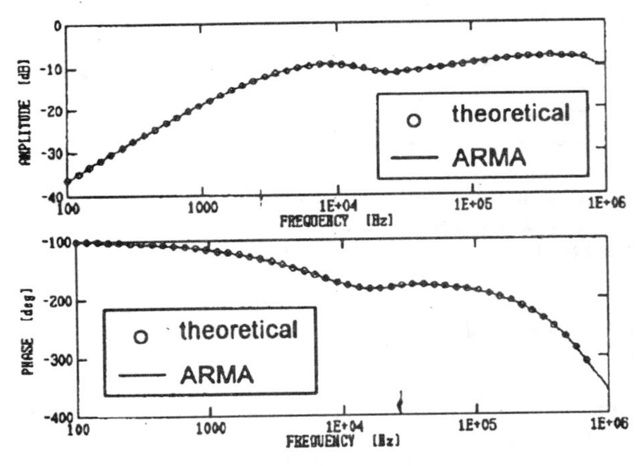

(b) Fitting results in the frequency domain

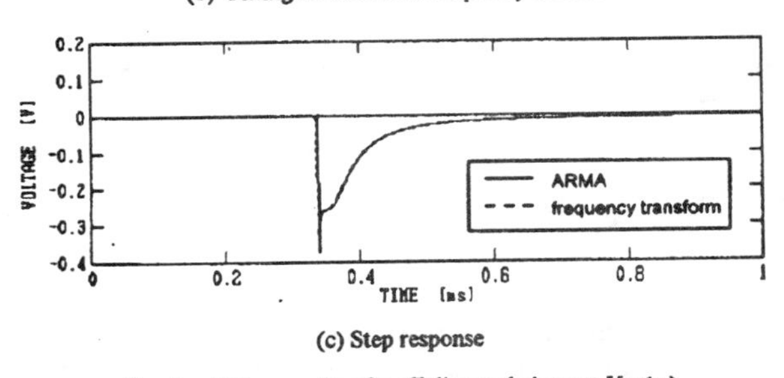

(c) Step response

Fig. A Fitting results of a off-diagonal element $H_{56}(\omega)$

Best Copy Available

TABLE A Coefficients of $H_{56}(\omega)$

Δt	0.5 μs
N	12 (order)
a_0	$1.419037627605408 \times 10^{-03}$
a_1	$-3.707989297151094 \times 10^{-01}$
a_2	1.817490407109451
a_3	-3.591784826682045
a_4	3.557198723072919
a_5	-1.763529792817600
a_6	$3.500055795163549 \times 10^{-01}$
a_{48}	$-3.073333098105003 \times 10^{-05}$
a_{49}	$-1.617214031477093 \times 10^{-04}$
a_{50}	$8.378241348680700 \times 10^{-04}$
a_{51}	$-1.270130798932017 \times 10^{-03}$
a_{52}	$8.236506522728622 \times 10^{-04}$
a_{53}	$-1.990873656694414 \times 10^{-04}$
b_1	-4.988118230362191
b_2	$1.011871989274092 \times 10^{+01}$
b_3	$-1.068433629862729 \times 10^{+01}$
b_4	6.332858647194881
b_5	-2.276268007811602
b_6	$6.209327978758516 \times 10^{-01}$
b_7	$-5.610115304335307 \times 10^{-02}$
b_8	$-1.923184241665601 \times 10^{-01}$
b_9	$2.204312368061560 \times 10^{-01}$
b_{10}	$-1.385472395859206 \times 10^{-01}$
b_{11}	$5.211864764644851 \times 10^{-02}$
b_{12}	$-9.371868655154460 \times 10^{-03}$

frequency-domain oscillating nature is inherently taken into account by the term z^{-qk} in (28).

The discussers suggest a fitting methodology, which performs the fitting for each mode. This methodology requires the convolutions due to modal transformation matrices. But it is pointed out that an element of the transformation matrices is not a causal function, which has a response in the negative time in ref. [A]. We doubt if it is possible to carry the convolution of a non-causal function in the time-domain simulation.

In the order determination process described in IV-D, the order is increased one by one until a best fit is found. Thus, an order evaluated to be unstable by Jury's method is not used, and the order is increased for the next fitting.

The influence using (22) instead of the original function is almost canceled by using a relative type of error evaluation described in section IV-C.

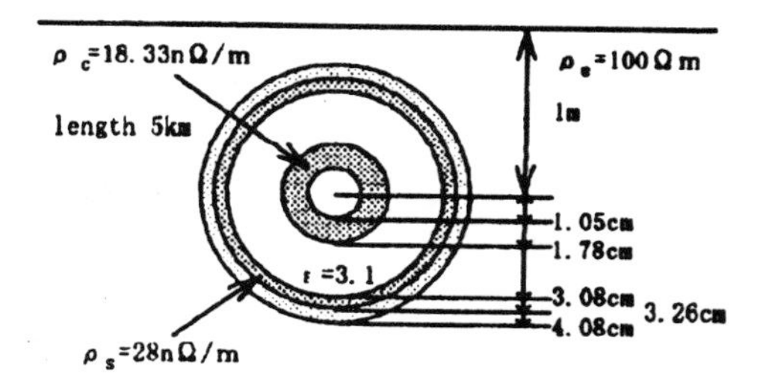

(a) An underground single-core cable

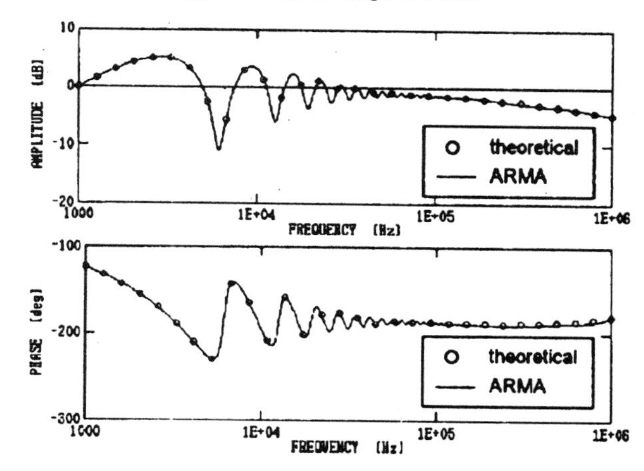

(b) Fitted results in the frequency-domain

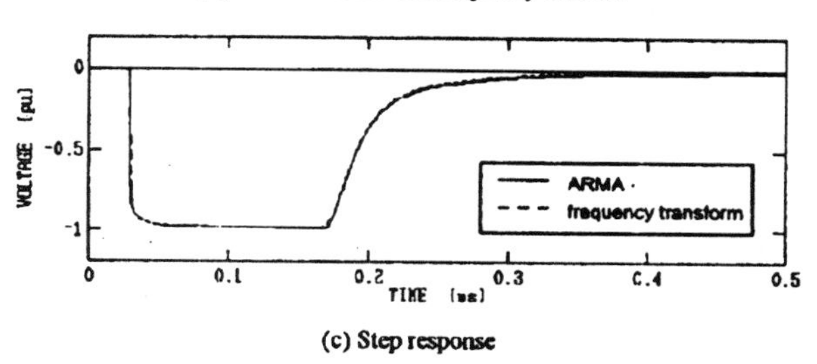

(c) Step response

Fig.B Fitted results of $H_{12}(\omega)$ of an underground single-core cable

[A] T.Ino and C.Uenosono, "An examination of highly accurate equivalent circuit for double-circuit DC transmission lines," *Trans. IEE Japan*, Vol. 114-B, No.5, 1994.

Manuscript received April 12, 1995.

EXPERIMENTAL EVALUATION OF A UHV TOWER MODEL FOR LIGHTNING SURGE ANALYSIS

T. Yamada, A. Mochizuki, J. Sawada, E. Zaima
Member
Tokyo Electric Power Company
Tokyo, Japan

T. Kawamura
Fellow
Shibaura Institute of Technology
Tokyo, Japan

A. Ametani
Fellow
Doshisha University
Kyoto, Japan

M. Ishii
Senior Member
University of Tokyo
Tokyo, Japan

S. Kato
Member
Toyo University
Saitama, Japan

ABSTRACT—An experimental investigation was performed on a UHV tower model for the EMTP multiconductor calculation of lightning overvoltage at substations associated with back-flashover at an adjacent transmission tower. The various lightning surge response characteristics were measured on an actual UHV tower, and parameters of a multistory transmission tower model that can reproduce voltages across the insulator strings, voltages of the crossarms, and voltages of the power lines were determined. A value of 120 Ω was determined as the surge impedance at each section of the multistory tower model, which closely agreed with the tower surge impedance measured for the UHV tower alone.

Key words: Lightning surge, EMTP, UHV, Multistory transmission tower model, Tower surge impedance, Back-flashover

INTRODUCTION

For economical insulation coordination in transmission and substation equipment, it is necessary to accurately predict the lightning surge overvoltage that occurs in an electric power system. Digital analysis techniques, such as the Electromagnetic Transients Program (EMTP)[1], have been used in recent years, and an important factor in such analysis is how to simulate transmission towers, transmission lines, substation equipment, and other components.

Experimental and theoretical research has been performed on the surge response characteristics of transmission towers and methods for modeling them, and there have been many reports on techniques for estimating tower surge impedance by analysis based on electromagnetic field theory[2]-[5]. Such research, however, approximates a tower as a cylinder or cone without specifying how to select the equivalent radius for the tower, and does not establish a method of calculating the surge impedance from the structure of the tower. Although a technique has been proposed by which the surge impedance can be calculated with a near-actual tower shape by numerical electromagnetic field analysis using antenna theory[6], it requires an enormous amount of computing time and is not suitable for practical surge analysis.

Thus, techniques have been proposed for measuring the surge characteristics of actual towers and modeling towers based on these measurements[7],[8]. The tower model study described here is also based on the results of actual measurements.

If the tower is represented as a traveling wave line, the basic parameters are its surge impedance, the traveling wave propagation velocity, and the attenuation and deformation characteristics of the traveling wave. Of these, the attenuation and deformation characteristics of the traveling wave determine the reflected wave from the tower base; that is, the wave tail after its peak in the tower's potential-rise waveform. It has been pointed out that if this is ignored and it is treated as a lossless line, like an equivalent circuit such as is standard in various countries, and if a nearly step-wave current flows into the tower, then, as soon as the reflected wave reaches the tower top, it is attenuated, and the wave tail does not change gently like the actually measured waveform[9]. When the tower is high, as is the case with a UHV transmission tower, a more precise representation is demanded in which the attenuation and deformation characteristics of the traveling wave are considered. Hence, what has been proposed are the multistory transmission tower model[8] and frequency-dependent tower model[10], both of which are based on actual measurements of the lightning surge response characteristics of 500 kV transmission towers[8]. In the former, the traveling wave line is represented as a parallel circuit of a resistance and an inductance connected to a lossless line, and in this model emphasis is mainly placed on reproduction of the voltages across the insulator strings[8]. In the latter, the tower, which has frequency characteristics in the traveling wave circuit elements themselves, is represented by a Semlyen model[10].

To predict the lightning surge overvoltage in UHV systems, which has great merit for economical insulation coordination, we measured the lightning surge response characteristics of an actual UHV tower and studied the multistory transmission tower model for a tower based on these measurements. The surge response characteristics that were modeled were voltages across the insulator strings, voltages of the crossarms, voltages of the power lines, and the voltage of the tower footing. Because a 1/70 μs ramp wave is used as the waveform of current injected onto the tower top in

94 WM 044-8 PWRD A paper recommended and approved by the IEEE Transmission and Distribution Committee of the IEEE Power Engineering Society for presentation at the IEEE/PES 1994 Winter Meeting, New York, New York, January 30 - February 3, 1994. Manuscript submitted July 30, 1993; made available for printing December 6, 1993.

lightning surge analysis[9], in considering the modeling of a tower, response waveforms of the 1/70 μs ramp-wave current calculated from the measured waveforms using Laplace transforms were taken to be fundamental. We also considered the effect of the positioning of wires to assist in taking measurements, and this was reflected in the tower modeling.

LIGHTNING SURGE CHARACTERISTICS OF UHV TRANSMISSION TOWER

Measurement method

Measurements were performed on the No.3 tower (tower height 140.5 m), equipped with ground wires and power lines, of the UHV Nishi-Gunma double-circuit line of TEPCO, as shown in Fig. 1. Fig. 2 shows the experimental setup for the measurements. For this purpose, a pulse generator installed at the top of the tower injected step-wave currents onto the top of the tower through a high resistance (2.5 kW), and voltages across the insulator, voltages of the crossarms, voltages of the power lines, and voltage of the tower footing were measured from the ground using optical transducers by the direct method. The current lead wire and the auxiliary potential wire were held up by balloons so as to be kept parallel to the ground and perpendicular to the transmission lines, thereby minimizing their inductive coupling with the tower and transmission lines. The voltage of the tower footing was also measured using an auxiliary potential wire positioned 1 m above the ground at a central position in the tower, with each tower footing crossed with copper wire. In taking the measurements, the power lines were grounded on the adjacent No.2 and No.5 towers, and the ends of the current lead wire and auxiliary potential wire were matched with a resistance (about 750 Ω, including the grounding resistance) equivalent to their surge impedance.

The pulse generator used for the measurements had a rise time of 20 ns. Copper wires having a radius of 1 mm were used for the current injection wire and auxiliary potential wire, the voltage were measured by high-resistance probes, and the current was measured by a current transformer. For the waveforms, a digital recorder capable of recording four channels simultaneously was used, and twenty-times averaging was applied to suppress the effects of noise and create the resulting waveforms.

Measurement results

Fig. 3(a) shows the measured waveforms of the injected current, voltages across the insulator strings, voltages of the crossarms, voltages of the power lines, and voltage of the tower footing.

Fig. 4 shows the maximum values of the voltages at each part of the tower and of voltages across the insulator string. The numerical values were calculated by using Laplace transforms to determine the response waveforms with respect to the step-wave current (Fig. 3(b) from the measured waveforms by a technique described in the next section. The voltage at each part of the tower showed a roughly linear relationship with the height. The voltage of the tower footing, indicating the surge characteristics of the grounding resistance, showed roughly flat characteristics.

Fig. 1 Tower for measurements (UHV Nishi-Gunma line No.3)

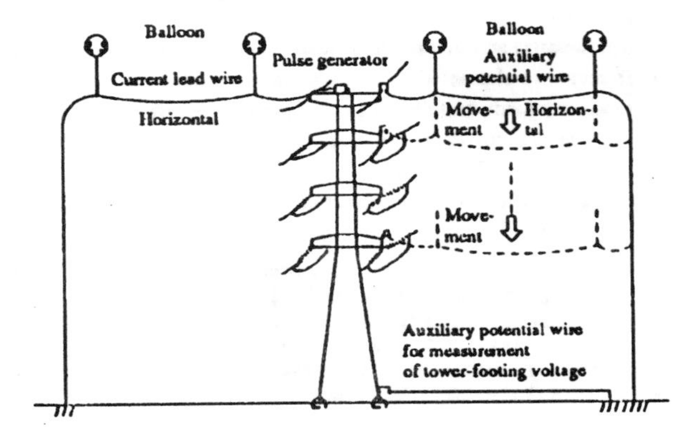

Fig. 2 Experimental setup for measurements

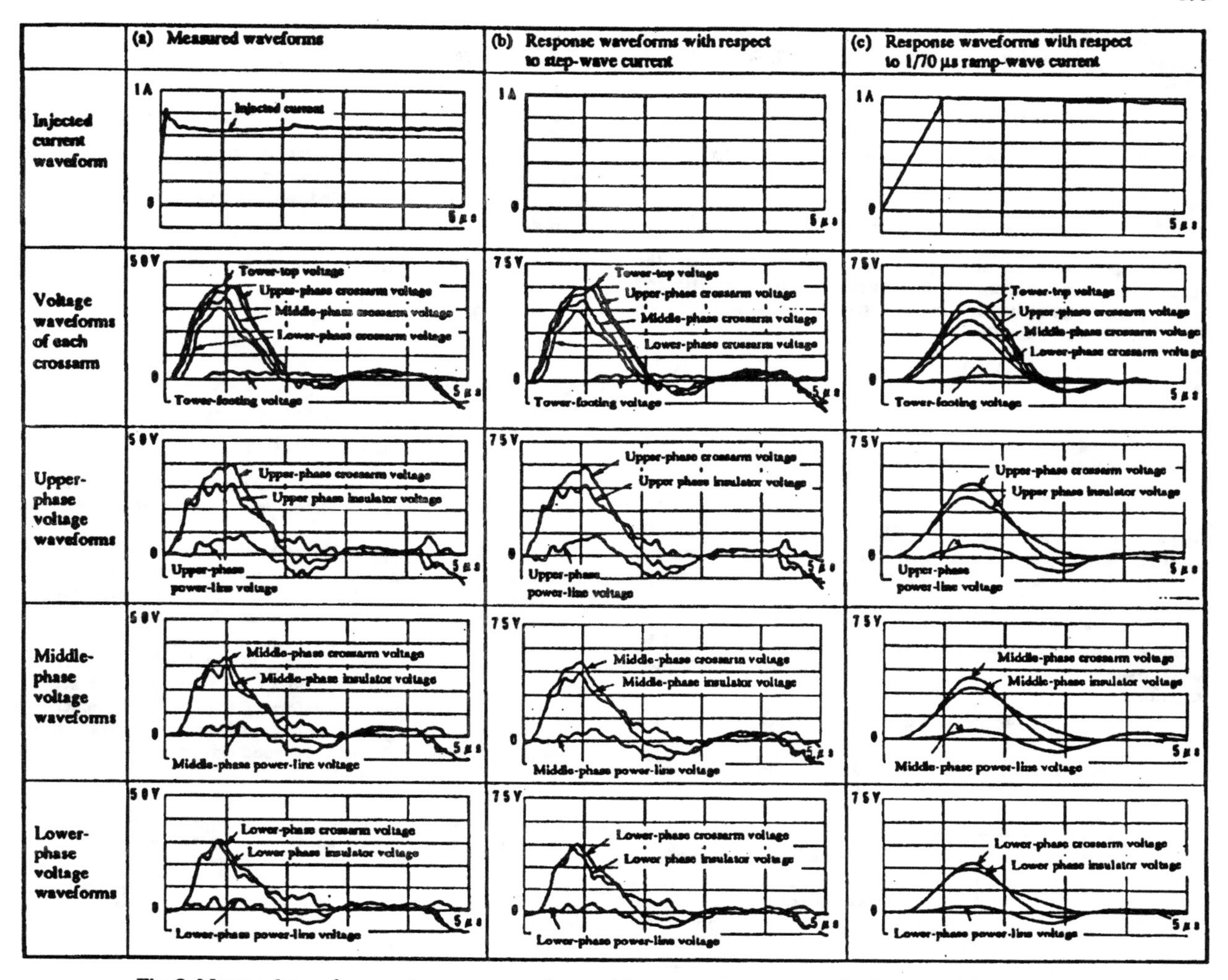

Fig. 3 Measured waveforms and response waveforms with respect to step-wave and 1/70 μs ramp-wave currents

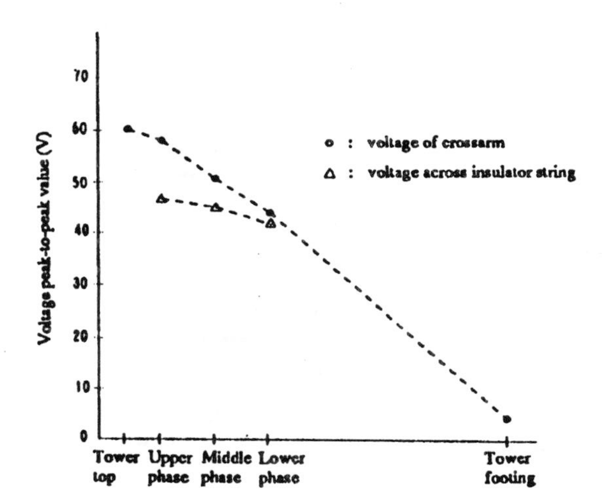

Fig. 4 Voltages at each part of tower

EVALUATION OF MULTISTORY TRANSMISSION TOWER MODEL

The current lead wire was arranged so as to be perpendicular to the tower and transmission line, but also had to be arranged in a straight line with the auxiliary potential wire used for measuring the voltages of the crossarms and power lines. Thus, due to the effects of induced voltage, the measured voltages of the crossarms and power lines were lower than the actual values.

We therefore studied a multistory transmission tower model that would enable the waveforms to agree with the voltages across the insulator string, eliminating the need for a voltage measurement auxiliary line, and the peak-to-peak value to agree with the voltages of the crossarms and power lines while taking the induced voltage into consideration.

Response waveforms with respect to 1/70 μs ramp-wave current

Because a 1/70 μs ramp wave is used as the injected current waveform in lightning surge analysis for UHV systems, the voltage response waveforms with respect to this current waveform were taken to be fundamental for the study of tower modeling as well, and these waveforms were calculated from the measured waveforms using Laplace transforms.

Letting $i_o(t)$ be the measured injected current waveform and $v_o(t)$ be the measured voltages waveform, and defining their frequency responses as $i_o(s)$ and $v_o(s)$, respectively, we have the following formulas.

$$v_o(s) = H(s)\, i_o(s) \quad (1)$$

where $i_o(s) = L\{i_o(t)\},\ v_o(s) = L\{v_o(t)\}$ (2)

and $s=\alpha+j\omega$: Laplace operator
L : Laplace transform
H(s) : transfer function

Now, representing an arbitrary current waveform as i(t) and the voltage response waveform for this current waveform as v(t), and defining their respective frequency responses as i(s) and v(s), we have the following formulas.

$$v(s) = H(s)\, i(s) \quad (3)$$

where $i(s) = L\{i(t)\},\ v(s) + L\{v(t)\}$ (4)

By applying an inverse Laplace transform to formula (3), we obtain the voltage response waveform i(t) by the following formula.

$$v(t) = L^{-1}\{H(s)\, i(s)\} \quad (5)$$

If the current waveform i(t) is made into a step wave and a 1/70 µs ramp wave, we can calculate voltage response waveforms with respect to the step-wave current and the 1/70 µs ramp-wave current, respectively.

The numerical Laplace transform was calculated by the following formula with a piece-wise linear Laplace transform[11], in which a piece-wise linear Fourier transform[12] is modified into a Laplace transform.

$$F(s)=L\{f(t)\}=\{f(t_o)-f(t_n)\exp(-s\cdot t_n)\}\cdot[1/s+\{1-\exp(s\cdot Dt)\}/(s_2 Dt)] + [\{\exp(-s\cdot\Delta t)+\exp(s\cdot\Delta t)-2\}/(s^2\Delta t)]\cdot \sum_{t=1}^{N} f(t_{i-1})\cdot \exp(-s\cdot t_{i-1}) \quad (6)$$

when $n = N+1,\ t_o=0,\ t_{n-1} = t_N = T_{max},\ \Delta t = T_{max}/N$

and N : sampling number
T_{max} : maximum calculation time

The final term in formula (6) can be calculated by a fast Fourier transform.

The numerical inverse Laplace transform was calculated with the following formula by a fast Laplace transform[13] that combines a Laplace transform with a window function and a fast Fourier transform.

$$f(n\cdot\Delta_t) = L^{-1}\{F(s)\} = Re[(1/\pi)\cdot\exp\{(a\cdot\Delta t+j\pi/N)\cdot n\}\cdot \sum_{k=1}^{N}\{F_k\cdot W^{(k-1)n}\}] \quad (7)$$

where $F\pi = F\{(2K-1)\Delta\omega-j\alpha\}\sigma_k\cdot 2\Delta\omega$,
$\sigma_k = \sin\{(2K^{-1})\pi/2N\},\ w = \exp(j2\pi/N)$,
$\Delta t = T_{max}/N,\ \Delta\omega = \pi/T_{max},\ \alpha = 2\pi/T_{mas}$

Fig. 3(c) shows the response waveforms with respect to the 1/70 µs ramp-wave current calculated from the actually measured waveforms of (a).

We verified the appropriateness of this Laplace transform technique by the voltage response waveforms corresponding to the measured 1/60 µs exponential function injected current. Based on transfer functions calculated from the actually measured waveforms with respect to the step-wave current, we determined the voltage response waveforms corresponding to the 1/60 µs exponential function injected current, then compared them with the actually measured voltage response waveforms. Fig. 5 shows comparisons of the tower-top voltage, and voltages of the upper phase. It was confirmed that these voltages show good agreement in terms of both peak-to-peak value and waveform.

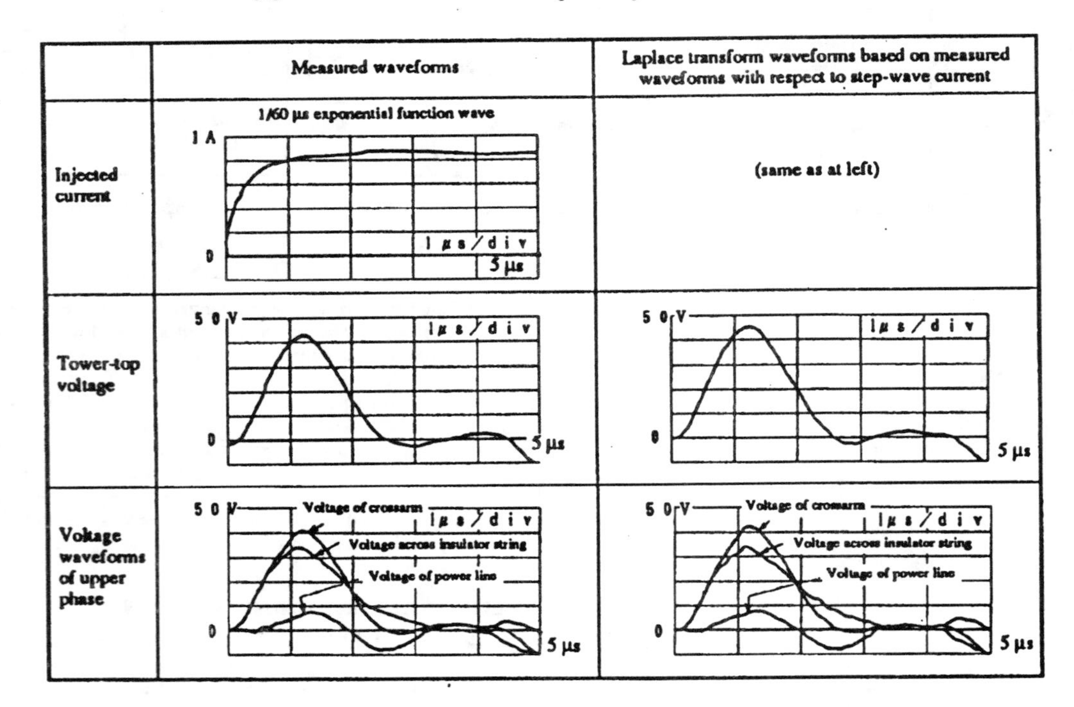

Fig. 5 Example of verification of appropriateness of Laplace transform

Effect of arrangement of current lead wire and auxiliary potential wire

Regarding the effect of the current lead wire and the auxiliary potential wire, it has been confirmed by experiments on cylindrical and cylindrical-cone models as well as by electromagnetic field analysis that there is an approximately 20 V/A difference in surge impedance between a perpendicular arrangement and straight arrangement[14].

We investigated the effect of the current lead wire and the auxiliary potential wire on a scale model. Fig. 6 shows the experimental setup of the scale model. The tower model was a scale model of a 500 kV transmission tower, with a height of 180 cm and a spacing between the tower base of 36.6 cm. A coaxial cable which was kept parallel to the ground was connected to the top of the scale model from a pulse generator on the ground. The voltage at the top of the tower was measured due to the position of the auxiliary potential wire varying from a perpendicular arrangement (90°) to a straight arrangement (0°) with respect to the current lead wire. Fig. 7 shows the injected current waveform, and Fig. 8 shows the tower-top voltage. The surge impedance was simply calculated by dividing the maximum voltage by the injected current at the moment when the voltage reached its maximum value. The surge impedance of the straight arrangement was approximately 16 V/A lower than that of the perpendicular arrangement. This is thought to be because, in the straight arrangement, the magnetic flux created by the injected current of the current lead wire interlinks with the auxiliary potential wire loop, and the resulting induced voltage reduces the measured voltage.

In addition, without ground wires, the upper-phase power line was arranged perpendicularly to the current lead wire and the power line voltage was measured in the perpendicular arrangement. In the absence of ground wires, almost no power line voltage should arise. However, induced voltage having a peak value of -14 V/A was measured, as shown in Fig. 9, and this is equivalent to the difference between the straight arrangement and the perpendicular arrangement.

Also, by electromagnetic field analysis using the moment method[6], we calculated how the difference between the straight and perpendicular arrangements varies with the arm position. The results of an analysis performed for a cylinder of 6 mm in diameter and 180 cm in height are shown in Table 1. The induced voltage was 15 V/A at the top of the tower, and tended to decrease with decreasing height.

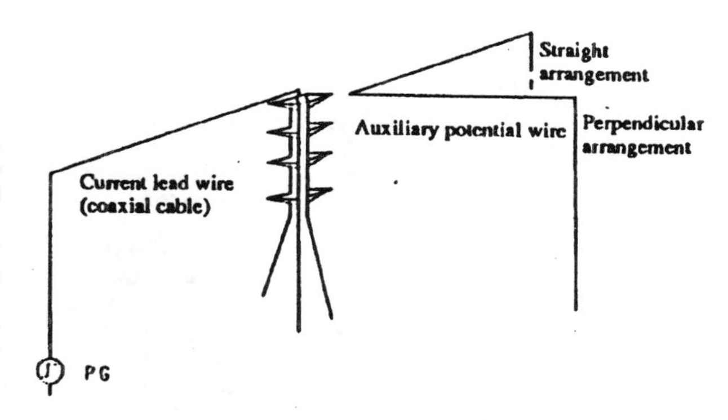

Fig. 6 Experimental setup of scale model

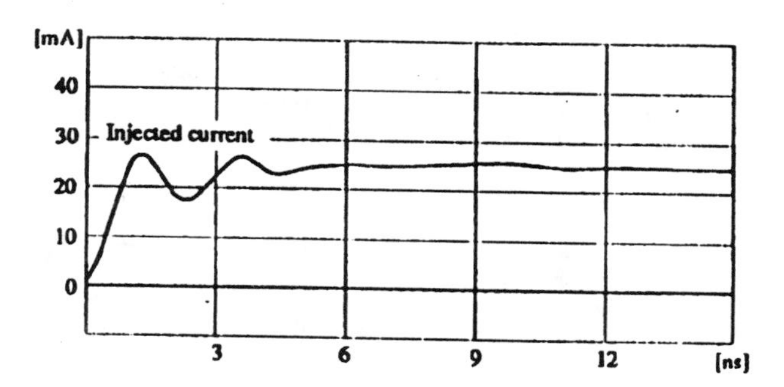

Fig. 7 Injected current waveform

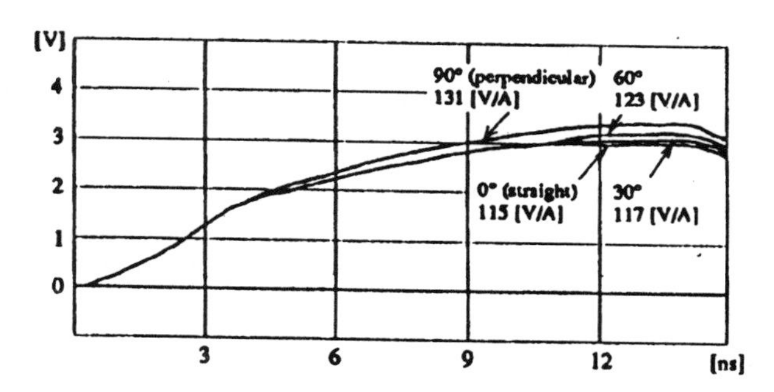

Fig. 8 Changes in tower-top voltage waveform according to arrangement of auxiliary potential wire

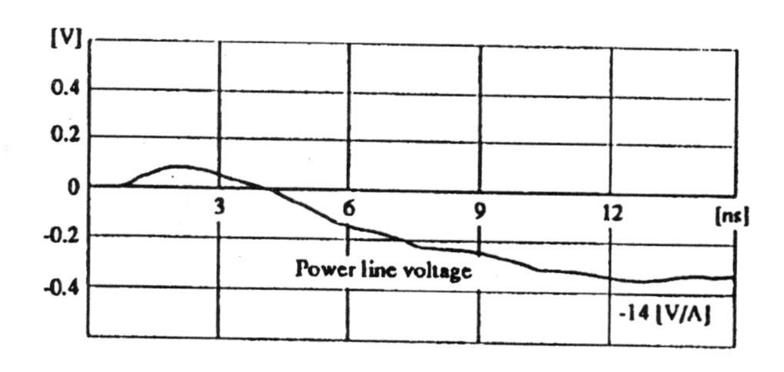

Fig. 9 Measured voltage waveform of power line

Table 1 Comparison of surge impedance between straight arrangement and perpendicular arrangement

(unit: V/A)

Height above ground [cm]	Straight arrangement	Perpendicular arrangement	Difference
180	345	330	15
150	330	320	10
120	315	305	10
90	300	290	10
60	280	266	14
30	220	215	5

EMTP analysis conditions

We studied the multistory transmission tower model for a UHV transmission tower in the multiconductor circuit shown in Fig. 10. The detailed analysis conditions are described below.

Tower model

Fig. 11 shows a multistory transmission tower model[8]. In the present study the tower arms have not been simulated. Because the surge response characteristic of the tower footing voltage shown in Fig. 3 was roughly flat, it was decided to simulate the grounding resistance with pure resistances, and resistance values measured by the voltage drop method (No. 2: 5.5, No. 3: 2.0, No. 4: 2.0, No. 5: 1.9 Ω) were set.

Transmission model

There are two types of EMTP line models, Semlyen model and J. Marti model, with frequency response characteristics. We adopted the J. Marti model for our calculation because this model can be easily used. The transmission conditions were as follows.

Conductor configuration	:	As shown in Fig. 12, the average height above the ground was chosen based on the undulations of the ground and the dip of the lines at 15°C with no wind blowing.
Ground wire	:	OPGW 500 mm^2, 2 wires
Power line	:	ACSR 810 mm^2, 8 conductors
Span between towers	:	No. 2-3: 854 m, No. 3-4: 613 m, No. 4-5: 627 m
Ground resistivity	:	30 Ωm
Calculation frequency	:	1 MHz
Ends	:	Power lines grounded at No. 2 and No. 5

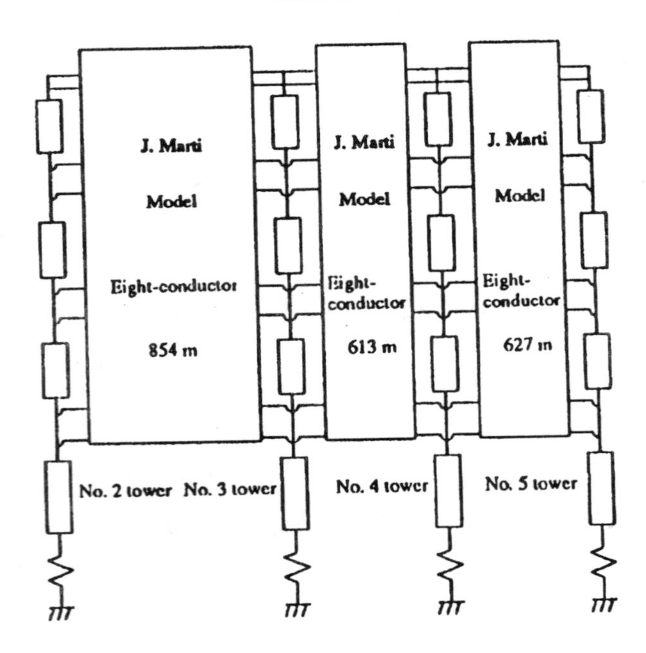

Fig. 10 Equivalent circuit for EMTP analysis

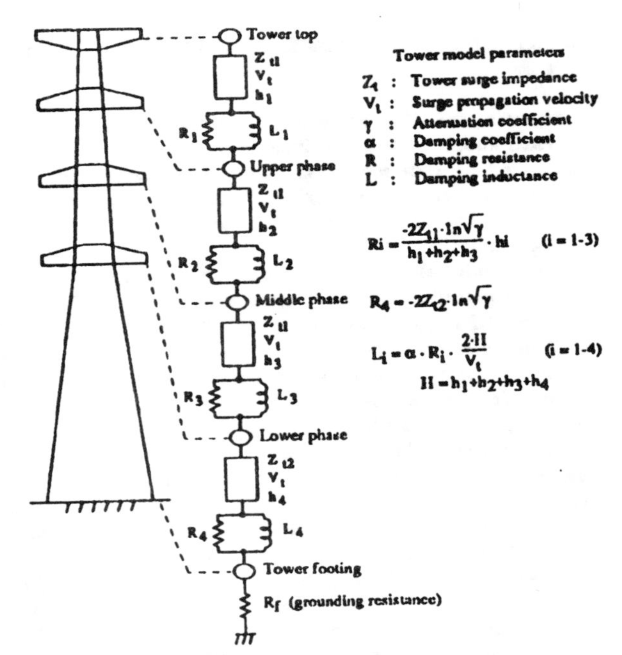

Fig. 11 Multistory transmission tower model

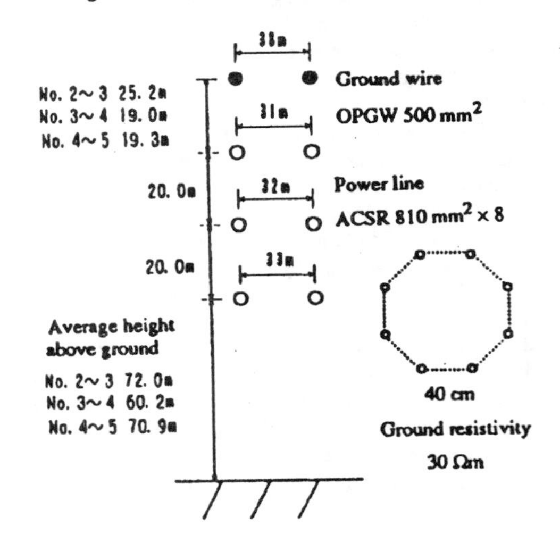

Fig. 12 Conductor configuration

Power source model

The experimental setup for the current lead wire is shown in Fig. 13, while Fig. 14(a) shows this circuit in the form of an analysis circuit for EMTP. The current source is determined by the measured current I. In the analysis, J was set so that I was a 1/70 µs ramp wave.

Because the point where the current is injected is the middle of the tower-top arm, there is static coupling between the current lead wire and the tower-top arm. To reflect this effect in the analysis, a capacitance was inserted in parallel with the resistance which was connected with the current wire, as shown in Fig. 14(b). The capacitance between the two conductors (circuit lead wire radius: 1 mm, tower-top arm equivalent radius: 1.0 m, spacing: 2.8 m, length: 20 m) was set to 150 pF.

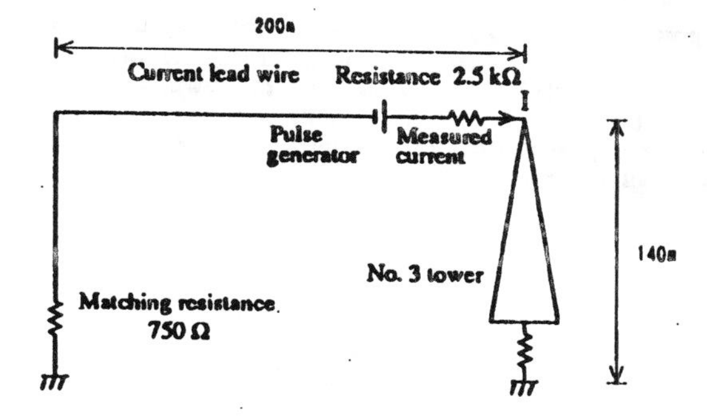

Fig. 13 Experimental setup for current lead wire

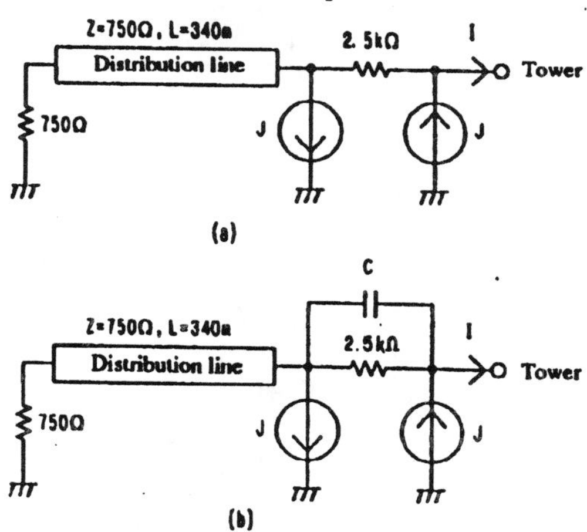

Fig. 14 Power source model

Parameters for multistory transmission tower model.

The parameters for the multistory transmission tower model were determined so as to fit the waveforms of the voltages across the insulator strings (crest value, virtual front time, virtual time to half value) and the crest values of the crossarm voltages and power line voltages taking into account the induced voltage of the auxiliary potential wire. As a result, the following parameters exhibited good agreement between the measured waveforms and the waveforms obtained by EMTP analysis, as shown in Fig. 15 and Table 2.

- Surge propagation speed : $V = 300$ m/μs
- Tower surge impedance : $Z_{t1} = Z_{t2} = 120\ \Omega$
- Attenuation coefficient : $\gamma = 0.7$
- Damping coefficient : $\alpha = 1$

The speed of light was taken as the surge propagation speed because (1), it is difficult to specify, from the measured waveform, the exact point of arrival of the reflected wave and (2), the physical speed of propagation of the current within a tower is the speed of light.

As Table 2 shows, the actual measured values for the voltages of the crossarms and power lines were 20-15 V/A less than the calculated values. This was due to the effect of the arrangement of the auxiliary potential wire mentioned above. It is thought that if this effect is taken into account, the results of measurement and of analysis will agree for the voltages of the crossarms and power lines as well.

Up to now, the following parameters have been proposed as multistory transmission parameters, based on actual measurements at a 500 kV transmission tower: $V=300$ m/μs, $Z_{t1}=Z_{t2}=220/150\ \Omega$, $\gamma=0.8$, $\alpha=1$[8]. If these parameters are used in calculations for a UHV tower, the tower-top voltage becomes quite high, as shown in Table 2. Because the actually measured valued of the surge impedance of the tower alone is 119 Ω, the above parameters slightly overstate the tower-top voltage. The multistory transmission tower model places great importance on reproducibility of the peak voltages across the insulator strings for injected current having a steep wavefront. Regarding the insulator voltages, there is not much difference between the two models. However, if we analyze the lightning overvoltage at substations associated with back-flashover at an adjacent transmission tower, the current dividing ratio between the tower and the ground wire is important and, in this sense, the proposed model is more reasonable.

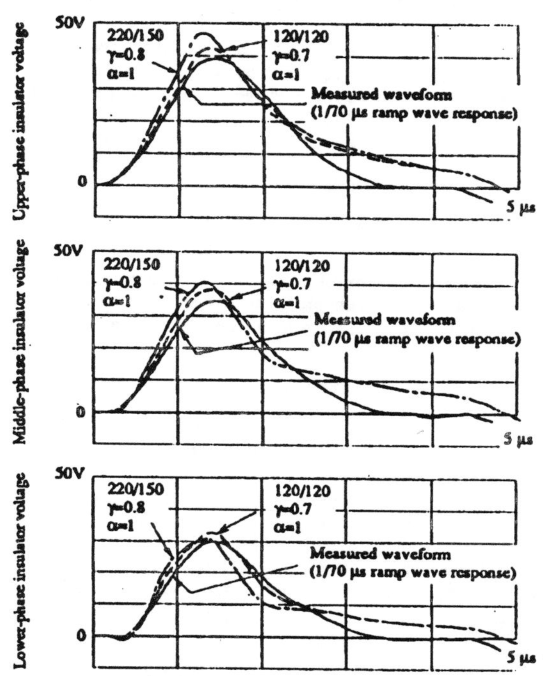

Fig. 15 Comparison of waveforms obtained by actual measurement and by analysis

Table 2 Comparison of values obtained by actual measurement and by analysis

(unit: V)

		Actually measured value	Zt1/Zt2= 120/120 Ω	Difference	Zt1/Zt2= 220/150 Ω
Tower-top voltage		53.3	72.5	-19.2	84.8
Upper phase	Crossarm voltage	48.3	67.4	-19.1	77.0
	Insulator voltage	39.2	41.6	- 2.4	46.9
	Power line voltage	8.6	25.9	-17.3	30.2
Middle phase	Crossarm voltage	41.1	58.0	-16.9	63.0
	Insulator voltage	34.9	39.1	- 4.2	41.0
	Power line voltage	5.3	19.1	-13.8	22.3
Lower phase	Crossarm voltage	34.1	47.4	-13.3	47.3
	Insulator voltage	30.6	33.7	- 3.1	31.5
	Power line voltage	4.2	13.9	- 9.7	16.2
Tower footing voltage		3.8	1.9	1.9	1.9

TOWER SURGE IMPEDANCE

Although it is different from measurement of a tower equipped with ground wires and power lines, we measured the surge characteristics of a tower alone using the No. 118 tower (tower height 120.5 m) of the UHV Nishi-Gunma double-circuit line, as shown in Fig. 16. Fig. 17(a) shows the injected current waveform and the tower-top voltage waveform. From these measurements we calculated the step response waveform by the Laplace transform technique mentioned earlier. The tower surge impedance was 126 Ω, as shown in Fig. 17(b). However, due to restrictions in taking the measurements, the angle between the current lead wire and the auxiliary potential wire was about 130°, so that from the results of the scale model experiments mentioned earlier, with a perpendicular arrangement it is estimated that this measurement would be approximately 130 Ω.

For the tower that we measured in the present study, we calculated the tower surge impedance using formulas proposed in recent years. The average surge impedance recommended by IEEE[15] and CIGRE[16] was calculated from the following formula, in which the tower is simulated by an inverted cone.

$$Z = 60 \cdot \log_e \cot\{0.5 \cdot \tan^{-1}(R/H)\} \qquad (8)$$

where R: equivalent radius of the tower
H: tower height

R in the above formula is calculated by dividing the tower into upper and lower truncated cones as shown in Fig. 18, and equivalently replacing them with a cone as defined by the following formula.

$$R = (r_1 h_2 + r_2 H + r_3 h_1)/H \qquad (9)$$

where r_1 : tower-top radius, m
r_2 : tower-midsection radius, m
r_3 : tower-base radius, m
h_1 : height from base to midsection, m
h_2 : height from midsection to top, m

For the tower that we measured (r_1=2.5 m, r_2=4.0 m, r_3=10.8 m, h_1=61.5 m, h_2=59 m), calculating the average surge impedance by formula (8) yielded Z=187 Ω (R=10.7 m), which is 60 Ω higher than the actually measured result.

A calculation formula has been proposed for determining the impedance of vertical conductors[17] based on Neumann's inductance formula and Deri's complex penetration depth[18]. This can be approximated by the following formula which is the same as Jordan's formula for surge impedance calculation[2].

$$Z = 60\{\log_e(H/R) - 1\} \quad (\text{where } R \ll H) \qquad (10)$$

The tower equivalent radius R in the above formula is determined by equivalently replacing the tower with a cylinder, as is done when the calculation is made by equivalently replacing the tower with a cone. R is defined by the following formula.

$$R = (r_1 h_2 + r_2 H + r_3 h_1)/(2 \cdot H) \qquad (11)$$

For the tower that we measured, calculating the surge impedance using formula (10) yields Z=126 Ω (R=5.4 m), which agrees closely with the actual measured value.

For the measured tower equipped with ground wires and power lines (tower height 140.5 m, r_1=2.5 m, r_2=4.0 m, r_3=13.0 m, h_1= 81.5 m, h_2=59 m), if we calculate the surge impedance using formula (10), which agrees closely with the actual measured value, we obtain Z=126 Ω (R=6.3 m), which agrees closely with the surge impedance of the tower as measured for the tower alone.

The surge impedance of a UHV transmission tower was approximately 130 Ω, which was almost equal to the surge impedance of the parameters of the multistory transmission model. This model shows good agreement for tower-top as well as insulator voltage, and is suitable for analyzing lightning overvoltage at substations associated with back-flashover at an adjacent transmission tower.

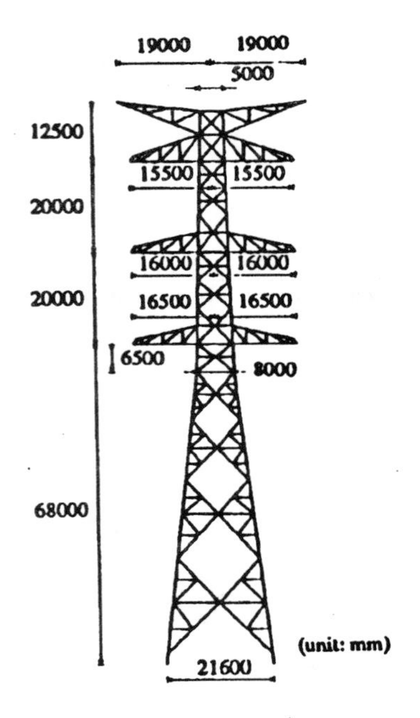

Fig. 16 Structural drawing of UHV tower (No. 118, Nishi-Gunma line)

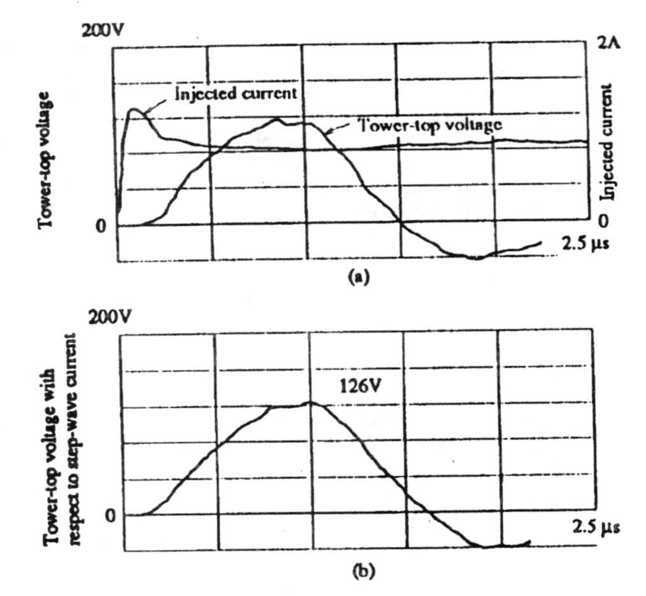

Fig. 17 Lightning surge response of UHV tower alone

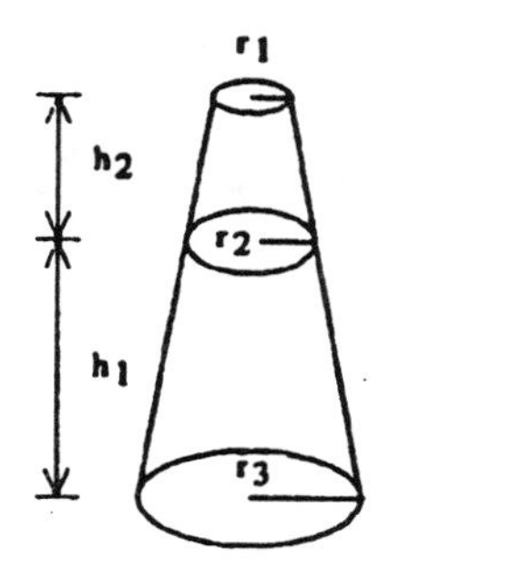

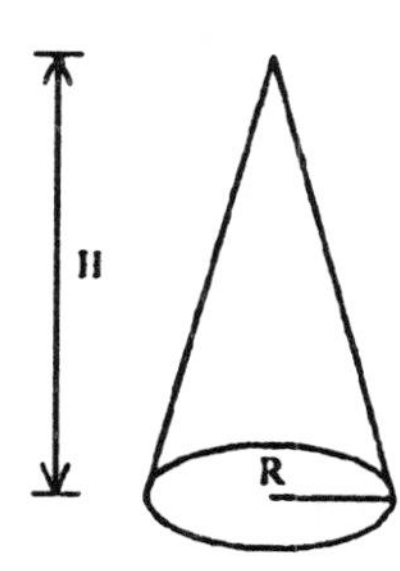

Fig. 18 Tower equivalent radius

CONCLUSIONS

This paper reports on the results of measurements of the lightning surge response characteristics of a UHV transmission tower, and the evaluation of a UHV tower model for lightning surge analysis based on these measurements.

(1) For a UHV transmission tower equipped with ground wires and power lines, we measured the lightning surge response characteristics of the voltages across the insulator strings, the voltages of the crossarms and power lines, and the tower footing voltage.
(2) From the measurements we calculated the response waveform corresponding to a 1/70 μs ramp-wave current using Laplace transforms, and used it as the basis for modeling.
(3) We considered how the arrangement of the auxiliary potential wire affects the voltages of the crossarms and power lines.
(4) The parameters for the multistory transmission tower model for a UHV tower based on the actually measured waveforms were $V=300$ m/μs, $Zt1=Zt2=120\ \Omega$, $\gamma=0.7$, $\alpha=1$.
(5) The surge impedance for a UHV tower was approximately 130 Ω, which was almost the same as the tower surge impedance of the parameters of the multistory transmission tower model.
(6) This model provides good agreement for tower-top voltage as well as voltages across the insulator strings, and is suitable for the analysis of lightning overvoltage at substations associated with back-flashover at an adjacent transmission tower.

REFERENCES

[1] Bonneville Power Administration, "Electromagnetic Transients Program (EMTP), Reference Manual, Mode 39," Portland, Oregon, May 1984
[2] C.A. Jordan, "Lightning Computations for Transmission Lines with Overhead Ground Wires Part II", G.E. Rev., 37, 1934
[3] R. Lundoholm, et al., "Calculation of Transmission Line Lightning Voltages by Field Concepts," AIEE Trans. Vol. 77, Feb. 1958, pp. 1271-1283
[4] C.F. Wagner and A.R. Hileman, "A New Approach to the Calculation of the Lightning Performance of Transmission Lines, III-A Simplified Method Stroke to Tower," AIEE Trans. Vol. 79, 1960, pp. 589-603
[5] M.A. Sargent and M. Darreniza, "Tower Surge Impedance," IEEE Trans. PAS-88, No. 5, 1969, pp. 680-687
[6] K. Kato et al., "Electric Field Analysis of Vertical Conductor for Steep Front Wave Voltage," 7th ISH 83.14, 1991, pp. 151-154
[7] M. Kawai, "Studies of Surge Response on a Transmission Line Tower," IEEE Trans., PAS-83, 1964, pp. 30-34
[8] M. Ishii et al., "Multistory Transmission Tower Model for Lightning Surge Analysis," IEEE Trans. PWRD Vol. 6, July 1991, pp. 1327-1335
[9] "Parameters for Lightning Surge Analysis in Electric Power Systems, and Their Effects," IEE Japan Technical Report (II), No. 301, 1989 (in Japanese)
[10] N. Nagaoka, "Development of a Frequency-Dependent Tower Model," IEE Japan B-111, 1991, pp. 51-56 (in Japanese)
[11] N. Nagaoka and A. Ametani, "A development of a generalized frequency-domain program—FTP," IEEE Trans. PWRD, Vol. 3, Oct. 1988, pp. 1996-2004
[12] L.M. Wedepohl and S.E.T. Mohamed, "Transient analysis of multiconductor transmission lines with special reference to nonlinear problem," Proc. IEE, Vol. 117, 1970, pp. 979-987
[13] A. Ametani, "The application of the fast Fourier transform to electrical transient phenomena," Int. J. Elect. Eng. Educ., Vol. 10, 1973, pp. 277-281
[14] "New Evaluation Method for Lightning Surges at Substations," IEE Japan Technical Report (II), No. 446, 1992 (in Japanese)
[15] "Estimating Lightning Performance of Transmission Lines II—Updates to Analytical Models," IEEE Working Group Report, 92 SM 453-1 PWRD
[16] "Guide to Procedure for Estimating the Lightning Performance of Transmission Lines," CIGRE SC33-WG01 Report, Oct. 1991
[17] A. Ametani et al., "A Frequency-Dependent Impedance of Vertical Conductors and a Multi-Conductor Tower Model," IEE GTD 93-2019, 1993
[18] A. Deri et al., "The complex around return plane: a simplified model for homogeneous and multi-layer earth return," IEEE Trans. PAS-100, No. 8, 1981, pp. 3686-3693

ACKNOWLEDGMENT

We are grateful to Messrs. Matsubara and Motoyama of the Central Research Institute of the Electric Power Industry for their cooperation in taking UHV tower surge measurements.

BIOGRAPHIES

Takeshi Yamada was born in Hokkaido, Japan on February 20, 1962. He received his B.S. and M.S. degrees from Hokkaido University in 1984 and 1986, respectively. He joined the Tokyo Electric Power Company in 1986, and has been engaged in design and research related to transmission lines.

Azuma Mochizuki was born in Yamanashi Prefecture, Japan on June 24, 1959. He received his B.S. and M.S. degrees from Tokyo Science University in 1982 and 1984, respectively. He joined the Tokyo Electric Power Company in 1984, and has been engaged mainly in substation engineering and insulation coordination.

Jun Sawada was born in Hokkaido, Japan on May 14, 1955. He received his B.S. and M.S. degrees from Nihon University in 1979 and 1981, respectively. He joined the Tokyo Electric Power Company in 1981, and has been engaged in design and research related to transmission lines.

Eiichi Zaima (M'90) was born in Kumamoto Prefecture, Japan on October 28, 1949. He received his B.S. degree in 1974 from the University of Tokyo and his M.S. degree in 1990 from the Massachusetts Institute of Technology. He joined the Tokyo Electric Power Company in 1974, and has been engaged mainly in substation engineering and insulation coordination.

Tatsuo Kawamura (M'59, SM'89, F'91) was born in Tokyo, Japan on August 16, 1930. He received his B.S., M.S. and D.Eng. degrees from the University of Tokyo in 1954, 1956 and 1959, respectively. He is a professor at the Shibaura Institute of Technology and also Professor Emeritus at the University of Tokyo. He has been involved mainly in the fields of HV and electric power engineering.

Akihiro Ametani (M'71, SM'83, F'92) was born in Nagasaki Prefecture, Japan on February 14, 1944. He received his B.S. and M.S. degrees from Doshisha University in 1966 and 1968, respectively, and his Ph.D. degree from the University of Manchester in 1973. He is a professor at Doshisha University. He has been involved mainly in transient analysis and computer analysis.

Masaru Ishii (SM'87) was born in Tokyo, Japan on March 11, 1949. He received his B.S., M.S. and D.Eng. degrees from the University of Tokyo in 1971, 1973 and 1976, respectively. He is a professor of the Institute of Industrial Science, University of Tokyo. He has been involved mainly in the field of HV.

Shohei Kato (M'77) was born in Aichi Prefecture, Japan on September 12, 1948. He received his B.S. degree from Yokohama National University in 1971, and his M.S. and Ph.D. degrees from the University of Tokyo in 1973 and 1976, respectively. He is a professor at Toyo University. He has been involved mainly in the fields of HV and numerical analysis of electromagnetic fields.

Discussion

W.A. CHISHOLM (Ontario Hydro Technologies): The discusser compliments the authors on an interesting experimental program. He wishes to raise the following points regarding measurement technique, and also regarding interpretation of the tower surge impedance.

MEASUREMENT TECHNIQUE

The authors are reporting measurements of tower-top, crossarm and phase-conductor voltages that have a significant defect. The Newi [A] probe technique for high-voltage measurement is used. A high-resistance arm is connected in series with an auxiliary potential wire (APW), which is expected to have a constant impedance over ground. The expected output voltage would be given by the ratio of APW surge impedance to the sum of APW impedance and probe resistance. However, the results provided by a Newi probe are invalid for approximately two tower travel times. The electromagnetic fields from the APW have no way of "knowing" that the ground plane and the resulting image APW are present, so the initial impedance of the APW will be its antenna impedance. This will increase to the surge impedance value within two travel times, and the result will be a slowly-increasing probe output that eventually reaches the correct value.

This defect of the Newi probe technique can be easily verified in the authors' model studies. The APW should be applied to a single conductor over a ground plane, and a step current then injected. The current and voltage waveform risetimes should be identical. Measurements with the APW, however, will give the illusion of a degraded voltage risetime. This illusion will be similar to the voltage waveform found in the author's Figure 8, in the presence of a tower. Measurements of potential difference from the driven to an adjacent, undriven conductor, will not have this problem. For this reason, potential difference from tower to phase conductor, converted to an optical signal [8] or routed to avoid coupling [B], will provide satisfactory estimates of insulator stress. As the authors point out in their work on probe routing, under transient conditions, the measured voltage depends on the path of measurement, while potential differences will remain unique.

INTERPRETATION

The authors make several statements about tower surge impedance, based on the observed tower-top voltage using the Newi probe technique. The questions raised above suggest that the authors are under-estimating the tower-top voltage during the initial surge. However, the authors are also comparing the surge impedance estimated for a round tower, with no crossarms, to a surge impedance measured on a square tower, with crossarms. Reference [16] discusses the issue of round versus square towers, and recommends that the impedance be reduced by a factor of $\sqrt{\pi/4}$, based on the differences in surface area. Also, measurements on a tower with crossarms should give a lower surge impedance, combined with a longer tower travel time, according to previous work [C]. Could the authors report on any surge impedance estimates obtained on their model towers, without crossarms? Are these in closer agreement with the estimates from expressions in [15] and [16]?

[A] G. Newi, "A High-Impedance Nanosecond Rise Time Probe for Measuring High-Voltage Impulses", IEEE Trans. PAS-87 No.9, September 1968.

[B] P.G. Buchan and W.A. Chisholm, "Surge Impedance of HV and EHV Transmission Towers", Final Report for Canadian Electrical Association Contract 78-71.

[C] W.A. Chisholm, Y.L.Chow and K.D. Srivastava, "Travel Time of Transmission Towers", IEEE Trans. PAS-104 No.10, October 1985.

Manuscript received March 9, 1994.

T. Yamada, A. Mochizuki, J. Sawada, E. Zaima, T. Kawamura, A. Ametani, M. Ishii and S. Kato: The authors would like to thank Dr. W. A. Chisholm for his interest in our paper.

Measurement Technique

The method by Newi is equivalent to the measurement of the current flowing into the APW. On the other hand, we have measured the voltages between the APW and the crossarms by potential dividers and optical transducers. Then, the method we have used is different from the Newi's method.

According to the discusser's interpretation, the risetime of the voltage waveform is degraded by the antenna impedance of the APW within two travel times. The slow risetimes of the measured voltage waveforms, however, were also found in the case of the measurements of the insulator voltages without the APW in Fig. 3 of our paper. The slow risetime is thought to be caused by the surge impedance of the tower which is dependent on time initially.

Interpretation

The method we have used is not the Newi's method, so we believe the tower top voltage we have measured is not underestimated.

The equivalent radius of the tower was calculated following the formula (9) in our paper, and the radius at each part was defined as the radius of the circle which is inscribed in the cross section at each part, resulting in the surge impedance for the No. 118 UHV tower of 187 Ω (R = 10.7m). If the radius at each part is defined as the radius of the circle which is circumscribed around the square section at each part, the tower surge impedance would be 166 Ω (R = 15.1m). The ratio of 166 versus 187 agrees closely with the reduction factor of $\sqrt{\pi/4}$ which you indicated. However, the tower surge impedance of 166 Ω is still higher than the actually measured result.

According to a model experiment [D], the wave shapes of the tower top voltage, measured in a similar way as ours, were different between pole towers with and without a crossarm, but the crest values were almost the same. Therefore, the effect of the crossarms on the tower surge impedance is not yet clear.

Reference

[D] T. Hara et al., "Lightning Surge Response of A Cylindrical Tower with A Crossarm", Sixth International Symposium on High Voltage Engineering, New Orleans, paper 27.12, August, 1989.

Manuscript received April 18, 1994.

IEEE Transactions on Power Delivery, Vol. 11, No. 2, April 1996

COMPUTATION OF ELECTROMAGNETIC TRANSIENTS ON NONUNIFORM TRANSMISSION LINES

M.T. Correia de Barros, Senior Member, IEEE

M.E. Almeida

IST-Universidade Tecnica de Lisboa / Instituto da Energia-INTERG
1096 Lisboa Codex, Portugal

Abstract **- A nonuniform transmission line model is presented, based on a finite-differences algorithm to solve the wave equations for any space variation of the line parameters. This line model is interfaced with EMTP and used for representing transmission line towers. Results are obtained for comparison with other solution methods, showing a good performance of the proposed model. Computer simulation of experimental tests is performed, and a good agreement between the simulation results and the published experimental results is found, the tower losses being considered as a frequency-constant nonuniform parameter.**

I. INTRODUCTION

An increasing interest is being paid to the transient analysis of power network elements representable by nonuniform transmission lines, in particular aiming at determining the lightning surge response of transmission line towers. Indeed, experimental results clearly show that the surge impedance of a transmission tower varies as the lightning surge travels along it [1-4], and the need for better tower models exists. Furthermore, other elements in a power network also present distributed parameters having space variation, such as the connection between an overhead line and an underground cable.

For evaluating the solution of the single-phase lossless nonuniform line, an adaptation of Bewley's reflection lattice method has been proposed by Menemenlis and Chun [5]. This adaptation is achieved by attributing distributed reflection and refraction coefficients to each point along the nonuniform transmission line.

The same problem has been solved by Saied et al using a Laplace transform based method [6, 7]. The single-phase lossless nonuniform line is modeled as a cascade of exponential transmission lines.

So far, to the authors' best knowledge, no solution has been proposed for including the transmission line losses, even when representing them as frequency-constant parameters. Furthermore, the need exists of having a nonuniform transmission line model suitable for use with general purpose programs, such as the Electro-Magnetic Transients Program - EMTP.

For modelling the transmission line tower in lightning performance studies, a multistory model has been proposed by Ishii et al [3, 4], suitable to be used in power system analysis by EMTP. The multistory tower is based on the individualization of the regions between crossarms, and consists of a series connection of lossless uniform lines with lumped R//L circuits. The nonuniform distributed parameters of the tower are not considered as such, but are approximated by attributing different values to the parameters of the different stories.

A particular tower model has been previously presented by the authors, corresponding to a lossless transmission line along which the surge impedance varies by discrete steps, following a geometric progression [8].

In this paper, a general nonuniform transmission line model is presented. Distributed nonuniform losses are considered. At the present stage, their frequency dependence is not taken into account.

II. TRANSMISSION LINE MODEL

The algorithm used to model the nonuniform transmission line is based on a finite-differences approximation to the partial derivatives in the wave propagation equations.

Considering a transmission line divided into N incremental segments, the partial differential equations describing wave propagation are represented by the following set of $2(N+1)$ equations in the time variable:

95 SM 397-0 PWRD A paper recommended and approved by the IEEE Transmission and Distribution Committee of the IEEE Power Engineering Society for presentation at the 1995 IEEE/PES Summer Meeting, July 23-27, 1995, Portland, OR. Manuscript submitted January 3, 1995; made available for printing June 7, 1995.

$$\begin{aligned} & h_0(u_0,i_0,u_0^{\cdot},i_0^{\cdot}) = 0 \\ & a_j = u_{j+1} - u_j + \Delta u_{j+1}(i_j,i_j^{\cdot}) = 0 \qquad j = 0,\ldots,N-1 \\ & b_j = i_j - i_{j-1} + \Delta i_j(u_j,u_j^{\cdot}) = 0 \qquad j = 1,\ldots,N \\ & h_N(u_N,i_N,u_N^{\cdot},i_N^{\cdot}) = 0 \end{aligned} \tag{1}$$

where $u^{\cdot}$ and $i^{\cdot}$ denote the time derivatives of u and i. The functions Δu_j and Δi_j (with $j=1,\ldots,n$) represent the longitudinal voltage drop and the transverse current on segment j, respectively, and contain in their expressions the nonuniformity of the line parameters. The boundary conditions are given by the implicit functions h_o and h_N. The segment equations are also kept in their general implicit form to keep the algorithm independent from the particular line parameter modelling techniques to be used in applications.

The time equations are solved using a multistep linear integration formula. The implemented algorithm gives to the user the option between the backward Euler and Gear's 2nd order methods. The very nature of the travelling wave problem implies that the system matrix has a band-diagonal structure, thus allowing fast recursive formulas to be used in evaluating its solution.

Typically, the discretization of wave propagation equations gives rise to system's supporting lightly damped oscillating modes at a high frequency determined by the spatial stepsize. In the Appendix it is shown that these can be smoothed out by the time integration algorithm by making an adequate choice of the integration formula, on one hand, and of the space and time steps, on the other. Trapezoidal integration, because of its lack of filtering, has proven inadequate for this purpose. Both the Backward Euler's and the Gear's 2nd order methods introduce an increasing damping for higher frequencies. However, accuracy is usually impaired by the introduction of damping. The diagrams in Appendix show that Gear's 2nd order, for a given bandwidth and a constant time-step, remains accurate enough over a wider range of frequencies. Therefore, among the considered integration methods, it is the most suitable to be used for the numerical filtering of the spurious oscillations introduced by the space-discretization of the transmission line equations.

Evaluation of the transmission line parameters is made by independent routines of any suitable type, which may implement simple equivalent circuits, as well as suitable physical models. The above described line model allows to consider any variation law of the line parameters. It also is adequate to include non-linear as well as frequency-dependent parameters. The latest by using, for instance, Foster circuits.

The boundary conditions may be evaluated by dedicated routines or by a general purpose program, such as EMTP.

III. RESULTS

The proposed nonuniform transmission line model has been interfaced with the ATP version of the EMTP, as a Type-94 element, using the simulation language MODELS. This approach enables considerable savings in computation time and memory requirements, as it allows the external program to take advantage of the sparsity of the distributed line equations matrix, and to use a local time step different from the time step of the circuit solution. Typically, for a 0.2 µs rise time surge, 0.2-meter space-discretization segments and a time step of 2 ns are adequate for modelling a transmission tower as a nonuniform line using the proposed algorithm.

In the applications presented in this paper, a simple representation by R_j, L_j and C_j elements has been considered for the nonuniform line parameters, the longitudinal voltage drop and transverse current per incremental line segment being given by:

$$\begin{aligned} & \Delta u_j = R_j i_j + L_j i_j^{\cdot} \\ & \Delta i_j = C_j u_j^{\cdot} \end{aligned} \tag{2}$$

and the space variation of the parameters being specified by the user.

Two examples have been selected for the validation of the model and its interface with the EMTP, illustrating the electromagnetic transients along a transmission tower hit by direct lightning.

A. Menemenlis and Chun test

The example presented in [5] by Menemenlis and Chun to illustrate the application of their travelling-wave method has also been used by Saied et al to validate their Laplace-transform based line model [6, 7]. Results obtained by the two methods agree, except in the wave tail of the voltage at the tower top, which shows negative values in the results presented by Menemenlis and Chun.

The 75-meter tower is modeled as a nonuniform lossless line with a characteristic impedance given by the equation:

$$Z_0(x) = k_1 + k_2\sqrt{x} \tag{3}$$

where $k_1 = 50\,\Omega$, $k_2 = 35\,\Omega m^{-1/2}$, and x is measured from the ground level. The tower footing is represented by a 10 Ω resistance. The lightning stroke current is considered a 0.2/25 µs impulse with 30 kA peak value. The surge impedance of the lightning channel is assumed to be 250 Ω. Menemenlis and Chun considered a triangular shape for the lightning impulse, and Saied et al a double-exponential shape.

The results shown in figs. 1 to 4 were obtained with the nonuniform transmission line model presented in this paper, for the triangular (figs. 1 and 2) and the double-exponential (figs. 3 and 4) lightning currents. The results in figs. 1 and 2 agree with those obtained by Menemenlis and Chun, except in the voltage wave tail. However, the authors consider that, according to the physics of the wave reflection at the tower bottom, no negative values should be expected in the voltage wave tail. The results in figs. 3 and 4 agree with those obtained by Saied et al applying the Laplace-transform to the highest number of cascaded exponential lines [7].

B. Ishii et al experiment

For simulating in EMTP the experimental tests reported by Ishii at al in [3], the circuit shown in fig. 5 is considered. The nonuniform transmission line losses are introduced by considering a damping resistance R (x). The equation giving this distributed parameter has been established according to the damping resistances considered in [3, 4]:

$$R(x) = -2\frac{\ln\sqrt{\gamma}}{H} Z_0(x) \qquad (4)$$

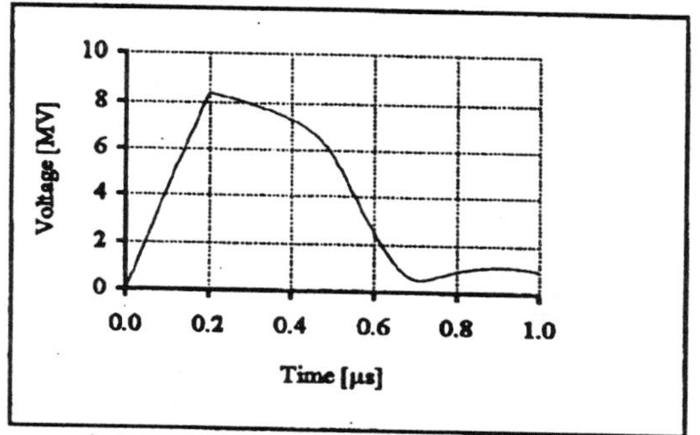

Fig. 1 Simulation results obtained for the Menemenlis and Chun test. Tower top voltage for a triangular 30 kA/ 0.2/25 μs lightning surge, as in [5].

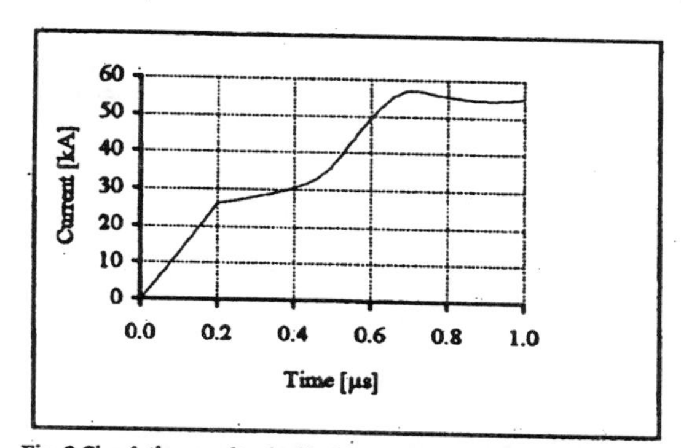

Fig. 2 Simulation results obtained for the Menemenlis and Chun test. Tower top current for a triangular 30 kA/ 0.2/25 μs lightning surge, as in [5].

H and γ being the tower height and the attenuation coefficient, respectively, and Z_0 corresponding to the lossless transmission line. The value $\gamma = 0.8$, as established in [3], was used. The lossless tower impedance Z_0 was evaluated according to equation (3), considering $k_1 = 150\,\Omega$, $k_2 = 8.83\,\Omega m^{-1/2}$, in order to fit the impedance values measured by Ishii et al [3]. The source current is represented by a 20-ns rise-time ramp-wave, with a peak value of 2 A. Results obtained are shown in figs. 6 to 9.

The simulation results closely reproduce the published experimental results [3], showing the adequacy of the proposed nonuniform line model for representing the transmission tower.

IV. CONCLUSIONS

A nonuniform transmission line model has been presented, suitable for use in power system electromagnetic transients analysis. In particular, the proposed model is adequate for representing the transmission line tower in lightning performance studies.

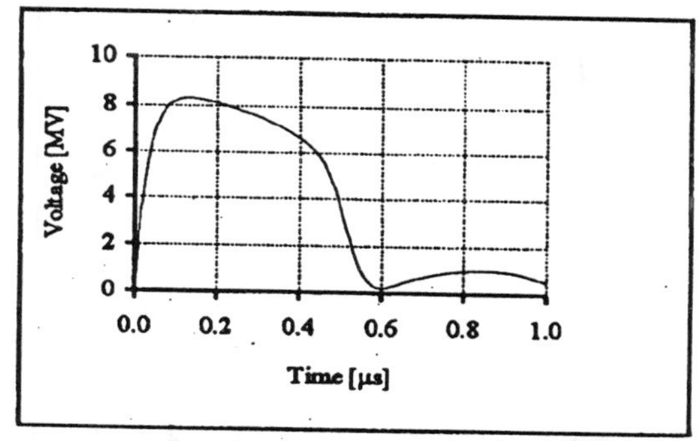

Fig. 3 Simulation results obtained for the Menemenlis and Chun test. Tower top voltage for a double-exponential 30 kA/ 0.2/25 μs lightning surge, as in [7].

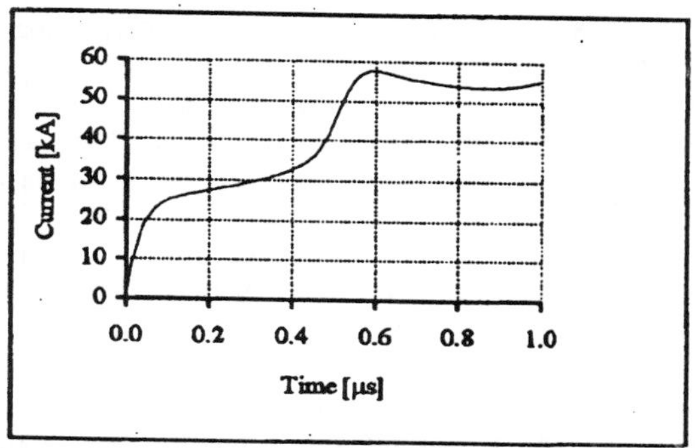

Fig. 4 Simulation results obtained for the Menemenlis and Chun test. Tower top current for a double-exponential 30 kA/ 0.2/25 μs lightning surge, as in [7].

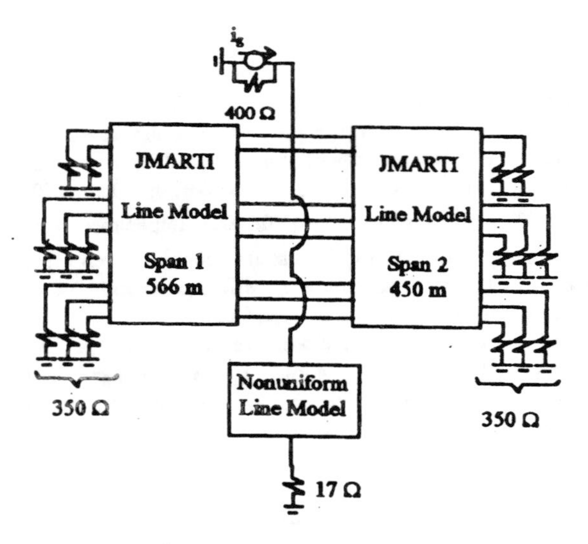

Fig. 5 Circuit for EMTP analysis of Ishii experimental test.

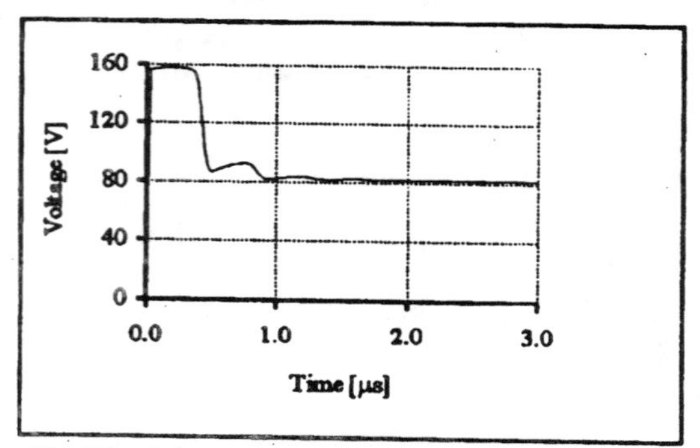

Fig. 6 Simulation results obtained for the Ishii et al experiment. Tower-top voltage.

The nonuniform line model is based on a finite differences solution of the wave propagation equations, and is able to take into account any space variation of the transmission line parameters.

The proposed model has been evaluated by comparison to other nonuniform transmission line models, and its adequacy to represent transmission towers by comparing simulation results and published experimental results. The power system analysis was carried out using EMTP, to which the nonuniform line model was interfaced.

One advantage of this nonuniform line model to represent a transmission tower, is its flexibility with regard to the line parameters variation law. It also allows eventual consideration of corona, as well as the frequency-dependence of the tower parameters. However, with this respect, modelling validation tests require that more experimental results are available.

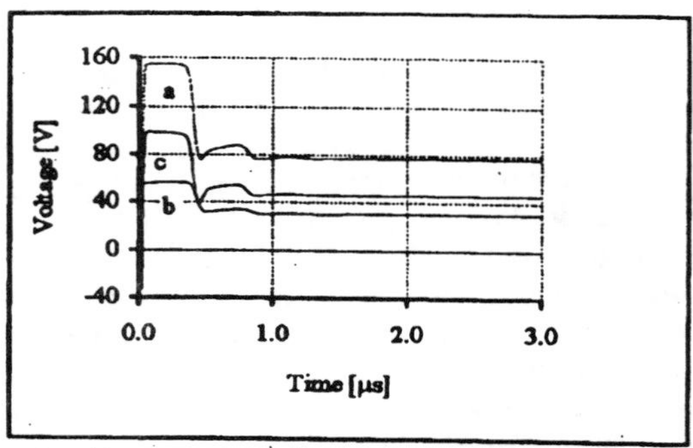

Fig. 7 Simulation results obtained for the Ishii et al experiment. Top-phase voltages.
a) -crossarm position. (b) - phase conductor. (c) - insulator.

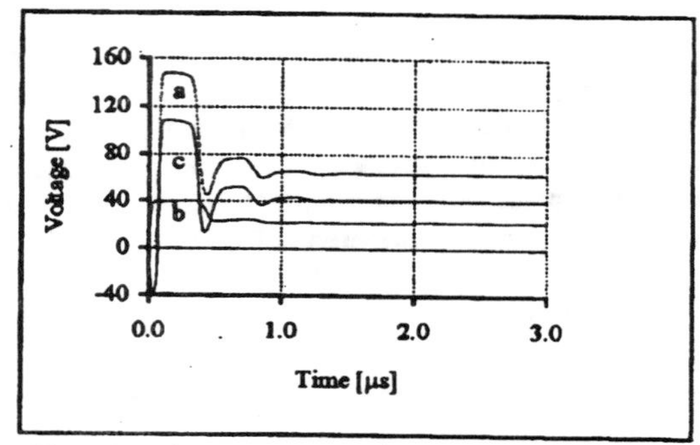

Fig. 8 Simulation results obtained for the Ishii et al experiment. Middle-phase voltages.
a) -crossarm position. (b) - phase conductor. (c) - insulator.

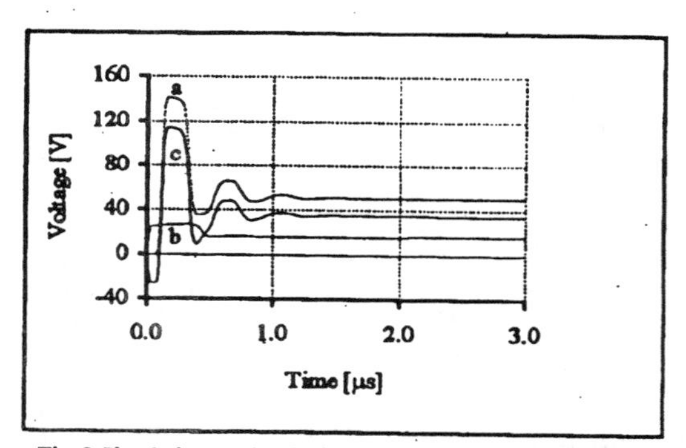

Fig. 9 Simulation results obtained for the Ishii et al experiment. Lower-phase voltages.
a) -crossarm position. (b) - phase conductor. (c) - insulator.

V. APPENDIX

STABILITY AND ACCURACY PROPERTIES OF LINEAR INTEGRATION FORMULAS

A. Background

Let us consider a p-step integration method being expressed by the general formula:

$$\sum_{i=0}^{p}\alpha_i y_{k-i} = h\sum_{i=0}^{p}\beta_i y'_{k-i} \qquad (A.1)$$

where index k denotes the current time-step (the formula applies for $k=p,p+1,...$, the values for $k=0,...,p-1$, being given or evaluated using a lower step formula), y' is the time derivative of y, h is the time step and the values of coefficients α_i, β_i, for $i=0,...,p$, identify each integration formula. An implicit formula will be identified by a non-zero value for β_0.

When a p-step integration formula is applied recursively to a set of linear ordinary differential equations:

$$y' = \lambda y \qquad (A.2)$$

it generates a set of y values corresponding to instants nh, n being an integer and h the chosen time step. These values can be expressed as:

$$y(nh) = \mu^n y_0 \qquad (A.3)$$

where y_0 is given by the initial conditions, and μ is a function of $h\lambda$. Since μ is related to α_i, β_i, the corresponding integration method is identified by expression (A.3). Using (A.2) and (A.3) in (A.1), we can get:

$$\sum_{i=0}^{p}(\alpha_i - h\lambda\beta_i)\mu^{-i} = 0 \qquad (A.4)$$

or, in the canonical form of a rational function of μ^{-i}:

$$h\lambda = \frac{\sum_{i=0}^{p}\alpha_i\mu^{-i}}{\sum_{i=0}^{p}\beta_i\mu^{-i}} \qquad (A.5)$$

The relationship between the μ-plane and the $h\lambda$-plane determines the accuracy and stability properties of the numerical integration method.

The stability region corresponds to the transformation, into the $h\lambda$-plane, of the circle defined by $|\mu|\le 1$ in the μ-plane.

The behaviour of each method regarding oscillations of different frequencies can be evaluated by transforming the imaginary axis of the $h\lambda$-plane into the μ-plane.

B. Backward Euler's method

Backward Euler is a one-step 1st-order method identified by the coefficients:

$$\alpha_0 = 1 \quad \alpha_1 = -1$$
$$\beta_0 = 1 \quad \beta_1 = 0 \qquad (A.6)$$

Taking into account equations (A.5) and (A.6), we derive:

$$h\lambda = \frac{\mu - 1}{\mu} \qquad (A.7)$$

and the results shown in Fig. A.1 are obtained, proving that this is an A-stable integration method and showing its frequency behaviour.

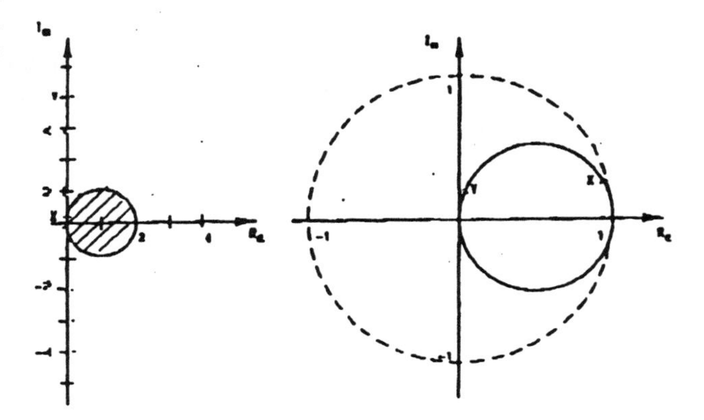

Fig. A.1 Stability and accuracy properties of the Backward Euler's integration method. The dashed area corresponds to $|\mu|>1$ (unstable region).

C. Trapezoidal integration method

The coefficients α_i, β_i, identifying the trapezoidal integration method are:

$$\alpha_0 = 1 \quad \alpha_1 = -1$$
$$\beta_0 = \frac{1}{2} \quad \beta_1 = \frac{1}{2} \qquad (A.8)$$

It is a one-step 2nd-order method.

Taking into account equations (A.5) and (A.7), we obtain:

$$h\lambda = 2\left(\frac{\mu - 1}{\mu + 1}\right) \qquad (A.9)$$

From this equation we find that the circumference $|\mu|=1$ is transformed into the imaginary axis in the $h\lambda$-plane (Fig. A.2). Besides showing that it is an A-stable integration method, these result clear indicates that the method is not selective in frequency, reproducing all the wave components with the same accuracy.

D. Gear's 2nd order method

Gear 2nd-order method is a two-step method defined by the coefficients:

$$\alpha_0 = 1 \quad \alpha_1 = -\frac{4}{3} \quad \alpha_2 = \frac{1}{3}$$
$$\beta_0 = \frac{2}{3} \quad \beta_1 = 0 \quad \beta_2 = 0 \tag{A.10}$$

Taking into account equations (A.5) and (A.8), we derive:

$$h\lambda = \frac{3}{2} - 2\frac{1}{\mu} + \frac{1}{2}\frac{1}{\mu^2} \tag{A.11}$$

and results shown in Fig. A.3 are obtained, proving that this is an A-stable integration method and showing its frequency selectivity and accuracy.

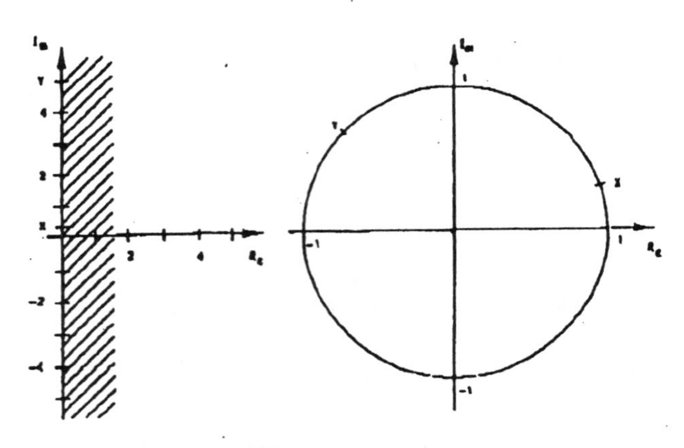

Fig. A.2 Stability and accuracy properties of the Trapezoidal integration method. The dashed area corresponds to $|\mu|>1$ (unstable region).

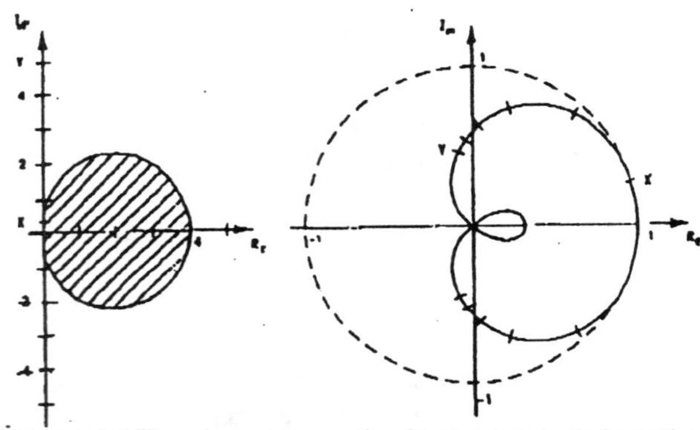

Fig. A.3 Stability and accuracy properties of the Gear's 2nd order integration method. The dashed area corresponds to $|\mu|>1$ (unstable region).

VI. ACKNOWLEDGEMENT

The authors acknowledge L. Dubé for his valuable support when writing the interface between the line model and the ATP version of the EMTP, using MODELS.

VII. REFERENCES

[1] M. Kawai, "Studies of the surge response of a transmission line tower", IEEE Trans. on Power Apparatus and Systems, Vol. PAS-83, pp. 30-34, January 1964.

[2] M.A. Sargent and M. Darveniza, "Tower Surge Impedance", IEEE Trans. on Power Apparatus and Systems, Vol. PAS-88, No. 5, pp. 680-687, May 1969.

[3] M. Ishii, T. Kawamura, T. Kouno, E. Ohsaki, K. Shiokawa, K. Murotani, T. Higuchi, "Multistory transmission tower model for lightning surge analysis", IEEE Trans. on Power Delivery, Vol. 6, No. 3, pp. 1327-1335, July 1991.

[4] T. Yamada, A. Mochizuki, J. Sawada, E. Zaima, T. Kawamura, A. Ametani, M. Ishii, S. Kato, "Experimental evaluation of a UHV tower model for lightning surge analysis", IEEE PES Winter Meeting Paper, No. 044-8 PWRD, 1994.

[5] C. Menemenlis, Z.T. Chun, "Wave propagation on nonuniform lines", IEEE Trans. on Power Apparatus and Systems, Vol. PAS-101, No. 4, pp. 833-839, April 1982.

[6] M. M. Saied, A.S. AlFuhaid, M.E. El-Shandwily, "s-Domain analysis of electromagnetic transients on nonuniform lines", IEEE Trans. on Power Delivery, Vol. 5, No. 4, pp. 2072-2083, November 1990.

[7] E.A. Oufi, A.S. AlFuhaid, M. M. Saied, "Transient analysis of lossless single-phase nonuniform transmission lines", IEEE Trans. on Power Delivery, Vol. 9, No. 3, pp. 1694-1701, July 1994.

[8] M.E. Almeida, M.T. Correia de Barros, "Tower modelling for lightning surge analysis usin EMTP", European EMTP User Group Meeting, Lyngby, Denmark, April 1994. To be published in Proc. IEE.

M. T. Correia de Barros was born in Lisbon, Portugal, in 1951, and received the Dipl. in Electrical Engineering in 1974 and the Doctor's Degree in 1985, both from IST - Technical University of Lisbon. She is currently an Associate Professor at the same University. Her main research interests are the fields of High Voltage Engineering and Electromagnetic Transients.

M.E. Almeida was born in Mozambique, in 1962, and received the Msc degree in Electrical Engineering in 1990, from IST-Technical University of Lisbon. She is currently a Research Assistant, and prepares a Ph D Thesis under the supervision of Prof. Correia de Barros.

Discussion

Adam Semlyen (University of Toronto): The authors have addressed the important and timely topic of modeling a nonuniform transmission line in the calculation of electromagnetic transients. Their approach consists in subdividing the line into segments in accordance with the spatial discretization of the underlying partial differential equations. The study is performed with great competence and care regarding the numerical approach chosen and its implementation. It is interesting to note that different time steps are used for simulation within the model and in the external system, a problem we have encountered in relation with transformer modeling [1,2].

One of the advantages of the method is its complete generality achieved by the fact that the identity of intermediate points along the line is preserved so that nonlinear effects, as for instance corona, can be included in the model. The authors deserve congratulation for their outstanding work.

In the following an alternative view is presented with the nonuniform transmission line being globally modeled from its terminals (or, perhaps, in relatively longer segments needed for the inclusion of intermediate, possibly nonlinear, elements). This terminal approach is an extension to the traditional modeling of uniform transmission lines with constant or frequency dependent parameters. It is of course only possible in the absence of nonlinearities. The *global, terminal results* can be derived for single or multi-phase lines, first in the frequency then in the time domain, as admittance matrices or with realization by *R,L,C* elements, or even for a representation in terms of traveling waves. The derivations are based on the *frequency domain* (vector-phasor) differential equations

$$\frac{d}{dx}\begin{bmatrix}\mathbf{v}\\ \mathbf{i}\end{bmatrix}=\begin{bmatrix}0 & \mathbf{Z}(x)\\ \mathbf{Y}(x) & 0\end{bmatrix}\begin{bmatrix}\mathbf{v}\\ \mathbf{i}\end{bmatrix},\quad \begin{bmatrix}\mathbf{v}(0)\\ \mathbf{i}(0)\end{bmatrix}=\begin{bmatrix}\mathbf{v}_0\\ \mathbf{i}_0\end{bmatrix} \qquad \text{(a)}$$

where x is the longitudinal space parameter (defined in the direction opposite to that for the current). We seek at $x=l$ a solution of the form

$$\begin{bmatrix}\mathbf{v}(l)\\ \mathbf{i}(l)\end{bmatrix}=\begin{bmatrix}\mathbf{A} & \mathbf{B}\\ \mathbf{C} & \mathbf{D}\end{bmatrix}\begin{bmatrix}\mathbf{v}(0)\\ \mathbf{i}(0)\end{bmatrix} \qquad \text{(b)}$$

The A,B,C,D-matrix in (b) can be readily identified by direct numerical integration of the (complex) ordinary differential equations (a), with the input vector taken sequentially equal to the columns of the unity matrix.

While its evaluation for a sufficient number of frequencies, as described above, requires repeated numerical integrations, the result obtained shows that an A,B,C,D-*matrix* and, consequently, an *admittance matrix* Y and its *π-equivalent* exist and can be calculated for the nonuniform line just as for a line with uniformly distributed parameters. Moreover, for the branches of the π-equivalent an *R,L,C-realization* (a Cauer or Foster circuit [3]) can be obtained for direct *time domain* calculation. This approach assures stability and the possibility of direct implementation in the EMTP or equivalent programs. A complete model of transmission tower with cross-arms could be assembled using this procedure.

The A,B,C,D matrix solution (b) permits to decompose the input $\{\mathbf{v},\mathbf{i}\}$ into an incident and a reflected wave at $x=0$. The former comes from, and the latter proceeds to, the $x=l$ end of the line but neither satisfies individually the continuity requirements for the wave voltages and currents at the "nodes" along the (discretized, nonuniform) line. This is consistent with the "local reflections" as the characteristic impedance varies along the line.

The resultant voltages and currents have of course a continuous variation. Thus, while *traveling waves* may be defined even in this case, the corresponding concepts appear to be more complex than in the case of uniform transmission lines.

In simple particular instances, the data for equation (a) may be available in analytical form. For example, in the case of the 75m tower with the characteristic impedance $Z_0(x)$ of equation (3) in the paper, we could likely take

$$Z(x)=\frac{j\omega}{c}Z_0(x),\quad Y(x)=\frac{j\omega}{c}\frac{1}{Z_0(x)} \qquad \text{(c)}$$

where c is the velocity of propagation ($\sim 3\times10^8$ m/s). Then, numerical integration of (a), as described above, identifies the 2×2 A,B,C,D-matrix for (b). We note that A and D are real while B and C are pure imaginary for a lossless, uniform or nonuniform, line. Rearranging the voltages and currents yields the 2×2 admittance matrix which must of course be symmetrical. Hence, equating its off-diagonal entries, we obtain the familiar relation

$$AD-BC=1 \qquad \text{(d)}$$

A non-symmetrical π-equivalent (since now $A\neq D$) can be readily obtained from the admittance matrix.

In order to obtain wave equations we note that a wave is characterized by a fixed relation between its voltage and current components and by a multiplying factor λ relating the waves at the two ends of the line. Consequently (b) becomes

$$\lambda\begin{bmatrix}V(0)\\ I(0)\end{bmatrix}=\begin{bmatrix}A & B\\ C & D\end{bmatrix}\begin{bmatrix}V(0)\\ I(0)\end{bmatrix} \qquad \text{(e)}$$

Thus λ is an eigenvalue of the A,B,C,D-matrix in (e). Taking (d) into account, we get the characteristic equation

$$\lambda^2-(A+D)\lambda+1=0 \qquad \text{(f)}$$

Its roots are of the form $\lambda=\exp(\pm j\beta l)$ (where βl results from $A+D=2\cos(\beta l)$), so that $|\lambda|=1$, as anticipated for a lossless line. The characteristic impedances Z_{0_1}, Z_{0_2} are obtained as the ratio of the voltage to current components of the corresponding eigenvectors.

We note that in the case of a lossless, *uniform* line we have $A=D=\cos(\sqrt{XB}\,l)$. Then (f) gives $\lambda=\exp(\pm j\sqrt{XB}\,l)$ and from the corresponding eigenvectors of the A,B,C,D-matrix we get the characteristic impedance $Z_0=\pm\sqrt{X/B}$, as expected. In the case of a *nonuniform* line, the characteristic impedances are complex and of equal magnitude, with Z_{0_2} the complex conjugate of $-Z_{0_1}$.

The views of the authors on the above and on related topics would be greatly appreciated.

[1] A. Semlyen and F. de Leon, "Computation of Electro-Magnetic Transients Using Dual or Multiple Time Steps", IEEE Transactions on Power Systems, Vol. 8, No.3, August 1993, pp. 1274-81.

[2] A. Semlyen and F. de Leon, "Complete Transformer Model For Electromagnetic Transients", IEEE Transactions on Power Delivery, Vol.9, No.1, January 1994, pp. 231-239.

[3] F. de Leon and A. Semlyen, "Time Domain Modeling of Eddy Current Effects for Transformer Transients", IEEE

Transactions on Power Delivery, Vol.8, No.1, January 1993, pp. 271-280.

Manuscript received August 15, 1995.

HUYEN V. NGUYEN (The University of British Columbia, Vancouver, B.C., Canada) The authors are complimented for an interesting paper. It would be helpful if the authors could comment on the following issues related to the Ishii et al experiment section of the paper:

1) Realizing the non-uniformity of the transmission tower structure and its associated surge impedance, the discusser has also been working on a non-uniform tower model since early of 1993. The conventional EMTP transmission line models only represent the TEM mode in the horizontal direction, but not the electromagnetic field propagation in the traverse direction. Therefore, a sudden voltage change at the tower top and ground wire connection will lead to sudden voltage changes in the phase conductors. This is why when the current source is injected into the tower top, the induced voltages appear instantaneously on the phase conductors, as shown in Figures 7-9. On the other hand, the tower model will show a time delay between the tower top and the crossarms. Thus, when the insulator voltages are calculated, negative spikes will appear initially on the waveforms (see Figures 7-9). However, these negative spikes do not exist in field measurements [3]. In order to avoid having these spikes one has

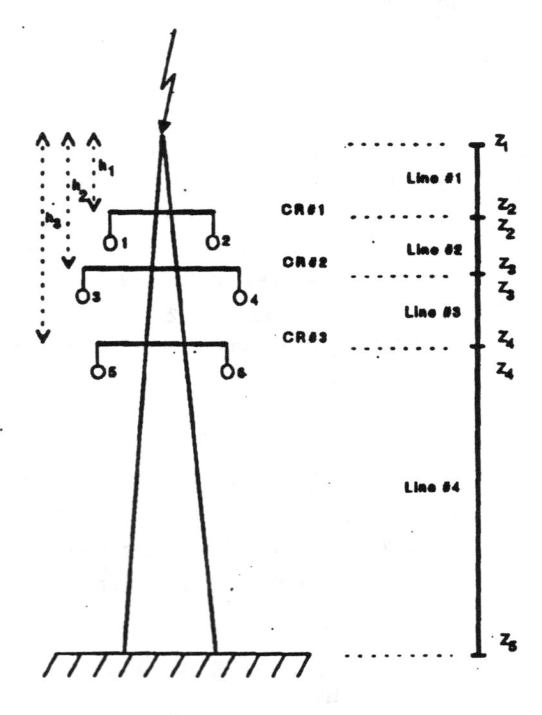

crossarm #1 insulator voltage = $v_{ca1}(t) - v_{p1}(t-\tau_1)$

crossarm #2 insulator voltage = $v_{ca2}(t) - v_{p2}(t-\tau_2)$

crossarm #3 insulator voltage = $v_{ca3}(t) - v_{p3}(t-\tau_3)$

where

$\tau_1 = h_1/c_1$, $\tau_2 = h_2/c_1$ and $\tau_3 = h_3/c_1$

Figure A1 - Insulator voltage calculations with time delays

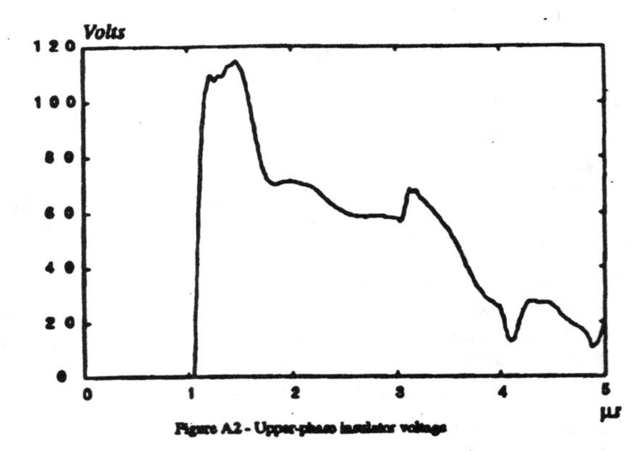

Figure A2 - Upper-phase insulator voltage

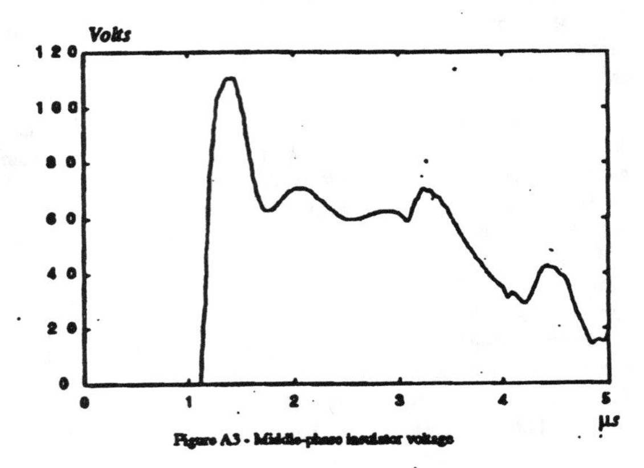

Figure A3 - Middle-phase insulator voltage

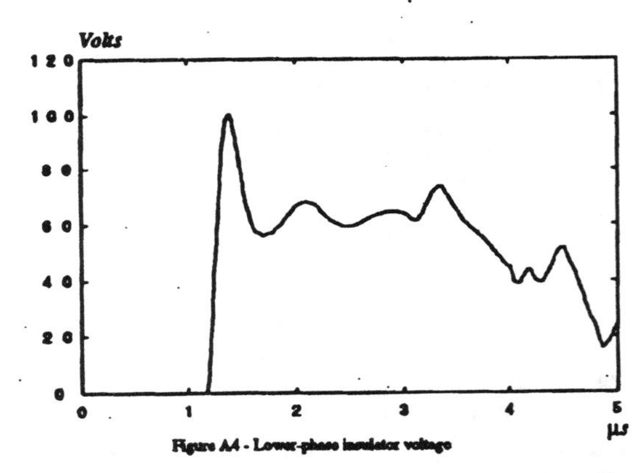

Figure A4 - Lower-phase insulator voltage

to take into account the traverse field retardation from the ground wires, or from the tower top, to the phase conductors. One way to approximate this is to ignore totally the induced voltages on the phase conductors from t = 0 up to the time when the injected current source reaches the crossarms. Another way is to include time delays in the insulator voltage calculations as shown in Figure A1.

2) Ishii et al mentioned that the tails of the insulator waveshapes are difficult to simulate [3]. Indeed, there is a distinct difference between the tails of the insulator voltage waveshapes shown in Figures 7-9 and those measured by Ishii et al. The insulator voltage waveforms in Figures 7-9 show a significant dip at the instant when the first reflection from the tower base comes back. On the other hand, field measurements show a gradual decreasing change. Furthermore, it is noticed that the order of these voltage magnitudes are in reverse as compared to field measurements.

The lower-phase insulator voltage shows the largest value. This is the opposite of what had been measured, which showed the upper-phase insulator voltage having the highest magnitude. For comparison purposes, the insulator voltages using the discusser's non-uniform line model are depicted in Figures A2-A4. The crossarms of the tower were ignored in these simulations. The tower was modelled as 4 transmission line sections, three uniform and one non-uniform. Note that the simulation results are very close compared to the measurements shown in Figure 7 of reference [3].

3) Have the authors used other types of space variations for the tower surge impedance besides that of equation (3)? One easy type would be the exponential variation. If this is used, would the simulated waveforms show significant differences?

4) Finally, do the tower crossarms have any impacts in the authors' simulation results?

Manuscript received August 17, 1995.

M.T. Correia de Barros and M.E. Almeida: We would like to thank the discussers for their interest in our paper and for their questions and comments. We will give our answers to each discusser separately.

Prof. Adam Semlyen:

We very much appreciate Prof. Semlyen's encouraging words about our work. His comments constitute a valuable contribution to the topic of the paper.

Prof. Semlyen points out the possibility of including non-linear effects in our model. Indeed our work on finite-difference line modelling targets to develop a general-purpose multi-phase transmission line model adequate for directly taking into account (a) distributed nonuniform parameters (b) distributed non-linear parameters and also (c) distributed sources. The latest feature is required for the calculation of lightning induced overvoltages.

Prof. Semlyen's alternative view for modelling globally a linear nonuniform transmission line is fully correct. Perhaps it is more adequate than the authors approach for modeling lines with frequency dependent parameters.

Inclusion of the frequency-dependence of the line parameters in a finite-difference line model increases significantly the required computation time. A comparison with Prof. Semlyen's approach will be very interesting.

Mr. H.V. Nguyen

We will address one by one the questions related to the Ishii et al experiment

1) We agree with the explanation given by Mr. Nguyen for the negative spikes showing in the simulation results, and which do not show in the experimental results.

Mr. Nguyen suggests two alternatives for eliminating these negative spikes. Both are external to the modelling techniques adopted for the tower and for the adjacent transmission line spans, and therefore, the resulting elimination of the negative spikes is somehow artificial. Figs. C1 to C3 show the simulation results obtained by considering a delay time for the voltages induced on the phase conductors, as suggested in fig. A1 of the discussion. By comparing these results with those shown in the paper (figs. 7-9) it is plain that the retardation of the induced voltages does not have much influence, except in eliminating the negative spikes.

In the paper, we chose to show the negative spikes in the simulation results, in order not to hide the inconsistency that really is incurred when considering the tower as a distributed parameter and, simultaneously, a TEM approach for modeling the adjacent transmission line spans.

2) Comparison of the frequency response of a transmission line model where losses are represented by simple resistances with the frequency response of a frequency-dependent parameters model shows that the first is more selective with respect to high frequencies. To the authors opinion this explains the dip at the instant corresponding to the first reflection from the tower base, and a frequency-dependent model will be adequate to reproduce the gradual

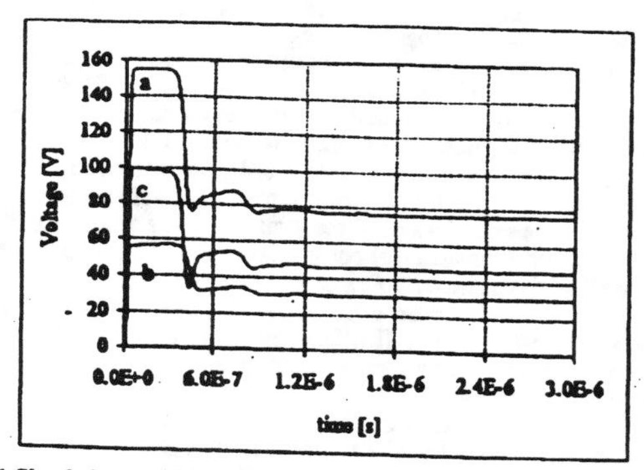

Fig. C1 Simulation results obtained for the Ishii et al experiment, considering a delayed induced voltage. Top-phase voltages.
a) crossarm position. b) phase conductor. c) insulator.

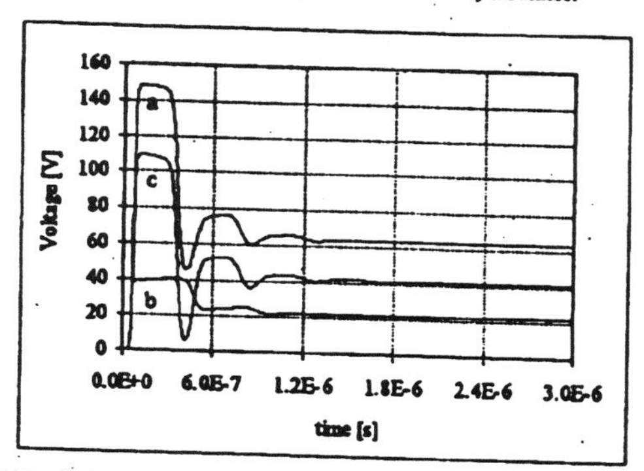

Fig. C2 Simulation results obtained for the Ishii et al experiment, considering a delayed induced voltage. Middle-phase voltages.
a) crossarm position. b) phase conductor. c) insulator.

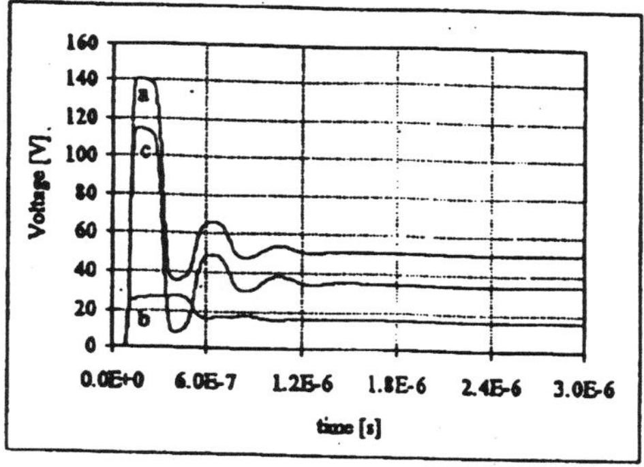

Fig. C3 Simulation results obtained for the Ishii et al experiment, considering a delayed induced voltage. Lower-phase voltages.
a) crossarm position. b) phase conductor. c) insulator.

decrease of the insulator overvoltage observed in the experimental results.

Reasoning about the insulator voltage magnitudes for the different phases, we must look at the order of magnitude of its components. The voltage induced on the phase conductor is always lower for the lowest placed conductor. If the voltages at the different crossarm positions would be equal, then the voltage at the insulators would increase from the top to the lower phase. Therefore, observation of an higher magnitude at the top phase can only be explained by the behavior of the tower itself.

Considering the height of the Ishii test tower, and the injected current rise-time, we conclude that reflections from the tower base do not influence the magnitude of the overvoltages at the crossarm positions.

Then, if the tower would behave as a uniform lossless transmission line, the overvoltage at the crossarms would all reach the same magnitude. As the tower behaves as non-uniform line, with the surge impedance decreasing from top to bottom, the crossarm voltage magnitude decreases from the upper to the lower phase. The tower losses have a similar effect.

Therefore, the values assumed for the tower parameters may show decisive in determining the order of the insulator voltage magnitudes.

For comparison purposes, it would be interesting to have more information about Mr. Nguyen's tower model and the values of the simulation parameters used for evaluating the voltages shown in figs. A2-A4. For a meaningful comparison, the voltages at the crossarms, as well as the induced voltages are also necessary.

3) We have also used other types of space variation for the tower surge impedance. See [8]. However, we consider that the existing knowledge about tower parameters does not allow a sustained choice of this variation law.

Due to the flexibility of the proposed nonuniform transmission line model with respect to the space variation of the line parameters, any variation law is as easy to handle as the exponential variation.

4) The crossarms were not considered in the tower model.

As a final remark we must add that a better fitting between simulation and experimental results can be obtained for the Ishii at al experiment by changing the tower parameters. However, we prefer not to follow the usual trial and error approach. The surge impedance variation law was chosen to allow comparison with previously developed nonuniform transmission line models [5-7], and was kept unchanged from one application to the other. The damping resistance equation proposed in [3,4] was respected.

We consider that further work on the evaluation of tower parameters is needed, which requires both theoretical and experimental work.

[8] M.E. Almeida, M.T. Correia de Barros, "Tower modelling for lightning surge analysis using EMTP", IEE Proc.-Gener. Transm. Distrib., Vol. 141, No. 6, pp. 637-639, November 1994.

Manuscript received October 23, 1995.

WIDE-BAND CORONA CIRCUIT MODEL FOR TRANSIENT SIMULATIONS

J. R. Martí (M) F. Castellanos (STM) N. Santiago

The University of British Columbia,
Vancouver, BC., Canada V6T 1Z4.

Abstract. **Corona in overhead transmission lines is a highly nonlinear and non-deterministic phenomenon. Circuit models have been developed to represent its behaviour, but the response of these models is usually limited to a narrow set of frequencies. The circuit model presented in this paper achieves a much wider frequency response than previous models: 1) by matching more closely the topology of the circuit, and 2) by using a second-order circuit response to match the high-order dynamic response of the phenomenon. The resulting model is valid for a wide range of frequencies and is able to represent waveshapes from switching to lightning surges. The model is applied to the EMTP program and simulations of q-v measurements and travelling surges are presented.**

Key words: Corona modelling, q-v loops, EMTP transient simulation.

1. Introduction

Corona in overhead transmission lines can be an important source of attenuation and distortion of fast surges and temporary overvoltages. A number of measurements have been taken and physical explanations have been formulated to better understand the corona characteristics (e.g., [1-11]). However, the physical complexities of the phenomenon have made it difficult to formulate simple models that can be used in the context of general-purpose system-level transients simulations, using for example the EMTP program.

Ideally, one should be able to derive some set of equations to describe the characteristics of the phenomenon from readily available line data and then formulate an EMTP-type circuit model. This is the case, for instance, in transmission line modelling [15], where relatively complicated equations (e.g., Carson's equations) have to be translated into simple and computationally efficient circuit models. For corona, no general analytical expressions from basic line data are yet available and most descriptions are based on experimentally measured charge versus voltage characteristics (q-v loops). However, regardless of whether the physical description is derived analytically or measured experimentally, it still has to be translated into an efficient and practical circuit model for general-purpose transmission system simulation with the EMTP.

A main drawback of conventional circuit models for corona is their limitations in representing the phenomenon for a wide range of waveforms using the same set of circuit parameters. The model presented in this paper overcomes this limited bandwidth problem by formulating a circuit topology that more closely matches the topology of the actual physical system and by including circuit components with a higher-order dynamic response.

The improved topology and dynamic response allow the model to match, with a single set of circuit parameters, a wide range of surge waveshapes, from slow switching surges to very fast lightning strokes. In addition, the model can correctly respond to the changes in shape and rise time as the surge propagates along the transmission line.

The paper expands on the work presented in [16] and is part of a new general transmission line model for the EMTP that takes into account the distributed nature of both the frequency dependence of the line parameters and the corona nonlinearity.

2. Q-V Curves and Conventional Models

Corona characteristics of transmission lines are usually obtained through experimental measurements. There are

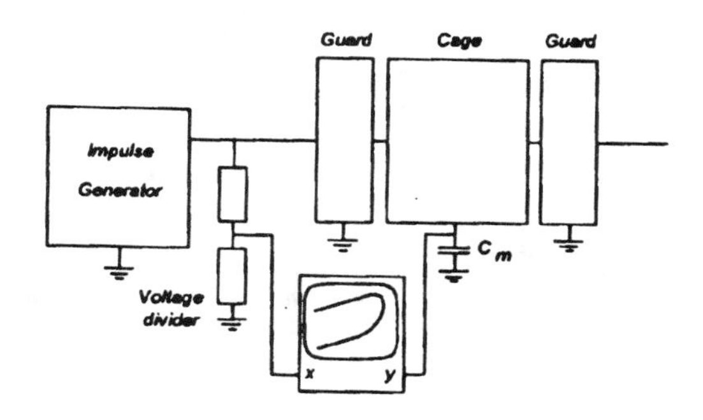

Figure 1 - Experimental set-up for q-v curves measurements.

94 SM 516-5 PWRS A paper recommended and approved by the IEEE Power System Engineering Committee of the IEEE Power Engineering Society for presentation at the IEEE/PES 1994 Summer Meeting, San Francisco, CA, July 24 - 28, 1994. Manuscript submitted July 30, 1993; made available for printing May 3, 1994.

two basic types of measurements: 1) q-v curves, which are plots of charge versus voltage, and 2) measurements of voltage surges propagating along a transmission line. The q-v curves are obtained in cage set-ups, as the one in Figure 1, where an overvoltage is injected into the conductor. The voltage is measured in a voltage divider, and the charge is obtained as the integral of the current in a probe capacitance connected to the return circuit.

Measurements of q-v curves for single conductors of standard dimensions and for thinner conductors for different voltage waveshapes are shown in Figure 2.

From observation of the q-v curve, it is possible to simulate the main characteristics of the loop by increasing the capacitance in the region between the knee point of the loop (apparent corona onset voltage E_{cor}) and the peak voltage, V_p. This type of representation can be achieved with the basic model shown in Figure 3(a), where C_g is the geometric capacitance of the conductor, and C_{cor} is the capacitance introduced by the corona effect. The q-v curve obtained with this model is shown in Figure 3(b). Comparing this idealized curve with the measured ones, it can be observed that the model is limited in its ability to follow the smooth transitions between curve regions and to represent the gradual increase of slope in the frontal lobe of the curve. Since the q-v curve for a given conductor changes according to the shape of the surge, the parameters of the simplified circuit of Figure 3 have to be recalculated for different surges and as the surge travels along the line.

The basic circuit of Figure 3 can be improved by connecting additional capacitances in parallel, additional corona branches in parallel, resistors in series or parallel, and other combinations. Comparative evaluations of some of these circuits using the EMTP were presented in [14] and [12].

Measurements of q-v curves for conductors in experimental cage arrangements are available from different references. Maruvada et al [6] presented a complete study

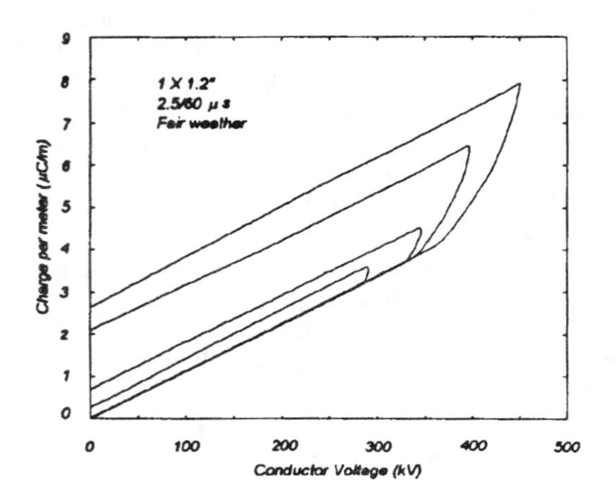

(a) Actual single conductor (diam. 1.2"), rise time 2.5 μs [6].

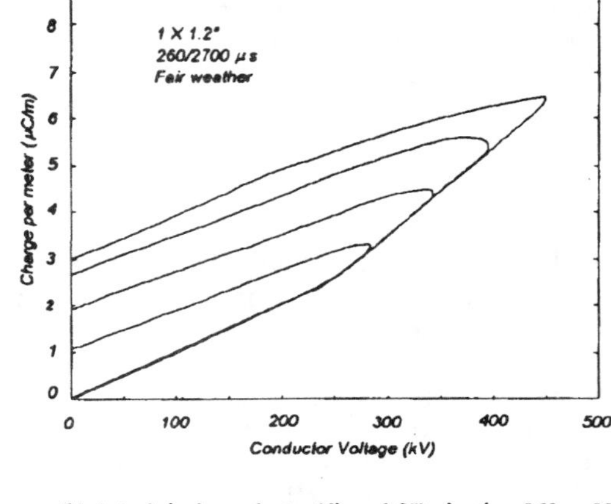

(b) Actual single conductor (diam. 1.2"), rise time 260 μs [6].

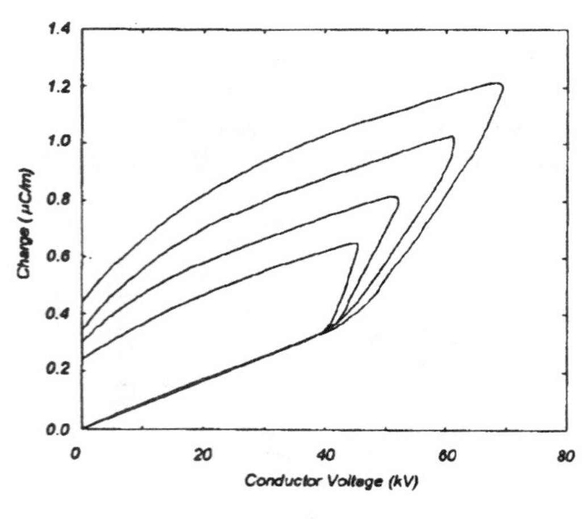

(c) Thin conductor (diam. 0.65 mm), rise time 0.12 μs [7].

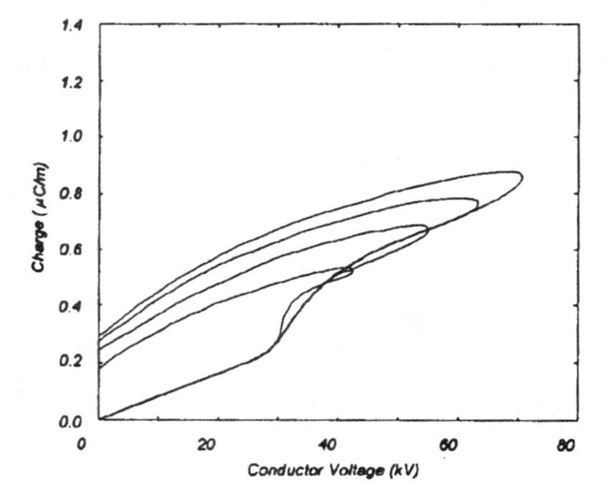

(d) Thin conductor (diam. 0.65 mm) rise time 1.2 μs [7].

Figure 2 - Typical q-v curves.

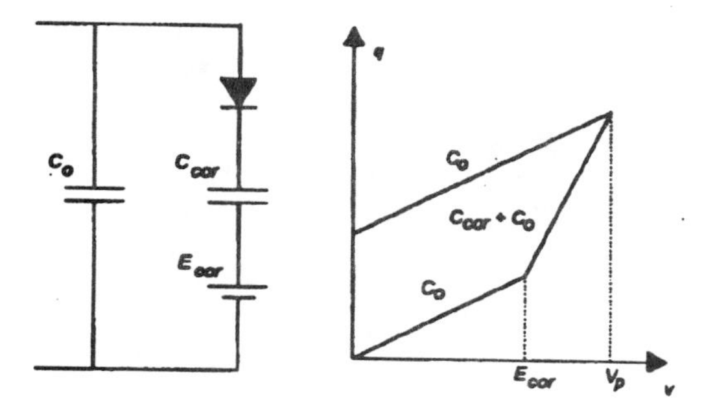

(a) Corona model [8]. (b) Typical response of the model.

Figure 3 - Basic corona model.

covering different conductors subjected to switching and lightning surges of different magnitudes. Figures 2(a) and 2(b) show some of these results. The following observations can be made with respect to these curves:

1) The value of the corona onset voltage increases with the derivative of the voltage with respect to time. This increase is more pronounced for faster rise times.
2) The q-v curves present a smooth transition around the onset voltage. This is more noticeable in the case of slower rise times.
3) The portion of the q-v curve corresponding to the front of the surge after the onset voltage increases its slope as a function of the derivative of the voltage. This effect is more pronounced for faster rise times.
4) The portion of the q-v curve just after the turnaround at the voltage peak presents a slope that decreases as a function of the voltage derivative. This effect is more pronounced for faster rise times.

The above observations suggest that the corona effect is probably of a higher order than the simple first-order system of Figure 3(a). The strong dynamics of the phenomenon are even more noticeable in thin conductors (curves in Figures 2(c) and (d)). The turning inwards of the slope in the characteristics of Figure 2(d) strongly suggests a system of order two or higher.

In order to reproduce the different behaviours of the q-v curves under different shapes of transient functions, many researchers have opted for changing the value of the model parameters according to the type of surge applied. This approach can give satisfactory results when replicating measured q-v curves by themselves, but it is not fully adequate for simulating travelling surges on transmission lines. During propagation, the shape of the surge gets distorted, and the rise time and frequency content change as the surge travels down the line.

A more general solution for a dynamic wide-band model of corona requires a higher-order circuit response and an improved topological representation of the geometry of the conductors system.

3. Proposed Equivalent Circuit

The proposed model achieves a wider frequency bandwidth than conventional models by matching more closely the topology of the equivalent circuit to the topology of the actual physical system.

The situation of a conductor above ground is shown in Figure 4. Under corona, the air surrounding the conductor is ionized from the conductor radius r to the border of the corona crown r_{cor}. The integral of the electric field from the conductor to ground, which equals the voltage applied between conductor and ground, is divided into two parts: from r to r_{cor}, under ionized air, and from r_{cor} to h, under normal air:

$$\int_r^h E\cdot dl = \int_r^{r_{cor}} E\cdot dl + \int_{r_{cor}}^h E\cdot dl$$

$$= f(r, r_{cor}, \varepsilon_{cor}, q') + \frac{q}{2\pi\varepsilon_{air}} \ln\frac{2h}{r_{cor}}$$

$$= V_{cor} + V_{air} = V$$

The particular form of $V_{cor} = f(r, r_{cor}, \varepsilon_{cor}, q')$ is not simple to define. Nonetheless, it can be seen from the expression above that there are actually two capacitances, C_{cor} and C_{air}, connected in series between conductor and ground (Figure 4(b)). As discussed next, this simple observation brings out an important topological difference between the physical system and the conventional circuit representation of Figure 3(a).

The proposed circuit representation of an overhead conductor under corona is shown in Figure 5. The upper part of this circuit represents the dynamic high-order processes of corona during the ascending and descending branches of the q-v loop characteristic. During the initial linear part of the q-v loop, before the source voltage $v(t)$ reaches the ideal corona onset voltage E_o, branches C_{cor}-L_h-R_h and R_g in the circuit of Figure 5 are open. The air capacitance C_{a1} (between the conductor radius r and the distance r_{cor}) and the air capacitance C_{a2} (between r_{cor} and ground) are connected in series, and their combined value equals the total geometric capacitance C_o between the conductor and ground with no corona.

When the surge voltage $v(t)$ reaches the ideal corona onset voltage E_o, the corona process starts, with a time delay given by the statistical time lag [11], which is the statistical time needed by a seed electron to start the corona discharge. After this delay, which depends on the rate of rise of the surge, the electron avalanche process proceeds very rapidly and the air surrounding the conductor becomes ionized. In the proposed circuit representation of Figure 5, when the voltage across the corona branch $v(t)$ reaches the DC source value E_o, the diode begins to conduct and the capacitive corona branch C_{cor}-L_h-R_h "clicks in". At the same time, the corona discharge branch R_g also "clicks in" through a spark gap with a breakdown voltage E_o.

With reference to the diagram of Figure 6 and the

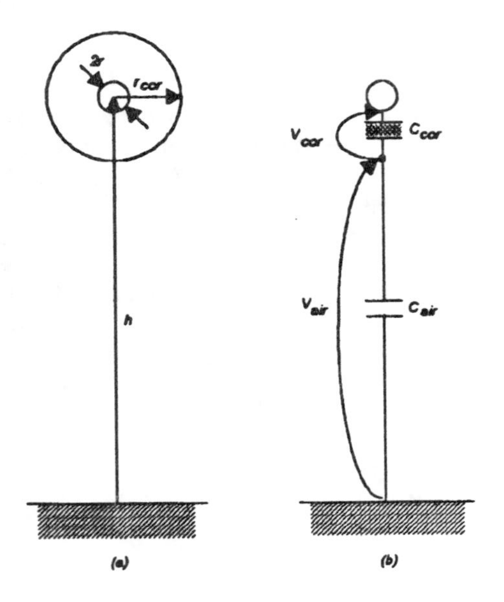

Figure 4. - (a) Conductor above ground.
(b) Associated capacitances.

equivalent circuit of Figure 5, C_{cor} represents the increase in capacitance due to the ionization of the air surrounding the conductor, R_h the additional conduction losses in the frontal lobe of the q-v loop, and R_g the additional losses in the tail of the loop. The combination L_h-R_h in the capacitive corona branch provides the time delay to simulate the statistical time lag observed in fast surges. The time constant of this branch has practically no effect on slow surges.

When the surge voltage $v(t)$ reaches the ideal corona onset voltage E_o, the two corona branches start conducting. Due to the much larger permittivity of the ionized air under corona, the capacitance C_{cor} becomes dominant and the circuit consists essentially of the capacitive corona branch C_{cor}-L_h-R_h in series with the normal air capacitance of the non-ionized region C_{a2}. (By contrast, in the conventional circuit of Figure 3(a), the capacitance C_{cor} is not limited to the ionized region but extends all the way from the conductor to ground.)

The diode in the proposed circuit of Figure 5, or in the simplified conception of Figure 3(a), stops conducting when the surge voltage reaches its peak value V_p. With the capacitive corona branch isolated from the circuit by the diode, the capacitance corresponding to the descending branch of the q-v loop is again the total air capacitance C_o, from conductor to ground, in both circuits. In the more realistic proposed circuit, at the moment the diode stops conducting the voltage split between capacitances C_{a1} and C_{a2} is not at their natural non-ionized air value (ratio of the capacitive voltage divider C_{a1}-C_{a2}) because capacitance C_{a2} was charged to a much higher value during the time the capacitive corona branch C_{cor}-L_h-R_h was connected by the diode.

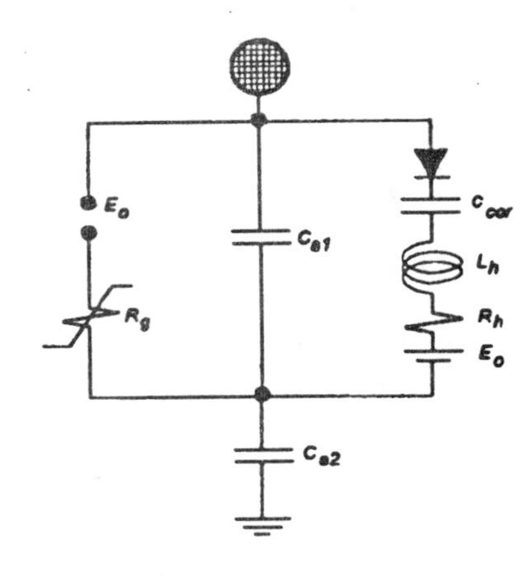

Figure 5 - Proposed corona model.

To restore the natural voltage ratio between capacitances C_{a1} and C_{a2}, a path must be provided for the charges to redistribute themselves. In the proposed circuit this path is provided by the resistance R_g. This resistance is switched into the circuit at the ideal inception voltage E_o by an air gap. Since the air gap does not cease to conduct until the applied voltage comes down to zero, R_g remains in the circuit during the entire descending part of the q-v characteristic. As the cycle proceeds along this descending branch, the voltages across capacitances C_{a1} and C_{a2} return to their natural ratio (values without corona) at a rate determined by the time constant of the circuit formed by C_{a1}, C_{a2} and R_g.

4. Equivalent Circuit Parameters

The process to derive the parameters of the equivalent circuit of Figure 5 can be explained with the help of the diagram of Figure 6. This diagram compares an idealized q-v loop with a real one. The real curve is built from the ideal curve by the effect of resistances R_h and R_g in shaping the lobe and the tail of the cycle, and by the time delay in the lobe of the curve provided by the time constant of the L_h-R_h combination.

The different effects of the various circuit parameters on the q-v curves of slow and fast surges provide asymptotic conditions to determine these parameters:

a) The loops for slow surges are not much affected by the statistical time lag, and their apparent corona inception voltage does not change. It is then easy to determine from these curves the value of E_{cor} and of the capacitances before and after corona. The DC source in the equivalent circuit is given by $E_o = C_{a2}/(C_{a1}+C_{a2})\cdot E_{cor}$. The initial slope of the q-v curve corresponds to the capacitance before corona, that is, $C_o = C_{a1}+C_{a2}$. By experience, it was found that the ratio $C_{a1}/C_{a2}\approx 1$, and therefore, $C_{a1}\approx C_{a2}\approx 2C_o$. The capacitance associated with the second slope of the curve includes the corona capacitance C_{cor} in combination with the air capacitances C_{a1} and C_{a2}

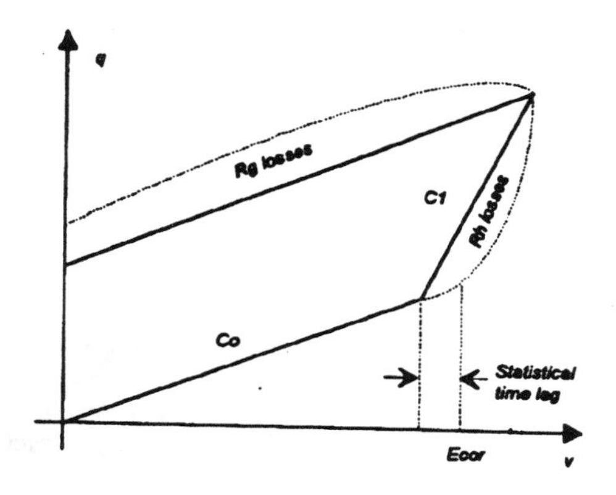

Figure 6 - Shaping of the q-v loop.

and is given by $C_1 = (C_{a1} + C_{cor}) \cdot C_{a2}/(C_{a1} + C_{a2} + C_{cor})$. Also, in the curves for switching surges, the influence of R_g is more noticeable than in the curves for lightning surges and R_g can be estimated from the additional losses in the tail of the loop in these curves.

b) The loops for fast surges are more sensitive to the values of R_h and L_h. The value of R_h can be estimated from the additional losses in the lobe of the loop, while the value of L_h/R_h is adjusted to match the apparent corona inception voltage E_{cor}.

In the opinion of the authors, the parameters of the proposed equivalent circuit could be determined more accurately if the measurements of corona included the current-time curves [14] in addition to the q-v curves, and thus provide a more complete picture of the phenomenon.

5. Simulation Results

Simulation of Q-V Curves

To validate the proposed model, simulations of q-v measurements were performed using the EMTP. The value of the parameters for the proposed circuit were obtained through observations of a large number of sets of q-v curves derived from the experimental measurements in [6] using the procedure explained above.

Figures 7 and 8 compare some of the q-v curves obtained with the proposed model of Figure 5 for surges varying from switching (260/2700 μs) to lightning (2.5/60 μs). The experimental curves of [6] were obtained for a single conductor of 3.04 cm of diameter.

All the simulated curves of Figures 7 and 8 for the proposed model were obtained with a fixed set of circuit parameters. The capacitances were C_{a1} = 21 pF, C_{a2} = 21 pF, and C_{cor} = 140 pF. Resistance R_g was a piecewise linear resistance with values of 90, 45 and 9 MΩ switched on at 0, 50 and 125 kV. The other parameters were L_h = 23 mH, R_h = 3 kΩ and E_o= 125 kV.

From the results of Figures 7 and 8, it can be seen that the main characteristics of the experimental tests, particularly in the ascending part of the q-v loops, are reproduced accurately by the proposed model with a fixed set of parameters.

The following general observations can be made in connection with the results of Figures 7 and 8:

1) The rise portion of the simulated q-v curves matches the experimental results very well for different types of surges.

2) The increase of onset voltage E_{cor} for faster surges is simulated in the proposed circuit by the dynamics of the R_h-L_h-C_{cor} branch using a constant value of DC source voltage E_o. The ability of the circuit to simulate this effect is more clearly observed in Figure 7.

3) The proposed circuit can reproduce the negative slopes observed at the tip of the loop in some of the measured q-v curves. This is achieved by the dynamic behaviour of the second-order circuit model and does not require the concept of a negative C in the circuit. This effect is more

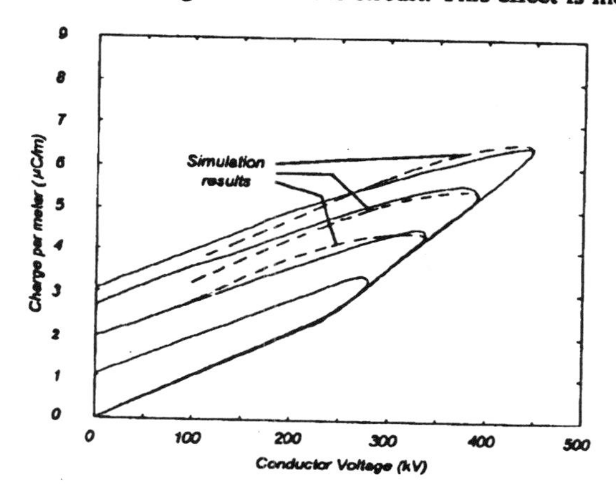

Figure 7 - Comparison between simulated and measured q-v curves for switching surges (1.2" conductor).

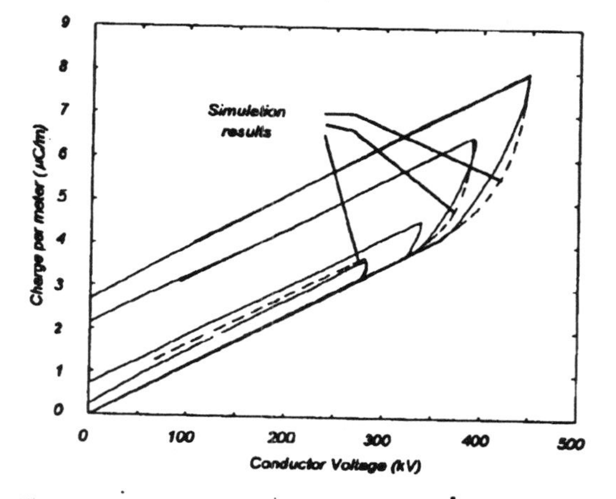

Figure 8 - Comparison between simulated and measured q-v curves for lightning surges (1.2" conductor).

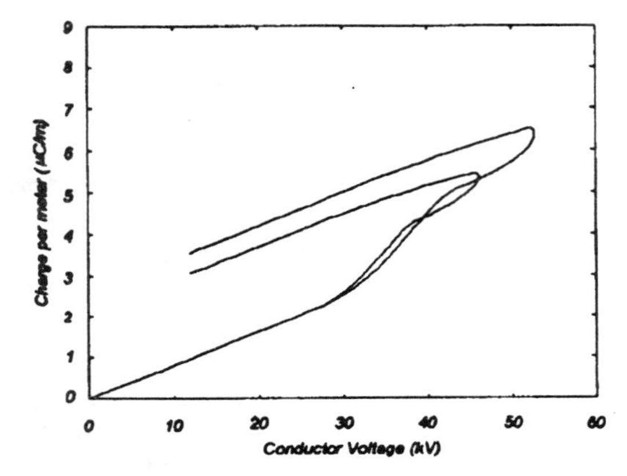

Figure 9 - Simulated q-v curves for thin conductor (0.65 mm) (rise time 1.2 µs).

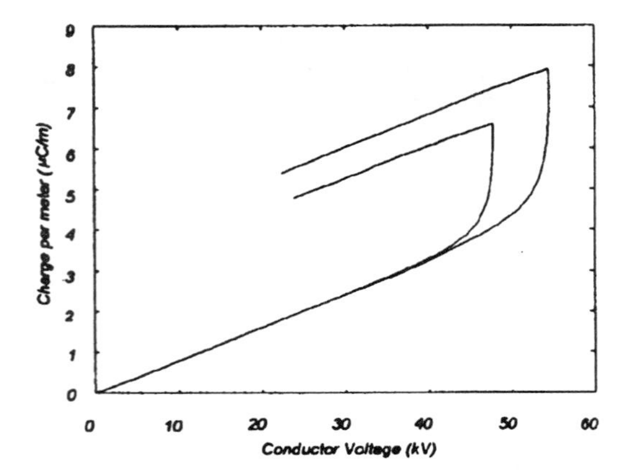

Figure 10 - Simulated q-v curves for thin conductor (0.65 mm) (rise time 0.12 µs).

noticeable in Figure 7.

4) The descending part of the simulated q-v curves does not fit exactly the measured ones. This effect is more noticeable in Figure 7.

Figures 9 and 10 show simulated q-v curves for the case of thin conductors. These simulations emphasize the capability of the model in reproducing the unusual dynamics observed in experimental measurements with thin conductors (Figures 2(c) and 2(d)). All the simulated curves were obtained with the same set of circuit parameters: $C_{a1} = 16$ pF, $C_{a2} = 16$ pF, and $C_{cor} = 100$ pF; R_g was a piecewise linear resistance of values 3 and 0.03 MΩ with switchovers at 0 and 117 kV; $L_h = 0.3$ mH, $R_h = 30$ Ω, and $E_o = 117$ kV.

Surge Propagation

A second set of simulations were performed for voltage surges propagating along a transmission line. In order to test the proposed model using standard EMTP components, the line is divided into short segments, with the corona model connected between them. In the case of surges with

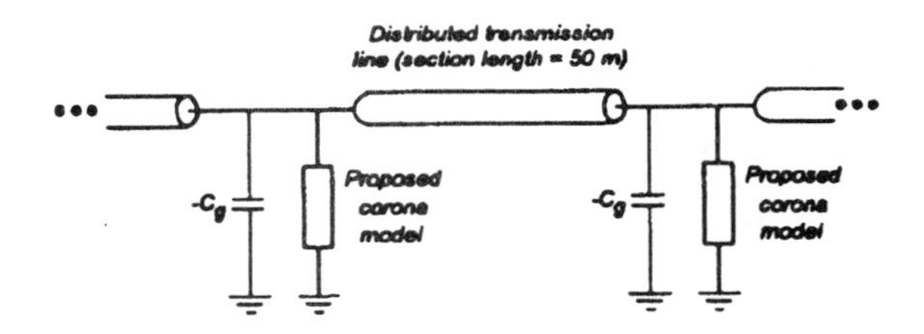

Figure 11 - Transmission line simulation.

rise times in the range of lightning, a segment length of 50 m is considered adequate [13]. Each line segment is represented using the distributed parameters model. Since the corona model already includes the geometric capacitance of the line, it is necessary to connect at each corona section a negative capacitance of the same magnitude, in order not to count it twice. Figure 11 shows a general scheme of one section of the modelled transmission line.

To test the dynamic behaviour of the proposed corona circuit and its ability to simulate a measured case of surge propagation, an equivalent circuit was set up to duplicate the results reported in [1] for the Tidd line. The experimental case has the following data:

1) Conductor type: ACSR 2.35 cm diameter.
2) Average conductor height: 18.9 m.
3) Applied impulse voltage: 1650 kV peak, 0.7 µs rise time.
4) Points of voltage measurement: 0, 0.62, 1.28 and 2.22 km.
5) Load impedance: 484 Ω.

A transmission line of 2.5 km was considered for the simulation. The line was terminated with a load impedance of 484 Ω. A comparison of the simulated and measured results is shown in Figure 12. The following observations can be made from these curves:

1) The surge's amplitude decreases as it travels.

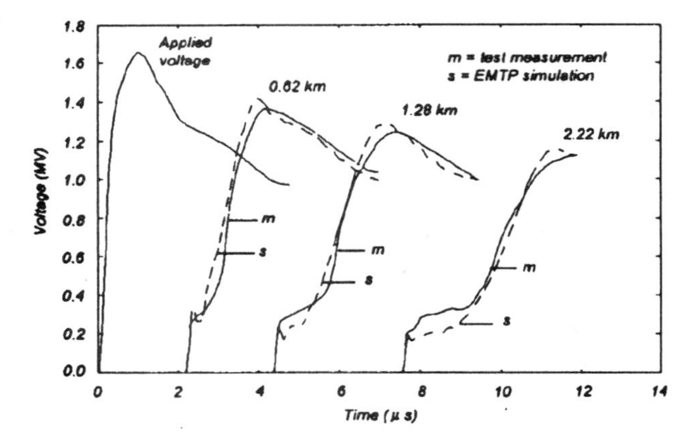

Figure 12 - Simulation and experimental results [11] for a travelling surge.

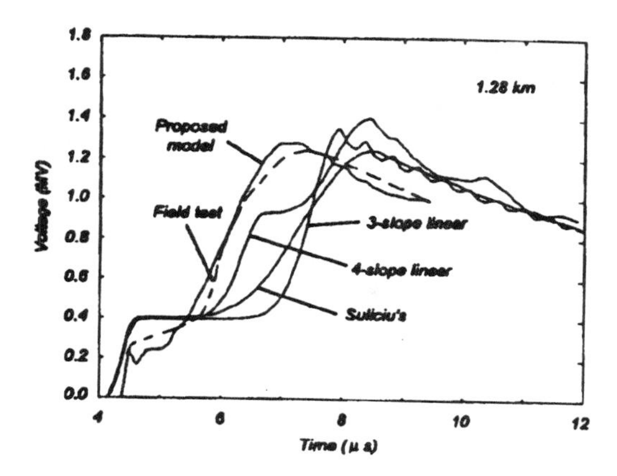

Figure 13 - Comparison between analytical methods [11] and the proposed model.

2) The apparent onset voltage decreases as the surge travels.

3) The steepness of the surge decreases as it travels.

Comparing the simulation and test results, it is seen that the proposed model is capable of following very well the changes in magnitude, apparent corona onset voltage, and steepness of the surge as it travels along the transmission line. This very good agreement is noticeable even for distances as long as 2.2 km from the surge injection point. This agreement was obtained despite the possible accumulated error of cascading more than forty line and corona sections in the EMTP model. The simulated results present small oscillations around the apparent onset value of voltage, which do not appear in the measured ones. These oscillations are due to the dynamics of the corona inception branch C_{cor}-L_h-R_h in Figure 5. Some of the results presented in the literature [1, 9] for other voltages and conductors have shown similar oscillations.

The results obtained with the proposed model were also compared with results obtained with other corona models in [13] for the Tidd line measurements. Figure 13 shows the waveforms at the relatively distant measuring point of 1.28 km for the field test, the proposed model, and the models reported in [13]. It can be seen that the proposed model matches much more closely the field measurement.

6. Conclusions

This paper proposes a corona model that can closely match, with a constant set of circuit parameters, the q-v characteristics of a wide range of fast and slow surges. Conventional models require different sets of parameters for different waveshapes. The proposed model achieves this wider frequency response by more closely matching the topology of the circuit to the topology of the actual conductor system and by using a circuit with a second-order dynamic response. The model is capable of simulating the apparent increase in the corona inception voltage for faster surges without changing the value of E_o in the equivalent circuit. To our knowledge, this is the first time in the reported literature that this can be achieved using an equivalent circuit with constant parameters. Due to its dynamic characteristics, the model can also reproduce the negative slope at the upper tip of q-v loop with normal positive values of the circuit parameters.

The q-v response of the model matches very closely the experimentally measured q-v response of the conductors in the ascending branch of the loop for a wide range of fast and slow surges. This region includes the statistical time lag needed to initiate the electron avalanche process.

Previous dynamic models, such as Suliciu's, are not directly realizable with simple circuit elements and require the solution of a system of differential equations to achieve a dynamic adaptation of the model to the changing q-v characteristic. The proposed model is much simpler than Suliciu's and is based on standard circuit components available in the EMTP.

The results obtained with the proposed model are not totally satisfactory in the representation of the descending branch of the q-v loops and further research is needed to better model this region.

The simulation of travelling surges for the Tidd line case shows that the proposed model can match the general behaviour of the phenomenon better than previously published comparisons. The Tidd line test, however, has a number of uncertainties regarding the accuracy of the data and of the measurements. A more thorough assessment of the proposed model in terms of surge propagation would probably require a fresh set of experiments with well-known line data and more accurate measurements.

Work is currently under progress at the University of British Columbia to implement the proposed corona model as part of a new multiphase transmission line model in the EMTP.

Acknowledgment

The authors wish to express their acknowledgement to the System Planning Department of B. C. Hydro and Power Authority, and to the Natural Sciences and Engineering Research Council of Canada for their financial support. Also gratefully acknowledged is the financial support from CAPES and Universidade Federal do Rio de Janeiro, Brasil during Dr. Nelson Santiago's work as Visiting Professor at the University of British Columbia.

References

[1] C. F. Wagner, I. W. Gross, B. L. Lloyd, "High Voltage Impulse Test on Transmission Lines," *AIEE Trans.*, Vol. 73 - Pt. III, pp. 196-210, April 1954.

[2] C. F. Wagner, B. L. Lloyd, "Effects of Corona on Travelling Waves," *AIEE Trans.*, Vol 74 - Pt. III, pp. 858-872, Oct. 1955.

[3] T. Giao, J. Jordan, "Modes of Corona Discharges in Air," *IEEE Trans. on PAS*, Vol. PAS-87, No. 5, pp. 1207-1215, May 1968.

[4] R. T. Waters, T. E. S. Rickard, and W. B. Stark, "Direct Measurement of Electric Field at Line Conductors During A. C. Corona," Proc. IEE, Vol. 119, No. 6, pp. 717-723, June 1972.

[5] IEEE, *EHV Transmission Line Corona Effects*, IEEE Tutorial Course 72 CH0644-5 PWR, New York, IEEE Press, 1972.

[6] P. S. Maruvada, H. Menemenlis, R. Malewski, "Corona Characteristic of Conductor Bundles Under Impulse Voltages," *IEEE Trans. on PAS*, Vol. PAS-96, No. 1, pp. 102-115, Jan/Feb 1977.

[7] M. M. Suliciu, I. Suliciu, "A Rate Type Constitutive Equation for the Description of the Corona Effect," *IEEE Trans. on PAS*, Vol. PAS-100, pp. 3681-3685, August 1981.

[8] EPRI. *Transmission Line Reference Book - 345 kV and Above*, USA, Electric Power Research Institute, 1982, pp. 169-203.

[9] C. Gary, D. Cristescu, G. Dragan, "Distortion and Attenuation of Travelling Waves Caused by Transient Corona," *CIGRE Report*, Paris, 1989.

[10] N. Santiago, A. Junqueira, C. Portela, A. Pinto, "Attenuation of Surges in Transmission Lines Due to Corona Effect - Three Phase Modelling," *Proc. 7th International Symposium on High Voltage Engineering*, Paper 84.06, Dresden, August 1991.

[11] N. Harid and R. T. Waters, "Statistical Study of Impulse Corona Inception Parameters on Line Conductors," IEE Proceedings-A, Vol. 138, No. 3, pp. 161-168, May 1991.

[12] S. Carneiro, "A Comparative Study of Some Corona Models and Their Implementation in the EMTP," *Trans. CEA, Engineering and Operating Division, Vol. 27, 1988.*

[13] S. Carneiro, J. Martí, "Evaluation of Corona and Line Models in Electromagnetic Transients Simulations," *IEEE Trans. on Power Delivery*, Vol. PWRD-6, No. 1, pp. 334-341, Jan. 1991.

[14] N. Santiago, F. Castellanos, "Physical Aspects of Corona Effect During Transient Overvoltages and Their Simulation with Circuit Models," *Proc. IASTED International Conference- Power Systems and Eng.*, Vancouver, pp. 9-14, Aug. 1992.

[15] J. R. Martí, "Accurate Modelling of Frequency-Dependent Transmission Lines in Electromagnetic Transient Simulations," IEEE Trans., vol. PAS-101, 1982.

[16] J.R. Martí, F. Castellanos, and N. Santiago, "A wide bandwidth corona model," Proceedings PSCC, 11th Power System Computation Conference, Avignon, France, Aug 30-Sept 3, 1993, pp. 899-905.

Biographies

José R. Martí (M'71) was born in Lérida, Spain in 1948. He received the degree of Electrical Engineer from Central University of Venezuela in 1971, the degree of M.E.E.P.E from Rensselaer Polytechnic Institute in 1974 and the Ph.D. degree from the University of British Columbia in 1981. In Venezuela, he worked for industry and taught at Central University of Venezuela. At present he teaches at the University of British Columbia, Canada. Dr. Martí has been involved for a number of years in the development of models and solution techniques for the transient analysis program EMTP.

Fernando Castellanos Trujillo. (M'82) was born in Bogotá, Colombia in 1961. He received the degree of Electrical Engineer and the M.Sc. degree from Universidad de Los Andes, Colombia, in 1983 and 1986, respectively. He has worked for industry, consulting companies and taught at Universidad de Los Andes. He is currently a Ph.D. candidate at the University of British Columbia, Vancouver, Canada.

Nelson Henrique Costa Santiago. was born in Brasil in 1950. He received the degree of Electrical Engineer from the Pontificia Universidade Católica do Rio de Janeiro, Brasil, in 1973, and the M.Sc. and D.Sc. degrees in Electrical Engineering from COPPE-Universidade Federal do Rio de Janeiro, Brasil, in 1982 and 1987, respectively. From 1975 to date, he has been with the Electrical Engineering Department of Universidade Federal do Rio de Janeiro. Since 1973, Dr. Santiago has been involved, as consultant of several Brazilian utilities, in a number of power system projects, including the well-known ITAIPU project. From January 1992 until June 1993 he was a Visiting Professor at the University of British Columbia, Vancouver, Canada.

Discussion

Adam Semlyen (University of Toronto): I would like to congratulate the authors for their interesting and practical contribution to the simulation of corona in the calculation of electromagnetic transients.

The physical phenomenon of corona is very complex but the authors have provided a simple approximation to the experimental data available for both lightning and switching surges. The parameters of the model have been skillfully tuned to closely match the measured q-v curves, including the differences related to the speed of the applied impulses. The essential improvement with respect to previously proposed models seems to be the presence of the inductance L_s (Figure 5) which cannot be related to any inductive effect in the corona phenomenon but is useful for simulating the observed dynamic effects. The approach is similar to that in the dynamic corona model of reference [A] where $v = f(q, \dot{q})$ (equation (10)). There too, the IREQ measurements of [6] have been used for fitting. Regarding the identification of the parameters, I was a bit puzzled by the authors' remark that current-time data would have been useful: indeed, since $v(t)$ is known from the description of the impulse (as a double exponential with specified time constants), $q(t)$ can be inferred and thus $i(t)$ becomes readily available.

The focus of the paper is on the propagation of unidirectional pulses. Especially in switching phenomena changes of polarity are however quite common and neither the measurements of reference [6] nor the simulation of the paper cover the case of the corona phenomenon under variable polarity inputs. How would the authors expand their corona model to permit the simulation of corona effects under general transient conditions? We have addressed this problem in reference [A] but the computational effort required for following the physical phenomena around the conductor under corona was higher than in the case of the present paper due to the complexity and history-dependence of the phenomena. We also have found it useful to limit the corona model by a cylindrical surface (as in Figure 4a) and to subtract the line capacitance (as in Figure 11 of the paper); see Figures 10 and 11 of [A]. As seen from the latter, this separation of the corona model and association to each phase-conductor facilitates the modeling for three phase transmission lines [B].

[A] A. Semlyen and W.-G. Huang, "Corona Modelling for the Calculation of Transients on Transmission Lines", IEEE Transactions on Power Delivery, Vol. PWRD-1, No. 3, July 1986, pp. 228-239.

[B] W.-G. Huang and A. Semlyen, "Computation of Electro-Magnetic Transients on Three-Phase Transmission Lines with Corona and Frequency Dependent Parameters", IEEE Transactions on Power Delivery, Vol. PWRD-2, No. 3, July 1987, pp. 887-898.

Manuscript received August 9, 1994.

M.T. Correia de Barros, IST / Technical University of Lisbon, Portugal:
The authors are congratulated for their valuable work on corona modeling using a circuit approach. This work may prove to be very useful for practical applications, in particular if its generalization to multiphase systems is successfully achieved.

The present discussion is composed of the following two parts:

(A) Corona modeling

The paper presents a very interesting circuit synthesis exercise for corona modeling, in particular as it allows the circuit parameters to remain unchanged for a given conductor, independently from the time constants of the applied surge. However, the discusser considers that the attempt to give physical meaning to the circuit parameters is in general not successful.

In deriving the parameters of the proposed equivalent circuit, the authors mention that, by experience, it was found that the ratio between the air capacitances (C_{a1}/C_{a2}) is approximately unitary. Is the fitting process for these circuit parameters based on the evaluation of r_{cor}? How is this radius evaluated? Is it independent from all the parameters of the applied voltage?

If the interpretation and separation of the R_h and R_g losses given in Fig. 6 is applied to the case of Fig 2 (c) - see Fig. below - how is it possible to explain the behavior of the charge during the wave tail? Is the developed circuit model able to reproduce this part of the experimental results?

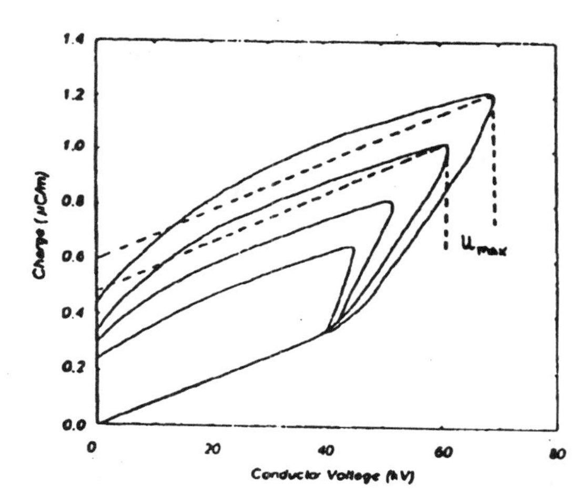

(c) Thin conductor (diam 0.65 mm), rise time 0.12 μs [7]

In the simulation of q-v curves, the authors give the values of the circuit parameters corresponding to figs. 7-8 and 9-10, respectively. It would be very interesting to have also the circuit parameters corresponding to the Tidd line conductor considered for the computation of surge propagation. To which type of surge correspond the experimental q-v curves that have been used for the circuit parameter fitting?

For practical applications, it is important to be able to use a corona model for bipolar transient overvoltages. Have the authors considered the possibility of generalizing the proposed model in order to allow such use?

(B) Computation of surge propagation

Corona is taken into account by inserting a corona circuit every 50 m. The proposed circuit is able to take into account the dependence of the charge-voltage curve on the voltage rate of rise. However, considering the values of the surge rise time and the length of the line segments, isn't the accuracy of the corona model being wasted?

The corona circuit model includes the line capacitance in the absence of corona (geometric capacitance), which is also included in the line segments model. How have the authors overcome this problem?

When comparing the results of surge propagation obtained from the proposed model with results obtained from other corona models, the authors claim that the proposed model allow a much better fitting of the experimental results. Couldn't a better choice of the modeling parameters used with the other models [13] give results closer to the experimental tests? It would be interesting to know which q-v curves were used for the corona simulation parameters fitting in [13] for the Tidd line.

Manuscript received August 22, 1994.

J. R. Martí, F. Castellanos, N. Santiago. We thank the discussers for their interest and valuable comments.

Corona understanding and modelling is a difficult problem that touches fundamental physical phenomena and stretches the capabilities of normal circuit simulation by combining frequency dependence, nonlinearity, and distributed parameters. We feel that after a relatively long time gap of research in this area much progress is being achieved recently, both in the understanding of the phenomena ([A], [B]) and in its modelling ([C], [D], [F]). Our paper's intended contribution is in the second area.

We will refer next to Part (A) of Dr. Correia de Barros' questions. We agree with her comments in the first paragraph of Part (A) regarding the difficulty of relating some of our proposed circuit-model components with the current knowledge of the physical corona phenomena. However, for as long as these phenomena can be described by differential equations, and we can find a combination of elementary circuit components that responds according to these equations, the model is valid from a network synthesis point of view. A danger that is always present in a mathematical description of a physical phenomena is the possible numerical stability of the model. Our proposed corona model includes only positive time constants and is numerically absolutely stable.

Dr. Correia de Barros' second question refers to the air capacitances C_{a1} and C_{a2} and their relationship to r_{cor}. A more detailed explanation of the procedure to calculate these capacitances on a test cage setup is given next.

The first parameter that can be easily calculated or measured from the q-v curves is the total geometric capacitance of the cage C_g. Next, the capacitances C_{a1} and C_{a2} should be estimated. Their value depends on the maximum corona radius, r_{cor}. Despite the complexity of the electric field surrounding the conductor, r_{cor} can be estimated under the following assumptions:

1) The electric field at the surface of the conductor is equal to Peek's critical field E_c.
2) Streamers (charge movement) develop nearby as long as the electrical field in front of them is not lower than a critical field E_{cri} [B].
3) The electric field inside the ionization cloud can be approximated by the electric field due to the conductor charges.

It then follows that the voltage peak V_p applied to a conductor of radius r at the centre of a cylindrical cage of radius R can be calculated as:

$$Vp = E_{cri} \cdot r_{cor}\left[\frac{E_c \cdot r}{E_{cri} \cdot r_{cor}} \ln\left(\frac{r_{cor}}{r}\right) + \ln\left(\frac{R}{r_{cor}}\right)\right]$$

To determine the maximum radius r_{cor} of the corona charge, this equation can be solved by a Newton Raphson method for a given voltage peak V_p. With this value the capacitances C_{a1} and C_{a2} can then be calculated.

If different values of peak voltage are used, average values for the parameters can be used and the model still gives good responses as long as the range of peak voltage values produces small variations in the values of capacitances C_{a1} and C_{a2} .(For the case presented in the paper the ratio of the average values was close to one.)

Dr. Correia de Barros' third question refers to the tail portions of the loops in Fig. 2 (c) of the paper and their relationship to the proposed circuit model. As mentioned in the Conclusions section of the paper, we are not totally happy with the fitting results for the descending part of the q-v loop and we are continuing our investigation of this aspect. This region's behaviour seems more of a problem in the case of thin conductors, where the dynamics of the corona phenomena seem to be more complicated.

We appreciate the opportunity given to us by Dr. Correia de Barros' fourth question to elaborate further on the procedure to obtain the proposed model parameters (Fig. 5 in the paper) from experimental data. Experimental q-v curves are usually obtained for cage setups. However, for as long as the calculated C_{a1} for the cage test is close to the calculated C_{a1} for the conductor above ground, the part of the circuit-model that corresponds to the ionized air (entire circuit less C_{a2}) will not be too different for the cage or for the overhead line. This observation is based on the fact that the electric field distribution between r and r_{cor} is very similar for both cases (for as long as r_{cor} << distance to ground). Under these conditions, to transfer the model from the cage to the line, the only change needed is in the value of the air capacitance C_{a2}.

In the case of the Tidd line, the correlation between cage and line cannot be accomplished easily due to the lack of a proper set of published q-v curves for the conductor. As just explained, however, a good approximation could still be obtained by using the q-v curves presented in the paper for a similar conductor if the capacitances C_{a1} proved to be close enough. Unfortunately, this was not the case using the Tidd line's published data. Given the limited available data for this case, an estimation of the circuit-model parameters was made based on the geometry of the line, the q-v curves of a similar conductor, and the measured waveshapes. This estimation resulted in a circuit-model that was able, as shown in the paper, of reproducing the measured waveshapes as they travel down the line better than previously published comparisons. These results prove that given an appropriate set of circuit parameters the proposed circuit-model can reproduce very well the dynamics of the wave distortion as the wave travels down the line. The parameters used for the Tidd line simulation were (for a 50 m section): C_{a1} = 0.81 nF, C_{a2} = 0.6 nF, C_{cor} = 5.1 nF, L_h = .04 mH, R_h = 200 Ω and E_o= 110 kV.

Dr. Correia de Barros' fifth question in Part (A) of her discussion refers to the generalization of the model for bipolar surges. We consider this an important issue and we are working on a generalization of the proposed model for asymmetrical bipolar behaviour.

In part (B) of her discussion, Dr. Correia de Barros expresses her concern about the length of the line sections (50 m) used in the simulations. In the Tidd line case, the applied source has a rise time of about 0.7 μs, which corresponds to about four sections per wavefront. In our experience this resolution is enough for a reasonable tracing of the wavefront. This was confirmed by the ability of the model to follow quite closely the various distortions of the waveshape as it travels down the line (Fig. 12). A problem of lumping the corona branches at finite-length sections is the reflections of the wave at the junctions between sections. These appear as ripple oscillations in the simulations (even though these ripples were not excessive in the presented simulations). We are in the process of developing a space-sectionalized model, following the principles in [C], which will alleviate the problem of reflections between line sections.

Regarding the second question on Part (B), Fig. 11 in the paper shows that a negative capacitance $-C_g$ is connected in parallel with the corona branch to avoid "counting twice" the line's geometric capacitance. This has been suggested by Dr. Semlyen in [D] and [E].

Regarding the comparison with previous Tidd line simulations in Fig. 13 of the paper, we would like to point out that even though a better choice of parameters in the other models could have improved the fitting of the waveshape at a particular line length from the source, it is the ability of the proposed model to adapt its response to the changing magnitude and frequency content of the travelling waveform that allows it to remain accurate at increasing distances from the source (Fig. 12 in the paper).

We address next Dr. Semlyen's comments. Indeed, the dynamic relation $v = f(q, q')$ in equation (10) of reference [D] achieves a similar effect to the "non-physical" inductance in our proposed corona circuit-model. We apologize for the omission of not directly referencing this excellent work in the main body of our paper.

As indicated in the paper, we do not consider the results obtained with the proposed circuit model in the simulation of the descending branch of the q-v loops to be yet satisfactory. It is believed, as discussed in reference [14], that a better understanding of the turnaround phenomena in the loop would be gained from direct measurements of the corona current. Due to the difficulties in establishing very accurate experimental setups for corona measurements, the "higher definition" required in the loop's turnaround region is hard to obtain taking $i = dq/dt$ from the measured data.

We concur with Dr. Semlyen (as also expressed by Dr. Correia de Barros) on the need for a bi-directional corona model. In connection with our circuit model of Fig. 5, we have been experimenting with a dual-branch model in which positive or negative branches conduct according to the polarity of the voltage. Using different circuit parameters for both branches, we can represent the different positive and negative corona polarization characteristics.

Once more we want to thank the discussers for the interesting and detailed comments.

References

[A]. C. de Jesus, M. T Correia de Barros, "Modelling of Corona Dynamics for Surge Propagation Studies", *IEEE/PES 1994 Winter Meeting,* 94WM043-0 PWRD, New York, February 1994.

[B]. R. T. Waters, D. M. German, A. E. Davies, N. Harid, H. SB. Eloyyan, "Twin Conductor Surge Corona", *Fifth International Symposium on High Voltage Engineering,* (4 pages), Federal Republic of Germany, 24-28 August 1987.

[C]. H. M. Barros, S. Carneiro, R. M. Azevedo, "An Efficient Recursive Scheme for The Simulation of Overvoltages on Multiphase Systems Under Corona", *IEEE/PES 1994 Summer Meeting,* 94SM468-9 PWRD, San Francisco, CA, July 1994.

[D]. A. Semlyen and W. G. Huang, "Corona Modelling for the Calculation of Transients on Transmission Lines", *IEEE Transactions on Power Delivery,* Vol. PWRD-1, No.3, July 1986, pp. 228-239.

[E]. W. G. Huang and A. Semlyen, "Computation of Electro-Magnetic Transients on Three-Phase Transmission Lines with Corona and Frequency Dependent Parameters", IEEE Transactions on Power Delivery, Vol. PWRD-2, No.3, July 1987, pp. 887-898.

Manuscript received November 23, 1994.

IEEE Transactions on Power Apparatus and Systems, Vol. PAS-99, No. 3 May/June 1980

A GENERAL FORMULATION OF IMPEDANCE AND ADMITTANCE OF CABLES

A. AMETANI
Doshisha University
Kyoto, Japan

ABSTRACT

Interest in the analysis of wave propagation characteristics and transients associated with cable systems has rapidly increased. In order to answer the need of the analyst, impedances and admittances of various cables have to be known. This paper describes a general formulation of impedances and admittances of single-core coaxial and pipe-type cables. The formulation presented here can handle a coaxial cable consisting of a core, sheath and armor, a pipe-type cable of which the pipe thickness is finite and an overhead cable, which has not been discussed in the literature heretofore.

Using the formulation presented in this paper, it now becomes possible to analyze wave propagation characteristics and transients on any type of cable system.

1. INTRODUCTION

The growing use of cable systems and the increasing levels of capacity makes the analysis of wave propagation characteristics and transients on cable systems an important task. The cases of underground single-core coaxial cables (SC cables) consisting of a core and sheath, and pipe-type cables (PT cables) of which the pipe thickness is assumed to be infinite have been well studied.[1-5] However, SC cables consisting of a core, sheath and armor, PT cables of which the pipe thickness is finite, and overhead cables have never been studied.

SC cables with a core, sheath and armor are quite often seen in the submarine cable case, and, in fact, this author has been asked about the possibility of calculating transients on such cables.[6] So far, pipe enclosures were assumed to act as complete shields, thus avoiding consideration of earth return currents. As far as wave propagation characteristics and transients on inner conductors of the pipe enclosure are concerned, the assumption of the infinite thickness of the pipe is quite acceptable. But once the wave propagation and transients on the pipe are to be included, all the previous studies are not applicable. Thus, we need a way of handling voltage and current on the pipe.

An analysis of overhead cables seemed to be overdue, although there is some need for it. This author has been asked to calculate transients in a gas insulated substation, where a bus and circuit breaker are enclosed in a pipe, and the pipe is overhead.[7] This can be considered to be an overhead cable.

Because of the situation explained above, a formulation of impedances and admittances, which is able to deal with an SC cable consisting of a core, sheath and armor, a PT cable having a finite thickness of the pipe, and an overhead cable, has been developed in the present paper. The formulation is carried out in a generalized manner so as to be able to handle all the above cases.

F 79 615-6 A paper recommended and approved by the IEEE Insulated Conductors Committee of the IEEE Power Engineering Society for presentation at the IEEE PES Summer Meeting, Vancouver, British Columbia, Canada, July 15-20, 1979. Manuscript submitted January 26, 1979; made available for printing May 1, 1979.

2. IMPEDANCE AND ADMITTANCE

The impedance and admittance of a cable system are defined in the two matrix equations.

$$d(V)/dx = -[Z]\cdot(I) \quad (1)$$

$$d(I)/dx = -[Y]\cdot(V) \quad (2)$$

where (V) and (I) are vectors of the voltages and currents at a distance x along the cable. [Z] and [Y] are square matrices of the impedance and admittance.

In general, the impedance and admittance matrices of a cable can be expressed in the following forms.

$$[Z] = [Z_i] + [Z_p] + [Z_c] + [Z_o] \quad (3)$$

$$\left.\begin{aligned} [Y] &= s\cdot[P]^{-1} \\ [P] &= [P_i] + [P_p] + [P_c] + [P_o] \end{aligned}\right\} \quad (4)$$

where [P] is a potential coefficient matrix, and $s=j\omega$.

In the above equations, the matrices with subscript "i" concern an SC cable and the matrices with subscripts "p" and "c" are related to a pipe enclosure. The matrices with subscript "o" concern cable outer media, i.e. air space and earth. When a cable has no pipe enclosure, there exists no matrix with subscripts "p" and "c".

In the formulation presented here, the following assumptions are made.

(1) The displacement currents and dielectric losses are negligible.
(2) Each conducting medium of a cable has constant permeability.
(3) The pipe thickness is greater than the penetration depth of the pipe wall for the PT cable case.

The details will be explained in the following sections.

2.1 Single-Core Coaxial Cable (SC cable)

2.1.1 Impedance

When an SC cable consists of a core, sheath and armor as shown in Fig.1 (a), the impedance is given in the following form based on the result of Appendix 1.

$$[Z] = [Z_i] + [Z_o] \quad (5)$$

where

$[Z_i]$ = SC cable internal impedance matrix

$$= \begin{bmatrix} [Z_{i1}] & [0] & \cdots & [0] \\ [0] & [Z_{i2}] & \cdots & [0] \\ \vdots & \vdots & \ddots & \vdots \\ [0] & [0] & \cdots & [Z_{in}] \end{bmatrix} \quad (6)$$

$[Z_o]$ = impedance matrix of the cable outer medium (earth return impedance)

$$= \begin{bmatrix} [Z_{o11}] & [Z_{o12}] & \cdots & [Z_{o1n}] \\ [Z_{o12}] & [Z_{o22}] & \cdots & [Z_{o2n}] \\ \vdots & \vdots & \ddots & \vdots \\ [Z_{o1n}] & [Z_{o2n}] & \cdots & [Z_{onn}] \end{bmatrix} \quad (7)$$

All the off-diagonal submatrices of $[Z_i]$ are zero.

0018-9510/80/0500-0902$00.75 © 1980 IEEE

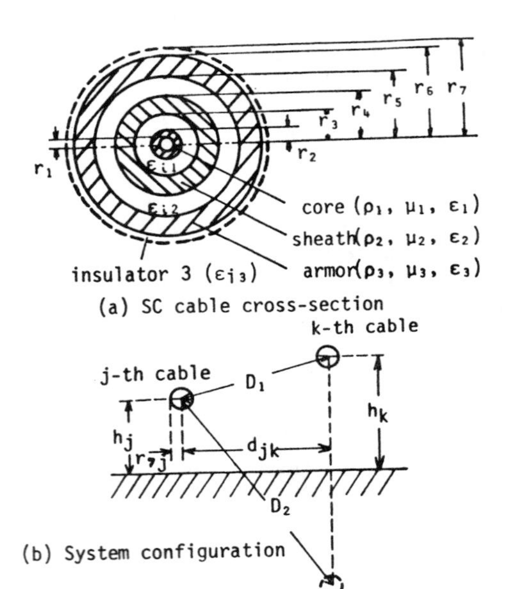

Fig. 1 An SC cable system

A diagonal submatrix expresses the self-impedance matrix of an SC cable. When the SC cable consists of a core, sheath and armor, the self-impedance matrix is given by:

$$[Z_{ij}] = \begin{bmatrix} Z_{ccj} & Z_{csj} & Z_{caj} \\ Z_{csj} & Z_{ssj} & Z_{saj} \\ Z_{caj} & Z_{saj} & Z_{aaj} \end{bmatrix} \quad (8)$$

where

Z_{ccj} = core self-impedance

$= z_{cs} + z_{sa} + z_{a4} - 2z_{2m} - 2z_{3m}$

Z_{ssj} = sheath self-impedance

$= z_{sa} + z_{a4} - 2z_{3m}$

Z_{aaj} = armor self-impedance $= z_{a4}$

Z_{csj} = mutual impedance between the core and sheath

$= z_{sa} + z_{a4} - z_{2m} - 2z_{3m}$

Z_{caj} = mutual impedance between the core and armor

$= z_{a4} - z_{3m}$

Z_{saj} = mutual impedance between the sheath and armor

$= Z_{caj}$ (9)

where

$$\left.\begin{aligned} z_{cs} &= z_{11} + z_{12} + z_{2i} \\ z_{sa} &= z_{20} + z_{23} + z_{3i} \\ z_{a4} &= z_{30} + z_{34} \end{aligned}\right\} \quad (10)$$

When the SC cable consists of a core and sheath, the matrix of eq.(8) is reduced to a 2 x 2 matrix.

$$[Z_{ij}] = \begin{bmatrix} Z_{ccj} & Z_{csj} \\ Z_{csj} & Z_{ssj} \end{bmatrix} \quad (11)$$

where

$$\left.\begin{aligned} Z_{ccj} &= z_{cs} + z_{s3} - 2z_{2m} \\ Z_{ssj} &= z_{s3} \\ Z_{csj} &= z_{s3} - z_{2m} \\ \text{and} \\ z_{s3} &= z_{20} + z_{23} \end{aligned}\right\} \quad (12)$$

If an SC cable consists only of a core, the submatrix is redued to one element.

$$[Z_{ij}] = Z_{ccj} = z_{11} + z_{12} \quad (13)$$

The component impedances per unit length in the above equations are given in the following equaions for an SC cable shown in Fig.1(a).[1,9,10]

(1) z_{11} : internal impedance of core outer surface

$$z_{11} = (s\mu_0\mu_1/2\pi)\cdot(1/x_2D_1)\cdot\{I_0(x_2)\cdot K_1(x_1) + K_0(x_2)\cdot I_1(x_1)\}$$

(2) z_{12} : core outer insulator impedance

$$z_{12} = (s\mu_0\mu_{i1}/2\pi)\cdot \ln(r_3/r_2)$$

(3) z_{2i} : internal impedance of sheath inner surface

$$z_{2i} = (s\mu_0\mu_2/2\pi)\cdot(1/x_3D_2)\cdot\{I_0(x_3)\cdot K_1(x_4) + K_0(x_3)\cdot I_1(x_4)\}$$

(4) z_{2m} : sheath mutual impedance

$$z_{2m} = \rho_2/2\pi r_3 r_4 D_2$$

(5) z_{20} : internal impedance of sheath outer surface

$$z_{20} = (s\mu_0\mu_2/2\pi)\cdot(1/x_4D_2)\cdot\{I_0(x_4)\cdot K_1(x_3) + K_0(x_4)\cdot I_1(x_3)\}$$

(6) z_{23} : sheath outer insulator impedance

$$z_{23} = (s\mu_0\mu_{i2}/2\pi)\cdot \ln(r_5/r_4)$$

(7) z_{3i} : internal impedance of armor inner surface

$$z_{3i} = (s\mu_0\mu_3/2\pi)\cdot(1/x_5D_3)\cdot\{I_0(x_5)\cdot K_1(x_6) + K_0(x_5)\cdot I_1(x_6)\}$$

(8) z_{3m} : armor mutual impedance

$$z_{3m} = \rho_3/2\pi r_5 r_6 D_3$$

(9) z_{30} : internal impedance of armor outer surface

$$z_{30} = (s\mu_0\mu_3/2\pi)\cdot(1/x_6D_3)\cdot\{I_0(x_6)\cdot K_1(x_5) + K_0(x_6)\cdot I_1(x_5)\}$$

(10) z_{34} : armor outer insulator impedance

$$z_{34} = (s\mu_0\mu_{i3}/2\pi)\cdot \ln(r_7/r_6)$$

where

$$D_1 = I_1(x_2)\cdot K_1(x_1) - I_1(x_1)\cdot K_1(x_2)$$

$$D_2 = I_1(x_4)\cdot K_1(x_3) - I_1(x_3)\cdot K_1(x_4)$$

$$D_3 = I_1(x_6)\cdot K_1(x_5) - I_1(x_5)\cdot K_1(x_6)$$

$$x_k = \beta_k\sqrt{s}\ ,\ \beta_2 = r_2\sqrt{\mu_0\mu_1/\rho_1}$$

$$\beta_3 = r_3\sqrt{\mu_0\mu_2/\rho_2}\ ,\quad \beta_4 = r_4\sqrt{\mu_0\mu_2/\rho_2}$$

$$\beta_5 = r_5\sqrt{\mu_0\mu_3/\rho_3}\ ,\quad \beta_6 = r_6\sqrt{\mu_0\mu_3/\rho_3}$$

A submatrix of the earth return impedance $[Z_o]$ in eq.(7) is given in the following form.

$$[Z_{ojk}] = \begin{bmatrix} Z_{ojk} & Z_{ojk} & Z_{ojk} \\ Z_{ojk} & Z_{ojk} & Z_{ojk} \\ Z_{ojk} & Z_{ojk} & Z_{ojk} \end{bmatrix} \quad (14)$$

When the SC cable consists of a core and sheath, the above matrix is reduced to:

$$[Z_{ojk}] = \begin{bmatrix} Z_{ojk} & Z_{ojk} \\ Z_{ojk} & Z_{ojk} \end{bmatrix} \quad (15)$$

If the SC cable consists only of a core, the matrix includes only one element.

$$[Z_{ojk}] = Z_{ojk} \quad (16)$$

Z_{ojk} in eqs.(14) to (16) is the earth return impedance between the j-th and k-th cables. When a cable system is overhead, the impedance is given by Carson![11] When a cable system is underground, the impedance given by Pollaczek[12] is used. If a cable is above a stratified earth, the earth return impedance developed by Nakagawa, et.al.[13] can be used.

2.1.2 Potential coefficient

The admittance matrix of a cable system is evaluated from the potential coefficient matrix as given in eq.(4). In the SC cable case, $[P_p]$ and $[P_c]$ are zero, and when the cable system is underground, $[P_o]$ is also zero. Thus, based on the result of Appendix 2,

(1) Overhead cable

$$[P] = [P_i] + [P_o] \quad (17)$$

(2) Underground cable

$$[P] = [P_i] \quad (18)$$

where

$[P_i]$ = cable internal potential coefficient matrix

$$= \begin{bmatrix} [P_{i1}] & & 0 \\ & [P_{i2}] & \\ 0 & & \ddots \ [P_{in}] \end{bmatrix} \quad (19)$$

$[P_o]$ = potential coefficient matrix of the system in air

$$= \begin{bmatrix} [P_{o11}] & [P_{o12}] & \cdots & [P_{o1n}] \\ [P_{o12}] & [P_{o22}] & \cdots & [P_{o2n}] \\ \vdots & \vdots & \ddots & \vdots \\ [P_{o1n}] & [P_{o2n}] & \cdots & [P_{onn}] \end{bmatrix} \quad (20)$$

All the off-diagonal submatrices of $[P_i]$ are zero. A diagonal submatrix expresses the potential coefficient matrix of an SC cable. When the SC cable consists of a core, sheath and armor as shown in Fig.1 (a), the diagonal submatrix is given in the following form. (See Appendix 2.)

$$[P_{ij}] = \begin{bmatrix} P_{cj}+P_{sj}+P_{aj} & P_{sj}+P_{aj} & P_{aj} \\ P_{sj}+P_{aj} & P_{sj}+P_{aj} & P_{aj} \\ P_{aj} & P_{aj} & P_{aj} \end{bmatrix} \quad (21)$$

where

$$\left.\begin{aligned} P_{cj} &= (1/2\pi\varepsilon_0\varepsilon_{i1})\cdot \ln(r_3/r_2) \\ P_{sj} &= (1/2\pi\varepsilon_0\varepsilon_{i2})\cdot \ln(r_5/r_4) \\ P_{aj} &= (1/2\pi\varepsilon_0\varepsilon_{i3})\cdot \ln(r_7/r_6) \end{aligned}\right\} \quad (22)$$

When the cable consists of a core and sheath, the above matrix is reduced to:

$$[P_{ij}] = \begin{bmatrix} P_{cj} + P_{sj} & P_{sj} \\ P_{sj} & P_{sj} \end{bmatrix} \quad (23)$$

If the cable consists only of a core, then $[P_{ij}]$ includes only one element.

$$[P_{ij}] = P_{cj} \quad (24)$$

The submatrices of $[P_o]$ are given in the following form.

$$[P_{ojk}] = \begin{bmatrix} P_{ojk} & P_{ojk} & P_{ojk} \\ P_{ojk} & P_{ojk} & P_{ojk} \\ P_{ojk} & P_{ojk} & P_{ojk} \end{bmatrix} \quad (25)$$

where P_{ojk} is the space potential coefficient and is given for the case of Fig.1 (b) by:

$$\left.\begin{aligned} P_{ojj} &= (1/2\pi\varepsilon_0)\cdot \ln(2h_j/r_{7j}) \\ P_{ojk} &= (1/2\pi\varepsilon_0)\cdot \ln(D_2/D_1) \end{aligned}\right\} \quad (26)$$

2.2 Pipe-Type Cable (PT Cable)

2.2.1 Impedance

The impedance matrix of a PT cable shown in Fig.2, where an inner conductor is assumed to be an SC cable, is given in the same manner as the SC cable case.[8,9]

(1) Pipe thickness assumed to be infinite

$$[Z] = [Z_i] + [Z_p] \quad (27)$$

(2) Pipe thickness being finite

$$[Z] = [Z_i] + [Z_p] + [Z_c] + [Z_o] \tag{28}$$

where

$[Z_i]$ = SC cable internal impedance matrix

$$= \begin{bmatrix} [Z_{i1}] & [\ 0\ ] & \cdots & [\ 0\ ] & 0 \\ [\ 0\ ] & [Z_{i2}] & \cdots & [\ 0\ ] & 0 \\ \vdots & \vdots & \ddots & \vdots & \vdots \\ [\ 0\ ] & [\ 0\ ] & \cdots & [Z_{in}] & 0 \\ 0 & 0 & \cdots & 0 & 0 \end{bmatrix} \tag{29}$$

$[Z_p]$ = pipe internal impedance matrix

$$= \begin{bmatrix} [Z_{p11}] & [Z_{p12}] & \cdots & [Z_{p1n}] & 0 \\ [Z_{p12}] & [Z_{p22}] & \cdots & [Z_{p2n}] & 0 \\ \vdots & \vdots & \ddots & \vdots & \vdots \\ [Z_{p1n}] & [Z_{p2n}] & \cdots & [Z_{pnn}] & 0 \\ 0 & 0 & \cdots & 0 & 0 \end{bmatrix} \tag{30}$$

$[Z_c]$ = connection impedance matrix between pipe inner and outer surfaces

$$= \begin{bmatrix} [Z_{c1}] & [Z_{c1}] & \cdots & [Z_{c1}] & Z_{c2} \\ [Z_{c1}] & [Z_{c1}] & \cdots & [Z_{c1}] & Z_{c2} \\ \vdots & \vdots & \ddots & \vdots & \vdots \\ [Z_{c1}] & [Z_{c1}] & \cdots & [Z_{c1}] & Z_{c2} \\ Z_{c2} & Z_{c2} & \cdots & Z_{c2} & Z_{c3} \end{bmatrix} \tag{31}$$

$[Z_o]$ = earth return impedance matrix

$$= \begin{bmatrix} [Z_o] & [Z_o] & \cdots & [Z_o] & Z_o \\ [Z_o] & [Z_o] & \cdots & [Z_o] & Z_o \\ \vdots & \vdots & \ddots & \vdots & \vdots \\ [Z_o] & [Z_o] & \cdots & [Z_o] & Z_o \\ Z_o & Z_o & \cdots & Z_o & Z_o \end{bmatrix} \tag{32}$$

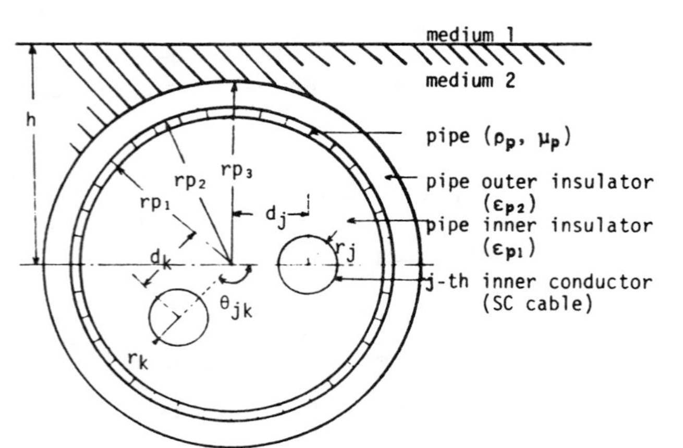

Fig. 2 A PT cable

In eqs.(29) and(30), the last column and row correspond to the pipe conductor. Thus, these should be omitted when the pipe thickness being assumed infinite. A diagonal submatrix of $[Z_i]$, i.e.eq.(29), is given in eq.(8). A submatrix of $[Z_p]$, eq.(30), is given in the following form.

$$[Z_{pjk}] = \begin{bmatrix} Z_{pjk} & Z_{pjk} & Z_{pjk} \\ Z_{pjk} & Z_{pjk} & Z_{pjk} \\ Z_{pjk} & Z_{pjk} & Z_{pjk} \end{bmatrix} \tag{33}$$

When an inner conductor consists of a core and sheath, eq.(33) is reduced to 2x2 matrix, and when the inner conductor consists only of a core, eq.(33) is further reduced to a column matrix in the same manner as explained in the case of $[Z_i]$. (See eqs.(8), (11) and (13).) This is the same for all other impedance and admittance matrices explained in this section.

Z_{pjk} in eq.(33) is the impedance between the j-th and k-th inner conductors with respect to the pipe inner surface, and is given by[3,8]:

$$Z_{pjk} = (s\mu_0/2\pi)\cdot[\mu_p K_0(x_1)/\{x_1 K_1(x_1)\} + Q_{jk} + 2\mu_p \sum_{n=1} C_n/\{n(1+\mu_p) + x_1 K_{n-1}(x_1)/K_n(x_1)\}] \tag{34}$$

where

$$Q_{jj} = \ln[(r_{p1}/r_j)\cdot\{1-(d_j/r_{p1})^2\}]$$

$$Q_{jk} = \ln[r_{p1}/\sqrt{d_j^2 + d_k^2 - 2d_j d_k \cos\theta_{jk}}] - \sum_{n=1}^{\infty} C_n/n$$

$$C_n = (d_j d_k/r_{p1}^2)^n \cdot \cos(n\theta_{jk}) \tag{35}$$

and

$$x_1 = \beta_1\sqrt{s}\ , \quad \beta_1 = r_{p1}\sqrt{\mu_0\mu_p/\rho_p} \tag{36}$$

A submatrix and the last row and column elements of $[Z_c]$ in eq.(31) are given in the following form[8,14]

$$\left.\begin{aligned} [Z_{c1}] &= \begin{bmatrix} Z_{c1} & Z_{c1} & Z_{c1} \\ Z_{c1} & Z_{c1} & Z_{c1} \\ Z_{c1} & Z_{c1} & Z_{c1} \end{bmatrix} \\ Z_{c1} &= Z_{c3} - 2z_{pm} \\ Z_{c2} &= Z_{c3} - z_{pm} \\ Z_{c3} &= z_{p0} + z_{p3} \end{aligned}\right\} \tag{37}$$

where

$$\left.\begin{aligned} z_{pm} &= \rho_p/(2\pi r_{p1} r_{p2} D_p) \\ z_{p0} &= (s\mu_0\mu_p/2\pi x_2 D_p)\cdot\{I_0(x_2)\cdot K_1(x_1) + K_0(x_2)\cdot I_1(x_1)\} \\ z_{p3} &= (s\mu_0/2\pi)\cdot\ln(r_{p3}/r_{p2}) \end{aligned}\right\} \tag{38}$$

and

$$\left.\begin{aligned} D_p &= I_1(x_2)\cdot K_1(x_1) - I_1(x_1)\cdot K_1(x_2) \\ &x_1 \text{ is given in eq.(36).} \\ x_2 &= \beta_2\sqrt{s}\ , \quad \beta_2 = r_{p2}\sqrt{\mu_0\mu_p/\rho_p} \end{aligned}\right\} \tag{39}$$

A diagonal submatrix of $[Z_o]$ in eq.(32) is given by:

$$[Z_o] = \begin{bmatrix} Z_o & Z_o & Z_o \\ Z_o & Z_o & Z_o \\ Z_o & Z_o & Z_o \end{bmatrix} \quad (40)$$

where Z_o in the above matrix is the self earth return impedance of the pipe.

2.2.2 Potential coefficient

The potential coeffidient matrix of a PT cable shown in Fig.2 is given in the following form.[8,9]

(1) Pipe thickness assumed to be infinite

$$[P] = [P_i] + [P_p] \quad (41)$$

(2) Pipe thickness being finite

(a) Underground cable

$$[P] = [P_i] + [P_p] + [P_c] \quad (42)$$

(b) Overhead cable

$$[P] = [P_i] + [P_p] + [P_c] + [P_o] \quad (43)$$

where

$[P_i]$ = SC cable internal potential coefficient matrix

$$= \begin{bmatrix} [P_{i1}] & [\,0\,] & \cdots & [\,0\,] & 0 \\ [\,0\,] & [P_{i2}] & \cdots & [\,0\,] & 0 \\ \vdots & \vdots & \ddots & \vdots & \vdots \\ [\,0\,] & [\,0\,] & \cdots & [P_{in}] & 0 \\ 0 & 0 & \cdots & 0 & 0 \end{bmatrix} \quad (44)$$

$[P_p]$ = pipe internal potential coefficient matrix

$$= \begin{bmatrix} [P_{p11}] & [P_{p12}] & \cdots & [P_{p1n}] & 0 \\ [P_{p12}] & [P_{p22}] & \cdots & [P_{p2n}] & 0 \\ \vdots & \vdots & \ddots & \vdots & \vdots \\ [P_{p1n}] & [P_{p2n}] & \cdots & [P_{pnn}] & 0 \\ 0 & 0 & \cdots & 0 & 0 \end{bmatrix} \quad (45)$$

$[P_c]$ = potential coefficient matrix between pipe inner and outer surfaces

$$= \begin{bmatrix} [P_c] & [P_c] & \cdots & [P_c] & P_c \\ [P_c] & [P_c] & \cdots & [P_c] & P_c \\ \vdots & \vdots & \ddots & \vdots & \vdots \\ [P_c] & [P_c] & \cdots & [P_c] & P_c \\ P_c & P_c & \cdots & P_c & P_c \end{bmatrix} \quad (46)$$

$[P_o]$ = potential coefficient matrix of the pipe in air

$$= \begin{bmatrix} [P_o] & [P_o] & \cdots & [P_o] & P_o \\ [P_o] & [P_o] & \cdots & [P_o] & P_o \\ \vdots & \vdots & \ddots & \vdots & \vdots \\ [P_o] & [P_o] & \cdots & [P_o] & P_o \\ P_o & P_o & \cdots & P_o & P_o \end{bmatrix} \quad (47)$$

In eqs.(44) and(45), the last column and row corresponding to the pipe conductor,these should be omitted when the pipe thickness being assumed infinite. A diagonal submatrix of $[P_i]$ in eq.(44) is given in eq.(21). Submatrix $[P_{pjk}]$ of $[P_p]$, eq. (45), is given in the following form.

$$[P_{pjk}] = \begin{bmatrix} P_{pjk} & P_{pjk} & P_{pjk} \\ P_{pjk} & P_{pjk} & P_{pjk} \\ P_{pjk} & P_{pjk} & P_{pjk} \end{bmatrix} \quad (48)$$

P_{pjk} in the above equation is the potential coefficient between the j-th and k-th inner conductors with respect to the pipe inner surface, and is given in the following equation using Q of eq.(35).

$$P_{pjj} = Q_{jj}/2\pi\varepsilon_{p1}\varepsilon_0 \;, \quad P_{pjk} = Q_{jk}/2\pi\varepsilon_{p1}\varepsilon_0 \quad (49)$$

A submatrix and the last column and row elements of $[P_c]$ in eq.(46) are given by:

$$\left.\begin{aligned} [P_c] &= \begin{bmatrix} P_c & P_c & P_c \\ P_c & P_c & P_c \\ P_c & P_c & P_c \end{bmatrix} \\ P_c &= (1/2\pi\varepsilon_{p2}\varepsilon_0)\cdot\ln(r_{p3}/r_{p2}) \end{aligned}\right\} \quad (50)$$

A submatrix and the last row and column elements of the space potential coefficient matrix$[P_o]$ is given in the following form.

$$\left.\begin{aligned} [P_o] &= \begin{bmatrix} P_o & P_o & P_o \\ P_o & P_o & P_o \\ P_o & P_o & P_o \end{bmatrix} \\ P_o &= (1/2\pi\varepsilon_0)\cdot\ln(2h/r_{p3}) \end{aligned}\right\} \quad (51)$$

3. DISCUSSION

The formulation of impedances and admittances of various cables given in the previous section includes some approximations. It may be important to discuss these approximations so as to make the limit of applicability clear when the formulation is used.

First of all, the major assumptions made for the formulation of impedances and admittances (on page 1 of the paper) should be discussed. The first assumption is constant permeability. Quite offten, a pipe and armor are ferromagnetic. It, however, seems to be rather unusual to have high currents to cause saturation of the pipe or armor. Thus, in most cases, the saturation of the pipe or armor can be neglected. When one needs to take the saturation into account, methods proposed in references (3) and (4) can be used. In regard to the second assumption, displacement currents are negligible as far as low frequencies (less than about 1MHz) are concerned. In the analysis of transients and wave propagation on a cable system, the frequency of interest is, in most cases, less than 1 MHz. The dielectric losses are small in comparison with the losses in conducting media of cables and earth. Thus, the assumption is valid. The third assumption will be discussed later.

No approximation is made for the impedances and admittances of an SC cable as far as Carson's and Pollaczek's earth impedances and Scheikunoff's cylindrical conductor impedance are concerned. One should pay attention to the fact that Carson's and Pollaczek's formulas of the earth return impedance are not applicable at frequencies higher than about 1MHz because the effect of displacement currents is not included in the formulas.[13] Thus, the formulation of the impedances of both SC and PT cables is correct only upto about 1MHz.

One can easily find that the formulation of the impedances and admittances of an SC cable given in this paper is identical to that given in reference (1) for the case of a coaxial cable consisting of a core and sheath.

Two assumptoins are included in the PT cable case. The first one is that the eccentric cable positions within the pipe do not affect the internal impedances and admittances of the inner conductors (SC cable) and the impedances and admittances between the inner and outer surfaces of the pipe. Thus, the inner conductor impedance and admittance of a PT cable become the same as those of an SC cable. The same assumption has been made in references (3) and (4). If one needs to take into account the effect of the eccentricity on the inner conductor impedance, the formula of the outer surface impedance of the inner conductor given in reference (5) can be used.

The second assumption concerns the case of finite pipe thickness. It is assumed that the pipe thickness will be greater than the penetration depth in the pipe wall. If the pipe thickness is smaller than the penetration depth, the formulas of the pipe internal impedance given in eqs.(34) and (35) and potential coefficient given in eq.(49) can not be used. In that case, accurate formulas of the impedance and potential coefficient can be derived based on the work done by Tegopoulos and Kriezis.[15] Since these formulas are too complicated for practical usage, the assumption of infinite pipe wall thickness may be used, but only to calculate the impedance and potential coefficients of the pipe. Note that earth return currents are not neglected and that complete shielding is not assumed. This assumption introduces negligible error for actual PT cables and for frequencies above 10Hz. Fig.3 shows a comparison of the pipe impedances for the cases of the pipe thickness being finite and infinite. It is clear that the impedance for the finite pipe thickness case approaches that for the infinite thickness case, at the frequency of 1kHz. When the pipe thickness is 4mm, which is nearly equivalent to the penetration depth at 10 Hz, its impedance is almost identical to that for the infinite thickness case in the frequency range shown in the figure. The pipe thickness is, in most cases, greater than the penetration depth. Thus, the assumption is valid.

Calculated results of admittances of a single-phase SC cable are shown in Fig.4. From the results, it is clear that the admittance of an underground cable are much greater than those of an overhead cable. The impedance shows not a significant difference between underground and overhead cables. Thus, it should be expected that the attenuation of the undeground cable is much higher than that of the overhead cable, and the propagation velocity is lower in the underground case. Similar results are obtained for the PT cable case.

The internal impedances of SC cables are shown in Fig.5. Significant differences are observed for the cases of SC cables consisting only of a core, of core and sheath, and of core, sheath and armor.

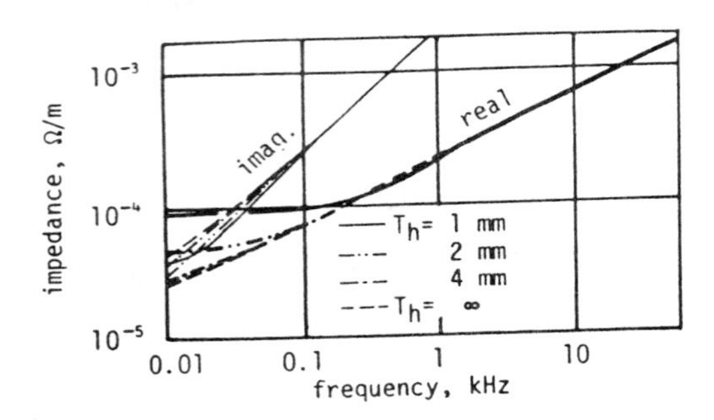

Fig. 3 Effects of pipe thickness on pipe inner surface impedance

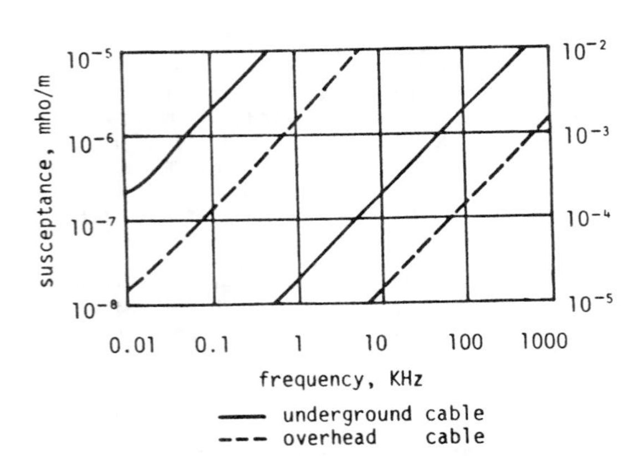

Fig. 4 Susceptances (imag.Y_{22}) of SC cables

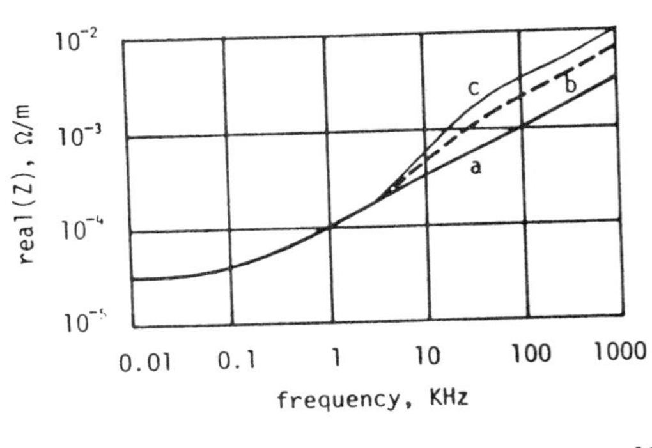

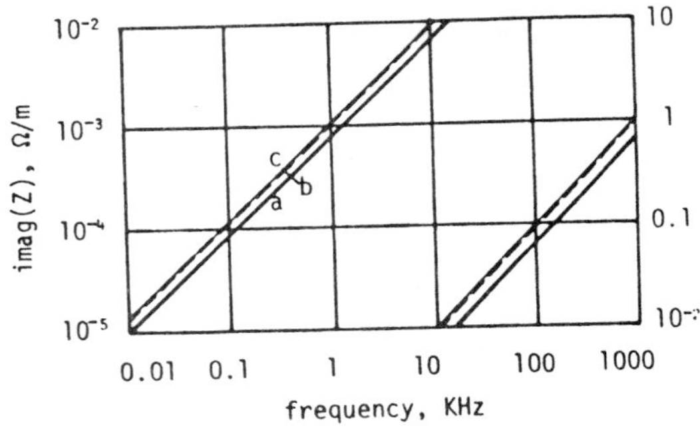

Fig. 5 Internal impedances Z_{cc} of SC cables

(a) Core and its outer insulator

(b) Core, sheath and its outer insulator

(c) Core, sheath, armor and its outer insulator

4. CONCLUSION

A general formulation of the impedances and admittances of single-core coaxial cables and pipe-type cables is given. The formulation presented in this paper can handle a coaxial cable consisting of a core, sheath and/or armor, a pipe-type cable of which the pipe thickness is either infinite or finite, and an overhead cable. Numerical results based on this formulation are readily available using BPA's computer program EMTP with subroutine CABLE CONSTANTS.

ACKNOWLEDGEMENTS

The author would like to thank Prof. K. Tominaga for his encouragement, and Prof. R. Schinzinger of University of Californua for his helpful discussion and critical reading of the manuscript. The author also wishes to express his appreciation for financial support by Bonneville Power Administration.

REFERENCES

1) L. M. Wedepohl and D. J. Wilcox:"Transient analysis of underground power-transmission systems",Proc.IEE vol.120, pp.253-260 (1973)

2) L. M. Wedepohl and D. J. Wilcox: " Estimations of transient sheath overvoltages in power cable transmission systems", ibid., vol.120, pp.877-882 (1973)

3) G. W. Brown and R. G. Rocamora: "Surge propagation in three-phase pipe-type cables, Part I-Unsaturated pipe", IEEE Trans. on Power App. & Syst., PAS-95, pp.89-95 (1976)

4) R. C. Dugan, et. al.: " Surge propagation in three phase pipe-type cables, Part II - Duplication of field tests including the effects of neutral wires and pipe saturation", ibid.,PAS-96,pp.826-833(1977)

5) R. Schinzinger and A. Ametani: " Surge propagation characteristics of pipe enclosed underground cables", ibid., PAS-97, pp.1680-1687 (1978)

6) B. Dixon: Private correspondence (1977.9)

7) B. P. A.: Private correspondence (1977.8)

8) A. Ametani: "Generalized program for line and cable constants", Bonneville Power Administration, Purchase Order No.70249, Report No.2 (1977.10)

9) A. Ametani: "Extension of generalized program for line and cable constants in EMTP", Bonneville Power Administration, Contract No.EW-78-C-80-1500, Report No.1 (1978.7)

10) S. A. Schelkunoff: " The electromagnetic theory of coaxial transmission line and cylindrical shields", Bell Syst. Tech. J., vol.13, pp.532-579 (1934)

11) J. R. Carson: " Wave propagation in overhead wires with ground return", ibid., vol.5, pp.539-554(1926)

12) F. Pollaczek: "Über das Feld einer unendlich langen wechsel stromdurchflossenen Einfachleitung",E.N.T., Band 3 (Heft 9), pp.339-360 (1926)

13) M. Nakagawa, et,al.: " Further studies on wave propagation in overhead lines with ground return ", Proc.IEE, vol.120, pp.1521-1528 (1973)

14) A. Ametani and T. Ono: " Wave propagation characteristics on a pipe-type cable, III - Consideration of pipe thickness", IEE Japan,Proceedings of Annual Meeting, Paper No.840 (1978)

15) J. A. Tegopoulos and E. E. Kriezis: "Eddy current distribution in cylindrical shells of infinite length due to axial currents, Part II - Shells of finite thickness", IEEE Trans. on Power App. & Syst., PAS-90, pp.1287-1294 (1971)

APPENDICES

Appendix 1 Impedance of an SC cable consisting of a core, sheath and armor

In the case of an SC cable with core, sheath and armor, an equivalent circuit for impedances is given in Fig.A-1.

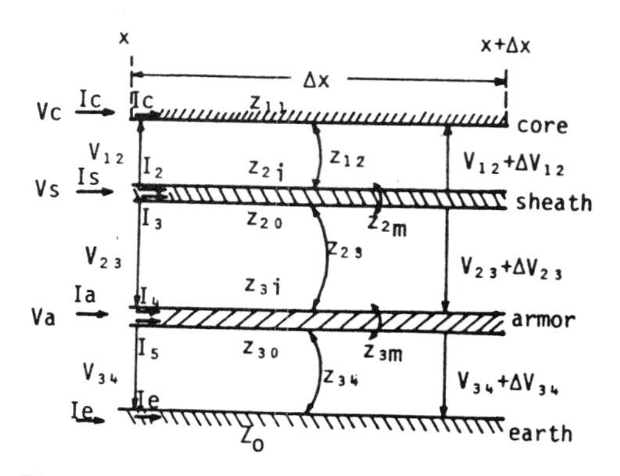

Fig. A-1 An equivalent circuit for impedances of an SC cable

Define currents flowing into the core, sheath, armor and outer medium (earth) by I_c, I_s, I_a and I_e at x. Also inner and outer surface currents of the sheath and the armor are I_2, I_3, I_4 and I_5 as shown in Fig.A-1. Voltages between the core, sheath, armor and outer medium are V_{12}, V_{23} and V_{34} at x, and are $V_{12}+\Delta V_{12}$, $V_{23}+\Delta V_{23}$, and $V_{34}+\Delta V_{34}$ at $x = x + \Delta x$.

Then, the following relation for currents are obtained.

$$I_2 = -I_c\ ,\quad I_3 = -I_4\ ,\quad I_5 = -I_e \qquad \text{(A-1)}$$

$$\left.\begin{aligned} I_s &= I_2 + I_3 = -(I_c + I_4) \\ I_a &= I_4 + I_5 = I_4 - I_e \end{aligned}\right\} \qquad \text{(A-2)}$$

From the above equations,

$$\left.\begin{aligned} I_4 &= -(I_c + I_s) \\ I_e &= -(I_c + I_s + I_a) \end{aligned}\right\} \qquad \text{(A-3)}$$

For voltage V_{12} between the core and the sheath,

$$V_{12} = z_{11}\Delta x I_c - z_{12}\Delta x I_2 - z_{2i}\Delta x I_2 - z_{2m}\Delta x I_3 + V_{12} + \Delta V_{12}$$

$$\therefore\ -\Delta V_{12}/\Delta x = (z_{11} + z_{12} + z_{2i})I_c + z_{2m}I_4$$

Define z_{cs} by:

$$z_{cs} = z_{11} + z_{12} + z_{2i} \qquad \text{(A-4)} = \text{eq.(10)}$$

Using the above equation,

$$-\Delta V_{12}/\Delta x = z_{cs}I_c + z_{2m}I_4 \quad \text{(A-5)}$$

For voltage V_{23},

$$-\Delta V_{23}/\Delta x = (z_{20} + z_{23} + z_{3i})I_4 + z_{2m}I_c - z_{3m}I_e$$

Define z_{sa} by:

$$z_{sa} = z_{20} + z_{23} + z_{3i} \quad \text{(A-6) = eq.(10)}$$

Then,

$$-\Delta V_{23}/\Delta x = z_{sa}I_4 + z_{2m}I_c - z_{3m}I_e \quad \text{(A-7)}$$

For voltage V_{34},

$$-\Delta V_{34}/\Delta x = (z_{a4} + Z_o)I_e - z_{3m}I_4 \quad \text{(A-8)}$$

where

$$z_{a4} = z_{30} + z_{34} \quad \text{(A-9) = eq.(10)}$$

Take the earth voltage of zero potential as reference,

$$V_a = -V_{34}$$
$$V_s = -(V_{23} + V_{34}) = V_a - V_{23} \quad \text{(A-10)}$$
$$V_c = V_{12} + V_s$$

Substituting eqs.(A-3) and (A-10) into eq.(A-8).

$$-\Delta V_a/\Delta x = (z_{a4} - z_{3m} + Z_o)\cdot(I_c + I_s) + (z_{a4} + Z_o)I_a \quad \text{(A-11)}$$

Substitute eqs.(A-3), (A-10) and (A-11) into eq.(A-7).

$$-\Delta V_s/\Delta x = (z_{sa} + z_{a4} - z_{2m} - 2z_{3m} + Z_o)I_c + (z_{sa} + z_{a4} - 2z_{3m} + Z_o)I_s + (z_{a4} - z_{3m} + Z_o)I_a \quad \text{(A-12)}$$

In the same manner,

$$-\Delta V_c/\Delta x = (z_{cs} + z_{sa} + z_{a4} - 2z_{2m} - 2z_{3m} + Z_o)I_c + (z_{sa} + z_{a4} - z_{2m} - 2z_{3m} + Z_o)I_s + (z_{a4} - z_{3m} + Z_o)I_a \quad \text{(A-13)}$$

Finally frpm eqs.(A-11), (A-12) and (A-13) with $x \to 0$,

$$d(V)/dx = -[Z]\cdot(I) \quad \text{(A-14)}$$

where [Z] is given by:

$$[Z] = [Z_i] + [Z_o] \quad \text{(A-15) = eq.(5)}$$

and

$$[Z_i] = \begin{bmatrix} Z_{cc} & Z_{cs} & Z_{ca} \\ Z_{cs} & Z_{ss} & Z_{sa} \\ Z_{ca} & Z_{sa} & Z_{aa} \end{bmatrix}, \quad [Z_o] = \begin{bmatrix} Z_o & Z_o & Z_o \\ Z_o & Z_o & Z_o \\ Z_o & Z_o & Z_o \end{bmatrix} \quad \text{(A-16) = eqs.(8) and (14)}$$

where

$$\left.\begin{aligned} Z_{cc} &= z_{cs} + z_{sa} + z_{a4} - 2(z_{2m} + z_{3m}) \\ Z_{ss} &= z_{sa} + z_{a4} - 2z_{3m}, \quad Z_{aa} = z_{a4} \\ Z_{cs} &= z_{sa} + z_{a4} - z_{2m} - 2z_{3m} \\ Z_{sa} &= z_{a4} - z_{3m} \end{aligned}\right\} \quad \text{(A-17) = eq.(9)}$$

Appendix 2 Potential coefficient

An equivalent circuit for the admittance of underground SC cable with a core, sheath and armor is shown in Fig.A-2. From the figure,

$$I_c = y_{cs}\Delta x(V_c - V_s) + I_c + \Delta I_c$$
$$I_s = y_{cs}\Delta x(V_s - V_c) + y_{sa}\Delta x(V_s - V_a) + I_s + \Delta I_s$$
$$I_a = y_{sa}\Delta x(V_a - V_s) + y_{a4}\Delta x V_a + I_a + \Delta I_a \quad \text{(A-18)}$$

Rewriting the above equations,

$$-\Delta I_c/\Delta x = y_{cs}V_c - y_{cs}V_s$$
$$-\Delta I_s/\Delta x = -y_{cs}V_c + (y_{cs} + y_{sa})V_s - y_{sa}V_a$$
$$-\Delta I_a/\Delta x = -y_{sa}V_s + (y_{sa} + y_{a4})V_a \quad \text{(A-19)}$$

Put $x \to 0$ in the above equations,

$$\frac{d}{dx}\begin{bmatrix} I_c \\ I_s \\ I_a \end{bmatrix} = -\begin{bmatrix} y_{cs} & & 0 \\ -y_{cs} & (y_{cs}+y_{sa}) & -y_{sa} \\ 0 & -y_{sa} & (y_{sa}+y_{a4}) \end{bmatrix}\cdot\begin{bmatrix} V_c \\ V_s \\ V_a \end{bmatrix} = -[Y_i]\cdot(V) \quad \text{(A-20)}$$

where

$$\left.\begin{aligned} y_{cs} &= s2\pi\varepsilon_0\varepsilon_1/\ln(r_3/r_2) \\ y_{sa} &= s2\pi\varepsilon_0\varepsilon_2/\ln(r_5/r_4) \\ y_{a4} &= s2\pi\varepsilon_0\varepsilon_3/\ln(r_7/r_6) \end{aligned}\right\} \quad \text{(A-21)}$$

Potential coefficients being inversly related to admittances,

$$[P_i] = \begin{bmatrix} P_c+P_s+P_a & P_s+P_a & P_a \\ P_s+P_a & P_s+P_a & P_a \\ P_a & P_a & P_a \end{bmatrix} \quad \text{(A-22) = eq.(21)}$$

$$P_c = s/y_{cs}, \; P_s = s/y_{sa}, \; P_a = s/y_{a4} \quad \text{(A-23) = eq.(22)}$$

When a cable is overhead, considering a space admittance being connected in series to y_{a4} in Fig.A-2, the potential coefficient matrix is derived in the same manner as the underground cable case.

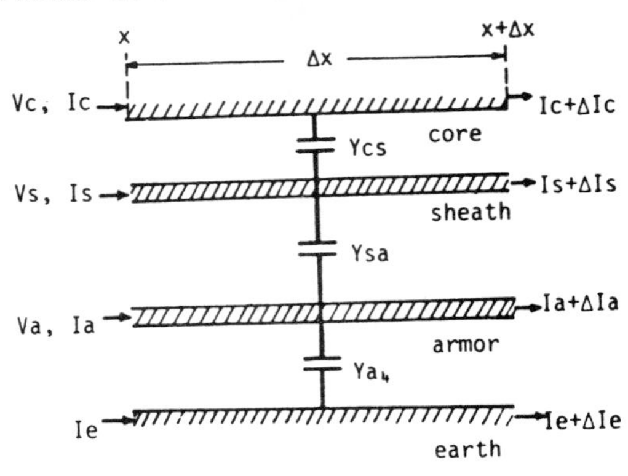

Fig. A-2 An equivalent circuit for admittances of an SC cable

Discussion

Adam Semlyen (University of Toronto, Toronto, Ontario, Canada): Dr. Ametani's paper on cable impedances and admittances is based on the assumption that such parameters are available between components of the cable. The contribution consists in assembling the basic data into matrices defined in (1) and (2). The complexity of cable layouts tends to obscure the analysis of basic phenomena and, therefore, a systematic matrix formulation is useful. Could the author indicate the reference which provides details for the calculation of the cable parameters needed for the computation of the impedance and admittance matrices?

The author's remark that pipe enclosures of finite thickness do not provide a complete shielding is theoretically correct, but, as shown in reference [A], the ground return current is actually quite small and, therefore, the ground path can be neglected.

Among the basic assumptions listed by the author, we find that displacement currents are negligible. It is probably in conductors and not in dielectrics where this assumption is considered, since all capacitive effects are related to displacement currents.

Clarifications concerning both problems discussed above would be welcome.

Reference

[A] A. Semlyen and D. Kiguel, "Phase Parameters of Pipe Type Cables", Paper No. A 78 001-0, presented at the 1978 IEEE PES Winter Meeting, New York City.

Manuscript received July 30, 1979.

A. Ametani: The author would like to thank the discussor for his interest in this paper.

In reply to his first comment, the author is not sure what the discussor meant by his question. If he asked the derivation of the component impedances and admittances, references 1 and 10 could be the answer for a coaxial cable and reference 3 for a pipe-type cable. If the discussor asked formulas of each component impedance and admittance, these are given in detail in the present paper. Only the formula of the earth return impedance is not shown in this paper. This, however, is well known and can be found in references 11 to 13.

Concerning the second comment, I agree with Prof. Semlyen's comment that the earth return current is actually quite small, and therefore, the earth return path can be neglected if one concerns only the propagation modes within the pipe. But, if it is the case that the propagation mode between the pipe and the earth, namely the earth return mode, becomes significant, for instance if one wants to know the surface voltage of a gas insulated transmission line or bus which is overhead, we need to include the effect of the earth return path. For such a case, the earth return impedance is to be included in the pipe-type cable case, though in most cases it can be neglected.

In reply to the third comment, the displacement currents mentioned in the paper is related to the conductor as Prof. Semlyen pointed correctly. The assumption of neglecting the displacement currents is concerned with the displacement currents between the cable and the earth, in other words, it concerns with the earth return impedance. As far as Carson's or Pollaczek's earth return impedance is adopted, we can not deal with the displacement currents between a conductor and earth.

Manuscript received October 22, 1979.

SIMULATION OF TRANSIENTS IN UNDERGROUND CABLES WITH FREQUENCY-DEPENDENT MODAL TRANSFORMATION MATRICES

L. Marti, Member IEEE
The University of British Columbia
Vancouver, Canada

Abstract

This paper presents a new mathematical model for the simulation of electromagnetic transients in underground high voltage cables. The solution is carried out in the time domain; therefore, this model is compatible with time domain solution algorithms, such as the one used in the EMTP. The frequency dependence of the cable parameters and of the modal transformation matrices is accurately taken into account. Comparisons with analytical and measured results are also presented.

1. Introduction

Accurate modelling of underground cables and transmission lines plays an important part in the simulation of transient phenomena in power systems. A number of models have been proposed to date. These models can be classified into two major groups, according to the solution techniques used in their host programs:

a) Time domain models.

In this class of models, the solution is carried out in the time domain without explicit use of inverse (Fourier or Laplace) transforms. Within this group, two types of models deserve attention:

i) Lumped-parameters models: The transmission system is represented by lumped elements (usually by several cascaded π-sections) evaluated at a single frequency [1]. A more sophisticated form of this type of model includes the representation of the ground return impedance using a suitable combination of several R-L branches. This representation is widely used in transient network analyzers [2]. The validity of these models is restricted to relatively short lines or cables and, in general, their frequency response is only good in the neighbourhood of the frequency at which the parameters are evaluated.

ii) Distributed-parameters, frequency-dependent models: The solution is performed in the modal domain. The frequency dependence and the distributed nature of the line or cable parameters are taken into account [3]. The validity of these models is restricted because the modal transformation matrices are assumed to be constant. This assumption can lead to poor results in many cases of unbalanced overhead transmission lines (especially multiple-circuit) when the simulation involves a wide range of frequencies. In the case of underground cables, the modal transformation matrices depend strongly on frequency, and constant-transformation-matrix models generally produce very poor results.

87 WM 154-8 A paper recommended and approved by the IEEE Transmission and Distribution Committee of the IEEE Power Engineering Society for presentation at the IEEE/PES 1987 Winter Meeting, New Orleans, Louisiana, February 1 - 6, 1987. Manuscript submitted Janaury 30, 1986; made available for printing December 2, 1986.

b) Frequency domain models.

In this class of models, the response of the transmission system is evaluated in the frequency domain. The time domain solution is then found using inverse transformation algorithms such as the FFT (Fast Fourier Transform) [4].

The frequency dependence of the line or cable parameters and of the modal transformation matrices is taken into account. Even though inherent numerical problems such as aliasing and Gibbs' oscillations have been alleviated (using windows and other oscillation-suppressing techniques), the applicability of these models is restricted by the limitations of their host programs.

There are no general-purpose transient analysis programs in the frequency domain with the overall simulation capabilities of time domain programs such as the EMTP [5]. Sudden changes in the network configuration (such as faults, opening and closing of circuit breakers, etc.), and the modelling of non-linear elements cannot be handled easily with frequency-domain solution methods.

The model presented in this paper belongs to the class of time-domain, frequency-dependent models. It overcomes the main limitation of existing time-domain line models; that is, it takes into account the frequency dependence of the modal transformation matrices. Also, by being compatible with the solution algorithm of the EMTP, it enjoys all the implicit advantages of a versatile host program.

The new cable model is accurate, as demonstrated with comparisons with analytical results. From a computational point of view, the speed of this model is comparable to the speed of frequency-dependent models with constant transformation matrices.

All numerical results presented in this paper refer to the simulation of underground high voltage cables. However, the model itself is general, and given its computational speed, it should also be very useful in the simulation of unbalanced and multiple-circuit overhead transmission lines.

2. Description of the model

The following conventions in notation will be used in this paper:

- Upper case letters indicate "frequency domain" quantities, whereas lower case letters are used to denote their "time domain" counterparts (e.g., V in the frequency domain, and v(t) in the time domain).
- Primed symbols refer to modal quantities; otherwise, all quantities represent phase components (e.g., v′(t) for modal voltage, and v(t) for its phase domain equivalent).

Consider an underground cable consisting of n conductors of length l. In the frequency domain, the relationship between voltages and currents at sending and receiving ends can be expressed as

$$Y_c V_m + I_m = F_m = A F_k \quad (1)$$

$$Y_c V_k - I_k = B_k = A B_m \quad (2)$$

where

0885-8977/88/0700-1099$01.00©1988 IEEE

$$Y_c = \sqrt{(YZ)^{-1}}\,Y$$

$$A = \exp(-\sqrt{YZ}\, l)$$

Yc is the characteristic admittance matrix; A is defined as the propagation matrix; Y and Z are the shunt admittance and series impedance matrices per unit length, respectively; V and I are voltage and current vectors of dimension n; F and B are intermediate vector functions which, in the time domain, can be interpreted as waves travelling in forward and backward directions, respectively. All matrices are dimensioned nxn, and subscripts "k" and "m" are used to indicate sending and receiving end quantities.

Equations (1) and (2) describe any transmission system, whether it is an overhead line or an underground cable. Only the values of Yc and A, as functions of frequency, determine their respective behaviours.

Let Q be the eigenvector or modal transformation matrix which diagonalizes YZ. Equations (1) and (2) can then be transformed into the "modal" domain

$$Y'_c\, V'_m + I'_m = F'_m = A' F'_k \tag{3}$$

$$Y'_c\, V'_k - I'_k = B'_k = A' B'_m \tag{4}$$

where $Y'c$ and A' are diagonal matrices, and

$$\begin{aligned} V &= Q^{-T} V' \\ I &= Q\, I' \\ F &= Q\, F' \\ B &= Q\, B' \\ A' &= Q^{-1} A\, Q \\ Y'_c &= Q^{-1} Y_c\, Q^{-T} \end{aligned} \tag{5}$$

Note that the elements of $Y'c$, A', and Q are complex-valued functions of frequency. Also note, that Q^T is the transpose of Q, and Q^{-T} indicates the transpose of Q^{-1}.

Transforming equations (3) and (4) into the time domain gives

$$y'_c(t) * v'_m(t) + i'_m(t) = f'_m(t) = a'(t) * f'_k(t) \tag{6}$$

$$y'_c(t) * v'_k(t) - i'_k(t) = b'_k(t) = a'(t) * b'_m(t) \tag{7}$$

where the voltages and currents, in modal components, are given by

$$\begin{aligned} v'(t) &= q^T(t) * v(t) \\ i'(t) &= q^{-1}(t) * i(t) \end{aligned} \tag{8}$$

Note that the symbol "*" is used to indicate matrix-vector convolutions. If the elements of matrices $Y'c$, A' and Q are synthesized using rational functions, then $y'c(t)$, $a'(t)$ and $q(t)$ become matrices whose elements are finite sums of exponentials (see Appendix I). Therefore, the convolutions in equations (6) to (8) can be evaluated numerically using well-known recursive techniques [6].

Algebraic manipulation of these equations finally leads to

$$y_{eq}\, v_m(t) + i_m(t) = h_m(t) \tag{9}$$

$$y_{eq}\, v_k(t) - i_k(t) = h_k(t) \tag{10}$$

where yeq is a real, constant, symmetric matrix; $hm(t)$ and $hk(t)$ are defined as equivalent history current sources. At time t, these vectors are completely defined in terms of variables already known from previous time steps (see Appendix I).

Equations (9) and (10) can be represented by the equivalent circuit shown in Figure 1. This equivalent circuit is compatible with the solution algorithm of the EMTP. In fact, the EMTP representation of lossless lines and of frequency-dependent lines with constant Q also share the same form. Only the updating of the history current sources at each time step, and the value of yeq are different; therefore, taking into account the variation with frequency of the modal transformation matrices becomes transparent to the main core of the EMTP.

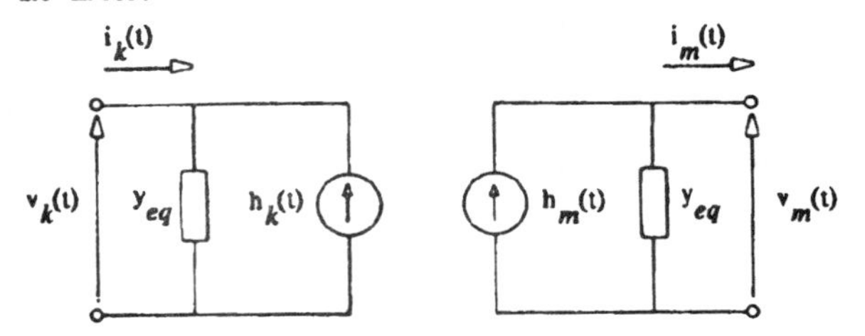

Fig. 1: *Equivalent circuit in the time domain.*

3. Synthesis of input data functions

Recursive convolution can be used to update the history current sources in equations (9) and (10) because the elements of $Y'c$, A' and Q can be expressed in terms of rational functions. The fitting algorithm used to synthesize these input data functions is a refined version of the one used in the frequency-dependent line model of the EMTP. The original algorithm is described in [3].

The elements of Q can be synthesized with rational functions when the following conditions are met:

a) The columns of Q (i.e., the eigenvectors of YZ) are scaled so that one of their elements becomes real and constant throughout the entire frequency range. With this normalization scheme, all the elements the eigenvectors become minimum-phase-shift functions.

b) The eigenvectors of YZ are continuous functions of frequency.

This last requirement is met when a very stable eigenvalue/eigenvector algorithm, such as Jacobi [7] is used. Note that the Jacobi algorithm is designed for symmetric matrices, and the product YZ is not symmetric. However, the unsymmetric eigenvalue/eigenvector problem $YZx=\lambda x$ can be converted into a symmetric one if $Zx=\lambda Y^{-1}x$ is solved instead [8]. The convergency rate and stability of the Jacobi algorithm can be improved considerably if YZ at a given frequency is pre and post multiplied by the transformation matrix obtained in the preceding frequency step. If Q is normalized so that its elements are minimum-phase-shift functions, the elements of $Y'c$ also become minimum-phase-shift, and can be likewise synthesized with rational functions.

The elements of A' can be synthesized with rational functions multiplied by $\exp(-j\omega\tau)$ [3]. The time delay constant τ depends on the difference between the phase angle of a given element of A' and the phase angle of its approximation by rational functions; also, τ is numerically close to the travel time of the fastest frequency component of a wave propagating on a given mode. Note that the propagation matrix A (in the phase domain) cannot be synthesized with rational functions because a single time delay cannot be associated with each of its elements.

Figure 2 shows the magnitude of the elements of eigenvector 3 for the 230 kV, three-phase underground cable

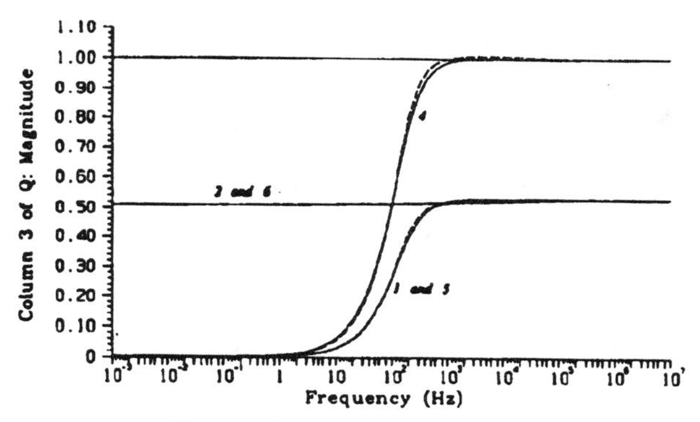

Fig. 2: *Magnitude of the elements of eigenvector 3 of YZ:*
------ *Exact function.*
——— *Synthesized function.*

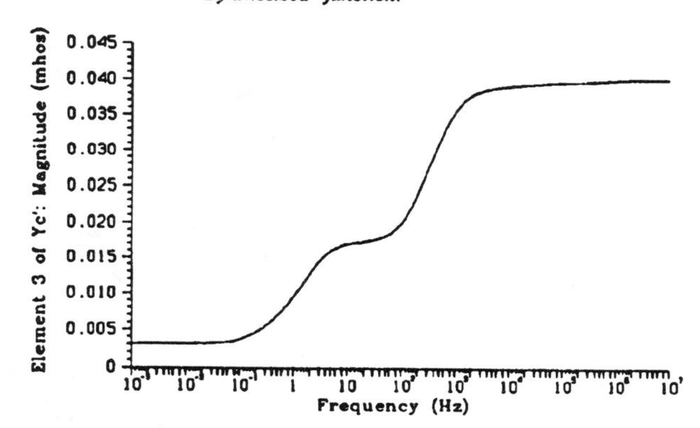

Fig. 3: *Magnitude of element 3 of Y'c:*
------ *Exact function.*
——— *Synthesized function.*

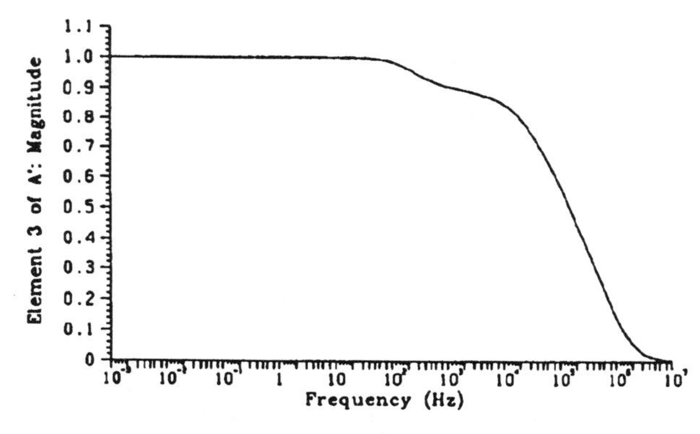

Fig. 4: *Magnitude of element 3 of A':*
------ *Exact function.*
——— *Synthesized function.*

described in Appendix II. Note that there are twelve curves superimposed on this plot, corresponding to the magnitudes of real and synthesized elements of eigenvector 3. Also note that this eigenvector contains only four distinct elements; that is, the magnitudes of elements 1 and 2 are identical to the magnitudes of elements 5 and 6, respectively. Figures 3 and 4 show Y'c and A' for mode 3. Their respective synthesized functions are also superimposed. Note that the synthesized functions match the original ones very closely. Errors are typically under 2% within the 0 to 1 MHz range.

The maximum number of functions that must be approximated for a cable with n conductors is n(n+1). For n=6, for example, the maximum number of elements to be synthesized would be 42. In practice, however, there is a considerable amount of symmetry within the elements of Q. For the 230 kV cable shown in Appendix II, there are only 14 distinct elements in Q. The total number of distinct functions that had to be synthesized in this case was 26.

The order of a given approximation depends on the shape of the curve. For example, in the case of the cable mentioned above, the average number of terms needed to approximate Y'c, A' and Q from 0 to 1 MHz were 21, 15 and 7, respectively.

4. Numerical results

In order to establish the accuracy of the new cable model, the analytical response of a three-phase underground cable is compared with the results obtained using the implementation of the new model in UBC's version of the EMTP. The measured impulse response of a crossbonded cable is also compared with its transient simulation. The effects of taking into account the frequency dependence of Q are illustrated with the simulation of a single-phase line to ground fault on a crossbonded cable.

4.1 Comparison with analytical results

Consider the 230 kV, three-phase underground cable shown in Figure 5, where a voltage source $v_s(t)$ is connected to the core of the first conductor (physical data for this cable, and the characteristics of its series impedance matrix, are shown in Appendix II).

Fig. 5: *Representation of a three-phase underground cable*

If $v_s(t)$ is a sinusoidal source of frequency ωo, the response will also be sinusoidal, and it can be found analytically (e.g., by solving equations (1) and (2) evaluated at $\omega = \omega o$). This is the steady-state response at $\omega = \omega o$.

If $v_s(t)$ is not sinusoidal, it can still be represented as an infinite sum of sine and cosine functions

$$v_s(t) = f(t) = a_o + \sum_{n=1}^{\infty} [a_n \cos(n\omega o t) + b_n \sin(n\omega o t)]$$

where $\omega o = 2\pi/T$, and T is the period over which the original function is represented by its series equivalent.

If a sufficiently large number of terms N is taken, the resulting series represents a reasonable approximation of the original input function. The exact response to this input series can be obtained by superimposing N steady-state responses evaluated at $\omega = n\omega o$, for n=1,2,...,N. This analytical solution can then be compared with the numerical solution obtained with the new model.

In the following simulation, $v_s(t)$ represents the Fourier series approximation of a square wave of unit amplitude. This approximation contains 1000 terms, and the period T is 40 ms (see Figure 6). The length of the cable is assumed to be 10 km.

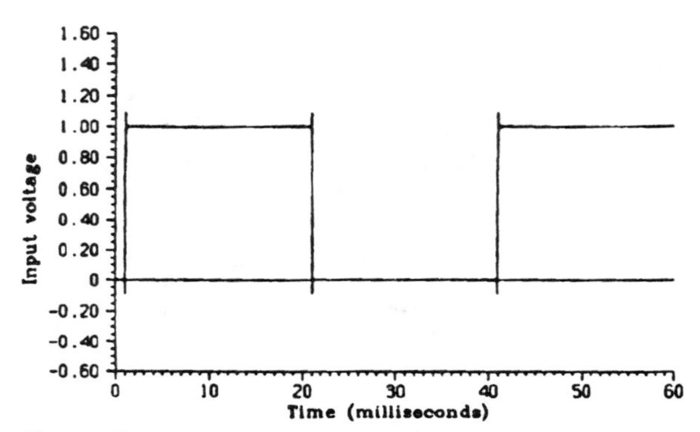

Fig. 6: *Fourier series approximation of a square wave. N=1000, T=40 ms.*

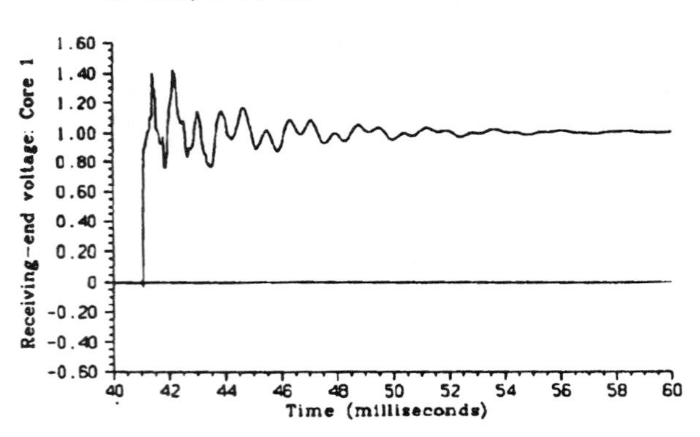

Fig. 7: *Square-wave response, receiving-end voltage. Core 1:*
------ *Analytical response.*
——— *New model.*

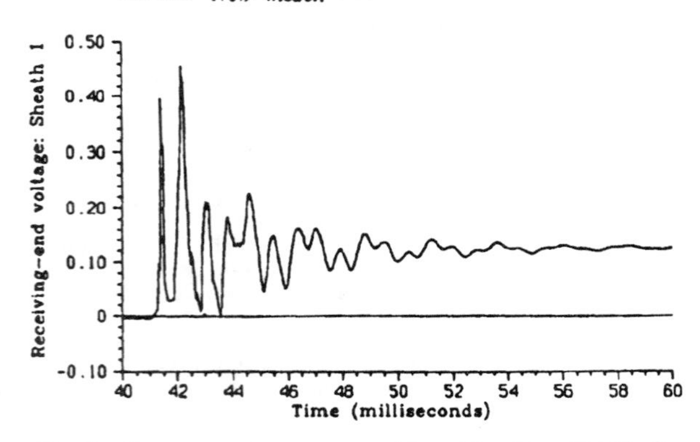

Fig. 8: *Square-wave response, receiving-end voltage. Sheath 1:*
------ *Analytical response.*
——— *New model.*

Figure 7 shows the voltage at the receiving end of core 1, and Figure 8 shows the voltage at the receiving end of sheath 1. The analytical solution is indicated with a dashed line, and the EMTP simulation is shown in solid trace. Figure 9 shows the relative differences between the two curves plotted in Figure 7.

From these results it can be seen that the agreement between analytical and transient solutions is very good (the superimposed curves in Figures 7 and 8 are practically indistinguishable). Note that the largest errors (less than 2%) occur in the first sharp peaks of the response. Also note, that

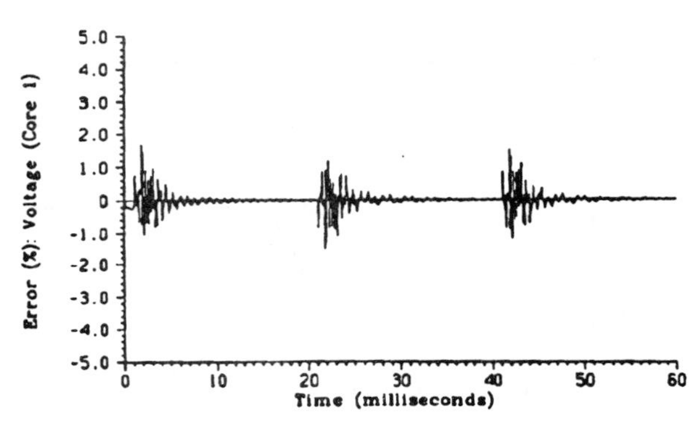

Fig. 9: *Relative error function, receiving-end voltage. Core 1:*
------ *Analytical response.*
——— *New model.*

these errors do not accumulate with time. In fact, this simulation was allowed to proceed up to 120 ms (12000 time steps) and no deterioration in the transient response was observed.

4.2. Comparison with a field test

The field test presented here has been reproduced from [1], which in turn quotes [9] as the original source. Figure 10 shows the circuit diagram of the test, and Appendix II summarizes the physical data of the crossbonded cable used.

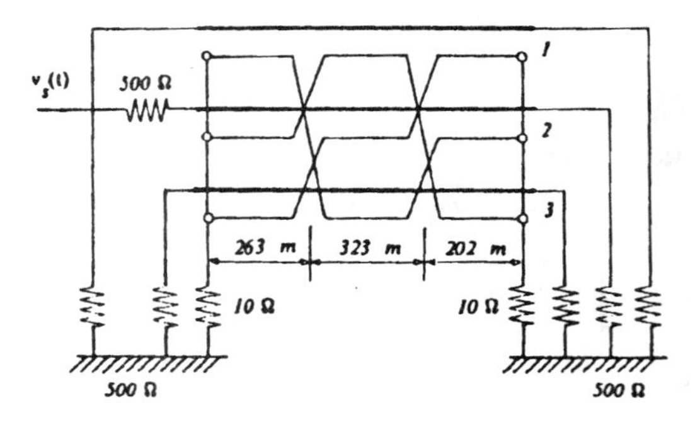

Fig. 10: *Field test connection diagram.*

An impulse of waveshape 0x40μs (i.e., negligible front time) with a peak magnitude of 7.3 kV was applied between the centre core and earth. The sheaths are connected together and grounded through a 10 Ω resistance at the sending and receiving ends.

Figure 11 shows the voltage at the sending end of core 2, and Figure 12 shows the voltage of sheath 2 measured at the second crossbonding point. Simulation results are shown in solid trace, while the measured response is shown in dashed trace.

The response obtained with the new model agrees well with the simulation results shown in [1]. The authors of reference [1] indicate that the available test data regarding complex permittivity, earth resistivity, etc., was not well known. Also, it appears that the description of the input waveform may have been oversimplified or loosely described in [9]. When the time-to-half-value of the input voltage impulse is changed from 40μs to 45μs, the agreement between calculated and measured results improves considerably (see Figure 13). The sheath voltages, however, are not affected much by this change in the input waveform.

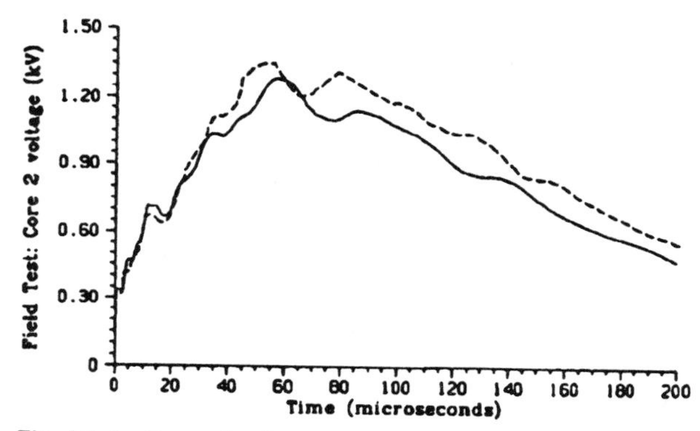

Fig. 11: *Sending-end voltage. Core 2:*
------ *Field test.*
——— *New model.*

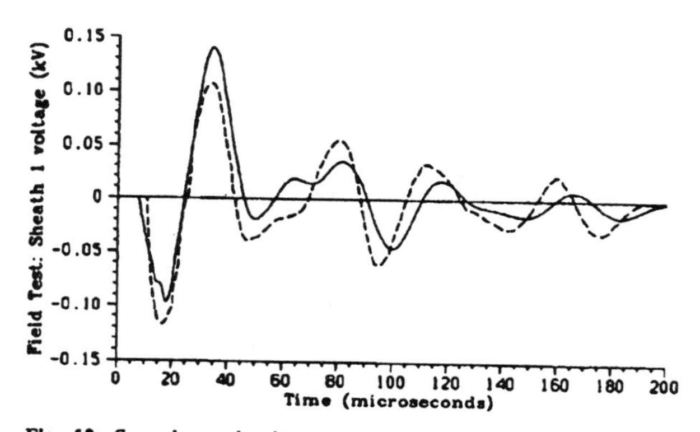

Fig. 12: *Second crossbonding-point voltage. Sheath 1:*
------ *Field test.*
——— *New model.*

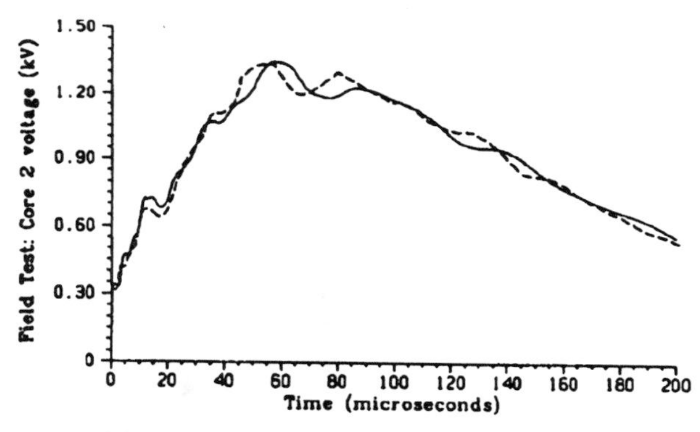

Fig. 13: *Sending-end voltage (waveshape 0x45µs).* Core 2:
------ Field test.
——— New model.

Overall, the agreement between the waveshapes of calculated and simulated results are reasonably good, considering the many uncertainties involved in such a test (e.g., earth not homogeneous, grounding resistances not well known, etc.).

4.3 Simulation of a single-phase, line to ground fault

The effects of taking into account the variation with frequency of the modal transformation matrix Q will be illustrated with the simulation of a line to ground fault. The connection diagram for this simulation is shown in Figure 14.

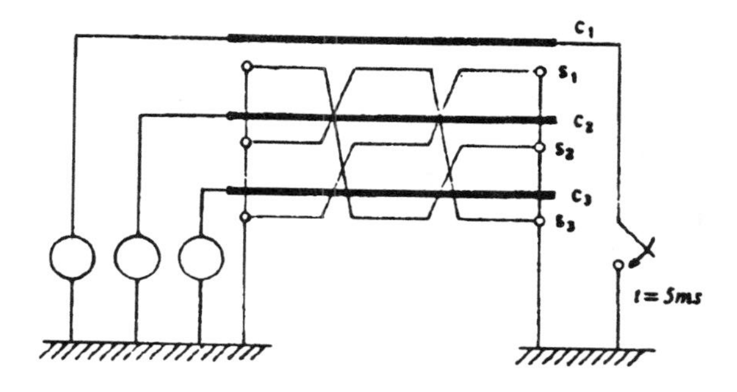

Fig. 14: *Simulation of a line to ground fault.*

The peak magnitude of the voltage sources is 1.0 p.u., and their phase angles are set 120° apart (i.e., −120°, 0° and +120°). The cores of this 230 kV underground cable are open at the receiving end; the sheaths have been crossbonded and grounded at the sending and receiving ends; the length of the cable is 3 km, and the crossbonding points are evenly spaced (i.e., each minor section is 1 km long). The simulation starts from 60 Hz steady-state initial conditions, and the receiving end of core 1 is connected to ground at t = 5 ms.

Figure 15 shows the receiving-end voltage of core 2, and Figure 16 shows the fault current (receiving end of core 1). The response obtained with the new cable model is shown in solid trace. The response obtained when Q is assumed to be constant (evaluated at 5 kHz) is shown in dashed trace.

These results clearly indicate that the frequency dependence of Q has a very significant effect on transient simulations of this type. Note that during the first 5 ms of the simulation, both solutions coincide. However, after the occurrence of the fault, the differences in the first peak of the voltage response are approximately 30%, while the differences in the first peak of the fault current exceed 60%.

Figure 17 shows the receiving-end voltage of core 2 when the simulation starts from zero initial conditions at t = 0. The response when Q is constant is quite good during the first 2 ms of the simulation. As the high frequency transients attenuate, the differences increase. After the fault occurs, the dominant frequency component is 60 Hz, and the response when Q is constant deteriorates considerably.

It has been found that the accuracy of frequency-dependent models with constant Q, depends on the type of transient situation being simulated, as well as the frequency at which Q is evaluated. For example, reasonably good

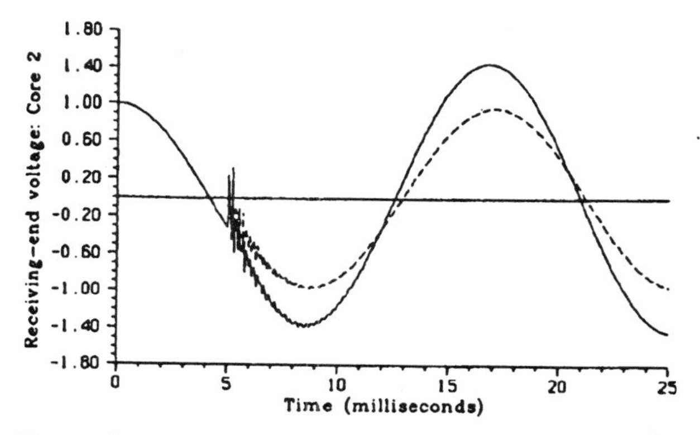

Fig. 15: *Receiving-end voltage. Core 2:*
------ *Frequency-dependent model with constant Q.*
——— *New model.*

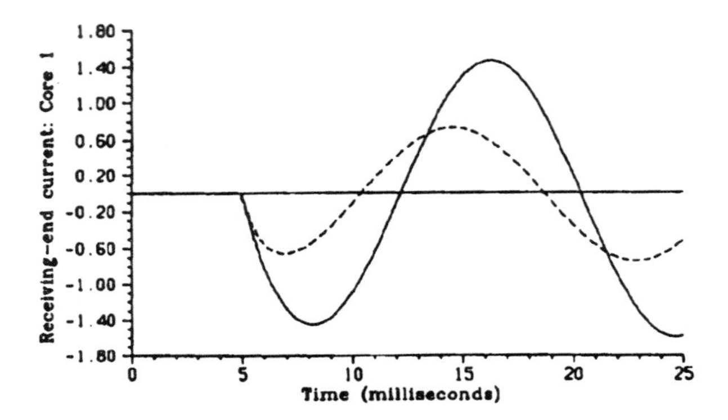

Fig. 16: *Receiving-end current. Core 1:*
------ *Frequency-dependent model with constant Q.*
——— *New model.*

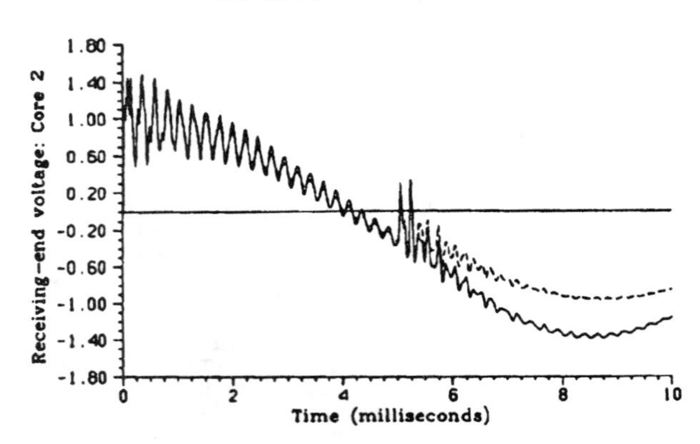

Fig. 17: *Receiving-end voltage (zero initial conditions). Core 2:*
------ *Frequency-dependent model with constant Q.*
——— *New model.*

results can be obtained during simulations where the currents flowing in the cable are very small. Also, good results can be obtained in transient simulations where the frequency range of interest is relatively high (e.g., above 1 kHz), or relatively low (e.g., below 1 Hz), and Q is evaluated at a frequency within this range (as long as the columns of Q can be normalized so that their imaginary parts are very small).

5. Computational speed

The elapsed CPU time in the time-step loop of UBC's EMTP was measured for a case where the energization of a three-phase underground cable was simulated. Using a constant-parameters model as a reference, the relative speeds of the new cable model and the frequency-dependent model with constant Q were determined. The order of the rational function approximations used was the same for both models. The results are shown in Table I.

Model	Relative timing
Constant-parameters	1.0
Frequency-dependent	6.0
New Model	6.14

Table I: *Relative CPU times. EMTP time-step loop.*

It can be seen from these results that the computational speed of the new cable model is comparable to that of the frequency-dependent model with constant Q.

Although these results are probably influenced by differences in the programming of the algorithms, they indicate that the additional computational effort required to take into account the frequency dependence of Q is relatively low.

Work is currently being carried out at UBC to improve the speed of frequency-dependent models by reducing the order of the synthesized functions at running time. The maximum frequency that can be reproduced in a numerical solution is limited by the sampling rate Δt ($f_{max}=1/(2\Delta t)$); therefore, the synthesized functions need not reproduce the original ones above this frequency. If poles above f_{max} are dropped in an appropiate manner at running time, the accuracy of the simulation is not affected, and considerable savings in computer time can be obtained.

6. Conclusions

The underground cable model presented here, accurately takes into account the variation with frequency of the parameters and of the modal transformation matrices. This overcomes the main limitation of existing frequency-dependent line models for time-domain solution algorithms.

The accuracy of the synthesized functions $Y\hat{c}$, A' and Q is very high, and generally higher than the accuracy with which earth resistivity, dielectric losses, and other cable parameters can be estimated. Therefore, from a practical point of view, the accuracy of the new cable model is only limited by the accuracy of the input data and by the simulation capabilities of the host program.

Computational speed is very high, and of the same order of magnitude as that of frequency-dependent models with constant transformation matrices. High computational speed makes the detailed simulation of crossbonded cables attractive. Also, given that minor sections are usually of the same length, only the parameters of a single section need to be synthesized. Furthermore, by modelling each section separately (making the crossbonding connections explicitly), non-linear voltage limiters at the crossbonding points can be easily taken into account.

The model is numerically stable. Relatively long simulations have been made (12000 time steps) and there has been no deviation from the correct answers.

The model is general. Its application to multiple-circuit overhead transmission lines should be of considerable practical importance.

Acknowledgements

The author would like to thank Professor H. W. Dommel for his encouragement and advice; Professor J. R. Marti for providing the rational-functions fitting routines used in the development of the new cable model, and for many stimulating discussions; Professor A. Ametani for providing the cable data used in the comparison with the field test; the reviewers of this paper, for their valuable comments and suggestions. Also, the financial support of Bonneville Power Administration, British Columbia Hydro and Power Authority, and the University of British Columbia is gratefully acknowledged.

References

[1] Nagaoka, N. and Ametani, A., "Transient Calculations on Crossbonded Cables." IEEE *Transactions on Power Apparatus and Systems*, April 1983, pp. 779-787.

[2] CIGRE Working Group 13.05, "The Calculation of Switching Surges: II.- Network Representation for Energization and Re-Energization Studies on Line Fed by an Inductive Source". *Electra*, No. 32, 1974, pp. 17-42.

[3] Marti, J. R., "Accurate Modelling of Frequency-Dependent Transmission Lines in Electromagnetic Transient Calculations." IEEE *Transactions on Power Apparatus and Systems*, January 1982, pp. 147-157.

[4] Wedepohl L. M. and Indulkar, C. S., "Switching Overvoltages in Long Crossbonded Cable Systems Using the Fourier Transform." IEEE *Transactions on Power Apparatus and Systems*, July/August 1979, pp. 1476-1480.

[5] Dommel H. W. and Meyer W. S., "Computations of Electromagnetic Transients". IEEE *Proceedings*, vol. 62(7), pp. 983-993, July 1974.

[6] Semlyen A. and Dabuleanu A., "Fast and Accurate Switching Transient Calculations on Transmission Lines with Ground Return using Recursive Convolutions." IEEE *Transactions on Power Apparatus and Systems*, March/April 1975, pp. 561-571.

[7] Strang, G., *Linear Algebra and its Applications*. New York: Academic Press, 1976, pp. 283-288.

[8] Hornbeck R. W., *Numerical Methods*. Quantum Publishers INC, 1985, pp. 229-231.

[9] Shinozaki, H., et al., "Abnormal Voltages of a Core at a Crossbonding Point.", *J. Tech. Lab. Chugoku Electric Power Co.*, vol 39, 1971, pp. 175-198.

[10] Wedepohl L. M. and Wilcox D. J., "Transient Analysis of Underground Power-Transmission Systems". IEE *Proceedings*, Vol. 120, No. 2, February 1973.

APPENDIX I

The elements of Q and Y'_c, when normalized as minimum-phase-shift functions, can be approximated by rational functions $P(\omega)$ of the form,

$$P(\omega) = k_o + \sum_{i=1}^{m} \frac{k_i}{j\omega + p_i} \tag{I.1}$$

where k_o, k_i and p_i are real constants.

In the time domain this equation becomes

$$p(t) = k_o \ \delta(t) + \sum_{i=1}^{m} k_i \exp(-p_i t) \ u(t) \tag{I.2}$$

where $\delta(t)$ is the Dirac impulse function, and $u(t)$ is the unit step function.

The convolution of $p(t)$ with a given time function $f(t)$ can be expressed as

$$g(t) = f(t) * p(t) = d \ f(t) + h(t) \tag{I.3}$$

where d is a real constant, and $h(t)$ depends on past history of $g(t)$ and $f(t)$ [3].

On the other hand, the elements of A' can be approximated by rational functions of the form

$$R(\omega) = P(\omega) \exp(-j\omega\tau) \tag{I.4}$$

where $P(\omega)$ is a rational function of the same form shown in (I.1) with k_o=0, and τ is a real constant associated to modal time delay. Equation (I.4), in the time domain, becomes

$$r(t) = \sum_{i=1}^{m} k_i \exp(-p_i(t-\tau)) \ u(t-\tau) \tag{I.5}$$

Unlike the numerical convolution $f(t)*p(t)$, the convolution of $f(t)$ with $r(t)$ is given by

$$b(t) = r(t) * f(t) = h(t) \tag{I.6}$$

where $h(t)$ only depends on past history values of $b(t)$ and $f(t)$.

Let us now consider equation (6).

$$y'_c(t) * v'_m(t) + i'_m(t) = f'_m(t) = a'(t) * f'_k(t)$$

where the elements of matrices $a'(t)$ and $y'_c(t)$ are sums of exponentials, as indicated in equations (I.2) and (I.5). Note that matrix-vector convolutions, rather that scalar convolutions, are now being indicated by the symbol "*".

The numerical convolution of $y'_c(t)$ with $v'_m(t)$ can then be expressed as

$$y'_c(t) * v'_m(t) = g'_m(t) = y'_{co} \ v'_m(t) + h'_{m_1}(t) \tag{I.7}$$

where y'_{co} is a real, diagonal matrix, and $h'_{m_1}(t)$ is a function of past history values of $g'_m(t)$ and $v'_m(t)$ (see equation (I.3)).

To obtain the phase voltages,

$$v'_m(t) = q^T(t) * v_m(t)$$

$$v'_m(t) = q_o^T \ v_m(t) + h'_{m_2}(t) \tag{I.8}$$

Similarly, the phase currents will be given by

$$i_m(t) = q(t) * i'_m(t)$$

$$i_m(t) = q_o \ i'_m(t) + h'_{m_3}(t)$$

$$i'_m(t) = q_o^{-1} \ i_m(t) - q_o^{-1} \ h'_{m_3}(t) \tag{I.9}$$

where q_o is a real, nxn matrix; $h'_{m_2}(t)$ is a function of past history terms of v_m and v'_m; $h'_{m_3}(t)$ is a function of past history terms of i_m and i'_m.

After introducing (I.7), (I.8), and (I.9) into (6), algebraic manipulation leads to

$$y_{eq} \ v_m(t) + i_m(t) = h_m(t) \tag{I.10}$$

where,

$$y_{eq} = q_o \ y'_{co} \ q_o^T$$

$$h_m(t) = q_o \ [f'_m(t) - y'_{co} \ h'_{m_2}(t) - h'_{m_1}(t)] + h'_{m_3}(t)$$

Since $f'_m(t)$ depends on past history values of f'_m and f'_k only, $h_m(t)$ is completely determined at time t.

Following a similar procedure for equation (7)

$$y_{eq} \ v_k(t) - i_k(t) = h_k(t) \tag{I.11}$$

Equations (I.10) and (I.11) represent an n-conductor transmission system in the time domain, when the transient is solved at discrete time steps. Equivalent history sources $h_m(t)$ and $h_k(t)$ are updated continuously throughout the solution.

APPENDIX II

Physical data for the 230 kV underground cable used in the analytical comparisons of section 4.1 and 4.3 (Cable 1), and for the 110 kV crossbonded cable used in the field test of section 4.2 (Cable 2) are given below.

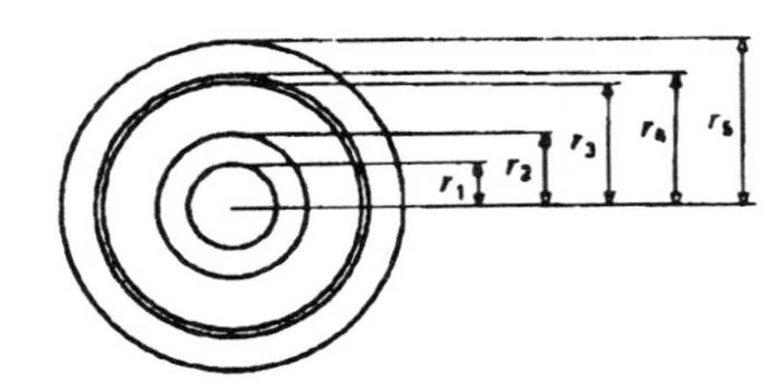

	Cable 1	Cable 2
r_1 (cm)	0.00	1.05
r_2 (cm)	2.34	1.78
r_3 (cm)	3.85	3.08
r_4 (cm)	4.13	3.26
r_5 (cm)	4.84	4.08
Core Resistivity (Ωm)	$0.0170\ 10^{-6}$	$0.0183\ 10^{-6}$
Sheath Resistivity (Ωm)	$0.2100\ 10^{-6}$	$0.0280\ 10^{-6}$
Inner insulation $\tan\delta$	0.001	0.040
Outer insulation $\tan\delta$	0.001	0.100
Inner insulation ϵ_r	3.5	3.3
Outer insulation ϵ_r	8.0	3.8
Earth resistivity (Ωm)	50	100

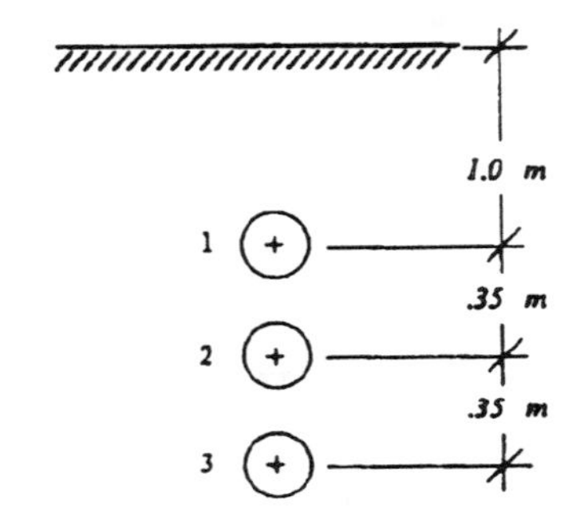

Fig. II.1: *Cable 1 configuration.*

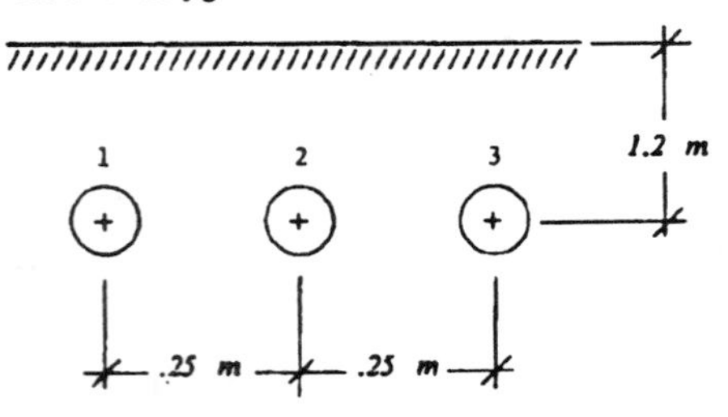

Fig. II.2: *Cable 2 configuration.*

Figures II.3 and II.4 show some of the elements of the impedance matrix for Cable 1. Note that these impedances are expressed as loop quantities; therefore, they do not depend on bonding or grounding connections [10]. Also note, that zero and positive sequence impedances can be derived directly from these loop impedances. In the captions of Figures II.3 and II.4, "core-sheath" corresponds to the loop formed by the core with return through the sheath; "sheath-ground" corresponds to the loop formed by the sheath with return through the ground; "sheath-ground-sheath" corresponds to the mutual impedance between two sheath-ground loops. There is also a mutual impedance between the core-sheath and sheath-ground loops, which is shown here.

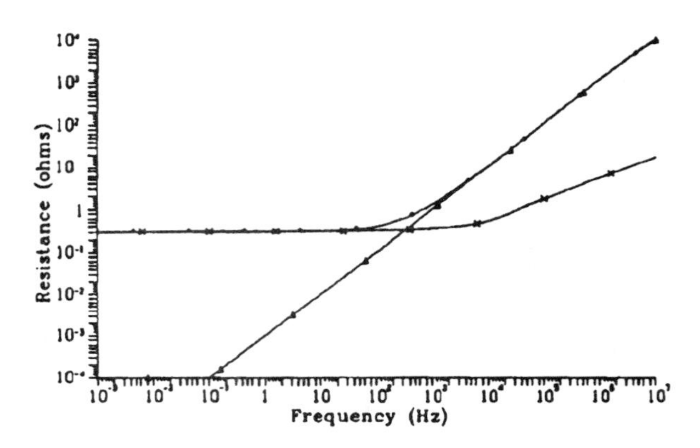

Fig. II.3: *Series impedance matrix in loop quantities. Resistance:*
—×— *Core-sheath.*
—●— *Sheath-ground.*
—△— *Sheath-ground-sheath.*

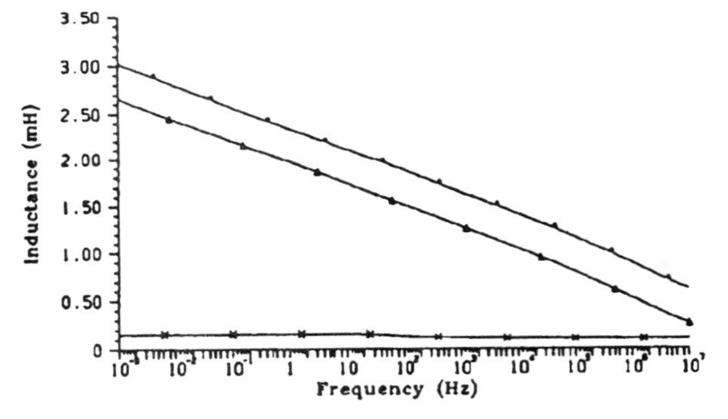

Fig. II.4: *Series impedance matrix in loop quantities. Inductance:*
—×— *Core-sheath.*
—●— *Sheath-ground.*
—△— *Sheath-ground-sheath.*

Discussion

A. Ametani and **N. Nagaoka** (Doshisha University, Kyoto, Japan): The author is commended for presenting an interesting paper. Inclusion of a frequency-dependent (FD) transformation matrix (Q matrix) into a transient calculation by a time-domain technique such as the EMTP has been one of the important problems to be solved to achieve a higher accuracy of the calculation. To overcome this problem, at least two different approaches have been proposed. One is to apply a real-time convolution to take into account the FD effect of the Q matrix at every step of transforming phasor components to modal components and vice versa within a modal framework [A]. The other one is to carry the transient calculation in the actual phase domain so as to avoid a usage of the FD Q matrix between the phase domain and the modal domain [B]. Because the EMTP is widely used all over the world as a transient analysis program and is structured to carry the transient calculation within a modal framework, the former approach has a practical significance.

The method to deal with the FD Q matrix in the paper is based on the former approach, of which the original idea was proposed by one of the discussers in [A] more than 10 years ago. (Eq. (8) in the present paper is same as (19) in [A].) Because there was no idea of recursive convolution at that time (and thus the modal approach requires a large computation time, as shown in Table 2 of [A]) the approach was not regarded practical at all. The author has completed the approach in a sophisticated manner for the case of a single-core coaxial (SC) cable. The discussers believe that the work makes an important contribution to the field of a transient analysis in an electric power system.

However, there are several questions which are not clear in the paper. The author's comments on the following questions are appreciated.

1) Eigenvalues and vectors calculation (Sec. 3(b) and Fig. 2)

i) The discussers have proposed a similar approach to improve the accuracy and stability of eigenvalues and vectors calculations and also to avoid a mode crossing or exchange phenomenon appearing during a numerical evaluation of the eigenvalues and vectors [C]. To confirm the accuracy of the calculated eigenvectors (Q matrix), it is necessary to re-diagonalize the original YZ matrix by the given Q matrix and measure the error of the re-diagonalization. This process is always carried out in the eigen-calculation program developed by the discussers. It is interesting to see the re-diagonalization error of the Q matrix, of which the third column elements are shown in Fig. 2, calculated by the author. For a reference, the re-diagonalization error of the Q matrix calculated by the discussers was 0.045 percent at maximum for the same cable as that in the present paper (Cable 1 in Appendix II).

ii) In general, the admittance matrix of an SC cable, which is a main concern of the present paper, is of a very simple form with many zero elements, and thus it is easily diagonalized. The admittance matrices of a pipe-type (PT) cable, a cable which is a multiphase SC cable with thin sheaths enclosed within a pipe, and an untransposed vertical overhead line are of a more complicated form. Thus, it is a matter of question if the author's method is applicable to the PT cable and the untransposed overhead line.

2) Fourier series approximation (Sec. 4.1, Figs. 6 to 9)

As a usual practice of a transient analyst, numerical Laplace transform with a weighting function is adopted to handle a transient with an ac source or a time varying source, to avoid a numerical instability due to poles along imaginary ($j\omega$) axis and well-known Gibb's oscillation. The discussers wonder why the author adopted Fourier series approximation rather than Laplace transform. Figures 6 to 8 clearly show Gibb's oscillation, which decreases the accuracy of the calculated results.

3) Comparison with a field test (Sec. 4.2, Figs. 11 to 13)

Because the dominant frequency of the transient on the crossbonded cable with 1-km length is quite high, the transformation matrix in this frequency region is almost constant and does not affect a transient calculation. In other words, a constant transformation matrix can give almost the same result as that with the frequency-dependent transformation matrix. Therefore, it is hard to discuss the accuracy or appropriateness of the proposed method from the results.

The source waveform (originally 40-μs wave tail, less than 0.1-μs wavefront) was deformed to have a 45-μs wave tail in Fig. 13. The change results in the decrease of the peak voltage and the phase shift of the waveform after the peak, and thus the calculated result shows a better agreement with the field test result. It is interesting to see if the change also results in a better agreement with the field test on the sheath voltages, for example, on the result corresponding to Fig. 12.

To clarify the inaccuracy of the source waveform, the impulse generator circuit which was connected to the left-hand side of the 500-Ω resistance in the field test is illustrated in Fig. A. It would be appreciated if the author would calculate again using this circuit and show the results.

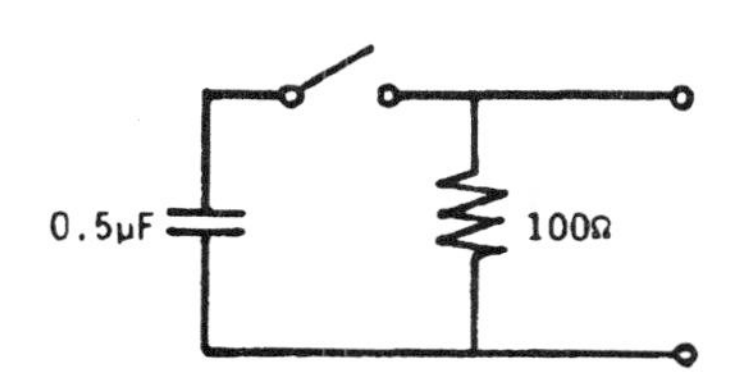

Fig. A. The source circuit in the field test.

4) Fault calculation (Sec. 4.3, Figs. 15 to 17)

Why are the calculated results by two methods almost identical for time smaller than 5 ms in Fig. 15? Since the transformation matrices used for the methods are different, the steady-state solutions seem to be different.

Also, it is a question why the calculated result only by constant Q matrix at 5 kHz is shown. To observe a steady state of a fault, Q matrix at 60 Hz should be used, and for a transient, Q matrix at a dominant transient frequency (maybe 5 kHz) should be used. In other words, if Q matrix at 60 Hz is used, a better agreement may be obtained by the constant Q matrix method for Fig. 15 to 17. However, the accuracy of a transient solution by the constant Q matrix at 60 Hz is, in general, poor. It might be better to show calculated results by the constant Q matrix at 60 Hz for comparison.

Once again, the author is commended for his interesting and timely paper.

References

[A] A. Ametani, "Refraction Coefficient Method for Switching-Surge Calculations on Untransposed Transmission Lines—Accurate and Approximate Inclusion of Frequency-Dependence," presented at IEEE 1973 PES Summer Meeting, C73-444-7, 1973.

[B] N. Nakanishi and A. Ametani, "Transient Calculation of a Transmission Line Using Superposition Law," *IEE Proc.*, vol. 133, Pt. C(5), pp. 263-269, 1986.

[C] N. Nagaoka, M. Yamamoto, and A. Ametani, "Surge Propagation Characteristics of a POF Cable," *Electr. Eng. Jpn.* (*USA*), vol. 105(5), pp. 67-75, 1985.

Manuscript received February 9, 1987.

Adam Semlyen and **H. Hamadanizadeh** (University of Toronto): We would like to commend the author for having implemented the repres ıta-tion of the frequency dependence of transformation matrices in time do ain calculations. The application shown is for cable transients but, as the author correctly points out, the effect of frequency dependence of the transformation matrix may be significant in the case of multicircuit overhead lines. Could the author provide some results related to this application?

The nice thing about the procedure is that an important part of the calculations, i.e., fitting the transformation matrix with rational functions, is performed in a preprocessing mode. Only the convolutions and updating of the past history vector in (I.8) and (I.9) requires additional computing time when calculating a transient, although this time may not be as insignificant as indicated by the author (increase of 2.3 percent from 6.00 to 6.14 in CPU time). According to the figures given in the paper for the average number of terms in the fitted Y_c', A', and Q, the number of first order differential equations to be solved for 6-modal Y_c' and A' are $6 \times (21 + 15) = 216$ and for 30 elements of Q this number is $30 \times 7 = 210$. Since at each end of the line the modal quantities have to be transformed to phase quantities for the solution of node voltages and then back to modal quantities again for updating the history vector, this nonconstant transformation matrix will increase the number of equations from 216 to 636. Does this increase not require much more computation time than 2.3 percent? If it does, then should one not try to model Q with a lower order approximation that will not result in a significant decrease in the accuracy of the model?

Manuscript received February 17, 1987.

Olov Einarsson (Asea Research): The author should be congratulated for a clearly written paper describing an efficient method for handling the frequency-dependent modal transformation matrices in time-domain modeling of cables and overhead transmission lines. I have the following questions and comments.

1) A property of the Jacobi eigensystem algorithm is that the eigenvectors are nearly orthogonal (using the metric defined by Y^{-1}) also for eigenvectors belonging to the same eigenvalue. It seems likely that this property is vital in order to meet the condition that the eigenvectors of YZ are continuous functions of frequency.

2) In the comparison with analytical results of section 4.1, the Fourier series expansion of the square wave of Fig. 6 exhibits the well-known overshot due to Gibb's phenomenon. Is this overshot also included in the input time function of the EMTP simulation using the new model, or can some part of the relative error shown in Fig. 9 be attributed to the overshot?

3) Are Figs. 15 and 16 correct? Fig. 15 seems to indicate that there is a longitudinal voltage of about 0.5 pu over core 2 in the steady state, or related to a transient having a time constant much longer than 25 ms. According to Fig. 16, this longitudinal voltage is roughly in phase with the current of core 1. Consequently the voltage cannot be induced by magnetic coupling between the two cable cores.

Manuscript received February 19, 1987.

B. R. Shperling (New York Power Authority, New York): A new transmission line time-domain model, which takes into account frequency dependent parameters, was developed by the author. From the transient analysis point of view, the suggested model represents a logical development of L. M. Wedepohl's results on the modal transformation matrices for multiphase systems and their application to the BPA's Electromagnetic Transient Analysis Program.

The author should be complimented for achieving very high accuracy in simulating system transients using the developed cable model. Indeed, the system frequency characteristics obtained with the help of the suggested method practically coincide with analytical results. At the same time, underground or submarine cables represent only one element of a conventional network. Thus, for example, in a simplest case of an underground cable energization, a quite precise representation of a sending end system, including its frequency response characteristics, is required to take full advantage of the suggested model. In addition, presence of nonlinear elements might also influence the results. Keeping these factors in mind, and realizing the inherent inaccuracy of system data, what are the accuracy requirements for the calculations of transient overvoltages in networks with underground or submarine cables?

Analysis of the transients during single phase-to-ground faults, which are illustrated by Figs. 15 and 16, needs some clarification. According to these figures, before fault application the voltages and currents calculated with the help of the new model and the frequency dependent model with constant Q at 5 kHz, are identical. After attenuation of the high frequency transients from fault application, the steady-state voltages and currents are significantly different for both methods. It is obvious that different cable parameters play different roles before and after fault application. Thus after a fault application, cable inductances, with their frequency dependencies, become dominant for transient and steady-state processes. At the same time, for cases without a line fault, the cable currents are relatively small, and precise simulation of the inductance frequency dependency becomes less important. Is it possible to generalize the discussed comparison and conclude that for cable switching operations without line faults a frequency dependent model with constant Q is sufficient?

Manuscript received February 19, 1987.

L. Marti: The author would like to thank the discussers for their interesting and useful comments. First of all, I would like to address the opening remarks made by Messrs. Ametani and Nagaoka. The importance of taking into account the frequency dependence of the modal transformation matrices of transmission systems has been recognized for a number of years. One of the first attempts to handle the problem within the framework of time-domain solution algorithms, was proposed in 1973 (reference [A] by the above discussers). In this reference, the equations describing a single-circuit untransposed transmission line are solved using modal analysis. The transformation between phase and modal voltages and currents in the time domain is obtained using direct numerical convolution between $q(t)$ and the corresponding voltage and current vectors (where $q(t)$ is obtained from $Q(\omega)$ using FFT techniques); that is,

$$i_{phase}(t) = q(t) * i_{mode}(t)$$

$$v_{mode}(t) = q'(t) * v_{phase}(t).$$

The methods proposed in [A] and the cable model proposed in this paper use this basic relationship between modal and phase quantities. The cable model in this paper differs from those proposed in [A], however, in the way it is interfaced with its host program (e.g., the EMTP), and in the way in which the convolutions are evaluated numerically. Essentially, the interface is done directly in phase quantities by splitting the convolutions into a part containing variables at instant t only, and into another part containing known history. This splitting leads to

$$i_{phase}(t) = q_o i_{mode}(t) + \text{hist}_{i\text{-phase}},$$

where

$$i_{mode}(t) = y_{co\text{-mode}} v_{mode}(t) + \text{hist}_{i\text{-mode}},$$

and

$$v_{mode}(t) = q_o' v_{phase}(t) + \text{hist}_{v\text{-mode}},$$

which, after some algebraic manipulation, produces the final form

$$i_{phase}(t) = [q_o y_{co\text{-mode}} q_o'] v_{phase} + \{ q_o [\text{hist}_{i\text{-mode}} + y_{co\text{-mode}} \text{hist}_{v\text{-mode}}] + \text{hist}_{i\text{-phase}} \}.$$

In this form, the matrix $[q_o y_{co\text{-mode}} q_o']$ is real and constant and the second term is known history which is evaluated by recursive convolution (the history term h_{m3} in Appendix I should not have been primed because it is in phase rather than in mode quantities).

The remaining questions posed by Messrs. Ametani and Nagaoka will now be addressed in the same order in which they appear in their discussion.

1) Regarding the calculation of eigenvalues and eigenvectors.

i) The approach proposed by the discussers in reference [C] is based on an eigenvalue-separation procedure proposed in 1964 [C1], where

$$P^{-1}[ZY - j\omega/cU]P$$

is diagonalized instead of ZY; c is the speed of light; and matrix P is given by

$$\begin{bmatrix} 1 & 1 & -1 & -1/2 \\ 1 & 1 & 0 & 1 \\ 1 & 1 & 1 & -1/2 \\ 1 & 0 & 0 & 0 \end{bmatrix}.$$

As the discussers correctly point out in [C], this indirect approach increases the accuracy of the calculation of the eigenvalues of YZ by increasing the numerical separation of almost coalescing eigenvalues. However, it is unclear why the discussers feel that this approach is similar to the use of seeding in combination with the modified Jacobi method presented in this paper, and why this approach prevents the eigenvector switchover phenomenon encountered with standard eigenvalue/eigenvector methods.

The accuracy of the eigenvalues obtained using seeded Jacobi is only limited by the machine accuracy, given that the method is completely stable. Default settings in the eigenvalue/eigenvector routines written by the author force the re-diagonalization errors to be less than 10^{-12}; that is, the Euclidean norm of the off-diagonal terms of $Q^{-1}YZQ$ is 10^{-12} times smaller than the norm of its diagonal entries.

ii) It is unclear to the author why the discussers feel that the particular form of the admittance matrix is a concern of this paper, since it is the matrix product YZ rather than Y alone that must be diagonalized in order to evaluate A', Y_c', and Q. The software needed to process the parameters of pipe-type cables and transmission lines (in a form compatible with the new model) will be written in the near future as part of the ongoing implementation of the new cable model in the DCG/EPRI version of the EMTP (preliminary results seem to indicate that the EMTP support routines CABLE CONSTANTS and LINE CONSTANTS do not produce the smooth eigenvector functions required by the cable model). Therefore, it will soon be possible to verify the applicability of the new model to other forms of transmission systems such as pipe-type cables and overhead transmission lines.

2) Regarding the comparison with analytical results shown in section 4.1, it was not the purpose of the simulation to calculate the exact response of a square wave. As indicated in the paper, only a comparison with an exact analytical answer was intended. Thus the use of a Fourier series expansion. The use of either Laplace or Fourier transforms would not have produced analytically exact answers, since numerical transformations are always subjected to a certain amount of error, and the comparisons would not have been rigorous from a mathematical standpoint.

3) With regards to the comparison with the field test shown in section 4.2, it is true that this test alone may not establish the accuracy of the model

completely. The author intends to do more comparisons with field tests, as they become available.

By changing the time-to-half value of the input waveform in the comparison with the field test, the peak value of the voltages at the sending end of the cable increased, thus giving a better agreement with the experimental results. However, as indicated in section 4.2, this change in the input waveform does not affect significantly the sheath voltages.

Figure C.1 below shows the sending-end voltage of core 2 when the equivalent circuit supplied by the discussers is used instead (in absence of additional data it was assumed that the 0.5-μF capacitor was charged to 7.3 kV). It can be seen from this plot that the response obtained with the new source representation is actually worse than the responses calculated with the 0 × 40 μs and 0 × 45 μs waveforms.

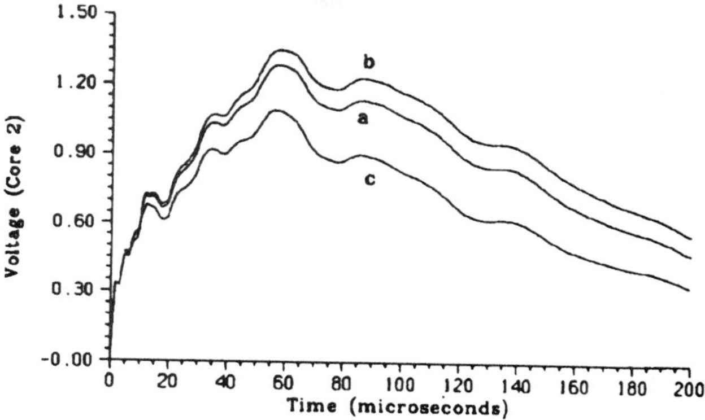

Fig. C.1. Comparison of Core 2 voltages for the field test simulation shown in section 4.2. a) 0 × 40 μs input function. b) 0 × 45 μs input function. c) New source representation.

4) In the type of cables studied in this paper, it has been observed that when the currents are relatively small, the use of a constant transformation matrix evaluated at frequencies above 1 kHz produces very good results. As soon as the currents are significant (i.e., after the occurrence of the line-to-ground fault) the response with Q constant deteriorates considerably, as shown in Figs. 15 to 17. Had Q been evaluated at 60 Hz, then the open-circuit response of the cable would have been poorer than the one shown in the paper because the magnitude of the imaginary part of Q at 60 Hz is not negligible, and the frequency-dependent line models currently used in the EMTP require that Q be real.

The comments made by Mr. Einarsson will be addressed next.

1) The use of the modified Jacobi method to solve the symmetric eigenproblem $Zx = \lambda Y^{-1}x$ is indeed an important factor in the calculation of smooth eigenvector functions of frequency. However, it should be pointed out that pre- and post-multiplication by the transformation matrix obtained in the previous frequency step (or seeding) is just as important. In fact, the use of the modified Jacobi method without seeding does not produce smooth eigenvector functions. Similarly, the use of seeding in combination with algorithms such as the QR decomposition, also fail to produce smooth eigenvector functions.

2) In the comparison with the analytical results shown in section 4.1, the input voltage used in the EMTP simulation is the Fourier series approximation of a square wave. Therefore, the overshoot due to Gibb's phenomenon is included in the EMTP results. The discrepancies between the analytical and numerical results shown in Fig. 9 seem to be consistent with the errors in the approximations by rational functions of Y_c', A', and Q.

3) The results obtained in the simulation of the single-phase line-to-ground fault may seem unusual, but they are indeed correct. The apparent contradiction pointed out by Mr. Einarsson can be explained using phasor analysis, assuming that the three-phase cable system is balanced and lossless. If the system is balanced the voltages induced in the unfaulted phases are directly proportional to the ratio Z_m/Z_s, where

$$Z_m = (Z_0 - Z_1)/3$$

$$Z_s = (Z_0 + 2Z_1)/3.$$

In the simulation shown in section 4.3, the voltage in core 2 (V_b in the phasor diagram shown in Fig. C.2) will be given by

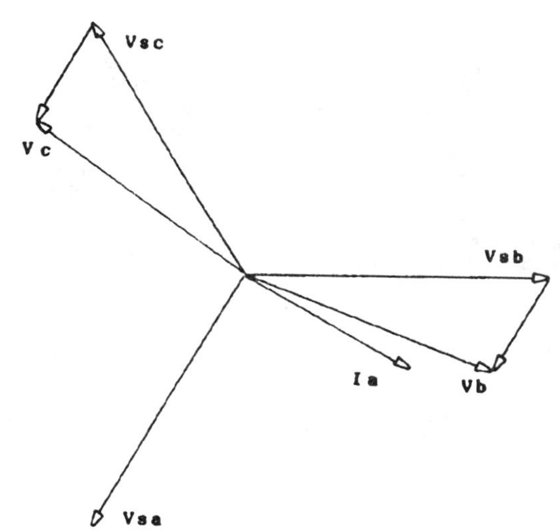

Fig. C.2. Steady-state voltages on a lossless three-phase system during a line-to-ground fault when $Z_0 < Z_1$.

$$V_b = V_{sb} - (Z_m/Z_s)V_{sa}$$

where V_{sa}, V_{sb}, and V_{sc} are the source voltages.

As illustrated in the phasor diagram shown below, if Z_m/Z_s is negative, then V_b will be roughly in phase with the fault current I_a. The series impedance matrix Z per unit length (evaluated at 60 Hz) for the underground cable under consideration is shown below. The sheaths have been eliminated because in this example they are grounded.

$$Z = \begin{bmatrix} (0.1600+j0.1530) & (0.0827-j0.0198) & (0.0581-j0.0354) \\ (0.0827-j0.0198) & (0.1430+j0.1370) & (0.0827-j0.0198) \\ (0.0581-j0.0354) & (0.0827-j0.0198) & (0.1600+j0.1530) \end{bmatrix} \text{ ohms/km}$$

The steady-state voltages and currents calculated using the correct value of Z also agree well with the results of the transient simulation shown in section 4.3.

The comments made by Messrs. Semlyen and Hamadanizadeh will be addressed next. The cable model presented in this paper should be applicable to the case of multiple-circuit transmission lines, where the modal transformation matrix Q is known to depend on frequency. At this point in time, however, the software needed to generate Q as a smooth function of frequency has not yet been adapted to existing line constants programs.

As indicated in section 5, the relative computational speed of the new cable model was measured using UBC's version of the EMTP. These results are strongly influenced by the programming techniques used in the implementation of the new cable model at UBC. The additional computational burden of taking into account the frequency dependence of Q is not directly proportional to the number of recursive convolutions evaluated. However, with programming considerations being equal, the differences in CPU time should be higher than those reflected in Table 1. This has been verified by forcing the approximations by rational functions of Q to be of order zero; that is, by assuming Q constant, but using the solution algorithm of the new cable model. The relative CPU time in this case is 3.6.

These results suggest that reducing the order of the approximations of the elements of Q would result in considerable reductions in computational time, as Messrs. Semlyen and Hamadanizadeh suggest. However, the largest savings that could be expected would be of the order of 40 percent in the limiting case when Q is constant. It is not clear, however, if computational savings of this order of magnitude would always justify the potential loss of accuracy.

Mr. Shperling raises an interesting question: If the system data and the models which represent other network components are only accurate to, for example, 10 or 15 percent, is it justifiable to try to model an underground cable within a 2-percent accuracy range?.

It is difficult to give a definite answer to this question because accuracy requirements depend on the type of simulation, and on the configuration of the system itself. There are studies where it is sufficient to model an underground cable as a pi-circuit or even as a simple shunt capacitance. When the main concern of a simulation is the transient response of the cable itself, it is probably best to model the cable with the highest accuracy possible, even if the basic data such as earth resistivity, dielectric permittivity, and other factors are not known very accurately. Knowing that a model is nearly as accurate as the input data available removes one source of uncertainty when the results of a transient simulation are interpreted.

Recent studies on the validity of the constant transformation matrix assumption in the case of overhead transmission lines [C2], [C3] seem to indicate that a constant transformation matrix Q evaluated at frequencies between 500 Hz and 5 kHz produces good results when the currents are relatively small. The same studies also indicate that significant loss of accuracy may occur in cases of strong asymmetry (e.g., when two circuits of different voltage levels share the same tower).

In the case of the type of underground cables studied at UBC, it appears that Q can also be assumed to be constant in situations where the currents are very small. It may not be wise to generalize these observations before further study of a larger variety of cable constructions and configurations is made. It might be better to say that in absence of a better model, the best answers can be obtained when Q is real, and evaluated at high frequencies, and when studies involve small currents.

References

[C1] R. H. Galloway, W. B. Shorrocks, and L. M. Wedepohl, "Calculation of Electrical Parameters for Short and Long Polyphase Transmission Lines," *Proc. IEE*, vol. 111, pp. 2051–2059, December 1964.

[C2] J. R. Marti, "Validation of Transmission Line Models in the EMTP," Presented to the Power System Planning and Operating Section of the Canadian Electric Association at Vancouver, BC, March 1987.

[C3] J. R. Marti, H. W. Dommel, L. Marti, and V. Brandwajn, "Approximate Transformation Matrices for Unbalanced Transmission Lines," to be published in the *Proc. Ninth Power Systems Computation Conference*, Lisbon, August 30–September 4, 1987.

Manuscript received April 10, 1987.

MATRIX REPRESENTATION OF THREE-PHASE N-WINDING TRANSFORMERS FOR STEADY-STATE AND TRANSIENT STUDIES

V. Brandwajn, Member, IEEE
Ontario Hydro
Toronto, Ontario, Canada

H.W. Dommel, Fellow, IEEE
The University of British Columbia
Vancouver, B.C., Canada

I.I. Dommel
Vancouver, B.C., Canada

ABSTRACT

Detailed transformer representations are needed in the analysis of electromagnetic transients and in the analysis of unbalanced steady-state conditions. This paper describes the derivation of models for three-phase and single-phase N-winding transformers in the form of branch impedance or admittance matrices, which can be calculated from available test data of positive and zero sequence short-circuit and excitation tests. The models can be used for many types of studies as long as the frequencies are low enough so that capacitances in the transformer can be ignored. The inclusion of saturation effects is briefly discussed.

1. INTRODUCTION

The representation of single-phase N-winding transformers for steady-state and transient studies is reasonably straightforward [1]. Representing three-phase transformers, on the other hand, has always been more difficult, on transient network analyzers as well as in digital computer studies. The most common approach has been the addition of an extra delta-connected winding to single-phase units, in order to approximate the magnetic coupling among the three cores. It is not always easy, however, to relate the data of this extra winding to the available test data. For instance, no such general relationship can be derived for a three-phase three-winding transformer.

The three-phase transformer models described here are based on the physical concept of representing windings as mutually coupled coils; with this approach, a three-phase two-winding transformer simply becomes a system of 6 coupled coils. The impedance or admittance matrix for the coupled coils can easily be derived from commonly available test data. The method is also valid for single-phase units.

2. BASIC CONCEPT FOR SINGLE-PHASE TRANSFORMERS

To explain the concept, a single-phase N-winding transformer will be considered first, which can be described by the following steady-state phasor equations:

$$\begin{bmatrix} V_1 \\ V_2 \\ \vdots \\ V_N \end{bmatrix} = \begin{bmatrix} Z_{11} & Z_{12} & \cdots & Z_{1N} \\ Z_{21} & Z_{22} & \cdots & Z_{2N} \\ \cdots & \cdots & \cdots & \cdots \\ Z_{N1} & Z_{N2} & \cdots & Z_{NN} \end{bmatrix} \begin{bmatrix} I_1 \\ I_2 \\ \vdots \\ I_N \end{bmatrix} \quad (1)$$

The matrix in Eq.(1) is symmetric. Its elements could theoretically be measured in excitation (no-load) tests: If coil k is energized, and all other coils are open-circuited, then the measured values for I_k and $V_1, \ldots V_N$ produce column k of the [Z]-matrix,

$$Z_{ik} = V_i/I_k \quad (2)$$

Unfortunately, the short-circuit input impedances, which describe the more important transfer characteristics of the transformer, get lost in such excitation measurements. The short-circuit input impedance Z_{ik}^{short} between energized coil i and short-circuited coil k is

$$Z_{ik}^{short} = Z_{ii} - \frac{Z_{ik}Z_{ki}}{Z_{kk}}, \quad (3a)$$

or

$$\frac{Z_{ik}^{short}}{Z_{ii}} = 1 - k^2, \quad (3b)$$

with the coupling coefficient

$$k = \sqrt{\frac{Z_{ik}Z_{ki}}{Z_{ii}Z_{kk}}} \quad (3c)$$

Large power transformers are tightly coupled, with k close to 1.0. For a typical short-circuit input impedance of 10% and for a typical exciting current of 0.4%, the values Z_{ii}, Z_{ik} and Z_{kk} would have to be measured with an accuracy of 0.001% to obtain the value of Z_{ik}^{short} to within ±10%. This is clearly impossible. It is, therefore, necessary to find [Z] or its inverse [Y] in a different way, namely by calculation as described later. The need for high accuracy still remains, however. In the example cited above, at least 6 significant digits would be needed for the elements of [Z], which could be close to the limit of computers with a short word length. The alternate representation of section 6 with [Y] avoids this precision problem.

For the analysis of electromagnetic transients, Eq.(1) is rewritten as a differential equation,

$$\begin{bmatrix} v_1 \\ v_2 \\ \vdots \\ v_N \end{bmatrix} = \begin{bmatrix} R_{11} & R_{12} & \cdots & R_{1N} \\ R_{21} & R_{22} & \cdots & R_{2N} \\ \cdots & \cdots & \cdots & \cdots \\ R_{N1} & R_{N2} & \cdots & R_{NN} \end{bmatrix} \begin{bmatrix} i_1 \\ i_2 \\ \vdots \\ i_N \end{bmatrix} + \begin{bmatrix} L_{11} & L_{12} & \cdots & L_{1N} \\ L_{21} & L_{22} & \cdots & L_{2N} \\ \cdots & \cdots & \cdots & \cdots \\ L_{N1} & L_{N2} & \cdots & L_{NN} \end{bmatrix} \frac{d}{dt} \begin{bmatrix} i_1 \\ i_2 \\ \vdots \\ i_N \end{bmatrix} \quad (4)$$

with [R] being the real part of [Z], and [L] being the imaginary part of [Z] divided by ω. This model is directly accepted by all versions of the Electromagnetic Transients Program of the Bonneville Power Administration. Getting from Eq.(1) to Eq.(4) implies a series connection of [R] and [L], which is a reasonable as-

81 SM 429-0 A paper recommended and approved by the IEEE Power System Engineering Committee of the IEEE Power Engineering Society for presentation at the IEEE PES Summer Meeting, Portland, Oregon, July 26-31, 1981. Manuscript submitted January 28, 1981; made available for printing May 4, 1981.

sumption.

3. EXTENSION OF BASIC CONCEPT TO THREE-PHASE TRANSFORMERS

The extension of Eq.(1) to three-phase transformers is conceptually easy. Each winding in Eq.(1) no longer consists of a single coil, but of three coils for the three phases or core legs. This means that a matrix element of [Z] becomes a 3x3 submatrix

$$\begin{bmatrix} Z_S & Z_M & Z_M \\ Z_M & Z_S & Z_M \\ Z_M & Z_M & Z_S \end{bmatrix} \tag{5}$$

where Z_S is the self impedance of a phase or leg and Z_M is the mutual impedance among the three phases or legs. As in any other three-phase power system component, these self and mutual impedances are related to the positive and zero sequence values Z_1 and Z_o by

$$Z_S = \frac{1}{3}(Z_o + 2Z_1), \tag{6a}$$

$$Z_M = \frac{1}{3}(Z_o - Z_1). \tag{6b}$$

Replacing an element of [Z] by the 3x3 submatrix of Eq. (5) and relating the diagonal and off-diogonal elements Z_S, Z_M to positive and zero sequence values is all that is needed to extend the methods developed for single-phase transformers to three-phase transformers.

Since the 3x3 submatrices contain only 2 distinct values Z_S and Z_M, it is not necessary to actually work with 3x3 submatrices, but only with pairs (Z_S, Z_M). D. Hedman derived a special "balanced-matrix algebra" in [2] for the multiplication, inversion and addition of such pairs.

4. DIRECT CALCULATION OF IMPEDANCE MATRIX

Recall that Eq.(1) is valid for three-phase N-winding transformers if it is understood that each element is replaced by the 3x3 submatrix of Eq.(5). These submatrices can be directly found from test data as long as the exciting current is not neglected.

First, calculate the imaginary parts of the diagonal element pairs (X_{S-ii}, X_{M-ii}) from the exciting current of the positive and zero sequence excitation tests. If excitation losses are ignored, and if "i" is the excited winding, then it follows from Eq.(2) that X_{ii} in per unit is simply the reciprocal of the per-unit exciting current. With the positive and zero sequence values X_{1-ii} and X_{o-ii} thus known, the pair values are simply obtained from Eq.(6),

$$X_{S-ii} = \frac{1}{3}(X_{o-ii} + 2X_{1-ii}) \tag{7a}$$

$$X_{M-ii} = \frac{1}{3}(X_{o-ii} - X_{1-ii}) \tag{7b}$$

For the other windings, it is reasonable to assume that the p.u. reactances are practically the same as in Eq. (7), since these open-circuit reactances are much larger than the short-circuit input impedances. If it is known, however, that a particular winding has very little stray flux (e.g., the tertiary winding no. 3 of a three-winding transformer with cylindrical coil construction), then one could imagine an equivalent circuit where the magnetizing reactance is connected across that winding [1], e.g., from 3 to neutral in Fig. 1 (following page). In that case, the p.u. impedances (Z_{S-11}, Z_{M-11}) and (Z_{S-22}, Z_{M-22}) of the primary and secondary windings would differ from the p.u. impedances (Z_{S-33}, Z_{M-33}) by the value of the short-circuit impedance 1-2 or 1-3, respectively, and the diagonal elements could be corrected for that difference accordingly.

If the winding resistances are known, they are added to the self impedance Z_{S-ii} of the diagonal element pairs (Z_{S-ii}, Z_{M-ii}). If they are not known, but if load losses from short-circuit tests are given, they could be calculated from the load losses. For two-winding transformers, one could assume $R_1 = R_2$ in p.u. in this calculation, while 3 equations in 3 unknowns could be used in the case of a three-winding transformer. Calculating winding resistances from load losses is not exact because these losses contain stray losses as well, but is probably better than simply setting winding resistances to zero if they are not known.

If excitation losses are known, they must not be included in the calculation of Eq.(7) because that would imply that they are modelled as a resistance in series with the magnetizing reactance. Instead, shunt resistances should be added across one or more windings to reproduce the excitation losses. These shunt resistances are additional branches which cannot be included in the impedance matrix representation of Eq.(1). Strictly speaking, the p.u. reactances X_{1-ii} and X_{o-ii} are then no longer the reciprocals of the p.u. exciting current, but the reciprocals of the imaginary part of the p.u. exciting current,

$$I_m = \sqrt{I_{exc}^2 - P_{exc}^2} \quad \text{in p.u. values} \tag{8}$$

with I_m = p.u. magnetizing current (imaginary part of p.u. exciting current)
I_{exc} = p.u. exciting current
P_{exc} = p.u. excitation loss.

In practice, I_m and I_{exc} differ so little that the value of I_{exc} can usually be used for I_m.

With the diagonal element pairs known, the off-diagonal element pairs (Z_{S-ik}, Z_{M-ik}) are calculated from the short-circuit input impedances with Eq.(3a). First find

$$Z_{ik} = Z_{ki} = \sqrt{(Z_{ii} - Z_{ik}^{short})\, Z_{kk}} \tag{9}$$

separately for positive and zero sequence, and then convert the values to the pair values with Eq.(6). Z_{ik}^{short} in Eq.(9) is the complex short-circuit impedance. In p.u., its real part are the p.u. load losses, and its imaginary part can be calculated from Eq.(21).

As already pointed out in section 2, the elements of [Z] must be calculated with high accuracy; otherwise, the short-circuit input impedance gets lost in the open-circuit impedances. The lower the exciting current is, the more equal the p.u. impedances Z_{ii}, Z_{kk} and Z_{ik} become among themselves in Eq.(3a). Experience has shown that the positive sequence exciting current should not be much smaller than 1% for a single-precision solution on a UNIVAC computer (word length of 36 bits) to avoid numerical problems. On computers with higher precision, the value could obviously be lower. On large, modern transformers, exciting currents of less than 1% are common, but this value can usually be increased for the analysis without influencing the results.

5. MODIFICATIONS IN ZERO SEQUENCE FOR DELTA-CONNECTED WINDINGS

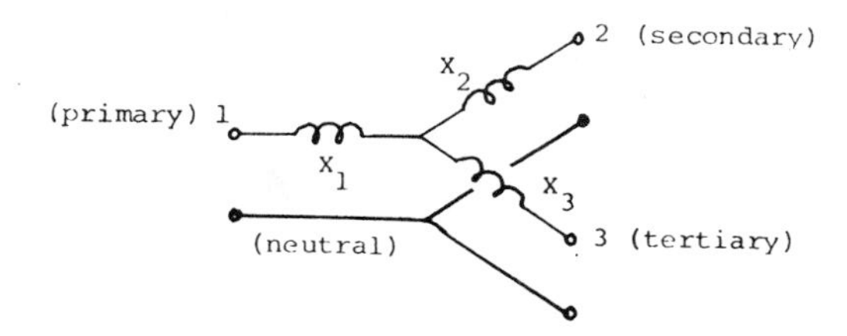

Fig. 1. Equivalent star circuit for zero-sequence short-circuit tests of a three-winding transformer (all reactances must be in p.u.)

Eq.(9) cannot be used directly for the zero sequence calculation of transformers with three or more windings if one or more of them are delta-connected. Assume that a three-winding transformer has wye-connected primary and secondary windings, with their neutrals grounded, and a delta-connected tertiary winding. In this case, the zero-sequence short-circuit test between the primary and secondary windings will not only have the secondary winding shorted but the tertiary winding as well, since a closed delta connection provides a short-circuit path for zero-sequence currents. This special situation can be handled in a number of ways, e.g., by modifying the short-circuit data for an open delta so that Eq.(9) can again be used. With the well-known equivalent star circuit of Fig. 1, the three test values supplied by the manufacturer are

$$X_{12}^{closed\ \Delta} = X_1 + \frac{X_2 X_3}{X_2 + X_3}, \qquad (10a)$$

$$X_{13} = X_1 + X_3, \quad \text{in p.u. values} \qquad (10b)$$

$$X_{23} = X_2 + X_3, \qquad (10c)$$

which can be solved for X_1, X_2, X_3:

$$X_1 = X_{13} - \sqrt{X_{23}X_{13} - X_{12}^{closed\ \Delta} X_{23}}, \qquad (11a)$$

$$X_2 = X_{23} - X_{13} + X_1 \quad \text{in p.u. values} \qquad (11b)$$

$$X_3 = X_{13} - X_1 \qquad (11c)$$

After this modification, the values $X_1 + X_2$, $X_1 + X_3$ and $X_2 + X_3$ are used as input data, with winding 3 no longer being shorted in the test between 1 and 2. The modification scheme becomes more complicated if resistances are included. For instance, Eq.(10a) becomes

$$\left| Z_{12}^{closed\ \Delta} \right| = \left| R_1 + jX_1 + \frac{(R_2 + jX_2)(R_3 + jX_3)}{(R_2 + R_3) + j(X_2 + X_3)} \right| \qquad (12)$$

in p.u. values

with $|Z_{12}^{closed\ \Delta}|$ being the value supplied by the manufacturer, and R_1, R_2, R_3 being the winding resistances. This leads to a system of nonlinear equations, which have been solved by Newton's method in a computer program based on the methods of this paper. This computer program can handle three-winding transformers with wye/wye/delta- and with wye/delta/delta- connections.

Fig. 1 is only used to derive the three reactances X_1, X_2, X_3. If it were used as a zero-sequence model, which it is not, terminal 3 would have to be shorted to the zero-sequence neutral bus to represent the closed delta. What is used here instead is the impedance matrix of Eq.(1) which is written in phase quantities rather than in symmetrical components. Eq.(1) is a system of branch equations, and as such does not contain any information about the winding connections. These connections must be established by the user through proper assignment of node names to the branch terminals. For the example of section 13, assume that the node names for the tertiary terminals are TA, TB and TC. To establish the proper delta connection, branch '3-A' must go from node TA to TB, branch '3-B' from TB to TC, and branch '3-C' from TC to TA. With these node assignments, the tertiary will provide a short-circuit for zero-sequence currents, and the phase shift between wye- and delta-connected windings in positive and negative sequence will be correct as well.

To explain the phase shift, assume that a load is connected to 3, with 2 open. If positive sequence currents I_{1pos} are fed into the high side, then the branch currents in the three high voltage windings are I_{1pos}, $a^2 I_{1pos}$ and aI_{1pos}, and in the three tertiary windings

$$I_{TB\ to\ TA} = \frac{230/\sqrt{3}}{50} I_{1pos}$$

$$I_{TC\ to\ TB} = \frac{230/\sqrt{3}}{50} a^2 I_{1pos}$$

$$I_{TA\ to\ TC} = \frac{230/\sqrt{3}}{50} a\, I_{1pos}$$

Since the line current I_{TA} on the tertiary side is

$$I_{TA} = I_{TB\ to\ TA} - I_{TA\ to\ TC},$$

it follows that

$$I_{TA} = \frac{230/\sqrt{3}}{50} (1-a) I_{1pos}$$

where the factor in parenthesis not only contains the correct factor of $\sqrt{3}$, but also the phase shift of -30°. For negative sequence currents, this factor would be $(1-a^2)$ which contains the phase shift of $+30^o$.

6. ADMITTANCE MATRIX REPRESENTATION

As mentioned in section 2, the elements of [Z] must be calculated with high accuracy, especially if the exciting current is low. If the exciting current is totally ignored, then [Z] cannot be used at all. In such cases, an alternative representation can be used in the form of

$$[I] = [Y][V], \qquad (13)$$

which is the inverse relationship of Eq.(1). Even though [Z] becomes infinite for zero exciting current, [Y] does exist, and is in fact the well-known representation of transformers used in power flow studies.

Let the transfer characteristics be expressed as voltage drops between winding i and the last winding N,

$$\begin{bmatrix} V_1 - V_N \\ V_2 - V_N \\ \vdots \\ V_{N-1} - V_N \end{bmatrix} = \begin{bmatrix} Z_{11}^{reduced} & Z_{12}^{reduced} & \ldots Z_{1,N-1}^{reduced} \\ Z_{21}^{reduced} & Z_{22}^{reduced} & \ldots Z_{2,N-1}^{reduced} \\ \ldots & \ldots & \ldots \\ Z_{N-1,1}^{reduced} & Z_{N-1,2}^{reduced} & \ldots Z_{N-1,N-1}^{reduced} \end{bmatrix} \begin{bmatrix} I_1 \\ I_2 \\ \vdots \\ I_{N-1} \end{bmatrix} \qquad (14)$$

in p.u. values

with $[Z^{reduced}]$ again being symmetric. Also, let the exciting current be ignored, which implies that

$$\sum_{k=1}^{N} I_k = 0 \quad \text{in p.u. values} \qquad (15)$$

The elements of the reduced matrix in Eq.(14) can then be found directly from the short-circuit test data, as shown by Shipley [3]. For a short-circuit test between i and N, only I_i in Eq.(14) is nonzero, and therefore

$$Z_{ii}^{reduced} = Z_{iN}^{short} \tag{16}$$

The off-diagonal element $Z_{ik}^{reduced}$ is found by relating Eq.(14) to the short-circuit test between i and k, where $I_k=-I_i$ and all other currents are zero. Then

$$V_i - V_N = (Z_{ii}^{reduced} - Z_{ik}^{reduced})\ I_i\ , \tag{17a}$$

$$V_k - V_N = (Z_{ki}^{reduced} - Z_{kk}^{reduced})\ I_i\ , \tag{17b}$$

or

$$V_i - V_k = (Z_{ii}^{reduced} + Z_{kk}^{reduced} - 2\ Z_{ik}^{reduced})\ I_i \tag{17c}$$

in p.u. values

By definition, the expression in parenthesis of Eq.(17c) must be the short-circuit input impedance Z_{ik}^{short}, or

$$Z_{ik}^{reduced} = \frac{1}{2}\ (Z_{iN}^{short} + Z_{kN}^{short} - Z_{ik}^{short}) \tag{18}$$

in p.u. values

Eq.(14) cannot be expanded to include all N windings, since all matrix elements would become infinite with the exciting current being ignored. Therefore, an admittance matrix formulation must be used. First, Eq. (14) is inverted, with

$$[Y^{reduced}] = [Z^{reduced}]^{-1}\ , \tag{19}$$

and then a row is added for I_N by using Eq.(15). This results in

$$\begin{bmatrix} I_1 \\ I_2 \\ \vdots \\ I_N \end{bmatrix} = \begin{bmatrix} Y_{11} & Y_{12} & \cdots & Y_{1N} \\ Y_{21} & Y_{22} & \cdots & Y_{2N} \\ \cdots & \cdots & \cdots & \cdots \\ Y_{N1} & Y_{N2} & \cdots & Y_{NN} \end{bmatrix} \begin{bmatrix} V_1 \\ V_2 \\ \vdots \\ V_N \end{bmatrix} \tag{20}$$

with

$$Y_{ik} = Y_{ik}^{reduced} \quad \text{from Eq.(19) for } i,k \leq N-1,$$

$$Y_{iN} = Y_{Ni} = -\sum_{k=1}^{N-1} Y_{ik}^{reduced} \quad \text{for } i \neq N,$$

and

$$Y_{NN} = -\sum_{i=1}^{N-1} Y_{iN}\ .$$

in p.u. values

For transient studies, [Y] must be split up into resistive and inductive components. This is best done by ignoring the resistances in building $[Z^{reduced}]$ of of Eq.(14) from the reactance part of the short-circuit test data,

$$X_{ik}^{short} = \sqrt{|Z_{ik}^{short}|^2 - (R_i + R_k)^2} \tag{21}$$

in p.u. values

with $|Z_{ik}^{short}|$ = p.u. short-circuit input impedance (magnitude),

$R_i + R_k$ = either p.u. load losses in short-circuit test between i and k, or p.u. winding resistances.

Then,

$$[L]^{-1} = j\omega[Y], \tag{22}$$

The winding resistances (either known, or calculated from load losses as indicated in section 4) then form a diagonal matrix [R], and the transformer is finally described by the equation

$$[\frac{di}{dt}] = [L]^{-1}[v] - [L]^{-1}[R][i] \tag{23}$$

The model of Eq.(23) has been implemented in the Electromagnetic Transients Program of the Bonneville Power Administration in 1974. With this model, the accuracy problems of the [Z]-matrix representation do not exist.

The extension from single-phase to three-phase is again achieved according to section 3. Specifically, Eq.(16) and (18) are solved separately for positive and zero sequence values, and these values are then converted to pairs (Z_S, Z_M) with Eq.(6). The inversion process in Eq.(19) again uses "balanced matrix algebra" with pairs, that is, an NxN matrix with element pairs is inverted rather than a (3N)x(3N) matrix. Similarly, element pairs replace single elements in Eq. (20).

The zero sequence short-circuit test data for closed delta connections is modified in the same way as discussed in section 5, before the calculation process begins with Eq.(16) and (18).

For three-phase transformers with three-legged core construction, the exciting current in the zero sequence test is fairly high (e.g., 100%). In this case, and in all other cases in which one would rather not ignore the exciting current, the model of Eq.(20) is modified as follows: From the p.u. exciting current I_{exc-1} and I_{exc-o} in the positive and zero sequence test, calculate the pair of p.u. shunt admittances

$$Y_S = -j\ \frac{1}{3}\ (I_{exc-o} + 2I_{exc-1}), \tag{24a}$$

in p.u. values

$$Y_M = -j\ \frac{1}{3}\ (I_{exc-o} - I_{exc-1})\ , \tag{24b}$$

and connect them either across one winding, or (1/N)-th of it across all N windings. In terms of Eq.(20), this means the addition of (Y_S, Y_M) to one diagonal element pair or the addition of (1/N)-th of it to all diagonal element pairs. If the excitation test on a three-winding transformer were made from winding 1, and if the magnetizing reactances are best connected across winding 3 because of negligible stray flux in that winding, as discussed in section 4 after Eq.(7), then (Y_S, Y_M) of Eq.(24) should be corrected in such a way that its inverse in series with the p.u. short-circuit impedance between 1 and 3 equals the reciprocals of the p.u. exciting currents.

If excitation losses are known, they are treated in the same way as discussed in section 4, with I_{exc} being replaced by I_m in Eq.(24) for increased accuracy.

7. IMPEDANCE MATRIX DERIVED FROM ADMITTANCE MATRIX

After [Y] has been modified with the shunt admittances of Eq.(24), it is no longer singular, though possibly ill-conditioned in cases of very low exciting current, and can therefore be inverted to produce an impedance matrix. Again, high accuracy is needed for the elements of this matrix, in contrast to the elements of [Y].

8. COMPARISON BETWEEN TWO IMPEDANCE MATRIX REPRESENTATIONS

The impedance matrix found from the inversion of [Y] will be practically identical with [Z] of section 4. Minor differences, if any, would arise from variations in obtaining the resistance matrix. If $[Z^{reduced}]$ of Eq.(14) is built from complex values Z_{ik}^{short}, with the p.u. real part being the p.u. load losses, then [R] becomes a full matrix in which the diagonal elements would in general not agree with specified winding resistances. This procedure was followed in a support routine in the Electromagnetic Transients Program of the Bonneville Power Administration for obtaining impedance matrices of single-phase transformers; since the user of that routine was only allowed to specify load losses but not winding resistances, this conflict never arose. On the other hand, if $[Z^{reduced}]$ is built from the X-values of Eq.(21) as suggested for transient studies, then [R] is a diagonal matrix, with its elements either being equal to the winding resistances or duplicating the load losses, but not both. The direct calculation of [Z] in section 4 allows the representation of winding resistances and load losses at the same time. If the load losses differ from the I^2R-losses, then [R] will again have off-diagonal elements. As the frequency approaches zero, off-diagonal elements in [R] imply that a dc-current in one phase would induce voltages in the other phases. Since R<<X in [Z], this erroneous effect would only show up at extremely low frequencies.

9. EXCITING CURRENT IN ZERO SEQUENCE TEST

If the transformer has delta-connected windings, the delta connections should be opened for the zero sequence excitation test. Otherwise, the test really becomes a short-circuit test between the excited winding and the delta-connected winding. On the other hand, if the delta is always closed in studies, any reasonable value can be used for the zero sequence exciting current (e.g., equal to positive sequence exciting current), because its influence is unlikely to show up with the delta-connected winding providing a short-circuit path for zero sequence currents.

If the zero sequence exciting current is not given by the manufacturer, a reasonable value can be found as follows: Imagine that one phase of a winding (A in Fig. 2) is excited, and estimate from physical reasoning how much voltage will be induced in the other two phases (B and C in Fig. 2). For the three-legged core design of Fig. 2, almost one half of flux ψ_A returns through phases B and C, which means that the induced voltages V_B and V_C will be close to 0.5 V_A (with reversed polarity). If k is used for this factor 0.5, then

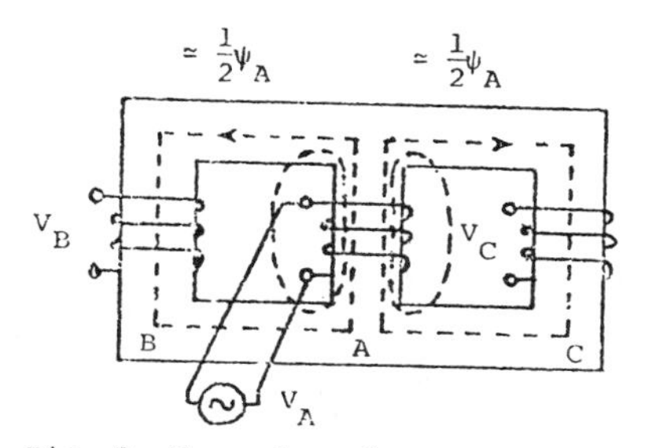

Fig. 2. Three-legged core-type design

$$\frac{I_{exc-o}}{I_{exc-1}} = \frac{1+k}{1-2k} \tag{25}$$

Eq.(25) is derived from

$$V_A = Z_S\, I_A \tag{26a}$$

$$V_B = V_C = Z_M\, I_A\,, \tag{26b}$$

with Z_S, Z_M being the self and mutual magnetizing impedances of the three phases of the excited winding. With

$$V_B = V_C = \frac{Z_M}{Z_S} V_A = \frac{Z_o - Z_1}{Z_o + 2Z_1} V_A\,, \tag{27}$$

and Z_1, Z_o inversely proportional to I_{exc-1}, I_{exc-o}, Eq.(25) follows.

Obviously, k cannot be exactly 0.5, because this would lead to an infinite zero sequence exciting current. A reasonable value for I_{exc-o} in a three-legged core design might be 100%. If I_{exc-1} were 0.5%, k would become 0.496, which comes close to the theoretical limit of 0.5. Exciting the winding on one leg (phase) with 100 kV would then induce voltages of 49.6 kV (with reversed polarity) in the other two legs.

10. SINGLE-PHASE TRANSFORMERS AND AUTOTRANSFORMERS

If a program is written for three-phase transformers, it will automatically produce correct results for three-phase banks consisting of single-phase transformers if all zero sequence values are simply set equal to their respective positive sequence values. In printing the matrices, the program should recognize, however, that the 9x9 matrix of a three-phase three-winding transformer actually consists of three separate 3x3 matrices for the three single-phase three-winding transformers.

If an autotransformer is treated the same way as a regular transformer, that is, if the details of the internal connections are ignored, the impedance or admittance matrix models will probably produce reasonably accurate results. It is possible, however, to develop more accurate models by modifying the short-circuit data, as explained in the appendix.

11. INCLUSION OF SATURATION EFFECTS

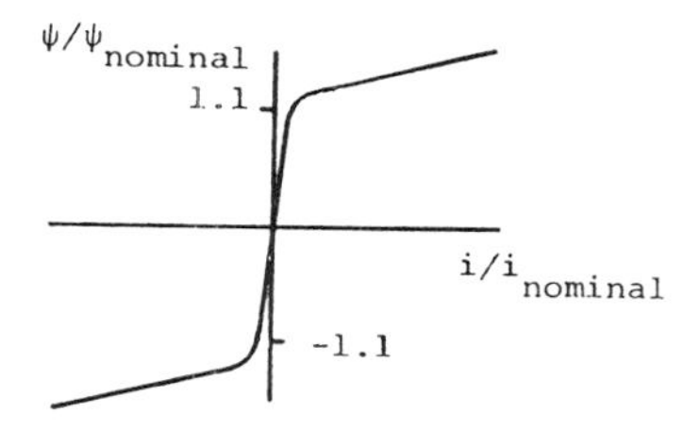

Fig. 3. Typical saturation curve

For transient studies with inrush currents, ferroresonance phenomena, etc., it is clearly necessary to include saturation effects. Modern high-voltage transformers with grain-oriented steel cores saturate typically somewhere above 1.10 to 1.20 times nominal flux [6], with a sharply defined knee (Fig. 3). Often, a two-slope piecewise linear inductance is sufficient to model such curves. The slope in the saturated region above the knee is the air-core inductance; it is almost linear and fairly low compared with the slope in the unsaturated region (typically twice the value of the short-circuit inductance [4]). Because of its low value, it may make a difference where the nonlinear inductance is added. It is best to put it across the winding which is closest to the core, at least in designs with cylindrical winding construction, which is usually the tertiary winding in three-winding transformers. Supporting evidence may be found in [5], [6] and [7].

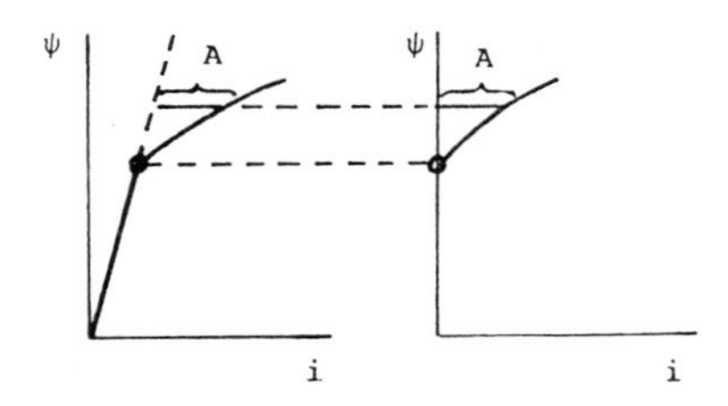

Fig. 4. Subtraction of unsaturated part in saturation curve (value of A equal in both curves)

If the $[L]^{-1}$-model of Eq.(23) without corrections for exciting current is used, then the nonlinear inductance of Fig. 3 is simply added across the winding closest to the core. In the case of three-phase transformers, usually only positive sequence saturation curves will be known. Then it is best to connect this nonlinear inductance across each one of the three phases, which implies that the zero sequence values are the same as the positive sequence values. If [Z]-models or [Y]-models are used which already contain the exciting current at rated voltage, then a modified nonlinear inductance must be added in which the unsaturated part has been subtracted out, as indicated in Fig. 4.

Frequently, the saturation curve is not defined as $\psi=f(i)$, but as root-mean-square characteristics $V_{RMS}=f(I_{RMS})$. A simple conversion technique is described in [1], which is based on the assumption that the influence of hysteresis and eddy current losses and of winding resistances in the saturation curve can be ignored. The Electromagnetic Transients Program of the Bonneville Power Administration contains a support routine with that method.

12. PER UNIT VALUES VERSUS PHYSICAL VALUES

Some equations in this paper are only correct for p.u. values on a common MVA base, as indicated by "in p.u. values" after the equation. In writing a program for the methods discussed here, it is best to do all calculations in p.u. on a common MVA base. Once [Z] or [Y] has been obtained in p.u., converting to physical values is straightforward. For instance, to convert [Z] in p.u. to physical values, simply multiply each element $Z_{ik-p.u.}$ with base values as follows:

$$Z_{ik\text{-physical}} = Z_{ik\text{-p.u.}} \frac{3 \cdot V_{i\text{-rating}} \cdot V_{k\text{-rating}}}{S_{rating}} \qquad (28)$$

with $V_{i\text{-rating}}$, $V_{k\text{-rating}}$ = rated voltage of windings i,k (line-to-ground for wye-connections, line-to-line for delta-connections),

S_{rating} = three-phase power rating used as base for p.u. values.

This conversion will automatically contain the correct turns ratios.

The authors of this paper prefer to use physical values in transient and multiphase steady-state studies, because they are less confusing when various system elements are connected together.

13. EXAMPLE

Assume that a three-phase three-winding transformer has the winding and short-circuit data of Tables 1 and 2. The positive sequence exciting current was given

Table 1. Winding data

Winding	Connection	Voltage rating (kV, line-to-line)	winding resist. (Ω)
primary (1)	wye (grounded)	230	0.2054666
secondary (2)	wye (grounded)	109.8	0.0742333
tertiary (3)	delta (closed)	50	0.0822

Table 2. Short-circuit data *)

between	Positive sequence impedance	Zero sequence impedance
1-2	8.74% based on 300 MVA	7.34% based on 300 MVA
1-3	8.68% based on 76 MVA	26.26% based on 300 MVA
2-3	5.31% based on 76 MVA	18.55% based on 300 MVA

*) MVA-base is three-phase

as 0.428% based on 300 MVA (three-phase), and the positive sequence excitation loss as 135.73 kW. Since zero sequence excitation data was not available, it was simply set equal to the positive sequence data. Though this may be unrealistic, it is immaterial because any zero sequence test will be a short-circuit test anyhow in the presence of a closed delta.

In the calculation, it was assumed that the excitation test was made from winding 1, but that the magnetizing impedance should be connected across winding 3. Using the method of sections 6 and 7, the following impedance matrix representation in physical units was obtained on a computer with a 64-bit double precision word length (each box represents a 3x3 submatrix with diagonal element Z_S and off-diagonal element Z_M):

[Z] =	1-A,1-B,1-C	2-A,2-B,2-C	3-A,3-B,3-C
1-A, 1-B, 1-C	Z_{S11} Z_{M11}	Z_{S12} Z_{M12}	Z_{S13} Z_{M13}
2-A, 2-B, 2-C	Z_{S21} Z_{M21}	Z_{S22} Z_{M22}	Z_{S23} Z_{M23}
3-A, 3-B, 3-C	Z_{S31} Z_{M31}	Z_{S32} Z_{M32}	Z_{S33} Z_{M33}

Z_{S11} = 0.2054666001E+00 + j0.4143208074E+05

Z_{M11} = -j0.5416527496E-01

Z_{S12} = Z_{S21} = j0.1977101943E+05

Z_{M12} = Z_{M21} = j0.9553782211E+00

Z_{S22} = 0.7423330004E-01 + j0.9437875034E+04

Z_{M22} = j0.7368306052E+00

$$Z_{S13} = Z_{S31} = j0.1557956118E+05$$
$$Z_{M13} = Z_{M31} = j0.1751039871E+01$$

$$Z_{S23} = Z_{S32} = j0.7437547077E+04$$
$$Z_{M23} = Z_{M32} = j0.8359311542E+00$$

$$Z_{S33} = 0.8220000005E-01 + j0.5866215636E+04$$
$$Z_{M33} = j0.6593239066E+00$$

In addition, a 3x3 shunt resistance matrix was calculated which has to be connected across winding 3 for the representation of excitation losses,

$$R_S = 0.551108E+05\ (\Omega)$$
$$R_M = 0.125332E+02\ (\Omega)$$

Using above data as input in the Electromagnetic Transients Program, and simulating the excitation and short-circuit tests with the steady-state option of the program, produced results which agreed with the original test data to within at least 4 significant digits.

CONCLUSIONS

This paper discusses the derivation of branch impedance and admittance matrices which model three-phase N-winding transformers simply as 3N coupled branches. It is shown that the elements of these matrices can be calculated from available test data of the positive and zero sequence short-circuit and excitation tests. The methods work for single-phase transformers as well, by simply setting zero-sequence values equal to positive sequence values.

For three-phase steady-state studies, either the branch admittance or branch impedance matrix can be used. For electromagnetic transients studies, the transformer is either represented by a series connection of [R] and [L]-branches, or of [R] and $[L]^{-1}$-branches.

For some types of transients studies, saturation effects are important. Ferroresonance phenomena are a typical example. It is shown how to add nonlinear inductances across one or more windings in such cases.

The matrix representations of this paper for three-phase and single-phase transformers are reasonably accurate up to moderate frequencies of approximately 1 kHz. For higher frequencies, capacitances between windings or layers of windings and between windings and and the tank would have to be represented. This may require further research along the lines suggested by R. C. Degeneff [8].

ACKNOWLEDGEMENTS

A major incentive for developing these three-phase transformer models came from discussions in Electromagnetic Transients Program Workshops at the University of Wisconsin, Madison, Wisconsin. The authors are grateful to the participants of these workshops, and to Dr. W.F. Long who organized them.

The authors are also indebted to Mr. T. Lou, who did part of the programming and testing at the University of British Columbia.

REFERENCES

[1] H.W. Dommel, "Transformer Models in the Simulation of Electromagnetic Transients." Proc. 5th Power Systems Computation Conference, Cambridge (England), September 1-5, 1975. Paper 3.1/4.

[2] D.E. Hedman, "Theoretical Evaluation of Multiphase propagation." IEEE Trans. Power App. Syst., vol. 90, pp. 2460-2471, November/December 1971.

[3] R.B. Shipley, D. Coleman and C.F. Watts, "Transformer Circuits for Digital Studies." AIEE Trans., pt. III, vol. 81, pp. 1028-1031, February 1963.

[4] D. Povh and W. Schulz, "Analysis of Overvoltages Caused by Transformer Magnetizing Inrush Current." IEEE Trans. Power App. Syst., vol. PAS-97, pp. 1355-1365, July/August 1978.

[5] K. Schlosser, "An Equivalent Circuit for N-Winding Transformers Derived from a Physical Basis." Brown Boveri-Nachrichten, vol. 45, pp. 107-132, March 1963 and "Application of the Equivalent Circuit of an N-Winding Transformer." Brown Boveri-Nachrichten, vol. 45, pp. 318-333, June 1963 (in German).

[6] M. Kh. Zikherman, "Magnetizing Characteristics of Large Power Transformers." Elektrichestvo No. 3, pp. 79-82, 1972 (in Russian).

[7] E.P. Dick and W. Watson, "Transformer Models for Transient Studies Based on Field Measurements." Paper F 80-244-4, presented at IEEE PES Winter Meeting, New York, N.Y., February 3-8, 1980.

[8] R.C. Degeneff, "A Method for Constructing Terminal Models for Single-Phase N-Winding Transformers." Paper A 78 539-9, presented at IEEE PES Summer Meeting, Los Angeles, California, July 16-21, 1978.

APPENDIX. AUTOTRANSFORMERS

For a more accurate representation of autotransformers, the high and low voltage terminals should be represented with the actual common winding II and series winding I, as shown in Fig. 5 for an autotransformer with a tertiary winding.

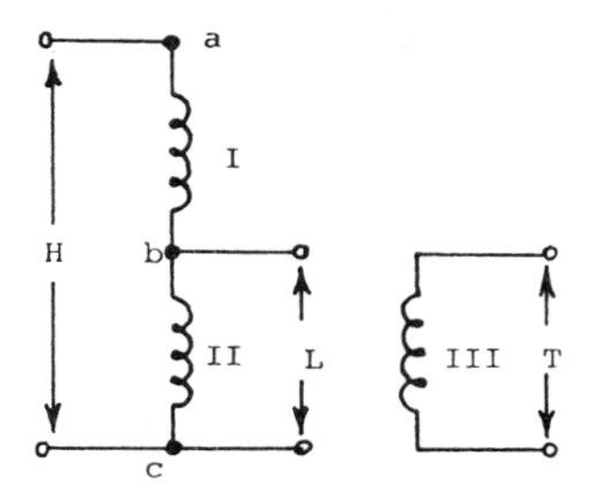

Fig. 5. Autotransformer with tertiary winding

This requires a re-definition of the short-circuit data in terms of windings I, II, III, with their voltage ratings

$$V_I = V_H - V_L,\quad V_{II} = V_L,\quad V_{III} = V_T. \tag{29}$$

The test between H and L is already the correct test between I and II, since II is shorted and the voltage is applied across I with b and c being at the same potential through the short-circuit connection. Therefore, simply change Z_{HL} to the new voltage base V_I,

$$Z_{I,II} = Z_{HL}\left(\frac{V_H}{V_H - V_L}\right)^2 \quad \text{in p.u. values} \tag{30}$$

No modifications are needed for the test between II and III,

$$Z_{II,III} = Z_{LT} \quad \text{in p.u. values} \tag{31}$$

For the test between H and T, the modification can best be explained in terms of the equivalent star-circuit of Fig. 1, with the impedances being Z_I, Z_{II}, Z_{III}, based on V_I, V_{II}, V_{III}, in this case. With III short-circuited, 1 p.u. current (based on $V_{III} = V_T$) will flow through Z_{III}. This current will also flow through I and II as 1 p.u. based on V_H, or converted to bases V_I, V_{II}, $I_I = (V_H - V_L)/V_H$ and $I_{II} = V_L/V_H$. With these currents, the p.u. voltages become

$$V_I = Z_I \frac{V_H - V_L}{V_H} + Z_{III}, \tag{32}$$

in p.u. values

$$V_{II} = Z_{II} \frac{V_L}{V_H} + Z_{III}. \tag{33}$$

Converting V_I and V_{II} to physical units by multiplying Eq.(32) with $(V_H - V_L)$ and Eq.(33) with V_L, adding them up, and converting the sum back to a p.u. value based on V_H produces the measured p.u. value

$$Z_{HT} = Z_I\left(\frac{V_H - V_L}{V_H}\right)^2 + Z_{II}\left(\frac{V_L}{V_H}\right)^2 + Z_{III} \tag{34}$$

in p.u. values

Eqs.(30), (31) and (34) can be solved for Z_I, Z_{II}, Z_{III} since $Z_{I,II} = Z_I + Z_{II}$ and $Z_{II,III} = Z_{II} + Z_{III}$, which produces

$$Z_{I,III} = Z_{HL}\frac{V_H V_L}{(V_H - V_L)^2} + Z_{HT}\frac{V_H}{V_H - V_L} - Z_{LT}\frac{V_L}{V_H - V_L}$$

in p.u. values (35)

The autotransformer of Fig. 5 can therefore be treated as a transformer with 3 windings I, II, III by simply re-defining the short-circuit input impedances with Eqs.(30), (31) and (35). This must be done for the positive sequence tests as well as for the zero sequence tests. If the transformer has a closed delta, then the zero sequence data must be further modified as explained in section 5.

This modification was tested using the example of section 13, with the assumption that the primary and secondary windings have an autotransformer connection. The resulting matrix again duplicated the original test data to within at least 4 significant digits. This case also verified that the modifications for closed deltas can be done after the re-definition of impedances with Eqs.(30), (31) and (35).

Discussions

Jaime Avila-Rosales (National Polytechnic Institute of Mexico, and Electric Power Research Institute of Mexico, Mexico City.):

The authors must be congratulated for their efforts in developing models of power system components for electromagnetic transient studies.

I would like to address the problem of modeling the magnetizing branch of the transformer. At The University of Wisconsin-Madison, we developed a frequency dependent transformer model[1] that could be implemented in a transients program like the BPA'S EMTP.

Briefly, the method used for developing the frequency dependent model was based on the analysis of the electromagnetic field distribution within the laminations of the transformer core. A one -- dimensional linear diffusion was analyzed, and a simple linear - equivalent circuit of the Norton type was synthetized. The circuit is an exact representation of the transformer core for any type of excitation at any frequency lower than the frequency "f_c" at which the electromagnetic field propagates mainly in the winding insulation. Saturation and hysteresis are included by modifying the equivalent linear circuit.

The resultant model is comparable to models for synchronous - machines recently obtained from field tests.

The linear frequency dependent model for the core of the transformer is shown in figure 1.

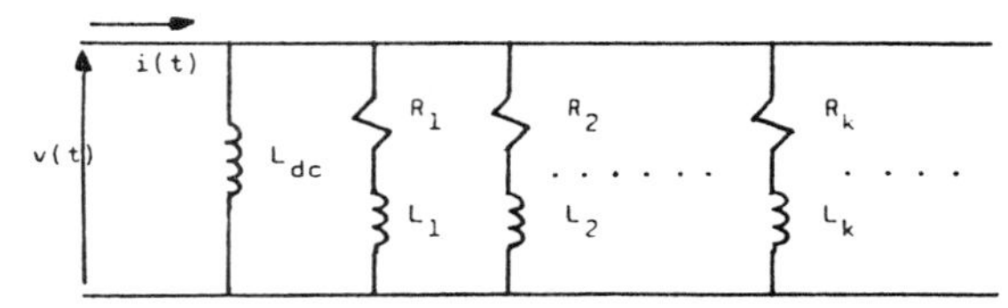

Fig. 1 Linear frequency dependent model for the core of the transformer.

Details of the development of the above model will be reported in a forthcoming paper.

I would appreciate the authors' opinion about the following:

a) Have the authors contemplated the analysis of the frequency dependency in transformers? If so, what approach have they used?
b) In what type of studies a frequency dependent transformer model will be necessary?

The discussor would be pleased to know the authors opinion on the frequency dependency matter for transformers.

REFERENCE

[1.] Jaime Avila-Rosales, "Modeling of the Power Transformer for Electromagnetic Transient Studies in Power Systems", Ph.D. Thesis, The University of Wisconsin-Madison, 1980.

Manuscript received August 18, 1981.

M. Owen (CEGB, London, England): There are several points in this paper upon which I should like to comment and amplify. I will use an example a typical 3 winding auto transformer of wye-wye-delta construction. The basis of my analysis refers to figures 1 and 2.

Leakage impedances of windings 1, 2, 3 and 4 are represented by Z_1, Z_2, Z_3 and Z_4 respectively. The number of turns per winding are N_1, N_2, N_3 and N_4 for which N_4 corresponds to a separate tap chaning winding. Fluxes have been chosen such that mutual flux coupling between phases is clearly indicated. Subscripts r, y and b are taken to reference the red, yellow and blue phases respectively.

The basic flux equation for the transformer operation on a 3 phase basis, assuming winding 3 to be delta connected is:

$$\frac{v_{3rb}}{N_3} = \frac{d}{dt}\left(\phi_r - \frac{\phi_y}{2} - \frac{\phi_b}{2} + \phi_o\right) \tag{1}$$

Which may be represented on a steady state basis as:

$$\frac{v_{3rb}}{N_3} = \left(i'_{3rb} - \frac{i'_{3yr}}{2} - \frac{i'_{3by}}{2}\right).N_3.\ Z_m + i'_{3rb}.N_3.\ Zo \tag{2}$$

Also since this flux links all 4 windings:

$$\frac{v_{1r}}{N_1} = \frac{v_{2r}}{N_2} = \frac{v_{3rb}}{N_3} = \frac{v_{4r}}{N_4} \tag{3}$$

Using the ampere turns balance across the perfect transformers in the equivalent circuit, then:

$$i_{1r}N_1 + (i_{1r} + i_{2r})\ N_2 + i_{2r}N_4 + (i_{3rb} - i'_{3rb})\ N_3 = 0 \tag{4}$$

Therefore:

$$i'_{3rb} = i_{1r}\left(\frac{N_1+N_2}{N_3}\right) + i_{2r}\left(\frac{N_2+N_4}{N_3}\right) + i_{3rb} \qquad (5)$$

$$i'_{3yr} = i_{1y}\left(\frac{N_1+N_2}{N_3}\right) + i_{2y}\left(\frac{N_2+N_4}{N_3}\right) + i_{3yr} \qquad (6)$$

$$i'_{3by} = i_{3rb}\left(\frac{N_1+N_2}{N_3}\right) + i_{3yr}\left(\frac{N_2+N_4}{N_3}\right) + i_{3by} \qquad (7)$$

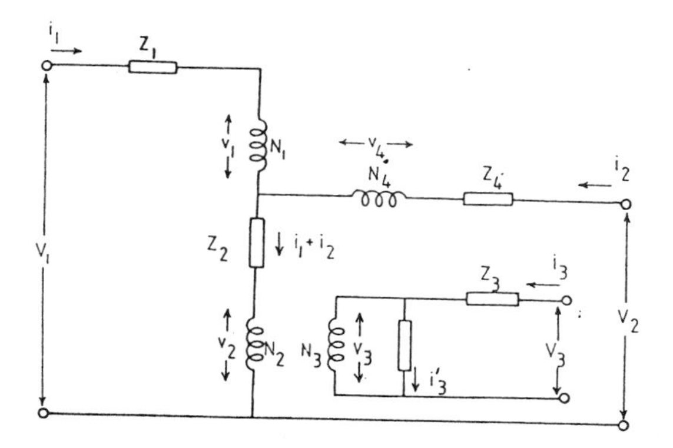

Figure 1: Single Phase Auto Transformer Equivalent Circuit

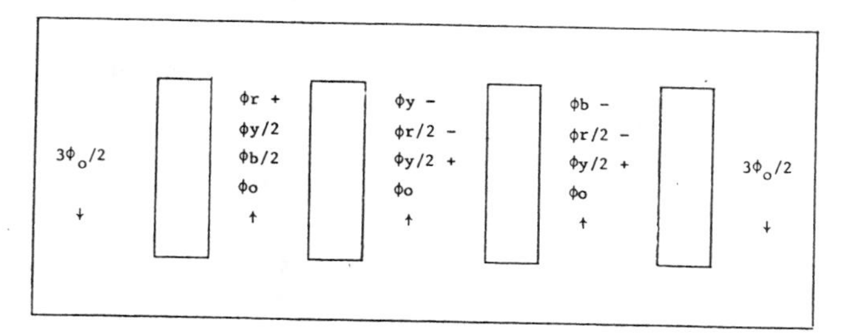

Figure 2: Flux Distribution in A 3 Phase Transformer Core

The network equations for the equivalent circuit may now be constructed, and for the red phase are shown to be:

V_{1r}	=	Z_1+Z_2	Z_2		$(Z_o+Z_m).(N_1+N_2).N_3$	$-Z_m/2.(N_1+N_2).N_3$	$-Z_m/2.(N_1+N_2).N_3$	i_{1r}
V_{2r}		Z_2	Z_2+Z_4		$(Z_o+Z_m).(N_2+N_4).N_3$	$-Z_m/2.(N_2+N_4).N_3$	$-Z_m/2.(N_2+N_4).N_3$	i_{2r}
V_{3rb}				Z_3	$N_3^2.(Z_o+Z_m)$	$N_3^2.-Z_m/2$	$N_3^2.-Z_m/2$	i_{3rb}
								i'_{3rb}
								i'_{3yr}
								i'_{3by}

(8)

Substitution for i'_{3rb}, i'_{3yr} and i'_{3by} may be made from equations 5, 6 and 7 to form:

	i_{1r}	i_{2r}	i_{3rb}	i_{1y}	i_{2y}	i_{3yr}	i_{1b}	i_{2b}	i_{3by}
V_{1r} =	$Z_1+Z_2+(Z_o+Z_m).(N_1+N_2)^2$	$Z_2+(Z_o+Z_m).(N_1+N_2)(N_2+N_4)$	$(Z_o+Z_m).(N_1+N_2).N_3$	$-Z_m/2\ (N_1+N_2)^2$	$-Z_m/2.(N_1+N_2)(N_2+N_4)$	$-Z_m/2.(N_1+N_2).N_3$	$-Z_m/2.(N_1+N_2)^2$	$-Z_m/2.(N_1+N_2)(N_2+N_4)$	$-Z_m/2.(N_1+N_2).N_3$
V_{2r} =	$Z_2+(Z_o+Z_m).(N_1+N_2)(N_2+N_4)$	$Z_2+Z_4+(Z_o+Z_m).(N_2+N_4)^2$	$(Z_o+Z_m).(N_2+N_4).N_3$	$-Z_m/2.(N_1+N_2)(N_2+N_4)$	$-Z_m/2.(N_2+N_4)^2$	$-Z_m/2.(N_2+N_4).N_3$	$-Z_m/2.(N_1+N_2)(N_2+N_4)$	$-Z_m/2.(N_2+N_4)^2$	$-Z_m/2.(N_2+N_4).N_3$
V_{3rb} =	$(Z_o+Z_m).(N_1+N_2).N_3$	$(Z_o+Z_m).(N_2+N_4).N_3$	$Z_3+(Z_o+Z_m)N_3^2$	$-Z_m/2.(N_1+N_2).N_3$	$-Z_m/2.(N_2+N_4).N_3$	$-Z_m.\frac{N_3^2}{2}$	$-Z_m/2.(N_1+N_2).N_3$	$-Z_m/2.(N_2+N_4).N_3$	$-Z_m.\frac{N_3^2}{2}$

(9)

Having developed such equations for all 3 phases the symetrical component transformation can be applied to transform from phase to sequence co-ordinates.
The voltage transformation is:

V_{1r}		1			1			1	
V_{2r}			1			1			1
V_{3rb}				$1-a$			$1-a^2$		
V_{1y}		a^2			a			1	
V_{2y}	=		a^2			a			1
V_{3yr}				a^2-1			$a-1$		
V_{1b}		a			a^2			1	
V_{2b}			a			a^2			1
V_{3by}				$a-a^2$			a^2-a		

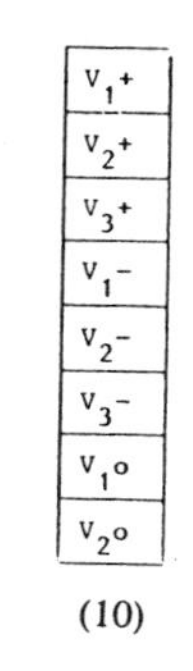

(10)

Which finally produces the sequence equations:

Positive sequence

$$\begin{bmatrix} V_1^+ \\ V_2^+ \\ \sqrt{3}\,V_3^+ \angle 30^\circ \end{bmatrix} = \begin{bmatrix} Z_1+Z_2+Z.N1^2 & Z_2.\frac{N1}{N2}+Z.N2N1 & \sqrt{3}.Z.N1.N3\angle 30^\circ \\ Z_2+Z.N1N2 & Z_2+Z_4+Z.N2^2 & \sqrt{3}.Z.N2.N3\angle 30^\circ \\ Z.N1.N3 & Z.N2.N3 & \sqrt{3}.Z3\angle 30^\circ + \sqrt{3}.Z.N3^2\angle 30^\circ \end{bmatrix} \begin{bmatrix} I_1^+ \\ I_2^+ \\ I_3^+ \end{bmatrix} \quad (11)$$

Negative sequence

$$\begin{bmatrix} V_1^- \\ V_2^- \\ \sqrt{3}.V_3^- \angle -30^\circ \end{bmatrix} = \begin{bmatrix} Z_1+Z_2+Z.N1^2 & Z_2.\frac{N1}{N2}+Z.N2.N1 & \sqrt{3}.Z.N1\angle -30^\circ \\ Z_2+Z.N1.N2 & Z_2+Z_4+Z.N2^2 & \sqrt{3}.Z.N2\angle -30^\circ \\ Z.N1.N3 & Z.N2N3 & \sqrt{3}.Z_3\angle -30^\circ + \sqrt{3}.Z.N3\angle -30^\circ \end{bmatrix} \begin{bmatrix} I_1^- \\ I_2^- \\ I_3^- \end{bmatrix} \quad (12)$$

Zero sequence (The complete matrix is obtained by inspection of equation 9)

$$\begin{bmatrix} V_1^o \\ V_2^o \\ 0 \end{bmatrix} = \begin{bmatrix} Z_1+Z_2+Z_o.N1^2 & Z_2+Z_o.N2N1 & Z_oN1 \\ Z_2+Z_oN1N2 & Z_2+Z_4+Z_oN2^2 & Z_oN2 \\ Z_o.N1N3 & Z_o.N2N3 & Z_3+Z_oN3^2 \end{bmatrix} \begin{bmatrix} I_1^o \\ I_2^o \\ I_{3\Delta}^o \end{bmatrix} \quad (13)$$

Where:

$$N1 = N_1 + N_2$$

$$N2 = N_2 + N_4$$

$$N3 = N_3$$

$$Z = \frac{3Zm}{2} + Z_o$$

The per unit equivalent circuit derived from the positive sequence impedance matrix equation 11 is shown in figure 3.

My following comments relate to the preceeding equations:

1. I would consider it essential that any such analysis should attempt to include impedance variation effects due to tap changing operations. For example a typical 400/132 kV, 240 MVA transformer on the CEGB system decreases its Z_{HL} impedance by a factor of 0.76 between its nominal and maximum tap positions. Such effects can actually be catered for by curve fitting the leakage impedances to a function of the tap position.
2. I would disagree with the authors and state that excitation resistances can be included in the impedance matrix representation. These are incorporated by Z_o and Z_m in the analysis presented with this discussion. For transient analysis of course, one should use the flux equation ((1) in the discussion) which caters for the losses from the hysteresis characteristic of the core.
3. It would seem that the auto-transformer leakage impedances derived in the appendix relate to the impedance matrix of equation (5) as diagonal elements when the transformations of equation (6) and (7) are implemented. From my analysis I find that the transformer impedance matrix contains off diagonal elements incorporating leakage impedance terms. Can the authors please comment.

Positive sequence

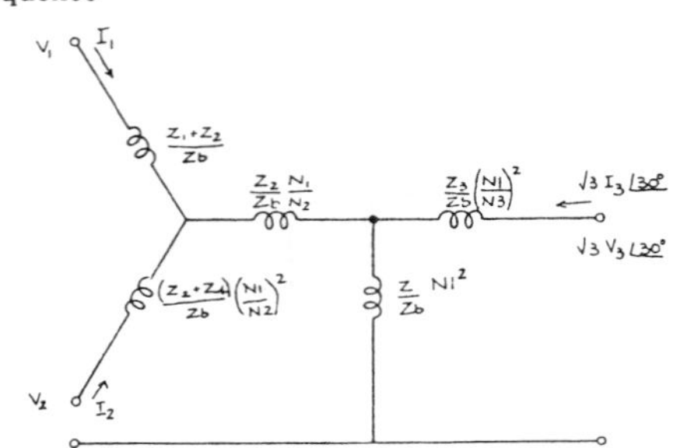

All currents and voltages are shown in per unit.

Also:

$$Zb = \frac{V_1^2}{\text{MVA base}} \quad (\text{nominal kV})$$

Fig. 3.
Three phase three winding auto-transformer equivalent circuit

Manuscript received August 19, 1981.

V. Brandwajn, H. W. Dommel and **I. I. Dommel:** The authors thank Dr. J. Avila-Rosales and Dr. M. Owen for their interest in the paper, and for their valuable comments.

The work of Dr. Avila-Rosales on the frequency-dependence of the magnetizing branch is very interesting, and the authors look forward to seeing more details in his forthcoming paper. While the authors have been aware of the frequency-dependence, they have never modelled it yet, and therefore find it somewhat difficult to judge in what types of studies it is important. From curves published by CIGRE Working Group 13.05 [A], it appears that the frequency-dependence may not only be important for the magnetizing impedance but for the short-circuit impedances as well.

Dr. Owen shows an alternative way of deriving transformer equations from leakage impedance data, while the authors chose to derive their model from short-circuit and excitation test data which are normally more readily available. The authors fully agree with Dr. Owen's first comment about impedance variations for different tap positions. Since the authors have used their model only for steady-state and electromagnetic transient studies in which tap positions did not change during the time span of the simulation, they have assumed a fixed tap position in their model, with test data given for that particular position. Dr. Owen's inclusion of the tapped part as a fourth winding is very interesting, and could be done with the authors' model as well provided that the six independent short-circuit impedances of the four-winding transformer [B] are known. Dr. Owen's model seems to require only four impedance values.

Dr. Owen's second comment is correct. For steady-state analysis, the excitation resistances can directly be included in the diagonal elements of [Y], and [Z] can be modified as well, though not quite as simply, to include them. The authors made the statement with transient analysis using [R]- and[L]-matrices in mind, because these matrices are directly accepted as input by the Electromagnetic Transients Program of the Bonneville Power Administration.

Since Dr. Owen's last question may result from some misunderstanding, the authors will attempt to summarize the derivation of the impedance matrix for an autotransformer: The equations in APPENDIX are used only to redefine the short-circuit impedances from those given among windings H, L, T to those for windings I, II, III. With the method of section 4, the diagonal elements are then found from Eq. (7) and the off-diagonal elements from Eq.(9). If it is a single-phase transformer, then the coupling impedance Z_M between phases is indeed zero in Eq.(5) and (6). The authors' program recognizes this and produces a 3×3 impedance matrix (rather than a 9×9 matrix) in case of single-phase units. This 3×3 matrix has nonzero off-diagonal elements, however. Three phase banks with single-phase units would be modelled by three such 3×3 impedance matrices, each representing a group of three coupled branches. The coupling between phases through the effect of the delta connection is not represented in the branch impedance matrices, but by proper node connections (e.g., the tertiary of the first unit could be connected from nodes TA to TB, of the second unit from TB to TC, and of the third unit from TC to TA). A three-phase unit would be represented by a 9×9 matrix with nonzero values for Z_m, as shown in the example of section 13.

REFERENCES

[A] CIGRE Working Group 13.05, "The Calculation of Switching Surges. II - Network Representation for Energization and Re-energization Studies on Lines Fed by an Inductive Source," ELECTRA, No. 32, pp. 17-42, 1974.

[B] L. F. Blume, A. Boyajian, G. Camilli, T. C. Lennox, S. Minneci and V. M. Montsinger, *Transformer Engineering*. Second Edition, John Wiley and Sons, New York, 1951, p. 117.

Manuscript received September 23, 1981.

IEEE Transactions on Power Delivery, Vol. 6, No. 1, January 1991

TRANSIENT SIMULATION AND ANALYSIS OF A THREE-PHASE FIVE-LIMB STEP-UP TRANSFORMER FOLLOWING AN OUT-OF-PHASE SYNCHRONIZATION

by
C.M.Arturi, Member, IEEE

Dipartimento di Elettrotecnica
Politecnico di Milano
Piazza Leonardo da Vinci 32 - 20133 Milano (Italy)

ABSTRACT

This paper presents the theoretical and the experimental analysis of the electromagnetic transient following the out-of-phase synchronisation of a three-phase five-limb step-up transformer; this means an abnormal condition where the angle between phasors representing the generated voltages and those representing the power network voltages at the instant of closure of the connecting circuit breaker is not near zero, as normal, but may be as much as 180°(phase opposition). When this happens, the peak values of the transient currents in the windings of the transformer might be sensibly higher than those of the failure currents estimated in a conventional way; in addition they correspond to unbalanced magnetomotive forces (MMF) in the primary and secondary windings of each phase of the machine.

The currents and fluxes during the transient are computed by a non-conventional circuital non-linear model of the transformer simulated by the ElectroMagnetic Transient Program (EMTP). The results of an experimental validation made on a specially built 100 kVA three-phase five-limb transformer are also reported.

Key Words: Three-phase 5-limb transformer models, Out-of-phase synchronization, Magnetic networks, Duality.

1. INTRODUCTION

The operation of generator step-up transformers following short-circuit events is, from the electromechanical stresses point of view, rather more secure than that of interconnection transformers and autotransformers. For the latter, in fact, a short circuit, whether on the high-voltage (HV) or on the medium-voltage (MV) network, is supplied from large powers and the resulting currents are very close to that determined by the only impedance of the machine. On the contrary, in step-up transformers, in the case of short circuit on the HV network, the currents are limited by the direct-axis subtransient impedance of the synchronous generator, which is relatively high, whereas a short circuit at the low voltage (LV) terminals is extremely unlikely, bearing in mind the type of the connection between synchronous generator and transformer, usually made with segregated bars.

Nevertheless it is not rare, unfortunately, that step-up transformers undergo an out-of-phase synchronization operation, that is an abnormal condition where the angle between phasors representing the generated voltages and those representing the power network voltages at the instant of closure of the connecting circuit breaker is not zero, as normal, but may be as much as 180 (phase opposition).

Although there are not many papers published on the analysis and the effects of the out-of-phase synchronization operation of transformers [1-2], this problem is very important in consideration of both the high repair cost of large power transformers and synchronuos machines added to the long outage time. Therefore a model which allows the prediction of the values of the currents in the windings during an out-of-phase synchronization event with suitable accuracy is very useful. From these currents, the magnitude of the related electromechanical stresses can be determined.

The axial electromechanical stresses due to a out-of-phase synchronisation transient might lead to the failure of the windings, particularly of the HV coils connected to the power network. Fig.1 shows a photograph of an HV coil of a large power three-phase five-limb transformer which has undergone a wrong parallel operation with a network the power of which was nearly 70 times greater than the rating of the transformer.

The parallel connection of a transformer, energized by a generator with the same power, to a network having a power 50-100 times greater, leads to an electromagnetic transient with different and very high saturation level of the various iron branches of the magnetic circuit, much heavier than that corresponding to the rated flux operation. In fact, if it is assumed

Fig.1 - HV coil of a large power step-up transformer which has undergone a failure caused by excessive axial electromechanical stresses due to an out-of-phase synchronization operation.

90 WM 225-3 PWRD A paper recommended and approved by the IEEE Transformers Committee of the IEEE Power Engineering Society for presentation at the IEEE/PES 1990 Winter Meeting, Atlanta, Georgia, February 4 - 8, 1990. Manuscript submitted July 31, 1989; made available for printing December 6, 1989.

that one phase voltage is zero at the instant of the synchronization with 180° phase-error, the corresponding flux linkage is maximum. Further, since this voltage gets on a half wave with the same sign of that just described, the linkage flux with the HV coil goes on to increase until a peak value not much less than 3 times the rated flux is reached. Consequently, the iron branches of the magnetic circuit undergo a very high saturation, analogous to that which would happen in an inrush operation with a residual flux density equal to the rated one. It follows that the MMFs on the iron branches are by no means negligible and the corresponding magnetizing currents are of the order of the rated ones. The MMFs associated to the iron branches combine with those associated to the leakage branches and give, as a result, the MMFs of the primary and secondary windings which, because of the contribution of the iron branches, are considerably unbalanced. In other words, a short-circuit transient adds to an inrush transient with maximum initial flux.

It is clear that, for a correct analysis of the outlined phenomenon, the transformer model has to consider a subdivision of the magnetic core in more branches with non-linear characteristics. Further, it is also evident that the conventional models of the three-phase transformer, set up with the leakage inductances and one saturable element per phase, are completely inadequate for the present analysis.

Details are given below of how to obtain a circuital model of the transformer by the procedure of the duality, starting from the magnetic network associated to the real configuration of the core and the windings. The obtained model for a large power transformer (370 MVA) is simulated with different impedance of the power network. An experimental validation of the theoretical results was carried out on a specially built 100 kVA transformer.

The simulation of the model was made by the ElectroMagnetic Transient Program (EMTP)[3].

2. METHOD OF ANALYSIS

2.1- Circuital model of the three-phase five-limb transformer.

The structure of a three-phase five-limb transformer is represented in Fig.2a. The windings are assumed to be made with three couples of concentric coils: three LV internal coils and three HV external ones. The core is made up of three wound limbs of equal sections, four intermediate yokes A-C, C-E, B-D, D-F and two lateral branches, composed of the lateral limbs and the related yokes. The lateral limbs and yokes normally have a smaller section than the wound limbs. Usually the intermediate yokes and the lateral limbs may have the same cross section (about 0.57-0.58 times the cross section of the wound limbs) or the lateral yokes and limbs may be smaller (nearly 0.4) and the intermediate yokes bigger (0.7 times the cross section of the wound limbs) accordingly the required reduction of the total core height.

The adopted circuital model comes from the magnetic network associated with the real configuration of the transformer by duality [4-8]. The magnetic network is set up by the nodes and the most significant flux tubes which approximate the real map of the magnetic field. The number of the nodes and branches to consider are to be selected in order to approximate the field satisfactorily, without complicating the magnetic network excessively.

The concept of magnetic network is based on the description of the field by means of the scalar magnetic potential [9], the definition of which implies rotH=0 in the considered space. Hence it is necessary to outline an essential difference between the flux tubes of the real magnetic field and that considered to set up the magnetic network. Whereas the first can also exist in regions where rotH is not zero, i.e. where there are currents, the second ones cannot. In other words, the field is assumed to be exclusively confined into the flux tubes which form the magnetic network, and they must be totally linked with the coils. Therefore, the radial thickness of the coils has not been considered. Further, the eddy-current loss of the magnetic core is assumed to be negligible.

At first, the nodes of the magnetic network have to be outlined. It is reasonable, in the present case, to select three nodes on the upper yoke (A, C, E) and three on the lower one (B, D, F). These nodes subdivide the core into nine branches outlined by the corresponding reluctances: three for the wound limbs, two for the lateral yokes and limbs, four for the intermediate yokes - two on the top and two at the bottom - which are in series and can be reduced to two. As a consequence, the network is simplified and the previous three nodes in the lower yokes can be put together and represented by just one.

Parallel to each iron branch there is an air branch, which is not usually considered in the conventional models of transformers but is indispensable for the analysis of the heavy saturated behaviour. Finally, between each top node and the corresponding lower node there is the leakage reluctance associated with each couple of concentric coils.

The sources of the magnetic network are represented by the MMFs of the six coils. These sources have to be concentrated out of the space where the field is supposed to be confined.

The magnetic network described is reported in Fig.2b with a linear or non-linear reluctance (the non-linear ones are shown in black) for each branch and a MMF source for each coil. In conclusion, the magnetic network includes seven non-linear reluctances and as many parallel linear reluctances, three leakage reluctances and six MMFs sources.

The corresponding dual electric network can be obtained from the magnetic network, assuming a common reference number of turns N (Fig.2c). An inductance ($L=N^2/R$) corresponds to a reluctance (R), a current source (i=F/N) corresponds to a MMF source (F), a node corresponds to a loop, series elements correspond to dual elements in parallel and so on.

The analysis of the electric network also gives all the information concerning the magnetic network of the transformer; in fact, on the base of the duality, there is a correspondance between the voltage v of a branch of the electric network and the flux Φ of the corresponding dual branch of the magnetic network ($\Phi=(\int v \cdot dt)/N+\Phi_0$) as well as between the current i of a branch of the electric network and the MMF F on the terminals of the corresponding dual branch of the magnetic network ($F=N \cdot i$).

The model so obtained has to be improved with six ideal transformers, both to take into account the actual number of turns of the coils and to allow for the star or delta connection of the windings. So far, no dissipation parameters have been considered and the radial thickness of the coils has been neglected. If the resistances of the windings and both the synchronous machine and the power network models are included, the equivalent network becomes that of Fig.2d. This is, finally, the circuital model used for the simulation of the wrong synchronizing transient.

As it is the first few periods of the transient just after the starting of the wrong parallel that are of interest, it is not necessary to adopt a dynamic model for the synchronuos machine. A three-phase voltage source in series with the direct-axis subtransient impedance is deemed to be suitable. The power network model is made by a three-phase voltage source in series with the impedance of the network.

Both the HV network and the HV transformer side have earthed neutrals.

A possible further improvement of the model of the

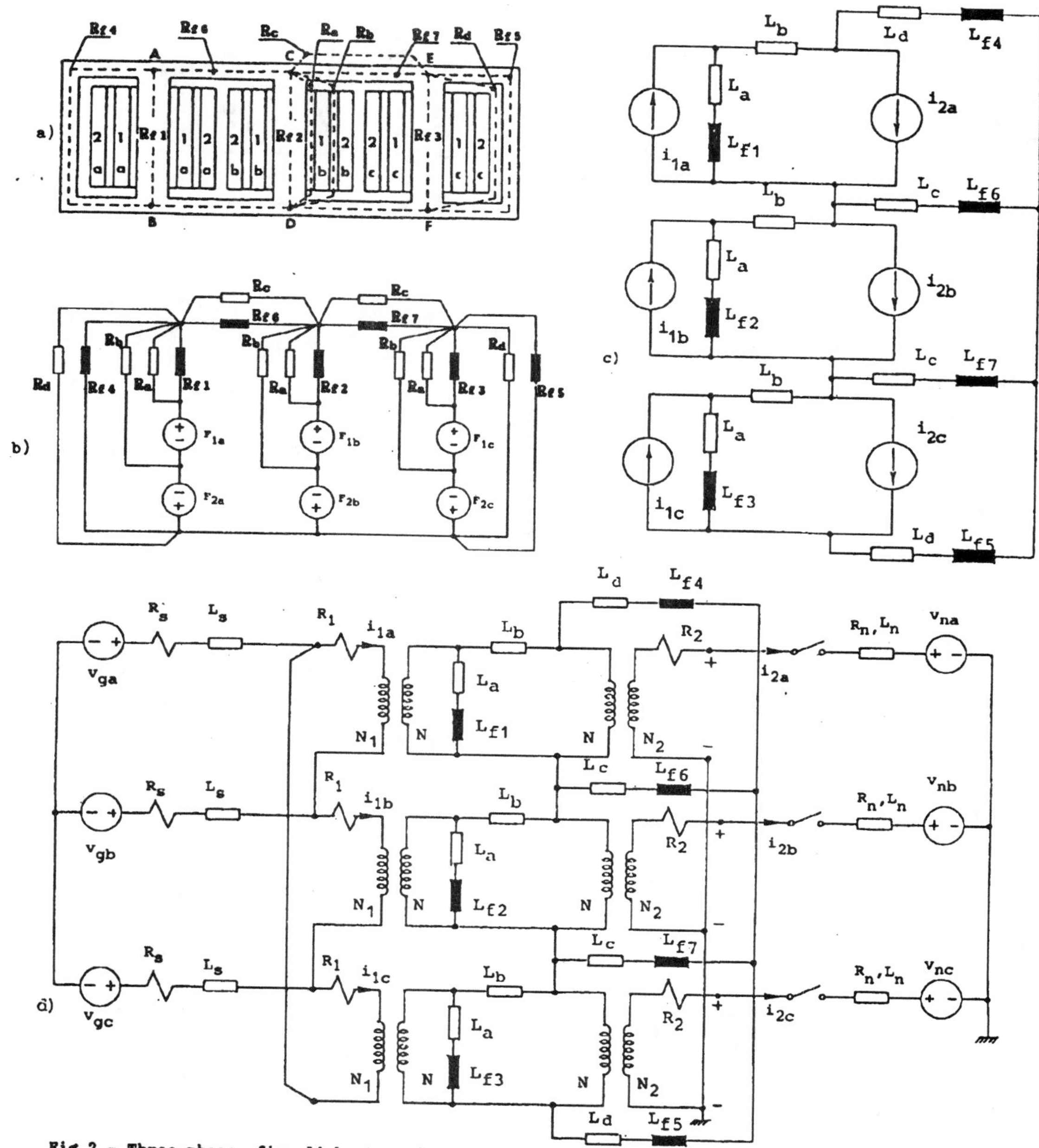

Fig.2 - Three-phase five-limb transformer: a)structure and main flux tubes; b)magnetic network; c)equivalent electric network; d) complete equivalent electric network of the transformer delta-star connected with neutral earthed, of the synchronous machine and of the power network.

transformer should consider the thickness of the coils. They could be taken into account by means of six negative inductances, each in series with the terminals of an ideal transformer. Each negative inductance corresponds to a negative air flux tube totally linked with each coil, i.e. in parallel to each MMF source, and take into account the influence of the partially linked phisical flux tubes [6].

2.2- Determination of the parameters of the equivalent network of a three-phase five-limb transformer.

The equivalent network of a three-phase five-limb transformer shown in Fig.2c has seven non linear inductances and ten linear ones.

The flux-current curve of the non-linear inductances is determined by the B(H) curve of the magnetic steel used to build the core, the section and the length of each iron branch. This is enough for computing the currents with very high saturation level but gives values which are well lower than the real ones when the flux is near to the nominal value, since the equivalent air gaps of the core have not been considered. Therefore, if the model must also be accurate for these last conditions, a linear inductance in parallel to each iron inductance is needed in the equivalent network. Such inductances are not visualized in the Fig.2c.

The linear inductances of the transformer model include three leakage inductances, which are assumed to have the same value and can be computed or determined by the short circuit test, and seven inductances associated to the air flux tubes in parallel to those in iron. Due to the symmetry, the three inductances in series with the non-linear inductances of the wound limbs are equal as well as the two inductances for the intermediate yokes and those for the lateral branches. Consequently, there are only three linear inductances to be determined: L_a, L_c and L_d (Fig.2c).

The calculation of L_a, L_c and L_d can be made by the following criterium: assuming that the equivalent network of the transformer is supplied from the terminals of a coil in order to be in a very high saturation condition, the inductance seen from those terminals is imposed to be equal to the air-core inductance of that coil. This value is a minumum physical limit which can be computed with certainty on the base of the geometrical dimensions of the coil. In this procedure, the vacuum permeability is used for computing the saturated inductances of the iron branches.
The inductance of the network as seen from three different couples of terminals, let us say, for example, those of the internal and the external coils of the central limb and those of a coil of a lateral limb, are expressed by a system of three algebraic equations which gives the three unknown inductances L_a, L_c and L_d.

The values of the air inductances are reported in App.A for the transformers analysed in this paper.

2.3 - Model simulation

The analyis of the out-of-phase synchronization operation of a three-phase five-limb transformer has been made with a phase error of 180°, that is in phase opposition.

First, the transformer is taken on no-load operation, and then the circuit breacker closes at the instant in which the voltage of the phase named 'a' is zero. The power of the network is assumed to be equal to 40 and 12.5 GVA, that is almost 108 and 34 times the rating of the transformer analysed, which is assumed equal to 370MVA.

The parameters of the equivalent network are listed in detail in App.A. The direct-axis subtransient reactance of the synchronuos machine is 22.7%; the leakage reactance of the transformer is 12.8% and the network reactance is 0.91% or 3%, respectively, for the two powers selected. The resistance is almost 1% for the generator, 0.2% for the transformer and 0.1% for the power network. The reference value is the nominal impedance of the transformer.

The results of the simulation of the transient operation of the three-phase five-limb transformer, with phase-error of 180°, are presented in Fig.3: phase voltages, flux linkages with the coils, phase currents just before and after the operation of the circuit breaker are shown. The maximum values of the various quantities, in their first peak, for both the values of the reactances of the network (0.91% and 3%), computed with the non-linear model (NLM) of Fig.2d are reported below and compared with those obtained not taking into account the magnetizing currents of the iron branches, i.e. by the corresponding linear conventional model (LM). All the computed quantities are expressed in per unit and refer to the corresponding nominal values of the transformer.

Phase currents (reference value: 755.3 A, peak value; 370MVA, 400kV, HV side):

	0.91%		3%	
phase	NLM	LM	NLM	LM
a1	-9.8	-10.5	-9.57	-9.94
a2	-11.5	-10.5	-10.59	-9.94
b1	8.0	7.9	7.51	7.52
b2	8.4	7.9	7.79	7.52
c1	8.2	8.0	7.66	7.62
c2	8.7	8.0	8.06	7.62

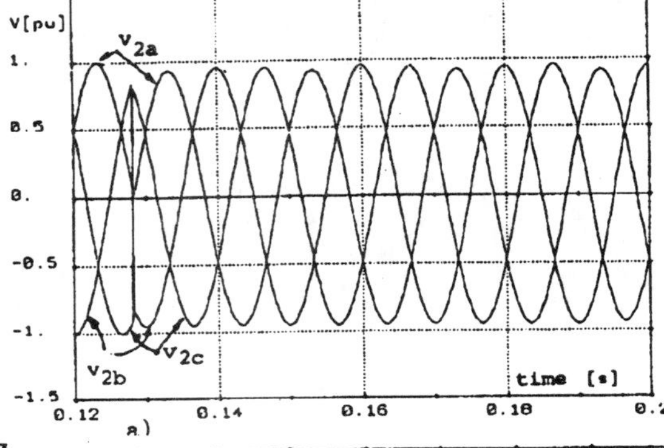

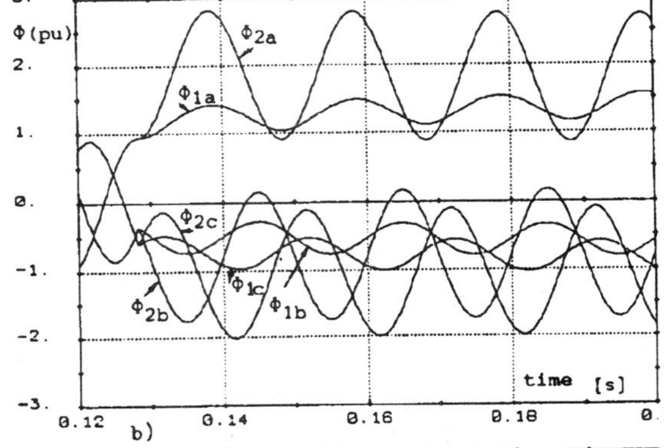

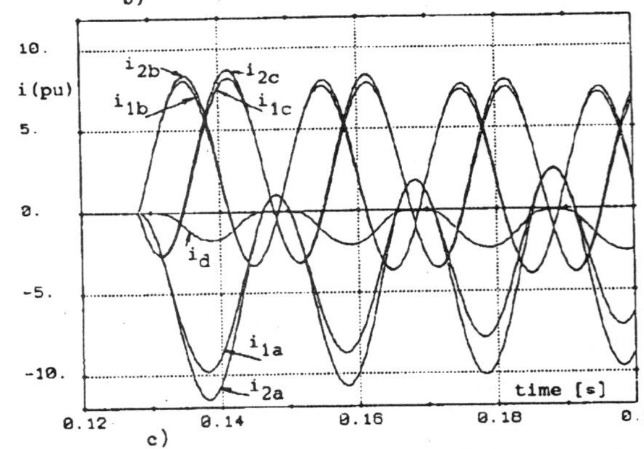

Fig.3- Computed transient of an out-of-phase synchronization operation of a three-phase 5-limb transformer with 180° phase-error: short circuit inductance 12.8%, direct-axis subtransient inductance 22.7%, power network inductance 0.91%; a) secondary phase voltages in per unit before and after the instant of the wrong parallel; b)flux linkages with the coils in per unit; reference flux linkage: 1040 Wb; c)phase currents and differential current $i_d=i_{a2}-i_{a1}$ of the most stressed phase in per unit; reference current of H.V.side: 755.3 A, peak value.

Flux linkages with the coils:

phase	0.91% NLM	0.91% LM	3% NLM	3% LM
a1	1.57	1.71	1.48	1.60
a2	2.83	2.83	2.61	2.63
b1	1.15	1.15	1.14	1.15
b2	1.75	1.15	1.61	1.14
c1	1.10	1.10	1.09	1.10
c2	2.00	2.00	1.85	1.87

Reference value: 1040 Wb (2.09 Wb·497 turns).

Total fluxes in the air and iron reluctances connected in parallel in the magnetic network of Fig.2b (for shortness, only the index number of the iron element is pointed out):

branch	0.91% NLM	3% NLM
4 e 5	0.95	0.902
6	1.88	1.72
7	1.11	0.996

Fluxes in the iron reluctances of the magnetic network:

branch	0.91% NLM	3% NLM
R_{f1}	1.41	1.36
R_{f2}	1.14	1.14
R_{f3}	1.09	1.09
R_{f4} e R_{f5}	0.707	0.702
R_{f6}	0.73	0.725
R_{f7}	0.70	0.692

Magnetizing currents of the iron branches:

branch	0.91% NLM	3% NLM
R_{f1}	2.07	1.55
R_{f2}	0.044	0.045
R_{f3}	0.014	0.014
R_{f4} e R_{f5}	0.548	0.454
R_{f6}	0.998	0.851
R_{f7}	0.350	0.261

The previous results, when the reactance of the power network is 0.91%, are commented below.

The ratio between the current peak values of phase 'a', which is the most stressed one, is 11.5/9.8=1.173. In comparison with the per unit current evaluated with the linear model (10.5), i.e. following the conventional procedure, the non-linear model gives a current 6.7% less in the internal coil and 9.5% more in the external one.

It is to be noted that the maximum value of the difference between the istantaneous value of the currents reaches the value of 3.1 p.u., which is greater than the difference between the peak values (1.7 p.u.), as shown in Fig.3c.

Still referring to phase 'a', the maximum flux linkage with the external coils reaches the value of 2.83 p.u. whereas the maximum flux linkage with the internal coil is 1.57 p.u.; the difference between them is equal to the leakage flux, as is evident from the magnetic network of Fig.2b. The flux linkage with the external coil is also equal to the sum of the total flux in the air-iron parallel of the adjacent lateral branches (0.95, with index number 4) and that of the air-iron parallel of the adjacent yoke (1.88 with index 6). In the lateral branch, 74% (0.707/0.95) of the flux goes in the iron path and gives a flux density of 2.11 T and a magnetizing current of 0.548 p.u. (it should be noted that the section of both the lateral limbs and yokes is 0.57 times that of the wound limbs and the nominal flux density is assumed to be equal to 1.7 T; therefore 0.707·1.7/0.57=2.11 T).

In the parallel of the adjacent yoke, only 39% of the flux goes in the iron path; it gives a flux density of 2.16 T and a magnetizing current of 0.998 p.u.

In the parallel of the other yoke (with index number 7), 63% of the flux goes in the iron path, i.e. more than in the yoke number 6, since the saturation level is lower. This flux gives a flux density of 2.07 T and a magnetizing current of 0.35 p.u.

In the parallel of the wound limb of phase 'a', 90% of the flux goes in the iron path; this gives a flux density of 2.41 T and a magnetizing current of 2.07 p.u., which is the greatest contribution due to the saturation of the iron branches.

It is important to compare the flux linkages with the internal and the external coils (1.57 and 2.83 p.u.) and magnetizing current associated to the respective iron branches (2.07 as compared with 0.548+0.989=1.546). It can be seen that the iron path of the wound limb has a saturation level (2.41 T) higher than that of both the iron of the lateral iron path (2.11 T) and the iron of the adjacent yoke (2.16 T) although the flux of the internal coil is nearly half the flux of the external coil. This is because the air inductance L_a, in parallel with the wound limb, is nearly 1/6th of the air inductance L_d, in parallel with the lateral limb and nearly 1/15th of of L_c, in parallel with the yoke. This clarifies the importance of the air branches in parallel with the iron branches in the magnetic network of a transformer operating with very high saturation level. On the contrary, if the model were to be used to simulate a nearly rated flux operation, the above mentioned air branches could be completely neglected.

It should be noted that the two lateral limbs have nearly the same fluxes with opposite sign because both the impedance of the power network is very small and its voltage system has a zero value resultant.

The second case, with 3% value for the power network impedance, is obviously less heavy than the previous one but substantially similar.

The maximum values of the unbalanced MMFs of the most stressed phase can be used for computing the magnetic field in the space occupied by the coils, by means of numerical methods, in order to evaluate the distribution of the axial electromechanical stress. A further study will consider this analysis.

3 - EXPERIMENTAL VERIFICATION

The experimental verification of the theoretical results was carried out on a three-phase five-limb transformer rated 100 kVA, specially built and proportioned in order to have a per unit short-circuit voltage nearly equal to that of a 370MVA power transformer, like that simulated in the previous paragraph. The characteristics of the test transformer are listed in detail in App.A. It has the same number of turns on both the primary and the secondary windings which are delta-star connected. A series of EMF search coils were mounted both on the coils and the core to measure, by integration, the corresponding fluxes.

The test circuit is schematically represented in Fig.4a and a photograph of the test equipment is shown in Fig.4b. The test circuit includes a source, represented by a synchronous machine (SM) rated 17MVA, a three-winding transformer (TWT) rated 17 MVA, the test transformer (TT), two impedances Z_a and Z_b and a synchronized circuit breaker (SW). The three-winding transformer (TWT) is connected delta-delta-star: the primary is connected to the synchronous generator, the secondary delta-connected winding supplies the primary side of the test transformer, through the Z_b impedance, and the secondary star-connected winding supplies the secondary side of the test transformer through the Z_a impedance and the synchronized circuit breaker.

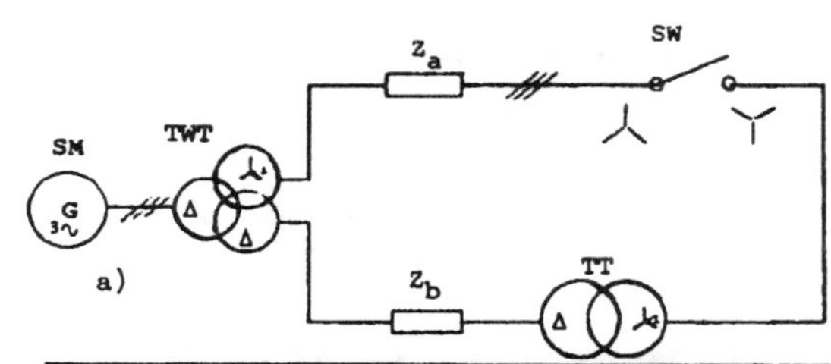

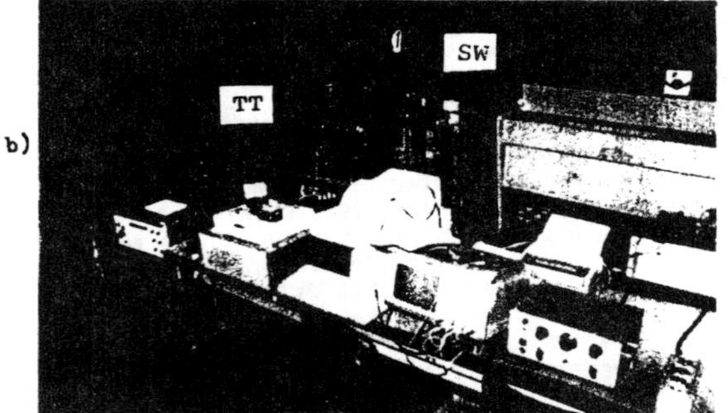

Fig.4 - a) Test circuit for the experimental verification of the out-of-phase synchronization operation of a three-phase 5-limb transformer rated 100 kVA. SM: synchronous machine rated 17 MVA; TWT: three-winding transformer rated 17 MVA; TT: test transformer rated 100 kVA; Z_a and Z_b impedances: 0.653{84.5°} Ω and 2.352{89.3°} Ω respectively; SW: synchronized circuit breaker. b) test equipment.

The connection of the two secondary windings of the TWT transformer is made in order to have phase opposition of the two voltage systems.

The synchronous machine SM and the TWT transformer are both rated 17 MVA, that is 170 times the rating of the test transformer. Therefore their impedance are very small in comparison with those that have to be included in order to simulate the per unit impedance of a real power system experimentally. Hence the need for the impedances Z_a and Z_b, the value of which are selected to represent the impedance of the power network and the direct-axis subtransient impedance of the SM, respectively.

The paralleling test with 180° of phase-error is made after the TT transformer has been supplied and is operating on no-load condition, by closing the synchronized circuit breaker at the instant when the voltage of a lateral phase, named 'a' is zero. The opening of the circuit breaker follows after nearly 0.5 s.

A multi-channel acquisition system, with sampling frequency of 33kHz (Multiprogrammer HP 6942), controlled by the computer HP 9826S, permits the acquisition and the processing of signals proportional to the quantities of interest. The sampled quantities are: phase voltage 'a', primary and secondary currents of the same phase, EMFs of the search coils mounted in order to give, by integration, the flux linkages with both coils of the examined phase or the flux in the various branches of the core.

Measured voltage of phase 'a', currents and flux linkages of both coils of the same phase are reported in Fig.5. The test transformer was supplied with the rated voltage and the power network impedance was 2.41%.

The computed quantities for phase 'a', corresponding to those of Fig.5, are reported in Fig.6. They were computed by the circuital model of Fig.2d and the parameters listed in detail in App.A.

The comparison among the computed and the measured values is reported below:

Measured		Computed		Difference	
i_{1a}	i_{2a}	i_{1a}	i_{2a}	i_{1a}	i_{2a}
[A]		[A]		%	
336.7	551.2	351.8	521.7	+4.5	-5.4
[p.u.]		[p.u.]			
7.14	11.69	7.46	11.07		

The reference current is the rated one: 47.14 A (peak value).

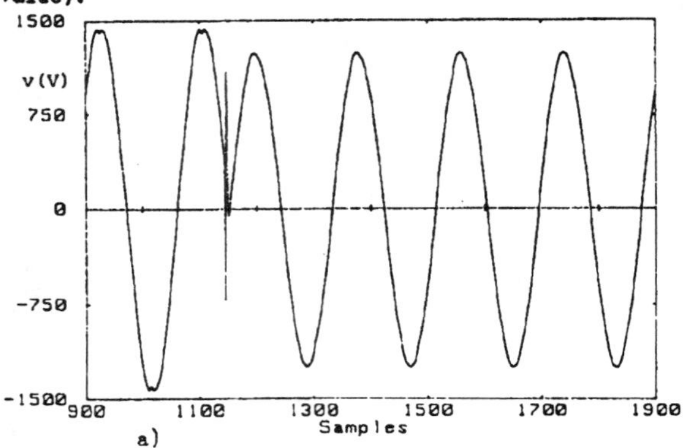

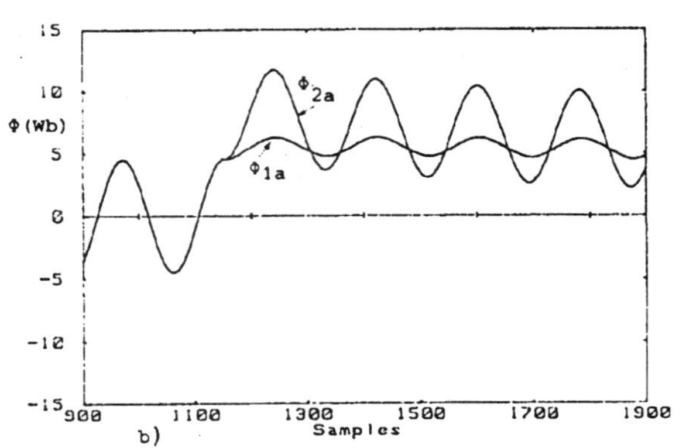

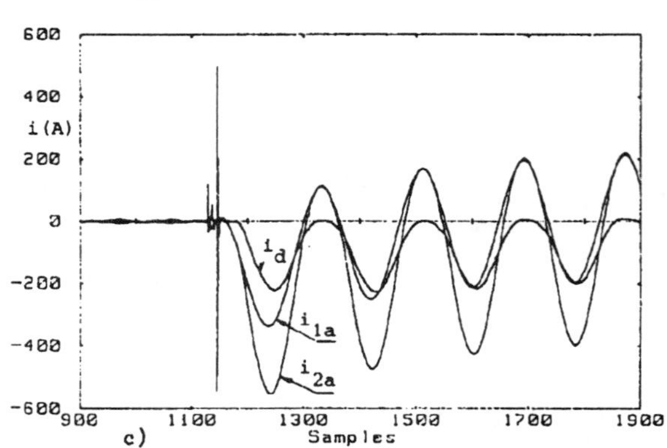

Fig.5- Measured transient of an out-of-phase synchronization of a three-phase 5-limb 100 kVA transformer with 180° phase-error: short circuit inductance 12.7%, direct-axis subtransient inductance 22.8%, power network inductance 2.41%; a) secondary line-to-neutral voltage of phase 'a'before and after the instant of the wrong parallel; b)flux linkages with the coils of phase 'a'; rated flux: 4.5 Wb; c)phase currents and differential current $i_d=i_{a2}-i_{a1}$ of the most stressed phase; rated current: 47.13A, peak value.

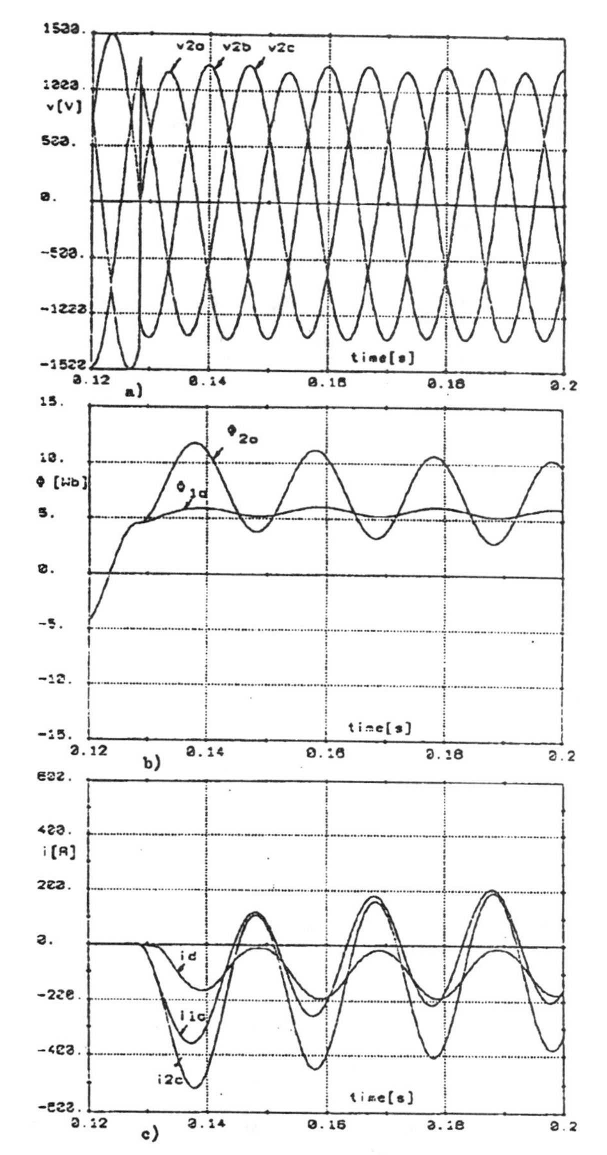

Fig.6- Computed transient of a out-of-phase synchronization of a three-phase 5-limb transformer rated 100 kVA with 180° phase-error, corresponding to the measured transient shown in Fig.5 with the same parameters.

The measured maximum unbalance between i_{2a} and i_{1a}, which is proportional to the MMFs unbalance, is of 226.9 A (i.e. 4.8 p.u.) and the computed one is 203.9A (4.3 p.u.). This maximun unbalance take place around the second peak of the currents and is greater than the difference between their peak values because the current peaks are not simultaneous.

It should be noted that the conventional evaluation of the currents, neglecting the magnetizing currents of the iron branches of the model of Fig.2d, would give a peak value $i_{a(conv)}$=415.5 A. Hence:

$$i_{1a}/i_{a(conv)} = 0.8467$$

and

$$i_{2a}/i_{a(conv)} = 1.256.$$

with the computed values.

The flux linkages with the internal coil (Φ_{1a}) and the external one (Φ_{2a}) are:

Measured		Computed		Difference	
Φ_{1a}	Φ_{2a}	Φ_{1a}	Φ_{2a}	Φ_{1a}	Φ_{2a}
[p.u.]		[p.u.]		%	
1.39	2.66	1.30	2.62	-6.5	-1.5

referred to the rated flux linkage, which is 4.5 Wb.

When the unbalance of the MMFs in the test transformer (1.48) is compared with that in the high power transformer simulated above (1.17), a remarkable differenze is evident. This is mainly because the per unit value of air inductance L_a of the 100kVA transformer is much less than that of the 370MVA transformer (0.03284 p.u. as compared with 0.4346 p.u.). This air inductance, associated to the air flux tube in parallel to the iron flux tube of the wound limbs, has the important role of limiting the corresponding magnetising current during the operation with very high saturation level.

The agreement among the computed and the experimentally measured results is satisfactory. Consequently, the validity of both the proposed model and the methodology adopted for the analysis can be regarded as positive. Therefore, this methodology can be applied for the study of the electromagnetic behaviour of large power three-phase five-limb transformers during the out-of-phase synchronization operation.

4.- CONCLUSIONS

This paper presents the analysis of the electromagnetic transient of a large power three-phase five-limb step-up transformer, during a 180° phase-error synchronization operation. The analysis has been made by a non-conventional circuital model for the transformer, to take into account the very high saturation level of some branches of its magnetic core. The simulation of the model was carried out by the ElectroMagnetic Transient Program (EMTP).

The obtained results have shown that:

(1) the peak value of the MMF in the HV coil of the most stressed phase is 10-20% greater than that evaluated in a conventional way, that is not taking the magnetizing current of the iron branches of the magnetic network of the transformer into account;

(2) there is a notable unbalance between the primary and the secondary MMFs of the same phase, due to the demand of MMFs on the very high saturated iron branches. The unbalance depends on the geometry of both the coils and the magnetic core of the transformer. Further, it increases as the power of the network increases. With a network having a power 100 times the rating of the transformer, the unbalance of the MMFs of the same phase is of the order of 3 p.u.

The experimental check of the validity of the proposed model has been carried out on a specially built 100 kVA three-phase five-limb transformer. The agreement between computed and measured results was satisfactory.

The result of the present work is the preliminary important step for the subsequent evaluation of the axial electromechanical stresses on the coils, which are expected to be notably higher than those conventionally evaluated with balanced MMFs. This will be the object of a future study.

ACKNOWLEDGEMENTS

This research was carried out in collaboration with the Societa' Nazionale delle Officine di Savigliano of

Turin (Italy). The author wishes to express to Mr.G.Sgorbati the recognition for his intuition concerning the greater magnitude of the actual stresses, in comparison with those predictable by conventional evaluations, in step-up transformers during wrong parallel operations. Further, the author wishes to thank both Mr.M.Borsani and Mr.M.Amprimo for their technical and experimental support as well as their valuable suggestions during the development of this research.

REFERENCES

[1] B.M. PASTERNACK, J.H. PROVENZANA, L.B. WAGENAAR, "Analysis of a generator step-up transformer failure following faulty synchronization", IEEE Transactions on Power Delivery, Vol.3, No.3, July 1988.

[2] C.M.ARTURI : "Analysis and simulation of the faulty synchronizing transient of single and three-phase transformers", L'ENERGIA ELETTRICA, N.2, 1989, pag.73-84 (in Italian).

[3] H.W. Dommel "Digital Computer Solution of Electromagnetic Transients in single and multiphase networks" IEEE Trans., Vol. PAS 88, pp. 389-399, April 1969.

[4] C.M.ARTURI, E.OLGIATI, M.UBALDINI, "Equivalent network of heavily saturated transformers and reactors", L'ENERGIA ELETTRICA, N.11, 1985, pag.476-484 (in Italian).

[5] G.FIORIO :"Magnetic networks and equivalent circuits of transformers", L'ENERGIA ELETTRICA, N.1, 1962, pag.29-44 (in Italian).

[6] A.BOSSI, G.CAPRIO, S.CREPAZ, G.PESCU', M.UBALDINI:"Equivalent networks of multi-winding transformers", L'ELETTROTECNICA, N.4, VOL.LXVI, 1979, pag.321-336 (in Italian).

[7] C.M.ARTURI, M.UBALDINI :"Equivalent networks of heavily saturated electromagnetic devices", ICEM '86, Munchen, Fed. Rep. of Germany, 8-9-10 Sept. 1986.

[8] C.M.ARTURI : "Equivalent magnetic networks of multi-windings electromagnetic static devices", Parte I - Theoretical aspects, L'ENERGIA ELETTRICA, N.1, 1989, pag.1-10 (in italian).

[9] W.H.HAYT,Jr.:"Engineering Electromagnetics", book, Mc Graw Hill, International Student Edition, fourth edition, 1985, pag.271-274.

APPENDIX A

Values of the parameters of the circuital model of Fig.2d for both a large power plant (in first column) and the test circuit (in second column).

Five-limb transformers

The nominal characteristics are:

Rated power	370 MVA	100kVA
Nominal voltages	20 kV/400 kV	1kV/√3kV

Connection: delta on the LV side and star with neutral earthed on the HV side.

All the parameter values refer to the number of turns of the HV coils which is 497 for the 370MVA transformer and 237 for the 100 kVA transformer.

The magnetization curve B(H) of the iron branches is that of a typical cold-rolled grain-oriented steel. The length (in m) of the various iron branches of the magnetic circuit are:

	370MVA	100kVA
$l_{f1}=l_{f2}=l_{f3}$	2.790	0.462
$l_{f4}=l_{f5}$	6.030	0.954
$l_{f6}=l_{f7}$	4.690	0.830

and the corresponding cross sections (in m²) are:

wound limbs	1.220	0.01099
yokes and lateral limbs	0.695	0.006375

Knowing the saturated inductances (in mH):

wound limbs	$L_{f1}=L_{f2}=L_{f3}$	135.9	1.679
lateral limbs	$L_{f4}=L_{f5}$	36.15	0.4717
intermediate yokes	$L_{f6}=L_{f7}$	46.48	0.5421
leakage	L_b	176.7(12.8%)	12.09(12.66%)

and the air inductances (in mH):

internal coils L_{a1}	206.2	4.533
external coils L_{a2}	314.7	13.43

the following air inductance values are obtained(in mH):

L_a	104	3.136
L_c	1602	95.26
L_d	609	32.20

The phase resistances (in Ω) are:

primary	R_1	0.3773	0.1208
secondary	R_2	0.5104	0.2104

Synchronous machines

Resistance	R_s (Ω)	1.44 (.999%)	0.2946(2.95%)
Direct-axis subtransient inductance	L_s(mH)	104.1 (22.69%)	7.565 (23.77%)

Power networks

Resistance	R_n(Ω)	0.43 (0.1%)	0.078 (0.26%)
Inductance	L_n (mH)	12.52 (0.91%) or 41.29 (3.0%)	2.301 (2.41%)

Cesare Mario Arturi (M'88)was born in Italy in 1950. He received his Electrical Engineering degree from the Politechnic of Milan (Italy) in 1975. From 1975 to 1985 he was Assistent Professor at the Electrical Department of the Politechnic of Milan for the Electric Machines course. Since 1985 he has been Associate Professor of Electrical Engineering at the Politechnic of Milan where he teaches Fundamentals of Electrical Engineering. His special fields of interest are the theory of the parametric transformer and the problems related to the thermal and electromagnetic modellization of large power transformers and reactors.

He is Member of the IEEE Power Engineering Society and Expert of the Technical Committee n.14 on Power Transformers of the Italian Electrotechical Commission.

Discussion

Adam Semlyen and Francisco de Leon (University of Toronto): We wish to congratulate Professor Arturi for his interesting and well written paper. Our discussion will focus on the derivation of the equivalent circuit, as shown in Fig.4.

While we fully agree with the final result of Fig.4d, we wonder whether the explicit representation of MMF sources in Fig.4b and of current sources in Fig.4c is necessary or useful? The reason for this question is that the symbol of a reluctance in Fig.4b could imply the existence of both a flux through it and of the related magnetic drop (negative MMF), provided that in all meshes with windings we assume the existence of the MMF excitation of the windings. Consequently, as an alternative to the author's procedure, we indicate a direct, graphical derivation of the equivalent circuit[1,2,3], based on the graph of the magnetic circuit drawn in thin lines in Fig.A. This part of Fig.A corresponds to the reluctances of Fig.4b, with two lines shown for those reluctances which are in parallel, without windings in between. The next step is to put nodes, represented by •, in all meshes with windings, and to draw inductance branches (strong lines in Fig.A) to cut all reluctance branches (the thin lines). The "external node" is represented by a dashed line. Inductance branches in parallel have been lumped together. The nodes with MMF sources (•) will be used to connect the corresponding ideal transformers (after the negative inductances have been added[3], as pointed out by Professor Arturi) and all the rest of the elements shown in Fig.4d. In Fig.A braces P_a, P_b, and P_c indicate the connection of the ideal transformers for the primary windings and S_a, S_b, and S_c for the secondary windings. Rotation by 90° of the electric part of the graph of Fig.A permits to verify that it is, in essence, identical to the central part of Fig.4d.

[1] E. Colin Cherry, "The Duality between Interlinked Electric and Magnetic Circuits and the Formation of Transformer Equivalent Circuits", Proc. of the Physical Society, Vol. (B) 62, Feb. 1949, pp. 101-111.

[2] G.R. Slemon, "Equivalent Circuits for Transformers and Machines Including Nonlinear Effects", Proc. IEE, Part IV, Vol. 100, 1953.

[3] H. Edelmann, "Anschauliche Ermittlung von Transformator-Ersatzschaltbildern", Arch. elektr. Übertragung, Vol. 13, 1959, pp. 253-261.

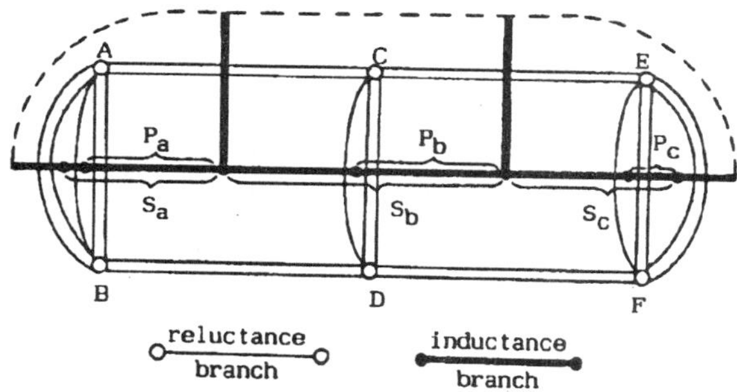

Fig.A Direct implementation of duality using the graph of the magnetic paths (thin lines) to obtain the graph of the inductances (strong lines) of the electric equivalent of the transformer. The external connections are via ideal transformers connected to P_a, P_b, P_c (primary) and S_a, S_b, S_c (secondary).

Manuscript received February 28, 1990.

Robert J. Meredith (New York Power Authority, White Plains, New York): Discusser agrees with the author on the necessity of developing saturable transformer models from physical parameters and commends him for his correct evolution of an equivalent electrical circuit from an initial magnetic circuit. However, the author's demonstration of winding currents only 10% greater than found with a linear model indicates that his test case is not very sensitive to the nonlinear and saturated linear parameters of his models. Consequently, his model for a saturated transformer cannot be validated very well, even against the physical model he developed. In the context of mechanical winding failure, discusser also believes that direct analysis of relative winding forces, being the product of current and an appropriate leakage flux density, would have provided more insight than a discussion of winding current differences, whose significance is unstated. A comparison of the test case against a design-basis transformer fault simulation would have been welcomed.

The author's iron flux path model appears appropriate, except for the omission of the steel tank. When the end limbs of the transformer saturate, flux is forced into air paths outside the core. The tank steel, being thousands of times better at conducting flux than air, cannot be ignored, except at great error to the external air-flux paths modeled. There also appear to be inconsistencies in the tabulations of iron fluxes. The smaller sized limbs all display saturated flux densities just above the 2-Tesla capability of transformer steels, but the author's identification of a 2.41-Tesla level within the wound core cannot be so reconciled. The discrepancy potentially causes large undercalculation of excitation currents, leakage fluxes and winding forces.

The use of the dimensions of core members to calculate "saturated inductances" cannot be recommended. The incremental reluctance, of what amounts to an air gap where a saturated core member exists, depends only on the geometry of adjacent unsaturated iron members or coils; it is unrelated to the cross-sectional area of the saturated limb. For wound limbs the "saturated inductance" is related to coil dimensions, rather than core dimensions. Even more to the point, the "saturated inductance" must be determined for each unique saturation condition of concern, in order to formulate an equivalent with either mutual couplings or additive terms reflecting a known sequence of limb saturation. Unlike the unsaturated iron-core counterpart, the closed loop air-flux path cannot be assumed to be composed of a summation of independent series segments whose reluctances can be individually varied without affecting the shape of the overall flux path.

Figure A displays a suggested air-flux reluctance model for the author's transformer, which would overcome many of these objections. The five iron nonlinearities at the right of the diagram are a rough attempt to model the tank as a type of nonlinear mesh. Only such a model can express the unsaturated tank's influence as a magnetic node above the core and as a shared low-reluctance path for flux in excess of fully-saturated core capacity. After tank and yoke saturation, all flux coupling between phases via air paths is assumed to disappear, approximating the very low levels which would actually exist. This is in sharp contrast to the low reluctance yoke air-path connections of the author's model. The air-flux reluctance

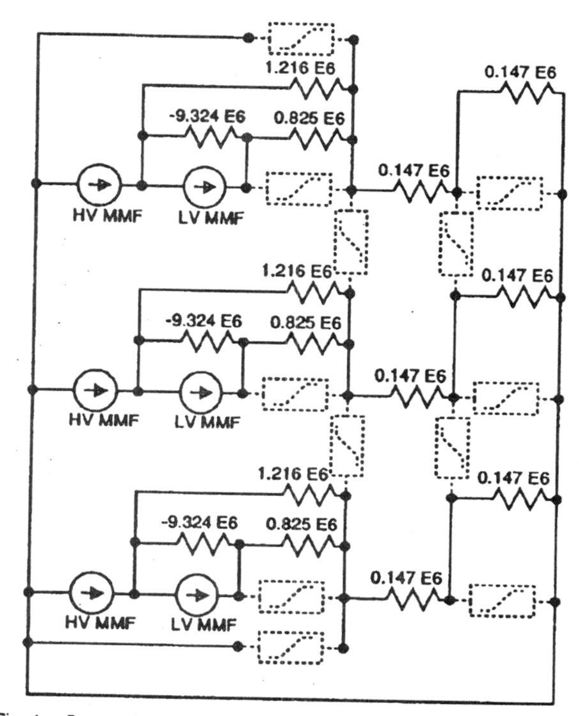

Fig. A. Suggested air-flux reluctance model for author's transformer

equivalent of Figure A closely duplicates air-core coupling results for the estimated dimensions of the coils on the author's transformer (60% of HV flux coupled to open circuit LV coil; 91% of LV flux coupled to open circuit HV coil). In addition it displays: 1) appropriate leakage reluctance between HV and LV coils; 2) stated air-core reluctances of both coils; 3) upper and lower magnetic node placement closely coinciding with that of the iron flux path nodes; 4) modeling to accommodate wound-leg saturation (solenoidal reluctances) followed optionally by return path saturation. Saturation onset in the reverse order would not necessarily be modeled as accurately.

It should be noted that any one-to-one relationship between iron- and air-flux paths is only coincidental. Phase models of the same form would be used for core-form transformers having any number of legs, including single-phase designs. The model is derived from coil configuration and is independent of core design.

Manuscript received February 27, 1990.

A. Narang (Ontario Hydro Research, Toronto, Canada): The author is to be congratulated for employing a model which seems particularly appropriate for this study. A physically based electric circuit model, displaying one-to-one correspondence with flux in the non-linear magnetic circuit, seems well equipped for simulating disturbances which may produce widely differing saturation states in parts of the core. The author has clearly recognized this need and opted for an "unconventional" model. Though uncommon, similar models based on this approach have been proposed earlier/A–E/, and are completely general and rigorously valid. I would welcome the author's response to the following:

1. The author recognizes that a further improvement to the model would include negative inductances to take account of coil thickness. However, these were apparently not included in the reported study. Would their inclusion not alter calculated flux levels and MMFs, and consequently also the computed axial electromechanical stresses appreciably? Can the author give approximate values for these inductances in relation to air-inductances La1 and La2 (Appendix A), or provide winding dimensions necessary for their calculation.
2. Results are presented for a "linear conventional model" and compared with the detailed model. This may be an unfair comparison since better models can be accomodated using existing EMTP components. For example, one might use three single phase, 3-winding transformers, with the innermost winding connected in open-corner delta and a non-linear inductance connected across the open-delta to model the influence of outer limbs for zero-sequence mode. Non-linear inductors would also be connected at the innermost winding terminals to model phase magnetizing impedances. Could such models produce comparable results for this study?

References

A. E. C. Cherry, "The Duality between Interlinked Electric and Magnetic Circuits and the Formulation of Transformer Equivalent Circuits", Proceedings of the Physical Society, Part 62, February 1949, pp. 101–110.

B. G. R. Slemon, "Equivalent Circuits for Transformers and Machines Including Non-linear Effects", Proc. IEE, Vol. 100, Part IV, July 1953, p. 129.

C. H. Edelman, "Anschauliche Ermittlung von Transformator Ersatzschaltbildern", Archiv Electr. Ubertrag, 1959, pp. 253–261 (In German).

D. K. Schlosser, "Eine auf Physikalischer Grundlage ermittelte Ersatzschaltung fur Transformatoren mit mehresen Wicklungen", BBC-Nachrichten, March 1963, pp. 107–132.

E. E. P. Dick, W. Watson, "Transformer Models for Transient Studies Based on Field Measurements", IEEE Transactions PAS, Vol. PAS-100, No. 1, January 1981, pp. 409–419.

I very much appreciate the broad interest shown in this paper and I thank all the discussors for their comments and objections stimulated by the accurate evaluation of the material I have presented. Their observations point out some aspects of the analysis I have made and permit the discussion of some points which have perhaps been stated in too concise a way.

I reply to Prof.A.Semlyen and Dr.F. de Leon, as regards the method of obtaining the equivalent network of the three-phase five-limb transformer, shown in Fig.2. The field sources, either the M.M.F. sources F, associated to the coil currents (F=NI), or the flux sources Φ, associated to the voltage applied at the coil terminals ($\Phi=(\int vdt)/N$), are modelled by two-terminal sources which could reasonably be included expressly in the magnetic network. Their localization is necessary since it is not possible to state a magnetic network with only reluctances and noy sources. The analysis of the behaviour of an electromagnetic device could also be made directly in terms of magnetic network, without considering the corresponding equivalent electric network. I therefore think that it is not necessary to wait for the electric network to point out the sources. It is also evident that the current sources of the electric network are the direct consequence of the M.M.F. sources of the magnetic network.

It should be noted that any duality procedure always presents two aspects: a topological one and a physical one. The topological aspect is given by the mesh-node correspondence whereas the physical aspect, which expresses the nature of the corresponding dual circuital elements, is better expressed by the analytical procedure which transforms the magnetic equations of the independent meshes in the corresponding electric equations of the independent nodes. In other words, starting from a planar magnetic network, the behaviour of which is expressed, in matrix terms, by the magnetic equation:

$$|R|\cdot|\Phi|=|F|,$$

one gets the electric matrix equation:

$$1/(d/dt)\cdot|1/L|\cdot|v|=|i|,$$

which expresses the behaviour of the dual equivalent electric network. The magnetic network is supplied by the M.M.F. sources $|F|$ just as the dual electric network is supplied by the source currents $|i|$.

In response to Mr.A.Narang as regards the negative inductances of the equivalent electric network and their influence on the obtained results, I give here the coil dimensions of the 370MVA rated transformer examined in the paper:

	L.V.coils	H.V.coils
internal diameter[mm]	1392	1842
external diameter[mm]	1634	2192
height [mm]	1834	1914

The negative inductances L_1 and L_2, which take into account the thickness of the coils, are to be connected in series to the terminals of each ideal transformer of the equivalent network. To compute them, it is necessary to consider two fictitious coils concentric to the actual ones: one inside the L.V. coil and the other outside the H.V. one. In this way one has a 4-winding transformer for which the binary short-circuit inductances are to be computed. From the relationships between the inductances of the equivalent network of the 4-winding transformers and the short-circuit inductances, the following inductances can be obtained:

$$L_1=-11.41 \text{ mH}, \quad L_2=-27.05 \text{ mH}, \quad L_b=215.2 \text{ mH}.$$

Obviously, the value of the inductance Lb of the equivalent network now has a different value in comparison with that seen in absence of negative inductances. The sum of L_1, L_2 and L_b must in fact always be equal to the leakage inductance, which in this case is 176.7 mH. With these air-inductances and the saturated-inductances of the iron paths one has to compute the air inductances L_a, L_c and L_d, associated to the air fluxes in parallel to the iron fluxes of the wound limbs, of the intermediate yokes and the lateral limbs. The obtained values are:

$$L_a=119.8 \text{ mH}, \; L_c=1401 \text{ mH e } L_d=586.9 \text{ mH}.$$

The simulation of the improved model, including the previous inductances, gives the following results, for the most stressed phase:

$$i'_{a1}=9.69 \text{ p.u.} \quad \text{e} \quad i'_{a2}=11.65 \text{ p.u.}$$

which are to be compared with

$$i_{a1}=9.8 \text{ p.u.} \quad \text{e} \quad i_{a2}=11.5 \text{ p.u.}$$

already obtained neglecting the negative inductances. As one can see, the unbalance between the M.M.F.s increases slightly (11.65/9.69=1.20 instead of 1.17) but the conclusions already obtained by neglecting the negative inductances are substantially the same.

The second point asked by Mr. Narang refers to the use of three single-phase 3-winding transformers in order to simulate by the EMTP program the equivalent network of the three-phase 5-limb transformer. My response is that only one non-linear inductance connected across the open-delta of the innermost windings does not seem enough to me to take into account the heavy and differing saturation of yokes and lateral limbs.

One should also consider that the EMTP program permits the equivalent network of Fig.2d to be easily implemented, as it is, without forced simplification, by means of seven type-98 pseudo non-linear reactors, six ideal transformers and a number of linear inductances and resistances. Therefore, I do not see the convenience of using a network configuration different from that which has been judged suitable for modelling the physical behaviour of the device under examination. This rule is always valid, for any structure of the magnetic network or number of windings.

Finally I reply to Mr.R.J.Meredith, who raised several objections. He proposed, in turn, a magnetic network for the three-phase 5-limb transformer, which should take into account the influence of the tank when highly saturated.

The 370MVA transformer examined in this paper and the corresponding wrong parallel operation represent a real matter (no a test case) for the analysis of which a circuital model has been set up. The circuit model has been also verified experimentally on a 100kVA test-transformer and has given satisfactorily results, as reported in the paper.

The effect of saturation is more or less evident in relation to the characteristics of the considered machine. As a matter of fact, the influence of saturation on the results is more notable in the 100kVA test-transformer than in the 370MVA transformer.

As regards the suggestion of Mr.Meredith for a direct analysis of the relative winding forces by means of "an appropriate leakage flux density", it is to be noted that the radial component of the flux density in the space occupied by the windings, which is responsible for the axial mechanical stress, varies considerably. With the F.M.M. reported in the paper for the most stressed phase, it varies from about 1.0T and zero, between the end and the middle of the winding. For the calculation of the mechanical forces one has therefore to evaluate the magnetic field by a numerical method imposing the unbalanced F.M.M. computed with the proposed circuital model and then evaluate the stress tensor on the surface of the coils in the axial and radial directions.

As regards the tank, I believe we can neglect its influence in the case of a three-phase 5-limb transformer. As a matter of fact, when the lateral limbs and the yokes are in high saturation conditions, the air-inductances Lc and Ld are computed in such a way to consider that the global behaviour of the devices has to lead to the value of the air-core inductance of the supplied coils. Therefore, the inductance La, Lc, Ld are associated to equivalent flux paths which may not have simple geometrical configurations.

It should also be considered that the tank has a thickness of almost 10 mm and an apparent cross-section of some percent of the cross-section of lateral limb. Further it has a skin depth of the order of 1 mm and therefore one should not neglect eddy currents. In other words, the tank behaves more like a magnetic shield than a magnetic shunt. In any case, I believe it is not worth while to complicate the model to take the tank into account. However, even when it is necessary to take the tank into account , as in the case of three-phase 3-limb transformer without delta-connected winding, one should not model the tank by a simple network of reluctances.

Referring to the flux density values reported in the paper for the wound limb and the lateral limbs and yokes, I have to observe that there are no inconsistencies in the fact that iron reluctances with less cross-section, like the yoke and the lateral limbs, have a less flux density (almost 2T) than the wound limb (2.41T) of the most stressed phase, since neither the flux nor the magnetic potential drop have been imposed. It is to be noted, in passing, that the lateral limbs have an average length of 6.03m against an average length of the wound limbs equal to 2.78m. The intermediate yokes have a mean length of 2.345m, which, multiplied per two, gives 4.69m, considering the series between the top and lower yoke. Although the B(H) characteristic is the same for all the iron paths, the ratio cross-section/length (S/l) is considerably different. Further it is to be considered that each iron path has air permeances in parallel with different values:

in	ratio S/l	permeance parallel
· wound limb	1.2196/2.78 = 0.4387	Λ_a=0.421 μH
· intermediate yoke	0.57*1.2196/4.69 = 0.1482	Λ_c=15.4 Λ_a
· lateral limb	0.57*1.2196/6.03 = 0.1153	Λ_d=5.86 Λ_a

As regards the "saturated inductances" used for the calculation of the air-inductances L_a, L_c end L_d, they do not represent the inductance of the coils in high saturation levels, as observed by Mr.Meredith. Indeed they only model the behaviour of the iron-path in high saturation and are in parallel with air-inductances. In the case of a wound limb, for instance, the "saturated inductance" is in parallel with La, which takes the air flux of the space between the coil and the iron-core into account.

Finally, referring to the magnetic network suggested by Mr.Meredith, in addition to the comments already made about the tank, I would observe that the numerical values given for the parameters do not correspond to the 370MVA transformer of this paper, as can easily be checked by the air inductances reported in Appendix A.

Furthermore, if the negative reluctance in parallel with each L.V. M.M.F. has to model the thickness of the corresponding coil I wonder why those of the H.V. coils are not considered as well.

In conclusion I hope that the interest stimulated by this paper will encourage other experts to consider the usefulness of the physical circuital models in the analysis of transient with high saturation level of the magnetic circuit, for electromagnetic devices in which it is possible to know, with good approximation, the distribution of the most significative magnetic flux tubes. It is also to be stressed that we should not over-estimate the field of applicability of the magnetic networks method and keep in mind their limits clearly, in order to avoid an improper or arbitrary use of this wonderful method of analysis.

Manuscript received April 9, 1990.

IEEE Transactions on Power Systems, Vol. 9, No. 3, August 1994

TOPOLOGY BASED MAGNETIC MODEL FOR STEADY-STATE AND TRANSIENT STUDIES FOR THREE-PHASE CORE TYPE TRANSFORMERS

Arun Narang, Member
Research Division

Russell H. Brierley, Member
System Planning Division

Ontario Hydro
Toronto, Ontario, CANADA

ABSTRACT - Existing matrix models for transformers reproduce specified short-circuit tests, portraying leakage flux in air, but do not explicitly model the core. Therefore, a precise relation between core flux and air flux is not clearly established. Core representation is defined simply on the basis of a specified excitation current (positive- and zero-sequence), permitting linear magnetizing branches to be incorporated in the matrix. Non-linear inductances are connected separately at winding terminals to introduce core non-linearities, however these are not identified physically with individual core limbs, making precise accommodation of hysteresis, saturation and eddy current effects difficult. Published models based on the duality existing between magnetic and electric circuits identify a means for correctly interfacing core flux with air flux, permitting core nonlinearities to be incorporated on a physical basis. However practical application of these models has been hampered by a difficulty in deriving the required model parameters. A formulation is presented here to build a topological model based on normally available test data. Results of short-circuit tests are presented demonstrating that existing matrix models can introduce errors.

Keywords: Transformers, Magnetic Model, Core Model, Core Non-Linearity, Transients, EMTP, Duality.

I. INTRODUCTION

Transformer models originating in the Bonneville Power Administration's (BPA) ElectroMagnetic Transients Program (EMTP) compute an admittance or impedance matrix reproducing specified short-circuit tests [1,2]. Core topology does not explicitly enter into the formulation, since results of short-circuit tests are dominated by leakage fields in air, outside the core. Non-linearities due to the iron core are incorporated separately by connecting magnetizing branches at winding terminals. However these are not rigorously identified with individual core limbs, making it difficult to relate the variation of magnetizing current with the magnetic state.

Published models [3-7] based on the principle of duality for magnetic and electric circuits [8] define a topologically correct interface between air flux and flux in individual core limbs. Core non-linearities are incorporated on a physical basis, permitting rigorous accommodation of core hysteresis, saturation and eddy current effects, all of which may play a role in transient studies. Existing models establish this interface implicitly, and are sufficiently general that a correct interface can be established neglecting non-linearities. To do so, however, requires due care and user manipulation of the specified data based on the findings presented here. The role of zero-sequence magnetizing inductance in conventional short-circuit tests (for 3-limbed units) is not recognized, sometimes causing modelled results to differ from specified values. It will be shown here that this role can be exploited to produce physically consistent models, enhancing modelling accuracy.

The reported formulation is part of a larger project aimed at improving transformer models. It builds on existing formulations, available in all versions of the EMTP which have evolved from the original BPA software. Sufficient details are also provided to benefit a wider audience. The model accommodates magnetic effects due to the core and bulk winding assemblies using normally available test data. It extends existing matrix methods, therefore its implementation utilizes certain elements of the existing BCTRAN code with minor changes, including the input/output user interface and eventual computation of the coupling matrix defining winding leakages. However it differs in its manipulation of the specified test data, incorporation of the core model, and the provision of facilities to produce physically consistent models. The resulting model differs from existing models in some respects, as illustrated by test cases and comparison with measurements.

II. REVIEW OF MATRIX MODELS

Transformer models in all versions of the EMTP having roots in the original BPA program use an impedance or admittance formulation relating terminal voltages and currents [1,2].

$$[v] = [R][i] + [L]\,d[i]/dt \qquad (1a)$$

$$d[i]/dt = [\Gamma][v] - [\Gamma][R][i] \qquad (1b)$$

Matrices [R], and [L] or its inverse [Γ], are accepted by the EMTP directly as input for time domain simulations. The topology based model utilizes the latter formulation, which is needed if magnetizing current is totally ignored (modelled separately), since the N x N impedance matrix [Z] (for N-winding single-phase transformers) is undefined and the admittance matrix [Y] is singular. In this event, a reduced impedance matrix [Z] of order (N-1) is computed first

93 SM 509-0 PWRS A paper recommended and approved by the IEEE Power System Engineering Committee of the IEEE Power Engineering Society for presentation at the IEEE/PES 1993 Summer Meeting, Vancouver, B.C., Canada, July 18-22, 1993. Manuscript submitted Aug. 27, 1992; made available for printing May 18, 1993.

PRINTED IN USA

(excluding winding resistances), taking winding N as reference,

$$[v_k - v_N] = [Z]^{reduced} [i_k] \quad (2)$$

where the diagonal elements correspond to short-circuit tests involving winding N (excluding a contribution due to winding resistances), and the off-diagonal elements (in pu values),

$$z_{ik}^{reduced} = (z_{iN}^{short\text{-}cct} + z_{kN}^{short\text{-}cct} - z_{ik}^{short\text{-}cct}) / 2 \quad (3)$$

$[Z]^{reduced}$ is inverted, and an additional row and column is generated such that the resulting N x N [Y] matrix is singular, or equivalently [Γ] related by frequency ω, since resistances are neglected,

$$[\Gamma] = j\omega[Y] \quad (4)$$

For three-phase transformers, first $[Z]^{reduced}$ is built separately from positive- and zero-sequence short-circuit quantities (z^+, z^o). Then the corresponding element pairs are transformed to self and mutual impedances (z_s, z_m), defining magnetic coupling in the phase domain,

$$z_s = (z^o + 2z^+) / 3$$
$$z_m = (z^o - z^+) / 3 \quad (5)$$

Each scalar value in (2) is therefore replaced by a balanced 3 x 3 sub-matrix. The resulting 3-phase $[Z]^{reduced}$ is inverted using "balanced matrix algebra" [9], operating on (z_s, z_m) pairs, and subsequently expanded to obtain a singular 3-phase [Y]. Winding resistances form a diagonal [R], completing the series [R], [Γ] formulation implemented by (1b).

III. THE DUALITY MODEL

Transformer models accounting for core topology are based on a correspondence between electric and magnetic circuits, as expressed by the principle of duality [8]. Voltage, current and inductance in electric circuits correspond to flux, MMF and reluctance respectively in magnetic circuits. Loops and nodes are topological duals. On this basis, an equivalent electric network is derived by identifying the principal flux paths in the magnetic circuit formed by the core and the windings [3-7].

Figure 1a illustrates a typical flux distribution presumed for 3-limbed core transformers, neglecting magnetic coupling among phases through air. The corresponding electric equivalent (Figure 1b) mirrors the magnetic circuit. Non-linear inductances correspond to iron flux paths in the magnetic circuit, permitting each core limb to be modelled individually. Hysteresis, saturation and eddy current effects are therefore accommodated on a physical basis [10,11]. Each L_k represents a top and bottom pair of horizontal limbs (yokes), and each L_b represents a wound limb. Inductances L_o represent the return flux path depicted in air, outside the core and around the windings. In 3-limbed transformers, the return path includes the transformer tank; in 5-limbed transformers, it is largely confined to the outer limbs. Finally, the ladder network of linear inductors sandwiched between L_o and L_b on each phase models

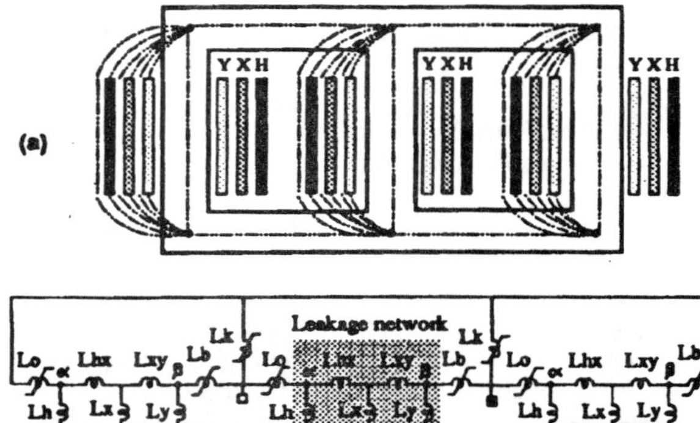

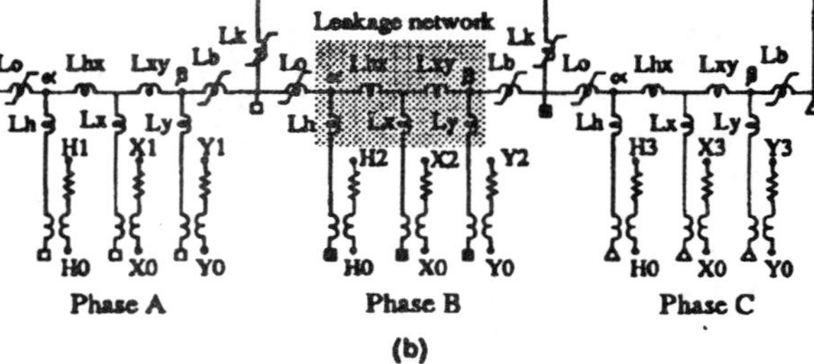

Figure 1: (a) Identification of magnetic flux paths for 3-limbed core transformer; (b) Electric equivalent portraying magnetic coupling in 3-limbed and 5-limbed transformers.

winding leakages through air, and is derived on the premise that leakage flux is axial, that windings are concentric around a central limb, and that the turns comprising each phase winding are contiguous (i.e. interleaving among windings is neglected). We shall refer to this network as the "leakage network" or "coupling matrix" interchangeably for convenience.

The leakage network is connected to the remainder of the model at internal nodes α and β, establishing an interface between leakage flux and magnetizing flux. The nodes are magnetically displaced from the respective winding terminals by inductances L_h and L_y, which account for unequal flux linkage among winding turns due to the finite winding radial build. They are therefore small in relation to L_o and L_b respectively. Published values based on detailed measurements on a 3-limbed unit [6] place L_h at about 1%, and L_y as about 2.5% of the saturated values of L_o and L_b respectively. Consequently, at a modest sacrifice in accuracy, we connect the respective magnetizing branches L_o and L_b magnetically at the respective winding terminals. This permits the order of the coupling matrix, and hence the number of parameters needed to model the transformer, to be reduced. We note the following, keeping in mind that L_b and L_k are typically orders of magnitude larger than the remaining inductances:

(a) Since the leakage network excludes all magnetizing branches, it requires a (singular) admittance matrix formulation; the impedance matrix is undefined.

(b) Individual leakage networks on each phase are effectively decoupled during short-circuit tests (positive-sequence and zero-sequence) in accordance with the flux distribution presumed in the derivation.

(c) During positive-sequence short-circuit tests, negligible current flows in the magnetizing branches (neglecting limb reluctance), confirming that all applied MMF is associated with leakage flux. The coupling matrix is therefore fully defined directly by N(N-1)/2 positive-sequence short-circuit tests normally provided by manufacturers. It can be computed in accordance with existing matrix methods for single-phase transformers.

(d) For zero-sequence short-circuit tests, the model yields three decoupled leakage networks, each with L_0 in shunt at the outermost winding. The applied MMF is therefore associated not just with winding leakage flux, but also magnetizing flux. If the leakage network is determined from positive-sequence tests, only L_0 is needed to complete the characterization of all air flux components. This requires just one zero-sequence test. By implication, all other tests are related, and are redundant according to the derivation. For 5-limbed units, L_0 is several orders of magnitude larger than the leakage inductances (being associated with the outer core limbs), hence zero-sequence short-circuit tests are no different than positive-sequence tests.

(e) For positive-sequence short-circuit tests, the measured per-unit impedance is the same regardless of which winding is excited. Hence, $z_{km(+)} = z_{mk(+)}$, where the subscripts denote the excited and short-circuited windings respectively. This is associated with the coupling matrix being singular. We shall term this a "reciprocal" test, though distinguishing this usage from the conventional term "reciprocity", which denotes a transfer measurement. Finally, we note that while positive-sequence tests are reciprocal, zero-sequence tests are not, at least for 3-limbed units, since L_0 provides a shunt path to ground at the outermost winding that cannot be neglected (Figure 3). Thus, $z_{km(o)} \neq z_{mk(o)}$. In conclusion, evidently N(N-1)/2 distinct measurements are possible in positive-sequence tests, and up to twice as many in zero-sequence tests.

IV. MODEL IMPLEMENTATION

Short-circuit tests normally performed on power transformers characterize positive-sequence and zero-sequence leakage impedances between windings, characterizing air flux outside the core. The MMF associated with core flux is negligible. It is therefore convenient to build a leakage model based on these tests, accepting that this implies assuming phase symmetry. While the foregoing duality based model requires just N(N-1)/2 + 1 short-circuit tests, and considers the remaining to be redundant, it may neglect factors likely to be accounted for in measurements. It is therefore prudent that the modelling formulation be more general. Existing matrix models permit matching N(N-1)/2 positive, and a like number of zero-sequence tests by building leakage models separately for each set of tests. Coupling among phases is then established by (5). We adopt this approach, replacing the leakage networks of Figure 1b by a full 3-phase matrix, incorporating symmetry among phases. The resulting model is depicted in Figure 2, with L_s and L_m denoting balanced self and mutual coupling among phases. Since the matrix representation is completely general and its elements are computed to match measurements,

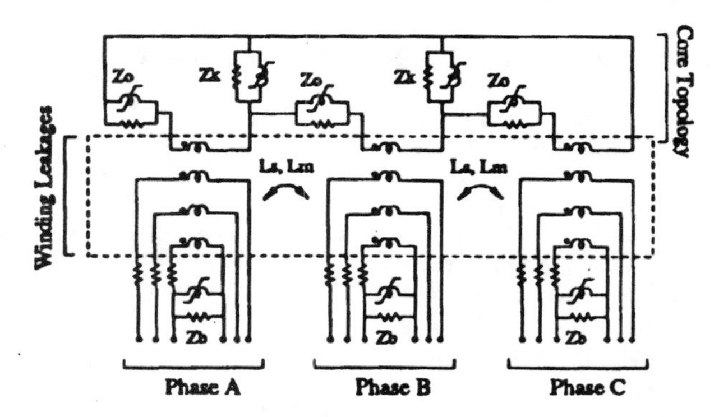

Figure 2: Schematic representation of model for 3-phase, 3-winding core type transformer.

the leakage model can accommodate arbitrary winding arrangements, including non-concentric and interleaved winding sections.

Computation of the coupling matrix [Γ] from positive-sequence short-circuit tests is straightforward, based on (3), since the tests do not involve any magnetizing currents. However, zero-sequence tests include an MMF contribution due to Z_0 (L_0 in parallel with a resistive branch accounting for excitation losses) which must be removed before the model can be implemented as depicted. This requires the specified data to be pre-processed as described in Section IV-A. An alternative appears possible based on initially incorporating Z_0 in an impedance coupling matrix $[Z] = [R] + j\omega[L]$, in accordance with the existing impedance formulation [1,2] implemented by (1a). However this approach produces a full [R] with negative off-diagonal terms, which may not be meaningful at very low frequencies (e.g. in studies involving core saturation due to Geomagnetic Induced Currents).

Finally, the leakage model is interfaced with magnetizing branches through a fictitious, uncoupled, 3-phase winding for electrical isolation (Figure 2). The additional winding is incorporated in an enlarged [Γ], as described in Section IV-B. The network of magnetizing branches formed by Z_b, Z_k and Z_0 mirrors the core structure. It exhibits centre-phase symmetry, and permits non-linearities to be incorporated on a physical basis. The established interface is topologically correct, according to the duality derivation, provided the innermost and outermost windings (α and β) are concentric, fully covering the wound-limb. While the present discussion centres on low-frequency magnetic effects, the model permits incorporation of frequency dependence due to eddy currents in all elements, including the winding resistances [10,11].

The model for 5-limbed units is just a special case of the above formulation. The leakage model is built directly from positive-sequence short-circuit tests; zero-sequence tests are redundant. The core model is topologically identical, differing only in its physical interpretation. Z_0 now corresponds to the outermost (non-wound) core limbs, hence it can be omitted from the centre phase, or perhaps replaced by a linear inductance matching its saturated value.

A. Pre-Processing of Zero-Sequence Short-Circuit Data

Processing is only needed for 3-limbed units, and is simple in the absence of a delta connected winding or if tests are performed with the delta connection opened (Class "A" formulation). If, on the other hand, tests are performed with the delta connection closed (Class "B" formulation), this really becomes an additional short-circuited winding whose effect must be removed before $[Z]^{reduced}$ can be computed as described. Short-circuit tests performed with more than one delta winding remaining closed are not accommodated. Processing is carried out using normalized quantities (i.e. pu values), and care is taken to distinguish between excited and short-circuited windings since tests are not reciprocal. For simplicity, Z_b is presumed sufficiently large (relative to the leakage impedances) that it plays no role in short-circuit tests.

1) CLASS A - No Delta Connected Winding

In the absence of a delta connected winding, the only processing needed is to remove the effect of Z_o and winding resistances. This is done by expressing the measured impedance as a function of the relevant network parameters, and solving the resulting nonlinear equation by Newton's method. Magnetizing branch elements L_o, R_o, and winding resistances are presumed known at this time, as specified or determined elsewhere. Tests are processed sequentially, in arbitrary order within each identified sub-class, as follows.

A-1: Compute $L_{\alpha k}$, denoting $L_\alpha+L_k$ or the loop inductance between nodes α and *k*, in tests involving the outermost winding α (Figure 3a).

A-2: Compute L_{km} (denoting L_k+L_m) for all remaining short-circuit tests, viz. those involving windings *k* and *m* (Figure 3b). Note that $L_{\alpha k}$ and $L_{\alpha m}$ are known from the previous step at this time.

2) CLASS B - Tests Involving a Closed Delta Winding

For generality, no restriction is imposed on the relative position of the delta connected winding around the wound limb. If it is the outermost winding, positive and zero-sequence tests should be indistinguishable since Z_o is short-circuited by the delta connection. In this case, no processing is needed since the coupling matrix can be computed using only positive-sequence tests in the same manner as for single-phase units or for 5-limbed cores. For a delta winding positioned anywhere else on the limb, processing is carried out sequentially by sub-class, as follows:

B-1: Compute $L_{\alpha\Delta}$ from the test involving nodes α and the delta connected winding Δ, as per Type A-1 tests (Figure 3a, with k denoting Δ).

B-2: Compute $L_{\alpha k}$, as per Type A-1 tests, from tests involving α and any other winding *k*, provided winding Δ does not remain closed during this test (Figure 3a).

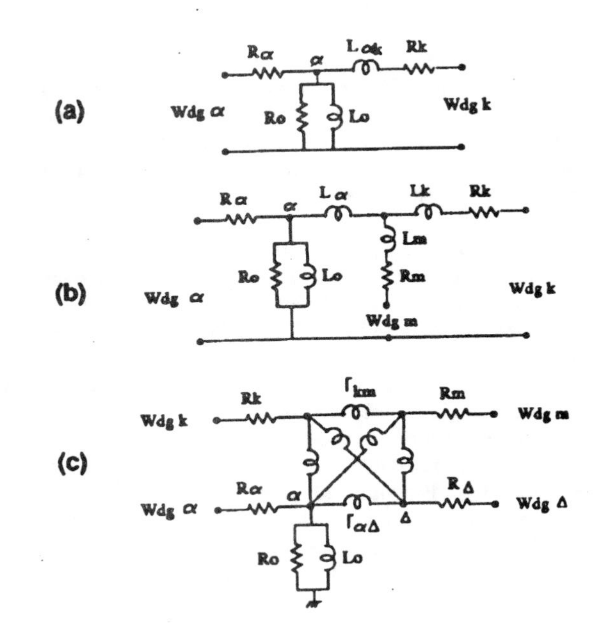

Figure 3: Equivalent networks for processing specified zero sequence short circuit impedances.

B-3: Tests involving winding *m* ($m \neq \alpha$) and winding Δ are represented as in Figure 3b (with k denoting Δ). With $L_{\alpha\Delta}$ already known from step B-1, solve for $L_{\alpha m}$ and $L_{m\Delta}$ by Newton's method, and perform Gauss-Seidel iterations to satisfy the specified test values $z^o_{\alpha m}$ (including delta connection on β) and $z^o_{m\Delta}$ concurrently.

B-4: The previous steps accommodate transformers with up to 3 windings. For transformers with 4 or more phase windings, short-circuit impedances involving windings *k* and *m* (distinct from α and Δ) are processed to compute the loop inductance L_{km}. If the delta connection on Δ remains intact, the solution is obtained by Gauss-Seidel iterations based on a nodal equivalent, as depicted in Figure 3c. Otherwise, Figure 3b applies, and is solved by Newton's method.

3) Remarks

The processing sequence is structured such that at most two test values are processed at any step (as in Step B-3). Iterations are therefore performed at most in 2-space (ie. for two parameters) rather than in N(N-1)/2 dimensions simultaneously (representing the number of zero-sequence tests). Processed results are used to compute $[\Gamma]^*$ (excluding the fictitious winding), as per the existing BCTRAN formulation [2]. Consequently, the leakage model differs from BCTRAN in that it incorporates a correction in the coupling matrix to take account of the zero-sequence magnetizing branch.

B. Topological Interface

The core model is electrically isolated from the leakage

network to ensure that disturbances in the external network do not disrupt its topology. Isolation is provided by means of 3 uncoupled windings η incorporated in $[\Gamma]$, each of which is tightly coupled to the respective α winding. The leakage impedance between each winding pair (α and η) is electrically in series with Z_0 when viewed from the external network. Therefore it must be small in relation to the saturated value of Z_0. This is accomplished by computing the off-diagonal elements for $[\Gamma]$ (for windings on the same phase) based on the respective diagonal element $\Gamma^*_{\alpha\alpha}$ (using normalized values),

$$\Gamma_{\eta\alpha} = \Gamma_{\alpha\eta} = -\delta\,\Gamma^*_{\alpha\alpha} \tag{6}$$

Here, the constant δ is a measure of the coupling between the two windings relative to that between winding α and all others. It is set to 10^3 by default, physically implying that on transformers with concentric windings, the leakage impedance between windings α and η is 0.001 times that between the outermost pair of windings. Finally, the respective diagonal elements $\Gamma_{\alpha\alpha}$ and $\Gamma_{\eta\eta}$ reflect coupling to the fictitious winding, and maintain matrix singularity,

$$\begin{aligned}\Gamma_{\alpha\alpha} &= (1+\delta)\,\Gamma^*_{\alpha\alpha} \\ \Gamma_{\eta\eta} &= \delta\,\Gamma^*_{\alpha\alpha}\end{aligned} \tag{7}$$

The network consisting of Z_0 and Z_k is formed at the terminals of windings η, as depicted in Figure 2. Magnetizing branches Z_b modelling the wound limbs are connected directly across the innermost winding, which is strictly valid only if winding resistance is neglected (or it is excluded from [R] and connected externally). This approximation avoids having to introduce yet another fictitious winding.

C. Excitation Current in Zero- and Positive-Sequence Tests

Zero-sequence excitation current is presumed to be the same in each phase for 3-limbed units, governed by Z_0 in the model. Its significance according to the duality model is the relation it establishes between positive- and zero-sequence short-circuit impedances, by providing an additional shunt path for zero sequence current at node α. Thus its value is easily calculated from specified short-circuit tests, performed from winding α to any other short-circuited winding "m",

$$Z_0 = \left(\frac{1}{\overset{o}{z}_{\alpha m}} - \frac{1}{\overset{+}{z}_{\alpha m}}\right)^{-1} \tag{8}$$

This is useful since a measured value for Z_0 is usually not available. Equation (8) presumes that no other windings remain connected in delta during the zero-sequence test. On this basis, the model permits zero-sequence excitation current to be established by default.

Positive-sequence excitation current for 3-limbed units is governed by Z_b and Z_k, since Z_0 and the winding leakage impedances are small in comparison. The model exhibits centre-phase symmetry due to symmetrically arranged yoke impedances Z_k, predicting equal current in the outer phases. The current in the centre phase is somewhat lower, depending on the relative values of Z_b and Z_k. Since Z_b and Z_k model limb reluctance (and shunt losses), their values are assumed to be related by $\gamma = l_k / l_b$, denoting the ratio of yoke length to wound limb length (set to 0.5 by default). Noting that each Z_k models an upper and lower yoke section,

$$Z_b = 2\gamma\,Z_k \tag{9}$$

Phase currents may be calculated analytically (see Appendix), yielding for the centre phase,

$$i_{ctr}(t) = \left(\frac{1}{3} + \frac{1}{2\gamma}\right)\frac{v_{ctr}(t)}{Z_k} \tag{10}$$

neglecting Z_0 and the winding leakage impedances. This permits Z_b and Z_k to be computed using (9) and (10) to produce a specified excitation current in the centre phase (neglecting non-linearities). The current in the outer phases relative to the centre phase,

$$\frac{|i_{outer}|}{|i_{ctr}|} = \frac{\sqrt{28\gamma^2 + 30\gamma + 9}}{2\gamma + 3} \tag{11}$$

is about 39% higher for $\gamma = 0.5$.

For 5-limbed units, the computation of Z_0, Z_b and Z_k is more complex due to differences in limb cross-sections, but they are estimated on a similar basis.

D. Modelling Facilities and Limitations

The formulation has been implemented for transformers with wye- or delta-connected windings, taking as input the same data needed for existing models, excluding non-linearities. For single-phase, and 3-phase 5-limbed units, this includes a set of N(N-1)/2 short-circuit impedances for only positive-sequence excitation (since zero-sequence tests are identical), producing an Isolated Leakage (Network) Model, as described. For 3-limbed units, supplementary data is needed to define zero-sequence short-circuit impedances, yielding two alternate versions of the model, as follows.

(i) Coupled Leakage Model: A set of N(N-1)/2 zero-sequence short-circuit impedances may be specified if available, and if it is desired that the model match these precisely. In this event, the formulation provides for leakage networks to be magnetically coupled, as defined by (5), producing [Y] of order 3(N+1). Pre-processing of the specified data (Section IV-A) is needed to build [Y], requiring that no more than one winding be connected in delta during tests.

(ii) Isolated Leakage Model: Only one zero-sequence test value may be specified, if a full set is not available. This determines Z_0 in accordance with (8), producing a physically consistent model. Leakage networks for each phase are decoupled (or isolated). The resulting model consists of three identical [Y], of order (N+1); one for each phase. Zero-sequence short-circuit impedances are established by default, based on Z_0 and its topological connection in the core model. Since no pre-processing of test data is needed, there is no restriction involving delta connected windings in zero-sequence tests.

Alternatively, only the zero-sequence excitation current may be specified. Since this corresponds to Z_o, the resulting model is comparable to the above.

If excitation losses are specified for positive and zero-sequence tests, suitable resistances are inserted in shunt with the magnetizing reactances. Winding resistances may be specified directly, or for 3-winding transformers they can be computed from specified load losses. Turns ratios are incorporated in [Γ] based on specified winding voltage ratings. The computed model comprises [R], [Γ], and a topologically connected network of (at present) linear magnetizing branches in a form accepted directly by the EMTP. Provision for generating suitable non-linear branches is contemplated.

Finally, it should be emphasized that the established topological interface is correct only if the innermost and outermost windings are concentric, fully covering the wound limb. For transformers not meeting this requirement, such as units with vertically displaced windings innermost on the core, or units with the tapped section of a high-voltage winding innermost on the core, an additional wye-connected winding should probably be introduced at the interface if core non-linearities matter.

V. VALIDATION & COMPARISON WITH EXISTING EMTP MODELS

Table I displays measured short-circuit impedances for a 3-phase core transformer having concentric HV and LV disk windings, and a tertiary (delta connected) layer winding innermost on the core. Zero-sequence excitation current is estimated as 34% on rating based on specified positive- and zero-sequence short-circuit test values. Measured positive-sequence excitation current is 0.03% on rating. EMTP Models [2] for this transformer were derived using TRELEG, BCTRAN, and the topology based model (TOPMAG). This requires specifying 3 positive-sequence, and 3 of the 6 available zero-sequence short-circuit impedances. Winding resistances were neglected. Modelled results are compared with measurements in Table II for 3 test cases, as discussed. Only zero-sequence test results are presented, since positive-sequence results are about the same.

Test Case 1 demonstrates that comparable accuracy is possible with TOPMAG (Coupled Leakage Model) and TRELEG in the absence of nonlinearities. Although short-circuit impedances are specified for only three of the six available tests, modelled results are in agreement with all measurements, confirming that the additional tests are redundant. Note that since positive-sequence short-circuit impedance $z^{\Delta}_{hx(+)}$ was not measured explicitly, its value was taken to equal $z^{\Delta}_{xh(+)}$. If a full complement of zero-sequence tests were not available, the Isolated Leakage Network version of the TOPMAG model would provide reasonable results. The displayed results, in this case, correspond to a model derived by computing I^o_{excit} from specified short-circuit tests $z_{hy(+)}$ and $z_{hy(o)}$, and allowing the remaining impedances to be established by default. Finally, since BCTRAN does not correct for the shunting effect of the zero-sequence magnetizing branch, simulated short-circuit impedances are always lower than specified. For best results using BCTRAN, the magnetizing branch should only be specified at the innermost winding terminal.

TABLE I
Factory Measured Short Circuit Impedances
500/240/28-kV, 750-MVA Autotransformer

Test Configuration $z_{excited, shorted}$	Measured Impedance (%) Positive Seq.	Zero Sequence
z^{Δ}_{hx} (Tertiary closed)	Not available	12.82
z_{hy}	33.30	29.89
z_{xy}	18.00	17.21
z^{Δ}_{xh} (Tertiary closed)	13.78	7.36
z_{hx} (Tertiary open)	Not available	12.82
z_{xh} (Tertiary open)	Not available	13.63

Test Case 2 supports the premise that Z_o, or equivalently the zero-sequence excitation current I^o_{excit} (which is the specified input for TOPMAG), is not an independent parameter. It establishes a relation between positive- and zero-sequence short-circuit tests. Experience with a limited number of units suggests that I^o_{excit} can range to 300% of rated current. A somewhat arbitrary specification of 100% in this example is shown to reduce the overall modelling accuracy. This error can matter in transient studies, even if the transformer is operated with a normally closed tertiary, since the zero-sequence leakage between the HV and LV windings is incorrect. For example, if a single line-to-ground fault were simulated at HV winding terminals, with balanced excitation on LV windings, Case #1 predicts a circulating current in the tertiary of 870 A, and Case #2 predicts 2.7 kA (using TOPMAG or TRELEG). Evidently, since the "x" to "h" leakage impedance is unduly large for Case #2, zero-sequence current favours the tertiary winding. Simplified calculation by symmetrical components predicts about 810 A.

Test Case 3 demonstrates that, TRELEG does not correctly accommodate some short-circuit test connections at this time. Zero-sequence tests are considered reciprocal, which appears to be an assumption in its implementation rather than a limitation in the formulation. A consequence is that modelled winding leakages can be in error, varying in magnitude depending on the value of Z_o relative to the leakages. For correct results at this time, specified zero-sequence tests must correspond to a short-circuited inner winding (i.e. the winding to which a higher number is assigned for data entry). In the cited example, specified test value $z^{\Delta}_{xh(o)}$ is reproduced in the $z^{\Delta}_{hx(o)}$ test, causing both to be incorrect. BCTRAN is less restrictive, making the correct distinction provided additional delta connected winding(s) are present during the specified test. Otherwise, a test value specified for $z_{xh(o)}$ is interpreted as $z_{hx(o)}$ instead (not shown in Table II).

Finally, if positive and zero-sequence short-circuit tests are related by Z_o, then the zero-sequence short-circuit data, after processing by TOPMAG, should correspond closely with the respective positive-sequence values. This is confirmed by Column 7 in Table II (under TOPMAG), showing good agreement for Case #1 and Case #3 (compare with corresponding values in Table I). In comparison, the results for Case #2 differ considerably, which is a consequence of an inappropriate choice for I^o_{excit}. Similar disagreement is expected if specified positive and zero-sequence short-circuit impedances are inconsistent, serving as a check on data integrity.

For completeness, we include one case study involving core nonlinearities. Figure 4 illustrates magnetizing currents computed using TOPMAG (Case #1, Table II) and BCTRAN. For physical consistency using BCTRAN, the non-linearity associated with the outer phases was scaled in accordance with the longer magnetic path due to the wound-limb and yoke sections. The delta connection on the tertiary winding was opened, since otherwise it would mask the 3-limbed core model being examined here. Current waveshapes predicted using the BCTRAN model are similar among phases, scaled in proportion to the magnetic path length. In comparison, phase currents predicted by TOPMAG exhibit greater interaction due to coupling provided by the core. Computed neutral currents exhibit these differences more clearly. Both models predict an appreciable third harmonic current, but BCTRAN also predicts a significant 60 Hz component, which is at odds with TOPMAG. Such differences may matter in harmonic studies,

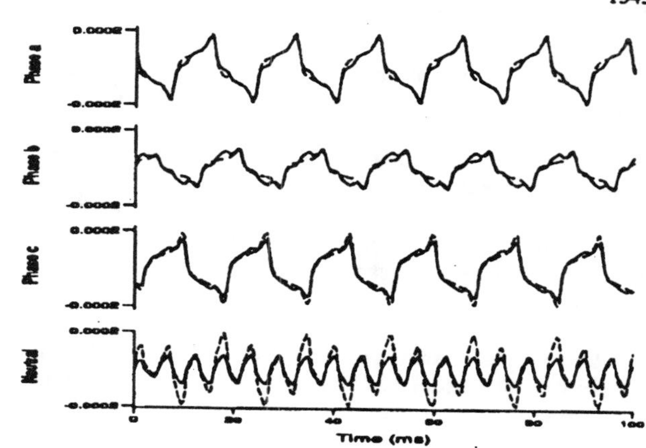

Figure 4: Simulated magnetizing current (in pu) for 1.0 pu excitation on 500-kV windings: TOPMAG —— BCTRAN - - -

for example involving Geomagnetic Induced Currents (GIC).

VI. CONCLUSIONS

A modelling formulation is described for 3-phase core type transformers which permits iron flux and air flux components to be modelled and interfaced on a topologically correct basis. It therefore accommodates core hysteresis, saturation, and frequency dependence due to eddy-current effects on a physical basis for each core limb. The approach is suitable for practical

TABLE II
Comparison of Modelled and Measured Zero-Sequence Short-Circuit Impedances

Test Case #	Short Circuit Test	Specified Short-Cct Impedance (Zero Seq)	Normalized Error (%) BCTRAN		TRELEG†	TOPMAG Coupled Leakage Model		Isolated Leakage Model
Specified Location of Magnetizing Branch -->			HV	Tertiary	HV	Processed Test Value	Normalized Error (%)	Normalized Error (%)
(1)	Z^{Δ}_{hx}	12.82	-4.2	0.00	0.00	13.4351	0.00	2.5
	Z_{hy}	29.89	-9.2	0.00	0.00	33.2690	0.00	0.1
	Z_{xy}	17.21	-5.3	0.00	0.00	18.3211	0.00	-1.7
$I^o_{excit.}$ = 34%	Z^{Δ}_{xh}	—	0.3	0.3	0.3	--	0.3	1.0
	Z_{hx}	--	-4.2	0.00	0.2	--	0.2	2.7
	Z_{xh}	—	-5.9	-10.0	-1.3	--	-1.4	1.1
(2)	Z^{Δ}_{hx}	12.82	-11.4	0.00	0.00	15.1384	0.00	-5.7
	Z_{hy}	29.89	-23.0	0.00	0.00	42.6294	0.00	0.3
	Z_{xy}	17.21	-13.1	0.00	0.00	21.3990	0.00	-10.8
$I^o_{excit.}$ = 100%	Z^{Δ}_{xh}	--	0.3	0.3	0.3	--	0.3	1.0
	Z_{hx}	--	-11.4	0.00	2.4	--	2.6	-5.5
	Z_{xh}	--	-5.9	-17.9	9.5	--	11.1	1.1
(3)	Z^{Δ}_{hx}	--	-4.4	-0.3	-42.6	--	-0.3	Same as Case #1
	Z_{hy}	29.89	-9.2	0.00	0.00	33.2690	0.00	
	Z_{xy}	17.21	-5.3	0.00	0.00	18.3235	0.00	
$I^o_{excit.}$ = 34%	Z^{Δ}_{xh}	7.36	0.00	0.00	-42.4	13.3959	0.00	
	Z_{hx}	--	-4.4	-0.3	-38.0	--	-0.1	
	Z_{xh}	--	-6.2	-10.3	-38.9	--	-1.7	

† Errors produced by TRELEG in Case #3 may be temporary, being caused at this time by an assumption that zero-sequence tests are reciprocal. Future versions of the model may correct this problem.

simulation of power system transients since it utilizes conventional short-circuit test data to produce a terminal model. A limitation is that zero-sequence short-circuit tests performed with more than one delta winding remaining closed are not always accommodated. It is also necessary that the innermost and outermost windings be concentric, fully covering the wound limb, if core non-linearities matter. The modelling approach appears suitable for other 3-phase core configurations (eg. shell type), though this is not discussed here. Among other findings, it is concluded that:

(1) Zero-sequence excitation current, which is specified as an input to the existing EMTP models, is not an independent parameter. It should not be specified arbitrarily.

(2) Positive- and zero-sequence short-circuit impedances are related by the zero-sequence excitation current. This permits (a) the latter to be estimated based on specified short-circuit test data, producing a more accurate model; and (b) a physically consistent model to be produced even in the absence of a full complement of zero-sequence short-circuit tests.

(3) Zero-sequence short-circuit tests are not reciprocal, hence care is required when specifying test data for building a model. Existing EMTP models do not always make this distinction, introducing errors without notice at this time if this limitation is overlooked.

ACKNOWLEDGEMENTS

This work was carried out as part of CEA Project 175 T 331G: Transformer Models for Electromagnetic Transient Studies, under funding from the EMTP Development Coordination Group (DCG).

REFERENCES

1) V. Brandwajn, H.W. Dommel, and I.I. Dommel, "Matrix Representation of Three-Phase N-Winding Transformers for Steady-State and Transient Studies", IEEE Transactions PAS 101, No. 6, June 1982, pp. 1369-1378.
2) H.W. Dommel, *Electromagnetic Transients Program Reference Manual (EMTP Theory Book)*, Bonneville Power Administration, Portland, Oregon, August 1986.
3) E.C. Cherry, "The Duality between Interlinked Electric and Magnetic Circuits and the Formulation of Transformer Equivalent Circuits", Proceedings of the Physical Society, Part 62, February 1949, pp 101-110.
4) H. Edelman, "Anschauliche Ermittlung von Transformator Ersatzschaltbildern" (*Demonstrative derivation of transformer equivalents*), Archiv. Electr. Übertrag, 1959, pp 253-261.
5) K. Schlosser, "Eine auf Physikalischer Grundlage ermittelte Ersatzschaltung für Transformatoren mit mehreren Wicklungen" (*An equivalent circuit for N-winding transformers derived from a physical basis*), BBC - Nachrichten, Vol. 45, March 1963, pp. 107-132.
6) E.P Dick and W. Watson, "Transformer Models for Transient Studies Based on Field Measurements", IEEE Transactions PAS, Vol. PAS-100, No. 1, January 1981, pp 409-419.
7) F. de Leon and A. Semlyen, "Reduced Order Model for Transformer Transients", IEEE Transactions on Power Delivery, Vol. 7, No. 1, January 1992, pp 361-369.
8) G.R. Slemon, Magnetoelectric Devices, Transducers, Transformers and Machines, John Wiley & Sons, 1966.
9) D.E. Hedman, "Theoretical Evaluation of Multiphase Propagation", IEEE Transactions PAS 90, No. 6, pp. 2460-2471, November/December 1971.
10) E.J. Tarasiewicz, A.S. Morched, A. Narang, and E.P. Dick, "Frequency Dependent Eddy Current Models for Nonlinear Iron Cores", IEEE Power Engineering Society Paper 92 WM 177-6-PWRS.
11) F. de Leon and A. Semlyen, "Time Domain Modelling of Eddy Current Effects For Transformer Transients", IEEE Power Engineering Society Paper 92 WM 251-9-PWRD.

BIOGRAPHIES

Arun Narang (M'75) received his B.A.Sc from the University of Waterloo in 1977. He has been at Ontario Hydro Research since 1979 studying power system transients, including high frequency measurement and modelling of windings on rotating machines and transformers. He is the project leader for CEA 175 T 331G: Transformer Models for ElectroMagnetic Transient Studies.

Russell H. Brierley (M'73) received his electrical engineering degree from Queens University in Kingston, Ontario in 1953. He has worked in various positions in Ontario Hydro System Planning Division, and has been conducting studies using the EMTP since 1969. He is a member of the Canadian National Committee of IEC TC28 on Insulation Coordination, and a co-author of CSA publication on the Principles and Practice of Insulation Coordination - C308.

APPENDIX

We use superposition to calculate the positive-sequence excitation current in each phase for 3-limbed units. Neglecting Z_o and the winding leakage impedances, the model (Figures 1, 2) reduces to a balanced set of wound-limb impedances Z_b, and an unbalanced set of yoke impedances Z_k. Excitation current in each phase due to Z_b is simply $v(t)/Z_b$. The current due to Z_k is obtained from the loop impedance matrix [v] = [Z] [i] relating source voltages and currents. Denoting the outer phases as "a" & "b", the matrix equation is inverted in closed form to yield source currents due to Z_k for balanced excitation,

$$\begin{aligned} i_a(t) &= \frac{v_a(t)}{Z_k} + \frac{v_b(t)}{3Z_k} \\ i_b(t) &= \frac{v_b(t)}{3Z_k} \\ i_c(t) &= \frac{v_c(t)}{Z_k} + \frac{v_b(t)}{3Z_k} \end{aligned} \tag{A-1}$$

which superimpose on those due to Z_b.

Discussion

B.C. PAPADIAS, N.D. HATZIARGYRIOU, J.M. PROUSALIDIS (National Technical University of Athens, Greece): We wish to congratulate the authors for their comprehensive paper on a timely subject. The development of three-phase transformer models suitable for transient studies has been the subject of many recent investigations including [1-3]. A number of three-phase transformer models, other than those reported in the paper, suitable for transient studies have been suggested (see ref. in [3] below). Inductive switching in particular, provides a major area of transient phenomena and is a suitable field of testing transient transformer models. It would be interesting to simulate such phenomena using the suggested model.

The authors have achieved in providing a considerable enhancement of BCTRAN-model which takes into account zero-sequence impedance and asymmetrical mutual phase coupling. The representation of zero-sequence impedance and asymmetries of the core structure is important in transient studies like switching overvoltages [1-3]. In addition the effect of winding capacitances is critical. Could the authors suggest where and how winding capacitances should be placed? Moreover, we would appreciate their comments on the following points:

1) In order to represent saturation of the outer limbs in the 5-limbed core case, it is suggested in the paper to replace Z_0 by a linear inductance matching its saturated value. Would it not be more precise if a non-linear inductance was used instead?

2) Could the authors clarify the following:

i) In step B-3 of B-class case, an iterative method is used to calculate L_{am} and $L_{m\Delta}$. What is the difference between L_{am} and L_{ak} (for k=m) already obtained in step B-2?

ii) Although it is claimed that the core network is electrically isolated from the leakage network, the magnetizing impedance Z_b appears directly connected to the leakage network of the innermost network (Figure 2).

The authors are strongly encouraged to continue their work and validate their model with measured results.

REFERENCES

[1] B.C. PAPADIAS, N.D. HATZIARGYRIOY, J.A. BAKOPOULOS: "The effect of transformer parameters in the switching overvoltages of reactor loaded transformers", Colloquium of CIGRE SC13, Serajevo, May 1989.

[2] N.D. HATZIARGYRIOU, J.M. PROUSALIDIS, B.C. PAPADIAS, :"A Generalized Transformer Model Based on the Analysis of its Magnetic Circuit", IEE Proceedings-C Generation, Transmission & Distribution, July 1993.

[3] B.C. PAPADIAS, N.D. HATZIARGYRIOY, J.A. BAKOPOULOS, J.M. PROUSALIDIS : "Transformer Models For Fast Electromagnetic Transients", paper No 93 SM 396-2 PWRD, IEEE 1993 Summer Meeting, Vancouver (Canada).

Manuscript received August 9, 1993.

HERMANN W. DOMMEL (The University of British Columbia, Vancouver, B.C., Canada) and IRMGARD I. DOMMEL (Power Systems Consultants, Vancouver, B.C., Canada): As the developers of the support routine BCTRAN, we appreciate it very much that the authors clearly identified a problem related to the zero sequence magnetizing impedance. There is no doubt that the short-circuit impedances must be corrected for the loading effect of the magnetizing impedance Z_0 in the zero-sequence circuit, because Z_0 is of the same order of magnitude as the short-circuit impedances.

In case of data redundancy, there is more than one way of exploiting it to arrive at the final model. If the "non-reciprocal" values from the zero-sequence short-circuit tests Z_{hx} (tertiary open) and Z_{xh} (tertiary open) are used as the starting point, then the short-circuit impedances can be modified in a slightly different way, as explained in the following steps (resistances ignored):

1) Assume that the three-winding transformer is described by the three-branch star-circuit of Figure A, with Z_0 connected to the high side. This star-circuit is also used in BCTRAN to make corrections for closed delta connections.

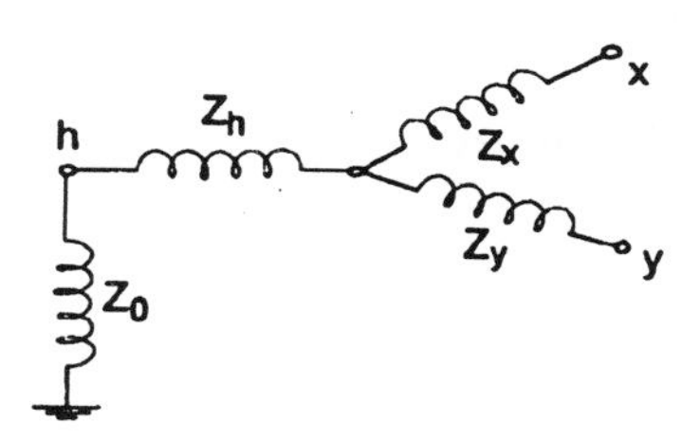

Figure A. Star-circuit for zero-sequence parameters.

2) With the tertiary open, Z_{xh} is equal to $Z_x + Z_h = 13.63$ %, and Z_{hx} will be the parallel combination of Z_0 and $(Z_h + Z_x)$, or
$$1/Z_0 + 1/13.63 = 1/12.82.$$
This produces $Z_0 = 215.7242$ %, or $I^0_{exci} = 46.35548$ %. This value is then used as input to BCTRAN. Note that it differs somewhat from the 34 % value derived in the paper from comparisons between positive and zero sequence test values.

3) Knowing Z_0, $Z_h + Z_y$ can easily be found for input into BCTRAN, because $1/Z_0 + 1/(Z_h + Z_y) = 1/29.89$, or
$$Z_h + Z_y = 34.697576 \text{ \%}.$$
This value is then used as input for the short-circuit impedance between h and y.

4) The correction of $Z_x + Z_y$, to account for the loading effect of Z_0, can be done as follows: since Z_{hx} is identical with and without the tertiary closed, according to the test data, Z_x must be close to zero, because then Z_y does not contribute anything to Z_{hx}. If $Z_x \approx 0$, then $Z_h \approx 13.63$, or
$$1/(13.63+215.7242)+1/Z_y = 1/17.21.$$
This produces a value of $Z_y \approx 18.606$, which we used as input to BCTRAN for the short-circuit impedance between x and y (assuming $Z_x \approx 0$).

5) After running BCTRAN, we calculated the short-circuit impedance between x and y from the [Z]-matrix, which turned out to be 17.20036 instead of 17.21 (already fairly close to correct value). We then iteratively increased the input value from 18.606 to 18.79644, until we practically obtained 17.21. This iteration was done "by hand", but it could easily be automated inside the program.

With the final [Z]-matrix, the following four measured values are matched exactly:

$$Z_{hx} = 12.82 \text{ (zero error)},$$

$Z_{hy} = 29.89$ (zero error),
$Z_{xy} = 17.21$ (zero error),
$Z_{xh} = 13.63$ (zero error).

The two remaining impedances Z^{Δ}_{hx} *and* Z^{Δ}_{x} will not be completely correct, but will have small errors. To calculate these impedances, one can either find Z_h, Z_x, Z_y from the three input values first,

$Z_h = 14.765568$ %
$Z_x = -1.135568$ %
$Z_y = 19.932008$ %

and then derive the two impedances from them, or read the two impedances directly from the admittance matrix output, with

$Z^{\Delta}_{hx} = 1/Y_{11} = 0.127592$,
$Z^{\Delta}_{xh} = 1/Y_{22} = 0.073465$.

The error in Z^{Δ}_{hx} found either way is -0.47 %, and the error in Z^{Δ}_{xh} is -0.18 %, which are both very small. Z^{Δ}_{hx} and Z_{hx} cannot be completely identical anyhow, unless Z_x is exactly zero in the equivalent circuit.

Instead of steps (4) and (5), one could also have used Z^{Δ}_{xh} as the input value, because this value does not have to be corrected for Z_0. In that case Z_{xy} would have been slightly off from the measured value.

Manuscript received August 10, 1993.

Francisco de León (Instituto Politécnico Nacional, México). The authors are to be congratulated for their very interesting transformer model which can be constructed from normally available data. This is a very useful feature that most models lack. Another important characteristic of the model is that it properly represents the interface between the iron and air fluxes which is a flaw in the models currently present in the EMTP.

I would like to raise a question on the frequency range of validity of the transformer model developed in the paper. The reasons for my concern is that: the parameters of the model are calculated from tests at 60 Hz; and, the currents in the windings are considered the same in all points (no internal capacitances are included). I estimate that the model, as it is, will not give reliable results for transients beyond a few kilohertz. It will not be able to predict the variation of the transformer's terminal impedance over a wide range of frequencies. Although the inductances representing the windings could be divided into several sections to insert the internal capacitances, the calculation of the latter will require some knowledge of the transformer design geometry. The authors comments on this matter would be very much appreciated.

Another point of concern is the incorporation of the frequency dependency to the parameters of the transformer model. The methodology for the representation of eddy current effects given in references [10] and [11] will be relatively easy to include in the iron core inductances L_o and L_k of Figure 1. The calculation of the parameters for the equivalent circuits needs the laminations data but a good estimation can be made. However, the proper modeling of eddy currents effects in the windings is a much more complicated problem. It requires the correct evaluation of the skin and proximity effects which vary from point to point within a winding. There exist several equations [A] for simplified geometries and low frequency performance which cannot be used for high frequency transients [11]. There is a methodology to include the effects of eddy currents in the winding [B] which requires the design parameters. Will the authors kindly comment on the modeling of the damping due to eddy currents in the windings and in the iron core?

In spite of the above remarks, the model presented in the paper should give very reliable results for low frequency transients. The authors should be commended for this achievement.

[A] M. Perry, "Low Frequency Electromagnetic Design", Marcel Dekker, Inc., 1985.

[B] F. de León and A. Semlyen, "Detailed Modeling of Eddy Current Effects for Transformer Transients", paper No. 93 WM 395-4 PWRD presented at the 1993 IEEE/PES Summer Meeting.

Manuscript received August 10, 1993.

Cesare M. Arturi (Politecnico di Milano, Italy). I wish to congratulate the authors for their contribution to the still partially unsolved problem of transformer modelling with the EMTP program.

The paper presents a transformer model, based on normally available experimental data, which takes into account the core structure. The three-phase core-type transformer is specifically considered and the model should be suitable for simulation of both steady-state and slow transients, i.e., with capacitive effects neglected.

In the majority of the practical cases, the transformers have three windings at most and the test data normally available are represented by the positive-sequence binary short-circuit tests and the no-load excitation test. Some time, the zero-sequence short-circuit and excitation test results are also available. With these test data, obtained considering the three-winding terminals, only a model with a 3x3 leakage inductance matrix or its inverse is possible. Thus, in spite of the generality outlined in the paper, where a number of N windings is considered, a simpler case with N=3 is usually of interest.

I would also add that, in my opinion, when a transformer model is to be made, the first attention should be placed on the equivalent inductive network, which represents the behaviour of the magnetic field. Secondly, the improvement of the model is possible to take into account the loss effects in the windings and the core. However, the two steps should be taken separated.

The inductive network can be obtained either considering the actual geometry of both the windings and the core (topological model) or considering the transformer as a black-box, by using the test data made on the winding terminals (external-effects equivalent model). The duality procedure gives a topological model for the leakage magnetic field (in air) and for the coupling magnetic field (confined in the iron core). Instead, the matrix of the inverse of the nodal inductances $[\Gamma]$, obtained from the binary short-circuit test data, represents a black-box model of the leakage magnetic field.

The authors have chosen a hybrid model, where the topological model of the coupling magnetic field (in the core) is connected to the black-box model of the leakage magnetic field (in air). Furthermore, for the three-phase model, eq.(5) expresses the link between the positive and zero sequence impedances and the self and mutual impedances. I would observe that eq.(5) are only valid for three-phase *linear* component and, in order to avoid confusion with loss effects, *self and mutual inductances* and *not impedances* should always be considered. If the component is highly non linear, as in the case of transformers, the incremental or the differential inductances should be used to express the relationship between the linkage flux and the excitation current. Therefore, it seems not appropriate the use of the impedance concept in non linear component. Since eq.(5) are here used for a transformer, the authors could indicate which linear inductance has been chosen and why.

I would observe that the magnetic network used for the duality procedure consider a core without loss. This is implicit in the definition of the scalar magnetic potential and the reluctance of a magnetic branch. Consequently, it seems improper to substitute the non linear inductances Lo, Lk, Lb of the equivalent network of Fig.1b with the impedances Zo, Zk, Zb of the circuit of Fig.2, including both loss and non-linearity.

I would also observe that the inductance Lo of the core-type transformer, associated with the flux reclosing outside the external winding, partially in the air and partially in the tank is non linear. Also for this inductances a linear value is apparently used in the paper. Would the authors justify the reason of their choice?.

Finally, I would note that if the presented model has to simulate correctly the transient behaviour in very high saturation conditions, then it seems that some linear inductances should be connected in parallel to the non linear inductances associated to the wound limbs and yokes [1], in order to consider the notable air flux around the iron branches when high saturation occurs. Could the authors comment this point?

Finally, a better degree of details would be appreciated, in order to allow the calculation of the numerical results presented to validate the proposed approach.

[1] C.M.Arturi :"Transient simulation and analysis of a three-phase five limb step-up transformer following an out-of-phase synchronisation", IEEE Trans. on Power Delivery, Jan. 1991, pag.196.

Manuscript received August 11, 1993.

XUSHENG CHEN (Department of Electrical Engineering, Seattle University, Seattle, WA 98122-4460, USA). The authors are to be commended for their research and for the informative results presented. this discussor has been following the authors' work for several years, and has learned a lot from their papers. However, I would like to make some comments and ask some questions.

1. The basis for this transformer model is duality between a magnetic circuit and an electric circuit. while duality is valid for linear magnetic circuits (at low frequency), its extension to the nonlinear magnetic circuits has never been rigorously proved. this discussor has met divergence problems of duality derived models for single-phase modern power transformers.

2. The neglect of the excitation current of a transformer will cause large errors in simulating switching and ferroresonance overvoltages of an unleaded transformer. The L-matrix can be formed directly by the formulas shown in Reference A. The BPA EMTP can accurately invert an L-matrix of a transformer with excitation current as low as .002%.

3. My work shows that the L-matrix for a typical three-phase three-legged transformer have the following feature: $L_{AA} = L_{CC} = .75\ L_{BB}$, and $L_{AB} = 2\ L_{AC}$. Therefore, to assume $L_{AB} = L_{AC} = L_m$ as is done in all versions of EMTP is not valid for unloaded transformers. Strictly speaking, symmetrical components is not applicable to unloaded transformers.

4. I incorporated my three-phase, two-winding, five-legged core transformer model (named SEATTLE TRANSFORMER) into BPA's EMTP in April 1993. the model is topology based [A]. It accepts terminal voltages from EMTP and returns EMTP with an up-dated L-matrix. The difference between this paper and my model is that I choose flux linkages of the windings as the state variables and solve the currents at each time step from the flux linkages using the magnetic circuit. Preliminary simulation results show that the model is very efficient and robust. since my model represents another approach for multi-legged transformer modeling, the comments from the authors will be greatly appreciated.

It is hoped that these questions and comments are useful to the authors and the others.

References:

[A] X.S. Chen and Paul Neudorfer, *A Digital Model for Transient Studies of a Three-Phase, Five-Legged Transformer*, IEE Proceedings-C, Generation, Transmission, and Distribution Vol. 139, No. 4 (July 1992), pp. 351-358.

Manuscript received August 16, 1993.

ARUN NARANG: We would like to thank the discussers for their interest in the paper as expressed by their thoughtful discussions. Our response to each follows, in the order received.

Profs. Papadias and colleagues point out the need for introducing capacitive effects for studies involving reactive switching. The reported formulation permits capacitive effects to be introduced in the same manner as existing models, not offering a novel alternative. The conventional approach of connecting terminal capacitances appears reasonable for frequencies approaching the first winding resonance frequency. For greater accuracy, and higher frequencies, more detailed models seem necessary [A, B]. In response to the numbered questions:

(1) We suggest only that Z_0 corresponding to the centre limb be eliminated in the equivalent network (Figure 2), since

it does not correspond to a physical limb; the remaining two "Z_0" in the network represent the outer limbs, and are therefore modelled by non-linear inductances.

(2) i) Processing of the specified zero-sequence test data is structured in a particular sequence, aimed at minimizing the number of "unknowns" at each step. This is intended to make the procedure more robust, and less likely to fail due to divergence in the iterative schemes. In our limited experience, the solution has failed to converge only in instances where the specified data is itself inconsistent. With this perspective, step B-2 applies to those specified tests which do not involve a closed delta-connected winding. Tests that do involve a closed delta are processed in step B-3; not at B-2. Thus a particular test is processed at either B-2 or B-3, but not at both.

ii) Magnetizing impedance Z_b is connected to the leakage network as an approximation, as indicated in Section IV-B of the paper. This nevertheless leaves the core network (top half of Figure 2) electrically isolated, meaning that the governing equations relating voltage and current for Z_b, Z_0 and Z_m are not altered by external disturbances.

We appreciate comments from Drs. Dommel, whose pioneering work culminating in BCTRAN is the basis of TOPMAG. The new formulation merely manipulates a different set of algebraic equations to process the specified test data based on some new physical insights. These insights yield additional benefits, permitting physically consistent models to be assembled by exploiting implied data redundancy. The sample calculation presented by Drs. Dommel illustrates several additional points, including that (a) L_0, as defined, cannot be determined uniquely; and (b) a relatively large error in its value can be tolerated without affecting appreciably the overall "fit" between specified test data and the model. A more precise calculation for L_0 is therefore not warranted.

We agree with Dr. de Leon completely, and can offer little additional insight. The presented model accommodates magnetic effects by establishing a physically meaningful interface between iron flux and air flux. Model elements Z_m, Z_k, Z_0 and winding resistances (Figure 2) can be viewed as specialized branches, incorporating algorithms mirroring the relevant physical mechanisms. In this way, the model provides a facility to accommodate eddy current effects in iron, and proximity effects in the windings. Dr. de Leon's recent publications on these topics are noteworthy, and seem suitable for interfacing directly with the reported formulation.

We thank Prof. Arturi for providing additional insight, and respond as follows:

(i) Equation (5) in the paper is used in reference to short-circuit impedances, relating applied MMF and leakage flux in air. It is therefore used in connection with (linear) leakage inductances.

(ii) We suggest a less restrictive interpretation of the duality derivation, whereby a branch element "Z", numerically a complex value, relates voltage and current in the electrical network. It corresponds physically to the relation existing between flux in the iron and the MMF required to establish this flux. To the extent that the applied MMF and the derivative of flux (corresponding to voltage) are not necessarily in quadrature, "Z" does not restrict voltage and current to be related by an inductive element.

(iii) Inductance L_0 is indeed non-linear in general, as displayed in Figure 2. For 3-limbed units, it is probably adequate to define it as a two-section, piecewise characteristic (e.g. see Reference [6] of the paper). The lower section corresponds to flux levels below those which cause the tank walls to saturate. This also represents the condition existing during short-circuit tests, which are the basis for our estimate. Consequently the quoted value for L_0 represents its unsaturated value, which is all that matters for most practical studies. Zero-sequence flux is unlikely to reach levels needed to saturate the tank walls in almost all practical applications, with the possible exception of studies involving large inrush currents on transformers not equipped with a delta connected winding.

(iv) The proposed linear inductance is implicit in the specified non-linearity, giving the curve a finite slope in saturation.

(v) While we have provided all the test data needed to duplicate illustrated results, and the steps involved in processing the specified test data, the algebraic equations involved in the iterative procedures have been omitted due to space constraints. In this respect, the sample calculation presented by Drs. Dommel should serve as a valuable guide.

Dr. Chen's first comment appears to echo Dr. Arturi's concern regarding the validity of duality for nonlinear circuits. Without responding rigorously here, we suggest simply that the duality principle yields an electric circuit which corresponds precisely to the presumed magnetic circuit. Thus, for instance, the correspondence between Figure 1(a) and 1(b) is rigorous and exact. No error is introduced here by applying duality. The approximation that the discussers imply is actually associated with an earlier step in the formulation, requiring conceptually discretizing the magnetic circuit formed by the transformer assembly into the equivalent network displayed in Figure 1(a). This step involves presumption of a particular field distribution, consistent with the intended application of the resulting model. Our own assumptions have been provided in the paper. In response to the subsequent numbered comments:

(2) We note for clarification, that excitation current is neglected only in the computation of the "leakage matrix". It is accommodated separately by the core model, comprising Z_m, Z_k and Z_0 (Figure 2).

(3) The relative inductance values quoted here are properly interpreted as unsaturated values governing magnetizing current. These depend on the relative limb dimensions, as

allowed for by Equation (9) in the paper. The TOPMAG formulation computes Z_m and Z_k (and Z_o for 5-limbed units) based on specified (or default) values of relative limb dimensions and excitation current.

(4) The referenced model employs a complete matrix formulation using magnetic state variables, therefore not invoking duality explicitly. It is rigorous, completely valid, and an analogous alternative to our formulation. Our choice among the two alternatives was based on convenience, since the dual electric network is easily introduced into the EMTP time-step loop using existing facilities, viz. as coupled, multi-phase branch elements.

Once again, we thank the discussers for their valued contributions, enhancing our presentation.

REFERENCES

[A] A. Morched, L. Marti, J. Ottevangers, "A High Frequency Transformer Model for EMTP", IEEE 1992 Summer Power Meeting, Paper 92 SM 359-0-PWRD, Seattle, WA, July 1992.

[B] M. Gutierrez, R.C. Degeneff, P.J. McKenny and J.M. Schneider, "Linear, Lumped Parameter Transformer Model Reduction Technique, IEEE 1993 Summer Power Meeting, Paper 93 SM 394-7 PWRD, Vancouver, Canada, July 1993.

Manuscript received October 1, 1993.

A High Frequency Transformer Model for the EMTP

A. Morched (SM) L. Martí (M) J. Ottevangers

Ontario Hydro, Canada

Abstract - A model to simulate the high frequency behaviour of a power transformer is presented. This model is based on the frequency characteristics of the transformer admittance matrix between its terminals over a given range of frequencies. The transformer admittance characteristics can be obtained from measurements or from detailed internal models based on the physical layout of the transformer. The elements of the nodal admittance matrix are approximated with rational functions consisting of real as well as complex conjugate poles and zeroes. These approximations are realized in the form of an RLC network in a format suitable for direct use with EMTP. The high frequency transformer model can be used as a stand-alone linear model or as an add-on module of a more comprehensive model where iron core nonlinearities are represented in detail.

Keywords - Transformer, High frequency, Frequency dependence, Electromagnetic transients, EMTP.

1. INTRODUCTION

The transformer is probably one of the most familiar components of a power system, but it is also one of the most difficult to model accurately. A recent survey comparing EMTP simulations with field measurements indicates that studies where transformer behaviour has the greatest influence on the results are those where EMTP simulations tend to be the least accurate[1].

To model a transformer in a transient simulation, nonlinear behaviour as well as frequency-dependent effects must be taken into account. Standard EMTP transformer models such as BCTRAN and TRELEG[2] can accurately reproduce the response of a transformer at the frequency at which the short-circuit and open-circuit tests are made; namely, at power frequency. However, these models do not account for the frequency dependence of copper and iron losses, or the effect of stray capacitances.

The behaviour of the transformer at higher frequencies can be approximated, to some extent, by modelling the distributed stray capacitances along the windings with lumped capacitances connected across the terminals of the transformer. This type of representation cannot reproduce the behaviour of the transformer beyond the first resonance frequencies. The calculation of the capacitances is not straightforward, and it is difficult to obtain accurate values, except for simple transformer designs[3,4].

92 SM 359-0 PWRD A paper recommended and approved by the IEEE Transformers Committee of the IEEE Power Engineering Society for presentation at the IEEE/PES 1992 Summer Meeting, Seattle, WA, July 12-16, 1992. Manuscript submitted February 3, 1992; made available for printing May 1, 1992.

A substantial number of transformer models have been proposed to date. While it is probably inaccurate to categorize all work done in high frequency transformer modelling, it is convenient to identify two broad trends to describe the model presented in this paper within the context of earlier work.

1) Detailed internal winding models. This type of model consists of large networks of capacitances and coupled inductances obtained from the discretization of distributed self and mutual winding inductances and capacitances[5,6]. The calculation of these parameters involves the solution of complex field problems and requires information on the physical layout and construction details of the transformer. This information is not generally available as it is considered proprietary by transformer manufacturers. These models have the advantage of allowing access to internal points along the winding, making it possible to assess internal winding stresses. In general, internal winding models can predict transformer resonances but cannot reproduce the associated damping. This makes this class of models suitable for the calculation of initial voltage distribution along a winding due to impulse excitation, but unsuitable for the calculation of transients involving the interaction between the system and the transformer. Furthermore, the size of the matrices involved (typically 100 x 100 or larger) makes this kind of representation impractical for EMTP system studies.

2) Terminal models. Models belonging to this class are based on the simulation of the frequency and/or time domain characteristics at the terminals of the transformer by means of complex equivalent circuits or other closed-form representations[7-11]. These "terminal" models have had varying degrees of success in reproducing the frequency behaviour of single-phase transformers accurately. The main drawback of the methods proposed to date appears to be that they are not sufficiently general to be applicable to three-phase transformers.

The high-frequency transformer model described here belongs to the class of models where the frequency dependent response at the terminals of the transformer is reproduced by means of equivalent networks. Unlike earlier frequency-dependent transformer models, the new model can simulate any type of multi-phase, multi-winding transformer as long as its frequency characteristics are known either from measurements or from calculations based on the physical layout of the transformer. The generation of the parameters for the model is automatic, and it does not require special skills on the part of the user. This model has been developed and implemented at Ontario Hydro as part of a new and comprehensive transformer model sponsored by the EMTP Development Coordination Group - DCG for the DCG/EPRI version of the EMTP. Although originally developed as a high-frequency representation, this model can also be used as a stand-alone linear model, if the frequency characteristics of the transformer are known over a sufficiently broad frequency range.

2. OVERVIEW

Consider a multi-phase, multi-winding transformer. The nodal equations which relate the voltages and currents at the accessible terminals of the transformer can be expressed as

$$[Y][V] = [I] \tag{1}$$

where the nodal admittance matrix [Y] is complex, symmetric, and frequency dependent. In a three-phase system, equation (1) can be expressed as

$$\begin{bmatrix} Y_{11} & Y_{12} & \cdots & Y_{1m} \\ Y_{21} & Y_{22} & \cdots & Y_{2m} \\ \vdots & \vdots & & \vdots \\ Y_{m1} & Y_{m2} & \cdots & Y_{mm} \end{bmatrix} \begin{bmatrix} V_1 \\ V_2 \\ \vdots \\ V_m \end{bmatrix} = \begin{bmatrix} I_1 \\ I_2 \\ \vdots \\ I_m \end{bmatrix} \tag{2}$$

$$[Y_{ij}] = \begin{bmatrix} y_{ij,aa} & y_{ij,ab} & y_{ij,ac} \\ y_{ij,ba} & y_{ij,bb} & y_{ij,bc} \\ y_{ij,ca} & y_{ij,cb} & y_{ij,cc} \end{bmatrix} \tag{3}$$

where $[Y_{ij}]$ is a 3×3 sub-matrix and m is the number of three-phase terminals under consideration. For example, the [Y] matrix for a two-winding, three-phase, Y-Y transformer with grounded neutrals would be of order six, with $3m\cdot(3m+1)/2 = 21$ distinct elements. The elements of the nodal admittance matrix can be obtained from measurements, or they can be calculated from a detailed winding model over a given frequency range.

The basic idea behind the new transformer model is to produce an equivalent network whose nodal admittance matrix matches the nodal admittance matrix of the original transformer over the frequency range of interest. Such representation would correctly reproduce the transient response of the transformer at its terminals. Consider then the multi-phase network shown in Figure 1. This network will be referred to as a multi-terminal π-equivalent. The parameters of this circuit can be calculated from its nodal admittance matrix using the well-known relationships

$$[Y_{ii,\pi}] = \sum_{j=1}^{m} [Y_{ij}] \tag{4}$$

and

$$[Y_{ij,\pi}] = -[Y_{ij}] \tag{5}$$

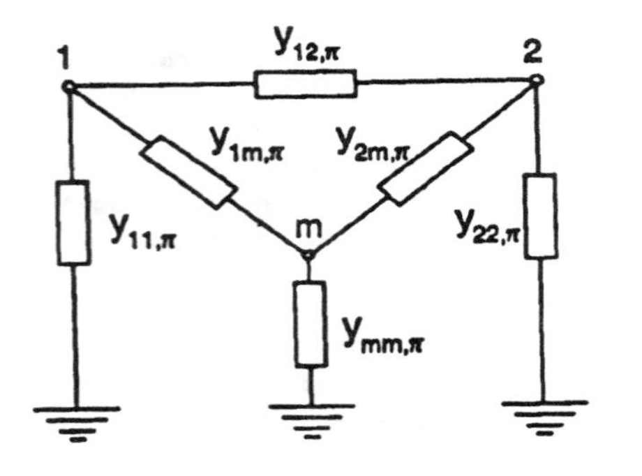

Fig. 1: *Single-line diagram of a multi-terminal π-equivalent.*

The elements of $[Y_{ij,\pi}]$ are approximated with rational functions which contain real as well as complex conjugate poles and zeroes. The rational functions can then be realized with RLC networks which can be combined using (4) and (5) to produce the parameters of the equivalent π-circuit.

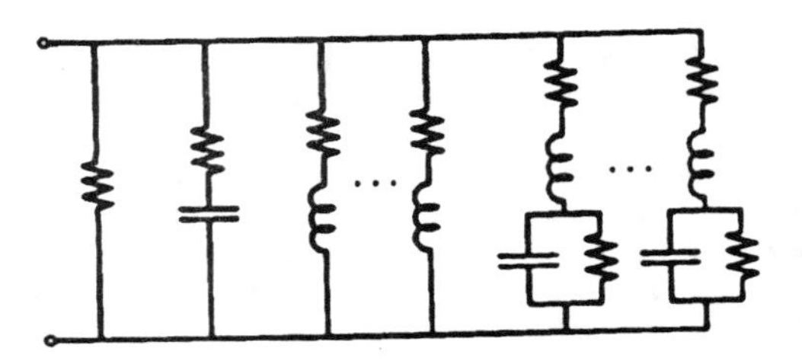

Fig: 2: *Structure of an RLC module.*

A typical RLC network used in the approximation of the elements of $[Y_{ij,\pi}]$ is shown in Figure 2. The general structure of these RLC networks reflects the known frequency characteristics of the admittance functions of a transformer:

- Inductive behaviour at low frequencies which includes frequency dependent effects due to skin effect in the windings and iron core eddy current losses. These are simulated by the RL branches.
- Series and parallel resonances from mid to high frequencies caused by winding-to-winding and winding-to-ground stray capacitances. These are reproduced by the RLC branches.
- Predominantly capacitive behaviour at high frequencies, represented by the single RC branch.

3. PRACTICAL CONSIDERATIONS

The transformer model must be sufficiently robust to produce consistent and numerically stable equivalent networks even when data are obtained from noisy or inconsistent measurements. To this effect, the following steps are taken:

- Fitting the elements of $[Y_{ij}]$ and calculating $[Y_{ii,\pi}]$ by adding the fitted functions using (4), instead of fitting $[Y_{ii,\pi}]$ directly.
- Averaging the diagonal and off-diagonal elements of $[Y_{ij}]$ so that $[Y_{ij}]$ become balanced matrices.

The explicit approximation of the elements $[Y_{ii,\pi}]$, would result in models of lower order than those obtained by first approximating $[Y_{ij}]$ and then adding the results. However, it has been found that models obtained by fitting $[Y_{ii,\pi}]$ directly may be numerically unstable.

Averaging the diagonal and off-diagonal elements of $[Y_{ij}]$ results in

$$[Y_{ij}] = \begin{bmatrix} y_{ij,aa} & y_{ij,ab} & y_{ij,ac} \\ y_{ij,ba} & y_{ij,bb} & y_{ij,bc} \\ y_{ij,ca} & y_{ij,cb} & y_{ij,cc} \end{bmatrix} = \begin{bmatrix} y_{ij,s} & y_{ij,m} & y_{ij,m} \\ y_{ij,m} & y_{ij,s} & y_{ij,m} \\ y_{ij,m} & y_{ij,m} & y_{ij,s} \end{bmatrix} \tag{6}$$

If the sub-matrices of [Y] are balanced, they can be diagonalized by a constant transformation [Q] such that

$$[Q]^{-1}\,[Y_{ij}]\,[Q] = [Y_{ij,mode}]$$

$$[Y_{ij,mode}] = \begin{bmatrix} y_{ij,0} & 0 & 0 \\ 0 & y_{ij,1} & 0 \\ 0 & 0 & y_{ij,1} \end{bmatrix} \tag{7}$$

Subscripts s and m in (6) stand for "self" and "mutual", respectively; subscripts "o" and "1" in (7) stand for the familiar zero and positive sequence components. Matrix [Q] could be any of a number of transformation matrices which diagonalize a balanced matrix. Note that even though each sub-matrix in [Y] can be diagonalized, [Y] itself will not be diagonal. For example, for a three-phase, two-winding transformer

$$[Y_{mode}] = \begin{bmatrix} Q^{-1} & 0 \\ 0 & Q^{-1} \end{bmatrix} \begin{bmatrix} Y_{HH} & Y_{HL} \\ Y_{LH} & Y_{LL} \end{bmatrix} \begin{bmatrix} Q & 0 \\ 0 & Q \end{bmatrix} = \begin{bmatrix} Y_{HH,mode} & Y_{HL,mode} \\ Y_{LH,mode} & Y_{LL,mode} \end{bmatrix}$$

$$[Y_{mode}] = \begin{bmatrix} y_{HH,o} & 0 & 0 & y_{HL,o} & 0 & 0 \\ 0 & y_{HH,1} & 0 & 0 & y_{HL,1} & 0 \\ 0 & 0 & y_{HH,1} & 0 & 0 & y_{HL,1} \\ y_{LH,o} & 0 & 0 & y_{LL,o} & 0 & 0 \\ 0 & y_{LH,1} & 0 & 0 & y_{LL,1} & 0 \\ 0 & 0 & y_{LH,1} & 0 & 0 & y_{LL,1} \end{bmatrix} \tag{8}$$

Introducing into equation (2) we finally obtain

$$\begin{bmatrix} Y_{HH,mode} & Y_{HL,mode} \\ Y_{LH,mode} & Y_{LL,mode} \end{bmatrix} \begin{bmatrix} V_{H,mode} \\ V_{L,mode} \end{bmatrix} = \begin{bmatrix} I_{H,mode} \\ I_{L,mode} \end{bmatrix} \tag{9}$$

where

$$[Q]^{-1} V_H = V_{H,mode} \; ; \qquad [Q]^{-1} I_H = I_{H,mode}$$

$$[Q]^{-1} V_L = V_{L,mode} \; ; \qquad [Q]^{-1} I_L = I_{L,mode}$$

The admittances to be approximated with rational functions are now the elements of $[Y_{mode}]$, instead of that the elements of $[Y_{ij}]$. Since the parameters of the positive and negative sequence networks are identical, the problem reduces to the fitting of only $m\cdot(m+1)$ distinct admittance elements, rather than $3m\cdot(3m+1)/2$.

Averaging the elements of $[Y_{ij}]$ to produce balanced matrices has obvious merits from the point of view of computational speed. For example, for a two-winding, three-phase transformer, 6 rather than 21 distinct functions would have to be approximated with RLC networks. Also, the time-step loop calculations in the EMTP are also substantially reduced. Averaging the elements of $[Y_{ij}]$ also adds some robustness and consistency to the raw measurements, and contributes further to the numerical stability of the model. While averaging may, in some instances, mask the effect of legitimate asymmetries in the transformer, the differences observed in the transformers studied appear to be relatively small (see Figure 3).

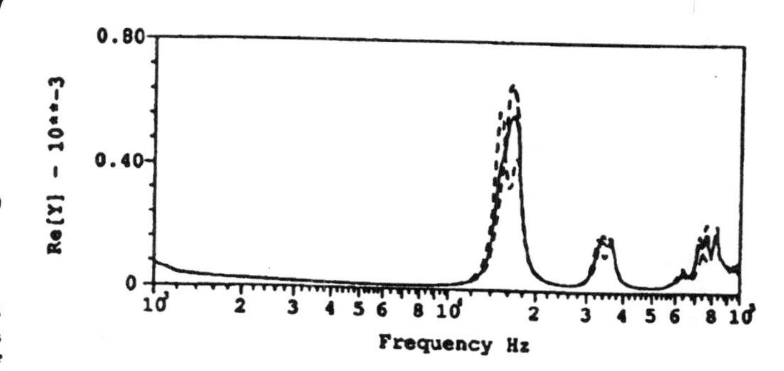

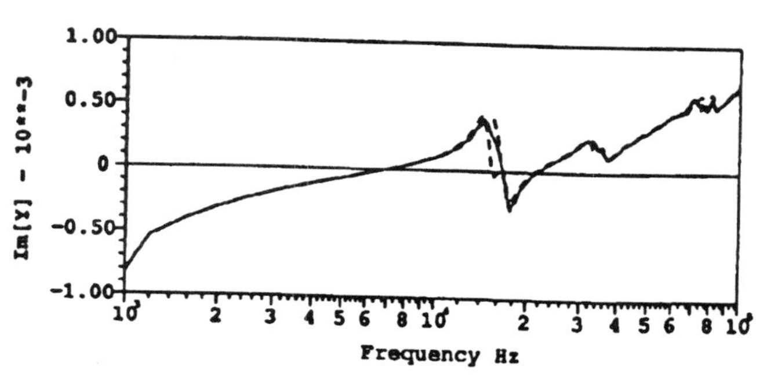

Fig. 3: *Diagonal elements of $[Y_{HH}]$. Solid trace: $y_{HH,s}$; Dashed traces: $y_{HH,aa}$, $y_{HH,bb}$, $y_{HH,cc}$*

To validate the transformer model, frequency domain as well as time domain measurements were conducted on a 125 MVA, 215/44 kV, three-limbed core-type transformer. The transformer is YY-connected with grounded neutrals at the high and low voltage sides. The transformer has a delta-connected tertiary winding with no accessible terminals.

The effect of averaging the elements of $[Y_{ij}]$ for the transformer indicated above, is illustrated in Figure 3, where the solid trace corresponds to the averaged functions, and the dashed traces correspond to the raw measurements.

4. FITTING PROCESS

The elements of $[Y_{mode}]$ in (8) are approximated with rational functions given by

$$Y(s) = Y_a(s) = Y_{RL}(s) + Y_{RC}(s) + Y_{RLC}(s) \tag{10}$$

$$Y_{RL}(s) = k_o + \sum_{j=1}^{NR} \frac{k_{RL,j}}{s - p_{RL,j}} \tag{11}$$

$$Y_{RC}(s) = \frac{s\,k_{RC}}{s - p_{RC}} \tag{12}$$

$$Y_{RLC}(s) = \frac{k_{RLC}\,(s - \gamma_o)}{(s - p_{NC})\,(s - p^*_{NC})} \cdot \prod_{i=1}^{NC-1} \left(\frac{(s - z_i)\,(s - z^*_i)}{(s - p_i)\,(s - p^*_i)} \right) \tag{13}$$

where k_o, $k_{RL,j}$, k_{RC} and k_{RLC} are real constants; p_{RL} and p_{RC} are real poles; γ_o is a real zero; p_i and z_i are complex poles and zeroes, and p^*_i and z^*_i are their respective complex conjugates. For the practical example given, the number of real poles NR and the number of complex conjugate poles NC in (11) and (13) are typically 6 and 15, respectively. All poles are confined to the left hand side of the complex plane and $s=j\omega$. $Y_a(s)$ can be described with the equivalent circuit shown in Figure 2, where Y_{RL} corresponds to the RL branches, Y_{RC} corresponds to the RC branch and Y_{RLC} corresponds to the RLC branches. The single resistive branch comes from k_o in equation (11), and its conductance is normally very small.

Let us now define f_{cut} as the frequency where the first parallel resonance of Y(s) occurs (see Figure 4). At frequencies below f_{cut}, the admittance functions behave as combinations of RL branches without resonances. At frequencies above f_{cut}, stray capacitances come into play and a number of resonances are present. Therefore, for an initial estimate of $F_a(s)$, it is assumed that the region between the first measured data point f_{min} and f_{cut} contains real poles only, while the region from f_{cut} to the last measured point f_{max} contains complex conjugate pairs only.

The steps followed in the approximation of Y(s) are:

1) Numerical noise in Y(s) is removed. Peaks whose magnitude fall below a user-controlled percentage of the largest peak in Y(s) are dismissed.

2) Initialize Y_{RC}. The response of an RC branch is

$$Re\{Y_{RC}(s)\} = \frac{R(\omega C)^2}{1+(\omega RC)^2}$$

$$Im\{Y_{RC}(s)\} = \frac{\omega C}{1+(\omega RC)^2}$$

The RC branch represents the asymptotic behaviour of the transformer at very high frequency. To calculate R and C, it is assumed that for the frequency range of interest $\omega RC << 1$ and that the imaginary part of $Y_{RC} \approx \omega C$. The value of C is found by

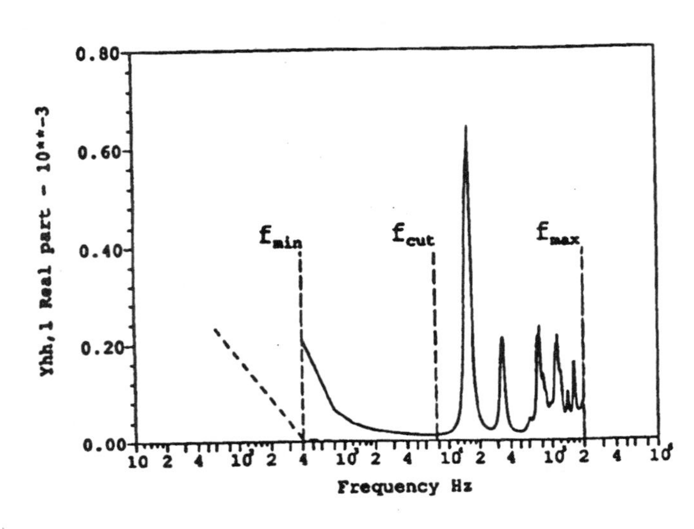

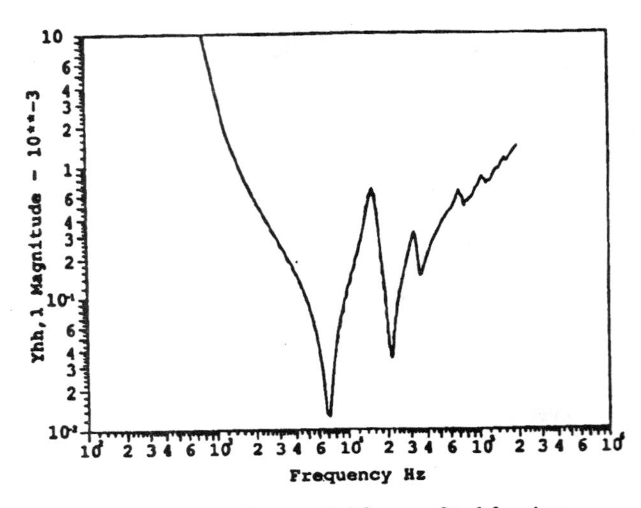

Fig. 5: *Approximation of $y_{HH,1}$. Solid trace: fitted function; Dashed trace: raw data.*

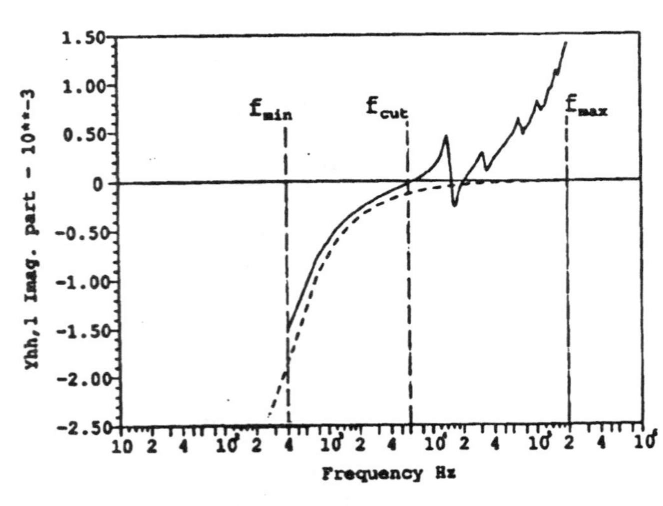

Fig. 4: *Element $y_{HH,1}$. Solid trace: raw data; Dashed trace: low frequency model.*

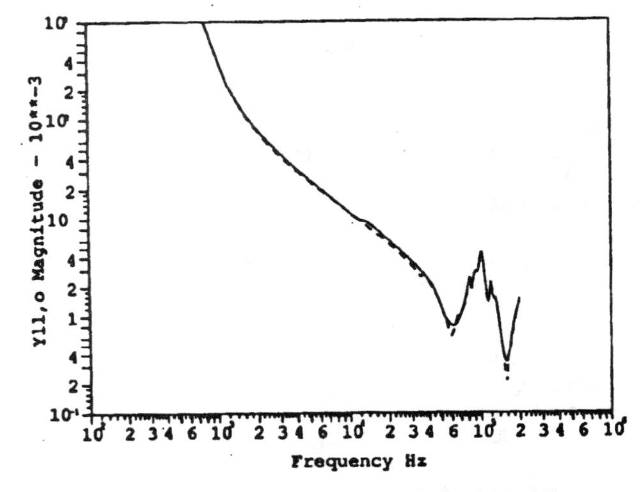

Fig. 6: *Approximation of $y_{LL,o}$. Solid trace: fitted function; Dashed trace: raw data.*

fitting the imaginary part of Y(s) with ωC in the least squares sense. With C known, R is found by matching Y_{RC} to a point on the lower envelope of the real part of Y(s). This technique is very simple but surprisingly effective: optimization seldom changes this initial estimate by more than five percent.

3) Initialize Y_{RLC} by identifying the local maxima and minima of the magnitude of the real part of Y(s). Each local maximum or peak corresponds to a complex conjugate pole $p_i=\alpha_i+j\beta_i$, and each local minimum or valley corresponds to a complex conjugate zero $z_i=\gamma_i+j\delta_i$. The angular frequency at which a maximum and minimum occurs determines β_i and δ_i ($\beta_i=2\pi f_{peak}$, $\delta_i=2\pi f_{valley}$). The real parts are arbitrarily initialized to 2.5% of their corresponding imaginary parts. The real zero γ_o is initialized to $\gamma_o=2\pi f_{cut}$. The number of poles and zeroes assigned is determined by the shape of Y(s) and by the tolerance that determines which peaks are considered meaningful.

3) Optimize the initial guess of $Y_a(s)=Y_{RC}+Y_{RLC}$, assuming $Y_{RL}=0$, over the frequency range from f_{cut} to f_{max} using a modified Marquardt algorithm[12]. The error function is defined as the magnitude of the difference function Y(s)-Y_a(s).

4) Initialize Y_{RL}(s) from 60 Hz to f_{cut} using the asymptotic fitting procedure described in the Appendix. This initialization algorithm does not require that the function to be approximated be a minimum phase-shift function in order to produce an accurate initial fit. The default number of poles is four, but this number is under user's control.

5) Optimize the entire function Y_a(s) from 60 Hz to f_{max} using the same optimization algorithm indicated in item 3 above.

During the optimization process, all poles are confined to the left hand side of the complex plane. Zeroes are not so constrained. This often leads to the realization of branches with negative values of R, L and C. Nevertheless, because of the constraints indicated above, these branches still have a positively damped response.

The entire fitting process is fully automatic and no user input or special skills are necessary to initialize it. Some of the fitting parameters can be overridden by the user to control the number of poles and zeroes of Y_a(s) and to control the desirable error levels in the approximations.

5. USAGE AS AN ADD-ON MODULE

When the high frequency model is used as an add-on module of a more complex representation with linear and nonlinear components, the response of the low frequency components Y_{low} must be subtracted from the measured response Y_{raw} before it is approximated. In a typical application, the frequency response of the low frequency nameplate model (e.g., BCTRAN or TRELEG), including iron core losses, is subtracted.

The subtraction of the effect of the low frequency models introduces some complications if the difference between their low frequency response and the measurements is not negligible in the transition region around the lowest measured frequency f_{min} (see Figure 4). Ideally, an add-on high frequency model should have no effect on the response at power frequency. If the difference at f_{min} is too large, no causal rational function will be able to approximate the transition region and still produce a negligible contribution at 60 Hz. In these cases, the approximation in the neighbourhood of f_{min} will be somewhat degraded.

When the response of the low frequency model is subtracted from the measured data, the region between 60 Hz and f_{min} defines a transition area where the magnitude of Y(s)=Y_{raw}-Y_{low} is determined by the consistency between measured data and the low frequency model and where the magnitude of Y(s) at 60 Hz should be zero. If the low frequency model were accurate from 60 Hz to f_{min} and the measurements were error free, then the magnitude of F(s) would be very small at f_{min}. The smaller the difference the better the fit around f_{cut}. If the difference is very large, then is not possible to approximate this transition region accurately and some compromises are necessary, namely, a larger error between f_{min} and f_{cut} and a relatively large Y_a(s).

A high frequency model was developed for the measured transformer as an add-on module to a TRELEG model of the same transformer. Figures 5 and 6 show the approximation of the elements $Y_{HH,o}$, and $Y_{LL,1}$ produced by the combined model. These illustrate the best and the worst fits, respectively, obtained for this particular transformer.

6. TRANSIENT RESPONSE

With the elements of [Y] available in a closed form, inclusion of the high frequency model in the EMTP is conceptually straightforward. Each branch in the equivalent network shown in Figure 1 is represented by a constant conductance matrix in parallel with a past history current source. During a transient simulation, modal voltages and currents are calculated from terminal voltages and the uncoupled sequence networks are solved to produce an updated set of modal history current sources. These current sources are transformed into phase quantities and used by the EMTP for the solution of the system in the next time step.

The approximations generated with the techniques described above, are ultimately combined using equations (1) and (2) to produce the branches of the equivalent sequence representation of the transformer. These networks can be simulated in the EMTP by means of the new FDB (Frequency Dependent Branch) model. This model was designed as a general-purpose tool to simulate multi-phase coupled RLC networks in the EMTP. The type of networks which can be modelled with the FDB model are more general than those required by the high frequency transformer model. In fact, even the EMTP implementation of the FDNE (Frequency dependent Network Equivalent) is a sub-set of the FDB model.

It is not within the scope of this paper to describe the implementation of the FDB model in the EMTP or Ontario Hydro's experience with the new high frequency transformer model. Due to space limitations these topics will have to be part of a separate paper.

To verify the EMTP implementation and to validate the developed model, a comparison between simulated versus measured transients on the same 125 MVA, 215/44 kV unit used earlier was conducted. Figure 7 shows the measured response of the transformer measured on phase 1 of the high voltage terminals when a step voltage is applied on phase 3 of the high voltage terminals. All other terminals are grounded. Figure 8 shows the results of the corresponding EMTP

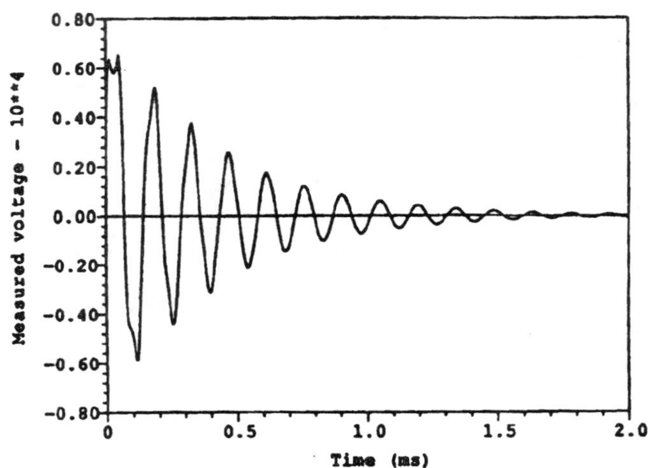

Fig. 7: *Step response. Field test.*

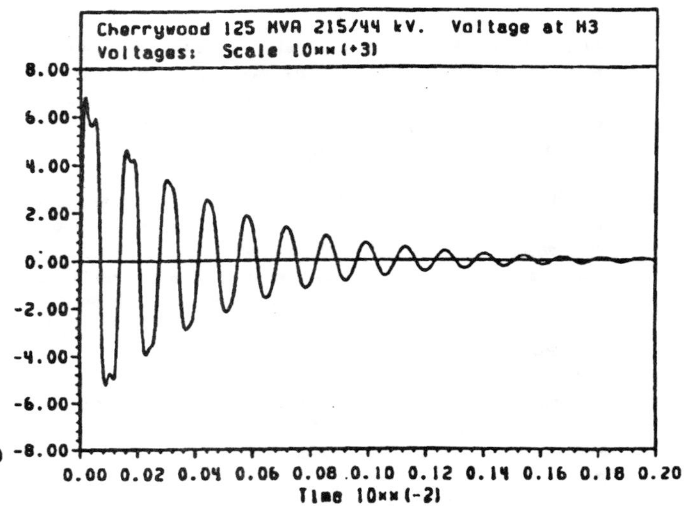

Fig. 8: *Step response. EMTP simulation.*

transient simulation. Numerical stability has been verified for this and other similar tests by allowing the simulation to run for extremely long times (several seconds).

7. CONCLUSIONS

This paper presents a model to simulate the behaviour of a multi-phase, multi-winding transformer over a wide frequency range. This model reproduces the behaviour of the transformer by means of combinations of RLC networks that match the frequency response of the transformer at its terminals. The frequency response of the transformer is assumed to be known from measurements, or from calculations with models based on geometry and construction details. Its most important features are:

1) It can be used to model multi-winding, multi-phase transformers for which the frequency response is known.

2) It can be used as an add-on module for a more complex transformer representation. It can also be used as a stand-alone linear model if the frequency response of the transformer is known over a sufficiently wide frequency range.

3) The fitting techniques developed to approximate the admittance functions of the transformer produce approximations of exceptional quality.

4) Validation tests performed indicate that the models produced are accurate and numerically stable.

5) The process to generate parameters for the model is completely automatic: no special skills or experience are required from the user.

Acknowledgements

The authors would like to acknowledge the use of the Marquardt optimization routine from the Harwell Subroutine Library. Also we would like to thank A. Narang from Ontario Hydro's Electrical Research Division for providing the measurements for the transformer used in the numerical examples. Thanks are also due to CEA for permission, on behalf of DCG, to publish this work. Funding for this project was provided by DCG.

REFERENCES

[1] J. Skliutas, and J. Panek, "Electromagnetic Transients Program (EMTP) - Field Test Comparisons". *EPRI EL-6768*, March 1990.

[2] H. W. Dommel, *Electromagnetic Transients Program Reference Manual (EMTP THEORY BOOK)*, Printed by The University of British Columbia, Vancouver B.C., Canada, August 1986, pp. 6-62 - 6-63.

[3] R. C Degeneff, "A Method for Calculating Terminal Models of Single Phase n-winding transformers". Paper No. A 78 539-9 presented at the IEEE PES Summer meeting in Los Angeles, July 1978.

[4] T. Adielson, A. Carlson, H. B. Margolis, and J. A. Hallady, "Resonant Overvoltages in EHV Transformers - Modelling and Application", *IEEE Transactions on Power Apparatus and Systems*, vol. PAS-100, pp. 3563-3572, July 1981.

[5] P. I. Fergerstad and T. Henriksen, "Inductances for the Calculation of Transient Oscillations in Transformers", *IEEE Transactions on Power Apparatus and Systems*, vol. PAS-93, No. 2, pp. 500-509, March/April 1974.

[6] R. C. Degeneff, "A General Method for Determining Resonances in Transformer Windings", *IEEE Transactions on Power Apparatus and Systems*, vol. PAS-96, No., pp. 423-430, March/April 1977.

[7] R. C. Degeneff, W. S. McNult, W. Neugebauer, J. Panek, M. E. McCallum, and C. C. Honey, "Transformer Response to System Switching Voltages", *IEEE Transactions on Power Apparatus and Systems*, vol. PAS, No. 6, pp. 1457-1470, June 1982.

[8] P. T. M. Vaessen, "Transformer Model for High Frequencies", *IEEE Transactions on Power Delivery*, vol. 3, No. 4, pp. 1761-1768, October 1988.

[9] Q. Su, R. E. James, and D. Sutanto, "A Z-Transform Model of Transformers for the Study of Electromagnetic Transients in Power Systems", *IEEE Transactions on Power Systems*, vol. 5, No. 1, pp. 27-33, February 1990.

[10] A. Keyhani, H. Tesai, and A. Abur, "Maximum Likelyhood Estimation of High Frequency Machine and Transformer Winding Parameters", *IEEE Transactions on Power Systems*, vol. 5, No. 1, pp. 212-219, January 1990.

[11] A. Keyhani, S. Chua, and S. Sebo, "Maximum Likelyhood Estimation of Transformer High Frequency Parameters from Test Data", *IEEE Transactions on Power Delivery*, vol. 6, No. 2, pp. 858-865, April 1991.

[12] D. W. Marquardt, "An Algorithm for Least-Square Estimation of Nonlinear Parameters", *J. Soc. Indust. Appl. Math*, vol. 11, No. 2, pp. 431-441, June 1963.

[13] J. R. Martí, "Accurate Modelling of Frequency-Dependent Transmission Lines in Electromagnetic Transient Calculations". *IEEE Transactions on Power Apparatus and Systems*, pp. 147-157, January 1982.

[14] L. Martí, "Low-order approximation of Transmission Line Parameters for Frequency-DependentModels". *IEEE Transactions on Power Apparatus and Systems*, pp. 3584-3589, November 1983.

APPENDIX

To approximate a minimum phase-shift function H(s) with a rational function P(s) that contains only real poles and zeroes which lie in the left hand side of the complex plane, it is sufficient to match the magnitude functions of H(s) and P(s). This is possible because the phase angle of a minimum phase-shift function is uniquely determined by its magnitude function: if |H(s)| and |P(s)| match, their phase angles will also match.

$$H(s) = P(s) = k_o \prod_{i=1}^{N} \frac{(s - z_i)}{(s - p_i)}$$

A very effective technique to match the magnitude of a minimum phase-shift function is suggested in [13] and [14]:

1) Subdivide the magnitude function into N equally-spaced segments. These segments define the location of the horizontal asymptotes h_i, $(i=1,...,N+1)$ of |P(s)|.

2) Place the corresponding vertical asymptotes at the frequency where |H(s)| equals the geometric mean of two adjacent horizontal asymptotes.

3) The initial location of poles and zeroes is defined by the intersection of vertical and horizontal asymptotes.

4) Optimize the initial location of the poles and zeroes by reducing the error function in the least squares sense.

This technique can be extended to approximate any analytical, non-minimum phase-shift function. The basic premise is that the imaginary part of an analytical function is uniquely determined by its real part. The modified method proceeds as follows:

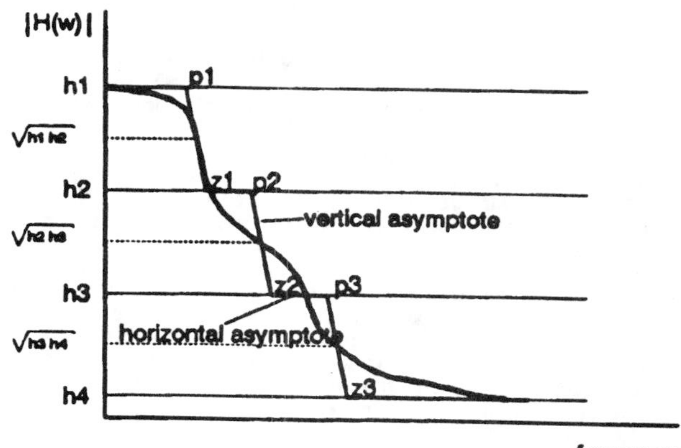

1) Calculate R(s), where

$$R(s) = \sqrt{Re\{H(s)\} - C}$$

where C is an arbitrary constant such that $Re\{H(s)\} > 0 \; \forall \; \omega \geq 0$

2) Use the fitting technique described above to approximate R(s) with R'(s).

3) Calculate the partial fraction expansion of the rational function R'(s)

$$R'(s) = k_o + \sum_{i=1}^{N} \frac{k_i}{(s - p_i)}$$

4) The approximation of H(s) is then given by

$$H(s) = k'_o + \sum_{i=1}^{N} \frac{k_i/p_i}{(s - p_i)}$$

where $k'_o = k_o + C$

BIOGRAPHIES

Atef S. Morched (M'77-SM'90) received a B.Sc. in Electrical Engineering from Cairo University in 1964, a Ph.D. and a D.Sc. from the Norwegian Institute of Technology in Trodheim in 1970 and 1972. He has been with Ontario Hydro since 1975 where he currently holds the position of Section Head - Electromagnetic Transients in the Power System Planning Division.

Luis Martí (M'79) received an undergraduate degree in Electrical Engineering from the Central University of Venezuela in 1979, MASc and PhD degrees in Electrical Engineering in 1983 and 1987, respectively, from The University of British Columbia. He did post-doctoral work in cable modelling in 1987-1988, and joined Ontario Hydro in 1989, where he is currently working in the Analytical Methods & Specialized Studies Department of the Power System Planning Division.

Jan H. Ottevangers received an MSc. in Electrical Engineering from the Delft Institute of Technology in 1956. He has been with Ontario Hydro since 1967 where he is currently working in the Analytical Methods and Specialized Studies Department of the Power System Planning Division.

Discussion

Q. Su (Monash University, Clayton, Australia): The authors are to be congratulated in having presented a comprehensive high frequency transformer model of equivalent networks. In electrical power systems, the transient overvoltages of high voltage power transformers, either at the terminals or inside the windings, are of great importance for the reliability of electricity supply. The model developed by the authors will be useful for the study of system transients in which the high frequency characteristics of transformer winding are to be considered.

Obviously, the model represented by a number of R, L. C components can easily fit in EMTP programs. For a detailed internal winding model, several hundred components may be used resulting in a large size of matrix impractical for EMTP system studies, as mentioned in the paper. The authors' RLC module in Figure 2 consists of at least 15 components and 12 such modules are used to represent a three-phase, two-winding transformer. It is therefore necessary to simulate each transformer of interest in power system with a network of 180 or more RLC components. Would this be a problem with EMTP system studies?

In my previous papers [1, 2], a closed-form transformer high frequency model was presented, as shown in Figure A(a). Extended to the mode form in Figure A(b), the model has also been used for three phase transformers. From my experience, the computing time for system transient studies increases significantly for a transformer represented by RLC networks rather than closed-form models.

Another question concerns the higher frequency response of a transformer under step voltages. The functions in Figures 5 and 6 fit measured data up to about 200 kHz and the calculated step voltage response in Figure 8 agrees with the measured in Figure 7. This confirms the fitting accuracy of the authors' method. Could the authors indicate the rise time of the step voltage and the time step interval used for the calculations of the step voltage responses?

References

[1] Q. Su, R. E. James and D. Sutanto, "A Z-transform Model of Transformers for the Study of Electromagnetic Transients in Power Systems", Co-authored with R. E. James and D. Sutanto, *IEEE Transactions on Power System*, No. 1, Vol. 5, 1990, pp. 27–33.

[2] Q. Su and T. Blackburn, "Application of Z-Transform Method for Study of Lightning Protection in Electrical Power Systems," *Proceedings of the 7th International Symposium on High Voltage Engineering*, Dresden, Germany, Aug. 26–30, 1991, pp. 139–142.

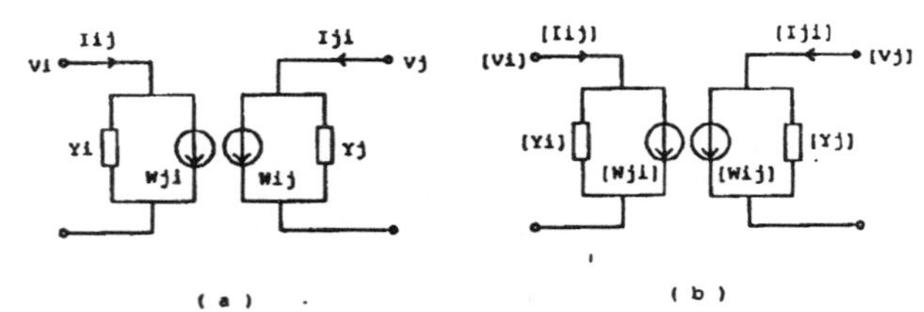

Fig. A A close-form high frequency model for two-winding, (a) single-phase and (b) three-phase transformers.

Adam Semlyen (University of Toronto): This is an interesting and useful paper as it solves the problem of providing a realistic model for multi-phase transformers for the purpose of EMTP simulations. It has benefited of the authors' expertise and experience in fitting stable circuit models to frequency domain data [A]. One of its outstanding features is the modal decomposition they have used: it has not only simplified the problem of fitting but, more importantly, it has reduced the dynamic size of the model to that of a minimal realization. The following remarks and questions are mainly related to the problem of modal decomposition.

I note that the authors assume that the off-diagonal block $[Y_{ij}]$ can be adjusted to become a balanced matrix. This, of course, implies an approximation. (The central phase may, for instance, have somewhat different parameters than the other two phases.) Then, as a result of the balancing, a real, constant, transformation $[Q]$ can be used to obtain the desired modes. After fitting, the modal approximations are transformed back to the original phase domain. They now correspond to the given matrix $[Y]$ of frequency domain measurements. My first question is whether a final refinement of the fitted results is performed or contemplated for obtaining a best match with the original set of data, in order to compensate for the approximation made by the initial balancing process?

The transformer connection used in the paper is Y-Y, with the particular feature that the $[Y_{ij}]$ block can in fact be balanced by a small adjustment. This is so because the connection does not produce an internal phase shift. When this is not the case, for instance in the important class of Y-Δ or Y-Z (zig-zag) connected transformers, the off-diagonal block $[Y_{ij}]$ has cyclic symmetry. For instance, in one particular Y-Δ connection (with an admittance y associated to each phase of the Y-connected winding), we have

$$[Y_{12}] = y\begin{bmatrix} 0 & 1 & -1 \\ -1 & 0 & 1 \\ 1 & -1 & 0 \end{bmatrix}, \qquad [Y_{21}] = [Y_{12}]^T$$

Eigenanalysis of this matrix leads to the symmetrical component transformation matrix with three, rather two, decoupled modes. Balancing would yield the zero matrix. An α-type input gives a β-type output and vice versa (as expected, see for instance [B]; thus the (real) Clarke transformation does not result in modal decoupling). Positive or negative sequence voltages result in currents of the same sequence with the expected phase rotation. In the modal domain there is of course no strict symmetry, as reciprocity now implies a rotation in the opposite direction if the voltages are applied to the secondary rather than to the primary winding.

Clearly, transformers with internal phase shifting effects pose more complex problems. Could the authors please elaborate on their thoughts regarding the solution of these problems?

Finally, I wish to reassert my appreciation regarding the merits of this paper and would like to congratulate the authors for their fine contribution.

[A] A.S. Morched, J.H. Ottevangers, and L. Marti, "Multi-Port Frequency Dependent Network Equivalents for the EMTP", IEEE paper no. 92 SM 461-4 PWRD, presented at the 1992 IEEE/PES Summer Meeting, in Seattle, WA.

[B] Edith Clarke, "Circuit Analysis of A-C Power Systems, Volume I: Symmetrical and Related Components", John Wiley & Sons, Inc., New York, 1943.

Manuscript received July 27, 1992.

X. Chen (Department of Electrical Engineering, Seattle University, Seattle, WA): This paper is very impressive in scope and in detail. The authors and their organization must be commended for their contributions to the accurate modeling of the high frequency behavior of multi-winding, multi-phase transformers. The fitting techniques to approximate the admittance functions of a transformer is both novel and practical. This discussor has learned a lot from their paper and the authors' earlier papers on transformer modeling. I would appreciate the authors' comments on the following questions:

(a) I have developed a computer program which can form the inductance matrix for a two-winding, three-phase, multi-legged transformer.

The inductance matrix for the primary winding of an unsaturated three-phase three-legged transformer computed by BCTRAN (pages XIX-C-15 to 20, ATP Rule Book, 1987–1992, BPA) is shown in Eqn. (A).

$$\begin{pmatrix} L_{aa} & L_{ab} & L_{ac} \\ L_{ba} & L_{bb} & L_{bc} \\ L_{ca} & L_{cb} & L_{cc} \end{pmatrix} = \begin{pmatrix} 879.72 & -438.02 & -483.02 \\ -438.02 & 879.72 & -438.02 \\ -438.02 & -438.02 & 879.72 \end{pmatrix} \text{ Henry (A)}$$

The inductance matrix computed by my program is shown in Eqn. (B).

$$\begin{pmatrix} L_{aa} & L_{ab} & L_{ac} \\ L_{ba} & L_{bb} & L_{bc} \\ L_{ca} & L_{cb} & L_{cc} \end{pmatrix} = \begin{pmatrix} 879.80 & -584.68 & -292.14 \\ -584.68 & 1172.06 & -584.68 \\ -292.14 & -584.68 & 879.80 \end{pmatrix} \text{ Henry (B)}$$

It is striking to note that L_{ab} is two times greater than L_{ac}, and L_{bb} is 1.33 times greater than L_{aa} and L_{cc}. Because of the asymmetry of the iron core of a three-legged, core-type transformer, my work is very possibly correct. If this is the case, then Eqs. (6) to (8) of the authors' paper might not be valid. It is common practice to represent a transformer by its sequence impedances for short circuit analysis. To apply the symmetrical components method to an unloaded three-phase core-type transformer is not always valid, even if there is no saturation involved.

(b) Figures 7 and 8 of the paper showed the comparison between the computed and measured step response of a 125 MVA, 215/44 kV transformer. The applied step voltage is much lower than the rated voltage of the high voltage terminals. The main objective of developing high frequency transformer models is to study transformer overvoltages caused by switching and lightning. Although many researchers claimed that magnetic saturation of the core has minor influence on fast transients, and therefore can be disregarded, this discussor is interested in knowing if the authors have compared the results of their model to the field test or measurements for overvoltages caused by a lightning surge on a transmission line which is connected to a transformer and operating at rated voltage. This discussor has a strong opinion that harmonic analysis is valid for linear and slightly nonlinear systems. Wherever severe nonlinearity is involved, differential equations should be used and nothing else.

Again, the authors are to be congratulated for their effort in developing a comprehensive transformer model.

Manuscript received August 7, 1992.

H. M. Beides and A. P. Sakis Meliopoulos (Georgia Institute of Technology). The authors should be commended for revisiting the problem of power transformer modeling. As it is widely known, a comprehensive and generally acceptable transformer model for transient simulation does not exist. One of the reasons is that transformers come in different designs and configurations and with different parameters of parasitic capacitances, etc. We would appreciate the authors response to the following comments and questions:

Has the proposed modeling method been tested using transformers with tertiary windings? If yes, the authors' comments on the accuracy and performance of the derived models will be appreciated. What are the effects of hysteresis losses and skin effect on the accuracy of the estimated resistive components of the transformer model?

The method requires measuring the frequency response of the transformers. Is the frequency response dependent on the design of the transformer alone (i.e. two transformers of the same manufacturer and type will have identical frequency response)? If this is not the case, it appears to us that it will be necessary to measure the frequency response of each transformer to be modeled.

Manuscript received August 11, 1992.

R. Malewski (Westmount, Quebec, Canada): This study can serve as an excellent example of successful and realistic approach to modeling of a large HV power transformer complex internal circuit. The authors recognize a necessity of taking measurements of the examined transformer characteristics in order to develop the EMTP model. As an alternative, they refer to the transformer design parameters; these however, are considered proprietary by the manufacturer and not accessible to the utility engineers.

The paper title includes the mention of high frequency, and at the end of paragraph #2 a statement is made on the predominantly capacitive behavior of the winding at high frequencies. This is correct if the f_{max} is set at some 200 kHz, as indicated in Fig. 1. After all, it

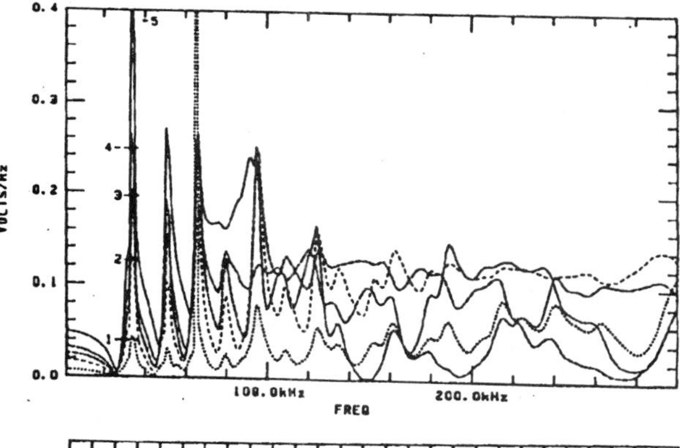

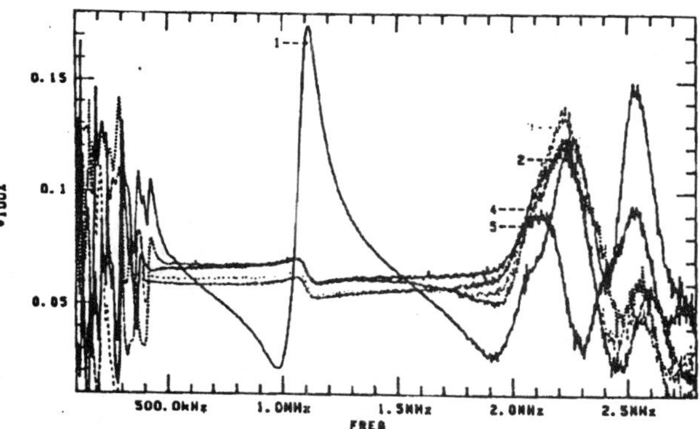

Fig. 1. Transfer function of five first discs of an interleaved HV transformer winding presented at three frequency scales. These transfer functions were deconvoluted in frequency domain from transients recorded on the untanked winding.

has been known since the time of Wagner [1] that the "standing wave" type of resonant frequencies is confined to a few hundred kilohertz interval.

A distinctly different behavior of typical HV transformer windings starts some 2 to 5 MHz, where the internal disc resonances come to play. It may be of interest to inspect a typical winding transfer function spanning all the four frequency intervals. Such a graph was obtained from an impulse voltage distribution measured along the discs on an untanked medium voltage unit. The transfer function was deconvoluted from the digitally recorded transients and the applied (low voltage) impulse. First five disc characteristic is shown in Fig. 1 at three frequency scales: 125 kHz, 300 kHz and 2.8 MHz. It can be seen that the interval from some 500 kHz to nearly 2 MHz can be modeled by a real pole circuit, but beyond that limit a different representation is required.

Clearly, this paper does not address the issue of very high frequency phenomena, although they are of practical importance for transformers directly connected to SF6 insulated bus bars [2].

Practical implementation of the EMTP model presented in this paper calls for measurements of the transformer transfer function in the frequency range of at least 200 kHz. Such measurements can not be easily taken on a large unit in substation, but the required measured characteristics can be obtained from an industrial laboratory performing the acceptance test of new transformers. At present, many laboratories use a digital recorder for monitoring the impulse test [3, 4]. The obtained records are usually processed in order to enhance the efficiency of fault detection. The processing often includes calculation of the frequency spectrum of the output and input impulses, and finding the transformer transfer function as quotient of these two spectra.

An analysis of the transfer function required for the dielectric fault detection, is not pertinent to the study presented by the authors. However, at a reduced voltage level, additional records can be taken during the impulse test, if requested by the utility purchasing the transformer. Such additional measurements can be included in the test program, on demand of the utility system planning department. An incremental cost of the additional measurement is negligible, since the impulse generator and recording system are anyhow prepared for the acceptance test.

The algorithms for measuring the HV to LV transfer function, and for retrieving the parameters required for modeling can be implemented on existing commercial digital impulse recorders, or a specialized recording and signal processing system can be developed using the accumulated experience in high frequency measurement of transformer winding characteristics.

References

[1] Wagner, K. "Das Eindringen einer elektromagnetischen Welle in eine Spule mit Windungkapazitat," Elektrotechnik und Maschinenbau, 1915, p. 89.

[2] Müller, W. "Fast Transients in Transformers," CIGRE SC12, WG12.11 Report presented at the Transformer Colloquim in Graz, 1990.

[3] Malewski, R., Poulin, B., "Impulse Testing of Power Transformers using the Transfer Function Method," IEEE Trans. Vol. PWRD-3, 1988, p. 476.

[4] Malewski, R., Gockenbach, E., Maier, R., Fellmann, K. H., Claudi, A., "Five Years of Monitoring the Impulse Test of Power Transformers with Digital Recorders and the Transfer Function Method", CIGRE Paper 12-201, 1992.

Manuscript received August 21, 1992.

A. Keyhani and **T. Tsai** (The Ohio State University, Electrical Engr., Columbus, OH): We would like to commend the authors for a well-written paper and for their efforts to develop a practical high frequency transformer model for the EMTP.

The essential ingredient of high frequency transformer modeling is to represent the transformer admittances as frequency dependent nonlinear functions. In general, these transfer functions are non-minimum phase system. Therefore both magnitude and phase of the transfer function are needed to uniquely identify the transfer function model. In this paper, since only the magnitude data were used for the transfer function identification, the transfer function model had to be modified into a minimum-phase plant. It is our belief that such practice may not be necessary and the phase data of the measured transformer admittances should be used for the transfer function estimation, because by including the phase data in the estimation process does not increase the number of unknown parameters which determines the size of the estimation problem. Furthermore it adds an important constraint on the variation of the estimated parameters.

Another important aspect of the high frequency transfer function estimation is the numerical stiffness problem. This problem becomes more severe if the resonant points are spreaded in a wide frequency range. In general, this problem can be resolved if a proper scaling scheme is adopted during the curve fitting. It would be interesting to know if any frequency scaling was performed or needed for this particular study.

The authors have provided the power industry with a valuable and practical technique for modeling the transformer high frequency dynamics for the EMTP. We would appreciate the authors' comments concerning the questions and issues raised in this discussion.

Manuscript received October 16, 1992.

A. S. Morched, L. Martí, and J. Ottevangers: We would like to thank the discussers for their interest and their many relevant questions presented.

Regarding Dr. Su's questions, we would like to make the following comments: After the admittance functions are approximated with rational functions, they become closed-form representations of the original functions. Expressing the admittance functions in terms of RLC modules is a convenient form of visualization and it does not imply that a number of RLC branches have to be connected explicitly in the EMTP. Inside the EMTP, the fitted functions are modeled with FDB modules. Each n-phase FDB module consists of a constant conductance matrix and a set of past history current sources. Figure I illustrates the EMTP representation of a two-winding transformer using FDB modules.

Therefore, the presence of a high frequency transformer (HFT) in the EMTP does not increase the size of the nodal admittance matrix of the system modeled, and it only adds n entries to the EMTP branch tables for each n-phase FDB module. For example, the transformer shown in Figure I only adds 9 branches to the EMTP branch tables. Additional storage is needed to keep track of the updating of the past history current sources. In broad terms, this additional storage amounts to 2 cells for each complex conjugate pole in (13) and one cell for each real pole in (11) and (12).

The computational burden of Dr. Su's transformer model should be comparable with that of the HFT model. For a single-phase, two-winding transformer Dr. Su's model requires the approximation of three functions and four numerical convolutions per time step of a transient solution. The HFT model, also requires the approximation of three distinct functions, but only three numerical convolutions per time step are needed. A comparison of the performance of both representations will depend largely on the number of terms needed in the fitting process.

The time step used in the EMTP simulation of the step response shown in Figures 7 and 8 of the paper, was the sampling rate used in the field measurement, i.e., 0.5 μs. An EMTP step function was used in the simulation (Δt rise time). In the field test, the input step reached 95% of its peak value in 1 μs.

We agree with Messrs Keyhani and Tsai when they indicate that a non-minimum phase shift function cannot be described uniquely with the magnitude function alone. However, any causal function is uniquely defined if its real part is known. In the identification/optimization process described in the paper, both real and imaginary parts are used to compensate for possible inconsistencies in measured data. Fre-

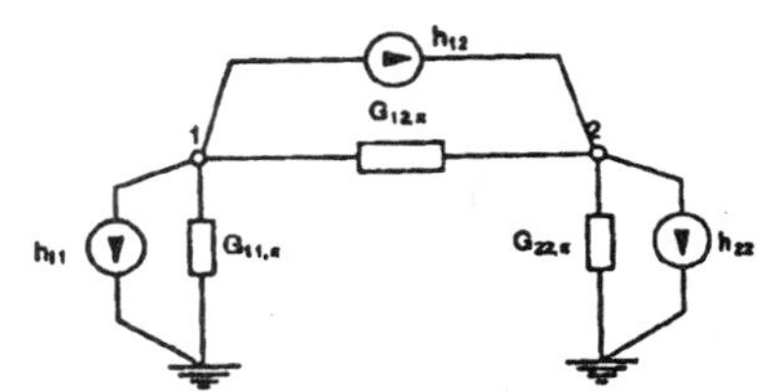

Fig. I. Single-line diagram representation of a two-winding transformer with the FDB model.

quency scaling is commonly used in the solution of least squares optimization problems. However, its use would not be advantageous within the context of the modified Marquardt optimization algorithm.

Messrs Beides and Meliopoulos ask whether the frequency response of a transformer depends on its design alone. We have observed that transformers of the same design and make show essentially the same frequency behaviour. It is unclear to us, at this point in time, how far to generalize these observations. In absence of actual measurements, it might be better to use the HFT model of a similar transformer than to use no high frequency model at all. While the transformer used in the paper has a buried delta winding, we have not yet modeled a delta-connected tertiary winding explicitly. This requires some special considerations which will be explained in more detail in our response to Prof. Semlyen's questions. Hysteresis and eddy current effects are normally taken into account by dedicated models (e.g., [i]) when the HFT is used as an add-on module. In this case, the HFT model will match the difference between the measured data and the frequency response of the linear portions of these models.

We will now address Dr. Malewski's comments. The choice of 200 kHz as the maximum frequency was based on the simulation needs of the transient simulations for which the transformer model was required. Figure II shows the magnitude of $y_{h_1 h_1}$ from 400 Hz to 1 MHz. Other than additional poles and zeros and the added computational burden, we do not feel that the extended frequency range presents a problem that the HFT model cannot handle. The graphs shown by Dr. Malewski also suggest to us that frequency range beyond 1 MHz does not pose any special problems either.

With regard to the techniques used to obtain the frequency responses, we feel that direct, low voltage frequency domain admittance measurements are probably simpler, cheaper and more reliable for the purposes of the HFT model. This type of measurements can be made with relative ease in the field. It would probably be difficult to persuade manufacturers to perform all the full scale chopped-wave tests required to obtain all the data required by the HFT model. Nevertheless, Dr. Malewski's measurement techniques could provide an alternative way to obtain data for the HFT model since some of them are normally done in acceptance tests anyway.

Professor Semlyen suggests an adjustment of the final fitted functions to account for phase asymmetries. It is not clear to us how this adjustment could be made after the fitted functions are obtained. It should be possible, however, to choose a real constant transformation matrix other than α, β, o to account for unbalances. This would be roughly the same type of approximation used to model frequency dependent unbalanced lines. Whether this constant transformation matrix would also give acceptable answers at higher frequencies is probably a subject of further research. Another possibility is to approximate each element of the Y matrix. This would not rely on any assumptions of symmetry, but the additional computational burden would be substantial.

Professor Semlyen correctly points out that in the case of Y-D or Y-Z-connections, the off-diagonal $[Y_{ij}]$ sub-matrices do not lend themselves to be approximated by a balanced matrix. Depending on the type of delta connection and node numbering scheme, variations of a cyclic matrix can be obtained. For instance,

$$[Y_{ij}] = \begin{bmatrix} y_{ij,aa} & y_{ij,ab} & y_{ij,ac} \\ y_{ij,ba} & y_{ij,bb} & y_{ij,bc} \\ y_{ij,ca} & y_{ij,cb} & y_{ij,cc} \end{bmatrix}$$

$$\approx y_a(\omega) \cdot \begin{bmatrix} 0 & -1 & +1 \\ +1 & 0 & -1 \\ -1 & +1 & 0 \end{bmatrix} \qquad \text{(i)}$$

There are several ways in which this situation can be handled: The most obvious one is to fall back on the approximation of every element of $[Y]$. On the other hand, it might be more practical to use a constant transformation matrix $[Q]$ whose elements $q_{i,k}$ for a n-phase system are given by

$$q_{i,k} = \frac{1}{\sqrt{n}} e^{-j\frac{2\pi}{n}\cdot(i-1)(k-1)} \qquad \text{(ii)}$$

The well-known symmetrical components transformation matrix is just a special case of the matrix defined by equation (ii). The resulting modal admittance matrix only has two non-zero elements, and these

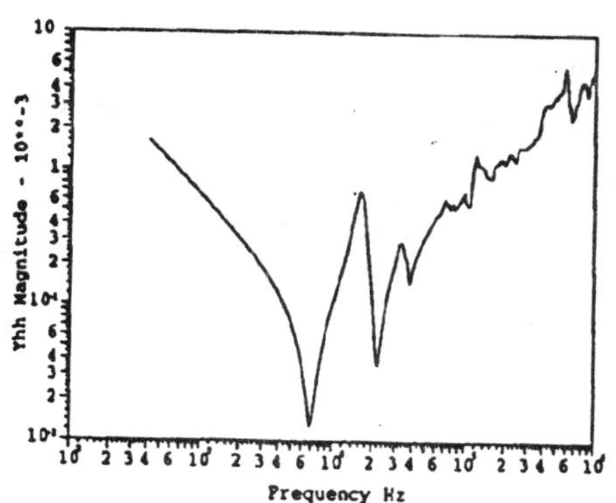

Fig. II. Magnitude of $y_{h_1 h_1}$.

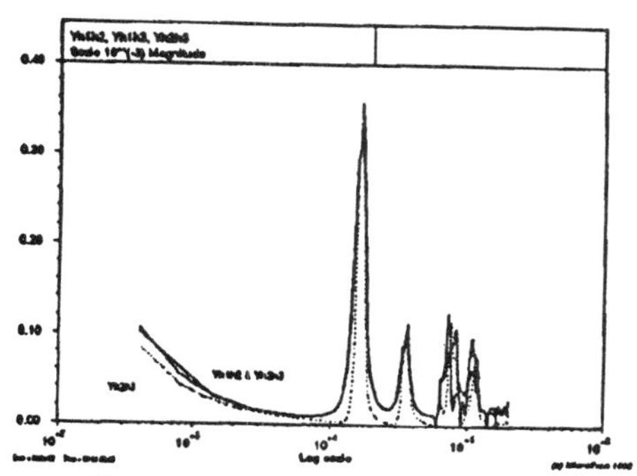
Fig. III. Off-diagonal elements of $[Y_H]$.

differ only by a constant. For example,

$$[Q]^{-1}[Y_{ij}][Q] = [Y_{modal}] \qquad \text{(iii)}$$

$$[Y_{modal}] = y_a(\omega) \cdot j\sqrt{3} \begin{bmatrix} 0 & 0 & 0 \\ 0 & 1 & 0 \\ 0 & 0 & -1 \end{bmatrix} \qquad \text{(iv)}$$

In this case, only one admittance function $y_a(\omega)$ has to be approximated. In the time-step loop of the EMTP the complex algebra does not present a problem because even if intermediate functions are nominally complex, the final phase voltages and currents are always real. In other words, the existing FDB model can easily be modified to account for cyclic symmetric modules.

Professor Chen correctly points out that a multi-legged transformer should show some asymmetry, which would degrade the accuracy of the assumption that the sub-matrices of $[Y]$ are balanced. However, the measurements available to us do not show the severe asymmetry indicated in Prof. Chen's calculations. Figure 3 in the paper shows that the diagonal elements of the high voltage winding block are nearly identical over a wide frequency range. Figure III below, shows the measured off-diagonal elements of the same sub-matrix.

From this plot it can be seen that while one element is indeed different, the unbalance ratio at low frequencies is in the order of 1.2 to 1.3 rather, 2.0 as Professor Chen's calculations suggest. Based on these measurements we are inclined to accept the balancing procedure as a reasonable simplification. It is clear, however, that the use of modal transformation matrix that accounts for center phase asymmetries would be desirable. Strictly speaking, this transformation matrix would also be frequency dependent. Therefore, further investigation would be needed to find what constant transformation matrix would represent an acceptable compromise over the entire frequency range of interest.

The question of the validity of superimposing linear high frequency behaviour on the nonlinear response due to saturation does not have a simple answer. Short of solving the nonlinear field problem with detailed knowledge of core and winding design, accounting for nonlinear and frequency dependent effects will always involve a certain degree of approximation. If the transformer is unsaturated, high frequency excitation cannot drive the transformer into saturation as the flux produced by a voltage input is inversely proportional to its frequency. If a transient of sufficient magnitude is impressed on a transformer which is already saturated or near saturation, then superposition is not strictly valid. On the other hand, situations where a transient is impressed on a transformer which is already in saturation may not be all that common. It is Ontario Hydro's practice not no operate transformers near saturation because of acoustic pollution requirements. This may contribute to the lack of field measurements that would validate the assumption of superposition under near-saturation conditions.

Reference

[i] E. Tarasiewicz, A. S. Morched, A. Narang and E. P. Dick, "Frequency Dependent Eddy Current Models for Nonlinear Iron Cores," Paper No. 92 WM 177-6 PWRS, Presented at the IEEE-PES Winter Meeting, New York, Feb. 1992.

Manuscript received October 16, 1992.

IEEE Transactions on Power Delivery, Vol. 10, No. 3, July 1995

Simplified Three-Phase Transformer Model for Electromagnetic Transient Studies

S. Chimklai

J. R. Martí (M)

The University of British Columbia
Vancouver, BC., Canada V6T 1Z4

Abstract: **This paper presents a simplified high-frequency model for three-phase, two- and three-winding transformers. The model is based on the classical 60-Hz equivalent circuit, extended to high frequencies by the addition of the winding capacitances and the synthesis of the frequency-dependent short-circuit branch by an RLC equivalent network. By retaining the T-form of the classical model, it is possible to separate the frequency-dependent series branch from the constant-valued shunt capacitances. Since the short-circuit branch can be synthesized by a minimum-phase-shift rational approximation, the mathematical complications of fitting mutual impedance or admittance functions are avoided and the model is guaranteed to be numerically absolutely stable. Experimental tests were performed on actual power transformers to determine the parameters of the model. EMTP simulation results are also presented.**

Key words: Transformer modelling, frequency dependence, electromagnetic transients, stray capacitance, leakage impedance.

INTRODUCTION

Figure 1 shows the measured short-circuit impedance of a 50 MVA 115/23 kV three-phase power transformer. Also shown in the figure is the short-circuit impedance $R_{60}+j\omega L_{60}$ of the 60-Hz constant-parameter model. The two responses begin to diverge for frequencies beyond 3 kHz or so.

A number of high-frequency transformer models, suitable for transient simulations, have been proposed in recent years (e.g., [1] - [8]). Even though some of these models can be very accurate, there are still a number of problems in terms of efficiency, numerical stability, and data acquisition requirements. Most full-frequency-dependent models (e.g., [7], [8]) are based on the fitting of the elements of a $[Y(\omega)]$ matrix that represents the transformer as seen from its external terminals. This matrix depends on the internal connection of the transformer windings (e.g., wye, or delta) and is generally obtained from frequency domain measurements of its self and mutual elements. The fitting of the self admittances of this matrix can be done with minimum-phase-shift rational approximations (i.e., both, poles and zeroes are located on the left side of the complex plane). The mutual admittances, though, are non-minimum-phase-shift functions, and very strict conditions regarding the location of the poles and zeroes of the approximating rational functions have to be satisfied to prevent numerical stability problems [9].

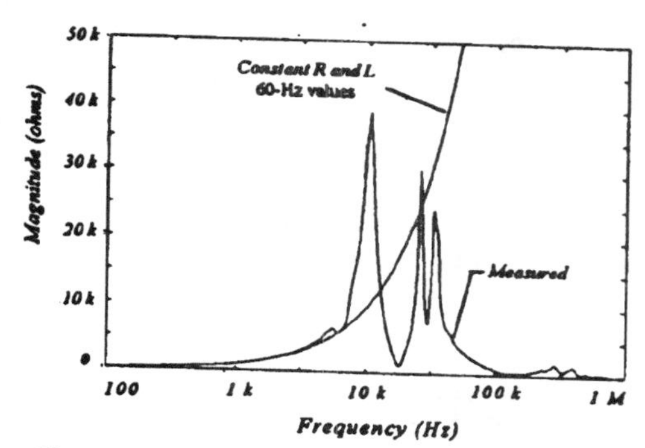

Figure 1. Measured short-circuit positive sequence impedance of a 50 MVA 115/23 kV transformer versus $(R_{60}+j\omega L_{60})$ model.

As in other frequency dependence modelling problems ([10], [11]), the relative complexity, accuracy, and numerical stability of the model is closely related to the choice of network representation. "Black box" models, without carefully defined topological structures, can result in very complicated functions to approximate and in numerical stability problems. The works in [7] and [8] represent important contributions towards simplification of the modelling and better numerical stability of the solution.

The work described in the present paper further simplifies the modelling problem by working at a more basic topological level: the coupled-coils level [12]. The classical 60-Hz T-circuit for single-phase, two- and three-winding transformers, is used to model the inductive and capacitive interaction among coils belonging to the same phase (coils mounted on the same leg of the core). For three-phase banks made up of single-phase units, this is all the modelling that is required. For common-core units (e.g., core-type and shell-type three-phase transformers), the mutual interaction among coils of different phases is eliminated through the use of a decoupling transformation matrix. Each decoupled mode can then be treated as a single-phase separate unit.

Modelling at the coil level presents two fundamental topological advantages over the $[Y(\omega)]$ approach: a) the

94 SM 410-1 PWRD A paper recommended and approved by the IEEE Transformers Committee of the IEEE Power Engineering Society for presentation at the IEEE/PES 1994 Summer Meeting, San Francisco, CA, July 24 - 28, 1994. Manuscript submitted December 30, 1993; made available for printing April 18, 1994.

modelling of the frequency-dependent series branch (short-circuit impedance) can be separated from the modelling of the shunt branches (constant stray capacitances and magnetizing branch), and b) the representation is independent of the particular external connection among windings (wye, delta, etc.). The particular connection is taken care of by node labelling at the EMTP level.

As a result of the simplified topology, the frequency dependence modelling problem is reduced to the fitting of single impedance functions (the short-circuit impedances). The functions to be fitted are fewer and simpler than with the [Y(ω)] formulation. This results in an appreciable reduction in the number of discrete-time terms, and, consequently, in the solution times during EMTP simulation. Since the short-circuit impedances are minimum-phase-shift functions, the mathematical and numerical problems of fitting the non-minimum-phase-shift mutual terms in the [Y(ω)] approach have been eliminated. The resulting model is accurate, efficient, and numerically stable.

Proposed Model

Figure 2 shows the proposed wide-band model for a pair of coupled coils of a transformer. The stray capacitances are connected at the terminals ("outside") of the circuit. In this way, the internal part of the model has exactly the same form as the conventional 60-Hz model.

Even though not a theoretical restriction, it is assumed in the present work that the interaction among phases in a common-core three-phase transformer unit is symmetrical. This permits the use of a balanced-system transformation matrix that can be taken as real and constant to decouple the system at all frequencies. It also facilitates the design of simple experiments to measure the parameters of the model, without having to open the internal connections among windings. In general, this assumption neglects the effect of uneven capacitive and leakage-inductance coupling among phases mounted in the outer and central limbs of a three-phase core-type transformer. As it has been discussed in [7], however, this balanced-phases assumption seems to be reasonable.

After decoupling the system, the circuit of Figure 2 can represent any of the decoupled modes, that is, positive, negative, and zero sequence networks for the balanced-system case. Figure 2(a) shows the physical conception of the model, while Figure 2(b) shows its simplified form.

The series impedance $Z_{winding}$ in Figure 2(b) includes the equivalent series resistance (current dependent losses) and the equivalent leakage inductance (self and mutual leakage flux) of the windings, in combination with part of the winding-to-winding capacitance. The frequency dependent response of this branch (Figures 8 and 9) can be obtained from short-circuit test measurements using a variable-frequency supply. The measured positive and zero sequence $Z_{winding}$ impedances can then be synthesized with a network of constant RLC elements (Figure 7). The rest of the transformer model in Figure 2(b) includes constant stray

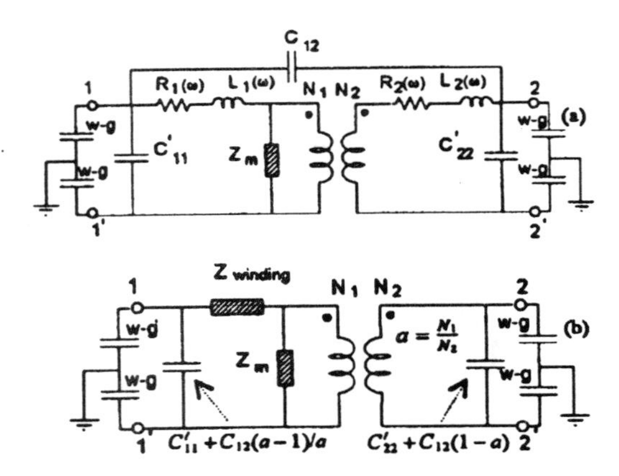

Figure 2. High-frequency model for a pair of coupled transformer coils. (a) Original Circuit. (b) Equivalent Circuit.

capacitances and the magnetizing branch. The value of the stray capacitances can be determined from experimental measurements.

For the magnetizing branch, it is assumed, as in the 60-Hz model, that it has little influence in the short-circuit test. This assumption is more accurate as the frequency becomes higher, since the flux required to induce a given voltage is inversely proportional to the frequency. Since the proposed model preserves the internal form of the conventional 60-Hz equivalent, any magnetizing branch model developed for the 60-Hz equivalent can be incorporated directly in the proposed high-frequency model. (The particular modelling of the magnetizing branch is beyond the scope of this paper.)

As already indicated, most previous wide-band transformer models are based on the fitting of the self and mutual elements of an admittance matrix [Y(ω)] that represents the transformer as seen from its terminals. Matrix [Y(ω)] is symmetrical, and for a three-phase M-winding transformer there are $3M\cdot(3M+1)/2$ distinct elements to be measured and approximated. For a three-phase two-winding transformer (3ϕ-2wdg) there are 21 distinct functions, while for a 3ϕ-3wdg transformer there are 45 distinct functions. By averaging the elements corresponding to phases *a*, *b*, and *c* (which is equivalent to assuming a balanced three-phase core arrangement), the number of functions to be fitted can be reduced to $M\cdot(M+1)$ [7], which gives six functions for a 3ϕ-2wdg transformer and 12 functions for a 3ϕ-3wdg transformer. The model proposed in the present paper requires the fitting of only the positive and zero sequence short-circuit impedances. In the case of a 3ϕ-2wdg transformer only two impedances have to be fitted, while for a 3ϕ-3wdg transformer six impedance functions need to be fitted. These impedances can be realized by minimum-phase-shift rational functions, thus guaranteeing the numerical stability of the model.

STRAY CAPACITANCES

Stray capacitance in transformers are physically distributed and very complicated to model in detail. For disturbances occurring in the power system outside of the transformer, a detailed distributed-parameter modelling of the inside of the transformer may not be necessary, and a simple terminal model for the capacitances is probably sufficient for most cases. The model proposed in this work considers four types of stray capacitances:

1) capacitance from winding to ground,

2) capacitance from winding to winding,

3) turn-to-turn capacitance and,

4) capacitance from the outside winding of one phase to the outside winding of the other phases.

These four types of stray capacitances are shown schematically in Figure 3. A number beside each capacitance identifies its type according to the four types indicated above.

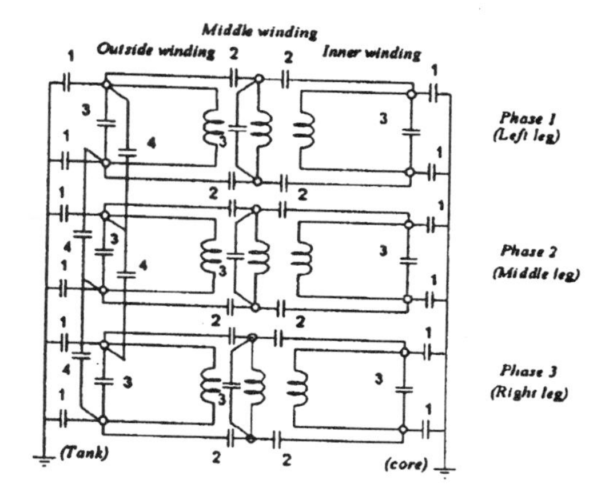

Figure 3. Stray capacitances in a three-phase three-winding transformer.

Physically, the coils corresponding to the same phase are mounted concentrically on the same leg of the core. The ground "facing" the outermost winding is the transformer tank, while the ground "facing" the innermost winding is the transformer core. Following the recommendation of [13], each type of capacitance, except the turn-to-turn capacitance, can be split into two equal parts, each part connected to one end of each coil.

All stray capacitances are assumed to be constant. This assumption was verified in the experimental tests. Figure 4 shows the combined capacitances to ground (plotted as measured impedance) for the tested 115/23 kV, 50 MVA transformer. The plotted capacitance is a combination of the capacitance to ground of the 23 kV winding and the capacitance between the 115 kV and 23 kV windings. As can be observed in these results, the capacitance is constant to about 100 kHz. The irregularities beyond 100 kHz are probably due to inaccuracies in the measurements.

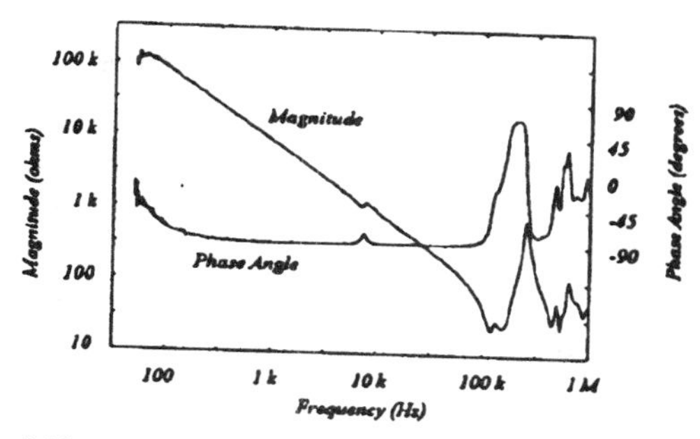

Figure 4. Measured stray capacitances ($Z_c = 1/j\omega C$) in the tested 115/23 kV transformer.

SHORT-CIRCUIT MEASUREMENTS

Short-Circuit Winding Impedance

Figure 5 shows the proposed model of Figure 2 for each sequence network under a short-circuit test. The short-circuit branch $Z_{winding}$ includes the resistance and leakage inductance of the windings, in combination with part of the winding-to-winding capacitance. The physical winding-to-winding capacitance (C_{12} in Figure 5(a)) is connected across the ideal coupling transformer. To combine this capacitance with the leakage branch, $R_{12}(\omega)$-$L_{12}(\omega)$, the terminal of C_{12} on the "wrong side" of the ideal coupling transformer can be transferred to the side of the leakage branch, as suggested in [14]. This results in a capacitance C_{12}' in parallel with the leakage branch plus additional capacitances across the primary and secondary terminals (Figure 5(b)):

$$C_{12}' = C_{12}/a \quad (1)$$

$$C_1' = C_1 + C_{12}(a-1)/a \quad (2)$$

$$C_2' = C_2 + C_{12}(1-a) \quad (3)$$

where $a = N_1/N_2$, and C_1 and C_2 include the winding-to-ground and the turn-to-turn capacitances. In addition, in the positive sequence equivalent circuit of common-core three-phase transformers, capacitance C_1 or C_2 (depending on which side the top winding, usually the high-voltage winding, is located) also includes the capacitance from top winding to top winding between phases (this capacitance does not exist in separate-core units).

Experimental measurements were performed to determine the frequency response of the $Z_{winding}$ branch. A variable-frequency power supply was used to perform measurements from a few Hz up to about 1 MHz. Details of the experimental set-up for these measurements are indicated later in the paper.

The results for the positive sequence (measured on the high-voltage side) and zero-sequence (measured on the low-voltage side) short-circuit responses of a 50 MVA 115/23 kV wye-delta transformer are shown in Figures 1, 8, and 9. The plots in these figures are typical of short-circuit frequency responses of power transformers, as found, for instance, in [15] and [16]. The multiple peaks are caused by

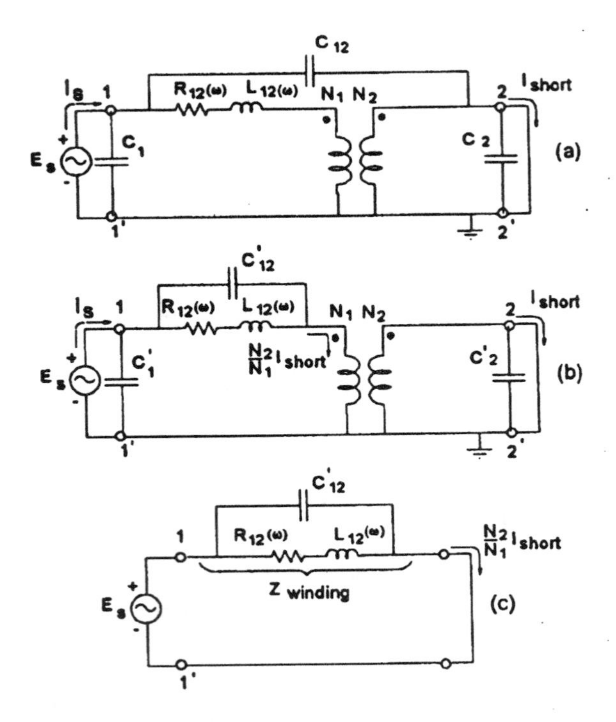

Figure 5. Short-circuit test. (a) Original circuit. (b) Transferring C_{12} to the primary side. (c) Equivalent $Z_{winding}$ impedance.

the non-uniformity of the windings, which may be produced by the presence of auxiliary tap changing windings [17] and voltage grading schemes [18].

Short-Circuit Impedance

Figure 5 shows the proposed equivalent circuit during a short-circuit test. Following reference [15], two experimental measurements are suggested to determine the series and shunt impedances of the circuit: a *transfer impedance* measurement and a *driving point* impedance measurement.

In the circuit of Figure 5(a), measuring the output current I_{short} and the applied voltage E_s, one can calculate (Figure 5(c)) the transfer impedance $Z_{winding} = aE_s/I_{short}$, where a is the turns ratio N_1/N_2. This impedance includes the winding's resistance and equivalent leakage inductance, in combination with part C_{12}/a (Equation 1) of the total winding-to-winding capacitance.

Terminal Capacitances

With reference to the circuits of Figure 5, measuring the driving-point impedance $Z_{driving} = E_s/I_s$ permits the calculation of the terminal capacitance C_1. In the circuit of Figure 5(b), as ω tends to infinity, $Z_{driving}$ tends to $j\omega C_{driving}$, where $C_{driving} = C_1' + C_{12}'$. Capacitance C_{12}' comes from the winding-to-winding capacitance C_{12} (Figure 5(a) and Equation 1), which can be measured separately with the set-up of Figure 6. Subtracting C_{12}' from $C_{driving}$ gives C_1', which together with Equation 2, gives the terminal capacitance C_1 in the circuit of Figure 5(a). Finally, subtracting the winding-to-ground capacitance, one obtains the terminal capacitance

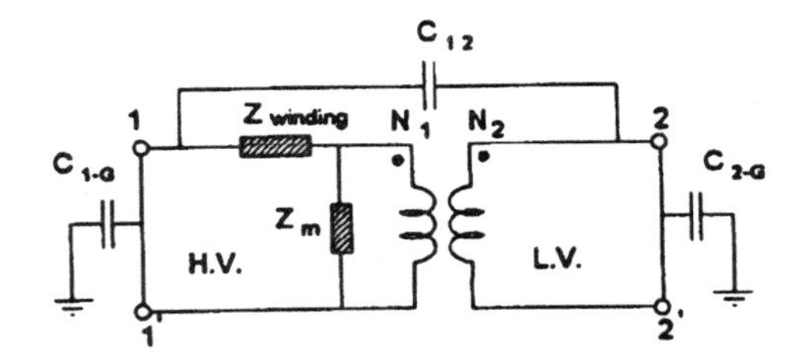

Figure 6. Windings connection for measurement of the winding-to-winding and winding-to-ground capacitances.

C_{11}' in the proposed equivalent circuit of Figure 2(b) for each decoupled mode of the transformer. The winding-to-ground capacitance can be measured experimentally in the set-up of Figure 6. To obtain the terminal capacitance C_{22}' on the other side of the equivalent circuit, the direction of the driving-point impedance measurements has to be reversed.

Due to limitations in the research facilities where the measurements were taken, the results presented in Figures 8 and 9 for $Z_{winding}$ were calculated from the $Z_{driving}$ measurements described above, subtracting the terminal capacitance C_1'. However, it is believed that a $Z_{transfer}$ test is a more direct and accurate way of obtaining $Z_{winding}$.

SYNTHESIS OF $Z_{WINDING}$

The next step in developing the proposed transformer model is to synthesize the short-circuit impedance $Z_{winding}$ with an equivalent network with constant R, L, and C elements (Figure 7).

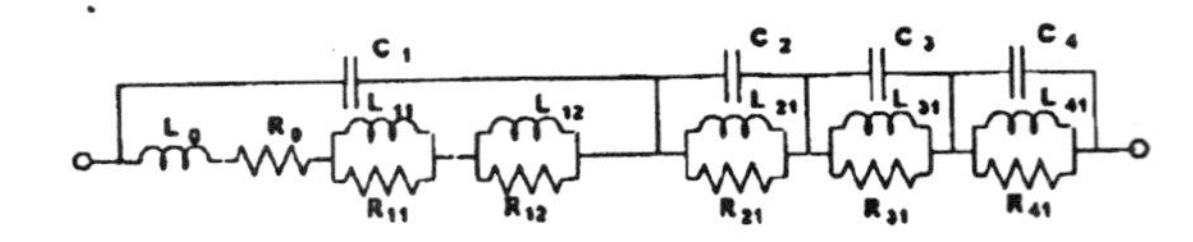

Figure 7. RLC synthesis network to approximate $Z_{winding}$.

The procedure to obtain the RLC synthesis network is described next. Since this network corresponds to a simple impedance function, it can be synthesized with a minimum-phase-shift rational approximation. Magnitude and phase or real and imaginary parts are, therefore, interlocked and only one of them needs to be matched. It was found in this work that a synthesis from the real part of the function leads to more accurate results than a synthesis from the magnitude of the function.

From the plots of Figures 8 and 9, it can be recognized that this frequency response can be produced by a parallel network of R, L and C components. As a first approximation, a number of simple RLC blocks (one R, one L, and one C) are used to match the peaks of the function. The values of R, L and C for each block can be found by applying a non-linear fitting routine, such as the one given in [19].

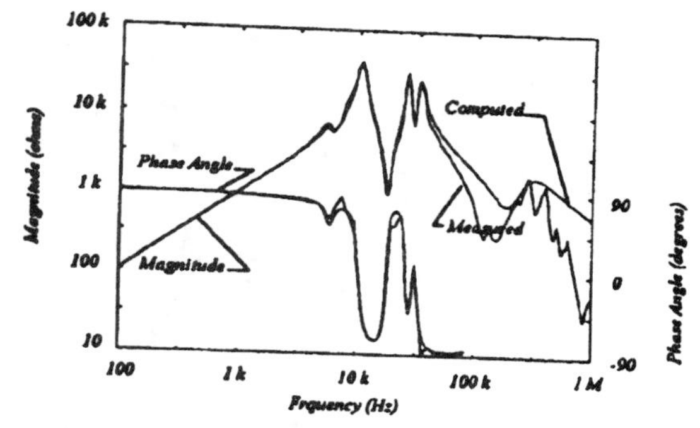

Figure 8 Comparison of measured and approximated $Z_{winding}$ positive-sequence.

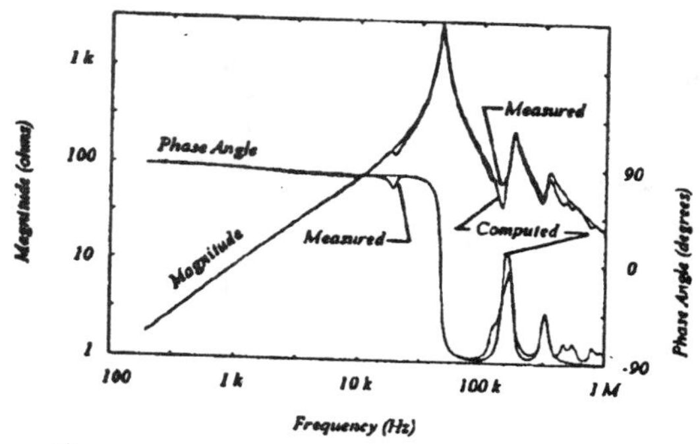

Figure 9 Comparison of measured and approximated $Z_{winding}$ zero-sequence.

Suppose, for example, that we want to match three peaks in the graph of Figure 8. Three simple RLC blocks are first calculated. On a second iteration, the real part of the function corresponding to the second and third blocks is subtracted from $Z_{winding}$. The left-over first block is now to be approximated in more detail. The impedance function after subtracting the secondary blocks has only one dominant peak. The C of the first block (main peak), found in the first iteration, is now subtracted from the left-over first block. What remains is now synthesized using the pole-zero asymptotic approximation procedure used in [10].

Figures 8 and 9 compare the measured and approximated $Z_{winding}$ functions for the positive and zero sequence modes of the tested transformer. Three RLC blocks were used to match the peaks and three RL sections were used for the low frequency region. The irregularities in the capacitance measurements of Figure 4 for frequencies beyond 100 kHz seem to indicate that, in general, the measured data may not be reliable beyond this frequency. For this reason, no special effort was made to more accurately fit this region with a higher number of RLC blocks. It should also be noticed that since the vertical scale in the plots of Figures 8 and 9 is logarithmic, the magnitude values in the irregular region beyond 100 or 200 kHz are much smaller than in the preceding region where the main magnitude peaks occur (this can be verified in the graph of Figure 1 in which the magnitude of the positive sequence short-circuit impedance is plotted in a linear scale).

Short-Circuit Tests for Three-Phase Two-Winding Transformers

Experimental tests to measure the short-circuit characteristics of a 115/23 kV wye-delta transformer were performed at the high-voltage laboratory of the Electricity Generating Authority of Thailand (EGAT). The short-circuit impedance at each frequency was measured with an HP Model 4192A LF Impedance Analyzer. This instrument has a rating of 5 Hz to 13 MHz with a maximum output voltage of 1.1 V_{RMS}. The four-terminal pair measuring technique ("Kelvin connection") was used in order to achieve a wide impedance range (1 mΩ to 10 MΩ) and to minimize measuring errors due to parasitic coupling with the test leads.

Two short-circuit tests were performed, one for the zero sequence equivalent circuit and one for the positive sequence equivalent circuit. Due to the difficulty of obtaining a variable-frequency three-phase power supply for high frequencies, both zero-sequence and positive-sequence tests were performed with a single power source (Figure 10). This is not an issue for the zero-sequence tests since in this case all phases should be connected together (Figure 10(b)). Interestingly, it was found that the positive-sequence test can also be performed with a single power source if one set of transformer windings is connected in delta. For this test (Figure 10(a)), one coil of the delta winding is short-circuited and the impedance analyzer is connected across another coil. Suppose the delta coils are named 1, 2, and 3. The following matrix equation can be written to relate the voltages across the delta coils to the currents in the coils:

$$\begin{bmatrix} I_1 \\ I_2 \\ I_3 \end{bmatrix} = \begin{bmatrix} Y_s & Y_m & Y_m \\ Y_m & Y_s & Y_m \\ Y_m & Y_m & Y_s \end{bmatrix} \begin{bmatrix} V_1 \\ V_2 \\ V_3 \end{bmatrix} \tag{4}$$

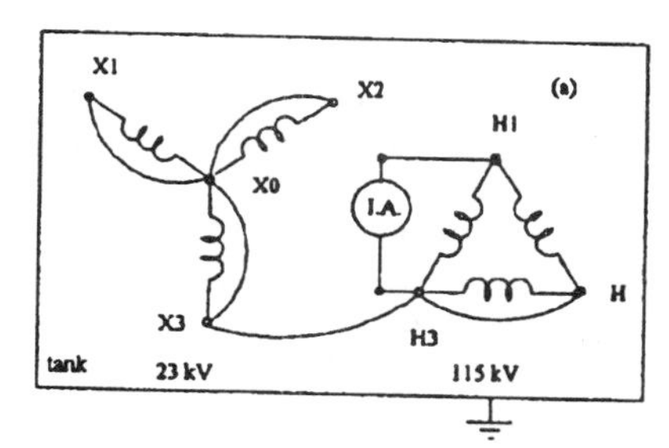

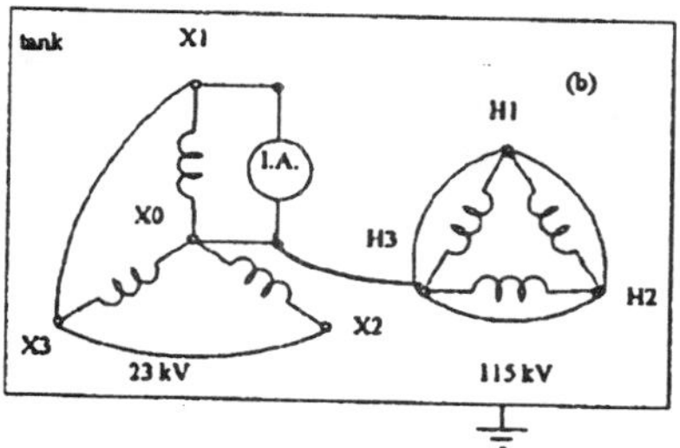

Figure 10. Short-circuit tests on a three-phase two-winding transformer. (a) Positive sequence test. (b) Zero sequence test.

With the connection of Figure 10(a), the voltage across one of the delta coils will be zero and the voltage across the other two coils will have equal value but opposite polarity. Taking, for example, $V_3 = 0$, $V_1 = V$ and $V_2 = -V$, and substituting these values into Equation 4 yields

$$I_1 = (Y_s - Y_m) \cdot V \quad (5)$$

$$I_2 = (Y_m - Y_s) \cdot V \quad (6)$$

The current flowing into the delta from the outside of the transformer is $I_1 - I_2$, which, from Equation 5 and 6, becomes $2 \cdot I_1$. The impedance read by the analyzer is then half the actual positive sequence impedance. To obtain an average of positive sequence results in unbalanced cores (e.g., core-type transformers), the test can now be repeated from the second and third coils of the delta.

Short-Circuit Test for Three-Phase Three-Winding Transformers

The measuring technique described above for the positive and zero sequence short-circuit tests on two-winding transformers can also be applied to three-winding transformers. The only difference is that in this case three tests are required to define the branches of the equivalent T-circuit [20] (Figure 11). Each of these tests involves two windings, with the third winding open-circuited. As in the case of the two-winding transformer, the zero sequence test can be performed directly with a single source. Also, as in the two-winding case, when one of the windings is connected in delta (as it is usually the case for the tertiary winding), the positive sequence test can performed with a single source.

Notice that for the short-circuit tests described above, it is not necessary to break apart the connections of the windings, and, therefore, can be easily performed on existing transformers.

EMTP Transformer Model

After obtaining the positive, negative, and zero sequence equivalent circuits, these circuits are discretized for time-domain analysis. After discretization, the RLC network to simulate $Z_{winding}$ (Figure 7) is reduced to just an equivalent current source in parallel with a constant equivalent resistance. Figure 11 shows the discrete-time model for one of the sequence networks for a three-phase three-winding transformer (capacitances not shown). The discrete-time positive, negative, and zero sequence equivalent circuits have to be transferred to phase coordinates in a form which is suitable for interfacing the transformer model with the rest of the power system solution. With reference to Figure 11 for a three-phase three-winding transformer:

$$i_1.R_1 + e_1 = v_1 + ih_1.R_1 \quad (7)$$

$$i_2.R_2 + e_2 = v_2 + ih_2.R_2 \quad (8)$$

$$i_3.R_3 + e_e = v_3 + ih_3.R_3 \quad (9)$$

$$n_1.i_1 + n_2.i_2 + n_3.i_3 = 0 \quad (10)$$

$$n_2.e_1 - n_1.e_2 = 0 \quad (11)$$

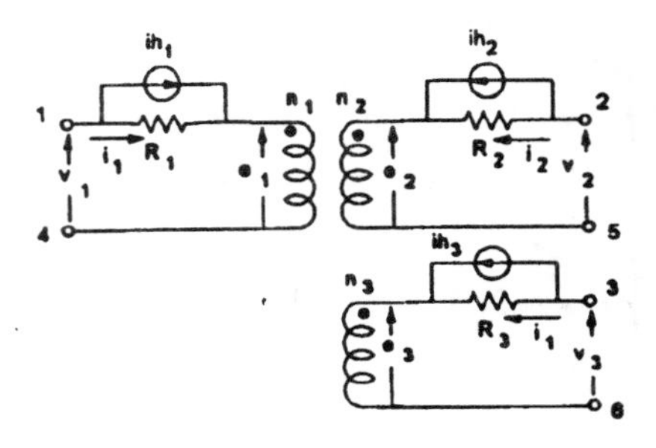

Figure 11. Equivalent circuit for a three-phase three-winding transformer in discrete-time form (capacitances not shown).

$$n_3.e_1 - n_1.e_3 = 0 \quad (12)$$

Equations 7 to 12 can be rewritten in symbolic form as

$$\begin{bmatrix} R & I \\ N_1 & N_2 \end{bmatrix} \begin{bmatrix} i(t) \\ e(t) \end{bmatrix} = \begin{bmatrix} I & R \\ \Phi & \Phi \end{bmatrix} \begin{bmatrix} v(t) \\ ih(t-\Delta t) \end{bmatrix} \quad (13)$$

where

$$[R] = \begin{bmatrix} R_1 & 0 & 0 \\ 0 & R_2 & 0 \\ 0 & 0 & R_3 \end{bmatrix} \quad (14)$$

$$[N_1] = \begin{bmatrix} n_1 & n_2 & n_3 \\ 0 & 0 & 0 \\ 0 & 0 & 0 \end{bmatrix} \quad (15)$$

$$[N_2] = \begin{bmatrix} 0 & 0 & 0 \\ n_2 & -n_1 & 0 \\ n_3 & 0 & -n_1 \end{bmatrix} \quad (16)$$

Submatrices [I] and [Φ] are the unity matrix and zero matrix, respectively.

If the transformer has only two windings per phase, the last rows and columns of the matrices in Equations 14, 15 and 16 must be eliminated. It should be noted that matrices [N_1] and [N_2] are singular. Equation 13 can be solved for the currents $i_1(t)$, $i_2(t)$ and $i_3(t)$ in windings 1, 2 and 3, respectively, in terms of the winding voltages and history terms. Solving Equation 13 for the currents,

$$[i(t)] = [N_2R - N_1]^{-1}N_2[v(t)] + [N_2R - N_1]^{-1}N_2R.[ih(t-\Delta t)] \quad (17)$$

or

$$[i(t)] = [Y][v(t)] + [I_{hist}(t-\Delta t)] \quad (18)$$

which is the [3×3] nodal admittance formulation for each sequence network.

Before transferring to phase coordinates, the [3×3] admittance matrices for each sequence network are combined into a single [9×9] block-diagonal $[Y]_{mode}$ matrix. A transformation matrix of the same size [9×9] is needed to transform the system from sequence to phase quantities. Using, for example, Clarke's transformation (Edith Clarke [21]), adapted

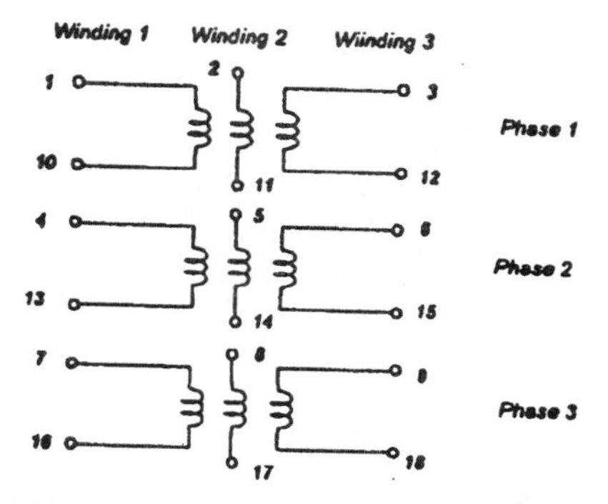

Figure 12. Nodes labelling for external network solution. Three-phase three-winding transformer.

to the present problem, results in the following [9×9] matrix for a three-phase three-winding transformer

$$[T]=\begin{bmatrix} \frac{1}{\sqrt{3}} & 0 & 0 & \frac{1}{\sqrt{2}} & 0 & 0 & \frac{1}{\sqrt{6}} & 0 & 0 \\ 0 & \frac{1}{\sqrt{3}} & 0 & 0 & \frac{1}{\sqrt{2}} & 0 & 0 & \frac{1}{\sqrt{6}} & 0 \\ 0 & 0 & \frac{1}{\sqrt{3}} & 0 & 0 & \frac{1}{\sqrt{2}} & 0 & 0 & \frac{1}{\sqrt{6}} \\ \frac{1}{\sqrt{3}} & 0 & 0 & -\frac{1}{\sqrt{2}} & 0 & 0 & \frac{1}{\sqrt{6}} & 0 & 0 \\ 0 & \frac{1}{\sqrt{3}} & 0 & 0 & -\frac{1}{\sqrt{2}} & 0 & 0 & \frac{1}{\sqrt{6}} & 0 \\ 0 & 0 & \frac{1}{\sqrt{3}} & 0 & 0 & -\frac{1}{\sqrt{2}} & 0 & 0 & \frac{1}{\sqrt{6}} \\ \frac{1}{\sqrt{3}} & 0 & 0 & 0 & 0 & 0 & -\frac{2}{\sqrt{6}} & 0 & 0 \\ 0 & \frac{1}{\sqrt{3}} & 0 & 0 & 0 & 0 & 0 & -\frac{2}{\sqrt{6}} & 0 \\ 0 & 0 & \frac{1}{\sqrt{3}} & 0 & 0 & 0 & 0 & 0 & -\frac{2}{\sqrt{6}} \end{bmatrix}$$

The voltages and currents on each of the nine coils of the transformer in phase coordinates are related to the corresponding sequence voltages and currents by

$$[V_{phase}] = [T][V_{sequence}] \tag{19}$$

$$[I_{phase}] = [T][I_{sequence}] \tag{20}$$

Using Equations 19 and 20 in Equation 17, the following equations are obtained in phase quantities,

$$[i_{phase}(t)] = [T][Y][T]^{-1}[v_{phase}(t)] + [T][I_{hist}(t-\Delta t)] \tag{21}$$

or

$$[i_{phase}(t)] = [Y_{phase}][v_{phase}(t)] + [I_{hist\text{-}phase}(t-\Delta t)] \tag{22}$$

Equation 22 is not yet in the right form to interface the transformer model with the rest of the network solution because the variables are still associated with the transformer coils and not with the external nodes. Labeling the nodes as shown in Figure 12, Equation 22 can now be transformed into node voltages and currents as follows:

$$[I_{node}(t)] = \begin{bmatrix} Y_{phase} & -Y_{phase} \\ -Y_{phase} & Y_{phase} \end{bmatrix}[V_{node}(t)] + \begin{bmatrix} I_{hist\text{-}phase}(t-\Delta t) \\ -I_{hist\text{-}phase}(t-\Delta t) \end{bmatrix} \tag{23}$$

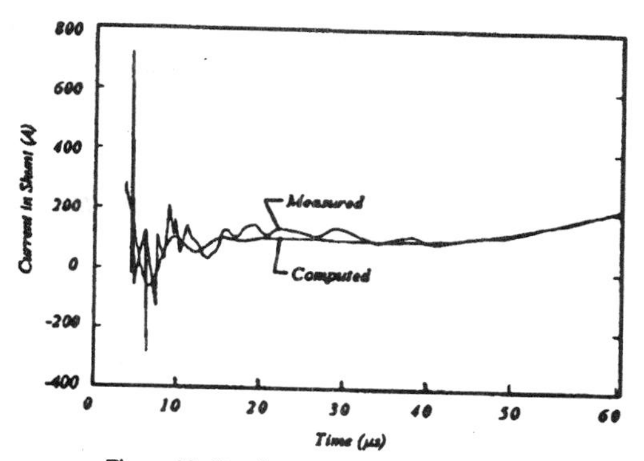

Figure 13. Step impulse transient. Comparison of measurement and simulation.

Equation 23 is the desired nodal equation and, in combination with the corresponding expression for the terminal capacitances, constitutes the complete frequency dependent model for the three-phase three-winding transformer.

TIME-DOMAIN SIMULATIONS

An impulse response test was performed on the tested 50 MVA 115/23 kV transformer. Figure 13 shows the measured current response to a 550 kV 1.2/50 μs voltage impulse applied to the delta winding, with the wye winding shorted. Considering the uncertainty of the measurements

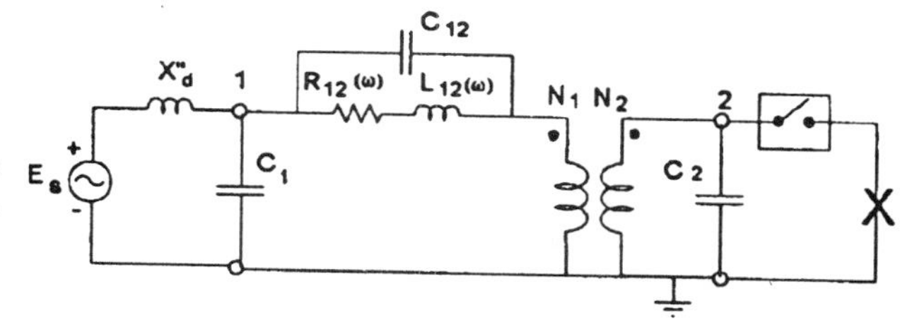

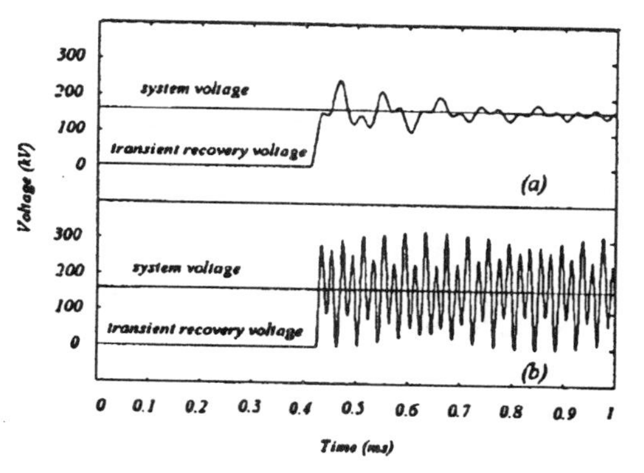

Figure 14. Transient recovery voltage in a circuit breaker after clearing a fault. (a) Frequency-dependent model. (b) Constant-parameter model

beyond 100 kHz (as discussed before), the simulated and measured test results are in good agreement.

To compare the proposed frequency dependent transformer model with the constant-parameter model, a fault interruption case (Figure 14) was simulated with the program MicroTran (UBC version of the EMTP). The plots in

Figure 14 compare the voltage across contacts of the circuit breaker obtained with the proposed frequency dependent model (with parameters derived from the tested 50 MVA 115/23 kV transformer) and with a constant parameters model in which the short-circuit impedance is represented with the 60-Hz resistance and inductance. The same external capacitances network is used for both models. The difference in the results illustrate the importance of more accurate transformer modelling in fast switching transients.

Conclusions

A wide-band general-purpose model has been developed for three-phase two- and three-winding power transformers. Instead of using a [Y(ω)] matrix formulation for the transformer as seen from its external terminals, the model uses the classical 60-Hz T-circuit to represent the electric and magnetic interaction among coils belonging to the same phase. For three-phase common-core units, the mutual interaction among different phases is decoupled through a modal transformation matrix. Even though the concept is general, a balanced-system transformation matrix is assumed in order to simplify the modelling and the test data requirements. The decoupled sequence networks consist of a frequency-dependent short-circuit branch and constant-valued terminal capacitances. These parameters were measured experimentally on a 50 MVA 115/23 kV three-phase core-type power transformer.

As a result of the simplified topology, the frequency dependence modelling problem is reduced to the fitting of simple minimum-phase-shift impedances (the short-circuit impedance). Therefore, the possible numerical stability problems associated with the synthesis of the mutual terms in the [Y(ω)] formulation have been eliminated. Also, the model has fewer and simpler frequency dependent functions to synthesize, making it much faster in time domain simulations.

Acknowledgement

The authors would like to thank Mrs. Supatra Phumiwat of the Electricity Generating Authority of Thailand (EGAT) for her great efforts in conducting transformer testing and providing the transformer data used in this paper. We are also indebted to the Canadian International Development Agency (CIDA) for their financial support of Mr. S. Chimklai's Ph.D. work at the University of British Columbia and to the Natural Sciences and Engineering Council of Canada (NSERC) for their financial support of our basic research work.

References

[1] R.C. Degeneff, "A General Method for Determining Resonances in Transformer Windings," IEEE Trans., vol. PAS-96, pp. 423-430, 1977.

[2] M. D'Amore and M. Salerno, "Simplified Models for simulating Transformer Windings Subject to Impulse Voltage," Paper A 79431-8, presented at IEEE PES Summer Meeting, Vancouver, British Columbia, Canada, July 15-20, 1979.

[3] P.T.M. Vaessen, "Transformer Model for High Frequencies," IEEE Trans. on Power Delivery, vol. 3, No. 4, pp. 1761-1768, October 1988.

[4] R.C. Degeneff, "A Method for Constructing Terminal Models for Single-Phase n-Winding Transformers, Paper A 78 539-9, IEEE PES Summer Meeting, Los Angeles, Calif., July 16-21, 1978.

[5] T. Adielson, A. Carlson, H.B. Margolis, J.A. Halladay, "Resonant Overvoltages in EHV Transformers-Modelling and Application," IEEE Trans., vol. PAS-100, pp. 3563-3572, 1981

[6] Francisco de Leon and Adam Semlyen, "Reduced Order Model for Transformer Transients," IEEE Trans. on Power Delivery, vol. 7, No. 1, pp. 361-369, January 1992.

[7] A. Morched, L. Martí, and J. Ottevangers, "A High-Frequency Transformer Model for the EMTP," IEEE Trans. on Power Delivery, vol. 8, No. 3, pp. 1615-1626, July 1993.

[8] F. de León and A. Semlyen, "Complete Transformer Model for Electromagnetic Transients," IEEE PES Winter Meeting, 93 WM 053-9 PWRD, 1993.

[9] H. Baher, *Synthesis of Electric Networks*, John Wiley & Sons, New York, 1984.

[10] J. R. Martí, "Accurate Modelling of Frequency-Dependent Transmission Lines in Electromagnetic Transient Simulations," IEEE Trans., vol. PAS-101, 1982.

[11] J. R. Martí, F. Castellanos, and N. Santiago, "A Wide Bandwidth Corona Model," Proceedings PSCC, 11th Power Computation Conference, Avignon, Aug. 30 - Sept 3, 1993, pp. 899-905.

[12] J. R. Martí, *Modelling of Power Transformers* (in Spanish), Central University of Venezuela, 1975.

[13] Allan Greenwood, *Electrical Transients in Power Systems, 2nd Ed.*, John Wiley & Sons, 1991.

[14] G. R. Slemon and A. Straughen, *Electric Machines*, Addison-Wesley, 1980, pp. 137-142.

[15] Ross Caldecot, Yilu Liu and Selwyn E. Wright, "Measurement of the Frequency Dependent Impedance of Major Station Equipment," IEEE Trans. on Power Delivery, vol. 5, No. 1, pp. 474-480, January 1990.

[16] Yilu Liu, Stephen A. Sebo and Selwyn E. Wright, "Power Transformer Resonance - Measurements and Prediction," IEEE Trans. on Power Delivery, vol. 7, No. 1, pp. 245-253, January 1992.

[17] P.T.M. Vaessen and E. Hanique, "A New Frequency Response Analysis Method for Power Transformers," IEEE Trans. on Power Delivery, vol. 7, No. 1, pp. 384-391, January 1992.

[18] Richard L. Bean, Nicholas Chackan, Jr., Harold R. Moore, Edward C. Wentz, *Transformers for the Electric Power Industry*, McGraw-Hill, 1959.

[19] William T. Vetterling, Saul A. Teukolsky, Willam H. Press and Brian P. Flannery, *Numerical Recipes-Example Book [C], 2nd Ed.*, Cambridge University Press.

[20] L. F. Blume, A. Bayajian, G. Camilli, T. C. Lennox, S. Minneci, V. M. Montsinger, *Transformer Engineering, 2nd ed.*, John Wiley & Sons, New York, 1951.

[21] Edith Clarke, *Circuit Analysis of A-C Power Systems, Volume I: Symmetrical and Related Components*, John Wiley & Sons, New York, 1943.

Biographies

Suthep Chimklai was born in Bangkok, Thailand in 1955. He received his Bachelor degree of Electrical Engineering in 1977 from Chulalongkorn University, Thailand. He has been working for the Electricity Generating Authority of Thailand since he graduated. In 1984, he received his Master degree in Electrical Engineering from Carnegie Mellon University, Pittsburgh, USA. He is currently a Ph.D. candidate at the University of British Columbia, Vancouver, Canada.

José R. Martí (M'71) was born in Lérida, Spain in 1948. He received the degree of Electrical Engineer from Central University of Venezuela in 1971, the degree of M.E.E.P.E from Rensselaer Polytechnic Institute in 1974 and the Ph.D. degree from the University of British Columbia in 1981. In Venezuela, he worked for industry and taught at Central University of Venezuela. At present he teaches at the University of British Columbia, Canada. Dr. Martí has been involved for a number of years in the development of models and solution techniques for the transient analysis program EMTP.

Discussion

Francisco de León (Instituto Politécnico Nacional, Mexico). The authors are congratulated for their very interesting and clearly written paper. They have developed a new, low order, model for the transformer windings. The model parameters are calculated from a set of short circuit terminal measurements. The authors' comments to the following questions and concerns would be greatly appreciated:

1) Can the authors kindly describe how the capacitance measurements were performed? Without this information, one could think that the measurements are correct for the whole frequency range. The oscillations at frequencies above 100 kHz. could be the result of the many internal resonances (series and parallel) in the winding, and not due to large inaccuracies in the measurements as suggested in the paper.

2) All models based on measurements are only applicable for transformers already built. Although, some general trends can be inferred from the tests, according to design, size, manufacturer, etc., accurate predictions for non-tested transformers cannot be assured. Therefore, the necessary data for the model might not be available at the planning stage. Will the authors comment on this?

3) It is stated in the paper, second paragraph of the Introduction, that the transformer model of reference [8] (or reference [A]) is based on the fitting of a $[Y(\omega)]$ matrix that represents the transformer at its terminal. Although, it is true that this model requires the fitting of a $[Z(\omega)]$ matrix, such a matrix is not calculated at the transformer terminals. On the contrary, it is based on *turn-to-turn* frequency dependent information (resistances and inductances) from where we later obtain coils, windings, and perform transformer connections. The details of the derivation of the model can be found in reference [B].

[A] F. de León and A. Semlyen, "Complete Transformer Model for Electromagnetic Transients," IEEE Transactions on Power Delivery, Vol. 9, No. 1, January 1994, pp. 231-239.

[B] F. de León and A. Semlyen, "Detailed Modeling of Eddy Current Effects for Transformer Transients," IEEE Transactions on Power Delivery, Vol. 9. No. 2, April 1994, pp. 1143-1150.

Manuscript received August 9, 1994.

B.C. Papadias, N.D. Hatziargyriou, J.M. Prousalidis (National Technical University of Athens, Athens, Greece): The discussers wish to congratulate the authors on their timely and interesting work. They would appreciate the authors' comments on the following points:

- It is shown in measurements that the stray capacitances are constant to about 100 kHz. However in CIGRE WG.13.02 reports [A,B], it is argued that **the effective capacitance**, with which transformer participates in switching phenomena, is in the order of 60% - 80% the value at the operating frequency (50/60 Hz). The discussers would appreciate further discussion on this subject.

- The asymmetries of transformer core magnetic circuit, as stated by the authors, play no significant role [C] in conditions where the contribution of shunt magnetizing term is not important, e.g. in short circuits, or in full load conditions. On the other hand, in no load conditions [C] or in highly unbalanced conditions [D], this contribution could be considerable. Could the authors comment on a possible extension of their model taking into account the frequency dependence of the magnetizing term and the unbalanced mutual phase coupling?

- The interphase capacitance (between outside windings of different legs), is considered to be the same for all phases, disregarding the fact that the distance between the outer limbs is approximately double the one between the central and an outer limb. In general, these capacitances are considered negligible [A], therefore no important information is lost by this assumption of the authors. However, it would be interesting to know if the authors have had any available measurements on these capacitances indicating a typical value of these terms or the discrepancies between different pairs of legs.

References

[A] CIGRE W.G. 13.02:" Chapter 5: switching of unloaded transformers; Part 1: Basic theory and single phase transformer interruption without reignitions", Electra No 133, pp.78-97

[B] CIGRE W.G. 13.02:" Chapter 5: switching of unloaded transformers; Part 2: Three phase transformer interruption, reignition phenomena, test result and conclusions", Electra No 134, pp. 23-134.

[C] J.M. Prousalidis, N.D. Hatziargyriou, B.C. Papadias: "The effect of mutual phase coupling on three-phase transformer models", EPST '93, Lisbon (Portugal), June 1993.

[D] B.C. Papadias, N.D. Hatziargyriou, J.A. Bakopoulos, J.M. Prousalidis:"Three-Phase Transformer Modelling for Fast Electromagnetic Transients", paper No 93 SM PWRD presented at 1993 IEEE PES Summer Meeting, Vancouver (Canada), July 1993.

Manuscript received August 24, 1994.

S. Chimklai and J.R. Martí (University of British Columbia, B.C., Canada): We thank the discussers for their kind comments and interesting questions. Dr. Papadias et al.'s and Dr. de León's first questions refer to the capacitance measurements shown in Figure 4 of the paper. As indicated in Figure 6 of the paper, these capacitance measurements were performed with the coils short circuited. Under these conditions, the impedances $Z_{winding}$ and Z_m in the equivalent circuit are shorted and only the C's (interwinding and winding to ground capacitances) are measured. That is, for as long as the proposed lumped-parameter circuit is valid, the expected measured impedance would be a constant capacitance. The measurement of Figure 4 suggests that this assumption is correct up to about 100 kHz. Beyond this frequency, it is unclear whether the oscillations in the measured impedance are due to instrumentation errors, or, as pointed out by Dr. de León, to local resonances in the actual distributed-parameter nature of the windings. In connection with Dr. Papadias et al.'s comments concerning CIGRE's reports, we are not familiar with the experimental context underlying these reports, but impulse-wave measurements, for example, would not separate capacitance from inductance as "cleanly" as our experimental set up, which short circuits the coils, and could give a lower value for the capacitances at switching frequencies.

We next answer the remaining questions of the discussers. We start with Dr. Papadias et al.'s second and third questions.

One of the advantages of the proposed circuit model is that it separates the identity of the various component parts: short-circuit impedance, capacitances, and magnetization branch. The "separate" magnetization branch in the circuit can then be modeled with as much detail as desired and incorporate, for instance, frequency dependence and saturation.

We concur with the comments in Dr. Papadias et al.'s third question in that the phase-to-phase capacitances are probably small compared to the other capacitances in the model and, therefore, their different values do not significantly affect the balanced model assumption. Our tests, however, do not directly measure these capacitances (Figures 5 and 10 of the paper) but a combined "balanced" capacitance for the proposed equivalent circuit. It is not possible from these measurements to isolate the phase-to-phase capacitance values or their differences between external and central phases.

It may be interesting to mention, nonetheless, that in the "simulated positive sequence" short-circuit tests (Figure 10(a) in the paper) conducted on a two-winding transformer, there were no significant differences in the derived capacitance values when the coil shorted was the coil in the middle leg of the core or the coil on an outer leg. Since the other capacitances involved in the test are more "balanced" than the phase-to-phase capacitances, these tests seem to verify the assumption that the phase-to-phase capacitances have less importance than the other transformer capacitances.

We address now Dr. de León's second and third questions. We agree with Dr. de León in that modeling at the planning stage requires data from design dimensions. Reference [A] in Dr. de León's discussion (reference [8] in the paper) is an excellent work in this connection. We also hope that given the importance of more accurate frequency dependent transformer modeling, even in common switching and fault transient studies, transformer manufacturers will include frequency responses as part of their standard transformer data.

Regarding Dr. de León's third comment, we mention in the Introduction to our paper that the formulations in [7] and [8] ([A] in the discussion) require the fitting of the self and mutual elements in the $[Y(\omega)]$ (or $[Z(\omega)]$) matrix, while our formulation only involves the fitting of a single $Z_{winding}$ impedance (short-circuit impedance) for the positive and zero sequence modes. We apologize for not making it clear that in the case of reference [8] ([A]) the elements of the matrix are not obtained from experimental measurements but built up from more basic data ([B]).

Manuscript received November 23, 1994.

A SIMPLE REPRESENTATION OF DYNAMIC HYSTERESIS LOSSES IN POWER TRANSFORMERS

Francisco de León
Instituto Politécnico Nacional - E.S.I.M.E.
Edificio No. 5, 3er Piso
07738 - México, D.F., México

Adam Semlyen
Department of Electrical and Computer Engineering
University of Toronto
Toronto, Ontario, Canada, M5S 1A4

Abstract - The paper describes a procedure for the representation of hysteresis in the laminations of power transformers in the simulation of electromagnetic transient phenomena. The model is based on the recognition that in today's iron cores the hysteresis loops are narrow and therefore the modeling details are only important in relation to the incurred losses and the associated attenuation effects. The resultant model produces losses proportional to the square of the flux density, as expected from measurement data. It is formulated as a simple, linear relationship between the variation $B-B_{rev}$ of the magnetic flux density B after a reversal point B_{rev} and the resulting additional field intensity H_{hyst}. This idea can be easily implemented in existing transformer models with or without frequency dependent modeling of eddy currents in the laminations. It has been found that in many simulation tests the representation of hysteresis is not necessary and those situations have been described where the modeling of hysteresis appears to be more meaningful.

Keywords: Transformer modeling, Electromagnetic transients, Hysteresis, Ferroresonance.

INTRODUCTION

Transformer modeling for the simulation of electromagnetic transients has made significant advances in the last decade. A fairly complete list of references in this field can be found in [1]. This reference summarizes our contributions to the field with the remark that it covers all the major phenomena that are relevant for transformer modeling with the exception, however, of the dynamic representation of hysteresis in the iron core. The main reason why this has been left out is the complexity of the phenomena, where the nonlinearity of saturation is coupled with the complicated dependence of the magnetic field intensity on the present and past values of the flux density, characteristic to hysteresis. Numerous studies exist, however, related to hysteresis [2]-[53] and successful achievements have been reported in the implementation of some models in the representation of transformers. Because of their significance, we present a fairly extensive overview of these models in the Appendix at the end of the paper.

A general characteristic of most existing hysteresis models is their sophistication and complexity. This may slow down the computer simulation of transients. A careful examination of the rationale for the representation of hysteresis in transformer models and of the plots showing results of measurements of dynamic hysteresis has lead us to the conclusion that a very simple hysteresis model could be adequate for achieving the correct representation of the attenuation of transients that can be attributed to hysteresis. We are thus in the position of presenting a simple and efficient hysteresis model to supplement the fairly complete transformer model we have previously described [1].

We make from the outset the following clarification regarding the terminology we use: the word "dynamic" in relation to hysteresis is used to indicate and emphasize that the phenomena are history dependent, rather than to include – as done in many classical texts – the effects of eddy currents in the laminations.

94 SM 407-7 PWRD A paper recommended and approved by the IEEE Transformers Committee of the IEEE Power Engineering Society for presentation at the IEEE/PES 1994 Summer Meeting, San Francisco, CA, July 24 - 28, 1994. Manuscript submitted August 20, 1993; made available for printing April 20, 1994.

Review of Existing Models

There are basically three types of vaguely defined approaches and originators in modeling of hysteresis in ferromagnetic materials. In the first group we have the physicists. They primarily look at the physical properties of the material, i.e., domain alignments, wall movements, spin rotations, etc. In the second group are those working in machine designing based on electromagnetic fields. They prefer a macroscopic description of hysteresis using mathematical models to predict the $B-H$ curve but without completely neglecting the physics of the material. In the third group we have power system engineers. They need equivalent circuits to be introduced in existing computer programs. Their base for modeling is the $B-H$ curve obtained by tests. The circuits should predict the losses in transient and steady state conditions. The purpose of the paper is to contribute to this last approach.

The bibliographic review presented in the Appendix is mainly devoted to models of the second two groups and especially to the last one, after 1970. Most of the publications pre-1970 can be found in the references of [3].

In the following we introduce, justify, and describe the new hysteresis model. Then we show its effects on different types of transformer transients.

DYNAMIC HYSTERESIS MODELING

Fundamental Remarks

Hysteresis is a very complex phenomenon. Curves showing the dynamic relation between B and H illustrate that the hysteresis related component H_{hyst} of the magnetic field intensity H is strongly dependent on the magnetization history. In figure 1 we show a measured hysteresis characteristic showing minor loops taken from reference [26], Figure 7. It is not our purpose to analyze or describe this problem. We make however two basic observations relative to the problem of hysteresis as it applies to power transformers:

• As a result of technological improvements, the iron core laminations have at present much reduced losses compared to past constructions. These are generally only a fraction of one percent (based on transformer rating). Therefore, the figures that describe hysteresis should be viewed as having increased scales for H in order to exhibit the details; when, however, the magnetization curve is displayed with a sufficient portion of the saturated branches, then the hysteresis loops narrow down to a very thin strip so that their details become immaterial and only the associated losses and attenuation remain relevant for the simulation of transients (see Figure 2). Figure 6 of reference [26] presents a measured full cycle that shows the described features (very narrow cycle). Therefore, in what follows, we shall focus primarily on an adequate reproduction of the hysteresis related losses and give preference to simplicity over precision as the latter has only negligible influence on the magnitude of the magnetizing current.

• Magnetizing curves have branches with asymptotically finite slopes at increasing flux densities. It may therefore appear that hyperbolic approximations [12] would be the most appropriate for their fitting. While they have been examined in great detail and implemented for the modeling of hysteresis loops [54], they do not have a flexibility comparable to polynomial approximations for improved fitting of the magnetization characteristic. Since, as discussed above, the precise representation of hysteresis loops is not of primordial importance, full freedom remains for the representation of the basic magnetization curve, including polynomial fitting.

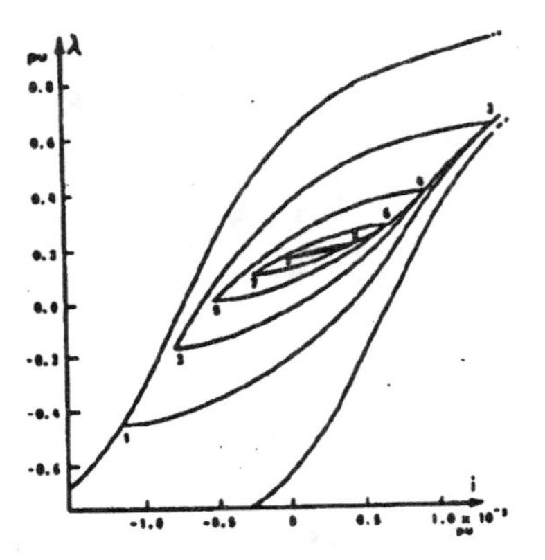

Figure 1. Measured magnetization curve [26] showing minor loops

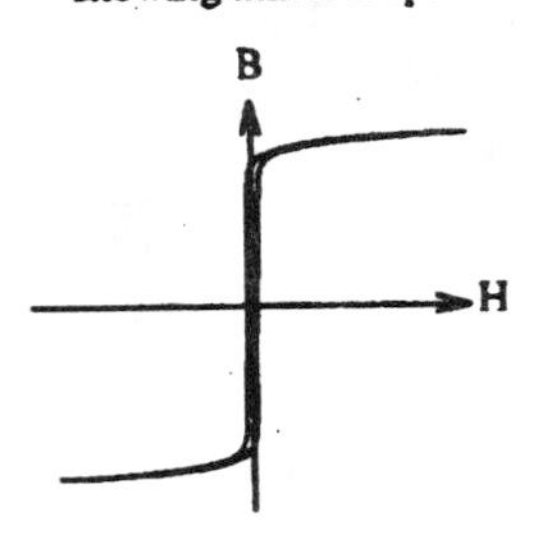

Figure 2. Narrow hysteresis main cycle

The New Dynamic Hysteresis Model

In order to build our hysteresis model we have found it convenient to postulate the existence of a "basis" magnetization curve

$$H_{basis} = f_0(B) \tag{1a}$$

This curve is related to the standard magnetization curve for the real magnetic material (i.e., in the presence of hysteresis) through the hysteresis losses, as reflected by the model described below. It should not be identified with the magnetization curve for the idealized behavior of the same magnetic material without hysteresis. The term "basis" simply reflects the fact that in our model hysteresis effects are assumed to originate and to end on this curve. It is a "reset" curve for hysteresis before any reversal of B.

As the hysteresis loop is very thin (as mentioned above), we will use a polynomial approximation for the basis curve with a very steep initial slope

$$H_{basis} = K_{basis} B + K_{n_1} B^{n_1} + K_{n_2} B^{n_2} \tag{1b}$$

In Figure 3 we show a basis curve for $n_1 = 17$ and $n_2 = 21$ with $K_{basis} = 0$, $K_{17} = 0.2181$, and $K_{21} = 0.1353$ (for S.I. units; see [47]).

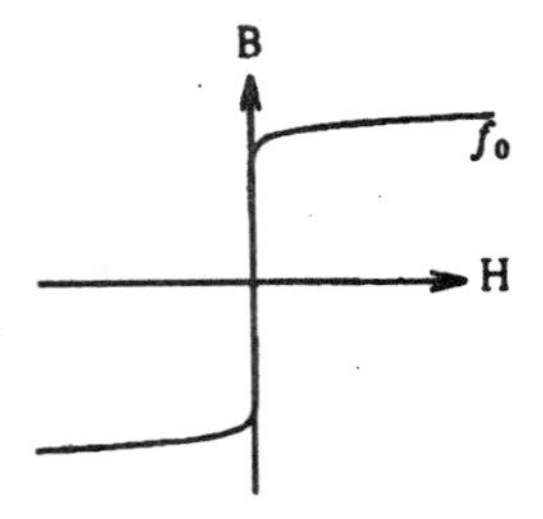

Figure 3. Basis curve

In our model we assume that there exists a hysteresis related field intensity, proportional to the change in B from the previous reversal point:

$$H_{hyst} = K_{hyst} (B - B_{rev}) \tag{2a}$$

Therefore,

$$H = H_{basis} + K_{hyst} (B - B_{rev}) \tag{2b}$$

(see Figure 4; here, for better illustration the basis curve is different from that of Figure 3). Reversal means that the time derivative of B changes sign.

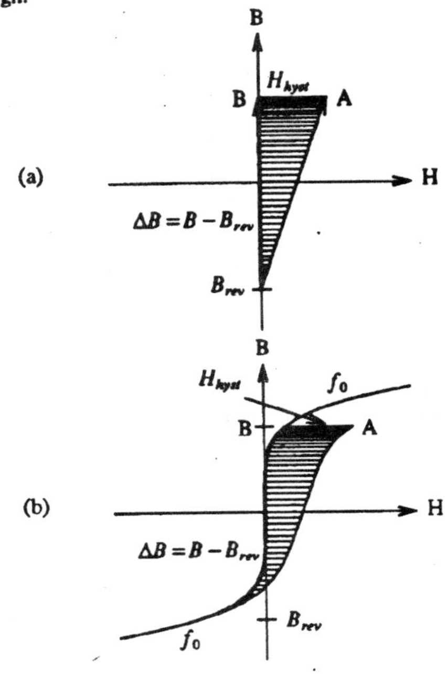

Figure 4. Basic idea of hysteresis modeling
(a) H_{hyst} component to be added, equation (2a)
(b) H_{hyst} added to basis curve, equation (2b)

If at the point A there is again a reversal, we return to the basis curve f_0. If the process is now duplicated with descending B, we have the loop shown in Figure 5. The resulting area of the loop is

$$AREA_{loop} = H_{hyst} (B - B_{rev}) = K_{hyst} (B - B_{rev})^2 \tag{3}$$

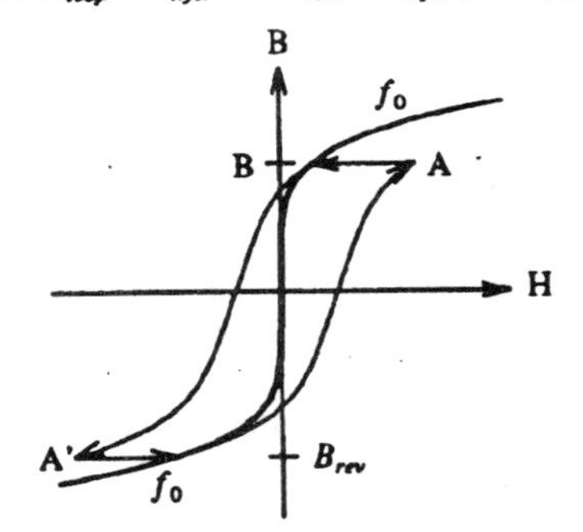

Figure 5. Asymmetrical loop

This indicates that even in minor loops the losses are proportional to ΔB^2, see reference [55]. In 1892 Steinmetz [56] proposed an empirical equation that relates the hysteresis losses to frequency and flux density:

$$P_{hyst} = K_{loss} f B^n \tag{4}$$

Steinmetz computed an exponent $n = 1.6$ which, however, for modern steels used in transformers varies between 1.5 and 2.5 and may not be constant [55]. Although an expression of the form (4) is not fully accurate for general use, as an approximation we have selected an exponent equal to two.

Consider now the symmetrical loop of Figure 6. Then

$$B_{rev} = -B \tag{5}$$

and (3) yields

$$AREA_{sym} = K_{loss} B^2 \tag{6}$$

where

$$K_{loss} = 4 K_{hyst} \tag{7}$$

The area of (6) corresponds to the hysteresis losses with symmetrical magnetization. Values of K_{loss} are available from measurements. Thus we also know

$$K_{hyst} = \frac{K_{loss}}{4} \tag{8}$$

Figure 6. Symmetrical loop

If symmetrical loops of different amplitude are repeated, we get the picture of Figure 7. This appears to be a generalization of idealized hysteresis loops presented in the literature. In [57], for example, straight line loops are proposed (Fig. 2.21), similar to those in the central, unsaturated part of Figure 7. Such idealized, symmetrical hysteresis loops shown in the literature are, however, the starting point for simulations, while in the approach of the paper they are the result of a more general, dynamic model (equation 2a) valid for any type of transient and not restricted to linear magnetization curves.

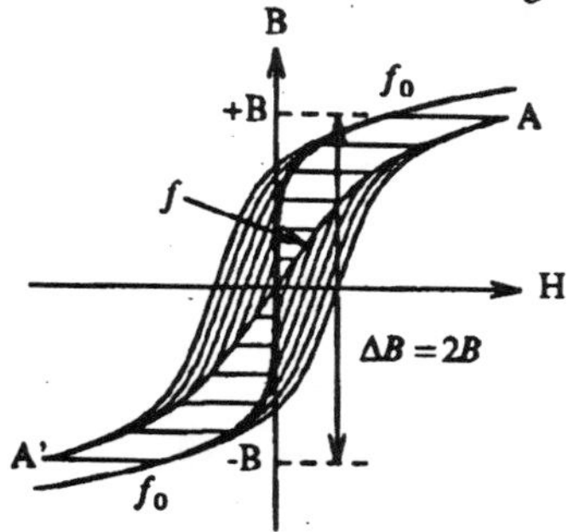

Figure 7. Symmetrical loops

By definition, the magnetization curve

$$H = f(B) \tag{9}$$

is the locus of the return points A and A' in Figure 7. It is obtained from f_0 of (1) by adding H_{hyst} of (2a) corresponding to $2B$, according to (5):

$$H_{hyst} = K_{hyst} \times 2B \tag{10}$$

This yields

$$H = H_{basis} + K_{hyst} 2B = K_1 B + K_{n_1} B^{n_1} + K_{n_2} B^{n_2} \tag{11}$$

where, by (7),

$$K_1 = K_{basis} + 2K_{hyst} = K_{basis} + \frac{K_{loss}}{2} \tag{12}$$

Accordingly,

$$K_{basis} = K_1 - \frac{K_{loss}}{2} \tag{13}$$

IMPLEMENTATION OF THE HYSTERESIS MODEL

Due to the model's simplicity, the computer implementation is straightforward. It is based on equation (2b). We start by computing K_{loss} from hysteresis loss measurements. Then, for a given approximation (11) of the magnetization curve, we use (13) to obtain K_{basis}. For time simulations one only needs to keep track of B for the present time and the two previous integration steps. Using a very simple logic (only one *if* statement) one can control the program flow. If the direction of the change in B is unchanged (i.e., the point where we are is not a reversal point), we continue using equation (2b). When a reversal point is encountered, then we first reset H of (2b) to H_{basis} and continue with equation (2b). What we do at a reversal point is, geometrically speaking, displacing the operating point horizontally to the basis curve (see Figure 5). The displacement at reversal points is horizontal in our case, consistently with the asymptotes of the polynomial describing the saturation characteristics, but primarily for simplicity. In a transformer model the above procedure is implemented in all magnetic branches, including those possibly used in the discretized representation of the laminations for the purpose of eddy current modeling [1]. However, the focus of the following simulations is on the effect of hysteresis itself.

In order to illustrate how the model works, we present in Figure 8 the simulation of an iron core driven by a sinusoidal voltage source with increasing amplitude. Figure 8a shows the excitation voltage and Figure 8b shows the response when the iron core has no remanent magnetization.

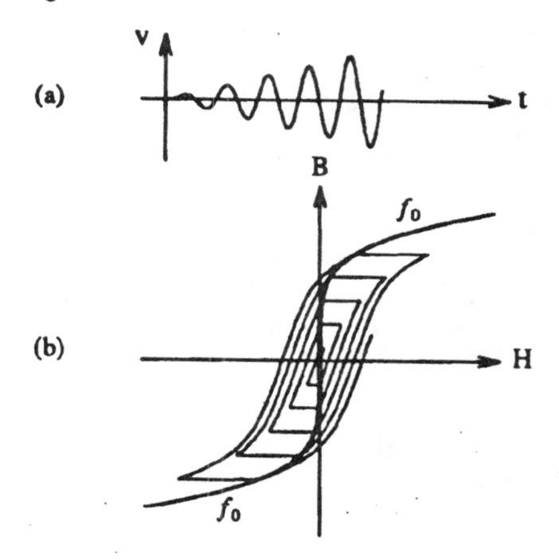

Figure 8. Dynamic hysteresis model
(a) Excitation voltage; (b) Model response

SIMULATION RESULTS

In this section we perform a number of transient studies to find the effect of including hysteresis in the simulations. To start, we note however that only transients involving a single winding may lead to dominance of iron core phenomena. This excludes all "longitudinal" transients. Moreover, even in open circuit, at high frequencies the magnetizing flux will be small and, therefore, modeling of both saturation and hysteresis becomes unimportant. Consequently, only a few situations which are suspected to remain of significance (although there may be more) are analyzed below.

Inrush Currents

When a transformer is energized, a large (inrush) current may be drawn from the source. There are a great number of references dealing with this problem; see, for example, the book by Greenwood [58] or reference [59]. To illustrate the effect of hysteresis in the inrush current, we use the simplest representation for the source, i.e., an ideal sinusoidal voltage source with constant amplitude. In Figure 9 we present the simulation of the inrush current for phase C of the three-phase three-legged transformer presented in reference [1]. The figure actually shows two cases: with and without hysteresis in the simulation. We note that there is no difference between the two cases in the magnitude and damping of the inrush current.

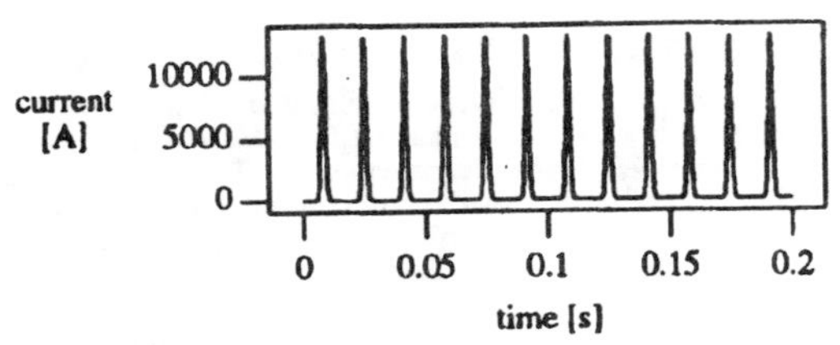

Figure 9. Inrush current with and without hysteresis

We believe, the explanation of this negative result is as follows: if we imagine the circuit representation of hysteresis as a resistance in parallel with the inductor, then the voltage source will absorb directly the losses caused by hysteresis. Since the magnitude of the inrush current is only dependent of how much the material becomes saturated, the effect of hysteresis is important solely in establishing the point from where the flux starts building up.

From our simulations we conclude:

- Hysteresis does not add noticeable damping to the inrush current.
- Hysteresis only affects the magnitude of the inrush current when there is remanent magnetization (which sets the initial condition).

Magnetizing Current Chopping

The chopping of magnetizing currents may lead to large transient overvoltages. This subject has also received very much attention in the literature; see for example [58], [60] and [61]. In this section we analyze the effects that hysteresis has in the disconnection of a transformer. The magnetizing current is abruptly chopped by a circuit breaker before its zero crossing, leaving a capacitance $C(=10^{-10}\,F)$ connected at the terminals of the transformer of [1]. In Figure 10 we show the transient voltage (without restrikes) when the starting point is well into saturation. Figure 10a corresponds to the simulation with the hysteresis model presented in this paper. We can observe that as the amplitude decreases, the frequency of the oscillations becomes smaller, as expected. If the damping due to hysteresis is represented by a shunt resistance calculated from the losses at 60 Hz, then the transient is excessively damped. If, on the other hand, we increase the value of the shunt resistance to give the correct damping at the highest frequency of the transient, then the damping is insufficient at lower frequencies; the results are shown in Figure 10b.

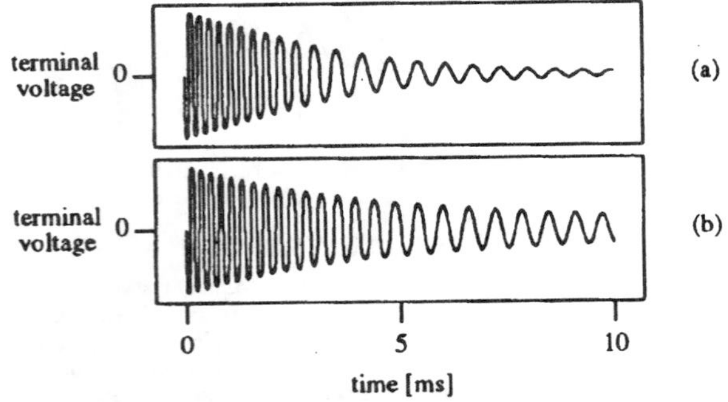

Figure 10. Transient following magnetizing current chopping
(a) Damping due to hysteresis model
(b) Damping due to high frequency equivalent resistance

Discussion

The frequency variation of the transient voltage is due to the fact that the effective inductance of the transformer varies with the saturation conditions. In saturation the inductance is smaller giving faster transient oscillations. Figure 11 gives the phase portrait (flux versus voltage) for the transient of Figure 10a. We note that the external contours, corresponding to the beginning of the transient, reflect the distortion due to saturation. As the transient attenuates, the contours describe more perfect ellipses, corresponding to non-saturated conditions.

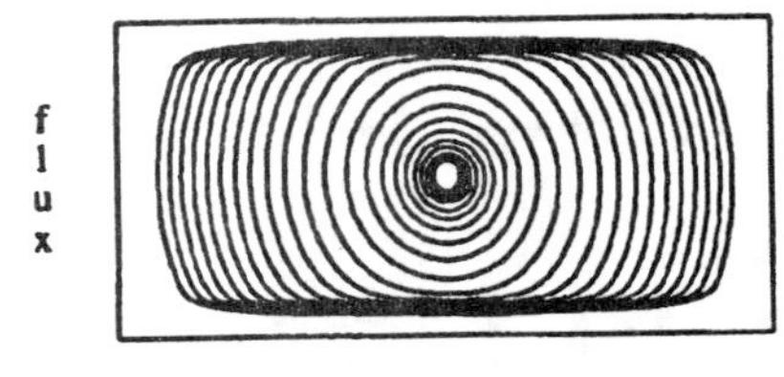

Figure 11. Flux versus voltage (of Figure 10a)

The differences in the amount of damping seen in the simulations of Figure 10 can be explained by examining a periodic oscillation. Then the voltage (gradient) is related to the flux density by

$$E = \omega B \tag{14}$$

Our model predicts the correct losses, proportional to B^2 and f. Thus, according to (4),

$$P_{hyst} = K'_{loss}\,\omega B^2 \tag{15}$$

(where $K'_{loss} = K_{loss}/(2\pi)$). From (14) and (15) we have

$$P_{hyst} = \frac{K'_{loss}}{\omega} E^2 \tag{16}$$

The losses in a constant resistance are

$$P_R = \frac{1}{R} E^2 \tag{17}$$

From (16) and (17) we see that for a given sinusoidal voltage the hysteresis losses vary inversely proportional with frequency, while for a shunt resistance the losses do not depend on frequency. To properly represent the hysteresis losses the equivalent resistance should be

$$R_{equiv} = \frac{\omega}{K'_{loss}} \tag{18}$$

From our observations we conclude:

- Hysteresis has a significant effect in the damping of transients due to magnetizing current chopping.
- For the calculation of electromagnetic transients, hysteresis losses cannot be adequately represented by a constant resistance connected in parallel with the nonlinear inductance.

Ferroresonance

The phenomenon is a series resonance between the nonlinear inductor of an iron core transformer and the capacitance of the cable connected to it [58]. A very large voltage can appear across the inductor or capacitor even if the applied voltage is within reasonable bounds. There has been a considerable amount of work in this area; see for example the recent publications [62]-[64]. The circuit for the analysis is shown in Figure 12.

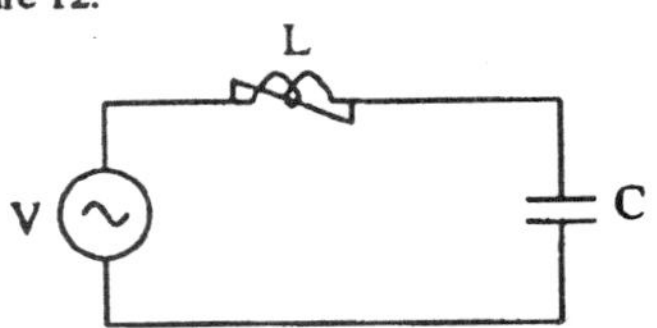

Figure 12. Circuit for the study of ferroresonance

Often, the situation in which ferroresonance occurs [62], [64] is when one or two phases of the feeder to an open-circuited transformer are disconnected from the supply source so that the capacitance to ground of these conductors appears in series with the magnetizing inductance of the transformer. The winding involved may have any connection, but if it has a star point, it should be isolated from ground. Any load or loss-producing element may prevent the appearance of a resonance condition.

Since the circuit of Figure 12 has no unique natural frequency (see Figure 10), one cannot analyze the phenomenon of ferroresonance using the simple concepts and approaches applied to the examination of resonance in linear circuits. In particular, one cannot separate a transient part from a steady state solution. The latter may not even exist and when it does it may and will often take many seconds (real, not simulation time), or even longer, to reach it. This long duration dynamic is very complex and extremely sensitive to small variations of all parameters of the problem. These include: C, V, the initial conditions (which can be considered as contributors to the initial stored energy in the system), and last but not least, the damping due to resistances and hysteresis.

While all our simulations eventually converged to a periodic steady state, with or without subharmonics, the long duration transient has often shown significant overvoltages, multiples of the peak of the source voltage. Therefore, the reliable simulation of ferroresonance is of great practical importance. Since we are dealing with a transient, it is inaccurate to assign a single frequency to it. Often subharmonic oscillations of different orders have been noticed in our simulations; see, for example, Figure 13. These results were obtained with the transformer used above for $v=100\sin(\omega t)$ and $C=10^{-7}$ F. However, due to space limitations we can only show a single sample of the many interesting results we have obtained. Figure 14 represents the peak values of the voltage oscillations across the capacitance in Figure 12, as a function of time. It corresponds to the transient shown in Figure 13 for a ten times longer time. A subharmonic of order 2 is clearly visible during most of the transient. This type of display is similar to those in [62] and is also related to the ideas of Poincaré sections, except that here we show maximum values (rather than periodically taken samples) because of their practical importance. We emphasize the significance of the special display used in our analysis: it gives directly the maximum values relevant for insulation coordination while any regular patterns can also be distinguished.

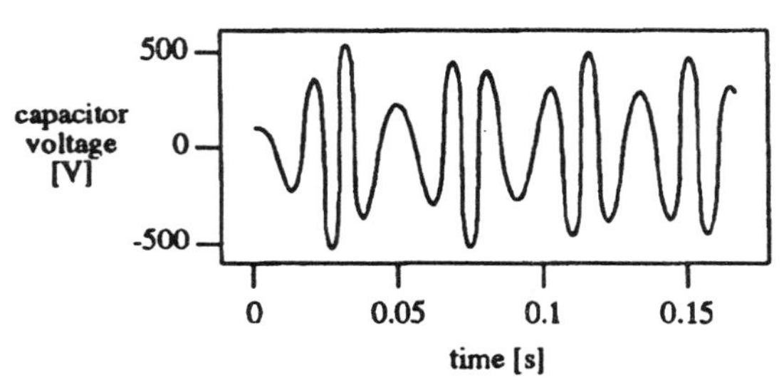

Figure 13. Ferroresonant voltage

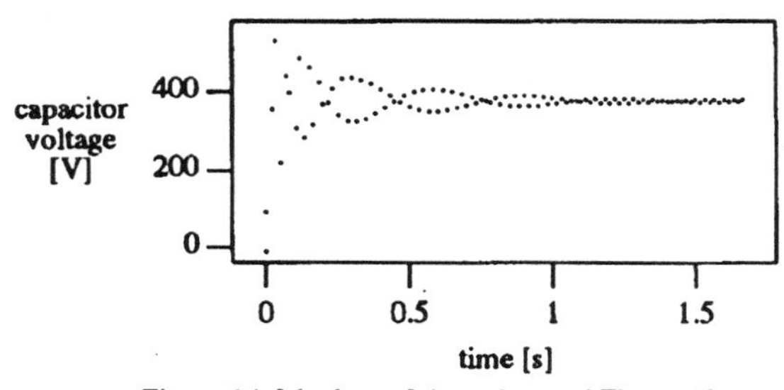

Figure 14. Maxima of the voltage of Figure 13

Since under transient conditions an equivalent resistance for the representation of hysteresis losses, chosen for the single frequency of 60 Hz, is inadequate, it was suspected that the simulation of ferroresonance, under otherwise equal conditions, would give different results with a hysteresis model than with an "equivalent" resistance. Figure 15a shows that indeed with hysteresis damping the simulation converges to a voltage of 60 Hz base frequency, while with an equivalent resistance (for 60 Hz), the steady state voltage, shown in Figure 15b, has a base frequency of 30 Hz (subharmonic of order 2). These simulations were obtained using the same transformer, as above, with $C=10^{-7}$ and $v=245\sin(\omega t)$.

We conclude by noting that even a small change in the hysteresis loss coefficient K_{loss} may significantly change the results. For practical purposes it is therefore useful to condense the results of simulations as in Figures 14 and 15a with K_{loss} varied over a reasonable range, by displaying only the maxima of the peak values, i.e., their upper envelope.

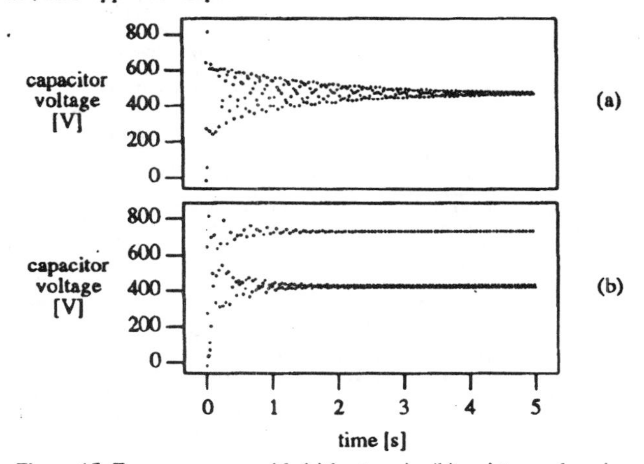

Figure 15. Ferroresonance with (a) hysteresis, (b) resistance damping

CONCLUSIONS

The paper describes a simple procedure for the representation of hysteresis in the laminations of power transformers for the simulation of electromagnetic transients. The model produces losses proportional to frequency and to the square of the flux density, as expected from measurements.

The main characteristics of the model are, besides its simplicity, the fact that it is dynamic (i.e., it is not restricted to symmetrical hysteresis loops or, in fact, any closed loops at all), and that it can be applied to any magnetization characteristic (described by polynomial, hyperbolic, or other types of functions). It deforms insignificantly the magnetization characteristic and affects a transient only through the incurred damping.

While the damping can be obtained by an equivalent parallel resistance, the frequency dependence of the two is different. Therefore, in cases where the dynamics of the phenomena is very sensitive to the losses and to speed and frequency, as in studies of ferroresonance and magnetizing current chopping, the dynamic modeling of hysteresis appears to be particularly important. We have found, however, that hysteresis does not add damping to the inrush currents. This indicates that the mere existence of losses may be of no practical importance in situations where they are directly covered from the power source.

ACKNOWLEDGEMENTS

Financial support by the Natural Sciences and Engineering Research Council of Canada is gratefully acknowledged. The first author wishes to express his gratitude to the Instituto Politécnico Nacional, Mexico, for the study leave at the University of Toronto.

APPENDIX

Review of Existing Models

References [2] to [53] represent a list of the most cited (or used) publications. It is of course not exhaustive. The first classic model for the prediction of hysteresis was by Preisach [2] in the 1930's. In this model materials are comprised by a number of magnetic dipoles each one exhibiting a square loop. Many researchers have followed Preisach's approach. It is the preferred approach of those developing finite elements programs. In 1970 Chua and Stromsmoe [3] presented the first attempt for the computer modeling of hysteresis with an electric circuit. Their model consists of a nonlinear resistor in parallel with a nonlinear inductor following a series of complicated function compositions. In 1971 Swift [4] states that eddy current losses are much more important than hysteresis losses for power transformers. Bouc [5] in 1971 presents a form of functional to give a mathematical description of hysteresis. Chua and Bass [6] improve on the model of reference [3] in 1972 to account for the d.c. loop with a still more complicated model. In 1974 Germay et al. [7] present static, dynamic and transient models based on the theory of Preisach [2]. The static model is obtained from the magnetization curve and the largest static hysteresis loop they could measure. The dynamic model is suitable for steady state a.c. conditions. Although, there are no details for their transient model, the authors admit that while it works very accurately for steady state simulations, transient conditions are calculated with limited accuracy. Nakra and Barton [8] in the same year proposed a transformer model where the hysteresis is represented as a number of closely spaced trajectories experimentally determined. Teape et al. [9] proposed in 1974 a model derived from experiments. Charap [10], also in 1974, proposed a two branch circuit with linear resistances and nonlinear inductances. This model is derived from movement of domain walls at 180° and 90°. In the same year, Wright and Carneiro [11] developed a hysteresis model for current transformers.

In 1975 Semlyen and Castro [12] proposed a hyperbolic model for saturation that, when shifted in two directions (parallel to the saturated slope), gives a representation of hysteresis. Also in 1975 Yamashita et al. [13] decompose the hysteresis cycle into several backlash elements with saturation. This model is suitable for the representation of minor loops but it is rather complicated since every segment is represented by a different equation. In the transformer model for the simulation of transients developed by Dommel [16] in 1975 the hysteresis characteristics are represented by a fundamental frequency cycle. Talukdar and Bailey [17] in 1976 propose a transient model for hysteresis following two families of curves "uppers" and "downers". Jufer and Apostolides [18], in 1976, obtain a transient and steady state model for saturation with a rational function of order one. The hysteresis model is obtained by displacing the curve and modifying the constants. Some more complicated functions are used for curved regions. In reference [19] O'Kelly (1977) presents an exponential function for the modeling of the hysteresis loop. Janssens [20], also in 1977, uses a family of curves fitted to experimental results. Coulson et al. [21] presented in 1977 another exponential model (more complicated) derived from the theory of Preisach [2]. Newbury [22] and [23], in 1978 and 1979, presents a hysteresis loop model for 3% grain-oriented silicon steel with square shape. His model includes minor loops and is relatively simple. Hannalla and Macdonald [24], in 1980, propose a formula for the prediction of hysteresis which gives good results in the calculation of inrush currents. Del Vecchio [25], in 1980, developed a complicated model for hysteresis based on the theory of Preisach [2] for non-oriented electrical steel laminations.

The model of Talukdar and Bailey [17] was improved by Dick and Watson [26], in 1981. They propose two models based on observations derived from tests. One model is based on hyperbolic functions and the other one uses several branches in series with current sources. Rivas et al. [27], in 1981, present a model as a rational function and get two curve families (one going up and another going down). Also in 1981, Stein [28] presents a transformer model for the calculation of transients using differential permeability. Ivanoff [29], in the same year, developed a model for hysteresis based on circulation rules related to four different paths of the major hysteresis loop. In reference [30], Raham et al. present a hysteresis model for hard magnetic materials. Del Vecchio [31], in 1982, presents a model of hysteresis that is suitable for calculation of fields using finite elements. In 1982, Saito [32] shows three-dimensional finite elements field calculations using a hysteresis model similar to Chua's [3]. Savini [33] used a piecewise linear representation of hysteresis in finite element calculations. Burais and Grellet [34], in 1982, present a rational function for the hysteresis modeling and its application for finite differences and finite elements.

Frame et al. [35] in 1982, present a hysteresis model that is useful for transient calculations and the model is included into the EMTP. This model is a modification of the model by Talukdar [17]. Avila and Alvarado [36], also in 1982, present a transient model for the iron core using the model proposed by Dick and Watson [26] consisting of current sources connected in series with the branches of a Foster circuit for the representation of hysteresis. Zaher and Shobeir [37] present a model for analog computer simulations which is similar to Talukdar's [17]. Jiles and Atherton [38], in 1983, present a model derived from the physical properties, i.e., magnetic dipoles, wall motions, domain rotations, etc. Saito et al. [39] present more results of their previous work [32]. Jiles and Atherton [40], in 1984, present a more detailed paper of their previous work [38]. In 1985, Mahmoud and Whitehead [42] perform a piecewise curve fitting of the magnetization characteristics. In reference [41] Prusty and Rao, in 1984, present an analytical expression obtained from the saturation curve and the no-load test. Udpa and Lord [43] use a Fourier descriptor for the representation of hysteresis. Ewart [44], in 1986, proposed a transformer transient model where the hysteresis is composed by infinitesimal dead-bands with saturation as hard limits. Green and Gross [45], in 1988, present a model for hysteresis in the study of harmonics which consists of four exponentials. Rajakovic and Semlyen [47], in 1989, use a polynomial for the representation of hysteresis in the harmonic domain. Joosten et al. [48], in 1990, show a very simple model consisting of straight lines with two slopes. The authors claim that in spite of its simplicity, more elaborate models have not shown better correlations with measurements. In 1989, Lin et al. [46] present a model with (so called) consuming functions. Their steady state representation of hysteresis is presented in reference [51] in 1991. The most recent attempt for hysteresis modeling is by Dolinar et al. [52] in 1993. They use Ewart's approach [44] to construct a transformer transient model together with a polynomial approximation for hysteresis.

Finally, in a recent paper [53] Marcki, Nistri and Zecca, in 1993, review from a mathematical point of view the existing models for hysteresis in various areas of engeneering, physics and mathematics. They provide a complementary list of references starting from 1897.

REFERENCES

[1] F de León and A. Semlyen, "Complete Transformer Model for Electromagnetic Transients", paper No. 93 WM 053-9 PWRD presented at the 1993 IEEE/PES Winter Meeting.

[2] F. Preisach, "Über die magnetische Nachwirkung", Zeitschrift der Physik 1935, 94, pp. 277-302.

[3] L.O. Chua and K.A. Stromsmoe, "Lumped-Circuit Models for Nonlinear Inductors Exhibiting Hysteresis Loops", IEEE Transactions on Circuit Theory, Vol. CT-17, No. 4, November, 1970, pp. 564-574.

[4] G.W. Swift, "Power Transformer Core Behavior Under Transient Conditions", IEEE Transactions on Power Apparatus and Systems, Vol. PAS-90, No. 5, September/October, 1971, pp. 2206-2210.

[5] R. Bouc, "Modèle Mathématique d'Hysteresis", Acustica, 1971 (24), pp. 16-25.

[6] L.O. Chua and S.C. Bass, "A Generalized Hysteresis Model", IEEE Transactions on Circuit Theory, Vol. CT-19, No. 1, January, 1972, pp. 36-48.

[7] N. Germay, S. Maestero, and J. Vroman, "Review of Ferroresonance Phenomena in High-Voltage Power Systems and Presentation of a Voltage Transformer Model for Predetermining Them", CIGRE, 1974 Session Paper 33-18.

[8] H.L. Nakra, T.H. Barton, "Three Phase Transformer Transients", IEEE Transactions on Power Apparatus and Systems, Vol. PAS-93, No. 6, November/December, 1974, pp. 1810-1819.

[9] J.W. Teape, R.R.S. Simpson, R.D. Slater, and W.S. Wood, "Representation of Magnetic Characteristic, Including Hysteresis, by Exponential Series", IEE Proceedings, Vol. 121, No. 9, September 1974, pp. 1019-1020.

[10] S.H. Charap, "Magnetic Hysteresis Model", IEEE Transactions on Magnetics, Vol. MAG-10, No. 4, December 1974, pp. 1091-1096.

[11] A. Wright and S. Carneiro, "Analysis of Circuits Containing Components with Cores of Ferromagnetic Material", IEE Proceedings, Vol. 121, No. 12, December 1974, pp. 1579-1581.

[12] A. Semlyen and A. Castro, "A Digital Transformer Model for Switching Transient Calculations in Three-Phase Systems", Proc. 9th PICA Conference, New Orleans, June 2-4, 1975, pp. 121-126.

[13] H. Yamashita, E. Nakamae, M.S.A.A. Hammam, and K. Wakisho, "A Program to Analyze Transient Phenomena of Circuits Including Precisely Represented Transformers", IEEE/PES Summer Meeting, July, 1975, paper No. A 75 403-6.

[14] A.M. Davis, "Synthesis and Simulation of Systems Possessing Ferromagnetic Hysteresis", Simulation, Journal of the Soc. of Computer Simulation, November 1975, pp. 153-157.

[15] J.L. Hay, R.I. Chaplin, "Dynamic Simulation of Ferromagnetic Hysteretic Behavior by Digital Computer", Simulation, Journal of the Soc. of Computer Simulation, December 1975, pp. 185-191.

[16] H.W. Dommel, "Transformer Models in the Simulation of Electromagnetic Transients", 5th Power System Computation Conference, Cambridge (England), September 1-5, 1975 Paper No. 3.1/4.

[17] S.N. Talukdar and J.R. Bailey, "Hysteresis Models for System Studies", IEEE Transactions on Power Apparatus and Systems, Vol. PAS-95, No. 4, July/August, 1976, pp. 1429-1434.

[18] M. Jufer and A. Apostolides, "An Analysis of Eddy Current and Hysteresis Losses in Solid Iron Based upon Simulation of Saturation and Hysteresis Characteristics", IEEE Transactions on Power Apparatus and Systems, Vol. PAS-95, No. 6, November/December, 1976, pp. 1786-1794.

[19] D. O'Kelly, "Simulation of Transient and Steady-State Magnetisation Characteristics with Hysteresis", IEE Proceedings, Vol 124, No. 6, June 1977, pp. 578-582.

[20] N. Janssens, "Static Models of Magnetic Hysteresis", IEEE Transactions on Magnetics, Vol. MAG-13, No. 5, September 1977, pp. 1379-1381.

[21] M.A. Coulson, R.D. Slater, and R.R.S. Simpson, "Representation of Magnetic Characteristics, including Hysteresis, using Preisach's Theory", IEE Proceedings, Vol 124, No. 10, October 1977, pp. 895-898.

[22] R.A. Newbury, "Prediction of Magnetization Hysteresis for Oriented Silicon Steel", Electric Power Applications, May 1978, Vol. 1, No. 2, pp. 60-64.

[23] R.A. Newbury, "Prediction of Magnetization Hysteresis Loops for Distorted Wave Forms", Electric Power Applications, April 1979, Vol. 2, No. 2, pp. 46-50.

[24] A.Y. Hannalla and D.C. Macdonald, "Representation of Soft Magnetic Materials", IEE Proceedings, Vol 127, Pt. A, No. 6, July 1980, pp. 386-391.

[25] R.M. Del Vecchio, "An Efficient Procedure for Modeling Complex Hysteresis Processes in Ferromagnetic Materials", IEEE Transactions on Magnetics, Vol. MAG-16, No. 5, September 1980, pp. 809-811.

[26] E.P. Dick and W. Watson, "Transformer Models for Transient Studies Based on Field Measurements", IEEE Transactions on Power Apparatus and Systems, Vol. PAS-100, No. 1, January 1981, pp. 409-419.

[27] J. Rivas, J.M. Zamarro, E. Martin, and C. Pereira, "Simple Approximation for Magnetization Curves and Hysteresis Loops", IEEE Transactions on Magnetics, Vol. MAG-17, No. 4, July 1981, pp. 1498-1502.

[28] B. Stein, "A Transformer Model for Electromagnetic Transients Programs", Proc. 7th PSCC, pp. 900-904, July 1981.

[29] D. Ivanoff, "Simulation of Magnetic Hysteresis by Algorithms Based on Circulation Rules", Proc. 7th PSCC, pp. 1204-1208, July 1981.

[30] M.A. Raham, M. Poloujadoff, R.D. Jackson, J. Perard, and S.D. Gowda, "Improved Algorithms for Digital Simulation of Hysteresis Processes in Semi Hard Magnetic Materials", IEEE Transactions on Magnetics, Vol. MAG-17, No. 6, November 1981, pp. 3253-3255.

[31] R.M. Del Vecchio, "The Inclusion of Hysteresis Processes in a Special Class of Electromagnetic Finite Element Calculations", IEEE Transactions on Magnetics, Vol. MAG-18, No. 1, January 1982, pp. 275-284.

[32] Y. Saito, "Three-Dimensional Analysis of Magnetic Fields in Electromagnetic Devices Taken into Account the Dynamic Hysteresis Loop", IEEE Transactions on Magnetics, Vol. MAG-18, No. 2, March 1982, pp. 546-551.

[33] A. Savini, "Modelling Hysteresis Loops for Finite Element Magnetic Field Calculations", IEEE Transactions on Magnetics, Vol. MAG-18, No. 2, March 1982, pp. 552-557.

[34] N. Burais and G. Grellet, "Numerical Modelling of Iron Losses in Ferromagnetic Steel Plate", IEEE Transactions on Magnetics, Vol. MAG-18, No. 2, March 1982, pp. 558-562.

[35] J.G. Frame, N. Mohan, and T. Liu, "Hysteresis Modeling in an Electro-Magnetic Transient Program", IEEE Transactions on Power Apparatus and Systems, Vol. PAS-101, No. 9, September 1982, pp. 3403-3412.

[36] J. Avila-Rosales and F.L. Alvarado, "Nonlinear Frequency Dependent Transformer Model for Electromagnetic Transient Studies in Power Systems", IEEE Transactions on Power Apparatus and Systems, Vol. PAS-101, No. 11, November 1982, pp. 4281-4288.

[37] F.A.A. Zaher and A.I. Shobeir, "Analog Simulation of the Magnetic Hysteresis", IEEE Transactions on Power Apparatus and Systems, Vol. PAS-102, No. 5, May 1983, pp. 1235-1239.

[38] D.C. Jiles and D.L. Atherton, "Ferromagnetic Hysteresis", IEEE Transactions on Magnetics, Vol. MAG-19, No. 5, September 1983, pp. 2183-2185.

[39] Y. Saito, H. Saotome, S. Hayano, and T. Yamamura, "Modelling of Nonlinear Inductor Exhibiting Hysteresis Loops and its Application to the Single Phase parallel Inverters", IEEE Transactions on Magnetics, Vol. MAG-19, No. 5, September 1983, pp. 2189-2191.

[40] D.C. Jiles and D.L. Atherton, "Theory of Magnetization Process in Ferromagnets and Its Application to Magnetomechanical Effect", Journal of Physics D. Applied Physics, Vol. 17, No. 6, June 1984, pp. 1265-1281.

[41] S. Prusty and M.V.S. Rao, "A Novel Approach for Predetermination of Magnetization Characteristics of Transformers Including Hysteresis", IEEE Transactions on Magnetics, Vol. MAG-20, No. 4, July 1984, pp. 607-612.

[42] M.O. Mahmoud and R.W. Whitehead, "Piecewise Fitting Function for Magnetisation Characteristics", IEEE Transactions on Power Apparatus and Systems, Vol. PAS-104, No. 7, July 1985, pp. 1822-1824.

[43] S.S. Udpa and W. Lord, "A Fourier Descriptor Model of Hysteresis Loop Phenomena", IEEE Transactions on Magnetics, Vol. MAG-21, No. 6, November 1985, pp. 2370-2373.

[44] D.N. Ewart, "Digital Computer Simulation Model of a Steel Core Transformer", IEEE Transactions on Power Delivery, Vol. PWRD-1, No. 3, July 1986, pp. 174-183.

[45] J.D. Greene and C.A. Gross, "Nonlinear Modeling of Transformers", IEEE Transactions on Industry Applications, Vol. 24, No. 3, May/June 1988, pp. 434-438.

[46] C.E. Lin, J.B. Wei, C.L. Huang, and C.J. Huang, "A New Method for Representation of Hysteresis Loops", IEEE Transactions on Power Delivery, Vol. 4, No. 1, January 1989, pp. 413-420.

[47] N. Rajakovic and A. Semlyen, "Harmonic Domain Analysis of Field Variables Related to Eddy Current and Hysteresis Losses in Saturated Laminations", IEEE Transactions on Power Delivery, Vol. 4, No. 2, April 1989, pp. 1111-1116.

[48] A.P.B. Joosten, J. Arrillaga, C.P. Arnold, and N.R. Watson, "Simulation of HVDC System Disturbances with References to the Magnetizing History of the Converter Transformers", IEEE Transactions on Power Delivery, Vol. 5, No. 1, January 1990, pp. 330-336.

[49] I. J. Binard and I.J. Maun, "Power Transformer Simulation Including Inrush Currents and Internal Faults", IMACS - TC1'90 Nancy, pp. 57-62.

[50] I. J. Binard and I.J. Maun, "Hysteresis Model for Power Transformer Transients Simulation Program", IMACS - TC1'90 Nancy, pp. 539-544.

[51] C.E. Lin, C.L. Chen, and C.L. Huang, "Hysteresis Characteristic Analysis of Transformer Under Different Excitations Using Real Time Measurements", IEEE Transactions on Power Delivery, Vol. 6, No. 2, April 1991, pp. 873-879.

[52] D. Dolinar, J. Philer, and B. Grcar, "Dynamic Model of a Three-Phase power Transformer", paper No. 93 WM 048-9 PWRD presented at the 1993 IEEE/PES Winter Meeting.

[53] J.W. Macki, P. Nistri, and P. Zecca, "Mathematical Models for Hysteresis", Society for Industrial and Applied Mathematics, SIAM Review Vol. 35, No. 1, March 1993, pp. 94-123.

[54] R.A.I. Roth, "Computer Model for Simulation of Transients in an Iron Core Inductor", M.A.Sc. Thesis, University of Toronto, 1979.

[55] M.I.T., "Magnetic Circuits and Transformers", The M.I.T. Press, 1943.

[56] C.P. Steinmetz, "On the Law of Hysteresis", AIEE Transactions, No. 9, 1892, pp. 3-51.

[57] V. Migulin, V. Medvedev, E. Mustel, and V. Parygin, "Basic Theory of Oscillations", Mir Publishers Moscow, 1978, pp. 66-67.

[58] A. Greenwood, "Electrical Transients in Power Systems", Second Edition, John Willey, 1991.

[59] D. Povh and W. Schultz, "Analysis of Overvoltages Caused by Transformer Magnetizing Inrush Currents", IEEE Transactions on Power Apparatus and Systems, Vol. PAS-97, No. 4, July/August 1978, pp. 1355-1365.

[60] E.J. Tuohy and J. Panek, "Chopping of Transformer Magnetizing Currents, Part I: Single-Phase Transformers", IEEE Transactions on Power Apparatus and Systems, Vol. PAS-102, No. 5, May 1983, pp. 1106-1114.

[61] S. Ihara, J. Panek, and E.J. Tuohy, "Chopping of Transformer Magnetizing Currents, Part II: Three-Phase Transformers", IEEE Transactions on Power Apparatus and Systems, Vol. PAS-102, No. 5, May 1983, pp. 1106-1114.

[62] B.A. Mork and D.L. Stuehm, "Application of Nonlinear Dynamics and Chaos to Ferroresonance in Distribution Systems", paper No. 93 SM 415-0 PWRD presented at the 1993 IEEE/PES Summer Meeting.

[63] R.A. Walling, K.D. Barker, T.M. Compton, and L.E. Zimmerman, "Ferroresonant Overvoltages in Grounded Wye-Wye Transformers with Low-Loss Five-Legged Silicon-Steel Cores", IEEE Transactions on Power Delivery, Vol. 8, No. 3, July 1993.

[64] A.E.A. Araujo, A.C. Soudack, and J.R. Marti, "Ferroresonance in Power Systems: Chaotic Behaviour", IEE Proceedings, Vol 140, Pt. C, No. 3, May 1993, pp. 237-240.

Francisco de León (M) was born in Mexico, in 1959. He received his B.Sc. degree and his M.Sc. degree (summa cum laude) from Instituto Politécnico Nacional (I.P.N.), Mexico, in 1983 and 1986, respectively. From 1984 to 1987 he worked as a lecturer at the same institute. He obtained the Ph.D. degree in 1991 at the University of Toronto. He continued his research there as a postdoctoral fellow until his return to the I.P.N., Mexico, in September 1992. He is now a professor in the Graduate Division of the School of Electrical and Mechanical Engeneering. His main research interests include electromagnetic fields and transients.

Adam Semlyen (F) was born and educated in Rumania where he obtained a Dipl. Ing. degree and his Ph.D. He started his career with an electric power utility and held an academic position at the Polytechnic Institute of Timisoara, Rumania. In 1969 he joined the University of Toronto where he is a professor in the Department of Electrical and Computer Engineering, emeritus since 1988. His research interests include the steady state and dynamic analysis of power systems, electromagnetic transients, and power system optimization.

FREQUENCY DEPENDENT EDDY CURRENT MODELS FOR NONLINEAR IRON CORES

E.J. Tarasiewicz (M) A.S. Morched (SM)
Ontario Hydro System Planning

A.Narang (M) E.P. Dick (SM)
Ontario Hydro Research Division

Abstract - *Frequency dependent representations of eddy currents in laminated cores of power transformers are developed. One representation is based on a continued fraction expansion for the frequency dependent magnetizing impedance while the other is based on a discretization of flux distribution within the lamination. Both models are low order and reproduce the frequency characteristics of the transformer magnetizing branch up to 200 kHz with less than 5% error. The importance of modeling the frequency dependence of eddy currents for the calculation of the transient recovery voltage across a low voltage transformer breaker interrupting a nearby fault is demonstrated.*

Key Words - *Eddy currents, transformer, EMTP, Frequency dependent eddy current models*

1. INTRODUCTION

Transformer modeling in power system transient studies still presents a substantial difficulty. A study was conducted recently by General Electric Company (GE) , under an Electric Power Research Institute (EPRI) contract, to assess the accuracy of Electro-Magnetic Transient Program (EMTP) simulations by comparing their results with field measurements. The study concluded that among the least accurate EMTP studies are those where the transformer is a critical component [1].

Difficulties in modeling transformers stem from the fact that some of the transformer parameters are both nonlinear and frequency dependent. Iron core losses and inductances are nonlinear due to saturation and hysteresis. They, also, are frequency dependent due to eddy currents in the laminations.

Models of varying complexity have been implemented in the EMTP to simulate transient behavior of transformers. The parameters of these models are extracted from power frequency measurements. Therefore, the models are valid only for a narrow frequency band. The EMTP transformer models assume complete symmetry among the phases and do not reflect the actual iron core structure of the transformer.

The need for a model that accurately reproduces the transformer nonlinear and frequency dependent characteristics, was recognized by the EMTP Development Coordination Group (DCG) who sponsored the development of such a model.

92 WM 177-6 PWRS A paper recommended and approved by the IEEE Power System Engineering Committee of the IEEE Power Engineering Society for presentation at the IEEE/PES 1992 Winter Meeting, New York, New York, January 26 - 30, 1992. Manuscript submitted September 3, 1991; made available for printing November 25, 1991.

This paper describes eddy current representation developed for the new transformer model. Results of a transient recovery voltage simulation demonstrating the importance of accurate modeling of eddy currents on the computed transients are presented.

2.0 TRANSFORMER MODEL STRUCTURE

The new EMTP transformer model developed reflects the iron core structure and the winding arrangements around the limbs. Its derivation is based on the duality between magnetic and electric circuits [2, 3, 4]. By applying this principle to a three phase, three winding, core type transformer with concentric windings an equivalent electrical network is derived, where each limb of the magnetic core can be accommodated separately. The resulting equivalent circuit is shown in Figure 1. As shown in the figure, flux paths in iron members are represented by nonlinear impedances in the dual electric circuit. The values of these impedances change with flux magnitude due to saturation and hysteresis, and with frequency due to eddy currents.

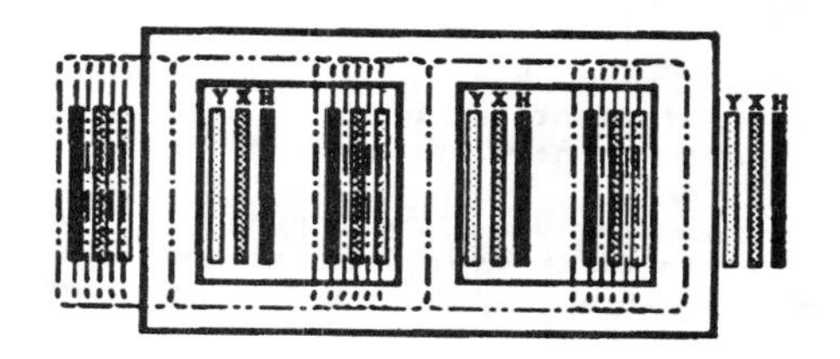

(a) Magnetic Flux Paths for 3-Limbed Core Transformer

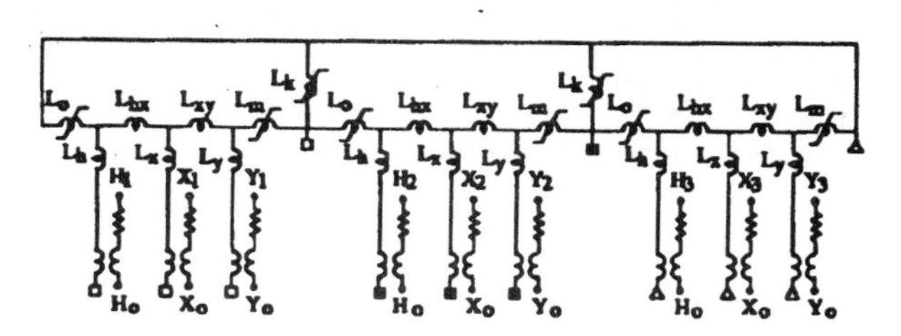

(b) Model for Magnetic Coupling in 3-Limbed and 5-Limbed Transformers

Fig.1 Equivalent network for 3 phase 3 winding core type transformer

3.0 EDDY CURRENTS AND THEIR REPRESENTATION

Eddy currents are induced in transformer core laminations by the alternating flux in the core. As frequency changes, flux distribution in the iron core lamination changes. For high frequencies the flux will be confined to a thin layer close to the lamination surface, whose thickness decreases as the frequency increases. This indicates that inductances representing iron path magnetization and resistances representing eddy current losses are frequency dependent.

Transformer models used for system transient studies have traditionally represented the magnetic induction in the core by a nonlinear inductance reflecting the saturation of iron. Hysteresis and eddy current losses are normally represented by a constant resistance in parallel with the magnetizing inductance. In more refined simulations hysteresis losses are represented by modified nonlinear saturation characteristics to reflect the dynamic BH curve adequately. Eddy currents, however, continue to be modeled using a constant resistance across the hysteresis model. The parameters of these models were extracted from open circuit tests at power frequency. Therefore, these models cannot adequately represent the frequency dependent eddy current effects over a wide frequency range.

The significance of frequency dependence of the eddy current effects on the transformer transient simulation has been emphasized in previously published work [5]. Models have been developed to simulate frequency dependence of the magnetizing inductance as well as losses due to eddy currents [6,7]. However, these models are computationally inefficient, and therefore are not suitable for implementation in transient simulation of systems involving several transformers.

The previously developed eddy current models belong to two main categories:

a) series expansion models - obtained by the realization of the analytical expression for the magnetizing impedance as a function of frequency, and

b) discretized models - obtained by subdivision of the lamination into a number of sublaminations and generation of their electrical equivalents.

These models reproduce the frequency dependence of the equivalent impedance for an unsaturated iron core. They have to be modified to account for the nonlinear iron core behavior.

3.1 The Series Expansion Model

An expression for the equivalent impedance of a coil wound around a laminated iron core limb was derived by solving Maxwell's equations with the assumption that the electromagnetic field distribution is identical in all laminations [6]. The obtained frequency dependent equivalent impedance can be written as:

$$Z(j\omega) = \frac{4N^2 Ax}{ld^2 \gamma} \tanh x \qquad (1)$$

where:
$x = d\sqrt{j\omega \mu \gamma}/2$
μ - magnetic permeability of the steel lamination
γ - electric conductivity of the steel lamination
A - total cross-sectional area of all laminations
d - thickness of the lamination.
l - length of core limb
N - number of coil turns

Substitute the partial fraction expansion of the hyperbolic tangent

$$\tanh x = 2x \sum_{k=1}^{\infty} \frac{1}{x^2 + [\pi (2k-1)/2]^2}$$

in equation (1). The resulting expansion for the equivalent impedance $Z(j\omega)$ can be synthesized using a series connection of parallel RL branches (Fig.2a). A dual circuit (Fig.2b) can be generated using the partial fraction expansion for the hyperbolic cotangent in the expression for the equivalent admittance. These representations are interchangeable and are referred to, in the literature, as the Foster series and parallel equivalents. The accuracy of either of these representations over a defined frequency range depends on the number of terms retained in the partial fraction expansion, and therefore the number of sections in either of them.

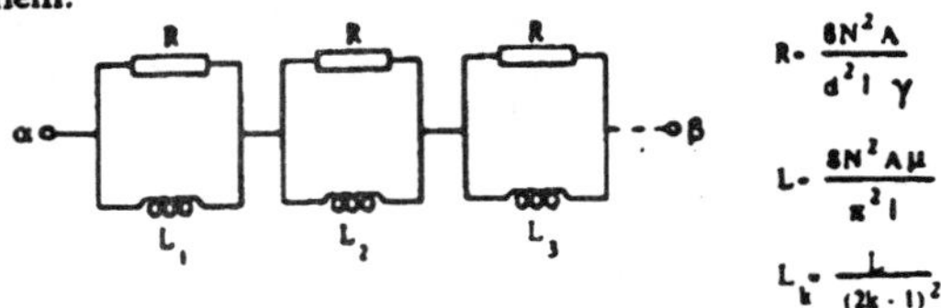

(a) Series Foster equivalent

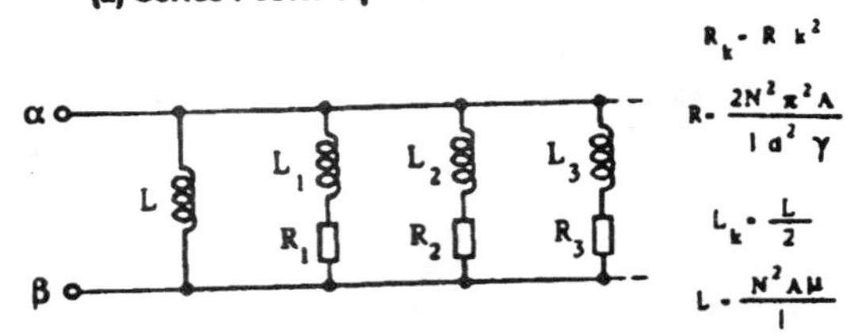

(b) Parallel Foster equivalent

Fig.2 Series expansion model.

3.2 The Uniformly Discretized Model

An alternative approach to model eddy current effects has been based on subdividing each lamination into a number of sublaminations which are sufficiently narrow so that a uniform flux distribution within each sublamination can be assumed [7]. The solution of Maxwell's equations for equal thickness sublaminations results in an equivalent representation with a longitudinal inductance L and transversal resistance R given by:

$$L = \frac{N^2 A \mu}{n l} \quad \text{and,}$$

$$R = \frac{4 N^2 A n}{l d^2 \gamma} .$$

where:
n - number of sublaminations

The corresponding equivalent circuit representation of the iron core limb is obtained by connecting "n" of these sections in a cascade as shown in Figure 3. The accuracy of the eddy current model over a given frequency range depends on the thickness of the sublamination, and therefore, on the number of ladder sections in the resulting circuit representation.

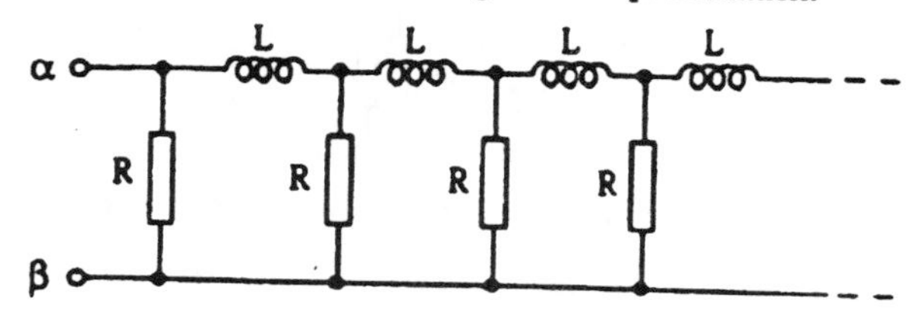

Fig.3 Uniformly discretized model

4.0 THE ORDER OF THE EDDY CURRENT MODELS

The described eddy current models, require an infinite number of sections in order to exactly represent the analytical solution. However, to achieve a given accuracy over a finite frequency range, only a finite number of sections needs to be retained in any of these models. The higher the accuracy and/or the wider the frequency range the more sections will have to be retained.

Considering the type of studies for which the eddy current model is required it was decided that the model should reproduce the frequency characteristics of the iron core limb with less than 5% error up to 200 kHz. To achieve this accuracy either of the two Foster equivalent models requires 72 sections while the uniformly discretized model requires 30 sections. The convergence characteristics of these models are shown in Figure 4 - plots (a) and (c).

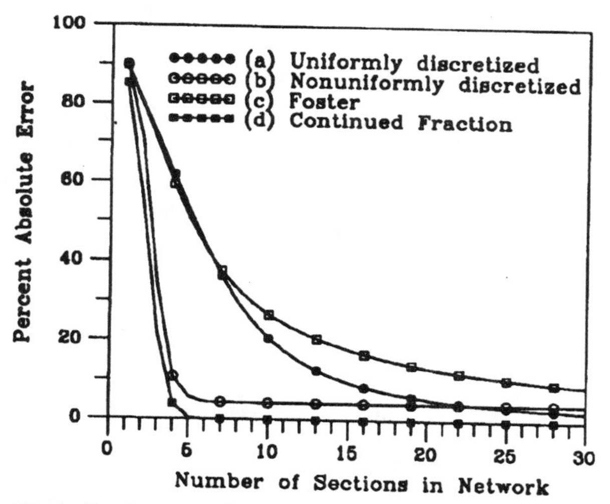

Fig.4 Absolute error in the equivalent impedance as a function of the number of sections.

The large number of sections necessary to achieve the required accuracy renders none of the discussed models suitable for inclusion in transient simulation programs, especially when we recognize that the transformer model shown in Figure 1 contains eight magnetizing branch models. Efficient models suitable for implementation in the EMTP without drastically increasing the computational burden had to be developed. The need to develop low order models becomes more compelling if we recognize that the inductive elements of the models will have to be nonlinear in order to represent the iron core saturation and hysteresis characteristics. The transient solution of a high order nonlinear model would be computationally prohibitive.

5.0 THE DEVELOPED MODELS

5.1 The Continued Fraction Model

The equivalent impedance expression (equation 1) can be expanded using the continued fraction, rather than series, expansion in order to reduce the number of sections, without changing the accuracy. The continued fraction expansion for the hyperbolic tangent, in equation (1), is given by [8]

$$\tanh x = \cfrac{x}{1 + \cfrac{x^2}{3 + \cfrac{x^2}{5 + \cfrac{x^2}{7 + \cfrac{x^2}{9 + \dots}}}}}$$

Equation 1 with this continued fraction expansion can then be realized in the form of the ladder network shown in Figure 5 with inductances

$$L_o = \frac{N^2 A \mu}{l} \qquad \text{and} \qquad L_k = \frac{L_o}{(4k-3)}$$

in the transversal branches, and resistances

$$R_o = \frac{4 N^2 A}{l\, d^2 \gamma} \qquad \text{and} \qquad R_k = R_o\,(4k-1)$$

in the longitudinal branches.

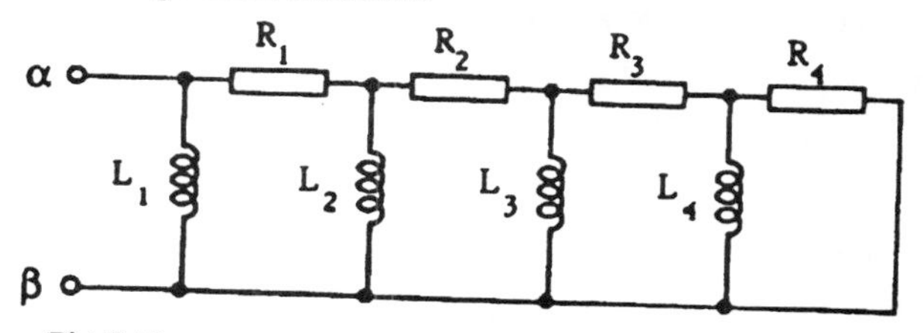

Fig.5 Four section continued fraction model

As in the previous models the more sections in the ladder the better the approximation. However, in order to represent the frequency range up to 200 kHz with error less than 5% only four terms need to be retained in continued fraction expansion. The convergence characteristic of this expansion is shown in Figure 4 (plot d). It can be shown (Fig.6) that the first section of the model governs its characteristics at frequencies up to a few kHz and each subsequent section comes into play as the frequency increases.

5.2 The Nonuniformly Discretized Model

The uniformly discretized model has an advantage of representing the spatial flux distribution over the iron lamination.

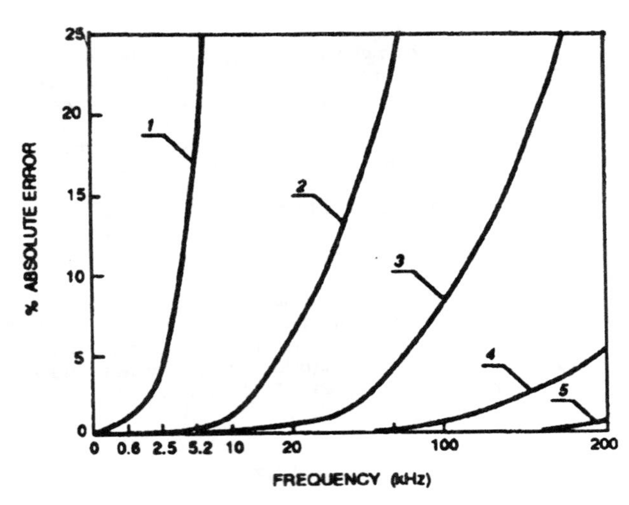

Fig.6 Absolute error, as a function of frequency, in the equivalent impedance of the Continued Fraction Expansion model with: (1) - one, (2) - two, (3) - three, (4) - four, and (5) - five sections.

This would permit the simulation of the saturation of different sublaminations separately. The large number of the resulting ladder sections can be reduced recognizing that the flux becomes confined to an increasingly thinner layer, near the lamination surface, as the frequency increases. Therefore, the sublamination thickness can be increased from the lamination surface towards its center resulting in fewer sublaminations [7].

Doubling the thickness of the sublaminations progressively from the surface to the center of the lamination results in a nonuniform ladder with longitudinal inductances and transversal resistances given by:

$$L_k = \frac{N^2 A \mu}{l} \frac{2^{(k-1)}}{\sum_{i=1}^{M} 2^{(i-1)}}$$

and

$$R_k = \frac{8 N^2 A}{l d^2 \gamma} \frac{\sum_{i=1}^{M} 2^{(i-1)}}{2^{(k-1)}}$$

where:

M - number of sections in a nonuniform ladder

k - sublamination number starting from the lamination surface

The resulting model is shown in Figure 7. It can be seen from the above expressions that the values of the inductance and resistance of any section depend on the number of sublaminations. This means that a lower order model cannot be obtained from a higher order model by simply leaving out some of the sections. New values of the parameters would have to be recalculated for every section of the reduced order model. With this scheme it is sufficient to use six sections to represent eddy current effects with less than 5% error up to 200 kHz. This brings the model within the realm of practical applications.

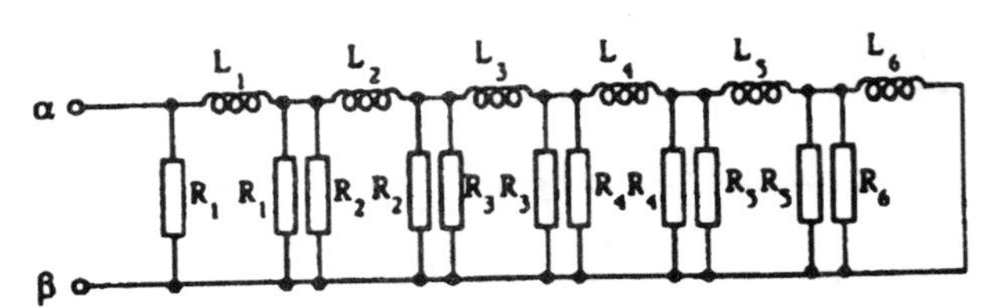

Fig.7 Six section nonuniformly discretized model

The surface sublamination is represented by the first section at the network terminals, the centre of the lamination corresponds to the last section. At low frequencies the values of the resistors are much higher than the values of the inductive reactances; current will flow through all inductances avoiding shunt resistances. This accounts for the fact that the low frequency flux penetrates deep into laminations. At high frequencies values of the inductive reactances become higher than the values of resistors. Resistors conduct more current as frequency increases and short-circuit inductors sequentially starting at the far end from the terminals. This represents the effect of limited penetration of flux into laminations at high frequencies.

A comparison of the convergence characteristic of this model (plot b) with other models is shown in Figure 4.

6.0 REPRESENTATION OF THE IRON CORE NONLINEARITIES

The eddy current models presented so far do not account for the iron core hysteresis and saturation effects. In order to account for these effects, inductive components of the models representing the magnetizing reactances have to be made nonlinear.

6.1 The Continued Fraction Model

Since the inductances and resistances in the continued fraction model do not represent any physical part of the iron lamination it is not immediately obvious how to incorporate the nonlinear effects. However, recognizing that the high frequency components do not contribute appreciably to the flux in the transformer core, it can be assumed that only low frequency components are responsible for driving the core into saturation. It may, therefore, be justifiable to represent as nonlinear only the first section of the continued fraction model which governs its behavior at low frequency. This assumption reduces the number of the modeled nonlinearities significantly.

6.2 The Nonuniformly Discretized Model

The inclusion of the iron nonlinear effects in the discretized nonuniform ladder model follows readily. In this case all the inductances representing different sublaminations have to be nonlinear. Since they represent sublaminations of different thicknesses, scaled hysteresis and saturation models are developed for each sublamination. Each eddy current model representing an iron core limb has six inductances and the transformer model contains eight eddy current models. Considerable computational burden is introduced to the overall transformer model when this model is used. The presence of a large number of nonlinearities in the model makes it difficult to initialize and results, in some cases, in a divergent iteration when the Newton-Raphson method is used for solving the nonlinear system.

6.3 Inclusion of Nonlinear Elements in the Transient Solution

Finding the transient solution using the EMTP type 96 pseudo-nonlinear hysteretic reactor involves moving along multi-slope piecewise linear segments. This algorithm necessitates the complete retriangularization of the nodal conductance matrix whenever the solution moves from one straight-line segment to another. In addition to the computer time involved, a piecewise linear representation has certain deficiencies (eg. overshooting) and a significant improvement can be obtained by representing the two halves of the major and minor hysteresis loops by continuous nonlinear functions (eg. hyperbolic equations).

To include the nonlinear elements in the transient solution the compensation method can be used to solve the network. In this method nonlinear elements are simulated as current injections, which are superimposed on the linear network after a solution without the nonlinear elements has been found [9].

The total network solution for the node voltages [v] is obtained by superposition of the value $[v_{open}]$ found without the nonlinear inductances and the contribution produced by nonlinear branch currents [i]:

$$[v] = [v_{open}] - [Z_{Thev}][i] \qquad (2)$$

The matrix $[Z_{Thev}]$ is the Thevenin equivalent impedance of the network as seen from the terminals of the nonlinear elements (Fig.8). $[V_{open}]$ and $[Z_{Thev}]$ are calculated by the EMTP at each time step.

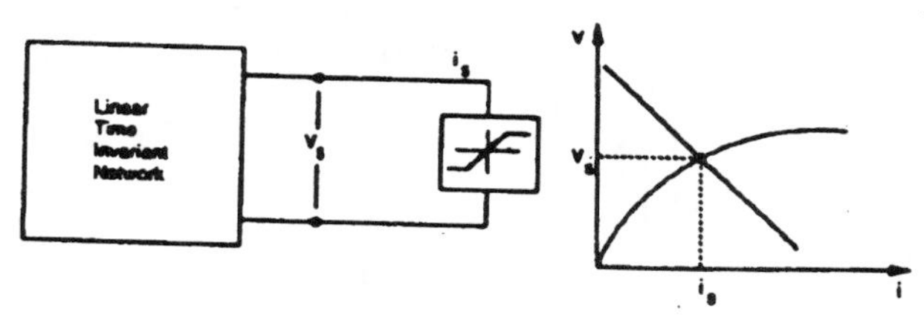

Fig.8 Superposition method for solving a network with a nonlinear inductance

The nonlinear v-i relationship between voltage and current for a saturable core can be obtained from the nonlinear instantaneous flux-current characteristic

$$i = f(\psi) \qquad (3)$$

Using the trapezoidal rule of integration the flux is expressed in terms of voltage as

$$\psi(t) = \frac{\Delta t}{2} v(t) + \psi(t - \Delta t) + \frac{\Delta t}{2} v(t - \Delta t) \qquad (4)$$

The Newton-Raphson iteration scheme is used to solve equations 2, 3 and 4 in every time step.

This method provides an accurate and efficient means of solving a system of nonlinear equations when an initial guess in the neighborhood of the solution can be identified. The computational cost of the iterative process is offset by eliminating the need for retriangularization of the modal conductance matrix.

6.4 Behavior of the Nonlinear Models

The behavior of the developed models with nonlinearities was tested using the Newton Raphson scheme outlined in Section 6.3 with a true nonlinear hysteresis/saturation model developed for the new EMTP transformer model. The new hysteresis/saturation model is the subject of another paper. Some testing had to be carried out using the existing EMTP pseudo-nonlinear hysteretic reactor type 96.

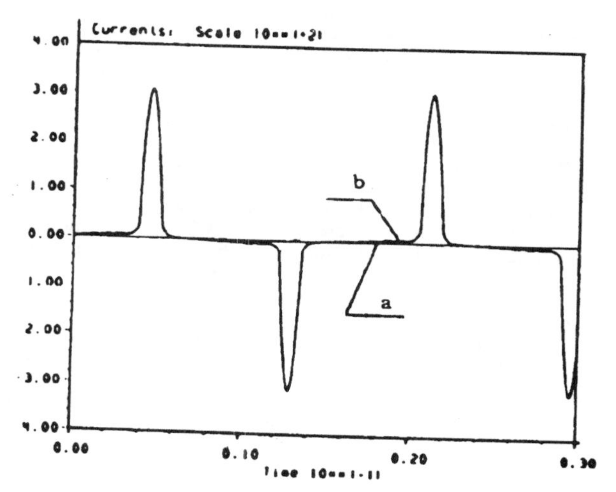

Fig.9 Terminal currents for the eddy current models energized with a 60 Hz voltage source
(a) the continued fraction eddy current model
(b) the nonuniformly discretized eddy current model

Figure 9 shows the currents drawn by each of the two models when energized with a 60 Hz voltage source of peak value corresponding to 1.15 the knee point of the saturation characteristic. These results were obtained with the new hysteresis model using the Newton-Raphson iteration scheme. As can be seen from Figure 9 the nonlinear behavior of the two models is identical for 60 Hz input voltages. However, the CPU time for the continued fraction model was one order of magnitude shorter than for the nonuniformly discretized model.

Examining the model behavior for a combined 60 Hz and a high frequency input voltage using the Newton-Raphson scheme was possible only with the continued fraction model. The series connection of the highly dissimilar highly nonlinear inductances in the nonuniformly discretized model, coupled with the Newton-Raphson method requirement of initial conditions close to the final solution, resulted in divergent iterations.

It is possible to obtain a solution, for input voltage containing high and low frequency components, using the EMTP type 96 hysteresis model for both eddy current models. However, extra caution has to be exercised in selecting the initial conditions in case of the nonuniformly discretized model in order to ensure that initial fluxes are within the hysteresis loops for all ladder sections. The behavior of the models under conditions involving both high frequency and nonlinearities is described in the next section in conjunction with a practical application example.

7.0 PRACTICAL APPLICATIONS

The importance of representing the nonlinear behavior of iron cores in transformer models used in system transient studies has been recognized and techniques of modeling the associated phenomena were developed. The importance of representing the frequency dependence of the magnetizing impedance due to eddy currents has yet to be demonstrated. The conventional model of a nonlinear inductance in parallel with a constant resistance is accurate to within 5% up to about 3 kHz (Fig.6).

The importance of modeling the frequency dependence of the magnetizing impedance becomes apparent for transients involving frequencies in the range of tens of kHz. Transients directly involving the core will be low frequency due to the high magnetizing inductances. High frequency phenomena involve the resonance of the transformer leakage reactances with small capacitances such as bushing and bus capacitances. Although the magnetizing branch may not be directly involved; the presence of the core losses will provide a major damping mechanism under these conditions. Therefore, the accurate representation of the core losses at high frequencies can be crucial to the accuracy of the calculated transients.

An example of this situation arises when a low voltage transformer breaker attempts to clear a nearby fault. Recovery voltages appearing between breaker contacts under these conditions can impose a severe duty on the breaker [10]. Comparisons with field measurements indicate that the use of a constant resistance to model eddy currents results in underestimating the peak value of the resulting TRV and overestimating its damping [11].

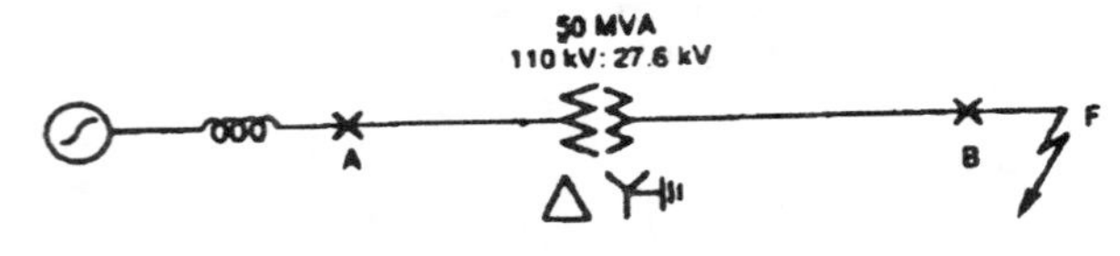

Fig.10 Single line diagram of the test system.

The transient recovery voltage appearing across the contacts of breaker B shown in Figure 10 for different types of faults at location F were calculated using the following models:

a) the continued fraction eddy current model;
b) the nonuniformly discretized eddy current model;
c) 60 Hz non-frequency dependent model (constant resistance in parallel with the magnetizing inductance)

Hysteresis and saturation were included in all three models using the EMTP type 96 nonlinear inductor. The simulation results obtained using the EMTP, for an ungrounded three phase fault are shown in Fig. 11 for the three models.

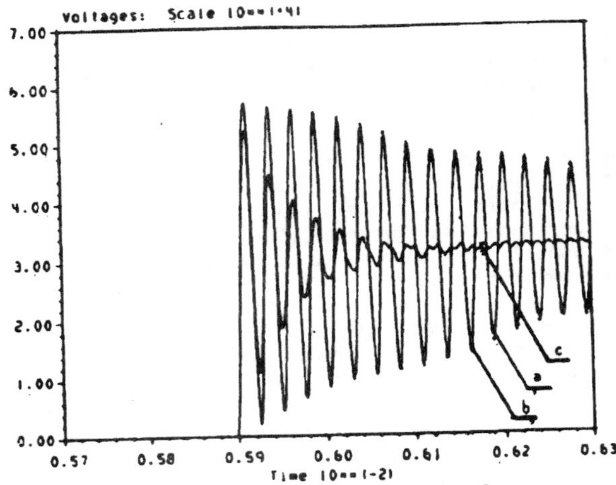

Fig.11 Simulated TRV in a low voltage transformer breaker using:
(a) continued fraction eddy current model
(b) nonuniformly discretized eddy current model
(c) 60 Hz non-frequency dependent model

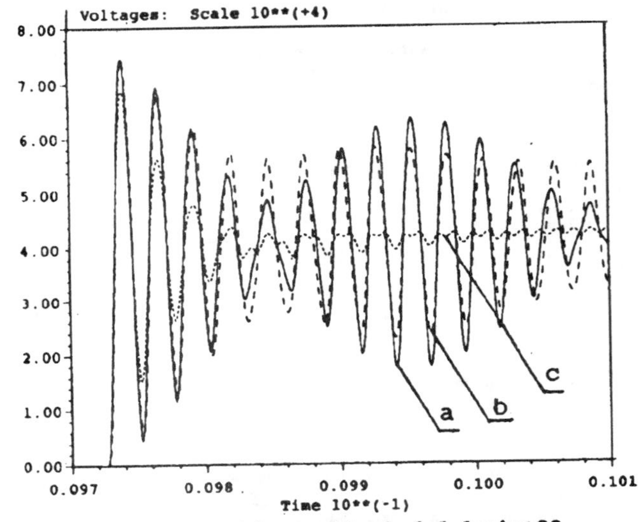

Fig.12 TRV results with saturation included using 96 nonlinear inductance and:
(a) continued fraction eddy current model
(b) nonuniformly discretized eddy current model
(c) 60 Hz non-frequency dependent model

The transformer leakage inductance and effective capacitance play the dominant role in determining the transient recovery voltage. The natural frequency of the recovery voltage oscillations is 37.5 kHz and the damping due to eddy currents, at this frequency, is much lower than at 60 Hz. Comparison of the results in Figure 11 shows that both eddy currents models produce identical results while the conventional model produces a TRV with a reduced peak and more damped transient. The effects of saturation were not apparent since the source voltage was well below the knee point of the saturation characteristic of the transformer for all three cases.

To examine the behavior of the frequency dependent models under saturation conditions, the TRVs were recalculated using a source voltage of peak value equal to 1.15 of the saturation curve knee point. The results obtained using each of the frequency dependent eddy current models, as well as the conventional model, with hysteresis and saturation represented using the EMTP type 96 nonlinear inductances are shown in Figure 12. The results shown indicate that while there is a reasonable agreement between the TRVs calculated using the frequency dependent eddy current models, the TRV calculated using the conventional model deviates considerably.

8.0 CONCLUSIONS

1. Computationally efficient and accurate eddy current models are developed for use in representing transformer magnetizing branches in electromagnetic transient studies. Two models based on the solution of the field equations within steel laminations are presented. One model is based on a frequency domain solution of the field problem within a lamination (continued fraction model) while the other is based on the division of the laminations into sublaminations with constant flux (nonuniformly discretized model).

2. The two models are capable of representing the frequency dependence of the transformer magnetizing impedances as well as their nonlinear characteristics. The nonuniformly discretized model represents the spatial distribution of the flux within a lamination and, therefore, is expected to produce more accurate results when nonlinearities are involved. However, this model is prone to numerical instability and is computationally less efficient than the continued fraction model which produces comparable results.

3. The importance of modeling the frequency dependence of eddy currents for the calculation of the transient recovery voltage across a low voltage transformer breaker interrupting a nearby fault is demonstrated. The example shown indicates that neglecting the frequency dependence of eddy current losses results in overestimating their magnitudes at high frequencies. This, in turn, leads to overestimation of damping of high frequency components, and therefore, to the underestimation of peaks and rates of rise of calculated transients.

ACKNOWLEDGMENT

The authors like to thank Professor A.Semlyen of the University of Toronto and Messers R.H. Brierely and L. Marti of Ontario Hydro for many useful discussions and suggestions. They also like to thank the CEA for the permission to publish this work on behalf of the EMTP-DCG.

REFERENCES

[1] Electromagnetic Transient Program (EMTP) - Field Test Comparisons, EPRI EL-6768, EMTP Development Coordination Group, Prepared by GE, Schenectady, New York, 1990.

[2] G.R. Slemon, Magnetoelectric Devices - Transducers, Transformers and Machines, John Wiley & Sons Inc., 1966.

[3] E.P.Dick, W. Watson," Transformer Models For Transient Studies Based on Field Measurements", IEEE Trans. on PAS, Vol. PAS-100, No.1, Jan. 1981.

[4] L.Krahenbuhl, B.Kulicke, A.Webs,"Simulationsmodell eines Mehrwicklungstransformators zur Untersuchung von Sattigungsvorgangen", Siemens Forschung and Entwicklungs Bericht, Vol 12 (1983), No.4

[5] E.J. Tuhoy, J. Panek, "Chopping of Transformer Magnetizing Currents - Part1: Single Phase Transformer", IEEE Trans. on PAS, Vol. PAS-97, No. 1, Jan/Feb 1978.

[6] J. Avila-Rosales, F.L. Alvarado," Nonlinear Frequency Dependent Transformer Model for Electromagnetic Transient Studies in Power Systems", IEEE Trans. on PAS, Vol. PAS-101, No.11, Nov.1982.

[7] J. Avila-Rosales, A. Semlyen,"Iron Core Modeling for Electrical Transients", IEEE Trans. on PAS, Vol. PAS-104, No.11, Nov.1985.

[8] M. Abramowitz, I.A. Stegun, Handbook of Mathematical Functions, Dover Publications Inc., New York, 1972.

[9] H.W. Dommel; Electromagnetic Transient Program, Reference Manual, (EMTP Theory Book), Aug. 1986

[10] W.E. Reid,"Effect of Transient Recovery Voltage (TRV) on Power System Interruption", McGraw-Edison Power Systems, Cooper Industries Inc.

[11] M.Rioual, R. Kouteynikoff,"Calculation of TRV of 20 kV Circuit Breakers Using EMTP", EMTP Review, Vol.2, No.1, Jan. 1988.

BIOGRAPHIES

Eva J. Tarasiewicz (M'85) received the B.Sc., M.Sc., and Ph.D. degrees in Electrical Engineering from Poznan University of Technology, Poland in 1977, 1978, and 1982, respectively. From 1982 to 1983 she worked as an Assistant Professor at the Poznan University of Technology. From 1983 to 1985 she continued her work on numerical methods in computation of electromagnetic field transients as a Post-Doctoral Fellow at the University of Manitoba. From 1985 to 1987 she worked at McMaster University as an Assistant Professor in the Department of Electrical and Computer Engineering. In 1987 she joined Ontario Hydro, where she is currently working in the Analytical Methods & Specialized Studies Department of the Power System Planning Division.

Atef S. Morched (M'77-SM'90) received a B.Sc. in Electrical Engineering from Cairo University in 1964, a Ph.D. and a D.Sc. from the Norwegian Institute of Technology in Trodheim in 1970 and 1972. He has been with Ontario Hydro since 1975 where he currently holds the position of Section Head - Electromagnetic Transients in the System Planning Division.

Arun Narang (M'75) received his B.A.Sc. from the University of Waterloo in 1977. He has been at Ontario Hydro Research since 1979 studying power system transients, including high-frequency modeling of machine windings and transformers.

Peter Dick (M'75-SM'90) graduated from the University of Waterloo (B.A.Sc.1971), and University of British Columbia (M.A.Sc. 1973). His first engineering position was at the Karlsruhe Nuclear Research Centre (Germany) on high energy switching (1974). Since then he has been at Ontario Hydro Research studying power system transients.

Discussion

R. J. Meredith (New York Power Authority, White Plains, NY): Having recently completed a transformer modeling project similar to that of the authors, I found their approaches to eddy loss/excitation modeling very thought-provoking, particularly where their approaches differed from my own.

Application of the continued fraction model is a valuable contribution to modeling planar linear materials. However, it is based on Equation 1, which applies only to linear materials, not to saturable iron. The Foster equivalents are equally limited, and clearly inferior to the uniformly descretized model. The authors correctly state that use of nonlinear inductances in the uniformly descretized model would allow it to model nonlinear performance, but fail to follow through. It is true that it is computationally more demanding, but discarding the only valid approach to nonlinear modeling—what I would term the nonlinear benchmark—leaves the appreviated models with no proof of accuracy. The problems with solving the benchmark model are not insuperable, as we have frequently solved pi-chains of 65 nonlinear series sections without problem and routinely incorporate an aggregate total of 30 sections in a low frequency transformer model (60 in a phase shifter).

The authors do present results of a non-uniformly descretized model with EMTP type 96 nonlinear elements; however such an approach is highly inaccurate if any saturation occurs. The reason it is highly inaccurate is that saturation drastically increases the penetration depth of high frequency signals, allowing them to reach the larger dimensioned sections which can not accurately model high frequency propagation. As a general rule, the thin laminations of the uniformly descretized model are mandated for any depth reachable by higher frequencies. Under saturation, that is almost anywhere within the lamination. This eliminates virtually all of the benefit of using graduated sections, unless the iron is constrained to a linear operating range.

The accuracy of what I refer to as finite section modeling, to place it in its correct niche between transmission line pi sections and finite elements, is directly related to the number of R-L-R pi sections per wavelength (as it is in transmission line modeling). Inasmuch as the penetration depth of any frequency is always 0.15915 wavelength, it is usually necessary to model one or more sections per penetration depth to obtain accuracy comparable to that obtained with 6 or more pi sections per wavelength in a transmission line model. Table 1 provides an indication of the error involved with modeling the surface impedance of a lamination in linear materials. All high frequency core modeling boils down to such a calculation, which in linear materials produces an R = X type impedance at every frequency whose penetration depth is much smaller than the lamination thickness.

The authors do not clearly state which of the three possible error statistics is referred to when they established their 5% accuracy criterion, but if it was impedance magnitude, the loss magnitudes are significantly more in error. It is clearly necessary to model more than one section per penetration depth to obtain reasonable accuracy. When graduated section sizes are permitted, consistent accuracy levels are obtained at all frequencies by starting at the center of the lamination and sizing the next outward finite section as X% of the remaining distance to the surface. The doubling series cited by the authors represents the case of X = 50, although greater or lesser figures can be used.

There is also a need to comment on the material parameters assumed for core steel. Our recently purchased transformers have steel with a linear relative permeability of about 50000 and incremental permeabilities significantly larger. A resistivity of about 15.E-8 ohm-meter provides reasonable agreement with measured eddy current losses in a nonlinear model. The penetration depth using such parameters is on the order of one-half of a 0.23 mm thick lamination. Consequently uniform flux density cannot be obtained without saturation (beginning above about 1.0 Tesla). It is apparent that the authors have not used parameters in this range when they state the accuracy of the (linear) continued fraction model with one and four sections. Using the above parameters the one section model deteriorates before 150 Hz rather than the 3 kHz stated and the four section model deteriorates before 15 kHz rather than 200 kHz. Either their models do not represent the permeability of "linear" steel or the steel that is modeled is not representative of that now being used.

The authors' assertion that saturation effects can be added to the continued fraction model by representing the first shunt inductance as nonlinear is not correct in terms of losses. Although it could tend to make the current drawn by the model look more reasonable at one frequency, it is apparent that it would have no effect on the eddy loss of the remaining linear model when excited at the same voltage (flux). The true effect of saturation can be seen in the uniformly descretized model where the series inductances are nonlinear. Saturation of the surface layers reduces the incremental inductance values (that seen by high frequencies) by as much as a factor of 50000 at full saturation. Such action effectively parallels several shunt resistances ahead of a remaining chain of lesser-saturated or unsaturated sections which would continue to present a high impedance to high frequencies. Thus high frequency signals see a qualitative effect of saturation in the addition of shunt resistances to a linear model, which increases high frequency eddy current losses.

The authors fail to point out that the surface impedance seen by a high frequency component is entirely determined by the biasing effect on the saturable series inductances caused by a superposed dc or low frequency signal. It is always the incremental inductance of the individual layers of the model which determines how far into the model the higher frequencies can propagate and create eddy currents. The incremental inductances vary cyclically under fundamental frequency excitation from very high values along the "linear" portion of the B-H curve to very low values at saturated levels. The high frequency eddy current paths depend entirely on what part of the B-H curve is being operated on when the high frequency (transient) occurs. In the absence of low frequency biasing, a third area of very low permeability below 0.2 Tesla, which is usually ignored in use of the term "linear slope", will dominate all high frequency behavior. The amount of incremental loss caused by the incremental high frequency eddy currents depends both on the magnitude of those currents in each layer and on what fundamental frequency eddy current is already flowing along the same paths. The latter effect makes damping of high frequencies a function of the point on the fundamental cycle, even in linear models.

Table 2 shows some representative eddy losses based on a comparison of the continued fraction model and a nonlinear model with 18 identical finite sections. An actual core B-H curve, forced to be linear below 1.2 Tesla and to match the slope of the continued fraction model, was used in the nonlinear model. Losses are for the same core sizes. One PU 60 Hz voltage is assumed to produce 1.6 Tesla flux density, a typical value. The superposed 10th harmonics signal represents about the largest magnitude transient which could be applied without arrester operation.

It may be seen that the incremental losses from superposition of frequencies is significantly larger when saturation is modeled, regardless of which signal is regarded as incremental. It may also be noted that even in a linear model the incremental loss due to a high

TABLE I
MODELING ERROR AS A FUNCTION OF UNIFORM SECTION SIZE

Sections/ Pen. Depth	Sections/ Wavelength	Z Magnitude Error %	Resistance Error %	Reactance Error %
0.75	4.7	−15.2	+8.6	−49.1
1.0	6.3	−5.4	+13.8	−29.7
1.5	9.4	−1.2	+9.0	−12.6
2.0	12.6	−0.4	+5.7	−6.8
3.0	18.8	−0.1	+2.7	−2.9
4.0	25.1	nil	+1.6	−1.6
5.0	31.4	nil	+0.7	−1.0

TABLE II
COMPARISON OF MODELED EDDY CURRENT LOSSES

Case	PU 60 HZ Voltage	PU 600 HZ Voltage	Finite Section Model Loss	Continued Fraction Model Loss
1	0.5	0.0	118.6	118.6
2	1.0	0.0	503.7	473.1
Ratio of Case 2 to Case 1 =			4.25	3.99
3	0.5	1.0	158.6	153.2
Case 3–Case 1 Increment			40.	34.6
4	1.0	1.0	923.0	699.4
Case 4–Case 2 Increment			419.3	226.3
Case 4–Case 3 Increment			764.4	546.2

frequency signal depends on the level of the fundamental signal. Although all the losses above were analyzed over a full cycle of the fundamental, the differences in losses between the two models occurred only while operating in the saturated region above 1.2 Tesla. The ratios of model losses occurring at such brief times would necessarily be much larger than the full cycle loss ratios indicated above.

It is the inability to show equivalence to the true nonlinear benchmark that most undermines the authors' continued fraction model. Without proof of equivalence or even use of the benchmark, the reference to nonlinearities in the paper's title is unwarranted. The reason proof cannot be given is that the continued fraction model unrealistically assumes a constant layer-to-layer and time-to-time permeability in its derivation. Even use of an average permeability is limited to the linear range since differences in average and incremental permeability, which differentiate fundamental frequency and high frequency responses, only appear with use of nonlinear series elements.

It might also be noted that the need for kilohertz level core modeling is essentially limited to cases with incremental low permeability effects, which cannot be modeled linearly. A high frequency transient occurring when the incremental permeability is high ("linear" portion of the B-H) is probably inconsequential for the following reasons. All of the (linear) higher order models produce a core surface impedance which rises essentially with the square root of the frequency. Thus by 1 kHz the (linear) magnetizing current is reduced by a factor of about four. Contemporary EHV transformers have net capacitive excitation currents when operating on their linear slopes (about 30-80% of rated voltage); only after saturation does the magnetizing current attain levels comparable to the capacitive current. That capacitive current can be expected to increase by a factor of 17 at 1 kHz. The capacitive component of the exciting current can be expected to exceed the (linear) magnetizing component of the exciting current by more than a factor of 70 at 1 kHz and above. Thus the magnetizing current effects can be readily ignored in comparison to capacitive current effects above 1 kHz, unless low-permeability effects are present and are appropriately modeled. Whether the core eddy current losses are really important to model at kilohertz frequencies, as asserted by the authors, would seem to depend on the relative magnitudes of those losses compared to those associated with capacitive currents, skin effects in conductors and other stray losses. Ensuring that eddy losses are not incorrectly modeled by simple shunt resistances may be more important than any of the details of a model at kilohertz frequencies.

The authors' Figure 3 contains some minor errors. Whether or not nonlinear inductances are included, the first resistance must be twice that of the remaining resistances, which effectively represent the ends of two pi sections. The preceding equation for resistance is for the value which occurs repetitively. The diagram also does not show that after "n" sections the chain is to be shorted. The value "n" does not actually represent the number of sublaminations, but half of the total, after the effects of symmetry in wave propagation from both sides of the lamination have been taken into account.

Manuscript received January 30, 1992.

A. Semlyen and F. de Leon (University of Toronto): We would like to congratulate the authors for their interesting contribution to the modeling of laminated iron cores. The two alternatives they have examined and compared with each other represent improvements of significant practical value over the Foster equivalents and the uniformly discretized models reported in the literature (figures 2 and 3).

It may seem that the continued fraction model of figure 5 is ideal because it is the only one that can be derived analytically from a continued fraction expansion. Other advantages of this model have also been identified in the paper. In the following we would like to present a few comments on both models.

Analytically derived models lose accuracy because of truncation: thus the impedance of a continued fraction model with a finite number of sections is always too high. On the other hand, a nonuniformly discretized model is only as good as its discretization permits. For example, a discretization in geometrical progression with ratio $\rho=2$, examined in the paper, is adequate for only a small number of sections; for more sections the ideal ratio should be smaller (and vice versa) since, otherwise, the errors remain significant in the medium frequency range, as reflected by curve *b* of figure 4. In fact, by the simple device of choosing a "best" ρ and by making the π-sections in figure 7 nonsymmetrical (with the resistance branches also reflecting the variation in geometric progression of the width of the sublaminations), the errors of the nonuniformly discretized model can be made smaller than those of the continued fraction model.

We see thus that the balance in the evaluation of the two models examined in the paper can easily tip one way or the other: it is therefore useful that both alternatives of ladder equivalents, with the inductances placed transversally or longitudinally, have been examined in the paper. In any case, both the problem of truncation inherent in the continued fraction model and of the need for improved discretization in the dual alternative can be easily resolved by iterative computational fitting with the methodology described in [A]. This leads to ladder equivalents of significantly lower order. For example, the nonuniformly discretized model has less than 1% error for frequencies up to 200 kHz with only 4 inductances, and up to 1 MHz with 5 inductances.

An advantage of the continued fraction model, suggested in the paper, is that only the leading inductance has to be nonlinear. This appears to be based on some heuristic reasoning: could the authors please elaborate on this topic? Also, in the case of the nonuniformly discretized model, could not all inductances, except the innermost one, be considered constant or assumed to vary in unison, so as to reflect the fact that the state of saturation of the lamination is essentially uniform, in accordance with the low frequency flux variations?

The requirement of only one nonlinear inductance in the lamination model becomes, in our experience [B], less stringent if, instead of the Newton-Raphson interface used in the paper, a simple iterative scheme is adopted, with voltage as input into the ladder network and current as output. The convergence is fast because these currents are small; see Appendix B of [C]. In addition, the representation of any number of nonlinear elements in a ladder equivalent is simple and straightforward.

[A] F. de Leon and A. Semlyen, "Time Domain Modeling of Eddy Current Effects for Transformer Transients", paper no. 92 WM 251-9-PWRD, presented at the 1992 IEEE/PES Winter Meeting.

[B] F. de Leon and A. Semlyen, "Reduced Order Model for Transformer Transients", IEEE Transactions on Power Delivery, Vol.7, No.1, January 1992, pp. 361-369.

[C] F. de Leon, "Transformer Model for the Study of Electromagnetic Transients", Ph.D. thesis, 1992, University of Toronto.

Manuscript received February 12, 1992.

E. J. Tarasiewicz: We welcome the comments of the discussers, elaborating further on the challenges in modeling eddy current effects including non-linearities. In particular, we are intrigued by the lengthy and insightful comments by Dr. Meredith.

The comments of Dr. Semlyen and Dr. de Leon, concerning further optimization of the nonuniformly discretized model are consistent with the findings of our CEA project. In particular, the optimum value of ρ varies with the physical parameters for laminations (thickness, conductivity, permeability) and the desired modeling bandwidth. We do favor a uniform ladder for studies requiring high accuracy including iron nonlinearity if the associated computational burden is not a concern. However, for practical power transmission studies, the continuous fraction model incorporating only one nonlinearity appears more efficient, though it sacrifices some accuracy at high frequencies in mild saturation. Maintaining the trailing ladder inductances constant at their respective unsaturated values, while ascribing core hysteresis and saturation for the entire lamination to the first inductor, reproduces the correct terminal response asymptotically for the two dominant magnetic states (i.e., unsaturated and saturated iron) as follows. For the unsaturated state, the ladder matches the analytical solution; in the saturated state, flux distribution is nearly uniform and is modeled accruately by the saturated value of L (Fig. 5; effectively short circuiting trailing ladder elements). In mild saturation, or in instances where the saturation state differs markedly through the lamination cross-section, modeling accuracy is indeed slightly compromised. To this extent (i.e., under mild saturation), we share Dr. Meredith's concerns. Nevertheless, and subject to the foregoing caveat, the continuous fraction model is an acceptable compromise between modeling accuracy and numerical efficiency for most power system transient studies.

In most studies, the applied power frequency excitation establishes an operating point on the saturation characteristic, and is associated with nearly uniform flux distribution in the laminations. The onset of a transient component merely causes a small flux perturbation around the operating point (or bias level). The latter causes traversals around minor hysteresis loops, associated with rather small incremental permeabilities. Dr. Meredith's discussion appears to overlook this mechanism, accounting largely for our different perspectives. Due to hysteresis, the effective incremental permeability of laminations for eddy current effects is not nearly as large as that in the unsaturated region.

The use of single valued function nonlinearities forces the effective permeability of the iron to be the same for the high frequency and low frequency components throughout the "linear range". This is contrary to the fact that minor hysteresis loops produced by the high frequency components have effectively lower permeability, whether the iron is in saturation or not. Highly accurate results can only be obtained if the high order uniform models are solved assuming hysteretic type nonlinearities with the correct minor loop representation. Thus the comparison of the results given by Dr. Meredith in Table 2 may not be meaningful, especially for the cases where more than one frequency component are involved, if the nonlinear models used did not account for hysteretic behavior of iron core. To illustrate using results of our measurements on a 230-kV, 125 MVA transformer, the incremental permeability at the steepest part of the major hysteresis loop (below 0.2 pu rated flux) is consistent with Dr. Meredith's quoted 50,000. Immediately upon a flux reversal the incremental permeability is closer to 1000 and rises gradually to a peak value of about 2000 for 0.1 pu flux excursion. We note that a 0.1 pu flux excursion corresponds to a progressively larger voltage swing at higher frequencies. On this basis we favoured a relatively low choice for the relative permeability for the purpose of comparing errors among the models. The remaining material parameters for laminations are 16.7E-8 ohm-meter and 0.33 mm, about the same as Dr. Meredith.

There is not disputing that the uniform ladder represents the ultimate in accuracy for any conceptual magnetization state. However, the model is simply impractical for most power system transient studies. To illustrate, consider an 18 section uniform ladder for frequencies to 600 Hz, as suggested for Table 2 of the discussion. An exact time domain solution for this ladder is obtained by iterations at each time step. For an EMTP study involving a 3-phase transformer, each wound-limb and yoke-pair on the core must be modeled by one such ladder, since the magnetic state of each limb may differ. Consequently, each transformer model incorporates 5 such ladder networks, or 90 nonlinear inductors, for modeling eddy currents accurately to just 600 Hz. Incorporating hysteresis into the nonlinearity only compounds the enormous challenge for each time step solution.

The nonuniformly discretized ladder permits a significant improvement in numerical efficiency. Nevertheless, it requires about 20 nonlinear inductors per transformer (for an optimized ladder) for 200 kHz modeling bandwidth (compared to almost 200 for the uniform ladder). We agree that saturation drastically increases penetration depth of high frequency signals, but this also allows more uniform flux distribution inside laminations, so that a coarser discretization is adequate.

Dr. Semlyen and Dr. de Leon favor seeking an efficiency improvement for the nonuniform ladder on a heuristic basis, suggesting that perhaps all but the innermost inductor could be made constant. The values for the fixed inductances would presumably correspond to the saturated case, in order for the unsaturated inductance to remain accurate. Conceptually, this implies a very fine discretization of laminations near the surface, and a correspondingly coarser subdivision towards the centre. As such, the model appears equivalent to the classical frequency-independent eddy current model, involving a single resistor in parallel with the magnetizing branch. The second proposal of Dr. Semlyen and Dr. de Leon requiring all inductors to vary in unison has greater merit as it appears equivalent to the assumption of a uniform bias. Further investigations may be warranted to assess whether this process is feasible and how it would affect the computational burden associated with iterations in the time-step loop.

Finally, the corrections noted by Dr. Meredith in the last paragraph of his discussion, concerning Figure 3 of the paper, are well founded. We regret this oversight.

Manuscript received April 16, 1992.

Transactions on Power Delivery, Vol. 7 No.1, January 1992

MODELING OF METAL OXIDE SURGE ARRESTERS

IEEE WORKING GROUP 3.4.11
APPLICATION OF SURGE PROTECTIVE DEVICES SUBCOMMITTEE
SURGE PROTECTIVE DEVICES COMMITTEE

Abstract: Working Group 3.4.11 on Surge Arrester Modeling has reviewed a number of ways to model metal-oxide arresters. Lab test data of metal-oxide arrester discharge voltage and currents available to the working group have indicated that metal-oxide arresters have dynamic characteristics that are significant for studies involving lightning and other fast-front surges. A model is described which will give an appropriate voltage response for a current surge which has a time-to-crest anywhere in the range of 0.5 us to 45 us. The paper presents a method for generating the parameters of the model from published manufacturer's data.

Key Words: Modeling, Metal-Oxide Arrester, Lightning, Frequency-Dependent Model, Insulation Coordination

INTRODUCTION

Working Group 3.4.11 on Surge Arrester Modeling Techniques was formed in 1971 by the IEEE Surge Protective Devices Committee's subcommittee on the Application of Surge Protective Devices. Initial efforts of the working group were directed toward obtaining a model of gap-type silicon-carbide arresters valid for switching-surge studies. Results of this work were presented at the 1981 Power Engineering Society Winter Meeting [1]. That paper suggested that an active-gap silicon-carbide arrester could be modeled as a time-dependent voltage, initiated by gap sparkover, in series with a non-linear voltage versus current (V-I) characteristic. The paper also suggested a logical study procedure for determining if insulation levels are protected satisfactorily.

Since 1981 the working group has been gathering data on the characteristics of metal-oxide surge arresters. Analysis of this data suggested that switching-surge studies could be performed by representing metal-oxide arresters only with their non-linear V-I characteristics. However, the test data that the working group has been able to obtain indicates that metal-oxide arresters have dynamic (or frequency-dependent) characteristics that are significant for lightning and other fast wavefront surges. The significant dynamic characteristics are that the voltage across a metal-oxide arrester increases as the time to crest of the arrester current decreases and that the arrester voltage reaches a peak before the arrester current reaches its peak. This would not be the case if the metal-oxide valve element performed strictly as a non-linear resistance. The working group believes that the dynamic effects are significant condiderations for surge arrester location and insulation coordination studies.

91 WM 012-5 PWRD A paper recommended and approved by the IEEE Surge Protective Devices Committee of the IEEE Power Engineering Society for presentation at the IEEE/PES 1991 Winter Meeting, New York, New York, February 3-7, 1991. Manuscript September 4, 1990; made available for printing January 22, 1991.

The goal of the group has been to find a mathematical model that will adequately produce these effects without requiring so much computing time that other aspects of system modeling would have to be sacrificed or compromised. The effort has been primarily directed toward station class arresters; however, the concepts would be the same for other classes of metal-oxide arresters. Also, the working group has concentrated on modeling gapless metal-oxide arresters and not arresters with a shunt or series gap. This emphasis was due to the increasing popularity of gapless arresters.

The dynamic effects tend to become significant for current waves which peak in the range of 8 us and faster. Data available to the working group and relevant to the tasks undertaken contain current waves which peak as fast as 0.5 us. Nanosecond data has not been sought or considered because of the difficulties in making reliable measurements in this time frame [2-5]. For currents with times to crest in the range of 0.5 us to 4 us or less, any stray inductance in the measurement circuit can result in the measurement of higher arrester discharge voltages than would actually be produced by the arrester. This type of measurement error tends to make the dynamic effects mentioned above more pronounced. The most recent data collected by the working group (see Appendix A) indicates that the arrester discharge voltage for a given discharge current magnitude is increased by only approximately 6% as the time to crest of the current is reduced from 8 us to 1.3 us. Earlier data indicated an increase of approximately 12%.

Another phenomenon which occurs when the time-to-crest of arrester discharge currents becomes shorter than approximately 4 us is that of voltage spikes. These are voltage peaks on the front of the arrester discharge voltage waveform which can sometimes exceed the subsequent discharge voltage level for the arrester. The magnitudes of these spikes are increased by any stray inductance in the measuring loops. The spike can be minimized by placing the voltage divider used in the measurement circuit in a coaxial arrangement within the metal-oxide block [2]. There is no consistent agreement in recent technical literature [2-5] concerning the realistic magnitude of these spikes. There is some evidence, however, that when careful measurements are made, the spike does not exceed the subsequent discharge voltage for currents cresting in 0.5 us or more [2]. Therefore, the working group has not tried to obtain a model which will reproduce the spike.

The effort by the working group was limited to modeling metal-oxide arresters only. Other system parameters such as arrester leads and separation distances are also important in studies with lightning and other fast-wavefront surges. The users of the arrester models described in this paper also need to take these other parameters into account.

MODELING EFFORTS

When specifying a metal-oxide arrester model for a study, the accuracy of the simulation is improved when the arrester characteristic is chosen to be consistent with the frequency or time-to-crest of the voltage and current expected during system perturbations. This concept is key to modeling because as mentioned previously, metal-oxide arresters are frequency-dependent devices, i.e., the voltage across the arrester is a function of both the rate of rise and the magnitude of the current conducted by the arrester. In order to obtain metal-oxide arrester characteristics for a wide range of waveshapes (lightning to temporary overvoltages), a number of different current test waveshapes have been used. Several test waves have been defined by ANSI C62.11 while others have not. The use of a simple nonlinear V-I characteristic which is derived from test data with appropriate times to crest would be adequate in the absence of a frequency-dependent model.

In searching for a frequency-dependent model, the working group tried to account for the variation of arrester voltage with time to crest of arrester current by adding an inductance in series with a non-linear characteristic. This approach had some merit because the voltage across the inductance, and hence across the arrester, would increase as the time to crest of the current decreased. A variation of this model included the addition of a shunt capacitance to the non-linear resistance. However, the shunt capacitance had a negligible effect on the results of the model. This type of model had some success in matching a particular test result. For example, an inductance could be chosen for the model such that it gave a reasonably good match of the voltage magnitude and waveshape for an arrester discharge current which reached its crest in 8 us. However, when the same inductance and other model parameters were used for an arrester discharge current which reached its crest in 2 us, the voltage magnitude was in error by a significant amount. Different parameters could be chosen for the model such that good results could be obtained for the voltage corresponding to an arrester current reaching its crest in 2 us. However, if the time to crest of the current differed very much from 2 us, the resulting voltage was in error. Hence, it became apparent that a more sophisticated model would be required if it were to be used to represent the response of an arrester to currents with a wide range of times to crest.

The next model considered by the working group is referred to as the frequency-dependent model [6]. For this model the non-linear V-I characteristic of an arrester is represented with two sections of nonlinear resistance designated A_0 and A_1 as shown in Figure 1. The two sections are separated by an R-L filter. For slow-front surges, this R-L filter has very little impedance and the two non-linear sections of the model are essentially in parallel. For fast-front surges the impedance of the R-L filter becomes more significant. This results in more current in the non-linear section designated A_0 than in the section designated A_1. Since characteristic A_0 has a higher voltage for a given current than A_1 (see Figure 2), the result is that the

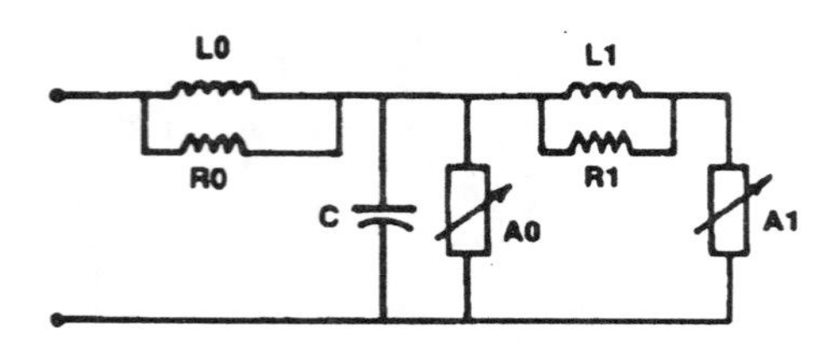

Figure 1 - Frequency-Dependent Model

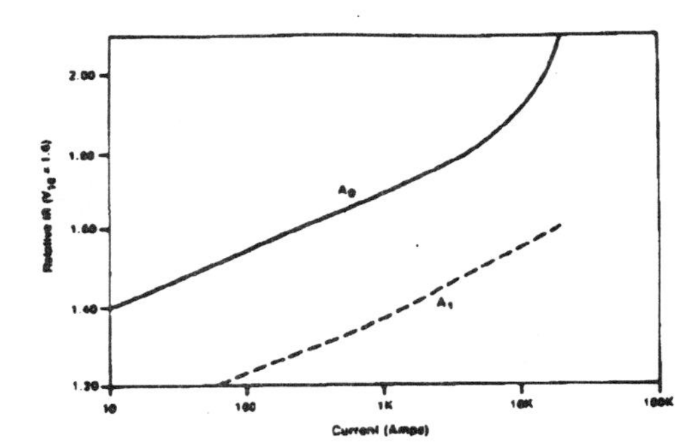

Figure 2 - V-I Relationships for Non-Linear Resistor Part of Model

arrester model generates a higher voltage. Since metal-oxide arresters have a higher discharge voltage for fast-front surges, the model matches the overall behavior of a metal-oxide arrester. More sophisticated versions of this model can be made by adding more sections of non-linear resistance separated by R-L filters. However, only the two-section model was investigated by the working group, because it gave good correlation with lab-test data. The primary difficulty with this type of model lies in choosing the parameters of the model. This aspect will be discussed later in the paper.

RECOMMENDED MODEL FOR TEMPORARY OVERVOLTAGE AND SWITCHING-SURGE STUDIES

One objective of a transient study is to evaluate the performance of metal-oxide arresters during temporary and switching-surge overvoltages on the system. A metal-oxide arrester model suitable for such studies would be a nonlinear resistance with characteristics which can be derived from a low frequency test wave consisting of a half sinusoid with a 1 ms time to crest. This test wave is designated as the "1 ms wavefront". An example of a 1 ms wavefront characteristic for a metal-oxide disk is shown in Figure 3 and should be used in system simulations involving temporary overvoltages and slow switching surges (e.g., switching a large capacitor bank in the absence of other banks). In this context, temporary overvoltages refer to overvoltages with durations of one second or less which produce arrester currents of one ampere or more. If arrester currents and overvoltage durations are outside of this range, the model may not be adequate. The characteristics shown in Figure 3 are for example only. Data should be obtained from manufacturers for use in modeling.

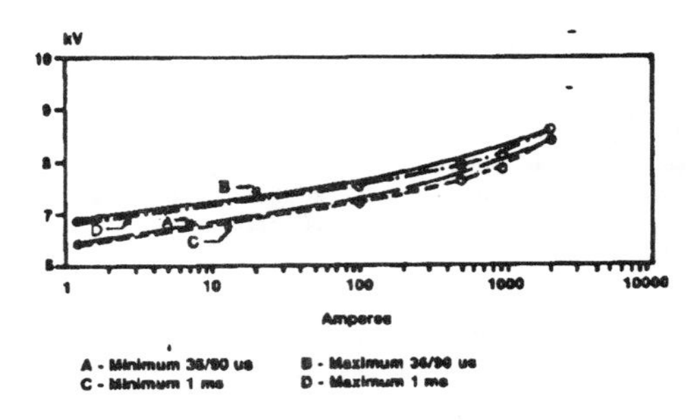

Figure 3 - Examples of Metal Oxide Disk Characteristics Including Manufacturing Tolerances

Similarly, the switching-surge characteristic of metal oxide arresters is obtained by applying a current wave with an actual time to crest between 45 to 60 us. Figure 3 also shows the voltage-current relationship for a 36x90 us current wave and is appropriate for analyzing arrester operations which involve the switching of transmission lines. Note that the discharge voltage of the arrester is higher for switching-surge situations as compared to the characteristic for temporary overvoltages.

An additional consideration when evaluating the performance of metal-oxide arresters is to recognize that there are manufacturing tolerances associated with the actual construction of the arrester. In critical arrester applications, the effects of manufacturing tolerances should be addressed. The arrester characteristic with the maximum voltage for a given current should be used in the computation of protective ratios because such a model yields the most conservative assessment of the protective ratio. On the other hand, the lower voltage-current curve should be considered for situations where the discharge energy duty of the arrester needs to be maximized.

The final consideration for arrester modeling in switching studies is to define the arrester model over a range of currents expected in the simulation. For example, transmission line discharge currents are typically less than 2000 A while currents associated with restrikes on large capacitor banks may be as high as 10 kA. If the current range defined in the model is insufficient, the arrester discharge voltage predicted by the model will be too high and arrester duty too low.

In summary, for temporary overvoltages and switching-surge studies, an arrester can be modeled by choosing an appropriate nonlinear V-I characteristic. There is no need for a frequency-dependent model.

RECOMMENDED MODEL FOR LIGHTNING STUDIES

The time to crest for surges used in lightning studies can range from 0.5 us to several us. For a given current magnitude in an arrester, the voltage developed across the arrester can increase by approximately 6% as the time to crest of the current is decreased from 8 us to 1.3 us. One approach for an arrester model for lightning studies would be to use a simple non-linear V-I characteristic based on 0.5 us discharge voltage. This would give conservative results (higher voltages) for surges with slower times to crest. However, this level of conservatism may be unnecessary. The frequency-dependent model described previously will give good results for current surges with times to crest from 0.5 us to 40 us.

Much data has been collected by the working group concerning the discharge voltages for arrester valve blocks for various current levels and times to crest. This data covered several manufacturers' valve blocks. Some data was also obtained for the discharge voltages of full scale arresters (588kV and 96kV duty-cycle ratings) for various current levels and times to crest. The frequency-dependent model was successfully used each time to match this data. The main problem with using the model was how to choose the parameters of the model.

Reference 6 gives some suggested formulas for choosing the parameters of the model based on an estimated height of an arrester and the number of parallel columns of metal-oxide disks. The inductance L_1 and the resistance R_1 of the model comprise the filter between the two nonlinear resistances. The formulas for these two parameters are:

L_1 = 15d/n microhenries
R_1 = 65d/n ohms

Where d is the estimated height of the arrester in meters (use overall dimensions from catalog data)
n is the number of parallel columns of metal oxide in the arrester.

The inductance L_o in the model represents the inductance associated with magnetic fields in the immediate vicinity of the arrester. The resistor R_o is used to stabilize the numerical integration when the model is implemented on a digital computer program. The capacitance C represents the terminal-to-terminal capacitance of the arrester.

L_o = 0.2d/n microhenries
R_o = 100d/n ohms
C = 100n/d picofarads

The non-linear V-I characteristics A_0 and A_1 can be estimated from the per unitized curves given in Figure 2.

The efforts of the working group in trying to match model results to the laboratory test data have indicated that these formulas do not always give the best parameters for the frequency-dependent model. However, they do provide a good starting point for picking the parameters. An investigation was made by the working group into which parameters of the model had the most impact on the results. To accomplish this, a model based on the preceding formulas was set up on the Electromagnetic Transients Program (EMTP M39 version).

The model represented one 76 mm diameter metal-oxide valve block. Each of the parameters of the model (L_o, R_o, L_1, R_1, and C) were varied from one tenth of the base case value to ten times the base case value. These results indicated that parameter L_1 has the most impact while the other parameters varied had little impact. Figure 4 illustrates the impact of varying L_1. This work led to the following procedure for choosing the parameters of the frequency-dependent model.

1. Use the previously given formulas to derive initial values for L_0, R_0, L_1, R_1, C, and the non-linear characteristics A_0 and A_1.

2. Adjust the per unit value on the curves for characteristics A_0 and A_1 to get a good match for the published discharge voltages associated with switching surge discharge currents (time-to-crest of approximately 45 us).

3. Adjust the value of L_1 to get a good match of

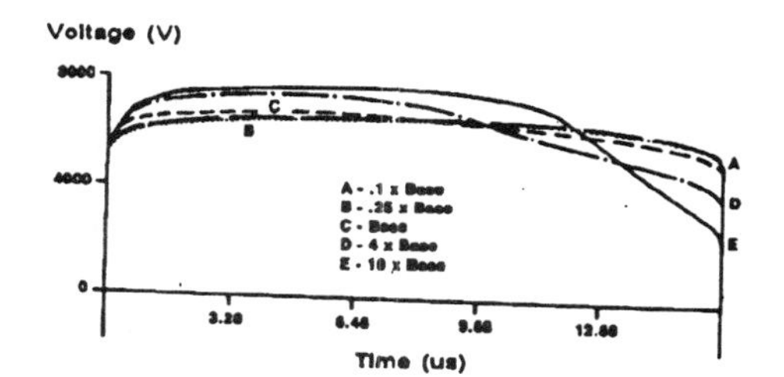

Figure 4 - Variation of Parameter L1 for Surges with 8 us Front (results for 75mm Disks)

published arrester discharge voltages for 8 x 20 us discharge currents.

The resulting parameters will give good results for surges with times to crest in the range of 0.5 us to 45 us. Appendix B gives an example of the use of this procedure.

Table 1 gives an example of model results obtained by the use of this procedure. Laboratory tests were made of the discharge voltage of one metal-oxide valve block for discharge currents ranging from 1kA to 10kA and times to crest ranging from 30 us to 2 us. The model results were within 5% of the test data as shown by the discharge voltage magnitudes in Table 1. Figures 5-8 illustrate how well the discharge waveshapes given by the model match the waveshapes obtained in the tests.

In summary the frequency-dependent model has been shown to give good results for arrester discharge voltage when the discharge current has a time to crest in the range of 0.5 to 45 us. Therefore, this model is recommended for lightning studies.

CONCLUSIONS

1. The voltage developed across a metal-oxide arrester for a given discharge current increases as the time to crest of the current decreases. The voltage will increase by approximately 6% as the time to crest is decreased from 8 us to 1.3 us.

2. For discharge currents cresting in 8 us or less, the discharge voltage reaches crest prior to the crest of the discharge currents.

3. Simple arrester models can be constructed consisting of a non-linear V-I characteristic and an inductance in series with it which can represent the two effects in Conclusions 1 and 2 for a narrow range of time to crest. Such a model would not be accurate for times to crest outside that range.

4. A model was chosen which can give satisfactory results for discharge currents with a range of times to crest for 0.5 us to 45 us. A procedure for choosing the parameters of the model is given in this paper. Comparisons of modeling results and test results were given to demonstrate the accuracy of the model.

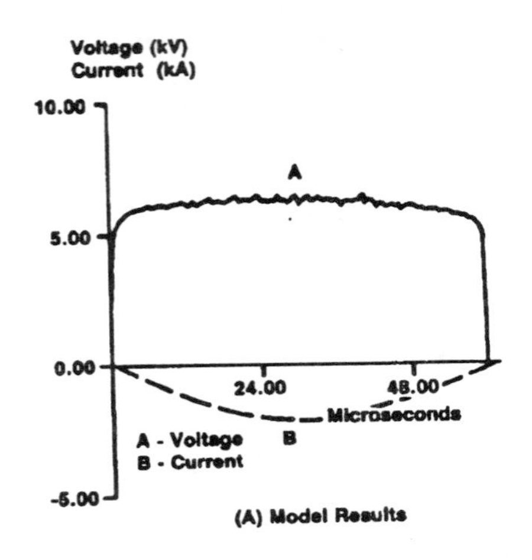

(A) Model Results

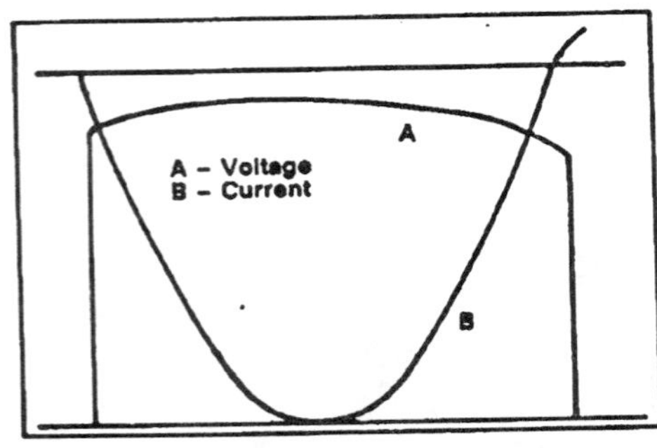

(B) Test Results (See Table 1 for Magnitudes)

Figure 5 - Comparison of Model Results with Test Results for 30 us Time-to-Crest, 2 kA Surge. Polarity of Current Reversed for Clarity.

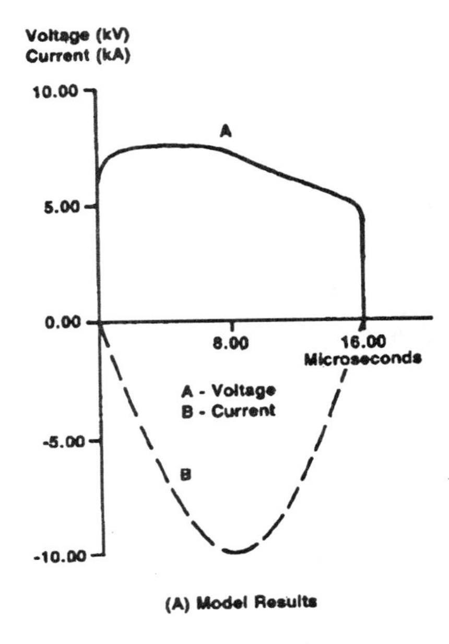

(A) Model Results

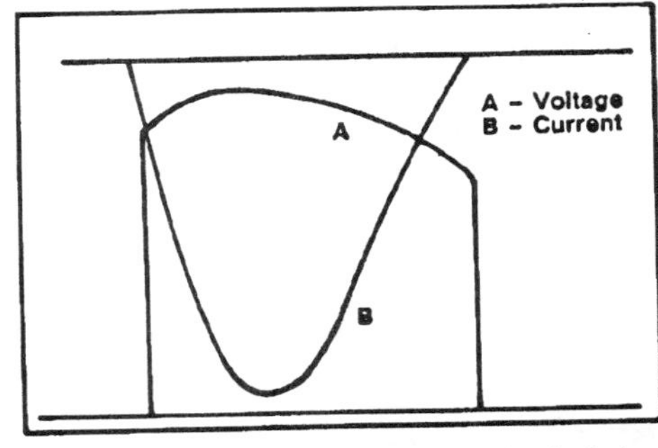

(B) Test Results (See Table 1 for Magnitudes)

Figure 6 - Comparison of Model Results with Test Results for 8 us Time-to-Crest, 10 kA Surge. Polarity of Current Reversed for Clarity.

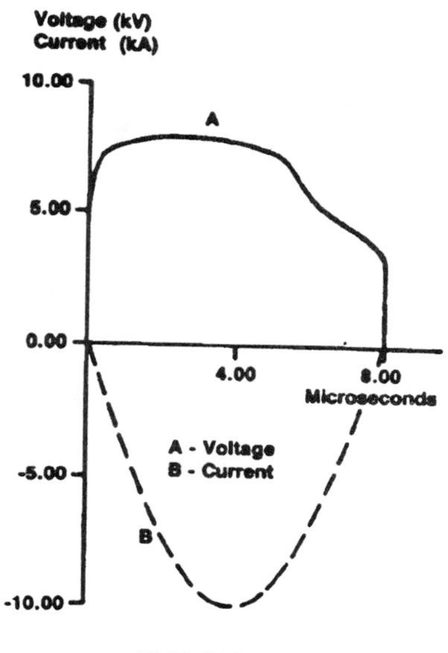

(A) Model Results

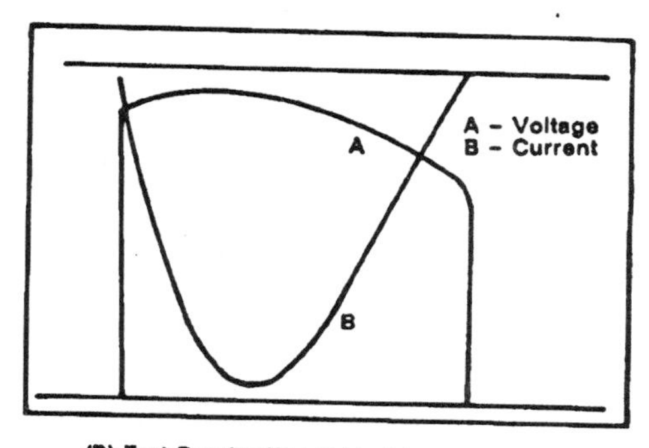

(B) Test Results (See Table 1 for Magnitudes)

Figure 7 - Comparision of Model Results with Test Results for 4 us Time-to-Crest, 10 kA Surge. Polarity of Current Reversed for Clarity.

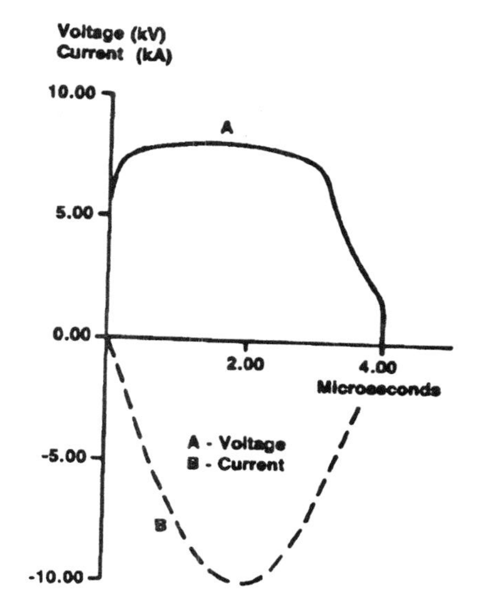

(A) Model Results

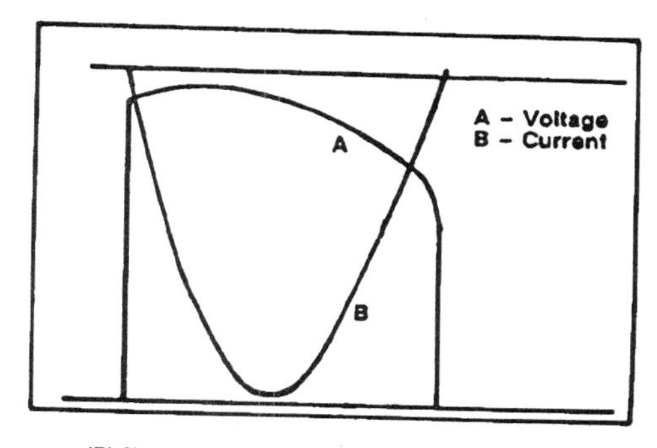

(B) Test Results (See Table 1 for Magnitudes)

Figure 8 - Comparision of Model Results with Test Results for 2 µs Time-to-Crest, 10 kA Surge. Polarity of Current Reversed for Clarity.

TABLE 1 - Model Results Compared to Test Data For One Valve Block

Surge Front	Test Current (kA)	Peak Test Voltage* Minimum	Maximum	Average	Peak Model Voltage	Error Between Model & Average
30 us	1	6218	6366	6301	6186	-1.8%
	2	6458	6629	6561	6504	- .9%
8 us	5	7010	7191	7098	6968	-1.8%
	10	7466	7647	7562	7558	0%
4 us	5	7101	7281	7194	7332	1.9%
	10	7609	7805	7704	7892	2.4%
2 us	5	7175	7359	7265	7621	4.9%
	10	7805	7987	7892	8105	2.7%

* Note: A total of 9 different valve blocks were tested. These all came from one surge arrester.

REFERENCES

1. IEEE Working Group of Surge Protective Devices Committee, "Modeling of Current-Limiting Surge Arresters", IEEE Transactions on Power Apparatus and Systems, Vol. PAS-100, pp. 4033-4040, August 1981.

2. W. Schmidt, J. Meppelink, B. Richter, K. Feser, L. Kehl, D. Qui, "Behavior of MO-Surge-Arrester Blocks to Fast Transients," IEEE Transactions on Power Delivery, Vol. 4, No. 1, pp. 292-300, January, 1989.

3. W. Breilmann, "Protective Characteristics of Complete Zinc-Oxide Arresters and of Single Elements

for Fast Surges," Paper 82.04, Fifth International Symposium on High Voltage Engineering, Braunschweig, Federal Republic of Germany, August, 1987.

4. A Bargigia, M.deNigris, A. Pigini, "The Response of Metal Oxide Resistors for Surge Arresters to Steep Front Current Impulses," Paper 82.01, Fifth International Symposium on High Voltage Engineering, Braunschweig, Federal Republic of Germany, August, 1987.

5. C. Dang, T. Parnell, P. Price, "The Response of Metal Oxide Surge Arresters to Steep Front Current Impulses," IEEE Transactions on Power Delivery, Vol. PWRD-1, No. 1, pp. 157-163, January, 1986.

6. D. W. Durbak, "Zinc-oxide Arrester Model For Fast Surges", EMTP Newsletter, Vol. 5 No. 1, January, 1985.

Acknowledgments

The membership of W.G. 3.4.11 consists of the following individuals:

R. A. Jones - Chairman

P. R. Clifton
Hieu Huynh
Glenn Grotz
Mike Lat
Frank Lembo *
D. J. Melvold
Olaf Nigol
J. P. Skiutas
Andrew Sweetana
Jon Woodworth
D. F. Goodwin
J. L. Koepfinger
D. W. Lenk
Yves Latour
R. T. Leskovich
Yasin Musa
Javad Adinzh
Keith Stump *
E. R. Taylor

* Past Chairman of working group

Appendix A

Test Data Gathered

Much test data was collected by the working group to determine the discharge voltage waveform for a metal-oxide arrester when subjected to discharge currents of varying magnitudes and times to crest. Some of the older data was supplied by arrester manufacturers for their own arresters. Some of the newer data came from three test labs which performed tests on metal-oxide blocks from another manufacturer. There is not sufficient space to publish all of this data. Therefore, only a summary table will be given here to give an idea of the data that the working group used.

COMPARISON OF COLLECTED ARRESTER DATA

	Time	1kA	1.5kA	2kA	3kA	5kA	10kA	20kA
I. NEW DATA								
LAB A	36 us	.833		.868				
LAB A	8 us					.939	1.00	
LAB A	4 us					.951	1.019	
LAB A	2 us					.961	1.044	
LAB A	1.3 us					-	-	
LAB B	29 us	.84		.87				
LAB B	8 us					.94	1.00	
LAB B	4 us					.96	1.02	
LAB B	2 us					.98	1.05	
LAB B	1.3 us					1.00	1.06	
LAB C	30 us						.974	
LAB C	8 us						1.00	
LAB C	2 us						1.05	
II. OLD DATA								
MFG. A 95mm	45 us		.867		.918			
MFG. A 95mm	10 us		.877		.933	.973	1.00	1.055
MFG. A 95mm	2 us		.904		.947	1.00	1.043	1.179
MFG. A 95mm	1.1 us		.892		.933	1.00	1.055	1.25
MFG. B 50mm	45 us		.766		.816			
MFG. B 50mm	8 us		.809		.868	.912	1.00	1.12
MFG. B 50mm	4 us		.824		.890	.941	1.04	1.16
MFG. B 50mm	1.3 us						1.10	1.26
MFG. B 75mm	45 us		.82		.85			
MFG. B 75mm	8 us		.85		.90	.95	1.0	1.11
MFG. B 75mm	4 us		.87		.92	.98		
MFG. B 75mm	1.3 us						1.10	1.29
MFG. C 75mm	47 us		.836		.874		.989	
MFG. C 75mm	12 us					.925	1.00	1.08
MFG. C 75mm	1 us					1.04	1.12	1.26

Note: Voltages are expressed in per unit of 10kA ,8X20 us voltage

Transactions on Power Delivery, Vol. 7 No.1, January 1992

Appendix B

Example of Parameter Selection for a Metal Oxide Arrester Model

This section details the determination of the metal-oxide model parameters for a one column arrester with an overall length of 1.45 meters. The discharge voltage, V_{10}, for this arrester is 248kV and the switching-surge discharge voltage, V_{ss}, is 225kV for a 3kA, 300 x 1000us current waveshape.

B1.0 Arrester Information Required

- d - length of arrester column in meters (use overall dimensions from catalog data)
- n - number of parallel columns of metal-oxide disks
- V_{10} - discharge voltage for a 10kA, 8 x 20uS current, in kV
- V_{ss} - switching-surge discharge voltage for an associated switching-surge current, in kV

B2.0 Determining the Initial Parameters

B2.1 Lumped Parameter Elements

Using the equations presented previously in this paper, the initial values for L_o, R_o, L_1, R_1 & C are determined as follows:

L_1 = (15d)/n uH = (15*1.45)/1 = 21.75uH

R_1 = (65d)/n ohm = (65*1.45)/ = 94.25 ohm

L_o = (0.2d)/n uH = (0.2*1.45)/1 = 0.29uH

R_o = (100d)/n ohm = (100*1.45)/1 = 145 ohm

C = (100n)/d pF = (100*1)/1.45 = 68.97pF

B2.2 Nonlinear Resistors

The nonlinear resistors, A_o & A_1, can be modeled in the EMTP as a piecewise linear V-I curve with characteristics defined point by point. The number of points selected to represent the nonlinear resistance depends on the smoothness desired. In this example, approximately a dozen points ranging from 10A to 20kA were selected. The nonlinear resistors could also be modeled in EMTP with equations of the form $I=BV^{\alpha}$

Figure 2 in the paper is used to determine the initial characteristics of the nonlinear resistors A_o and A_1. Each of the V-I points for the nonlinear resistors is found by selecting a current point and then reading the relative IR in pu from the plot. This value is then multiplied by (V_{10}/1.6) to determine the model discharge voltage in kV for the associated current. This scaling from pu to actual voltage is done by the application of the following formula to the "Relative IR" pu voltage found for that current as given in Figure 2:

For A_o, the

Discharge kV = [Relative IR in pu for A_o(i)]*[V_{10}/ 1.6].

Likewise for A_1, the

Discharge kV = [Relative IR in pu for A_1(i)*[V_{10}/ 1.6]

For the above arrester, the associated V-I voltage for a 10kA current for the nonlinear resistor, A_o is determined by reading the "Relative IR" for a 10kA current from Figure 2. Examination of the plot shows that the "Relative IR" for a 10 kA current is 1.9pu. Therefore the discharge kV for A_o associated with 10kA is:

Discharge kV = [1.9]*[155]
= 294.5kV

Other V-I points for the nonlinear characteristics are determined in a similar manner. See Table B0 for the V-I points for the nonlinear resistor A_o and Table B1 for the V-I points for the nonlinear resistor A_1. Figure B1 gives a schematic of the initial model.

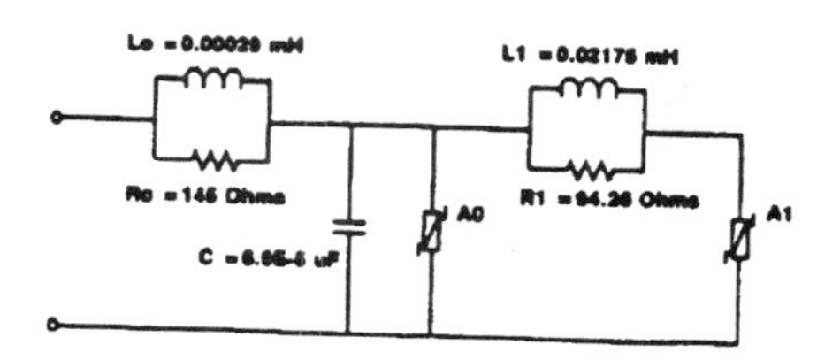

Figure B1 - Initial Arrester Model

Table B0
V-I characteristics for A_o

kA	V(pu)	V(kV)
0.01	1.40	217.0
0.1	1.54	238.7
1	1.68	260.4
2	1.74	269.7
4	1.80	279.0
6	1.82	282.1
8	1.87	289.9
10	1.90	294.5
12	1.93	299.1
14	1.97	305.3
16	2.00	310.0
18	2.05	317.7
20	2.10	325.5

Table B1
V-I characteristics for A_1

kA	V(pu)	V(kV)
0.1	1.23	190.50
1	1.36	210.80
2	1.43	221.65
4	1.48	229.40
6	1.50	232.50
8	1.53	237.15
10	1.55	240.25
12	1.56	241.85
14	1.58	244.95
16	1.59	246.45
18	1.60	248.00
20	1.61	249.55

0885-8977/91/$3.00©1992 IEEE

B3.0 Adjustment of A_0 and A1 to Match Switching-Surge Voltages

The metal-oxide arrester model with the initial parameters is tested for a match with the switching-surge current and voltage. Figure B2 is the circuit used to inject switching-surge current into the initial model. The current injected should be of the same magnitude and waveshape as the current used by the manufacturer to determine the switching-surge discharge voltage. Inject switching-surge test current magnitude and waveshape and examine resulting peak voltage.

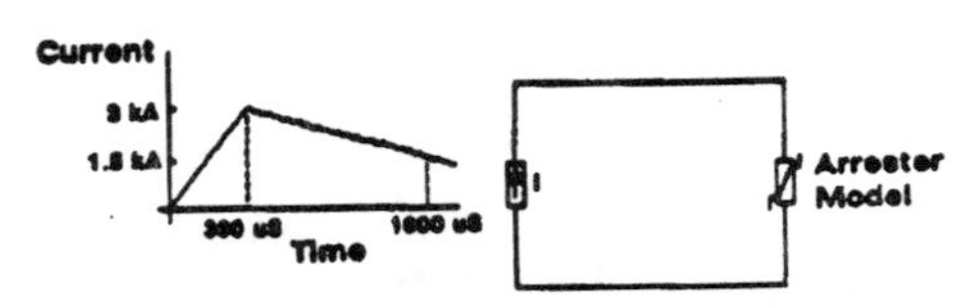

Figure B2 - Switching Surge Test Circuit

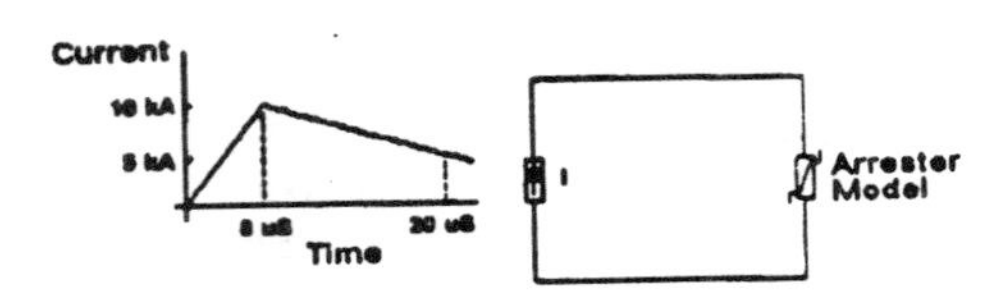

Figure B3 - V_{10} Surge Test

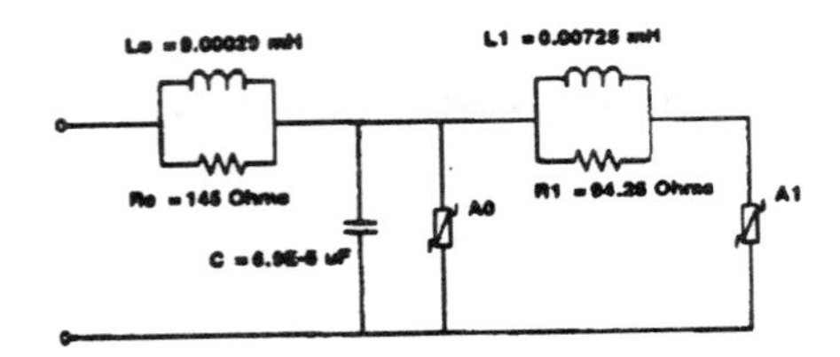

Figure B4 - Final Arrester Model

The nonlinear resistances, A_0 & A_1, are used to adjust the model for correct switching-surge voltages. This is done by adjusting the "Relative IR" pu voltages found in section B2.2 until there is a good match between the manufacturer's switching-surge voltage and current and the model test results.

In the example presented, injecting a 3kA, 300 x 1000 us current waveshape into the intital model resulted in a 225.6kV voltage peak that was in good agreement with the manufacturer's value of 225 kV for the same current. No adjustment of the nonlinear resistances, A_0 & A_1, was necessary in this step. Use the values shown in Tables B0 and B1 for the characteristics of the nonlinear resistances, A_0 & A_1 in the revised model.

B4.0 Adjustment of L_1 to Match V_{10} Voltages

The model with the correct nonlinear resistances, A_0 & A_1, is now tested to obtain a good match between the manufacturers' data and the model discharge voltages for an 8 x 20us current. Figure B3 is a schematic of the circuit used to inject an 8 x 20us current into the revised model. Examine the resulting peak voltage and adjust L_1 in the model until it produces a good match with the manufacturers' V_{10} voltage.

For the arrester in this example, the value of L_1 was determined by trial and error as shown in Table B2. Use L_1 = 0.00725mH in the final model. See schematic in Figure B4.

TABLE B2

Run No.	L_1 Magnitude mH	V_{10} kV	difference from 248kV %	L_1 for next trial mH
1	0.02175	262.61	5.6	0.02175/2
2	0.010875	252.34	1.7	0.010875/2
3	0.0054375	246.37	0.7	-
4	0.00725	248.37	0.15	-

DISCUSSION

David W. Jackson, R. W. Beck and Associates, Waltham, MA. This paper and the associated paper 91 WM 011-7 PWRD, Computationally Efficient MOV Model for Series Compensation Studies, provide an excellent basis for preparing an IEEE Application Guide or a Recommended practice. I encourage the Working Group to submit a Standards Project Authorization Request, and to prepare such a standard for general use by surge protection application engineers.

IEEE Transactions on Power Delivery, Vol. 11, No. 2, April 1996

STUDY OF ZnO ARRESTER MODEL FOR STEEP FRONT WAVE

Ikmo Kim, Member
Toshihisa Funabashi, Member
Haruo Sasaki, Member
Toyohisa Hagiwara, Non-Member
Misao Kobayashi Senior Member
Meidensha Corporation
Tokyo, Japan

Abstract - **Measurements were performed to get the response of arrester block to steep front impulse current(1/2μ sec), 4/10μsec impulse current and standard impulse (8/20μsec). As the steep front wave response model, the conventional model, IEEE model, and nonlinear inductance model have been studied.**

The method of building arrester block model was explained in this paper. The nonlinear inductance model proposed in this paper has accuracy with about 1% error in calculations of steep front impulse in block model study. This model can be built easily and simply.

Keywords - **Insulation coordination, Steep front impulse, residual voltage characteristics, Arrester block, IEEE recommended Model, Nonlinear inductance model**

1. Introduction

ZnO arrester generally has the characteristics that its residual voltage rises for current impulse having more steep front than standard impulse current(8/20μsec)[1]~[4]. Rising rate of residual voltage depends on the time to crest, peak value of current and kinds of block[6]. These characteristics become important in considering the insulation coordination. For examples, the insulation coordination of Gas-Insulated Switchgears, must be designed more rationally, taking account of the protection performance of surge arresters. Many GIS have been constructed and now in operation because of their compactness and environmental merits. This paper presents a ZnO arrester model considering steep front wave effect, which can be used in EMTP which is a powerful tool to analyze the power system.

A conventional model which is represented by nonlinear resistance and does not consider steep front wave effect, and IEEE recommended model which was proposed recently[7], have been selected from various models proposed in past studies. We also propose a nonlinear inductance(L) model in this paper, and have studied these three different types of model.

For this study, we measured the response characteristics of arrester block to steep front impulse current(1/2μsec), 4/10μsec impulse current, and standard impulse current(8/20μsec). The block level models were built and compared.

95 SM 369-9 PWRD A paper recommended and approved by the IEEE Surge Protective Devices Committee of the IEEE Power Engineering Society for presentation at the 1995 IEEE/PES Summer Meeting, July 23-27, 1995, Portland, OR. Manuscript submitted December 28, 1993; made available for printing May 11, 1995.

IEEE recommended model can be used in the range of current surge wave front time from 0.5 to 45μsec. Therefore, this model has to be a frequency dependent model which is complex. This approach has some deficiencies. It needs detailed information of arrester block to build models and also needs some calculations. It was clarified that this model has the voltage characteristics having too much drop in tail of voltage curve in block level studies. But the drop in tail of voltage curve is not important to the insulation coordination study.

Nonlinear inductance model consists of a nonlinear resistance and a nonlinear inductance in series. The nonlinear inductance model proposed in this paper has a good response characteristic to steep front wave impulse calculation. But the error of calculation was increased in standard wave impulse case. That case is not so important in considering insulation coordination. This model can be built easily. This model is built from the two pieces information which are the v-i characteristics of 8/20μsec impulse(conventional model) and the rising rate of steep front impulse voltage to standard impulse. Those data are normally available test data which are found in catalogs. The model calculation algorithm was proposed in this paper. Using the program, EMTP input formatted data(nonlinear inductance model) can be built very simply. This model was developed to cope with the steep front wave study. That is very important in considering insulation coordination. In point of model building time, the nonlinear inductance model proposed in this paper is much easier than IEEE model which needs several calculations to adjust the parameters of the model circuit.

2. Measurement of ZnO block

Measurement circuit is represented in Fig. 1. After charging the capacitance C, the air gap is discharged and current flows through arrester block. The voltage and current have been measured through the voltage divider(V.D.) and coaxial shunt(C.S.), respectively.

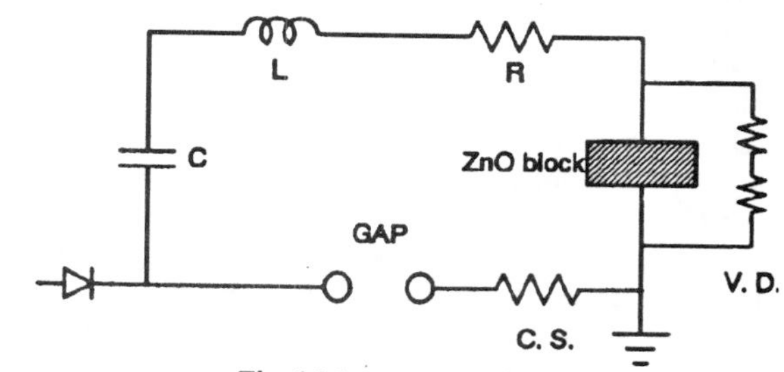

Fig. 1 Measurement circuit

Measurements were performed for three types of impulse. Those are standard impulse (its wave front time is about 8 μsec), steep front wave impulse(its wave front time is about 1 μsec) and 4/10μsec impulse currents. The peak current values were changed in each impulse test.

The results of measurement are represented in Fig. 2. This graph is based on the standard impulse test of which peak current value is 10kA. The voltage of 10kA impulse current is 1 PU. The x axis value of the graph is current peak value. The residual voltage characteristics depends on the kind of block. But the tendencies that the residual voltage of a steep front impulse is higher than that of a standard impulse at the same current are almost the same in the measurements.

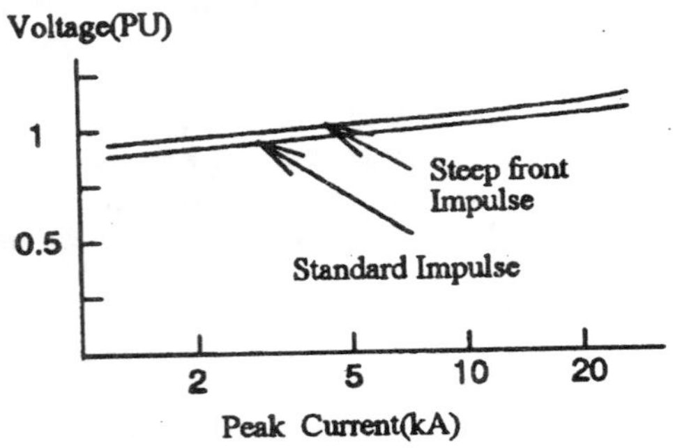

Fig. 2 The results of measurement

3. IEEE recommended model

The IEEE recommended model was presented by IEEE WG 3.4.11. This model can be used in range of front wave time from 0.5 μs to 45μs. The circuit of this model is represented in Fig. 3.

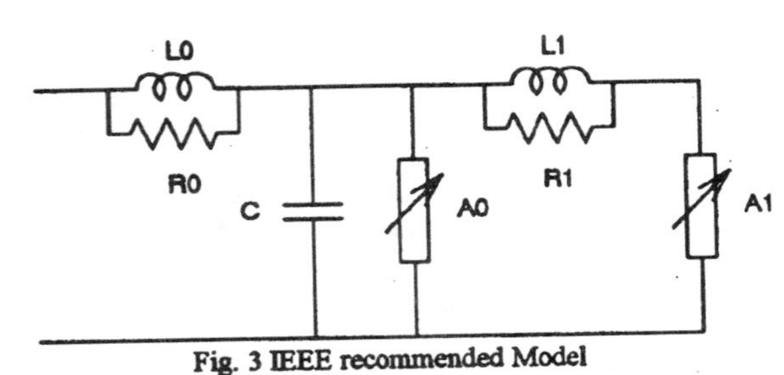

Fig. 3 IEEE recommended Model

The nonlinear resistor A0 and A1 were connected through the RL filter. The RL filter consists of R1 and L1 in parallel. For slow front surges, this RL filter has very little impedance and the two nonlinear sections of the model are essentially in parallel. For fast front surges the impedance of the RL filter became more significant. This results in more current in the nonlinear section designated A0 than in the section designated A1. Since characteristic A0 has a higher voltage for a given current than A1, the arrester model generates a higher voltage. To determine the circuit parameters A0 and A1, switching surge test results were used. The parameter L1 of RL filter was determined using the standard impulse test results. The other parameters were determined using the dimension of block. The IEEE Working Group 3.4.11 suggested formulas for choosing the parameter[7].

4. Nonlinear inductance model

ZnO arrester block has dynamic(or frequency dependent) characteristics that the voltage across the block increases as the time to crest of the arrester current decreases and that the arrester voltage reaches a peak before the arrester current reaches its peak[4][5]. Nonlinear resistance is used conventionally as the arrester block model. Since nonlinear resistance can not produce above dynamic characteristics, a nonlinear L was simply connected in series with nonlinear resistance. That circuit is represented in Fig. 4-(a). The conventional model can be represented as a v-i curve. The v-i curve was constructed from typical v-t and i-t characteristics for the standard impulse. A hysteresis loop can be constructed with one impulse test. The point v-i characteristic is determined as crossing point of peak current line and peak voltage line in Fig. 4-(b). To calculate the nonlinear inductance of nonlinear inductance model, first it is necessary to construct a hysteresis loop. The calculation method of nonlinear inductance and nonlinear resistance are proposed and the calculation algorithm of nonlinear inductance model is explained.

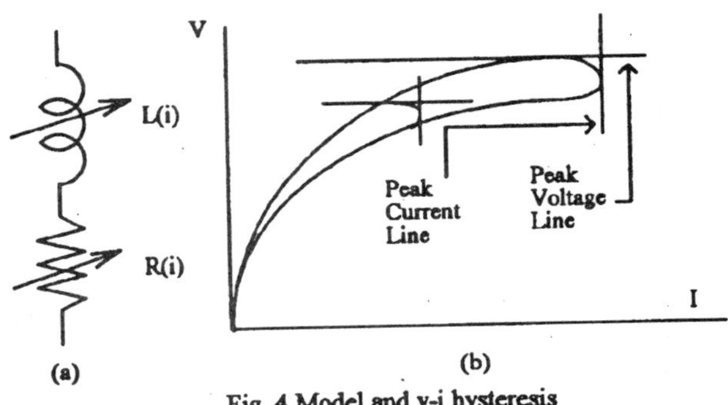

Fig. 4 Model and v-i hysteresis

4.1 Construction of hysteresis loop

A hysteresis loop was constructed based on the v-i curve. Nonlinear inductance and nonlinear resistance were calculated from the hysteresis loop.

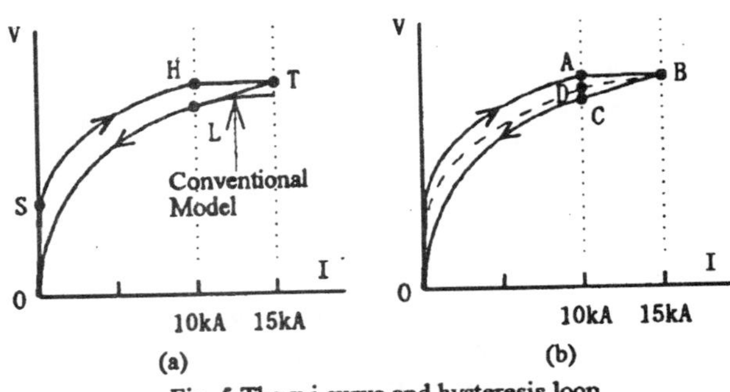

Fig. 5 The v-i curve and hysteresis loop

The nonlinear inductance model which is proposed in this paper can be derived from the conventional model. Thus, first the hysteresis loop should be built as follows.

The rising rate(RR) of voltage of the steep front wave to standard impulse test(each current peak value is about 10kA) is calculated by equation(1).

$$RR = \frac{PV1 - PV8}{PV8} \times 100 \qquad (1)$$

RR: The rising rate of steep front impulse to standard impulse[%]
PV1: The peak voltage of steep front impulse[V]
PV8: The peak voltage of standard front impulse[V]

The hysteresis loop in Fig. 5-(a) consists of the rising curve (OSHT) and falling portion(TLO). The voltage of point S is the residual voltage at 1mA. The falling portion from 10kA to zero(curve LO) is the v-i characteristic line which is the conventional model. The rising portion curve SH is constructed by raising the curve OL along the y axis. The degree of raising portion is simulated many times. The value (PV1-PV8) which means the voltage difference between steep front impulse and standard impulse, can be calculated from the RR value in equation(1). The value of raising portion is determined to be the value earned from multiplying the value (PV1-PV8) by 1.5. The top point(T) is constructed by raising a half of the above raising portion(1.5/2) from the conventional model voltage at 15kA. The point H and T is connected straight and also point T and L are connected the same way. The above values of 1.5, 1.5/2 are selected from a number of simulations. Thus, the hysteresis loop(OSHTLO) is completed.

4.2 Nonlinear inductance calculation method

Nonlinear inductance was calculated from the v-i hysteresis characteristic which are represented in Fig. 5-(b). In order to calculate the nonlinear inductance, the duration time of the hysteresis loop is selected as the time of steep front impulse.

The voltage across arrester block v(t) is applied to nonlinear resistance R(i) and nonlinear inductance L(i). And then equation (2) is formed.

$$v(t) = R(i) \times i(t) + L(i) \times \frac{di}{dt} \qquad (2)$$

As the voltage applied to nonlinear resistance, an average value (D point of Fig. 5-(b)) of the voltage of rising curve(A point of curve OAB) and the voltage of falling curve(C point of curve BCO) was adopted. Thus the nonlinear resistance V-I characteristic becomes the curve ODB in Fig. 5-(b). The voltage applied to nonlinear L(i) equals to the voltage which is derived by subtracting the voltage of the nonlinear resistance from the voltage across arrester. And then nonlinear inductance is calculated by equation (3).

$$L(i) = \frac{v(t) - R(i) \times i(t)}{\frac{di}{dt}} \qquad (3)$$

The nonlinear inductance must be represented with current and flux in the input format of EMTP, therefore the flux(ϕ) was calculated by equation (4).

$$\phi = \int L(i)di \qquad (4)$$

4.3 Nonlinear inductance calculation program

The nonlinear inductance calculation program was made by using the calculation method explained in 4.2. At least 20 voltage and current points as the conventional model are needed to derive the nonlinear inductance model. The maximum number of points which is limited in program is 100. In this program, the complete arrester model also can be calculated on the assumption that the residual voltage of the complete arrester is calculated by multiplying the one block residual voltage by the number of blocks of complete arrester. The flow chart of the program is represented in Fig. 6.

4.4 Comparison of model building methods

The conventional model is a nonlinear v-i characteristic which is derived from standard impulse test. The model building is done very easily by editing v-i points. The nonlinear inductance model can be derived from the above v-i points and the rising rate of steep front impulse voltage to standard impulse by executing the program explained in 4.3 once.

In building IEEE model, first, there was a program execution to determine the initial parameters. After that, to adjust A0 and A1 in Fig. 3, there was 2 times of the program execution and to adjust L1 in Fig. 3, there was 4 times of the program execution. The total program execution was 6 times.

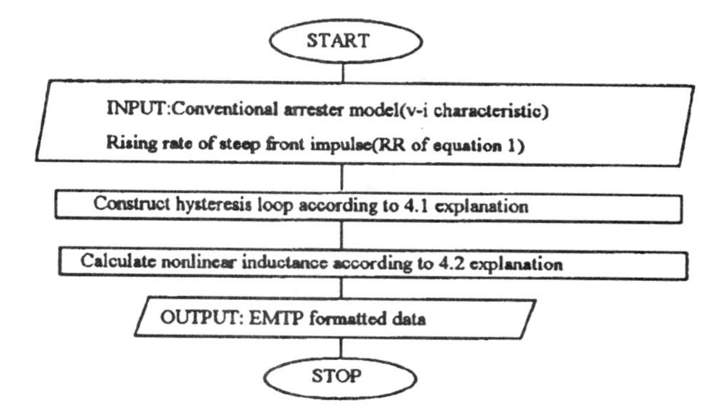

Fig. 6 Flow chart of the program

5. Comparative study of arrester block models

Comparative studies of conventional model, IEEE model, and nonlinear inductance model were performed at 2kA, 5kA, 10kA current peak value cases. These models are built to an arrester block. The current waveform of measured impulse is based in calculation. The 1/2μs impulse calculation results are represented in Fig. 7 and Fig. 8. The peak voltages of model calculation for the 1/2 μs impulse are presented in Table 1. The calculation results for the 4/10μs impulse were presented in Table 2. The model calculations to the 8/20μs impulse are presented in Table 3. In the tables, "P.C." means peak current of the impulse and "P.V." means peak voltage. The "Non. L" means the nonlinear inductance model. The measured and calculated waveforms of 8/20μs impulse are presented in Fig. 9 and Fig. 10.

In cases of steep impulse calculation, the conventional model produced a greatest error about 6%. The IEEE model has a good agreement, but the tail of voltage wave drops too sharply.

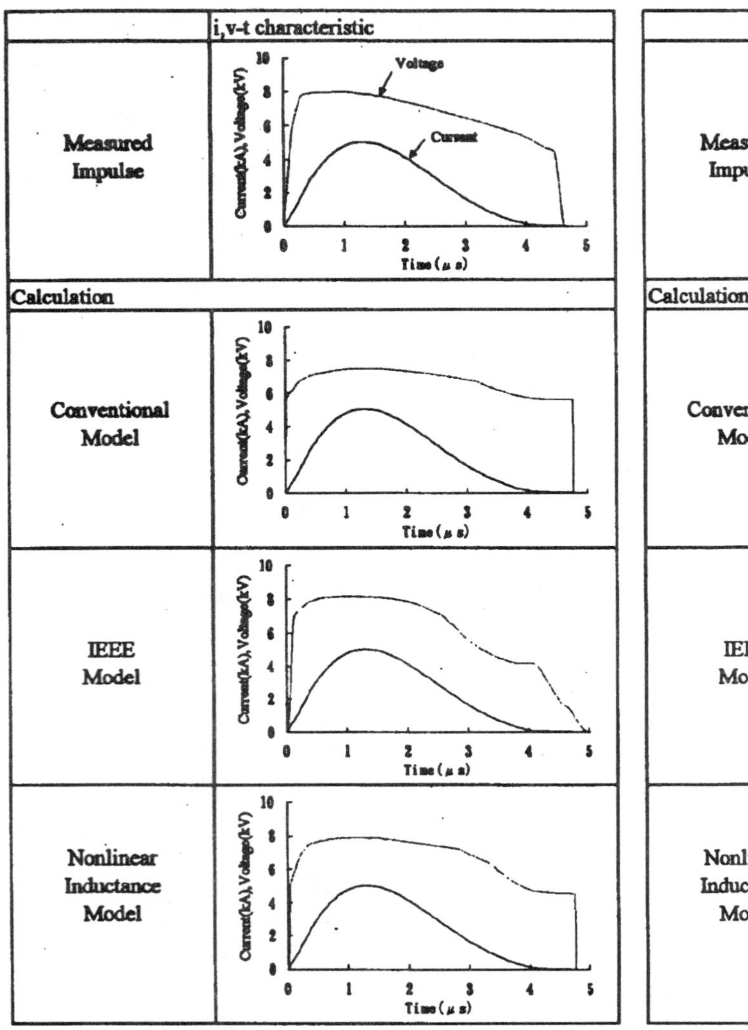

Fig. 7. 1/2μs Impulse waveforms(5.04kA)

Fig. 8. 1/2μs Impulse waveforms(10.48kA)

Table 1. 1/2μs Impulse

Measured Impulse	Calculation		
	Models	Voltage(kV)	Error(%)
P.C.:2.58kA P.V.:7.46kV	Conventional	7.05	-5.5
	IEEE	7.68	2.9
	Non. L	7.39	-0.9
P.C.:5.04kA P.V.:7.97kV	Conventional	7.50	-5.9
	IEEE	8.15	2.6
	Non. L	7.89	-1.0
P.C.:10.48kA P.V.:8.62kV	Conventional	8.14	-5.6
	IEEE	8.61	-0.1
	Non. L	8.62	0.0

Table 2. 4/10μs Impulse

Measured Impulse	Calculation		
	Models	Voltage(kV)	Error(%)
P.C.:2.95kA P.V.:7.26kV	Conventional	7.12	-1.93
	IEEE	7.32	0.83
	Non. L	7.47	2.89
P.C.:4.98kA P.V.:7.62kV	Conventional	7.49	-1.71
	IEEE	7.81	2.49
	Non. L	7.86	3.15
P.C.:10.42kA P.V.:8.22kV	Conventional	8.13	-1.09
	IEEE	8.35	1.58
	Non. L	8.55	4.01

The greatest error of nonlinear L model is about 1% in case of steep front impulse calculation. At 2.5kA standard impulse calculation the error is 7%. But for the error of standard impulse calculation, 7% is not so important in considering insulation coordination.

In steep front impulse measurements, the time to maximum voltage is shorter than the time to maximum current by 0.5μs. In conventional model calculation, the time to maximum voltage is the same as the time to maximum current since the conventional model only consists of the nonlinear resistance.

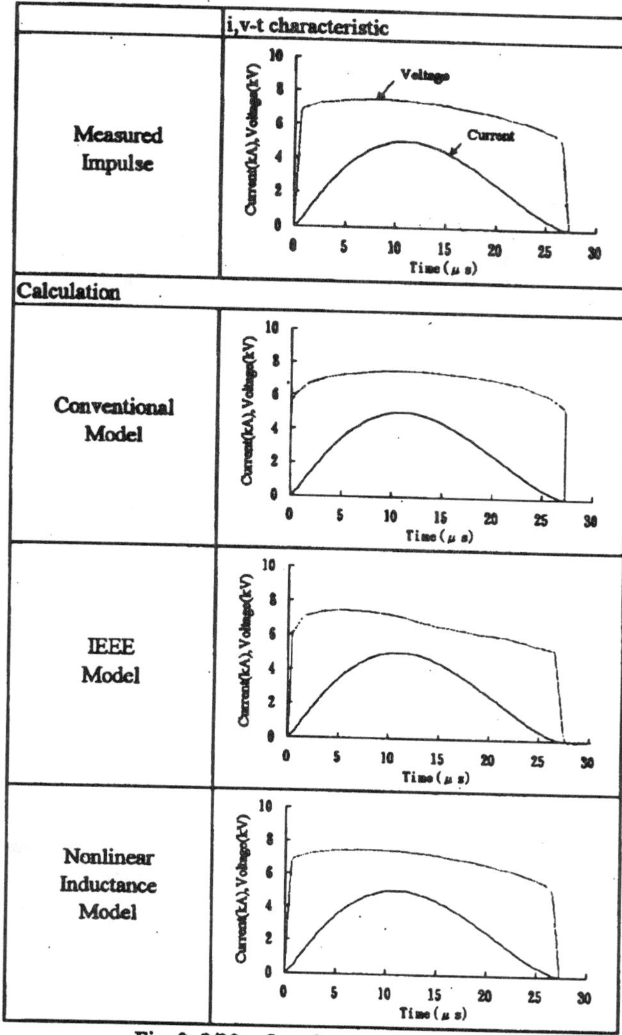

Fig. 9. 8/20μs Impulse waveforms(5.01kA)

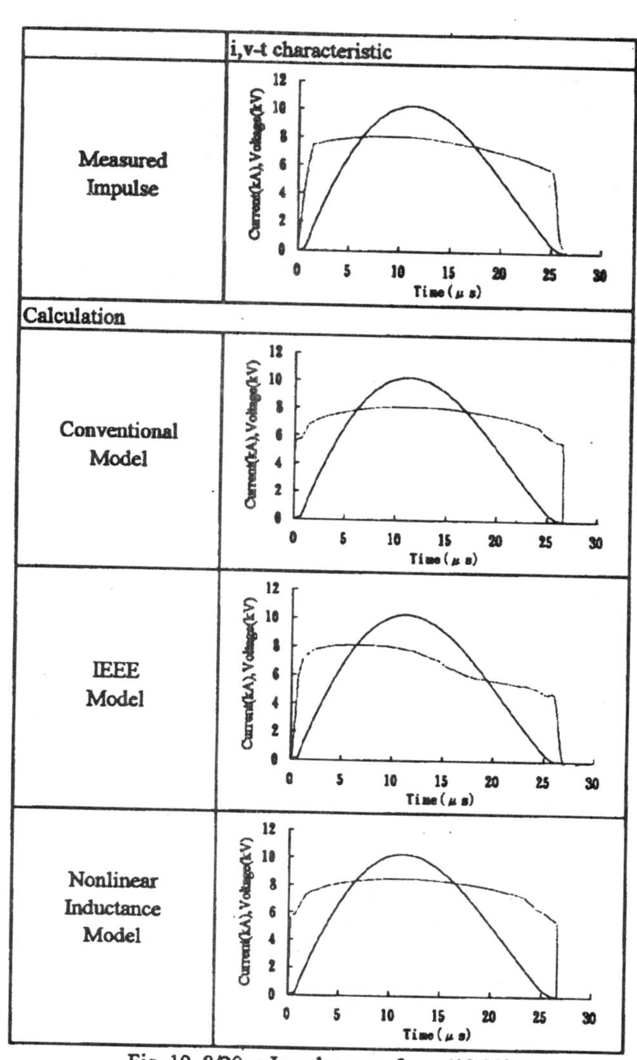

Fig. 10. 8/20μs Impulse waveforms(10.23kA)

Table 3. 8/20μs Impulse

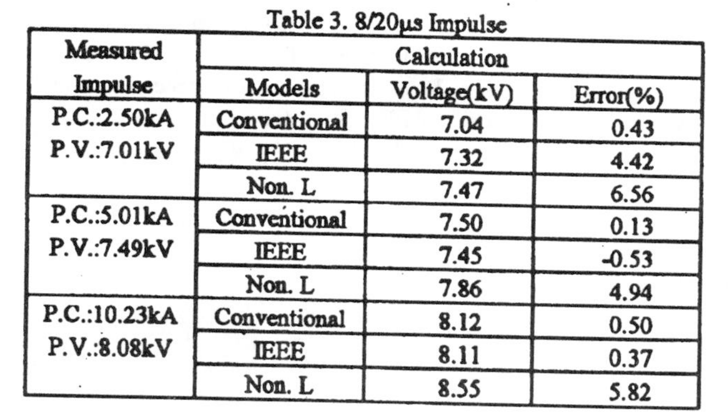

Measured Impulse	Calculation		
	Models	Voltage(kV)	Error(%)
P.C.:2.50kA P.V.:7.01kV	Conventional	7.04	0.43
	IEEE	7.32	4.42
	Non. L	7.47	6.56
P.C.:5.01kA P.V.:7.49kV	Conventional	7.50	0.13
	IEEE	7.45	-0.53
	Non. L	7.86	4.94
P.C.:10.23kA P.V.:8.08kV	Conventional	8.12	0.50
	IEEE	8.11	0.37
	Non. L	8.55	5.82

In the IEEE model and the nonlinear inductance model calculation, the time to maximum voltage is shorter than the time to maximum current by 0.38μs in the steep front impulse case. The two models are satisfactory in comparison with the measurements.

6. Conclusions

The nonlinear inductance model proposed in this paper has the accuracy of about 1% error in calculations of steep front impulse. The IEEE model has good accuracy over all cases considered in this paper, but the tail of voltage waveform dropped too sharply in block level study. But the drop in tail of voltage curve is not important to an insulation coordination study. The greatest error of the IEEE model was 4.42% at 2.5kA standard impulse calculation. In the case of steep front impulse calculation, the nonlinear inductance model has less error than the other models. In the IEEE model and the nonlinear inductance model calculation, the time to maximum voltage is shorter than the times to maximum current. The two models are satisfactory in comparison with the measurements. It is useful that the nonlinear inductance model is used in the steep front impulse calculation.

The effort of our study was limited to modeling the metal oxide arrester only. Other system parameters such as arrester leads and separation distances are also important in studies with lightning and other fast waveform surges. This study has concentrated on modeling gapless metal oxide arresters.

Simple arrester models can be constructed consisting of a nonlinear v-i characteristic(nonlinear resistance) and an inductance in series with it which can represent the dynamic characteristic for a narrow range of time to crest. Such a model would not be accurate for time to crest outside that range and outside a peak current value at which the inductance was calculated. The nonlinear inductance model proposed here would be accurate for a narrow range of time to crest. And this model would also be accurate for any magnitude of peak current value for a narrow range of time to crest. In this study the nonlinear inductance model has been proved to be accurate for 2kA to 10kA impulse peak current range. It is considered that the steep front impulse study is very important in considering insulation coordination. And using the nonlinear inductance model is reasonable in the steep front impulse calculations. The model can be developed simply by executing calculation program once, while IEEE model is developed by several times calculation.

References

[1] Study Committee of Insulation co-ordination against Surge, "Several parameters and their effects in lightning surge analysis for the power stations and the substations", Technical Report, IEE Japan, No. II-301, June 1989, pp77-79 (in Japanese)

[2] Study Committee of Insulation Design Rationalization, "Rationalization of Insulation Design", Electricity Cooperated Study Committee Report, Vol. 44, No.3, Dec. 1988, pp60-65 (in Japanese)

[3] Study Committee of Insulation Technology of GIS, "Steep front surge and GIS insulation problem", Technical Report, IEE Japan, No. II-324, Feb. 1990, pp44-52 (in Japanese)

[4] S.Yoshimura and K.Nakano, "Protection characteristics of ZnO arrester to steep front ", Papers of Switching and Protecting engineering of IEE Japan, SP-92-3, March, 1992 (in Japanese)

[5] N. Nagaoka, A. Ametani and K. Inaba, "Numerical Modeling of a ZnO arrester for surges with steep wave fronts", *Advances in Varistor Technology*, Ceramic Transactions Vol. 3, 1989, pp266-273

[6] W. Schmidt, J. Meppelink, B. Richter, K. Feser, L. Kehl and D. Qiu, "Behaviour of MO-Surge-Arrester Transients", *IEEE Transactions on Power Delivery*, Vol. 4, No. 1, Jan. 1989, pp292-300

[7] IEEE Working group 3.4.11, "Modeling of Metal Oxide Surge Arresters", *IEEE Transactions on Power Delivery*, Vol.7, No.1, Jan. 1992, pp302-309

Biographies

Ikmo Kim(M'89)

He was born in Kyungshangbukdo, Korea, on March 14 1954. He received the BS and MS degrees from Yon Sei University, Korea, in 1976 and 1981, respectively. he worked at the second naval academy as an instructor from 1976 to 1979. And then he was employed by You Han Technical college from 1980 to 1988. He joined Meidensha Corporation in 1988 and has worked on power system analysis. Now, He is a Chief staff of System Analysis Engineering Section, Meidensha Corporation. Mr. Kim is a member of IEEE, IEE of Korea and IEE of Japan.

Toshihisa Funabashi(M'90)

He was born in Aichi, Japan, on March 25, 1951. He received the BS degree from University of Nagoya, Aichi, Japan, in 1975 and has worked on power system analysis. Currently, he is a Manager of System Analysis Engineering Section, Meidensha Corporation. Mr. Funabashi is a member of IEEE and IEE of Japan.

Haruo Sasaki(M'90)

He was born in Hokkaido, Japan, on March 9, 1945. He received the BS degree from Nagoya Institute of Technology, Aichi, Japan, in 1968 and has worked on power system analysis. Currently, he is a Chief Engineer of System Development Division, Meidensha Corporation. Mr. Sasaki is a member of IEEE and IEE of Japan.

Toyohisa Hagiwara

He was born in Kanagawa, Japan, on Februrary 25, 1957. He received the BS degree from University of Ikutoku, Kanagawa, Japan in 1979 and has worked on research and development of surge arrester. He is a Assistant Manager of Engineering Section, Surge Arrester Factory, Meidensha Corporation. Mr. Hagiwara is a member of IEE of Japan.

Misao Kobayashi(M'89, SM'93)

He was born in Tokyo, Japan, on August 24, 1931. He received the BS degree from Tokyo Institude of Technology, Tokyo, Japan, in 1954 and has worked on surge arrester engineering and high voltage technology. He is a Chief Engineer of Production Headquarters, Meidensha Corporation. Mr. Kobayashi is a senior member of IEEE and a member of CIGRE and IEE of Japan.

Discussion

A. Haddad and R.T. Waters
(Electrical Division, University of Wales Cardiff, Wales, UK) :

We wish to congratulate the authors upon a novel and interesting contribution to the evaluation of the response of ZnO arrester elements to steep-fronted non-standard current impulses.

We seek some points of clarification and at the same time report briefly upon some recent steep-front test results with complete 15kV distribution polymeric ZnO surge arresters.

In Figure 5.a of the authors' paper, it appears that the line OL is obtained from tests with 8/20μs current impulses, using measured values of peak voltage and current. This would inevitably include induction effects in the element branch. We therefore have some difficulty in reconciling the broken curve ODB in Figure 5.b with the resistive characteristic as mentioned in section 4.2 of the paper.

The adoption by the authors of a constant shift to construct a hysteresis curve will produce a discontinuity at point S if the shift differs from V(1mA). Will this be important?

Figure 10 in the paper shows peak voltage and current to occur simultaneously in the non-linear inductance model, unlike the measured impulse. Does this mean that no hysteresis was assumed in the model? Have the authors checked this constructed hysteresis loop against a dynamically measured v-i loop?

We have recently conducted steep-front tests on 15kV rated surge arresters in a specially constructed test module[1]. Figure A shows voltage and current oscillograms for a current peak of 3.3kA and about 0.6μs risetime. We also show the associated measured dynamic hysteresis curve (Figure B). The shape of this curve is not well represented by the model in the present paper. In fact, it is the constant-shift construction adopted by the authors which makes necessary the use of a non-linear inductance in their model.

[1] Haddad A., Naylor P., German D.M. and Waters R.T. : 'A fast transient test module for ZnO surge arresters', Measurement Science Technology, Vol.6, No.5, pp.560-570, 1995.

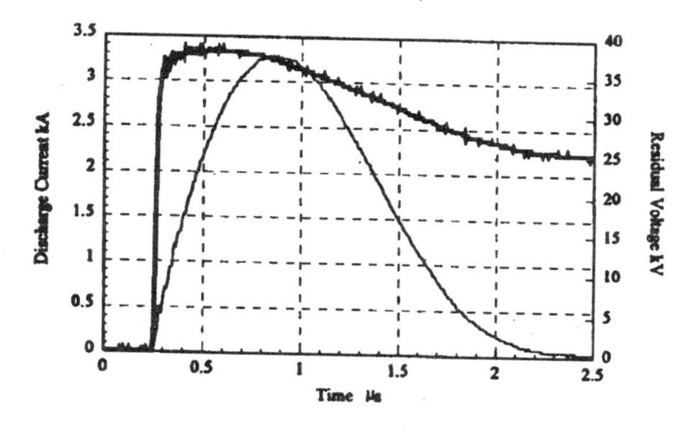

Figure A: Measured fast voltage and current traces on a 15kV rated Zno surge arresters.

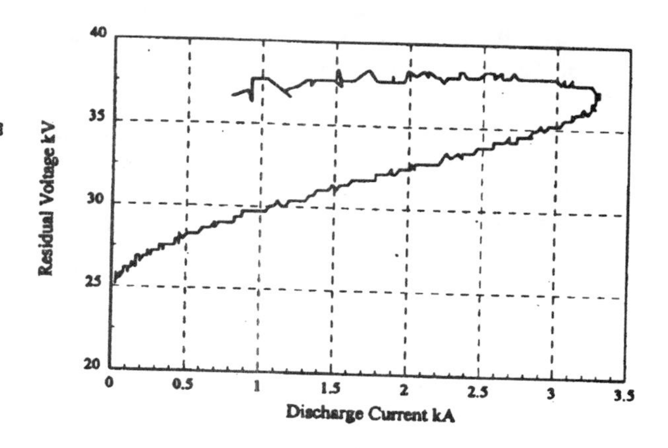

Figure B: Associated dynamic V-I curve of traces shown on Figure A above

Manuscript received August 21, 1995.

Ikmo Kim, Toshihisa Funabashi, Haruo Sasaki, Toyohisa Hagiwara and Misao Kobayashi
(Meidensha Corporation System Technology Division Riverside Building 36-2, Nihonbashi Hakozakicho Chuo-Ku, Tokyo, Japan):

The authors thank Messrs. Haddad and Waters for their discussion of our paper. The discussors state that in Fig. 5.b, the broken curve ODB would inevitably include induction effects in the element branch. We agree with their opinion in point of 8/20μs impulse case, but that curve is used to make an arrester model for steep front wave as an intermediate process. The curve ODB is used as the V-I characteristic which can be represented as the nonlinear resistance at making the model. That is an assumption to make the nonlinear inductance model. Therefore, we suppose that curve would not include induction effects in a steep front wave case.

The discontinuity will occur in any case, because the V-I characteristic of arrester has nonlinearity. The voltage at 1mA will not play an important role in the maximum voltage calculation of the analyzed network. But the voltage at 1mA is necessary because that model will be used in insulation coordination. For example in the lightning surge calculation, if that point were not used to construct the model of arrester, this produces a problem that insulation characteristic of arrester would be lost in small current range.

In Fig. 10, the peak of voltage and that of current had occurred simultaneously. We checked the hysteresis loop of that case, and the area of hysteresis loop is small , but the 8/20μs impulse is not so important target to making the our model. We focus on steep front wave model that is very important in considering insulation coordination.

The nonlinear inductance is connected to nonlinear resistor in series, and that makes the peak of voltage lead the peak of current in a steep front wave case. We focus on making the simple model. First, we connected an inductance with the nonlinear resistance. But the calculation results of that model did not match the measurements. And then we changed the inductance to the nonlinear inductance. The Fig. 5.a and 5.b are conceptual graphs. Those are not results of measurement.

Manuscript received October 16, 1995.

Multi-Port Frequency Dependent Network Equivalents for the EMTP

A. S. Morched
Senior member

J. H. Ottevangers

L. Martí
Member

Ontario Hydro. Ontario, Canada

Abstract - *A method is developed to reduce large power systems to single and multi-port frequency dependent equivalents. These equivalents consist of simple RLC modules that faithfully reproduce the frequency characteristics of the network. The method is implemented in the EMTP and has been extensively tested at Ontario Hydro. The implementation involves a pre-processor program to generate the model: the Frequency Dependent Equivalent (FDNE), and a EMTP time step loop module to calculate the transient response. The use of the FDNE results in major reductions in computer time and is especially beneficial for multi-case statistical EMTP studies. An example showing the accuracy and efficiency of the FDNE when used to reduce a large 500 kV network is presented.*

Keywords - Network equivalent, Multi-port, Frequency dependence, Electromagnetic transients, EMTP.

1. INTRODUCTION

The study of Electromagnetic transients often requires the detailed modelling of complex transmission networks. However, detailed representations may require prohibitive amounts of computer time, especially when statistical analysis is involved. It is, therefore, a common practice to represent in detail only a small portion of the system and to model the rest using equivalent networks.

Until very recently, only simplified equivalents have been used in transient studies. The most common representation consists of simple inductances derived from the short circuit impedances at the terminal buses evaluated at power frequency. A better representation, in terms of frequency response, can be obtained by shunting the power frequency impedances with the equivalent surge impedances of the lines attached to the buses. This produces improved first reflections, but degraded low frequency behaviour, and incorrect steady-state solutions.

Since these representations are only adequate for very simple transient studies, it has been necessary to represent in detail large portions of the system behind the terminal buses. The correct assessment of how much of the system should be modelled explicitly, is as much an art as it is the result of experience.

92 SM 461-4 PWRD A paper recommended and approved by the IEEE Transmission and Distribution Committee of the IEEE Power Engineering Society for presentation at the IEEE/PES 1992 Summer Meeting, Seattle, WA, July 12-16, 1992. Manuscript submitted January 31, 1992; made available for printing May 15, 1992.

Efforts to establish rules for the size of the networks to be represented have been made. Notably, the work of the CIGRE Working Group 13.05[1], which recommends the use of detailed representation of the system up to two buses behind the terminal buses. However, in dense networks with many transmission lines, even a two-bus explicit representation can be computationally prohibitive. If these dense networks also contain many short lines, the two-bus rule may not guarantee the accuracy of the results, and the detailed modelling of larger portions of the system could be required.

The use of the frequency dependent equivalents for power networks goes as far back as the late sixties and early seventies. Pioneering work in this area was conducted by N. Hingorani[2] and A. Clerici[3]. More recent work[4,5,6] attempted to establish systematic procedures to generate frequency dependent network equivalents. Morched and Brandwajn[4] proposed an approach to produce single-port equivalents with models that only matched the network admittances at the series resonant frequencies. Do and Gavrilovic[5,6] proposed a procedure where the component modules used to represent each of the admittance series resonances are selected by inspection, and a least squares method is used to match the network admittances over a range of frequencies. While they presented methods for generating multi-port equivalents, the treatment was not sufficiently general.

This paper describes an efficient method to calculate network equivalent admittances as seen from one or more ports. It describes the extension of the concepts developed in [4] to multi-port equivalents and the improvement of the fitting technique to match the admittances over a wide range of frequencies.

The application of these techniques has resulted in the development of the Frequency Dependent Network Equivalent (FDNE) program and its subsequent implementation in the EMTP. The FDNE can simulate any type of network, and the generation of its parameters is automatic and does not require special skills on the part of the user.

The FDNE is a stand-alone program which uses a description of the network similar to the one used by the EMTP. It evaluates the nodal admittance matrix seen from multi-port terminals over a user-specified frequency range. On output, the FDNE produces a number of modules consisting of RLC branches (in EMTP-compatible format) whose frequency response matches that of the system seen from the terminals.

The FDNE allows the modelling of large portions of the system at a small fraction of the computational cost required to model them explicitly. This results in more reliable simulations as the need to estimate what portion of the system should be modelled in detail becomes less crucial.

2. NETWORK COMPONENT MODELS

The first step in the creation of a network equivalent is to build the nodal admittance matrix of the portion of the system to be represented over a given frequency range. Rather than using the EMTP itself to produce frequency scans of the entire network, the FDNE calculates the admittance matrix of each system component independently according to its mathematical description. This is more efficient from a computational point of view, and it does not tie the FDNE to a given version of the EMTP. The representations used in the FDNE to model the major system components are described below.

2.1 Transmission Lines

The representation of transmission lines used in the FDNE is based on the following assumptions:

1) The line parameters are distributed and their dependence on frequency due to skin effect and finite earth resistivity is taken into account.

2) A transmission line consisting of only one circuit is forced to be balanced by averaging the diagonal and off-diagonal elements of the series impedance matrix [Z] and the shunt admittance matrix [Y] per unit length.

3) In the case of a multi-circuit line, or of several single-circuit lines sharing the same right-of-way, the zero sequence coupling between circuits is taken into account.

The relationships between voltages and currents at both ends of the line, in nodal admittance matrix form, are given in the Appendix. In its most general form, voltages and currents at sending and receiving ends of the line can be described with equation (A.4); namely,

$$\begin{bmatrix} \bar{I}_i \\ \bar{I}_j \end{bmatrix} = \begin{bmatrix} Y_{ii} & Y_{ij} \\ Y_{ji} & Y_{jj} \end{bmatrix} \begin{bmatrix} \bar{V}_i \\ \bar{V}_j \end{bmatrix}$$

The evaluation of $[Y_{ii}]$ and $[Y_{ij}]$ requires the calculation of the eigenvalues and the associated eigenvector matrix of [Z][Y] for each right-of-way at each frequency. This can result in excessive computation times, especially for crowded right-of-ways.

Since the area of interest in a transient simulation is normally modelled in detail and not included in a network equivalent, it is possible to assume symmetry among the phases of the circuits forming the equivalent without significant loss of accuracy.

As can be seen from the Appendix, this assumption results in two identical uncoupled networks (positive and negative sequence) and a third network (zero sequence), which would be coupled in the case of multi-circuit right-of-ways, and uncoupled in the case of isolated single-circuit lines.

For each right-of-way, the FDNE calculates the positive sequence parameters for each circuit using the single-conductor equations (A.2) and (A.3). The only frequency dependent parameter in these equations is the transmission coefficient λ. Given that λ is a very smooth function of frequency, it is only calculated explicitly at 10 frequencies over the entire frequency range; intermediate values of λ are calculated by interpolation.

The zero sequence admittance matrices are calculated using (A.5) and (A.6), where the transmission coefficients as well as the modal transformation matrix $[T_v]$ are frequency dependent. For the purposes of the FDNE, the eigenvector matrix $[T_{vo}]$ evaluated at 10 kHz is used to diagonalize $[\Lambda]^2$ over the entire frequency range with acceptable accuracy. As in the case of the positive sequence calculations, $[\Lambda]^2$ is evaluated explicitly at only 10 frequency points. The zero sequence admittance matrices are calculated with (A.5) and (A.6) with $[\Lambda]$ obtained using interpolation. The use of $[T_{vo}]$ instead of solving the eigenvalue problem at each frequency results in significant savings in computer time and in preserving the continuity of the eigenvalue functions.

2.2 Linear branches

Power system components other than transmission lines are normally represented as concentrated-parameter, single or multi-phase elements. For the purposes of the FDNE, symmetry among the phases of three-phase components is assumed.

Series Elements

In power networks, series elements appear as capacitive compensation of long lines, current limiting reactors or transformers (in their simplest representation).

The admittance $[Y_{ij}]$ of a series element can be calculated from its equivalent circuit, or it can be measured at the required frequencies. The nodal admittance matrix of a series element connected between buses i and j is given by

$$\begin{bmatrix} \bar{I}_i \\ \bar{I}_j \end{bmatrix} = \begin{bmatrix} Y_{ij} & -Y_{ij} \\ -Y_{ij} & Y_{ij} \end{bmatrix} \begin{bmatrix} \bar{V}_i \\ \bar{V}_j \end{bmatrix} \quad (1)$$

Shunt Elements

Shunt elements appear as generators, loads, and inductive or capacitive shunt compensation. The admittance matrix $[Y_{ii}]$ which describes the shunt element connected at bus i can be calculated from equivalent circuits or can be measured at the required frequencies.

3. NETWORK REDUCTION

The nodal admittance matrices of the network have to be calculated at every frequency, and decomposed into positive and zero sequence networks using the well-known Karrenbauer modal transformation matrix. In the case of multi-circuit right-of-ways, the contribution to the zero sequence network will be a matrix, rather that a scalar. The sequence admittance matrices are then reduced between a given number of terminals, or ports, producing a family of frequency-dependent reduced-order admittance matrices.

In the FDNE, the admittance matrices of the entire network are never fully calculated: only partial matrices including two bus layers beyond the reference ports are formed. In this process, the network is divided in layers. All transmission circuits and linear branches connected to the first reference port are counted in layer 1. The reference port is considered to be the 'from' bus. Corresponding 'to' buses are designated as layer 1 buses. All circuits of a right-of-way must be assigned to the same layer in order to account for the ground mode coupling between the circuits. The second layer consists of all remaining circuits connected to layer 1 buses.

The admittance matrix for the partial network consisting of the first two layers, and complemented by all shunts added to layer 1 and layer 2 buses, can now be reduced by eliminating the rows and columns corresponding to layer 1 buses. Any reference ports that might appear in layer 1 are retained.

The next stage is to extend the reduced admittance matrix by including buses belonging to a new layer, e.g., layer 3. This new layer consists of all remaining circuits connected between layer 2 and layer 3 buses, and circuits between layer 3 buses, including shunts. Matrix reduction will eliminate the now redundant layer 2 buses, with the exception of reference ports. This process of adding layers, followed by reducing the matrix is repeated until all circuits and shunts are included. In this context, linear branches are treated the same way as transmission circuits.

The network admittance matrix is kept to a relatively small size during this process and the final reduction produces an N x N matrix for a network with N reference ports. The matrix is symmetrical with N(N+1)/2 distinct elements. The admittance matrix is calculated over a range of frequencies. Each element of the matrix produces a frequency response curve, to be matched by a multi-branch linear network.

The described process is more efficient than the direct reduction of the admittance matrix of the full system because it confines the operations to low order, almost full, matrices. It also eliminates the need to store large complex matrices and would allow the program to run on small computers with limited memory regardless of the size of the network modelled.

4. FREQUENCY DEPENDENT EQUIVALENTS

Once the equivalent positive and zero sequence admittance matrices are calculated as functions of frequency, their elements are fitted using modules of the type shown in Figure 1. Each module consists of a number of parallel RLC branches.

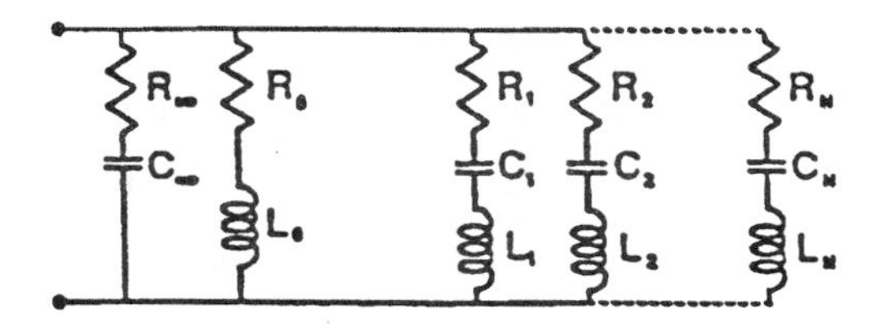

Fig. 1: *Structure of an FDNE module.*

Before proceeding with the description of the fitting procedure, the main differences between single and multi-port equivalents should be pointed out.

4.1 Single-Port Equivalents

The realization of the single-port equivalent is straightforward. The stability of the model is not an issue, provided that each of the fitted RLC branches has a positively-damped response.

Reduced-order models are obtained by the rejection of less significant admittance peaks as compared to the largest peak of the fitted frequency characteristic.

4.2 Multi-Port Equivalents

The multi-port equivalent is realized in the form of the multi-terminal π-equivalent shown in Figure 2. The components of this π-equivalent are easily calculated from the nodal admittance matrix of the reduced network. The admittance $y_{ii,\pi}$ of the component connected between node i and ground is given by the sum of the elements of the row i of the nodal admittance matrix.

$$y_{ii,\pi} = \sum_{j=1}^{N} y_{ij} \tag{2}$$

The component connected between nodes i and j of the π-circuit is given by the negative of the corresponding off-diagonal element of the reduced-order nodal admittance matrix.

$$y_{ij,\pi} = -y_{ij} \tag{3}$$

Each component of the positive and zero sequence π-equivalents has to be fitted to an RLC module for both positive and zero sequence networks.

A number of factors have to be taken into consideration during the realization of the multi-port equivalent:

a. The real part of the equivalent transfer admittances $y_{ij,\pi}$ have negative peaks. These can only be reproduced by negative RLC branches. This is not a problem, provided that each RLC branch has a positively damped response.

b. The shunt components of the π-equivalent can be produced by fitting them directly, or by fitting each of

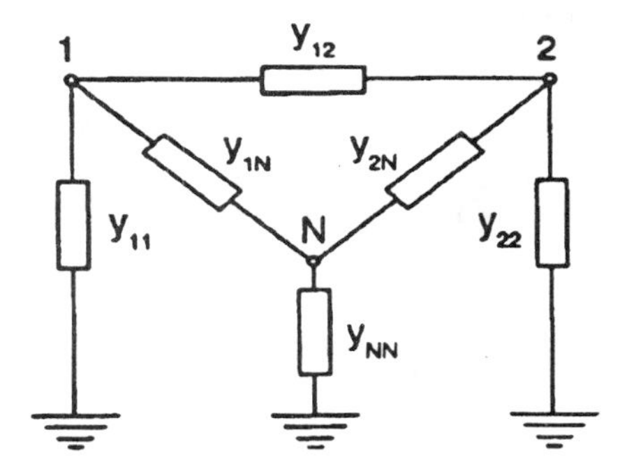

Fig. 2: *Equivalent π-representation*

the elements of the reduced nodal admittance matrix and then adding the approximations according to equation (2). The former procedure produces a lower order model but it may not be numerically stable.

These factors are taken into consideration when multi-port equivalents are produced. The process starts with scanning the frequency characteristics of each of the diagonal and off-diagonal elements of the reduced matrices. Series and parallel resonance frequencies and the corresponding admittances are identified. Peaks whose magnitudes are small compared to the magnitude of the largest resonance peak in all the elements are rejected. A list of all retained resonance frequencies is prepared. A second scan is made through each of the matrix elements, and peaks corresponding to any of the listed frequencies are retained regardless of their magnitude. This process is carried out on the positive and zero sequence equivalent networks.

5. CALCULATION OF THE RLC MODULES

Each matrix element is approximated by an RLC module whose frequency response matches that of the matrix element. An RLC module consists of one RC branch (R_∞, C_∞), one RL branch (R_o, L_o), and a number of RLC branches (R_k, L_k, C_k, k=1,...,M) connected in parallel (see Figure 1). The method described in [4] is used to generate a module that matches the frequency characteristic at series resonance frequencies only. This is then used as an initial guess for an optimization process to minimize the mean square of the error over the entire frequency range. An iterative procedure is followed where the parameters of the module are adjusted one branch at a time. The process is based on a steepest decent approach with the following steps:

a. The partial derivatives of the square of the error function with respect to L_k and the resonance frequency $\omega_{o,k}$ of each RLC branch are calculated over a limited frequency range. This establishes the direction in which to change each parameter in order to reduce the error. The influence of R_k, L_k, and C_k on the value of the admittance at frequencies which are not close to the resonance frequency of branch *k* is small. Therefore, the frequency range used in the optimization of branch *k*, extends only to the neighbouring parallel resonance frequencies.

b. The inductance of branch *k* is changed first. The change is carried out in two steps. On each step, L_k is changed by 20% of the value obtained in the previous iteration. The second step is carried out only if the first step results in a lower error. A linear search using the Fibonacci's ratio is conducted to locate the value of L_k that produces the lowest error within the frequency range.

c. The resonance frequency $\omega_{o,k}$ of branch *k* is the second parameter to be adjusted, and the change is limited to 1% of the resonance frequency calculated in the initialization process. A linear search is then carried out to establish the resonance frequency corresponding to the minimum error within the frequency range. The capacitance of the branch is updated using the new values of L_k and $\omega_{o,k}$.

d. The resistance of branch *k* is calculated by matching the equivalent system admittance at the series resonance frequency.

e. The value of C_∞ is adjusted by minimizing the square of the error of over the whole frequency range, and R_∞ is selected by matching the power system admittance at a very high frequency (default value 500 kHz).

f. The values of R_o and L_o are selected to match the admittance of the system at power frequency.

These steps are repeated until the error falls within acceptable limits, or until the maximum number of permissible iterations is reached. During this optimization process, the parameters of each branch are constrained to produce positively damped responses.

6. THE FDNE PROGRAM

The implementation of the Frequency Dependent Network Equivalents in the EMTP involves:

1) An auxiliary routine to calculate the model parameters.

2) A time step loop module to calculate the transient response.

6.1 The FDNE Auxiliary Routine

The FDNE auxiliary routine carries out the network reduction and the fitting procedures described above. The description of the network is similar to the one used in the EMTP and the resulting model can be included directly into an EMTP data case.

There are no intrinsic constraints regarding the size of the system that can be modelled. With default dimensions, a system may contain: 60 right-of-ways with 20 circuits each; 250 circuits connected between 250 busbars; 100 series elements, and 50 shunt elements. The default number of ports is 5. The number of frequencies for which the matching is attempted can be as high as 1000, either linearly or logarithmically spaced. These defaults can easily be changed as required.

6.2 The Time Step Loop Module

The time step loop module implemented in the EMTP as part of the work described in [4] can also be used to model multi-port equivalents, given that the original implementation allowed the connection of an RLC module between any two buses (rather than just between a bus and ground). Only minor coding changes were required to allow modules with negative R, L, and C.

7. EXAMPLE

The 500 kV system shown in Figure 3 is used to illustrate use of the FDNE. It consists of 41 transmission circuits connected between 20 busbars, with many circuits sharing the same tower or right-of-way. The 60 Hz self and transfer impedances of the underlying 230 kV system are connected to the terminal buses (not shown in Figure 3).

A two-port equivalent between busbars ES and HN was produced with the FDNE. Comparisons between system and model admittance characteristics are shown in Figures 4, 5 and 6. The figures illustrate the good agreement between the model and system characteristics.

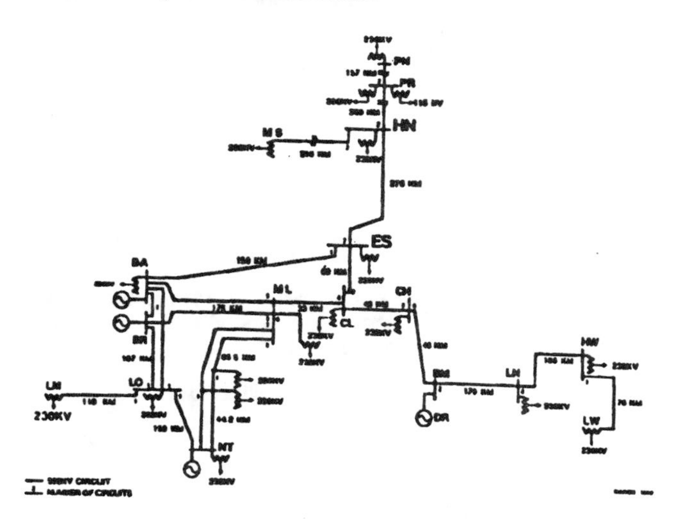

Fig. 3: *500 kV test system.*

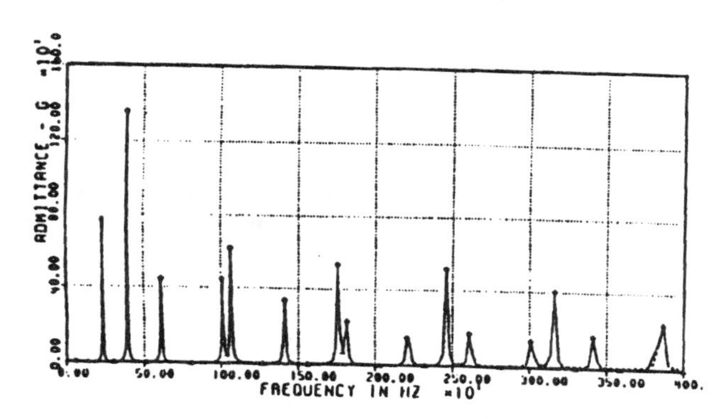

(a)

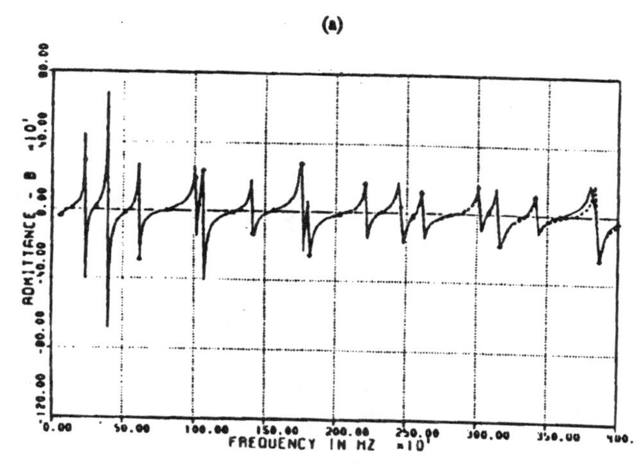

Fig. 4: Positive sequence admittance from bus HN to ground.
(a) Real part.
(b) Imaginary part.

The effect of different network representations on the switching surges caused by the energization of one circuit of the double-circuit line between buses HN and ES was examined. Figure 7 shows the voltages at bus ES calculated with the full system representation, and with the two-port equivalent between the buses ES and HN.

The full system simulation requires 30 minutes of CPU time on a VAX 8600, whereas the FDNE simulation requires only 4 minutes. Preparation time to generate the FDNE equivalent is 23 minutes, while the time required to generate 41 JMARTI line models is 2½ hours.

The results obtained using a simple 60 Hz equivalent (with and without surge impedance in parallel) are shown in Figure 8. CPU time for each of these runs was slightly under 4 minutes.

In the assessment of switching surge levels at Ontario Hydro, it is a standard procedure to carry out statistical studies based on a set of 100 switching operations. Considering the results shown, it would be impractical to use the full system representation and it would be inaccurate to use the simple equivalents. The use of the FDNE solves this problem.

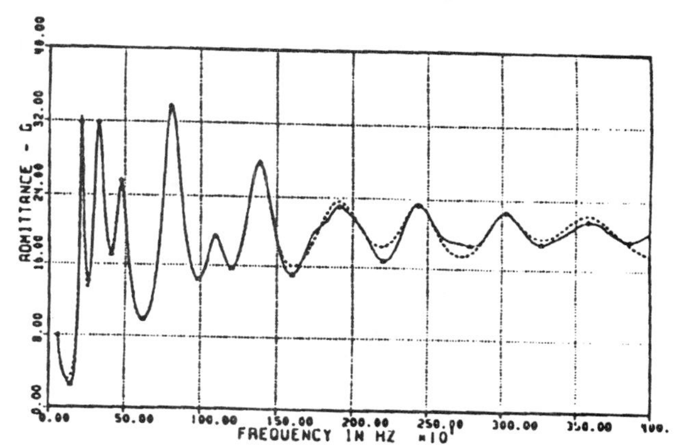

Fig. 5: *Zero sequence admittance from bus HN to ground (real part only).*

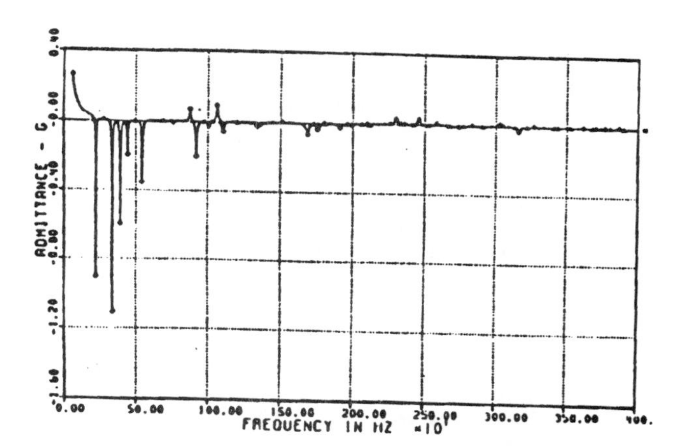

Fig. 6: *Positive sequence admittance between buses HN and ES (real part only).*

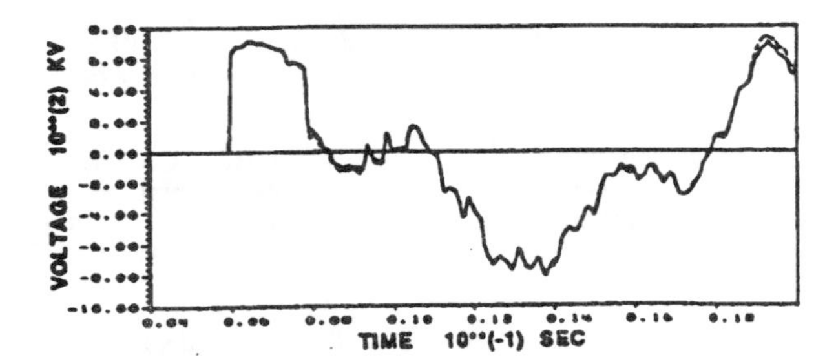

Fig. 7: *Transient simulation.*
Solid trace: Full system
Dashed trace: FDNE equivalent.

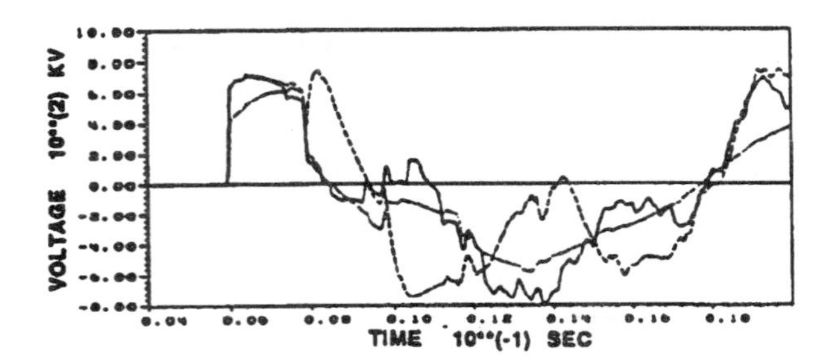

Fig. 8: *Results with conventional equivalents.*
Solid trace: Full system;
Dashed trace: 60 Hz equivalent;
Dotted trace: 60 Hz equivalent and surge impedance.

8. CONCLUSIONS

1. This paper describes the procedures developed to produce equivalent networks whose frequency response matches that of a much larger portion of a power system as seen from one or more terminals. These network equivalents make it unnecessary to use detailed but computationally expensive models to represent those parts of the system which are not the prime concern of a given transient simulation. Furthermore, since the FDNE does not use travelling wave representations of transmission lines, the time step of an EMTP simulation does not have to be smaller than the travel time of the shortest line in the equivalenced system. These computational advantages are specially beneficial for multi-case EMTP statistical studies.

2. The FDNE uses robust analytical techniques and has been extensively tested at Ontario Hydro. Input data is in the form of a network description similar to the one used in the EMTP. On output, the FDNE produces a series of RLC modules which can be included directly into an EMTP data case.

3. From the point of view of EMTP users, the FDNE may, in many instances, eliminate the need to re-dimension the EMTP to accommodate a large system. While re-dimensioning the EMTP is not difficult, it is a task which may involve system administrators, access privileges, and other time consuming details. Also, the FDNE makes possible the simulation of very large systems using computers with relatively modest memory and CPU resources.

REFERENCES

[1] GIGRE WG 13-05 III. "Transmission Line Representation for Energization and Re-energization Studies with Complex Feeding Networks", Electra Vol. 62, pp. 45-78, Jan. 1979.

[2] N.G. Hingorani, and M.F. Burbery, "Simulation of AC System Impedance in HVDC System Studies", IEEE Trans. Power App. Syst., Vol. PAS-89, PP. 820-28, May/June 1970.

[3] A. Clerici, and L. Marzio, "Coordinated use of TNA and Digital Computer for Switching Surge Studies: Transient Equivalent of a Complex Network", IEEE Trans., Power App. Syst., Vol. PAS-89, PP. 1717-26, Nov./Dec. 1970.

[4] A.S. Morched, and V. Brandwajn, "Transmission Network Equivalents for Electromagnetic Transients Studies", IEEE Trans., Power App. Syst., Vol. PAS-102, PP. 2984-90, Sept. 1983.

[5] V.Q. Do, and M.M. Gavrilovic, "An Interactive Pole-Removal Method for Synthesis of Power System Equivalent Networks", IEEE Trans. Power App. Syst., Vol. PAS-103, PP. 2065-70, August 1984.

[6] V.Q. Do, and M.M. Gavrilovic, "A Synthesis Method for One Port and Multi-Port Equivalent Networks for Analysis of Power System Transients", T-PWRD, Vol. 1, PP. 103-11, Apr. 1985.

APPENDIX

LINE EQUATIONS IN ADMITTANCE MATRIX FORM

A.1 Single Conductor Relationships

The relationship between voltages and currents at both ends i and j of a single conductor transmission line of length l at any frequency is given by

$$\begin{bmatrix} I_i \\ I_j \end{bmatrix} = \begin{bmatrix} y_{ii} & y_{ij} \\ y_{ji} & y_{jj} \end{bmatrix} \begin{bmatrix} V_i \\ V_j \end{bmatrix} \tag{A.1}$$

where

$$y_{ii} = y_{jj} = \lambda / y \cdot \coth(\lambda l) \tag{A.2}$$

$$y_{ij} = y_{ji} = \lambda / y \cdot csch(\lambda l) \tag{A.3}$$

z and y are the longitudinal impedance and transverse admittance of the line per unit length, and the transmission coefficient λ is the square root of zy.

A.2 Multi-Conductor Lines

The admittance matrix of a multi-conductor line is similar to that given in (A.1) where scalars are replaced with matrices.

$$\begin{bmatrix} \vec{I}_i \\ \vec{I}_j \end{bmatrix} = \begin{bmatrix} Y_{ii} & Y_{ij} \\ Y_{ji} & Y_{jj} \end{bmatrix} \begin{bmatrix} \vec{V}_i \\ \vec{V}_j \end{bmatrix} \qquad \text{(A.4)}$$

where

$$[Y_{ii}] = [Y_{jj}] = [Y][T_v][\Lambda]^{-1} \coth([\Lambda]\, l)[T_v]^{-1} \qquad \text{(A.5)}$$

$$[Y_{ij}] = [Y_{ji}] = -[Y][T_v][\Lambda]^{-1} \operatorname{csch}([\Lambda]\, l)[T_v]^{-1} \qquad \text{(A.6)}$$

$$[\Lambda]^2 = [T_v]^{-1}[Z][Y][T_v] \qquad \text{(A.7)}$$

[Z] and [Y] are the longitudinal impedance and transverse admittance matrices of the multi-conductor line per unit length. Λ is a diagonal matrix containing the modal transmission coefficients of the line; that is, the square root of the eigenvalues of the product [Z][Y]. Matrix $[T_v]$ is the corresponding eigenvector matrix. Both $[\Lambda]$ and [Q] are complex and frequency dependent.

A.3 Single-Circuit Lines

Matrices [Z], [Y], and [Z][Y] associated with a balanced three-phase single-circuit line can be diagonalized by a family of constant transformation matrices. Symmetrical components, Clarke, and Karrenbauer transformations are examples of these. For a balanced three-phase line, the following relationships hold:

$$[\Lambda]^2 = [H]^{-1}[Z][Y][H] = \begin{bmatrix} \lambda_o^2 & 0 & 0 \\ 0 & \lambda_1^2 & 0 \\ 0 & 0 & \lambda_2^2 \end{bmatrix} \qquad \text{(A.8)}$$

where

$$[Z] = \begin{bmatrix} z_s & z_m & z_m \\ z_m & z_s & z_m \\ z_m & z_m & z_s \end{bmatrix} \qquad [Y] = \begin{bmatrix} y_s & y_m & y_m \\ y_m & y_s & y_m \\ y_m & y_m & y_s \end{bmatrix} \qquad \text{(A.9)}$$

$$\lambda_o^2 = z_o y_o = (z_s + 2z_m)(y_s + 2y_m)$$

$$\lambda_1^2 = \lambda_2^2 = z_1 y_1 = (z_s - z_m)(y_s - y_m)$$

The transformation matrix [H] in (A.8) reduces the coupled relationships of a three-phase balanced line to three uncoupled relationships among the transformed quantities. The transformed quantities are referred to as a ground mode (zero sequence) and two identical sky modes (positive and negative sequence). Single-conductor relationships given in (A.1) are applicable for each of these modes.

A.4 Multi-Circuit Lines (or right-of-ways)

Consider two three-phase circuits sharing the same right-of-way; then

$$\begin{bmatrix} \Lambda_1^2 & \Lambda_c^2 \\ \Lambda_c^2 & \Lambda_2^2 \end{bmatrix} = \begin{bmatrix} H^{-1} & 0 \\ 0 & H^{-1} \end{bmatrix} \begin{bmatrix} Z_{11} & Z_{12} \\ Z_{21} & Z_{22} \end{bmatrix} \begin{bmatrix} Y_{11} & Y_{12} \\ Y_{21} & Y_{22} \end{bmatrix} \begin{bmatrix} H & O \\ O & H \end{bmatrix} \qquad \text{(A.10)}$$

$[Z_{11}]$ and $[Y_{11}]$ are the longitudinal impedance and transverse admittance matrices per unit length of circuit 1. Similarly $[Z_{22}]$ and $[Y_{22}]$ are the longitudinal impedance and transverse admittance matrices per unit length of circuit 2. If each circuit is assumed to be balanced,

$$[Z_{11}] = \begin{bmatrix} z_{s1} & z_{m1} & z_{m1} \\ z_{m1} & z_{s1} & z_{m1} \\ z_{m1} & z_{m1} & z_{s1} \end{bmatrix}; \quad \textit{and} \quad [Y_{11}] = \begin{bmatrix} y_{s1} & y_{m1} & y_{m1} \\ y_{m1} & y_{s1} & y_{m1} \\ y_{m1} & y_{m1} & y_{s1} \end{bmatrix} \qquad \text{(A.11)}$$

$[Z_{12}] = [Z_{21}]$ and $[Y_{12}] = [Y_{21}]$ are the longitudinal impedance and transverse admittance matrices per unit length representing the coupling between circuits 1 and 2. If the coupling between the phases of both circuits is assumed to be constant

$$[Z_{12}] = \begin{bmatrix} z_c & z_c & z_c \\ z_c & z_c & z_c \\ z_c & z_c & z_c \end{bmatrix}; \quad \textit{and} \quad [Y_{12}] = \begin{bmatrix} y_c & y_c & y_c \\ y_c & y_c & y_c \\ y_c & y_c & y_c \end{bmatrix} \qquad \text{(A.12)}$$

Introducing (A.11) and (A.12) into (A.10) gives

$$[\Lambda]^2 = \begin{bmatrix} \Lambda_1^2 & \Lambda_c^2 \\ \Lambda_c^2 & \Lambda_2^2 \end{bmatrix} = \begin{bmatrix} \lambda_{1o}^2 & 0 & 0 & \lambda_c^2 & 0 & 0 \\ 0 & \lambda_{11}^2 & 0 & 0 & 0 & 0 \\ 0 & 0 & \lambda_{12}^2 & 0 & 0 & 0 \\ \lambda_c^2 & 0 & 0 & \lambda_{2o}^2 & 0 & 0 \\ 0 & 0 & 0 & 0 & \lambda_{21}^2 & 0 \\ 0 & 0 & 0 & 0 & 0 & \lambda_{22}^2 \end{bmatrix} \qquad \text{(A.13)}$$

where

$$\lambda_{1o}^2 = (z_{s1} + 2z_{m1})(y_{s1} + 2y_{m1})$$

$$\lambda_{1o}^2 = \lambda_{12}^2 = (z_{s1} - z_{m1})(y_{s1} - y_{m1})$$

$$\lambda_c^2 = 3z_c y_c$$

For the uncoupled sky modes (A.1) is applicable, while for the coupled zero sequence mode (A.4) should be used instead.

BIOGRAPHIES

Atef S. Morched (M'77-SM'90) received a B.Sc. in Electrical Engineering from Cairo University in 1964, a Ph.D. and a D.Sc. from the Norwegian Institute of Technology in Trodheim in 1970 and 1972. He has been with Ontario Hydro since 1975 where he currently holds the position of Section Head - Electromagnetic Transients in the Power System Planning Division.

Jan H. Ottevangers received an MSc. in Electrical Engineering from the Delft Institute of Technology in 1956. He has been with Ontario Hydro since 1967 where he is currently working in the Analytical Methods and Specialized Studies Department of the Power System Planning Division.

Luis Martí (M'79) received an undergraduate degree in Electrical Engineering from the Central University of Venezuela in 1979, MASc and PhD degrees in Electrical Engineering in 1983 and 1987, respectively, from The University of British Columbia. He did post-doctoral work in cable modelling in 1987-1988, and joined Ontario Hydro in 1989, where he is currently working in the Analytical Methods & Specialized Studies Department of the Power System Planning Division.

Discussion

Adam Semlyen (University of Toronto): This is a well written, interesting and useful paper. It gives a comprehensive description of a methodology for obtaining external system equivalents for the EMTP for the general case when there are several connecting buses between the study zone and the external system.

One particular procedure has caught my attention: it is the building of the reduced external system in the frequency domain in a gradual, sequential manner. I believe the method represents a useful alternative to sparsity techniques and would appreciate it if the authors would provide a more detailed description of the procedure.

In closing, I wish to commend the authors for their fine paper.

Manuscript received July 27, 1992.

N. R. Watson (University of Canterbury, New Zealand): I would like to congratulate the authors on a very useful and interesting paper. I was interested to note that the matching was performed on the admittance matrix elements, whereas we have tended to invert this and match the elements of the impedance matrix. Also we use a state variable implementation (TCS) [1, 2], which is computationally more expensive than the EMTP approach, therefore comparison of the computation times using the different equivalents cannot be directly made. The reason for using state variable approach is the accuracy obtained when modelling HVDC systems as it does not suffer from the same numerical oscillation problems that electromagnetic transient programs experience, due to frequent switchings.

I would be grateful if the authors would comment on the following questions:

1. At how many frequencies is the error between the equivalent and system evaluated when calculating the mean square error over a limited frequency range? Is there a fixed interval between each error evaluation or a fixed number of evaluations?
2. What error function is used, i.e., is the error in the admittance magnitude minimized, or a combination of magnitude and angle (or equivalently conductance and susceptance)?
3. Typically how many iterations (of steps a, b, c, d and e) are required to reach convergence and what is the convergence criteria?
4. Please clarify what is meant by *constrained to produce positively damped response* and how does this relate to the networks representing mutual terms (which will have some negative R, L and C values).
5. The Fibonacci's search requires the two initial points to bracket the minimum and this is achieved by the stepping procedure (by 20% in inductance) prior to the search. However, there is the possibility of local minimum being found rather than global being found. This is probably more likely when an optimization is performed over the full frequency range (such as performed for C_x) rather than limited frequency range (as for L_k). Has any such problem been experienced?
6. Can you comment on why 500 kHz was chosen as the very high frequency to determine R_x?
7. Although a multi-port π representation is shown, it is not clear how this is actually implemented since the aerial and ground mode admittances are used for the matching process. The use of a Y-D transformer is made when one port equivalent is used, however, how is this extended to multi-port, and in particular the incorporation of the mutual admittances between ports?

References

[1] Watson N. R. and Arrillaga J., "Frequency-Dependent A. C. System Equivalents for Harmonic Studies and Transient Convertor Simulation," *IEEE Trans. on Power Delivery*, Vol. 3, No. 3, July 1988, pp. 1196–1203.

[2] Watson N. R., Arrillaga J., and Arnold C. P. "Simulation of HV DC System Disturbances with Frequency-Dependent AC-System Models *IEE Proc.*, Vol. 136, Pt. C, No. 1, January 1989, pp. 9–14.

Manuscript received August 6, 1992.

M. C. Kieny (Électricité de France, France): I would like first to congratulate the authors for their detailed presentation and excellent results obtained with a method both efficient and rigorous. I also would like to point out the usefulness of the routines implemented on EMTP.

My questions are:

- The structure of the RLC network to model an FDNE branch is slightly different from the ones you use to model the High Frequency Transformer Model described in [A]. Can you comment on that? Also, is your fitting method appropriate for highly resistive networks which exhibit, for instance, many real poles?
- How do you model rotating machines? I guess you use X''_d in series with a resistance and a voltage source. But how would you represent a machine in a longer simulation involving X'_d and/or including 50 Hz (60 Hz) steady state? I am thinking of load rejection or line tripping studies.
- I am interested in using your equivalent to study nonlinear phenomena such as ferroresonance. How would you deal with the case where one or several nonlinear elements are connected to nodes inside the reduced part of the network? Could you define one or more equivalent nonlinear elements connected to the remaining nodes?

Reference

[A] A. S. Morched, L. Martí, J. H. Ottevagers, "A High Frequency Transformer Model for The EMTP," Presented to IEEE/PES 1992 Summer Meeting, Seattle, WA, July 12–16, 1992.

Manuscript received August 10, 1992.

HARI SINGH, Texas A & M University, College Station, Texas; The authors have presented a systematic method of obtaining multi-port frequency-dependent network equivalents useful for studying switching transients in electric power systems. The method is implemented as FDNE program in the DCG/EPRI EMTP and is a useful tool, particularly because it requires minimal intervention and judgmental decisions by the user. However, the user will always have to specify the desired frequency range (bandwidth) for which the equivalent produced is valid and, therefore, it is under user control in FDNE. Since the method involves obtaining the frequency response of the network's driving-point and transfer admittances in the specified bandwidth, which is highly oscillatory in nature, the frequency resolution, Δf, used to obtain the response is a very important parameter. The FDNE provides user control over this by allowing upto a maximum of 1000 points within the bandwidth of interest. Since switching transients in power systems have frequencies in the range of 3 Hz – 30 kHz [A], equivalents which are valid (atleast) uptil 30 kHz will usually be required. This constrains the best possible resolution for this typical equivalencing problem to $\Delta f = 30Hz$.

The frequency response of the admittances consists of multiple series and parallel resonances (poles and zeros) and is influenced by the topology (sparse or dense) and the elements (low-loss or high-loss) of the network to be equivalenced. Therefore, whether a particular value of frequency resolution is "good" or "bad" will depend on the characteristic of the network. Specifically, a worst-case situation will be one where the resonance peaks have a high Q-factor (low-loss network) and/or the response consists of many closely-spaced series/parallel resonance frequencies. In that case, even a Δf of 30 Hz may not be good enough since there is a high likelihood that the true resonance frequencies of the network will not be detected unless most of them are an integral multiple of Δf. If this error occurs for many of the dominant resonance frequencies, it will result in "capturing" a frequency response very different from the true response of the network. Consequently, FDNE will generate an inaccurate equivalent.

In view of the above, the authors' comments on the following questions will help in providing better guidelines for use of FDNE.

- During their extensive testing of FDNE, have the authors encountered any networks whose frequency response approaches the worst-case described above? If not, is there anything inherent in the toplogy and elements of physical power system networks which precludes such a case from arising?

- As a practical guideline for FDNE use, can the authors suggest some criteria for *a priori* selection of frequency resolution that will ensure accurate equivalencing? Also, in the same vein, can the authors comment on the factor(s) influencing the choice of logarithmic or linear spacing of frequency points in the bandwidth of interest? In the limited experience provided by a few studies using FDNE, little difference in results was observed by us using either selection!

[A] EMTP Application Guide, EPRI Publication No. EL-4650, Chap. 1, pp. 7.

Manuscript received August 17, 1992.

A.S. Morched, J. Ottevangers, L. Martí: We would like to thank the discussers for their interest and their many thought-provoking questions presented.

We agree with Professor Watson when he indicates that a direct comparison between the FDNE and network equivalencing methods implemented in different platforms is indeed very difficult to make. We have not had the opportunity to evaluate the state variable approach mentioned by Prof. Watson and its advantages in terms of speed and accuracy as compared to the FDNE. However, we would like to point out that while it is true that in the past the EMTP has had problems in the simulation of HVDC and power electronics because of numerical oscillations, these restrictions have been overcome in recent years by different means. The Critical Damping Adjustment (CDA) procedure [a], developed at the University of British Columbia, is an example of a successful implementation of a numerical oscillation suppression method in the EMTP. The CDA procedure has already been implemented in some versions of the EMTP [b].

In the fitting of an admittance function in the FDNE, the error function is defined as

$$\varepsilon^2(\omega) = \sum_{n=1}^{N} [Re\{Y(\omega_n)\} - Re\{Y_a(\omega_n)\}]^2 + [Im\{Y(\omega_n)\} - Im\{Y_a(\omega_n)\}]^2$$

where $Y(\omega)$ and $Y_a(\omega)$ are the actual and approximating admittance functions, respectively. The error function is calculated at all data points within a given frequency range. The number of data points is controlled by the user. The present implementation allows a maximum of 1000 points in either linear or logarithmic scales. A typical application will use 500 points in a linear scale, and convergence generally occurs within 3 to 8 iterations. Convergence is assumed when the error function decreases less than 0.5 % between two consecutive iterations (maximum number of iterations allowed is 30). As indicated in the paper, the fitting algorithm is constrained to produce a positively damped response. This is achieved by forcing the R, L and C elements of a given RLC branch to have the same sign (an RLC branch where R, L, and C have the same sign is always stable). The choice of 500 kHz as the "high frequency" for asymptotic calculations is an anachronism from early FDNE days. It was dictated by the largest argument of the hyperbolic tangent that the compiler allowed in those days.

The Fibonacci search used in the optimization process can converge to a local rather than to a global minimum in a given iteration. In practice, obvious occurrences of this situation seldom happen. Perhaps this is the case because the initial guess is relatively close to the final solution. If the optimization process does become locked in a local minimum, and a satisfactory fit cannot be obtained, the best alternative is to disturb the problem slightly, by either using a different number of points per function or by changing the distribution of the data points (e.g., using a logarithmic, rather than a linear scale).

The last question posed by Prof. Watson can best be answered by an example. Assume a two-port equivalent between buses 1 and 2. The nodal admittance matrix will be given by

$$[Y] = \begin{bmatrix} Y_{1,1} & Y_{1,2} \\ Y_{2,1} & Y_{2,2} \end{bmatrix}$$

where each Y_{ij} represents a 3x3 matrix. The "one-line diagram" of the multi-port π for this matrix is shown in Figure I.a.. Each "element" of this circuit is a 3-phase sub-network. The nodal admittance matrix of each of these sub-networks is then diagonalized with a modal transformation matrix. The resulting zero and positive sequence admittances are then fitted in the FDNE. In the EMTP, these modal networks are not modelled explicitly with series RLC branches and transformers, but with a dedicated model for each sub-network. This model consists of a constant conductance matrix and past history current sources, as illustrated in Figure I.b. These current sources are updated at each time step of the transient simulation based on the sequence networks fitted outside the EMTP.

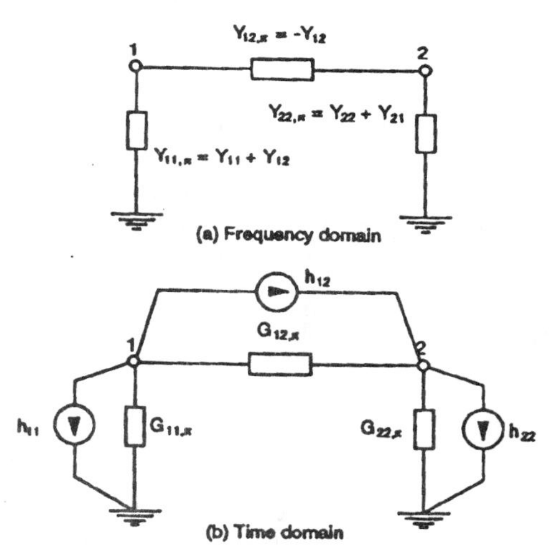

Fig I: *Single-line diagram, multiphase π-circuit*

We will now address the questions asked by Mr. Kieny. The structure of an FDNE branch is shown in Figure 1 of the paper. The structure of a branch used in the High Frequency

Transformer (HFT) model to which Mr. Kieny refers is shown in Figure II below.

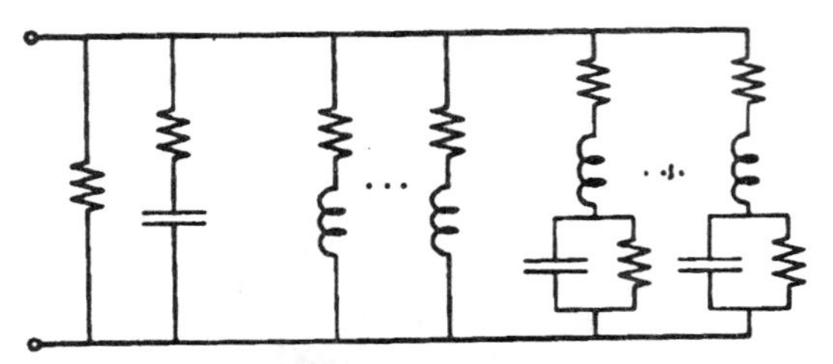

Fig II: *RLC module of the HFT model.*

The reason for the differences in structure is that in the development of the HFT model it was felt that a larger degree of freedom in the fitting process would be beneficial in the approximation of measured data. To achieve this additional flexibility, the following steps were taken: (a) adding a resistance across the capacitance (since this is the most general realization of a complex conjugate pole); (b) optimizing poles and zeros (rather than the R, L, and C components of each branch); (c) adding more than one RL branch. In the FDNE, only two real poles (the RL and RC branches) must provide the dynamics that cannot be accounted for by using complex conjugate poles alone. While this provides good overall answers over a wide frequency range, we feel that quality of the approximations at lower frequencies (e.g., between 60 Hz and 1 kHz) would improve with the addition of extra real poles and zeros. Part of the ongoing FDNE development is concerned with the addition of these extra poles and zeros.

- If the behaviour of a network element like a machine is important to the outcome of a transient simulation, the best approach is not to include it inside the network equivalent. As Mr. Kieny correctly points out, there is no simple way to account for the behaviour of a rotating machine over a broad range of operating conditions by using only an impedance behind an infinite bus. Likewise, if the nonlinear behaviour of a transformer is important, it should not be included inside the equivalent. There are, however, no restrictions regarding what type or how many elements are connected to the external nodes of the FDNE equivalent, or to the number of FDNE equivalents in a given simulation.

Mr. Singh's concerns regarding the number of points required to obtain reasonable resolution for fitting purposes are legitimate, although his conclusions require further examination. The limit of 1000 points indicated in the paper is strictly a programming limit and it does not reflect any inherent FDNE limit. This number was chosen to give ample leeway in the resolution required in all the practical cases examined. The highest Q we have observed in most practical positive sequence networks lies between 80 and 100, which, depending on the resonant frequency, results in a bandwidth between 70 and 150 Hz. A 30 Hz resolution would be sufficient in these cases since to identify and to fit a complex conjugate pole/zero, very few points are really necessary. If a peak is missed because a 30 Hz resolution is not sufficient to identify them, chances are that their effect in a transient simulation would be minimal: closely-packed peaks generally result from transmission lines with small differences in length, and the resulting differences in a transient simulation would only become evident after a relatively long time.

If the FDNE approximations are not acceptable, there are two courses of action: reduce the tolerance in the identification of resonances and/or increase the number of frequency points. Experience, however, suggests that the number of frequency points is less critical than the tolerance in the initial identification of resonant points. A linear scale effectively increases the resolution at high frequencies while a logarithmic scale increases the resolution at lower frequencies. The fact that linear/logarithmic options yielded similar answers in Mr. Sigh's tests, suggests that the high frequency poles were well separated and that the selected number of frequency points specified was sufficient.

Professor Semlyen's interest in the network reduction technique used in the FDNE is gratifying. We did not place too much emphasis in the description because we felt that in today's world of vast computer resources such an algorithm would not be as important as it was when it was first developed. Perhaps the best way to clarify the process is by means of a small example: Consider the network shown in Figure III. Layer 1 consists of all branches connected between layer 1 buses and/or between layer 1 buses and the reference bus. Layer 2 consists of all remaining branches connected between layer 1 and layer 2 buses. The nodal admittance matrix involving these nodes is formed. Once formed, the buses corresponding to layer 1 can be eliminated. Any reference buses that might appear in layer 1 are retained. Next, the reduced admittance matrix is extended to include layer 3 buses, as shown in Figure III. This new layer consists of all remaining circuits connected between layer 2 and layer 3 buses, and circuits between layer 3 buses. Again, matrix reduction will eliminate the now redundant layer 2 buses, with the exception of reference ports. This process of adding layers, followed by matrix reduction is repeated until all buses are included.

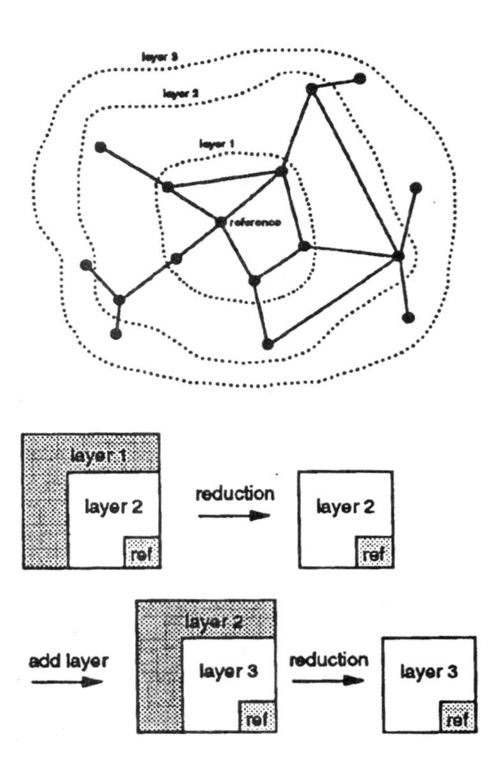

Fig III: *Network reduction*

Some sparsity techniques are also used throughout the process, such as keeping track of zero elements. However, the reduced matrix rapidly becomes full and sparsity techniques are only effective when a new layer is added.

References

[a] J. R. Martí and Jimin Lin, "Suppression of Numerical Oscillations in the EMTP". *IEEE Transactions on Power Systems*, pp. 739-747, May 1989.

[b] Jimin Lin and J. R. Martí, "Implementation of the CDA Procedure in the EMTP". *IEEE Transactions on Power Systems*, Vol. 5, No. 2, pp. 394-402, May 1990.

Manuscript received September 28, 1992.

PART 3

SIMULATION TOOLS

PART 3
SIMULATION TOOLS

Based on Dommel's pioneering works, the development of the simulation tool designated as Electro Magnetic Transients Program (EMTP) was started in the late 1960s at Bonneville Power Administration (BPA). Since then, very intense activity has been dedicated to the development and improvement of electromagnetic transients programs (EMTPs). Several simulation tools (EMTP, ATP, EMTDC, MICROTRAN, NETOMAC) based on the trapezoidal rule are currently available. Other tools, such as PSPICE and SABER, using a different integration rule, can also be applied to the simulation of some transient phenomena in power systems.

A general purpose program is usually made of three parts:

- a pre-processor or graphical user interface (GUI) aimed at creating the input data file using a drafting facility
- a processor dedicated to performing calculations needed to obtain the solution of a transient process
- a post-processor to provide capabilities for analysis and documentation of the simulation results

The most widely used EMTPs are made of several programs dedicated to the tasks described above. The four papers included in this section of the book give an overview of the current status.

The first paper presents PSCAD (Power Systems Computer Aided Design) [1], a GUI integrated by three user-friendly modules that support EMTDC, an EMTP initially developed for simulation of AC/DC links.

The second paper illustrates some capabilities of DCG/EPRI EMTP. The document presents the application of this tool to several cases and the validation of simulation results by comparison to field test measurements.

The third paper introduces the general-purpose program MODELS [3]. This tool was interfaced to the original BPA EMTP and to the ATP (Alternative Transients Program) in 1990. It was initially developed to replace the TACS program, see Part 1, and to overcome some of its limitations. The main purpose was to develop a new tool for the representation of control systems. However, MODELS is a high-level language that can be used for other purposes and as an autonomous program for transient analysis.

The fourth paper describes the implementation of an EMTP using general-purpose software as a computational engine. The creation of MatEMTP, as the new program is known, is the future trend in the development/creation of transients program using sophisticated software tools.

The additional bibliography includes references to other tools developed to perform some or all of the tasks above described.

REFERENCES

Reprinted Papers

[1] O. Nayak, G. Irwin and A. Neufeld, "GUI enhances electromagnetic transients simulation tools", *IEEE Computer Applications in Power*, vol. 8, no. 1, pp. 17-22, January 1995.

[2] W. Long et al. "EMTP. A powerful tool for analyzing power system transients", *IEEE Computer Applications in Power*, vol. 3, no. 3, pp. 36-41, July 1990.

[3] L. Dubé and I. Bonfanti, "MODELS : A new simulation tool in the EMTP", *European Transactions on Electrical Power Engineering*, vol. 2, no. 1, pp. 45-50, January/February 1992.

[4] J. Mahseredjian and F. Alvarado, "Creating an electromagnetic transients program in MATLAB : MatEMTP", Paper 96 WM 098-4 PWRD, *1996 IEEE/PES Winter Meeting*, January 21-25, Baltimore.

Additional References

[A1] K. Carlsen, E.H. Lenfest and J.J. LaForest, "MANTRAP : Machine and network transients program", *Proc. of IEEE PICA*, pp. 144-151, 1975.

[A2] S.N. Talukdar, "METAP - A modular and expandable program for simulating power system transients", *IEEE Trans. on Power Apparatus and Systems*, vol. 95, no. 6, pp. 1882-1891, November/December 1976.

[A3] F.L. Alvarado, R.H. Lasseter and Y. Liu, "An integrated engineering simulation environment", *IEEE Trans. on Power Systems*, vol. 3, no. 1, pp. 245-253, February 1988.

[A4] F.L. Alvarado and Y. Liu, "General purpose symbolic simulation tools for electric networks", *IEEE Trans. on Power Systems*, vol. 3, no. 2, pp. 689-697, May 1988.

[A5] P. Bornard, P. Erhard and P. Fauquembergue, "MORGAT : A data processing program for testing transmission line protective relays", *IEEE Trans. on Power Delivery*, vol. 3, no. 4, pp. 1419-1426, October 1988.

[A6] E. Gunther, T. Grebe, R. Adapa and D. Mader, "Running EMTP on PCs", *IEEE Computer Applications in Power*, vol. 6, no. 1, pp. 33-38, January 1993.

[A7] H.K. Hoidalen, "*ATPDRAW - Reference Manual*", Portland, 1994.

[A8] B. Kulicke, "Simulation program NETOMAC : Difference conductance method for continuous and discontinuous systems", *Siemens Research and Development Reports*, vol. 10, no. 5, pp. 299-302, 1981.

[A9] P. Lehn, J. Rittiger and B. Kulicke, "Comparison of the ATP version of the EMTP and the NETOMAC program for simulation of HVDC systems", *IEEE Trans. on Power Delivery*, vol. 10, no. 4, pp. 2048-2053, October 1995.

GUI Enhances Electromagnetic Transients Simulation Tools

Omprakash Nayak, Garth Irwin*, and Arthur Neufeld**

For more than two decades, power systems engineers have been using electromagnetic transients simulation programs (generically designated herein as *emtp*) for planning, designing, and operating power systems. While several of these programs are already known for their technical competence, their user interfaces are severely lacking compared to the advanced interfaces used in other fields. Although user-friendliness is the most apparent advantage of a graphical user interface (GUI), there are other benefits of a more essential nature.

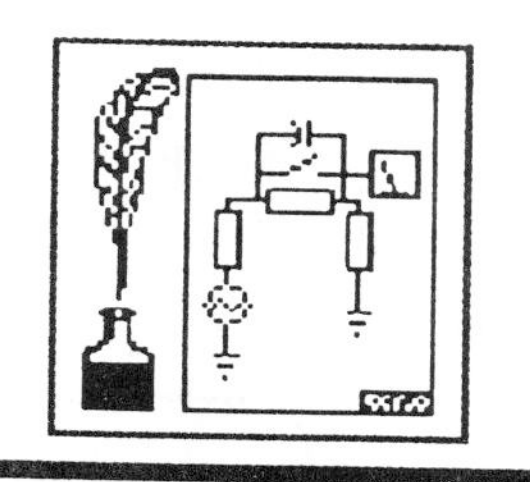

Graphical user interfaces increase productivity by enabling swift assembly of circuits, fast interactive simulations, and quick and easy analysis of results

The most compelling reason to have a GUI for an emtp is the resulting increase in productivity. All electromagnetic transients programs need input data files, which are hand assembled for most programs. This can be a tedious and time-consuming process. A computer-aided drafting facility can create these data files in a fraction of the time. Moreover, it can also help to document the simulation by producing high-quality drawings of the system being studied. Graphical interfaces for running and interacting with a simulation as well as for analysis of results can result in further productivity. Also, the learning period for new users is much shorter with a GUI. It is easier to understand and build a circuit with a graphical representation instead of entering numbers in a data file.

There are additional benefits of a GUI for emtp. Normally, a user needs to have a certain degree of computer expertise to use a conventional emtp. Through a GUI, a power systems engineer can easily use the program with little or no computer expertise, thus making the program more accessible and less intimidating for novice users.

At present, input files such as data files are not compatible among different emtps, which makes it difficult to simulate the same circuit using different programs, to compare the results to gain confidence in the simulations, or to simply exchange the simulation case with another user who happens to have a different program. This is possible if the drafting facility can create data files for different emtp from the same circuit drawing. This is a very useful feature but, in practice, it is not easy to achieve.

Currently, some graphical user interfaces are available, but many are still in infant stages of development and are not viable for commercial studies. This article describes a revolutionary graphical user interface, PSCAD™. The description of this GUI is incomplete without discussing the capabilities, scope, and reliability of the simulation program it was designed for, mainly because the simulation results obtained using a GUI are only as good as and limited by the simulation program it uses. At present, the interface fully supports two electromagnetic transients simulation tools:

- EMTDC™, a general-purpose, high-performance emtp (EMTDC is described briefly in this article.)
- RTDS™, a general purpose, real-time, electromagnetic-transients digital simulator.

User Interface

The specifications of a GUI are unique for each application. The criteria used in the design of the GUI program developed by the Manitoba HVDC Research Centre are as follows:

- **Easy-to-Use:** The GUI should have sufficient CAD features at every stage and should be easy-to-use.
- **Networks:** The GUI should take full advantage of network facilities to improve efficiency.
- **Standards:** The GUI should follow industry standards such as X/Windows and Windows to avoid any special hardware or software requirements.
- **User Models:** It should be easy to add user models and icons to the interface.
- **Online Help:** Context-sensitive help with topic links should be available to assist beginners as well as experts.
- **Diagnostics:** The GUI should provide useful diagnostic messages throughout every stage of the study.

* Manitoba HVDC Research Centre

- **Interchangeable:** The GUI should support data interchangeability among different types of emtp.
- **Interactive:** The GUI should facilitate dynamic interaction with the simulation in progress.

The result (after 10 man-years of effort) is Power Systems Computer Aided Design (PSCAD), which meets most of these criteria. The GUI, which is a collection of well-integrated modules (DRAFT, RUNTIME, and MULTIPLOT), forms a complete graphical environment for power systems simulation. The use of these modules in a typical study is illustrated in Figure 1. The design and operation of the GUI is presented in the following sections through the description of its constituent modules.

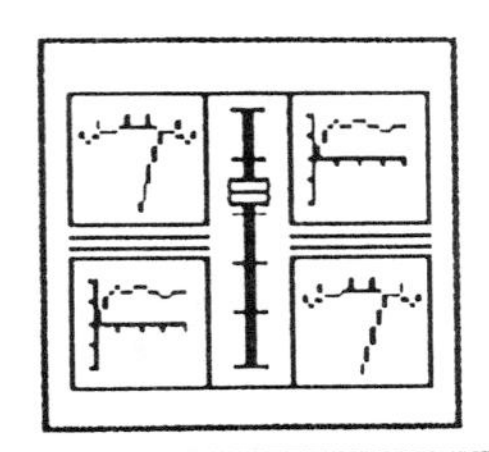

Manipulation of operator-controlled inputs (sliders, push buttons, switches, and rotary dials) dynamically affects the simulation in progress

Preparation of Data Files

The first step in a simulation study, after gathering the necessary data of network and controls, is to prepare the data files in a form suitable to the simulation program. This task is easily performed using the GUI's DRAFT graphical circuit design module.

Graphical Assembly of Circuit

Figure 2 shows a sample screen with an HVDC system. The right side of the screen is a library area and the left side is a drawing canvas. Both are scrollable and adjustable. A library is merely a collection of spatially laid-out components or circuits, which can be loaded into the library area. The libraries contain user-written components as well as master components supplied with the GUI. Users can create and organize their own libraries according to their needs.

The user can draw a picture of the circuit by selecting the graphical components from the library and arranging them on the canvas. A pop-up edit window with default parameter values for each component accepts new values and conducts a *sanity check*. The components can be rotated or mirrored to fit into the circuit. In addition, wires can be resized. Modularity of the control circuits can be maintained by arranging them on multiple pages. Electric circuits can span across pages with transmission lines or cables connecting them.

Meters (voltmeter, ammeter, power meter, etc.) are available for measuring electrical quantities. Measured quantities are assigned to output channels by connecting plot symbols at the required points in the circuit. The module allows the use of manually controlled inputs through its slider, push button, switch, and rotary-dial components. They can be controlled interactively during a simulation run.

The module is complete with essential computer-aided design features such as move, copy, cut, and paste, as well as advanced features such as select, deselect, mirror, rotate, undo, group, select all, find, and keyboard shortcuts, etc., which help to speed-up the circuit building process. To facilitate easy navigation through the circuit and the library, scroll and slide features and miniature pallets (at the bottom corners of the screen) are provided. File commands include save, save as, and load. The print feature supports laser printer or plotter output of the circuit drawing on different paper sizes and is ideal for documenting the circuit drawings. Online help is available for all components.

When asked to compile, the module conducts a *sanity check* on the circuit and creates input files (data files, etc.) required for running the chosen simulation program. Diagnostic messages generated during this stage are very useful in correcting errors.

DRAFT — For Quick Assembly of Circuits

RUNTIME — For Interactive Running of Simulations

MULTIPLOT — For Advanced Analysis/Printing of Results

Figure 1. Sequence of tasks in a simulation study

Custom Components

Master libraries supplied with the GUI may not be sufficient to meet a user's individual needs in spite of the libraries being very exhaustive. This module provides an easy-to-use facility for creating and adding custom components. To create a new component, users simply select a menu item that brings up an editor with a generic graphics description and input/output connections. The graphics, described with simple text, can be easily customized. Users can define the component's function either by typing the Fortran statements using this editor or by entering the name of the file containing the Fortran code. The module's component graphics are open to users to view and copy. Typically, users copy a master component into their own library, rename it, and perform simple modifications to suit their needs.

There are two common situations for which users tend to write their own components. The first situation is when users cannot find a component in the master library to meet a specific functional requirement. They may then choose to create their own component. Users routinely write specific custom components despite the availability of an exhaustive default library. The second reason for writing custom components is when users (normally organizations) already have their own graphical user interface on their product (such as a programmable logic controller) and want to maintain the look and feel of their interface when the circuit is modeled on PSCAD. In this situation, users may choose to rewrite the components even though they may be functionally identical to master components. This may result in a completely customized component library. One such example is Asea Brown Boveri's HIDRAW controls library in DRAFT (available from NESA AS, a Danish utility), which is used to model ABB's HVDC controls. Another example of a custom library is the SIMADYN D controls library of Siemens AG. These libraries contain features such as the capability to run with multiple time steps which duplicate how digital controls function in the real system.

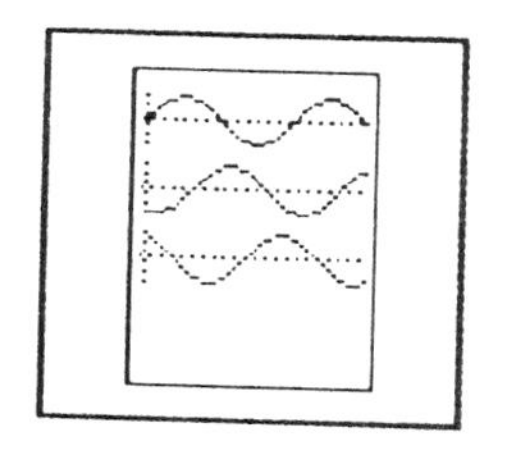

A collection of well-integrated modules forms a complete graphical environment for power systems simulation

Running the Simulation

After completing the assembly of the circuit, i.e., preparation of data files, the next step for the user is to run the simulation. The RUNTIME module provides a convenient graphical user interface for this task.

Figure 3 shows a sample screen for running a simulation. The module gets all the necessary information about the circuit from the batch file created by the graphical circuit design module. After loading the batch file, the simulation is started by pressing on a single menu button. Once started, the simulation can be paused, single stepped, and resumed. A maximum of ten simulations can be run simultaneously on any computer on the network from each system console.

Flexible Layout

The user has to set up the console the first time a study is run to display the desired graphs and controls. In a large-scale study, it is natural to have many control variables and output quantities that are accessed through the module for running the simulation, either for controlling or for monitoring purposes. Monitored outputs are displayed either in plots or in meters, and they are often used for a quick analysis through visual inspection. Inputs to the simulation that are controlled from the console are represented as sliders, push buttons, switches, or rotary dials. The module allows the user to customize the console by providing many useful features that can be used during and after the run. Plots, meters, sliders, etc., can be created through a pull-down menu, and these can be arranged on the palette at the user's discretion and can also be

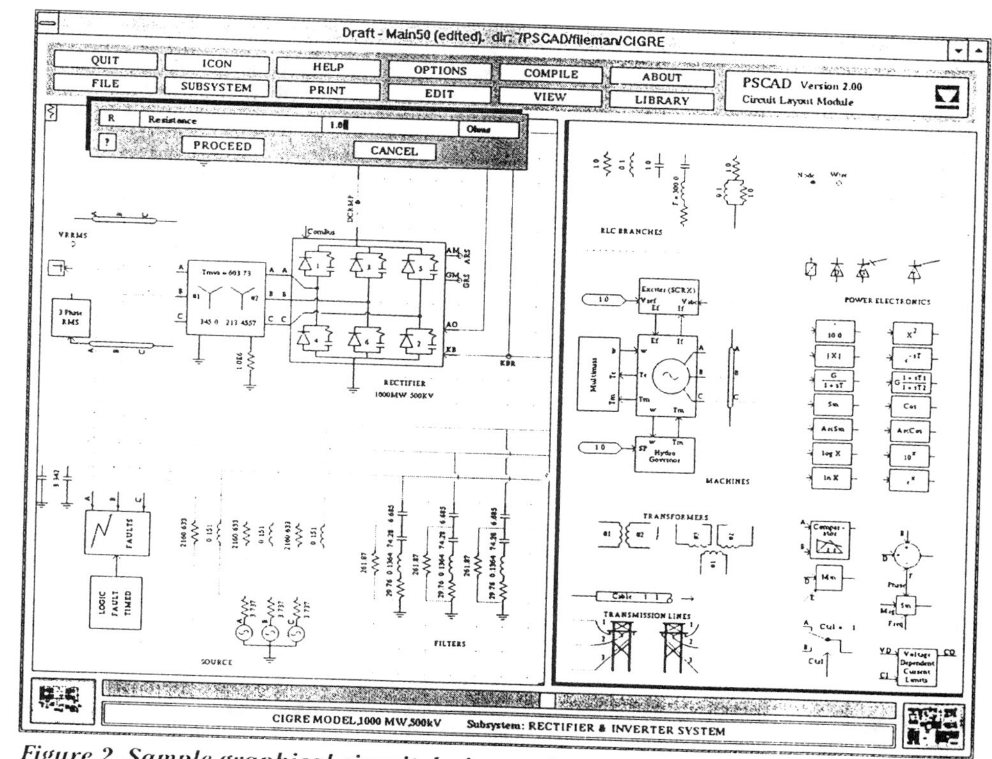

Figure 2. Sample graphical circuit design module screen showing drawing canvas, component library, and edit window

grouped and temporarily hidden. Plots can be zoomed, scaled, and resized. Cross-hairs can be used to find the exact value at a point on the graph, while meters provide both analog and digital displays. Multiple curves can be plotted on a single graph and multiple graphs can be aligned in a group. The user can save either a selected group of curves or the entire output to a text file for postprocessing. Run messages, such as status and diagnostics, are conveniently displayed in a window whenever requested. A console layout can be saved in a batch file for later use in studies involving the same circuit.

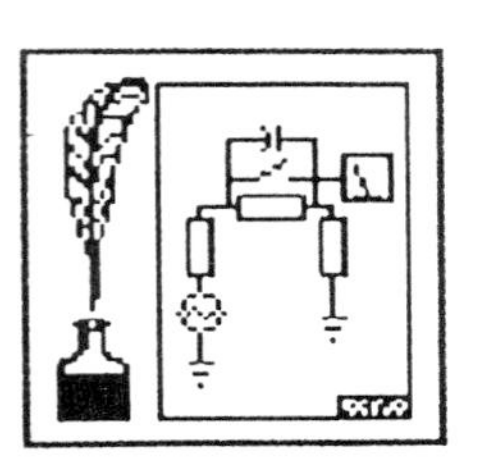

A graphical circuit design module prepares data files in a form suitable to the simulation program

Interactive Runs

A simulation run with a conventional emtp is performed in a batch mode, which means that the user does not have any control over the simulation when it is running. In contrast, this GUI run-time console is interactive in a way very similar to an actual operator console in a control room. Manipulation of operator-controlled inputs (sliders, push buttons, switches, and rotary dials) dynamically affects the simulation in progress, and the results can be observed on the screen interactively in a meter or a plot. Thus, a user can push a button to trip a generator or create a fault, move a slider to change the power order, turn a dial to set the transformer tap, or toggle a switch to change the controller topology while the simulation is in progress. The snapshot feature allows the user to save the state of the system at any point during the run from which point a later run can be started. Note that the GUI must have communication links with the simulation program it is running in order to achieve these interactive features, hence, may require modifications to the emtp to incorporate them in the GUI.

Multiple Runs

For studies that need repeated simulations of the same system with one or two parameters changed, the multiple-run feature is convenient. For example, it can be used to determine optimum gains of a controller or to find point-on-wave to determine a worst overvoltage due to a fault. Multiple runs let you change one or two parameters in a random or sequential order over a specified range and plot the objective function contour.

Analysis and Documentation of Results

An important last step in a simulation study is the analysis and documentation of results. The module for running the simulation can be used for initial online analysis of the results. For advanced analysis and customized documentation of results, MULTIPLOT provides additional useful capabilities, some of which are discussed in this article. A sample analysis screen layout is shown in Figure 4.

Fourier and harmonic distortion analysis can be used to determine the harmonic content of a single waveform or of three phase waveforms.

A waveform calculator enables many useful operations, such as addition, multiplication, or scaling of waveforms and even generation of new functions.

A flexible layout, graph formatting, and labeling facilities allow easy customizing of the printout, which can be printed directly on a laser printer or stored in a form suitable for inclusion in a word processor to generate reports.

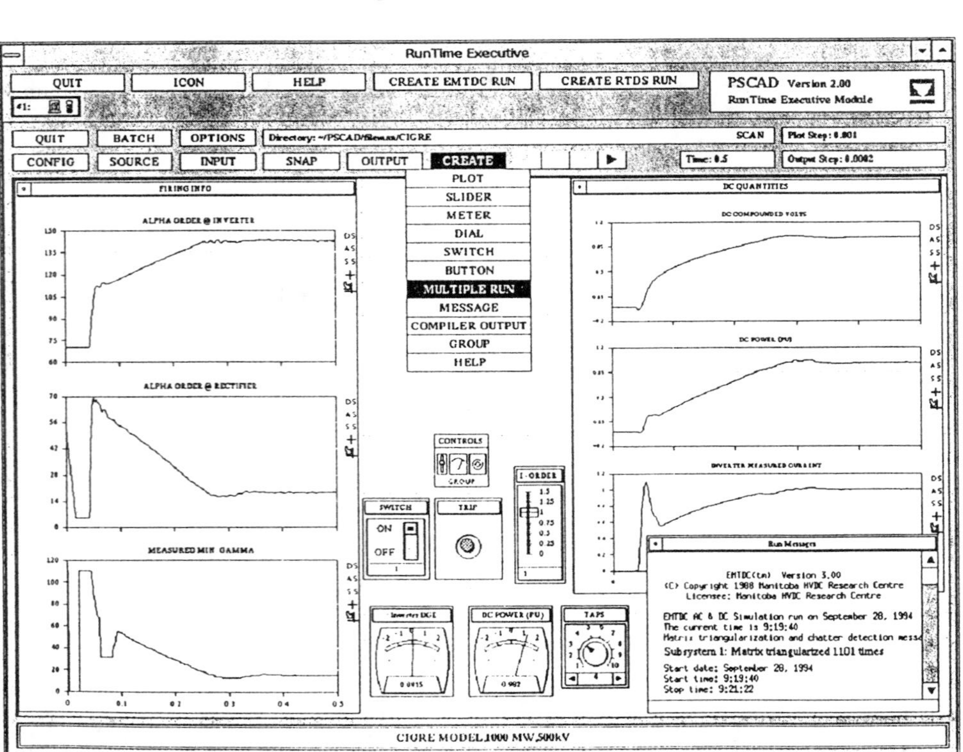

Figure 3. Sample user console of a simulation

It is very common in large-scale studies to generate printouts of all the results only to realize that one of the parameters is incorrect and that the entire simulation must be repeated to generate new results. The update feature of the analysis module lets you print the new results effortlessly.

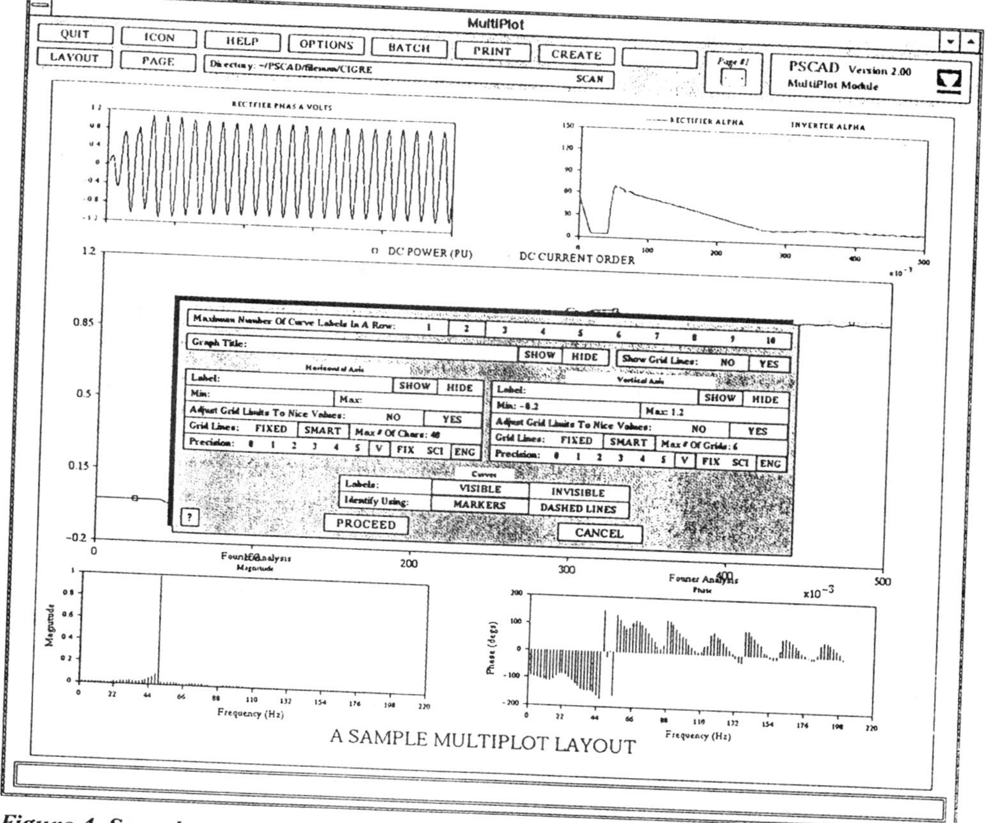

Figure 4. Sample analysis screen

Simulation Program

As a result of the pioneering work by Dr. Hermann Dommel, the first electromagnetic transients simulation program, specifically designated as EMTP, was made available in early 1970s. Encouraged by the potential of such programs for studying the dynamic behavior of power systems, many people worldwide started using EMTP. EMTDC was initially developed at Manitoba Hydro in the late 1970s to study ac/dc systems, since EMTP at that time was not suitable for such studies. It is based on the same trapezoidal integration algorithm developed

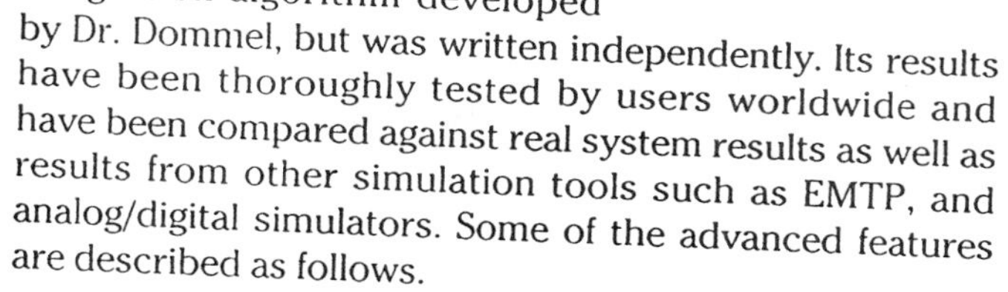

by Dr. Dommel, but was written independently. Its results have been thoroughly tested by users worldwide and have been compared against real system results as well as results from other simulation tools such as EMTP, and analog/digital simulators. Some of the advanced features are described as follows.

Interpolation

In a basic emtp, events such as switching of a thyristor occur at discrete time steps. If the switching criterion is based on an event such as a voltage zero crossing (for example), the required switching instant may not coincide with the simulation time step. The switching would then occur at the next time step, resulting in a small error. The effect of this error is negligible in most situations if the simulation time step is small enough and the switching is infrequent. However, in circuits with many fast switching power electronic devices, this error could be significant. Manitoba Hydro's simulation program can interpolate the switching events of diodes, thyristors, GTOs, surge arresters, and breakers to precisely the correct instant, thus eliminating switching errors. Another important feature is the removal of spurious chatter from inductive node voltages and capacitive currents. This problem exists in all emtps using trapezoidal integration, unless special measures are taken. The chatter removal procedure is implemented using a half-step interpolation.

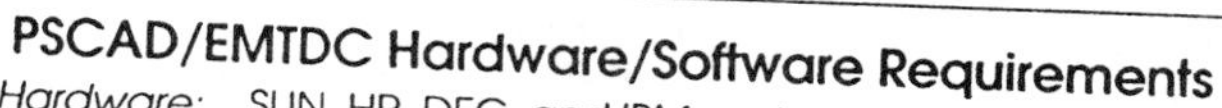

PSCAD/EMTDC Hardware/Software Requirements

Hardware: SUN, HP, DEC, and IBM workstations
Software: Unix operating system, X11 graphics, and Fortran 77 compiler
Memory: 16 MB internal RAM
Storage: 10 MB disk space

Initialization

Initializing the network and control system to steady-state conditions is a major effort in many emtps. The snapshot feature of this simulation program eliminates the need for an elaborate initialization process. It allows the state of the system to be saved at any point during the run so a later run can be started from that point. At first, the system is usually started from standstill, just the way an actual system is started. For example, for an HVDC system the current order is gradually ramped up to the rated value. A snapshot is taken when the system reaches steady state. Subsequent runs can be started from this snapshot from an initialized state. The snapshot eliminates the need for time-consuming initialization and makes it possible to model very complex systems with multiple nonlinearities, which would be difficult if not impossible to initialize directly. However, in cases with many machines, special start-up procedures may be necessary.

Fortran Interface
An easy-to-use Fortran interface is used by all models in EMTDC, whether user written or included with the program. A flexible graphics language and a general Fortran interface make a powerful combination in a simulation program. Advanced users may choose to write every one of their models, be it a simple control block or a detailed machine model.

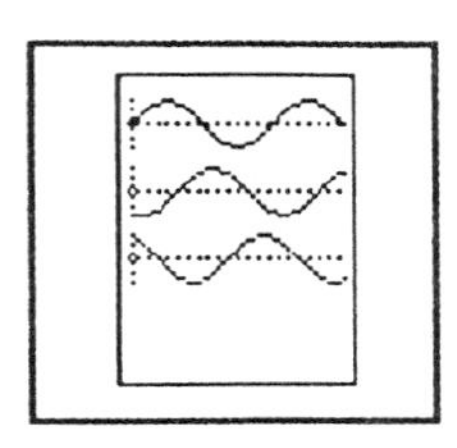

The MULTIPLOT module enables advanced analysis and customized documentation of results

Computational Accuracy
In digital computers, the accuracy of a solution can be affected by the precision of variables used in the program. Double-precision representation is more accurate than single-precision representation, but requires more computation time. The GUI/simulation package selectively uses double precision computation only in critical areas to maintain accuracy. A full double-precision version can also be invoked if necessary.

Computational Speed
In spite of the computational burden of additional features (such as interpolation), the simulation program still runs faster than most other emtps. This is made possible by using the concept of subsystems to decouple the network, sparsity techniques, and optimal node ordering in addition to efficient programming. The modeling of network dynamics and controls using the general Fortran interface (as opposed to internal matrix representations) also greatly speeds the solution.

Simulation Scope

The scope of simulation that can be performed using an emtp is limited by the range and capabilities of models available. Manitoba Hydro's GUI/simulation system provides a wide range of component models in the form of master libraries to deal with most situations. The following is a partial list of these models:

- **Network Components:** passive RLC components; transformers with saturation; frequency dependant transmission lines and cables; synchronous machines and induction motors with exciters, governers, and multimass models; faults; breakers; surge arrestors; current and voltage sources; multiple harmonic injectors
- **Power Electronics:** thyristors, diodes and GTOs; HVDC valve groups, SVS, and other FACTS devices
- **Control Blocks:** derivative, delay, differential lag, integrator, limit, complex pole, real pole, lead lag, filters, amplifier, switch, and Boolean functions
- **Meters:** voltmeters, ammeters, real and reactive power, peak detector, phase angle, frequency, and rms.

The types of studies routinely performed by users worldwide cover all general power systems electromagnetic transients studies such as dynamic performance, overvoltage, relay coordination, and harmonic interaction. Modeling of FACTS devices and electrical drives is gaining popularity among users, mainly due to the ease of modeling power electronic circuits and its controls.

Future Developments

The Manitoba HVDC Research Centre is expanding the capabilities of the GUI to support other electromagnetic transients simulation tools.

At present, the GUI can run only on Unix workstations with X/Windows graphics. A version for personal computers with Microsoft Windows is currently under development.

Acknowledgments

PSCAD and EMTDC are the creation of a closely knit team of electrical engineers and computer scientists at the Manitoba HVDC Research Centre, Manitoba Hydro, and the University of Manitoba. The ongoing developments would not be possible without their creative and cooperative efforts.

For Further Reading

P.G. McLaren, R. Kuffel, R. Wierckx, J. Giesbrecht, L. Arendt, "A Real Time Digital Simulator for Testing Relays," *IEEE Transactions on Power Delivery*, Volume 7, Number 1, January 1992, pages 207-213.

P. Kuffel, K.L. Kent, G.B. Mazor, M.A. Weekes, "Development and Validation of Detailed Controls Models of the Nelson River Bipole 1 HVdc System," *IEEE Transactions on Power Delivery*, Volume 8, Number 1, January 1993, pages 351-358.

Biographies

Omprakash Nayak received his BE (hons.) from Mysore University, India in 1984, after which he worked with Bosch for 1 year and with Siemens as a project engineer for 2 years. He received MSc and PhD degrees from the University of Manitoba in 1990 and 1993, respectively. Dr. Nayak is a research engineer at the Manitoba HVDC Research Centre, working on EMTDC developments.

Garth Irwin received a BSEE from the University of Manitoba in 1987 and is currently pursuing a MS degree. He has been employed by the Manitoba HVDC Research Centre since 1987, where he is now in charge of the development of PSCAD/EMTDC.

Arthur Neufeld received his BS degree in computer engineering from the University of Manitoba in 1988 and is currently pursuing a MS degree. He started work at the Manitoba HVDC Research Centre in 1988 as a research engineer, specializing in the design of graphic programs.

EMTP A Powerful Tool for Analyzing Power System Transients

Willis Long[1], David Cotcher[2], Dan Ruiu[2], Philippe Adam[3], Sang Lee[4], and Rambabu Adapa[5]

The Electromagnetic Transients Program (EMTP) is a general purpose computer program for simulating high-speed transient effects in electric power systems. The program features an extremely wide variety of modeling capabilities encompassing electromagnetic and electromechanical oscillations ranging in duration from microseconds to seconds. Examples of its use include switching and lightning surge analysis, insulation coordination, shaft torsional oscillations, ferroresonance, and HVDC converter control and operation.

In the late 1960s Hermann Dommel developed the EMTP at Bonneville Power Administration (BPA), which considered the program to be the digital computer replacement for the transient network analyzer. The program initially comprised about 5000 lines of code, and was useful primarily for transmission line switching studies. As more uses for the program became apparent, BPA coordinated many improvements to the program. As the program grew in versatility and in size, it likewise became more unwieldy and difficult to use. One had to be an EMTP afficionado to take advantage of its capabilities.

Cooperative EMTP Development

To address the problems that accompanied this code expansion, the EMTP Development Coordination Group (DCG) was formed in 1982. The Electric Power Research Institute (EPRI) joined the activity in 1984. The first result of this cooperation, DCG/EPRI EMTP Version 1.0, appeared in 1987. It included improved documentation (including four Workbooks), a user hotline, and an effort directed at identifying and removing program errors. Version 2.0 appeared in mid-1989, and it included a number of enhancements to the base code. Among these were multi-frequency initialization, multi-port frequency dependent network equivalents, a circuit-breaker arc model, a voltage-current controlled dc converter model, and new air-gap and arrester models.

To meet growing industry demands, EPRI and DCG have taken on the major task of developing an EMTP PC Workstation, which will be available in Spring, 1990. This workstation, which is based on the 0S/2 operating system, will provide the user with a powerful interactive environment that includes on-line help, graphic presentation of the EMTP output in a window format, and several other features. Plans are also underway to develop the EMTP for UNIX-based workstations.

Version 3.0, the next planned DCG/EPRI release, is expected to be available in Spring, 1992. Enhancements will include a high-frequency transformer model, a complete dc converter bridge model, an SVC model, a three-phase load flow module, improved line and cable models, protection system models, corona models, and revisions to TACS, the control system subroutine.

To illustrate the capabilities of the EMTP, three examples of its use have been selected. These are excerpted from previous issues of the *EMTP Review* (see For Further Reading), a newsletter for users of the DCG/EPRI EMTP. These examples are TNA-type problems that have been solved using EMTP V 1.0. In two of the examples, comparisons between the simulations and actual field tests are shown. The accuracy of replication of the results is good.

Long Transmission Line Transients

The operation of the Athabasca Electric System in northern Saskatchewan, Canada, started when SaskPower completed and commissioned the new Beaverlodge-Rabbit Lake electric line in October 1988. This new 355 km, 115 kV line connected three existing hydroelectric plants and their substations located near Uranium City to several northern communities

[1] University of Wisconsin-Madison, Madison, Wisconsin
[2] SaskPower, Regina, Saskatchewan
[3] Electricite de France, Paris, France
[4] ABB Power Systems, Inc., New Berlin, Wisconsin
[5] Electric Power Research Institute, Palo Alto, California

0895-0156/90/0700-0036$1.00 © 1990 IEEE

and a mine at Rabbit Lake. Figure 1 shows a single-line diagram of this transmission system.

Since the system comprised a rather long 115 kV line connecting small hydro units and tap-off substations energized with the line, harmonic overvoltages were anticipated. Therefore, as part of the planning activities for this project, EMTP studies were done to provide recommendations on equipment characteristics and operating procedures for line and transformer energization. These studies were carried out using the DCG/EPRI EMTP Version 1.0 running on SaskPower's PRIME 750 computer.

The EMTP studies included:

- Simulations of line and transformer energization,load rejection and fault clearing, which were done to help determine the required transformer, circuit breaker and other equipment characteristics.
- Additional simulations, which were used to develop recommended procedures for line and transformer energization.
- Comparisons of several recorded waveforms with the results of EMTP simulations of the same events. Prior to the commissioning of the 115 kV line, digital transient fault data recorders (TFDRs) were installed at the Beaverlodge and Stony Rapids stations. These TFDRs record the voltages and currents for faults, breaker switching, or other transients.

Line Energization

The EMTP studies of the Beaverlodge to Stony Rapids 115 kV line energization were an important part of the planning activities for this project. The system has been designed so that the two Fond du Lac 115-25 kV transformers, two Stony Rapids 115-25 kV transformers, and one Stony Rapids reactor are energized together with the 115 kV line. The EMTP frequency scan was used to determine the resonant frequencies of the system with the hydro units and the 115 kV line up to Stony Rapids. Depending on the number of generators operating, there is a peak harmonic impedance near the second or third harmonic. Therefore, harmonic overvoltages were anticipated when energizing the transformers together with the 115 kV line.

EMTP simulations were done for several alternatives to determine the equipment characteristics and operating procedures required to energize the Beaverlodge to Stony Rapids line with acceptable peak overvoltages. The resulting study recommendation was to reduce the Beaverlodge voltage to 0.8 p.u. before closing the Beaverlodge 115 kV breaker to energize the system up to Stony Rapids. A 66-115 kV transformer with an on-load tapchanger was installed at Wellington.

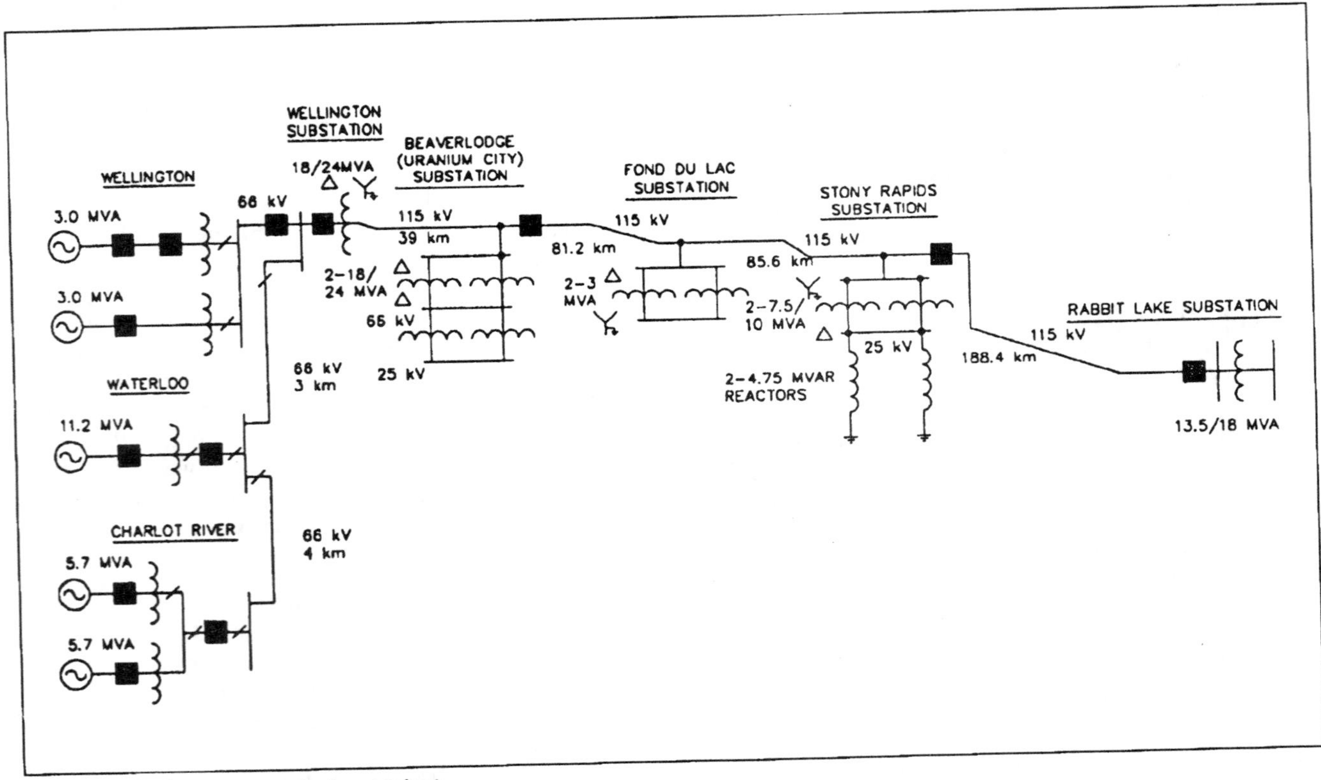

Figure 1. Athabasca transmission project

The selected standard operating procedure was to reduce the 115 kV voltage to 0.8 p.u. at Beaverlodge by lowering the Wellington transformer tap to the bottom of its range.

Figure 2 shows a plot of the Stony Rapids 115 kV voltage from an EMTP simulation of this energization using the recommended procedure. This is a worst case with residual flux in the transformers and a breaker closing instant that gives maximum inrush current. The peak overvoltage shown is 138 kV or 1.47 p.u. The voltage exceeds 1.4 p.u. for only three cycles, which was judged to be acceptable. EMTP runs were done with the full range of generation units on-line from the two Wellington units up to all five hydro units. The study concluded that the system must never be energized with only the two Wellington units connected because, in this case, the overvoltage could exceed 2.0 p.u. with a second harmonic resonance.

Field Test Comparisons

When the Beaverlodge to Stony Rapids line was energized as part of the system commissioning on October 20, 1988, the line and transformer voltages and currents were recorded by the Beaverlodge TFDR. These recorded waveforms were then compared with EMTP simulations of the same event. Prior to the energization, the two Charlot River generators were operating with the line in-service up to Beaverlodge and the 115 kV voltage reduced to 0.8 p.u. Figure 3 shows the comparison of the Beaverlodge voltage for the actual TFDR recording and the EMTP simulation. This simulated overvoltage is lower than the one shown in Figure 2, because there is no residual flux in the transformers. The line and transformers had previously been de-energized with no load connected. A

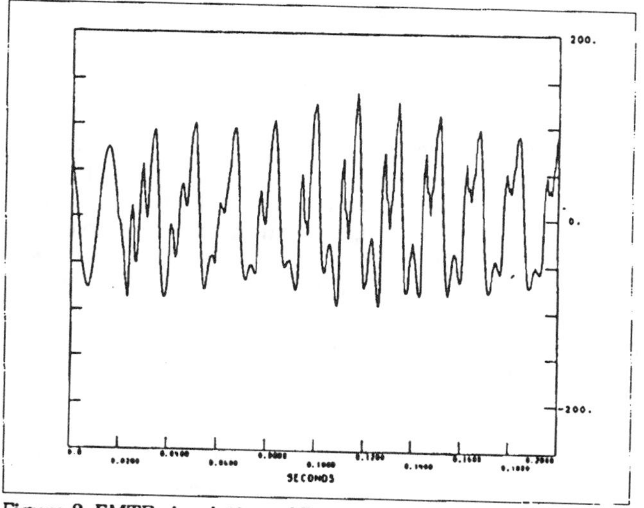

Figure 2. EMTP simulation of Beaverlodge-Stony Rapids line and transformer energization. Transformers with residual flux. Beaverlodge 115kV voltage (1p.u.=93.9kV)

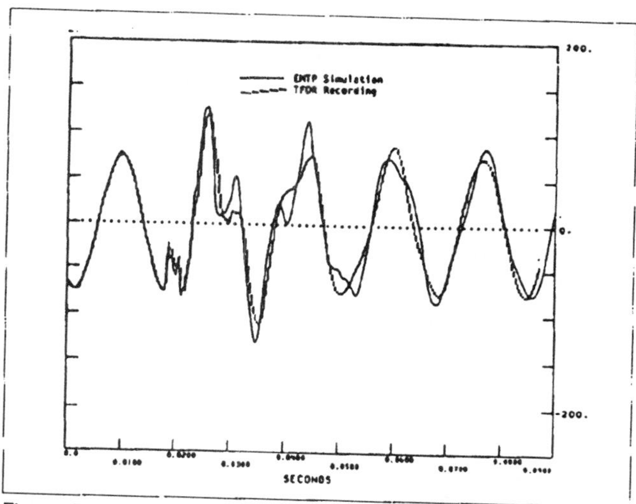

Figure 3. Energization of Beaverlodge-Stony Rapids line and transformers. Beverlodge 115 kV voltage (1p.u. = 93.9kV)

decaying natural frequency oscillation at 70 Hz was shown as predicted by the EMTP simulation. Therefore, the simulation for energization shown in Figure 3 was run with no residual flux in the hysteresis model used to model transformer saturation.

The EMTP simulations were run with different models to check the sensitivity of the results. Table 1 shows the comparison of the peak overvoltages for the TFDR recording and the EMTP simulations. The Stony Rapids TFDR was not yet in service at that time. The EMTP simulation results shown in Figure 3 correspond to Case #2 in Table 1, which was the base model for these studies.

The simulation Case #1 presents the results obtained with the original design transformer data used in the earlier planning studies. After the transformers were built and tested and additional information was obtained from the manufacturer for the saturation characteristic, the transformer modelling was revised.

	Peak Phase-Ground Voltages (1 pu = 93.9 kV)			
	Beaverlodge		Stony Rapids	
	kV	pu	kV	pu
TFDR Recording	125.5	1.34		
EMTP Simulations:				
1. Source impedance at Wellington 66 kV. Transformers with design data. Constant distributed parameter lines.	123	1.31	125	1.33
2. Source impedance at Wellington 66 kV. Transformers with test data. Constant distributed parameter lines.	127	1.35	131	1.40
3. Source impedance at Wellington 66 kV. Transformers with test data. Frequency dependent distributed parameter lines (JMARTI).	132	1.41	136	1.45
4. Synchronous machine models. Transformers with test data. Constant distributed parameter lines.	125	1.33	125	1.33

Table 1. Peak overvoltages for TFDR recording and EMTPs simulations

The constant distributed parameter line model was used in most studies, but some runs using frequency-dependent line models (JMARTI) showed very little difference with the low frequency harmonics in this case.

EMTP simulation Case #4 in Table 1 used the synchronous machine models for individual hydro units. The base models used one voltage source and an equivalent source impedance at the Wellington 66 kV bus. The simulations with the synchronous machine models gave results very close to the same as for the equivalent source. The difference in EMTP results using different models are small. All are within five percent of the recorded peak voltage at Beaverlodge. All of the EMTP simulations produced waveshapes which were very close to the waveshape shown in Figure 3. The harmonics appear to damp out in the attached TFDR recording sooner than in the simulation.

In conclusion, the EMTP program was a valuable tool to help determine the required equipment characteristics and acceptable operating procedures for the Athabasca System. The actual TFDR recordings were extremely valuable for verifying the validity of the EMTP models and results. In the Athabasca Study, the EMTP results compared quite well with the actual recordings.

Converter Transformer Switching Transients

The 2000 MW HVDC link between France and Great Britain has been in commercial operation since October 1986, following extensive commissioning tests that were carried out by Electricite de France (EdF) and Central Electricity Generating Board (CEGB). The first part of these tests consisted of the energization of all the EHV equipment of the first 1000 MW bipole, which includes two 618 MVA converter transformers and four 160 MVAR harmonic filter banks at the 400 kV busbar of Les Mandarins substation (Figure 4.)

The purpose of the simulation exercise was to check the protection settings and calculate the transient disturbances due to the switching-in of equipment or fault clearing near the station. Extreme configurations of the 400 kV network have been investigated on site. Short-circuit levels from about 4 GVA up to 14 GVA were available during the testing sequences, thus simultaneous energization of transformers and filter banks was possible.

Modeling with the EMTP

The converter transformers were modeled with the EMTP saturated transformer model. The representation of the surge arresters on the 400 kV busbars, on the valves, and on some filters components used the pseudo non-linear resistors available in EMTP. The ac system was represented in detail to simulate the field test as precisely as possible. The major difficulty was to give correct damping to the ac system impedance for a wide range of frequencies. Correct representation is necessary from dc (to damp the dc component of transformers inrush currents) to the natural frequencies of the ac system, which are repeatedly excited every time a transformer phase is saturated or desaturated. The number of parameters that can affect the waveshapes, with the switching instant and the remanent flux in transformers being the two most important, presents another difficulty in the validation task.

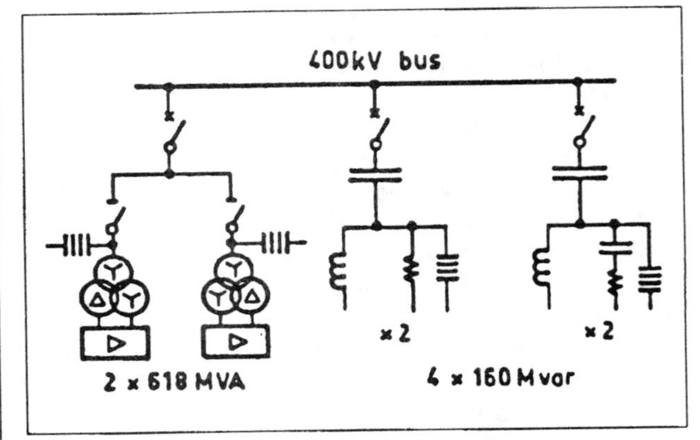

Figure 4. Converter station arrangement

Comparison Between Field Tests and Simulations

Figure 5 is an example of a comparison between the field test and the EMTP results for a sequence of a simultaneous energization of one filter bank and one converter transformer on a weak ac system. Simulations have also been carried out for 12 other switching sequences. Most of them give satisfactory results with maximum errors on peak values of voltages and currents of 10 percent compared with values recorded on site.

Capacitor Switching Study

This study simulated the back-to-back switching of new 115 kV 300 MVAr grounded-wye capacitor banks to be installed at four close locations in a metropolitan area. The study investigated various concerns in connection with the new installations: capacitor energizing transient overvoltage, propagation and amplification of the overvoltage at remote substations, transient recovery voltage, surge arrester energy duty, inrush and outrush currents.

EMTP Simulations

The study took full advantage of EMTP's modular capability. EMTP data modules describing various parts of the system were stored in a modular library. Diverse data cases could be quickly set up by referring

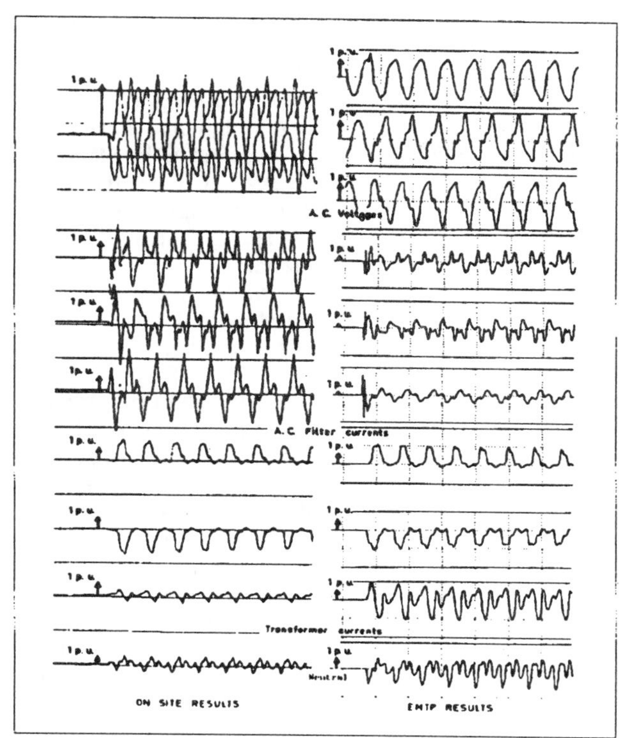

Figure 5. Energization of converter transformer and filter bank with weak ac system.

to these data modules. The represented system included 230 kV and 115 kV buses, 230-115 kV autotransformers, 115 kV lines, 96 kV metal-oxide arresters, 115 kV capacitor banks, current limiting reactors and capacitor switching devices. The model required positive and zero sequence data.

Statistical Switching

Capacitor bank energization was modeled with the STATISTICS switch of EMTP. Magnitudes of energizing transients depend on actual closing times of capacitor switching devices. The actual closing time of a breaker contact is somewhat different from the aiming point. This difference and the aiming point on the voltage waveform are of random nature. Figure 6 shows a summary page produced by the program for a statistical energization study. The summary page includes the plot of the cumulative probability function for the voltage of a switched capacitor bank. The plot shows that the probability of the voltage exceeding two per unit is about eight percent.

Energy duties of arresters near the 115 kV capacitor banks are determined by the energy dissipated when a restrike occurs at the peak of the transient recovery voltage. Figure 7 shows voltage, current, and energy of a 96 kV metal-oxide arrester located at a capacitor bank terminal following a

CAPACITOR BANK SWITCHING TRANSIENT

Back-to-back energization at COWLITZ
Switched bank 40 MVAR, fixed bank 20 MVAR

SWITCHED CAPACITOR BANK VOLTAGE
STATISTICAL SUMMARY OF 100 ENERGIZATIONS:

Mean:	1.882 P.U.
Standard Deviation:	0.138 P.U.
Maximum:	2.013 P.U.
Minimum:	1.369 P.U.

CUMULATIVE DISTRIBUTION:

PROBABILITY (%): 0.00, 20.00, 40.00, 60.00, 80.00, 100.00
PER UNIT VOLTAGE: 1.20, 1.40, 1.60, 1.80, 2.00, 2.20

Figure 6. Statistical summary produced for a capacitor bank voltage

restrike. The energy absorbed is within the arrester's design capability.

Acknowledgement

The EMTP Development Coordinator Group (DCG) presently comprises the Canadian Electrical Association (utility members), Hydro Quebec, Ontario Hydro, Western Area Power Administration, United States Bureau of Reclamation, ABB Power Transmission (associate member), and Central Research Institute, Electric Power Industry-Japan (associate member). In addition to the Electric Power Research Institute, both Electricite de France and the American Electric Power Service Corporation participate in EMTP development as EPRI associate members. All of the above organizations contribute significantly to this EMTP development financially, and/or with human resources.

For Further Reading

John Douglas, "Designing for Disaster," *EPRI Journal*, October-November, 1989.

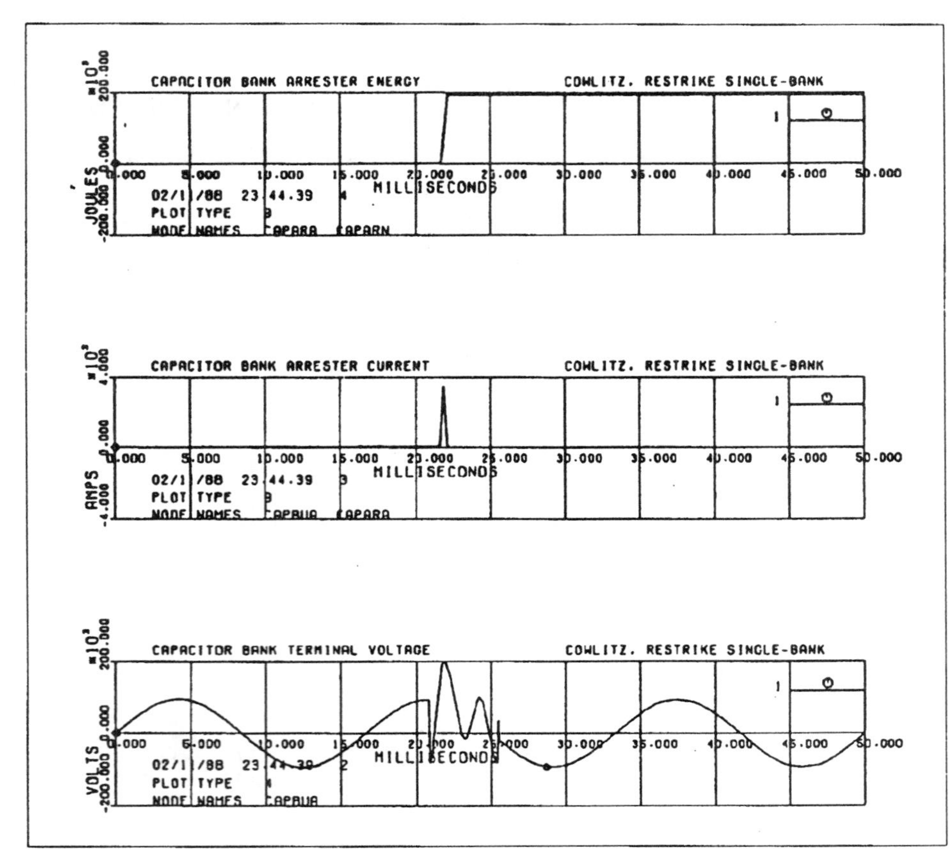

Figure 7. Arrester voltage, current, and energy after restrike

David Cotcher and Dan Ruiu, "EMTP Studies for the Athabasca Transmission Project," *EMTP Review*, January, 1990.

Ph. Adam, "Transients When Switching In Large Converter Transformers and Filter Banks," *EMTP Review*, July, 1988.

Sang Lee, "Capacitor Switching Transient Study Using EMTP EMPT REVIEW, October, 1988.

(The *EMTP Review* is a quarterly newsletter providing information on current developments in the DCG/EPRI EMTP. Subscription information is available from Professor Willis Long, Editor, University of Wisconsin-Madison, 432 N. Lake Street, Madison, WI 53706.)

Biographies

Willis Long is Professor, Departments of Engineering Professional Development and Electrical and Computer Engineering, University of Wisconsin-Madison. His research and teaching activities focus on the computer analysis of power systems, especially EMTP simulation of high-voltage direct-current systems. He is a Fellow of IEEE and a Registered Professional Engineer in Wisconsin.

David Cotcher obtained a degree in Electrical Engineering at the University of Saskatchewan. He started employment at SaskPower in 1976 in control design, and since 1979 has worked in the Planning Group on reliability studies, system stability, EMTP, and transmission planning. At present, he is Senior Engineer in Transmission and Interconnection Planning.

Dan Ruiu obtained a degree in Electrical Engineering, followed by post-graduate studies and degrees in power engineering and economics. He has worked with utilities, engineering consulting companies, government, and international organizations in Europe, the United States, and Canada. At present, he is Manager, System and Development Planning in SaskPower.

Philippe Adam received his degree of Engineer at the Ecole Centrale des Arts et Manufactures. As a research engineer at Electricite de France he has been involved in the studies and commissioning of the 2000 MW France-England dc link, and is now chief of the HVDC Transmission Studies Group. He is a member of IEEE, and convenes CIGRE Working Group 14-03.

Sang Lee did his graduate studies in Electrical Engineering at the University of Saskatchewan and the University of Wisconsin-Madison. He joined ASEA Power Systems Center in 1985 and has conducted various transient and harmonic studies for power transmission and distribution systems using the EMTP. He is now employed by Advanced Systems Technology, ABB Power Systems Inc.

Rambabu Adapa is a Project Manager at the Electric Power Research Institute, where among other activities he manages EMTP development and maintenance. He received his Ph.D. in Electrical Engineering from the University of Waterloo, Ontario, Canada. He is a member of IEEE and a Registered Professional Engineer in Wisconsin.

MODELS: A New Simulation Tool in the EMTP

L. Dubé, I. Bonfanti

Abstract

MODELS is an autonomous general-purpose simulation program recently interfaced to two versions of the Electromagnetic Transients Program (EMTP). This interfacing enables the description of components that could not be represented easily until now using the electrical elements of the EMTP and the set of control blocks available in its TACS (Transient Analysis of Control Systems) section. This paper evaluates the usage of MODELS in the EMTP by stressing the differences in approach and use as compared to TACS. An application example is provided and discussed.

1 Introduction

The TACS program [1], installed in the EMTP [2] in 1976 at BPA (Bonneville Power Administration, USA), had been designed for the representation of simple control systems associated with the operation of synchronous machines and HVDC converters in the EMTP at that time. It has been then employed by EMTP users for applications beyond its original scope, rapidly demonstrating its usefulness as well as its limitations in terms of model representation, model solution, and model initialisation. After a number of attempts to increase the usefulness of TACS in later years, mainly through the addition of more components and revisions of the sequencing and initialization algorithms, some ideas emerged and led to the design of a completely new program, based on a different approach for all three aspects of representing, solving, and initialising the modeled components.

The development of the MODELS language and its associated simulation solver was started in 1985 by Dubé, the author of TACS. An evaluation of the language was then conducted jointly with a group of EMTP users in 1988. This was followed by interfacing MODELS to BPA's EMTP in 1989, and to the royalty-free ATP [3] version of the EMTP in 1990 [4]. Although MODELS had been identified originally as "the new TACS", it has become apparent that it does not replace TACS. Rather, MODELS provides an alternate approach to the representation and simulation of new components and of control systems in the EMTP, with TACS continuing to play a useful role for the representation of simple control systems which can be described using the existing TACS building blocks.

The main characteristics of the MODELS approach are introduced from a user's point-of-view, with comparisons to the approach used in TACS. The three aspects of model description, model solution, and model initialization in MODELS are covered in detail. This is followed by a description of how MODELS and EMTP interact with each other. An application example is provided and discussed, illustrating the use of MODELS in the EMTP.

2 Background of Modeling in the EMTP

2.1 Simulation in the EMTP without Dynamic Control

An electric system is depicted in the EMTP as an arbitrarily-configured network of individual multi-terminal components (R-L-C branches, transmission lines, etc.). Each component describes the relation between the values of voltage and current present at its terminals over the time of the simulation. Although the components are described separately from each other, the shared values of voltages and currents at the components' terminals are determined simultaneously and globally at the successive instants of the simulation.

Variations in the operation of the system are represented either by variations in the circuit's topology (using the opening and closing of switches and the activation and deactivation of sources), or by variations in the operating function of certain components (using variable voltage and current sources, and variable impedance functions). The types of variations associated with these components are fixed and pre-programmed; only their parameters can be determined by the user. In addition, when not using TACS or MODELS, many of these parameters must be specified with fixed values prior to the simulation (for example, the closing time of a switch), as opposed to being made dynamically dependent on the ongoing results of the simulation.

The initialization of the variables of the network is program-calculated based on a steady-state operation, a completely-linear circuit (the non-linearities are excluded from the circuit), and a unique local frequency of the sources (no superimposed harmonics).

2.2 Simulation in the EMTP with Dynamic Control

The addition of dynamic control to the EMTP (using TACS or MODELS) gives the user the possibility to specify dynamic variations in the operation of the electrical components of the system according to arbitrary rules. The variations (switch opening/closing, operating values of sources, control of machines, and values of non-linear elements) can be determined directly by using control circuits (in TACS) or functional algorithms (in MODELS), which can themselves be made dependent on values of voltages and currents and on status of switches and machines observed in the electric system during the simulation.

One limitation to the application of this form of dynamic control to the electrical components is the non-iterative solution method of the EMTP itself. Because the linear part of the electric system is solved without iteration as a simultaneous set of equations, any control action based on the state of the electric system observed at one instant of the simulation will take effect only when evaluating the state of the electric system at the following instant of the simulation. In practice, this limitation has not been an obstacle to the use of dynamic control in the EMTP, with the time constants of the control loop generally being much larger than the one-step delay artificially introduced in the operation of a switch or machine or in the variation of a resistance or inductance value. More caution has to be exercised, however, when directly controlling the value of a voltage or current source, a situation equivalent to attempting to simulate the operation of an individual electrical component (rather than a control component) as a non-linear or non-simultaneous element without the benefit of iteration, in effect using prediction without immediate correction.

A more serious difficulty is the exclusion of all non-linearities by the initialization algorithm of the electrical part of the EMTP. Because this exclusion also applies to all dynamic control signals received from TACS and MODELS, the equivalent effect of the controls prior to the beginning of the simulation must be simulated by other means explicitly by the user in order to correctly initialize the electric network. This divided initialization of the electric components and the control system very often requires an adjustment period in the first instants of the simulation prior to obtaining a true representation of the steady-state condition of the complete system from which the simulation of the transient phenomena of interest can then be started.

2.3 Simulation in the EMTP with TACS

The mode of representation offered by TACS is that of a user-defined block-diagram arrangement of simple input-output components. A set of pre-defined types of building components are available, the parameters of their operation being specified by the user when configuring the system. The set of outputs from all the assembled components constitutes the complete set of variables of the control system, each variable being available for use as input to other components of the control system and as a control signal to components of the electric system.

The system represented in TACS is assumed to be mostly linear and mostly simultaneous in its operation, with non-linear elements and sequencing of elements treated as isolated exceptions. Similarly to the electrical side, an overall matrix of linear algebraic equations is formed for the simultaneous solution of all linear components. Unlike the treatment of the non-linear electrical components (solved together by iteration using a Thevenin-equivalent of the linear part of the circuit), the non-linear TACS functions (for example any of the supplemental devices) are assumed to be few and isolated from the linear part of the control system. They are solved sequentially without iteration, using as input the values of variables calculated either at the present or at the previous instant of the simulation, as determined by the sequencing algorithm responsible for the solution of the non-simultaneous part of the TACS system.

Similarly to the initialization of the electrical side, the initialization algorithm of the TACS program attempts to calculate the values of the simultaneous variables of the linear part of the TACS system at the initial time of the simulation, and requires the user to manually initialize the non-linear components. The algorithm can only represent DC and/or single-frequency steady-state operation.

2.4 Simulation in the EMTP with MODELS

In MODELS, all aspects of the representation, solution, and initialisation of the modeled component or control system are completely under user control.

The user represents a system by explicitly specifying the procedural algorithm which describes its operation. This is done by having the possibility of specifying how the components or sub-systems are interrelated structurally (using a hierarchy of models and sub-models), and how the structure of this composition can itself vary during the system's operation.

The user represents the operation of a system by defining a set of numerical and logical variables and by formulating the expressions, functions, sub-models, and algorithms which define their values dynamically during the simulation of the system's operation, with the possibility of explicitly stating the conditions regulating both assignment of the values and sequence of the procedures.

The initialization of a system and its sub-systems also can be specified by means of complete algorithms (using variables, conditions, etc.), allowing the user to build procedures that define the state of the system prior to the starting instant of the simulation with as detailed a past history as required.

3 Advantages of MODELS over TACS

3.1 Modeling Assumptions

TACS was designed for representing systems that were mostly linear, tightly coupled, and of fixed topology. Non-linear elements were considered exceptions and solved outside the overall set of simultaneous equations. Experience has shown that most systems are non-linear and loosely coupled, with operation discontinuities, and variable topology. The tightly-coupled elements, if any, are found in well-defined isolated sub-systems. These are the basic assumptions on which MODELS is based.

3.2 Composition

In TACS, all parts of the system are at the same level in the description space, creating difficulties in naming variables and in separating, testing, modifying, and re-using sub-components. MODELS provides a mechanism for easily structuring a system into separate sub-components. This allows decoupling of each subfunction with local internal naming, for easier construction, testing, modification, and re-use. It also permits to easily allocate the assembly of a large model among groups of developers.

3.3 Representation

In addition to the block diagram approach of TACS, MODELS accepts component descriptions in terms of procedures, functions and algorithms. The user is not limited to a pre-defined set of components, but can build libraries of components and sub-models as required by each application. Similarly, the user is not limited to an input-output signal flow representation, but can also combine this with numerical and logical manipulation of variables inside symbolic algorithms.

3.4 Description and Use of a Model

A clear distinction is made in MODELS between the description of a model and its use. The description contains a declaration part (specifying inputs, variables, sub-models, ...) and an execution part defining how the model operates. Once a model has been declared, it can be used as many times as required, each occurrence having its own set of inputs, its own history, and even its own solution step size. In addition, one can also apply the "use" mechanism of MODELS to models written in a different language (e. g. Fortran, C, ...).

3.5 Initialisation

MODELS gives powerful control of the initialisation of the state of a model to the user. The aspects of the initialisation that are not usage-dependent can be pre-built in the model description, in the form of a general procedure and of history-function assignments. The aspects that are dependent on the usage context can be defined at that stage. This offers the possibility to initialise the model differently depending on the many possible situations that can occur during the simulation process. This also allows a satisfactory solution to the initialisation problem when taking the model from a library and implementing it into an "a priori" unknown electrical network. In contrast with the partial calculation of initial values in TACS, MODELS does a normal execution of a model at initialisation, generating the complete definition of its state.

3.6 Self-Diagnosis

Self-diagnosis is especially important when large models have to be set up, debugged, and fine-tuned. Detection of the existence of potential mis-operation can be built directly into the model by the user, forcing an erroneous simulation to be halted via the ERROR statement with clear indication of the problem that occurred. Warnings and other observations can also be recorded during the simulation for later analysis.

3.7 Tools

There are explicit tools for easing the construction of a model. A high-level language is available for describing the operation of the model in a manner very close to how it is understood in the user's mind. This allows the user, together with the freedom of the format rules, to concentrate himself on the logic of the model without being distracted by the difficulties of implementing the thinking using lower-level tools. There are tools for the logic definition of the flow of operations (IF, WHILE, FOR, ...) and for specifying non-trivial mathematical expressions using differential equations (DIFFEQ), integrals (INTEGRAL), polynomials of derivatives (DERIVPOL), and Laplace functions (LAPLACE) among the many others.

3.8 Arrays

Among the many tools available in MODELS, an important one from the user's point of view is the possibility of handling variables by arrays, avoiding the repetition of the same procedure applied to the "*n*" variables of interest. A typical case is a three-phase electrical component that can be easily represented in a compact and easily readable form.

3.9 No Format Rules

As opposed to TACS, MODELS does not rely on identifier codes, fixed columnar formats and cryptic description rules for representing individual components and their interconnection. This facilitates readability and understanding. Only few explicit key words grouping homogeneous functions (INPUT, OUTPUT, VAR, DATA, ...) are needed. The free format of the description and the possibility of using arbitrarily-long names facilitates self-documentation, valuable especially when large systems and team work are of concern. Comments and illustrations can be introduced anywhere in a model description for clearer documentation.

```
BEGIN NEW DATA CASE
  5.E-08   4.E-4      50.
    100        1         1         1         1                              1
C ==========================================================================
C Example of a circuit-breaker with variable arc resistance.
C ==========================================================================
C Opening of a short-circuit current of 25 kA rms, V=15 kVrms, Q factor =70,
C capacitance in parallel to the breaker=1 nF, Zc at 1st discontinuity=798 Ohm
C ==========================================================================
MODELS                        -- begin the MODELS section in the EMTP datacase
INPUT                         -- inputs from the EMTP to MODELS
  curr1     {i(intmon)}          -- measured current through the series switch
  breakstat{switch(intmon)} -- status, open or close, of the series switch
OUTPUT                        -- outputs from MODELS to the EMTP
  resist                         -- controls the value of the arc resistance
  signal                         -- control signal to open/close the series switch
$INCLUDE ARCRESIS.MDL
USE arcresistance AS arc      -- use model "arcresistance" directly in the EMTP
  DATA                           -- assigning values to the model's data
    tau:=4.E-7, power:=50000, imax:=35355
  INPUT                          -- assigning values to the model's inputs
    current:=curr1, status:=breakstat, opencommand:=(t>0)
  OUTPUT                         -- assigning the model's outputs to MODELS outputs
    resist:=arcresis, signal:=openclose
ENDUSE                           -- end of this USE statement
RECORD                           -- to be recorded as EMTP printed/plotted values
  arc.conductance AS conduc
ENDMODELS                        -- end of the MODELS section in the EMTP datacase
C BRANCHES =================================================================
C <----><----><----><----><---R><---L><---C>
  SOURCEMACHIN              5.7E-3   0.4
  MACHINSOURCE              50930.
  MACHIN                      2.5        1.E-02
  MACHINREARCO              2.9E-3   0.2
  REARCOMACHIN              25465.
  REARCO                      25.        1.E-03
  INTVAL                    1.E-05
C CONTROLLED RESISTANCE
91REARCOINTMONTACS  RESIST                                                 2
BLANK CARD ENDING BRANCHES
C SWITCHES =================================================================
C <----><----><---------><---------><---------><---------><---------><----><----><-->
13INTMONINTVAL                                         CLOSED          SIGNAL  13
  REARCOINTVAL        -1.     5.E-08      50000.                                 2
BLANK CARD ENDING SWITCHES
C SOURCES==================================================================
C <---->  <----AMPL><----FREQ><---------><---------><---------><--TSTART><---TSTOP>
14SOURCE      21213.2       50.       178.                           -1.        1.
BLANK CARD ENDING SOURCES
BLANK CARD ENDING NODE VOLTAGE REQUEST
BLANK CARD ENDING PLOT
BEGIN NEW DATA CASE
BLANK
```

Listing 1. EMTP data case

4 Use of MODELS in the EMTP

4.1 General

The interaction between MODELS and the electrical side of the EMTP is identical to the interaction between TACS and the electric system in all aspects. Any of the quantities which can be observed in the EMTP by TACS can also be used as inputs to MODELS (node voltages, switch currents, status of switches, and machine variables). Any output variables from MODELS can be used in the EMTP in the same manner as TACS variables are used, for the control of voltage and current sources, the operation of switches, the operation of machines, and in the ATP version, the control of non-linear elements. Finally, any variable in MODELS can be recorded as a printed/plotted value in the EMTP, when specified as such in a RECORD declaration in MODELS.

4.2 Example of Application

The example illustrates the use of a model that simulates the opening and closing of a circuit breaker using a representation of the variable conductance of the arc during the opening operation. The accompanying circuit diagram (**Fig. 1**) shows a circuit breaker ("REARCO" to "INTVAL") represented by a MODELS-controlled type-13 switch ("INTMON" to "INTVAL") in series with the arc resistance represented by a MODELS-controlled type-91 variable resistance ("REARCO" to "INTMON"). The circuit breaker is initially in a closed position, and is then made to interrupt the current flowing through the short-circuit resistance placed between "INTVAL" and ground, after which time it remains open until the end of the simulation.

The closed circuit breaker is represented with its switch initially in a closed position and with a very low value applied to its series resistance. Another closed switch is placed in parallel with the two series elements of the circuit breaker in order to represent the condition of the circuit breaker during the steady-state initialization of the electric circuit (the EMTP does not include type-91 resistances in the steady-state initialization). This parallel

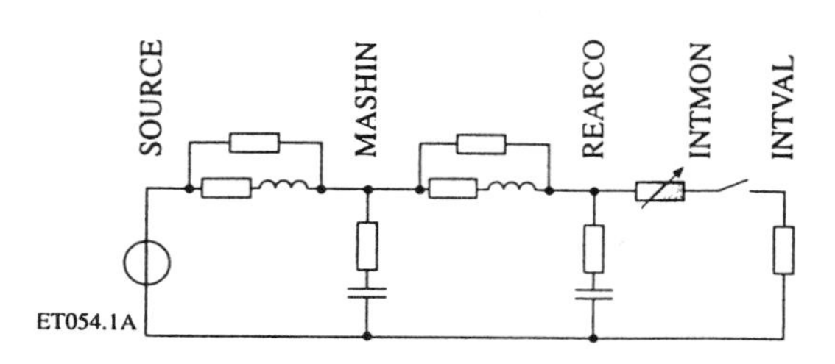

Fig. 1. Circuit diagram of the data case

switch is opened immediately at the first time-step of the simulation, and remains open through the rest of the simulation, monitoring at the same time the transient recovery voltage (TRV) across the breaker for plotting purposes. The circuit breaker model is made to open at the first time-step of the simulation by applying the logical value "$t > 0$" to its input signal "open-command". The application of the command initiates the simulated separation of the breaker's contacts and the building of the arc.

The variation in the arc conductance during the opening is represented by a differential equation,

$$g + \tau(\mathrm{d}g/\mathrm{d}t) = i^2/p$$

with

- g arc conductance,
- τ time constant of the arc,
- i current in the circuit breaker and
- p steady-state power loss of the arc.

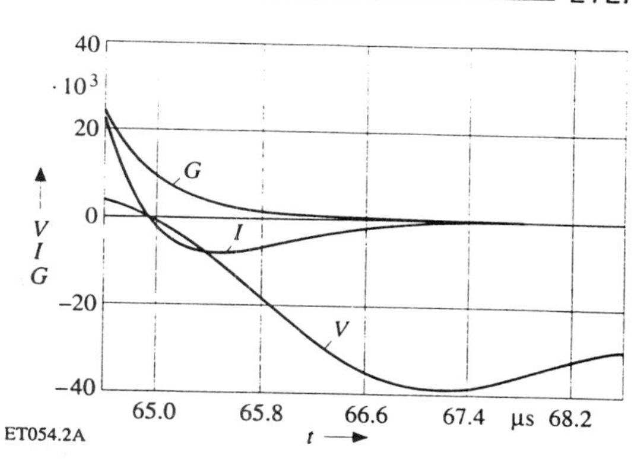

Fig. 2. Simulation results:
V voltage across the arc (multiplying factor: 3)
I current through the arc (multiplying factor: $2 \cdot 10^4$)
G conductance of the arc (multiplying factor: $3 \cdot 10^7$)

Plots of the voltage, current, and conductance of the arc (**Fig. 2**) during the opening operation show how the decreasing arc conductance fully interrupts the current in the circuit breaker without requiring the operation of the series switch, which is opened eventually only after the value of the conductance has become negligible (when "conductance ≤ minconduc", here at $t = 68.4$ μs).

The ATP data case corresponding to the above described example is presented in **Listing 1**. The items of major importance are the following:

- the inputs from the electric circuit into MODELS are "curr1" and "breakstat";
- the outputs from MODELS to the electric circuit are "resist" and "signal";
- one variable of MODELS is to be recorded as an EMTP printed/plotted value, under the name "conduc";
- the "arcresistance" model is used, defining its DATA parameters, and local inputs and outputs;
- the model is left outside of the data case; it is called from a library of models via an EMTP $INCLUDE command.

The definition of the used model, written in the MODELS language, is shown in **Listing 2**. The items of major importance are the following:

```
MODEL arcresistance           -- begin the description of the model
CONST                          -- constants of the model
  minconduc {VAL:1.E-7}          -- minimum conductance value
DATA                           -- data to be assigned when using the model
  power  {DFLT:60000}            -- steady-state arc power loss
  imax   {DFLT:30000}            -- peak of the sine wave current in the breaker
  tau    {DFLT:5.E-6}            -- arc time constant
INPUT                          -- inputs to be assigned when using the model
  current                        -- arc current
  status                         -- status of the series switch
  closecommand {DFLT:false}      -- optional command to immediately close the breaker
  opencommand  {DFLT:false}      -- optional command to start opening the breaker
VAR                            -- local variables of the model
  conductance                    -- arc conductance
  arcresis                       -- arc resistance (the inverse of conductance)
  driving                        -- driving function
  openclose                      -- signal to open/close the series switch
  isopening                      -- flag showing that the breaker is opening
OUTPUT                         -- outputs available when using the model
  arcresis
  openclose
TIMESTEP MAX: tau/8            -- limiting the timestep to not greater than tau/8

INIT ---------------------- the initialization procedure of the model ---------
  IF status=open THEN openclose:=open;   conductance:=minconduc
                 ELSE openclose:=closed; conductance:=imax**2/power
  ENDIF
  arcresis:=recip(conductance); isopening:=false
ENDINIT

EXEC ---------------------- the execution procedure of the model ---------------
  IF status=open AND closecommand THEN ---------------------- close immediately
    openclose:=closed; conductance:=imax**2/power; arcresis:=recip(conductance)
  ELSIF status=closed AND opencommand THEN ---------------------- start opening
    isopening:=true
  ENDIF
  IF isopening THEN ------------- continue opening until conductance<=minconduc
    driving:=current**2/power
    DIFFEQ( 1.0|D0 + tau|D1 )|conductance:=driving
    IF conductance<=minconduc THEN ----------------------- opening is completed
      openclose:=open; conductance:=minconduc; isopening:=false
    ENDIF
    arcresis:=recip(conductance)
  ENDIF
ENDEXEC
ENDMODEL                          -- end of the description of this model
```

Listing 2. Circuit-breaker model using the MODELS language.

- the constant "minconduc" is defined and assigned a value;
- data parameters are defined and assigned default values; these values may be redefined in the USE of the model, as shown in Listing 1;
- four inputs are defined, two of which being assigned default values; notice that only three inputs are defined in the USE statement, the fourth one keeping its default value;
- variables used in the procedure must be declared; the names are chosen to be representative of their function;
- of all variables of the model, only two are designated as outputs, and used in the USE statement;
- a maximum is specified for the solution step of the model, limiting its value to a fraction of the data parameter "tau" that is defined dynamically when the model is used;
- an initialization algorithm is defined in the model as a procedure distinct from the execution procedure, illustrating the possibility of using conditional assignments during initialisation;
- the functional operation of the model is given in the EXEC procedure; it calculates the value of the variable "arcresis", using some of the tools available in the MODELS language;
- the key words of the language are shown in the example in capital letters;
- the specific format and the placement of the comments are chosen arbitrarily to suit the purpose of this example.

5 Conclusions

The usage of MODELS in the EMTP has been evaluated, stressing the differences in approach and usage as compared to TACS. A simple example has been presented and discussed, illustrating the main aspects of the definition of a model and of its use in the EMTP. It is believed that MODELS is a more powerful and general tool and offers more flexibility than TACS in representing the type of components and control systems needed in EMTP simulation.

References

[1] *Dubé, L.; Dommel, H. W.*: Simulation of Control Systems in an Electromagnetic Transients Program with TACS. IEEE Power Engng. Soc. (PES) Power Ind. Comput. Appl. (PICA) 1977, Conf.-rec. vol. 10, pp. 266–271

[2] *Phadke, A. G. (Ed.)*: Digital Simulation of Electrical Transient Phenomena. IEEE Tutorial Course (1981) no. 81 EHO173-5-PWR

[3] ATP Rule Book. Leuven: EMTP Center, 1990

[4] *Dubé, L.; Bortoni, G.; Bonfanti, I.*: The EMTP's New TACS. 16th Eur. EMTP Meeting Leuven (EMTP Center) 1989, Proc. pp. 89–11

[5] *Martinez Velasco, J. A.; Capolino, G. A.*: TACS and MODELS: Drive Simulation Languages in a General Purpose Program. 3rd IEEE Workshop MCED'91, 1991, Proc.

Manuscript received on October 22, 1991

The Authors

Laurent Dubé (1949) received his B.A. and his Electrical Engineering degree from Sherbrooke University, Québec, in 1967 and 1972, and his M.Appl.Sc. from Ecole Polytechnique de Montréal in 1973. From 1973 to 1976, he studied at the University of British Columbia and developed the TACS program, which he installed in the EMTP in 1976. He spent the next ten years pursuing independent studies in psychology, microcomputer architecture, and software engineering, while working for the Canadian Coast Guard as a lighthouse-keeper. From 1985 to 1988, he developed the MODELS language and program, which he added to BPA's EMTP in 1989, and to the ATP in 1990. He is currently working as a consultant on simulation projects with CESI and BPA, while continuing the development of MODELS. He is also an ongoing contributor to the development of the ATP with the Leuven EMTP Center. He is a member of the IEEE Power Engineering Society, the IEEE Computer Society, the Society for Computer Simulation, and the Association for Computing Machinery. (6442 S.W. Barnes Road, Portland/Oregon 97221, USA, T + 15 03/2 92 65 29, Fax + 15 03/2 74 03 41)

Ivano Bonfanti (1958) received his Electrical Engineering doctor degree in 1983 from Polytechnic of Milano, Italy. In the Network Study Division of CESI since 1982, he has been involved in system studies in general, with emphasis on electromagnetic transients. He is CESI's representative to LEC (Leuven EMTP Center) and was a faculty member for the 1987, 1989 and 1991 EMTP Summer courses held in Leuven. In High Power Division of CESI since 1989, he has also been involved in various test and study activities related to AC-DC converters, SVC and circuit-breakers. He is a member of the CIGRE 13/14.08 Working Group, CIGRE 14.01.02 TF, Italian member of IEC 22F WG 7 and IEC 22F WG6, and is involved in the activity of CIGRE 14.01 TF 03. He is also secretary of SC 22F and member of SC 17A of CEI, the Italian Electrotechnical Committee. He currently works in the High Power Division of CESI in the position of Departmental R&D Assistant. (CESI, via Rubattino 54, I-20134 Milano, T + 3 92/2 12 52 95, Fax + 3 92/2 12 54 40)

Creating an Electromagnetic Transients Program in MATLAB: MatEMTP

Jean Mahseredjian (IEEE member)
Institut de Recherche d'Hydro-Québec (IREQ)
1800 Montée Ste-Julie
Varennes, Québec, Canada J3X 1S1

Fernando Alvarado (IEEE fellow member)
University of Wisconsin-Madison
Electrical & Computer Engineering
1415 Johnson Drive, Madison, WI 53706, USA

ABSTRACT: The traditional method for developing electric network analysis computer programs is based on coding using a conventional computer language: FORTRAN, C or Pascal. The programming language of the EMTP (Electromagnetic Transients Program) is FORTRAN-77. Such a program has a closed architecture and uses a large number of code lines to satisfy requirements ranging from low level data manipulation to the actual solution mathematics which eventually become diluted and almost impossible to visualize. This paper proposes a new design idea suitable for EMTP re-development in a high level programming context. It presents the creation of the transient analysis numerical *simulator* MatEMTP in the *computational engine* frame of MATLAB. This new approach to software engineering can afford a dramatic coding simplification for sophisticated algorithmic structures.

Keywords: EMTP, MATLAB, time-domain network analysis, software engineering

1. INTRODUCTION

In a conventional electric network *simulator* design, everything is based on line-by-line coding. Every component is implemented this way, as is the network analysis algorithm and any minor details of the overall computation and data manipulation process. The actual network model equations and network matrix operations are diluted in a large number of cryptic code lines created by specialized and experienced developers. Moreover, old-fashioned and historically supported programming techniques inhibit modularity and are geared towards memory conservation. Models for any one component appear in more than one place in the code. This is the case of the EMTP [1] (Electromagnetic Transients Program) code. The low level design methodology of such a code explains its low renewal and enhancement rate. It is also prohibitive to experiment with modern algorithmic ideas for eliminating solution limitations or for improving the computational speed on changing computer architectures.

Most network solution and modelling methods are simple to visualize and support mathematically, but their translation into an actual large scale working code is complex. Commonly used programming languages are ill-suited to human abilities for dealing with complexity. Software built using such languages is often inadequate. Some other new languages such as ADA, C++ and FORTRAN-90, provide powerful features for the formulation of appropriate abstractions [2] for the desired application. But programming is always easier if a specialized language is already available for the creation of similar applications. Specialized applications should use dedicated *computational engines* where the developer can build and compose with high level constructs. In addition to defining a new library of functions and overloading existing operators, such an engine must provide a minimal number of portable graphical data visualization and manipulation functions. It is obvious that programming a *computational engine* from scratch is a major effort.

This paper proposes to use a widely used general purpose program available on most popular computer platforms as a *computational engine*: MATLAB [3]. MATLAB has a large number of built-in functions and constructs covering a wide range of EMTP development needs and is expandable by means of optional toolboxes. The recent implementation of sparse matrix manipulation capabilities eliminates a major feasibility barrier.

This paper presents the creation of MatEMTP: a transient analysis program in MATLAB M-files. It is based on a new formulation of the main system of network equations, designed to eliminate several topological data restrictions and capable of handling arbitrary switch interconnections. The existing EMTP is used for validation and as a reference for solution timings.

96 WM 098-4 PWRD A paper recommended and approved by the IEEE Transmission and Distribution Committee of the IEEE Power Engineering Society for presentation at the 1996 IEEE/PES Winter Meeting, January 21-25, 1996, Baltimore, MD. Manuscript submitted August 1, 1995; made available for printing December 5, 1995.

2. SOLUTION METHOD

2.a Fundamental principles

The basic time-domain solution method implemented in MatEMTP is similar to the existing EMTP approach [4]. A large set of algebraic-differential equations is first transformed into a discrete algebraic equivalent and then solved over the requested interval $[0, t_{max}]$. The solution is available at discrete time-points $(0, t_1, t_2, \ldots, t_{max})$. The design utilizes a fixed integration time-step Δt as is the case for EMTP.

The high level matrix manipulation capabilities of MATLAB stimulate algorithmic ideas based on matrix computations. MatEMTP uses matrices and vectors for coding and solving network equations, closely replicating the underlying mathematics of network theory. The core code operates by defining a larger and more general matrix to represent network equations than is customary.

2.b Network equations: the core code

The network component interconnecting equations constitute the core code equations and must be defined before accordingly programming the individual component models. The following augmented sparse formulation is used:

$$\begin{bmatrix} \mathbf{Y_n} & \mathbf{V_a}^t & \mathbf{S_a}^t \\ \mathbf{V_a} & \mathbf{0}_{V_s} & \mathbf{0}_{V_sS} \\ \mathbf{S_a} & \mathbf{0}^t_{V_sS} & \mathbf{S_0} \end{bmatrix} \begin{bmatrix} \mathbf{V_n} \\ \mathbf{I}_{V_s} \\ \mathbf{I_S} \end{bmatrix} = \begin{bmatrix} \mathbf{I_n} \\ \mathbf{V_s} \\ 0 \end{bmatrix} \quad (1)$$

where $\mathbf{Y_n}$ is the standard $n \times n$ nodal admittance matrix excluding switches, $\mathbf{V_a}$ is the $nV_s \times n$ node incidence matrix of voltage sources, $\mathbf{S_a}$ is the $n_S \times n$ node incidence matrix of closed switches, $\mathbf{0}_{V_s}$ is an $nV_s \times nV_s$ null matrix, $\mathbf{0}_{V_sS}$ is an $nV_s \times n_S$ null matrix, $\mathbf{S_0}$ is an $n_S \times n_S$ sparse binary matrix used to nullify open switch currents, $\mathbf{V_n}$ is the vector of unknown node voltages, $\mathbf{I}_{V_s}$ holds the unknown voltage source currents, $\mathbf{I_S}$ holds unknown switch currents, $\mathbf{I_n}$ holds known nodal current injections and $\mathbf{V_s}$ stands for known source voltages.

This new formulation is less restrictive than the standard EMTP nodal analysis. It expands modified nodal analysis [5] by including explicitly the switch equations. Equation (1) is used in both steady-state and time-domain solutions. The node incidence switch matrix $\mathbf{S_a}$ is modified to avoid the reformulation of $\mathbf{Y_n}$ when the topology changes. MatEMTP can model voltage sources not connected to ground, floating switch nodes and branch to branch relations. All switch currents are automatically calculated and the explicit switch matrix $\mathbf{S_a}$ usage simplifies the detection of illegal switch loops. A switch loop creates linearly dependent rows in the switch matrix $\mathbf{S_a}$. This dependency is deleted by removing redundant closed switches.

The steady-state solution is a frequency domain solution. Its objective is to initialize the time-domain solution when steady-state conditions exist before transient analysis. MatEMTP can handle a fundamental frequency and harmonic initialization [6].

2.c Component models

Network models consist of an interconnection of component models. Component models interact with the core code by inserting their frequency domain and time-domain equations into (1). Node incidence matrices are used for formulating the interconnection of component model equations.

For a passive component, the frequency domain requires a complex admittance matrix at each solution frequency. Active components insert their voltage or current phasors in the right hand side of (1).

The time-domain solution is based on the discretization of the component models. Although trapezoidal integration is the default discretization method, other integration methods such as Backward Euler are applied in individual model equations, as long as compliance with core code requests exists. In addition to handling discontinuities [7], Backward Euler integration is useful for startup from user-defined initial conditions.

Several components of the same type (same model) usually exist in a given network. Fig. 1 shows the ith element of a multiphase coupled component model. The following equation can be written for this component type during the time-domain solution:

$$\mathbf{I_x} = \mathbf{Y_x}\mathbf{V}_{km} + \mathbf{I}_{x_h} \quad (2)$$

Bold characters denote matrices and vectors. Subscript h stands for history terms. Matrix $\mathbf{Y_x}$ is a sparse block-diagonal admittance matrix containing individual matrices $\mathbf{Y}^i_x$.

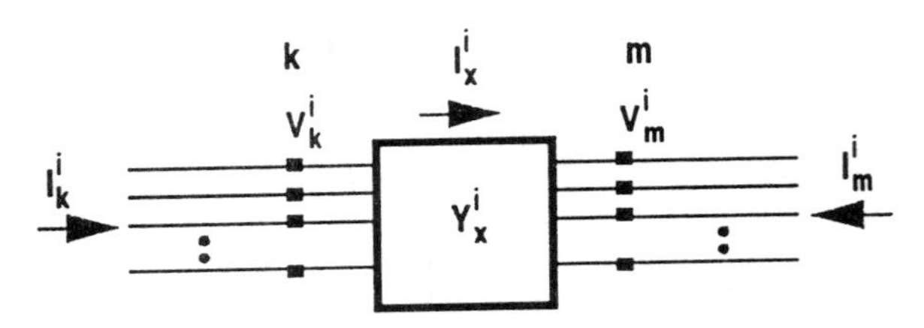

Figure 1: A coupled multiphase component model

If all component types possess their own sparse node-incidence matrix $\mathbf{M_a}$, equation (2) is inserted into (1) using the following formulas:

$$\mathbf{Y_n}^{after} = \mathbf{Y_n}^{before} + \mathbf{M}^t_a\mathbf{Y_x}\mathbf{M_a} \quad (3)$$

$$\mathbf{I_n}^{after} = \mathbf{I_n}^{before} - \mathbf{M}^t_a\mathbf{I}_{x_h} \quad (4)$$

3. THE MatEMTP CODE

3.a Main structure

The objective is to program MatEMTP using only MATLAB M-files [3]. These files include standard MATLAB statements and may also refer to other M-files. An M-file is an

ASCII script or function file. Since these files are run directly in the MATLAB environment and there is no requested compilation stage, MatEMTP inherits an open source code.

In inexperienced hands, the large number of available MATLAB building functions and constructs, can result in inefficient and cryptic code. Some experience is needed for programming with a minimal number of code lines and for minimal CPU time. The key to minimal CPU time is the vectorization of the solution algorithms. Other important rules to follow for increased efficiency are: avoid extreme modularity; use function files instead of script files; minimize the number of logical statements for model and option selections; minimize data initialization; avoid data storage pointers; preallocate vectors and matrices of predictable size. Except for memory preallocation, vectorization and the above outlined rules actually improve code readability and simplicity. Blind usage of dynamic memory allocation simplifies programming but places a heavy burden on the MATLAB interpreter.

By programming through matrix and vector operations the MatEMTP code is naturally vectorized. To eliminate useless testing, initialization procedures and repetitive dead code executions, the ready-to-run structure of Fig. 2 is proposed. This data adaptable structure relies on the *input processor* to interconnect the M-files. The *input processor* is a separate program (also written with M-files) that decodes standard EMTP data files [1] and creates the *case.m* file. This file is a processed file of network data created from the external *case data* format.

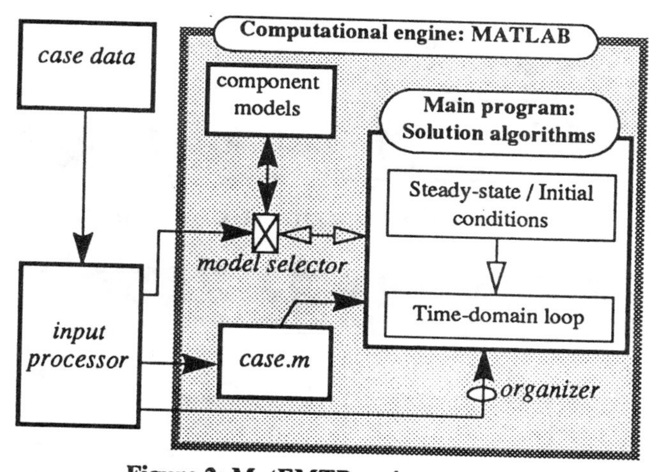

Figure 2: MatEMTP main structure

The *model selector* is a set of M-files created by the *input processor* for connecting required *case.m* models to the main program. All models are programmed in separate M-files that obey to a set of predefined core code requests. A typical request for a component model is "*provide admittance matrix*" or "*update history*". The creation of any new model is as simple as programming a new M-file which is automatically recognized and inserted into the appropriate code location by the *model selector*.

The *organizer* is another M-file created by the *input processor* that calls solution M-files according to selected options and overall solution needs. Thus, MatEMTP is based on a data dependent interconnection of individual code modules. Here is a valid sequence of files called in by the *organizer* for solving a typical case *case.m*:

1❐ *matemtp.m*: program startup and request for data case
2❐ *case.m*: the actual case file, any name can be used
3❐ *start.m*: initial setups, initial conditions, initialization of the time-domain solution
4❐ *timeloop.m*: the time-domain loop for the simulation

The *start.m* script file initializes all network variables (including automatic frequency domain initialization for any subnetwork where active sources exist at $t < 0$).

Appendix A shows a section of code called from *start.m* for linear harmonic initialization. The listing for *timeloop.m* shown in Appendix B demonstrates the advantages of programming within the *computational engine* frame of MATLAB.

Since all component models appear hidden to the main MatEMTP code, the *model selector* can only communicate through 3 built-in generic function files: *msouroe.m*, *mbranch.m* and *mswitch.m*. A file *mglobal.m* is used to transfer data from the main code to model function files.

The simple test circuit of Fig. 3 demonstrates the above outlined functionality. The contents of automatically generated *test2iwh.m* (this is now *case.m*) are listed in Appendix C. The names of model M-files used in this circuit are available in an *input processor* library. The *model selector* consists of the following files created by the *input processor*:

mglobal.m: (called in from *matemtp.m*)

```
gvsine; %sinusoidal voltage source data
gisine; %sinusoidal current source data
rlcglob;        %RLC model data
sw0glob;        %ordinary switch model data
```

msource.m:

```
function msource(ido)
vsine(ido);    %sinusoidal voltage source
isine(ido);    %sinusoidal current source
```

mbranch.m:

```
function mbranch(ido)
rlcmod(ido);   %RLC model
```

mswitch.m:

```
function mswitch(ido)
sw0(ido); %ordinary switch model
```

As an example of model data connection file, here are the contents of *gvsine.m*:

```
global Vadj Vsinein Vmag Vstart Vstop Vphi Vw;
```

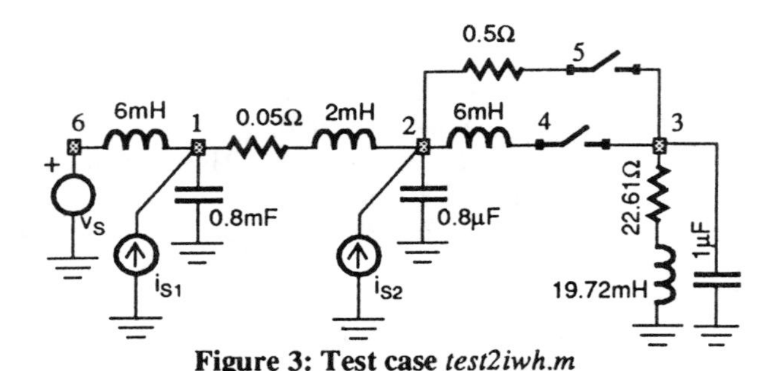

Figure 3: Test case ***test2iwh.m***

3.b Programming the component models

Every component model is located in a separate M-file and responds to a standardized number of *ido* values sent to it by the core code. As an example, an *ido=2* requests the insertion of the component model admittance matrix into $\mathbf{Y_n}$.

To illustrate the simplicity of programming, here is a portion of code from *rlcmod.m*:

```
if ido==2  %insert into Yn for steady-state
  Yn=Yn+RLCadj'*sparse(1:nRLC,1:nRLC,
    1./(RLCR+jz*(w*RLCL - RLCC/w ) ) )*RLCadj;
elseif ido==5 %insert into Yn in time-domain
  GRLC=sparse(1:nRLC,1:nRLC,
    1./(RLCR+(2/Dt).*RLCL + (Dt/2).*RLCC));
  Yn=Yn+RLCadj'*GRLC*RLCadj;
elseif ...
```

The programming of a transmission line model [4] usually requires the implementation of pointers for holding and updating history. MatEMTP avoids this complexity by using a single two dimensional sparse array for holding history and a sparse rotation vector for extracting and storing history terms at each time-point. This is best demonstrated by the following self explanatory code lines taken from the lossless single phase transmission line model (*tlmod.m*):

```
elseif ido == 6 %insert into In in time-domain
 Tikh=(1-Tinter).*Tikhist(:,1)+
    Tinter.*Tikhist(:,2); %k side history
 Timh=(1-Tinter).*Timhist(:,1)+
    Tinter.*Timhist(:,2); %m side history
 In=In+Tadjk'*Tikh; %contribute to In
 In=In+Tadjm'*Timh;
elseif ido == 7 %update history
  Tikh=Tadjk*Vn./TZc-Tikh;     %ik=vk/Zc-ikh
  Timh=Tadjm*Vn./TZc-Timh;     %im=vm/Zc-imh
  Tikhx=Tadjm*Vn./TZc+Timh;    %ikh=vm/Zc+im
  Timhx=Tadjk*Vn./TZc+Tikh;    %imh=vk/Zc+ik
  Tikhist=Tikhist*Trotate+
   spconvert([(1:nT)', TNhist, Tikhx]); %store
  Timhist=Timhist*Trotate+
   spconvert([(1:nT)', TNhist, Timhx]); %store
elseif ...
```

The transmission line is connected between nodes k and m. The following arrays and variables are calculated for *ido=1* in *tlmod.m*: `nT` is the total number of single phase lossless transmission lines, `Tinter` is an interpolation vector according to the propagation delay of each line, `TNhist` is a vector holding the number of history cells required for each line, `Trotate` is the sparse rotation matrix, `Tadjk` and `Tadjm` are sparse node incidence matrices found from the main node incidence matrix `Tadj` of this line model. Only minor modifications are needed to incorporate lumped resistances for losses [4].

Since equations (2) to (4) are applicable to any number of phases, programming of multiphase component models is based on matrix manipulations similar to single phase models.

3.c User interface

MATLAB provides high level functions that enable a portable programming of a graphical user interface (GUI) for MatEMTP. Fig. 4, for example, shows the GUI appearing during the initial program startup procedure. It is used to modify basic simulation data and options. The menu item Schematic opens the schematic capture GUI shown in Fig. 5.

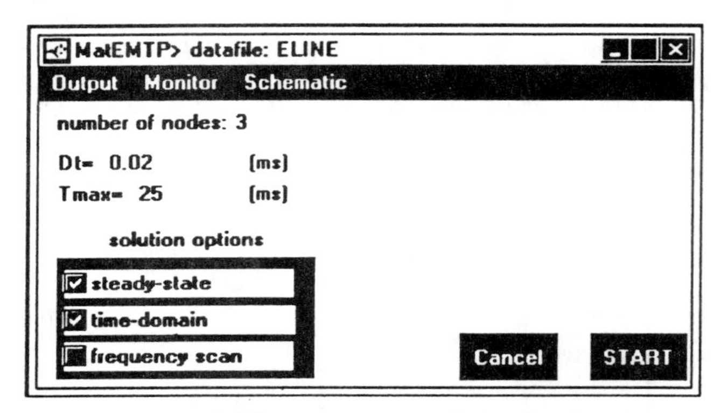

Figure 4: The initial data capture GUI of MatEMTP

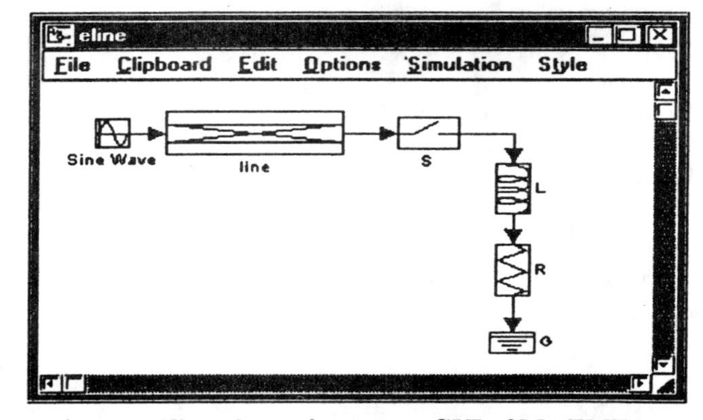

Figure 5: The schematic capture GUI of MatEMTP

The GUI of Fig. 5 can be used for creating and modifying

4/8

an arbitrary multiphase circuit diagram. Clicking on a given component opens the corresponding data capture panel. The programming of this GUI is based on the MATLAB-SIMULINK [8] toolbox. Any new component icon or subcircuits can be created through block masking [8]. Available network list generation functions allow the translation of a circuit diagram into the actual data case M-file. It must be remarked that the SIMULINK GUI was originally created for assembling control circuits and its usage for circuit diagrams suffers from visual limitations such as obligatory arrows and boxed blocks.

4. TEST CASES

4.a Case 1

The circuit diagram of this test case is shown in Fig. 3. The standard EMTP cannot handle steady-state initialization with different source frequencies in the same subnetwork, and according to *test2iwh.m* (see Appendix C) the harmonic current sources i_{S1} and i_{S2} are connected for $t < 0$. Thus, EMTP starts with wrong initial conditions (both current sources disconnected in the 60Hz initialization) and enters almost perfect steady-state only after 8s of simulation time. MatEMTP solves this case directly through its initialization algorithm (Appendix A). Fig. 6 superimposes an EMTP waveform delayed by 7.95s to the MatEMTP waveform. Both solutions are undistinguishable.

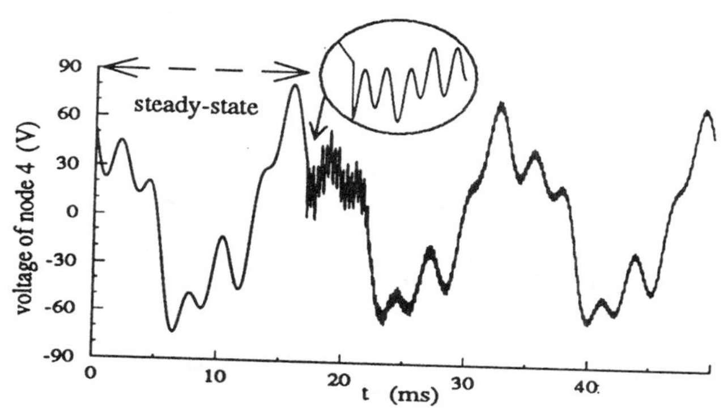

Figure 6: MatEMTP and shifted EMTP solutions, Case 1

4.b Case 2

This test case is taken from an EMTP Workbook [9]. It simulates the energization of a three-phase 15 mile 230kV cable. The circuit diagram for phase a is shown in Fig. 7. The cables are represented using 9 two-phase pi-sections. The sheath is grounded at the sending end and at each pi-section. A partial listing for *pi.m* is given in Appendix D. The simulation results from EMTP and MatEMTP shown in Fig. 8 are perfectly identical.

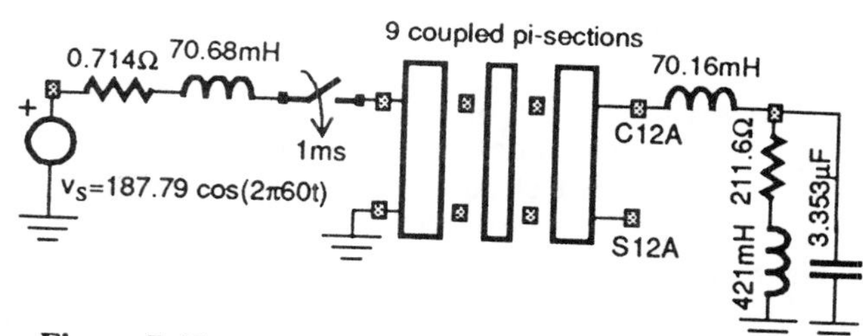

Figure 7: Test case *pi.m*: energization of cable phase a.

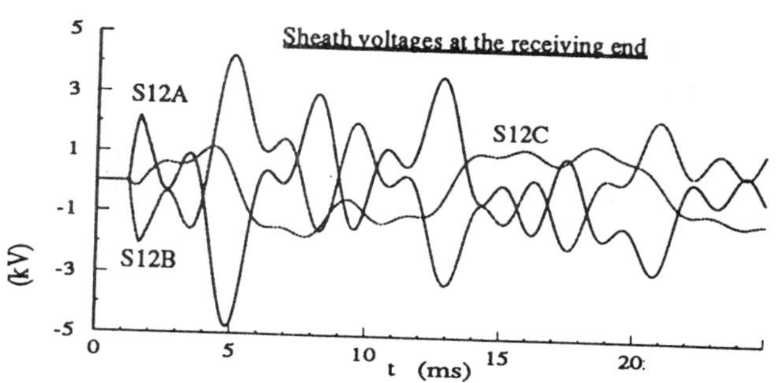

Figure 8: MatEMTP and EMTP solutions, Case 2

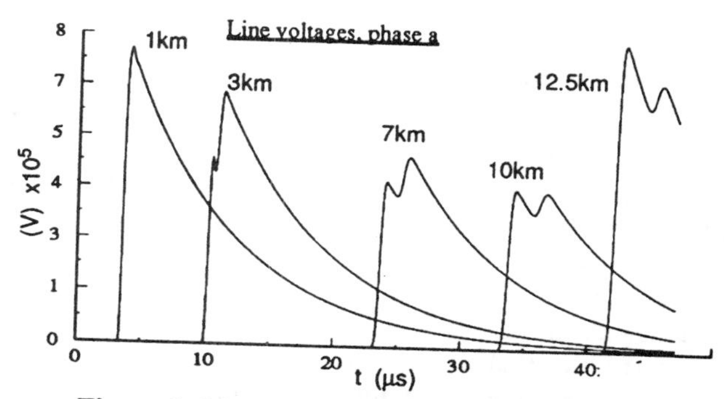

Figure 9: MatEMTP and EMTP solutions, Case 3

4.c Case 3

The objective of this case is to validate and test the performance of MatEMTP distributed parameter line modelling where a sparse matrix based history maintenance method has been proposed. The line setup taken from [10] is used for corona modelling, each phase is subdivided into 500 sections and a total of 1503 nodes is created. Only phase a is energized with a surge voltage function [10]. Simulation results are shown in Fig. 9. Since MatEMTP does not yet possess a corona model, corona branches are disconnected in EMTP and a constant distributed parameter line model is used. Zooming on these waveforms will show a minor Δt delay between EMTP and MatEMTP, related to the programming of the surge function in EMTP-TACS [1], MatEMTP is actually more precise.

5. DISCUSSION

The short length of Appendices A and B indicates that only a small number of code lines is needed to express solution procedures and elaborate sparse matrix manipulations through readily available MATLAB functions and constructs. This is a dramatic improvement over conventional coding for performing similar tasks. Data output and plotting are easily handled through available MATLAB functions.

The next step in this paper is to compare MatEMTP computational performance against EMTP. The *eratio* is defined as total MatEMTP elapsed execution time over EMTP execution time.

For Case 1 EMTP needs a much longer simulation time, and the found *eratio* of $\cong 0.25$ is in favor of MatEMTP. This *eratio* is achieved only after modifying EMTP [11] to disable plot data storage before 7.95 seconds.

If the EMTP initilization time is excluded, the *eratio* becomes 2.5 for $\Delta t = 50\mu s$ and 10 for $\Delta t = 10\mu s$. MatEMTP CPU time is an almost linear function of the total number of solution steps.

Case 2 has an *eratio* of $\cong 7$. The case 3 *eratio* is $\cong 6$. If the number of line sections is dropped to 500 (by deleting phases b and c) then *eratio* $\cong 5.7$. This relative insensitivity to network dimensions is the result of both methodologies using sparse matrix techniques.

A detailed analysis of MatEMTP CPU usage in the time-step loop for the typical case of Fig. 3, shows the following disposition: less than 15% for LU factorization and triangular solution, close to 60% for updating the right hand side of equation (1) and the remaining is for individual model updates. Half of that 60% is drained by the source function *msource.m*. A promising possibility is the replacement of such functions by compiled C language MEX-files [12], but this should be applied only at the last stage of programming. Another possibility is to resort to an automatic M-file compiler.

CONCLUSIONS

This paper has demonstrated an implementation of a comprehensive electromagnetic transients analysis program using MATLAB as a *computational engine*.

Used algorithms provide results identical to those from the EMTP. However, the proposed environment is implemented in very few lines of code, is easily expandable, modifiable and highly portable. It also eliminates EMTP modelling limitations through a less restrictive formulation of main network equations.

Although in a few cases the new environment is faster, in general studies show that the conventional coding retains a speed advantage ranging from 2.5:1 in the best case to 10:1 in the worst case. Ideas for reducing this ratio have been proposed.

The ultimate contribution of this paper is a dramatic illustration of the possibilities afforded by this new approach to software development.

APPENDIX A

MatEMTP linear initialization module: *steadylin.m*

The following is a listing of *steadylin.m*:

```
Wall=[];
msource(5); %put all source ws in Wall
Wall=sort(Wall);
Vn_init=zeros(n,1);%preallocate n node voltages
IVs=zeros(nVs,1);      %preallocate IVs
IS=zeros(nS,1);        %preallocate IS
nfreq=size(Wall,1);    %the number of ws to do
ifreq=1; wdone=[];
while ifreq <= nfreq
 w=Wall(ifreq);
 if w ~= wdone
   steady1; %(see code below)
   wdone=w;
   mbranch(3); %accumulate steady-state at t=0
 end
 ifreq=ifreq+1;
end
Vn=Vn_init;  %solution at t=0
```

The following lines are from *steady1.m*:

```
%steady-state module step for the frequency w
Yn=sparse(n,n); %Build Yn
mbranch(2); %contribution to Yn by branch models
Ytmp=[Yn Vadj'; Vadj sparse(nVs,nVs)];
Stmp=sparse(1:nS,1:nS,Sactive)*
   [Sadj  sparse(nS,nVs)]; %active switches
Sz  =sparse(1:nS,1:nS,~Sactive);
Yaug=[Ytmp Stmp'; Stmp Sz ];
%
In=zeros(n,1);     %n is the number of nodes
msource(3); %put sources in Vs and In for w
Itmp=[In; Vs];
Iaug=[Itmp; zeros(nS,1)]; %account for switches
Vaug=Yaug\Iaug;            %compute unknown phasors
Vn=Vaug(1:n);              %nodal phasor voltages
Vn_init=real(Vn)+Vn_init;       %at t=0 accumulate
IVs=real(Vaug(n+1:n+nVs))+IVs; %v source currents
IS=real(Vaug(n+nVs+1:n+nVs+nS))+IS; %switch currents
```

APPENDIX B

MatEMTP time-domain solution module: *timeloop.m*

```
Yn=sparse(n,n);    %Initialize the conductance matrix
mbranch(5);       %Contributions to Yn
Ytmp=[Yn Vadj'; Vadj sparse(nVs,nVs)];
Saug=[Sadj sparse(nS,nVs)]; %account for switches
reBuild=1;
Vs=zeros(nVs,1); %eliminate alloc functions
%
for itime=1:tmax %start of main loop
   ... printing and plotting functions ...
  t=t+Dt;
  In=zeros(n,1);          %currents may add
  msource(4);             %contribution to Vs and In
```

```
    mbranch(6);  %contribution to In from history
    Iaug=[In; Vs; zeros(nS,1)];
    %
    if (reBuild) %process switches and LU if rebuild
      Stmp=sparse(1:nS,1:nS,Sactive)*Saug;
      Sz=sparse(1:nS,1:nS,~Sactive);
      Yaug=[Ytmp Stmp'; Stmp Sz];
      [LL,UU]=lu(Yaug);
    end;
    tmp=LL\Iaug; Vaug=UU\tmp;
    Vn=Vaug(1:n);          %extract nodal voltages
    IVs=Vaug(n+1:n+nVs); %voltage source currents
    IS=Vaug(n+nVs+1:n+nVs+nS);  %switch currents
    %
    mbranch(7);  %update history terms
    mswitch(7);  %update switch status, signal reBuild
end %of main loop
```

APPENDIX C

MatEMTP data file for the test case of Fig. 3

The following is a listing of *test2iwh.m*, manual comments have been added for readability:

```
Dt=10e-06; tmax=ceil(0.05/Dt); %this is 50ms
storedata=0;    %indicates hard disk store when 1
steadystate=1;  %request for steady state when 1
%
n=6; %number of nodes
%
BUS=['BUS1   ';'BUS12 ';'BUS13L';'BUS13S';'BUS1S
';'SRC    ';]; %node names
%
%RLC model
RLCadj=sparse(8,n);
RLCadj(1,6)=1; RLCadj(1,1)=-1;
RLCadj(2,1)=1; RLCadj(2,2)=-1;
....
RLCadj(8,3)=1;
%
RLCout=[5;]; %current output for 5
RLCR=[0;0.05;0;0;0;22.61;0.5;0;];
RLCL=[0.006;0.002;0;0;0.006; 0.01972;0;0;];
RLCC=[0;0;8e-07; 8e-07;0;0;0;1e-06; ];
%
%Sine current source model
Isineadj=sparse(2,n);
Isineadj(1,2)=1; Isineadj(2,1)=1;
Imagn=[2.001; 1.1; ]; Iphi=[10.0; 5.0; ];
Istart=[-1.0; -1.0; ]; Istop=[Inf; Inf; ];
Iw=2*pi*[180.0; 360.0; ];
%
%Ordinary switch model
Sadj=sparse(2,n);
Sadj(1,3)=-1; Sadj(1,4)=1;
Sadj(2,3)=-1; Sadj(2,5)=1;
Sclose=[17.E-3; 22.E-3; ];
Seps=[0; 0; ]; Sopen=[Inf; Inf; ];
%
%
%Sine voltage source model
Vadj=sparse(1,n);
Vadj(1,6)=1;
Vsinein=[1;]; Vmag=[56.34; ]; Vphi=[0.0; ];
Vstart=[-1.0; ]; Vstop=[Inf; ]; Vw=2*pi*[60.0;];
%
Vnout=[2; 3; 4;]; %output request of node voltages
```

APPENDIX D

Partial listing for the test case: *pi.m*

```
...
%PI section model
PIadj=sparse(54,n); %node incidence matrix
PIadj(1,3)=1;  PIadj(1,4)=-1; %first pi-section
               PIadj(2,5)=-1;
...
PIR=sparse(54,54);  %resistance matrix
PIL=sparse(54,54);  %inductance matrix
PIC=sparse(54,54);  %capacitance matrix
%
for k=1:2:54
PIR(k:k+1,k:k+1)=[.25387 .10212; .10212 .69831];
PIL(k:k+1,k:k+1)=[.56461 .13758;
                 .13758 .13139]*1e-03;
PIC(k:k+1,k:k+1)=[.7268 -.7268
                 -.7268 3.4012]*1e-06;
end
...
```

REFERENCES

[1] Electric Power Research Institute, EMTP Development Coordination Group, EPRI EL-6412-L: Electromagnetic Transients Program Rule Book, Version 2

[2] G. Bray and D. Pokrass: Understanding Ada, A Software Engineering Approach. John Wiley & Sons, 1985

[3] MATLAB, High-Performance Numeric Computation and Visualization Software. The MathWorks, Inc. MATLAB User's guide, August 1992

[4] H. W. Dommel: Electromagnetic Transients Program reference manual (EMTP Theory Book). Bonneville Power Administration, August 1986.

[5] C. W. Ho, A. E. Ruehli and P. A. Brennan: The modified nodal approach to network analysis. Proc. 1974 International symposium on circuits and systems, San Francisco, pp. 505-509, April 1974

[6] X. Lombard, J. Mahseredjian, S. Lefebvre and C. Kieny: Implementation of a new harmonic initialization method in the EMTP. IEEE Trans. on Power Systems, Summer Meeting 94, paper 94 SM 438-2 PWRD

[7] B. Kullicke: Simulation program Netomac, Difference conductance method for continuous and discontinuous systems. Siemens Research and Development Reports, Vol. 10, pp. 299-302, 1981, no. 5

[8] SIMULINK, Dynamic System Simulation Software. The MathWorks, Inc. (April 1993

[9] F. Alvarado: EMTP Workbook II. University of Wisconsin at Madison. EL4651, Volume 2, June 1989

[10] C. Gary, A. Timotin, D. Critescu: Prediction of surge

propagation influenced by corona and skin effect. Proc. IEE, 130-A, pp. 264-272, July 1983.

[11] J. Mahseredjian: The EMTP SUN and CRAY UNIX versions. Rapport IREQ-93-065, March 1993, Hydro-Québec

[12] MATLAB, High-Performance Numeric Computation and Visualization Software. The MathWorks, Inc. External Interface guide, January 1993

BIOGRAPHIES

Jean Mahseredjian (M) received the B.Sc.A., M.Sc.A. and Ph.D. in Electrical Engineering from Ecole Polytechnique de Montréal (Canada) in 1982, 1985 and 1990 respectively. At present he is a researcher at Institut de Recherche d'Hydro-Québec and an associate-professor at Ecole Polytechnique de Montréal.

Fernando L. Alvarado (F) was born in Lima, Peru in 1945. He received the BEE and PE degrees from the National University of Engineering in Lima, Peru, the MS degree from Clarkson College (now Clarkson University) in Potsdam, New York, and the Ph. D. degree from the University of Michigan in 1972. Since 1975 he has been with the University of Wisconsin in Madison, where he is currently a Professor of Electrical and Computer Engineering.

PART 4

APPLICATIONS AND CASE STUDIES

PART 4
APPLICATIONS AND CASE STUDIES

INTRODUCTION

EMTP was developed at BPA to have a tool that could duplicate TNA results. The first versions of the program had a limited number of applications; EMTP studies were mainly devoted to overvoltage calculations and insulation coordination. Several capabilities were added during 1970s. They expanded EMTP applications, among others, to HVDC link simulations and subsynchronous resonance (SSR) studies.

The implementation of advanced models for some components and some refinements for others already implemented increased the scope of applications and improved the accuracy of simulation results during the subsequent years.

Significant improvements were made during the late 1980s and the 1990s. New frequency-dependent models for lines, cables and transformers were developed, and MODELS language was embedded in some EMTPs.

However, some of the most important changes have not been related to modeling improvements. Advances in hardware and software have made possible the development of efficient EMTP versions for personal computers and user-friendly GUIs.

In addition, an important change has taken place in the main applications. Studies performed in the early years were related to switching overvoltages. Although they still form an important part of the current applications, the most significant efforts are now devoted to the improvement of lightning simulations, the performance analysis of FACTS (Flexible AC Transmission Systems) devices and power quality studies.

The collection of papers included in this section provide an overview of the most important applications. The papers have been divided into three areas:

- overvoltage calculations
- simulation of power electronics equipment (HVDC links and FACTS devices)
- power quality studies.

This set of papers is just a small sample of the applications for which EMTPs have been used. The additional bibliography complements the list of applications.

OVERVOLTAGE CALCULATIONS AND INSULATION COORDINATION STUDIES

Overvoltages in power systems can be produced by a wide variety of factors, such as faults, switching operations and lightning strokes. It is not advisable to design power equipment to withstand all types of overvoltages. The power engineer usually seeks a compromise between insulation or protection level and economics. The goals of insulation coordination studies are to select appropriate insulation levels for equipment and protection devices so as to minimize damage and interruptions.

Overvoltages that can occur in a power system are classified according to their duration and frequency range:

1) Temporary overvoltages are long-duration power- frequency oscillations. Conditions that lead to this type of overvoltages are faults, load rejection, linear resonance, ferroresonance and open conductors.

2) Switching overvoltages result from the operation of switching devices, either during normal conditions or as the result of fault clearings. These transients have a duration from tens to thousands of microseconds. They belong to the category of slow-front transients. Main operations that can produce switching overvoltages are line energization and re-energization, capacitor and inductor switchings, occurrence of faults and breaker openings.

3) Lightning overvoltages are fast-front transients, in the order of microseconds. They are either caused by direct strokes to phase conductors and backflashovers, or by strokes to earth close to the line.

4) Disconnector operations or faults in gas-insulated substations (GIS) can originate very-fast-front overvoltages, in a range higher than 1 MHZ.

Although a correspondence between the causes of overvoltages and the type of transient phenomena that they produce is well established and accepted, there are some exceptions. For instance, distant lightning strokes can induce slow-front overvoltages, while too fast arc quenchings can result in very-fast-front current chopping overvoltages.

Insulation coordination studies and simulations performed with a digital computer can be classified into two categories: deterministic and statistical. A deterministic procedure is performed preselecting all parameters involved in calculations. Due to the random nature of many overvoltages, a statistical procedure is frequently advisable. Goals of these studies are the determination of the Line FlashOver Rate (LFOR) for overhead lines, or of the Mean-Time-Between-Failures (MTBF) for substations.

The digital simulation of every type of overvoltage, using either a deterministic or a statistical procedure, requires that a significant number of capabilities be available in an EMTP. Since the early versions, most EMTPs have been used for simulation of switching transients, although capabilities for simulation of other overvoltages were available. However, only recently has an increasing interest in simulation of lightning overvoltages raised the development of frequency-dependent models for some components, such as transformers and arresters.

Presently, most types of transient overvoltages are successfully simulated using some EMTPs. Reprinted papers and additional references of this part cover the majority of overvoltage and insulation coordination studies. They are:

- temporary overvoltages [1], [2], [A1] - [A6]
- computation of switching overvoltages and TRV studies [3] - [8], [A7]-[A18]
- lightning overvoltage studies [10] - [12], [A19]- [A24]
- simulation of overvoltages in GIS [14], [15], [A26] - [A34]

Extensive insulation coordination studies, dealing with several types of overvoltages, are presented in [9], [A25]. EMTP modeling of grounding systems for lightning protection and electromagnetic compatibility studies are presented in [13].

SIMULATION OF POWER ELECTRONICS EQUIPMENT

Power electronics applications are currently present at all voltage levels. Main applications include HVDC links, adjustable speed drives, static var compensators, energy storage, renewable energy sources and the new generation of FACTS and CPS (Custom Power Systems) devices.

There are two main objectives of digital time-domain simulations of power electronics systems. The first one covers steady-state calculations where the goal is to evaluate harmonics generated by power electronics equipment, their propagation and their effects on the power system performance. The second covers the dynamic performance. The goals here are the evaluation of control strategies and the identification of potential problems produced during a transient phenomenon. Common to both types of studies are the verification of a design and the prediction of its performance.

The simulation of HVDC converters and the associated control strategy is one of the most important applications of EMTPs. A very extensive bibliography has been produced during the last twenty years. Three reprinted papers on HVDC links present both modeling and validation of control systems [16] - [18]. Each paper includes the study of an actual case, using a different EMTP, and comparisons between simulation results and field test measurements. Other studies related to this application have been covered in the recent literature [A35] - [A45].

Static VAR compensation was one of the first applications of power electronics in transmission networks. Advantages of static VAR compensators (SVC) are many, and their applications cover all voltage levels. Pioneering work on the simulation of SVCs using an EMTP was performed at the early 1980s [A46]. Digital simulation of SVCs has been useful, among others, for design considerations [A47], to illustrate how SVC equipment can be used to increase power system damping [A48], or to show how to balance low voltage networks [A49].

The papers on SVC simulation present three different cases. The first describes the development of an SVC model and its interface to the EMTDC program [19]. The second provides a detailed modeling of an actual SVC system and compares digital and TNA simulation results [20]. The third presents the development and application of an SVC model using the data module concept [21]. Additional works on SVC modeling are presented in [A50] - [A53].

The application to the simulation of new FACTS devices is shown in a paper presenting the study of a TCSC (Thyristor Controlled Series Capacitor) for SSR mitigation [22].

Subsynchronous resonance studies are not covered here. However, a very important activity on the application of EMTPs to SSR studies has been performed since the late 1970s [A54] - [A68].

An important bibliography on the simulation of FACTS devices is presently available: series compensation [A69] - [A73], IPC (Interphase Power Controller) [A74], STATCON (advanced static var compensator) [A75], [A76], UPFC (Unified Power Flow Controller) [A77], [A78], SPS [Static Phase Shifter] [A79], active filters [A80], [A81], custom power applications [A82], fault current limiters with series compensation [A83]. Design considerations of fast controllers using EMTPs are discussed in [A84].

The last paper in this section presents a detailed description of a SMES (superconducting magnetic storage) system [23]. The study was performed to demonstrate the feasibility of the SMES system for various purposes. Other works showing the application of EMTPs to SMES studies are in [A85], [A86].

Semiconductor devices are usually represented by means of built-in switches. A power converter is modeled using switch and TACS capabilities, and assembled as any other part of the network. A different approach based on a separately programmed module for simulation of power converters was presented in [A87]. The module is interfaced to the EMTP and uses its own solution method.

POWER QUALITY STUDIES

The term power quality is applied to a wide variety of electromagnetic phenomena and includes a broad range of concerns. According to the definition proposed in [A88], a power quality problem is a problem "manifested in voltage, current or frequency deviations that results in failure or misoperation of customer equipment".

Phenomena causing electromagnetic disturbances are many. In fact any transient phenomenon causing overvoltages should be considered a power quality problem. The main distinction between insulation coordination and power quality studies is related to voltage level: power quality studies involve customer equipment, and they are generally restricted to medium and low voltage networks. A common ground to both types of studies is overvoltage calculation and protection.

Topics included in power quality studies are, among others, harmonics and power frequency variations, voltage fluctuations, unbalances and interruptions.

Capabilities available in most EMTPs cannot be efficiently used to analyze many of these phenomena. However a great variety of studies and simulations can be performed. Four papers related to this subject have been reprinted. The first analyzes voltage magnification on costumers systems produced by switched capacitors [24]. The second presents an arc-furnace model developed to evaluate flicker effects [25]. The third paper analyzes the harmonic impact of compact fluorescent lamps [26]. The last paper evaluates voltage notching problems caused by adjustable speed drives [27].

Other power quality studies performed with the help of an EMTP are presented in [A89] - [A96].

OTHER APPLICATIONS

The areas for which EMTPs can be used have been expanded over the years. A sample with some additional applications follows : secondary arc studies [A97] - [A101], bus transfer [A102], analysis of internal failures [A103], six-phase transmission [A104], induced overvoltages in parallel lines [A105], reverse metering analysis [A106], geomagnetic disturbances [A107], [A108], wind energy conversion systems [A109], photovoltaic systems [A110], adjustable motor drives [A111], [A112], magnetic levitation train coils [A113], protection systems [A114].

REFERENCES

Reprinted Papers

[1] D.A.N. Jacobson, D.R. Swatek and R.W. Mazur, "Mitigating potential transformer ferroresonance in a 230 kV converter station", *1996 IEEE Transmission and Distribution Conference Proceedings*, Los Angeles, September 15-20.

[2] O. Bourgault and G. Morin, "Analysis of harmonic overvoltage due to transformer saturation following load shedding on Hydro-Quebec - NYPA 765 kV interconnection", *IEEE Trans. on Power Delivery*, vol. 5, no. 1, pp. 397-405, January 1990.

[3] A.C. Legate, J.H. Brunke, J.J. Ray and E.J. Yasuda, "Elimination of closing resistor on EHV circuit breakers", *IEEE Trans. on Power Delivery*, vol. 3, no. 1, pp. 223-231, January 1988.

[4] K.C. Lee and K.P. Poon, "Statistical switching overvoltage analysis of the first B.C. Hydro phase shifting transformer using the Electromagnetic Transients Program", *IEEE Trans. on Power Systems*, vol. 5, no. 4, pp. 1054-1060, November 1990.

[5] R.S. Bayless et al., "Capacitor switching and transformer transients", *IEEE Trans. on Power Delivery*, vol. 3, no. 1, pp. 349-357, January 1988.

[6] B.C. Furumasu and R.M. Hasibar, "Design and installations of 500-kV back-to-back shunt capacitor banks", *IEEE Trans. on Power Delivery*, vol. 7, no. 2, pp. 539-545, April 1992.

[7] B. Bhargava et al., "Effectiveness of pre-insertion inductors for mitigating remote overvoltages due to shunt capacitor energization", *IEEE Trans. on Power Delivery*, vol. 8, no. 3, pp. 1226-1238, July 1993.

[8] N. Kolcio et al., "Transient overvoltages and overcurrents on 12.47 kV distribution lines : Computer modeling results", *IEEE Trans. on Power Delivery*, vol. 8, no. 1, pp. 359-366, January 1993.

[9] Q. Bui-Van et al., "Overvoltage studies for the St-Lawrence River 500-kV DC cable crossing", *IEEE Trans. on Power Delivery*, vol. 6, no. 3, pp. 1205-1215, July 1991.

[10] A. Inoue and S. Kanao, "Observation and analysis of multi-phase grounding faults caused by lightning", *IEEE Trans. on Power Delivery*, vol. 11, no. 1, pp. 353-360, January 1996.

[11] Y. Matsumoto et al., "Measurement of lightning surges on test transmission line equipped with arresters struck by natural and triggered lightning", *IEEE Trans. on Power Delivery*, vol. 11, no. 2, pp. 996-1002, April 1996.

[12] H. Elahi et al., "Lightning overvoltage protection of the Paddock 362-145 kV gas-insulated substation", *IEEE Trans. on Power Delivery*, vol. 5, no. 1, pp. 144-150, January 1990.

[13] F.E. Menter and L. Grcev, "EMTP-based model for grounding system analysis", *IEEE Trans. on Power*

Delivery, vol. 9, no. 4, pp. 1838-1849, October 1994.

[14] S. Yanabu et al., "Estimation of fast transient overvoltage in gas-insulated substation", *IEEE Trans. on Power Delivery*, vol. 5, no. 4, pp. 1875-1882, October 1990.

[15] Z. Haznadar, C. Carsimamovic and R. Mahmutcehajic, "More accurate modeling of gas insulated substation components in digital simulations of very fast electromagnetic transients", *IEEE Trans. on Power Delivery*, vol. 7, no. 1, pp. 434-441, January 1992.

[16] P. Kuffel et al., "Development and validation of detailed controls models of the Nelson River Bipole 1 HVDC system", *IEEE Trans. on Power Delivery*, vol. 8, no. 1, pp. 351-358, January 1993.

[17] A. Hammad et al., "Controls modelling and verification for the Pacific Intertie HVDC 4-terminal scheme", *IEEE Trans. on Power Delivery*, vol. 8, no. 1, pp. 367-375, January 1993.

[18] G. Morin et al., "Modeling of the Hydro-Quebec - New England HVDC system and digital control with EMTP", *IEEE Trans. on Power Delivery*, vol. 8, no. 2, pp. 559-566, April 1993.

[19] A.M. Gole and V.K. Sood, "A static compensator model for use with electromagnetic transients simulation programs", *IEEE Trans. on Power Delivery*, vol. 5, no. 3, pp. 1398-1407, July 1990.

[20] A.N. Vasconcelos et al. "Detailed modeling of an actual static var compensator for electromagnetic transients studies", *IEEE Trans. on Power Systems*, vol. 7, no. 1, pp. 11- 19, February 1992.

[21] S. Lefebvre and L. Gérin-Lajoie, "A static compensator model for the EMTP", *IEEE Trans. on Power Systems*, vol. 7, no. 2, pp. 477-486, April 1992.

[22] W. Zhu et al., "An EMTP study of SSR mitigation using the Thyristor Controlled Series Capacitor", *IEEE Trans. on Power Delivery*, vol. 10, no. 3, pp. 1479-1485, July 1995.

[23] I.D. Hassan, R.M. Bucci and K.T. Swe, "400 MW SMES power conditioning system development and simulation", *IEEE Trans. on Power Electronics*, vol. 8, no. 3, pp. 237-249, July 1993.

[24] M.F. McGranaghan et al., "Impact of utility switched capacitors on customer systems - Magnification at low voltage capacitors", *IEEE Trans. on Power Delivery*, vol. 7, no. 2, pp. 862-868, April 1992.

[25] G.C. Montanari et al., "Arc-furnace model for the study of flicker compensation in electrical networks", *IEEE Trans. on Power Delivery*, vol. 9, no. 4, pp. 2026-2036, October 1994.

[26] R. Dwyer et al., "Evaluation of harmonic impacts from compact fluorescent lights on distribution systems", *IEEE Trans. on Power Systems*, vol. 10, no. 4, pp. 1772-1780, November 1995.

[27] L. Tang et al., "Voltage notching interaction caused by large adjustable speed drives on distribution systems with low short circuit capacities", *IEEE Trans. on Power Delivery*, vol. 11, no. 3, pp. 1444-1453, July 1996.

Additional References

[A1] T. Sakurai, K. Murotani and K. Oonishi, "Suppression of temporary overvoltages caused by transformer and AC filter inrush currents at the Shin-Shinano frequency converter station", *IEEE Trans. on Power Apparatus and Systems*, vol. 100, no. 4, pp. 1608-1613, April 1981.

[A2] T. Adielson, A. Carlson, H.B. Margolis and J.A. Halladay, "Resonant overvoltages in EHV transformers", *IEEE Trans. on Power Apparatus and Systems*, vol. 100, no. 7, pp. 3563-3572, July 1981.

[A3] F. Iliceto, E. Cinieri and A. Di Vita, "Overvoltages due to open-phase occurrence in reactor compensated EHV lines", *IEEE Trans. on Power Apparatus and Systems*, vol. 103, no. 3, pp. 474-482, March 1984.

[A4] J.A. Halladay and C.H. Shih, "Resonant overvoltage phenomena caused by transmission line fault", *IEEE Trans. on Power Apparatus and Systems*, vol. 104, no. 9, pp. 2531-2539, September 1985.

[A5] M.M. Adibi, R.W. Alexander and B. Avramovic (Power System Restoration WG), "Overvoltage control during restoration", *IEEE Trans. on Power Systems*, vol. 7, no. 4, pp. 1464-1470, November 1992.

[A6] Y. Rajotte, J. Fortin and G. Raymond, "Impedance of multigrounded neutrals on rural distribution systems", *IEEE Trans. on Power Delivery*, vol. 10, no. 3, pp. 1453-1459, July 1995.

[A7] B.C. Papadias, "The accuracy of statistical methods in evaluating the insulation of EHV systems", *IEEE Trans. on Power Apparatus and Systems*, vol. 98, no. 3, pp. 992-999, May/June 1979.

[A8] S.H. Sarkinen, G.G. Schockelt and J.H. Brunke, "High frequency switching surges in EHV shunt reactor installation with reduced insulation levels", *IEEE Trans. on Power Apparatus and Systems*, vol. 98, no. 3, pp. 1013-1021, May/June 1979.

[A9] D. Bhasavanich et al., "Digital simulation and field measurements of transients associated with large capacitor bank switching on distribution systems", *IEEE Trans. on Power Apparatus and Systems*, vol. 104, no. 8, pp. 2274-2282, August 1985.

[A10] J.R. Ribeiro and M.E. McCallum, "An application of metal oxide surge arresters in the elimination of need for closing resistors in EHV breakers", *IEEE Trans. on Power Delivery*, vol. 4, no. 1, pp. 282-291, January 1989.

[A11] E.J. Michelis, "Comparison between measurement and calculation of line switching with inductive and complex source", *IEEE Trans. on Power Delivery*, vol. 4, no. 2, pp. 1432-1440, April 1989.

[A12] K.H. Lee and J.M. Schneider, "Rockport transient voltage monitoring system : Analysis and simulation of recorded waveform", *IEEE Trans. on Power Delivery*, vol. 4, no. 3, pp. 1794-1805, July 1989.

[A13] G.T. Wrate et al., "Transient overvoltages on a three terminal DC transmission system due to monopolar ground faults", *IEEE Trans. on Power Delivery*, vol. 5, no. 2, pp. 1047-1053, April 1990.

[A14] F. Iliceto, F.M. Gatta, E. Cinieri and G. Asan, "TRVs across circuit breakers of series compensated lines. Status with present technology and analysis for the Turkish 420-kV grid", *IEEE Trans. on Power Delivery*, vol. 7, no. 2, pp. 757-766, April 1992.

[A15] A.K. McCabe et al., "Design and testing of a three-break 800 kV SF6 circuit breaker with ZnO varistors for shunt reactor switching", *IEEE Trans. on Power Delivery*, vol. 7, no. 2, pp. 853-861, April 1992.

[A16] R.G. Andrei, A.J.F. Keri, R.J. Albanese and P.B. Johnson, "Bridge capacitor bank installation concept reactive power generation in EHV systems", *IEEE Trans. on Power Systems*, vol. 8, no. 4, pp. 1463-1470, November 1993.

[A17] C.W. Taylor and A.L. Van Leuven, "CAPS : Improving power system stability using the time-overvoltage capability of large shunt capacitor banks", *IEEE Trans. on Power Delivery*, vol. 11, no. 2, pp. 783-792, April 1996.

[A18] D.F. Peelo et al., "Mitigation of circuit breaker transient recovery voltages associated with current limiting reactors", *IEEE Trans. on Power Delivery*, vol. 11, no. 2, pp. 865-871, April 1996.

[A19] J.J. Burke and E.C. Sakshaug, "The application of gapless arresters on underground distribution systems", *IEEE Trans. on Power Apparatus and Systems*, vol. 100, no. 3, pp. 1234-1243, March 1981.

[A20] F. Iliceto et al., "New concepts on MV distribution from insulated shield wires of HV lines", *IEEE Trans. on Power Delivery*, vol. 4, no. 4, pp. 2130-2144, October 1989.

[A21] J. Panek, M. Sublich and H. Elahi, "Criteria for phase-to-phase clearances of HV substations", *IEEE Trans. on Power Delivery*, vol. 5, no. 1, pp. 137-143, January 1990.

[A22] A.J.F. Keri, Y.I. Musa and J.A. Halladay, "Insulation coordination for delta connected transformers", *IEEE Trans. on Power Delivery*, vol. 9, no. 2, pp. 772-780, April 1994.

[A23] N. Fujiwara et al., "Development of a pin-post insulator with built-in metal oxide varistors for distribution lines", *IEEE Trans. on Power Delivery*, vol. 11, no. 2, pp. 824-833, April 1996.

[A24] R.J. Harrington and M. Mueen, "A simple approach to improve lightning performance of an uprated substation", *IEEE Trans. on Power Delivery*, vol. 11, no. 3, pp. 1633-1639, July 1996.

[A25] M. Sanders, G. Köppl and J. Kreuzer, "Insulation co-ordination aspects for power stations with generator circuit breakers", *IEEE Trans. on Power Delivery*, vol. 10, no. 3, July 1995.

[A26] S. Narimatsu et al., "Interrupting performance of capacitive current by disconnecting switch for gas insulated switchgear", *IEEE Trans. on Power Apparatus and Systems*, vol. 100, no. 6, pp. 2726-2732, June 1981.

[A27] S. Matsumara and T. Nitta, "Surge propagation in gas insulated substation", *IEEE Trans. on Power Apparatus and Systems*, vol. 100, no. 6, pp. 3047-3054, June 1981.

[A28] L. Blahous and T. Gysel, "Mathematical investigation of the transient overvoltages during disconnector switching in GIS", *IEEE Trans. on Power Apparatus and Systems*, vol. 102, no. 9, pp. 3088-3097, September 1983.

[A29] S. Ogawa et al., "Estimation of restriking transient overvoltage on disconnecting switch for GIS", *IEEE Trans. on Power Delivery*, vol. 1, no. 2, pp. 95-102, April 1986.

[A30] T. Yoshida at al., "Distribution of induced grounding current in large-capacity GIS using multipoint grounding system", *IEEE Trans. on Power Delivery*, vol. 1, no. 4, pp. 120-127, October 1986.

[A31] J. Ozawa et al., "Suppression of fast transient overvoltage during gas disconnector switching in GIS", *IEEE Trans. on Power Delivery*, vol. 1, no. 4, pp. 194-201, October 1986.

[A32] M. Rioual, "Measurements and computer simulation of fast transients through indoor and outdoor substations", *IEEE Trans. on Power Delivery*, vol. 5, no. 1, pp. 117-123, January 1990.

[A33] S. Okabe, M. Kan and T. Kouno, "Analysis of surges measured at 550 kV substations", *IEEE Trans. on Power Delivery*, vol. 6, no. 4, pp. 1462-1468, October 1991.

[A34] A. Ardito et al., "Accurate modeling of capacitively graded bushings for calculation of fast transient overvoltages in GIS", *IEEE Trans. on Power Delivery*, vol. 7, no. 3, pp. 1316-1327, July 1992.

[A35] J. Reeve and S.P. Chen, "Versatile interactive digital simulator based on EMTP for AC/DC power system transient studies", *IEEE Trans. on Power Apparatus and Systems*, vol. 103, no. 12, pp. 3625-3633, December 1984.

[A36] J. Reeve and S.P. Chen, "Digital simulation of a multiterminal HVDC transmission system", *IEEE Trans. on Power Apparatus and Systems*, vol. 103, no. 12, pp. 3634-3642, December 1984.

[A37] D.A. Woodford, "Validation of digital simulation of DC links", *IEEE Trans. on Power Apparatus and Systems*, vol. 104, no. 9, pp. 2588-2595, September 1985.

[A38] T. Ino et al., "Validation of digital simulation of DC links - Part II", *IEEE Trans. on Power Apparatus and Systems*, vol. 104, no. 9, pp. 2596-2603, September 1985.

[A39] S.Arabi, M.Z. Tarnawecky and M.R. Iravani, "Dynamic performance of an HVDC quasi 24-pulse series tapping station", *IEEE Trans. on Power Delivery*, vol. 3, no. 4, pp. 2112-2118, October 1988.

[A40] T. Shome et al., "Adjusting converter controls for paralleled DC converters using a digital transient simulation program", *IEEE Trans. on Power Systems*, vol. 5, no. 1, pp. 12-19, February 1990.

[A41] L.A.S. Pilotto, M. Szechtman and A.E. Hammad, "Transient AC voltage related phenomena for HVDC schemes connected to weak AC systems", *IEEE Trans. on Power Delivery*, vol. 7, no. 3, pp. 1396-1404, July 1992.

[A42] J. Reeve and M. Sultan, "Robust adaptive control of HVDC systems", *IEEE Trans. on Power Delivery*, vol. 9, no. 3, pp. 1487-1493, July 1994.

[A43] X. Jiang and A.M. Gole, "Energy recovery filter with variable quality factor", *IEEE Trans. on Power Delivery*, vol. 9, no. 3, pp. 1625-1631, July 1994.

[A44] R. Verdolin et al., "Induced overvoltages on an AC-DC hybrid transmission system", *IEEE Trans. on Power Delivery*, vol. 10, no. 3, pp. 1514-1524, July 1995.

[A45] A. Sarshar, M.R. Iravani and J. Li, "Calculation of HVDC converter noncharacteristic harmonics using digital time-domain simulation methods", *IEEE Trans. on Power Delivery*, vol. 11, no. 1, pp. 335-344, January 1996.

[A46] R.H. Lasseter and S.Y. Lee, "Digital simulation of static VAR system transients", *IEEE Trans. on Power Apparatus and Systems*, vol. 101, no. 10, pp. 4171-4177, October 1982.

[A47] H.K. Tyll et al., "Design considerations for the Eddy County static var compensator", *IEEE Trans. on Power Delivery*, vol. 9, no. 2, pp. 757-763, April 1994.

[A48] Q. Zhao and J. Jiang, "Robust SVC controller design for improving power system damping", *IEEE Trans. on Power Systems*, vol. 10, no. 4, pp. 1927-1932, November 1995.

[A49] J.A. Martínez, "EMTP simulation of a digitally-controlled static var system for optimal load compensation", *IEEE Trans. on Power Delivery*, vol. 10, no. 3, pp. 1408-1415, July 1995.

[A50] J. He and N. Mohan, "Switch-mode var compensator with minimized switching losses and energy storage elements", *IEEE Trans. on Power Systems*, vol. 5, no. 1, pp. 90-95, February 1990.

[A51] S.Y. Lee et al.,"Detailed modeling of static var compensators using the Electromagnetic Transients Program (EMTP)", *IEEE Trans. on Power Delivery*, vol. 7, no. 2, pp. 836-847, April 1992.

[A52] O.B. Nayak et al., "Dynamic performance of static and synchronous compensators at an HVDC inverter bus in a very weak AC system", *IEEE Trans. on Power Delivery*, vol. 9, no. 3, pp. 1350-1358, August 1994.

[A53] R.A. Kagalwala et al., "Transient analysis of distribution class adaptive var compensators : Simulation and field test results", *IEEE Trans. on Power Delivery*, vol. 10, no. 2, pp. 1119-1125, April 1995.

[A54] N.C. Abi-Smara et al., "Analysis of thyristor-controlled shunt SSR countermeasures", *IEEE Trans. on Power Apparatus and Systems*, vol. 104, no. 3, pp. 584-597, March 1985.

[A55] IEEE Committee Report, "Second benchmark model for computer simulation on subsynchronous resonance", *IEEE Trans. on Power Apparatus and Systems*, vol. 104, no. 5, pp. 1057-1066, May 1985.

[A56] B.L. Agrawal and R.G. Farmer, "Effective damping for SSR analysis of parallel turbine-generators", *IEEE Trans. on Power Systems*, vol. 3, no. 4, pp. 1441-1448, November 1988.

[A57] IEEE Subsynchronous Resonance WG, "Comparison of SSR calculations and test results", *IEEE Trans. on Power Systems*, vol. 4, no. 1, pp. 336-344, February 1989.

[A58] M.R. Iravani, "Coupling phenomenon of torsional modes", *IEEE Trans. on Power Systems*, vol. 4, no. 3, pp. 881-888, August 1989.

[A59] R.M. Hamouda, M.R. Iravani and R. Hackam, "Torsional oscillations of series capacitor compensated AC/DC systems", *IEEE Trans. on Power Systems*, vol. 4, no. 3, pp. 889-896, August 1989.

[A60] M.R. Iravani, "Torsional oscillations of unequally-loaded parallel identical turbine-generators", *IEEE Trans. on Power Systems*, vol. 4, no. 4, pp. 1514-1524, October 1989.

[A61] A. Edris, "Series compensated schemes reducing the potential of subsynchronous resonance", *IEEE Trans. on Power Systems*, vol. 5, no. 1, pp. 219-226, February 1990.

[A62] M.R. Iravani, "A method for reducing transient torsional stresses of turbine-generator shaft segments", *IEEE Trans. on Power Systems*, vol. 7, no. 1, pp. 20-27, February 1992.

[A63] M.R. Iravani and A. Semlyen, "Hopf bifurcations in torsional dynamics", *IEEE Trans. on Power Systems*, vol. 7, no. 1, pp. 28-36, February 1992.

[A64] A. Edris, "Subsynchronous resonance countermeasure using phase imbalance", *IEEE Trans. on Power Systems*, vol. 8, no. 4, pp. 1438-1447, November 1993.

[A65] W. Shi and M.R. Iravani, "Effect of HVDC line faults on transient torsional torques of turbine-generator shafts", *IEEE Trans. on Power Systems*, vol. 9, no. 3, pp. 1457-1464, August 1994.

[A66] M.K. Donnelly et al., "Control of a dynamic brake to reduce turbine-generator shaft transient torques", *IEEE Trans. on Power Systems*, vol. 8, no. 1, pp. 67-73, February 1993.

[A67] R.J. Piwko et al., "Subsynchronous resonance performance tests of the Slatt thyristor-controlled series capacitor", *IEEE Trans. on Power Delivery*, vol. 11, no. 2, pp. 1112-1119, April 1996.

[A68] R. Rajaraman et al., "Computing the damping of subsynchronous oscillations due to a thyristor controlled series capacitor", *IEEE Trans. on Power Delivery*, vol. 11, no. 2, pp. 1120-1127, April 1996.

[A69] B. Pilvelait, T.H. Ortmeyer and D. Maratukulam, "Advanced series compensation for transmission systems using a switched capacitor module", *IEEE Trans. on Power Delivery*, vol. 8, no. 2, pp. 584-590, April 1993.

[A70] S.G. Helbing and G.G. Karady, "Investigations of an advanced form of series compensation", *IEEE Trans. on Power Delivery*, vol. 9, no. 2, pp. 939-947, April 1994.

[A71] S.G. Jalali et al., "A stability model for the Advanced Series Compensator (ASC)", *IEEE Trans. on Power Delivery*, vol. 11, no. 2, pp. 1128-1137, April 1996.

[A72] T. Godart et al., "Feasibility of Thyristor Controlled Series Capacitor for distribution substation enhancements", *IEEE Trans. on Power Delivery*, vol. 10, no. 1, pp. 203-209, January 1995.

[A73] C.J. Hatziadoniu and A.T. Funk, "Development af a control scheme for a series-connected solid-state synchronous voltage source", *IEEE Trans. on Power Delivery*, vol. 11, no. 2, pp. 1138-1144, April 1996.

[A74] K. Habashi et al., "Design of a 200 MW Interphase Power Controller prototype", *IEEE Trans. on Power Delivery*, vol. 9, no. 2, pp. 1041-1048, April 1994.

[A75] R.W. Menzies and Y. Zhuang, "Advanced static compensation using a multilevel GTO thyristor inverter", *IEEE Trans. on Power Delivery*, vol. 10, no. 2, pp. 732-738, April 1995.

[A76] Y. Zhuang et al., "Dynamic performance of a STATCON at an HVDC inverter feeding a very weak AC system", *IEEE Trans. on Power Delivery*, vol. 11, no. 2, pp. 958-964, April 1996.

[A77] L. Gyugyi et al., "The Unified Power Flow Controller : A new approach to power transmission control", *IEEE Trans. on Power Delivery*, vol. 10, no. 2, pp. 1085-1097, April 1995.

[A78] R. Mihalic, P. Zunko and D. Povh, "Improvement of transient stability using Unified Power Flow Controller", *IEEE Trans. on Power Delivery*, vol. 11, no. 1, pp. 485-492, January 1996.

[A79] M.R. Iravani, P.L. Dandeno, K.H. Nguyen and D. Maratukulam, "Applications of static phase shifters in power systems", *IEEE Trans. on Power Delivery*, vol. 9, no. 3, pp. 1600-1608, July 1994.

[A80] C. Wong et al., "Feasibility study of AC- and DC-side active filters for HVDC converter terminals", *IEEE Trans. on Power Delivery*, vol. 4, no. 4, pp. 2067-2075, October 1989.

[A81] M. Aredes and E.H. Watanabe, "New control algorithms for series and shunt three-phase four-wire active power filters", *IEEE Trans. on Power Delivery*, vol. 10, no. 3, pp. 1649-1656, July 1995.

[A82] G. Venkataramanan, B.K. Johnson and A. Sundaram, "An AC-AC power converter for custom power applications", *IEEE Trans. on Power Delivery*, vol. 11, no. 3, pp. 1666-1671, July 1996.

[A83] S.Sugimoto et al., "Principle and characteristics of a fault current limiter with series compensation", *IEEE Trans. on Power Delivery*, vol. 11, no. 2, pp. 842-847, April 1996.

[A84] D. Woodford, "Electromagnetic design considerations for fast acting controllers", *IEEE Trans. on Power Delivery*, vol. 11, no. 3, pp. 1515-1521, July 1996.

[A85] R.H. Lasseter and S.G. Jalali, "Dynamic response of power conditioning systems for superconductive magnetic energy storage", *IEEE Trans. on Energy Conversion*, vol. 6, no. 3, pp. 388-393, September 1991.

[A86] B.M. Han and G.G. Karady, "A new power-conditioning system for superconducting magnetic energy storage", *IEEE Trans. on Energy Conversion*, vol. 8, no. 2, pp. 214-220, June 1993.

[A87] J. Mahseredjian, S. Lefebvre and D. Mukhedkar, "Power converter simulation module connected to the EMTP", *IEEE Trans. on Power Systems*, vol. 6, no. 2, pp. 501-510, May 1991.

[A88] R.C. Dugan, M.F. McGranaghan and H.W. Beaty, *Electrical Power Systems Quality*, McGraw-Hill, 1996, New York.

[A89] M.F. McGranaghan et al., "Impact of utility switched capacitors on customer systems - Part II : Adjustable-speed drive concerns", *IEEE Trans. on Power Delivery*, vol. 6, no. 4, pp. 1623-1628, October 1991.

[A90] G. Manchur and C.C. Erven, "Development of a model for predicting flicker from electric arc furnaces", *IEEE Trans. on Power Delivery*, vol. 7, no. 1, pp. 416-426, January 1992.

[A91] D.M. Dunsmore et al., "Magnification of transient voltages in multi-voltage-level, shunt-capacitor-compensated, circuits", *IEEE Trans. on Power Delivery*, vol. 7, no. 2, pp. 664-673, April 1992.

[A92] N. Mohan, M. Rastogi and R. Naik, "Analysis of a new power electronics interface with approximately sinusoidal 3-phase utility currents and a regulated DC output", *IEEE Trans. on Power Delivery*, vol. 8, no. 2, pp. 540-546, April 1993.

[A93] J. Lamoree et al., "Description of a Micro-SMES system for protection of critical customer facilities", *IEEE Trans. on Power Delivery*, vol. 9, no. 2, pp. 984-991, April 1994.

[A94] L. Tang et al., "Analysis of DC arc furnace operation and flicker caused by 187 Hz voltage distortion", *IEEE Trans. on Power Delivery*, vol. 9, no. 2, pp. 1098-1107, April 1994.

[A95] T.A. Bellei, R.P. O'Leary and E.H. Camm, "Evaluating capacitor- switching for preventing nuisance tripping of adjustable-speed drives due to voltage magnification", *IEEE Trans. on Power Delivery*, vol. 11, no. 3, pp. 1373-1378, July 1996.

[A96] S. Varadan, E.B. Makram and A.A. Girgis, "A new time domain voltage source model for an arc furnace using EMTP", *IEEE Trans. on Power Delivery*, vol. 11, no. 3, pp. 1685-1691, July 1996.

[A97] A.J. Fakheri et al., "Single-phase switching tests on the AEP 675 kV system - Extinction time for large secondary arc currents", *IEEE Trans. on Power Apparatus and Systems*, vol. 102, no. 8, pp. 2775-2783, August 1983.

[A98] S. Goldberg, W. Horton and D. Tziouvaras, "A computer model of the secondary arc in single-phase operation of transmission lines", *IEEE Trans. on Power Delivery*, vol. 4, no. 1, pp. 586-595, January 1989.

[A99] G. Thomann, S.R. Lambert and S. Phaloprakan, "Non-optimum compensation schemes for single pole reclosing on EHV double circuit transmission lines", *IEEE Trans. on Power Delivery*, vol. 8, no. 2, pp. 651-659, April 1993.

[A100] D. Woodford, "Secondary arc effects in AC/DC hybrid transmission", *IEEE Trans. on Power Delivery*, vol. 8, no. 2, pp. 704-711, April 1993.

[A101] M. Kizilcay and T. Pniok, "Digital simulation of fault arcs in power systems", *European Transactions on Electrical Power Engineering*, vol. 1, no. 1, pp. 55-60, January/February 1991.

[A102] I.D. Hassan et al., "Evaluating the transient performance of standby diesel-generator units by simulation", *IEEE Trans. on Energy Conversion*, vol. 7, no. 3, pp. 470-477, September 1992.

[A103] A.S. Morched et al., "Analysis of internal winding stresses in EHV generator step-up trasnformer failures", *IEEE Trans. on Power Delivery*, vol. 11, no. 2, pp. 888-894, April 1996.

[A104] J.R. Stewart et al., "Transformer winding selection associated with reconfiguration of existing double circuit line to six-phase operation", *IEEE Trans. on Power Delivery*, vol. 7, no. 2, pp. 979-985, April 1992.

[A105] P.K. Dwivedi et al., "Safety procedures for working on de-energized EHV lines sharing common right of way", *IEEE Trans. on Power Delivery*, vol. 7, no. 3, pp. 1371-1378, July 1992.

[A106] M.B. Hughes, "Revenue metering error caused by induced voltage from adjacent transmission lines", *IEEE Trans. on Power Delivery*, vol. 7, no. 2, pp. 741-745, April 1992.

[A107] R.A. Walling and A.H. Khan, "Characteristics of transformer exciting-current during geomagnetic disturbances", *IEEE Trans. on Power Delivery*, vol. 6, no. 4, pp. 1707-1714, October 1991.

[A108] M.A. Eiztmann et al., "Alternatives for blocking direct current in AC system neutrals at the Radisson/LG2 complex", *IEEE Trans. on Power Delivery*, vol. 7, no. 3, pp. 1328-1337, July 1992.

[A109] L. Tang and R. Zavadil, "Shunt capacitor failures due to windfarm induction generator self-excitation phenomenon", *IEEE Trans. on Energy Conversion*, vol. 8, no. 4, pp. 513-519, September 1993.

[A110] Y.Yao, P. Bustamante and R.S. Ramshaw, "Improvement of induction motor drive systems supplied by photovoltaic arrays with frequency control", *IEEE Trans. on Energy Conversion*, vol. 9, no. 2, pp. 256-262, June 1994.

[A111] Z. Daboussi and N. Mohan, "Digital simulation of field-oriented control of induction motor drives using EMTP", *IEEE Trans. on Energy Conversion*, vol. 3, no. 3, pp. 667-673, September 1988.

[A112] C.T. Liu, "An efficient EMTP compatible algorithm for modelling switch-controlled drive circuits", *IEEE Trans. on Power Delivery*, vol. 9, no. 4, pp. 2018-2025, October 1994.

[A113] A. Ametani et al., "A study of transient induced voltages on a maglev train coil system", *IEEE Trans. on Power Delivery*, vol. 10, no. 3, pp. 1657-1662, July 1995.

[A114] J.N. Peterson and R.W. Wall, "Interactive relay controlled power system modeling", *IEEE Trans. on Power Delivery*, vol. 6, no. 1, pp. 96-102, January 1991.

MITIGATING POTENTIAL TRANSFORMER FERRORESONANCE IN A 230 KV CONVERTER STATION

D. A. N. Jacobson
Member IEEE

D. R. Swatek
Member IEEE

R. W. Mazur
Member IEEE

Manitoba Hydro
Winnipeg, Manitoba, R3C 2P4

***ABSTRACT* - A wound potential transformer failed catastrophically recently on the Manitoba Hydro system. The failure was attributed to excessive current flow in the primary winding due to a sustained ferroresonance between the circuit breaker grading capacitance and nonlinear magnetizing inductance of the potential transformer. In order to prevent future occurrences, switching guidelines have been revised and permanently connected damping resistors were installed. Modern nonlinear techniques of analysis are evaluated for their suitability of defining safe and unsafe zones of operation. A new two-dimensional bifurcation diagram is proposed as a new visualization tool.**

***KEY WORDS* - Ferroresonance, nonlinear dynamics, chaos, bifurcation.**

I. INTRODUCTION

The Dorsey HVdc converter station 230 kV ac bus is comprised of four bus sections on which the converter valves and transmission lines are terminated. At 22:04, May 20, 1995, bus A2 (Fig. 1a) was removed from service to commission replacement breakers, current transformers and to perform disconnect maintenance and trip testing. At approximately 22:30, a potential transformer (PT) failed catastrophically causing damage to equipment up to 33 m away. The switching procedure resulted in the deenergized bus and the associated PTs being connected to the energized bus B2 through the grading capacitors (5061 pF) of nine open 230 kV circuit breakers. A station service transformer, which is normally connected to bus A2, had been previously disconnected. A ferroresonance condition caused the failure of the PT.

Ferroresonance is not a new phenomenon. One author has traced the origin of the word "ferroresonance" back to a 1920 paper [1]. Over the past eighty years, several methods have been used to help model, understand and analyze ferroresonance. Rudenberg's [2] graphical approach based on modified phasor theory shows the existence of multiple operating points but cannot be used to identify safe and unsafe operating regions nor does it address subharmonic ferroresonant states. Hayashi [3] and Janssens et. al. [4] have used the harmonic balance method to directly calculate the final periodic state. This method requires a prior knowledge of the harmonic content of the final state and does not consider the possibility of ferroresonance being switching surge induced. Another direct calculation method proposed by Swift [5] based on incremental describing functions is suitable for predicting spontaneous jumps to fundamental frequency ferroresonance but is not suitable when switching surges or subharmonics are considered. Fortunately, digital time-domain simulation permits the explicit representation of nonlinear circuit elements and thus provides a simple method of determining whether or not a particular operating configuration (or disturbance) will lead to a ferroresonant state. Digital simulation also demonstrates the chaotic nature of ferroresonance; that is, small variations in initial parameters can lead to drastic changes in operating states [1],[6]. This underlying chaotic nature necessitates a tremendous number of simulations in order to gain confidence in a particular mitigation scheme. Bifurcation diagrams have been borrowed from the chaos literature as a means of organizing this huge mass of output into a single coherent illustration of the operating state [1],[6],[10]. Unfortunately, the traditional one-dimensional bifurcation diagram cannot describe the global behavior of the ferroresonant circuit since, for any practical system having more than one variable parameter, an infinite number of such bifurcation diagrams are possible.

In this paper we present an EMTP model of the ferroresonant Dorsey bus outage, and the results of the associated mitigation study. Enhanced visualization of the "global" behavior of the ferroresonant bus is achieved through the use of a novel two-dimensional bifurcation diagram that represents a planar cross section of the complete n-dimensional bifurcation space. It is hoped that this visualization technique will benefit other investigators of ferroresonance.

II. EMTP MODEL

The EMTP (ElectroMagnetic Transients Program) was used for simulation of this ferroresonance phenomena. The program works well, however, other methods are more suitable for specific problems. A Runge Kutte state variable solution method is discussed later in the paper.

The model used in the EMTP ferroresonance investigation is shown in Fig. 1.

0-7803-3522-8/96 $5.00 © 1996 IEEE

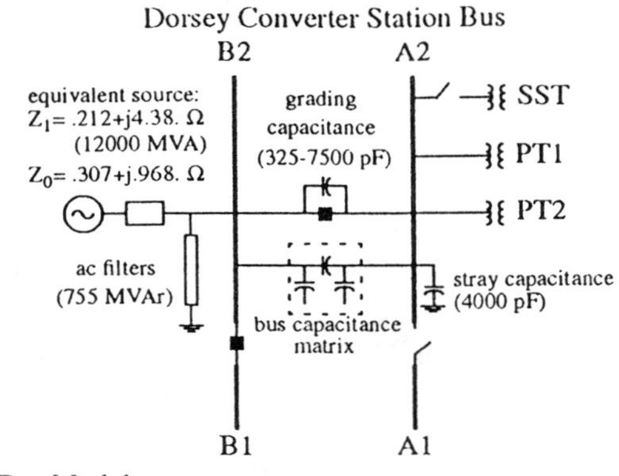

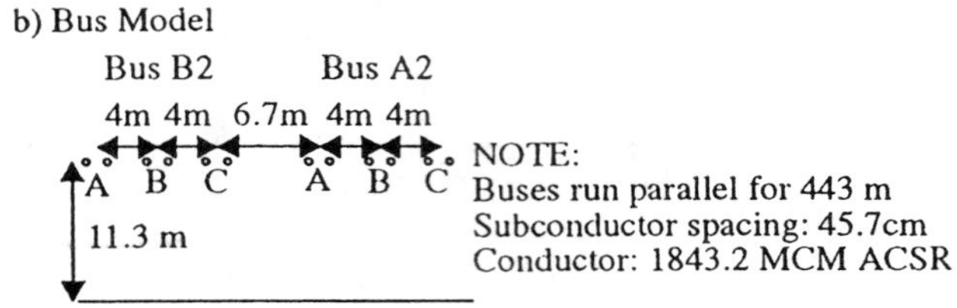

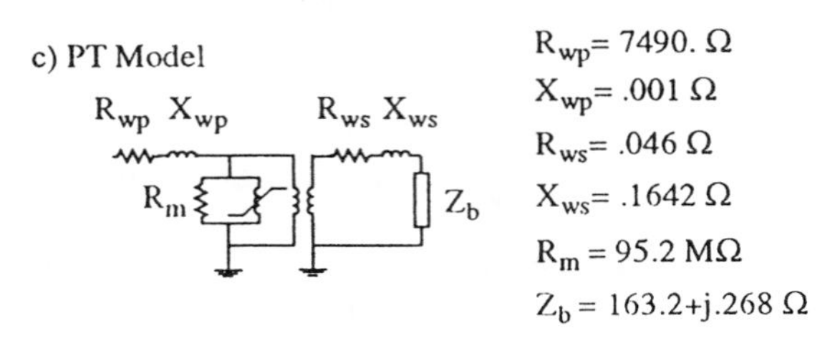

Fig. 1. EMTP model.

A. Equivalent Source

The short circuit strength at Dorsey does not influence the onset of ferroresonance because the impedance is not of the same order as the saturated potential transformer impedance. A strong equivalent impedance was used in all simulations.

B. Filters

The exact number of ac filters at Dorsey will not influence the onset of ferroresonance. Detailed models of each ac filter branch were included in the model however, to assess the effectiveness of mitigating ferroresonance by switching a filter onto the deenergized bus.

C. Bus Capacitance

An accurate model of the bus capacitance is required to simulate ferroresonant conditions reliably. A value can be estimated if the induced voltage on the deenergized busbar is measured and the grading capacitor value is known. The system cannot be in ferroresonance for an accurate estimate to be made.

When bus A2 is deenergized, it remains capacitively coupled to the parallel bus B2. A first order model of the capacitive coupling between conductors was created using the data given in Fig. 1b. The equivalent bus capacitance to ground value on each phase (6450 pF) is too low to match field measurements of the steady-state coupled voltage under non-ferroresonant conditions. An additional value of 4000pF to ground was added to account for stray capacitances due to equipment such as bushings, etc.

D. Grading Capacitors

Grading capacitors are necessary for obtaining the proper voltage distribution across multi-breaks as well as improving the short line fault TRV. The total amount of grading capacitance on the Dorsey bus has increased over the last two years as circuit breakers have been upgraded to a 63 kA standard interrupting rating. The magnitude of grading capacitance is a critical parameter for determining the probability and severity of ferroresonance.

A total of 12 circuit breakers can be connected to bus A2 following an outage. Seven breakers are SF6/CF4. Each has 2 breaks per phase with a 1500 pF capacitor across each break. Five breakers are minimum oil with 4 breaks per phase. Four of these five breakers are equipped with a 1350 pF capacitor across each outside break and 1250 pF across the inside breaks. The fifth breaker has 1350 pF across each break.

At the time of the bus deenergization, the disconnects on three circuit breakers were open leaving a total of 5061 pF connecting bus B2 with bus A2. The maximum grading capacitance of all 12 breakers is 6885.5 pF. Typical tolerances in grading capacitor values are ±2%. Therefore, considering possible circuit breaker upgrades, the maximum circuit breaker grading capacitance operating range is between 325 and 7500 pF.

E. Potential Transformers

The two wound potential transformers involved in the ferroresonant disturbance (PT1 and PT2) are both 4 kVA, 138000 Volt grounded wye-115V/69V 1200/2000:1 voltage transformers. Critical parameters for the wound PT include core losses, winding resistance and exciting current.

The losses measured in an excitation test are the iron-core losses which are due to the cyclic changes in flux. The iron-core losses are composed of both hysteresis and eddy current losses. Hysteresis losses are a nonlinear function of flux linkage (ϕ) and a linear function of frequency (f). They represent the work done in re-orienting the magnetic moments of the material as the magnetizing curve is traversed. Eddy current losses are proportional to flux linkage squared and frequency squared. The eddy current losses represent the losses in the core caused by circulating currents. A constant resistance was chosen to model the iron-core losses (1).

$$P_{iron-core} = \frac{V_{rms}^2}{R_m} = \frac{(\omega \cdot \phi)^2}{2 \cdot R_m} = \frac{2 \cdot \pi^2}{R_m} \cdot \phi^2 \cdot f^2 \qquad (1)$$

The approximation is valid if one considers that typically hysteresis losses are only one-third the eddy current losses[7]. Dick and Watson [11] have suggested that hysteresis may be safely neglected in cases of high-current ferroresonance. Detailed study of subharmonic ferroresonance may warrant the use of more accurate hysteresis models [8]. A value of 200 watts (no-load loss) per phase at nominal voltage (115 V) was used for each PT. The corresponding value for R_m reflected to the primary is 95.2 MΩ.

In regions of high saturation, the value of resistance (R_m) is no longer constant but is greatly reduced thus increasing the effective iron-core losses [9]. The change is due to the magnetic properties of the core and to supplementary losses. By ignoring these additional losses, an additional safety margin is ensured in a mitigation scheme.

The leakage inductance of the primary winding is assumed to be negligible and is set at .001 ohms. The remaining component values are listed in Fig. 1c. A burden of 75 VA is assumed to be the normal PT loading. The R (163.2 Ω) and X (.268 Ω) values for the PT burden were taken from CSA standard C13-M83.

Manufacturer's data was used to model the PT saturation characteristic. The fully saturated (air-core) inductance was not available. A value of 62 H was assumed as it produced a ferroresonant state which matched field recordings. The magnetization curve used in the EMTP model is shown in Fig. 5.

F. Station Service Transformer

The station service transformer (SST) is a 10 MVA, 230 kV grounded wye - 4.16 kV wye transformer. The secondary is grounded through a 2.4 ohm neutral resistor rated at 2400 volts and 1000 amps for 10 seconds. The positive sequence impedance is 12.2%. Core losses are 17 kW at nominal voltage. Manufacturer's data was used to model the saturation characteristic. The fully saturated (air-core) inductance was calculated by the manufacturer to be 884 ohms (2.34 H). The air-core inductance was used to calculate an extrapolated point on the magnetization (φ/i) curve shown in Fig. 5.

G. Benchmark Case Results

An oscillogram was located which shows the A2 Bus voltage for the Dorsey ferroresonant event of 95/05/20 at 22:04, after the last breaker cleared the bus. The trace is included in Fig. 2 for reference. The oscillogram shows phase A and B experienced chaotic oscillations for the first 700 ms before settling into a steady state fundamental frequency ferroresonant state. Phase C did not experience ferroresonance.

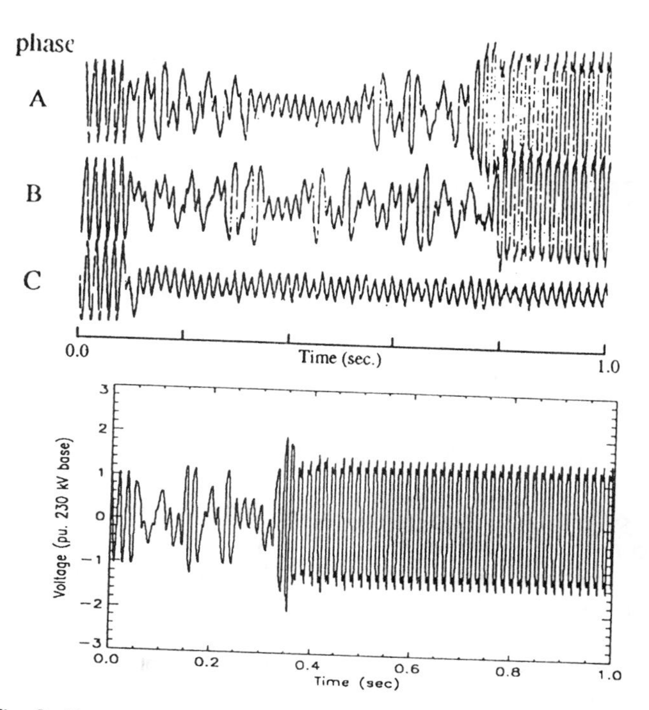

Fig. 2. Benchmark case results. Top: recording. Bottom: EMTP.

The particular chaotic pattern shown in the oscillograph is almost impossible to duplicate. It is a function of the breaker opening times, pre-switch voltage and the exact values of all parameters in each phase. Benchmark tests showed for example that by varying the time step either all phases could jump into the final ferroresonant state immediately or any phase or combination of phases could remain at low voltage. A typical EMTP simulation is shown at the bottom of Fig. 2.

III. EMTP Study Summary

The study recommended revising the manual bus clearing procedure, installing a permanently connected 200 ohm/phase load on the secondary of the station service transformers (SSTs) and to lock a filter on the bus in the event of a SST outage. The recommendations were based on the criterion that no ferroresonance is acceptable. A longer term solution which is being evaluated is the replacement of the PTs with capacitive voltage transformers (CVTs).

Revised switching guidelines reduce the risk of ferroresonance by minimizing the grading capacitance coupling the two buses. Disconnects on all but two breakers are opened thus limiting the maximum grading capacitance to 1500 pF. Damping resistors are not required in this case if the SST is in-service.

Several mitigation measures were evaluated for suitability of preventing ferroresonance following automatic bus clearing. Permanently connected damping resistors connected to the secondary of the SST were found to be the most suitable short term mitigation measure. Resistors were installed in September, 1995. Trip testing (Fig. 3) has shown the resistors prevent ferroresonance.

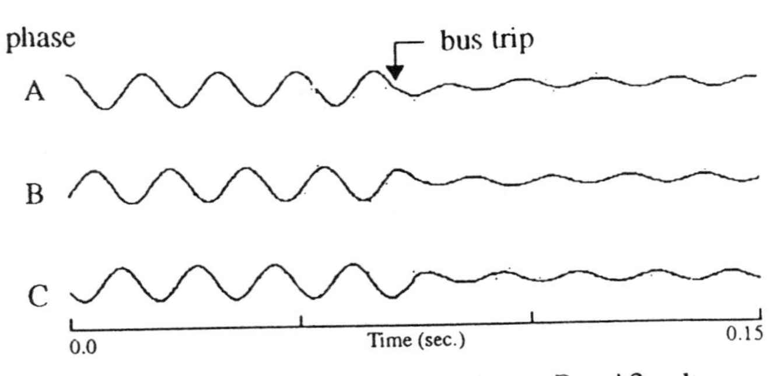

Fig. 3. Trip testing with 200 ohm resistors. Bus A2 voltage.

Other options that would also work include: a permanently bus connected 2 MVAr air-core reactor or filter (80-100 MVAr), automatic disconnection of PT/SST via motor operated disconnects or a three phase grounding switch. The decreased system security, reduced reliability, operating inflexibility and cost preclude the use of these options. Bus surge arresters are not feasible as large overvoltages did not occur.

Permanently connected resistors on the PT secondaries are not feasible since the minimum required per phase load is at least double the 4 kVA thermal rating. The continuous losses associated with a permanent resistor can be avoided by switching a resistive load upon detection of a bus isolation signal. The degraded reliability associated with switched elements makes this option unattractive. As well, the resistive load needs to be increased (up to 40%) in order to extinguish an established ferroresonant state. The damping resistor needs to dissipate energy faster than the system can supply energy in order for the ferroresonant state to collapse into a non-ferroresonant 60 Hz operating mode. If the rate of dissipation is too slow, the ferroresonance will stabilize at a new lower energy state.

IV. State Variable Modelling

As was previously mentioned, the exact replication of a chaotic system transient is next to impossible. Small variations in system parameters or in the nature of the disturbance can lead to a completely different final state. In order to gain necessary insight into the ferroresonance circuit, a simplified single-phase equivalent state variable model was constructed. This model has the flexibility needed to perform systematic evaluations such as bifurcation analysis.

A. Derivation of Equations

The essential features of the detailed EMTP model are retained and are shown in Fig. 4.

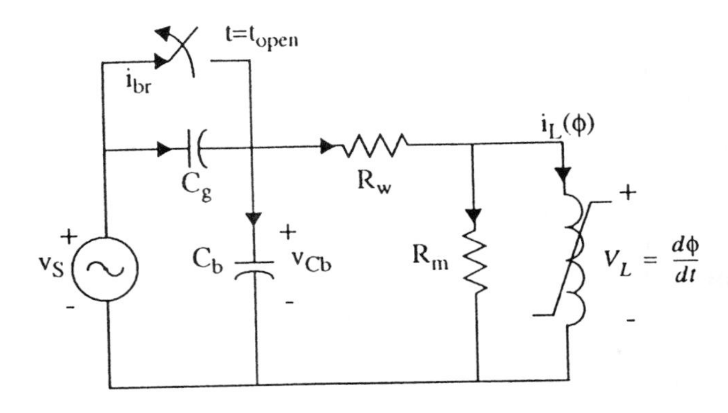

Fig. 4. Ferroresonant state variable model.

The equations describing the dynamics of the circuit may be derived.

Solving for flux linkage:

$$\frac{d\phi}{dt} = -\frac{R_m \cdot R_w}{R_m + R_w} \cdot i_L(\phi) + \frac{R_m}{R_m + R_w} \cdot V_{Cb} \quad (2)$$

and for bus voltage:

$$\frac{dV_{Cb}}{dt} = \frac{-(V_{Cb} + R_m \cdot i_L(\phi))}{(R_m + R_w) \cdot (C_g + C_b)} + \frac{C_g}{C_g + C_b} \cdot \frac{dV_S}{dt} \quad (3)$$

The nonlinear inductor is represented as a continuous curve of the form given by (4)[6] rather than a piecewise linear representation used by EMTP.

$$i_L(\phi) = a \cdot \phi + b \cdot \phi^n \quad (4)$$

If current and flux are per-unitized, the constants a=.1, b=.06 and n=15 provide a reasonable approximation to the nonlinear PT saturation characteristic (Fig. 5). The SST characteristic up to the kneepoint can be well represented if a=.001, b=.005 and n=9. Beyond the kneepoint, the magnetization curve is unknown except that the slope should asymptotically approach the air-core inductance. The flux contained by the iron core under full saturation (ϕ_s) can be obtained by linearly extrapolating the highly saturated values back to the flux linkage axis. Dick and Watson [11] have made some measurements and determined that ϕ_s lies in the range 1.18 to 1.345 times the nominal flux. It is expected that the true ϕ/i curve will lie somewhere between the polynomial and piecewise linear approximations. More experimental work in transformer core modelling during ferroresonance would be of benefit.

A set of initial state equations can be derived with the circuit breaker closed. Equation (2) remains unchanged, however, equation (3) is greatly simplified (5).

$$V_{Cb} = V_S = E \cdot \cos(\omega t) \quad (5)$$

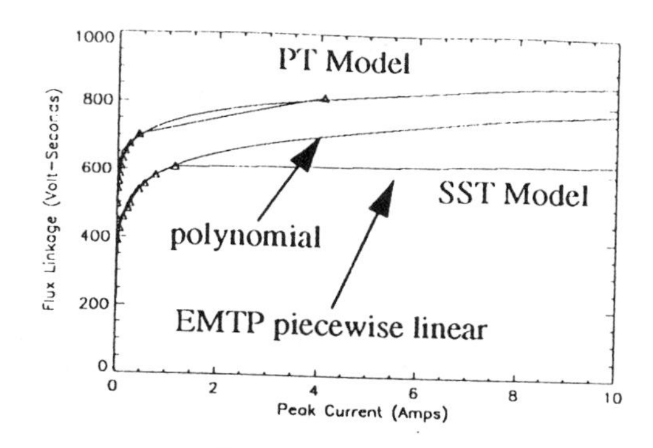

Fig. 5. Magnetization curves.

The transition from the initial to the final set of state equations occurs at a current zero. The exact time step where a current zero occurs is found by detecting a change in sign of (6).

$$i_{br} = C_b \cdot \frac{dV_{Cb}}{dt} + \frac{1}{R_m} \cdot \frac{d\phi}{dt} + i_L(\phi) \tag{6}$$

B. Solution Method

The set of first-order differential equations (2)-(3) were solved using a fourth order Runge-Kutte numerical integration technique. A fixed time step of 50 µsec was found to be a good compromise between accuracy and cpu time.

Comparisons were made between the detailed EMTP model and the simple single-phase model and were found to be in good agreement.

V. Nonlinear Dynamics Analysis Method

Concepts such as attractors, bifurcation and basins of attraction provide the best available framework for discussion and analysis of the ferroresonance phenomenon. These ideas and their applications are discussed below.

A. Characterization of Attractors

An attractor is a description of the final operating state of a dynamic system. The attractor is most clearly illustrated by plotting the trajectory of the system response in state-space. Attractors can be characterized as being single points, periodic orbits or complex ribbons (chaotic attractor) folding through state-space. Two examples of periodic orbits are given in Fig. 6.

Poincaré sections are a useful method of displaying or characterizing the attractor. In this paper, Poincaré sections refer to slices across the time axis of a two dimensional attractor (e.g. flux(ϕ) vs. voltage (V_{Cb})). The sections are calculated by sampling the state variables once per 60 Hz drive cycle at a fixed phase angle. The aggregate of all slices (i.e. Poincaré sections covering 0 to 360 degrees) is the full attractor.

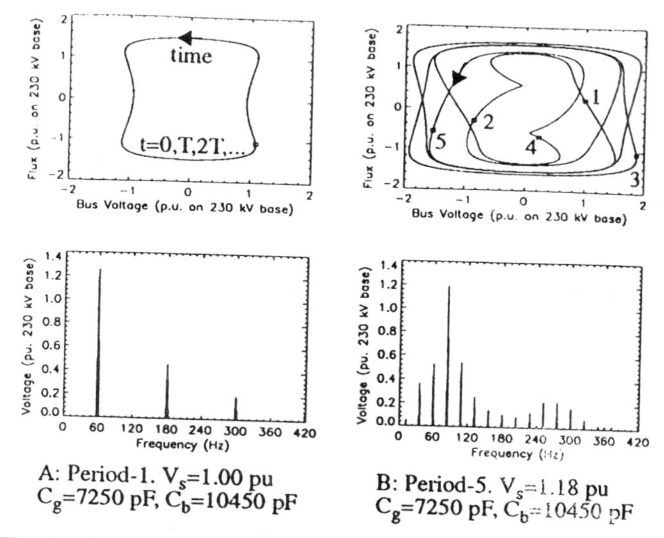

A: Period-1. V_s=1.00 pu
C_g=7250 pF, C_b=10450 pF

B: Period-5. V_s=1.18 pu
C_g=7250 pF, C_b=10450 pF

Fig. 6. Characterization of attractors. Top: state space trajectory (SST flux vs. V_{Cb}). Bottom: spectrum (DFT of V_{Cb}).

The ratio of the natural period of the dynamic system to the period of the drive is equal to the number of points in the Poincaré section. A single point indicates the frequency of the attractor is the same as the sampling frequency and is referred to as "period-1" or fundamental frequency ferroresonance.

A discrete Fourier transform (DFT) of the time series can also be used for characterization. The lowest frequency of a period -5 attractor, for example, is (60/5) or 12 Hz. The frequency spacing can correspond to integer or odd multiples of the lowest frequency.

B. Traditional Bifurcation and Basin of Attraction Analysis

A bifurcation is the abrupt change in the qualitative nature of the system's final operating state as a system parameter is quasi-statically varied. The type of bifurcation is of interest. Period doubling (or pitchfork) bifurcations refer to the emergence of a periodic orbit near another periodic orbit and has been shown to be a route to chaos [10]. Jump resonances refer to a sudden change in the solution of the differential equation without a change in its period.

The basin of attraction is normally displayed as the set of initial conditions leading to a particular attractor. Given a set of system parameters, a pair of initial conditions (e.g. ($\phi(0)$,$V_{Cb}(0)$)) can be chosen on the two-dimensional phase plane and the final trajectory calculated. At the coordinates of the initial conditions chosen, a symbol representing the final attractor is placed.

Practically, the initial conditions do not cover the entire phase plane. The initial conditions are prescribed at the instant the circuit breaker opens. Therefore, only two initial conditions (or six in the three phase model) are possible for a normal unfaulted bus clearing since the breaker opens at a current zero.

C. Hybrid Approach

Systematic analysis of our practical circuit requires the mapping of the final operating state as parameters are varied as in the traditional bifurcation diagram, yet each operating state must be derived from a rest state or pre-switched state as in a basin of attraction. Thus we have taken a hybrid approach by combining the traditional bifurcation and basin of attraction concepts.

A kick-initiated rather than a slow parameter varying approach was taken in the generation of bifurcation diagrams. The slow parameter varying approach exhibits hysteretic mode transitions depending on whether the parameter is being increased or decreased [1]. The "kick" is initiated at time t_{open}=.04, when the circuit breaker is ordered to open. After one second, the transients are assumed to have decayed leaving only the final attractor. A Poincaré section is then taken for thirty cycles, thus recording the essential features of the attractor (i.e. the periodicity). As an example, a bifurcation diagram is shown in Fig. 7 for the case of two parallel PTs with 10450 pF of bus capacitance and 5080 pF of grading capacitance. As the source voltage is varied between two extremes, the character of the bifurcation diagram changes greatly.

Classic period doubling bifurcations begin at 6.7 pu source voltage. At low values of source voltage, transitional chaos dominates. Two or three dominant attractors are fighting to determine which will be the system's final state.

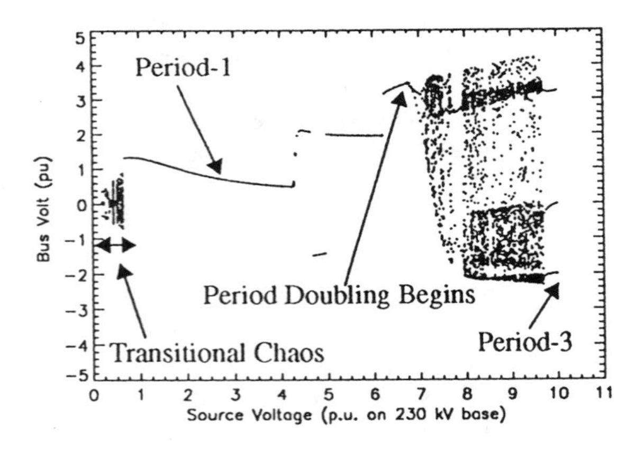

Fig. 7. Bif. diagram. Two PTs.C_g=5080 pF, C_b=10450 pF

The region of transitional chaos is the most interesting from a practical point of view. An alternative way of visualizing the dynamics in this region is to construct a two-dimensional (2D) bifurcation diagram. Two system parameters are selected to be varied and the final state of the system determined. A symbol is used to represent the final attractor's periodicity. For the ferroresonant circuit under study, the source voltage and grading capacitance may vary the most during normal system operation. Other parameters will remain fixed, however there may be some uncertainty in their values which would lead to an n-dimensional bifurcation diagram.

Two cases are shown to illustrate the improved visualization which can be realized by a 2D bifurcation diagram. The first case (Fig. 8) maps the ferroresonant regions of two parallel PTs. A parallel SST is included in the second case (Fig. 9).

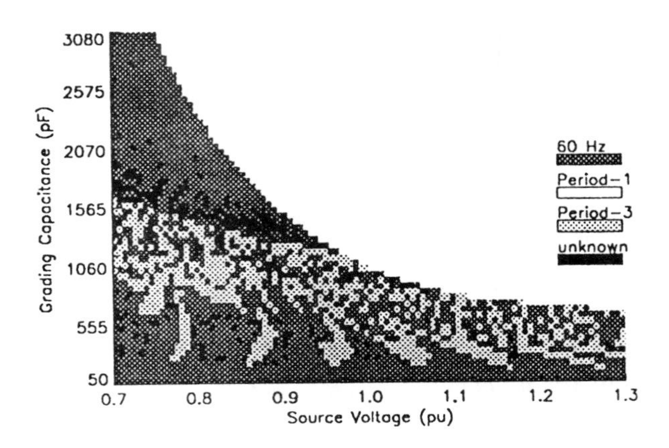

Fig. 8. 2D bifurcation diagram. Two parallel PTs.

The states marked unknown in Fig. 8 are still indeterminate after 15 seconds due to the low damping of this circuit.

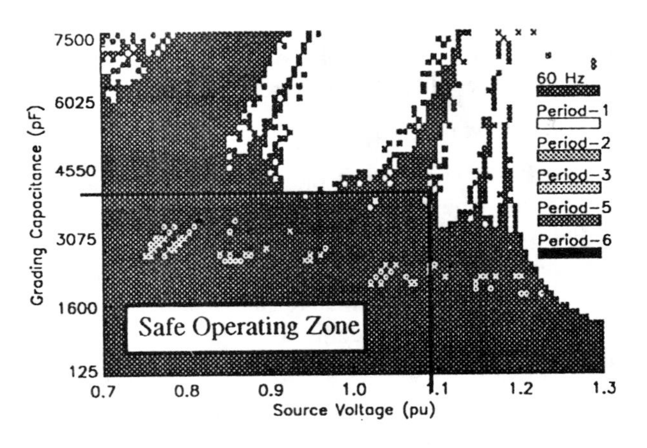

Fig. 9. 2D bifurcation diagram. Two parallel PTs and SST.

The main effect of including the SST is an increase in the steady state losses occurs which causes the first transition to fundamental frequency (period-1) ferroresonance to shift to a higher grading capacitance value. The transitional chaos region is not as complex as for the two PT case as fewer subharmonic states are sustained.

One possible "safe" operating zone is indicated in Fig. 9. If the source voltage is kept below 1.1 pu and the grading capacitance is kept below 4000 pF, the risk of ferroresonance is minimized. The maximum grading capacitance in our circuit could be as high as 7500 pF which requires mitigation. Inclusion of permanently connected 200 ohm/phase damping resistors on the secondary of the SST extends the safe operating zone such that the entire plane in Fig. 9 is covered.

VI. CONCLUSION

Digital simulation has been used to study ferroresonance in a 230 kV circuit. The study has shown that there are sufficient losses in a station service transformer to prevent ferroresonance up to a grading capacitance value of 4000 pF. Due to circuit breaker upgrades, the maximum grading capacitance could reach 7500 pF which requires mitigation. A 200 ohm/phase damping resistor installed on the secondary of the SST eliminates ferroresonance for faulted and unfaulted bus clearing. The resistor increases the effective core losses of the SST by a factor of six. If the SST is unavailable, ferroresonance will develop in wound potential transformers even at minimum grading capacitance.

A two dimensional bifurcation diagram is an ideal tool for visualizing the behavior of a ferroresonant circuit. The margin of safety between ferroresonant and non-ferroresonant states can clearly be seen. A kick-initiated approach should be used in the generation of bifurcation diagrams since it mirrors true system conditions and does not exhibit hysteretic bifurcations.

REFERENCES

[1] Mork, B.A. and Struehm, D.L.,"Application of Nonlinear Dynamics and Chaos to Ferroresonance in Distribution Systems", *IEEE Trans. on Power Delivery*, Vol. 9, No. 2, pp. 1009-1017, April 1994.

[2] Rudenberg, R., *Transient Performance of Electric Power Systems*, New York: McGraw Hill, pp.642-656, 1950.

[3] Hayashi,C., *Nonlinear Oscillations in Physical Systems*, New York: McGraw Hill, 1964.

[4] Janssens, N., et. al.,"Direct Calculation of the Stability Domains of Three-Phase Ferroresonance in Isolated Neutral Networks with Grounded-Neutral Voltage Transformers", presented at the *1995 IEEE/PES Summer Meeting*, 95 SM 420-0 PWRD.

[5] Swift, G.W.,"An Analytical Approach to Ferroresonance", *IEEE Trans. on Power Apparatus & Systems*, Vol. PAS-88, pp. 42-46, January 1969.

[6] Mozaffari,S., Henschel, S., and Soudack, A.C.,"Chaotic Ferroresonance in Power Transformers", *IEE Proc. Gen. Transm. Distr.*, Vol. 142, No. 3, pp.247-250, May 1995.

[7] Swift, G.W.,"Power Transformer Core Behavior Under Transient Conditions",*IEEE Trans. on Power Apparatus and Systems*, Vol. PAS-90, No.5, pp. 2206-2210, Sept./Oct. 1971.

[8] de León, F., and Semlyen, A., "A Simple Representation of Dynamic Hysteresis Losses in Power Transformers", *IEEE Trans. on Power Delivery*, Vol. 10, No. 1, pp.315-321, January 1995.

[9] Janssens,N.,et. al.,"Elimination of Temporary Overvoltages due to Ferroresonance of Voltage Transformers: Design and Testing of a Damping System",*CIGRE 1990*, Paper No. 33-204.

[10] Tan. C., et. al.,"Bifurcation, Chaos, and Voltage Collapse in Power Systems", *Proceedings of the IEEE*, Vol. 83, No. 11, pp. 1484-1496, November 1995.

[11] Dick, E.P. and Watson, W.,"Transformer Models for Transient Studies Based on Field Measurements", *IEEE Trans. on Power Apparatus and Systems*, Vol. PAS-100, No. 1, pp. 409-419, January 1981.

David A. N. Jacobson received the degrees of B.Sc. (E.E.) with distinction and M.Sc. from the University of Manitoba in 1988 and 1990 respectively. Currently, he is pursuing a Ph.D. From 1990 until the present, he has been a planning engineer with Manitoba Hydro. He was a visiting researcher with the Siemens Power System Planning group in 1994. Research interests include nonlinear dynamics, power system control and FACTS devices. Mr. Jacobson is a registered professional engineer in the province of Manitoba.

David R. Swatek obtained his B.Sc.(EE) from the University of Manitoba in 1988 and is presently engaged in Doctoral studies in the field of applied electromagnetics. He joined Manitoba Hydro in 1988 where he currently holds the position of Transmission Development Engineer, and is responsible for the study of novel transmission line technology and the modeling of electromagnetic phenomena. Mr. Swatek's main interests include the propagation and scattering of electromagnetic waves, and Russian existentialism. He has been a registered engineer with the association of Professional Engineers of the Province of Manitoba since 1990

Ronald W. Mazur obtained his B.Sc. degree in 1971 and his M.Sc. degree in 1989, both in Electrical Engineering from the University of Manitoba. He joined Manitoba Hydro in 1974 where he has worked in switching station design, system operation and system planning. He presently holds the position of Senior Planning Engineer, and is responsible for high voltage transmission development, grid supply transmission and interconnection planning. He has been a registered engineer with the Association of Professional Engineers of the Province of Manitoba since 1974.

ANALYSIS OF A HARMONIC OVERVOLTAGE DUE TO TRANSFORMER SATURATION FOLLOWING LOAD SHEDDING ON HYDRO-QUEBEC - NYPA 765 KV INTERCONNECTION

Omer Bourgault

Gaston Morin, Member, IEEE

Hydro-Quebec, Montreal, Canada

Abstract - The paper presents an analysis of the harmonic overvoltage recorded on the interconnection between Hydro-Quebec and the New York Power Authority (NYPA) in the Châteauguay area. The overvoltage was caused by transformer saturation, following tripping of the 765-kV transmission line linking both systems. EMTP simulation yielded good agreement with field measurements and was valuable in re-evaluating operating practices at the Châteauguay station.

INTRODUCTION

The Châteauguay-Beauharnois complex (Fig. 1) is mainly operated to export energy from the Hydro-Quebec power system to the United States. Energy from part of the radially operated Beauharnois generating station and if necessary, from the main 735-kV system (via 1000-MW converters at Châteauguay) is carried by line 7040 as far as Massena substation in New York State (NYPA).

Figure 1 - The Châteauguay-Beauharnois complex.

89 SM 602-4 PWRD A paper recommended and approved by the IEEE Transmission and Distribution Committee of the IEEE Power Engineering Society for presentation at the IEEE/PES 1989 Summer Meeting, Long Beach, California, July 9 - 14, 1989. Manuscript submitted January 17, 1989; made available for printing April 21, 1989.

This paper describes harmonic overvoltages observed following load-shedding on the interconnection in March 1988. When line 7040 tripped, it was carrying 240 MW from six units at Beauharnois (Fig. 2) via line 1363 and three 125/765-kV transformers (840 MVA ea.) towards the NYPA system. Converter 1 was blocked, while transformer T105 and the 120-kV buses were energized, and converter 2 was isolated. The power line carrier (PLC) filters connected to the 120-kV buses are designed to attenuate high-frequency noise generated by the converters. The overvoltages were recorded at the 765-kV Châteauguay bus by a continuous-monitoring system and by the potential transformers on the 120-kV buses. The locations of the measuring equipment are shown in Fig. 2.

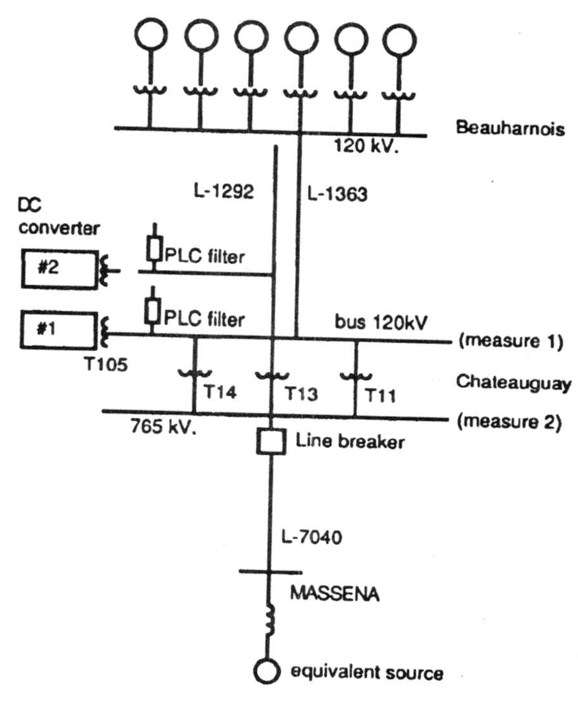

Figure 2 - System configuration before load shedding.

After the 765-kV line tripped, the Beauharnois generating units remained connected to the 120-kV equipment at Châteauguay substation. This configuration presents a state of parallel resonance at 350 Hz between the 120-kV bus PLC filters and the degraded system. This led to generator overspeed and overvoltages causing symmetrical saturation of the 125/765-kV transformers.

The voltage regulators, which are very slow, did not manage to counter the overvoltages, with the result that in saturating, the transformers acted as sources of harmonic currents, building up harmonic voltages through the system impedance [1]. The generator overspeeding increased the fundamental frequency of the voltage, so that the 5th-harmonic currents excited the system resonance, when system frequency reached 70 Hz (5 X 70 Hz = 350 Hz) creating overvoltages of up to 1.8 p.u. at 120 kV.

0885-8977/90/0100-0397$01.00 © 1989 IEEE

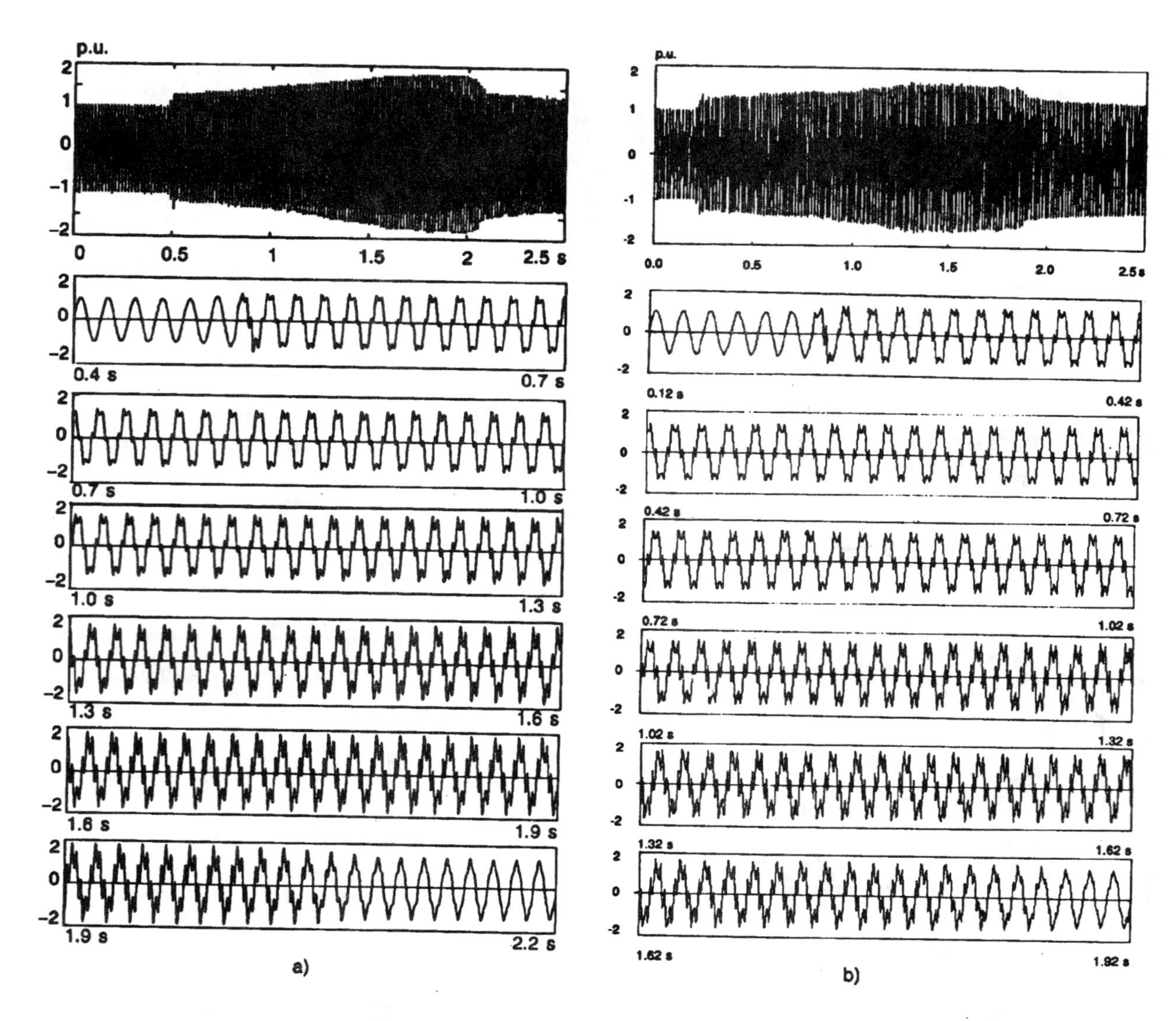

Figure 3 – 120–kV voltage after line 7040 tripped.
a) Measured b) Simulations

ANALYSIS OF HARMONIC CONTENT OF 120-kV VOLTAGE

Fig. 3a shows the 120-kV voltage (phase C, measured), following the tripping of line 7040, with the overvoltage reaching 1.8 p.u. after 1.5 s. Substantial 5th-harmonic distortion can be seen on the 120-kV voltage. Further analysis also reveals the presence of a 7th-harmonic component of much lower amplitude. The simulation (Fig. 3b) correlates well with the field measurements.

Fig. 4 shows the variation in amplitude of these harmonics as a function of time; the analysis was made over a 4-cycle window with a 2-cycle overlap. The amplitude of the fundamental frequency increases to 1.29 p.u. whereas the frequency is raised to 72 Hz in 1.7 s by the generator overspeed. The maximum amplitude of the 5th harmonic reaches 0.66 p.u. and 0.13 p.u. for the 7th harmonic.

The 120-kV lines and the transformers remained connected radially to the Beauharnois generators. This configuration presents a parallel-resonance condition at 350 Hz between the PLC-filters capacitance and the system inductance as shown in Fig. 5. Quality factor used in that calculation doesn't consider the effect of frequency dependence of losses.

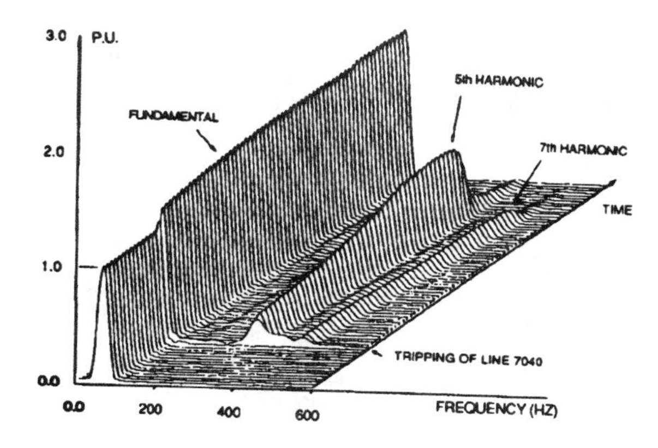

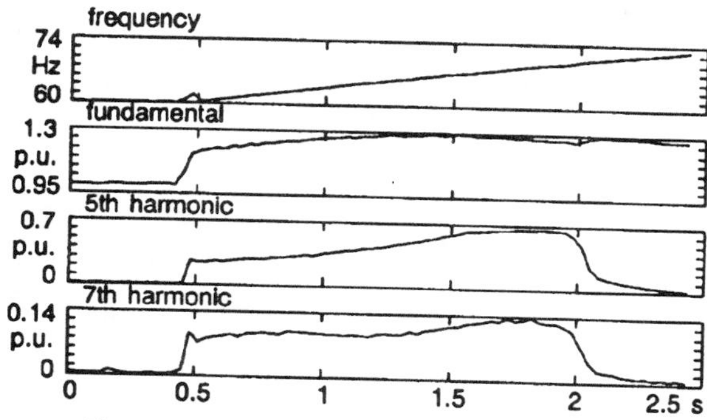

Figure 4 - Analysis of the harmonic content of the 120-kV Voltage.

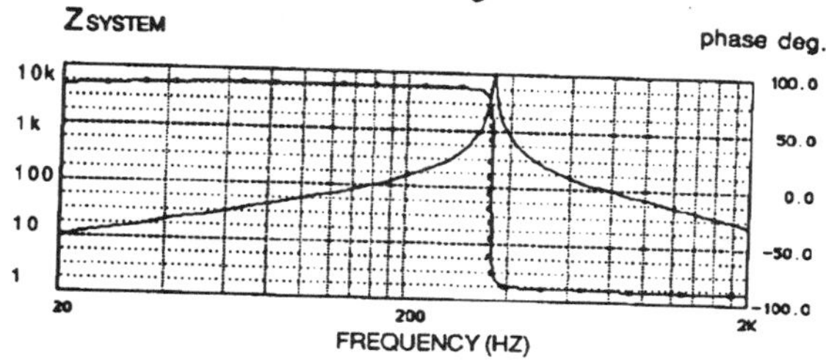

Figure 5 - Impedance of the degraded system following tripping of line 7040.

A theoretical analysis of harmonic currents generated by a transformer under symmetrical saturation is presented in the appendix. The simplified magnetization characteristic of a single-phase system transformer is represented by two slopes in a flux-current diagram, when losses are not considered: a quasi-infinite slope under unsaturated conditions, the other representing the impedance ωL_s of the saturated transformer for excitation beyond its saturation knee (see Fig. A-1 in the appendix). For a sinusoidal overvoltage applied to the transformer, its current has the same shape as the thyristor-controlled-reactor (TCR) branch current in a static compensator. Fourier analysis of such currents reveals that it consists exclusively of odd harmonics, whose amplitude varies with the amplitude of over-excitation. In addition, the phase of these odd harmonics in relation to the fundamental component of the magnetization current is either 0° or 180°, also depending on the amplitude of over-excitation. This stems from the fact that the coefficients a_n as well as the even coefficients b_n in the general Fourier series expression of the magnetizing current, are zero. A harmonic in phase with its fundamental is defined, when one zero crossing going positive occurs simultaneously with that of the fundamental. It should be recalled that the Fourier series of a 2π periodic function is expressed by the equation:

$$f(t) = \frac{a_o}{2} + \sum_{n=1}^{\infty} (a_n \cos n\omega t + b_n \sin n\omega t)$$

with
$$a_n = \frac{1}{\pi} \int_0^{2\pi} f(t) \cos n\omega t \; d(\omega t)$$

$$b_n = \frac{1}{\pi} \int_0^{2\pi} f(t) \sin n\omega t \; d(\omega t)$$

At the start of the overspeed ($5f < 350$ Hz), the 5th-harmonic currents passed through a system of inductive equivalent impedance, building up 5th-harmonic voltages leading by $\pi/2$. It should be stressed that these currents are in phase with the fundamental component of the magnetizing current, which itself lags the fundamental voltage by $\pi/2$ as shown in Fig. 6. The vector diagram of 5th-harmonic voltages and currents in this figure turns counter-clockwise five times faster than the 60-Hz diagrams. The vectors are shown at the point where the zero crossing (going positive) of the fundamental component of the magnetizing current is simultaneous to that of its 5th-harmonic component. Assuming that the transformer magnetizing current is defined as positive when it enters at the "polarity mark", the 5th-harmonic current produces a voltage drop through the system impedance, which is subtracted from the fundamental-frequency voltage. This corresponds to superimposing a 5th-harmonic voltage lagging fundamental-frequency voltage by π.

Approximately 1.5 s later, the 5th-harmonic component of the magnetizing current reaches 350 Hz ($5f = 350$ Hz $= f_o$), by which time it has built up an in-phase 5th-harmonic voltage in the system, whose equivalent impedance is resistive. This corresponds to superimposing a 5th-harmonic voltage lagging the fundamental-frequency voltage by $\pi/2$.

Finally, when the 5th harmonic of the fundamental frequency of the degraded system exceeds resonance ($5f > 350$ Hz), the 5th-harmonic current injected into a capacitive equivalent impedance system, builds up 5th-harmonic voltages lagging by $\pi/2$. This corresponds to superimposing a 5th-harmonic voltage in phase with the fundamental-frequency voltage.

From the 120-kV voltage (phase C) signal in Fig. 3, the overvoltages are observed to be much higher before resonance and to diminish very rapidly afterwards (that is, when the 5th-harmonic voltage is added in phase with the fundamental). This is explained by the increase in the amplitude of the fundamental component of the 120-kV voltage, for example, from 1.2 p.u. to 1.29 p.u., while the frequency rises from 61 Hz to 70 Hz. The transformer flux linkage, which is the integral of the applied voltage, consequently decreases from 1.18 p.u. to 1.10 p.u. In this way, as the frequency increases, transformer over-excitation decreases, gradually reaching an unsaturated state (below the saturation knee). This strongly reduces the magnetizing current and its 5th-harmonic component.

It should be emphasized that the mathematical development of the magnetizing current of a transformer in a state of sinusoidal over-excitation (appendix) is based on a two-slope representation, although total accuracy cannot be achieved for over-excitation values slightly above the saturation knee.

Measurements taken on large system transformers [2] show that the knee is generally located around 1.1 p.u. and that the slope goes gradually from a very high value without saturation to the saturation slope (L_s), reflecting a more rounded knee than the characteristic shown in Fig. A-1.

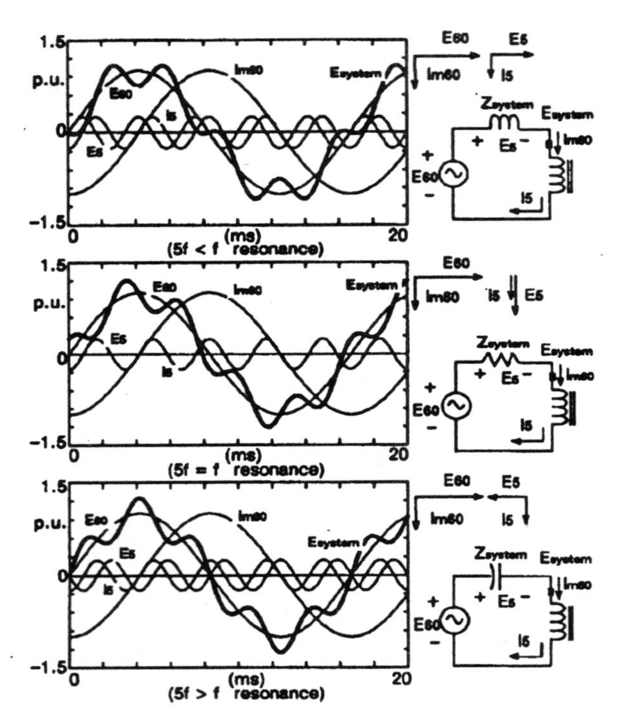

Figure 6 - 5th-harmonic phase shift vs system resonance.

SIMULATION MODELS

The simulated system (Fig. 2) basically comprises hydraulic generating units at Beauharnois (6 X 48 MVA), two 120-kV lines (1292, 1363), three 125/765-kV transformers (840 MVA ea.) and two PLC circuits at Châteauguay substation, a 765-kV line (7040) and a 60-Hz equivalent system at Massena substation. The simulations were carried out on the Electro Magnetic Transient Program (EMTP) [3]. Models are shown schematically in Fig. 7.

Source

The Beauharnois generating units were represented using a complete model of synchronous machines using Park's equations, disregarding saturation. In addition, it has been shown that voltage regulator (IEEE type 1 [4]) of these machines are very slow. The voltage is regulated as follows: when the error is less than 5% the regulator is switched to transfer function 1/(1 + 20 s); when more than 5%, the regulator is switched to a faster function, namely 1/(1 + 1 s). This excitation system was simulated using Transient Analysis of Control Systems (TACS) functions [3].

The NYPA system was represented using a 60-Hz Thevenin equivalent with a short-circuit level of 5000 MVA at Massena 765-kV substation, which resulted in a contribution of about 3000 MVA at Châteauguay substation (120-kV side).

Lines

The total contribution of the capacitive effect of lines 1292 and 1363 is approximately 0.26 μF. Theoretically, the line parameters vary as a function of frequency. As far as the positive sequence resistance R and inductance L are concerned, these parameters may be assumed to be constant, at least up to 1 kHz. On the other hand, zero sequence R and L are highly dependent on frequency [5] but this dependence was not represented and the short (11 km) 120-kV lines were simulated using a single π section.

Line 7040 (90 km) is represented simply using a π model since only the steady state preceding the switching operation need be thoroughly simulated.

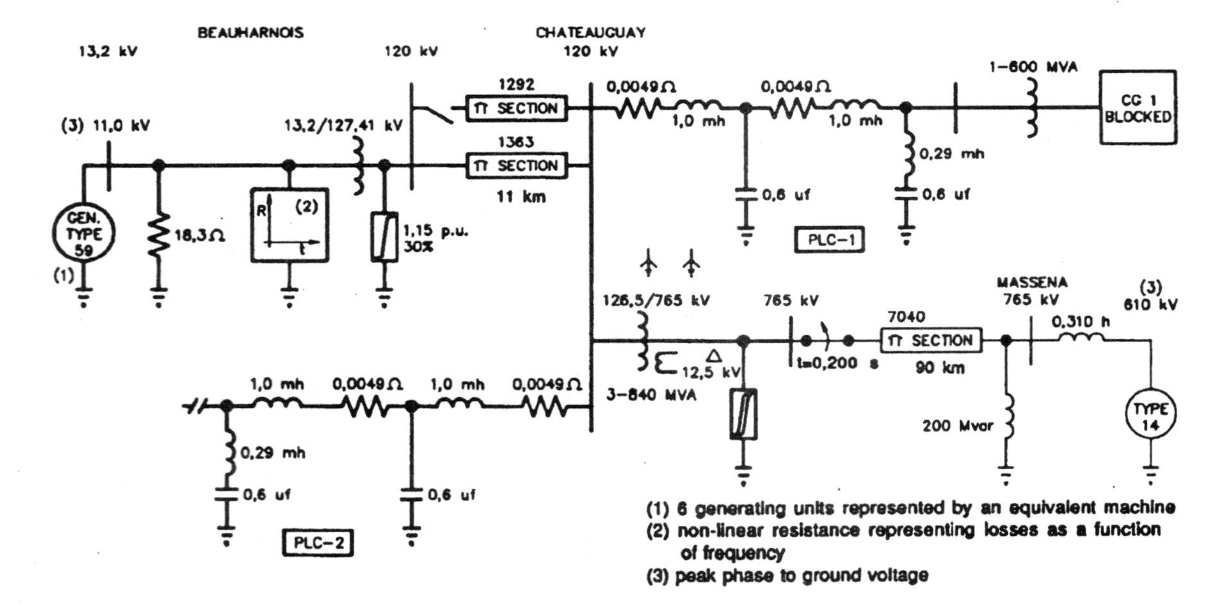

Figure 7 - Beauharnois - NYPA simulated system.

Transformers

Transformer modeling constitutes a major aspect of this type of study. Since the transformer is a source of harmonics, its magnetizing characteristic must be represented accurately. The most important transformers are the 125/765-kV units at Châteauguay totaling 2520 MVA. Since the hysteresis of one of these units had already been measured [2], it served as a model for this bank of transformers.

For the Beauharnois transformers, typical values were used, namely a saturation knee of 1.15 p.u. and a saturation slope of 30%. Regarding the transformer of converter 1, the manufacturer evaluates the saturation knee between 1.3 and 1.4 p.u., making its effect negligible in the context of this study.

Representation of losses vs frequency

The main difficulty in simulating resonant systems is that of properly representing losses; these determine the impedance and overvoltages amplitudes at resonance. According to Hydro-Quebec data, the Beauharnois generators, have losses amounting to 3.7% of the full load at 60 Hz. these comprise:

- ventilation and friction
- no-load core losses
- nominal-load losses
- exciter losses
- rotor losses
- stator losses (0.4%)

It should be noted that for load shedding involving an increase in fundamental frequency (through an acceleration of the rotor), an increase in generator losses in general can be expected. Stator losses, represented by the armature resistance (Ra) of the machine, total 0.4%. A shunt resistance of 18.3 Ω connected to the generator terminals represents the remaining losses (3.3%).

Measurements of the operational impedance [6, 7] of a generator at Beauharnois showed that the generator may be represented by a complex vector, whose real part increases with frequency. This reflects a quality factor decreasing with frequency. The curve in Fig. 8 suggests that the quality factor decreases rapidly at frequencies close to 60 Hz, reaching a minimum around 250 Hz.

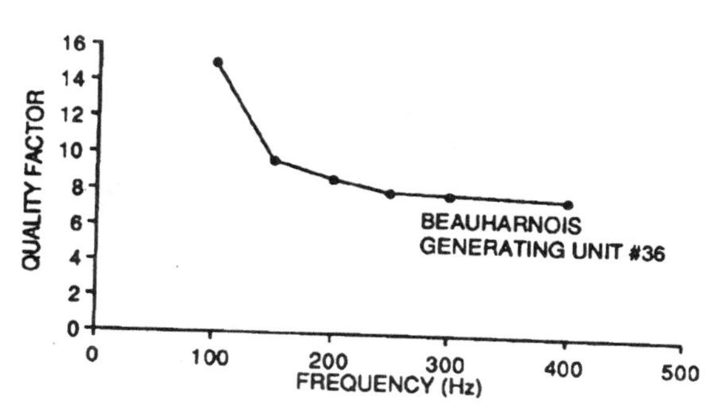

Figure 8 - Quality factor of a Beauharnois generating unit as a function of frequency.

The 60-Hz core losses of power transformers, were represented by a resistance parallel to the magnetizing branch in the case of Beauharnois and the transformer of converter 1 at Châteauguay. The Châteauguay 125/765-kV transformers, for their part, have a magnetizing branch represented by a hysteresis curve [2] ; the core losses are proportional to the included area in the hysteresis and rise with the fundamental frequency and its harmonics. Copper losses are represented at 60 Hz.

Research reveals that modeling losses vs frequency on the different types of equipment involved except for lines, is not treated fully in the literature. However representation of the losses is so significant in simulating this system, in view of the harmonics present, that a single 60-Hz representation produces solution in which the overvoltages are very high (> 3 p.u.) which interrupts the simulation. It was therefore decided to add a shunt resistor connected to the generator terminals, decreasing as a function of time and corresponding to a system-quality factor around 6 at 350 Hz. This is an attempt to represent the increase in losses with the rise in fundamental frequency and the presence of harmonics.

VALIDATION OF THE RESULTS

A 240-MW load shedding following the tripping of line 7040 produces a 60-Hz overvoltage followed by acceleration of the generators. The overvoltage excites the transformers connected to the bus whose magnetizing current immediately increases (Fig. 3 and 9) and the odd harmonics develop 5th-harmonic voltages in the system. The transformer over-excitation decreases with the rise in frequency to the point of dropping below the saturation knee. The magnetizing current then drops significantly simultaneously causing the harmonic overvoltages in the system to disappear.

The envelope of the magnetizing-current curve in Fig. 9 increases as the transformer excitation decreases owing to the fact that the harmonic distortions of the voltage applied to the transformers also distorts their flux linkage($\phi = \int V dt$) and affect their magnetizing current via their saturation slope. The phase of the magnetizing-current harmonic distortions thus shifts 180° just like the 5th harmonic of the voltage (Fig. 6) as the fundamental system frequency increases to scan the resonance. The system goes from an inductive (5f < fo) to a capacitive (5f > fo) equivalent impedance.

The generator terminal voltage, rectified and filtered, varying with 0, 1 or 2 PLC filters and rapid voltage regulator is shown with system frequency, in Fig. 10.

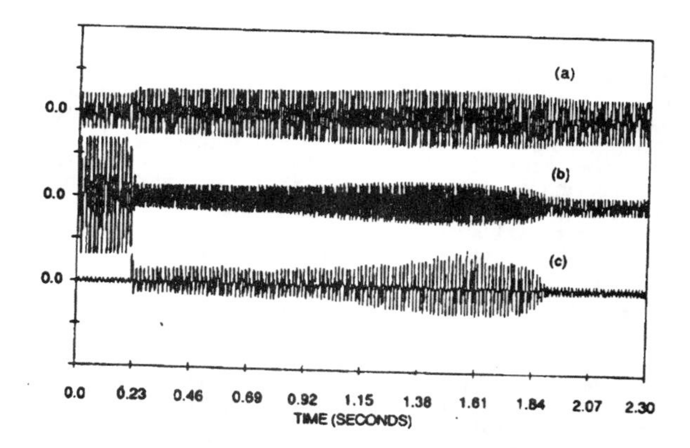

Figure 9 - Simulation results.
a) Châteauguay 120-kV phase C voltage
b) Beauharnois phase C stator current
c) 125/765-kV transformer magnetizing current phase C

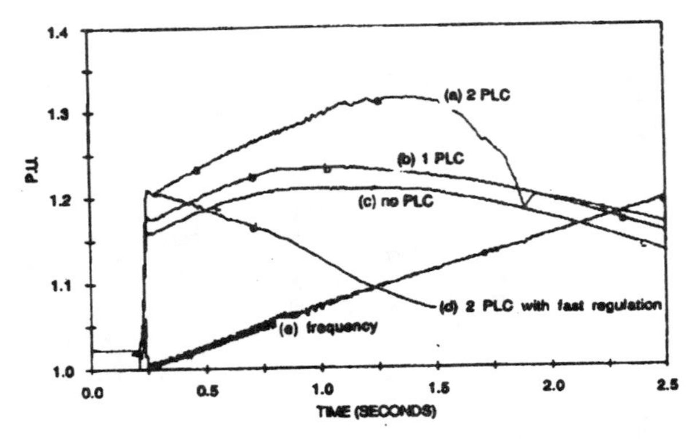

Figure 10 - Effect of the number of PLC filters in service and speed of regulation on Beauharnois generators terminal voltage (rectified and filtered).

EXTRAPOLATION

One PLC filter in service

With one PLC filter in service, a 42% reduction in the capacitance C shifts the resonance to about 450 Hz, which means that the fundamental should reach 90 Hz (5 X 90 Hz = 450 Hz) to scan this resonance with the 5th harmonic. This allows the voltage regulator to react and reduce the transformer over-excitation. The combined effect of transformers magnetizing currents of 5th and 7th harmonic (at a lower amplitude) acting out of phase each other, for over-excitation less than 1.25 p.u. (Fig.A-2) result in an overall reduction in 120-kV overvoltages.

No PLC in service

Disconnection of both PLC filters leaves a very low capacitance of about 0.4 µF, producing a first pole at about 830 Hz. For all practical purposes, overvoltages are developed at fundamental frequency (Fig. 10).

Note that the voltage increase at the machine terminals is similar to the case of a constant voltage behind X"d because the voltage regulator is so slow. Consequently the fundamental voltage increase when line 7040 trips, is function of initial-loading conditions and proportional to the capacitance connected on 120-kV buses (Fig. 10).

Effect of a fast voltage regulator on harmonic overvoltages

To verify the effect of the voltage regulator response time on overvoltages, its time constant was reduced from 1.0 to 0.05 s, to investigate cases where the system voltage is more than 105% of the reference voltage, but without altering the gain in any way. The 5th-harmonic overvoltages are still present, albeit limited by the drop in fundamental voltage (Fig. 10), reducing transformer over-excitation and, at the same time, the generation of 5th-harmonic currents. The voltages reached are less than 1.3 p.u. and fall below 1.2 p.u. on 120-kV buses approximately 700 ms after line 7040 trips.

Effect of the number of 125/765-kV Transformers at Châteauguay

The duration of 5th-harmonic overvoltages after line 7040 trips is related to the duration of transformer over-excitation. Transformers absorb reactive power in proportion to the fundamental component of their magnetizing current. This slightly lowers the fundamental-frequency overvoltage as the transformers increase in number. However, their rise in number increases the 5th-harmonic current and voltage components, with the result that the overvoltage amplitude (Fig. 11) increases from 1.5 to 1.75 p.u. with two, three or four 840-MVA transformers at Châteauguay.

Addition of transformers connected to the 120-kV bus at Châteauguay has the effect of rounding the saturation knee, and reducing the total saturation impedance, so that the overall flux linkage has to drop to a lower value for the same magnetizing current value. This explains the slightly higher duration of the over-excitation (200 ms) when the number of transformers rises from two to four (Fig. 11 and 12).

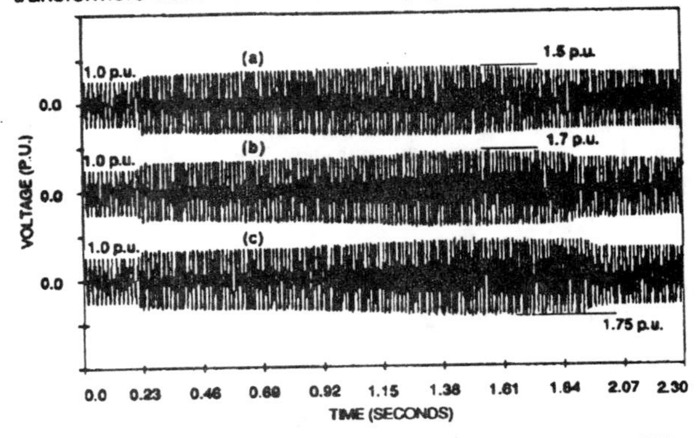

Figure 11 - Effect of number of Châteauguay 125/765-kV transformers on overvoltages. Simulation results phase C.
(a) 2 transformers (b) 3 transformers
(c) 4 transformers

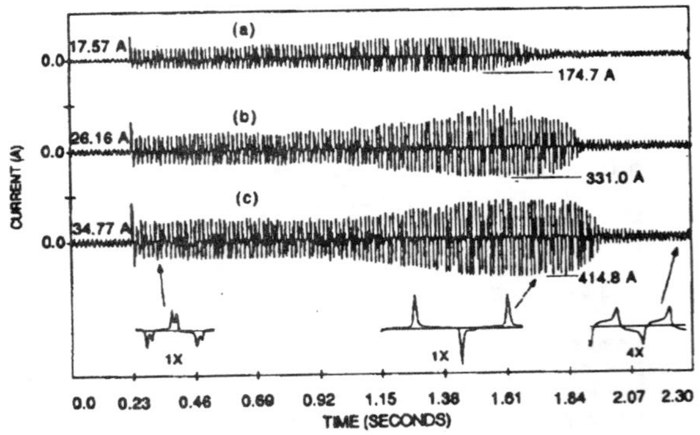

Figure 12 - Effect of number of Châteauguay 125/765-kV transformers on total magnetizing current. Simulation results phase C.
(a) 2 transformers (b) 3 transformers
(c) 4 transformers

CONCLUSIONS

This study made it possible to determine the cause and type of overvoltage recorded following the tripping of the Hydro-Quebec-NYPA interconnection at Châteauguay. It shows that symmetrical over-excitation of the transformers generated odd harmonics, of which the fifth excited the degraded system at resonance frequency while generator overspeed scanned that frequency (5 X 70 Hz). This produced an amplification of 5th-harmonic voltages, giving rise to overvoltages of 1.8 p.u. at 120 kV.

The Beauharnois voltage regulation systems were shown to be slow; faster regulators would have controlled overvoltages to lower amplitudes and shorter durations. It was also shown that keeping unnecessary equipment in service (PLC filters and power transformers) increases the risk of overvoltages in unexpected situations.

The problems encountered in simulating the overvoltage phenomenon brought out the need to conduct further research on representing losses as a function of the frequency for the different types of power system equipment.

APPENDIX

The magnetizing current of a single-phase (lossless) transformer under sinusoidal-overvoltage conditions is similar to the TCR branch current of a static compensator. If the transformer has the simplified magnetic characteristic shown in Fig. A-1, it is possible to develop the mathematical expression for its inrush current when it is energized, as well as for its magnetizing current under over-excitation conditions.

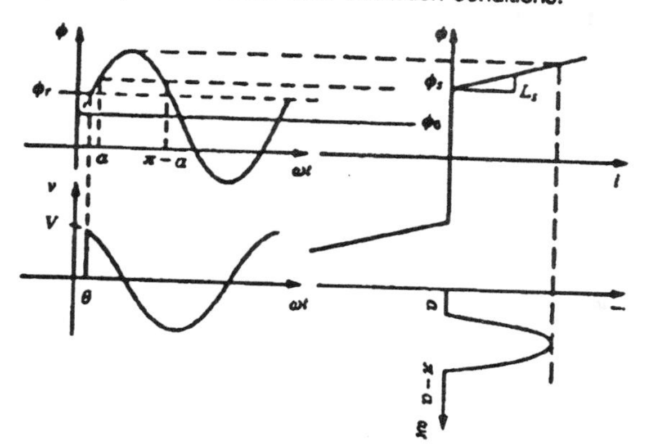

Figure A-1 – Simplified magnetization characteristic of a lossless transformer.

If the voltage applied to the transformer is $v = V \cos \omega t$ at time "t_0" for which $\omega t_0 = \theta$, the transformer flux linkage is given by the integral of the applied voltage, namely:

$$\phi = \frac{V}{\omega} \sin \omega t + \phi_0 \qquad \text{(A.1)}$$

The transient flux "ϕ_0" is equal to the remanent flux "ϕ_r" plus the initial flux condition,namely: $-(V/\omega) \sin\theta$

$$\phi_0 = \phi_r - \frac{V}{\omega} \sin\theta \qquad \text{(A.2)}$$

Assuming that the voltage source is not disturbed by the transformer saturation current, the inrush current will be expressed by:

$$i = \frac{1}{L_s}(\phi - \phi_s) = \frac{1}{L_s}\left(\frac{V}{\omega} \sin \omega t + \phi_0 - \phi_s\right) \qquad \text{(A.4)}$$

If the current is normalized in relation to the saturation voltage V_s and saturation impedance ωL_s, the inrush current will be written:

$$\frac{i}{I_s} = \frac{i}{\frac{V_s}{\omega L_s}} = \frac{\frac{V \sin \omega t}{V_s\, \omega L_s}}{\omega L_s} + \frac{\frac{\phi_0}{V_s\, L_s}}{\omega L_s} - \frac{\frac{\phi_s}{L_s\, V_s}}{\omega L_s} \qquad \text{(A.5)}$$

$$\frac{i}{I_s} = \frac{V \sin \omega t}{V_s} + \frac{\phi_0}{\phi_s} - 1$$

$$\frac{i}{I_s} = \frac{V}{V_s} \sin \omega t + S - 1 \quad \text{for } a \le \omega t \le \pi - a \qquad \text{(A.6)}$$

$$\frac{i}{I_s} = 0 \qquad \text{elsewhere}$$

$$\text{with } S = \frac{\phi_0}{\phi_s}, \quad a = \sin^{-1}\left[(1 - S)\frac{V_s}{V}\right], \quad I_s = \frac{V_s}{\omega L_s}$$

When the overvoltage applied to the transformer is balanced (positive sequence only) and the transformer saturation is symmetrical, this is only a special case of asymmetrical saturation for which $\phi_0 = 0$ or $S = 0$

The symmetrical saturation current under overvoltage conditions will therefore be written:

$$\frac{i}{I_s} = \frac{V}{V_s} \sin \omega t - 1 \qquad a \le \omega t \le \pi - a$$

$$\frac{i}{I_s} = \frac{V}{V_s} \sin \omega t + 1 \qquad \pi + a \le \omega t \le 2\pi - a \qquad \text{(A.7)}$$

$$\frac{i}{I_s} = 0 \qquad \text{elsewhere}$$

and will have the following theoretical form:

$\frac{i}{I_s} \Rightarrow$ [waveform sketch with marks 0, a, $\pi - a$, π, $\pi + a$, $2\pi - a$, 2π]

Fourier analysis of this current is expressed by the equation:

$$\frac{i}{I_s}(t) = \frac{a_0}{2} + \sum_{n=1}^{\infty}\left(a_n \cos n\omega t + b_n \sin n\omega t\right)$$

$$a_n = \frac{1}{\pi}\int_0^{2\pi} \frac{i}{I_s}(t) \cos n\omega t \, d\omega t \qquad \text{(A.8)}$$

$$b_n = \frac{1}{\pi}\int_0^{2\pi} \frac{i}{I_s}(t) \sin n\omega t \, d\omega t$$

Demonstration

Considering that $\sin a = \frac{V_s}{V} \rightarrow V_s = V \sin a$, the function f(t) may also be written:

$$\frac{i}{I_s} = \frac{V}{V_s}\left(\sin \omega t - \sin a\right) \quad \text{for} \quad a \le \omega t \le \pi - a$$

$$= \frac{V}{V_s}\left(\sin \omega t + \sin a\right) \quad \text{for} \quad \pi + a \le \omega t \le 2\pi - a$$

$$= 0 \qquad \text{elsewhere} \qquad \text{(A.9)}$$

Calculation of the coefficient a_1

Since there is no DC component, $a_0 = 0$ and the coefficient a_1 is given by the equation:

$$a_1 = \frac{1}{\pi}\int_0^{2\pi} f(t) \cos \omega t \, d(\omega t)$$

hence:

$$a_1 = \frac{1}{\pi}\frac{V}{V_s}\left[\int_a^{\pi - a} (\sin \omega t - \sin a) \cos \omega t \, d(\omega t) + \int_{\pi + a}^{2\pi - a} (\sin \omega t + \sin a) \cos \omega t \, d(\omega t)\right] \qquad \text{(A.10)}$$

knowing that: $\int \sin Ax \cos Ax \, dx = \frac{\sin^2 Ax}{2A}$ (A.11)

therefore: $a_1 = 0$

Calculation of coefficients a_n

The coefficients a_n are given by the equation:

$$a_n = \frac{1}{\pi}\int_0^{2\pi} f(t)\ \cos n\omega t\ d(\omega t)$$

hence:

$$a_n = \frac{1}{\pi}\frac{V}{V_s}\left[\int_{\alpha}^{\pi-\alpha}(\sin\omega t - \sin\alpha)\cos n\omega t\ d(\omega t) + \int_{\pi+\alpha}^{2\pi-\alpha}(\sin\omega t + \sin\alpha)\cos n\omega t\ d(\omega t)\right] \qquad (A.12)$$

knowing that:
$$\int \sin px \cos qx\ dx = \frac{-\cos(p-q)x}{2(p-q)} - \frac{\cos(p+q)x}{2(p+q)} \qquad (A.13)$$

therefore: $a_n = 0$

Calculation of the coefficient b_1

The coefficient b_1 is given by the equation:

$$b_1 = \frac{1}{\pi}\int_0^{2\pi} f(t)\ \sin\omega t\ d(\omega t)$$

hence:

$$b_1 = \frac{1}{\pi}\frac{V}{Vs}\left[\int_{\alpha}^{\pi-\alpha}(\sin\omega t - \sin\alpha)\sin\omega t\ d(\omega t) + \int_{\pi+\alpha}^{2\pi-\alpha}(\sin\omega t + \sin\alpha)\sin\omega t\ d(\omega t)\right] \qquad (A.14)$$

knowing that:
$$\int \sin^2 ax\,dx = \frac{x}{2} - \frac{\sin 2ax}{4a} \qquad (A.15)$$

therefore:

$$b_1 = \frac{1}{\pi}\frac{V}{Vs}\left[\pi - 2\alpha - \sin 2\alpha\right] \qquad (A.16)$$

The calculation of coefficients b_n

The coefficients b_n are given by the equation:

$$b_n = \frac{1}{\pi}\int_0^{2\pi} f(t)\sin n\omega t\ d(\omega t)$$

hence:

$$b_n = \frac{1}{\pi}\frac{V}{Vs}\left[\int_{\alpha}^{\pi-\alpha}(\sin\omega t \sin n\omega t - \sin\alpha\sin n\omega t)\ d(\omega t) + \int_{\pi+\alpha}^{2\pi-\alpha}(\sin\omega t\sin n\omega t + \sin\alpha\sin n\omega t)\ d(\omega t)\right] \qquad (A.17)$$

knowing that:
$$\int \sin px \sin qx\,dx = \frac{\sin(p-q)x}{2(p-q)} - \frac{\sin(p+q)x}{2(p+q)} \qquad (A.18)$$

therefore:

$$b_n = \frac{1}{\pi}\frac{V}{Vs}\left[+\frac{\sin(1-n)(\pi-\alpha)}{(1-n)} - \frac{\sin(1+n)(\pi-\alpha)}{(1+n)} + \frac{2\sin\alpha\cos n(\pi-\alpha)}{n} - \frac{\sin(1-n)\alpha}{(1-n)} + \frac{\sin(1+n)\alpha}{(1+n)} - \frac{2\sin\alpha\cos n\alpha}{n}\right] \qquad (A.19)$$

b_n is equal to zero for all even values of n, so that the *sine* and *cosine* terms cancel each other out. Only the odd terms of b_n (namely b_{2n+1}) are different from zero; replacing n with $2n+1$ in the equation, the coefficients $b_{(2n+1)}$ will be:

$$b_{(2n+1)} = \frac{1}{\pi}\frac{V}{Vs}\left[\frac{\sin[1-(2n+1)](\pi-\alpha)}{[1-(2n+1)]} - \frac{\sin[1+(2n+1)](\pi-\alpha)}{[1+(2n+1)]} + \frac{2\sin\alpha\cos(2n+1)(\pi-\alpha)}{(2n+1)} - \frac{\sin[1-(2n+1)]\alpha}{[1-(2n+1)]} + \frac{\sin[1+(2n+1)]\alpha}{[1+(2n+1)]} - \frac{2\sin\alpha\cos(2n+1)\alpha}{(2n+1)}\right] \qquad (A.20)$$

$$b_{(2n+1)} = \frac{-1}{\pi}\frac{V}{Vs}\frac{1}{(2n+1)}\left[\frac{\sin 2(n+1)\alpha}{n+1} + \frac{\sin 2n\alpha}{n}\right] \qquad (A.21)$$

The Fourier series for the function i/Is described in equation A.9, therefore equals:

$$\frac{i}{Is} = \frac{1}{\pi}\frac{V}{Vs}\left[\pi - 2\alpha - \sin 2\alpha\right]\sin\omega t - \sum_{n=1}^{\infty}\frac{1}{\pi(2n+1)}\frac{V}{Vs}\left[\frac{\sin 2(n+1)\alpha}{n+1} + \frac{\sin 2n\alpha}{n}\right]\sin(2n+1)\omega t \qquad (A.22)$$

The crest amplitude of each harmonic therefore depends on the severity of the sinusoidal over-excitation of the transformer V/Vs. To clearly illustrate these results, Fig. A-2 shows the amplitude of each normalized harmonic in relation to the saturation current, as a function of the crest voltage applied in over-excitation conditions.

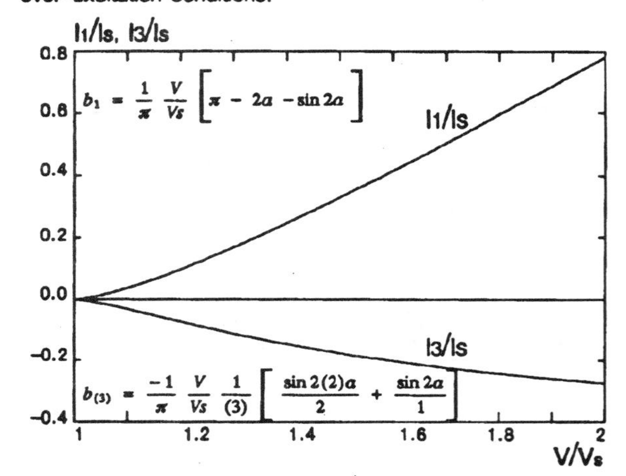

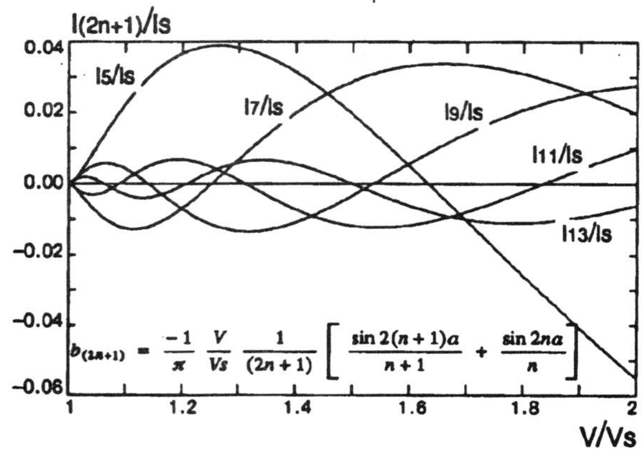

Figure A-2 – Amplitude of each normalized harmonic in relation to the saturation current, as a function of the crest voltage applied in over-excitation conditions

ACKNOWLEDGMENTS

The authors wish to thank Claude Loyer, Robert Parent and Michel Guilbault of Hydro-Quebec for their respective contributions in taking system measurements and in processing the data. The invaluable assistance provided by Georges Desrochers and Serge Lefebvre of IREQ in the harmonic analysis and support for the EMTP software, respectively, was also greatly appreciated. Technical comments from Jean Bélanger of IREQ during preparation of this paper have been very helpful. A final word of thanks goes to Sylvie Humbert of Hydro-Quebec for editing help.

REFERENCES

[1] G. Sybille, M.M. Gavrilovic, J. Bélanger, V.Q. Do, *Transformer saturation effects on EHV system overvoltages*, IEEE PAS-104, no. 3, March 1986.

[2] O. Bourgault, M. Tessier *Mesure des paramètres magnétiques du transformateur #14 du poste Châteauguay*, Hydro-Quebec report no. 88019, 1988.

[3] *Electro-Magnetic Transient Program, rule book*, EPRI version 1.0, April 1986.

[4] *Computer representation of excitation systems, IEEE Committee report PAS*, June 1968.

[5] *EMTP theory book*, Bonneville Power Administration, Portland, August 1986.

[6] G. Crôteau, O. Bourgault *Identification of generator parameters by P.R.B.S. signal injection through a static compensator*, CEA presentation March 1986, Hydro-Quebec report no. 8660.

[7] G. Crôteau *Mesure de l'impédance opérationnelle d'un alternateur de la centrale LG-2*, Hydro-Quebec report no. 8571, 1985.

BIOGRAPHY

Omer Bourgault was born in Quebec area, Canada in 1948. He received a B.Sc.A degree in Electrical engineering from Laval University of Quebec in 1972.

He is with Hydro-Quebec since 1972, working in Power System Testing and Technical Expertise department. His main fields of activities have been power system transients and acceptance tests on SVC and HVDC.

He is author of publications related to transients on 735-kV reactor switching, transformer saturation impedance measurement, on-line transfer function measurement of power system equipment with correlation techniques, instability of SVC and HVDC controls in presence of harmonics and finally dynamic stability margin measurement for SVC and HVDC control optimization.

Gaston Morin (M'85) was born in Quebec area, Canada, in 1954. He received the B.Sc.A. degree in Electrical engineering from University of Sherbrooke, and the M.Sc.A. degree from Ecole Polytechnique of Montreal in 1978 and 1983 respectively.

He is with Hydro-Quebec since 1978 working in Power System Operation department.

His main interests concern harmonics, DC Systems and power system transients. He is author of publications related to harmonic overvoltages during power system restoration. He is currently member of the IEEE Power System Restoration Task Force and IEEE working group 15.05.02, Dynamic Performance and Modeling of DC systems.

IEEE Transactions on Power Delivery, Vol. 3, No. 1, January 1988

ELIMINATION OF CLOSING RESISTORS ON EHV CIRCUIT BREAKERS

A. C. Legate, Member
J. H. Brunke, Member
J. J. Ray, Member
E. J. Yasuda, Senior Member

Bonneville Power Administration
Portland, Oregon

Abstract

Closing resistors may be eliminated from EHV power circuit breakers when staggered (pole) closing or polarity closing is used on transmission lines that have metal oxide surge arresters (MOSA) only at the terminals. This paper reports the results of computer studies and field tests that validate the concept and the techniques for statistical studies of specific cases. The investigation concludes that 500 kV lines up to 320 km may be energized successfully without the use of closing resistors.

INTRODUCTION

EHV power circuit breakers have used one-, two-, and even three-step closing resistors, sometimes in combination with control of the timing of contact closing and resistor shunting to reduce switching surges. To implement reduced switching surge design methods on existing power systems, much of the earlier developments focused on the power circuit breakers in terms of opening/closing resistors and controlling the time (the instant on voltage wave) at which the contacts close. Other surge reduction methods were available but not commonly applied because the power system switching surge design levels were generally attainable with resistor equipped breakers. These complex circuit breakers are expensive to purchase and maintain. Bonneville Power Administration's (BPA) experience with these complex breakers shows mechanical malfunctions as the most common cause of circuit breaker failures. Reduction of mechanical complexity should greatly improve reliability. As an alternative to closing resistors, BPA has investigated staggered and polarity closing and equipping each line terminal with MOSA. (Staggered closing means the instant of main contact closing in individual poles is separated approximately one cycle. Polarity closing means closing each phase when the source voltage and trapped charge of the respective phases are the same polarity).

The development of MOSA is having significant impact on the power industry. The ability of MOSA to passively limit overvoltages, absorbing only the energy in the overvoltage without power follow, has opened a variety of new applications. One such application is the limiting of surges when switching 500-kV transmission lines using power circuit breakers without closing resistors.

Since the introduction of MOSA in the late 1960's, extensive operating experience has been gained to demonstrate their thermal stability, lower switching surge protective level (1.5 - 1.7 pu), higher energy discharge capability, and overall reliability.

86 SM 373-4 A paper recommended and approved by the IEEE Switchgear Committee of the IEEE Power Engineering Society for presentation at the IEEE/PES 1986 Summer Meeting, Mexico City, Mexico, July 20 - 25, 1986. Manuscript submitted January 31, 1986; made available for printing April 23, 1986.

Printed in the U.S.A.

Furthermore, MOSA designs have progressed where parallel columns now permit still lower protective levels and still higher energy discharge capabilities.

In the design of power systems, equipment reliability is a major concern. MOSA are static devices which do not depend on mechanical moving parts to perform their functions. As such, they provide the additional advantage of simplicity which in turn provides higher degrees of reliability. BPA's experience has substantiated the high reliability of MOSA when applied to reactors and transformers that are connected to the terminals of transmission lines and subjected to frequent line switching. This is in addition to the high reliability of MOSA applied on reactors and transformers connected to the station bus.

The concept of breakers without closing resistors and MOSA on line terminals as a means of reducing and controlling line switching surges extends the range of potential applications and benefits on power systems. Some of these applications and benefits include:

1. Reduce cost of new lines by reducing the line switching surge overvoltages.
2. Simplify new breaker designs without resistors and still maintain or lower the line switching surge design level and recovery voltages.
3. Voltage uprating of existing lines without major line modifications.
4. Increase existing line reliability against switching transients.

BPA's consideration for the elimination of breaker resistors and the application of MOSA on line terminals of existing 500-kV lines resulted in extensive transient overvoltage studies to determine the application guidelines for limiting switching overvoltages at and below the design levels.

Background

The Bonneville Power Administration (BPA) was a pioneer in the application of 500-kV circuit breakers with closing resistors and controlled closing to limit switching surges to 1.5 pu for 98 percent of all line energizations [1, 2, 3]. The purpose was to provide an assured margin for BPA's first 500-kV reduced insulation lines [4] and serve as a measure of feasibility for 1.5 pu design for 1100 kV. The 500-kV reduced insulation (1.7 pu) line design, based on the NESC alternate (switching surge) method for clearances, has had an excellent performance record at BPA. The cost savings have been in the order of 10-20 percent. While this method has not seen widespread application in the power industry, the demonstrated performance and cost savings for the line, along with simplified switching surge control methods indicated in this paper, should cause more interest and future applications of optimized power line designs based on the NESC alternate method for computing clearances.

Until 1980, BPA policy was to purchase 500-kV breakers designed to limit switching surges to 1.5 pu with two-step closing resistors or single resistor breakers with a closing time consistency of ± 3 ms. For lines exceeding 129 km, controlled closing devices were also required. Experience over a 2-year period (1978-1980) with application of these breakers on the 1.7 pu line design indicated the circuit breaker design was conservative and needed to be re-evaluated. No switching surge flashovers had been recorded and the breaker design was more complex than necessary.

As a result of this re-evaluation, including digital model studies, the policy was modified in 1980 to purchase 1.7 pu breakers with single closing resistor only. The need for closing consistency with controlled closing devices, or two-step closing resistors, was eliminated. For 1.7 pu lines exceeding 160 km without shunt reactors, staggered closing is required and magnetic potential transformers (MPT) were applied on the line terminals to reduce trapped charges. Lines less than 160 km without shunt reactors did not require MPT's. It should be noted that the 1980 policy did not consider single pole reclosing. The overvoltages are considerably reduced for single pole reclosing which has been adopted by BPA for most new 500-kV lines.

The IEEE Working Group on Switching Surges has reported [5] that MOSA can not protect an entire line when located only at the terminals. While the voltage in the middle of the line is not limited to the arrester protective level, the purpose of this paper is to demonstrate that it can be limited to the switching surge design level per NESC for 98 percent of the energizations. The Working Group also reported that the switching surge magnitudes may be controlled by reducing the pole closing span between the first and last pole to close. However, this is not practical. Very short pole closing spans (less than one travel time of transient on the line) are difficult to achieve and maintain due to prestriking and minor timing differences between poles. Staggered closing approaches the same reduction of switching surge magnitude as that for zero pole closing span. With staggered closing, the next pole is not closed until the transient voltages from the previous pole closing have damped out. A one- to two-cycle delay is not significant for completion of closing or reclosing operations.

STAGED TEST ON POWER SYSTEM

The purpose of the staged system test was to provide data to validate the digital model studies and to demonstrate the feasibility of switching lines with breakers without closing resistors. The test was performed in conjunction with another test for reasons of economy, and as a result the circuit was more complicated than necesary for the validation. The transformer high side jumpers were dropped on the 500/230-kV autotransformer at BPA's Satsop Substation, leaving the MOSA connected to the line terminals (Figure 1).

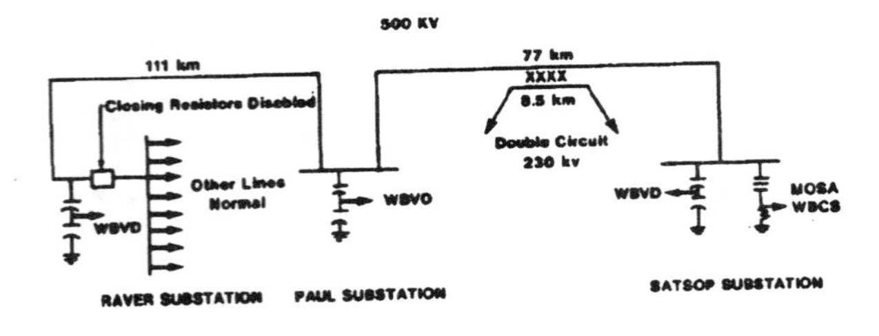

Fig. 1 Configuration for the staged system test and measurment locations.
WBVD - Wideband Voltage Divider
WBCS - Wideband Current Shunt

The Satsop line runs 77 km to Paul Substation where it was connected through the bus to the 111 km long Paul-Raver 500 kV line. This bus and these two lines were isolated from all other lines at Paul Substation. At Raver Substation, the closing resistor contacts on an air blast power circuit breaker were disabled so no closing resistor contacts would be operated. The total line length was 188 km (117 miles). An 8.5-km section approximately in the middle of the Paul-Satsop line is double circuit; sharing the structures with the Olympia-Satsop 230-kV line No. 2. (see Figure 1)

Wideband voltage dividers were installed on all three phases at Raver, Paul and Satsop Substations to measure line voltages. At Satsop Substation, wideband shunts were used to measure arrester currents. At all three locations the data was recorded on an analog instrumentation tape recorder with 80 kHz bandwidth. The overall measurement system bandwidth on all measurements was 80 kHz.

A total of 17 three-phase switching tests were performed. Each test consisted of a three-phase trip followed by a reclose after approximately 30 cycles. The poles closed consecutively, staggered by 1-cycle each. A sequence controller was used to control both the opening and closing instant of each pole. Closing angles were adjusted so the closing occurred at the same and opposite polarity as the trapped charge.

In the 17 trip/reclose test operations the highest overvoltage recorded was 1.9 pu. A summary of the test data is provided in Table I. The tabulated overvoltages in this table are not random statistical data since the breaker was controlled by a sequence controller and was adjusted to produce various closing angles. Figure 3 shows the transient voltage measured at Paul Substation (near the midpoint of the line) and the arrester transient voltage and current measured at the Satsop Substation (the open end of the line). The peak arrester current was 1700 amperes. The voltage and current data from Satsop was digitized and arrester energy calculated at 1.0 MJ for this transient event. The transient voltage at Paul crested 1.9 pu in 1.05 ms. The risetime (10-90%) was 0.7 ms. The steepest portions of the wavefront had rates of rise of more than 8 kV/microsecond.

Table I
Measured Per Unit Overvoltages at Paul Substation

Test	A-phase	B-phase	C-phase
1	1.1	1.5	1.3
2	1.9	1.4	1.6
3	1.1	1.2	1.4
4	1.1	1.2	1.2
5	1.5	1.6	1.1
6	1.1	1.1	1.1
7	1.1	1.2	1.2
8	1.1	1.2	1.2
9	1.1	1.5	1.2
10	1.1	1.3	1.3
11	1.1	1.7	1.2
12	1.1	1.9	1.4
13	1.1	1.1	1.7
14	1.1	1.1	1.9
15	1.1	1.1	1.9
16	1.5	1.3	1.1
17	1.1	1.4	1.1

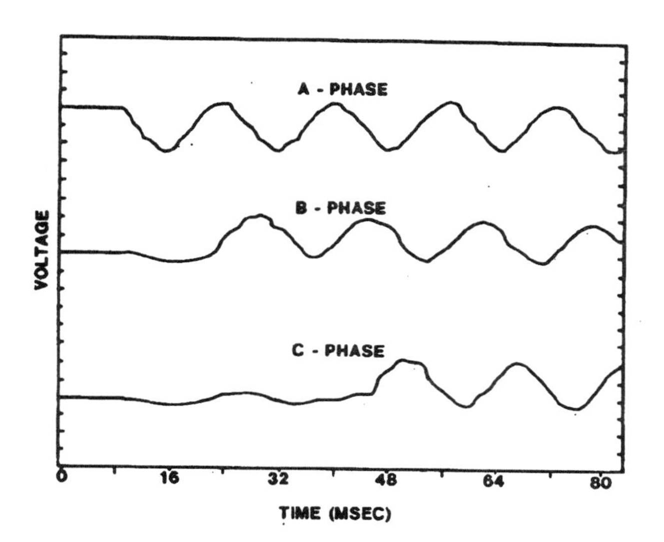

Fig. 2 Staged system test voltage waveforms recorded at Paul Substation with polarity closing of breaker at Raver (refer to Fig. 1). Maximum overvoltage did not exceed 1.25 pu (compare with Fig. 3E)

As part of the staged system test, the breaker at Raver Substation was closed with polarity closing. Figure 2 shows the voltages recorded at Paul Substation for one case of polarity closing and illustrates the effectiveness of this closing technique in controlling switching surges. Compared to the voltage of Figure 3E, the overvoltages in Figure 2 are insignificant.

RESULTS OF MODEL STUDIES

Digital model studies, made with the BPA Electromagnetic Transients Program (EMTP) [6, 7], were initiated to determine methods of limiting switching surge line-to-ground voltages to 1.7 pu (1 pu = 450 kV peak), with full trapped charges as an initial condition, and with power circuit breakers without closing resistors.

Results of the initial statistical studies with staggered and polarity closing warranted the performance of a staged test on the power system, previously described, to validate the computer model. In the validation of the computer model, the parameters and conditions representing the actual staged tests were represented. A brief description of the major components simulating the power system follows.

Transmission Line

The 500-kV lines of Figure 1 were modeled with the EMTP using a frequency dependent option, which represents frequency dependence in both the single and double-circuit sections. [8, 9]

Except for the 8.5-km double-circuit portion, the Raver-Paul-Satsop line was modeled in 17-km sections connected in tandem.

Arresters

Shunt gapped 396-kV station class MOSA connected to the open end of the line at Satsop were modeled with E - I characteristics supplied by the manufacturer. These characteristics, obtained from laboratory tests in which 45 x 90 microsecond current surges were applied to the arresters, were simulated by a relatively new multi-expotential method described in the EMTP manual. [10] Details of the arrester (MOSA) model and its application will be reported in a future paper planned for the IEEE 1987 Winter Power Meeting.

Source Configuration

In addition to the switched line, 8 separate 3-phase 500-kV lines terminate on the Raver bus. The 8 unswitched lines, ranging in length from 16 km to 280 km, were modeled as distributed parameter lines back to the next bus away from Raver. Since staged test results show that the major transient voltages of interest occur within 1-2 line travel-times after switching, no attempt was made to duplicate waveforms resulting from the longer period reflections originating from the lines connected to these remote buses, which were modelled simply as lumped source impedances.

Staged Test Results versus Model Results

The staged tests, previously discussed, produced a maximum voltage at Paul of 1.9 pu. One specific staged test reclosure which produced the 1.9 pu maximum voltage was selected for duplication with the EMTP model. Since the 1.9 pu voltage occurred only on A-phase, the waveforms on this phase were chosen for duplication and subsequent comparison. Only one phase was compared for approximately 8-10 travel times following reclosure of the Raver breaker with staggered closing. The longer period reflections did not contribute to the highest overvoltages.

In the simulation of the staged test, the phases of the breaker at Raver were closed at the same times, with respect to the source voltage waveforms, as were the respective phases of the breaker during the staged test. Trapped charges identical to those in the test were simulated in the model and approximately 2.0 pu voltage existed across the breaker contacts at the instant of contact closure.

Figure 3 compares the staged test and digital study results. As shown, good correlation was obtained in terms of waveshape, magnitudes and rise times for the voltage across the Satsop arrester, arrester current, and the voltage at Paul Substation. The minor differences in the magnitudes of the test and model results are attributed to a possible difference in line parameters since the model parameters were computed with a ground resistivity of 100 ohm-meters.

Arrester discharge energy was determined for both the staged test and model results in Figure 3. The model simulation produced an arrester discharge energy of 1.13 MJ which agrees with the staged test discharge energy of 1.0 MJ. These levels are well below the discharge energy limit of 5 MJ for standard available MOSA.

Statistical Studies

Results of statistical studies made with the model of the staged system test configuration (refer to Figure 1) are shown in Figure 4. Each statistical curve represents the results of 100 separate 3-phase reclosures. Curves (1) and (2) show the effects of arrester protective levels of 1.7 and 1.5 pu in combination with staggered closing. Curves (3) and (4) show the effectiveness of polarity closing with closing time variations of + 4 ms (3-sigma) about the optimum closing time for each phase.

Curves (1) and (2) illustrate that staggered closing with MOSA is an effective combination for controlling surge voltages to 2.0 pu or less with MOSA

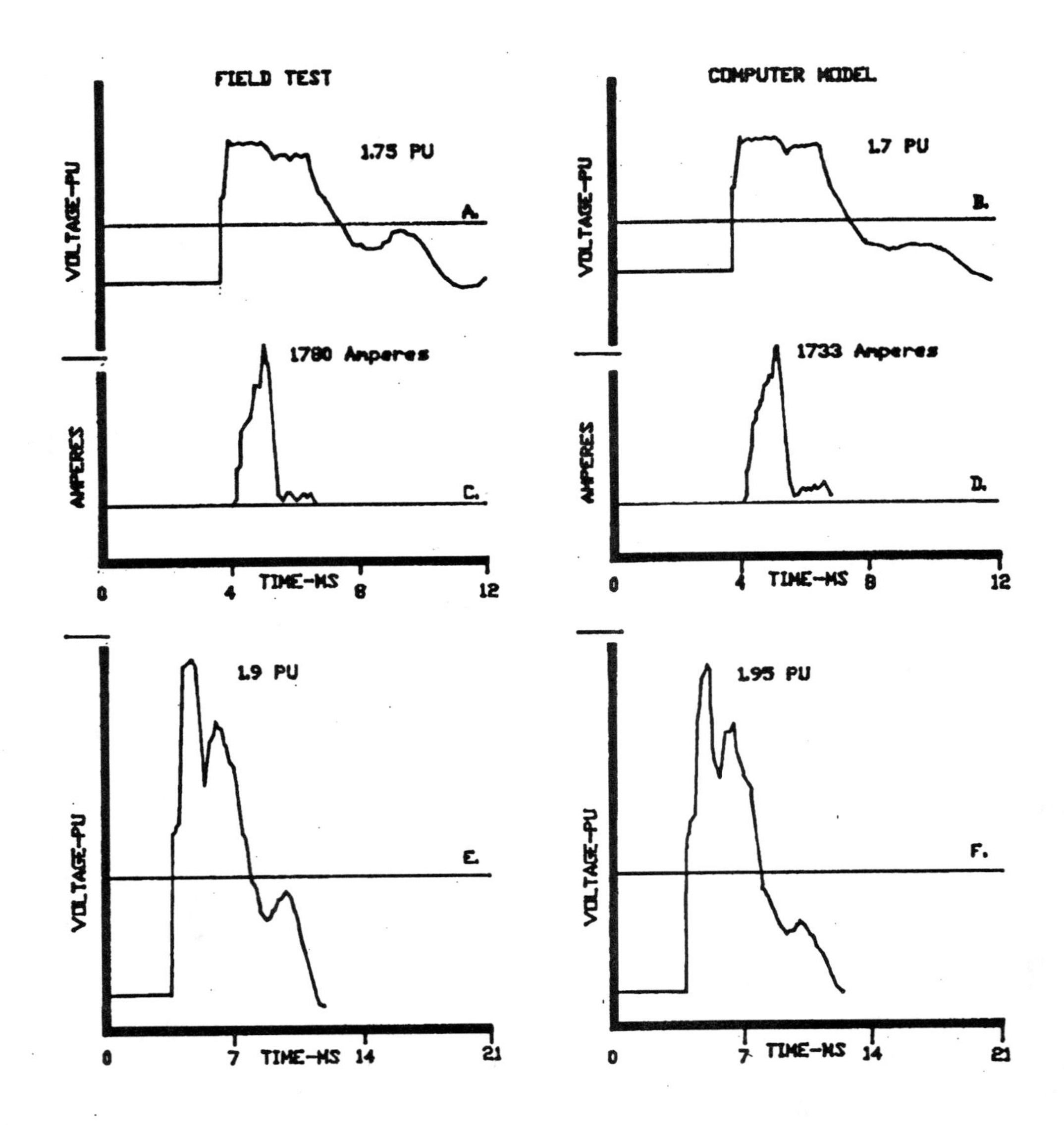

Fig. 3 Comparison of staged system test measurements and model results.

WAVEFORMS A AND B - LINE-TO-GROUND VOLTAGE ACROSS ARRESTER AT SATSOP

WAVEFORMS C AND D - CURRENT THROUGH ARRESTER AT SATSOP

WAVEFORMS E AND F - LINE-TO-GROUND VOLTAGE AT PAUL

at the receiving end. Similarly, curves (3) and (4) show that polarity closing and MOSA are equally effective for limiting the voltages to 1.7 pu or less. With arresters at the receiving end of the line, the highest overvoltages will occur in the middle 30-50 percent of the line.

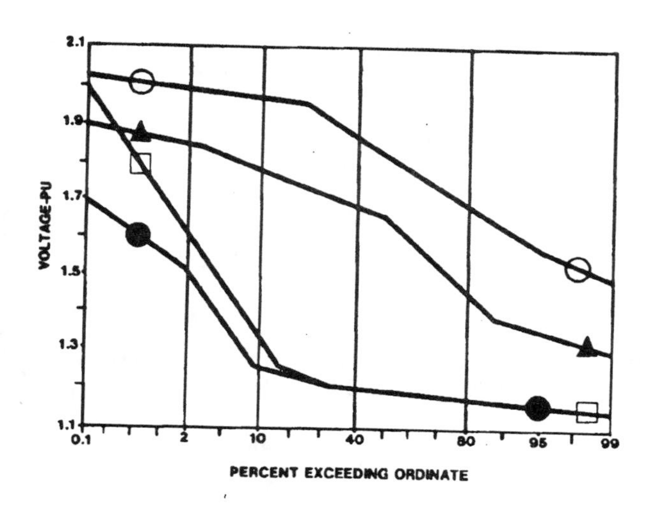

Fig. 4 Results of EMTP statistical studies on the staged system test line of Fig. 1
(1) O - Random staggered closing of the Raver breaker with the standard available arresters at Satsop and with random trapped sharges on the line
(2) ▲ - Same as above except with 1.5 pu arresters at Satsop
(3) □ - Polarity closing of the Raver breaker with the standard available arresters at Satsop
(4) ● - Same as above (polarity closing) except with 1.5 pu arresters at Satsop

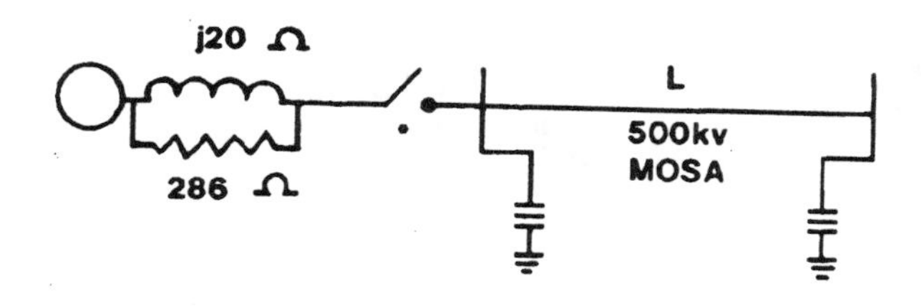

Fig. 5 Circuit configuration used to produce results shown in Fig. 6 and 7

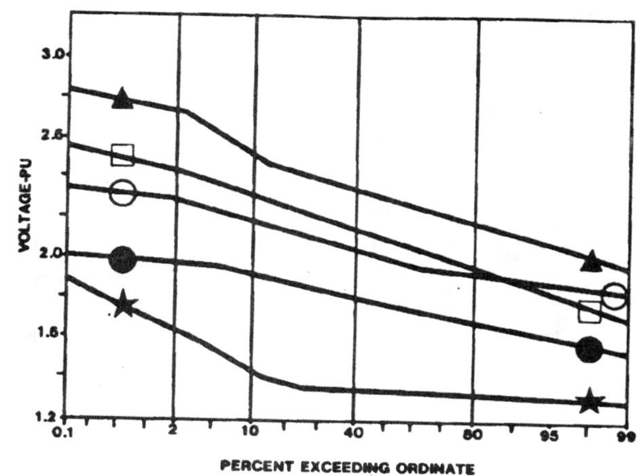

Fig. 6 Results of EMTP statistical studies with system of Fig. 5
- Line length (L) = 320 km
(1) ▲- Random closing of the breaker with 1.7 pu arresters at the line terminals but without staggered closing. Random trapped charges were represented
(2) □ - Same as above except with 1.5 pu arresters
(3) O - Random staggered closing of the breaker with 1.7 pu arresters and with random trapped charges
(4) ● - Same as above (staggered closing) except with 1.5 pu arresters
(5) * - Polarity closing with 1.5 pu arresters.

The results shown in Figure 6 were made with the circuit configuration of Figure 5 and a line length of 320 km. Each curve of Figure 6 represents 200 separate 3-phase reclosures. In each case, arresters were connected to both ends of the line since breaker reclosure can be initiated at either end.

Curves (1) and (2) show that neither 1.7 nor 1.5 pu arresters will limit surge voltages to the design level of 2.0 pu without staggered closing. Curve (3) shows that 1.7 pu arresters also will not limit the voltages to 2.0 pu even with staggered closing. Curve (4) shows, however, that with staggered closing, 1.5 pu arresters are quite effective in controlling surge voltages to 2.0 pu for a 320-km line.

For a 1.7 pu line design of this length, however, Curve (5) shows that controlled polarity closing and 1.5 pu arresters are necessary. Figure 2 under Staged Test on Power System shows recorded voltages that illustrate the effectiveness of these measures in controlling switching surges. The data on curve (5) was obtained with a breaker closing time variation of $\pm$ 5 ms (3-sigma).

The maximum energy absorbed by an arrester was 1.85 MJ and 2.6 MJ for arresters with 1.7 and 1.5 pu protective levels, respectively. These values are well within the MOSA discharge capability.

The results in Figure 7 were obtained with the system of Figure 5 but with a line length of 113 km. Staggered closing is still required if the voltages are to be limited to 2.0 pu with either 1.7 or 1.5 pu arresters.

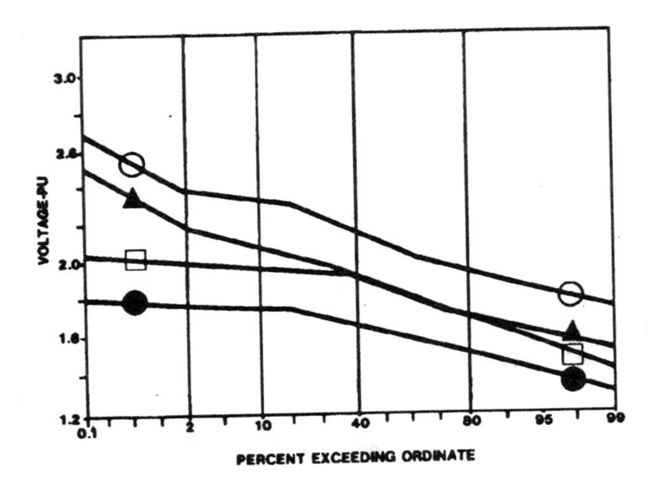

Fig. 7 Same as Fig. 6 except that L = 113 km.
(1) O - Random closing of PCB with 1.7 pu arresters and with random trapped charges but without staggered closing
(2) ▲ - Same as above except with 1.5 pu arresters
(3) ● - Random staggered closing of PCB with 1.5 pu arresters and with random trapped charges
(4) □ - Same as above (staggered closing) except with 1.7 pu arresters

CONCLUSION

The described staged system tests have verified the results of computer studies and thereby validated the technique for controlling switching surges. These investigations also validated the method to be used for computer studies of other system configurations employing breakers without closing resistors and using staggered closing with MOSA at line terminals to reduce switching surges. This technique will allow lower cost breakers and line designs as well as simpler breakers that will require less maintenance.

For BPA's 500-kV designs, the studies conclude that:

1. Staggered closing, when combined with MOSA on the line terminals, provides a 15 to 28 percent reduction in switching surge levels for breakers without closing resistors.

2. Lines up to 320 km that are designed for 2.0 pu (98%) switching surge voltage do not require closing resistors when 1.5 pu MOSA are applied at both line terminals along with staggered closing. For 2.0 pu lines, less than 113 km, 1.7 pu MOSA can be used.

3. Lines up to 80 km that are designed for 1.7 pu (98%) switching surge voltage do not require closing resistors when 1.5 pu MOSA are applied at the line terminals along with staggered closing. For 1.7 pu line lengths between 80 km and 320 km, polarity close will be required with a closing time consistency of ± 5 ms (3-sigma). The polarity close will not be required where single-pole reclosing is used. Also, additional studies not reported in this paper conclude that where single pole reclosing is used, MOSA and closing resistors are not required for 1.7 pu lines that are less than 35 km, and possibly longer.

4. Additional studies not reported in this paper conclude that for short 2.0 pu lines, in the order of 2 km, MOSA are not required when staggered closing is used and there are no closing resistors.

5. The switching surge energy dissipation in the MOSA for all of the above cases is well within the capabilities of standard available MOSA.

6. The installed cost of MOSA on both 500-kV line terminals is less than the cost of closing resistors for 3 breakers (1 1/2 breaker scheme).

The above conclusions should not be considered as an application guide for all cases of line connected apparatus. For example, statistical studies have not been made for reactor or transformer terminated lines.

REFERENCES

1. G. E. Stemler, "BPA's Field Test Evaluation of 500 kV PCB's Rated to limit Line switching Overvoltages to 1.5 per unit", IEEE Transactions on Power Apparatus and Systems, vol. PAS-95, pp.352-361, February 1976

2. H. E. Konkel, A. C. Legate, and H. C. Ramberg, "Limiting Switching Surge Overvoltages with Conventional Power Circuit Breakers", IEEE Transactions on Power Apparatus and Systems, vol. PAS-96, No. 2, pp. 535-542, March/April 1977.

3. A. C. Legate, G. E. Stemler, K. Reichert, N. P. Cuk, "Limitation of Phase-to-Phase and Phase-to-Ground Switching Surges, Field Tests in Bonneville Power Administration 550 kV System", CIGRE Paper, 33-06, 1976.

4. E. J. Yasuda and F. B. Dewey, "BPA's New Generation of 500 kV Lines", IEEE Transactions on Power Apparatus and Systems, vol. PAS-99, No. 2, March/April 1980, pp. 616-624

5. IEEE Working Group on Switching Surges, "Switching Surges Part IV--Control and Reduction on AC Transmission Lines", IEEE Transactions on Power Apparatus and Systems, vol. PAS-101, No. 8, August 1982.

6. H. W. Dommel, "Digital Computer Solution of Electromagnetic Transients in Single and Multi-phase Networks", IEEE Transactions on Power Apparatus and Systems, Vol. PAS-88, pp. 388-399, April 1969.

7. H. W. Dommel, W. S. Meyer, "Computation of Electromagnetic Transients", Proceedings of the IEEE, Vol. 62, pp. 983-993, 1974.

8. J. F. Hauer, "Power System Identification by Fitting Structured Models to Measured Frequency Response", IEEE Transactions on Power Apparatus and Systems, Vol. PAS-101, No. 4, pp. 915-923, April 1982.

9. J. F. Hauer, "State-space Modeling of Transmission Line Dynamics via Nonlinear Optimization", IEEE Transactions on Power Apparatus and Systems, Vol. PAS-100, No. 12, pp. 4918-4925, December 1981.

10. "Electromagnetic Transients Program (EMTP) Rule Book", Bonneville Power Administration, Revised June 1984.

A. C. Legate (M'63) graduated from the University of Portland, Portland, Oregon, in 1949. His experience includes working at the Bureau of Standards, Naval Ordinance Laboratory and the Bonneville Power Administration where he specialized in switching surge investigation.

For approximately 20 years he has served with the Working Group on Switching Surges of the IEEE Transmission and Distribution Committee. He has authored and co-authored 5 IEEE papers on switching surge overvoltages and participated in the preparation of 2 Working Group IEEE Committee reports.

John H. Brunke (S'72, M'75) was born in Portland, Oregon, on February 16, 1950. After serving with the U.S. Navy he received his B.S. in Engineering and Applied Science in 1974 and M.S. in Applied Science in 1980.

Since 1975 he has been employed by the Bonneville Power Administration as an Electrical Engineer. His work has centered on the development and testing of high voltage equipment, especially switchgear. He is presently the Chief of the Test and Development Section in the Division of System Engineering.

Mr. Brunke is a registered Professional Engineer in the State of Oregon.

J. J. Ray (S'56, M'65) was born in Austin, Minnesota, on December 18, 1939. He received the B.S. degree in Electrical Engineering from the University of Notre Dame in 1961. He did his graduate work at the University of Colorado, Denver University, and Colorado State University.

His professional career began with the Bureau of Reclamation, Division of Design, in 1961, with primary responsibilities in design and application of hydro powerplant and hydro pumping plant electrical systems and equipment, and in underground high-voltage and EHV cables. In 1973, he joined Bonneville Power Administration where he has worked in both Transmission Engineering and System Engineering.

Mr. Ray has served as Chairman of the Cable Supply Systems Subcommittee of the Insulated Conductors Committee and is active in the Power Engineering Society. He is a registered Professional Engineer in the State of Colorado.

Edward J. Yasuda (M'65, SM'85) was born in Pahoa, Hawaii, March 11, 1936. He received his B.S. Degree in Electrical Engineering from the University of Missouri in 1959 and B.S. Degree in Meteorology from Pennsylvania State University in 1960.

In 1962, Mr. Yasuda joined Bonneville Power Adminstration and is now the Chief of the High Voltage Section of the Division of System Engineering.

Mr. Yasuda is a senior member of IEEE and an active member of the IEEE PES Surge Protective Devices and the Transformers Committee. He is also an active member of ASC C62 Committee on Surge Arresters and a registered Professional Engineer for the State of Oregon.

Discussion

Fred Schaufelberger (Consultant, Portland, OR): The authors have presented data on an interesting scheme of staggered or polarity closing to limit line closing overvoltages which utilizes protective devices (MOSA) and eliminates breaker closing resistors. With the elimination of closing resistors, the need to limit line switching overvoltages has come full-circle, since it was the development of 500-kV systems that required closing resistors. It is good to see applications that take advantage of new technology, especially when lower costs and reduced complexity can be realized.

It is stated that the installed cost of MOSA on both 500-kV line terminals is less than the cost of closing resistors for three breakers (1-1/2 breaker scheme). However, unless this concept is adopted by the utility industry, the elimination of a closing resistor could conceivably make the circuit breaker "special." Depending on the manufacturer and specific design, a "special" rather than a "standard" breaker might not result in any cost savings. Has the expected cost savings been confirmed by several manufacturers?

The curves of Figs. 6 and 7 indicate that some fairly high overvoltages (2.2 pu to 2.4 pu, 2 percent of the time) can be expected even with 1.5 pu arresters if the closing device should fail and random closing occur. Does the scheme have a fail-safe provision to prevent breaker closing and line energization if the device controlling staggered closing should fail? Would overvoltages of this magnitude be within the switching surge energy dissipation capability of the MOSA?

No details are given on the control device used for staggered closing. Is closing completely random or is closing initiated at a specific point on wave of the first phase to close? If the latter is true, then it appears that the random staggered closing curves of Figs. 6 and 7 are based on various breaker closing times rather than the fixed closing time of a specific breaker. Would the authors comment on this?

Field test data (staged tests) enhance a technical paper and it is particularly valuable when model studies are compared with field measurements. Fig. 3 in the paper shows excellent correlation. It is stated that for the field test, closing occurred at the same and opposite polarity as the trapped charge. It would be helpful if Table 1 indicated if reclosing was of the same or opposite polarity as the trapped charge.

Fig. 2 shows the effectiveness of polarity closing to reduce switching overvoltages. The oscillogram indicates that closing is also staggered. Was this just a test condition or will the control device also provide staggered closing along with polarity closing?

Manuscript received July 21, 1986.

John E. Harder (Westinghouse Electric Corporation, Bloomington, IN): This timely paper is very interesting. The authors have done a nice job of presenting the results of their investigation.

With the former gapped silicon carbide arresters, line entrance arresters often required higher ratings than arresters at the transformer because of the remoteness of the ground source with the local breaker open. This higher rating resulted in higher switching surge protective levels. With the application of metal oxide arresters based on continuous operating voltage, the line entrance arrester may now have a lower voltage rating, resulting in much more attractive switching surge protective levels. Further, the high energy capability of modern metal oxide arresters gives substantial margin for the regular operation of these arresters to limit switching overvoltages.

There are several items on which the authors' comments would be appreciated.

1) Did the authors evaluate phase-to-phase switching overvoltages, comparing resistors with arresters? While phase-to-phase voltages may not be critical on a line which is not transformer terminated, a "feel" for the relative effectiveness of the two methods of limiting phase-to-phase switching overvoltages would be of interest.
2) The resistors tend to limit overvoltages to a certain per unit of actual operating voltage. The arresters limit voltages based on their protective level. It would appear that for the same maximum switching surge voltage, the distribution of voltages might be somewhat higher on the average for the arrester limited system, since the system will be operated at maximum voltage only part of the time. Since the probability of flashover of the insulation increases as the voltage goes up, the probability of flashover for the arrester protected system might be higher than for the insertion resistor protected system. Have the authors considered this effect and can they comment?
3) The authors mention the use of 1.5 per unit and 1.7 per unit switching surge protective levels. A quick survey of several manufacturers' standard catalog information indicates that standard 396-kV arresters have a switching surge protective level slightly above 1.7 per unit of MCOV. Special low protective level arresters were produced with a 1.5 per unit protective level under an EPRI R&D project. These low protective level arresters may require substantially more metal oxide than standard units if the design and application margins are not to be compromised. The authors' conclusion No. 6 "The installed costs of MOSA on both 500 kV line terminals is less than the cost of closing resistors for 3 breakers (1½ breaker scheme)." Was this conclusion for both protective levels of arrester?

IEEE Transactions on Power Systems, Vol. 5, No. 4, November 1990

STATISTICAL SWITCHING OVERVOLTAGE ANALYSIS OF THE FIRST B. C. HYDRO PHASE SHIFTING TRANSFORMER USING THE ELECTROMAGNETIC TRANSIENTS PROGRAM

K. C. Lee, Senior Member

K. P. Poon

B. C. Hydro
Vancouver, B. C.
Canada V6Z 1Y3

Abstract: A 400 MVA 230 kV phase shifting transformer (PST) is to be installed in the Nelway Substation of the British Columbia Hydro (B. C. Hydro) transmission network in 1990. This will be the first PST in the B. C. Hydro system. The Nelway PST is to mitigate the problems associated with the inadvertent loop-flows through the interconnections between the B. C. Hydro and the Bonneville Power Administration (BPA) systems. Phase shifting transformers for similar purposes are being planned or installed throughout the Western Systems Coordinating Council (WSCC).

Due to the unique design and construction of the PST, a detailed analysis of the switching overvoltages for insulation coordination is required. This analysis employs the application of the Electromagnetic Transients Program (EMTP). For accurate representation of the distributed and untransposed nature of overhead transmission lines in the vicinity of the Nelway Substation, the untransposed line model developed for the EMTP is used. Statistical switching studies are also performed using the EMTP. Only with recent advancement in computer technology are statistical studies of larger sample sizes feasible. The scanned maximum overvoltages obtained from the statistical runs are used to determine the energy requirement of the metal oxide surge arresters. The overvoltage protection level of the PST and the station equipment is also investigated with statistical analysis.

Keywords: Statistical analysis, transmission line modelling, EMTP switching simulations, insulation coordination.

1. INTRODUCTION

The major B. C. Hydro transmission system with the respective tie-lines is shown in Figure 1. Excessive loop flows through the B. C. Hydro/BPA systems via the tie-lines have been previously documented.[1] In recent years, with the additions of generation and transmission in the Northeastern region of the Pacific Northwest, this loop flow is becoming very heavy. The effects of this are an increase in transmission losses and a decrease in transfer capabilities for the B. C. Hydro system. At times, some circuits have to be opened due to overloading conditions by this loop flow. Studies done by the B. C. Hydro System Transmission Planning Department had identified the solution to this problem by the installation of a 400 MVA, 230 kV, ±40 degree PST at the Nelway Substation to control the flow in the 230 kV tie-line with the BPA network.[2]

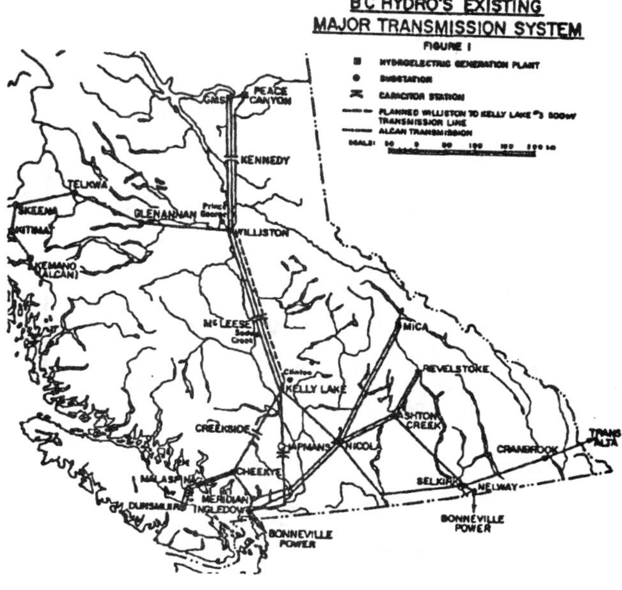

Fig. 1 Simplified B.C.Hydro Transmission network

90 WM 256-8 PWRS A paper recommended and approved by the IEEE Power System Engineering Committee of the IEEE Power Engineering Society for presentation at the IEEE/PES 1990 Winter Meeting, Atlanta, Georgia, February 4 - 8, 1990. Manuscript submitted August 31, 1989; made available for printing January 16, 1990.

PST's are rarely installed on power systems until recently. Documentation of transient overvoltages due to switching of PST's is scarce. Although the design of the individual windings in a PST is similar to that of an ordinary power transformer, the circuit configuration is completely different. Therefore, a detailed analysis of switching transient overvoltages for insulation co-ordination of the PST is necessary. In this analysis, the requirement of metal oxide arresters (MOA) in terms of locations, rating and energy requirements is investigated. The statistical switching overvoltage distributions for different switching operations are also documented.

2. SWITCHING OVERVOLTAGE STUDY BY THE EMTP

A. Network Modelling

For efficient analysis of switching overvoltages at the Nelway PST, only the 230 kV networks around the Nelway PST station are represented in detail. The reduced network arrangement for EMTP simulations is shown in Figure 2.

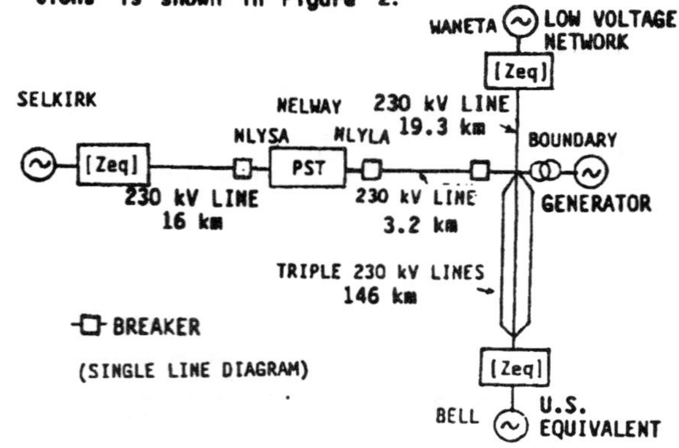

Fig. 2 Reduced 230kV network for EMTP simulation

The details of the models are discussed in the following:

1. Phase Shifting Transformer - The electrical network connection and the voltage vector diagram of the PST are shown in Figures 3 & 4. The PST under consideration is of the grounded 'Y' network construction.[3,4] This type of PST consists of a series unit and an exciting unit with an on load tap changer for controlling the amount of shift in the phase angle.

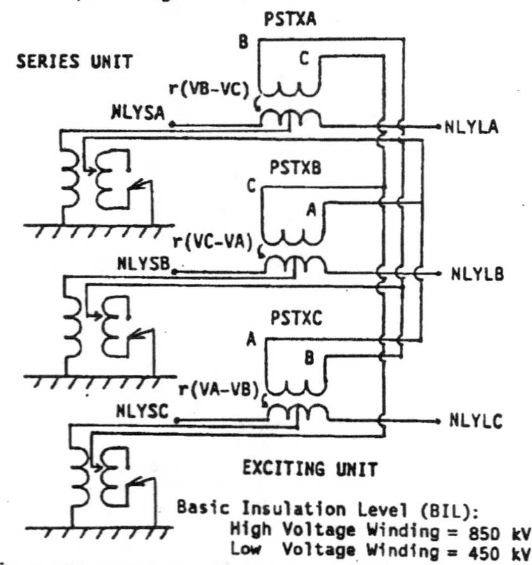

Fig. 3 Detailed 3 phase circuit diagram for the Nelway 230 kV phase shifting transformer.

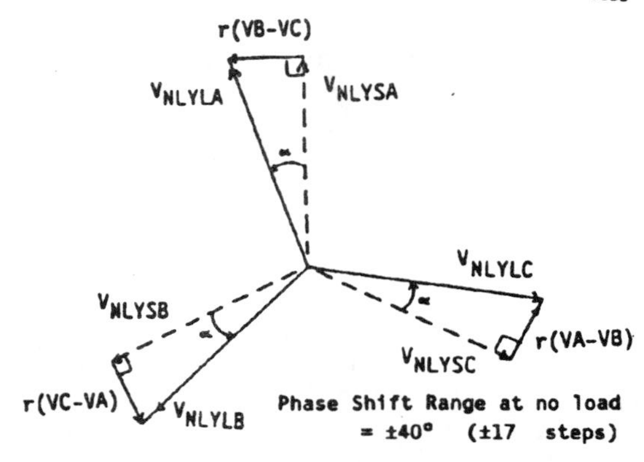

Fig. 4 Three phase voltage vector diagram for PST source and load side terminals.

The EMTP matrix representations of the series and exciting unit of the PST are initially derived by the auxiliary EMTP Transformer programs. The required input data to this Transformer program are the magnetizing current, load loss, short-circuit impedance and voltage ratios of the windings. The initial studies showed that the effect of PST saturation in the overvoltage results is small, the non-linear PST model is thus not employed in the statistical switching studies. The bushing and winding capacitances are represented by lumped elements. The windings are connected as shown in Figure 3. Later on, an equivalent impedance matrix provided by the manufacturer representing the PST externally are also used.

2. Overhead Transmission Line - For some short lines where the electromagnetic wave travelling times are small compared to the EMTP simulation time step, they were represented by the lumped pi-circuits. For longer lines with larger wave travelling times, the distributed parameter line model should be used. Due to the untransposed nature of the 230 kV lines, the untransposed distributed line model is applied. This untransposed line model will provide better accuracy than the pi model. The line parameters are evaluated at the dominant wave frequency of the worst overvoltage condition in order to account for frequency dependence effect of the transmission line.[5] The details of the mathematical derivation with eigen-analysis and a special rotational scheme for this untransposed line model is shown in Appendix A.

3. Circuit breakers - The breakers are modelled as simple time controlled switches in the EMTP. For statistical switching simulations, the characteristics of the breaker mechanical operating parts are modelled by a random pole spread having a standard deviation of 1.5 ms for the 3-phase breaker operating times. Randomly generated switch closing times varying from 0 to 16.66 ms are also used to account for random operation at any point on the 60 Hz power frequency waveform.[6,7,8]

4. Transformers - The EMTP models for various switching and generating station transformers are derived by the auxiliary EMTP Transformer programs. The required data are available in the recorded test results from the manufacturers.

5. Equivalent network - Network equivalencing is employed to reduce the rest of the B. C. Hydro 500 kV/230 kV network external to the Nelway PST study area. The complete external network will be reduced to a voltage source behind a Thevenin equivalent matrix of [Zeq]. The [Zeq] is obtained by the conventional method by initializing a 3-phase fault at the respective bus and obtaining the sequence impedances at the faulted bus. All this can be handled readily by the Fault Study Program with available fault data.

After the network equivalencing scheme is completed, the number of nodes in the EMTP network model is reduced from 675 to 69, and the number of branches is reduced from 981 to 93. The corresponding EMTP Central Processing Unit (CPU) simulation time is also reduced by a factor of 11.

The magnitudes and phase angles of all the voltage sources are found from a Load Flow program. Available data from the existing databank for a heavy load system condition has been chosen.

B. Accuracy of Reduced Network Representation

Previous EMTP simulations using more detailed B. C. Hydro network representation have been compared well with respective field test results.[9,10] However, for efficient computation especially for large number of statistical simulations, network reduction is desired. To verify the accuracy of the network reduction scheme, a 3-phase energization simulation, with both full and reduced network representations are compared. The respective waveforms for the phase to ground and phase to phase overvoltages are shown in Figures 5 and 6. The overvoltage spikes are due to the random closing of breakers at both sides of the PST during 3-phase energization process.

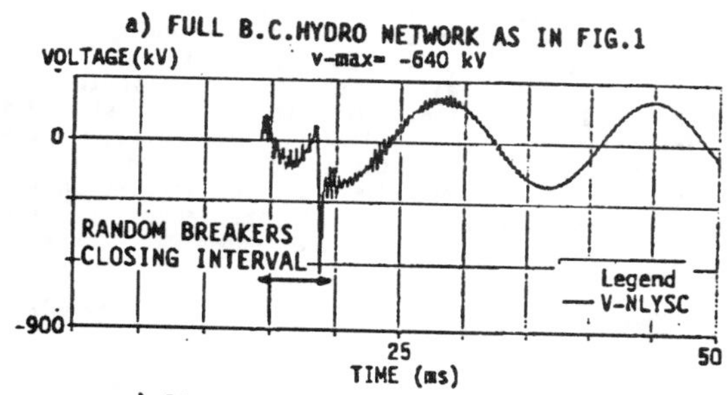

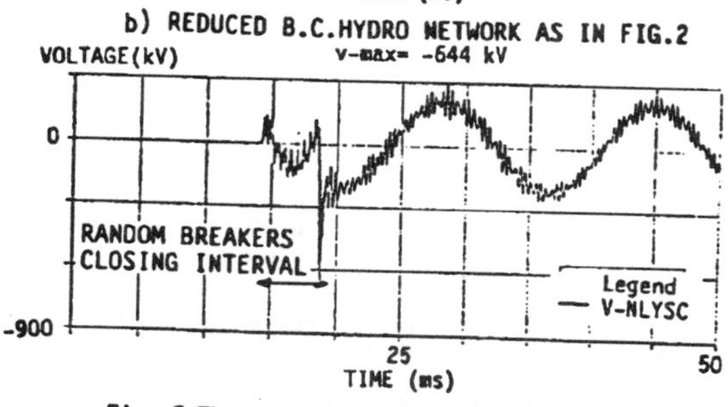

Fig. 5 Three phase energization: phase-ground voltage by full and reduced network representation. (dt=50 μs)

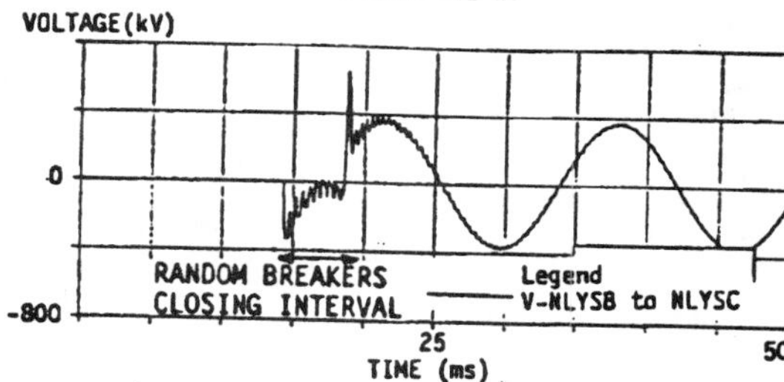

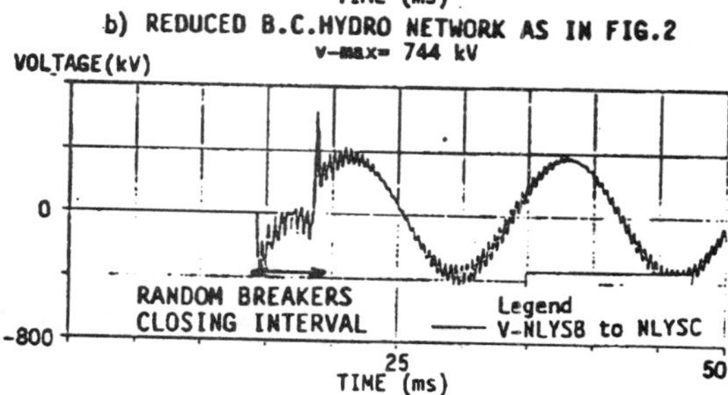

Fig. 6 Three phase energization: phase-phase voltage by full and reduced network representation. (dt=50 μs)

The following observations are obtained for this comparison:

1. The maximum phase to ground voltage for both full and reduced network representation agrees well (e.g. 1% deviation, 640 kV vs. 644 kV). The maximum phase-to-phase overvoltages deviate more (e.g. 4% deviation, 712 kV vs. 744 kV), but these typical deviations in result are still acceptable.

2. The amount of damping in the reduced network is less than in the full network representation. However, this does not affect the maximum overvoltage value which occurs soon after the switching operation.

As shown in Figs. 5a and 6a, the transients will be damped out within 2 to 3 cycles after network changes. Therefore, studies of subsequent circuit breaker operations can assume that a new steady state condition has been reached.

Due to the presence of lumped inductances and capacitances in the PST model, the switching output overvoltage results are also checked for the inherent numerical oscillation as caused by the trapezordal rule used in the EMTP. These numerical oscillations are characterized by a period of 2 simulation time steps. If required, they can be eliminated readily either by the backward Euler Critical Damping Adjustment Scheme[11,18] or the averaging of the output results at successive time steps.[12] Thus, the numerical oscillations of the EMTP will not cause accuracy problems in this investigation.

C. Switching Studies

After the accuracy of the reduced network representation in the EMTP is confirmed, a series of switching transient studies is performed. The following switching cases are investigated:

1. Fault initialization (at PST terminals for worst overvoltage conditions):
   a. Single phase-to-ground fault (φ-g);
   b. Phase-to-phase fault (φ-φ);
   c. Double phase-to-ground fault (2φ-g);
   d. Three phase-to-ground fault (3φ-g).
2. PST bypassing (BP).
3. Three phase energization (3φ).
4. Fault removal/PST de-energization:
   a. Single phase-to-ground fault removal (φ-g);
   b. Phase-to-phase fault removal (φ-φ);
   c. Double phase-to-ground fault removal (2φ-g).
5. Reclose into fault:
   a. Reclosed into single phase-to-ground fault (φ-g);
   b. Reclosed into phase-to-phase fault (φ-φ);
   c. Reclosed into double phase-to-ground fault (2φ-g);
   d. Reclosed into three phase-to-ground fault (3φ-g).

Each switching case is performed for 200 statistical simulations.[15] The breaker or fault initialization times are varied according to the sequences of generated random numbers.[16] The maximum overvoltages obtained by scanning the 200 simulations among the PST terminals of the series and exciting units and the CPU time used in each 200-runs case on the SUN 4/280 micro-computer systems are tabulated in Table 1.

			Type	Max Vφ-g (pu)	CPU Time (hr)
1.	Fault Initialization	a.	φ-g	1.48	0.82
		b.	φ-φ	1.22	0.81
		c.	2φ-g	1.67	0.86
		d.	3φ-g	1.61	0.83
2.	By-pass		BP	1.15	0.86
3.	Energization		3φ	3.75	0.80
4.	Fault removal	a.	φ-g	5.12	0.80
		b.	φ-φ	7.89	0.81
		c.	2φ-g	6.12	1.02
			*(φ-φ	7.95	9.7)*
5.	Reclosed into fault	a.	φ-g	3.33	0.84
		b.	φ-φ	4.53	0.86
		c.	2φ-g	3.53	0.85
		d.	3φ-g	2.16	0.82

(1 pu = 193 kV)

* Full B. C. Hydro network representation.

Table 1. Scanned maximum overvoltages from 200 switching runs for the Nelway PST without MOA.

As shown in Table 1, the switching overvoltages on the PST can be exceedingly high. The designed Basic Insulation Level (BIL) for the 230 kV PST is 850 kV. Applying a Switching Insulation Level (SIL) of 85% of the BIL and allowing an extra 20% margin for insulation aging during equipment service life span, the maximum allowable switching transient overvoltage must be lower than 578 kV or 3 pu. Without the installation of any surge arresters, the maximum switching overvoltages at the PST terminal during a faulted condition was found to be as high as 8.0 pu.

D. Surge Arrester Requirement

It is normal practice in B. C. Hydro to have lightning arrester installed to protect station equipments from damages by lightning. In view of the fact that extremely high switching transient overvoltages are observed in some of the EMTP statistical simulations, further studies are needed to evaluate the performance of the surge arresters for the protective level and energy requirement. The MOA rated at a maximum continuous operating voltage (MCOV) of 152 kV[14] are planned.

The MOA are initially modelled by a group of fitting exponentials for the nonlinear v-i discharge characteristics. The resulting non-linear equations in the EMTP are then solved by the Newton-Raphson iteration scheme.[8] It has been found that this MOA exponentials model causes numerical convergence problems in certain network conditions of the EMTP statistical simulation. In order to guarantee the successful completion of all the statistical simulations, the piece-wise linear model is applied. The accuracy of the piece-wise model has also been confirmed by comparing the respective results with the exponential models for the MOA.

After the numerical convergence problem for MOA non-linear model have been solved, a series of switching operations are simulated using the EMTP. It has been found that the installation of MOA at only one side of the PST is not sufficient to protect the other side during switching operations. Thus, the MOA must be installed at the source side as well as the load side of the PST for full protection. With this double side MOA installation, the switching overvoltages are lowered to an acceptable level. The maximum overvoltages scanned over the 200 simulation and the respective statistical parameter for the severe switching cases are shown in Table 2.

Type	Vmax (pu)	μ Mean (pu)	σ St. dev. (pu)	(Vmax-μ)/σ
Energization				
3 φ	1.26	0.67	0.19	3.1
Fault removal				
φ-g	1.28	1.00	0.08	3.5
φ-φ	1.70	0.86	0.20	4.2
2φ-g	1.46	1.02	0.11	4.0
Reclosed into fault				
φ-g	1.37	1.20	0.05	3.4
φ-φ	1.44	1.18	0.06	4.3
2φ-g	1.49	1.31	0.05	3.6
(φ-φ	1.69	1.25	0.10	4.4)*

* Worst initial trapped charge condition

Table 2. Scanned maximum overvoltages for the Nelway PST external nodes with MOA installed at the source and load side.

For illustration purpose, a typical most severe overvoltage condition of the phase to phase fault removal case without MOA installed is shown in Figure 7. This overvoltage far exceeds the SIL of the PST. However, these overvoltages are lowered to the acceptable level with the MOA installed at both sides of the PST. As shown in Figure 8, the overvoltages at the source and load sides with MOA installed will be clipped at about 220 kV (1.17 pu). The maximum discharge energy of the MOA is found to be 0.96 MJ (see Figure 9) which is well within the rated capabilities of the MOA's in this class (1.37 MJ).

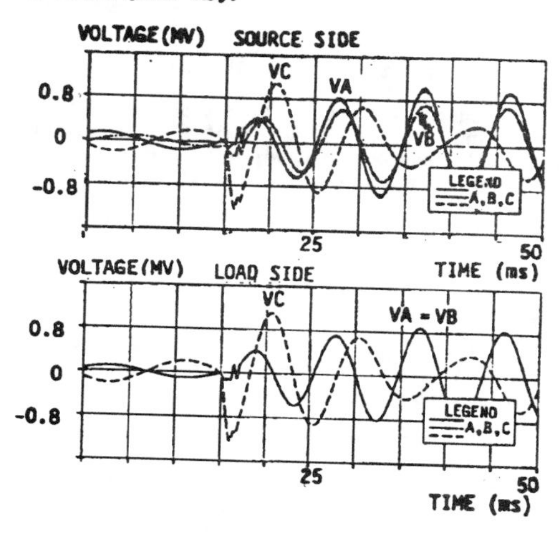

Fig. 7 Typical waveform of phase-ground voltage for the phase to phase fault removal without MOA.

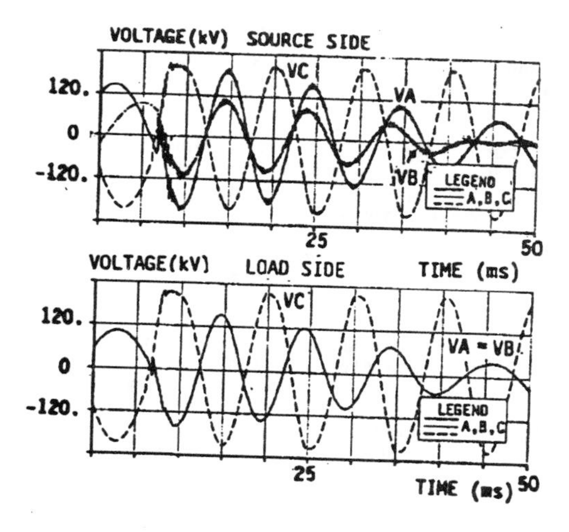

Fig. 8 Typical waveform of phase-ground voltage for the phase to phase fault removal with MOA.

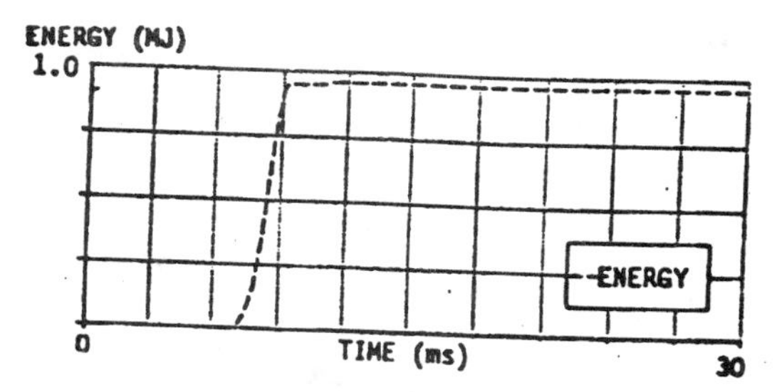

Fig. 9 Typical waveform of metal oxide arrester discharged energy for the worst condition.

E. Statistical Distribution of Overvoltages

The statistical distributions for the different switching overvoltages are also analysed. The overall maximum overvoltages from the various sets of simulations are illustrated by the normal distribution. For all cases, the scanned maximum overvoltages are usually within 3.0 to 4.5 standard deviations above the mean value.

The cumulative probabilities and histograms for a typical breaker closing scheme are shown in Figures 10 and 11. These statistical simulations involve random breaker closing operations. The overvoltage distribution for 200 runs can be described by the normal distribution with the mean and standard deviation calculated directly from EMTP results. For checking purpose, extended statistical simulations of 2000 runs are performed. However, no significant improvements are found. The cumulative probability plots for both the 200 and 2000 runs can be described by the same normal distribution. The statistical parameters of the mean, standard deviation and scanned maximum overvoltage for the 200 and 2000 runs are (0.67, 0.19, 1.26 pu) and (0.67, 0.20, 1.32 pu), respectively.

However, for statistical simulations involving random breaker opening operations, hypothetical normal distribution from the calculated mean and standard deviations of the data will not be sufficient. This is because the breaker would open randomly but the circuit would not be broken immediately until the switch current is below a certain current margin depending on the arc quenching mechanism. This usually results in a higher switching overvoltage. A typical breaker opening simulation is shown in Figures 12 and 13. A larger percentage of overvoltages occurs at the high end around the 2% cumulative probability level for both the 200 and 2000 runs. Since the scanned maximum overvoltage level is of major concern for the PST internal insulation requirement, a normal distribution fit can be performed beyond the 2% level. This may require more data points than is available from 200 runs. The statistical distribution of overvoltages from both the 200 and 2000 runs essentially also agree well. The normal distribution fitted for the high end is facilated by more data points from the 2000 run especially around the 0.1% region. The corresponding statistical parameters of the mean, standard deviation, and scanned maximum overvoltage for the 200 and 2000 runs of this breaker opening case are {0.94, 0.26, 2.43 pu} and {0.94, 0.30, 2.46 pu} respectively.

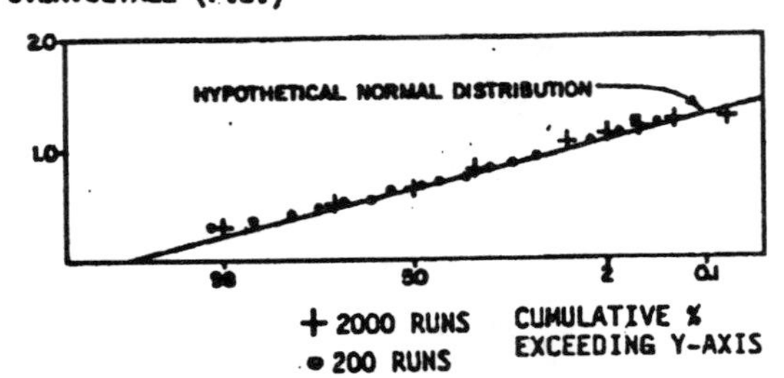

Fig. 10 Typical cumulative probability plot of overvoltage distribution for breaker closing operation.

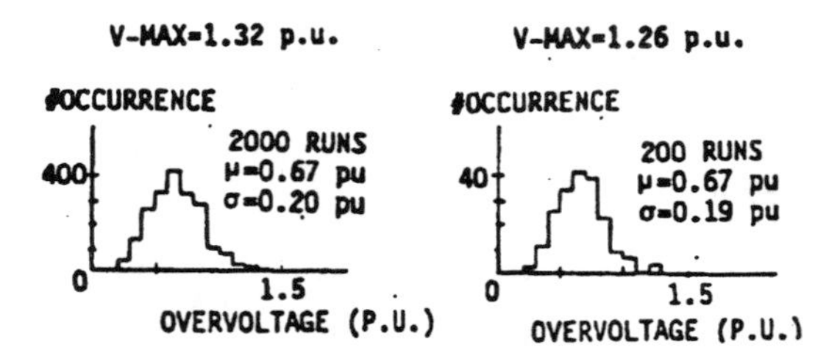

Fig. 11 Typical histogram of overvoltage distribution for breaker closing operation.

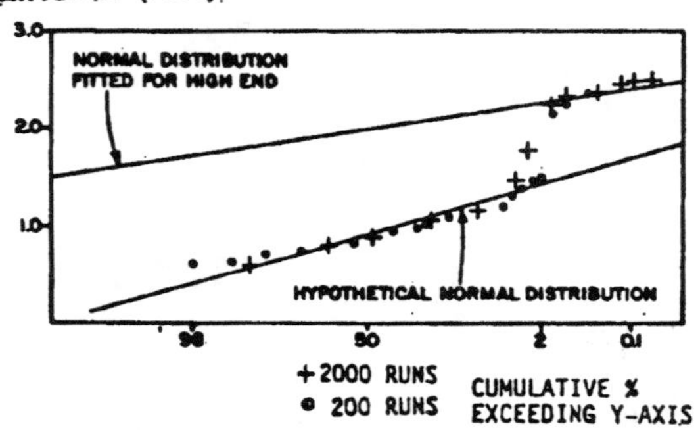

Fig. 12 Typical cumulative probability plot of overvoltage distribution for breaker opening operation.

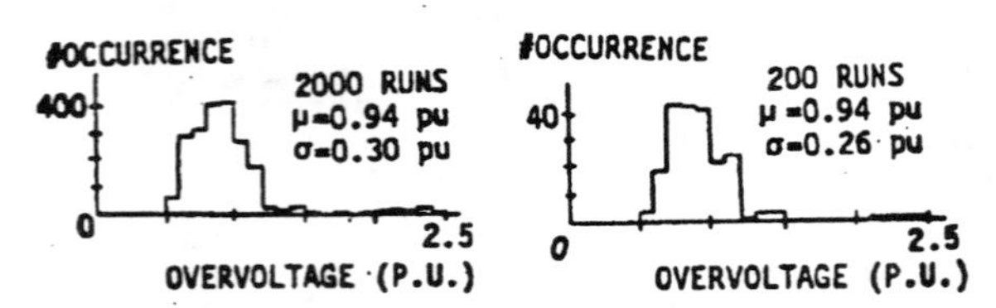

Fig. 13 Typical histogram of overvoltage distribution for breaker opening operation.

The maximum overvoltage distribution on the low voltage windings of the PST is also examined. The mean, standard deviation and scanned maximum overvoltages for the worst phase-to-phase fault removal case are (0.33, 0.07, 0.60 pu). This is acceptable for the designed level of 1.59 pu.

CONCLUSION

Switching transient overvoltages on the PST have been simulated successfully by the EMTP. The overvoltages at the PST could be unacceptably high without MOA connected at both sides of the PST. The highest overvoltages at both the primary and secondary windings of the series and exciting units of the PST are within the allowed level with proper MOA installation.

Statistical simulations using the EMTP have been shown to be useful for insulation co-ordination studies of the PST. Surge arrester applications have been analysed by using the most severe switching operation which results in maximum overvoltage conditions as obtained from the EMTP statistical simulations.

Experiences with suppression of possible numerical oscillations by different approaches and effect of modelling of non-linear metal oxide arrester on numerical convergence of the EMTP has also been investigated.

Validity of representation of overvoltage distribution by normal distribution for different breaker opening and closing operations has also been examined.

ACKNOWLEDGEMENT

The authors are grateful to Profs. H. W. Dommel and J. R. Marti for consultation through the B. C. Hydro/ Univ. of B. C. Research Agreement. The authors would also like to thank Messrs. B. J. Mills, M. Scott, J. H. Sawada, J. Polvi and B. L. Avent for discussions and support.

REFERENCE

1. K. P. Poon, "System Application: Nelway PST", B. C. Hydro Report, 1 June 1988.
2. K. P. Poon, "BCH-US Loop Flow as Affected by New System Additions in the U.S.", B. C. Hydro Report, September, 1984.
3. P. R. Russman "Design and Application of High-Voltage Phase Angle Regulating Transformers", presented to Canadian Electrical Association Meeting, Montreal, March 1988.
4. B. K. Patel et al "Application of Phase Shifting Transformers for Daniel-McKnight 500 kV Interconnection", IEEE PWD p. 167-172, July 1986.
5. K. C. Lee, J. H. Sawada, L. Marti "Comparison of Various EMTP Transmission Line Models, Part I, II" EMTP NEWSLETTER, Vol. 2-4, p. 58-69, May 1982.
6. S. F. Mauser, T. E. McDermott "EMTP Primer", EPRI Report EL-4202, RP2149-1, Westinghouse, Sept. 1985.
7. S. F. Mauser, T. F. McDermott "EMTP Application Guide" EL-4650, RP2149-1, Westinghouse Electric Corporation, Nov. 1986.
8. V. Brandwajn, "EMTP Version 1 Revised Rule Book", EPRI Report EL-4541, RP2149, SCI, April 1986.

9. Y. Mansour, T. G. Martinich, J. E. Drakos "B. C. Hydro Series Capacitor Bank Staged Fault Test", IEEE PAS p. 1960-1969, July 1983.

10. M. B. Hughes, R. W. Leonard, T. G. Martinich "Measurement of Power System Driving Point Impedance", IEEE PAS p. 619-629, March 1984.

11. J. R. Marti, J. Lin "Suppression of Numerical Oscillations in the EMTP", IEEE PWRS, p. 739-747 May 1989.

12. H. W. Dommel, I. I. Dommel, "Transients Program User's Manual", UBC Publication, Aug. 1988.

13. H. W. Dommel et al "EMTP Theory Book", Bonneville Power Administration Contract, August 1986.

14. "Application Guide: Dyna Var Metal-Oxide Surge Arresters", Ohio Brass Company, 1982.

15. A. R. Hileman, P. R. Leblanc, G. W. Brown "Estimating the Switching Performance of Transmission lines", IEEE PAS p. 1455-1469, Sept. 1970.

16. L. Nyhoff, S. Leestma, "Fortran 77 for Engineers and Scientists", MacMillan Publishing Co., 1988.

17. K. C. Lee, H. W. Dommel "Addition of modal analysis to the UBC Line Constants Program" UBC/BCH Report, Jan. 1980.

APPENDIX A

Theory of Untransposed Line Representation in the EMTP

The EMTP Line Constants Program produce the distributed line parameters of any line configuration in the form of a series impedance matrix $[Z^{phase}]$ and a shunt admittance matrix $[Y^{phase}]$ at any frequency. These matrices are used to describe the change in voltages and currents along the line by N coupled differential equations. These can then be transformed into N decoupled equations.

$$\left[\frac{d^2 V^{mode}}{dx^2}\right] = [T_v]^{-1} [Z^{phase}] [Y^{phase}] [T_v] [V^{mode}]$$

$$= [\Lambda] [V^{mode}] \qquad (B1)$$

$$\left[\frac{d^2 I^{mode}}{dx^2}\right] = [T_i]^{-1} [Y^{phase}] [Z^{phase}] [T_i] [I^{mode}]$$

$$= [\Lambda] [I^{mode}] \qquad (B2)$$

where $[T_i]$ = matrix of eigenvectors of $[Y^{phase}] [Z^{phase}]$.

$$\text{and } [T_v] = ([T_i]^t)^{-1} \qquad (B3)$$

We can obtain the modal $[Z^{mode}]$ and $[Y^{mode}]$ as:

$$[Z^{mode}] = [T_i]^t [Z^{phase}] [T_i] \qquad (B4)$$

$$\text{and } [Y^{mode}] = [T_i]^{-1} [Y^{phase}] ([T_i]^t)^{-1} \qquad (B5)$$

It has to be noted that the eigenvectors (columns of $[T_i]$ or $[T_v]$) are only determined to within a multiplicative constant. Each eigenvector can, therefore, be multiplied with any non-zero complex scalar, and it will produce zero modal conductances. A rotation scheme is used which makes the modal admittance matrix $[Y^{mode}]$ purely imaginary,[17]

$$[Y^{mode}]_{rotate} = [0] + j[B^{mode}]_{rotate} \qquad (B6)$$

This rotation is equivalent to dividing the i-th eigenvector (i-th column of $[T_i]$) by a factor D_i. First, find the angle θ_i of Y_i^{mode} with:

$$D_i = e^{j\frac{90^\circ-\theta_i}{2}} \qquad (B7)$$

With all D_i's forming a diagonal matrix [D], the modified matrix of eigenvectors becomes

$$[T_i]_{rotate} = [T_i] [D] \qquad (B8)$$

$$\text{Then, } [Y^{mode}]_{rotate} = [D] [T_i]^{-1} [Y^{phase}] ([T_i]^t)^{-1} [D]$$

$$= [D] [Y^{mode}] [D] \qquad (B9)$$

$$\text{and } [Z^{mode}]_{rotate} = [\Lambda] [Y^{mode}]^{-1}_{rotate}$$

BIBLIOGRAPHY

Kai-Chung Lee was born in Kwungton, China in 1951. He received the BSc degree from the University of Wisconsin in 1973 and the MSc, MASc and PhD degrees from the University of British Columbia in 1975, 1977 and 1980.

He was a post-doctoral fellow in UBC from 1980 to 1981, and joined B. C. Hydro in 1981. He is an Analytical Studies Engineer in the System Planning Division. He is senior member IEEE, member IEE, Registered Professional Engineer in the Province of B. C. and Chartered Engineer in the U.K.

Kenneth Keng-Por Poon was born in Hong Kong in 1939. He received his Diploma in Electrical Engineering from the Hong Kong Technical College in 1962.

He joined the Transmission Planning Department of B. C. Hydro in 1974 and has been a member of the WSCC System Review Work Group since 1980. In this capacity, he has contributed to the various WSCC technical activities. He is a Registered P.Eng. in the Province of B. C.

IEEE Transactions on Power Delivery, Vol. 3, No. 1, January 1988

CAPACITOR SWITCHING AND TRANSFORMER TRANSIENTS

BY

R.S. BAYLESS
Member, IEEE
Pacific Power and Light Co.
Portland, OR

J.D. SELMAN
Senior Member, IEEE
Tri-State G & T Assn., Inc.
Denver, CO

D.E. TRUAX
Senior Member, IEEE

W. E. REID
Member, IEEE
McGraw-Edison Company
Power Systems Division
Canonsburg, PA

ABSTRACT

The wide use of shunt capacitor banks on high voltage and extra high voltage systems in recent years has resulted in significant economic savings. These shunt capacitor applications have also resulted in some new application considerations. In this paper the impact of capacitor switching on transformer transients is evaluated. Two specific transformer failure events are described. Each of these failures coincided with the switching of a capacitor bank some distance away from the transformer. The causes of these failures are evaluated, and field test transient voltage waveforms are duplicated by computer simulations. In addition, the simulation capability is extended to determine system conditions which are susceptible to these transient voltages as well as the means to minimize them.

INTRODUCTION

Shunt capacitor banks are being applied with increasing frequency at virtually all system voltage levels. These capacitors are being applied to increase power transfer capability, reduce equipment loading, reduce energy costs, and control system voltage. The economic benefits are usually substantial. One of the significant characteristics of shunt capacitor banks is that they are switched quite often, typically on the order of twice a day, to react to changing system load conditions. As a result, capacitor switching is the most prolific generator of transient voltages on many utility systems. Historically, lightning strokes and the switching of lines and transformers have been the major sources of transients upon which the insulation coordination philosophies of utility systems have been based. These types of transients typically occur less frequently than do capacitor switching transients. Consequently, methods for minimizing capacitor switching transient voltages have been determined and are applied as needed. [1,2,3]

Several instances of transformer failure coincident with capacitor switching have occurred in recent years. In this paper two of these events are documented and evaluated so that the potential high overvoltage concerns are clearly illustrated.

In one case the overvoltages of concern were high phase-to-phase transients (Pacific Power and Light). In the second case the concern was high overvoltages internal to the transformer resulting from a resonance condition (Tri-State G and T). In each case the transformers were at the end of a radial line coming from the substation where the switched capacitor bank was located.

86 SM 419-6 A paper recommended and approved by the IEEE Transmission and Distribution Committee of the IEEE Power Engineering Society for presentation at the IEEE/PES 1986 Summer Meeting, Mexico City, Mexico, July 20 - 25, 1986. Manuscript submitted January 31, 1986; made available for printing April 23, 1986.

Printed in the U.S.A.

In this paper field measurements of transient voltages recorded on the PP&L system are compared with simulated waveforms. The duplication of the field waveforms allows a more confident extension of this analysis to identify potential problem areas as well as solutions for those conditions.

PACIFIC POWER & LIGHT EXPERIENCE

Prior to May 1980, PP&L had installed a 230 kV phase shifting transformer to control flows from the 500 kV Pacific Intertie into the parallel 230 kV system at Malin Substation in southern Oregon (see Figure 1). A 50 MVAR capacitor bank is located 34.5 miles away at the Klamath Falls 230 kV substation. The capacitor bank was switched as a unit with circuit switchers which were not equipped with preinsertion resistors. The capacitor bank is required for voltage support during peak load periods and during certain outage conditions. Consequently, it is switched daily during peak winter load seasons. The phase shifter was protected by 180 kV surge arresters.

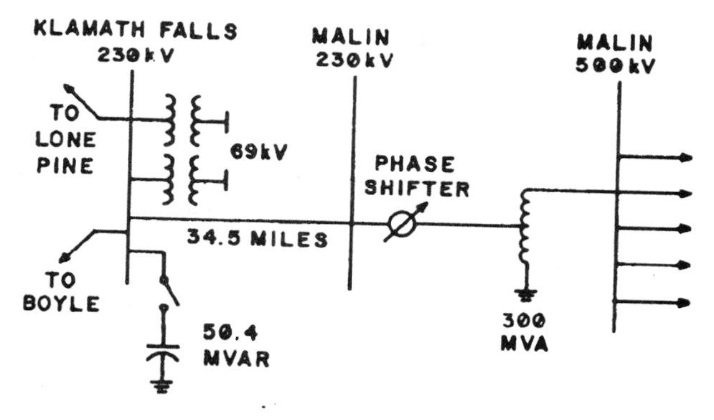

Figure 1.
P P & L System Diagram

On May 27, 1980 the phase shifting transformer failed from what appeared to be a phase-to-phase insulation failure between B and C phases. Later investigations indicated that the failure was coincident with the switching of the Klamath Falls capacitor bank. From analyzing arrester count records, it did not appear that transients at Malin had been high enough to spark over phase to ground protective surge arresters on the 230 kV bus at Malin.

Following the failure of the Malin phase shifter, Transient Network Analyzer (TNA) studies were conducted to investigate the characteristics of the transients produced by switching the Klamath Falls capacitor bank. These results indicated that high phase-to-phase transient could occur. These studies and later tests indicated that the transient, produced at Klamath Falls when the 50 MVAR capacitor bank is energized, is most pronounced when the capacitor switch is closed near crest phase voltage. The circuit switcher at Klamath Falls assures that the individual phases of the capacitor bank are

0885-8977/88/0100-0349$01.00©1988 IEEE

consistently energized near peak phase voltage because the contacts close slowly relative to 60 Hz.

When two phases close together, transients of opposite polarity occur in the two phases. These transients propagate from Klamath Falls to Malin. Transformer termination at Malin appears as virtually an open circuit to the high rate-of-rise transient. Because of the high refraction coefficient, the voltage transient nearly doubles at the Malin bus. The transient is then reflected back to Klamath Falls where it sees the capacitor bank as a short circuit, changes polarity, and is again reflected back to Malin. The transient oscillates back and forth across the 34.5 mile, 230 kV Malin to Klamath Falls line until it is damped out. The phase-to-phase voltage at Malin is highest when the Klamath Falls capacitor switch energizes two phases nearly simultaneously and near peak voltage on each of the two phases. This causes the alignment of high magnitude, opposite polarity surges on each phase and results in a high phase-to-phase surge voltage.

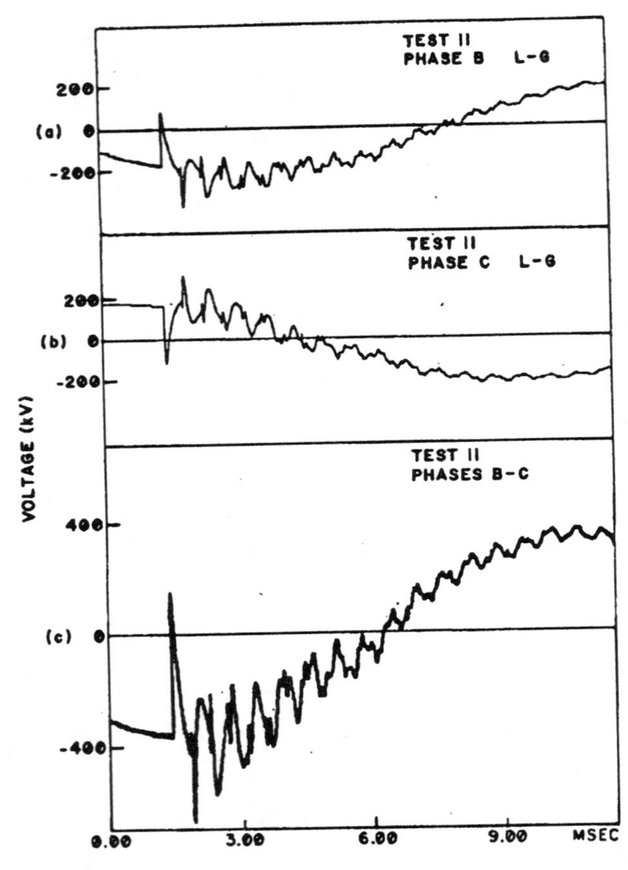

Figure 2.
Malin Field Measurements (shot 11)

A series of tests were conducted by PP&L and the Bonneville Power Administration (BPA). Using BPA's fiber optic instrumentation and digital high frequency recorders, these tests showed transients of the characteristic waveshape illustrated in Figure 2. This transient voltage was the highest recorded in thirteen energizations of the Klamath Falls capacitor bank (shot 11). The resulting phase-to-phase and phase-to-ground voltage transients at the Malin 230 kV bus are shown. The results of shot 11 show that phases B and C of the capacitor switch close very close together. At the Malin 230 kV bus, 648 kV peak was recorded from phase B to phase C with 352 kV peak from B phase to ground and 300 kV peak from C phase to ground. Results of the field tests show that switching sequences which were similar to shot 11 occurred twelve out of thirteen test shots, and because of the mechanics of the switchers, it is now assumed that this sequence could occur during most switching operations.

A simulation corresponding to the field test condition in Figure 2 is given in Figure 3 (when the per unit system is used, both the line-to ground and the line-to-line voltages are given in per unit of the line-to ground peak voltage, which is 188 kV). This simulation was carried out using the Electromagnetic Transients Program (EMTP). To accomplish a duplication of this accuracy required a good model of the 500 kV system as well as the 230 kV system. This good duplication allows an evaluation of the effect of other parameters on the voltage transient characteristics. This is carried out later in this paper.

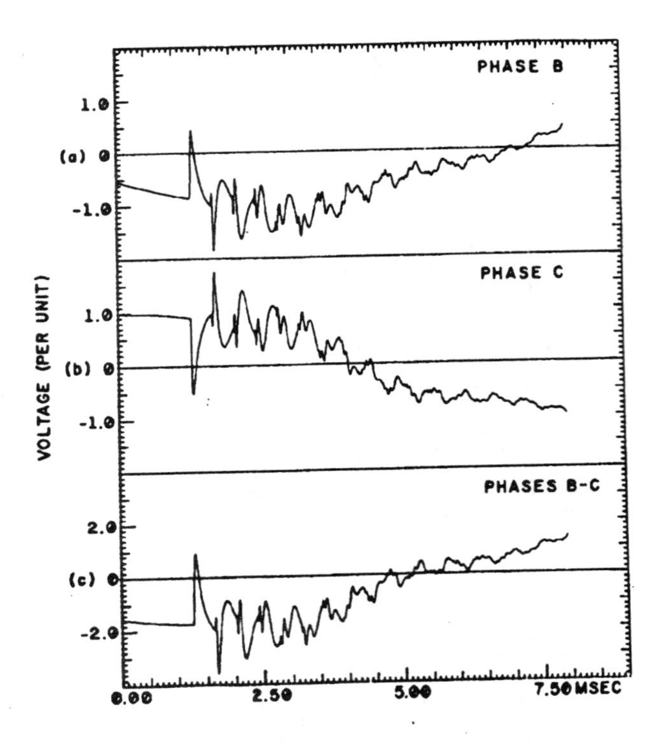

Figure 3.
Malin Simulations

Several options were studied as possible solutions to minimize the transients at Malin. These options included:

1. Preinsertion resistors in the Klamath Falls capacitor switches,
2. Staggered phase closing of the capacitor switch,
3. Placement of a capacitor bank at the Malin 230 kV bus,

4. Zero voltage controlled closing of the capacitor switches at Klamath Falls and,

5. Surge reactors in series with the Klamath Falls capacitor bank.

Due to the economics and practicality of the application, the use of reactors in series with the capacitor bank at Klamath Falls was implemented to limit the phase-to-phase transients at Malin.

TRI-STATE G & T EXPERIENCE

The Tri-State Generation and Transmission Association is a power cooperative with interconnections to numerous utilities in the western U.S. One interconnection point is at the Riverton Substation by means of a 230/115/13.8 kV, 60 MVA 3-phase autotransformer. Another interconnection is at the Pilot Butte 115 kV substation (See Figure 4). The 115 kV line connecting Riverton to Pilot Butte is 29 miles long, and is tapped 4.2 miles from Riverton at the Wind River Substation.

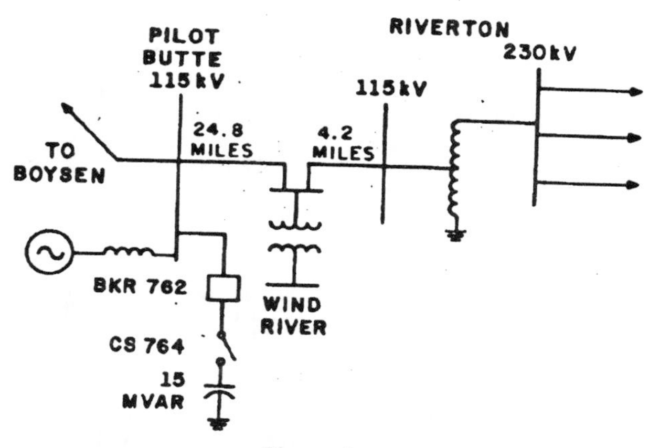

Figure 4.
Tri-State System Diagram

On March 29, 1984, the 15 MVAR capacitor bank at Pilot Butte was commissioned. Breaker 762 was closed prior to energization. Upon closing circuit switcher 764 and energizing the capacitor bank, breaker 762 tripped via the fault bus protection. Upon investigation, it was determined that the fault bus differential relay was set too sensitively and likely picked up due to the capacitor bank inrush current. Coincident with the capacitor bank switching, the Riverton transformer failed.

Inspection of the transformer indicated a turn-to-turn failure in the B-phase series winding group closest to the 115 kV (X) bushing. The transformer manufacturer concluded that the turn-to-turn breakdown in the winding insulation was caused by a transient overvoltage condition. In addition to the transformer failure, evidence of an overvoltage condition was indicated by a failed internal protective gap on the B-phase 115 kV Bus CCVT (capacitive-coupled voltage transformer) at Riverton.

It was discovered that the oil circuit breaker (762) used for fault interruption on the capacitor bank was not rated for capacitor switching. Consequently, restriking was suspected on the March 29th incident; however, there was no available oscillograph record to confirm this.

A switching transient study of the Riverton Pilot Butte area was conducted with the following objectives:

1. Determine the transient voltages which could occur at the Riverton 115 kV bus due to energizing and de-energizing the Pilot Butte capacitor bank.

2. Determine if internal transformer resonance is possible at Riverton due to the switching of the Pilot Butte capacitor bank.

The physical arrangement of the Riverton transformer windings is illustrated in Figure 5. The failure occurred approximately midway between the X (115 kV) and H (230 kV) terminals. Tests on this transformer indicated a natural frequency near 8 kHz. A detailed model of the transformer was developed to evaluate the possible transients at the point of failure. The results of a simulation of a switch restrike are illustrated in Figure 6. The simulated voltage at the 115 kV bus of the transformer is illustrated in Figure 6(a). The characteristic of the transient is very similar to that shown in Figures 2 and 3 for the PP&L system. This is to be expected since the system configurations are very similar.

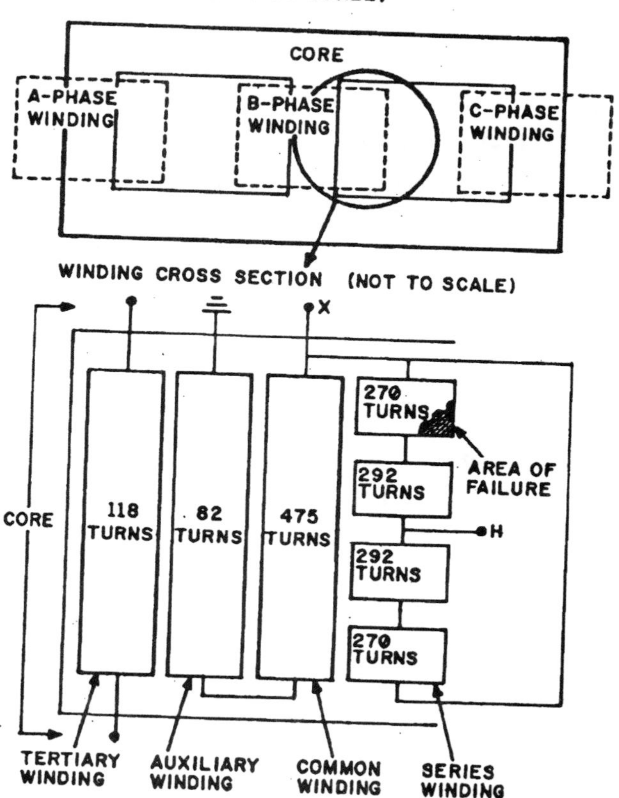

Figure 5.
General Physical Arrangement of Riverton Transformer (not to scale)

In Figure 6(b) the calculated voltage at the point of failure is shown. This voltage has an increasing oscillatory wave shape which is usually characteristic of transformer part-winding resonance conditions. The part-winding resonance condition has been discussed in numerous papers [4, 5,]. This has historically been associated with transmission line fault initiation where the transient reflects between the transformer and the point of fault setting up a semi-oscillatory transient voltage at the terminal of the transformer which excites the internal natural frequency of the transformer. The capacitor switching transient is similar to the fault initiation transient in that multiple reflections occur at the transformer resulting in a potentially significant internal oscillatory transient. If the timing of the reflections corresponds to the natural frequency of the transformer, a build up in the transient voltage within the transformer windings can occur as is illustrated in Figure 6.

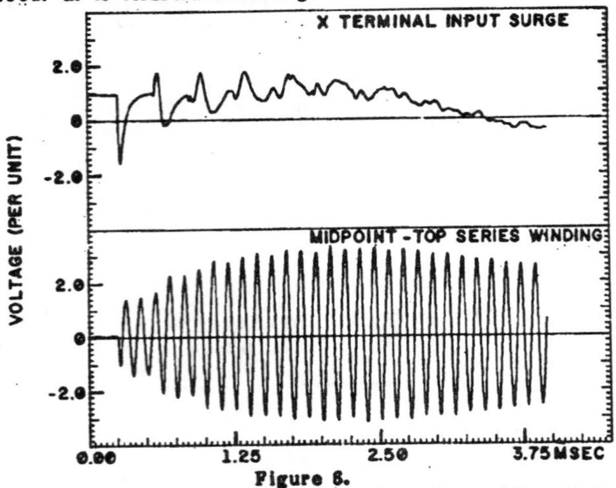

Figure 6.
Riverton Transients for a Single Restrike at Pilot Butte

In Figure 7 the results of a similar case are illustrated where the transformer capacitance has been increased by 15%. The transformer natural frequency has been changed enough to significantly reduce the magnitude of the transient. Clearly then even if the system configuration illustrated in both the PP&L and Tri-State systems occurs, the line length and the transformer natural frequency must correspond to result in a voltage build up similar to that illustrated in Figure 6 for the internal resonance condition.

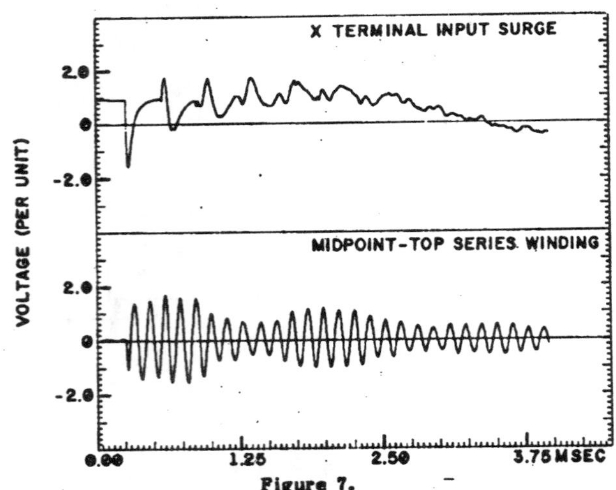

Figure 7.
Riverton Transients for a Single Restrike at Pilot Butte (15% Added to Transformer Capacitance)

The analysis of the Tri-State system resulted in the following conclusions:

- Even though normal surge protection is used, high internal transient voltages can occur in the Riverton transformer. This happens because the system parameters are such as to generate transients which excite the internal natural frequency of the transformer.
- These transients could occur for normal closing and for restrikes upon opening. Multiple restrikes would result in repetitive high surges and potentially higher magnitudes.
- Transformers are more prone to this type of an internal resonance when they are at the end of a line which emanates from a bus where a capacitor bank is switched.
- It is possible to control such transients so as to minimize their effect by using capacitor reactors or switch closing resistors; closing resistors, however, do not help the restrike condition. The effectiveness of these methods is dependent on system parameters and the natural frequency or frequencies of the transformer.

TRANSIENT CHARACTERISTICS

The characteristics of the transient voltage waveforms which occur at the transformers in both the PP&L and Tri-State cases are very similar. The initial portion of the PP&L waveform of Figure 3 is expanded in Figure 8 to illustrate its key characteristics, indicated by the circled numbers explained as follows:

(1) At Klamath Falls the switches on phases B and C close simultaneously. The bus voltages go to zero nearly instantaneously.

(2) The transient arrives at Malin approximately 183 μsec after switch closing. If the Malin terminal was open circuited the transient would double. However, at Malin there is a 300 MVA transformer which connects to a strong 500 kV system. The result is that the Malin terminal appears nearly as the short circuit inductance of the transformer to the surge. This low inductive termination results in sloping back the surge as it arrives at Malin and tailing off as illustrated in Figure 8.

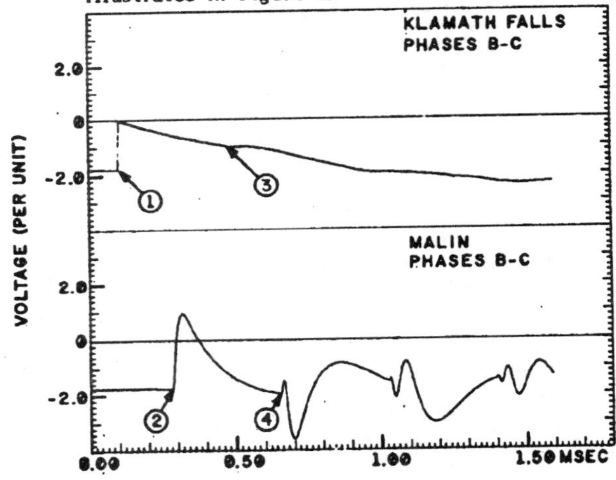

Figure 8.
Base Case, Expansion of Figure 3

(3) The surge returns to Klamath Falls and is barely discernible due to the very low surge impedance of the shunt capacitor bank.

(4) The surge returns to Malin giving a transient surge in the opposite direction and resulting in the peak transient.

From this analysis of the waveforms the key system parameters which determine the wave shape can be divided into the following three categories:

1. Transient characteristic at the capacitor
2. Transformer termination characteristic
3. Line length

Each of these categories and their effect on the waveshape is discussed below.

Transient Characteristic at the Capacitor

In Figure 9 three different conditions at Klamath Falls are simulated to illustrate how the transient may vary. The waveform in Figure 9(b) corresponds to the basic system as did the waveforms in Figures 2, 3, and 8 for the energization of a 50 MVAR capacitor bank. In Figure 9(c) the transient is for the energization of a 200 MVAR capacitor. In the first 183 μsec after closing the transient changes at a slower rate. At Klamath Falls, where there is virtually no local inductance source, the initial transient is determined by the R-C circuit of the line surge impedances charging up the capacitor. Consequently, for the larger capacitor the effective charging time is longer and the change in voltage, as illustrated in Figure 9(c), is less for the first 183 μsec.

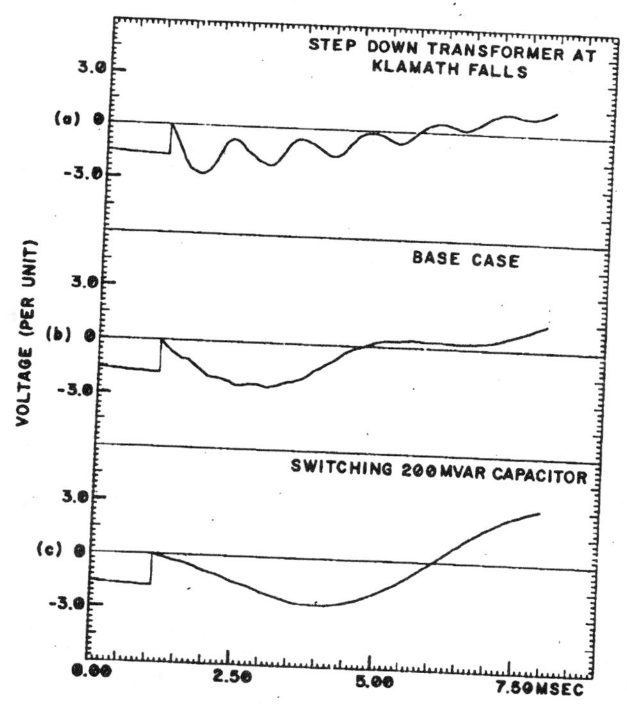

Figure 9.
Phases B-C Voltages at Klamath Falls

For Figure 9(a) a step down transformer is added at Klamath Falls which adds the equivalent of 26.6 kA of fault current at that bus. This strong inductive source tends to dominate the line surge impedance and results in an oscillatory transient in the first 183 μsec after switch closing. This results in a transient which is changing at more than twice the rate after closing than is the base case in Figure 9(b).

For the three cases described in Figure 9, the voltages at the Malin transformer are given in Figure 10. The strong source at Klamath Falls results in the highest transient at Malin while energizing the large 200 MVAR capacitor results in the lowest transient. These results indicate that the highest transients at the transformer are likely to occur for switching a small capacitor bank on a strong system.

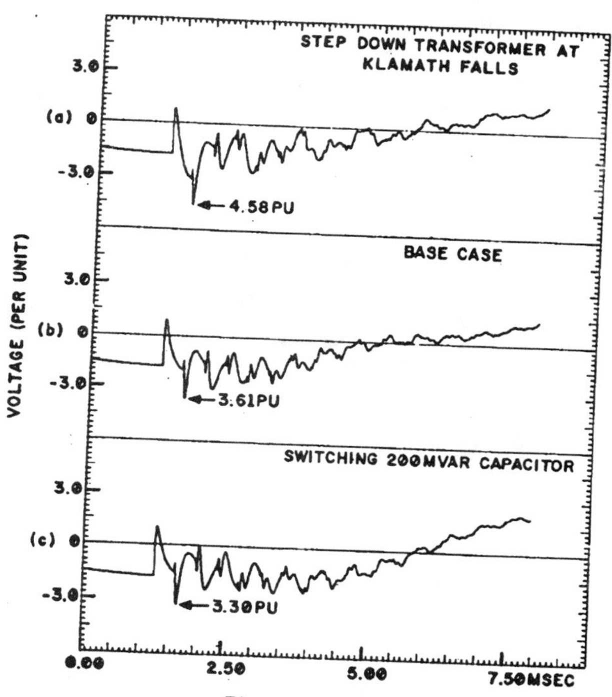

Figure 10.
Phases B-C Voltages at Malin

Transformer Termination Characteristics

For the field test at Malin the large 300 MVA transformer has a significant influence on the waveform. In other applications the radial line from the capacitor switching substation may feed a much smaller transformer which steps down to a lower voltage system. In such cases the termination may nearly approach an open circuit for the transient surge. To evaluate this condition the same three cases used for figures 9 and 10 were reevaluated assuming that the line was open ended at Malin. These results are given in Figure 11.

Comparing Figures 10 and 11, it is apparent that in Figure 11 the transient nearly doubles for the first reflection. This helps to contribute to making a transient that is more pronounced and longer lasting than was shown in Figure 10. In Figure 11(a) the transient reaches 6.00 per unit compared to 4.58 per unit in Figure 10(a).

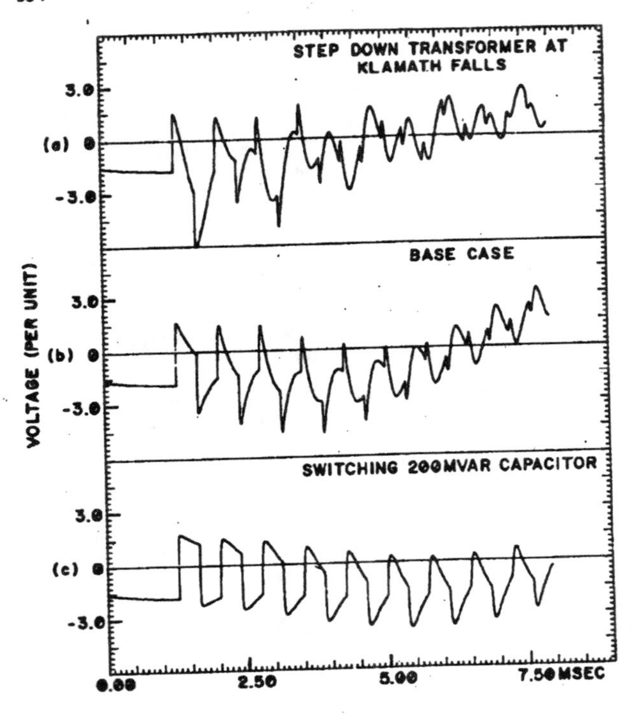

Figure 11.
Phases B-C Voltages at Malin; Malin Open-ended

In Figures 11(b) and 11(c) the peak transient does not occur at the second reflection as it did in Figures 10(b) and 10(c). The peak tends to occur significantly later in the transient and at a higher magnitude. These transients are similar to those discussed by Jones and Fortson [6].

Line Length

The length of the line between the point of capacitor switching and the transformer terminated line end is also an important factor in determining the transient characteristic. Referring back to Figure 10 a shorter line length would tend to reduce the maximum peak of these transients while a longer line would tend to increase them.

REDUCING THESE VOLTAGE TRANSIENTS AND THEIR EFFECTS

In the previous section the effect of system conditions on the transformer transients was discussed. In this section a number of possible steps are discussed which can reduce these transients.

Closing Resistors

Closing resistors in the capacitor switching device will help to reduce the transient. An example of using a 60 ohm closing resistor for the PP&L case is illustrated in Figure 12. At Klamath Falls the voltage change on the bus is reduced with the resistor as is the transient at Malin. The effect of using a 1000 ohm closing resistor is illustrated in Figure 13. The initial transient is very small, and the major transient actual occurs when the resistor is bypassed. In this particular case the resistor size which gives the lowest transient is in the range of 200 to 400 ohms.

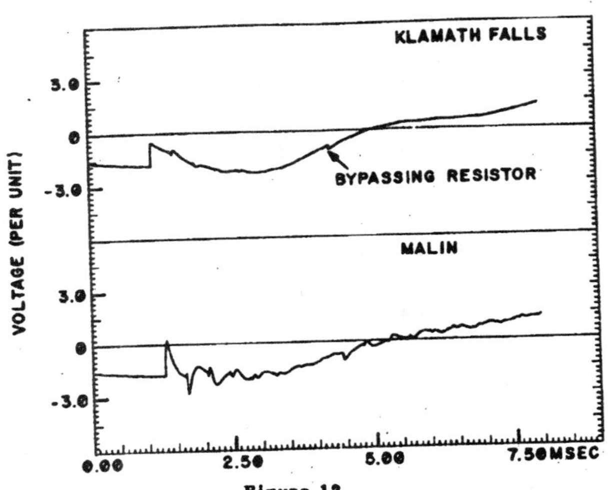

Figure 12.
Phases B-C Voltages;
Base Case with 60 ohm Closing Resistor

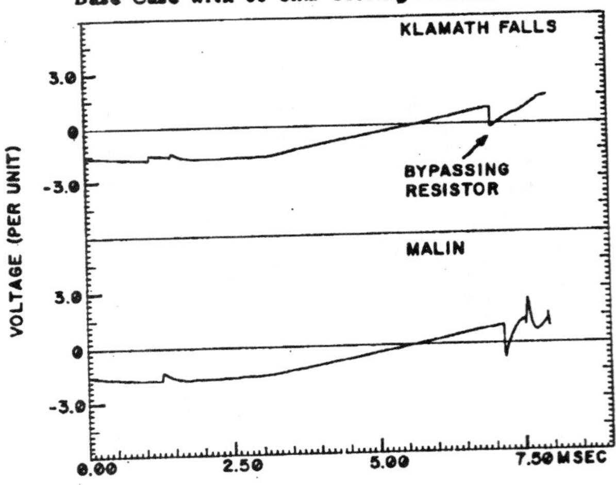

Figure 13.
Phases B-C Voltages;
Base Case with 1000 ohm Closing Resistor

Controlled Closing

In some switching devices it is possible to control the point of closing of the switch so as to minimize the transient voltage. [7] If the contacts close when the voltage across them is nearly zero, the transient will be greatly reduced.

Staggered Closing

In grounded wye capacitor banks it is possible to reduce the phase-to-phase transient by staggering the closing times to insure that no two phases close together. This does not help for ungrounded wye banks and also does not reduce the internal resonance condition.

Capacitor Bank Reactors

A reactor in each phase of the capacitor bank can also act to reduce the transient voltage. This is illustrated in Figure 14 for the PP&L case with a 2.5 ohm reactor. This is the solution that PP&L implemented on their capacitor bank at Klamath Falls.

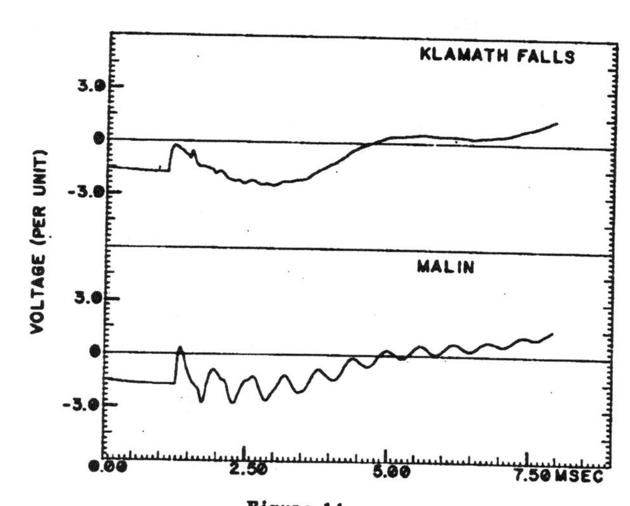

Figure 14.
Phase B-C Voltages;
Base Case with 2.5 ohm Capacitor Reactor

Surge Arresters

Surge arresters at the transformer will also act to reduce the transient. Using the lowest rated MOV arresters, the L-G voltage on each phase can be limited to nearly 2.0 per unit. For the worst case phase-to-phase condition of +2.0 per unit on one phase and -2.0 per unit on another phase, the phase-to-phase voltage could reach 4.0 per unit.

If arresters of the lowest rating were applied phase-to-phase, they could at least theoretically provide protection which is 87% of that for phase-to-ground arresters. This would reduce the maximum transient from 4.0 per unit to 3.5 per unit, but it is dependent upon being able to purchase the proper arrester rating.

Capacitor Bank Connection

For switching devices which are truly random in their closing times, ungrounded-wye capacitor banks will tend to have a high percentage of high phase-to-phase voltages while grounded-wye banks will have a significantly lower percentage. For a grounded wye bank the transients on each phase are nearly independent of each other. A high phase-to-phase voltage will occur on a grounded wye bank when two phases close nearly coincident with each other.

On an ungrounded wye bank it is not necessary for two phases to close together. Since the bank is ungrounded, the first phase to close does not generate a transient except through the stray capacitance of the equipment. The second phase then will always generate a significant phase-to-phase voltage, but the magnitude will depend upon the exact closing time. This characteristic is why ungrounded wye banks will tend to result in a higher percentage of high phase-to-phase voltages than grounded banks.

It should be noted that, in general, slow moving devices which close in air will tend to make near the peak of the voltage waves consistently resulting in maximum transients. For the PP&L case, a device closing in air was used on a grounded wye bank. In the field tests two phases consistently closed together near the peak voltage resulting in high transients on 12 of 13 closings.

Restrikes

When deenergizing capacitor banks, switch restrikes may occur occasionally. They will tend to result in more severe transients than normal closing. In addition, some precautions which have been taken to minimize closing transients may be ineffective for the contingency condition of a switch restrike. These would include closing resistors, controlled closing, and staggered closing. Opening resistors and surge arresters are the most effective methods in minimizing restrike transients. Reactors would also tend to help somewhat.

TRANSFORMER STANDARDS

Phase-to-Phase Transients

In the early 1970s, papers were written [8,9] which suggested that EHV transformers should have phase-to-phase switching impulse insulation levels of 3.4 times the maximum system peak 60 Hz voltage to ground. This recommendation came from system studies and measurements which indicated that this would likely be the maximum value encountered on normal EHV systems and that value would occur very rarely. These recommendations were implemented in IEEE Trial Use Standard 262B-1977 [10] and are summarized in Table 1. These values do not change for different line-to-ground BIL ratings.

TABLE I

(From IEEE Standard 262B -1977)

Nominal System Voltage (kV)	Phase-to-Phase Switching Impulse Insulation Level (kV) Crest
345	1050
500	1550
765	2300

In the special application described in this paper it is possible to see phase-to-phase transients in excess of 4.0 per unit. Typically, surge arresters would limit the surges on each phase to 2.0 per unit to ground of nominal system voltage and, therefore, to 4.0 per unit phase-to-phase. Due to the high number of capacitor switching events, these high transients may occur quite frequently for a transformer in this type of application. In normal insulation coordination, at least a 15% margin is applied between the arrester protective level and the equipment basic switching impulse insulation level (BSL). Applying this traditional 15% margin (it is not clear that the traditional 15% margin would be adequate for these highly frequent events) and using standard insulation levels would result in the BSL's given in Table 2 for HV and EHV systems.

It should also be noted that in many cases control of the transients, as was discussed above, will result in keeping the transient voltages below 3.4 per unit.

Internal Resonance Transients

A Working Group on Resonant Overvoltages within the IEEE Transformer Committee worked on this topic from the early 1970s to the mid 1980s. This group studied the phenomena extensively and attempted to define a dielectric test which would identify potential problems with resonant overvoltages. Although much was learned, no such test was defined. The basic problem is that

TABLE II

(For Special Capacitor Switching Applications)

Nominal System Voltage (kV)	Calculated Phase-to-Phase BSL (kV Crest)
115	450
138	550
161	650
230	900
345	1300
500	1925
765	2925

every transformer has internal natural frequencies, and these may be excited if the corresponding transient is applied at its terminal. This problem has a relatively low probability of occurring; however, the best available way at this time to limit its possibility, even more, is to control the transient at the source, as was discussed above.

CONCLUSIONS

Based upon the analysis summarized in this paper, the following conclusions are made:

1. Both PP&L and Tri-State experienced transformer failures coincident with capacitor switching. The PP&L failure was attributed to a phase-to-phase overvoltage and the Tri-State failure to an internal resonance oscillation. Both were initiated by capacitor bank switching.

2. Field tests, conducted on the PP&L system, were duplicated using a digital transients computer program. The parameters used in these simulations were varied to evaluate their effect on the transformer transient voltages.

3. The transients discussed in this paper are highest for a transformer which is located at the end of a radial line coming from the substation where the switched capacitor bank is located.

4. Capacitor switching transients can be reduced significantly by using switch closing resistors, switch opening resistors, controlled closing, and/or capacitor bank reactors. The capacitor bank connection and the system configuration also have a significant influence on the transient voltage characteristics.

5. Transformers which are exposed to capacitor switching generated transient voltages may require higher phase-to-phase basic switching impulse insulation levels (BSLs) than are currently given in IEEE Trial-Use Standard 262B-1977. A set of calculated BSL's is given in Table 2 based on using typical surge arresters with a 2.0 per unit protective level and applying the traditional 15% margin. This higher BSL may be needed when capacitor switching transients are not limited at the capacitor bank.

6. Historically, system insulation levels have been based on lightning and switching surge transients which tended to occur rather infrequently. Conversely, capacitor switching is often done on a daily basis, and may be the most prolific generator of transient voltages on many utility systems. This high frequency of occurrence may be a significant factor in determining adequate equipment insulation levels in the future, and higher than traditional margins may be required.

ACKNOWLEDGEMENTS

A number of persons were involved in the evaluation of this analysis over the last several years. The authors would like to extend their appreciation to all who were involved. A special note of recognition is made to the BPA engineers who were involved in the PP&L field tests and some of the early simulations.

REFERENCES

1. M.F. McGranaghan, W.E. Reid, S.W. Law, and D.W.Graham, "Overvoltage Protection of Shunt Capacitor Banks Using MOV Arresters", IEEE Transactions, PAS, Vol. 104, No. 8, pp. 2326-2336, August 1984.

2. S.S. Mikhail and M.F. McGranaghan, "Evaluation of Switching Concerns Associated with 345 kV Shunt Capacitor Applications," presented at the IEEE 1985 Summer Power Meeting, Vancouver, B.C.

3. C.G. Troedsson, E.F. Gramlich, R.F. Gustin, and M.F. McGranaghan, "Magnification of Switching Surges as a Result of Capacitor Switching on a 34.5-kV Distribution System", Proceedings of the American Power Conference, Vol. 46, pp. 513-517, 1984.

4. R.J. Musil, G. Preininger, E. Schopper and S. Wenger, "Voltage Stresses Produced by Aperiodic and Oscillating System Overvoltages in Transformer Windings", IEEE Transactions, PAS, Vol. 100, No. 1, pp. 431-441 January 1981.

5. R.C. Degeneff, W.J. McNutt, W. Neugebauer, J. Panek, M.E. McCallum, and C.C. Honey, "Transformer Response to System Switching Voltages", IEEE Transactions, PAS, Vol. 101, No.6, pp. 1457-1470, June 1982.

6. R.A. Jones and H.S. Fortson, Jr., "Consideration of Phase-to-Phase Surges in the Application of Capacitor Banks", Presented at the IEEE 1985 Summer Power Meeting, Vancouver, B.C.

7. R.W. Alexander, "Synchronous Closing Control for Shunt Capacitors", IEEE Transactions, PAS, Vol.104, No. 9, pp. 2619-2626, September 1985.

8. IEEE Committee Report, "Coordination of External Insulation for EHV Transformers", IEEE Transactions, PAS, Vol. 90, No. 5, pp. 2321-2329, September/October 1971.

9. IEEE Committee Report, "Dielectric Tests and Test Procedures for EHV Transformers Protected by Modern Surge Arresters and Operated on Effectively Grounded Systems 345 kV through 765 kV", IEEE Transactions, PAS, Vol. 92, No. 5, pp. 1752-1762, September/October 1973.

10. "IEEE Trial-Use Standard Dielectric Test Requirements for Power Transformers for Operation on Effectively Grounded Systems 345 kV and Above", IEEE Standard 262B-1977.

Discussion

Raymond P. O'Leary (S&C Electric Company, Chicago, IL): The authors are to be commended for a clearly written paper dealing with an increasingly important subject.

I would like to direct my discussion to the Pacific Power & Light test results. The fact that two poles of the switching device in the PP&L/BPA test experienced simultaneous closings 12 out of 13 times is of particular interest. Because both the source and the capacitor bank neutrals were grounded, one would expect the three phases of a switching device to close independently. However, as reported, this was not the case, primarily due to the nature of the switching transient, and not to the characteristics of the switching device. During the field test, B phase tended to be the first pole to close, probably because of the mechanical adjustment of this switch. When B phase closed near peak voltage, the bus voltage on B phase collapsed to zero, a change in voltage of approximately 175 kV. Assuming a coupling to A and C phases of 20 percent to 30 percent, the other phases would experience transient voltages on the order of 50 kV. In each case, these transients increased the voltage across the open switching-device poles. Since at this instant, A phase voltage was near zero, the added transient voltage had little effect. However, C phase voltage was near its peak value and the addition of the coupled transient voltage was sufficient to cause C phase to strike, resulting in the high probability of simultaneous closing on B and C phases.

The probability of simultaneous closing is dependent upon the magnitude of the coupled voltage and adjustment of the switching device, be it a circuit-switcher or circuit-breaker. Further, switching devices equipped with preinsertion resistors will have a lower probability of simultaneous closing. In general, the use of preinsertion resistors will limit the extent to which bus voltage will collapse during the energization of the capacitor bank. This reduction in the instantaneous change in bus voltage will be manifested in a reduction in the coupled voltages on the adjacent phases. In the case of the switching device at Klamath Falls, this would result in a significant lowering of the probability of simultaneous closing of B and C phases. Consequently, one would experience a twofold benefit from the use of preinsertion resistors — the first being the obvious reduction in the magnitude of the transients involved, and the second being a reduction of the transient voltages coupled to the other poles of the switch, such that simultaneous closings are less likely.

The authors note that series inductors were utilized at the Klamath Falls Bank to minimize the transients at Malin. The use of series inductors provides a significant advantage over preinsertion resistors. During the energization of a capacitor bank through a series inductor, the extent to which bus voltage collapses is significantly reduced as compared to a series resistor, resulting in lower voltages coupled to adjacent phases during energization, such that the probability of simultaneous closing will be further reduced, as compared to preinsertion resistors. In addition, the rate of change of the drop in bus voltage is significantly lower. Although the magnitude of the voltage transients to be expected using either preinsertion resistors or series inductors may be approximately equal, there is a significant benefit in that the transient developed with the series inductor is of a much lower frequency, as compared to a resistor or to no impedance at all. The impedance of a remote transformer, or similar high impedance device, at this much lower frequency will be much less and, therefore, the voltage of the transient will not double when it reaches the remote device, as is the case with a preinsertion resistor or no impedance. Also, because the transient is of a much lower frequency, the ability of the transformer winding to distribute this voltage more uniformly across the winding will be improved and a significant reduction in stress concentrations at the ends of the winding will be experienced.

S&C has recently introduced a new device as an alternative to preinsertion resistors or series inductors. The device is a preinsertion inductor which can be mounted on the S&C Circuit-Switcher in much the same fashion as the previously available preinsertion resistor. The conversion to preinsertion inductors was prompted by thermal limitations imposed by preinsertion resistors, due to the high energy that must be dissipated by the resistor during each closing operation. With preinsertion inductors, switching of larger-capacity banks is now permitted and, in some instances, more frequent switching of such capacitor banks is permitted.

Because preinsertion inductors are only temporarily energized during the closing operation, they offer significant advantages over fixed series inductors, which must carry current continuously. Preinsertion inductors are smaller, lighter, less costly, and typically have higher inductance (20 millihenries @ 230 kV) than series inductors. Further, since preinsertion inductors are installed directly on the Circuit-Switcher, no additional mounting structure is needed.

Manuscript received August 12, 1986.

R. S. Bayless, J. D. Selman, D. E. Truax, and W. E. Reid: We appreciate Mr. O'Leary's comments, especially his explanation of the simultaneous closings of two phases on the PP&L capacitor bank. We would agree that the use of either closing resistors or closing reactors will reduce the transients at both the capacitor bank and the transformer. The effectiveness of either method, however, is a function of the system parameters and the value of the resistors or the reactors. If the resistors or reactors are not within an appropriate range of values, they may have a very small effect on the transient. It is also possible that adding a reactor may change the transient frequency such as to coincide with the natural frequency of the transformer, thus making the transient more severe as in the Tri-State case discussed in the paper.

Most of the attention in the paper and in the discussion has been focused on minimizing the closing transients. However, as noted in the paper, when deenergizing capacitor banks, switch restrikes may occur occasionally. If they should occur, the transients could be significantly more severe than those associated with closing. Unfortunately, some precautions which have been taken to minimize closing transients may be ineffective for the contingency condition of a switch restrike, e.g., closing resistors or reactors, controlled closing, and staggered closing. Generally, restrike transients can be reduced by the use of devices which employ opening resistors as well as by the use of surge arresters. In addition, permanently installed reactors may also help to reduce the severity of the restrike transient.

Manuscript received September 9, 1986.

DESIGN AND INSTALLATION OF 500-KV BACK-TO-BACK SHUNT CAPACITOR BANKS

Brian C. Furumasu
Senior Member

Robert M. Hasibar
Senior Member

Bonneville Power Administration
Portland, Oregon

Key words: Shunt capacitors, back-to-back switching, current-limiting reactors, inrush current, transient stability control.

ABSTRACT

This paper describes Bonneville Power Administration's first 500-kV back-to-back shunt capacitor installation. The primary purpose of the capacitor banks is to support AC system voltage for transient stability following an outage of the HVDC Intertie line between Oregon and Southern California. Design studies and equipment requirements are detailed herein in addition to Electromagnetic Transients Program (EMTP) studies and subsequent field test results. Current-limiting reactors are used to limit inrush and outrush currents during capacitor switching. It is believed that this is the first 500-kV back-to-back capacitor installation in the United States.

INTRODUCTION

Bonneville Power Administration (BPA) determined the need for controlled shunt capacitors as a transient stability aid on the Pacific Northwest/Southwest (NW/SW) 500-kV AC Intertie following an HVDC Intertie line disturbance. These banks are used in conjunction with existing high-speed switched series compensation. Early investigations focused on static var systems. Further studies determined that two 500-kV mechanically-switched shunt capacitor banks located at Malin Substation, BPA's southern terminal of the AC Intertie, could be utilized for this purpose since post-disturbance damping control is not required (See Figure 1). Both banks are identically rated and can be energized separately or together, depending on system requirements.

Back-to-back switching requires evaluation of a number of concerns, including:

- Breaker interrupting capability and operating limitations.
- Type of device or method needed to limit inrush (and outrush) current.
- Insulation coordination.
- Capacitor bank trapped charge during high-speed switching.

91 WM 241-0 PWRD A paper recommended and approved by the IEEE Transmission and Distribution Committee of the IEEE Power Engineering Society for presentation at the IEEE/PES 1991 Winter Meeting, New York, New York, February 3-7, 1991. Manuscript submitted August 8, 1990; made available for printing January 3, 1991.

Described herein is the procedure used for developing design parameters and EMTP investigation of transient phenomena associated with capacitor bank switching. Field test results and study verification are included.

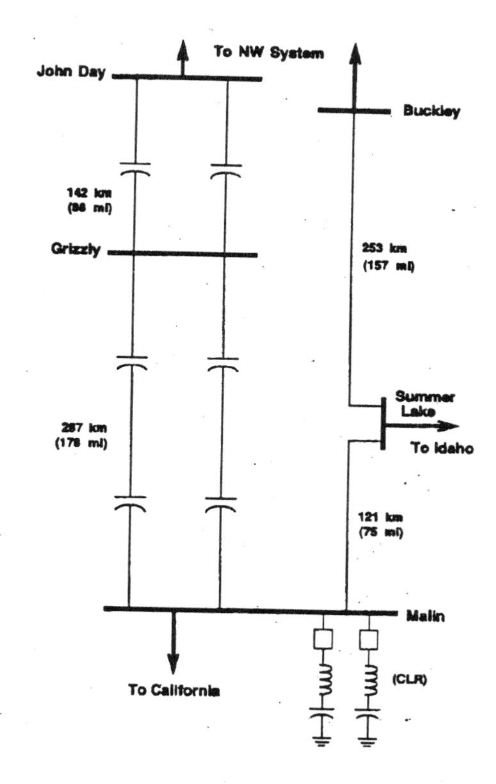

Figure 1. Northern section of Pacific NW/SW AC Interties showing 500-kV shunt capacitors at Malin.

EQUIPMENT DESIGN STUDIES

Design Considerations

Among the primary design considerations for these banks were the inrush current and the switching duty cycle specified for the capacitor banks. The switching duty (5 open-close operations -- wait 30 minutes — 5 open-close operations -- wait 5 hours, repeat the duty cycle) was based on possible future multiple operations to enhance the transient stability limit of the Pacific NW/SW AC Intertie. The open and close times of the breaker were specified at 5 cycles.

0885-8977/91/$3.00©1992 IEEE

The capacitor banks were placed on the North bus of a breaker and one-half bus arrangement at Malin Substation. The breaker was chosen for this application because of the switching duty requirements and the ability to interrupt the 23 000 A three-phase fault current. The decision to use current-limiting reactors (CLRs) over insertion resistors to limit the inrush current was based on the overall assessed reliability, plus the fact that the CLRs also provide the benefit of limiting the outrush current for faults external to the shunt capacitor bank. In addition, full BIL CLRs (terminal-terminal) were compared with a reduced BIL reactor protected by a metal oxide surge arrestor. The economics favored the full BIL CLR. The option of using a synchronous switching device was also considered, but a proven device was not available to meet the specifications at 500 kV.

Determination of the CLR inductance was an iterative process which took into consideration the following parameters:

1. Per unit overcurrent for capacitor units and fuses.
2. Inrush current capability of the SF6 puffer device.
3. Comparison with inrush current levels of successfully-operated, 230-kV back-to-back switched banks, measured during prior field tests.
4. Possible future expansion by adding two shunt capacitor banks of the same ratings, using the same value of CLR.

An EMTP model of the present and future shunt bank configurations was assembled to determine the CLR inductance. A value of 2 mH was chosen for each CLR. This value limited the inrush to 20 kA assuming maximum trapped charge on the fourth bank closing in at the opposite polarity of the bus voltage. This inductance limited the inrush current of an individual capacitor to approximately 50 pu of rated current. This inrush current value is conservative when compared to the transient overcurrent capability guidelines for capacitors in IEEE Standard 18, 1980 [1]. Magnitudes of inrush current for various scenarios are described in the following section.

This bank inrush current was comparable to 230-kV back-to-back banks operated successfully for many years, with satisfactory capacitor and fuse performance. The 500-kV banks are also designed with a peninsular grounding system [2]. This grounding system keeps the inrush current flow between the banks, with a single point connection to the main substation ground grid. This concept keeps the inrush current resulting from back-to-back switching contained within the area of the shunt banks and minimizes the possible electromagnetic interference (EMI) in the main substation yard.

STUDY RESULTS AND ANALYSIS

Transient stability control required the capacitors to be energized and deenergized over a short period of time. This would impose possible high stress on the breakers, CLR, and the capacitors themselves. Evaluation of this switching duty was done using the EMTP. The network shown in Figure 1 was extensively modelled and numerous studies performed.

One of the first investigations concerned the high inrush current experienced by back-to-back switching [3]. There were several options to minimize this inrush current:

(1) Insertion resistor in the breaker,

(2) Synchronous closing of the breaker [4,5],

(3) Ungrounded shunt banks, or

(4) Current-limiting reactor in series with the capacitor.

Since BPA has been moving towards eliminating breakers which use insertion resistors (reliability and safety concerns), the first option was not considered. Option (2) was not pursued due to the unavailability of a proven synchronous closing device at the time of installation. The third option is not BPA standard practice for control and protection. Choosing the current-limiting reactor was based on available technology and economic considerations.

Energizing any shunt capacitor bank at a time when the bus voltage is a maximum will produce the highest current in the capacitors. Also, due to the rapid switching required for this application, a trapped charge on the capacitors will remain and cause an even higher current (and CLR voltage) when closing at the opposite polarity of charge and bus voltage peak. These conditions were simulated with the EMTP to obtain the maximum inrush current and CLR voltage required for this application. From these results, specific technical requirements were developed for the shunt banks.

Choosing the CLR Inductance

As theory and standards describe [6], transient inrush currents can be reduced by inductance in the circuit between capacitor banks. An EMTP model of the system shown in Figure 1 was developed and the CLR inductance varied until a proper value was found which would limit the inrush current to 20 kA (peak). Figure 2 compares the maximum inrush current vs. CLR

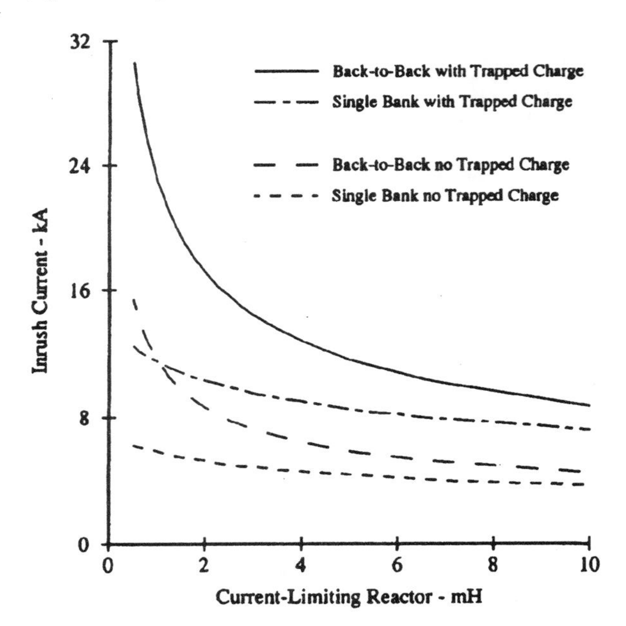

Figure 2. Maximum inrush current for shunt capacitor banks vs. CLR inductance.

inductance for single-bank and back-to-back energization. Figure 3 depicts the CLR peak voltage during bank energization in a similar manner to the inrush current plot. Based on these results, and considering a possibility of future additional capacitor banks, a CLR of 2 mH and 1800-kV BIL was chosen and found to be the most economical. The BIL of the reactor was coordinated with the station insulation of 1800-kV BIL.

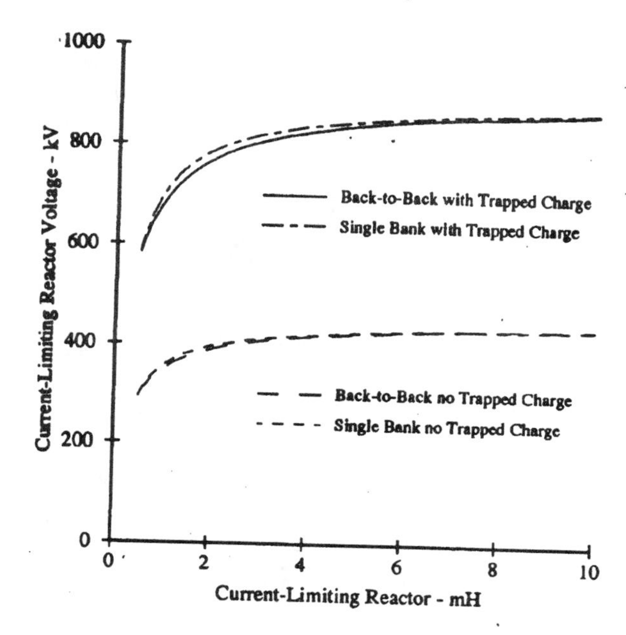

Figure 3. Maximum voltage across the CLR vs. CLR inductance of the shunt capacitor banks during energization.

Examples of EMTP capacitor inrush current waveforms for single-bank and back-to-back energization are shown in Figures 4(a) and (b). Figure 4(c) illustrates a typical CLR voltage during a back-to-back energization study. The ringing frequency shown in Figures 4(b) and (c) is the natural frequency of both shunt capacitors, CLR, and associated bus inductance (approximately 2500 Hz).

SYSTEM FIELD TESTS

Field tests were conducted at Malin in July 1989, to demonstrate the shunt capacitor operation and collect important data on:

(1) Switching capability for the duty cycle required for transient stability control.

(2) Outrush current flowing from an energized capacitor bank during a close-in single line-to-ground (SLG) fault.

(3) Inrush currents for single-bank and back-to-back energization.

Single-Bank and Back-to-Back Energization

Figure 5 is a plot of a field test recording of inrush current for one phase during single-bank energization.

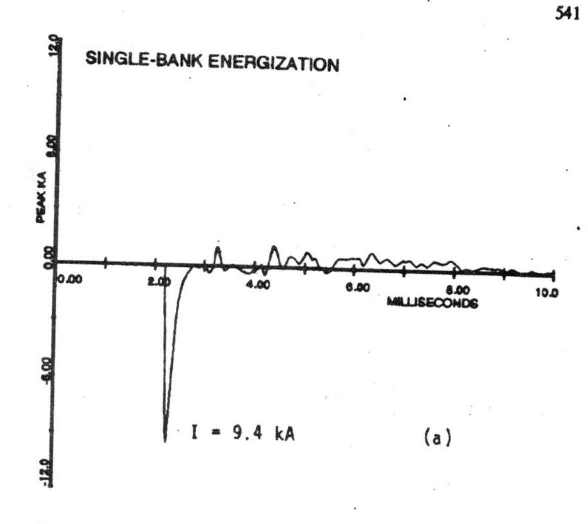

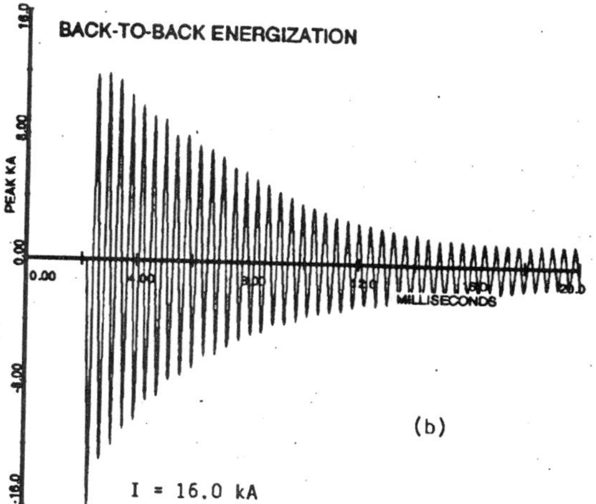

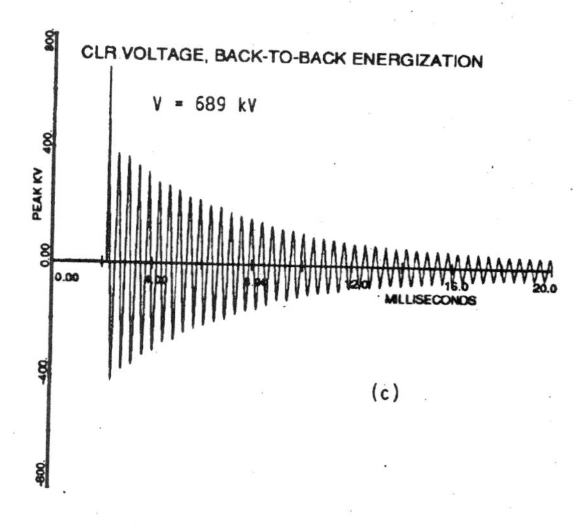

Figure 4. EMTP simulations of capacitor current during (a) single-bank energization, and (b) back-to-back energization. An EMTP plot of CLR voltage during back-to-back switching is shown in (c).

This compares with the EMTP plot of Figure 4(a). Differences in the current magnitude between simulation and test results is due to the closing time of the breaker with respect to the peak bus voltage.

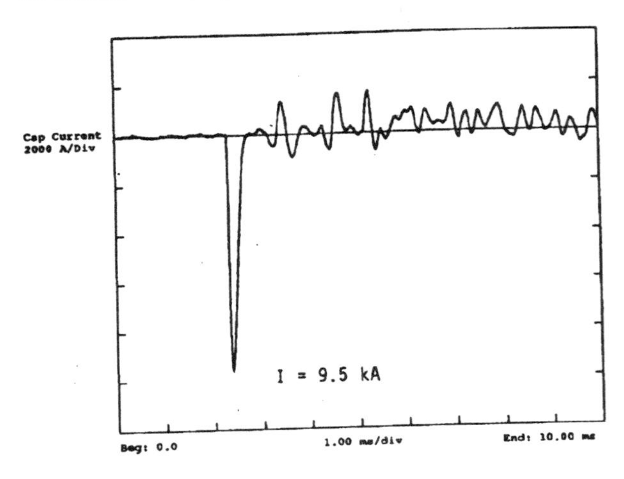

Figure 5. Field test recording of capacitor current during single-bank energization (with trapped charge on the capacitor).

For the stability application of these banks, trapped charge is a significant design parameter. This is the condition that exists for rapid switching for transient stability control. Since the trapped charge will not dissipate to zero during this multiple-switching sequence, it is necessary that the equipment be designed to withstand these high inrush currents.

A specific test was performed to demonstrate that the shunt capacitor banks could meet the 5 successive close/open operations with trapped charge on every shot. Each trip and close of the breaker was separated by 42 cycles (60 Hz) and resulted in a maximum current of 14 kA. This back-to-back switching imposed the most severe duties specified for this equipment which performed as designed.

Figure 6 is the test recording of back-to-back inrush current which compares with the EMTP simulation of Figure 4(b). Both this test and the simulation indicated an initial trapped charge on the capacitor of -.88 pu. Good agreement is shown between field results and simulation for current magnitude and oscillation frequency.

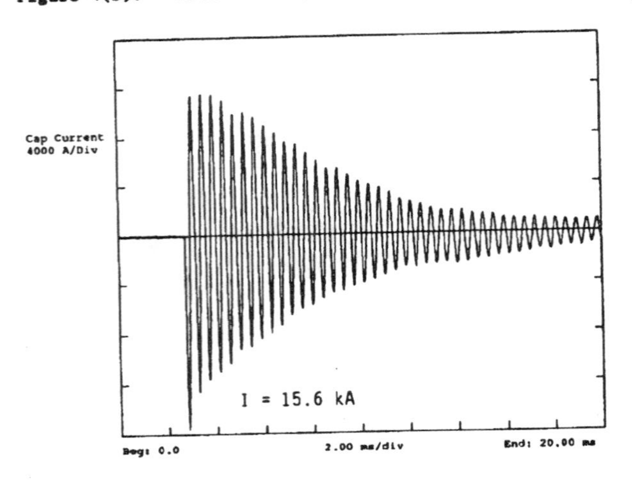

Figure 6. Field test recording of inrush current during back-to-back energization (with trapped charge on the capacitor).

Voltage measured across the capacitor bank is shown in Figure 7. A maximum of 800 kV across the bank is due to trapped charge and time of breaker closing.

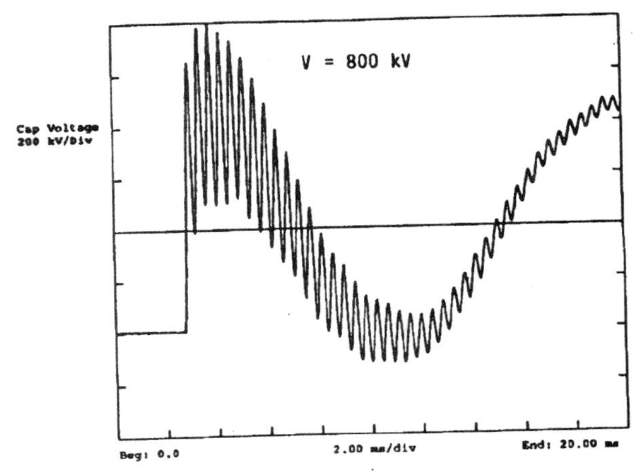

Figure 7. Field test recording of shunt capacitor voltage during back-to-back energization.

<u>Shunt Capacitor Current During a Close-in SLG Fault</u>

An SLG fault test was performed on one of the two 500-kV lines from Malin to California for the purpose of investigating single-pole switching on this line. The fault was applied on the line immediately outside Malin Substation. Important to the capacitor test was the current flowing in an energized capacitor bank during this fault. Figure 8 is the test recording of current in one energized bank during the fault. This recording shows a current of approximately 10 kA on the faulted phase in the capacitor bank (in this case, a B-phase fault). Subsequent EMTP simulations revealed the effects of the CLR in reducing this current magnitude. Removal of the CLR in the simulation resulted in capacitor currents that were four times larger than the test results.

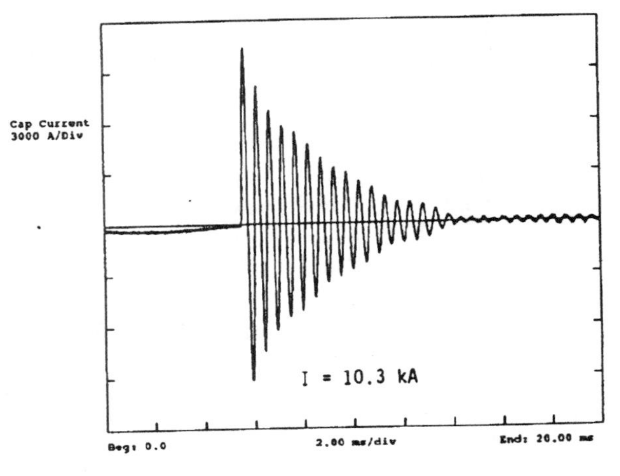

Figure 8. Field test recording of current in one phase of capacitor bank during a close-in SLG fault.

Peninsular-Ground Measurements

During all capacitor switching tests (a total of 37 tests), the ground potential of the peninsular ground was measured. Single-bank and back-to-back energization produced potentials of approximately 2 and 5 kV respectively. These measurements simply verified the effectiveness of the peninsular grounding system.

Conclusions

Back-to-back shunt capacitor switching for 500-kV transmission networks has been demonstrated to be feasible with existing technology. Furthermore, transient stability control is possible for the BPA system when utilizing these mechanically-switched capacitors. Use of this stability control method may be applicable on other systems as an option to various static var devices on an EHV network.

Current-limiting reactors in series with the shunt capacitors are capable of minimizing inrush and outrush currents when switching. Also, existing SF6 breaker technology is sufficient to withstand transient voltages during switching without other protective devices such as surge arresters across the switch.

BPA is planning to install two new back-to-back 500-kV shunt capacitor banks on its EHV system in the fall of 1990. These will provide voltage control during high power flows in the Seattle area. An existing bank will be retrofitted with the proper CLR and breaker.

ACKNOWLEDGMENT

The authors thank Mr. Randy Suhrbier for his assistance with the field test results. His valuable suggestions during and after this test are greatly appreciated.

References

[1] ANSI/IEEE Standard 18-1980, IEEE Standard for Shunt Power Capacitors Section 8 Guide for the Application and Operation of Power Capacitors.

[2] E.J. Rogers and D.A. Gillies, "Shunt Capacitor Switching EMI Voltages, Their Reduction in Bonneville Power Administration Substations," IEEE/PES WPM January/February 1984.

[3] K.L. Spurling, A.E. Poitras, M.F. McGranaghan, and J.H. Shaw, "Analysis and Operating Experience for Back-to-Back 115-kV Capacitor Banks," IEEE/PES 1986 Transmission and Distribution Conference, Anaheim, California, September 14-19, 1986, Paper No. 86 T&D 602-7.

[4] R.W. Alexander, "Synchronous Closing Control for Shunt Capacitor Banks," IEEE Transactions-PAS, Vol. 104, No. 9, pp. 2619-2626, September 1985.

[5] J.H. Brunke, G.G. Schockelt, "Synchronous Energization of Shunt Capacitors at 230-kV," Presented at 1978 Winter Power Meeting, New York, New York, January 29 to February 3, 1978, Abstract A78 148-9.

[6] ANSI/IEEE Standard C37.012-1979, AC High-Voltage Circuit Breakers Rated on a Symmetrical Current Basis, Section 4.7ff.

Brian C. Furumasu (M'75, SM'90) was born in Spokane, Washington, in 1953.

He received his BS and MS degrees in Electrical Engineering from Washington State University in 1975 and 1977, respectively.

Mr. Furumasu is currently the Section Chief of the High Voltage Section for the Bonneville Power Administration. In this position, he is responsible for BPA's high voltage policies and practices, electrical requirements for the transmission system and components, and R&D projects involving high voltage equipment.

Mr. Furumasu is active in IEEE Standards activities as secretary of the Capacitors Subcommittee and chairman of the Working Group to Rewrite the Series Capacitor Standard.

Robert M. Hasibar (M'78, SM'90) was born in Ketchikan, Alaska in 1942. He received his BSEE from Gonzaga University in Spokane, Washington in 1965. He did graduate work in Electrical Engineering at Santa Clara University, Santa Clara, California, in 1969-70.

Following two years in the U.S. Army after graduation, Mr. Hasibar joined the Bonneville Power Administration in November, 1967. His work at BPA has dealt mainly with EMTP simulation and many steady-state and transient problems as they apply to the transmission system. He was a member of the BPA engineering team which applied the High-Speed Grounding Switch (HSGS) to single-pole switched lines. He is presently Chief of the Power System Analysis Section at BPA.

Mr. Hasibar is a member of PES and a registered Professional Engineer in Oregon.

Discussion

J. E. Harder, (ABB Power T&D Company, Bloomington, Indiana): Compliments to the authors for an interesting paper illustrating how shunt capacitors may be switched in response to system dynamic conditions. There are a couple of questions concerning the reported results which are of interest:

What criteria was used in the selection of the Q of the inrush limiting reactors? It appears that the Q used for the study results was about 44. Is this a low value to improve damping? A high value to minimize losses during normal operation? An expected value for this particular installation?

One interesting possibility is to use a higher Q reactor (for low normal operating losses) paralleled by a damping assembly. The damping assembly is a low voltage MOV arrester in series with a linear resistor. The low voltage arrester has an MCOV above the normal reactor voltage so that the losses in the damping assembly are negligible during normal operation. During transients the linear resistor is sized for nearly critical damping. Such a damping assembly would minimize the duty to the capacitors and fuses and reduce the transient disturbance to the substation and system.

Figure 4a and 5 illustrate the current during single bank energization with trapped charge on the capacitor. The shape of the inrush current suggests some non-linearity. The normal energization of a capacitor bank would result in some ringing of the inrush current. Is it possible that an arrester operated during the operation (both in the simulation and the actual case)? If so, what was the arrester energy? Is the arrester capable of repeated operations with this energy? If it is not arrester operation, can the authors explain in qualitative terms the current waveshape noted in Figure 4a and 5? It would be interesting to see oscillograms of the substation phase-to-ground voltage associated with these two figures.

A very real value of this paper appears to be in the use of standard shunt capacitors to improve system dynamic performance. What are the plans for implementing this technique on a continuing basis?

Manuscript received February 26, 1991.

B. Bhargava and R. L. Ensign: The authors should be commended for writing this excellent paper. It is really surprising to note that the results of the study conducted on the EMTP and the field tests show such a close correlation. The problem of switching shunt capacitors is becoming more and more acute as more capacitors are being applied on Electric Power Systems to either enhance the system stability or to provide the needed var or voltage support.

The authors show the plots from the simulations, the field tests for the currents and the voltage across the series reactors. The voltage across the reactors is seen to reach as high as up to 880 KV. This is because of both the high frequency and the current magnitude that has to flow out/in to the capacitors during back to back switching. For normal shunt reactors and wave traps, the voltages are generally much lower because they are due to the fault currents at the power frequency. These kinds of high voltages across the reactor for capacitor switching could require extra turn to turn insulation across the reactor turns. Will the authors explain if some special kind of reactors were installed, or was this kind of voltage insulation withstand specified? Please also indicate the short circuit duty for which the reactors are designed.

Although the reactors reduce the capacitor switching inrush currents and frequencies caused by back to back switching, very high dv/dt or the rate rise of recovery voltage on the breakers can also result if a fault occurs after the reactors. Some manufacturers have expressed concern that their breakers may not be able to interrupt such faults. Did you have any such concerns? If so, how did you solve this problem?

The figures also show a very high rate of damping of the high frequency oscillations. The typical time constants for these oscillations can reach few cycles, but can increase further because of the high inductance added in the back to back circuit. Will the authors please explain if the damping shown in the figures is the natural damping of the buses and the series reactors, or was additional damping added to the circuit. Please also indicate the X/R ratio of the series reactors. Do the authors think that this high damping could be resulting from the skin effect of the busbar conductor?

SCE is planning to install two 230 KV shunt capacitor banks back to back at its Victor Substation. We are planning to install series reactors as well, to limit the inrush currents and frequencies to the levels specified in the ANSI Standards. Our concern is the large voltage that can appear across the reactors and also the low rate of damping of these high frequencies which reduces further, especially when the reactors are added. We are concerned that if the breaker is closed in a fault during back to back switching and has to open, the high frequencies could result in its failure to open. In view of these concerns, we are investigating the use of zinc oxide surge arresters in parallel with the series reactors to both limit the voltages across the reactors and also to provide additional damping. Did the authors look into this alternative? Have the capacitors or reactors had any problem with overvoltages?

Manuscript received February 20, 1991.

Brian C. Furumasu and Robert M. Hasibar: The authors appreciate the comments and interest provided by Mr. Harder and Messers Bhargava and Ensign.

Regarding Mr. Harder's questions, the primary concern BPA had during the design was to keep the i^2t in the fuses to a level that would not cause them to operate. This was especially so for this bank which was designed for multiple high speed back-to-back operations.

The current-limiting reactor (CLR) at Malin has a Q of 27 at 60 Hz. There was no additional damping added or specified for the CLR or other parts of the current loop. Some manufacturers have indicated that CLR designs are available where the Q at the discharge ringing frequency may be designed to be lower than the Q at the power frequency current. This would increase the circuit damping. However, the Malin banks have been switched back-to-back at high speed (with trapped charge) with excellent performance. To date only one capacitor unit has failed and nuisance fuse operations have not occurred.

Regarding the shape of the current during single-bank energization, for an EMTP simulation of the existing system, with all connected elements; i.e., lines, transformers, etc., the current is correct as shown in Figures 4a and 5. There was no arrester conduction either in the model or the field. (See the field test results below of the bus voltage, capacitor current, and capacitor voltage for this operation.) A simple EMTP model would produce some ringing, but the system representation has too much damping to develop appreciable oscillations.

BPA has no plans at this time to implement this shunt capacitor insertion scheme at other stations. It has, however, proven itself to be a visible alternative to the more expensive static var systems for system dynamic performance applied at Malin. Also, BPA's System Operations has automatic voltage control installed on these capacitor banks which has operated occasionally for improving the system voltage at Malin.

Messers Bhargava and Ensign indicate the large voltage appearing across the CLR during back-to-back switching. It was for this reason that the BIL across the CLRs are specified to the full line-to-ground insulation of the station. For this particular application the BIL across the CLRs is 1550 kV. In addition, the CLRs were specified to withstand the maximum three-phase fault duty at Malin (22 000 MVA).

As part of the field test, high-speed back-to-back switching operated successfully. This indicates that the breaker operated as designed when

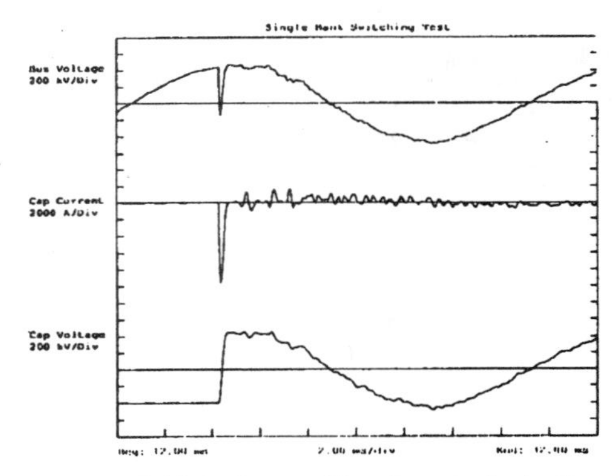

Field test recording for single-bank energization.

switching these capacitor banks. In addition, BPA has installed and successfully operated two other 500 kV back-to-back shunt capacitor banks. To date there have been no problems encountered with the SF_6 switches.

Applying zinc-oxide arresters across the CLR requires some further comment. During the EMTP study stages of the design an arrester was connected across the CLR to reduce the BIL. These results indicated that a full BIL CLR was more economic and a simpler solution. Conduction of the arrester resulted in some bypassing of the CLR.

Manuscript received September 16, 1991.

IEEE Transactions on Power Delivery, Vol. 8, No. 3, July 1993

EFFECTIVENESS OF PRE-INSERTION INDUCTORS FOR MITIGATING REMOTE OVERVOLTAGES DUE TO SHUNT CAPACITOR ENERGIZATION

Bharat Bhargava
Member
Southern California Edison
Rosemead, California

Aftab H. Khan, Member; Ali F. Imece, Member; Joseph DiPietro, Member
GE Industrial and Power Systems
Power Systems Engineering Department
Schenectady, New York

Abstract - Energization of shunt capacitor banks on power systems can potentially create significant overvoltages both at the location of the switched bank and at other locations in the system. Pre-insertion resistors have been used for years by the electric utility industry as one method of controlling these capacitor bank energization overvoltages. Recently, pre-insertion inductors have been introduced as a cost-effective alternative to pre-insertion resistors. The results of a system study investigating remote overvoltage mitigation methods for shunt capacitor switching have shown that pre-insertion inductors are not always as effective as pre-insertion resistors. A parametric study has been performed to determine the effectiveness of pre-insertion inductors for controlling remote overvoltages, such as those at the end of open-ended lines or transformer-terminated lines. The results of the parametric study show that the pre-insertion inductors used for controlling remote overvoltages may become completely ineffective for certain system conditions. This finding is explained in terms of the physical phenomena.

Key Words - Shunt capacitor energization, pre-insertion inductors, transient overvoltages, Transient Network Analyzer, Electromagnetic Transients Program.

INTRODUCTION

Energization of shunt capacitor banks can produce severe phase-to-ground and phase-to-phase overvoltages at the switched bus (i.e., capacitor bus), and also at remote locations. The overvoltages at remote locations, such as those at the end of open-ended lines or transformer-terminated lines are of special concern. The phase-to-ground overvoltages at the end of open lines can exceed the BSL of line end equipment, and cause air gap and insulator sparkovers. The high rate of rise of these overvoltages may further aggravate the situation. The limits on phase-to-phase equipment insulation withstand levels and clearances can also be exceeded by the phase-to-phase overvoltages due to capacitor energization. In the case of transformer-terminated lines, the phase-to-phase insulation of power transformers may be of special concern due to lack of specific standards for phase-to-phase switching surge withstand of power transformers [1].

Various methods of controlling overvoltages during capacitor energization have been investigated in the literature. These methods include the use of pre-insertion resistors, pre-insertion inductors, synchronous closing, and metal oxide surge arresters. Each has various advantages and disadvantages in terms of cost, reliability, and effectiveness [1-8].

Pre-insertion resistors are one of the most effective means of reducing the magnitude of the surges caused by switched capacitors. However, their reliability is a growing concern for some U.S. utilities and circuit breaker/switcher manufacturers [2,3].

Synchronous closing schemes use breaker control to synchronize the actual closing of contacts with the corresponding voltage zeros of the power-frequency waves, thus reducing voltage transients [4]. In practice, synchronous closing schemes have closing tolerances which may impact the effectiveness of this scheme.

Inductors, which are primarily used for limiting inrush currents during back-to-back energization, can also provide overvoltage control. A recent study suggests pre-insertion inductors as the preferred overvoltage control method over pre-insertion resistors and synchronous closing [2]. Note that pre-insertion inductors are more economical than the pre-insertion resistors.

Metal oxide varistors (MOVs) or surge arresters can effectively limit the overvoltages to the arresters protective level at the point of application [3]. However, for arresters placed at the capacitor bank bus, the reduction of overvoltages at remote locations during capacitor energization may not be adequate.

Overvoltages at remote locations during restrike of capacitor switching devices are of particular concern due to the ineffectiveness of most of the overvoltage mitigation options discussed above. Restrike may occur, for the breakers/switches not equipped with opening resistors, when the magnitude or rate of recovery voltage across the contacts exceeds specified limits after an opening operation. While SF6 breaker/switches are designated as restrike-free devices, restrike may occur if SF6 gas pressure falls below manufacturer specified limits. Restrike overvoltages cannot be protected by pre-insertion devices because the pre-insertion device is typically bypassed and disconnected at the time of breaker/switch opening.

The Southern California Edison Company (SCE) is in the process of installing 48.6 and 79.6 MVAR shunt capacitors on their 115 kV and 220 kV systems, respectively. Since SCE's normal operating strategy is to leave unused lines open-ended for var support and quick energization in case of an emergency, potential overvoltages at the end of the lines are of special concern during capacitor bank energization. To evaluate the impact associated with these capacitor installations, a switching study was performed on a Transient Network Analyzer (TNA). Various overvoltage mitigation methods were investigated in the TNA study. The results of the study indicated that commercially available pre-insertion inductors

92 SM 495-2 PWRD A paper recommended and approved by the IEEE Transmission and Distribution Committee of the IEEE Power Engineering Society for presentation at the IEEE/PES 1992 Summer Meeting, Seattle, WA, July 12-16, 1992. Manuscript submitted September 3, 1991; made available for printing May 15, 1992.

were not very effective in reducing the remote overvoltages for some system contingencies. This finding has not been analyzed in detail in the current literature.

A parametric study was performed using the BPA's Electromagnetic Transients Program (EMTP) to find the bounds of satisfactory performance for capacitor energization with pre-insertion inductors. This paper describes the results of both TNA and EMTP simulations, and illustrates the effects of system parameters.

SYSTEM STUDIES FOR SHUNT CAPACITOR BANK ENERGIZATION

System studies are required to determine the maximum overvoltages at the capacitor bank and remote locations due to shunt capacitor energization for normal and restrike conditions. These maximum overvoltages are best determined by means of a statistical analysis which considers the random point-of-wave closing during energization. The TNA overvoltage analysis utilizes a statistical maximization approach, performing the energization operation 300 times, randomly varying the breaker pole closings over the closing span with normal probability distribution. A 120° closing span between phases is considered in this TNA study. While manufacturer specifications of closing span can be as small as 32°, a larger closing span considers performance which may be a result of aging and/or improper maintenance.

Figure 1 shows the Victor-Kramer 220/115 kV network modeled for the Transient Network Analyzer Study. Each transmission line in the TNA model was represented by five π-sections to permit proper simulation of traveling waves. The frequency response of the TNA transmission line models used in this investigation have a roll-off frequency of approximately 10 kHz. The transformer models are also critical, and are modeled by an array of inductors and saturable elements adjusted to model all leakage impedances and air core impedances without any approximations. The external system beyond the area of interest was terminated by voltage sources behind Thevenin equivalent impedances representing system positive, negative, and zero sequence characteristics. The proposed capacitor bank installations are those shown at the Victor and Kramer 220 kV and 115 kV buses. For each location, a worst case scenario was developed considering various system contingencies. The worst case overvoltages occurred at the end of open lines while energizing a single capacitor bank. The maximum per unit phase-to-ground overvoltages at each location are shown as uncontrolled energization in Table 1. Figures 2.a and 2.b show typical voltage waveforms at the Kramer 220 kV bus (i.e., switched bus) and at the end of open-ended 48 mile long line (Kramer-Lugo 220 kV line in Figure 1) for 79.2 MVAR shunt capacitor switching at Kramer 220 kV bus, respectively. The corresponding statistical distribution curve is shown in Figure 3.

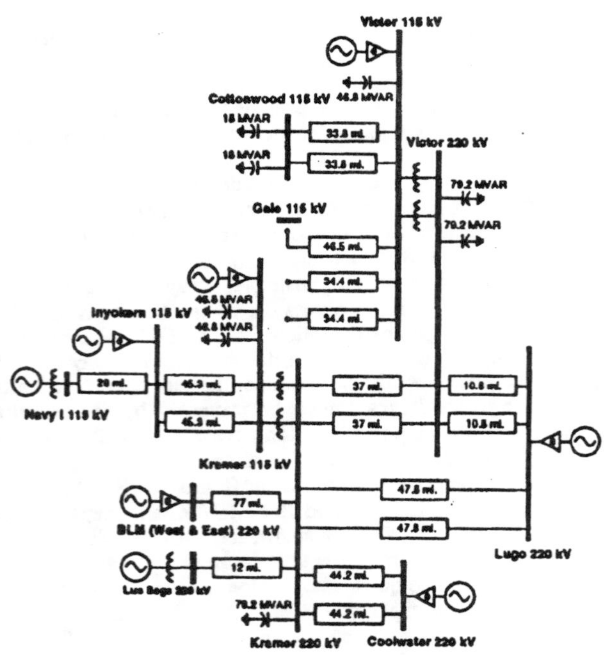

Figure 1. TNA Model for Victor-Kramer 220/115 kV Capacitor Switching Studies.

The maximum phase-phase overvoltage can be estimated from the statistical phase-to-ground overvoltages. The theoretical maximum of phase-to-phase overvoltage is two times the maximum phase-to-ground overvoltage. However, this is not usually the case, since if one phase-to-ground

Table 1
TNA Study Results - Overvoltages at the End of Open-Ended Transmission Lines
(1. pu = $\sqrt{2}/\sqrt{3}$ rated system voltage)

	Switched Capacitor Bank Location [1]			
	Victor 220 (79.2 MVAR)	Kramer 220 (79.2 MVAR)	Victor 115 (46.8 MVAR)	Kramer 115 (46.8 MVAR)
	Line End Voltages at:			
	Kramer 220[2] (37 Miles)	Lugo 220[3] (48 Miles)	Cottonwood 115[4] (34 Miles)	Inyokern 115[5] (45 Miles)
Uncontrolled Energization	2.66 u	2.28 pu	2.15 pu	2.58 pu
Pre-insertion Resistor				
100 Ohms	1.52 pu	-	-	-
200 Ohms	1.43 pu	1.38 pu	1.71 pu	1.68 pu
300 Ohm	1.65 pu	-	-	-
Pre-insertion Inductor				
10 mH	2.36 pu	2.39 pu	1.80 pu	1.74 pu
20 mH	1.87 pu	2.13 pu	-	-
40 mH	1.51 pu	1.83 pu	-	-
Synchronous Closing				
±1 ms	-	1.32 pu	-	-
±2 ms	-	1.88 pu	-	-
MOV//Inductor				
2 kV Vp/1.5 mH		2.51 pu	-	-
2 kV Vp/3.0 mH		2.30 pu	-	-
Surge Arrester				
180 kV, or 96 kV	2.20 pu	2.20 pu	2.15 pu	2.20 pu

1. "-" entries indicate that the mitigation method was not investigated at that particular location.
2. Kramer end of Victor-Kramer 220 kV line is open.
3. Lugo end of Kramer-Lugo 220 kV line is open.
4. Cottonwood end of Victor-Cottonwood 115 kV line is open.
5. Inyokern End of Kramer-Inyokern 115 kV line is open.

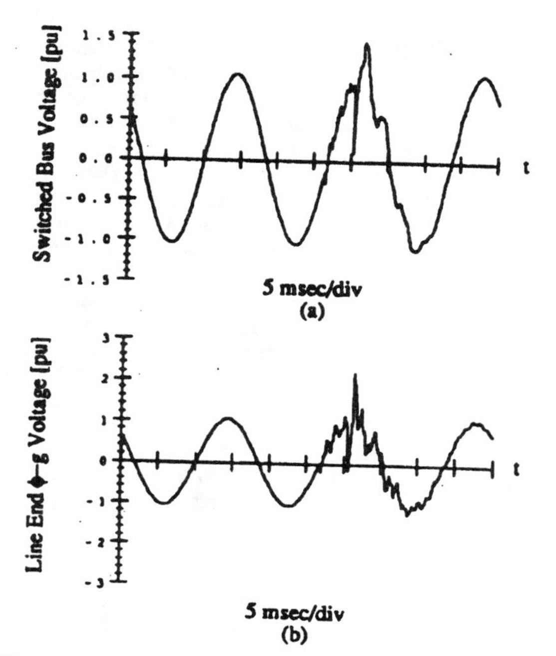

Figure 2. Line End Phase-to-Ground Voltage versus Line Length. (1. pu = $\sqrt{2}/\sqrt{3}$ rated system voltage)

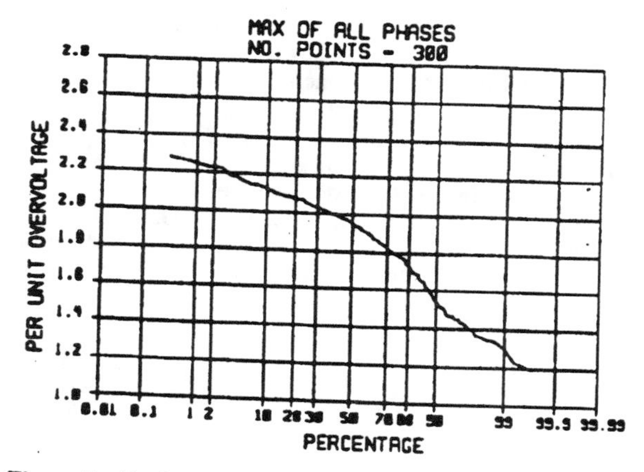

Figure 3. Statistical Distribution for Kramer-Lugo 220 kV Line End Uncontrolled Phase-to-Ground Voltage.

overvoltage is high, there is low probability that another phase voltage is of high magnitude and opposite sign. In practice, the ratio between phase-to-phase and maximum phase-to-ground overvoltages is around 1.5 [9].

OVERVOLTAGE MITIGATION

Various mitigation methods discussed previously were investigated for reducing remote overvoltages. In addition, a fixed overvoltage damping device was considered, which consists of an MOV and resistor series combination, in parallel with a current limiting reactor [8]. Table 1 summarizes the effectiveness of all the mitigation methods investigated.

Overvoltage transients associated with both energizing and bypassing of the pre-insertion devices were considered. The bypassing transients were found to be considerably less than energizing transients.

Of all the overvoltage mitigation devices considered, synchronous closing with a closing tolerance of ± 1 ms was the most effective. However, it may be difficult to obtain synchronous closing schemes with such a small closing tolerance.

The pre-insertion resistor was found to be consistently effective. A range of closing resistor values were investigated. The 200 Ohm closing resistor was found to be the optimum value for both 220 kV and 115 kV shunt capacitor installations.

Surge arresters placed at line ends can also effectively limit the overvoltages, including the overvoltages due to restrike. Only phase-to-ground arresters are considered. Arrester energies for arresters placed at line ends are found to be within the single-column energy limits for the worst case restrike condition. The arresters with standard ratings at the switched bus are not effective in reducing the transient overvoltages at the end of the open lines including the restrike conditions. Arrester with lower protective levels can be utilized to limit these overvoltages. However, multiple columns may be necessary to meet the energy dissipation requirements, thereby adding economic constraints.

As indicated in Table 1, the surge arrester in parallel with the fixed inductor was not effective in reducing the remote overvoltages. The exact selection of surge arrester protective level, the resistance in series with the surge arrester, and inductance requires an optimization of parameters considering the surge arrester conduction period, since this will determine the "effective" damping resistance. As the voltage across the current limiting reactor rises during energization, it is expected that the surge arrester protective level will be exceeded, inserting a resistance into the circuit and effectively acting as a pre-insertion resistor [8]. Experimentation was performed for two configurations of this device. Both configurations of the device considered a 3 kV protective level for the surge arrester and a 200 Ohm series resistor with either a 1.5 mH or 3 mH fixed inductor. Although both configurations were found to be ineffective, this device should not be ruled out as an overvoltage mitigation device, but requires further investigation which is beyond the scope of this study. Note that this device potentially offers protection against remote overvoltages during restrike.

Standard pre-insertion inductor values of 10 and 20 mH were considered. The pre-insertion inductor is attractive in terms of cost and reliability; however, the results indicated that the pre-insertion inductor is not effective for all system configurations. In particular, a 10 mH pre-insertion inductor caused a slightly higher line end overvoltage than the corresponding uncontrolled case for Kramer 220 kV capacitor bank switching. These controlled overvoltage waveforms are shown in Figures 4-a and 4-b for the switched bus and line end, respectively. Note that the switched bus voltage is not the same as the capacitor voltage when the overvoltage mitigation device (i.e., pre-insertion resistor, pre-insertion inductor, etc.) is in the circuit (see Figure 5). Results for the 115 kV network, however, have shown that a 10 mH pre-insertion inductance is almost as effective as pre-insertion

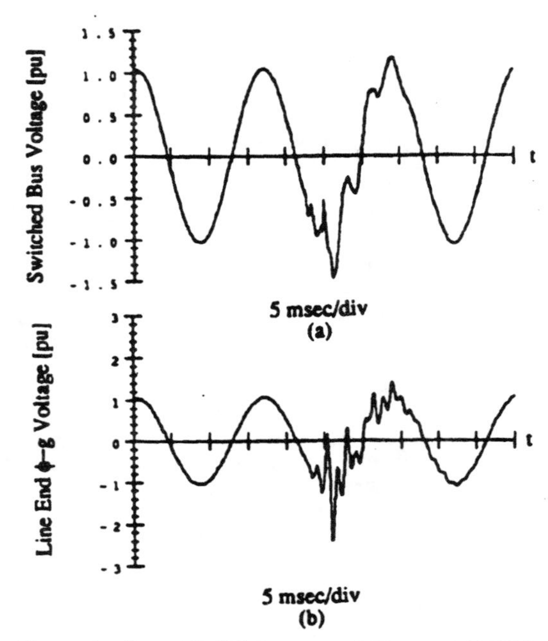

Figure 4. Controlled Voltages at a) Kramer 220 kV Switched Bus and b) Kramer-Lugo 220 kV Line End.

resistance. These conflicting observations of pre-insertion inductor effectiveness prompted a parametric study to better illustrate the influence of various system parameters.

PRE-INSERTION INDUCTOR PERFORMANCE EVALUATION

The generic 230 kV network shown in Figure 5 was used to investigate the influence of system strength, capacitor bank size, and line length on pre-insertion inductor effectiveness. Each of these system parameters was independently varied to determine the uncontrolled and controlled phase-to-ground overvoltages at the end of the open line and at the capacitor bank bus. In addition, the effect on phase-to-phase overvoltages at the end of the open line was considered. All overvoltages were statistically maximized using BPA's Electromagnetic Transient Program (EMTP). As in the TNA study, a 120° closing span was considered.

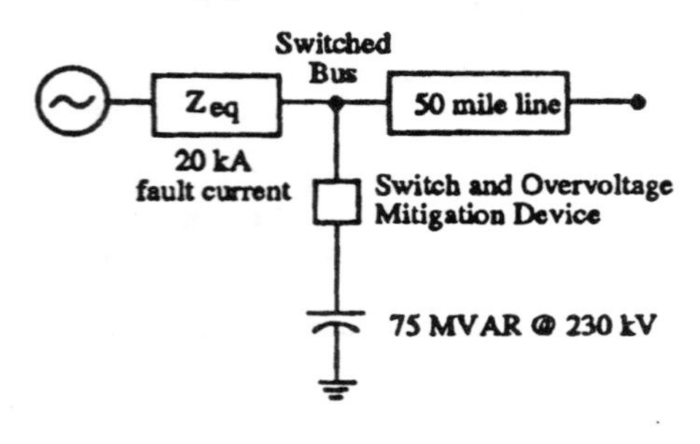

Figure 5. 230 kV Network Configuration Used for Parametric Studies.

The 230 kV base system used for the parametric variation consisted of a system strength of 20 kA available three-phase fault current at the bus, a 75 MVAR switched capacitor bank, and a 50 mile open-ended line. The source equivalent impedance (Z_{eq}) modeled as lumped-element, mutually-coupled R-L branches representing system positive, negative, and zero sequence characteristics. Fault current variations were performed by modifying the equivalent impedance to obtain the desired system strength while maintaining a system X_0/X_1 ratio of 3. A simple source representation was used over a more-detailed model to quantify the effects of basic system parameters. The transmission line was represented by a distributed parameter model (non-frequency dependent) adequate for the frequency ranges of interest. The results of the parametric study for the line length variations are provided in Figures 6 through 8 in terms of line end phase-to-ground, line end phase-to-phase, and switched bus voltage plots, respectively. Similarly, the results for capacitor bank size variations are shown in Figures 9 through 11, and the results for system strength variations are presented in Figures 12 through 14. Each of these figures includes curves for uncontrolled energization, and 10 mH and 20 mH pre-insertion inductors. In addition, the line length variation analysis of Figures 6 through 8 considers 50 and 200 Ohm pre-insertion resistors for comparison to pre-insertion inductors.

The line end phase-to-ground and phase-to-phase voltages versus line length demonstrate similar characteristics as shown in Figures 6 and 7. The phase-to-phase voltages are per-unitized by $\sqrt{2}/\sqrt{3}$ rated system voltage to allow direct comparison with the results shown in the literature [2]. Both uncontrolled overvoltages reach a peak near a 50 mile line length. The 200 Ohm pre-insertion resistor is the most effective in reducing the uncontrolled overvoltages. The 50 Ohm pre-insertion resistance and the 20 mH pre-insertion inductor are the next most effective overvoltage control methods. For certain line lengths, the 50 Ohm pre-insertion is more effective than the 20 mH pre-insertion, and for other line lengths the opposite occurs. The 10 mH pre-insertion

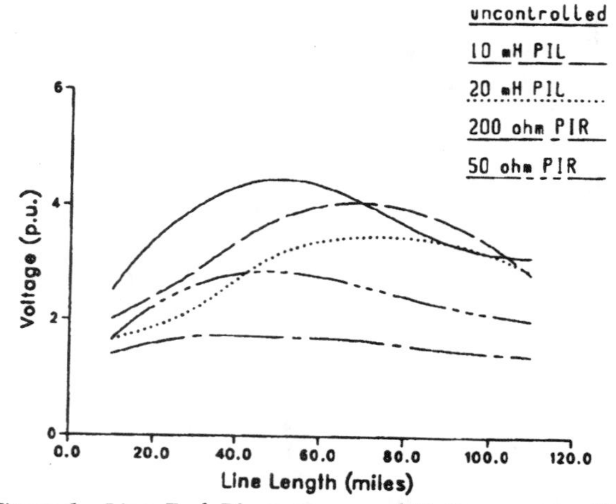

Figure 6. Line End Phase-to-ground Voltage versus Line Length.

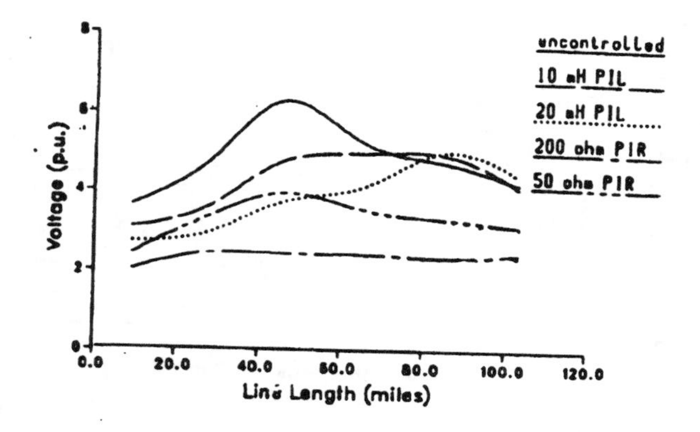

Figure 7. Line End Phase-to-phase Voltage versus Line Length.

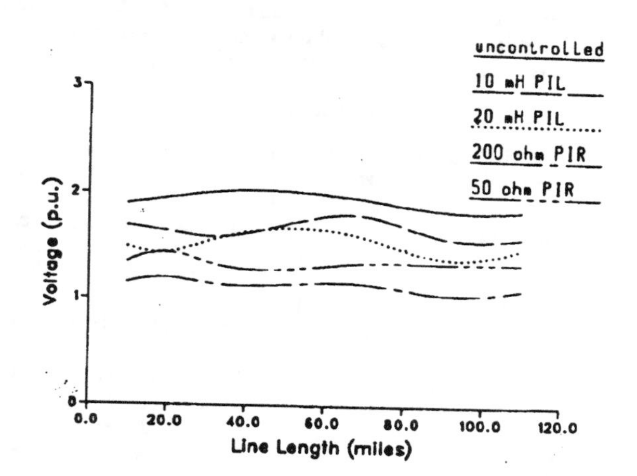

Figure 8. Switched Bus Voltage versus Line Length.

inductor is the least effective in reducing the uncontrolled overvoltage. While both pre-insertion resistor sizes always reduced the overvoltage below the uncontrolled level, the pre-insertion inductors did not exhibit this behavior. In fact, over a range of line lengths, both the 10 and 20 mH pre-insertion inductors exceed the uncontrolled level. This type of behavior is consistent with the TNA study results, but has not been observed previously [2].

Figures 9 and 10 show the line end phase-to-ground and phase-to-phase voltages versus capacitor size. Again, the controlled and uncontrolled overvoltages of these figures exhibit similar characteristics. For both phase-to-ground and phase-to-phase voltages, the uncontrolled overvoltage trend increases with capacitor bank size. The 10 and 20 mH pre-insertion inductors are either ineffective or slightly effective for smaller capacitor sizes, but become increasingly more effective as the capacitor size increases. The 20 mH pre-insertion inductor is generally more effective than the 10 mH pre-insertion inductor.

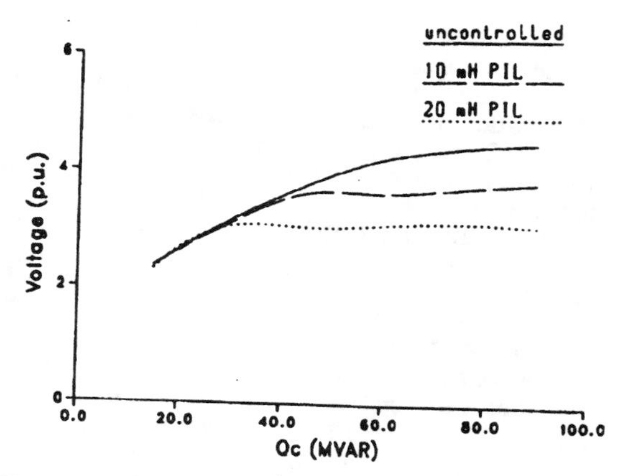

Figure 9. Line End Phase-to-ground Voltage versus Capacitor Size.

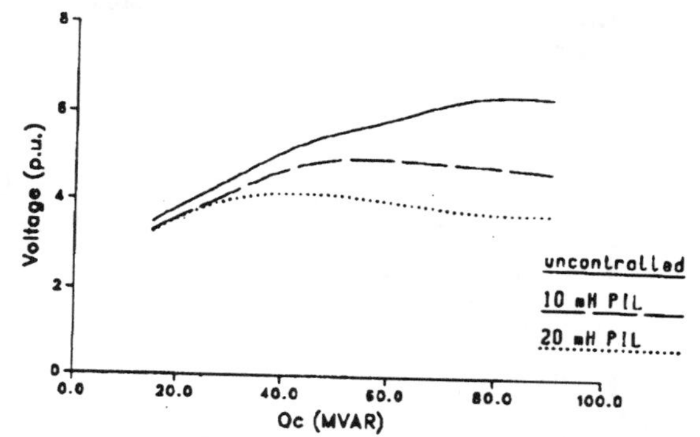

Figure 10 Line End Phase-to-phase Voltage versus Capacitor Size.

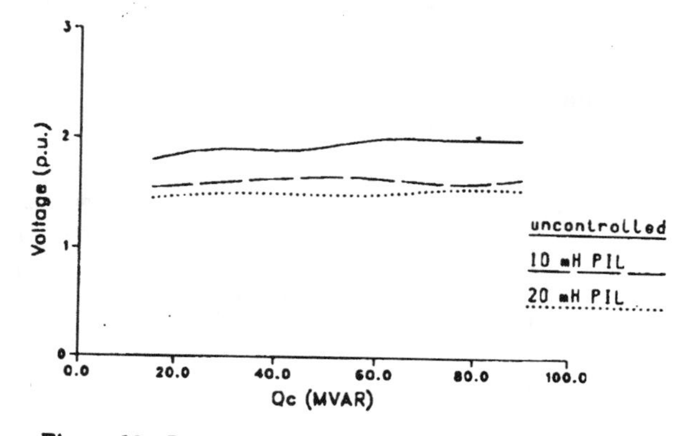

Figure 11. Switched Bus Voltage versus Capacitor Size.

The line end phase-to-ground and phase-to-phase overvoltages versus system strength are shown in Figures 12 and 13. Again, these figures exhibit similar trends. The uncontrolled overvoltages reach a peak and begin to decrease as the system strength is increased. Both pre-insertion inductor sizes are generally effective in reducing the overvoltage, but the 20 mH pre-insertion inductor is much more effective than the 10 mH pre-insertion inductor.

Figures 8, 11, and 14 show the switched bus voltage versus line length, capacitor size, and system strength, respectively. For all cases, the switched bus voltages with controlled energization are less than the corresponding uncontrolled voltage. The capacitor size and system strength variations of Figures 11 and 14 clearly indicate that the 20 mH pre-insertion inductor is more effective than the 10 mH pre-insertion inductor. The line length variation of Figure 8, generally shows this trend; however, for certain line lengths, the pre-insertion inductors are equally as effective. Also, Figure 8 indicates that both pre-insertion resistors are generally more effective than the pre-insertion inductors.

It should be noted that changes in the zero sequence network will effect the magnitudes of the observed overvoltages. The actual zero sequence effects are given not only by the system zero sequence impedances but also by the breaker closing span. While zero sequence effects will change overvoltage magnitudes, the relative effectiveness of the mitigation methods shown in this parametric study will remain unchanged.

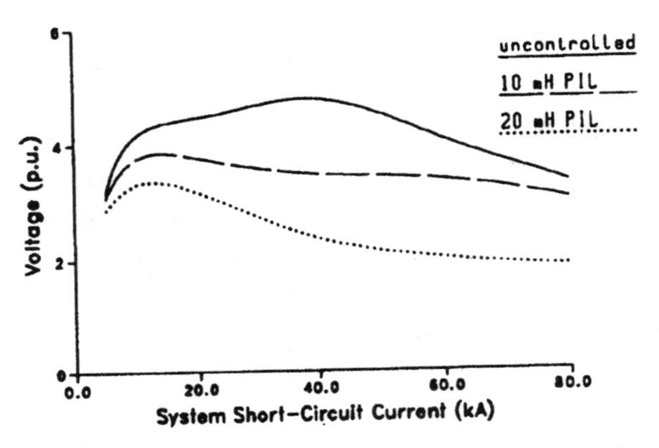

Figure 12. Line End Phase-to-Ground Voltage versus System Strength.

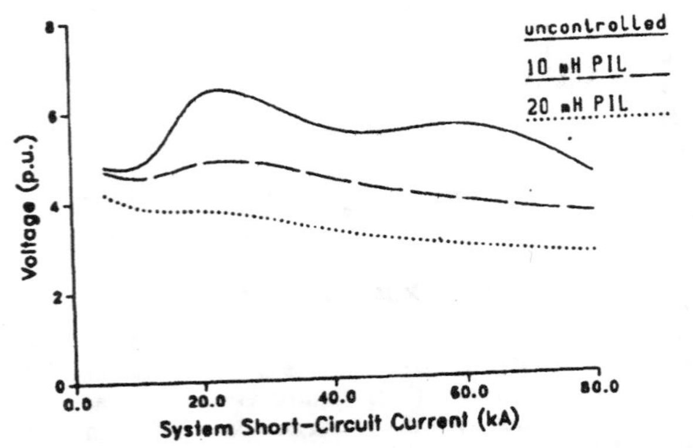

Figure 13. Line End Phase-to-Phase Voltage versus System Strength.

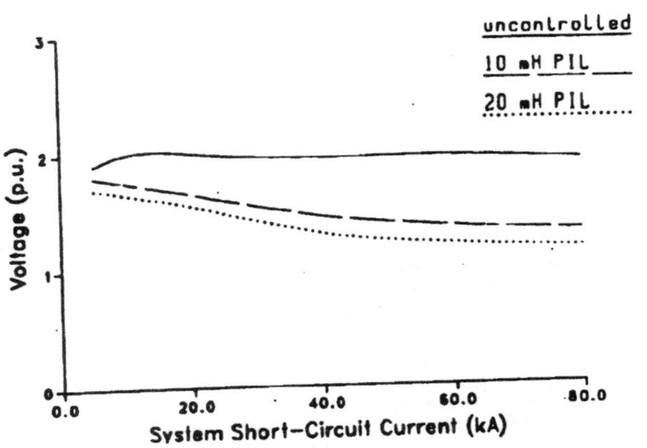

Figure 14. Switched Switched Bus Voltage versus System Strength.

The varying degrees of pre-insertion inductor effectiveness indicate an interaction between system parameters and pre-insertion inductance. This interaction may be understood by investigating the physical phenomena that causes high line end overvoltages during capacitor energization.

REMOTE OVERVOLTAGE PHENOMENA

A first step towards understanding the effectiveness and/or ineffectiveness of pre-insertion inductors is to understand the cause of remote overvoltage during capacitor bank energization. This discussion applies to overvoltages at the end of open lines, as considered for SCE, or transformer-terminated lines because transformers have a high surge impedance relative to transmission lines [1,5].

As discussed before, the simple circuit configuration shown in Figure 5 is designed to show relevant system parameters for the overvoltage phenomena. During normal capacitor bank energization, the switched bus voltage suddenly drops to zero, sending a negative 1.0 pu surge down the open line [5]. During this transient period, the switched bus voltage will recover and overshoot the nominal system voltage, reaching 2.0 pu in the theoretical limit. This in turn sends up to a 2.0 pu surge down the line, opposite in polarity to the original 1.0 pu surge. If the time for the bus voltage to rise from zero to 2.0 pu is equal to the round trip travel time for a surge on that line, the worst case overvoltage will occur at the end of the open line. The line end overvoltage can be 4.0 to 5.0 pu in the worst case; however, this is highly dependent on the line length and the time for the switched bus voltage to rise from zero to its peak. The time-to-peak for the switched bus voltage is equal to one half of the period associated with the damped resonant frequency of the network. This resonant frequency is governed by the system equivalent impedance, (i.e., short circuit MVA) capacitor bank size, and line surge impedances.

For open-ended lines, the first resonant frequency for line end voltage magnification depends on the line length, and corresponds to four travel times. The resonance repeats at integer multiples of the first resonant frequency. For example, the second resonant frequency is at two times the first resonant frequency, while the third one is at three times the first resonant frequency, and so forth. If the line length is such that the first resonant frequency is excited by the system's natural response, then worst case overvoltages occur at the end of the open line. In other words, if the system damped resonant frequency is close to the resonant frequencies of the line, there is potential for high overvoltages at the line end. This phenomenon is the reason for the peaks observed in maximum overvoltage versus line length curves of Figures 6 and 7 for uncontrolled and pre-insertion inductor energizations. As a quantitative example, consider the line-end overvoltage versus line length characteristics for uncontrolled energization shown in Figure 6 (solid line), which peaks around 50 miles. The resonant frequency of the system is determined by the switched capacitance and system strength. For 75 MVAR switched capacitance at 230 kV voltage level and 20 kA available fault current, the system resonant frequency corresponds to approximately 620 Hz. For a typical 230 kV line, the velocity of propagation for the traveling wave is around 175,000 miles/sec. The first resonant frequency of the line transfer function between the line end and switched bus voltages corresponds to four travel times along the line. The line length that will create a first resonant frequency of 620 Hz can be approximately calculated as 70 miles (175,000/(4*620 Hz) ≈ 70 miles), which should correspond to the peak of uncontrolled energization curve in Figure 6 (i.e., 50 miles). Note that this approximate calculation ignores the interaction between the system resonance and line resonance which would modify the resultant line length if considered, and is the source of discrepancy.

When a pre-insertion inductor is utilized, the total inductance value is effectively increased lowering the system resonant frequency. Therefore, the worst case line end overvoltages will occur for a line with lower resonant frequencies, which in turn corresponds to a longer line length. This explains why the peaks for pre-insertion inductor energization characteristics of Figures 6 and 7 occur at a longer line length than the corresponding peaks for uncontrolled energization.

It is also possible to have multiple peaks in the overvoltage performance curves in Figures 6 and 7, since the damped system resonant frequency may interact with the first resonant frequency of a line, and interact with a second or higher resonant frequency of a longer line. This trend would be observed in Figures 6 and 7 if longer line lengths were considered. Interactions with higher line resonant frequencies are expected to cause lower remote overvoltages due to the larger degree of damping at high frequencies.

In general, the behavior of the network shown in Figure 5 during uncontrolled or controlled capacitor energization can be explained by analyzing the interactions between the damped resonant frequency of the system and line resonant frequencies. The other curves in Figures 9 and 12 that characterize the overvoltage performance in terms of capacitor size and system strength further demonstrate this interaction.

CONCLUSION

Pre-insertion inductors should be carefully evaluated by means of detailed statistical system studies before being used for mitigating remote overvoltages due to capacitor energization. The results of a TNA study performed for SCE have shown that pre-insertion inductors may be ineffective for certain system conditions. This observation has been verified by a parametric study considering the influence of system parameters, such as line length, system strength, and capacitor size. The ineffectiveness of pre-insertion inductors has also been explained in terms of the physical phenomena. As a result of the TNA study for the 220 kV capacitor bank installations, SCE has decided not to use pre-insertion inductors at those locations. However, since the TNA study results have shown pre-insertion inductors to be effective for the 115 kV capacitor bank installations, SCE is considering pre-insertion inductors for these locations.

REFERENCES

[1] R.A. Jones, H.S. Fortson, "Consideration of Phase-to-Phase Surges in the Application of Capacitor Banks," **IEEE Transactions on Power Delivery**, Vol.PWRD-1, No.3, pp.240-244, July 1986.

[2] R.P. O'Leary, R.H. Harner, "Evaluation of Methods for Controlling the Overvoltages Produced by the Energization of Shunt Capacitor Banks," CIGRE Paris 1988, Paper 13-05.

[3] J.R.Ribeiro, M.E. McCallum, "An Application of Metal Oxide Surge Arresters in the Elimination of Need for Closing Resistors," **IEEE Transactions on Power Delivery**, Vol. 4, No. 1, January 1989.

[4] R.W. Alexander, "Synchronous Closing Control for Shunt Capacitors," **IEEE Transactions**, PAS, Vol.104, No.9, pp.2619-2626, September 1985.

[5] T.R. Sims, "Concerns for Switching EHV Capacitor Banks - Transient Voltage Considerations at a Remote Radial Transformer Termination," Internal Correspondence, Southern Company Services.

[6] H.M. Pflanz, G.N. Lester, "Control of Overvoltage on Energizing Capacitor Banks," **IEEE Transactions**, PAS, Vol.92, pp.907-915, May/June 1973.

[7] Sue S. Mikhail, Mark F. McGranaghan, "Evaluation of Switching Concerns Associated with 345 kV Shunt Capacitor Applications," IEEE Paper 85-SM, presented at the IEEE Power Engineering Society, July 1985.

[8] "Perfecting a Capacitor Bank Damping Device," Power Engineering Review, February 1991, p. 26

[9] **Transmission Line Reference Book, 345 kV and Above**, EPRI, 1982.

Bharat Bhargava was born on November 8, 1939. He received a B.E. (Electrical) degree from the University of Delhi, India in 1961 and a M.S. (Electrical Power) degree from Rensselaer Polytechnic Institute, Troy, New York, in 1976. Mr. Bhargava is presently working with the Southern California Edison Company, Rosemead, California. His fields of interest are Electrical Transients, Subsynchronous Resonance and System Dynamics. He is a member of IEEE and CIGRE.

Aftab H. Khan (M'87) was born in Chicago, Illinois, on December 18, 1965. He received his Bachelor of Science in Electrical Engineering from the University of Alaska, Fairbanks, and Master of Engineering in Electric Power Engineering from Rensselaer Polytechnic Institute in 1988 and 1989, respectively.

Mr. Khan joined the Power Systems Engineering Department of GE's Industrial and Power Systems as an Application Engineer in 1989, where he is currently responsible for performing harmonic and transient analysis of ac and HVDC transmission systems.

Mr. Khan is a member of IEEE, Tau Beta Pi, and Phi Kappa Phi.

Ali F. Imece (S'81-M'87) was born in Ankara, Turkey, on February 5, 1960. He received the B.S. degree with honors in Electrical Engineering from Middle East Technical University in 1981, the M.S. degree in Electrical Engineering from Michigan Technological University in 1983, and the Ph.D. degree in Electrical Engineering from Auburn University in 1987. Dr. Imece was a research assistant at Michigan Tech from 1981 to 1982, and at Auburn University from 1983 to 1987. He also worked at Sandia National Laboratories, Albuquerque, as a visiting researcher in 1984. Dr. Imece joined the Power Systems Engineering Department of GE in 1987. His current interests include electromagnetic transients, substation voltage uprating, and frequency dependent equivalent modeling of power systems.

Dr. Imece is a member of IEEE Power Engineering Society, National Society of Professional Engineers, Phi Kappa Phi and several IEEE Working Groups. He has contributed many papers and reports in the area of substation uprating, static var compensator analysis, and photovoltaics.

Joseph DiPietro (M'70) joined GE in 1948. Since 1964, he has been responsible for planning and conducting transient performance studies on GE's Transient Network Analyzer (TNA) in Schenectady, NY. He has participated in numerous studies over this period involving many of EHV transmission systems in service today. In addition, he has been responsible for a number of more complex and unique TNA studies relating to GE product design parameters to power systems. These studies have involved power circuit breakers, zinc oxide surge arresters, static var systems, and series capacitor protective devices. Mr. DiPietro is a member of IEEE.

DISCUSSION

R. P. O'LEARY, T. J. TOBIN, S&C Electric Company, Chicago, IL :

The authors have presented a stimulating paper on a subject which is not widely understood in the industry. The fact that they present conclusions contrary to those of their second reference is of particular interest to these writers.

Their implication that the pre-insertion inductor (PII) can yield transients higher than those created by an uncontrolled closing is particularly perplexing. Understanding these results requires further discussion of the nature of the transients, the nature of the switching device and its interaction with the system, and the analytical models used to evaluate performance.

- The transient network analyzer (TNA) model underestimates the overvoltages of uncontrolled energization and pre-insertion resistor (PIR) energization because the frequency response of this model is not fast enough to track the high-frequency transients associated with uncontrolled and PIR energizations. On the other hand, the frequency response of the model is adequate for the lower-frequency transients of the PII.

During an uncontrolled energization of a capacitor bank, the local bus voltage abruptly falls to zero in less than 1 micro-second, sending a steep-front wave down any lines connected to the bus. With a PIR, the magnitude of the transient is reduced but the steep-front nature of the transient is unchanged. When a PII is utilized, not only is the magnitude of the transient reduced, but the rate of voltage drop is also significantly reduced, thus lowering the transient frequency and eliminating the steep front wave traveling on the transmission lines. The steep rate-of-change obtained with the PIR and uncontrolled cases makes the frequency response of the model critical to the comparison of the different control means. In the TNA model used by the authors, the 10 kHz frequency response of the π-sections is not fast enough to track the megahertz-like transient voltages. This frequency response can, however, follow the slower transient of the PII. A model with an appropriate frequency response, would likely predict the uncontrolled phase-to-ground overvoltage to be closer to 3.5 PU, instead of the 2.2 to 2.7 PU reported by the paper. Such an enhanced model would also likely predict a considerably higher transient peak for the PIR, but not a significantly different peak for the PII.

- Both the TNA and Electro-Magnetic Transients Program (EMTP) models tend to favor PIR energization because they overestimate the transients for the PII energization. These models ignore system damping due to loads and/or surge impedances of transmission lines, which are essential to properly model the PII and uncontrolled energization.

The PIR has inherently high damping, and additional system damping has little effect on the transients. With uncontrolled energization or PII energization, there is little inherent damping, but transients are damped by losses in other parts of the system. However, loads (which are likely to be on the system, thus necessitating the energization of the capacitor bank) are not the only system elements providing damping. The damping due to the surge impedances of the interconnected transmission system at 220 kV can be as effective as 80 MVA of load. The equivalent source used in the EMTP model does not represent the surge impedances necessary for proper overvoltage calculations.

- The independent statistical pole closings of both the TNA and EMTP models tend to predict lower phase-to-phase overvoltages for uncontrolled and PIR energizations, due to an unsuitably low probability of simultaneous energizations.

When energizing an ungrounded capacitor bank, two poles of the switching device will close simultaneously for every energization, so that high phase-to-phase overvoltages are very likely. In practice, simultaneous closings of two phases are very likely in the case of uncontrolled energizations of grounded capacitor banks,[1] because the transient established by the first phase closing is coupled to the other two phases. The coupled voltage added to the normal phase-to-ground voltage is likely to induce a second pole to strike, thus "closing" two phases within micro-seconds of each other. Although the use of a pre-insertion device de-couples the system phases somewhat, a significant probability of simultaneous closings still exists.

The phase-to-phase overvoltages are underestimated for the uncontrolled and PIR energizations because coincident short-duration transients, with opposite polarity on two phases, are minimized by the independent statistical closings—where simultaneous closings are not likely. The lower-frequency PII transients have a greater likelihood of overlapping, reducing the criticality of the closing modeling. Since the mean closing time for simultaneous closings would occur near peak phase-to-phase voltage, the ratio of phase-to-phase to maximum phase-to-ground overvoltage would be closer to 1.73 than the 1.5 reported in the paper.

- The PII offers a level of control of the rate-of-change of overvoltage that is unavailable with PIRs.

The paper did not consider the fast rate-of-change of voltage associated with uncontrolled or PIR closings, which can be particularly distressful to system apparatus such as transformers[2] due to stress concentrations. The more moderately changing overvoltage associated with PII closings imposes much less severe duty on end-winding insulation because the voltage is more equally distributed along the transformer windings.

- PIRs sized for optimum overvoltage control do little to control inrush currents associated with back-to-back capacitor bank switching. PIIs limit inrush currents to 3.5 kA or less, depending on capacitor bank size.

When energizing a shunt capacitor bank with one or more additional banks already energized, inrush currents will flow between the banks. With only bus inductance to limit the current, magnitudes can easily reach 20 kA at frequencies of tens of kHz. These high-magnitude, high-frequency currents can induce dangerous step potentials and interfere with substation controls. A 200 Ω PIR optimized for voltage control at 115 kV on a 47 MVAR capacitor bank, produces 38 kV rms, or 0.58 PU, as continuous current flows through it. There is a high-frequency inrush current associated with bypassing or switching out the PIR that will be 58% of the uncontrolled energization, or about 12 kA. PIRs optimized to control back-to-back inrush currents to 3 or 4 kA are much smaller than PIRs optimized for voltage control but result in significantly higher overvoltages. 10 mH PIIs at 115 kV and 20 mH PIIs at 230 kV limit the back-to-back energization inrush current to less than 3.5 kA; and because of their low continuous-current voltage, the bypass current transient will typically be a few hundred amperes. An additional benefit of the PII is that the energization inrush current has a frequency of a few hundred Hz.

- 10 mH and 20 mH PIIs offer good overvoltage control. If needed, PII performance could be improved by increasing the inductance.

The overvoltage when bypassing the pre-insertion device is similar in nature to the uncontrolled energization response, except that its magnitude is driven by the voltage across the pre-insertion device. A pre-insertion device is optimum if its bypass overvoltage is equal to its magnitude of the energization transient. Again, as an example, a 200 Ω PIR applied at 115 kV on a 47 MVAR capacitor bank has a continuous-current voltage of 38 kV rms, or 0.58 PU. The PIR bypass overvoltage may thus be 58% of the uncontrolled case, or nearly equal to the PIR energization overvoltage. Under the same conditions with a 10 mH PII, the continuous-current voltage is 2 kV rms, or 0.02 PU. The PII bypass overvoltage will be barely noticeable.

The comparisons of 10 mH vs. 20 mH performance at 220 kV is interesting, since it shows a clear improvement with the latter. 20 mH inductors are offered for Circuit Switchers at 220 kV and larger values may be possible.

In summary, further consideration of switching device characteristics, pre-insertion device characteristics, system characteristics, as well as the nature of the transient overvoltages is needed when developing a model to predict a system's response to the energization of a shunt capacitor bank. The resulting enhanced model should yield a clear order of effectiveness in overvoltage control. Other potential problems associated with shunt capacitor bank energization might also influence the choice of control means. Relative strengths and weaknesses of several control schemes are outlined in the following table, where a value of 1 represents the most effective, and 4 the least.

Device	Peak Overvoltage	Rate-of-Change of Voltage	Back-to-Back Inrush Current
PIIs greater than 20 mH	1	1	1
10 mH to 20 mH PIIs	2	2	2
40 Ω to 80 Ω PIRs (optimized for current control)	3	4	3
100 Ω to 300 Ω PIRs (optimized for voltage control)	1	3	4

In summary, PII offers an adequate level of overvoltage control while minimizing the rate-of-change of voltage and the inrush currents on back-to-back switching. Cost, reliability, and practicality need to be evaluated against the electrical performance when selecting a pre-insertion device.

[1]R. P. O'Leary, Discussion of Bayless et al.

[2]R. S. Bayless, J. D. Selman, D. E. Truax, W. E. Reid, "Capacitor Switching and Transformer Transients," IEEE Transactions on Power Delivery, Volume No. 1, January, 1988, pp. 349-357.

A. Sabot (Member IEEE), **C. Morin** (Electicité de France, Direction des Etudes et Recherches, Les Renardières Electical Laboratories, France) : The authors should be commended for their interesting contribution to one increasing insulation co-ordination problem on the HV networks. It is always interesting to hear about different experience, approach or point of view on basic problems.

We would like to comment on the damping circuit composed of an inductor in parallel with a resistor in series with a Metal Oxide Varistor [L//(R↑MOV)] just mentioned in the paper. To our knowledge this damping circuit with MOV was first described in [1] and [2]. These descriptions surely raised the curiosity of Mr. J.E. Harder who asked B.C. Furumasu and R.M. Hasibar [3] many questions on the behavior of such a damping circuit.

The basic idea was to solve two problems with only one circuit :

- using an inductor for reducing the overvoltages (thanks to O'Leary and Harner [4]),
- using a damping resistor in parallel with the current-limiting inductor to avoid high frequency current zero in case of restrike of a line circuit breaker during fault clearing (thanks to Jansen and Van der Sluis [5,6]).

The details of this damping circuit were presented to the technical audience only in summer of 1992 [7]. So the inductance value of 3 mH (Table 1) chosen by the authors is too low to reduce the overvoltages efficiently. A higher inductance should be more suitable. In fact the L//(R↑MOV) damping circuit must be designed as follows :

- for overvoltage reduction the inductance value should be higher than the minimal short-circuit inductance (typically 120 to 150% of the equivalent short circuit inductance at the substation)
- for damping high frequency current with the resistor R in case of a line circuit breaker restrike due to excessive transient recovery voltage (TRV) not covered by the standard. A very high value of R does not damp sufficiently the high frequency current and allows high frequency current zero crossings, too low value of R short circuits the inductor reducing its efficiency in overvoltage reduction ; according to our studies a compromise is obtained with a time constant RC of 1 ms (R resistor value, C phase capacitor value)

In our study, the resistor R was introduced not for overvoltage damping but it nevertheless contributes to the overvoltage reduction. The MOV in series with the resistor allows to insert the resistor during transients and to disconnect it during steady state conditions (minimizing power frequency losses in the resistor).

Another advantage of the R↑MOV branch is that it limits the high frequency oscillations in the initial part of the TRV of the capacitor bank circuit breaker in case of a fault between the damping circuit and the capacitor, thus avoiding potential clearing problems due to the high value of the inductor.

According to the above rules the damping circuit investigated by the paper authors should have the following parameters :

$$L \approx 20 \text{ mH and } R \approx 266 \text{ ohms}$$

The Maximum Continuous Operating Voltage of the MOV must be higher than the voltage across the inductor including the possible effect of harmonic currents. For this reason the L value might be limited to avoid resonance with the capacitor : the resonance frequency should be lower than 5 to 6 times the power frequency of the network.

For different values of the line length of figure 5 of the paper, our EMTP overvoltage calculations for energization of the capacitor bank are presented in the following table :

line length in km	Maximum phase to ground overvoltage at the line open end in p.u.		Maximum phase to phase overvoltage at the line open end in p.u.	
	max	$S_{2\%}$	max	$S_{2\%}$
80	2.5	2.4	3.5	3:3
120	2.8	2.6	4	3.8
160	2.9	2.6	4	3.8
180	2.9	2.6	4.1	3.7
200	2.9	2.4	3.5	3.3
250	2.2	2.0	2.9	2.9
160 but with only 20 mH	3.2	3.1	5	4.8

According to these calculations, the damping circuit does reduce the overvoltages (especially phase to phase ones) to levels lower than those expected without any reduction circuits or with the pre-insertion inductor alone (fig 6 and 7 of the paper).

If the high frequency current in case of a line circuit breaker restrike during fault is not judged as a potential problem, a lower resistor value may achieve further overvoltage reduction but needs further investigation. In any case pre-insertion resistor seems more efficient in solving the overvoltage problem during capacitor energization ; however one must not forget that capacitor switching might create other problems not solved by the pre-insertion resistors.

Finally, for comparison purposes the 2% statistic overvoltage ($S_{2\%}$) is generally preferred. The $S_{2\%}$ calculated above are acceptable for the insulation levels of the EDF 245 kV network. What about the Southern California Edison 230 kV network ?

The discrepancy between the authors results and those of O'Leary and Harner could be caused by :

- three phase line modelling, since O'Leary and Harner results do not show any influence,
- line skin effect modelling.

The authors responses to our comments will be appreciated.

[1] A. Sabot, C. Morin : "Transient Conditions Created by Banks of Capacitor" : Perfecting a Damping Device, Lab Echo 3.90.

[2] "Perfecting a Capacitor Bank Damping Device", in Sparks of Power Engineering Review, February 1991, p 26.

[3] B.C. Furumasu, R.M. Hasibar :"Design and Installation of 500-kV back-to-back shunt capacitor banks" IEEE Transaction on Power Delivery, April 1992, vol 7, number 7, pp 539-545.

[4] R.P. O'Leary, R.H. Harner : "Evaluation of methods for controlling the overvoltages produced by the energization of a Shunt Capacitor Bank", 1988 CIGRE session, report 13-05.

[5] A. Janssen, L. Van Der Sluis : "Controlling the transient currents and overvoltages after the interruption of a fault near shunt capacitor banks", 1988 CIGRE session, report 13-13.

[6] A. Janssen, L. Van Der Sluis : "Clearing faults near shunt capacitor banks", IEEE Power Delivery, July 1990, vol 5, n° 3, pp 1346-1354.

[7] A. Sabot, C Morin and al : "A Unique Damping Circuit for shunt Capacitor Bank Switching", IEEE / PES Summer Meeting, Seattle, Washington, July 12-16 1992, paper 92 SM 605-6 PWRD.

A. S. Morched and **E. J. Tarasiewicz** (Ontario Hydro, Toronto, Canada): The authors have presented an interesting paper on the use of pre-insertion inductors for controlling capacitor energization overvoltages. An extensive study to examine the effectiveness of this measure as compared with other mitigating measures is described in the paper. We concur with the authors conclusions that pre-insertion reactors may not always be an effective way of controlling capacitor switching overvoltages. In the following we offer comments in support of the authors conclusions and raise some questions for which the authors' response will be appreciated.

1. As pointed out by the authors, as well as in references [1] and [2] of the paper: high remote overvoltages due to capacitor bank energization arise from the unfavorable combination of transient components due to line fundamental frequencies and capacitor/system oscillation frequency. Since the addition of pre-insertion inductor will change both the line fundamental frequency and the capacitor/system frequency; it is to be expected that this change could result in deteriorating the situation rather than improving it.
2. In many system configurations, a reasonably sized pre-insertion reactor resulting in the effective control of remote overvoltages can be identified. However, the normal growth of the system or the addition of transmission circuits to the bus may alter the resonance frequencies in such a manner that renders the pre-insertion reactor ineffective.
3. Energization of capacitor banks can create overvoltages at the remote end of a short feeder line of such a frequency that may interact with one of the internal resonances of a connected transformer. The pre-insertion inductors would reduce this frequency in a manner that may tune (or detune) the system to one of these frequencies and endanger the transformer.
4. As discussed above, and as pointed out in the paper, high remote end overvoltages are the outcome of a resonance phenomenon. Would this fact make it easier to identify system configurations which are candidates for potential problems? Could this make it possible to reduce the number of necessary simulation studies?
5. Capacitor bank energization can result in especially high phase-to-phase overvoltages at remote ends that can endanger connected transformers. Unfortunately there are limits to how these can be effectively controlled by the phase-to-ground surge arresters. In the paper the authors indicate that in their study only phase-to-ground arresters were considered. Did the authors consider phase-to-phase arresters? What would the pros and cons of such installations be? Could the use of these provide a means of eliminating the need for the high energy absorbing capability necessary for phase-to-ground arresters.
6. Our studies indicate that the presence of the capacitor bank on the bus can result in increasing the magnitude of remote end overvoltages due to line energization. The implementation of a pre-insertion inductor (or resistor) in the capacitor bank breaker would have no effect on these. Have the authors explored similar phenomena in their studies?

Manuscript received August 10, 1992.

B. BHARGAVA, A. H. KHAN, A. F. IMECE, and J. DIPIETRO: The authors thank Dr. Morched, Dr. Tarasiewicz, Mr. Sabot, Mr. Morin, Mr. O'Leary, and Mr. Tobin for their discussions which enhance the value of our paper. We will address the questions in the order that they appear in the discussions, beginning with Dr. Morched's and Dr. Tarasiewicz's discussion.

- As Dr. Morched and Dr. Tarasiewicz have pointed out, and as pointed out in the paper, the high overvoltages at the end of the open-ended lines are the result of a resonance phenomena involving the line natural frequencies and capacitor/system resonant frequency. We agree that this fact would make it easier to identify a range of system parameters for which performance of the pre-insertion inductor is satisfactory and this can be used as a screening tool to reduce the number of simulations. As a starting point, we suggest calculating a capacitor/system oscillation frequency based on some equivalent source impedance as seen from the capacitor bus and compare the time to peak of this oscillation to the round trip travel time of a surge on the open-ended or transformer terminated line in question. It should be noted, however, that this is an approximation and does not represent the true system behavior due to traveling wave effects lost in the system equivalent impedance.

- Phase-to-ground arresters at the line-ends of the SCE system were considered but were ruled out due to the large number of line-end terminations which need to be protected. The preferred approach was to mitigate the overvoltages at the source of disturbance, i.e., the capacitor bank. The discussors have raised an interesting question regarding the use of phase-to-phase arresters. We agree that phase-to-phase arresters could have a lower energy absorbing requirement than phase-to-ground arresters. Also, in the case of transformer terminated lines, where the phase-to-phase overvoltages are of concern due to lack of specific standards for the phase-to-phase switching surge withstand of transformers rated 230 kV and below [1], phase-to-phase arresters appear to provide a means of protection. However, it should be noted that phase-to-phase arresters are not commonly used, requiring detailed studies to insure proper application for various system conditions.

- We agree with the discussors that the presence of a capacitor bank on a bus can in some situations increase the line energization overvoltages, and that pre-insertion devices are ineffective under these conditions. If the presence of a capacitor bank poses a dangerous situation to line energization, a fixed inductance may provide overvoltage mitigation if properly applied considering the resonance phenomena identified in this paper. Alternatively, the inductor in parallel with a resistor in series with an MOV (L//(R-MOV)), described in a discussion to this paper by Mr. Sabot and Mr. Morin, may provide some benefit.

The following responds to Mr. Sabot's and Mr. Morin's discussion:

- The discussors have presented a novel concept for limiting overvoltages due to capacitor bank energization. As we pointed out in the paper, the L//(R-MOV) should not be ruled out as a result of studies performed for SCE but requires further investigation. The discussors have presented valuable information concerning the optimized parameters for this device and have shown preliminary calculations in support of its performance. While we see this concept as an improvement to the pre-insertion inductor by potentially mitigating overvoltages as well as inrush/outrush currents, detailed statistical studies may still be required to protect against the resonance phenomena concerns associated with the inductance portion of the device. Furthermore, the losses associated with the 20 mH fixed inductance that the discussors suggest may be of concern.

- For comparison of reduction means in the SCE system, the results for maximum statistical overvoltage ($S_2\%$) presented by the discussors would be acceptable for the SCE system.

- We agree that the discrepancy between our parametric analysis and a similar analysis presented in reference [2] may be due to modeling techniques used in their analysis. As the discussor's have pointed out, the waveforms presented in reference [2] do not indicate coupling between phases, and it is unclear whether high frequency effects (i.e. skin effect) have been modeled correctly. Another source of discrepancy may be the method of maximizing overvoltages. It appears in reference [2] that maximum phase-to-phase overvoltages are obtained by simultaneous pole closings. However, we do not believe that this necessarily corresponds to the maximum phase-to-phase or phase-to-ground overvoltage. The proper maximized overvoltage for a real system, containing coupling between phases, is obtained by performing a statistical analysis of breaker pole closings. Also, since the results of reference [2] present the results of parametric analysis for phase-to-phase overvoltage control and not for phase-to-ground overvoltage control, it is difficult to compare our analysis with reference [2] for phase-to-ground overvoltage control.

The following responds to Mr. O'Leary's and Mr. Tobin's discussion:

- The discussors contend that the frequency response of the TNA line model underestimates the overvoltages of the uncontrolled and pre-insertion resistor (PIR) energizations, while adequately capturing the lower frequency transients of the pre-insertion inductor (PIL). Our experience has shown that the TNA is capable of correctly capturing the transients associated with uncontrolled and PIR energizations. The TNA has been designed specifically for investigating switching surges similar to the remote overvoltage phenomena investigated in this paper. A majority of the EHV lines in the United States have been designed using TNA results for switching surge performance and numerous closing resistor applications have been analyzed using the TNA. Industry experience has consistently shown the TNA results to be conservative, indicating that uncontrolled and closing resistor energization operations predicted by the TNA tend to overestimate, rather than underestimate these transients. One of the reasons for good correlation between field results and TNA results is due to the roll-off of steep-front surges in real transmission systems. The fast front surges mentioned by the discussors is attenuated by frequency-dependent effects of transmission lines, as well as by corona, bringing the true system response within the TNA model bandwidth. The higher uncontrolled overvoltage results predicted by the discussors for fast front surges is not realistic considering these effects.

- The analysis and conclusions of the paper have not ignored the lines connected to the capacitor bank bus as the discussors suggest. We agree with the discussors that lines connected to the capacitor bank bus effect this phenomena and they have been considered in the TNA model. However, whether or not these lines provide damping is dependent on the length of the lines. Effective damping is provided by lines if the time for a surge to return from these lines is much greater than the period of the transient being investigated. Otherwise, returning reflection add complications to this phenomena, increasing or decreasing the resultant transient. The TNA model consisted of an extensive representation of the SCE system around the critical 220 kV capacitor bank locations as shown in Figure 1 of the paper. The EMTP parametric analysis was prompted by the TNA study results which have shown the ineffectiveness of the pre-insertion inductor under certain conditions. Also, at transmission system voltage levels, we do not believe that system loads provide significant damping to this phenomena as the discussors suggest.

System loads are typically located at lower voltage levels, involving several stages of transformation from the transmission level. Transformer leakage impedances tend to electrically isolate loads from transmission system transients. Load damping is therefore minimal for these transients and it is accepted practice to neglect loads for transient studies at transmission voltage levels. Furthermore, because the amount and type of load varies on utility systems, it is reasonable to expect any overvoltage mitigation device to perform under no load conditions.

- The statistical overvoltage analysis of the TNA is performed by analyzing the results of 300 repeated three-phase operations of the same switching event. While we agree with the discussors that the overvoltage distribution is not specifically weighted towards simultaneous pole closings, these are a subset of our statistical variation in breaker pole closings. Due to the large number of repeated switching operations performed (300), we are confident that our approach captures the transients associated with these events. Although simultaneous pole closings are more probable in practice due to coupling between phases, simultaneous pole closings do not necessarily relate to the maximum phase-to-phase or maximum phase-to-ground overvoltages. Since the investigation for SCE focused on controlling maximum phase-to-ground overvoltages at the open-ended lines, the statistical breaker operations used are justified.

- We agree with the authors that the PIL provides greater control over the rate of change of voltage than the PIR. This fast rate of change due to capacitor energization or any other switching operation, under certain conditions, can be distressful to system apparatus such as transformers. While such transformer failures have been documented, exposure of transformers to such switching surges may not be a reason for undue concern. Transformers are tested to withstand a chopped wave (very fast front), having a magnitude 15% greater than the BIL rating. Although the results of this test indicate that the transformer should withstand fast front switching surges, the impact of repeated exposure to such surges is undefined. Transformer failures may also be caused by switching surges which excite internal resonances [3]. The application of pre-insertion inductors can either tune or de-tune such a phenomena. Determination of transformer internal resonances involves detailed studies of individual transformers involving design details which are often difficult to obtain.

- The discussors point out the advantage of pre-insertion inductors for controlling back-to-back capacitor switching currents. However, it should be noted that proper application of capacitor banks requires protecting switchgear from all sources of outrush currents. A capacitor bank bus fault is one situation where pre-insertion inductors would be ineffective in limiting the high frequency outrush currents. This situation may require a fixed inductance for limiting the frequency of this current to within acceptable levels for the switchgear.

- The 10 and 20 mH pre-insertion inductors can provide overvoltage control if detailed system studies are performed to investigate the resonance phenomena identified in this paper. As the discussor's have mentioned, the cost, reliability, and practicality of the pre-insertion device must be weighed against the electrical performance. The intent of this paper has been to identify electrical performance concerns with the pre-insertion inductor. The results of the analysis presented in this paper indicate that peak overvoltage control by PIR's is less dependent on system resonant conditions (i.e. line length, system strength, and capacitor bank size) than are PIL's. For this reason, detailed statistical system studies are recommended prior to PIL application.

REFERENCES

1. R.A. Jones, H.S. Fortson, "Consideration of Phase-to-Phase Surges in the Application of Capacitor Banks," IEEE Transactions on Power Delivery, Vol. PWRD-1, No. 3, pp. 240-244, July 1986.
2. R.P. O'Leary, R.H. Harner, "Evaluation of Methods for Controlling the Overvoltages Produced by the Energization of Shunt Capacitor Banks," CIGRE Paris 1988, Paper 13-05.
3. R.C. Degeneff, W.J. McNutt, W. Neugebauer, J. Panek, M.E. McCallum, C.C. Honey, "Transformer Response to System Switching Voltages," IEEE Transactions on Power Apparatus and Systems", Vol. PAS-101, No. 6, June 1982.

Manuscript received October 2, 1992.

IEEE Transactions on Power Delivery, Vol. 8, No. 1, January 1993

TRANSIENT OVERVOLTAGES AND OVERCURRENTS ON 12.47 kV DISTRIBUTION LINES: COMPUTER MODELING RESULTS

N. Kolcio
Senior Member

J. A. Halladay
Member

G. D. Allen
Senior Member

E. N. Fromholtz
Non-Member

American Electric Power Service Corporation
Columbus, Ohio

Abstract - American Electric Power conducted field tests to study transient overvoltages and overcurrents on a 12.47 kV distribution line. A separate paper [1] has reported on the field test results. This paper covers the computer modeling and simulation of the field tests.

The paper describes the adaptation and refinement of the transformer, load and line models. Computer simulation of the field tests using the Electromagnetic Transients Program (EMTP) give results which are very close to the actual field tests. Also reported in the paper are the results from a parametric study which were utilized to develop a simplified computer model.

INTRODUCTION

In an attempt to study transient overvoltages and overcurrents on distribution lines, American Electric Power has conducted a series of tests on a one mile section of a 12.47 kV distribution line. A recently published paper on field test results [1], describes the data obtained from capacitor switching and load dropping tests. The aim of this paper is to describe the computer modeling and the simulation of the field tests.

The EMTP was used to perform the field simulation study. This program has been successfully used in Canada to simulate test results from switching large capacitor banks on distribution lines [2] and to analyze transient recovery voltages [3]. The success and accuracy of computer simulation of the field results also depend on the modeling of the test circuit. In this paper, particular attention was given to the refinement of the transformer, load and line models in order to determine the degree of refinement necessary to achieve parity with the field results. For example, the transformer's transient response was compared with the field tests to confirm the magnitude, with its associated rate of decay, and the dominant frequencies measured. Also, parametric studies were made to gain information on the influence of several system parameters on the magnitude of transient overvoltages at the measuring site. The parameters under review were: grounding resistance, earth resistivity values, type of line model (dependence of frequency) and size of load. Results of this study indicate that a simplified model of the test circuit can be achieved. Further evaluation of the simplified computer model has shown little loss in accuracy with a corresponding large decrease in computer run time.

The simulation of the test results using the EMTP program has been achieved with reasonable accuracy. The noted difference for most cases was less than 10%.

92 WM 273-3 PWRD A paper recommended and approved by the IEEE Transmission and Distribution Committee of the IEEE Power Engineering Society for presentation at the IEEE/PES 1992 Winter Meeting, New York, New York, January 26 - 30, 1992. Manuscript submitted March 25, 1991; made available for printing December 23, 1991.

Due to space limitation, the paper covers in some detail only a representative sample of the tests. A comparison of the results between field tests and computer simulations was limited to one example of line dropping and one of capacitor switching.

Maximum anticipated overvoltages were obtained from the simplified distribution model by varying the switching angles. A 200 shot statistical closing using Gaussian (normal) distribution were performed. Results were obtained for one case of capacitor switching and one for line dropping tests.

DESCRIPTION OF LINE SYSTEM MODELS USED IN EMTP COMPUTER SIMULATIONS

EMTP Distribution Line Model

The three-phase 12.47 kV distribution line was modeled in the EMTP using the Marti frequency dependent distributed parameter line model. The ground resistivity (RHO) used was 100 ohmmeters and the frequency at which the calculated current transformation matrix for converting between phase-modal domains was 5 kHz. The single-phase 7.2 kV (line to neutral) distribution lines connected to the three-phase line had the same Marti line model and parameters. The typical "TRIPLEX" service drop from the transformer pole to the residence was modeled as a continuously transposed three conductor bundle using the Marti frequency dependent line model with the above mentioned parameters. The physical line data is listed in Table 1.

Table 1: PHYSICAL 12.47 KV LINE AND HOUSE TRIPLEX DATA

1. Primary (7.2 kV) Line	2. Triplex (120/240 V) Service Drop
a. 3 Phase, 336 MCM ACSR Conductor	a. Phase and neutral conductor: 3 #2-AA Twisted Triplex
b. Phase average height: Outside: 31.33 ft.; Center: 32.58 ft.	b. Average height: 21.0 ft.
c. Neutral conductor: 3/0 ACSR	c. Conductor O.D.: 0.292 inch
d. Neutral average height: 28.0 ft.	d. Conductor resistance: 1.41 Ω/mi
e. Conductor O.D.: 0.710 inch	e. Conductor spacing: 0.337 inch
f. Conductor resistance: 0.2790 Ω/mi	
g. Neutral O.D.: 0.502 inch	
h. Neutral resistance: 0.5400 Ω/mi	
i. Phase to phase separation: 3.667 ft.	

EMTP Distribution Transformer Model

The transformer model was derived by its manufacturer for fast rise time transients [4]. The shell-form 25 kVA 7.2 kV-120/240V non-interlaced secondary winding transformer consists of a linear network of an impedance matrix and its interwinding and winding to ground capacitances. The detailed electromagnetic model is shown in Figure 1. The saturation characteristics for its magnetization (flux-current) curve were obtained from tests. The data obtained were the open circuited voltage and current at the transformer's terminals as a function of incremental (0.05 p.u.) step increases in terminal voltage. From this information, the appropriate hysteresis curve (flux-current) was derived for the EMTP program to correspond to the transformer's core material. The shape of the hysteresis loop for an inductor depends primarily on the material of the core. This transformer's core was assumed to consist of the common core material ARMCO M4 oriented silicon steel.

FIGURE 1
25 KVA DISTRIBUTION TRANSFORMER MODEL: 7.2 KV/120/240 V. NON-INTERLACED SECONDARY

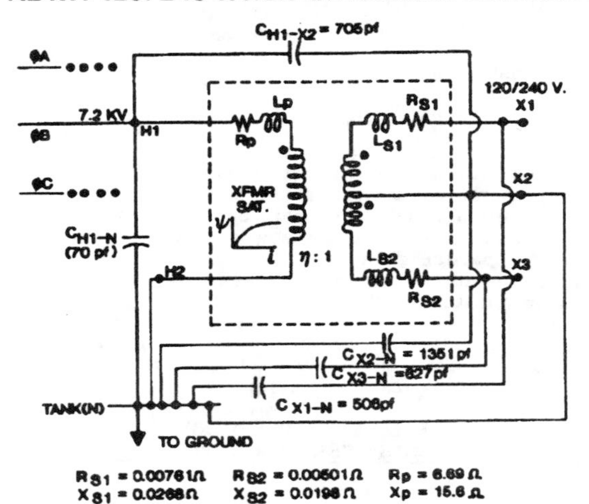

To verify the transformer model, a comparison was made to a specific field test. Test F1-K1 from Ref. [1], Field Test Results, provided the transformer's inrush current when its terminal voltage was near a voltage zero. The line was energized at Wallen Station with capacitor bank and transformer surge arresters disconnected. The transformer, on Phase B by itself, exhibited a typical inrush current characteristic. The first current peak is approximately nine times the 60 Hz steady state crest current, with the transient having a rapid decrement during the first few cycles and decaying more slowly thereafter. The damping coefficient, R/L is not constant because of the variation of the transformer inductance with saturation. During the first few current peaks, the degree of saturation of the iron is high, making the inductance low. The inductance of the transformer increases as saturation decreases, and thus the damping factor becomes smaller as the current decays. Figure 2 shows the computer simulation of the exciting-current inrush for a shell-form single-phase transformer energized at the zero voltage point on the waveform. Figure 3 is the actual field test waveform.

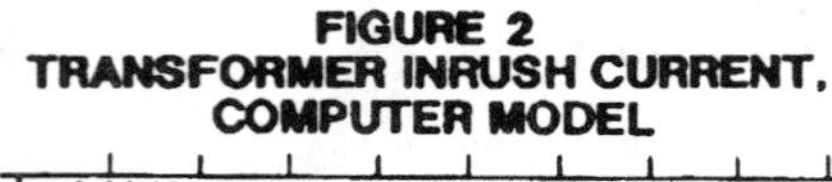
FIGURE 2
TRANSFORMER INRUSH CURRENT, COMPUTER MODEL

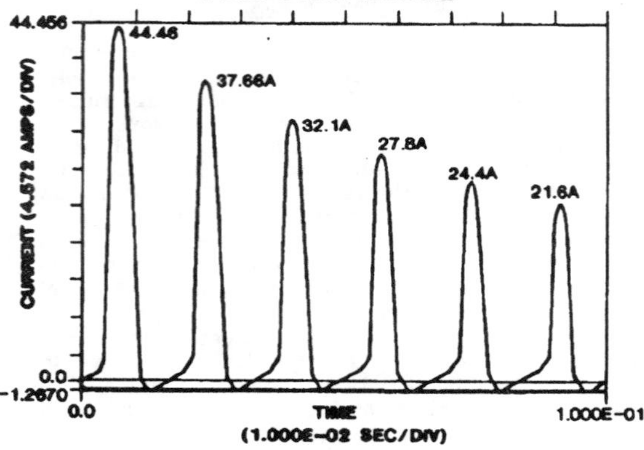

FIGURE 3
TRANSFORMER INRUSH CURRENT FIELD TEST F1-K1

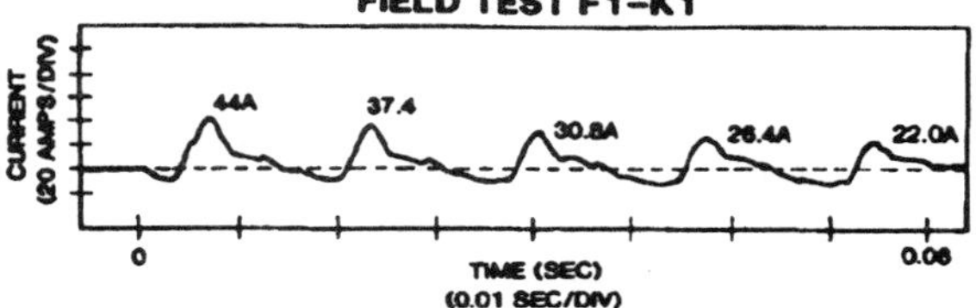

From the Table 2, it can be seen that the computer model agrees quite well when comparisons are made of the peak magnitudes for the first five current peaks and the corresponding rate of decay of the inrush current. The remaining 18 distribution transformers used the same model as it was assumed that they would be similar in design and core material.

Table 2: DISTRIBUTION TRANSFORMER INRUSH CURRENT: LINE ENERGIZATION AT NEAR ZERO VOLTAGE

Peak Current #	Ampere's (Crest)	% of 1st Point (Rate of Decay)
Test F1-K1, Channel No. 7		
1	44.0	1.00
2	37.4	0.85
3	30.8	0.70
4	26.4	0.60
5	22.0	0.50
Computer Model		
1	44.5	1.00
2	37.7	0.85
3	32.1	0.72
4	27.8	0.62
5	24.4	0.55

Ground Resistance Model

The transformer pole and house ground resistances were assumed to be 12.3 and 29.0 ohms, respectively, at each of the 19 distribution transformers. These values were measured values taken from the M1 transformer site as shown in Figure 5.

450 kVAR Capacitor Bank Model

The capacitor bank had three 7.675 uF capacitors connected to a common grounding wire, which was modeled by 15 uh inductor to represent a 7m (23 ft.) down lead and measured pole ground resistance of 4.0 ohms. The capacitor was shunted by a 5.72 megohm resistor to ground which would effectively discharge the capacitor to near zero voltage in approximately 5 minutes.

Table 3: HOUSE LOAD MODEL
Load Impedances in Ohms for 25 kVA Transformer

Load (% of 25 kVA)	5.0	20.0	40.0	60.0	80.0	100.0
Load (Watts)	1250	5000	10,000	15,000	20,000	25,000
H: R_{120} (Ohms)	92.2	23.0	11.5	7.68	5.76	4.61
S: R_{240} (Ohms)	61.4	15.4	7.68	5.12	3.84	3.07

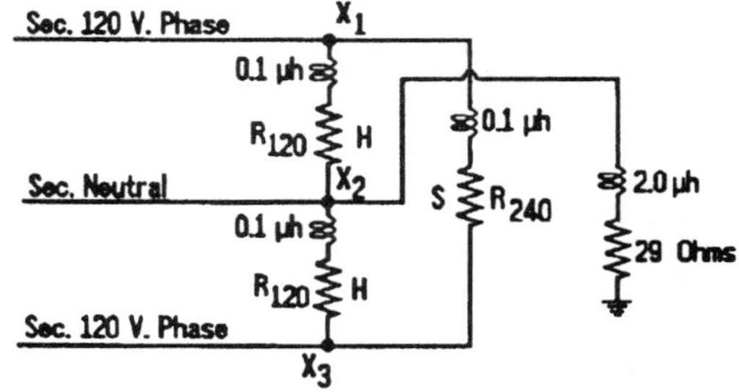

House Load Model

The load model for the house wiring system as shown in Table 3 was assumed initially to be a balanced phase-to-neutral load modeled by resistances R_{120}, and the phase-to-phase load modeled with resistance R_{240}.

The house wiring system consists of many radial branch circuits emanating from the service entrance panel. The larger loads such as water heaters, clothes dryers, ranges, central air conditioners, and electric furnaces are served phase-to-phase at 240V. Smaller loads are supplied through wall receptacles at 120V. Initially, the phase-to-phase load was assumed to be (3/4) three-quarters of the total load, with the remaining load, (1/4) one-quarter, being connected to the two 120 volt circuits (phase-to-neutral). The presence of 240 volt motors was ignored as the motor's high inductances makes them appear as an open circuit to surges. The assumption is in agreement with other studies [5]. Since the resistive loads were not located at the service entrance panel, the effect of the multiple branch circuits was represented with small inductances of 0.1 uh in series with the resistances.

Upon investigation of the field tests, it was noted that the M1 test site residence load was very light due to the fact that testing was done during the daylight hours on two consecutive midweek days (Tuesday and Wednesday). Specifically, the load varied between test shots, ranging from a low 2.3% to a high of 16.8% of full load (25 kW). Generally, the load was confined to one 120V circuit. The computer studies selected to compare with seven field test shots incorporate the measured loading at M1 just before the test shot was executed. It is expected that the light loading of the transformer will result in higher transient overvoltages, as heavy loading conditions at the measuring site should tend to minimize the transients measured. The other 18 transformers, for which their loads were not measured, were assumed to be loaded to 10% of their ratings for the computer study.

Wallen Station Model

The Wallen Station consisted of four 34.5 kV lines, a 10 kVA, 19.92 kV-120/240V single-phase station auxiliary load transformer, a 7.5 MVA, 34.5-12.47 (GR.-Y) kV transformer feeding three 12.47 kV distribution lines. Using a short circuit study, the station was equivalized per EMTP program to a R-L system equivalent. In addition, the 34.5 kV and 12.47 kV bus capacitances, and the transformer bushing and breaker capacitances were incorporated as lumped capacitances to ground at the station.

Distribution Surge Arrester Model

The surge arrester used on each transformer and at the capacitor bank was a metal-oxide varistor (MOV). The Distribution Class arrester was rated 9/10 kV rms. The corresponding protective characteristic was derived for EMTP using the manufacturer's supplied 8 x 20 us discharge wave data. The maximum switching surge protective level is 20.3 kV crest (2.0 p.u.) at 500 Amperes discharge current. The arrester's energy capability rating was 4 kJ/kV of rated rms voltage or 40 Kilojoules for a 1 minute period.

COMPUTER SIMULATION VS. FIELD TESTS

General Results

In the study, seven test shots were chosen to be simulated using the EMTP program. The detailed computer model, consisting of a large network of linear and nonlinear components, was used to simulate three capacitor switching tests and four line dropping tests. The six voltage parameters for each test shot that were reproduced from specific transient conditions, consistently came within ±10% of the corresponding measured value for the magnitude and the dominant frequency. The worst discrepancy occurred for a capacitor switching case in which the computed Phase C 7.2 kV transient overvoltage was approximately 15% smaller than measured. Also, the secondary 240V transient overvoltage was approximately 15% larger than the measured value. This could be due to accuracy in determining the exact point on the waveform that the switching of the capacitor occurs. The width of the trace line on the Gould graphical plots is approximately 0.15 milliseconds and it is this difference in switching times that could result in slightly different transient voltages being computed.

FIGURES 4A,4B,4C
CAPACITOR TEST: J-10
TRANSIENT SWITCHING CHARACTERISTICS

4A: M1 XFMR: 7.2 KV PRIMARY PHASE C

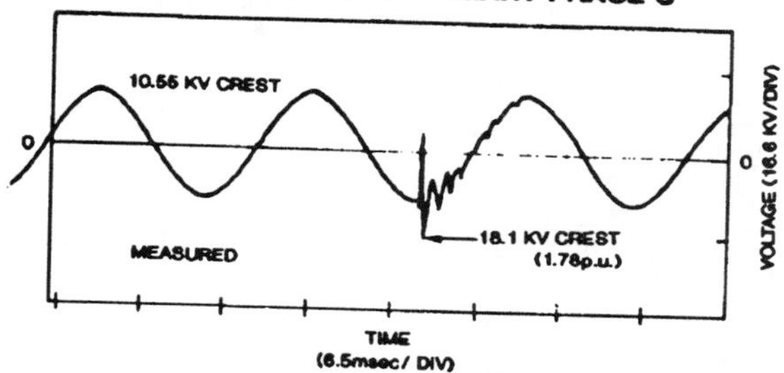

4B: EXPANDED TRANSIENT IN FIGURE 4A

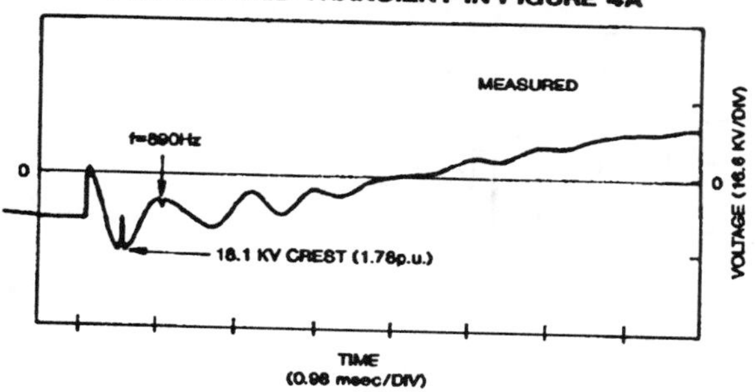

4C: M1: XFMR: 7.2 KV PRIMARY PHASE C

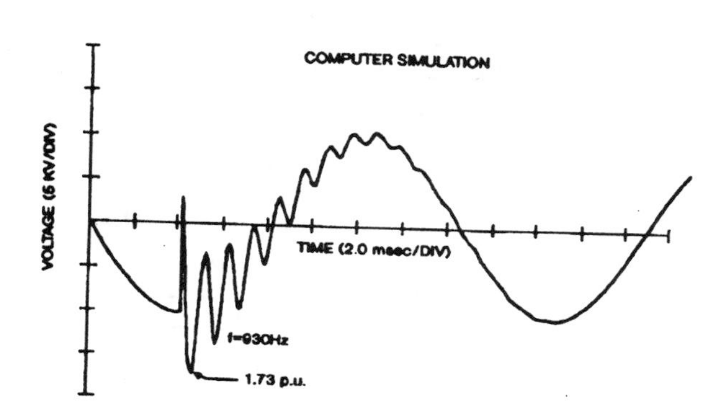

Results From Selected Tests

Due to space limitations, a comparison of the results between field tests and computer simulations was limited to one example of capacitor switching (J - 10) and one of line dropping (F2-K2). The selected test shots are the same as those used in the earlier paper [1] on field test results.

Table 4 and Figure 4 give the summary of capacitor switching tests. Figure 4 illustrates the transient switching characteristics of Phase C primary 7.2 kV voltage at M1 transformer. An expanded transient of Phase C is shown in Figure 4B, and Figure 4C gives the computer simulation results.

The line dropping test (F2-K2) consisted of a sequence of line deenergization with capacitor bank disconnected and line reenergization with no capacitors and no arresters connected. Table 5 shows the results of line reenergization which provided higher transient values.

The results from Tables 4 and 5 are representative of the overall tests. The difference between the field tests and the computer models for voltage parameters is less than 10%. For transient primary and secondary currents, this difference is greater and may be caused by inaccurate modeling of the house load.

PARAMETRIC STUDIES

Studies were completed to note the effects on transients when certain circuit parameters are changed. The parameters under review were: grounding resistance, load size, and type of line model. Table 6 illustrates the effect of changing the capacitor bank ground resistance on transients at various locations on the circuit. Definite variation of the transient magnitudes were observed from the measured value of 4.0 ohms. However, varying the grounding resistance at the transformer pole or at the house did not have much of an effect on the transient's magnitude.

The effect of the secondary load at the measurement transformer was calculated for a 21.8% load and 100% load (25 kW) for line energization transients with and without the capacitor bank connected. The transient overvoltages are approximately 17% lower with

Table 4: CAPACITOR SWITCHING TESTS:
Summary Of Voltage And Current Comparisons

PARAMETER		TEST: J-10 (Detailed Model)				LOAD: 7.8%		TEST: J-10 (Simplified Model)				LOAD: 7.8%	
		FIELD TEST		COMPUTER MODEL		PERCENTAGE DIFFERENCE (% of Field Test)		FIELD TEST		COMPUTER MODEL		PERCENTAGE DIFFERENCE (% of Field Test)	
		Voltage (p.u.crest)	Freq. (Hz.)	Voltage (p.u.crest)	Freq. (Hz.)	Voltage	Freq.	Voltage (p.u.crest)	Freq. (Hz.)	Voltage (p.u.crest)	Freq. (Hz.)	Voltage	Freq.
1. Primary Voltage 7.2kV	Phase A	1.15	1,000	1.14	960	-0.9	-4.0	1.15	1,000	1.14	960	-0.9	-4.0
	Phase B	1.18	900	1.22	930	3.4	3.3	1.18	900	1.20	900	1.7	0.0
	Phase C	1.78	890	1.73	930	-3.1	4.5	1.78	890	1.69	925	-5.1	3.9
2.Secondary Voltage	V_1 (120V)	1.62	1,000	1.65	960	1.9	-4.0	1.62	1,000	1.65	925	1.9	-7.5
	V_2 (120V)	1.71	1,000	1.65	960	-3.5	-4.0	1.71	1,000	1.65	925	-3.5	-7.5
	V_1-V_2 (240V)	1.65	1,000	1.67	960	1.2	-4.0	1.65	1,000	1.65	925	0.0	-7.5

PARAMETER		Steady State Amp crest	Tran-sient Amp crest	Freq. (Hz.)	Steady State Amp crest	Tran-sient Amp crest	Freq. (Hz.)	Steady State % Diff.	Tran-sient % Diff.	Freq. % Diff.	Steady State Amp crest	Tran-sient Amp crest	Freq. (Hz.)	Steady State Amp crest	Tran-sient Amp crest	Freq. (Hz.)	Steady State % Diff.	Tran-sient % Diff.	Freq. % Diff.
3. Primary Neutral Current	i_{TRN}	1.25	30.5	990	1.20	48.6	960	-4.0	59.1	-3.0	1.25	30.5	990	1.11	41.7	890	-11.2	36.7	-10.1
4.Secondary Currents	i_1 (120V)	3.6	4.0	1,000	3.6	6.0	960	0.0	50.0	-4.0	3.6	4.0	1,000	3.7	6.2	925	2.8	55.0	-7.5
	i_2 (120V)	19.2	25.2	900	19.2	26.0	960	0.0	3.2	6.7	19.2	25.2	900	19.0	36.7	925	-1.0	45.6	2.8
	$i_{neutral}$	15.7	35.4	900	15.6	32.0	960	-0.6	-9.6	6.7	15.7	35.4	900	15.3	30.9	925	-2.5	-12.7	2.8

Table 5: LINE DROPPING AND REENERGIZATION TESTS
Summary of Voltage and Current Comparisons

TEST: F2-K2 LOAD: 12.8%		FIELD TEST		COMPUTER MODEL		PERCENTAGE DIFFERENCE	
PARAMETER		Voltage (p.u. crest)	Frequency (Hz.)	Voltage (p.u. crest)	Frequency (Hz.)	(% of Field Test) Voltage	Frequency
1. Primary Voltage 7.2kV	Phase A	1.02	30,000/60	1.02	28,000/60	0.0	-6.7
	Phase B	1.76	30,000/60	1.69	28,000/60	-4.2	-6.7
	Phase C	1.09	30,000/60	1.01	28,000/60	-7.4	-6.7
2. Secondary Voltage	V_1 (120V)	2.13	30,000/60	2.17	28,000/60	1.8	-6.7
	V_2 (120V)	1.68	30,000/60	1.67	28,000/60	-0.5	-6.7
	V_1-V_2 (240V)	1.63	30,000/60	1.75	28,000/60	7.7	-6.7

PARAMETER		Steady-state (A.crest)	Tran-sient (A.crest)	Freq. (Hz.)	Steady-state (A.crest)	Tran-sient (A.crest)	Freq. (Hz.)	Steady-state % Diff.	Tran-sient % Diff.	Freq. % Diff.
3. Primary Neutral Current	i_{TRN}	1.05	2.31	60	0.72	2.32	60	-31.4	0.4	0.0
4. Secondary Currents	i_1 (120V)	< 0.5	< 0.5	---	0.1	0.7	60	---	---	---
	i_2 (120V)	18.5	47.2	60	38.2	43.7	60	107.	-7.4	0.0
	$i_{neutral}$	13.3	37.8	60	38.0	43.8	60	186.	15.9	0.0

the capacitor bank connected regardless of the transformer's secondary load. Increasing the load from 21.8% to 100%, with capacitors on, decreases the overvoltages by less than 1% and no decrease is noticed if the capacitor bank is disconnected. The values were computed for a house ground R = 29 ohms, transformer pole ground R = 12.3 ohms and a capacitor bank ground R = 4 ohms.

For switching surges generated by the line breaker at Wallen Station or the capacitor switches near the measurement site, the EMTP's frequency-dependent Marti line model provided the same results as the simpler frequency-independent Lee Line Model. The Marti phase-modal current transformation matrix is calculated at 5 kHz initially, as recommended by EMTP. Comparison of results calculated using line parameters for the Marti model derived at 1 kHz (near the dominant frequency when the 450 kVAR capacitor bank is connected to the system), reveal similar transient values. The EMTP literature and Line Constants Program Manual state that the frequency independent Lee line model calculation at 60 Hz is accurate for overhead lines for a frequency between 10 Hz and 10,000 Hz. Clearly, the simpler Lee Model should be used as it provides good results for switching surges. The number of data lines is reduced from 100 to 9 per section of distribution line model, and the program runs considerably faster on the computer when using the Lee frequency-independent line model.

Furthermore, computer calculations were completed for both the Lee (60 Hz) and Marti (@ 5 kHz and @ 1 kHz) line models for which the ground resistivity RHO was varied between 10 ohmmeters and 400 ohm-meters. Again, the results are similar. For switching surges on distribution lines, the transients do not depend significantly on the type of line model nor on the ground resistivity.

SIMPLIFIED COMPUTER MODEL RESULTS

Computer simulation of the test results using EMTP program has been achieved with reasonable accuracy. However, the use of EMTP has proven to be very time consuming and expensive. Parametric studies have led to a development of a simplified computer model which can reduce computation time by approximately a 50:1 ratio and yet provide acceptable accuracy. Further refinement and simplification of the EMTP model is possible. Table 4 illustrates the results of a second order of model simplification. A comparison is made in Table 4 between the field test results and the detailed model versus the simplified model. This case covers the capacitor switching test. In comparison to the field tests both models give accurate results for the magnitude and transient frequency of the over-voltages associated with the primary and secondary terminals of the distribution transformer. The detailed model provided transient magnitudes within 4% and dominant frequencies within 5%, when compared to the field tests. The simplified model yielded magnitudes within 5% and frequencies within 8% of the measured values. The steady-state primary and secondary load currents were reasonably accurate, with the detailed model having the greater precision.

Simplified Model

The one line diagram of the simplified second order model is shown in Figure 5. The 19 distributed loads are consolidated into 3 load sites situated at the midpoint and 2 terminal ends of the 1.70 km (1.07 mi.) three-phase distribution line. Load sites 1, 2, and 3 have the appropriate load connected to Phases A and C, along with the three-phase 450 kVAR capacitor

Table 6: CAPACITOR BANK SWITCHING
EFFECT OF CHANGING CAPACITOR BANK POLE GROUNDING RESISTANCE
TEST B-1 LOAD = 21.8%
(House Ground = 29 Ohms, Transformer Pole Ground = 12.3 Ohms)

PARAMETER (Maximum Crest Value)	CAPACITOR BANK GROUNDING RESISTANCE R = (OHMS)												
	0	1	2	4*	5	8	12	15	30	100	500	10,000	INF.
1. Primary 7.2kV Voltage to Ground (kV)	17.32	17.07	16.85	16.47	16.32	15.89	15.48	15.23	14.55	13.79	14.59	16.88	17.06
2. Secondary 120V. to Ground (V.)	318.1	293.6	278.7	259.9	252.9	236.2	220.1	211.0	185.7	184.3	185.2	205.3	208.4
3. Secondary Neutral Voltage to Ground (V.)	43.0	26.07	18.76	14.16	13.44	12.64	12.85	12.95	13.29	13.04	13.73	15.80	15.96
4. House Voltages a. 120V.	278.4	269.6	264.1	249.6	243.1	228.5	211.8	201.8	179.2	176.1	178.5	192.4	194.1
4. House Voltages b. 240V.	549.7	534.6	519.1	490.8	478.5	447.5	417.0	400.0	352.0	349.1	350.7	378.0	384.0
5. Capacitor Bank Neutral to Ground Voltage (V.)	0.0	413.2	796.5	1483.	1791.	2603.	3469.	4000.	5710.	7858.	9416.	10,050	10,200
6. XFMR Primary Current (A.)	4.97	3.02	2.17	1.63	1.55	1.46	1.48	1.50	1.53	1.50	1.58	1.82	1.84
7. Secondary Load Current (A.)	83.3	81.0	78.6	74.3	72.5	67.8	63.2	60.5	53.3	52.9	53.1	57.2	58.1
8. Secondary Neutral Current (A.)	62.7	60.5	58.5	55.2	53.8	51.7	48.1	45.5	40.4	40.0	40.4	43.7	44.1
9. Capacitor Ground Current (A.)	429.2	413.2	398.3	370.8	358.2	325.4	289.1	266.7	190.3	78.6	18.8	1.005	0.0
10. Phase A Capacitor Closing Current (A.) (ph.B to ph.A)	498.7	487.4	477.3	459.8	451.8	430.6	407.5	393.5	346.7	270.1	305.6	462.5	475.3
11. House Load Currents a. 120V.	72.3	70.3	68.2	64.5	62.9	58.8	54.8	52.6	46.3	45.9	46.1	49.7	50.5
11. House Load Currents b. 240V.	11.0	10.87	10.30	9.92	9.77	9.04	8.35	8.17	7.00	7.20	7.20	7.66	8.06
12. Wallen Station Phase B Current (A.)	431.1	415.3	400.3	372.8	360.2	327.2	304.9	305.1	297.8	254.0	297.1	464.4	477.5

* Capacitor Bank Grounding Resistance R = 4 ohms is the value measured in the Field Tests.
1. One Per Unit Voltages: Primary: 10.2 kV crest, Secondary: 169.7 V crest/339.4 V crest
2. INF. = Infinite Ohms

bank at load site 3. Thus, the original 19 transformers, loads, and line sections are reduced to 7 transformers and loads, with 2 three-phase line sections. In addition, the 19 triplex service drops from the transformers to the customers' homes are eliminated.

FIGURE 5
ONE-LINE DIAGRAM OF SIMPLIFIED MODEL

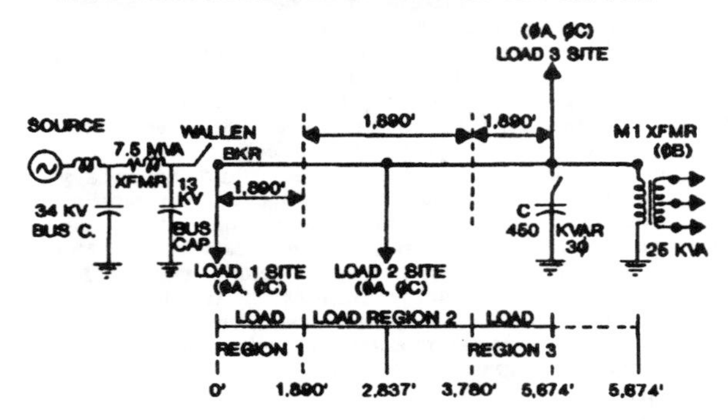

In the parametric studies, the transformer model was not simplified because preliminary investigations have indicated that modeling of hysteresis characteristics are desirable for accurate reproduction of the field test. Also, the transformer model used for the three load sites is the same as the M1 transformer which is connected to Phase B at load site 3. From the parametric studies using the detailed model, it was determined that transient overvoltages were not influenced greatly by the values of resistance for the house and transformer pole grounds. Hence, the measured values of the ground resistances at the M1 site were also used at the other load sites. Furthermore, the Lee Line Model was used assuming a ground resistivity of 100 ohmmeters. The cumulative effect of the model simplification as described above is to yield a 50 times reduction in computer run time while sacrificing little loss of accuracy.

STATISTICAL EVALUATION OF MAXIMUM VALUES

It is unlikely that the field tests with a limited number of switching operations have provided maximum values of overvoltages for each test condition. This type of information could be more readily obtained from computer statistical simulations of the test circuit. Such a study if accurately and reliably performed would then provide the maximum anticipated overvoltages which are needed for live line tool design and new designs of distribution line equipment such as arresters or circuit breakers.

Work was completed on an EMTP model to obtain the maximum overvoltages and overcurrent values by randomly varying the closing times for the capacitor switches and the Wallen Station breaker.

The reduced second order simplified distribution system was used to simulate the random 200 shot three-phase reenergization of the distribution line using the Wallen Station breaker. To achieve the worst case overvoltages and overcurrents, the distribution MOV surge arresters were not connected for these studies.

The statistical closing target time is the maximum Phase B (center) voltage at the capacitor bank or at the maximum Phase B breaker voltage at Wallen Station. The statistical characteristics were derived from the following assumptions and equations. The pole closing span for the switching operations was 6 msec [$3\sigma_P - (-3\sigma_P) = 6\sigma_P$ value]. The random closing of the three phases was allowed to vary over 1 cycle of 60 Hz (16.67 msec) using a Gaussian (normal) distribution. Given that Phase B closes at time $t = t_B$ such that:

$$t_{BP} - 3\sigma_B \le t_B \le t_{BP} + 3\sigma_B \quad (1)$$

where σ_B is the standard deviation of the Phase B random time to close, which was targeted at the peak of the Phase B breaker voltage. The closing times of Phases A and C about the Phase B closing time (t_B) were selected to have a normal distribution with a standard deviation of $\sigma_P = 1.0$ msec (range is 6.0 msec from above assumption), and were calculated as follows:

$$t_B - 3\sigma_P \le t_A \le t_B + 3\sigma_P \quad (2)$$

$$t_B - 3\sigma_P \le t_C \le t_B + 3\sigma_P \quad (3)$$

Calculating the specific values for σ_P, σ_B, t_A, t_B, t_C and t_{BP}, we have:

$$-(-3\sigma_P) + 3\sigma_P = 6\sigma_P = 6.00 \text{ msec} \quad (4)$$

$$\sigma_P = 1.00 \text{ msec} \quad (5)$$

$$-(-3\sigma_B) + 3\sigma_B = 6\sigma_B = (16.67-6.0) = 10.67 \text{ msec} \quad (6)$$

$$\sigma_B = 1.78 \text{ msec} \quad (7)$$

Thus, if we assume that the Phase B target closing time at the peak of Phase B breaker voltage occurred at time $t_{BP} = 14.00$ msec, we have from equations 1, 2 and 3:

$$8.67 \le t_B \le 19.34 \text{ msec} \quad (8)$$

$$8.67 - 3(1.00) \le t_A \le 19.34 + 3(1.00) \text{msec}$$
$$5.67 \le t_A \le 22.34 \text{ msec} \quad (9)$$

$$5.67 \le t_C \le 22.34 \text{ msec} \quad (10)$$

Thus, the window of the three phase closing of the breaker was between the boundaries of 360° of the power frequency fundamental.

The results are shown in Table 7 for the 200 shot statistical capacitor switching case (TEST:J) and for the 200 shot statistical line reenergization case (TEST:F-K). Test J was chosen since it had the largest measured overvoltage (2.29 p.u.) at the M1 site for the primary 7.2 kV line voltage. Test F-K, which consisted of line deenergization and reenergization with the Wallen breaker and without the 450 kVAR capacitor bank in service, had the highest secondary (120/240V) overvoltage of 2.13 p.u. at the M1 transformer.

Table 7 lists the measured field test value and compares it to the EMTP calculated maximum value for the 200 shot closing operations. The results provide good correlation for items A1 to A3 and A5 in Table 7. The secondary load current item A4 is dependent on the house load model. This load is modeled as a linear resistance corresponding to the steady state current in its respective circuit. Under transient conditions, the house load impedance would be modified by some additional percentage of nonlinear resistance and inductance. An additional complication is that these nonlinear impedances are also time varying (oscillatory rate of decay). Thus, developing a load model that is accurate for steady state and transient conditions is difficult at best. Therefore, it is expected that a discrepancy would appear for the secondary load current. In general, the greatest anticipated overvoltage to appear at the primary 7.2 kV line terminals at the M1 site is 2.29 p.u. due to three-phase 450 kVAR capacitor switching using vacuum switches. This agrees with the observation that

Table 7: STATISTICAL SWITCHING OF 450 KVAR CAPACITOR BANK AND WALLEN STATION 12.47 KV LINE BREAKER: 200 SHOTS

Parameter	Measured Maximum Test Value	EMTP Values		
		Maximum	Mean	Standard Deviation
A. J Test: Capacitor Bank Switching: Vacuum Switches, No Arresters				
1. M1 Site: 7.2 kV Line Voltage (p.u.)	2.29	2.15	1.56	0.28
2. M1 Site: 120/240 V. Secondary Voltage (p.u.)	1.73	1.80	1.29	0.20
3. M1 Transformer: Primary Current (A.)	30.50	28.00	17.50	4.51
4. M1 Site: 120 V. Secondary Load Current (A.)	84.00	61.88	44.28	8.82
5. Wallen: 7.2 kV Bus Voltage (p.u.)	2.00	1.96	1.45	0.23
B. F-K Test: Wallen Station Line Breaker Switching: No Capacitor Bank, No Arresters				
1. M1 Site: 7.2 kV Line Voltage (p.u.)	1.89	1.96	1.53	0.25
2. M1 Site: 120/240 V. Secondary Voltage (p.u.)	2.13	2.23	1.60	0.25
3. M1 Transformer: Primary Current (A.)	44.30	46.71	36.47	5.91
4. M1 Site: 120 V. Secondary Load Current (A.)	80.40	68.90	57.50	7.12
5. Wallen: 7.2 kV Bus Voltage (p.u.)	1.83	1.85	1.42	0.20

vacuum switches yielded greater overvoltages than oil switches.

The line reenergization case yielded a slightly lower primary line (7.2 kV) overvoltage maximum of 1.96 p.u. However, the secondary 120/240V voltage maximum of 2.23 p.u. at the residents' circuits occur for the line reenergization case with no capacitor bank connected. This is in agreement with the field test results which show that the line reenergization with the capacitor bank connected can significantly reduce secondary 120/240V overvoltages (Cases C-D,F-H vs. Cases C-G,F-K) as shown in Table 2 of Reference [1].

CONCLUSIONS

1. In comparison with field tests, good agreement was achieved in modeling the distribution transformer, lines, and associated equipment.

2. The capacitor bank ground resistance has a significant influence on the magnitude of transients caused by capacitor switching.

3. Transformer pole ground and house ground resistances do not influence significantly the magnitude of transients due to distribution switching surges.

4. The magnitude of the load connected to distribution transformer has little effect on the transient overvoltage due to distribution switching surges.

5. It was determined that the Lee Line model (frequency-independent) provided the same results for switching surges as the Marti (frequency-dependent) Line model. In addition, the results for both line models were not significantly affected by a change in the value of the ground resistivity, RHO.

6. Changing the capacitor bank size from 450 kVAR to 900 kVAR further reduces the dominant transient frequency from 1,000 Hz to 700 Hz. The frequency is inversely proportional to the square root of the relative change in the capacitance of the bank.

7. Spectrum analysis of the highly nonlinear transformer inrush and secondary load currents provided relatively large 3rd (67%) and 5th (44.8%) transient harmonic levels, as a percentage of a 1.0 p.u. 60 Hz fundamental component.

8. A simplified distribution system model utilizing the frequency-independent Lee Line model looks promising to provide the needed magnitude and frequency accuracy. The 19 transformers and their corresponding loads could be equivalized and lumped at 2 or 3 locations along the radial line to reduce the computer run time considerably. The cumulative effect of model simplification for this case was a 50 times reduction of computer execution time.

9. Maximum anticipated overvoltages were obtained from the simplified distribution model by varying the switching angles. For the 200 shot random closing of the capacitor vacuum switches, the maximum line overvoltage was 2.15 p.u., whereas the measured value was 2.29 p.u. Similarly, for a 200 shot random closing of the Wallen Station breaker, the maximum was 2.23 p.u., with the measured value being 2.13 p.u.

10. Further work is needed to develop a load model which could provide more accurate results during steady state and transient conditions.

REFERENCES

[1] N. Kolcio, J. A. Halladay, G. D. Allen, E. N. Fromholtz, "Transient Overvoltages and Overcurrents On 12.47 kV Distribution Lines: Field Test Results", IEEE-PES 1991 Winter Meeting, Paper 91 WM 097-6-PWRD, February, 1991.

[2] "Electrical Surges on Distribution Systems - 12 kV", British Columbia Hydro and Power Authority Canadian Electrical Assoc. Report 77-48B, April 20, 1981.

[3] "Electrical Surges on Distribution System", IREQ, Canadian Electrical Association Report, 77-48A, March 19, 1980.

[4] J. L. Puri, N. C. Abi-Samra, T. J. Dionise and D. R. Smith, "Lightning Induced Failures in Distribution Transformers", IEEE-PES 1987 Summer Meeting, Paper 87SM537-4, July, 1987.

[5] R. C. Dugan and S. D. Smith, "Low Voltage Side Current-Surge Phenomena in Single-Phase Distribution Transformer Systems", IEEE-PES T & D Conference and Exposition, Paper 86T&D553-2, September, 1986.

BIOGRAPHY

The authors biographies are listed in the paper entitled, "Transient Overvoltages and Overcurrents on 12.47 kV Distribution Lines: Field Test Results." IEEE-PES 1991 Winter Meeting, Paper 91 WM 097-6-PWRD, February, 1991.

Discussion of IEEE Paper 92 WM 273-3-PWRD

GEORGE GELA and ANDRE LUX, HVTRC, Lenox, MA: The authors should be congratulated on the thorough study of the application of the EMTP to a distribution line problem. The authors investigated several line models and concluded that a fairly simple representation of a distribution line suffices to provide sufficiently accurate computed results. Accuracy evaluations based on comparisons with experimental data (which is the only ultimate accuracy test) is always difficult, since many experimental uncertainties exist, and not all details of the real world can be easily simulated. The discussers agree that a discrepancy of less than 10% between computed and measured results, is a good state of affairs.

It is interesting to note that the good accuracy was obtained even with the frequency-independent Lee line model. Careful analysis of the distribution line problem shows that this finding should not be surprising.

The lack of sensitivity of the computed results to the values of the ground resistivity ρ suggests that the Carson's ground return impedance (the only line parameter that uses ρ), is not a very important contributor to the line model. This conclusion is very significant when one recognizes that the skin effect in the ground is typically far more important than the skin effect in the conductors of the overhead distribution line. As a result, one is led to conclude that skin effect is not an important aspect in calculation of transients on distribution lines. With skin effect "out of the picture", one is left with line parameters (resistance R, inductance L and capacitance C; conductance G is negligible for aerial lines) which are independent of frequency. Then, line series impedance Z and shunt admittance Y are linear function of frequency, and calculation of transients is simplified significantly. The authors' discussions in this regard are very valuable, since it not only demonstrates the applicability of the simpler line model, but also provides a high degree of assurance that the simpler model is appropriate for typical distribution lines.

The next level of simplification investigated by the authors, where the 19 transformers were consolidated into 3 load centers, is also a valuable exercise. Consolidating, or lumping transformers has the effect of neglecting the travel time of the transients between the individual transformers. As discussed previously in connection with Reference 1 of the paper, travel time for short distribution lines (1 mile in this case) is very small compared to dominant transient frequencies and waveform rise times. Therefore, one should be justified in neglecting the travel times and representing the line by lumped parameters. The authors' results seem to confirm this expectation. In view of this, however, an interesting question arises. Since the line can be approximated by lumped circuits, is it possible to use programs that solve lumped circuit problems, in order to calculate the transients at the line ends, or at some intermediate points? A popular circuit program is PSPICE, and the EMTP can also solve lumped circuits. Have the authors tried to use the simple lumped parameter line representation, and if so, what were the findings? Further, if the authors did perform a lumped parameter analysis, the answers to the following questions would be greatly appreciated.

1) What are the parameters which the authors used and what are typical values?

2) Based on a lumped parameter analysis: what would be typical expected risetimes, frequencies, etc.; what were calculated; and what was measured?

3) In the lumped parameter analysis, did the authors take into account coupling between the primary conductor and the neutral? What affect would the mutual impedance between the primary and neutral have on the propagation of the transients?

Manuscript received February 25, 1992.

N. Kolcio, J. A. Halladay, G. D. Allen and E. N. Fromholtz. The authors wish to thank Drs. G. GELA and A. LUX for their discussion and kind words regarding the paper's results. Specifically, the authors appreciate the valuable comments with respect to the parametric studies presented in the paper. Here, the discussers are in agreement with the paper's conclusion which states that for distribution systems, many load centers can be consolidated into several (2 or 3) load locations. A reduction of the computer run time is one of the advantages of this simplification of the system model. In addition, a simplified distribution line system eliminates the need to model very short line segments without sacrificing accuracy.

In the authors' opinion, programs that solve lumped circuit problems can be used to calculate transients at the line ends or at some intermediate points. This may be done for the switching phenomena in which the dominant frequency is below several kilohertz. A typical example is capacitor switching overvoltage of approximately 1,000 hz. It is envisioned that lumping all the distribution components would introduce a somewhat greater error in the transient calculations than the consolidation into two or three load centers. The error can be attributed to the PI-section line model L's and C's which could introduce an artificial "ringing" frequency. Thus, some limitations may occur for magnitude and waveshape simulation of the field tests. It is likely that modeling higher (above several tens of kilohertz) frequency disturbances, such as lightning or line dropping, with PI-section line models would introduce much greater errors and render this procedure impractical.

The authors have not investigated a total lumped parameter solution using EMTP. Perhaps, this work could be accomplished in the future. The authors would like to encourage further investigations of the total lumped parameters solutions.

Manuscript received April 13, 1992.

IEEE Transactions on Power Delivery ,Vol. 6, No. 3, July 1991

OVERVOLTAGE STUDIES FOR THE ST-LAWRENCE RIVER 500-kV DC CABLE CROSSING

Q. Bui-Van - G. Beaulieu - H. Huynh (Member)
Hydro-Québec
V.P. Planification du réseau
Montréal, Canada

Roger Rosenqvist (Member)
ABB Power Systems Inc.
Advanced Systems Technology
Pittsburgh, PA

ABSTRACT

An extensive insulation coordination study has been conducted for the 500-kV d.c. cables that will be installed in a tunnel under the St.-Lawrence river. These cables will be part of the Radisson-Nicolet section of the Quebec-New England HVDC line. This paper provides an overview of the studies and the phenomena that will cause decisive overvoltage stresses on the cable insulation. The basic assumptions and the simulations of these phenomena with the DCG/EPRI-released Electromagnetic Transients Program (EMTP) are also discussed. Finally, the results of the overvoltage studies and the specified insulation levels for the d.c. cables are presented.

INTRODUCTION

Phase II of the Quebec - New England HVDC system consists of a 1,500-km long bi-polar d.c. line and five converter stations - three in Quebec: Radisson, Nicolet, and Des Cantons; and two in New England: Comerford and Sandy Pond. The d.c. line between Radisson and Nicolet crosses the St.-Lawrence river at Grondines on the north shore and Bois-des-Hurons on the south shore. Figure 1 shows the geographical location of the multi-terminal d.c. link.

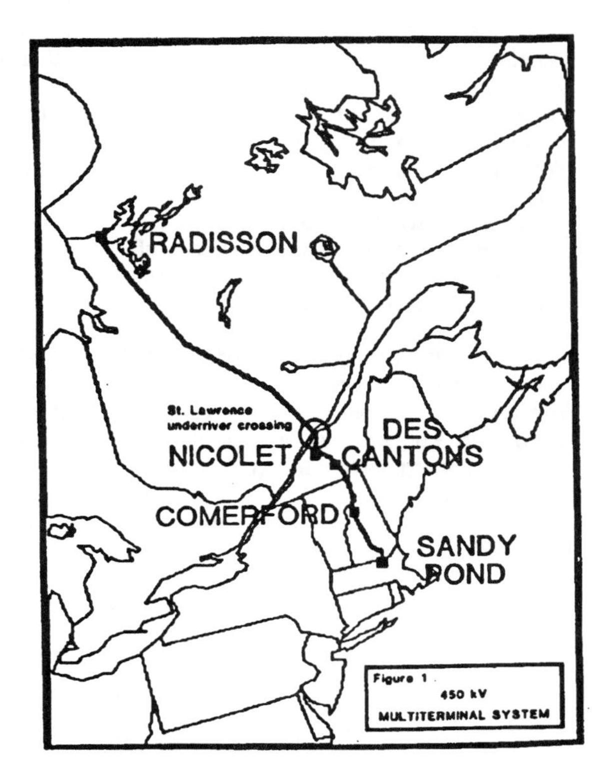

Figure 1. Quebec - New England HVDC Line

The approximately 4-km long cable section at the St.-Lawrence under-river crossing will be the first commercial d.c. cables operated at 500-kV. The integration of these relatively short cables requires special attention due to the unique characteristics of the application and the many possible operating conditions for the multi-terminal HVDC system. Due to the non-self-restoring characteristics of cable insulation, cable failures require long repair times. Therefore, extensive overvoltage studies were conducted to ensure an adequate insulation coordination of the d.c. cables for all practical operating conditions.

The following sections of this paper provide a summary of the studies that were performed with the

91 WM 121-4 PWRD A paper recommended and approved by the IEEE Transmission and Distribution Committee of the IEEE Power Engineering Society for presentation at the IEEE/PES 1991 Winter Meeting, New York, New York, February 3-7, 1991. Manuscript submitted August 28, 1990; made available for printing January 3, 1991.

0885-8977/91/0700-1205$01.00©1991 IEEE

DCG/EPRI-released EMTP to determine the maximum lightning and switching overvoltages on the cables. The required insulation levels for the cables are also provided.

LIGHTNING OVERVOLTAGES

For lightning strokes on the overhead line sections of a combined overhead line/cable transmission system, lightning surges will propagate to the cable terminals and transmit part of their discharge energy to the cables. In general, for long cables, since the surge impedance for a cable is much lower than for an overhead line, lightning impulse overvoltages will be significantly reduced and damped by the cable resistance as they penetrate into the cable. In the case of a short cable section such as the St.-Lawrence under-river crossing, however, multiple wave reflections at the line-cable junctions could result in high overvoltages stressing the cable insulation.

Lightning strokes that occur on the overhead line far away (i.e., several kilometers) from the cable terminals are less harmful to the cable insulation than close-in strokes since a large portion of the lightning discharge energy is lost along the overhead line.

Lightning surges are caused by the following phenomena:

- Back flashovers:

Lightning strokes of very high magnitudes (in the order of hundreds of kA) on overhead shield wires or towers may generate overvoltages of sufficient amplitudes to sparkover the pole conductor insulator strings. When this happens, part of the surge current will be transferred to the cables via the arc across the insulator string. Due to the d.c. voltages on the two poles of the line, back flashovers will occur only on the pole where the polarity of the d.c. voltage is opposite to that of the lightning stroke. Since the major part of the stroke current flows into the ground during back flashovers, the quality of the tower footing grounding has a major impact on the overvoltages on the cables.

- Direct lightning strokes:

Lightning strokes of low magnitudes (a few tens of kA) can bypass the overhead shield wires and strike directly on one of the two pole conductors. These direct lightning strokes will penetrate into the cables via the pole conductors. The worst case condition will occur when a direct stroke generates an overvoltage which is just below the flashover voltage for the pole conductor insulator string. In this case, the tower footing grounding will have no effect on the overvoltage stresses on the cables.

MODELING OF LIGHTNING SURGE PHENOMENA

The lightning surge studies were conducted on the equivalent system in Figure 2.

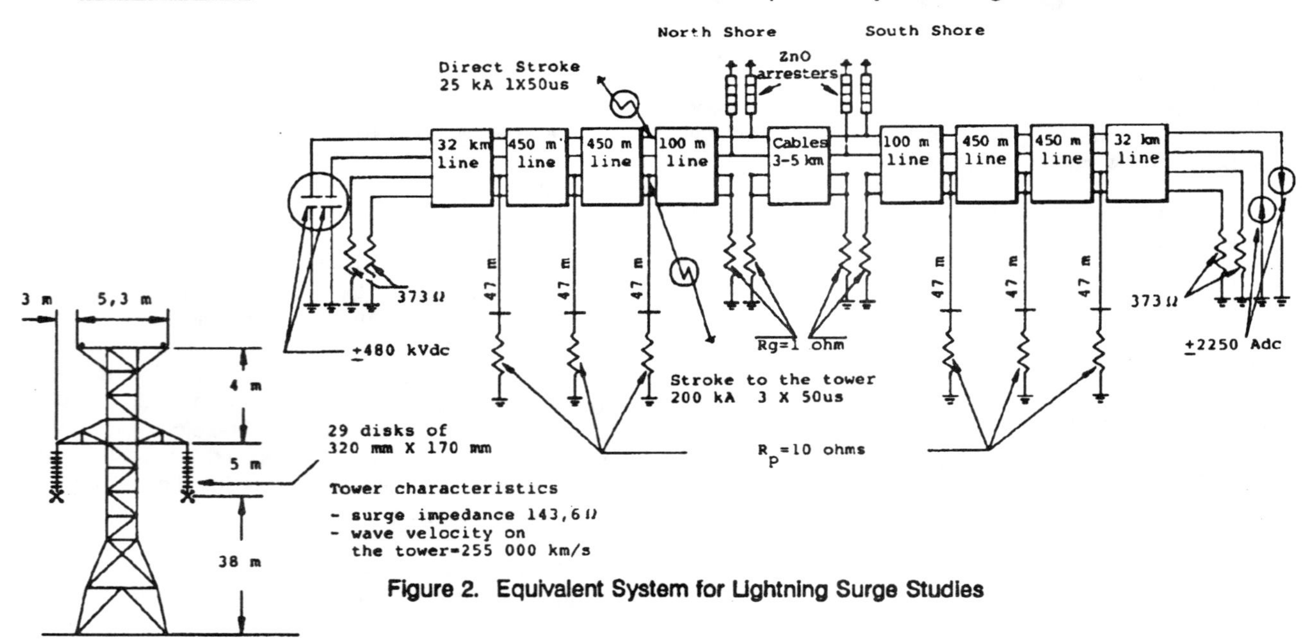

Figure 2. Equivalent System for Lightning Surge Studies

The equivalent system consists of the following elements:

The sections of the bi-polar overhead line (including overhead shield wires) connected to the cable terminals: These line sections were modeled by their frequency dependent parameters (J. Marti's model) [1]. The first three towers closest to the cable terminals were represented in detail, including the tower grounding.

The underwater d.c. cables: The cables were also represented by J. Marti's model. The preliminary cable dimensions are shown in Figure 3.

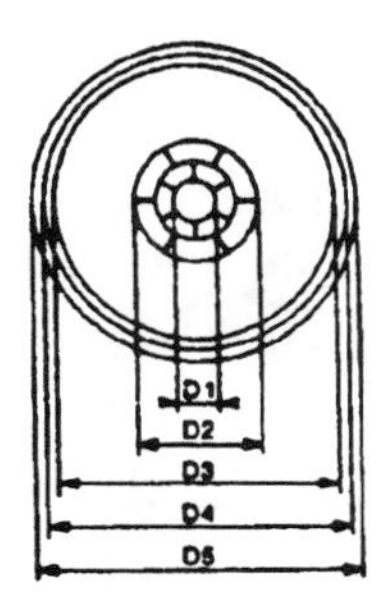

	D1 mm	D2 mm	D3 mm	D4 mm	D5 mm	S mm²	Zc Ω
Cable A	12	52	96	108	119	2000	19.1
Cable B	24	52	102	110	120	1600	21.0

S: Conductor cross-section
Z_c: Cable surge impedance

Figure 3. Preliminary Cable Dimensions

BACK FLASHOVERS

Basic Assumptions

The analysis of lightning overvoltages on the cables as a result of back flashovers was based on the following assumptions:

1. The amplitude of lightning strokes causing back flashovers on the d.c. line can be as high as 200-kA with a 3/50-μs waveform. This maximum stroke current was estimated from the results of the d.c. line lightning outage rate calculations, and the calculated back flashover rate was 0.06/100-km per year. The sparkover voltage of the insulator string for a 3-μs front was estimated to 2,650-kV.

2. The worst case cable overvoltage conditions will occur for back flashovers at the tower closest to the cable terminals.

3. Metal-oxide surge arresters with the following protective characteristics will be installed at the cable terminals:

   - Lightning impulse protective levels

     1,220-kV (10-kA, steep front surge)
     1,130-kV (10-kA, 8 x 20-μs)

   - Switching impulse protective level
     960-kV (3-kA, 30 x 60-μs)

4. A cable length of 4-km. The length of the under-river cable crossing could vary between 3 and 5-km depending on the final locations of the cable terminals. At the time when these studies were conducted, the final locations of the cable terminals were still under review. Therefore, an average cable length of 4-km was assumed.

5. The footing resistance of the towers closest to the cable terminals was estimated to be 10-ohms. This resistance was based on a preliminary investigation of the soil conditions in the area around the cable terminals.

6. Under normal operating conditions, there are two cables in parallel per pole. However, during contingency conditions, operation of the under-river cable crossing with one cable per pole at reduced pole current is foreseen.

Overvoltages on the Cables

The results of the back flashover simulations showed several rapid voltage polarity reversals and maximum positive and negative crest voltages of 980-kV and -800-kV respectively; see Figure 4. The overvoltage waveforms are characterized by an initial and rapid polarity reversal of about 10-μs followed by high frequency oscillations of about 10-kHz. This particular phenomenon is due to the short length of the cables and has not been dealt with in the C.I.G.R.E. WG21.01 recommendations for tests of power transmission d.c. cables [2]. Therefore, special tests will have to be performed to verify the performance of the St.-Lawrence under-river cable insulation for this type of stresses.

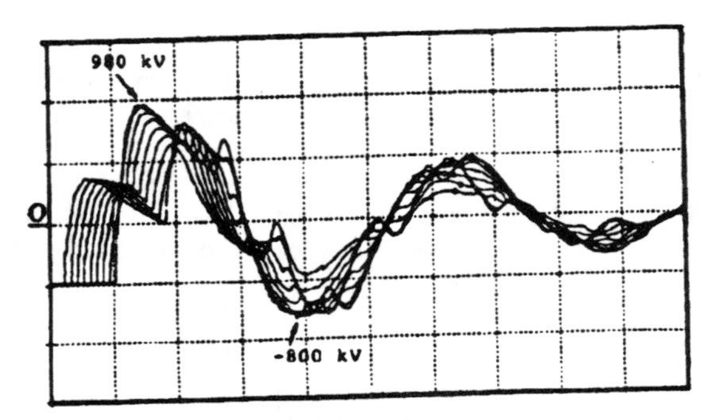

Figure 4

Overvoltages at every 500 m along the cable including both cable terminals. Back flashover at the tower closest to the cable terminal for base case conditions, i.e.:

- cable length = 4 km
- one cable "B" per pole
- tower footing resistance = 10 ohms

Vertical: 500 kV/div. - Horizontal:30 us/div.

Sensitivity Analysis

The effect on the overvoltages by variation of the system parameters was investigated.

1. Effect of cable dimensions

As shown in Figure 3, two preliminary cable designs were analyzed: cable "A" and cable "B" having conductor cross sections of 2,000-mm² and 1,600-mm² respectively. The overvoltage stresses on cable "B" were about 7% higher than those on cable "A".

2. Effect of the cable length

To study the effect of different cable lengths, simulations were carried out for 3, 5, 12 and 18-km long cables. Although the real length of the under-river cable crossing will be between 3 and 5-km, simulations were carried out with cable lengths of up to 18-km for analyzing purposes. The results in Figure 5 indicate that:

- the overvoltages on the cables decrease significantly when the cable length increases.
- the high frequency oscillation in the cable voltage is quickly damped with increasing cable length, thus, reducing the number of polarity reversals.

The results confirmed that the length of the cables has a significant impact on the insulation coordination. The St.-Lawrence cables are short compared to those of existing d.c. links. Therefore, the insulation coordination methods applied to previous d.c. cable projects are only in part applicable to the St.-Lawrence under-river cables.

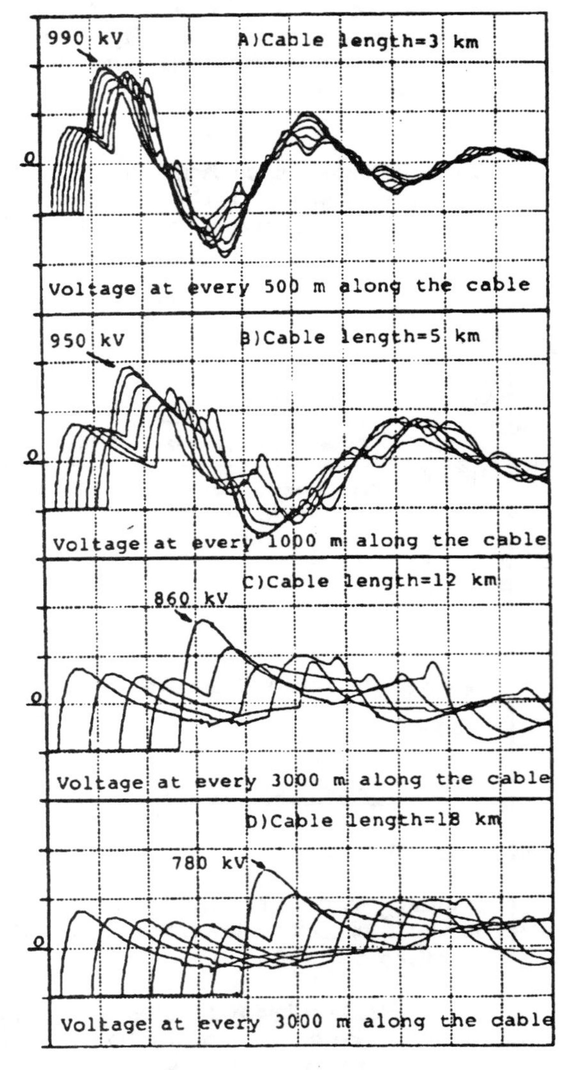

Figure 5

Effect of the cable length-One cable "B" per pole and tower footing resistance of 10 ohms

Vertical=500 kV/div. - Horizontal 30 us/div.

3. Effect of the distance between the back flashover and the cable terminals

To analyze the effect of the distance between the back flashover and the cable terminals, stroke currents were injected on the towers located 100, 550, 1,000, and 1,450 meters away from the cable terminals. The results indicate that increasing distance between the back flashover and the cable terminals significantly reduces the overvoltage stresses for distances of 1-km and above. Furthermore, when the back flashover is moved away from the cable terminals, the frequency of the voltage oscillations and the number of polarity reversals decrease significantly.

The results confirm the assumption that lightning strokes near the cable terminals cause the most severe stresses on the cable insulation.

4. Effect of tower footing grounding

During a back flashover, the major part of the stroke current flows into ground via the overhead shield wires and towers and, therefore, overvoltages on the cables can be reduced by using an effective tower footing grounding. Simulation results with different tower footing resistance showed that the overvoltage stresses on the cables can be reasonably controlled for tower footing resistances of 10 ohms or lower; see Table I. Since lightning strokes far away from the cable terminals are not as severe for the cable insulation, a low tower grounding resistance is required only for the four or five towers closest to the cable terminals.

5. Effect of parallel cables

The base case conditions for the lightning surge studies assumed that there is only one cable per pole. Under normal operating conditions, however, there will be two cables in parallel per pole and the reduced surge impedance of the cables in parallel is expected to limit the overvoltage stresses on the cables. Simulation results with two cables in parallel per pole confirmed this assumption; see Table I (cases 4, 5 and 6 versus 1, 2, and 3).

DIRECT LIGHTNING STROKES

The worst case conditions will occur when a direct stroke generates an overvoltage at the striking point which is just below the flashover voltage for the pole conductor. This case corresponds to the injection of a direct lightning stroke of 25-kA with a 1 μs front. The simulation results revealed that:

1. the cable overvoltages will have waveforms of lightning impulses superimposed on the initial d.c. voltage. The crest value can be as high as 1,200-kV: see Table II, case 1;
2. there is no high frequency oscillation in the voltage;

TABLE I
OVERVOLTAGES ON THE CABLES DURING BACK FLASHOVERS

Case No.	Length Of The Cables	Number Of Cables Per Pole	Cable Dimensions (2)	Tower Footing	Max. Positive Overvoltage (kV Crest)	Max. Negative Overvoltage (kV Crest)
1(1)	4-km	1	Cable B	Rp = 10 ohms	980	-800
2	4-km	1	Cable B	Rp = 25 ohms	1,325	-925
3	4-km	1	Cable B	2 counterpoise	1,300	-875
4	4-km	2	Cable B	Rp = 10 ohms	750	-720
5	4-km	2	Cable B	Rp = 25 ohms	1,000	-850
6	4-km	2	Cable B	2 counterpoise	1,000	-900

Notes 1 : basic system parameters.
2 : see Figure 3.

3. when direct lightning strokes are of opposite polarity to the d.c. voltage, polarity reversals are much less severe than those observed during back flashovers;

4. the overvoltages are less severe with two cables in parallel per pole: see Table II (cases 2 and 4 versus 1 and 3).

INTERNALLY GENERATED OVERVOLTAGES

Overvoltages can be generated by events within the transmission system itself, such as short circuits, switching, and converter malfunctions. Three types of internally generated overvoltages were considered in the St.-Lawrence under-river cable study, i.e. :

1. overvoltages resulting from pole-to-ground faults on the d.c. system;

2. overvoltages resulting from energization of an open ended line; and

3. overvoltages caused by injection of large a.c. voltages into the d.c. system.

POLE-TO-GROUND FAULTS

During pole-to-ground faults on a bipolar d.c. system, overvoltages will be induced on the healthy pole. The magnitudes of these overvoltages and their impact on the cable insulation, therefore, need to be studied. In addition, for a transmission system incorporating short cable sections, it also becomes necessary to study cable voltage polarity reversals on the faulted pole.

Polarity Reversals on the Faulted Pole

When a pole to ground fault occurs on the overhead line close to the cable terminals, the discharge of the cable will result in a high frequency oscillation and a rapid polarity reversal at the other cable terminal. The steady state d.c. voltage prior to the fault will not be higher than the rated voltage of the cable. However, a pole-to-ground fault may be connected to a temporary or transient overvoltage condition on the d.c. system and the voltage increase under such conditions, combined with a polarity reversal, can affect the requirements on the cable insulation.

The studies of the cable voltage polarity reversal assumed a pre-fault transient overvoltage of minus 775-kV (this value covers most types of overvoltage conditions on the d.c. system). The cables were simulated with a detailed model similar to the one used for the lightning overvoltage studies. The results show (Figure 6) a rapid polarity reversal and a voltage magnitude of +820-kV after the initial polarity reversal. The resulting stress on the d.c. cable insulation, however, is less severe than observed in the lightning back flashover studies.

Induced Overvoltages on the Healthy Pole

This type of overvoltage was studied with EMTP for two-terminal configurations in which Radisson was operating as rectifier (the only possible mode for this terminal).

TABLE II
OVERVOLTAGES ON THE CABLES DUE TO DIRECT LIGHTNING STROKES

Case No.	Length Of The Cables	Number Of Cables Per Pole	Cable Dimensions ([1])	Stroke Current Striking The Positive Pole	Max. Overvoltage On The Cable (kV Crest)
1	4-km	1	Cable B	+25-kA, 1 x 50-μs	1,200
2	4-km	2	Cable B	+25-kA, 1 x 50-μs	1,060
3	4-km	1	Cable B	-25-kA, 1 x 50-μs	-600
4	4-km	2	Cable B	-25-kA, 1 x 50-μs	-160

Note 1 : see Figure 3.

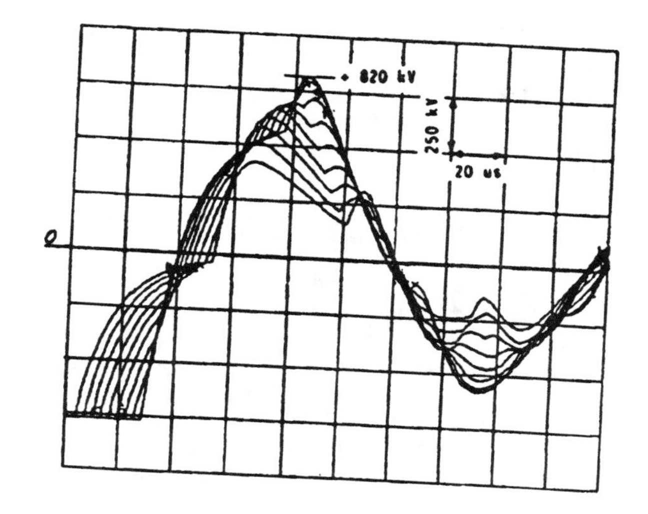

Figure 6.
Polarity Reversal Along the Faulted Pole Cable when Subjected to a Pre-Fault Transient Overvoltage

Figure 7 depicts the Radisson - Nicolet configuration. The bipolar line was represented by distributed parameter models and the d.c. cables were considered electromagnetically and electrostatically isolated from each other. The converters were modeled by a simplified equivalent using a current-controlled d.c. voltage source (EMTP type 16) to simulate the performance of the converter controls. Additional simulations were run with a more detailed converter model developed by IREQ to validate the results of the simplified model. Pole-to-ground faults were applied at several points along the line as well as at the cable terminals.

The highest induced overvoltages were found for faults near the cable terminals. Figure 8 summarizes results for the Radisson-Nicolet configuration. For the Radisson to Sandy Pond configuration, induced overvoltages on the healthy pole of the cable were slightly higher, reaching 1.55 p.u. (775-kV). Variation of the cable length from 3 to 5 km appeared to have no significant impact on the cable overvoltages.

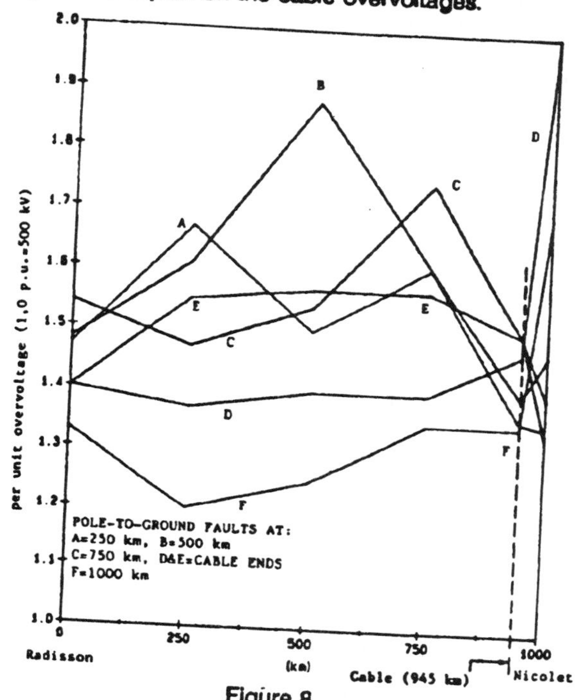

Figure 8.
Induced Overvoltage Profile on the Healthy Pole for Pole-to-Ground Faults in the Radisson-Nicolet Configuration

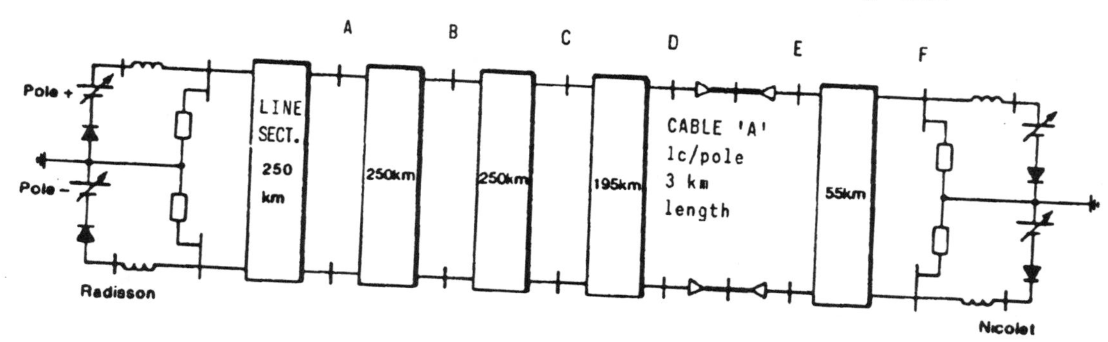

Figure 7.
Equivalent of the Radisson - Nicolet Configuration for Pole-to-Ground Fault Simulations

ENERGIZATION OF AN OPEN ENDED LINE

Switching of an Open Ended Line

A d.c. line is normally energized through a relatively slow rise of the converter voltage. However, the introduction of the sectionalizing switches in the Quebec - New England multi-terminal scheme provides a possibility of energizing the Radisson - Nicolet section of the line by closing the sectionalizing switch at one of the converter stations at the south portion of the system. In this case travelling wave phenomena will cause a transient component to be superimposed on the d.c. voltage, thus, creating an overvoltage condition.

A condition that may occur in practice is that one of the poles of the line-cable-line section is energized from the Des Cantons terminal, while Radisson at the north end of the line is blocked. During such conditions the line will be terminated by the d.c. filters only. TNA simulations conducted at IREQ for this configuration revealed overvoltages of 1.55 p.u. at the cable location. The pre-insertion resistors that will be installed on the d.c. switches reduced these overvoltages to 1.3 p.u. (300 ohms resistors and a pre-insertion time of 8 ms).

Deblocking of Radisson at Maximum Voltage

Deblocking of Radisson at maximum d.c. voltage with an unloaded line would generate higher overvoltages on the cable. However, with modern control systems, there are so many contingencies involved that the probability of such an event can be disregarded. Normal de-blocking of the rectifier is permitted by the control system only if an inverter is already connected to the system. At the deblocking of a converter, the control system will let the voltage rise gradually over a long period of time as compared to the natural frequency of the d.c. line, thus preventing any considerable overshoot in the voltage. If the inter-station telecommunication system is out of service and the system operators attempt to deblock the converters in the wrong order, other functions, such as the d.c. side voltage limiter, will act to limit the overvoltages.

A situation which would be more likely to occur is a restart of a rectifier against an open ended line due to protective blocking of the inverter without by-pass pairs (only for two-terminal configurations). If the inter-station telecommunication system is out of service at the same time, the rectifier will be retarded by the line protection and subsequently attempt to restart after a short delay. To prevent an overshoot of the voltage for such conditions, the overvoltage limiter will act to decrease the voltage. According to experience from HVDC systems already in operation, the maximum voltage trapped on the line and cable would be limited to about 1.4 p.u.

D.C. AND SUPERIMPOSED A.C. OVERVOLTAGES

Substantial alternating voltages may appear on the d.c. line and cable due to converter station faults and malfunctions. The most common type of disturbance is a line-to-ground fault on the a.c. system which produces unbalanced voltages at the converter a.c. bus. Due to the actions of the d.c. converters, the fundamental negative sequence a.c. voltage will be frequency transformed to 120 Hz on the d.c. side.

Large sustained 60 Hz voltages can be injected to the d.c. system due to converter malfunctions. For the purpose of determining the insulation requirements for the St.- Lawrence d.c. cables, a fault condition resulting in a non-conducting valve in one of the 6-pulse groups was considered.

These alternating voltage components can be amplified along the d.c. system if a resonance condition exists. With a multi-terminal d.c. system, it may not be possible to avoid resonance conditions near the fundamental frequency and the second harmonic because of the numerous operating modes.

The resonance conditions can be quantified by comparing the amplitude of the a.c. voltage at the cable location with the injected source voltage (amplification factor).

A detailed study to identify the resonance conditions and determine the highest a.c. voltage components that may appear on the St-Lawrence under-river cables was conducted with the DCG/EPRI-released EMTP. A summary of the study methodology and the results is presented in the following sections.

Study Methodology

To simulate large a.c. signal injection on the d.c. system, the converters were represented by a passive circuit using a.c. and d.c. voltage sources connected in series with the impedance of the a.c. system and the converter transformers as seen from the d.c. side during commutation. This simple representation of the converter and the a.c. system should give conservative results [3].

The sources representing the voltage injection at the converters were adjusted to reproduce conditions during sustained disturbances. For the 60 Hz injection, a missing valve in one 6-pulse group was assumed. For the 120 Hz, an a.c. single-line-to-ground fault close to the converter bus will give the highest injection.

Equivalent source voltages representing the disturbances were combined with different d.c. operating configurations. Resonance conditions at the d.c cable were investigated for five multi-terminal configurations and their three possible operating modes; i.e., bipolar, monopolar-ground-return and monopolar-metallic-return. Different d.c. filter outage conditions were also considered in order to find the maximum voltage amplification. Finally, the system resonance frequency was shifted to maximize the amplification factor by varying the short circuit impedance of the converter buses between the minimum and maximum values.

Results

1. <u>Overvoltages due to d.c. and superimposed 60 Hz voltages</u>

Simulation results showed that the amplification of the 60 Hz component is most severe for the Radisson-Nicolet configuration in monopolar-ground-return mode. The amplification factor is affected by the grounding status of the other pole, which was ungrounded for the worst case conditions. The most severe overvoltage stresses were revealed for 60 Hz injection at the Nicolet converter. Figure 9 provides the amplification factor and time simulation plots for this case. The maximum 60 Hz plus d.c. voltage at the cable can be expressed by the following equation:

$$U_{60}\ (kV) = 395 + 540 \cdot \sin(377 \cdot t)$$

The maximum duration of this overvoltage was determined based on the operating time for the slowest back-up protection (d.c. harmonic protection). This time is less than 600 ms.

2. <u>Overvoltages due to d.c. and superimposed 120 Hz voltages</u>

From simulation results it was found that the highest amplification of the 120 Hz component was obtained for the Radisson-Nicolet configuration in monopolar-metallic-return mode. The most severe overvoltage stresses were obtained for 120 Hz injection due to a single-phase-to-ground fault on the a.c system at Radisson.

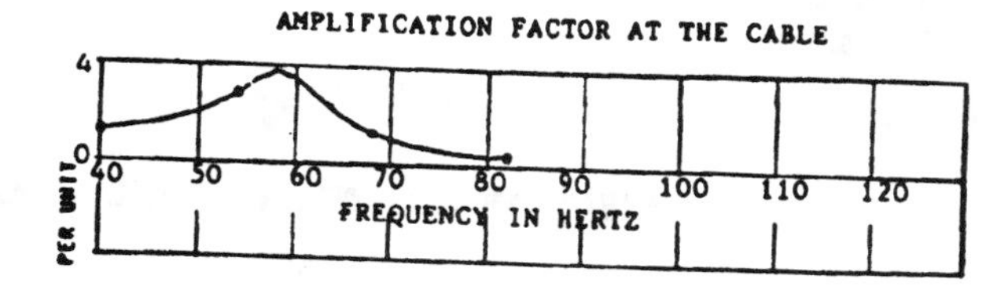

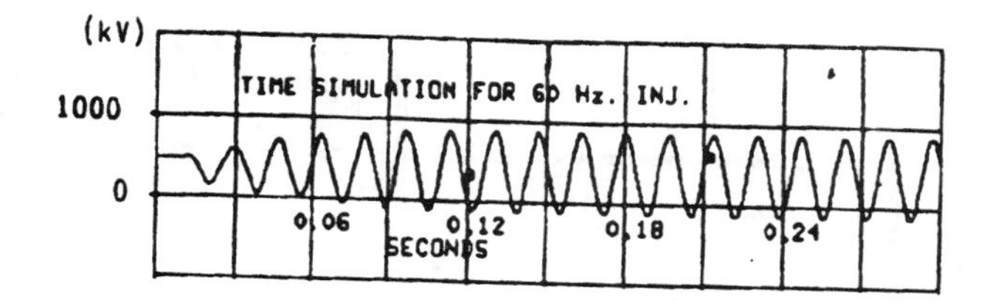

Figure 9.
Frequency Scan and Time Simulation for 60 Hz Injection

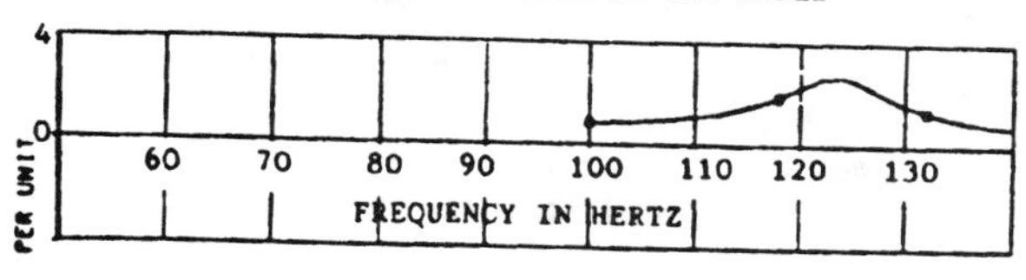

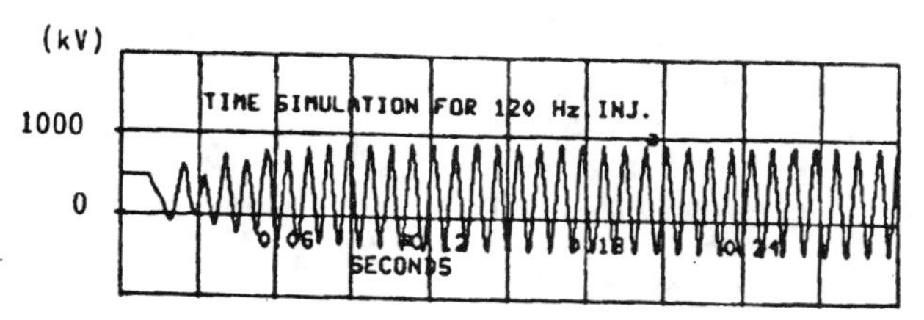

Figure 10.
Frequency Scan and Time Simulation for 120 Hz Injection

Frequency scan and time simulation plots for this case are provided in Figure 10. The maximum cable voltage due to d.c. with superimposed 120 Hz oscillations can be expressed as:

$$U_{120}\,(kV) = 257 + 685 \cdot \sin\,(754 \cdot t)$$

The maximum duration of this overvoltage would be the longest time to clear a single-phase-to-ground fault with a back-up breaker, which is less than 500 ms.

INSULATION LEVELS FOR THE D.C. CABLES

The results of these overvoltage studies, together with recommendations by C.I.G.R.E. WG21.01, have been used to select the insulation levels for the St-Lawrence under-river cables. The specified insulation levels are provided in Table III.

TABLE III
INSULATION LEVELS FOR THE ST.-LAWRENCE UNDER-RIVER D.C. CABLES

	Maximum Overvoltage (kV crest/ kV d.c.)	Insulation Levels (kV crest/ kV d.c.)	Margin (%)	
Loading cycles and polarity reversals test: (1) • without polarity reversal • with polarity reversals	 - -	 1,000 750	 - -	(1)According to CIGRE recommendations
Lightning withstand voltage (1.2/50-μs)	1,200 (2)	1,450	20.8	(2)Direct lightning stroke
Switching withstand voltage (250/2,500-μs)	960 (3)	1,175	22.4	(3)Protective level of surge arrester at 3 kA (30/60 μs)
Lightning impulse superimposed on a d.c. voltage of opposite polarity • crest value • initial d.c. voltage	 980 (4) 500	 1,225 500	 25.0 -	(4)First polarity reversal during back flashover
Lightning impulse followed by h. f. oscillations (5). • initial d.c. voltage • followed by 1st crest voltage of • followed by 2nd crest voltage of • followed by 3rd crest voltage of	 -500 +980 -800 +500	 -500 +1,225 -1,000 +625	 - 25.0 25.0 25.0	(5)Overvoltages during back flashover
D.c. and a.c. superimposed overvoltage (6) • initial d.c. voltage • followed by 60 Hz a.c. and d.c. (1 - 3 seconds): or • followed by 120 Hz a.c. and d.c. (1 - 3 seconds):	 - - -	 500 400+540sin(377t) 260+685sin(754t)	 - - -	(6)These overvoltage were determined based on very conservative assumptions. Therefore, no additional insulation margins were applied.

CONCLUSIONS

The results of the overvoltage studies for the St-Lawrence under-river d.c. cables have shown that the cables could be subjected to overvoltage stresses beyond the recommendations of C.I.G.R.E. WG 21.01 . These high stresses are due to the unusual configuration of a long overhead line section connected to a short cable.

The most severe overvoltage stresses will occur for direct lightning strokes on the line or back flashovers at the towers. The direct-stroke overvoltages can be limited to 1200-kV by metal-oxide surge arresters at the cable terminals. The magnitude of back-flashover overvoltages can be reduced to 980-kV by keeping the tower-footing resistance below 10 ohms for the five towers closest to the cable terminals.

High frequency oscillations with polarity reversals will occur during back flashovers and special tests have to be devised to verify the cable insulation for this type of stress.

The most severe internal overvoltages occur due to faults on the AC system or malfunctioning of the converters. These overvoltages do not reach the same order of magnitude as those caused by lightning. However, they have a longer duration (500 - 600-ms) and special tests have, therefore, been devised to verify the insulation strength also for this type of stress.

REFERENCES

[1] EMTP-version I, Rule Book, Published by the EMTP Development Coordination Group (EPRI), April 1986.

[2] Recommendations for tests of power transmission d.c. cables for a rated voltage up to 600 kV, Paper presented by W.G. 21-01 of C.I.G.R.E. Committee No. 72, October 1980, pp 105-144.

[3] D.C. System Resonance Analysis, M. Bahrman et al., IEEE Trans PWRD-2, No.1, January 1987, pp 156-164.

Germain Beaulieu received his B.Sc.A. and M.Sc. in electrical engineering degrees from the University of Sherbrooke and the Ecole Polytechnique de Montréal in 1978 and 1986 respectively. Mr. Beaulieu has been with Hydro-Québec since 1978. He spent most of his career in the Transmission System Planning Department with a few years in Distribution system planning. Mr. Beaulieu has been involved in insulation coordination studies, equipment specifications and harmonic analysis.

Mr. Beaulieu is a registered professional Engineer in Québec and member of a CIGRE joint WG on HVDC cables and a CSA technical committee on Electrical Coordination.

Hieu Huynh received his B.Sc.A. and M.Sc. in electrical engineering degrees from Ecole Polytechnique de Montréal in 1975 and 1979, respectively. In 1976, he joined the System Planning department of Hydro-Québec, where he has been involved in overvoltage studies, harmonic analysis and application studies including HVDC, Static VAr Compensators and series capacitors.

Mr. Hieu Huynh is a member of several working groups in the IEEE Surge Protective Devices Committee.

Que. Bui-Van was born in Saigon, Vietnam in 1952. He received his B.Sc.A in electrical engineering from Ecole Polytechnique de Montréal in 1975. He has been with Hydro-Quebec's System Planning Department since graduation in 1975. He has been involved in high voltage equipment specifications, lightning performance of transmission lines, insulation coordination of high voltage ac/dc power cables and GIS. He has also been involved in several system studies for implementation of special surge protective devices, static VAr compensators, series compensation and HVDC interconnections. Mr. Bui-Van is a registered professional engineer in the province of Québec.

Roger Rosenqvist received his M.Sc. in electrical engineering from the Chalmers University of Technology in Göteborg, Sweden, in 1980. In 1980, Mr. Rosenqvist joined the HVDC division of ABB Power Systems in Ludvika, Sweden, where he was involved in system design studies for several large HVDC transmission projects including Itaipu, Intermountain, Highgate and Rihand-Delhi. In 1987, he was appointed technical director for the Québec-New England HVDC project in Montréal, Canada. In 1989, Mr. Rosenqvist relocated to Pittsburgh, PA, where he became manager of the Transmission Center at Advanced Systems Technology.

IEEE Transactions on Power Delivery, Vol. 11, No. 1, January 1996

Observation and Analysis of Multiple-Phase Grounding Faults Caused by Lightning

Atsuyuki Inoue, Member, IEEE
Central Research Institute of Electric Power Industry
(CRIEPI)
Komae-shi, Tokyo, 201 Japan

Sei-ichi Kanao
Hokuriku Electric Power Co., Inc.
Toyama-shi, 930 Japan

Abstract—This paper describes four-phase and five-phase grounding faults caused by lightning on a 154-kV overhead transmission line. The authors measured insulator voltages and currents flowing along the ground wire and the tower. In addition, they photographed lightning strokes to the transmission line and flashovers between the arcing horns and examined the fault phases at substations. The paper analyzes insulator voltage waveforms using the Electromagnetic Transients Program (EMTP) and estimates the fault processes.

I. INTRODUCTION

Because of space restrictions, approximately 80% of all 500-kV transmission lines in Japan are double-circuit lines with vertical configurations. A recent survey on tripouts of 500-kV transmission lines indicates an annual lightning-induced tripout rate of 3.44 times per 100 km in the districts facing the Japan Sea, compared with 0.61 times per 100 km for other regions. The occurrence of double-circuit faults is 46% of the total in the coastal areas and 10% elsewhere [1]. Some 90% of the double-circuit faults in coastal areas occur in winter. Double-circuit faults can lead to severe power supply interruptions over a widespread area. Determining why double-circuit faults occur particularly frequently in winter is vital to establishing measures to counter the faults.

The most straightforward and reliable method of approaching this type of problem is to observe lightning flashes at an operating transmission line and simultaneously measure tower and groundwire currents and insulator voltages, then study the fault processes. For this purpose, the authors selected the 154-kV Tsuruga line, operated by the Hokuriku Electric Power Co., Inc., along the Japan Sea coast. The authors photographed lightning at a 4,273-m section between towers No. 24 and No. 35 and measured currents and voltages at tower No. 28.

In this paper, the authors describe the measurements for two multiple-phase grounding faults in November and December 1989, analyze the measured insulator voltage waveforms using the Electromagnetic Transients Program (EMTP) and estimate the fault processes.

95 WM 231-1 PWRD A paper recommended and approved by the IEEE Transmission and Distribution Committee of the IEEE Power Engineering Society for presentation at the 1995 IEEE/PES Winter Meeting, January 29, to February 2, 1995, New York, NY. Manuscript submitted August 1, 1994; made available for printing January 4, 1995.

II. OBSERVATION

A. Tsuruga Line

The Tsuruga line is an unbalanced-insulation double-circuit transmission line with one ground wire. Arcing horns in the higher insulation circuit (HIC) are separated by 1,650 mm and in the lower insulation circuit (LIC) by 1,050 mm. The phasing is ABC both in the HIC and LIC from top to bottom. The conductors are of type 240 mm^2 ACSR and the ground wire of type 70 mm^2 GSW. The structure of tower No. 28 is shown in Fig. 1(a), and the tower heights, span lengths and tower footing resistance values are listed in Table 1.

B. Lightning Channel Measurement

Several cameras (eight 35-mm multiple-stroke, two 35-mm light-integration, and two 8-mm framing and light-integration cameras) [2] were arranged to photograph lightning channels in the line section from two or more directions.

The multiple-stroke camera detects the initial lightning flash with a photodiode, opens the shutter after 20–30 ms, shoots the subsequent lightning channel and closes the shutter after 350 ms. The light-integration camera monitors the light exposure on the film at night, in wait of a lightning flash. When the exposure exceeds a threshold value, the camera automatically forwards the film, sets a new film and waits for a second lightning flash. The framing camera takes a photograph by forwarding frames at a rate of 18 frames/s when the photodiode detects a flash. Both 35-mm cameras can record shooting time.

C. Current Measurement

Fig. 1(b) shows the arrangement of measuring equipment. Current transformer CT_1 measured the lightning stroke currents flowing through the 1.2-m lightning rod installed at the top of the tower. Transformers CT_2

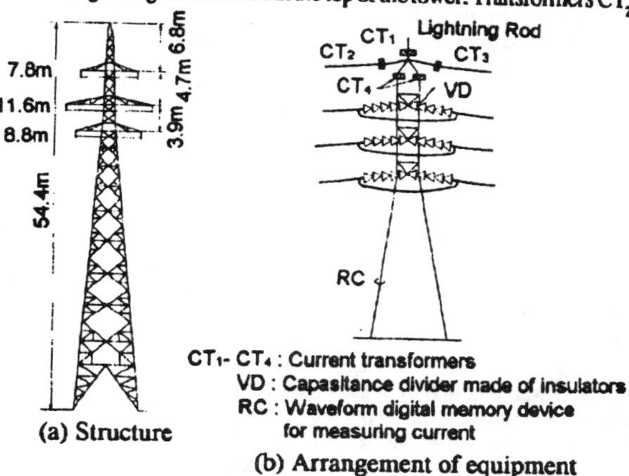

(a) Structure (b) Arrangement of equipment

Fig. 1. Tower No. 28.

TABLE I
MAIN VALUES USED FOR CALCULATION

Tower no.	Tower height (m)	Span length (m)	Tower footing resistance (Ω)
20	39.4	229	11
21	42.8	411	9
22	37.0	359	10
23	54.0	584	34
24	51.0	469	10
25	36.4	295	7
26	39.4	486	7
27	54.0	528	7
28	54.4	224	9
29	33.4	518	14
30	45.0	169	13
31	54.0	471	12
32	75.0	495	7
33	57.0	258	8
34	51.4	360	6
35	63.4		5

and CT_3 measured the current in the ground wire, and CT_4, a cluster of four transformers, measured the current flowing through the upper part of the tower. Each of these current transformers was installed on a main pole, and output voltages were summed. The risetime response of each transformer was less than 0.1 µs [3].

Output of each current transformer was recorded on a floppy disk through a five-unit transient recorder with the time it was recorded. The measuring system comprised electrical-to-optical transducers (E/O transducers), optical fiber cables and optical-to-electrical transducers (O/E transducers). The frequency bandwidth of each optical system was between 35 Hz and 3 MHz [3]. Each unit of the transient recorder had a memory capacity of 2 kb, which amplitude resolution of 10 bits. By selecting a 50-ns sampling interval, up to 100 µs of the waveform was recorded.

The relationship between the measured current and the output voltage of the O/E transducer was nonlinear in the low-current region, because the E/O transducer did not have a power supply, but was driven only by the detected electrical signal. Measured current waveforms were corrected to linearity using correction curves.

An RC waveform digital memory device was used to measure tower leg currents [4]. The device was installed 12 m above ground on the main pole (on the side of tower No. 27 tower and in the LIC). The device comprises a Rogowski coil, a battery-driven active integrator and a data storage device. The waveforms and the times were recorded on a memory card. The frequency bandwidth of the system was between 10 Hz and 350 kHz. Of the 24-kb storage device, a 20-kb portion was used to record the initial part of the waveform for a sampling interval of 0.2 µs, with the remainder used to record a sampling interval of 5 ms for the subsequent portion. The device was therefore able to record a 24-ms waveform. The RC's records were not synchronized with those taken by the current transformers. However, these records were associated according to the time they were taken.

D. Insulator Voltage Measurement

The voltage waveform was measured only on the upper phase at tower No. 28 (on the side of tower No. 29 and in the LIC) using a capacitor made from an insulator string. This capacitance divider measured voltage with a risetime response of less than 0.5 µs [3]. As with current measurements, signals were recorded on a floppy disk. The frequency bandwidth of this optical system was also 35 Hz to 3 MHz [3].

There was a nonlinear relationship between the measured voltage across the insulator string and the output voltage of the O/E transducer as shown in Fig. 2. The difference in the polarities was due to corona at the caps and the pins of the insulator strings, coupled with the polarity-dependent sensitivity difference in the optoelectronic circuit. The measured voltage waveform was corrected to linearity according to the relationship shown in Fig. 2. The voltage was synchronized with the currents measured by the transformers.

III. DATA

A. Record 89113004 (Five-Phase Grounding Fault)

The authors observed a five-phase grounding fault from the perturbo oscillographs taken at the substations at 16:27 on November 30, 1989. The faults were noted at the middle and lower phases of the HIC and at all the phases of the LIC. A 35-mm multiple-stroke camera photographed the lightning stroke at the top of tower No. 26 and recorded the flashovers between five arcing horns as shown in Fig. 3. The occurrence of a two-phase grounding fault at tower No. 28 (faults at the upper and lower LIC phases) was confirmed by the same photograph. The light intensity of the flashovers at tower No. 28 was much lower than that of flashovers at tower No. 26.

Fig. 4 shows the measured current and voltage waveforms. The groundwire current on the side of tower No. 27 could not be measured because of equipment failure. The leg current, shown in Fig. 4(b), had a wavefront time of 120 µs, with a peak value of +8 kA.

B. Record 89121413 (Four-Phase Grounding Fault)

At the substations, the authors observed a four-phase grounding fault at 18:09 on December 14, 1989, in the middle and lower phases of the HIC and the upper and middle phases of the LIC. A 35-mm multiple-stroke camera recorded a lightning stroke and several flashovers at the same time. It was inferred from other photos that lightning struck the ground wire between towers No. 26 and No. 27, slightly closer to tower No. 26. The photograph in Fig. 5 shows a four-phase grounding fault at tower No. 26 and two grounding faults at towers No. 24 and No. 25. However, the authors were unable to determine the phases of the latter faults. No flashovers were photographed at towers No. 27 and No. 28.

The light-integration framing camera took four frames of the flashover at tower No. 26. From these four frames the authors estimated that the fault continued between 167 ms and 222 ms.

Fig. 6 shows measured current and voltage waveforms. The groundwire current on the side of tower No. 27 could not be measured owing to equipment failure. The leg current, shown in Fig. 6(c), had negative polarity, a wavefront time of 30 µs and a peak value of –7.1 kA, while the peak value of the tower current, Fig. 6(b), measured by transformer CT_4 was –16.4 kA.

IV. ANALYSIS OF LIGHTNING FAULTS

Measurements of current and voltage waveforms caused by natural lightning on the transmission lines is extremely difficult because in general there is only a very slight chance of lightning striking a tower

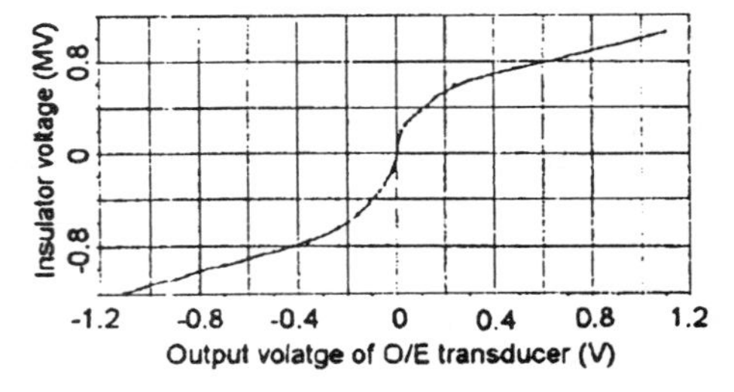

Fig. 2. Output voltage of O/E transducer vs. insulator voltage.

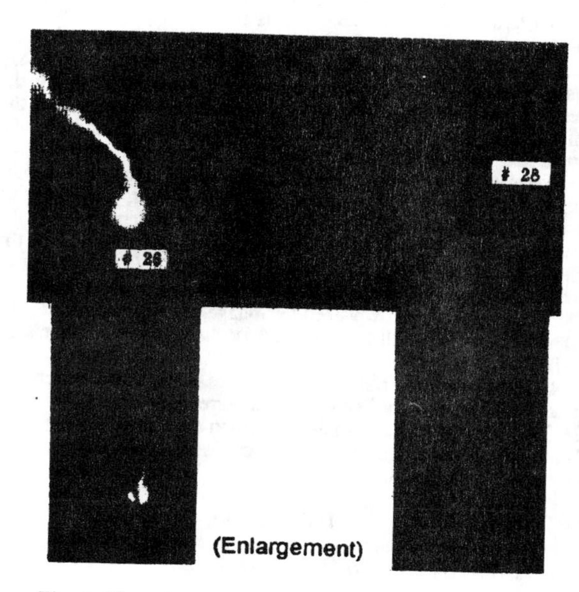

Fig. 3. Five-phase grounding fault at tower No. 26 and two-phase grounding fault at tower No. 28 (Record 89113004).

where measuring equipment has been installed. Such was true in this case, as well. Lightning stroke points were at the top of tower No. 26 and at the ground wire between towers No. 26 and No. 27. The equipment, however, was concentrated at tower No. 28. In such cases, comparison of calculated waveforms with measured ones is difficult. First, the calculations must be begun using assumed waveforms and amplitudes because the original lightning stroke current is not measured. Second, current waveforms of the tower and the ground wire at some distance from the stroke point are influenced considerably by the surge propagation characteristics of the towers and the ground wire, as well as by tower footing resistance characteristics, etc., thereby complicating the reconstruction of current waveforms. Third, insulator voltage waveforms vary considerably, according to when and where flashovers occur. The insulator voltage waveforms at towers where there is no flashover become more complex because they are produced from the tower potentials and the surge voltages the flashovers introduce in the conductors. Therefore, this paper's analysis aims principally to estimate the lightning-induced fault processes from limited records. The EMTP is used to aid the analysis.

A. Calculation Conditions for EMTP

The main EMTP calculation conditions are as follows:

1) The tower was divided into four sections between the top and the upper crossarms, the upper and the middle crossarms, the middle and the lower crossarms, and the lower crossarms and the tower foot. Each section was simulated by distortionless line modeling. Surge impedance, attentuation factors and propagation velocities were assumed 150 Ω, 0.85 per 50 m and 210 m/µs in each section, respectively.
2) K.C. Lee's model was used for the transmission line. A frequency of 350 kHz was used.
3) EMTP time-controlled switches were used to simulate flashovers between the arcing horns.
4) Groundwire and line conductor impedances were matched at towers No. 20 and No. 35.
5) The surge impedance of the lightning channel was assumed to be 400 Ω.
6) The instantaneous power-frequency voltages were ignored.

B. Calculated Results

1) Record 89113004: Figs. 7(b) through 7(d) show current and voltage waveforms calculated under the assumption that after the lightning stroke current, which had a peak value of +260 kA and is shown in Fig. 7(a), flowed into the top of tower No. 26, flashovers occurred in the sequence indicated in Table 2, where "re-flashover" means a flashover on an adjacent tower caused by a back-flashover on a tower. Each waveform is similar to that in Fig. 4.

2) Record 89121413: Figs. 8(b) through 8(d) show current and voltage waveforms calculated under the assumption that after the lightning stroke current, which had a peak value of –250 kA and is shown in Fig. 7(a), flowed into the ground wire at the midspan between towers No. 26 and No. 27, several flashovers occurred in the sequence listed in Table 3. Waveforms are similar to that shown in Fig. 6.

C. Interpretation of Voltage Waveforms of Record 89113004

The voltage waveforms of record 89113004 can be interpreted with the help of the EMTP calculations as follows:

1) Part A of Fig. 4(c) shows the upper crossarm voltage at tower No. 28 that was raised by a portion of the lightning stroke current prior to the first back-flashover at tower No. 26. It should be noted that the positive crossarm voltage raised by the positive tower current was measured as a negative voltage, since the reference point was the crossarm in the insulator voltage measurement.

2) Part B shows the difference between the conductor voltage induced at the upper phase in the LIC by the back-flashovers at the middle and lower phases of the HIC and the LIC at tower No. 26, and the upper crossarm voltage at tower No. 28.

3) Part C shows that the high voltage was introduced by the back-flashover into the upper phase of the LIC at tower No. 26. The V-shaped valley between parts B and C, which does not appear in Fig. 7(d), suggests the presence of predischarge phenomena.

4) The abrupt change from parts C to D shows a re-flashover and its extinction after a very short time, i.e. the recovery of air insulation, at the upper phase of the LIC at tower No. 28. The insulator voltage reached

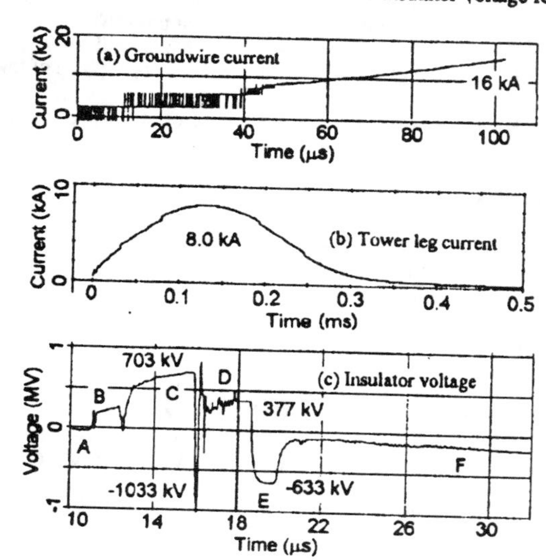

Fig. 4. Measured waveforms (Record 89113004).

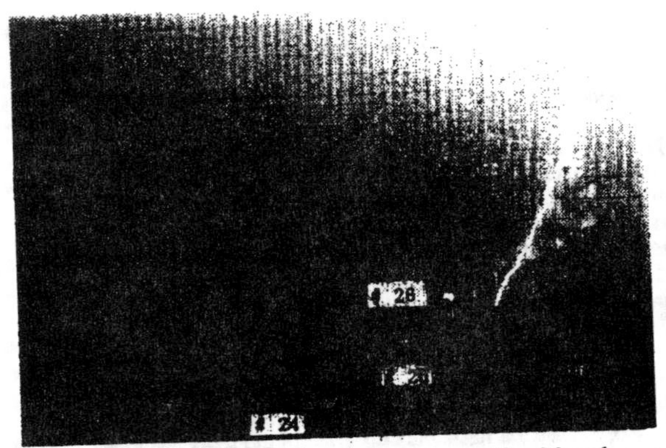

Fig. 5. Four-phase grounding fault at tower No. 26 and two grounding faults at towers No. 24 and No. 25 (Record 89121413).

+703 kV and caused a flashover. This re-flashover at tower No. 28 is not contradictory, because the 50% flashover voltage of the arcing horns on the side of tower No. 27 in the LIC is +658 kV for the positive standard lightning impulse, which is lower than the insulator voltage. It was estimated from the calculation that another re-flashover occurred simultaneously at the lower phase of tower No. 28 in the LIC, and this flashover extinguished the re-flashover at the upper phase. The lower peak value of part D compared that of part C may be attributed to the re-flashover having continued at the lower phase of tower No. 28. There is a considerable difference in part D between the calculated and the measured results. This suggests the tower model was insufficient.

5) Part E may be interpreted as follows. The re-flashover occurred at the upper phase of the LIC at tower No. 24 prior to the re-flashover at tower No. 28. As a result, a negative voltage was introduced in the upper conductor at tower No. 24. The pulse width, τ, of the voltage surge propagated to tower No. 28 according to the formula:

$$\tau = t_1 + t_2 - t_3$$

where t_1 is the extinction time of re-flashover at tower No. 24, t_2 is the propagation time of the surge between towers No. 24 and No. 26 and t_3 is the extinction time of the back-flashover at the upper phase of tower No. 26

Calculation confirmed that the voltage across the upper insulator at tower No. 24 was high enough to cause a re-flashover, and that a re-flashover could not occur at tower No. 25 because the coupling between the ground wire and the upper phase conductor was very large. The authors believe the reason no photographs were taken of the re-flashover at tower No. 24 is that this tower was covered by thundercloud.

6) Part F mainly shows the upper crossarm voltage, which was created by a portion of the lightning stroke current flowing into tower No. 28.

D. Interpretation of Voltage Waveform in Record 89121413

The voltage waveform in record 89121413 can be interpreted with the aid of calculations based on the EMTP as follows:

1) Part A in Fig. 6(d) does not exist in the calculated result of Fig. 8(d). This voltage was probably induced on the upper conductor by the thundercloud or the lightning channel.

2) Part B shows the upper crossarm voltage at tower No. 28 raised by a part of the lightning stroke current prior to the first back-flashover at tower No. 26.

3) Part C shows the difference between the high voltage introduced into the upper phase in the LIC by the back-flashover at the upper phase of the LIC at tower No. 26 and the upper crossarm voltage at tower No. 28. A few small spikes in part C may be attributed to the generation and extinction of the other back-flashovers at tower No. 26. In Fig. 8(d), they are shown as two small changes of P_A and P_B. The P_A change is due to the generation and extinction of the flashovers at the middle phases of the HIC and LIC at tower No. 26, and that of the successive flashover at the upper phase of the LIC at tower No. 26, while the P_B change is that of the flashover at the lower phase of the HIC at tower No. 26. Calculated results support the above-mentioned inference for those spikes. The peak value was clipped to −1,142 kV because of the excessive input level of the transient recorder. This value was sufficient for re-flashover because the 50% flashover voltage of the arcing horns is −799 kV for the negative standard lightning impulse. However, no flashover occurred at tower No. 28. The reason may be explained by voltage-time characteristics. The authors believe the small superimposed oscillation shows an unfinished re-flashover.

4) Part D primarily shows the upper crossarm voltage, which was generated by a part of the lightning stroke current flowing into tower No. 28. The authors estimate from the calculation that the change from C to D was caused by the extinction of the back-flashover at the upper phase of the LIC at tower No. 26.

V. DISCUSSION

A. Peak Values of Lightning Stroke Currents

In the EMTP calculation, the peak values of +260 kA and −250 kA were assumed to be lightning stroke currents. The occurrence probability of exceeding these values is extremely small. Upon referring to [5], [11], this stroke current magnitude in negative lightning flashes corresponds to a cumulative probability distribution of less than 0.1%. However, in Japan there are a few measured examples of lightning stroke currents with comparable peak values in winter ([6], [7]). Even taking

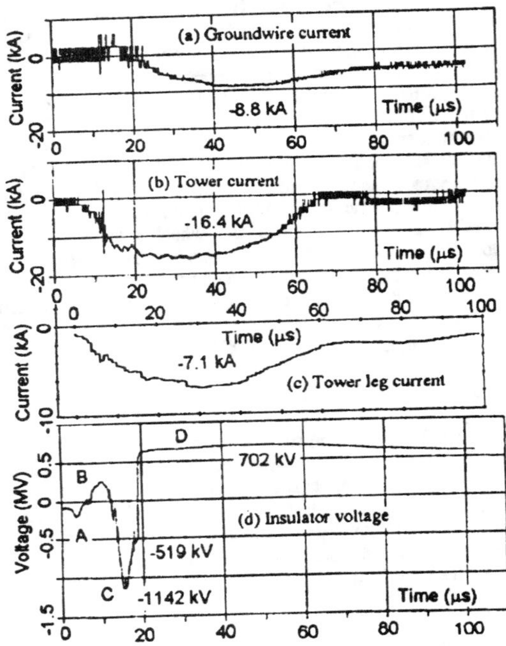

Fig. 6. Measured waveforms (Record 89121413).

into consideration that tower No. 28 is at some distance from the lightning stroke points, the wavefront times of approximately 120 µs in record 89113004 and 30 µs in record 89121413 are very long. Therefore, fairly large current peak values are required to cause multiple-phase grounding faults with such current waveforms.

B. Conversion of Tower Leg Current to Tower Current

In record 89121413, the tower current at the upper part of the tower was measured by current transformer CT_4, together with the tower leg current by the RC waveform digital memory device. The peak value of the tower current was −16.4 kA, while the tower leg current was −7 kA. Therefore, the conversion coefficient of the tower leg current into the tower current is 2.3 (16.4/7). This value is relatively small in comparison with the general conversion coefficient. The authors believe the main reason for this was that the four legs experienced a nonuniform current flow. The authors have noted during other experiments that when there are large differences between four tower footing resistance values, most of the tower current tends to rapidly flow to the least resistant footing as time progresses. One example is shown in the Appendix. It has also been reported that when lightning strikes a tower top, currents at the adjacent towers flow more easily along the tower legs on the side of the tower that is struck [8]. The leg where the RC device was installed is at the side of tower No. 27, i.e. at the side of the lightning stroke point—a situation the authors estimate made the conversion coefficient small.

In record 891113004, the wavefront steepness of the tower current was very low. Therefore, the amplitude of the current was extremely small in a 100-µs interval and as a result, the CT_4 transformer could not record the tower current. The peak value of the tower current became +18.4 kA when using the conversion coefficient of 2.3. The calculated value of +11.4 kA was 62% of the measured value, hence there was not a large difference.

In record 89121314, the calculated peak value of the tower current was −20.3 kA, or 124% of the measured value.

The following reasons may account for some of the difference between the measured and calculated waveforms:

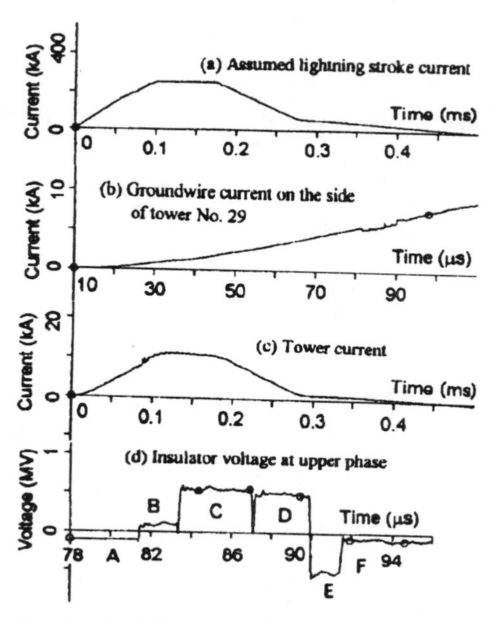

Fig. 7. Calculated waveforms (Record 89113004).

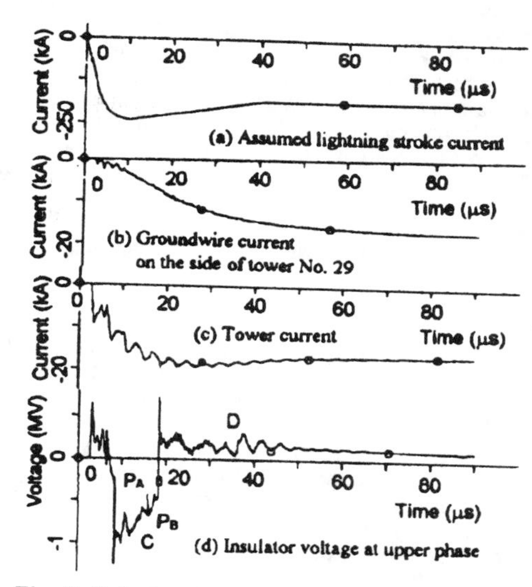

Fig. 8. Calculated waveforms (Record 89121413).

- The tower model is insufficient.
- The natural waveform of the lightning stroke current is not as simple as the one used for calculation.
- Predischarge phenomena for the flashovers were ignored.
- The corona attenuation and distortion were neglected for the surges on the tower, at the ground wire and at the line conductors.
- The non-linear and hysteresis effects were neglected for tower grounds.
- The electromagnetic coupling between the lightning channel and the conductors was ignored.

C. Flashovers and Calculated Voltage Waveforms

Flashover phenomena were simulated by closing time-controlled EMTP switches. When a flashover occurred, the voltage difference between arcing horns became zero abruptly and the conductor voltage was raised with a rise time of zero. This rapid change in the circuit added various spikes to the calculated waveforms. However, actual flashovers

TABLE 2
SEQUENCE OF FLASHOVERS IN RECORD 89113004

Tower no.	1L phase	2L phase	Generation time (µs)	Extinction time (µs)	Back-flashover or re-flashover
24		Upper	83.5	85.5	R.F.O.
26		Upper	80.0	86.5	B.F.O.
26		Middle	78.0	79.9	B.F.O.
26	Middle		78.0	79.9	B.F.O.
26	Lower		78.0	79.9	B.F.O.
26		Lower	78.0		B.F.O.
28		Upper	87.0	87.1	R.F.O.
28		Lower	87.0		R.F.O.

R.F.O.: Re-flashover B.F.O.: Back-flashover

TABLE 3
SEQUENCE OF FLASHOVERS IN RECORD 89121413

Tower no.	1L phase	2L phase	Generation time (µs)	Extinction time (µs)	Back-flashover or re-flashover
24		Upper	10.0	12.5	R.F.O.
26		Upper	5.0	15.0	B.F.O.
26		Middle	3.0	5.0	B.F.O.
26	Middle		3.0	5.0	B.F.O.
26	Lower		14.0	15.0	B.F.O.

R.F.O.: Re-flashover B.F.O.: Back-flashover

are completed through the streamer and leader discharge stages. The change in the voltage waveform, therefore, becomes slower than the change simulated by the simple switch. This is probably the reason for the difference between part C in Fig. 6(d) and part C in Fig. 8(d).

The corona discharges on the tower, at the ground wire and at the line conductors help suppress the rapid changes in various phenomena. Even if high-frequency pulses are generated, most of them are smoothed by the corona. The authors assume that this is the reason part D in Fig. 6(d) is smoother than part D in Fig. 8(d).

Though self-extinction of the flashovers was introduced into the calculations, this phenomenon has already been reported in [9].

D. Tower Modeling

Tower modeling for the purpose of lightning protection design has been investigated as a potential response of the tower to the current with a steep wavefront. For a lightning stroke current with a long wavefront time of 10–100 μs, the tower potential is created by the multiple forward and backward reflections of the currents along the towers and along the ground wire. In such cases, there is a relatively strong possibility of the small errors generating a large cumulative error between calculated and measured results. Any new tower modeling is especially important for lightning in winter, which frequently accompanies a current with a long wavefront time [10].

VI. CONCLUSIONS

1) The authors obtained and analyzed the various records about two multiple-phase grounding faults induced by lightning. Since this is the first time such research has been performed, it should contribute to improving the accuracy of lightning protection design.

2) The lightning stroke current for the five-phase grounding fault was estimated to have a peak value of +260 kA and a wavefront time of 100 μs, while these figures for the four-phase grounding fault were –260 kA and 10 μs. In Japan, lightning strokes with such large peak values and long wavefront times are thought to cause some multiple-phase grounding faults in winter.

3) At the 154-kV transmission line, the authors confirmed the occurrence of re-flashover both by measurement and through calculation. Moreover, they determined that the re-flashover was apt to occur at towers some distance away rather than at the tower nearest the lightning stroke point.

4) It was shown on the basis of the EMTP calculation that the complex insulator voltage waveforms could be interpreted by considering the generation and extinction of flashovers.

Future related research topics:

- It will be necessary to create a new tower model for a lightning stroke current having a long wavefront time.
- The mechanism of self-extinction in lightning flashovers should be investigated.
- Analysis of additional faults will be required to further clarify the processes of lightning-induced multiple-phase grounding faults. To accomplish this, current and voltage measuring equipment must be installed at many towers. The authors recommend measuring multiphase insulator voltages as well as currents in each of four tower legs. The instantaneous power-frequency voltages should also be measured.

VII. ACKNOWLEDGMENTS

The authors wish to express their thanks to Mr. M. Miki of CRIEPI and Mr. K. Shimizu of Hokuriku Electric Power Co., Inc., for their participation and their contributions to the observation.

The authors also thank Mr. Y. Katsuragi of Chubu Electric Power Co., Inc., Mr. F. Suzuki of Sankosha Co., Ltd., and Mr. M. Yokota of Hokkei Co., Ltd., for their support of the observation and Mr. Jeff Loucks for preparing this manuscript in English.

VIII. APPENDIX

Concentration of Current in One Tower Leg:

Figure A-1 shows the measured results of the currents at four tower footings when an impulse current was injected into the top of a 48.2-m high tower with a ground wire. The impulse generator was located 70 m from the tower at a right angle to the transmission line. The waveform was (1.7/50) μs, and the peak value was about 150 A. The tower footing resistance value at the leg C was sufficiently lower than the others, because three counterpoises were connected to leg C. The current was found to gather at leg C as time progressed. After 3 μs, the current peak at leg C was approximately four times that of leg B.

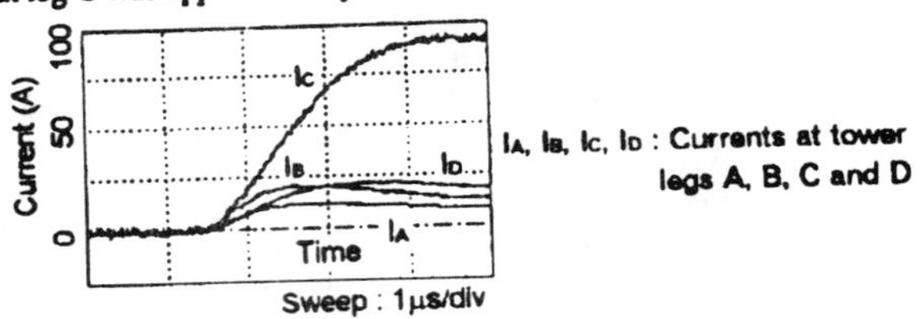

Fig. A-1. Tower footing currents.

IX. REFERENCES

[1] A. Inoue, "Lightning Outages and Countermeasures of Overhead Transmission Lines," *Electrical Review of Japan*, vol. 75, no. 6, June 1990, pp. 673–680.

[2] T. Tada and H. Fujinami, "Development of Automatic Lightning Path Photographing Equipment and Observation Results," *Trans. IEE of Japan*, vol. 104-B, no. 2, February 1984, pp. 85–92.

[3] Working Group on Lightning Protection for Transmission Systems, "Lightning Observations on Japan Sea Coast in Winter," *CRIEPI Report*, no. T10, January 1989.

[4] M. Arakane and Y. Katsuragi, "Development of Lightning Wave Memory and Its Application on Lightning Observation on Transmission Towers and Ground Wires," *Trans. IEE of Japan*, vol. 111-B, no. 1, January 1991, pp. 31–37.

[5] IEEE Working Group on Estimating Lightning Performance of Transmission Lines, "A Simplified Method for Estimating Lightning Performance of Transmission Lines, *IEEE Trans. Power Apparatus & Systems*, vol. 104, no. 4, April 1985, pp. 919–932.

[6] H. Narita, Y. Goto, H. Komuro and S. Sawada, "Bipolar Lightning in Winter at Maki, Japan," *J. Geophys. Res.*, no. 94, 1989, pp. 13191–13195.

[7] Y. Goto and Narita, "Observations of Winter Lightning to an Isolated Tower," *Res. Lett. Atmos. Electr.*, vol. 12, no. 1, 1992, pp. 57–60.

[8] Y. Katsuragi, M. Saiki, A. Inoue and M. Yokota, "Observations of Lightning Strokes to 275-kV Transmission Tower in Winter Thunderstorms," *IEE Tech. Report*, HV-92-66, October 1992, pp. 105–110.

[9] M. D. Perkins, S. F. Mauser, D. C. Mikell and S. L. Nilsson, "Summary of Transient Data Obtained from Long-Term Monitoring of a 138-kV and a 500-kV Transmission Line," *IEEE Trans. Power Apparatus & Systems*, vol. PAS-103, no. 8, August 1984, pp. 2290–2298.

[10] K. Miyake, T. Suzuki, K. Shinjou, "Characteristics of Winter Lightning Current on Japan Sea Coast," *IEEE Trans. Power Delivery*, vol. 7, no. 3, July 1992, pp. 1450–1457.

[11] Working Group 01 of Study Committee 33, "Guide to Procedures for Estimating the Lightning Performance of Transmission Lines," *CIGRE Technical Brochure 63*, October 1991.

Atsuyuki Inoue was born in Osaka Prefecture, Japan, on November 17, 1944. He received the B.S., M.S. and Ph.D. degrees from Kyoto University in 1967, 1969 and 1983, respectively.

He joined the Central Research Institute of Electric Power Industry, Tokyo, Japan, in 1969. Since that time he has been engaged in research of lightning protection design for power transmission lines. He is a Senior Research Fellow of the Electrophysics Department in the Komae Research Laboratory. Dr. Inoue is a member of the Institute of Electrical Engineers of Japan and the Institute of Atmospheric Electricity of Japan.

Sei-ichi Kanao was born in Toyama Prefecture, Japan, on January 8, 1951. He received the B.S. degree in electrical engineering from Toyama National College of Technology in 1971. He joined Hokuriku Electric Power Company Inc. in Toyama, Japan, in 1971. He is currently engaged in construction and maintenance of overhead power transmission lines at the Central Transmission and Substation Construction Office. Mr. Kanao is a member of the Institute of Electrical Engineers of Japan.

DISCUSSION

ABDUL M. MOUSA (British Columbia Hydro, Vancouver, B.C., Canada): I wish to congratulate Dr. Inoue and his co-author for a very interesting paper. This is one of the few papers which document observations in which the same lightning flash caused flashovers at more than one tower. It is interesting that a recent EPRI R&D project revealed similar incidents but the results may not have been published yet. The authors' response to the following questions and comments would be appreciated:

1. **The power frequency voltage was ignored in the calculations. This may be justified when the lightning currents are very large as in the cases reported in the paper (+260kA and -250kA). For smaller currents, on the other hand, the discusser suggests that the power frequency voltage needs to be taken into consideration to determine which phases may be involved in a flashover.**
2. **The phasing arrangement of the subject line is that of a "super bundle" (identical ABC phasing from top to bottom on each circuit). Would the incidence of simultaneous outage of both circuits decrease if phases A and C of one of the two circuits were interchanged to form the so-called "low reactance bundle"?**
3. **500 kV Double circuit (DC) outages are quoted in the paper to be 46% of the total for lines in the coastal areas and only 10% elsewhere. The tower footing resistances given in Table 1 of the paper are relatively low. If these are typical for Japanese power lines in coastal areas, then the excessive DC outage rate in those areas is apparently caused by an increase in the median current of the collected lightning strokes. The fact that 90% of the DC outages in coastal areas occur in winter, where positive polarity flashes with their higher amplitudes dominate, seems to support the above conclusion. The possible mitigative measures hence appear to be as follows: a) Minimizing the portion of the routes of future lines which is located along the coast if possible. b) Decreasing the grounding resistance of the towers further if possible. c) Carrying the shield wires on separate structures to minimize the insulator voltages generated by the collected lightning strokes.**
4. **Fig. A-1 of the paper shows the inequality of the currents flowing in the 4 legs of the tower based on measurements involving the injection of a total current equal to 150A. It should be noted that even larger non-uniformities occur when the current is large as in the case of lightning discharges [12]. Such non-uniformity is caused by the variation of the ionization gradient along the surfaces of the grounding electrodes as a result of the non-uniform distribution of the water within the soil.**

REFERENCE

[12] A.M. Mousa, "The Soil Ionization Gradient Associated with Discharge of High Currents into Concentrated Electrodes", IEEE Trans. on Power Delivery, Vol. 9, No. 3, pp. 1669-1677, July 1994.

Manuscript received March 1, 1995.

A. INOUE and S. KANAO: We would like to thank Dr. Mousa for his interest in this paper and offer the following replies to his questions and comments.

1. We agree with Dr. Mousa that to estimate flashovers precisely, the power frequency voltages should be taken into consideration. Therefore, in the future it will be necessary to base an analysis such as the one in this paper upon measuring surge voltages together with power frequency voltages across insulators at many towers.

2. According to our method of estimating lightning outage rates on transmission lines [13], in an unbalanced-insulation system there is little difference in double-circuit outage rates between "super bundles" and "low reactance bundles." This shows that the effect of gap length between arcing horns on flashovers is superior to that of power frequency voltages. On the other hand, in a normal-insulation system, the double-circuit outage rates become considerably lower in "low reactance bundles" than in "super bundles," which corresponds to Dr. Mousa's estimation. This shows that the transfer from single-phase outages to double-phase ones in the double circuit occurs more easily in "super bundles," owing to the phases of power frequency voltages.

3. The following four items are considered possible causes of high double-circuit outage (tripout) rates in the coastal areas of the Japan Sea.

   1) High median values of lightning currents in winter
   2) Decreases in flashover voltages under abnormal atmospheric conditions
   3) Simultaneous lightning strokes to various points of a transmission line
   4) High values of tower footing resistances in winter

The first possibility corresponds with Dr. Mousa's comment. We also consider this cause the most strongly supported, because in winter some records of high-current amplitudes [6],[7] have been obtained in addition to those estimated in this paper.

As for the second possibility, the cloud bases of winter thunderstorms sometimes lie at altitudes of only several hundred meters. In such cases, transmission lines in mountainous districts are encompassed by thunderclouds. According to one reference [14], the flashover voltages of insulator strings with arcing horns at 1-meter gap spacings decrease to approximately 80% in an electric field of 100 kV/m and a space charge density of 10^{-6} C/m^3.

The likelihood of the third possibility has not yet been confirmed. However, it seems probable, judging from numerous photographs of simultaneous flashes in winter.

The fourth possibility seems somewhat unlikely to be one of the causes, because the frost zone in the soil of these districts is relatively shallow. As to the discussor's comments on measures to counter lightning outages, we consider the following.

First, lightning stroke points are not uniformly distributed over the length of the transmission line. Therefore, a study should be initiated that provides insight into the correlation between frequency of lightning strokes and terrain. This information would be put to good use in selecting the routes of future lines.

Second, we have in practice decreased the grounding resistance of towers that have high resistance values through such methods as burying counterpoises. We are estimating the effects of these activities by investigating the lightning performance at these towers.

Under present conditions, the third countermeasure, which suggests carrying the shield wires on separate structures to minimize the insulator voltages generated by collected lightning strokes, could prove problematic because of Japan's narrow land mass. However, it would be practical to enact this countermeasure in sections where the frequency of lightning strokes is high. Therefore, this measure should be investigated closely in the future.

4. The decrease in grounding resistance due to the high current is particular to the thin wires, such as those of the counterpoises, and the thin rods, because corona streamers are generated easily and the rate of the equivalent increase in the surface area of such grounding electrodes is large. However, in a normal tower footing the effect of high currents on a decrease in resistance is not so large [15]. The reason will be attributable to the fact that the tower footing has a naturally wide contact area with soil. So, Dr. Mousa's comment is correct, but care must be taken not to overestimate this effect.

References

[13] A. Inoue, "A Method for Estimating Lightning Outage Rates of Transmission Lines," CRIEPI Rep. No. T87089, September 1988.

[14] H. Fujinami, Y. Aihara, "Reduction of Flashover Voltage of Insulator Strings under High Electric Field and Ion Flow Field," CRIEPI Rep. No. 185023, January 1986.

[15] T. Matsui, M. Adachi, "Measurement Results of the Various Grounding Electrodes' Surge Impedances with Large Current," 1995 National Convention Record IEE Japan, No. 1607, March 1995.

Manuscript received April 17, 1995.

IEEE Transactions on Power Delivery, Vol. 11, No. 2, April 1996

Measurement of Lightning Surges on Test Transmission Line Equipped with Arresters Struck by Natural and Triggered Lightning

Y. Matsumoto Non-Member, O. Sakuma Non-Member, K. Shinjo Non-Member, M. Saiki Non-Member, T. Wakai Non-Member, T. Sakai Non-Member, H. Nagasaka Non-Member
Hokuriku Electric Power Co.
Toyama, Japan

H.Motoyama Member
CRIEPI
Tokyo, Japan

M.Ishii Senior Member
University of Tokyo
Tokyo, Japan

Abstract - Measurement of lightning surges caused by natural or rocket-triggered lightning strokes has been carried out on a 275kV test transmission line since 1987.

In November 1993, lightning currents at the tower top, the ground wires, the tower legs and the transmission line arresters, along with the voltages across the insulator strings were measured simultaneously four times. One of the four flashes was natural lightning, and its peak value was +132kA. In this case, the arresters operated, and the insulator voltages where the arresters were equipped were controlled properly.

The result of a multi-phase analysis by EMTP agreed well with the measured waveforms.

I. INTRODUCTION

Most transmission line outages in Japan are caused by lightning. Especially the mountainous areas along the Sea of Japan are affected by winter lightning. Such lightning has often caused multi-phase outages including double circuit outages, and even strand fusing of overhead ground wires has occurred [1]. These outages cannot be explained by the conventional design concept of transmission lines against lightning [2], and the reason is that the characteristics of winter lightning and conventional summer lightning differ [3]. Therefore, it is very important to investigate the features of transmission line outages to improve the protection method against winter lightning. Installation of transmission line arresters has been reported to be quite useful to prevent double circuit outages [4][5][6].

Since 1987, measurement of lightning surges has been carried out at the 275kV Okushishiku Test Transmission Line in winter in cooperation with the study group of rocket-triggered lightning of Japan [7]. Fig. 1 shows the location of the 275kV Okushishiku Test Transmission Line. In November 1993, the authors succeeded in the simultaneous measurement of lightning currents at the tower top, and the currents of various parts of the transmission tower, along with the voltages across the insulator strings.

This paper reports the results of the measurement. Moreover, this paper also shows that the results of calculation using the method for lightning surge analysis by the Electromagnetic Transients Program (EMTP) [8] [9], agreed very well with the measured waveforms.

95 SM 383-0 PWRD A paper recommended and approved by the IEEE Transmission and Distribution Committee of the IEEE Power Engineering Society for presentation at the 1995 IEEE/PES Summer Meeting, July 23-27, 1995, Portland, OR. Manuscript submitted Dec. 28, 1994; made available for printing May 11, 1995.

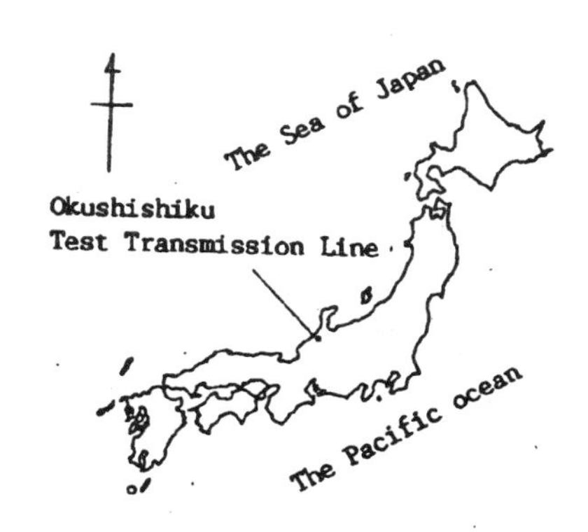

Fig. 1 Location of the 275kV Okushishiku Test Transmission Line.

II. TEST SET-UP

The Okushishiku Test Transmission Line is located about 10km inland from the coast of the Sea of Japan, and is an unenergized 275kV double circuit transmission line equipped with a single ground wire. The total length of the test line is 2.15km composed of 6 spans. Fig. 2 shows the outline of this transmission line, and Fig. 3 shows the dimensions of each tower. The terminals of all the conductors at the No.29 tower are terminated by 500Ω resistors to make the reflection of travelling waves on the conductors negligible. The terminals of all the conductors at the No.34 tower are directly connected to crossarms, as the reflection from this end, about 1.5km from the No.30 tower, does not affect the insulator voltage waveforms at the No.30 tower very much.

The No.30 tower is a suspension steel tower erected 926m above the sea level. Three transmission line arresters for a 154kV line are equipped on the circuit 2 at this tower as shown in Fig. 4. Each phase conductor is insulated by double insulator strings of 2,550mm in length.

A lightning rod of 4m in length is installed at the top of the No.30 tower. The lightning current to the tower top and currents flowing into the ground wires are measured by coaxial resistive shunts of 2mΩ, and the currents flowing into the tower legs and the arresters are measured by current transformers (CT). The voltages across the insulator strings are measured by resistive voltage dividers of 20kΩ. Fig. 4 shows the arrangement of measuring equipment, and Fig. 5 shows the equipment installed on the upper phase crossarm of the circuit 2. Table 1 shows the types and the specifications of the sensors. The arrester is a gapless type composed of zinc-oxide elements. The discharge current withstand capacity and the discharge voltage are 50kA for $2\mu s \times 20\mu s$ and 550kV, respectively. Fig. 6 shows the voltage-current characteristic of this transmission line arrester designed for a 154kV line.

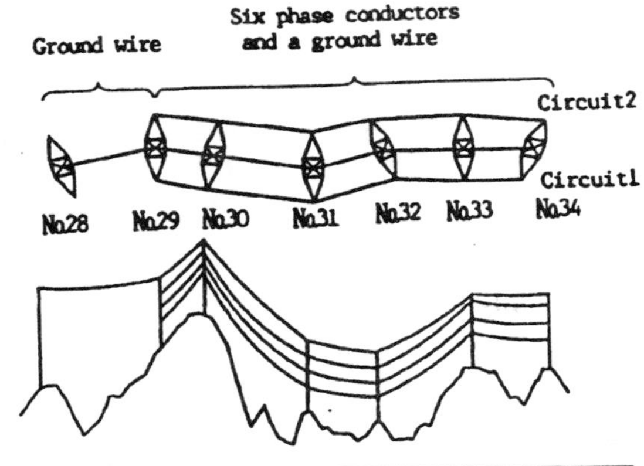

Tower No.	No28	No29	No30	No31	No32	No33	No34
Span length [m]	505	182	437	297	414	318	
Ground wire	HAS 150mm² (16.0mmφ)	OPGW 170mm² (17.5mmφ)		HAS 150mm² (16.0mmφ)			
Conductor	None	IACSR 610mm² (34.2mmφ)					
Average conductor height [m] upper	—	42	62	54	55	62	
Average conductor height [m] middle	—	31	51	41	42	49	
Average conductor height [m] lower	—	21	41	30	30	37	
Tower height [m]	85.3	59.4	59.4	59.4	59.2	61.3	62.4
Insulator string	None	320mm 13×2	280mm 15×2	320mm 13×2	320mm 13×2	320mm 13×2	320mm 13×2
Arcing horn gap [mm]	—	1,780	1,200	1,850	1,850	1,850	1,850
Grounding resistance [Ω]	8.7	2.3	5.0	4.5	3.2	8.0	9.1

OPGW : Composite fiber-optic ground wire
HAS : High aluminium steel wire
IACSR: I-aluminium alloy conductors steel-reinforced

Fig. 2 Outline of the Okushishiku Test Transmission Line.

Tower	No28	No29	No30	No31	No32	No33	No34
a	15.0	16.0	12.0	16.0	15.6	15.0	16.0
b	18.0	19.0	16.6	19.0	18.6	18.0	19.0
c	30.6	30.0	21.4	30.0	29.6	28.2	30.0
d	20.8	21.2	19.0	21.2	20.8	20.8	21.2
e	7.5	13.0	5.0	13.0	13.0	7.5	13.0
f	13.8	13.2	10.0	13.2	13.0	13.8	13.2
g	12.0	11.2	8.0	11.2	11.2	12.0	11.2
h	52.0	22.0	36.4	22.0	22.0	28.0	25.0
i	14.3	10.6	14.3	10.6	10.0	9.7	11.3

Fig. 3 Dimensions of each tower in meter.

Table 1 Types and specifications of the sensors.

CH	Measured quantity	Recording range	Sensor	Frequency range
1	Tower top current	± 180kA	Coaxial resistive shunt of 2mΩ	0 ~10MHz
2~ 3	Ground wire current	± 100kA		
4~ 7	Tower leg current	± 50kA	Current transformer	5Hz~ 2MHz
8~10	Arrester current	± 30kA	Current transformer	5Hz~ 2MHz
11~16	Insulator voltage	± 2,600kV	Resistive voltage divider of 20kΩ	0 ~10MHz

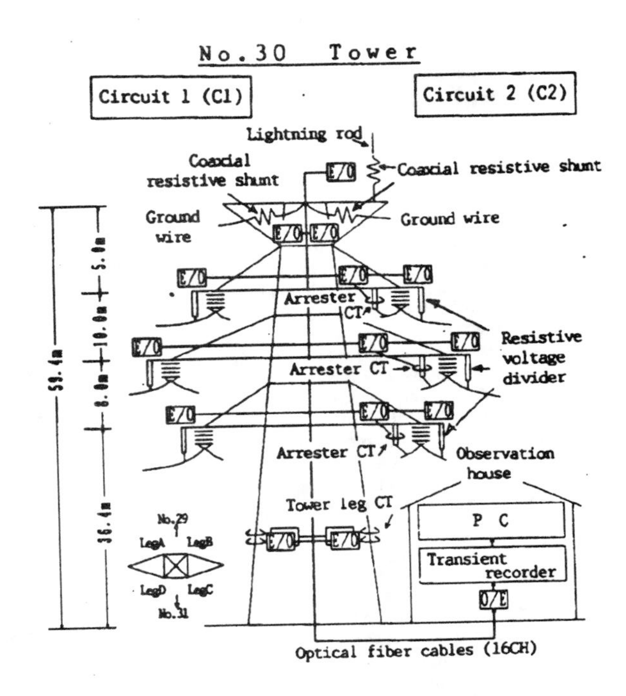

Fig. 4 Arrangement of measuring equipment.

Fig. 5 Equipment installed on the upper phase crossarm of circuit 2.

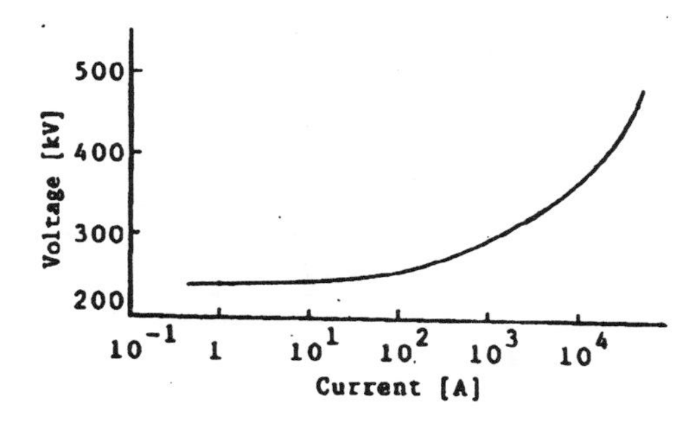

Fig. 6 Voltage-current characteristic of transmission line arrester designed for 154kV line.

The signals sensed by the sensors are converted into light signals and are transmitted through optical fiber cables to an observation house located about 30m away from the No.30 tower. In the observation house, the signals are converted back into electrical signals and are digitally recorded by a transient recorder with the resolution of 10 bits. The frequency range of the analog optical link is 10Hz~50MHz. The system consisting of 16 channels are triggered by a signal into either one of the selected 8 channels. The sampling interval of the transient recorder can be changed during operation. In this measurement, the steep-front current waveform is recorded for 920μs at the sampling interval of 100ns; after 920μs, the long-duration waveform is recorded at the sampling interval of 10μs. The total recording time is 72.6ms. Two diesel generators are used to feed power to the equipment and to keep off lightning surges from the power supply.

When a positive lightning current flows down along the lightning rod, all the polarities of measured currents are defined as positive. The polarities of insulator voltages at that time are defined as negative.

III. RESULTS

Four flashes struck the lightning rod of the No.30 tower during the period from November 5 to November 30, 1993. Three of these flashes were rocket-triggered lightning and one was natural lightning. When the natural lightning hit the No.30 tower, the arresters operated, and voltages and currents of various parts were measured. But the currents flowing into the tower legs B and C were not measured because of malfunction in the equipment.

Table 2 shows the peak values and the duration of lightning currents at the tower top. The duration is defined as the interval while the current is discernible.

A. Flash 1993-10 by Rocket-Triggered Lightning

Fig. 7 shows the lightning current waveform of Flash 1993-10, and Fig. 8 shows the photograph of Flash 1993-10. This flash was positive and had very long duration exceeding the maximum record length of 72.6ms. The peak value was +32.3kA and many pulses superposed in the beginning.

Table 2 Lightning currents at the tower top.

Flash No.	Lightning	Date Mon./Day	Time (J S T)	Peak value of lightning current [kA]	Duration [ms]
1993-10	Triggered	11/23	23:34:17	+ 32.3	72 <
1993-11	Triggered	11/23	23:53:10	+ 27.4	63
1993-12	Triggered	11/24	00:25:33	+ 53.8	69
1993-19	Natural	11/28	07:21:04	+131.8	35

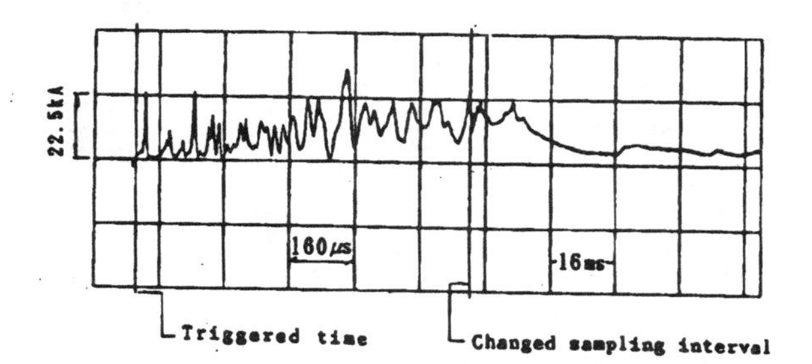

Fig. 7 Waveform of lightning current at tower top (Flash 1993-10).

Fig. 8 Photograph of rocket-triggered lightning to tower (Flash 1993-10).

Fig. 9 shows the measured waveform of each channel. The current pulses at the wavefront flowed mostly into the tower legs, and the currents of CH2 and CH3 on the ground wires were rather smooth. The DC-like current after 1ms mostly flowed into the ground wires. The duration of insulator voltages was short, and the maximum values appeared in the beginning where the current steepness was high. They did not appear after about 1ms reflecting the waveform of the current flowing into the tower.

B. Flash 1993-19 by Natural Lightning

Fig. 10 shows the lightning current waveform of Flash 1993-19, which struck 13 seconds after launching of a rocket, and can therefore be classified as natural lightning. Its peak value of 132kA was the highest one ever observed at the Okushishiku Test Transmission Line. The initially positive bipolar pulse lasted for 110μs, and after that, positive current of about 10kA followed for 35ms.

Fig. 11 shows the measured waveform of each channel. The waveforms of the currents at the tower legs were similar to that at the tower top in the beginning, and the arrester currents began to flow at about 17μs. All the three arresters operated properly, and the insulator voltages of the circuit 2 equipped with the arresters were suppressed as had been intended.

At Flash 1993-19, the peak value of the lightning current was +132kA, but the current flowing into the arrester located on the upper phase, which was the largest among the three phases, had a peak of only +3.0kA with the duration of 7μs. This was small enough compared with the withstand capacity of the arrester.

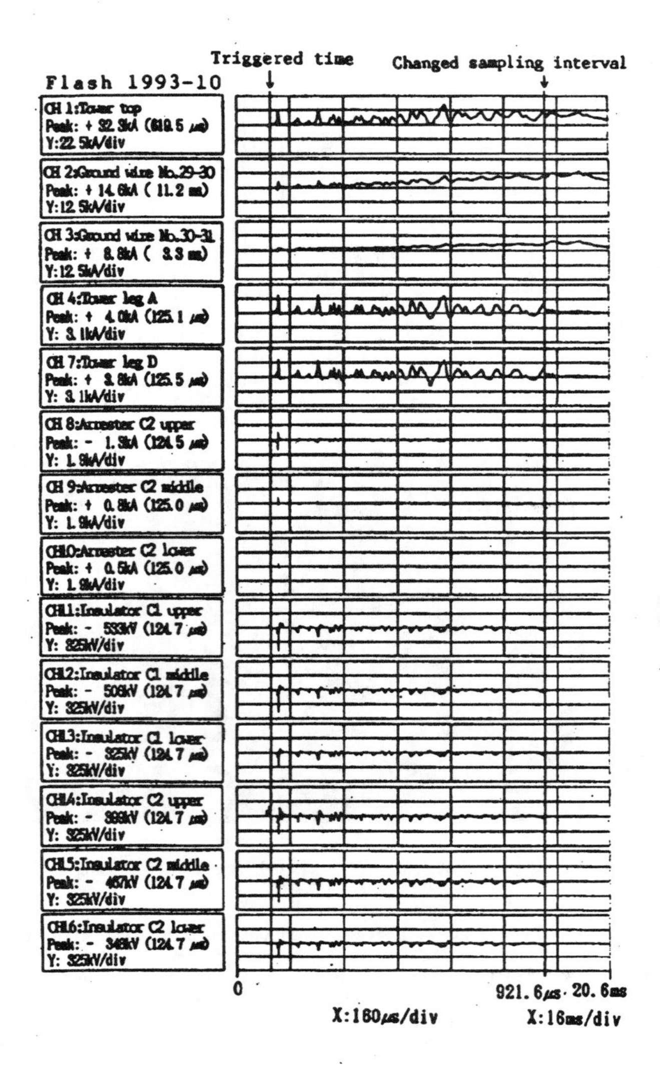

Fig. 9 Current and voltage waveforms (Flash 1993-10).

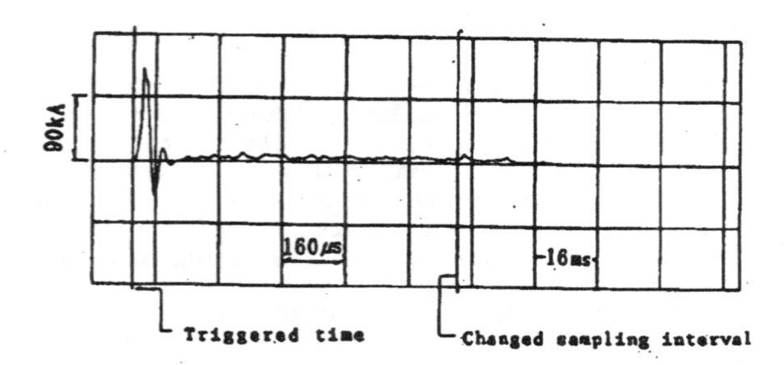

Fig. 10 Waveform of lightning current at tower top (Flash 1993-19).

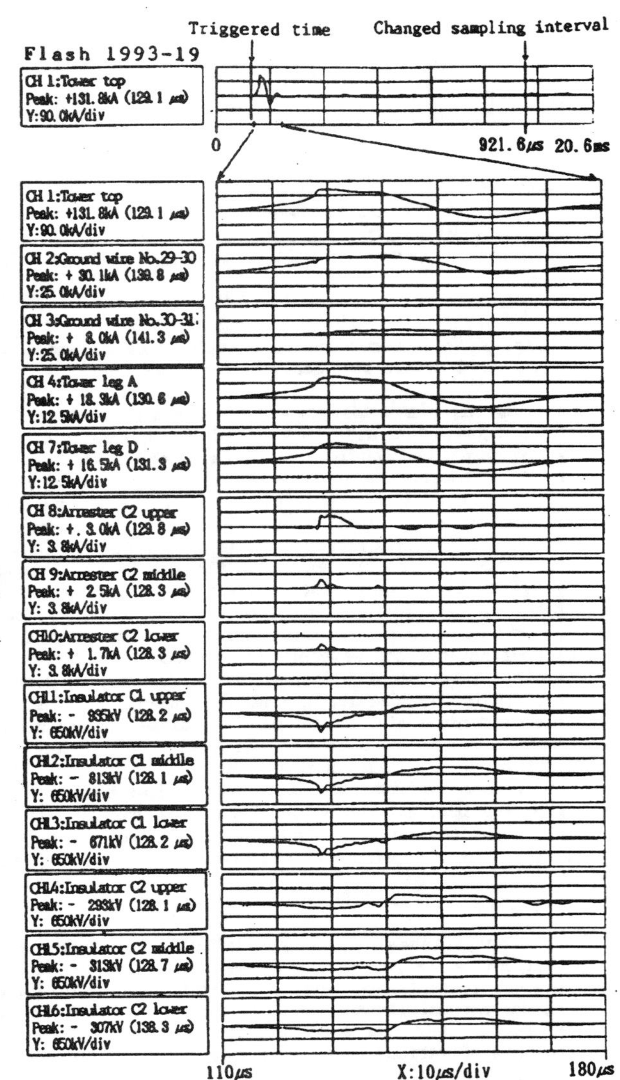

Fig. 11 Current and voltage waveforms (Flash 1993-19).

IV. EMTP ANALYSIS

Accuracy of the model proposed in [8] and [9] was tested by using the measured data of Flash 1993-19. The DCG/EPRI EMTP was used for this analysis.

A. Calculation Condition

Fig. 12 shows the model circuit employed in the EMTP calculation. The J. Marti model of 7 phases was used to simulate frequency dependent attenuation in a multi-phase transmission line. The transmission towers were simulated by using the multistory tower model proposed in [8].

The tower surge impedance Zt1 and Zt2 were supposed to be 220 and 150 Ω (case 1) or 120 and 120Ω (case 2) as had been proposed in [8] and [9], respectively. Cases for other combinations were also calculated to find the combination of Zt1 and Zt2 which gave the best fit between the measured and calculated results. As a result, a combination of 200 and 135Ω (case 3) was finally chosen. A case without arresters (case 4) was calculated for the circuit parameters of case 3 to confirm that the insulator voltages were actually controlled by the arresters.

The grounding resistance of each tower in the model circuit was pure resistance of the measured value. The arrester was modeled by non-linear resistance. The surge impedance of the lightning channel was assumed to be 400Ω. The waveform of the injected current in the EMTP calculation was piecewise linear approximation of the actually measured current waveform at the tower top. Table 3 shows the parameters in the model circuit for the representative four cases. Values of Ri and Li were determined from the attenuation coefficient γ and the length of each section of the tower [8].

B. Results and Discussion

Table 4 shows the measured peak values and the calculated ones. Fig. 13 shows the measured and calculated waveforms.

As shown in Table 4, the calculated peak value of the ground wire current from the No.30 tower to the No.29 tower was lower compared with the measured value, and that from the No.30 tower to the No.31 tower was higher than the measured value. This might have been caused by the induction from the lightning channel, however, the current into the tower was close to the measured value.

Table 3 Tower model constants.

		case 1 [8]	case 2 [9]	case3,4
Tower surge impedance	Zt1	220Ω	120Ω	200Ω
	Zt2	150Ω	120Ω	135Ω
Surge propagation velocity	Vt	300m/μs		
L/R		2T *		
Zt2/Zt1		0.68		
Attenuation coefficient	γ	0.8	0.7	0.8
Damping resistor (Example of No.30 tower)	R1	10.7Ω	9.30Ω	9.70Ω
	R2	21.4Ω	18.6Ω	19.4Ω
	R3	17.1Ω	14.9Ω	15.5Ω
	R4	33.5Ω	42.8Ω	30.1Ω
Damping inductance (Example of No.30 tower)	L1	4.23μH	3.68μH	3.84μH
	L2	8.45μH	7.37μH	7.69μH
	L3	6.76μH	5.89μH	6.15μH
	L4	13.3μH	16.9μH	11.9μH

* T=H/Vt : Tower travel time

Among the combinations of circuit parameters in Table 3, the case 3 gave the best fit for the peak values of measured and calculated voltages and currents. As shown in Fig. 13, the calculated waveforms of the case 3 correspond to the measured waveforms very well except that of CH8. However, the values of 200 and 135Ω for the tower surge impedance in the case 3 are obviously higher than the experimentally determined values for a UHV tower [9]. One of the causes of this difference is supposedly the difference in the orientation of the paths of current injection. In the experiment of the UHV tower, the current was injected from a horizontal wire, whereas the channel of the natural lightning, which is the case of this paper, is believed to be oriented more vertical; the induction from a vertical lightning channel will make the insulator voltage higher than from a horizontal channel, if the currents flowing the channels are the same.

By comparing the cases 3 and 4 in Table 4, equipped with and without arresters, it is obvious that the peak voltages stressing the insulator strings of the circuit 2 are greatly reduced by the arresters. If the arresters were not equipped, the peak voltage across the upper phase insulator strings would have risen higher than 1,150kV. The spacings of the arcing horn gaps are 1.2m, and the insulation strengths of these gaps for chopped impulse voltages having 1.5 μs and 2μs risetimes would be 1,109kV and 987kV, respectively, as per [10]. Therefore, a back flashover would likely have occurred for Flash 1993-19 if the arresters had not been equipped.

Table 4 Measured peak values and calculated peak values by EMTP.

Measured item		Measured peak value	Calculated peak value			
			case 1	case 2	case 3	case 4
Tower top	kA	+132	—	—	—	—
Ground wire No29-30	kA	+30.1	+24.2	+24.1	+24.1	+24.2
Ground wire No30-31	kA	+8.0	+11.6	+11.1	+11.5	+11.9
Tower body	kA	—	+109	+111	+110	+111
Arrester C2 upper	kA	+3.0	+2.9	+2.2	+2.7	—
Arrester C2 middle	kA	+2.5	+2.1	+1.8	+2.0	—
Arrester C2 lower	kA	+1.7	+1.7	+1.7	+1.6	—
Insulator C1 upper	kV	-935	-992	-823	-942	-1,174
Insulator C1 middle	kV	-813	-869	-785	-832	-1,061
Insulator C1 lower	kV	-671	-742	-729	-716	-932
Insulator C2 upper	kV	-293	-336	-323	-332	-1,174
Insulator C2 middle	kV	-313	-327	-319	-325	-1,061
Insulator C2 lower	kV	-307	-320	-317	-318	-932

case 4 : Calculation of the case 4 is without arresters.

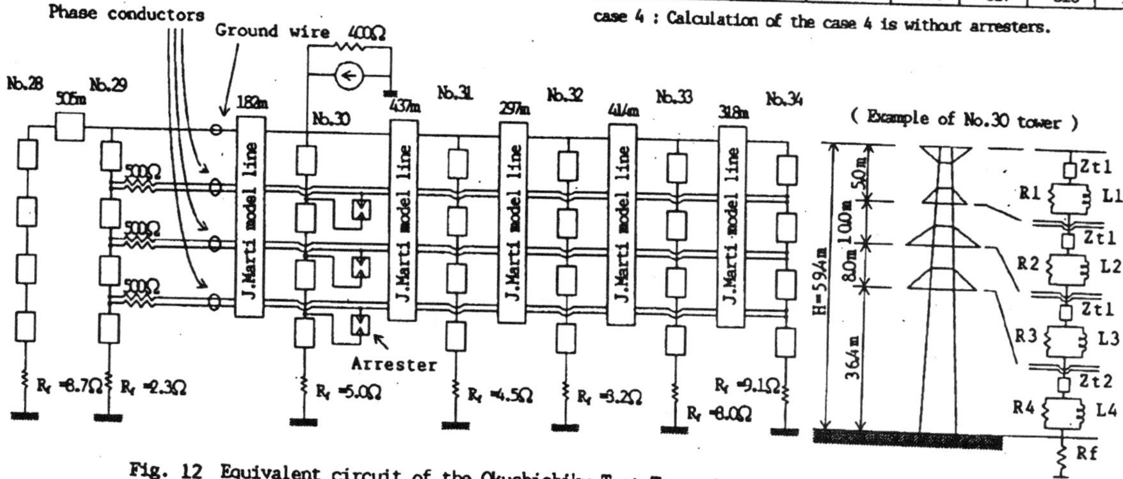

Fig. 12 Equivalent circuit of the Okushishiku Test Transmission Line for EMTP analysis.

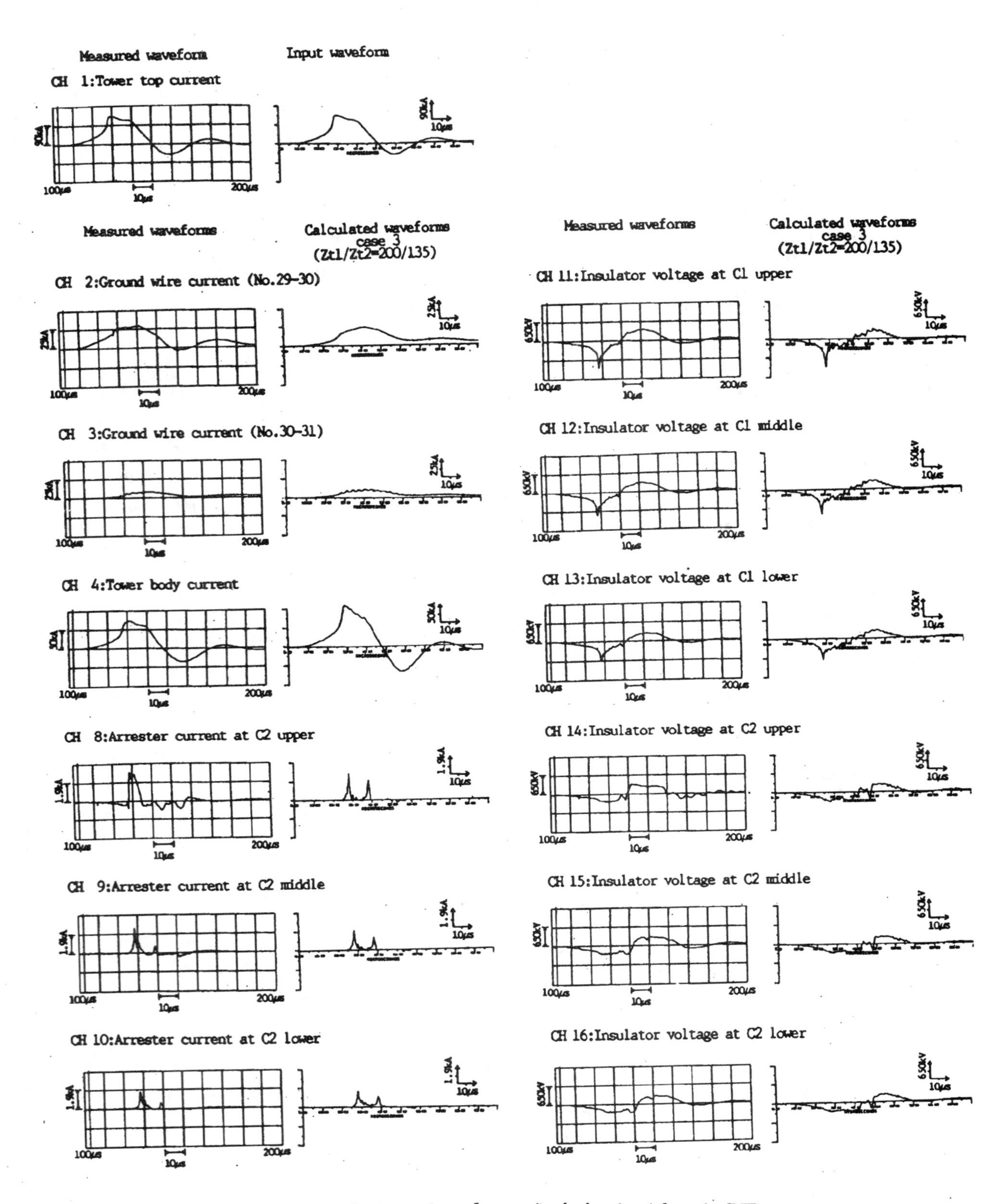

Fig.13 Measured waveforms and calculated waveforms by EMTP.

The current waveform of Flash 1993-19 has a slower risetime and much longer duration compared with typical negative lightning strokes. It is also reported that the average risetimes of lightning currents are slow in winter [3]. The slower risetime results in lower peak insulator voltages, however, at Flash 1993-19, the steep rise of the current just before the peak gave narrow and sharp peaks of the insulator voltages as shown in Fig. 13. Many double circuit outages experienced in winter also do not allow us to be too optimistic in protecting the system due to the observed many slow-front lightning current waveforms. The long duration of the current might be fatal for a transmission line arrester of a lower voltage line such as a 66kV line, if a ground wire is not equipped.

From the above result, it is proved that the EMTP model incorporating the multistory tower model is accurate enough to predict the performance of transmission line arresters. From the analysis of the "Flash 1993-19" data, it is evident that the transmission line arresters are effective in lowering the insulator voltages stressed by lightning surges.

V. CONCLUSIONS

Measurement of lightning surges was carried out at the 275kV Okushishiku Test Transmission Line. The following results were obtained.

1. When natural lightning with a peak value of +132kA struck the tower top equipped with arresters, the lightning currents flowing in various parts and the voltages across the insulator strings were measured simultaneously for the first time.

2. The ability of transmission line arresters to lower the insulator voltages was proved by a natural lightning stroke. Also it was confirmed that the arrester performed as designed.

3. The analysis by EMTP using the J. Marti model and the multistory transmission tower model could simulate actual waveforms and peak values of lightning surges accurately. Therefore, this method for estimation of the effect of installing transmission line arresters was proved to be useful.

VI. ACKNOWLEDGMENT

The authors are grateful to the university staffs who joined the rocket-triggered lightning project for their cooperation.

VII. REFERENCES

[1] The Working Group on Lightning Protection for Transmission Systems, "Lightning Observation on Japan Sea Coast in Winter", CRIEPI Report, T11, 1989 (in Japanese).

[2] IEEE Working Group Report, "Estimating Lightning Performance of Transmission Line II -Updates to Analytical Models", IEEE Trans., PWRD-8, No.3, pp.1254-1267, 1993.

[3] K. Miyake et al., "Characteristics of Winter Lightning Current on Japan Sea Coast", IEEE Trans., PWRD-7, No.3, pp.1450-1456, 1992.

[4] S. Furukawa et al., "Development and Application of Lightning Arresters for Transmission Lines", IEEE Trans., PWRD-4, No.4, pp.2121-2129, 1989.

[5] K. Ishida et al., "Development of a 500kV Transmission Line Arrester and its Characteristics", IEEE Trans., PWRD-7, No.3, pp.1265-1274, 1992.

[6] T. Yamada et al., "Development of Suspension-Type Arresters for Transmission Lines", IEEE Trans., PWRD-8, No.3, pp.1052-1060, 1993.

[7] K. Nakamura et al., "Artificially Triggered Lightning Experiments to an EHV Transmission Line", IEEE Trans., PWRD-6, No.3, pp.1311-1318, 1991.

[8] M. Ishii et al., "Multistory Transmission Tower Model for Lightning Surge Analysis", IEEE Trans., PWRD-6, No.3, pp.1327-1335, 1991.

[9] T. Yamada et al., "Experimental Evaluation of a UHV Tower Model for Lightning Surge Analysis", IEEE Winter Meeting, 94 WM 044-8 PWRD, 1994.

[10] J. G. Anderson, "Transmission Line Reference Book 345kV Above / Second Edition", EPRI, Palo Alto, California, pp. 563-564, 1982.

Yasuhiro Matsumoto was born in Toyama, Japan, on September 3, 1956. He received the B.S. degree in electrical engineering from Niigata University, Japan, in 1979. In the same year he joined Hokuriku Electric Power Co., Toyama, Japan, and now with the company's Engineering Research & Development Center.

Osamu Sakuma was born in Toyama, Japan, on October 29, 1959. He graduated from Toyama Technical High School, Japan, in 1978. In the same year he joined Hokuriku Electric Power Co., Toyama, Japan, and now with the company's Engineering Research & Development Center.

Kazuo Shinjo was born in Toyama, Japan, on August 7, 1960. He received the B.S. degree in electrical engineering from Kanazawa University, Japan, in 1983. In the same year he joined Hokuriku Electric Power Co., Toyama, Japan, and now with the company's Engineering Research & Development Center.

Masaki Saiki was born in Toyama, Japan, on July 5, 1954. He received the B.S. degree in electrical engineering from Nagoya Institute of Technology and the M. Eng. degree from Kanazawa University, Japan, in 1977 and 1979, respectively. In the same year he joined Hokuriku Electric Power Co., Toyama, Japan, and now with the company's Engineering Research & Development Center.

Takeo Wakai was born in Toyama, Japan, on December 18, 1941. He received the B.S. degree in electrical engineering from Kanazawa University, Japan, in 1964. In the same year he joined Hokuriku Electric Power Co., Toyama, Japan, and now is a Deputy General Manager of Engineering Research & Development Center.

Tsutomu Sakai was born in Toyama, Japan, on April 14, 1942. He graduated from Toyama Technical High School, Japan, in 1961. In the same year he joined Hokuriku Electric Power Co., Toyama, Japan, and now is an Assistant Manager of Engineering Research & Development Center.

Hideo Nagasaka was born in Toyama, Japan, on December 1, 1942. He received the B.S. degree in electrical engineering from Yokohama National University, Japan, in 1966. In the same year he joined Hokuriku Electric Power Co., Toyama, Japan, and now is the General Manager of Engineering Research & Development Center.

Hideki Motoyama (M'93) was born in Hokkaido, Japan, on October 8, 1961. He received the B.S. and the M. Eng. Degrees in electrical engineering from Doshisha University, Kyoto, Japan, in 1985 and 1987, respectively. In the same year he joined Central Research Institute of Electric Power Industry, Tokyo, Japan. He has been engaged in the insulation problems of electric power systems.

Masaru Ishii (SM'87) was born in Tokyo, Japan, on March 11, 1949. He received the B.S., the M.S. and the Dr. Eng. degrees from the University of Tokyo in 1971, 1973 and 1976, respectively. He is a professor of the Institute of Industrial Science, University of Tokyo, Japan. He has been involved in the field of high voltage engineering.

IEEE Transactions on Power Delivery, Vol. 9, No. 4, October 1994

EMTP-Based Model for Grounding System Analysis

Frank E. Menter, Member, IEEE
Technical University of Aachen
Aachen, Germany*)

Leonid Grcev, Member, IEEE
University of Skopje
Skopje, Republic of Macedonia

***Abstract* — EMC and lightning protection analyses of large power systems require the knowledge of the dynamic behavior of extended grounding systems. They cannot be regarded as equipotential planes, but must be treated as coupling paths for transient overvoltages. This contribution presents a model for linear earth conductors based on the transmission line approach and outlines its integration in the transients program EMTP. Validation of the presented model is achieved by comparison with field measurements and with a rigorous electromagnetic model. Overvoltages and electrical fields throughout electrical power systems thus can be computed.**

KEYWORDS - Grounding Systems, Buried Conductors, EMC, Lightning Protection, Computer Modellling, Electromag netic Fields, EMTP

I. INTRODUCTION

Grounding systems serve two purposes: Firstly, their task is to disperse fault currents into the earth, which can be evoked either by internal, unbalanced faults or by external sources, such as lightning. Secondly, extended grounding systems create a reference potential for all electric and electronic apparatus making up a large-scale system. Local inequalities of this reference potential and disturbances conducted across the grounding system reportedly are a source of malfunction and destruction of components in electrical connection with the grounding system. This contribution treats grounding systems from the point of view of the electromagnetic compatibility (EMC) and thus regards a grounding system as a path of coupling between a source of interference, usually being an impressed fault or lightning current, and a consequence, typically a transient overvoltage or field stress, at the location of a disturbed object.

94 WM 135-4 PWRD A paper recommended and approved by the IEEE Substations Committee of the IEEE Power Engineering Society for presentation at the IEEE/PES 1994 Winter Meeting, New York, New York, January 30 - February 3, 1994. Manuscript submitted July 29, 1993; made available for printing December 22, 1993.

Since in most cases EMC studies require the knowledge of the system's performance over a wide frequency range, the presented methodology extends the range of validity of the well-known models for grounding systems towards higher frequencies in the order of magnitude of some megahertz. An approach could be based on the three basic concepts developed thus far. These are:

— *The Network Approach*, which models an earth conductor as equivalent π-circuits made up of lumped *R-L-C* elements. The coupling of earth conductors can be taken into account by mutually coupled inductances. Among others, Velazquez and Mukhedkar describe the procedure [1].

— *The Transmission Line Approach* was brought to practical applicability by Sunde [2]. The topology of the network of interconnected linear ground conductors is treated by the travelling wave technique pioneered by Bergeron. Nowadays, Papalexopoulos and Meliopoulos are basing their work on this method (e.g., [3]).

— *The Electromagnetic Field Approach* exhibits the most rigorous theoretical background of all three approaches. Strictly based on the theorems of electromagnetism and with the least neglects possible, the problems are defined in terms of retarded potentials, and among the possible strategies for their solution, the method of moments proved to be most efficient. Dawalibi could translate the highly complex relationships into practical, engineering programs [4].

For a model valid in the envisaged frequency range, the network approach was assumed not to be appropriate. In contrast, this contribution shall point out that the transmission line approach is a suitable and practical choice for the determination of both the transient ground potential rise and the electric fields in the vicinity of the grounding system. An existing program based on the electromagnetic field approach could validate this statement.

A brief review of Sunde's transmission line theory and the evaluation of the transmission line characteristics for common earth electrodes introduces this paper, followed by the description of an interface to the ATP version [5] of the wide-spread *Electromagnetic Transients Program* (EMTP) [6]. A post processor which computes electric fields in the soil from the current distribution in the

*) At present with SIEMENS AG, Transportation Systems, Erlangen, Germany

grounding system is sketched next, preceding the presentation of field measurements and of comparisons with a rigorous electromagnetic model. Finally, a lightning protection study for a 123 kV substation underlines the great advantage of not having just a stand-alone grounding systems program, but a tool capable of analyzing specific grounding systems in conjunction with the entire electrical system.

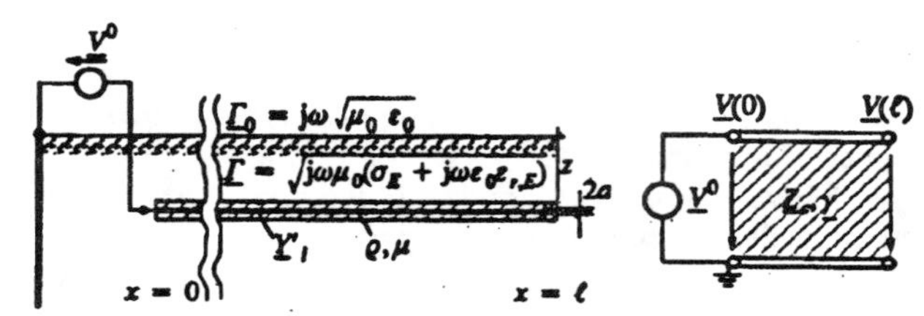

Fig. 1. Horizontal earth electrode and transition to an equivalent transmission line segment

II. TRANSMISSION LINE MODEL FOR BURIED CONDUCTORS

Fig. 1 illustrates the transition from a linear earth electrode to an equivalent transmission line with complex-valued, frequency dependent parameters $\underline{Z}_c(\omega)$ and $\underline{\gamma}(\omega)$, which denote the characteristic impedance and propagation function, respectively:

$$\underline{Z}_c(\omega) = \sqrt{\frac{\underline{Z}'(\omega)}{\underline{Y}'(\omega)}} \tag{1}$$

$$\underline{\gamma}(\omega) = \sqrt{\underline{Z}'(\omega)\cdot\underline{Y}'(\omega)} \tag{2}$$

The angular frequency, $\omega = 2\pi f$, uses f in Hz ($j = \sqrt{-1}$). This form corresponds to any two-conductor transmission line. Sunde derived equivalent expressions for a single conductor in contact to the soil, with the current returning through the earth. Both the longitudinal impedance per unit length, $\underline{Z}'$, and the transversal admittance per unit length, $\underline{Y}'$, of a horizontal conductor consist of an internal term and an earth return term in the following manner [2]:

$$\underline{Z}' \approx \underline{Z}_i' + \frac{j\omega\mu_0}{2\pi}\cdot\log\frac{1.85}{\sqrt{\underline{\gamma}^2+\underline{\Gamma}^2}\cdot\sqrt{2az}} \tag{3}$$

$$\underline{Y}'^{-1} \approx \underline{Y}_i'^{-1} + \frac{1}{\pi(\sigma_E + j\omega\varepsilon_0\varepsilon_{r,E})}\cdot\log\frac{1.12}{\underline{\gamma}\cdot\sqrt{2az}} \tag{4}$$

Here, $\underline{Z}_i'$ denotes the internal impedance of a conductor of radius, a, buried at a depth, z, which is mainly governed by the skin effect, whereas $\underline{Y}_i'$ stands for the insulation admittance of an eventual coating, through which the conductor is in contact to the surrounding medium. The latter is characterized by the earth's conductivity, σ_E, relative permeability, $\varepsilon_{r,E}$, and permittivity, μ_0. All three lead to a propagation function,

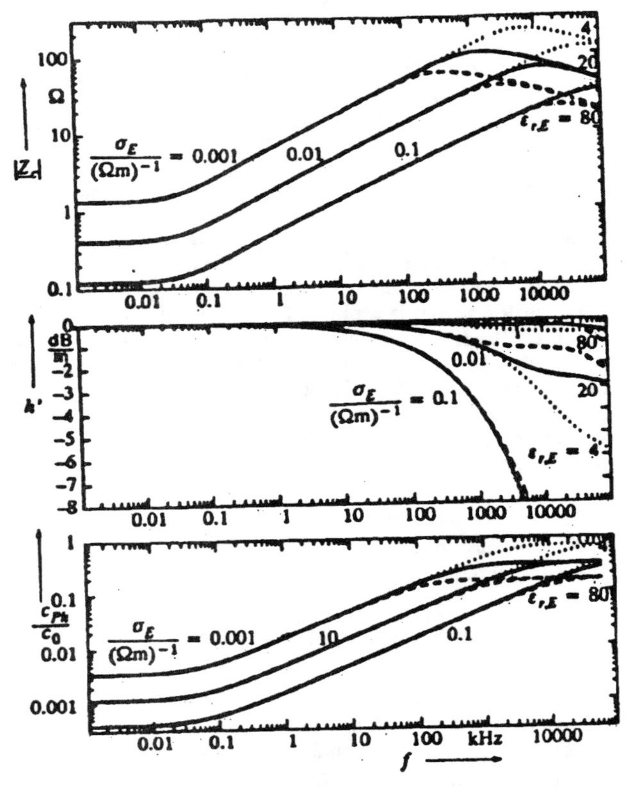

Fig. 2. Surge impedance, attenuation and phase velocity of a horizontal, bare copper wire (35 mm²) buried at a depth of 0.5 m

$$\underline{\Gamma}(\omega) = \sqrt{j\omega\mu_0\cdot(\sigma_E + j\omega\varepsilon_0\varepsilon_{r,E})} \quad , \tag{5}$$

which would govern the transmission of impulses along the conductor if it were imbedded in a homogeneous soil with these parameters. Plumey, Kouteynikoff et al. derived analogous expressions for vertical ground rods [7], reading

$$\underline{Z}' \approx \underline{Z}_i' + \frac{j\omega\mu_0}{2\pi}\cdot\log\frac{1.12}{\sqrt{\underline{\gamma}^2+\underline{\Gamma}^2}\cdot a} \tag{6}$$

$$\underline{Y}'^{-1} \approx \underline{Y}_i'^{-1} + \frac{1}{2\pi(\sigma_E + j\omega\varepsilon_0\varepsilon_{r,E})}\cdot\log\frac{\sqrt{\underline{\gamma}^2+\underline{\Gamma}^2}\cdot a}{3.56} \tag{7}$$

Equation (2) together with (3) and (4) leads to a complex-valued, transcendent equation for the propagation function

$$\underline{\gamma} = \sqrt{\underline{Z}'(\underline{\gamma})\cdot\underline{Y}'(\underline{\gamma})} \quad , \tag{8}$$

which is repeatedly solved for all frequencies under investigation by means of a standard IMSL routine [8,9].

Fig. 2 shows the resulting quantities as functions of frequency for a variety of soil parameters σ_E and $\varepsilon_{r,E}$. It is

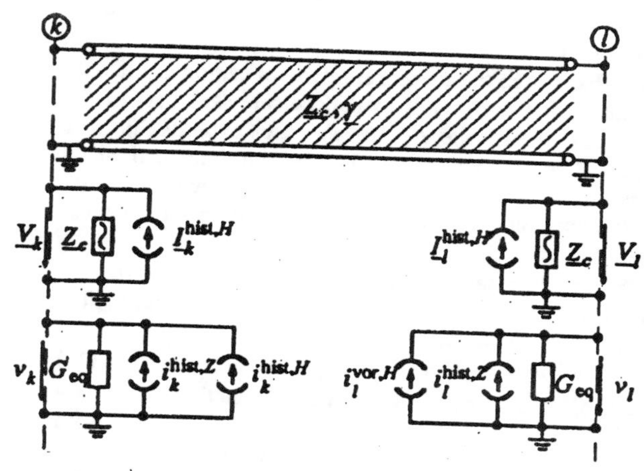

Fig. 3. Equivalent circuit for frequency-dependent transmission line

common engineering practice to express the propagation function in terms of transfer function, $\underline{H}=\exp\{-\underline{\gamma}\cdot\ell\}$, or attenuation per unit length, h', and phase velocity, c_{Ph}:

$$h'(\omega)=\frac{20\text{ dB}}{\ell}\cdot\log\left|\underline{H}(\omega)\right|\cong-8.686\text{ dB}\cdot\operatorname{Re}\{\underline{\gamma}(\omega)\} \quad (9)$$

$$c_{Ph}(\omega)=\frac{\omega}{\operatorname{Im}\{\underline{\gamma}(\omega)\}} \quad (10)$$

Only for lower frequencies, the properties of the ground are exclusively determined by the conductivity of the earth. Above some tens of kilohertz, its dielectric nature has to be taken into account.

III. Usage of EMTP for Grounding System Analysis

Once the characteristic impedance and the transfer function of linear earth conductors are known, any kind of extended grounding system can be modelled by a network of transmission line segments, provided that inductive and capacitive coupling between the different line segments can be neglected. At first sight, it is not evident that this assumption is permissible, but the purpose of this contribution is to show that for the grounding systems analyzed thus far, the resulting error is within acceptable limits. From an engineering point of view, the gained versatility and flexibility in modelling an actual earthing system together with live parts of the installation compensates the loss in accuracy.

A. Interface of the Ground Conductor Model to EMTP

Live and earthed parts of the electrical system are modelled by the Electromagnetic Transients Program. EMTP incorporates Bergeron's method for transmission lines, i. e. replacing the line by Thevenin equivalent circuits at both terminals, consisting of the constant and real-valued characteristic impedance and a current source for the history of the travelling waves with the travel time τ (Fig. 3, upper index/sign for forward travelling wave, lower index/sign for backward travelling wave):

$$i_{l/k}^{\text{hist}}(t)=\frac{1}{Z_c}\cdot v_{k/l}(t-\tau)\pm i_{k/l}(t-\tau) \quad (11)$$

In the frequency domain, the transition to complex-valued, frequency-dependent parameters necessary for earth conductors is simply,

$$\underline{I}_{l/k}^{\text{hist},H}=\underline{H}\cdot(\underline{Z}_c^{-1}\cdot\underline{V}_{k/l}\pm\underline{I}_{k/l}) \quad , \quad (12)$$

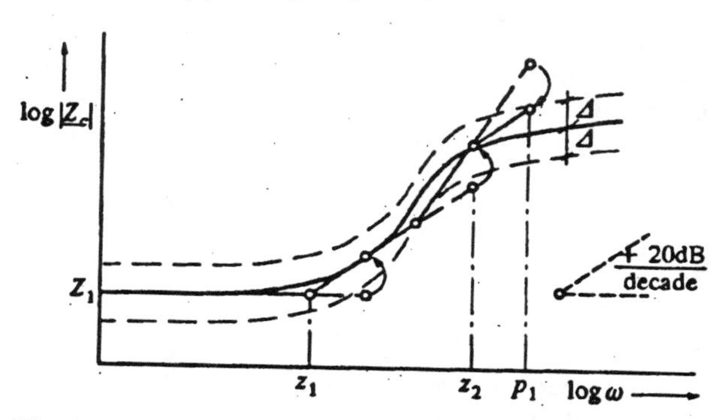

Fig. 4-a. Marti's strategy of approximation applied to the modulus of the characteristic impedance

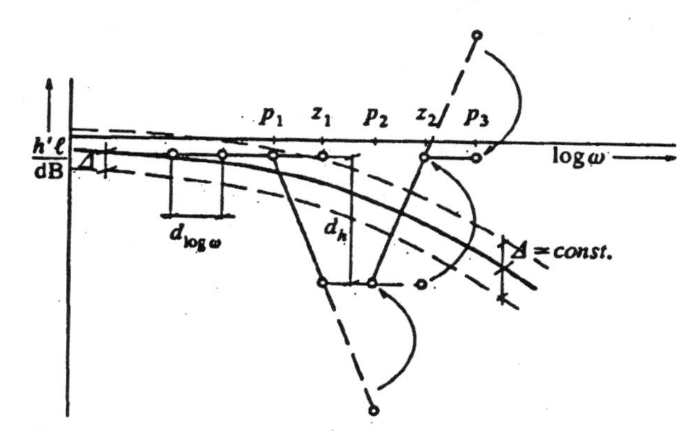

Fig. 4-b. Occurrence of numerical oscillations by Marti's strategy if applied to the attenuation

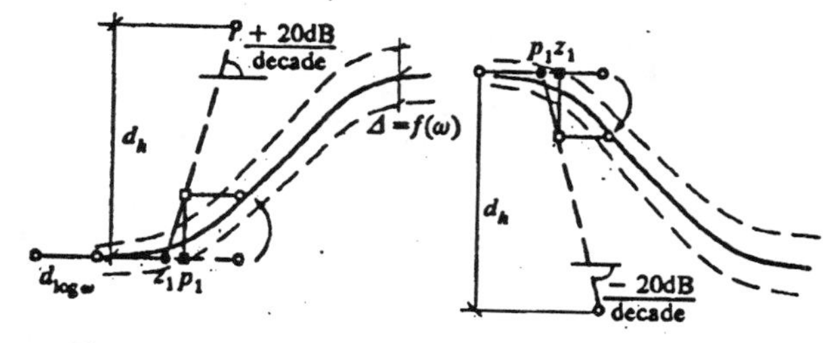

Fig. 4-c. Remedy by modified strategy (cf. text)

but since EMTP works in the time domain, the back transformation leads to two nested convolution integrals:

$$i_l^{hist,H}(t) = h_s(t) * [\mathcal{F}^{-1}\{\underline{Z}_c^{-1}(\omega)\} * v_k(t) \pm i_k(t)] \quad (13)$$

The numerical effort for the convolutions can be reduced drastically if the technique of the so-called recursive convolution can be applied [10]. For this purpose, it is necessary to approximate the characteristic impedance and the transfer function by rational functions, i. e. it is mandatory to identify their poles and zeroes. Martí's original strategy ([11], Fig. 4-a) of proceeding form lower to higher frequencies in a log-log-chart and placing a pole or zero wherever an asymptotic approximation leaves a tolerance strip of width 2Δ is not adequate in the present case: In contrast to its original field of application, observed frequency, characteristic impedance and attenuation cover several orders of magnitude, and thus frequently cause numerical oscillations (Fig. 4-b).

A remedy could be found by the introduction of a relative error criterion for the tolerance strip and by the possibility of placing closely adjacent pole/zero-pairs in order to jump exactly on the target function ([12], Fig. 4-c).

B. Modelling the Soil Ionization with EMTP

If large current densities emanate from the conductor into the soil, the critical field strength can be exceeded and breakdowns occur. Then, the conductor will be surrounded by a cylindrical corona-type discharge pattern, which augments the available surface for the transition of the current from the conductor to the earth. The breakdown strength, E_{crit}, of the soil is a function of its conductivity, measurements [13] lead to the empirical formula

$$E_{crit} = 241 \cdot \sigma_E^{-0.215} \quad (14)$$

for E_{crit} in kV/m and the soil conductivity, σ_E, in $(\Omega m)^{-1}$. The distance from the center of the earth conductor at which the ionization of the soil ceases, may be regarded as an effective radius. Ohm's law governs the relationship between current density at a distance, a, from the conductor and field in the soil,

$$\frac{\partial i / \partial x}{2\pi a} = \sigma_E \cdot E \quad . \quad (15)$$

The combination of the two latter expression for $a = a_{eff}$ and $E = E_{crit}$ yields

$$a_{eff} = \frac{1}{2\pi \cdot 241} \cdot \frac{\partial i}{\partial x} \cdot \sigma_E^{-0.785} \quad (16)$$

for the effective radius, a_{eff}, in m, and the leakage current per unit length, $\partial i/\partial x$, in kA/m. The seemingly increased circumference of the conductor will affect only the transversal admittance, $\underline{Y}'(\partial i/\partial x)$, whereas the by far most part of the longitudinal current will stay within the conductor: the longitudinal impedance, $\underline{Z}'$, remains unchanged by ionization phenomena.

The basic idea of incorporating an ionization model into the simulation network is to concentrate that part of the transversal admittance, $\Delta\underline{Y}'(\partial i/\partial x)$, which exceeds the value disregarding the ionization, $\underline{Y}_0'$, at the intersections of transmission line segments, which can be generated without the necessity of taking the soil ionization into account (Fig. 5).

The implementation of the model makes use of the observation that the additional transverse admittance, $\Delta\underline{Y}'(\omega, \partial i/\partial x)$, can be decomposed into

$$\Delta\underline{Y}'(\partial i / \partial x, \omega) \approx \kappa(\partial i / \partial x) \cdot \Delta\underline{Y}_1'(\omega) \quad , \quad (17)$$

where $\Delta\underline{Y}_1'$ is a specific fundamental function and κ is a multiplier, only depending on the leakage current. Both functions can be determined by formal variations of ω and $\partial i/\partial x$ before the simulation run. At the beginning of the run, $\Delta\underline{Y}_1'$ is similarly approximated by its poles and zeroes, which are passed to EMTP's *laplace xform* module for the representation of arbitrary, frequency dependent impedances. During the run, the additional transversal admittances are scaled by the multiplier κ, whose locally valid actual argument, $\partial i/\partial x(x)$, is determined at each time step from the difference of the longitudinal currents and the length, ℓ, of the corresponding transmission line segment. These run-time evaluations, the interpolation on the pointlist function $\kappa(\partial i/\partial x)$ and the scaling of all elements $\Delta\underline{Y}_1'$ is performed by the recently developed tool *models* of the ATP version of EMTP [14,15].

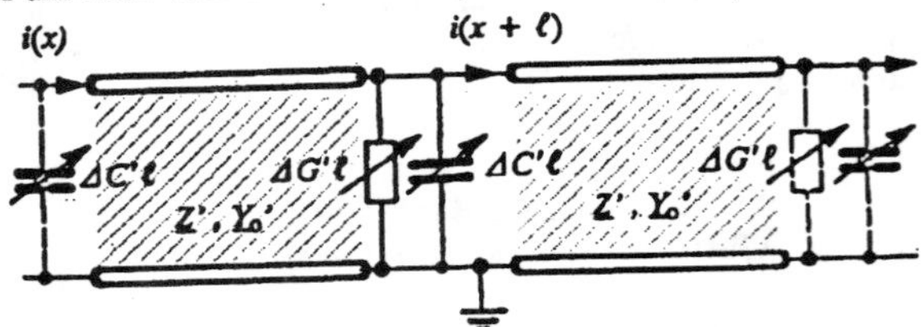

Fig. 5. Leakage current controlled lumped elements to model soil ionization

IV. Computation of the Electric Field Distribution

Considering the possibility to extend the application of the developed model to EMC studies, the electric field vector is chosen as a key quantity. The main reason for this is the well known fact that in the general case the voltage between points along specified path is not equal to the potential difference, but is defined as a line integral of the electric field vector along the path. In order to compute the electric field distribution in the vicinity of the grounding system, a post processor to EMTP has been developed. Since the grounding system is modelled as a network of arbitrarily connected or disconnected straight thin conductors, the electric field can be computed as a superposition of the contributions of field-generating currents on all straight conductors, which are readily available as EMTP results.

Usually, in order to avoid complications in the solution, only the electric charge distribution in the grounding system conductors is considered as a source of the electric field. This is equivalent to consider only the leakage currents from the conductors in the evaluation of the electric field and the voltages [16]. In other words, this simplification is based on the neglect of the time-varying longitudinal current in the ground conductors as an additional source of the electric field. But it has been shown [17] that such simplification can lead to extremely wrong results in the computation of voltages. The procedure used here takes into account both components of the electric field, due to the electric charges and longitudinal current distribution in all the ground conductors.

In this study, a frequency-domain approach is used, i.e. fields are computed from the steady-state current distribution for $f = 0 \ldots 1$ MHz. This yields the transfer function for the electric field at prescribed observation points in the vicinity of the grounding system. Such transfer functions subsequently can be Fourier transformed to obtain the time-domain response.

Grounding conductors are divided in a number of smaller segments in the EMTP and the longitudinal currents are determined only in the end or junction points of the segments. The current in all other points can be determined by interpolation. Among the many choices for the interpolating function, one is exceptionally attractive for our purposes: the so called piecewise sinusoidal approximation of the current distribution $\underline{I}(\ell)$ that can be expressed by:

$$\underline{I}(\ell) = \sum_{k=1}^{N} \frac{P_k(\ell)}{\sinh(\underline{\Gamma} d_k)} \left\{ \underline{I}_k^- \sinh\left[\underline{\Gamma}\left(\ell_k^+ - \ell\right)\right] + \underline{I}_k^+ \sinh\left[\underline{\Gamma}\left(\ell - \ell_k^-\right)\right] \right\} \tag{18}$$

$\underline{\Gamma}$ is explained above (5), and $P_k(\ell)$ are unit pulse functions with the values

$$P_k(\ell) = \begin{cases} 1, & \ell_k^- \le \ell \le \ell_k^+ \\ 0, & \text{elsewhere} \end{cases} \tag{19}$$

Here, ℓ denote points along the axis of the conductors, ℓ^-_k is the first, ℓ^+_k is the second end point and d_k is the length of the k-th segment. $\underline{I}^-_k$ and $\underline{I}^+_k$ are current phasors at the first and the second end point of the k-th segment, respectively.

The main reason for the approximation of the current with sinusoidal functions on the segments (18) is to exploit their properties revealed by Schelkunoff [18]. The sinusoidal line current source is probably the only finite source with simple and exact closed form expressions for the near fields [19]. In this study, the influence of the interface between the air and the earth is taken into account approximately by the modified image theory [16].

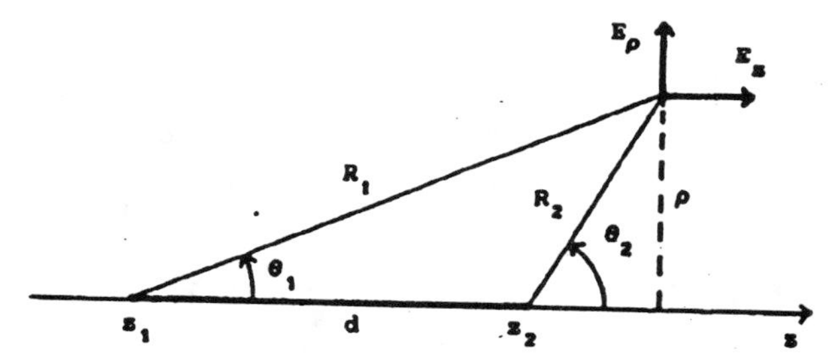

Fig. 6. Computation of the electric field distribution

By this way the electric field at a point in the earth due to the sinusoidal sources on the k-th segment is obtained by superposition of the field of the original and image sinusoidal sources both placed in the unbounded conducting medium. The exact expressions for the electric field in a local cylindrical co-ordinate system illustrated in Fig. 6, due to the current distribution given in (18) on the k-th segment at a near point are:

$$\underline{E}_{k\rho} = \frac{\underline{\eta}}{4\pi\rho\sinh(\underline{\Gamma}d_k)}\Big[\left(\underline{I}_k^- e^{-\underline{\Gamma}R_1} - \underline{I}_k^+ e^{-\underline{\Gamma}R_2}\right)\sinh\underline{\Gamma}d_k + \left(\underline{I}_k^- \cosh\underline{\Gamma}d_k - \underline{I}_k^+\right)e^{-\underline{\Gamma}R_1}\cos\theta_1 + \left(\underline{I}_k^+ \cosh\underline{\Gamma}d_k - \underline{I}_k^-\right)e^{-\underline{\Gamma}R_2}\cos\theta_2 \Big]$$

$$\underline{E}_{kz} = \frac{\underline{\eta}}{4\pi\sinh\underline{\Gamma}d_k}\left[\left(\underline{I}_k^- - \underline{I}_k^+\cosh\underline{\Gamma}d_k\right)\frac{e^{-\underline{\Gamma}R_2}}{R_2} + \left(\underline{I}_k^+ - \underline{I}_k^-\cosh\underline{\Gamma}d_k\right)\frac{e^{-\underline{\Gamma}R_1}}{R_1}\right] \tag{20}$$

where $\underline{\eta}$ is the intrinsic impedance of the medium:

$$\underline{\eta} = \sqrt{\frac{j\omega\mu_0}{j\omega\varepsilon_0\varepsilon_{r,E} - \sigma_E}} \tag{21}$$

The various geometrical quantities are illustrated in Fig. 6. The references concerning the derivation of (20) can be found elsewhere [19].

A brief description of the steps involved in the derivation are included in the Appendix for completeness.

V. Verification and Application

A. Comparison with Field Measurements by EDF

The Direction des Études et Recherches of the Électricité de France granted an insight in their recordings of extensive field measurements performed in the mid-80's [20]. Impulse currents with front times down to 0.2 µs have been fed into single- and multi-conductor earthing arrangements as used industrially. The resulting

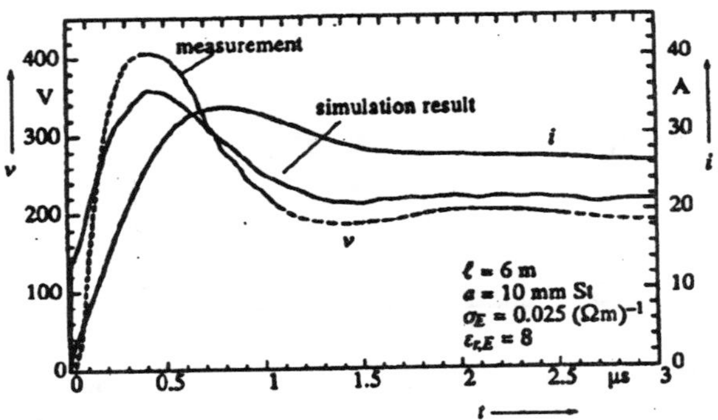

Fig. 7. Measured and calculated ground potential rise of a ground rod

transient ground potential rise has been measured by means of a 60 m long ohmic divider, still having a measuring bandwidth of 3 MHz.

Fig. 7 shows the oscillograms of voltage and current as recorded on a ground rod (steel, 20 mm in diameter, other parameters in the figure). Included in the figure is the simulation result of the voltage when a current according to the measured one is impressed in the corresponding transmission line segment. The peak of the measured voltage curve is always higher than the corresponding value of the simulation, which can be explained by some remaining inductive voltage drop during the wave front along the divider added to the actual potential rise at the clamp of the ground rod.

The effective grounding impedance (Fig. 8) has been determined by a division of the Fast Fourier Transforms of measured voltage and current on a 10 m long horizontal ground wire made of copper alloy. The corresponding simulation curve has been generated by EMTP's frequency scan option, requesting a repeated set of phasor solutions of the simulation network. The curves show good agreement up to frequencies of 1 MHz, the underestimation at higher frequencies may be explained by the above-mentioned effect.

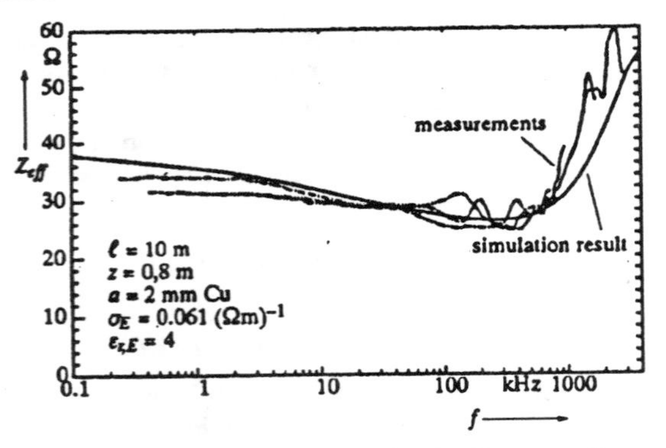

Fig. 8. Measured and calculated effective grounding impedance of a horizontal earth conductor

More complex grounding arrangements have been studied, too. Fig. 9 depicts the results in time domain of a tower footing, consisting of a set of four ground rods and two square loops (6 m × 6 m) in 1 m and 4 m depth, respectively. Except for a 200 ns long period during the wave front, the agreement between measurement and simulation is satisfactory.

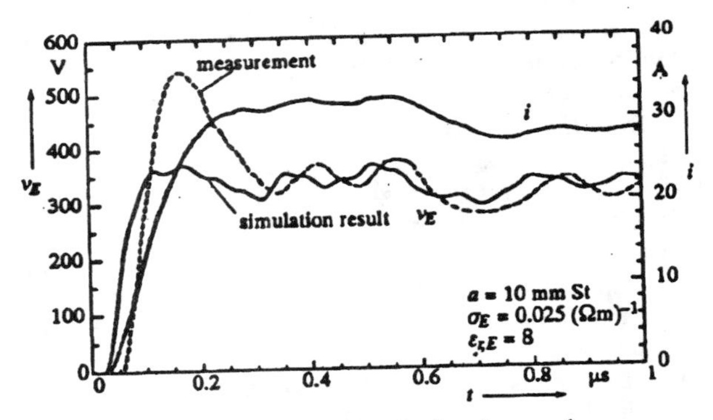

Fig. 9. Measured and calculated ground potential rise of a tower footing

B. Comparison with a Rigorous Electromagnetic Model

As an example, a grid-type grounding system of 10 m × 10 m with 3 × 3 meshes has been studied. A sinusoidal current of 1 kA and variable frequency is fed into the slightly asymmetric arrangement at its center ((x;y) = (5 m;5 m)), the first regular mesh branch being reached at (x;y) = (5 m;6.67 m). The conductors have a diameter of 10 mm, are assumed to be of ideal conductivity and are buried at a depth of 0.5 m in a soil with $\sigma_E = 0.01\ (\Omega m)^{-1}$ and $\varepsilon_{r,E} = 1$.

Since the local distribution of longitudinal and leakage currents within the grounding system is only an intermediate quantity in the computation of touch and step voltages above the grounding system, a comparison of the resulting electric fields at the earth's surface has been performed. Firstly, the current distribution in the grounding system according to the transmission line approach outlined above has been taken as input to the post processor described in section IV. The resulting field distribution is compared to a second, independent computation according to the rigorous electromagnetic approach developed by the second author [21]. The results are compared in Fig. 10.

The transmission line model slightly overestimates the results of the rigorous model, which is taken as a reference here. But still the interesting phenomena are modelled correctly, and the results are obtained within a fraction of the computation time required for the rigorous approach.

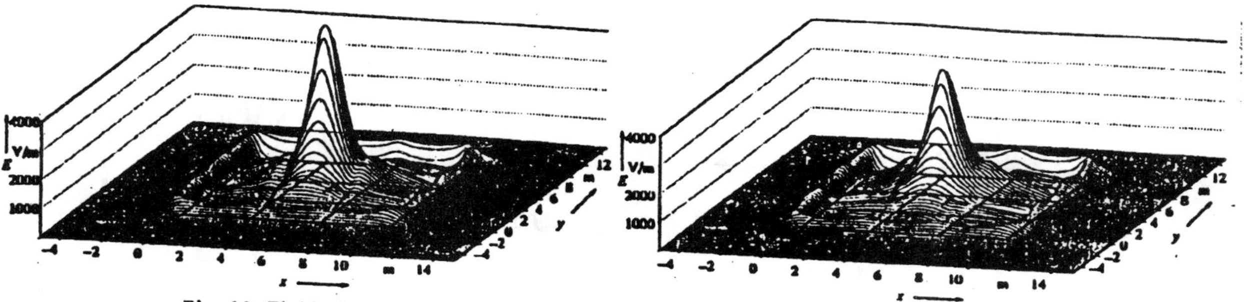

Fig. 10. Field distribution above a mesh-type grounding system. Comparison for $f = 1$ MHz of the transmission line approach (left) and the rigorous electromagnetic approach (right)

C. Lightning Protection study for a 123 kV substation

This application example sketches the procedure and results of a lightning protection study for a 123 kV substation. The lightning is assumed to hit the third tower of an overhead line (Fig. 11) and should have a crest value of 200 kA and an impulse shape as recommended in [22]. The EMTP data deck for the two-system, three-phase overhead line as well as the representation of the switchyard (voltage transformers, surge arresters, busbars and power transformer) has been made available by a CIGRÉ working group [23]. It has been combined with a detailed description of the tower footings and the meshed grounding system of the switchyard (70 m × 50 m) by means of the transmission line model.

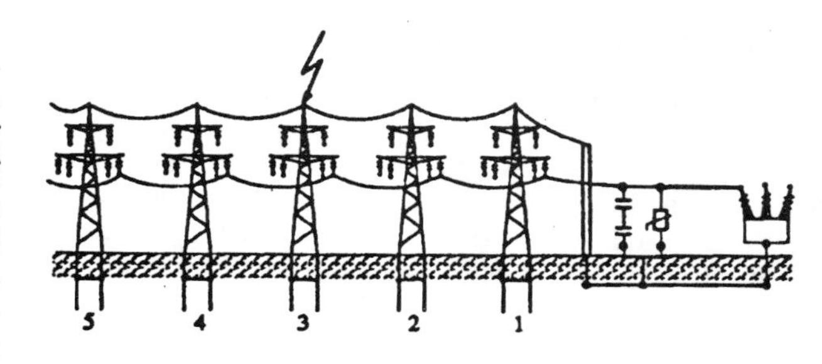

Fig. 11. Lightning-struck overhead line entering 123 kV substation

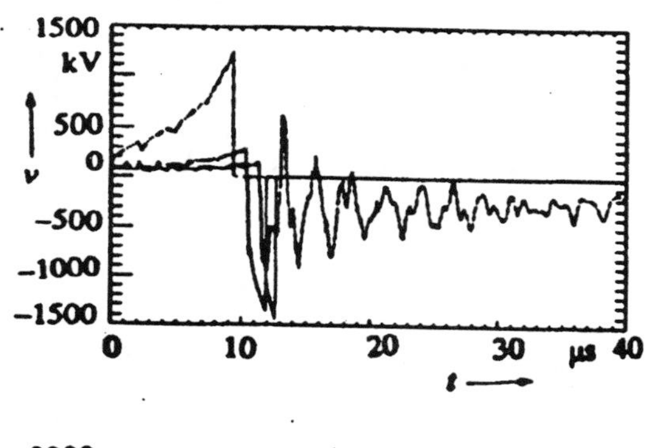

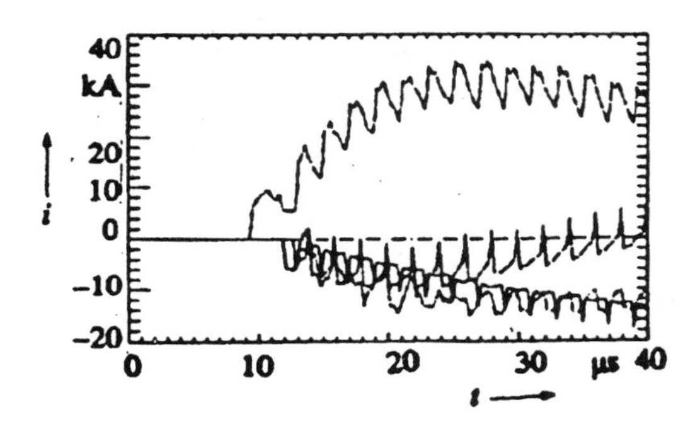

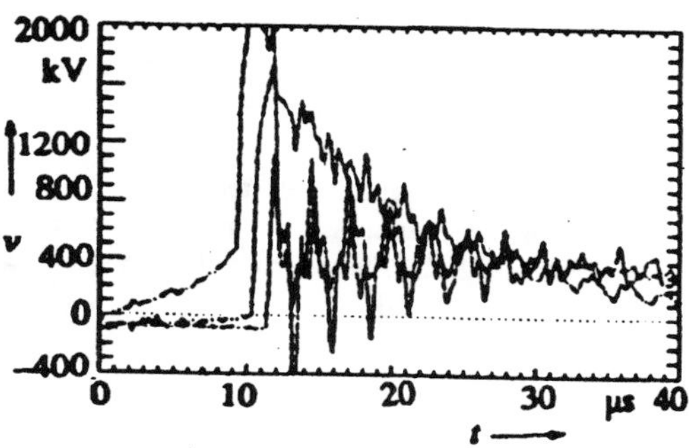

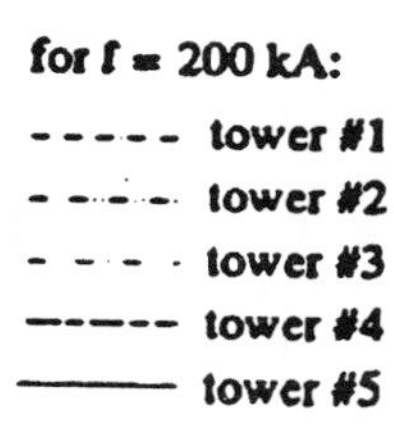

Fig. 12. Voltages and currents across insulators, and conductor potential at towers #1...#5 for phase A.

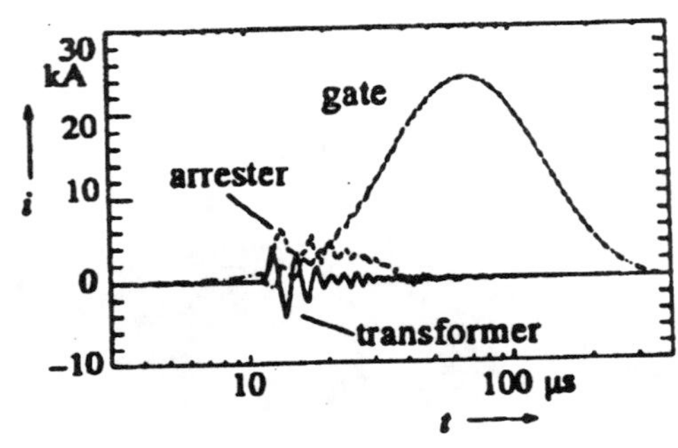

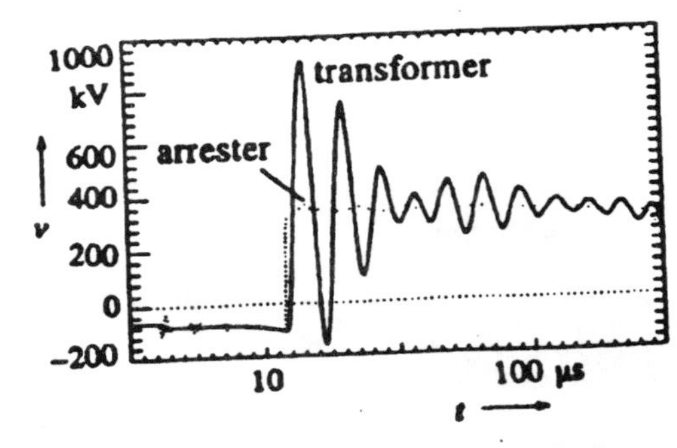

Fig. 13. Currents injected in the earthing system and voltages across transformer and arrester terminals,

The employed time-dependent flashover model leads to the fact that, as a consequence of the lightning stroke to tower #3, backflashovers occur at the towers #2, #4 and #5, whereas the flashover at tower #3 impresses a portion of the lightning current into the phase conductor (Fig. 12). Within the substation (Fig. 13), an oscillating overvoltage of nearly 9 p.u. very probably will destroy the power transformer windings. The metal oxide arrester (v_{max}=3.5 p.u.) has been placed at too large a distance from the transformer to provide sufficient protection. Under such fault conditions, currents are impressed at different locations into the grounding system, the major part of which through the foundation of the gate to which the neutral conductor of the overhead line is connected. This situation results in considerable local variations of the transient ground potential rise, temporarily reaching 100 kV. Such potential differences within the grounding system are able to cause EMC problems, if, for example, the shield of a measuring cable between voltage transformer and relaying room is grounded at both ends. Then, a rapidly varying current will flow on the cable shield and crosstalk will occur onto secondary circuits.

VI. Conclusion

Lightning protection and EMC analyses of power electric systems paying special attention to the grounding conditions require the knowledge of the dynamic properties of ground conductors and extended earthing systems. For this purpose, a methodology has been developed which is based on a transmission line approach for linear ground conductors. Thus, it is the possible to integrate the grounding system as a network of such conductors into EMTP. The necessary interface algorithm has been outlined as well as a post processor capable of deriving electrical fields in the vicinity of the grounding system. Validation of the method developed by the first author could be achieved by means of comparisons with field measurements by the EDF at Paris, France, and with computational results produced by the second author's code based on a rigorous electromagnetic field approach. As an example, some aspects of a lightning protection study for a 123 kV substation have been selected in order to underline the versatility and flexibility of the proposed method, capable of modelling not only the grounding system itself, but also the electrical apparatus connected to it.

VII. Acknowledgements

Financial support of the project had been granted by the Deutsche Forschungsgemeinschaft, Bonn, Germany, which is gratefully appreciated. Messrs. Hervé Rochereau of EDF, Paris, France and Vincent Vanderstockt of LABORELEC, Brussels, Belgium, amiably made available the measurement and substation data, respectively.

The fist author wishes to express his gratitude to his academic teacher, Prof. Dr. Klaus Möller. Under his supervision, the Ph.D. thesis could be realized upon which parts of the present contribution are based.

VIII. References

[1] R. Velazquez and D. Mukhedkar, "Analytical modelling of grounding electrodes transient behavior," *IEEE Trans. Power Apparatus and Systems*, vol. 103, 1984, pp. 1314-1322.

[2] E. D. Sunde, *Earth conduction effects in transmission systems*, 2nd ed., New York: Dover Publications, 1968.

[3] A. D. Papalexopoulos and A. P. Meliopoulos, "Frequency dependent characteristics of grounding systems," *IEEE Trans. Power Delivery*, vol. 2, October 1987, pp. 1073-1081.

[4] F. Dawalibi and A. Selbi, "Electromagnetic Fields of Energized Conductors," *IEEE/PES 1992 Summer Meeting*, Paper 92 SM 456-4 PWRD.

[5] D. van Dommelen [ed.], *Alternative Transients Program — Rule Book*, Leuven EMTP Center, Catholic University of Leuven, Belgium; July 1987.

[6] W. Long, D. Cotcher, D. Ruiu, P. Adam, S. Lee, and R. Adapa, "EMTP — A powerful tool for analyzing power system transients," *IEEE Trans. Computer Application in Power*, July 1990, pp. 36-41.

[7] J. P. Plumey, D. J. Robertou, J. M. Fontaine, and P. Kouteynikoff, "Impédance haute fréquence d'une antenne déposée dans un demi-espace conducteur," *Proc. Colloque sur la Compatibilité Électromagnétique*, paper C.1, Trégastel, France; January 1981.

[8] IMSL, Inc., *IMSL Reference manual*, 8th ed.; Houston, TX: IMSL, 1980.

[9] P. Wolfe, "The secant method for simultaneous nonlinear equations," *Communications of the ACM*, vol. 2, 1959.

[10] A. Semlyen and A. Dabuleanu, "Fast and accurate switching transients calculations on transmission lines with ground return using recursive convolutions," *IEEE Trans. Power Apparatus and Systems*, vol. 94, 1975, pp. 561-571.

[11] J. R. Martí, "Accurate modelling of frequency-dependent transmission lines in electromagnetic transient simulations," *IEEE Trans. Power Apparatus and Systems*, vol. 101, 1982, pp. 147-155.

[12] F. Menter, "Accurate modelling of conductors imbedded in earth with frequency-dependent distributed parameter lines," *Proceedings of the 21st EMTP User Group Meeting*, paper 92.09, Kolymbari, Greece; Athens: National Technical University, June 1992.

[13] E. E. Oettle, "A new general estimation curve for predicting the impulse impedance of concentrated earth electrodes," *IEEE/PES 1987 Winter Meeting*, Paper 87 WM 567-1 PWRD.

[14] L. Dubé and I. Bonfanti, "MODELS, a new simulation tool in EMTP," *European Transactions on Energy and Power (ETEP)*, vol. 2, 1992, pp. 45-50.

[15] F. Menter, *Computation of electromagnetic transients in extended grounding systems (in German)*, Ph.D. thesis, Technical University of Aachen; Aachen: Shaker 1993.

[16] T. Takashima, T. Nakae, and R. Ishibashi, "High Frequency Characteristics of Impedances to Ground and Field Distributions of Ground Electrodes," *IEEE Trans. Power Apparatus and Systems*, Vol. 100, 1980, pp. 1893-1900.

[17] L. Grcev, "Computation of Transient Voltages near Complex Grounding Systems Caused by Lightning Currents," *IEEE 1992 International Symposium on Electromagnetic Compatibility*; Anaheim, CA, pp. 393-400.

[18] S. A. Schelkunoff and H. T. Friis, *Antennas, Theory and Practice*, New York: Wiley, 1952, p. 401.

[19] D. V. Otto and J. H. Richmond, "Rigorous Field Expressions for Piecewise-Sinusoidal Line Sources," *IEEE Trans. Antennas and Propagation*, Vol. 17, 1969, p. 98.

[20] H. Rochereau, "Comportement des prises de terre localisées parcourues par des courants à front raide," *EDF Bulletin de la Direction des Études et Recherches*, série B, no. 2, 1988, pp. 13-22.

[21] L. Grcev and F. Dawalibi, "An electromagnetic model for transients in grounding systems," *IEEE Trans. Power Delivery*, vol. 5, 1990, pp. 1773-1781.

[22] CIGRÉ working group 33.01, *Guide to procedures for estimating the lightning performance of transmission lines*, Guide no. 63, Paris: CIGRÉ, October 1991.

[23] V. Vanderstockt, *Application procedures for station insulation co-ordination*, IWD7 SC 33-92 (WG11), Paris: CIGRÉ, May 1992

[24] J. H. Richmond, *Computer Analysis of Three-Dimensional Wire Antennas*, Technical Report 2708-4, ElectroScience Laboratory, Ohio State University, 1969.

APPENDIX

The derivation in this Appendix is based on [18] and [24]. Acccording to the fundamental analysis [24], the z-component of the electric field vector of the line source with arbitrary current distribution illustrated in Fig. 6 is given by:

$$\underline{E}_z = \frac{1}{j\omega\underline{\varepsilon}} \int_{z_1}^{z_2} \left(\frac{\partial^2 \underline{g}}{\partial^2 z} - \underline{\Gamma}^2 \underline{g} \right) \underline{I}(z')dz' \qquad \text{(A1)}$$

with

$$\underline{\varepsilon} = \varepsilon_0 \varepsilon_{r,E} + \frac{\sigma_E}{j\omega}, \quad \underline{g} = \frac{\exp(-\underline{\Gamma} r)}{4\pi r}, \quad r = \sqrt{\rho^2 + (z - z')^2}$$

Noting that

$$\frac{\partial \underline{g}}{\partial z} = -\frac{\partial \underline{g}}{\partial z'}, \qquad \frac{\partial^2 \underline{g}}{\partial z^2} = \frac{\partial^2 \underline{g}}{\partial z'^2} \qquad \text{(A2)}$$

and integrating the first term in (A1) by parts twice and substituting (A2) in (A1) we have:

$$\underline{E}_z = -\frac{1}{j\omega\underline{\varepsilon}} \left[\underline{I}'(z')\underline{g} + \underline{I}(z') \frac{\partial \underline{g}}{\partial z} \right]_{z'=z_1}^{z'=z_2} + \frac{1}{j\omega\underline{\varepsilon}} \int_{z_1}^{z_2} \left[\frac{d^2 \underline{I}(z')}{dz'^2} - \underline{\Gamma}^2 \underline{I}(z') \right] \underline{g}\, dz' \qquad \text{(A3)}$$

If the current between $z'=z_1$ and $z'=z_2$ is of form

$$\underline{I}_k(z) = A\cosh\underline{\Gamma}z + B\sinh\underline{\Gamma}z \quad , \tag{A4}$$

the bracketed expression in the integrand of (A3) vanishes. Hence:

$$\underline{E}_z = \frac{1}{4\pi j\omega\underline{\varepsilon}}\left[\frac{\underline{I}'(z_1)e^{-\underline{\Gamma}R_1}}{R_1} - \frac{\underline{I}'(z_2)e^{-\underline{\Gamma}R_2}}{R_2} + \underline{I}(z_2)\frac{\partial}{\partial z}\frac{e^{-\underline{\Gamma}R_2}}{R_2} - \underline{I}(z_2)\frac{\partial}{\partial z}\frac{e^{-\underline{\Gamma}R_1}}{R_1}\right] \tag{A5}$$

where $\underline{I}'(z_1)$ and $\underline{I}'(z_2)$ denote the derivatives of the current at the end points. The constants A and B can be eliminated to express the current distribution in terms of the endpoint currents $\underline{I}(z_1)$ and $\underline{I}(z_2)$ as follows:

$$\underline{I}(z) = \frac{1}{\sinh\underline{\Gamma}d_k}\left[\underline{I}(z_1)\sinh\underline{\Gamma}(z_2 - z) + \underline{I}(z_2)\sinh\underline{\Gamma}(z - z_1)\right] \tag{A6}$$

From (A6):

$$\underline{I}'(z_1) = \frac{\underline{\Gamma}}{\sinh\underline{\Gamma}d_k}\left[\underline{I}(z_2) - \underline{I}(z_1)\cosh\underline{\Gamma}d_k\right] \tag{A7}$$

$$\underline{I}'(z_2) = \frac{\underline{\Gamma}}{\sinh\underline{\Gamma}d_k}\left[\underline{I}(z_2)\cosh\underline{\Gamma}d_k - \underline{I}(z_1)\right] \tag{A8}$$

Equations (A5) to (A8) yield the z-component of the electric field in (18). This expression excludes the field contributions from the point charges at the endpoints of the line segments, since these charges disappear when two segments are connected. Such charges are also neglected in case of a segment with no connected end point. It is assumed that the longitudinal current at the open endpoint is zero. Since the point charges there are equal to $-\underline{I}/j\omega$, they are also zero. It should be noted that this assumption only applies on the longitudinal current at the segment's open endpoint and not on the radial or leakage current.

The same procedure can be repeated for the ρ-component of the electric field. This will yield:

$$\underline{E}_\rho = \frac{\eta}{4\pi\rho}\left\{\left[\underline{I}'(z_2)\cos\theta_2 - \underline{\Gamma}\underline{I}(z_1)\right]e^{-\underline{\Gamma}R_2} - \left[\underline{I}'(z_1)\cos\theta_1 - \underline{\Gamma}\underline{I}(z_2)\right]e^{-\underline{\Gamma}R_1}\right\} \tag{A9}$$

The ρ-component of the electric field in (18) follows from (A9) and (A6) to (A8).

Dr. Frank E. Menter (M 93) was born in Essen, Germany, on August 14, 1963. He graduated from the Rheinisch-Westfälische Technische Hochschule in Aachen, Germany, and received the Dipl.-Ing. degree. His practical experiences include various stays with Siemens AG, Nixdorf Computers and the Research Laboratory of Hitachi Cable, Inc. As a scientific assistant at the Institute for HighVoltage Engineering at Aachen University, he worked as a project co-ordinator of two larger projects, the first of which resulting in a lightning protection scheme for DC driven urban railways and the second treating EMC problems of high voltage switchgears. In May 1993, he received the doctoral Dr.-Ing. degree from the same University.

Currently, he is with Siemens Transportation Systems at Erlangen, Germany. In the Traction Power Supply Department, he is concerned with grounding and EMC problems. Dr. Menter is a member of the IEEE Power Engineering Society and the VDE.

Dr. Leonid Grcev (M 84) was born in Skopje, Macedonia on April 28, 1951. He received Dipl. Ing. degree from the University of Skopje, and the M.Sc. and Ph.D. degrees from the University of Zagreb, Croatia, all in Electrical Engineering, in 1978, 1982 and 1986, respectively. From 1978 to 1988, he held a position with the Electric Power Company of Macedonia, Skopje, working in the Telecommunication and Information Systems Departments. He worked on a number of projects involving radio systems, PLC and electromagnetic compatibility related problems and specialized computer software development. He was a Head of the Information System Department from 1985. In 1988, he joined the Faculty of Electrical Engineering at the University of Skopje, and is now an Associate Professor of Electrical Engineering. His present research interests are in computational electromagnetics applied to grounding and interference.

Dr. Grcev is a member of IEEE PE, AP, MTT, EMC, MAG and EI Societies. He is also a member of the Applied Computational Electromagnetic Society.

DISCUSSION

Abdul M. Mousa (B.C. Hydro, Vancouver, Canada. I wish to congratulate Dr. Menter and Dr. Grcev for an interesting paper. My comments pertain to soil ionization, a factor which is significant in case of concentrated electrodes, especially where soil resistivity is high. In this connection, it should be noted that Oettle's work [13] has been superseded by Mousa's recent paper [25]. The main findings as they pertain to this paper are as follows:

1. The mechanism of breakdown of the soil indicates that no direct correlation exists between the ionization gradient E_{crit} and the resistivity ρ (or the conductivity) of the soil. That fact is proven by the scatter in the relation between E_{crit} and ρ observed by Oettle [26]. Please see Fig. 14.

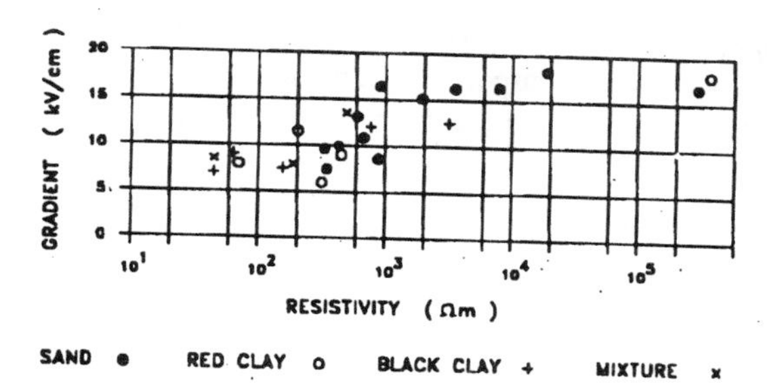

Fig. 14. Relation between the ionization gradient and resistivity of the soil according to Oettle.

2. E_{crit} is mainly governed by the water content of the soil.
3. Due to inhomogeneity of the soil, E_{crit} varies from point to point along the electrode. The **effective** value of E_{crit} is **not the average** of the subject values, but is rather **equal to the minimum** value encountered along the electrode.
4. For practical applications, the value of the ionization gradient should be taken equal to 300 kV/m. Based on the above, equations (14) and (16) should be replaced by:

$$E_{crit} = 300 \text{ kV/m} \quad \ldots(14.1)$$

$$a_{eff} = (\rho/600\pi)\,(\partial i/\partial x) \quad \ldots(16.1)$$

where $(\partial i/\partial x)$ is in kA/m, and ρ is in Ωm.

5. With typical water content, the value of the relative permittivity of the soil should be around 10 or even higher. The values shown in Figs. 7 and 8 are reasonable. On the other hand the value 1 used in part B of Section V is not realistic.
6. It should be noted that soil ionization was not present in the examples given in Figs. 7 and 9, because the amplitudes of the currents were too small. In the case of Fig. 7 in which a 6 m long rod is located in a soil having $p = 40$ Ωm, the minimum current needed to initiate ionization would be as follows:

a) About 1.4 KA if eqn. (14.1) is used.
b) About 2.5 KA if eqn. (14) is used.

The actual current, on the other hand, was only about 30 A.

REFERENCES

[25] A.M. Mousa, "The Soil Ionization Gradient Associated with Discharge of High Currents into Concentrated Electrodes", IEEE Paper No. 94 WM 078-6 PWRD.

[26] E.E. Oettle, "The Characteristics of Electrical Breakdown and Ionization Processes in Soil", Trans. of the South African IEE, pp. 63-70, December 1988.

Manuscript received February 22, 1994.

F.P. DAWALIBI, Safe Engineering Services & technologies ltd., Montreal, Quebec, Canada, H3M 1G4. The authors should be commended for an excellent paper illustrating the use of transmission line theory and its integration with the EMTP transient program. There is an urgent need for technical contributions in the area of transient performance of grounding systems in order to help refine the various analytical methods in use and determine the domain of their validity by comparing their computation results.

One major simplification of this transmission line approach is the neglect of the inductive, capacitive and perhaps the conductive coupling between the conductor segments of the ground network. According to the authors, this neglect does not lead to significant errors. This conclusion may not be valid at all frequencies and for all ground grid configurations. Indeed our own investigations [1,2] have revealed that coupling from aboveground conductors and loops may affect results even at low frequencies. Is the authors' conclusion based on extensive simulations or simply the result of a limited number of tests on simple ground configurations?

[1] F.P. Dawalibi, W.K. Daily, "Measurements and Computations of Electromagnetic Fields in Electric Power Substations," Paper No. 93 WM 220-0 PWRD T-PWRD, Presented at IEEE PES 1993 Winter Meeting in New York, New York.

[2] W. Xiong, F.P. Dawalibi, "Transient Performance of Substation Grounding Systems Subjected to Lightning and Similar Surge Currents," Paper No. 94 WM 139-6 PWRD, Presented at IEEE PES 1994 Winter Meeting in New York, New York.

Manuscript received February 22, 1994.

DR. FRANK E. MENTER, PROF. DR. LEONID GRCEV: The authors thank Messrs. A. M. Mousa and F. P. Dawalibi for their kind words. We are especially grateful to Mr. Abdul M. Mousa for his discussion, which enhanced the quality of our paper. His remarks on recent findings in the experimental analysis of breakdown phenomena are appreciated. Since we did not perform such analyses ourselves, we had to adopt one model available though the literature. So, we chose Oettle's model which lead to equations (14) and (16). Since the model Mr. Mousa now proposes (eqns. (14.1), (16.1)) is even less complicated, its introduction is feasible and we are delighted to incorporate his suggestion into our procedures.

Regarding the questions raised by Mr. Farid F. Dawalibi, it should be noted that the aim of this paper was to demonstrate that the transmission line approach for grounding system analysis integrated within the EMTP could lead to practically useful results. The authors' conclusion for the accuracy of the results is based on comparisons with field measurements performed by the EDF, Paris, France. An extensive set of experiments had been performed in Les Renardières in 1976-78 and in St-Brieuc in 1985. All experiments had been repeated to cover any seasonal and weather effects. Among the grounding arrangemets under study were vertical ground rods, horizontal earth electrodes, hemispheres, grids, star- and serpentine-shaped electrodes and tower footings. Further reference on the measurements can be found in [20]. The comparison has been performed in Paris in 1992/93 and has been documented in [15]. However, it is true that the investigated earthing structures do not include larger grids; the validation for such structures has been achieved by comparison with a separate program developed by the second author, implementing a rigorous electrodynamic approach. Additional checks and analyses of large structures with conductors above ground may be subject of a future paper.

Manuscript received May 3, 1994.

IEEE Transactions on Power Delivery, Vol. 5, No. 1, January 1990

LIGHTNING OVERVOLTAGE PROTECTION OF THE PADDOCK 362-145 kV GAS-INSULATED SUBSTATION

H. Elahi
Senior Member

M. Sublich
Member

GE Industry & Utility Sales
Schenectady, New York 12345

M.E. Anderson (P.E.)
Member

B.D. Nelson (P.E.)
Member

Wisconsin Power & Light Company
Madison, Wisconsin

Abstract - Backflashovers close to the Paddock 362-145 kV Gas-Insulated Substation (GIS) have been analyzed with the Electro-Magnetic Transient Program (EMTP) using a frequency dependent multi-conductor system. The severity of the lightning stroke currents were derived based on recent recordings in the eastern United States. Impacts of corona attenuation and distortion were accounted for using a shunt linear model approach. Turn-up effects of both line insulator flashover voltages and surge arrester protective characteristics were represented based on manufacturer's volt-time curves. Wave shaping effects of substation capacitances (i.e., PT's, transformers, CCPD's) were also modeled. Results show the importance of various modeling details in determining the overvoltages inside the GIS due to close backflashovers, which are caused by lightning strokes with varying intensity. These results are aimed at better evaluation of lightning protection requirements for GIS protected by metal-oxide surge arresters.

Key Words - Backflashover, Gas-Insulated Substation (GIS), Gas-Insulated Bus (GIB), Insulation Coordination, Corona, Critical Flashover Voltage (CFO), Capacitive Coupling Potential Device (CCPD), Metal-Oxide Surge Arrester

INTRODUCTION

Gas-Insulated Substations (GIS) are exposed to the same variety of overvoltages as air-insulated substations (i.e., lightning, switching, and temporary overvoltages). However, in air-insulated stations, primary concern is placed on the protection of transformers and some risk of failure across insulator strings or bushings is accepted since air insulation is self-restoring. In GIS, on the other hand, the entire gas insulated assembly (including enclosures, circuit breakers, disconnect and grounding switches) must be protected because the gas insulated system must be viewed as non-self-restoring.

In the insulation coordination design of the GIS, lightning overvoltages are found to be critically important. The most severe lightning overvoltages may be caused by close backflashovers. At the origin, where a high current lightning stroke intercepted by the tower (or shield wires) has caused the line insulators to flashover, very steep lightning surges with crests near, or above, the line critical flashover voltage (CFO) level are initiated. Voltage magnifications due to reflections of such lightning surges at various junctions within the GIS are often the determining criteria for selection of surge arrester rating and locations.

89 SM 801-2 PWRD A paper recommended and approved by the IEEE Substations Committee of the IEEE Power Engineering Society for presentation at the IEEE/PES 1989 Summer Meeting, Long Beach, California, July 9-14, 1989. Manuscript submitted January 31, 1989; made available for printing May 9, 1989.

The objective of this paper is to summarize the findings of a detailed backflashover surge analysis for Wisconsin Power and Light Company's (WP&L) Paddock 362-145 kV GIS. This paper focuses on studies performed for the 345 kV GIS. See Figure 1 for the system's one-line diagram, which shows the GIS configuration, including distances, starting from the gas-insulated bus (GIB) entrance. A surge impedance of 62.8 ohms was used for all gas-insulated buswork based on manufacturer's data. Lightning surges were assumed to be incoming on the line to Rockdale. (See the line between circuit breakers no. 1 and no. 2.)

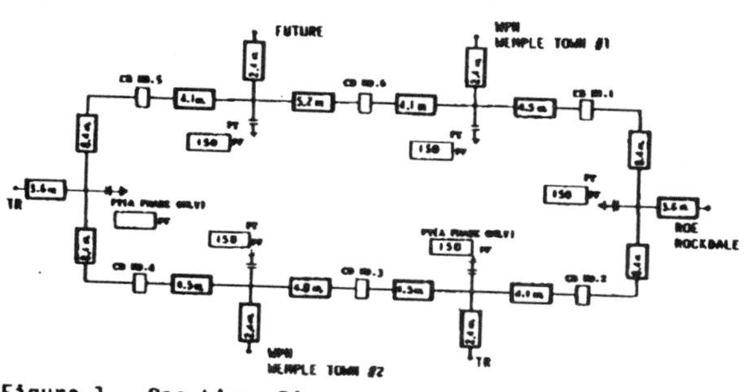

Figure 1. One-Line Diagram for the Paddock 345 kV, Six Breaker Gas-Insulated Substation

The analysis demonstrates the importance of various modeling details and techniques for achieving more economical and reliable application of metal-oxide surge arresters for GIS lightning protection.

BACKFLASHOVER SURGE ANALYSIS

Analysis of lightning backflashover overvoltages requires proper accounting of numerous phenomena, some of which are not reliably predictable. The following discussion summarizes this paper's approach in modeling the most important phenomena recognized by the industry:

- Parameters of the lightning stroke (i.e., crest, rise time, duration, frequency of occurrence): This has been traditionally the most uncertain part of the lightning surge analysis. In some studies, a direct stroke was simply modeled as a voltage source at the stroke location with a certain steepness [1]. In this study, the lightning stroke was represented as a current source in parallel with a large resistor representing the surge impedance of the stroke channel, see Figure 2. The crest of the lightning stroke was varied from 50 kA up to 200 kA based on recent recordings of over six million lightning ground flashes in the eastern United States [2,3], see Figure 3 for the number of

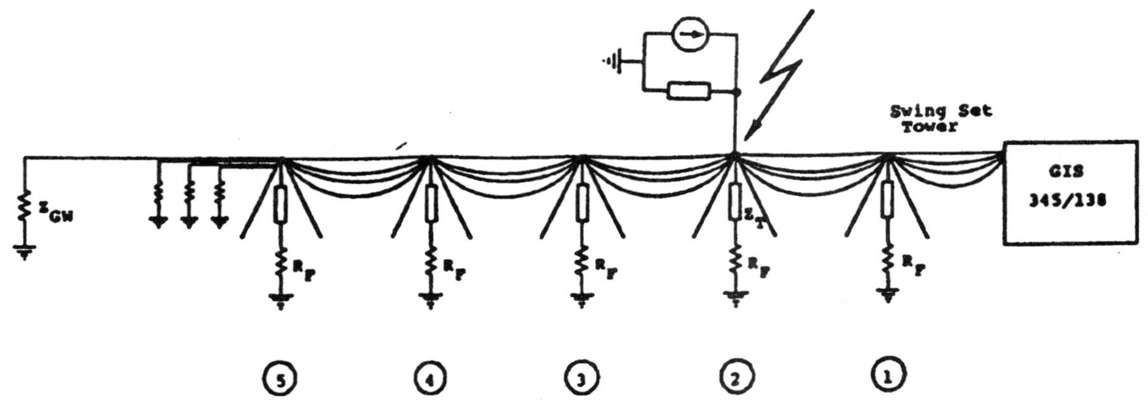

Figure 2. Stroke Current at Tower 2 Splits Between Shield Wires and Ground

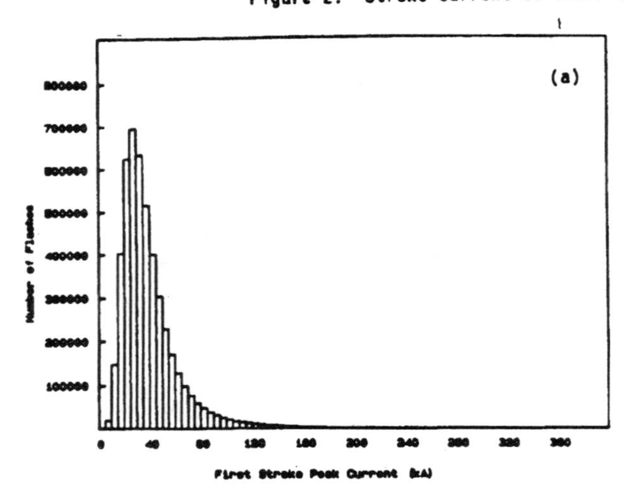

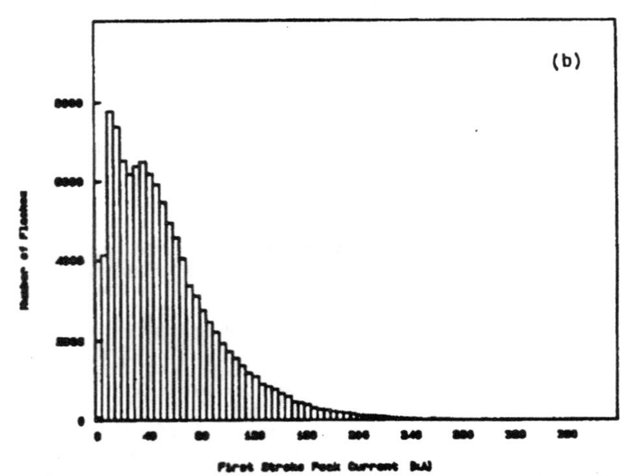

Figure 3. The Number of Flashes vs. First Stroke Peak Current for:

a) 4,785,203 Flashes Lowering Negative Charge to Ground in 1986

b) 106,145 Flashes Lowering Positive Charge to Ground in 1986 [3]

flashes versus peak current amplitudes lowering negative, and positive charges to ground. The steepness of the stroke current and its rate of decay were determined based on the guidelines given in References [4,5] and are given below:

Peak Current (kA):	50.	75.	100.	150.	200.
Time to Crest (μs):	1.5	1.75	2.0	2.25	2.5

- Turn-up effects of line insulator flashover voltage: It is well recognized that the insulator flashover voltage is dependent upon the steepness of the voltage stressing it. In fact, depending upon the waveshape of tower top voltage, line back-flashover may occur above the insulator CFO level. To account for this phenomena, the voltages across line insulators were monitored, and flashover voltage was updated based on volt-time data given in References [6,7]. The CFO at the struck tower for a 1.5 x 40 μs wave was obtained from Reference [7] as 1585 kV (i.e., for 18 insulator units). If the observed insulator voltage had a rise time to this CFO level (i.e., 1585 kV) of, for example, 1 microsecond, the CFO level was adjusted to 15% higher (i.e., 1823 kV) based on guidelines given in Reference [6].

- Corona effects: The attenuation and distortion of overvoltage waves due to corona effects are well recognized [8]. This was particularly important in this study since the incoming 345 kV overhead lines were not bundled, resulting in relatively lower corona starting voltage for the phase conductors (V_C = 490 kV). Note that this is well below the CFO level for typical 345 kV lines. In this study the corona effects were represented using a shunt linear model approach suggested in Reference [9] at each tower, as shown in Figure 4. Parameters of the corona model shown in Figure 4 were determined based on data in References [8,10].

- Turn-up effects of surge arresters: As steep lightning surges, modified by the corona effects, arrive at the GIS entrance, they are limited by the protective characteristics of the entrance arresters. However, the volt-ampere characteristics of the metal-oxide surge arresters are also a function of incoming surge steepness. This effect was represented in this study by monitoring the arrester's discharge voltage and updating its maximum non-linear v-i curve, shown in Table 1, based on the manufacturer's volt-time data, shown in Figure 5 [11].

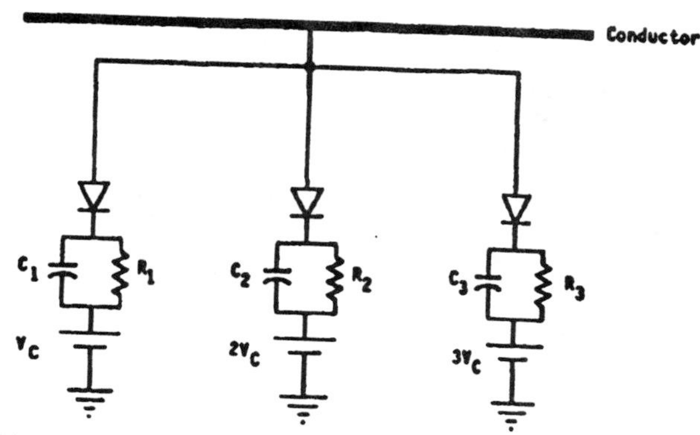

Figure 4. Linear Corona Model

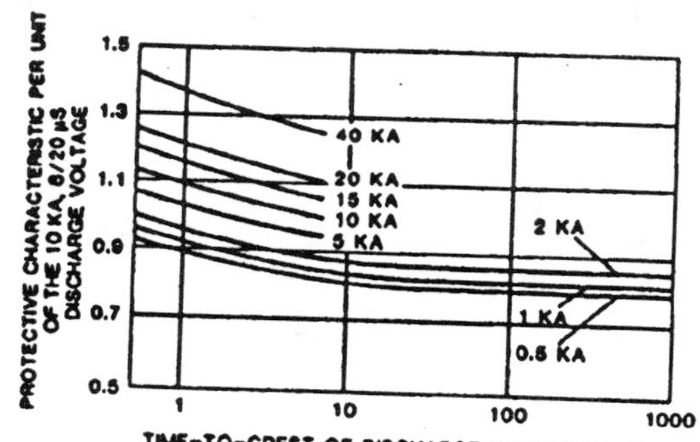

Figure 5. Protective Characteristic for Arrester Ratings 54 kV Through 360 kV (Crest Discharge Voltage vs. Time-to-Crest of Discharge Voltage) [11]

Table 1

v-i Characteristics for the Single-Column 276 kV Rated Arresters Used in the Study

Discharge Current (A)	8x20 μs Discharge Voltage (kV)
1	446
10	468
100	497
1,000	547
5,000	611
10,000	646
15,000	685
20,000	716
40,000	808

Note that the non-linear arrester v-i curve needs to be modeled up to at least 20 or 40 kA, since high current surges initiated by close backflash-overs can result in arreter discharge currents above 10 kA.

- Impact of substation capacitances: Surge wave-shape modifying effects of substation capacitances are also recognized as an important aspect of substation lightning insulation design [12]. WP&L plans to install CCPD's (2.1 nF) on two phases of the 345 kV bus for relaying purposes. The effects of CCPD's, and the stray capacitances of PT's and transformers, were modeled in this study as lumped shunt capacitances according to manufacturer's data. Note that in order to be conservative, the phase where backflashover occurred was assumed to be the one without the CCPD at the bus entrance.

The discussions presented above highlight the important modeling features of the backflashover surge analysis for the Paddock 345 kV GIS. Details of individual simulations and the findings of the study are presented next.

STUDY RESULTS

To make efficient use of the available space, and in order to make the results as general as possible, emphasis is placed on a simulation where the most severe lightning overvoltages were observed. This case corresponded to the situation where circuit breakers No. 1 and No. 2 were both open and lightning struck the second tower away from the GIS on the line to Rockdale (see Figure 1). Other simulations with different circuit breaker arrangements, and with the transformer connected, were not as severe. Also, in order to simulate the most unfavorable initial conditions (i.e., 60 Hz system voltage), the phase conductors were energized at 1.05 per unit of the nominal voltages with phase A (where backflashover occurred) at crest negative polarity to simulate the opposite polarity condition.

In the early stages of the study, the impact of various modeling details, as discussed above, under various stroke current levels were investigated. First, the effect of corona modeling was investigated. Figure 6 illustrates the observed lightning overvoltages at the GIB entrance and at the open circuit breakers 12 meters away. It was observed that:

- Backflashover occurred for stroke currents with peaks greater than 50 kA,
- With corona effects represented, observed peak overvoltages were up to 20% lower than the corresponding simulations without corona modeling. This pattern appeared to be more pronounced at higher stroke current magnitudes,
- Although stroke currents above 100 kA are rare (see Figure 3), they can impose much higher overvoltages inside the GIB. However, if gas-insulated systems are to be viewed as non-self restoring insulation systems, attention must be given to stroke currents in the range of 100 to 200 kA.
- As stated earlier, in all simulations the backflashover was assumed to occur on the phase without the CCPD (2.1 nF) at the GIB entrance. However, if this is not true, the surge modifying effects of the CCPD tend to reduce the steepness of the incoming lightning surges at the GIB entrance. (Note that the GIB entrance arrester will also operate at a lower protective level due to a less steep discharge voltage, see Figure 5.) Furthermore, at the open breaker the impact of the CCPD is less pronounced. However, depending on the distance to the open breaker, and the magnitude of the stroke current, slightly higher overvoltages with longer rise times were observed at the open circuit inside the GIS.

Figure 6 clearly illustrates the importance of corona modeling in GIS lightning analysis. Corona modeling becomes critically important issue if the incoming transmission line conductors are not bundled (i.e., corona starting voltages are much lower), as is the usual case in system voltages in the range of 230 kV and below.

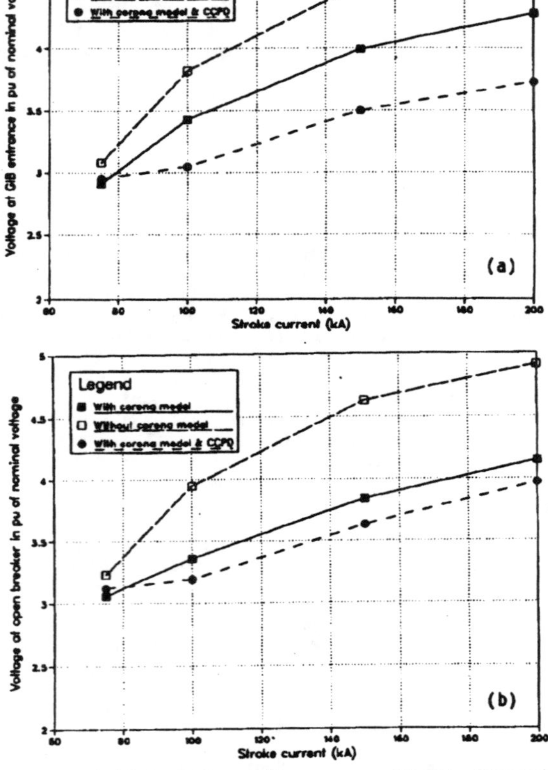

Figure 6. Lightning Overvoltage vs. Stroke Current, with Arrester and Insulator Turn-Up Effects Modeled, at:

a) GIB Entrance
b) Open Circuit Breaker

In the next set of simulations, impact of the arrester turn-up effects were studied. WP&L had decided to apply 276 kV rated metal-oxide arresters at every GIB line entrance. Originally, these arresters were planned to be installed approximately 30 feet away from the GIB entrance (i.e., from arrester terminal to GIB entrance bushing) with a total lead length of 32 feet (i.e., 16 feet each, from arrester to station ground, and from GIB entrance to station ground). The inductance of total lead length in the loop was represented by an inductor L in series with the arrester, per guidelines given in ANSI/IEEE Standards C62.22 [13]. The lightning backflashover scenario discussed earlier was studied next with and without arrester turn-up effects modeled. Results are shown in Figure 7. Simulations, with arrester turn-up modeled, were performed iteratively in order to update the arrester non-linear v-i curve according to the data in Figure 5. For example, in this study the steepest observed discharge voltage at the arrester terminal had a rise time of approximately 0.5 microseconds. Figure 5 shows that, compared to the 10 kA discharge voltage for an 8x20 μs wave (this data is often used in in lightning studies), the discharge voltage with a 0.5 μs rise time is nearly 13% higher. This is a sizeable increase in voltage. If neglected, it may result in an optimistic and unreliable arrester design. Note that Figure 7 shows the results without insulator turn-up effects.

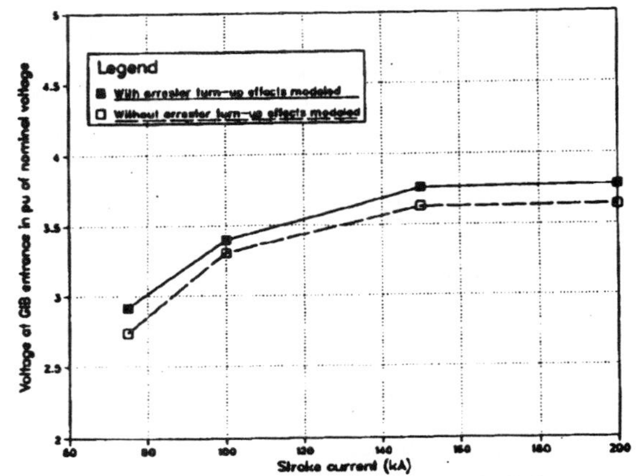

Figure 7. Lightning Overvoltage at GIB Entrance vs. Stroke Current, with Corona Model and Without Insulator Turn-Up Effects

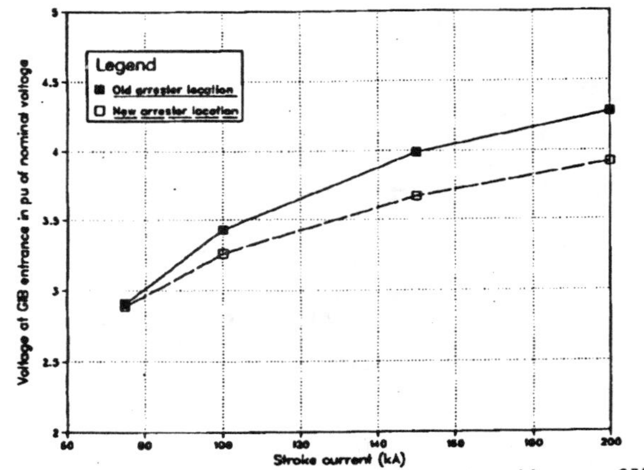

Figure 8. Impact of Revised Arrester Location on GIB Entrance Lightning Overvoltages

This allows illustration of the insulator turn-up effects by comparing the solid curves in Figures 6 and 7 (i.e., at 100 kA stroke current the difference is about 12%).

Following the earlier sensitivity studies discussed above, a decision was made to relocate the GIB entrance arresters closer to the GIB (i.e., approximately 14 feet instead of 30 feet). This decision was made since the gas-insulated buswork had a BIL of 1050 kV. Figure 8 shows the impact of the new arrester location on reducing the lightning overvoltages at the GIB entrance. Finally, to illustrate the backflashover phenomena simulated in this study, selected waveforms are shown in Figure 9. This simulation was performed assuming a 100 kA stroke current and GIB entrance arresters at their new location.

PROBABILITY OF CLOSE BACKFLASHOVERS

The following discussion is an attempt to estimate the probability of close backflashovers, resulting from lightning strokes with peak currents of 100 kA or above, at the Paddock GIS.

The average number of flashes collected by a transmission line, N_s, is approximately [14]:

$$N_s = N_g\ (28\ h^{.6} + b)/10$$

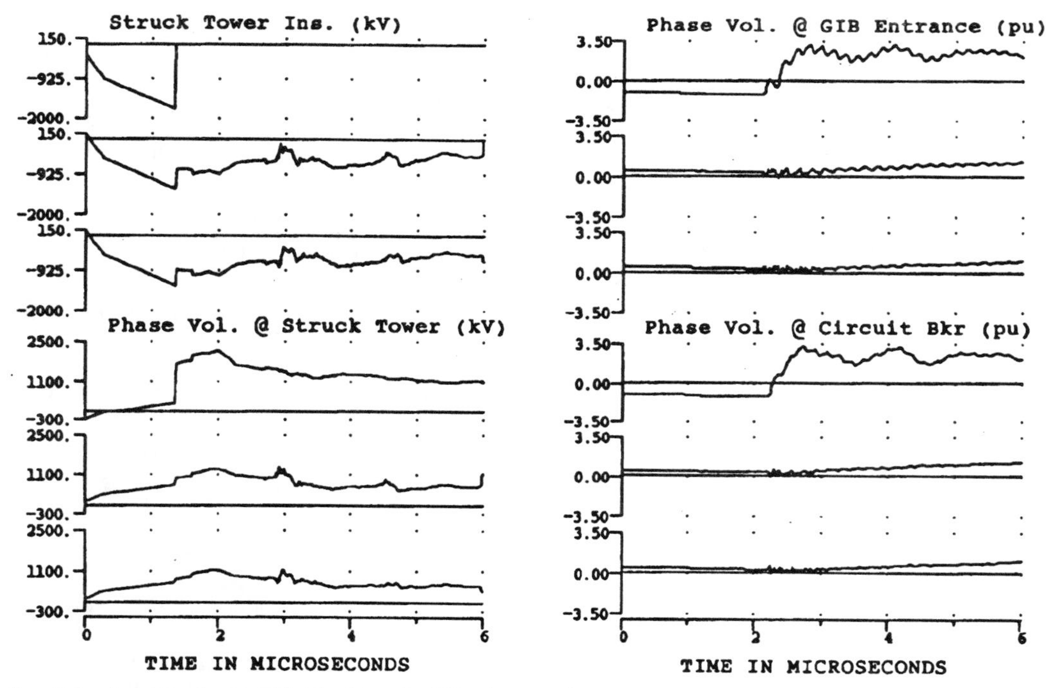

Figure 9. Selected Waveforms Illustrating Backflashover of Phase A at Tower 2 Due to a 100 kA Stroke Current

where:

h = tower height, m
b = overhead ground wire seperation distance, m
N_g = ground flash density, flashes/Km^2 /year
N_s = flashes/100 Km/year

Reference [14] also suggests that the value of N_g may be estimated using the following equation:

$$N_g = 0.04T^{1.25}$$

where T is the number of thunderstorm days per year or the Keraunic level.

Using a Keraunic level of 40, from Reference [4], and the above equations, the number of flashes for 345 kV and 138 kV lines are calculted as follows:

<u>345 kV Lines</u>

$$N_s = N_g\ (28 \times 28.1^{.6} + 10.97)/10$$
$$= 0.88 \text{ flashes}/K_m/\text{year}$$

<u>138 kV lines</u>

$$N_s = N_g\ (28 \times 21.3^{.6})/10$$
$$= 0.71 \text{ flashes}/K_m/\text{year}$$

Based on the electrogeometric theory [4, 15], shielding failures may occur with relatively low currents and the severely high currents considered here (i.e., stroke currents of 100 kA or above) may be assumed to be all terminated on the ground wires.

Using a span length of 260 meters, there are approximately four towers for each kilometer of the 345 kV lines and the number of flashes intercepted by each tower is approximately

$$N_{ST} = \frac{0.88}{4} = .22 \text{ flashes/tower/year}$$

Similarly, using a span length of 96 meters, there are approximately ten towers for each kilometer of 138 kV lines. Therefore,

$$N_{ST} = \frac{0.71}{10} = .07 \text{ flashes/tower/year}$$

In the final stage, the Paddock GIS will have four 345 kV lines and four 138 kV lines. Therefore, the number of flashes intercepted by the last towers closest to the station is approximated as:

$$N\ (\text{Total}) = 4\ (.22) + 4\ (.07) \approx 1.0 \text{ flash/year}$$

Figure 3 suggests that the probability of the first stroke current (both negative and positive) being at, or exceeding, 100 kA is approximately 8%. Also, it may be assumed that the circuit is disconnected from the station, and not grounded at the station end, for no more than 10 days a year. Therefore, the probability of occurrance of severe close backflashovers is approximately:

$$P = (10/365)(0.08)(1) = .0022, \text{ per year}$$
or once in 456 years.

Note that with the circuit in service, lightning surges both at the GIB entrance and within the station, will be low even for close backflashovers. This is because the high amplitude surges initiated by the first current component in the lightning flash would arrive at the station with the station breakers still closed. Hence, there will be a large negative reflection at the station. By the time the surge triggered by the subsequent components in the same flash arrive at the station, the terminal breakers are normally open. But these surges are of low magnitude, as the currents in the subsequent stroke components are typically well below 50 kA crest.

CONCLUSIONS

Findings of a detailed backflashover surge analysis have been summarized illustrating the impact of various modeling details. The primary objective of this study was to establish whether 276 kV rated arresters could effectively protect 1050 kV BIL gas-insulated buswork. Based on the findings of the study, surge arresters were relocated closer to the GIB entrance (i.e., approximately 14 feet separation distance). The criterion for acceptance of the new arrester location was based on backflashover at the second tower, away from the GIS, due to a 100 kA stroke current (see Figure 8). The severity of the various unfavorable conditions assumed in the study was justified by viewing the GIS as a non-self-restoring insulation system.

Recent recordings in the eastern United States [2, 3] suggest that stroke currents with peaks as high as 100 kA, or above, must be recognized for GIS lightning studies. With the coast-to-coast lightning detection network in place in North America by the end of this year, better understanding of the lightning discharge activities in specific areas is expected, and parameters of lightning strokes will be modeled more reliably.

The corona model suggested in Reference [9] is a simple and efficient method suitable for lightning surge analysis by EMTP. However, much care is needed in calculating the parameters of this model. It is suggested that users of this model consult the impulse test data given in Reference [10].

The response of metal-oxide surge arresters to steep lightning surges are also very important in backflashover surge studies. Close backflashovers could impose lightning surges at the GIS entrances with rise times as fast as 0.5 μs, where arrester protective levels are considerably higher than the 8x20 μs discharge levels typically used in lightning studies. Manufacturer's data, as shown in Reference [11], and guidelines in ANSI/IEEE Standards C62.22 [13] are suggested references.

Finally, the study techniques and references discussed in this paper may be utilized by engineers in lightning surge analysis of both gas-insulated and air-insulated substations.

REFERENCES

[1] J.C. Cronin, R.G. Colclaser, R.F. Lehman, "Transient Lightning Overvoltage Protection Requirements for a 500 kV Gas-Insulated Substation," IEEE Trans. PAS, Vol. 97, pp. 68-77, January/February 1978.

[2] EPRI Report EL-4729, Project 2431-1, "Lightning Flash Characteristics: 1985," Interim Report, August 1986.

[3] EPRI Report EL-5667, Project 2431-1, "Lightning Flash Characteristics: 1986," Interim Report, February 1988.

[4] J.G. Anderson, Chapter 12 of Transmission Line Reference Book 345 kV and Above/Second Edition, EPRI EL-2500, 1982.

[5] R.B. Anderson, A.J. Eriksson, "Lightning Parameters for Engineering Applications," Pretoria, South Africa, CSIR, June 1979, Report ELEK 170.

[6] M. Darveniza, F. Popolansky, E.R. Whitehead, "Lightning Protection of UHV Transmission Line," Electra, No. 41, July 1975, pp. 39-69.

[7] LAPP Insulators Catalog 10B, LeRoy, New York, 14482, U.S.A.

[8] K.C. Lee, "Non-Linear Corona Models is an Electromagnetic Transient Program," IEEE Trans., PAS, Vol. PAS-102(9), pp. 2936-2942, 1983.

[9] A. Ametani, H. Motoyama, "A Linear Corona Model," EMTP Newsletter 1987.

[10] C.F. Wagner, I.W. Gross, B.L. Lloyd, "High Voltage Impulse Tests on Transmission Lines," AIEE Transaction Pt. III, 73, pp. 196-210, 1954.

[11] Tranquell AC Station Surge Arresters, Product and Application Guide, GE Company, GET 6951, 100 Woodlawn Avenue, Pittsfield, Massachusetts, 01201.

[12] J.J. Minick, R.A. Hedin, K.W. Priest, "Substation Lightning Insulation Coordination - Impact of Substation Capacitance, Stroke Location and Power Frequency Voltage," IEEE paper A77 623-2, presented in IEEE PES Summer Meeting, July 17-22, 1977.

[13] ANSI/IEEE C62.22 1988, Guide for the Application of Metal Oxide Surge Arresters for AC Currents, April 1988.

[14] A.J. Eriksson, "The Incidence of Lightning Strikes to Power Lines," IEEE Transaction on Power Delivery, July 1987, pp. 859-870.

[15] F.S. Young, J.M. Clayton, A.R. Hileman," Shielding of Transmission Lines," AIEE Transactions on Power Apparatus and Systems, Special Supplement, Paper No. 63-640, pp. 132-154, 1963.

Hamid Elahi received B.S., M.S., and Ph.D from Iowa State University in 1977, 1979, and 1983, respectively.

Since joining General Electric's Systems Development and Engineering Department in the summer of 1983, Mr. Elahi has been responsible for conducting power system transients studies on various projects, including TNA and HVDC simulator studies.

Mr. Elahi is an active member of IEEE's Switching Surge Working Group, IEEE's Working Group on Estimating the Lightning Performance of Transmission Lines, and IEEE's Task Force on Voltage Stability.

Maribeth Sublich received her BEE from Pratt Institute in 1986. She is currently pursuing an advanced degree in Electric Power Engineering at Rensselaer Polytechnic Institute.

Ms. Sublich joined the General Electric Company in 1986 as an Application Engineer. Her current responsibilities include performing studies to determine the transient behavior of power transmission systems, including insulation coordination projects and series capacitor protection projects.

Ms. Sublich is a member of IEEE Power Engineering Society. She has coauthored three IEEE papers.

Michael E. Anderson graduated from the University of Illinois at Urbana-Champaign in 1967 with a B.S., in Electrical Engineering. He started with Wisconsin Power and Light Company as a distribution engineer.

He currently is a Substation and Transmission Line Engineer responsible for the design and construction of various projects. He is a member of the IEEE and a registered engineer in the state of Wisconsin.

Bradley D. Nelson received a B.S. in Electrical Engineering from the University of Illinois at Urbana-Champaign in 1976. He joined Wisconsin Power and Light Company as a transmission planning engineer.

Nelson currently is a system protection engineer, responsible for relaying and control projects. He is a member of IEEE and a registered professional engineer in the state of Wisconsin.

ESTIMATION OF FAST TRANSIENT OVERVOLTAGE IN GAS-INSULATED SUBSTATION

S. Yanabu, Senior Member
H. Murase,
H. Aoyagi,
H. Okubo, Member
Y. Kawaguchi Fellow

Toshiba Corporation
Kawasaki, Japan

Abstract – By using a commercial 550kV GIS to measure disconnector-induced FTO (fast transient overvoltages) on site, extensive data were obtained. The maximum FTO estimated from observation was 2.7 pu. Such a high FTO was observed infrequently and occurred only at the open end of bus bars.

Through a comparison between simulation and measurement by employing a 1-GHz surge sensor, the authors demonstrated that when estimating the level of FTO by EMTP, no large errors are likely to be involved even without strict simulation of such GIS components as spacers, disconnectors, and short bus branches. Thus, the estimated FTO levels analytically obtained agreed well with measured values within an error of 0.1 pu.

Keywords: GIS (gas-insulated switchgear) – Disconnector – Restrike – Prestrike – EMTP simulation – 1-GHz surge sensor – Spark collapse time – Short surge impedance discontinuity

INTRODUCTION

It is well known that FTO (fast transient overvoltages) appear when disconnectors in a GIS (gas-insulated switchgear) are operated [1]. A step-shaped traveling wave generated between disconnector contacts propagates in both ways, reflecting at the GIS ends and branches, finally resulting in a very complicated waveform. Pointed out as problems associated with FTO are not only the phenomenon of "flashover to ground" from the disconnector contacts [2][3], but also influences on other components such as transformers and bushings [4]. Other problems such as erroneous operation and failure of electronic control circuits connected to GIS units are subjected to electromagnetic interference by FTO. Furthermore, it has recently been reported that the dielectric strengths of non-uniform electric fields formed by the metallic particles involved are reduced when FTO are applied [5].

For these reasons, FTO generated in GIS units should be considered an important factor in the insulation design of not only GIS units but the entire substation, such FTO behavior being brought to the attention of researchers.

The levels and waveforms of such FTO are determined by the construction of each substation, the levels varying with each point of the substation. Thus, when designing a substation, it is important to accurately estimate the levels of FTO at each point.

To calculate the levels and waveforms of FTO, the EMTP (electromagnetic transient program) is commonly used. In EMTP calculation, it is necessary to simulate a substation by adopting a suitable equivalent circuit. There are several proposals not yet fully agreed upon as to the method for such simulation [6][7]. Accuracy of the simulation method is checked by comparing simulated waveforms with measured waveforms. Therefore, with the objective of establishing a simulation method for EMTP, it is necessary to measure and accumulate data on FTO in a commercial substation. However, there are few measured data of FTO of commercial substations [6][8].

88 SM 628-0 A paper recommended and approved by the IEEE Switchgear Committee of the IEEE Power Engineering Society for presentation at the IEEE/PES 1988 Summer Meeting, Portland, Oregon, July 24 - 29, 1988. Manuscript submitted January 28, 1988; made available for printing May 4, 1988.

The authors measured FTO at various points of a commercial 550kV GIS by operating energized disconnectors. From the results of these measurements, the possible levels of FTO at various points of the GIS and the probability of occurrence of large levels were discussed.

Laboratory tests were also performed by employing a newly developed 1-GHz surge sensor to check on the insignificance of high-frequency components caused by short-surge impedance discontinuities such as spacers, disconnectors, and short bus branches. Waveforms calculated by rough simulation that ignored the short surge impedance discontinuities were compared with measured waveforms at the commercial substation, proving that simulated FTO levels agreed well with measured values.

FTO MEASUREMENT AT A COMMERCIAL 550kV SUBSTATON

Outline of the Substation and Measuring Instruments

Figure 1 is an exterior view of the 550kV substation used for full-voltage measurement. Figure 2 is a one-line diagram of the substation. The transformers and the GIS are connected by power cables. The bus bars are single phase, with length of the main bus bars being about 50m.

Fig. 1 550kV substation where disconnector-induced FTO measurement was conducted

For measurement, a shield electrode installed in the spacer was utilized [8][9]. Two types of measuring instruments were used: one for observing FTO and the other for observing power frequency and step-like waveforms; both waveforms were observed simultaneously.

As to actual measurement, four types of typical GIS circuit diagrams were chosen, and four disconnectors were operated. From five to seven measuring points were chosen for each GIS circuit diagram, the number of the measuring points were 24 in total. The total number of on-and-off cycles of operation of the four disconnectors was 120. One of the GIS circuit diagrams, the corresponding operated disconnector, and the corresponding measuring points are also shown in Fig. 2.

Magnitudes of Restrike/Prestrike and Trapped Charge

Figure 3 shows the probability densities of the magnitude of trapped charge generated after opening operation of the disconnectors. The horizontal axis represents trapped charge in pu, normalized by the peak value of the ac voltage applied; the vertical axis represents the probability densities. As a high-speed type, the disconnector shows no trapped charge reduction effect caused by field strength

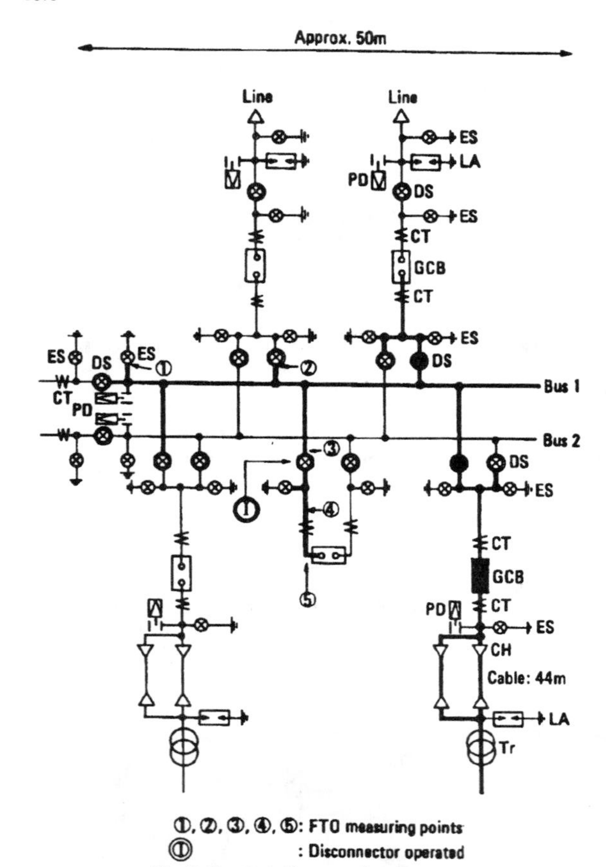

Fig. 2 One-line diagram of the substation

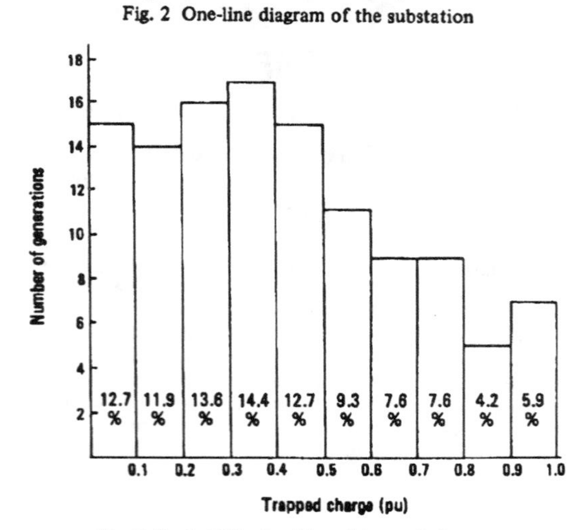

Fig. 3 Probability densities of trapped charges

inequality of the disconnector contacts, which is commonly recognized [1], thus showing a high probability of large trapped charges. However, very few restrikes appeared within one opening operation.

Figure 4 shows the probability densities of the magnitude of maximum restrike that appeared within one opening operation. The horizontal axis represents interpolar breakdown voltage of the maximum restrike in pu, normalized by the peak value of the ac voltage applied; the vertical axis represents the probability densities. The probability of an occurrence of maximum restrikes in the range of 0.5 – 1.2 pu is nearly 70%. The probability of occurrence of maximum restrikes in the range of 1.8 – 2.0 pu is as low as less than 1%, but actually does exist. Above 1.8 pu, a restrike of 1.86 pu was observed.

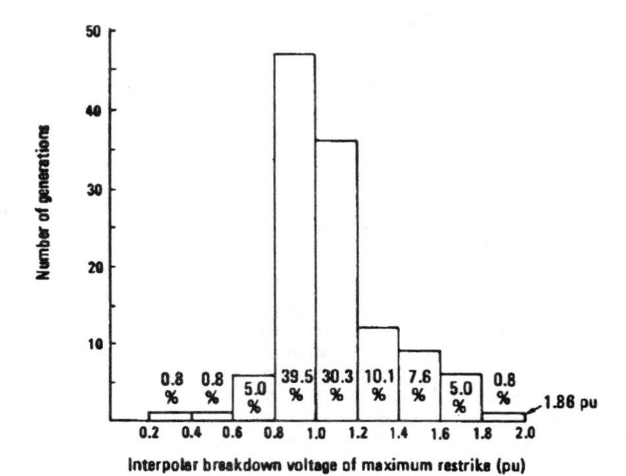

Fig. 4 Probability densities of interpolar breakdown voltages of maximum restrikes

Figure 5 shows the probability densities of the magnitude of maximum prestrike that appeared within one closing operation. Here again, the probability of the occurrence of prestrikes above 1.8 pu is as low as less than 1%, although a prestrike of 1.85 pu was actually observed.

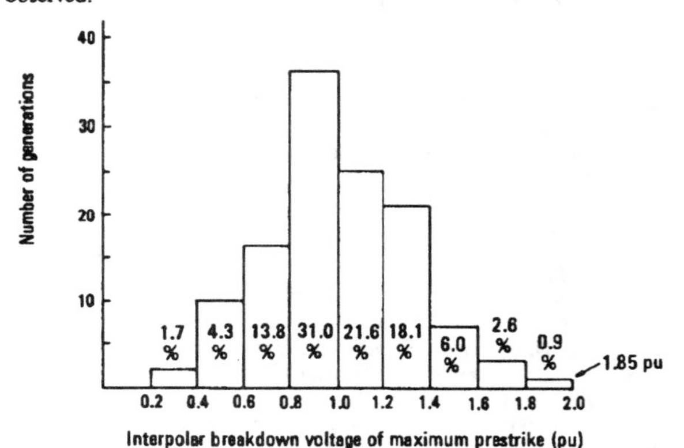

Fig. 5 Probability densities of interpolar breakdown voltages of maximum prestrikes

Results of FTO Measurement

Shown in Fig. 6 is a typical waveform of FTO obtained at measuring point ⑤ when disconnector Ⓘ shown in Fig. 2 was operated. Such measured FTO have various levels in accordance with interpolar breakdown voltages of restrikes/prestrikes. Since maximum restrike/prestrike occurs at an interpolar breakdown voltage of 2 pu, the possible FTO level can be estimated by proportional calculation with the amplitude ratio of the observed FTO waveform and the interpolar breakdown voltage of the corresponding restrike/prestrike.

Figure 7 shows the probability densities of the possible FTO level obtained from the 24 measuring points by using the above-mentioned processing. The measuring points were chosen by giving consideration to three categories of locations in the GIS—namely, near the disconnectors operated, at the open ends of the bus bars (at the end of the

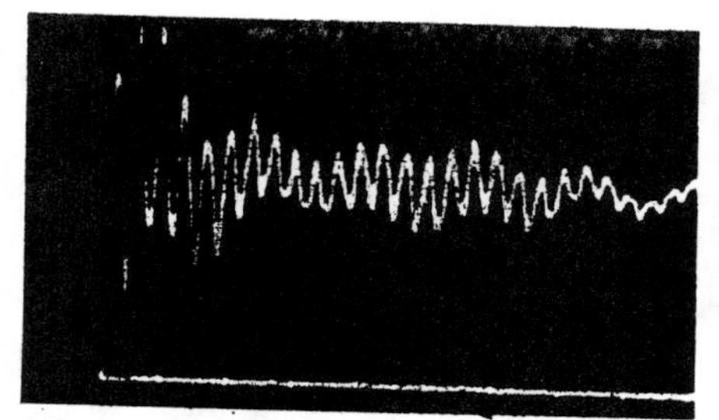

Fig. 6 Typical measured FTO waveform; measuring point ⑤ (180 kV/div, 1 μs/div)

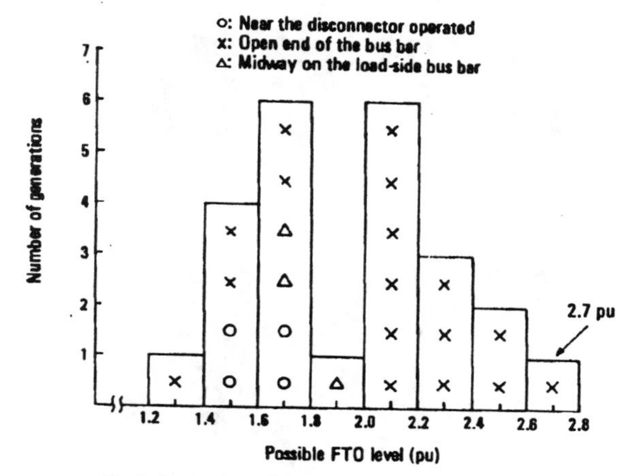

Fig. 7 Probability densities of possible FTO levels

main bus bar and near the open circuit breakers or disconnectors), and midway on the load-side bus bar. The possible FTO levels in the range of 1.3 – 2.7 pu were obtained by measurement. As shown in Fig. 7, the large FTO above 2.0 pu occurred consistently at open ends of the bus bars. In contrast, the FTO near the disconnectors operated were smaller, within the range of 1.5 – 1.7 pu. Smaller FTO at open ends of the bus bar are also found in Fig. 7. This is because these open ends were located on the opposite-side main bus bar, considerably away from the disconnector operated. Since the probability of an occurrence of a large restrike/prestrike, whose interpolar breakdown voltage lies within the range of 1.8 – 2.0 pu, is less than 1% (shown in Figs. 4

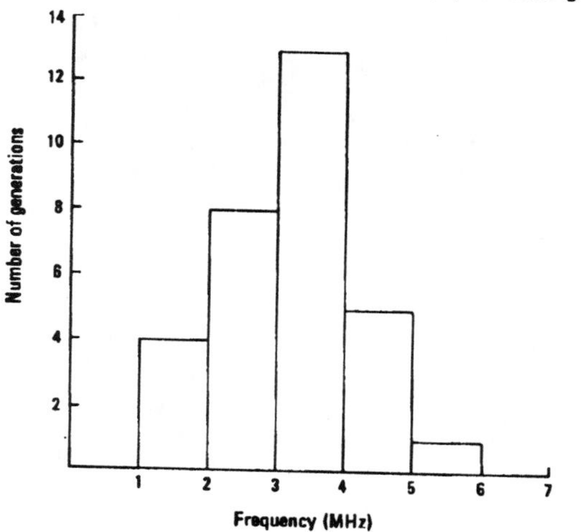

Fig. 8 Probability densities of basic frequencies of FTO

and 5), the probability of an occurrence of a large FTO is small.

Figure 8 shows the probability densities of basic frequencies of FTO obtained at the entire 24 measuring points. If the FTO contain typical two frequency components, both frequencies are plotted. There was no significant difference in frequency among the categories of measuring points.

LABORATORY TEST

The basic frequencies of the FTO are mainly determined by length and layout of the bus bars. However, higher frequency components up to 100 MHz arising from short surge impedance discontinuities such as spacers, disconnectors, and short bus branches may possibly be superposed on the basic frequencies. It is not known what degree of effect such high-frequency components have on the FTO levels. Also, to estimate FTO levels using an EMTP, opinions are divided as to whether or not to strictly simulate such surge impedance discontinuities.

When evaluating the results of measurement at the commercial substation mentioned in the previous section, it is very important to examine how strongly FTO levels are influenced by such high-frequency components.

Development of 1-GHz Surge Sensor

It is known that the risetime of step-shaped traveling waves caused by restrikes/prestrikes of disconnectors is within the range of 2 – 12 ns. The present study, therefore, necessitates a sensor that can respond to such risetime. The sensor must have a frequency characteristic of over several hundred MHz. To measure FTO in the laboratory, a capacitive potential divider consisting of a disk electrode installed in the GIS tank is commonly used [10][11]. Without exception, these sensors are designed to guide the potential at a point on the disk electrode to a coaxial cable via an impedance matching resistor. Here, it is important to eliminate the resonance mode in the low-voltage-side capacitor formed by the disk electrode and the GIS tank. In reference [10], rather than the junction of the coaxial cable and the disk electrode being located at the center of the disk electrode, it is positioned at the node of the standing wave of the resonance mode to prevent the resonance mode from propagating to the coaxial cable.

The authors developed a new type of sensor that was able to eliminate internal resonance modes. The structure of this surge sensor is shown in Fig. 9. The disk electrode is coupled to the coaxial cable

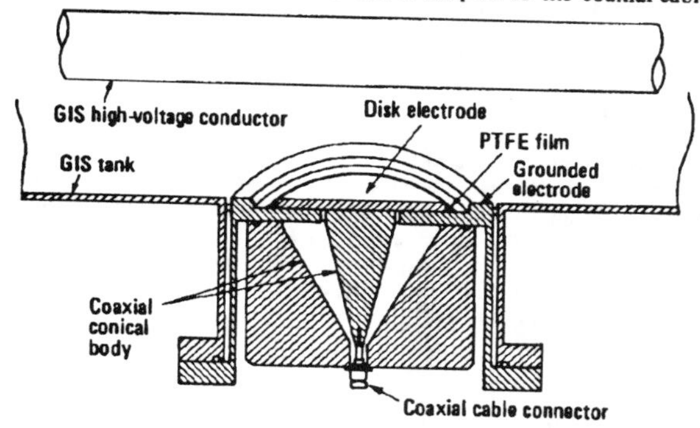

Fig. 9 Structure of 1-GHz surge sensor

by utilizing a coaxial conical body. This structure enables potentials of the disk electrode to be detected in a wide area, and therefore, to be detected by the mean potential of the disk electrode, thus permitting the internal resonance mode of the low-voltage-side capacitor to be eliminated. Diameter of the disk electrode was 15cm; the low-voltage-side capacitor, about 5 nF, was made from PTFE film 25μm thick.

Figure 10 shows the step response of this surge sensor. An oscilloscope having a frequency band width up to 1 GHz was used, together with a 1m-long coaxial cable to couple the oscilloscope with

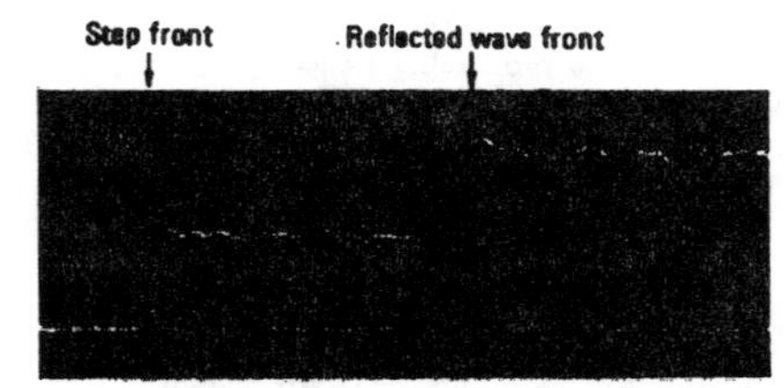

Fig. 10 Step response of 1-GHz surge sensor (2 ns/div)

the sensor. As shown in Fig. 10, a clear response waveform with few ripples was obtained, and step response time was about 350 ps. The second wave front in Fig. 10 is the reflected wave from the open end of the bus duct located about 1.3m away from the surge sensor. The response time corresponds to a frequency characteristic of about 1 GHz.

Experimental Apparatus

Figure 11 is a sketch of the experimental apparatus: the apparatus had a 275kV disconnector with an earthing switch, four disk-type spacers, a load-side bus bar about 10m long with three post-type spacers, and a 550kV gas bushing containing a stress capacitor. The

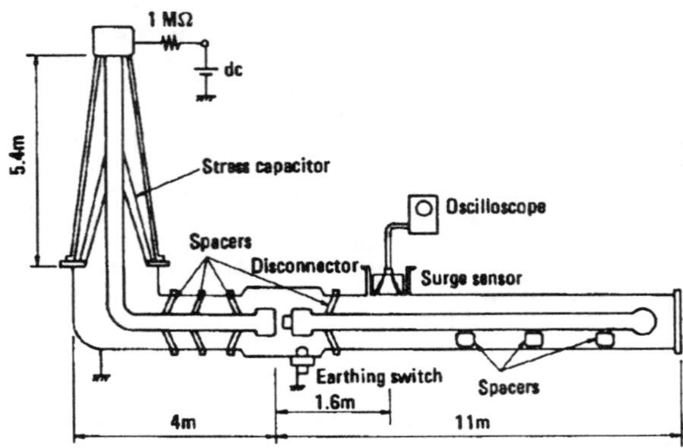

Fig. 11 Sketch of the experimental apparatus

1-GHz surge sensor mentioned above was located near the disconnector (placed 1.6m from the contacts) and connected to the oscilloscope with 1m-long coaxial cable. The oscilloscope was housed in a small shielded box connected to the tank with a copper plate; the coaxial cable was double shielded to eliminate noise caused by electromagnetic interference. Holding the load-side bus bar at zero potential, dc voltage was applied from the high-voltage dc power supply to the bushing via a 1 MΩ resistor, and the FTO waveforms generated by closing operation of the disconnector were observed. The dc voltage applied was positive, and the moving contact of the disconnector was located on the load side.

Results

Some observed waveforms are shown in Figs. 12 and 13. While the waveform in Fig. 12 (a) was observed when 10kV dc voltage was applied, the waveform in Fig. 12 (b) was observed when 300kV dc voltage was applied. The low-voltage test waveform in Fig. 13 was observed with the mercury switch connected in the shortest route between the disconnector contacts, and with a commercially available 350 MHz probe coupled to the tank and the high-voltage conductor at the surge sensor mounting location.

The waveforms in Figs. 12 (a) and 13 agree well. The waveform in Fig. 12 (b) differs from the other two, with a much smaller high-frequency component superposed on the basic frequency. To investigate the causes of this, the risetimes of FTO were observed by using

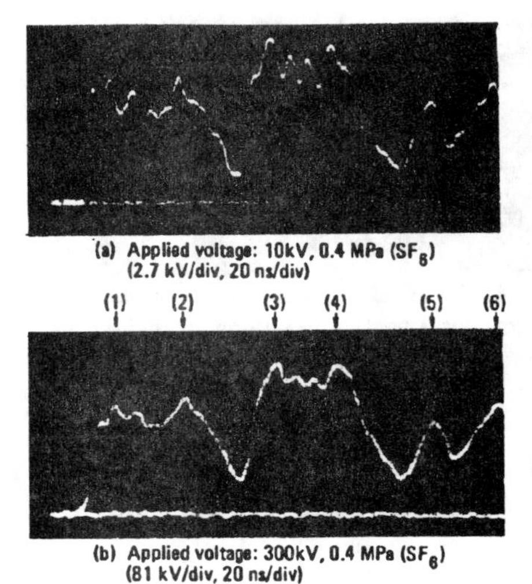

(a) Applied voltage: 10kV, 0.4 MPa (SF_6) (2.7 kV/div, 20 ns/div)

(b) Applied voltage: 300kV, 0.4 MPa (SF_6) (81 kV/div, 20 ns/div)

Fig. 12 FTO waveform measured by 1-GHz surge sensor

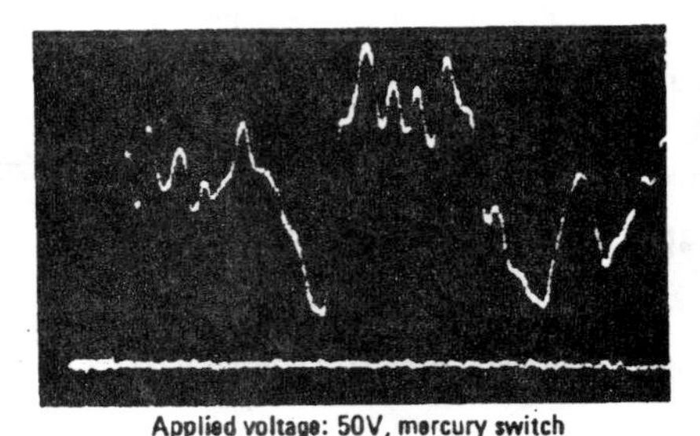

Applied voltage: 50V, mercury switch

Fig. 13 Low-voltage test waveform measured by 350 MHz commercially available probe (10 V/div, 20 ns/div)

various applied voltages as parameters. One of the waveforms observed is shown in Fig. 14. The results are shown in Fig. 15. The error bars denote the dispersion of twelve trials. It is evident that with a constant SF_6 gas pressure, a higher interpolar breakdown voltage causes a longer risetime. With the same voltage, a lower gas pressure also causes a longer risetime.

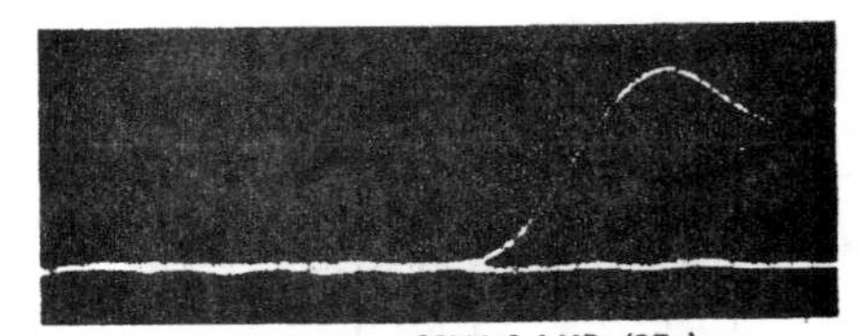

Applied voltage: 30kV, 0.4 MPa (SF_6)

Fig. 14 Typical waveform for risetime observation (2 ns/div)

Variation of the risetime is explained by the difference of spark collapse times (the times required for the formation of a conducting spark channel after a breakdown channel has been connected with the contacts). A longer spark length seems to cause the longer spark collapse time. Thus, in the waveform shown in Fig. 12 (b), the risetime was probably extended by the longer spark collapse time, leaving little room for a high-frequency component. For this reason, in the measured waveform upon application of 300kV, the high-frequency component is about one-tenth the amplitude of the basic frequency determined by length of the bus bars.

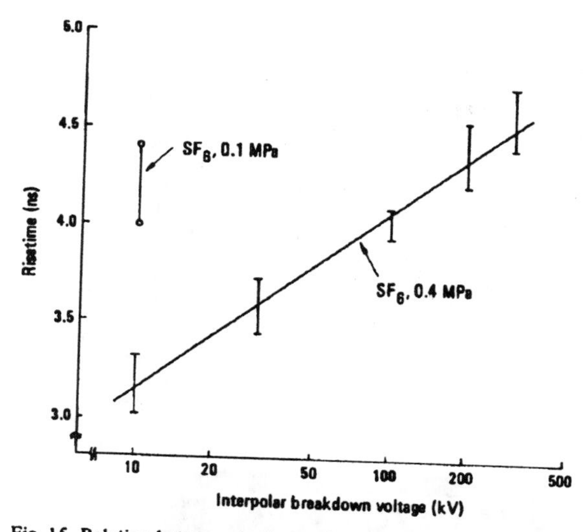

Fig. 15 Relation between risetime and interpolar breakdown voltage

Comparison with EMTP Waveforms

Waveforms by EMTP with strict simulation of short surge impedance discontinuities such as spacers, disconnectors, and short bus branches are shown in Figs. 16 (a) and (b). Spacers and the short bus branch of the earthing switch were simulated by lumped capacitors of 4 – 20 pF. The surge impedance of each bus duct is calculated from diameters of the inner and outer electrodes. Velocity of the traveling wave was given at the velocity of light. The time interval of the calculation was 40 ps. Stress capacitance installed in the bushing was divided into eight sections, and eight capacitances and eight lines of slightly changed surge impedances were adopted for the simulation. The conductor of the bushing was divided into five sections, and from 200Ω to 250Ω was given as surge impedance to the ground. Figure 16 (a) shows the waveform without a spark collapse time; Fig. 16 (b) shows the waveform with a spark collapse time. A spark collapse time can be given by time-variable resistance [1]. In the authors' simulation, the FTO front was approximated by a linearly increasing spark current: $i = (E/2Z)\cdot(t/tz)$. Here, Z, E, and tz represent surge impedance of the bus duct, applied voltage, and spark collapse time, respectively. The spark current i is also given by using a spark resistance r as $i = E/(2Z + r)$. Thus, the following equation is obtained.

$$i = \frac{E}{2Z + r} = \frac{E}{2Z} \cdot \frac{t}{tz} \quad (0 < t < tz) \tag{1}$$

From equation (1), spark resistance r is given by–

$$r = 2Z\left(\frac{tz}{t} - 1\right) \tag{2}$$

Measured rise time 4.5 ns shown in Fig. 15 was given for spark collapse time tz.

Figure 17 shows the waveform obtained from rough simulation ignoring short surge impedance discontinuities of less than 1.2m.

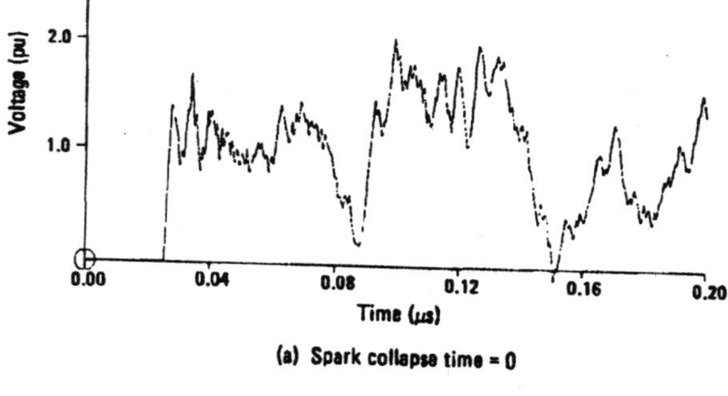

(a) Spark collapse time = 0

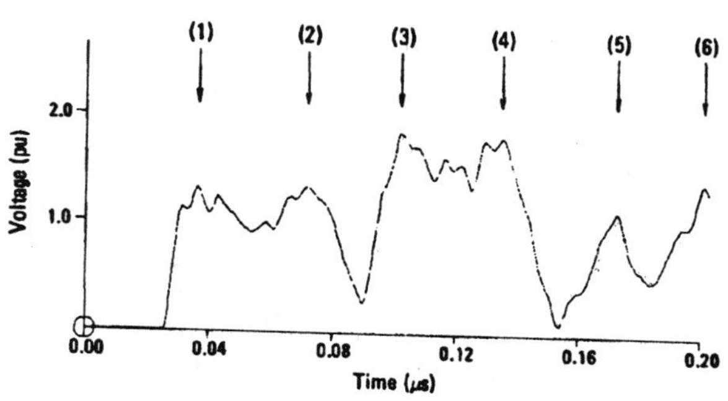

(b) Spark collapse time = 4.5 ns

Fig. 16 FTO waveforms by strict simulation

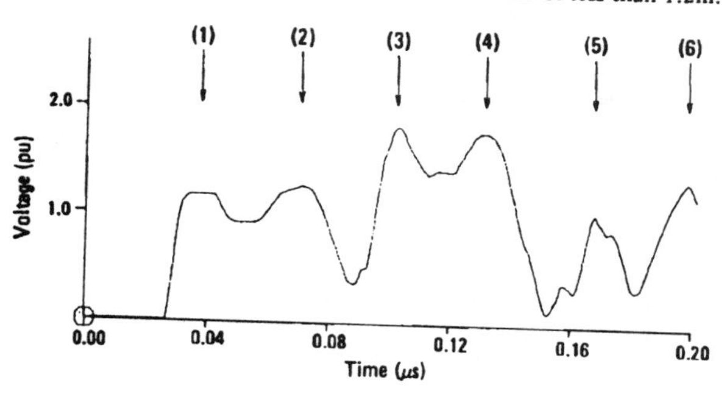

Spark collapse time = 4.5 ns

Fig. 17 FTO waveform by rough simulation

Surge impedance was rearranged by the mean value, and traveling wave velocity was less than light velocity (source-side bus bar: 250 m/μs, load-side bus bar: 290 m/μs). However, since length of the bushing was not negligible compared with those of the bus bars, the simulation method was not altered. The time interval of the calculation was also 40 ps.

If the spark collapse time is ignored, large high-frequency components of about 100 MHz are superposed on the basic frequency, as shown in Fig. 16 (a). On the contrary, as for a waveform with a spark collapse time of 4.5 ns, as shown in Fig. 16 (b), these high-frequency components are considerably damped, and the waveform reveals good agreement with the measured one shown in Fig. 12 (b). At the same time, it becomes similar to the roughly simulated waveform.

To evaluate the errors of FTO level estimations by EMTP, the values of the six peaks of the three waveforms (Figs. 12 (b), 16 (b), and 17) were compared. The results are shown in Table I. Here, the applied voltage of 300kV was divided by two (source side: +150kV, load side:-150kV), and 150kV was used as 1 pu. The maximum error 0.1 pu appears at peak (1) between measured and roughly simulated waveforms. At the maximum peak (peak (3)), however, the errors lie within 0.03 pu for both waveforms–strictly simulated and roughly

Table I Comparison of six peak values of three waveforms

Peak number in the figures	(1)	(2)	(3)	(4)	(5)	(6)
Measured waveform (Fig. 12 (b))	1.27	1.34	1.80	1.76	1.08	1.31
Strictly simulated waveform (Fig. 16 (b))	1.28	1.31	1.81	1.78	1.11	1.39
Roughly simulated waveform (Fig. 17)	1.17	1.25	1.83	1.77	1.04	1.36

Unit: pu (= 150kV)

simulated waveforms. Thus, it was proved that if the spark collapse time was taken into consideration, the FTO level estimation was performed within an error of 0.1 pu, even by rough simulation. The bushing, however, had length comparable with those of the bus bars, so strict simulation was necessary.

Many more spacers, disconnectors, and short bus branches are installed in a commercial GIS than in the experimental apparatus. These short surge impedance discontinuities cause the degradation in risetime of the traveling wave. Since the maximum peak usually appears after several peaks, the traveling waves must have passed a considerable number of the short surge impedance discontinuities. Thus, in a commercial GIS, risetime may be considerably degraded. For this reason, errors smaller than 0.1 pu are expected.

ESTIMATION OF FTO LEVELS OF THE SUBSTATION BY EMTP

EMTP simulation, easily performed with the help of a computer, can be an important means in estimating FTO levels of substations. In calculation, component units must be suitably simulated. The simulation method, considered as very important, has been widely studied by many researchers. The authors have demonstrated, as described in the previous section, that if risetime caused by spark collapse time is taken into account, FTO levels can be estimated with an error of less than 0.1 pu even without simulating short surge impedance discontinuities. Thus, through rough simulation, the authors conducted FTO waveform calculations of the commercial substation described in the previous section. As to the method for simulating component units such as transformers, circuit breakers, and capacitors between circuit breaker poles, refer to reference [6].

The waveform in Fig. 18 (a) is the one with 4.5 ns spark collapse time, which ignored attenuation caused by arc resistance. It is located at the same point as that used for the measured waveform in Fig. 6. Up to the fourth peak, the measured and the simulated waveforms agree well. In other parts from the fifth peak downward, the measured waveforms reveal considerably large attenuation. To simulate this, an arc resistance between 1Ω and 10Ω was assumed, but this was not sufficient to completely simulate attenuation of the measured waveform. Figure 18 (b) shows the waveform with an arc resistance of 2Ω. This suggests that a time-variable arc resistance must be considered.

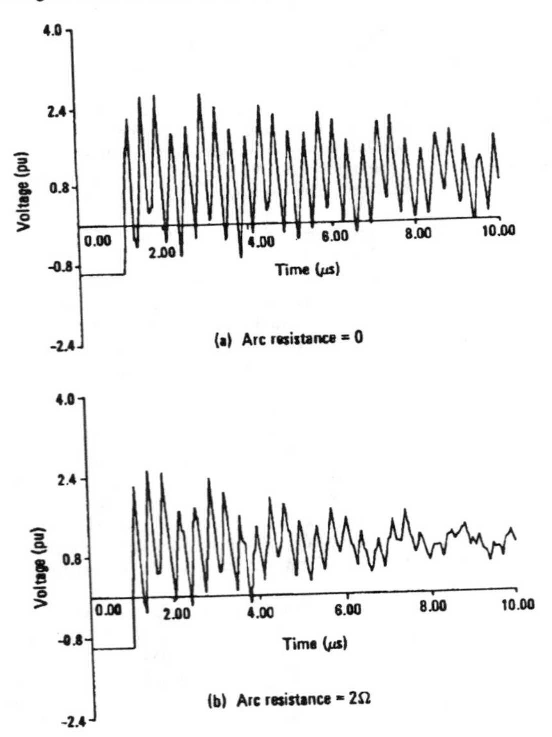

Fig. 18 Simulated FTO waveforms of the substation : measuring point ⑤

As shown in Figs. 19 (a) and (b), however, with FTO caused at the same measuring point, under the same conditions, but by restrike or prestrike, the degrees of attenuation are not always identical. Therefore, it does not appear to be very meaningful to simulate such attenuation precisely. Thus, we compared simulated FTO levels, without arc resistance, with measured levels at 24 measuring points. The results are shown in Fig. 20. The horizontal axis represents the time interval between the FTO front and the maximum peak given by simulated waveforms, and the vertical axis represents the difference of levels between the simulated and the measured waveforms. Almost all points are within errors of ±0.1 pu. With the time interval above 1 μs, errors are somewhat larger. If a simulated peak occurs in a time above 1 μs after the wave front, care should be taken in evaluating the result.

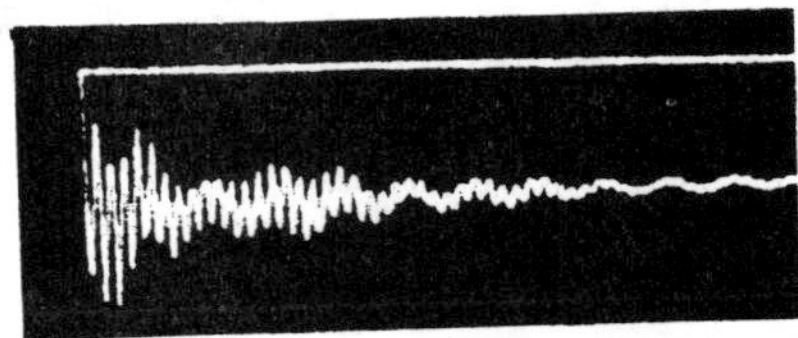

(a) FTO waveform caused by restrike: interpolar breakdown voltage was 1.4 pu (360 kV/div, 2 μs/div)

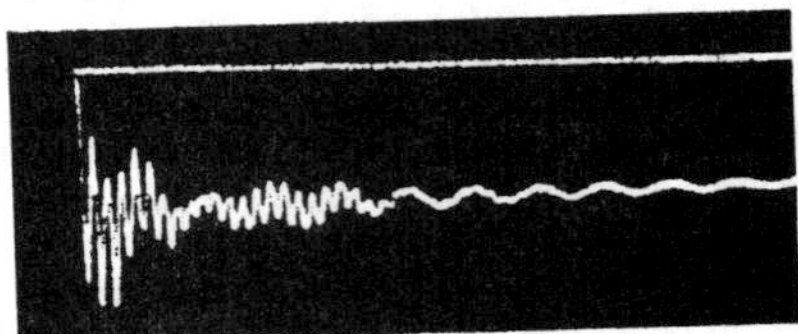

(b) FTO waveform caused by prestrike: interpolar breakdown voltage was 1.4 pu (360 kV/div, 2 μs/div)

Fig. 19 Two differently attenuated FTO waveforms measured at same measuring point (⑤), under same conditions

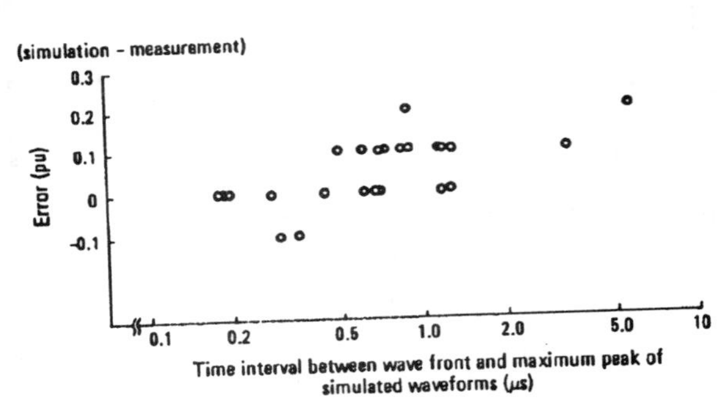

Fig. 20 Errors of FTO level between simulated and measured waveforms

CONCLUSIONS

By measuring disconnector-induced FTO at a commercial 550kV substation, a great deal of useful data were obtained. The simulation method for EMTP was also discussed. The concluded results are as follows:

(1) In case of a high-speed disconnector, the probability of an occurrence of restrike or prestrike caused by interpolar breakdown voltage of 1.8 – 2.0 pu is less than 1%.

(2) The estimated maximum FTO level resulting from measurement was 2.7 pu, and an FTO of more than 2.0 pu appeared only at the open end of the bus bar. On the contrary, FTO levels near the disconnectors operated were below 1.7 pu.
(3) A spark collapse time was correctly simulated by the time-variable resistor. By this spark collapse time, risetime of the FTO is extended, and the high-frequency component caused by short surge impedance discontinuities such as spacers, disconnectors, and short bus branches were fairly well damped.
(4) If errors of about 0.1 pu are permitted when estimating FTO levels, rough simulation ignoring the short surge impedance discontinuities and attenuation caused by arc resistance is sufficient. However, if the results of simulation show that the maximum peak appears from the wave front more than 1 μs later, care should be taken to evaluate these results.

The authors have obtained the same conclusions by measurement of FTO at a few other commercial substations with operating voltages above 275kV.

ACKNOWLEDGMENT

The authors express their sincere thanks to Dr. S. A. Boggs and Dr. F. Y. Chu, who offered them useful suggestions in their developing a surge sensor for fast transient overvoltages.

REFERENCES

[1] S. A. Boggs, F. Y. Chu, N. Fujimoto, A. Krenicky, A. Plessel, and D. Schlicht, "Disconnect Switch Induced Transients and Trapped Charge in Gas-insulated Substations," IEEE Transactions on Power Apparatus and Systems, vol. PAS-101, No. 10, pp. 3593–3602, 1982.

[2] S. Narimatsu, K. Yamaguchi, S. Nakano, and S. Murayama, "Interrupting Performance of Capacitive Current by Disconnecting Switch for Gas Insulated Switchgear," IEEE Transactions on Power Apparatus and Systems, vol. PAS-100, No. 6, pp. 2726–2732, 1981.

[3] S. Nishiwaki, Y. Kanno, S. Sato, E. Haginomori, S. Yamashita, and S. Yanabu, "Ground Fault by Restriking Surge of SF_6 Gas-insulated Disconnecting Switch and Its Synthetic Tests," IEEE Transactions on Power Apparatus and Systems, vol. PAS-102, No. 1, pp. 219–227, 1983.

[4] N. Fujimoto and S. A. Boggs, "Characteristics of GIS Disconnector-induced Short Risetime Transients Incident on Externally Connected Power System Components," IEEE 87 WM 185-2, New Orleans, Feb. 1987.

[5] W. Boeck and W. Taschner, "Insulating Behaviour of SF_6 with and without Solid Insulation in Case of Fast Transients," CIGRE Paper No. 15-07, Aug. 1986.

[6] S. Ogawa, E. Haginomori, S. Nishiwaki, T. Yoshida, and K. Terasaka, "Estimation of Restriking Transient Overvoltage on Disconnecting Switch for GIS," IEEE Transactions on Power Systems, vol. PWRD-1, No. 2, pp. 95–101, 1983.

[7] R. Witzman, "Fast Transients in Gas Insulated Substations (GIS) –Modeling of Different GIS Components," Fifth International Symposium on High Voltage Engineering, No. 12.06, 1987.

[8] H. Murase, I. Ohshima, H. Aoyagi, and I. Miwa, "Measurement of Transient Voltages Induced by Disconnect Switch Operation," IEEE Transactions on Power Apparatus and Systems, vol. PAS-104, No. 1, pp. 157–163, 1985.

[9] K. Nojima, S. Nishiwaki, H. Okubo, and S. Yanabu, "Measurement of Surge Current and Voltage Waveform Using Optical-transmission Techniques," IEE Proceedings, vol. 134, Pt. C, No. 6, pp. 415–424, 1987.

[10] S. A. Boggs and N. Fujimoto, "Techniques and Instrumentation for Measurement of Transients in Gas-insulated Switchgear," IEEE Transactions on Electrical Insulation, vol. EI-19, No. 2, pp. 87–92, 1984.

[11] J. Meppelink and P. Hofer, "Design and Calibration of a High Voltage Divider for Measurement of Very Fast Transients in Gas Insulated Switchgear," Fifth International Symposium on High Voltage Engineering, No. 71.08, 1987.

Satoru Yanabu (M'68-SM'81) was born in Shimane, Japan on July 15, 1941. He received his B.S. degree in electrical engineering from the University of Tokyo, Japan in 1964 and his Ph.D. degree from the Univeristy of Liverpool, England in 1971.

He joined Toshiba Corporation, Japan in 1964. From 1964 to 1969, he was in the High Voltage Laboratory, engaged in research on corona discharge detection. From 1969 to 1971, he had a leave of absence from Toshiba Corporation to join in arc research work at the University of Liverpool. From 1971 to 1982, he has been in the High Power Laboratory, Development Section, of Toshiba Corporation, engaged in the development of switchgear. He is presently senior manager of the High Voltage and High Power Laboratories.

Dr. Yanabu is a member of IEE of Japan and a senior member of IEEE.

Hiroshi Murase was born in Gifu Prefecture, Japan on July 9, 1952. He received his B.S., M.S., and Ph.D. degrees from Tokyo Institute of Technology, in 1975, 1977, and 1980 respectively, all in electrical engineering.

In 1980, he joined the High Voltage Laboratory of Toshiba Corporation, Japan. From 1984 to 1985, he had a leave of absence from Toshiba Corporation to join the plasma physics laboratory and high-voltage laboratory at the Technical University of Munich. Having returned to the High Voltage Laboratory of Toshiba Corporation, he has been engaged in the study of high-voltage insulation for SF_6 gas.

Dr. Murase is a member of IEE of Japan.

Hirokuni Aoyagi was born in Yokohama, Japan on August 12, 1941. He graduated from Toshiba Technical Junior College in 1963.

He is working in the High Voltage Laboratory of Toshiba Corporation, Japan, where he engages in the development of gas-insulated switchgear (GIS). He is presently involved in the study of a SF_6 gas insulation system for GIS.

Mr. Aoyagi is a member of IEE of Japan.

Hitoshi Okubo (M'82) was born in Nagoya, Japan on October 29, 1948. He received his B.S., M.S., and Ph.D. degrees in electrical engineering from Nagoya University, Nagoya, Japan in 1971, 1973, and 1984 respectively.

He joined Toshiba Corporation, Japan in 1973, and from 1976 to 1978 he was with the Technical University of Aachen and Technical University of Munich, West Germany where he studied high-voltage technique. Since 1985, he has been manager of the High Voltage Laboratory of Toshiba Corporation.

Dr. Okubo is a member of IEE of Japan and IEEE.

Yoshihiro Kawaguchi (M'58-SM'68-F'88) was born in Yokohama, Japan on January 15, 1932. He received his B.S. and Eng. Doctor degrees from Waseda University in 1953 and 1970 respectively.

He joined Toshiba Corporation, Japan in 1953, and has worked on the development and design of various high-voltage apparatus such as power transformers, surge arresters, and switchgear in the Heavy Apparatus Engineering Laboratory and Designing Department of Toshiba Corporation. He is presently fellow specialist in Heavy Apparatus Engineering Laboratory.

He is also a Japanese representative for IEC TC28-WG02 (Insulation Co-ordination) and for CIGRE SC33-WG04-TF06 (Artificial Pollution Testing of Metal-Oxide Surge Arresters).

Dr. Kawaguchi is a member of IEE of Japan and a fellow of IEEE.

Discussion

N. Fujimoto, (Ontario Hydro Research Toronto, Ontario, Canada): The authors present some interesting data concerning the characteristics of GIS fast transients measured in an actual commercial GIS. In addition, the authors give an indication of how factors such as risetime and arc resistance affect the transient waveform. Essentially, the results in the paper agree with other studies on the subject, including our own. However, some comments are offered for the authors' consideration.

The authors indicate in the paper that the probability of large FTO is small. We fully concur with this statement and add that the probability for slowing operating disconnectors (common in North America and Europe and as opposed to the fast operating switches common in Japan) would be even lower as there is less of a tendency for generating large trapped charges and high intercontact breakdown voltages. With statistical knowledge of the disconnector operating characteristics and knowledge of the normalized overvoltage factor, the probability of high FTO can be computed quantitatively. For instance, if C is the intercontact breakdown voltage and Q the momentary trapped potential which exists prior to intercontact breakdown of magnitude C, then (C, Q) represents, when treated as random variables, the statistical operating characteristics of the disconnector. These values could be obtained from the same data which generated figures 3, 4, 5 of the paper. The overvoltage factor K, which is the normalized overvoltage for which $(C, Q) = (1, 0)$, is also a statistical parameter which varies as a function of position within the GIS and operational configuration. These values can be obtained by computer simulation or direct measurement under controlled conditions. As a result, the statistical characteristics of the FTO, represented by the parameter V, is computed by the simple operation:

$$V = K \times C + Q$$

Since the statistical distributions of K, C and Q are known by measurement or simulation, the statistical distribution of the FTO, V can also be determined. This procedure was demonstrated in a previous publication [1] for a hypothetical operating configuration. Similar formulations for FTO have also been suggested by others and are found throughout the literature. From the data presented in the paper, the authors should also have sufficient information to perform this calculation.

Caution should be given to the emphasis on the absolute peak magnitude when evaluating FTO. Although this is a parameter which is in common use, evidence confirming that peak magnitude is the most important parameter is lacking. For instance, one school of thought suggests that high frequency oscillations (in the range of 100s of MHz) might have a strong influence on leader development [2]. The presence of such oscillations might have a greater influence on the severity of the transient (from an insulation point of view) than the simple consideration for the peak magnitude. Phenomena such as this are, in part, responsible for differing opinions regarding modeling for computer simulations, as the authors have mentioned in the paper.

From the work described in the paper and from other publications, the GIS fast transient phenomenon appears to be well understood. The remaining issues deal more with the consequences as opposed to the phenomenon itself. For instance, questions related to the influence of high frequency oscillations, the importance of peak magnitude and the amount of detail required in modeling, as discussed above all depend on the physics of breakdown under fast transient stress. However, the prevailing opinion is that well-designed GIS should not experience dielectric problems with fast transients and problems which occasionally occur are the result of defects or irregularities. Consequently, fast transient studies should now focus on understanding breakdown processes and the implications for the testing required to detect the defects sensitive to fast transients.

1. Boggs, S. A., N. Fujimoto, M. Collod and E. Thuries. "The modeling of statistical operating parameters and the computation of operation-induced surge waveforms for GIS disconnectors." CIGRE 1984, paper 13-15.
2. Luxa, G., E. Kynast, A. Pigini, A. Bargigia, W. Boeck, H. Hiesinger, S. Schlicht, N. Wiegart and L. Ullrich. "Recent Research Activity on the Dielectric Performance of SF_6 with Special Reference to Very Fast Transients." CIGRE 1988, paper 15-06.

Manuscript received August 17, 1988.

S. Yanabu, H. Murase, H. Aoyagi, H. Okubo and Y. Kawaguchi: We would like to thank Dr. N. Fujimoto for his constructive discussion of our paper.

We recognize well that the probability of large FTO for slowing operating disconnectors is very small or even nearly zero. On the contrary, high-speed types show no trapped charge reduction effect caused by field strength inequality, and the probability of large FTO becomes higher. In such a case, however, the probability of the re-/prestrike between 1.8pu and 2.0pu was less than 1%. We would like to stress that although the probability of large FTO is small, there are some possibilities, and re-/prestrike above 1.8pu was actually observed. Therefore, we have to take into consideration the severest case. From this reason, we discussed the possible FTO level calculated from the measured value, that is the severest case, in our paper.

The equation that gives the statistical distribution of FTO,

$$V = K \times C + Q$$

is very useful and interesting. It will be possible to calculate the statistical distribution for high-speed type disconnectors. Since the probability that (C, Q) becomes principal maximum value, that is (2, 1), is not zero as described above, we mainly discussed about such a case. Figure 7 in our paper represents the measured statistical distribution of V when (C, Q) = (2, 1) is supposed. These statistical distributions correspond to the distribution of K. It is beyond doubt that the probability of (C, Q) nearly equals (2, 1) for slowing operating disconnectors is almost zero, but our interest lies if it should be perfectly zero. Since the number of re-/prestrikes within one operation of slowing operating disconnectors is large, this point seems to be important.

We also have great interest in the influence of high frequency oscillations (in the range of 100s of MHz) on leader development. The strict simulation must be necessary for the laboratory study of this subject in order to grasp the exact FTO waveforms appearing in commercial substations. But, we are not sure how to evaluate the influence of such a high frequency component and the method is now under investigation. A hope arise from our study, however, that is one of the conclusions of our paper; "By a spark collapse time, risetime of the FTO is extended, and the high-frequency component caused by short surge impedance discontinuities were fairly well damped."

We agree that "fast transient studies should now focus on understanding breakdown processes and the implications for the testing required to detect the defects sensitive to fast transients." And we are now engaged in these studies.

Manuscript received September 21, 1988.

MORE ACCURATE MODELING OF GAS INSULATED SUBSTATION COMPONENTS IN DIGITAL SIMULATIONS OF VERY FAST ELECTROMAGNETIC TRANSIENTS

Z. Haznadar
University of Zagreb
Fac. of Electrical Eng.
Zagreb, Yugoslavia

S. Čaršimamović
"Energoinvest"
Electric Power Institute
Sarajevo, Yugoslavia

R. Mahmutćehajić
University of Osijek
Fac. of Electrical Eng.
Osijek, Yugoslavia

ABSTRACT

Very fast electromagnetic transients caused by switching operations in gas insulated substations (GIS) cannot be calculated if conventional techniques of modeling and simulations are used. Choice and adjustments of the most suitable models as well as determinations of their limitations in digital simulations of very fast transients in GIS are investigated in the paper. Results obtained from very extensive field tests as well as from digital simulations for different types of GIS have been used for development of more accurate models for GIS components and GIS as whole. A comparison between field test and calculation results enables development of a model which takes into account the most decisive physical phenomena inherent to the very fast transients in the GIS. This is a prerequisite for designing and manufacturing of power system components as well as for control and protection of the system.

1. INTRODUCTION

Coaxial conductors in gas insulated substations have higher specific capacitance to earth in comparison with the open air substations. Therefore, the capacitive currents of off-loaded bus in GIS are larger than the capacitive currents in open air substations. Further differences between the characteristics of these two types of substations are lower characteristic surge impedance and inductance as well as larger gradient of electric field between the prestrike and restrike arcs in sulfur - hexafluoride gas under pressure in reference to grounded enclosure. This causes very fast transient with waveshapes characterized by surge fronts of very short durations (about 10 ns). The disconnector contacts in GIS are moving slowly (in the order of 1 cm/s) causing numerous strikes and restrikes between contacts. When the contacts are closed, the capacitive charging flowing through the contacts is from 0.3 to 1 A *r.m.s.*, depending on the rated voltage and length of the bus which is switched. Strikes and restrikes occur as soon as the dielectric strength of the gas between contacts is exceeded by overvoltage. The overvoltage is defined by the distance between contacts, the contacts geometry and the pressure of the gas as well as by characteristics of the gas at the instant of strike. Every strike causes high-frequency currents tending to equalize potentials at the contacts. When the current is interrupted, the voltages at the source side and the loading side will oscillate independently. The source side will follow the power frequency while the loading side will remain at the trapped voltage. As soon as the voltage between contacts exceeds the dielectric strength of the gas between contacts defined by the distance between contacts, the restrike will occur, and so on. Successive strikes occurring during the closing and opening operations of off-loaded bus by disconnector are shown in Fig. 1, *a* and *b*, respectively.

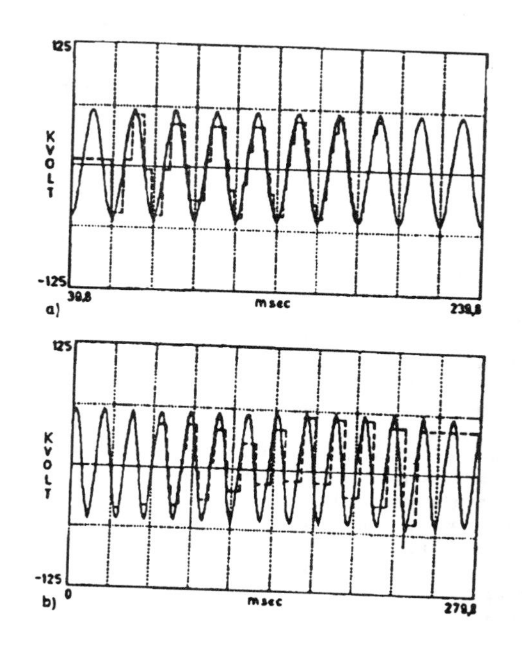

Fig. 1 The voltage due to the disconnector switching
a) disconnector opening
b) disconnector closing

At the transient beginning, the intervals between particular strikes are on the order of a millisecond, while just before the last strike, the period can reach about one half of cycle at power frequency (Fig. 1*a*). When closing takes place, the first strike will occur at the maximum value of the source voltage. Its values can be positive or negative. As the time passes a series of successive strikes will occur (Fig. 1*b*). The number of strikes during a switching operation depends on the speed of the contacts. The maximum values of voltage and the maximum values of the wave-front increasing will take place at the maximum distances between contacts. For the purpose of the investigation of the insulation strength, the most important are the first few strikes during the closing operation or the last few strikes during the opening operation. Each individual strike causes an oscillatory current with the basic frequency on the order of 100 kHz up to several Mhz and peak value up to a few kA. Very fast overvoltage transient due to the closing operation of the disconnector at the load side of the test circuit is shown in Fig2. The transient is recorded by a waveshape digitizer.

CH2948-8/91/0000-0260$1.00©1991 IEEE

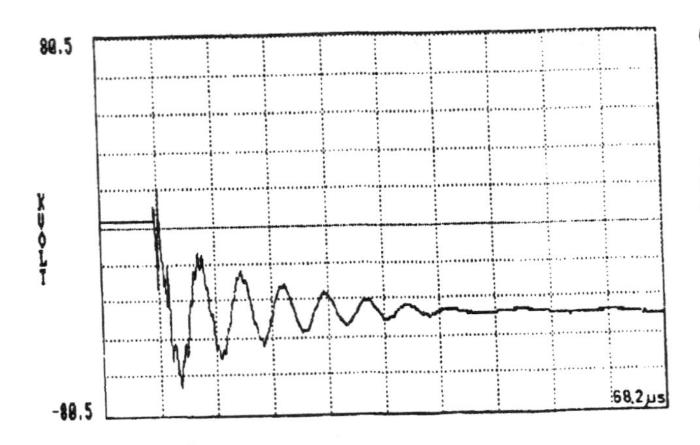

Fig. 2 Very fast transient overvoltage due to the closing at the load side

These transients can continue for several tenths of microseconds, i.e. up to the equalization of the overvoltage with the instant value of the power frequency voltage. The overvoltages due to the disconnector switching have electromagnetic wave nature and can be considered as a travelling wave propagating along particular elements of the substation: the characteristics of the overvoltage surges depend on the substation configuration and they are important factors in the insulation co-ordination. Depending on the voltage level, they can reach values which will cause a flashover from the bus to the grounded enclosure. Therefore, having the best possible insight into critical points in substation in which the peak value of overvoltage will occur is of key importance in the selection of substation equipment with the corresponding insulation levels.

The investigators are faced today with numerous and different approaches and methods of the calculation of very fast transient overvoltages (VFTO). Every substation element can be represented by different mathematical models suitable for various digital simulation approaches. Interfacing of the models in the totality which represents a substation can be made in different ways. Answering the question "How to select an optimal model for every individual element of a substation and how to make their connection in the totality?", represents a decisive task in modern engineering approaches to the area. This paper gives some answers. The investigations are based on very extensive field and computer testings of electromagnetic transients in GIS performed at a test section. Using the comparisons between the field test and computation results, a complete model for different types of substations is developed.

From the investigations performed up to now [1-12] it is possible to derive the following conclusions dealing with the problem of modeling and simulating of very fast transients in GIS:

(a) The elements of GIS can be represented using different models and, therefore, the simulation results depend on the selection of the models;

(b) Evaluation of the results from a validity point of view can be carried out by digital simulation of considered configuration response to the unit step impulse and consequent comparison with the results obtained from the field test for the same configuration:

(c) The performed investigations, when critically considered, have a disadvantage because they cannot give reliable answers on applicability as well as on correctness of the models applied in the cases of real substations:

(d) The simulation time in all available investigations is rather short and reliable conclusions on the model behavior in a longer time period cannot be derived from them.

These common properties of all available models point to the necessity of supplementing the investigations with two previously missing aspects: *(i)* selecting the most suitable models for individual elements of GIS which will be consistent among themselves, reliable as much as possible and accurate and, finally, efficient from the computation point of view, and *(ii)* considering the complete transient in a time period longer than the period encompassed in the approaches developed up to now. Answering these above mentioned tasks, in the following section of the paper we will discuss and resolve the most important aspects of the problem.

2. FIELD TESTS AND SIMULATIONS FOR LOW VOLTAGE RESPONSES

The waveshape of the overvoltage surge due to disconnector switching is affected by all GIS elements. Accordingly, the simulation of electromagnetic transients in GIS assumes an establishment of the models for bus, bushings, elbow, spherical terminations, transformers (power transformers, current and capacitive and inductive voltage transformers), surge arresters, breakers, disconnectors, spacers, reactors, enclosures and so on.

Accuracies of the models for each element have to be consistent. Furthermore, when the simulations are performed using programs based on the time-domain techniques, when the time step must be equal for all elements, it is necessary to have a deep insight into the properties of all individual elements, looking from the wave propagation point of view. On the other hand, this must be brought into accord with the duration of the considered transient.Since the wave propagates along rather short distances and is subdued to manifold refractions and reflections in different kinds of nodes, the consistency of the model as a whole is important as well as the accuracy of each model for individual elements.

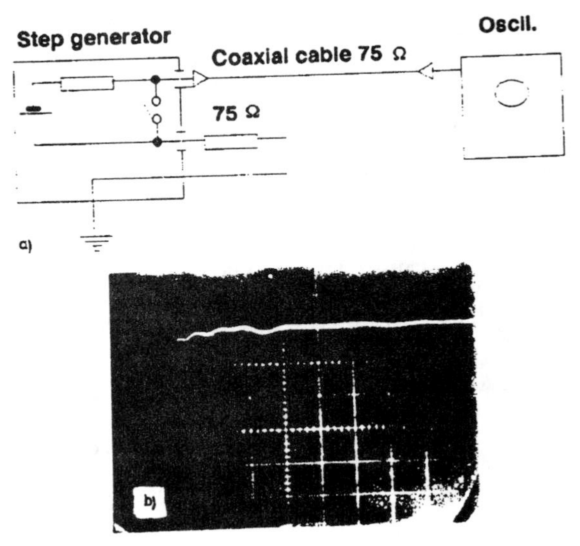

Fig. 3 Generator of the step voltage and the waveshape of the impulse

To resolve the above task, the following procedure is used: *(a)* for the configuration shown in Fig. 7 all its elements are modeled using one of the possible approaches, *(b)* a selected model of the configuration is tested by a series of simulations for the transient caused at different points when the sending-end is energized by a unit step impulse produced by a voltage generator. (The voltage generator produces a unit step voltage of 90 V with the time of the wave front rise of 4 ns, as shown in Fig. 3). The most suitable model is achieved by using results obtained from the simulation in the second step. Then, the simulation results are compared with field test results. Finally, such established criteria of modeling are applied to the modeling of an actual substation. Three configurations are considered. The first one is the case when the disconnector is in an open position. The second one is when the disconnector is in a closed position. And the third one is when the disconnector is in a closed position and has a capacitor connected at the receiving-end.

2.1 Model of test circuit with disconnector in open position

For the test circuit shown in Fig. 4*a*, when the disconnector is in an open position, the waveshape is recorded at point M_1 and it is given in Fig. 4*b*. The models of GIS elements as well as the model of measuring circuits are given in Fig. 4*c*. The comparison between the waveshapes obtained by simulation and field test is given in Fig. 4*d*.

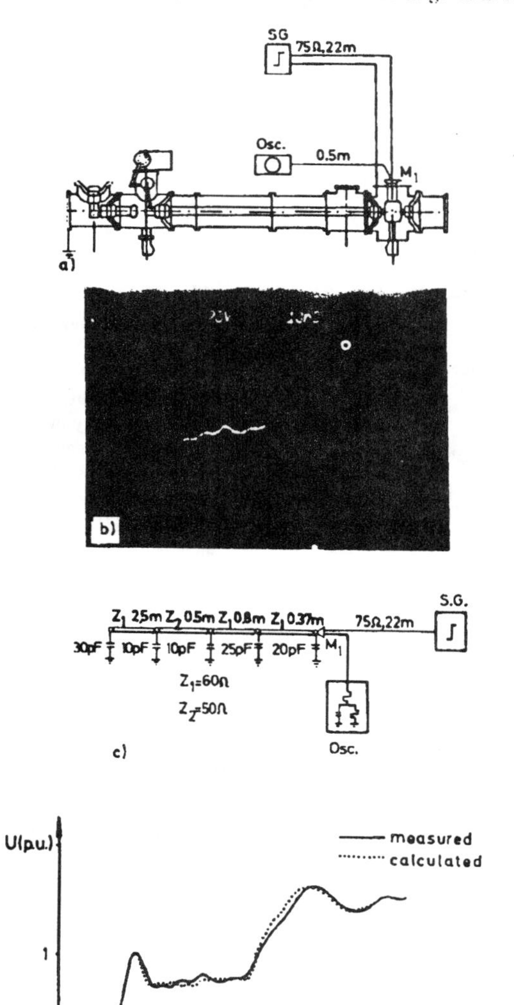

Fig. 4 The comparison between waveshapes obtained from simulation and field test

2.2 Model of test circuit with disconnector in closed position

The effect of the disconnector on the waveshape is measured for point M_1, as it is shown in Fig. 5*a*. The recorded waveshape is given in Fig. 5*b*. The model of GIS elements as well as the model of measurement circuit are given in Fig. 5*c*. The comparison between results obtained from simulation and field test is given in Fig. 5*d*.

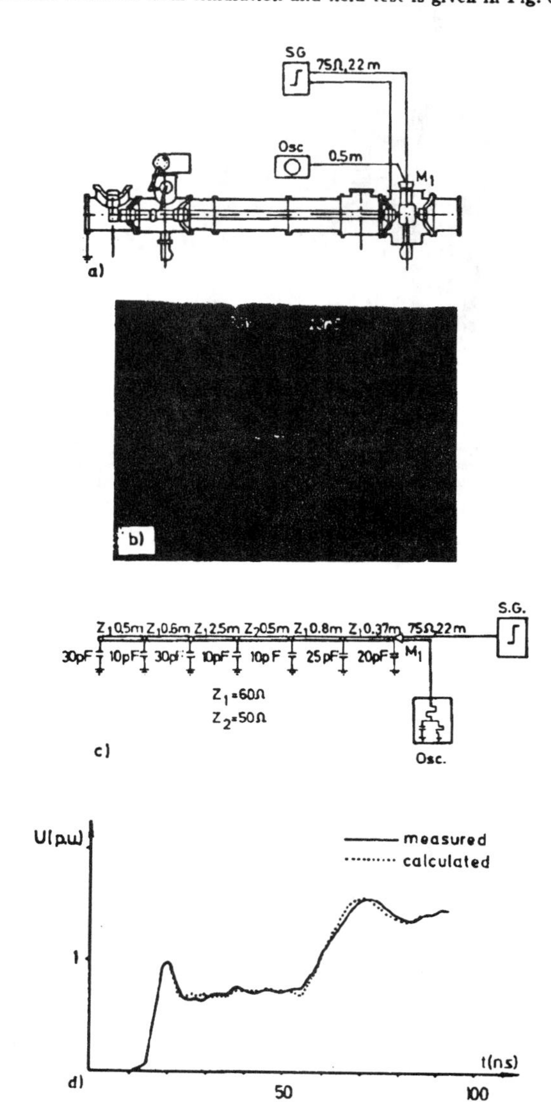

Fig. 5 Comparison between waveshapes obtained from simulation and field test

2.3 Model of test circuit with disconnector in closed position and a capacitor connected at the receiving end

The effect of the capacitor connected at the receiving end of the test circuit (Fig. 6*a*) is recorded at point M_1 as it is shown in Fig. 6*b*. The model of GIS elements as well as model of measurement circuit are given in Fig. 6*c*. The comparison between the waveshapes obtained from simulation and field test is given in Fig. 6*d*.

As can be observed from Figures 4 to 6, the simulated results match very well the results obtained from the field test.

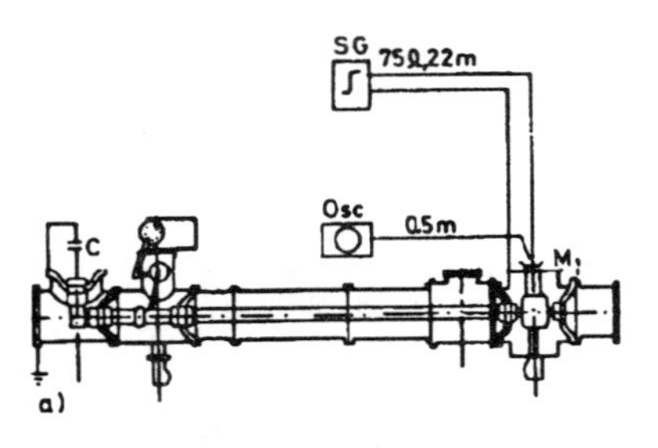

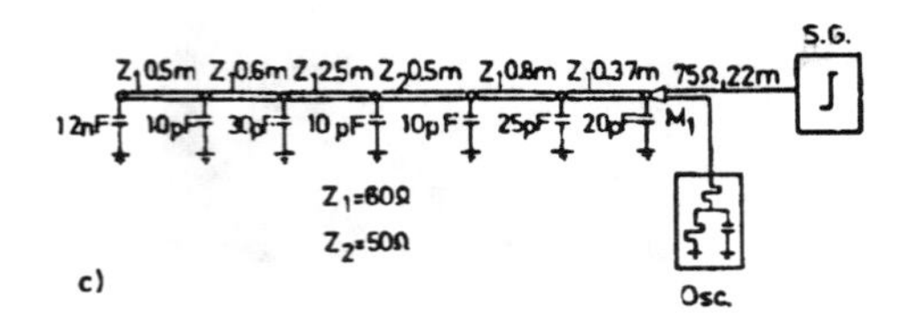

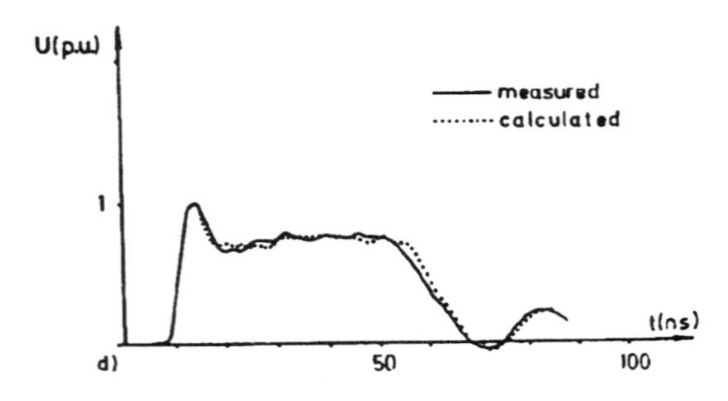

Fig. 6 Comparison between waveshapes obtained from simulation and field test

3. FIELD TESTS AND SIMULATIONS FOR HIGH VOLTAGE RESPONSES

The comparison between results obtained from simulation and field test will be carried out for the first strike when the disconnector is closing. The first strike will occur at the instant when the source voltage has its own peak value. Detailed illustration of the considered test circuit with all elements is given in Fig. 7.

The field tests are performed at the test circuit when the capacitors are connected to the source side C_1 as well as at the load side simulating the real operation condition of the disconnector for actual substation. The value of capacitor C_1 is varied from 40 to 60 nF, while the value of capacitor C_2 is equal to 7 nF. The latter value is taken to simulate equivalent length of the bus of 100 m. Thus, the measuring equipment fulfills the conditions of the frequency dependence characteristics of the considered surges. The rated current of the interruption was 0.3 A *r.m.s.* This value corresponds to the requirements for switching tests of metal enclosed switchgear issued by CIGRE Working Group 13.04 [18].

The measurements are performed by means of equipment consisting of the universal voltage diverters (type RC-Z) Haefely Co.

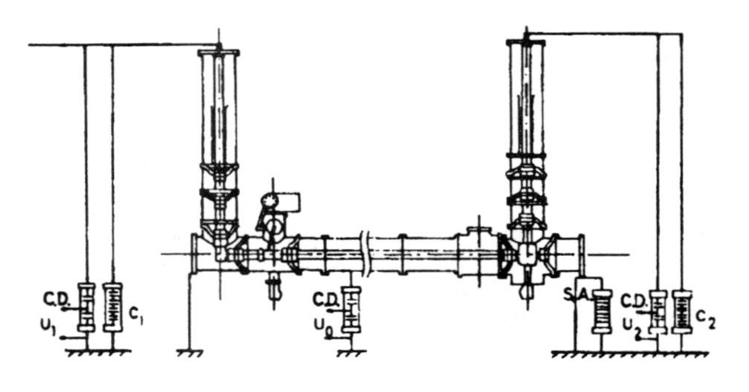

Fig. 7 The considered test circuit

The characteristics of the diverters are:

- the response of the measurement circuit is less than 19 ns;
- the partial response T is less than 35 ns;
- the time step rise from 10 to 90 percents T is less than 90 ns;
- the frequency range is from 0 to 7 MHz.

Very fast transients are recorded using a two-channel waveshape digitizer manufactured by Tektronix Co.; the frequency of recording for the digitizer is 60 MHz with an interface to a PC computer.

The recorded waveshape of the overvoltage at the load side when the capacitance of the source side C_1 is equal to 40 pF and for the ratio of capacitance of the source side and the capacitance of the load side is equal to 5.7 is shown in Fig. 8.

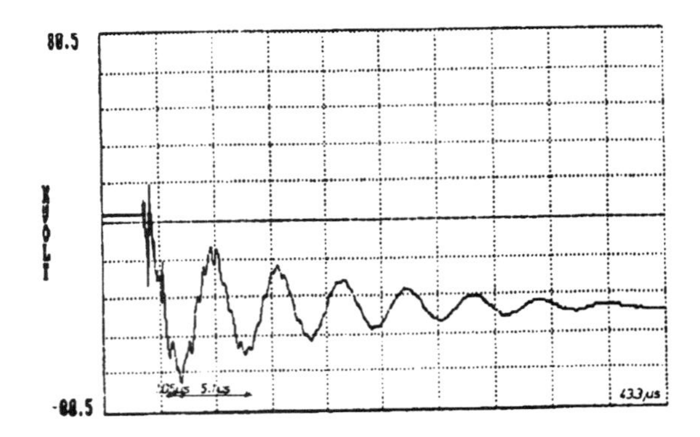

Fig. 8 Waveshape of the overvoltage at the load side

The overvoltage factor k is equal to 1.4 p.u. and the dominant frequency of considered transient f_d is equal to 195 kHz. To the dominant frequency an additional frequency f_{ad} is superposed. In this case the frequency is equal to 925 kHz.

The waveshape of the overvoltage at the load side when the capacitance at the source side (C_1) is equal to 60 pF and the ratio of the capacitances of the source side and the load side (C_1/C_2) is equal to 8.57 is shown in Fig. 9.

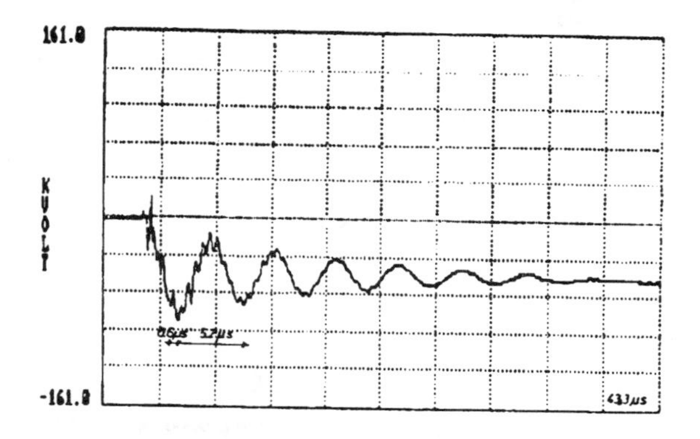

Fig. 9 The waveshape of the overvoltage at the load side

The overvoltage factor k is equal to 1.78 p.u. and the dominant frequency of the considered transient is equal to 192 kHz while the additional frequency is equal to 1.6 MHz.

4. MATHEMATICAL MODEL OF THE TEST CIRCUIT

The GIS elements which have to be mathematically modeled for the purpose of electromagnetic transient simulations are given in Fig. 10.

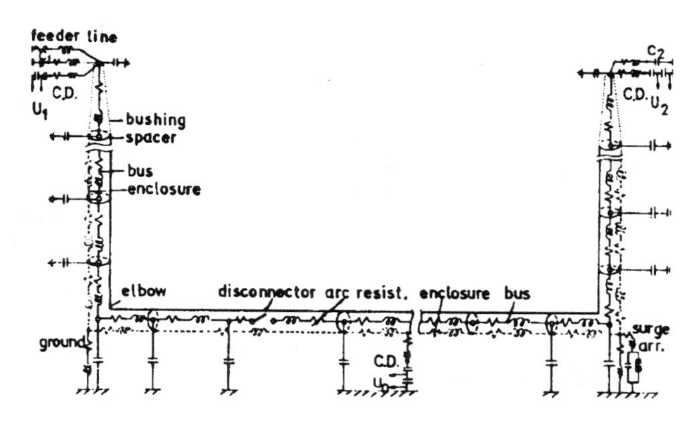

Fig. 10 The GIS elements which have to be modeled

The GIS elements from Fig. 10 are modeled by one of the available approaches given in Table I.

Table I - Models for GIS elements with corresponding assumptions and limitations of applicability

Element	Model	Characteristic	Limitation of applicability
Feeder line	Untransposed transmission line with distributed parameters [13]	Input data given by modal parameters (R, L, C); modal transformation matrix (T_i) calculated for the dominant frequency	Parameters R,L,C, are frequency independent for very high frequencies; elements of the transformation matrix are frequency independent for very high frequencies. Carson's and Pollaczek's correction factors are used (ρ_g = 250 Ωm) [14, 15]
Bushing	Untransposed transmission line with distributed parameters and capacitance added to the terminals	Parameters depending on the system topology	
Spacer	Lumped capacitance toward the ground	Value of the capacitance depending on the system topology	
Bus	Untransposed transmission line with distributed parameters	Input data given by modal parameters (R, L, C); modal transformation matrix (T_i) calculated for the dominant frequency	Coupling between conductor and enclosure is neglected at very high frequency. If $f_{cha} < f_{dom} < f_{cri}$ or if $f_{cri} < f_{dom}$ transformation matrix is assumed to be constant [16, 17]. If $f_{dom} < f_{cha}$ or $f_{dom} = f_{cri}$ frequency dependence of transformation matrix has to be taken into account. Carson's and Pollaczek's correction factors are used (ρ_g = 250 Ωm)
Enclosure	Condition as for bus	Condition as for bus	Condition as for bus
Elbow	Untransposed transmission line with distributed parameters and capacitance added to the terminals	Parameters depending on the ratio between conductor and enclosure radius. Value of the capacitance C depending on the system topology	

(cont. on the next page)

Element	Model	Characteristic	Limitation of applicability
Disconnector	Untransposed transmission line with distributed parameters and capacitance added to the terminals	Parameters depending on the ratio between conductor and enclosure radius. Value of capacitance C depending on the system topology	At the instant of strike, disconnector is assumed to be closed through arc resistance, while the disconnector in open position is modeled as a spherical termination
Arc resistance	Nonlinear function of time	If $t < 1\ \mu s$, $R = 0\ \Omega$; If $t > 1\ \mu s$ R varies from 0 to 5 Ω	If $t < 0$ R is assumed to be infinity
Surge arrester	Protection characteristic connected in parallel with arrester capacitance	In the case of very fast transient (0.5 μs) the protection characteristic is corrected in reference to the characteristic for the surge 8/20 μs. Inductance of the grounding connection is taken into account	Self inductance of the surge arrester is neglected
Power transformer	Lumped capacitance toward the ground	Value of the capacitance depending on the transformer mer type, voltage level, winding connection and winding types	Inductive branch toward the ground neglected due to a very high impedance at very high frequencies. Nonlinear behavior of the core is neglected.
Spherical termination	Untransposed transmission line with distributed parameters and capacitance added to the terminals	Parameters depending on the ratio between conductor and enclosure radius. Value of the capacitance C depending on the system topology	

The models given in Table I. are developed and selected in accordance with the investigated properties of the system. They are based on the travelling wave techniques applied on multiconductor transmission systems consisting of coaxial conductor and insulator structures. To evaluate accurate limits of applicability for each model, let us consider Fig. 11 illustrating the impedances of the multiconductor system.

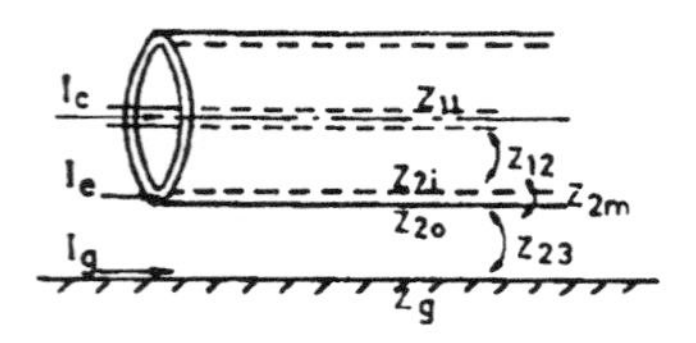

Fig. 11 Illustration of the impedances of multiconductor coaxial system

At frequencies higher than the critical frequency (f_c) no coupling between the bus core and the ground is observable. Therefore, the impedance of the core toward the ground is much larger than the one between the core and the enclosure, i.e.

$$Z_{cc} \gg Z_{12} \tag{1}$$

where

$$Z_{cc} = Z_{11} + Z_{12} + Z_{2i} + Z_{20} + Z_{23} - Z_{2m} + Z_g \tag{2}$$

$$Z_{12} = Z_{11} + Z_{12} + Z_{2i} \tag{3}$$

In equations (2) and (3) the meanings of the symbols used are as follows:

Z_{11} - inner impedance of outer surface of conductor
Z_{12} - impedance of outer conductor insulation
Z_{2i} - inner impedance of inner surface of enclosure
Z_{2m} - mutual impedance between outer and inner surface of enclosure
Z_{20} - inner impedance of outer surface of enclosure
Z_{23} - impedance of outer enclosure insulation
Z_g - impedance of the ground

From equations (1), (2) and (3) the following expression is obtained:

$$Z_{20} + Z_{23} - Z_{2m} + Z_g \gg Z_{2m} \tag{4}$$

From equation (4) the expression for the critical frequency can be obtained:

$$f_c = \frac{\rho_{cc}}{d^2 \pi \mu_0 \mu_{0k}} \tag{5}$$

where $d = \sqrt{2\rho/\omega\mu}$ = the penetration depth (equal to the thickness of the sheath).

When the frequency becomes higher, the penetration depth decreases and no current flows through the outer medium of the enclosure but complete current flows through enclosure as a return path.

When the simulation is performed using one of the programs which use Carson's or Pollaczek's formulae, the capacitive currents through the ground are assumed to be negligible. Furthermore, this assumes that the ground return path can be taken as a conductor. Thus, the applicability of these programs is limited to the cases in which the maximum frequency does not exceed an approximate value of 1 MHz.

The ground behavior changes from the conductor to the insulator when the frequency is equal to $1/2\pi\epsilon\rho$, i.e. for the value at which the density of the conductive current ($i_r = E/\rho$) becomes equal to the density of the capacitive current ($i_c = \omega\epsilon E$). From these conditions it is possible to give a graphical representation of the change frequency (at which the conductive behavior changes to the capacitive behavior) versus the earth resistivity. As it is shown in

Fig. 12, three regions can be detected. The first region corresponds to the capacitive behavior of the ground return path. The second one corresponds to the transient behavior, when the ground return path shows the conductive behavior as well as the capacitive behavior. The third one corresponds to the conductive behavior of the ground return path.

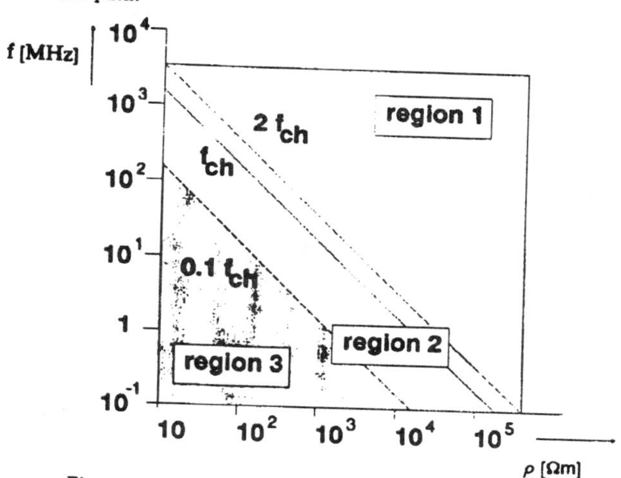

Fig. 12 The change frequency versus the earth resistivity

At very high frequencies ($f > 2f_{ch}$), conductive current through the ground is negligible and the ground has insulator properties (region 1). For the lower frequencies ($f < 0.1f_{ch}$), the capacitive current is negligible, and the ground have the conductor properties. Carson's approach assumes that the capacitive current can be neglected and, therefore, his formulae can be applied for region 3. Consequently, the limit frequency of applicability of Carson's and Pollaczek's formulae depends on the earth resistivity. If the earth resistivity is low enough, the limit frequency can be higher than 1 MHz. Theoretically, the frequency can be up to 100 MHz.

The waveshapes of recorded and simulated overvoltage surge at the load side, in the case when capacitance C_1 is equal to 40 nF, are given in Fig. 13.

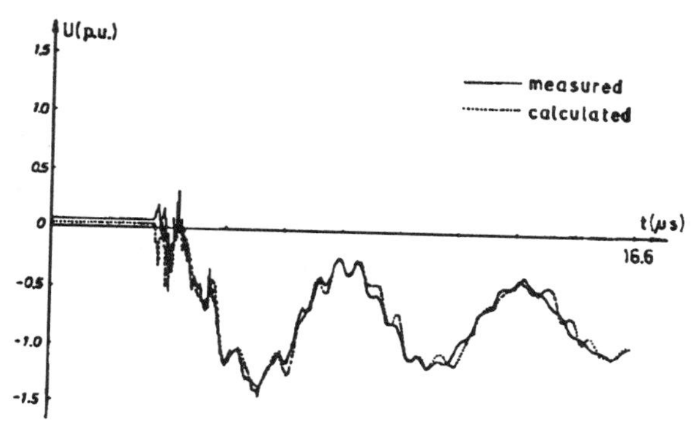

Fig. 13 Waveshapes of recorded and simulated overvoltage surge

As it can be seen from Fig. 13, the curves match each other very well. The peak values difference for measured and simulated overvoltage is 7 percent. The dominant frequency and the additional frequency match very well over all period of 16 μs during which the transient has been shown, to provide more detailed illustration.

The waveshapes of recorded and simulated overvoltage surge at the load side, in the case when capacitance C_1 is equal to 60 nF, are given in Fig. 14.

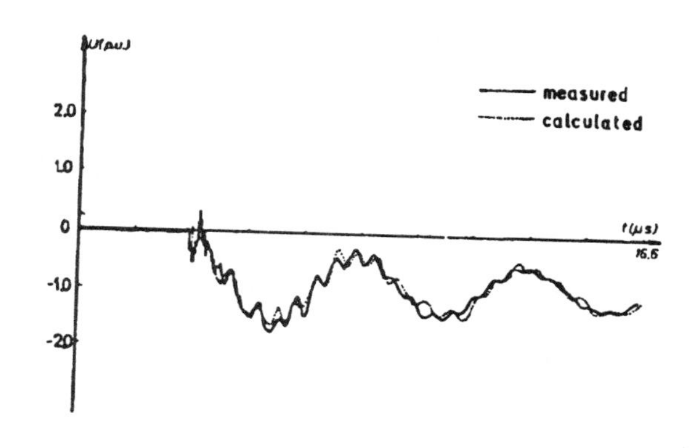

Fig. 14 Waveshapes of recorded and simulated overvoltage surge

The difference between magnitudes of measured and simulated overvoltages is 8 percent. The dominant frequency and the additional frequency match very well during all 16 μs, as it can be seen from Fig. 14.

5. CONCLUSION

An optimal model of gas insulated metal enclosed substations suitable for digital transient simulations can be developed by successive comparison between field test and simulation results obtained for a test circuit. So, a developed model can be used for investigation of electromagnetic transient in actual gas insulated substations. Performed computer investigations provide deeper insight in transient phenomena which are decisive for insulation coordination, protection techniques and optimal substation design. Of course, the model is limited by assumptions inherent to the different parts of theoretical background. Therefore, the model can be used only if the limits of applicability are evaluated in a proper manner. These limits are, in fact, crucial criteria in transient calculation in the systems of this kind and their establishment improves engineering practice. Using the developed GIS model and evaluating the limits of its applicability, digital simulation tests can give very instructive knowledge on physical phenomena in different GIS elements. And consequently, the techniques of overvoltage dumping in metal enclosure can be established. Furthermore, the approach proposed in the paper can be applied onto different transient phenomena occurring in gas insulated substations.

REFERENCES

[1] S. Narimatsu, K. Yamaguchi, S. Nakano, S. Maruyama, "Interrupting performance of capacitive current by disconnecting switch for gas insulated switchgear", *IEEE Transactions on Power Apparatus and Systems*, vol. PAS-100, No. 6, June 1981.

[2] S. Matsumura, T. Nitta, "Surge propagation in gas insulated substation", *IEEE Transactions on Power Apparatus and Systems*, vol. PAS-100, No. 6, June 1981.

[3] T. Yoshiyumi, S. Matsuda, T. Nitta, "Fast transient overvoltages in GIS caused by the operation of isolators", *Proceedings of Third Int. Symp. on Gaseous Diel.*, Knoxville, USA, 1982.

[4] K. Nakanishi, M. Ueda, "Charging current switching by disconnector in gas insulated switchgear and some laboratory test methods", *CIGRE Paper 13-06*, Paris 1984.

[5] S.A. Boggs, A. Plessl, N. Fujimoto, "Disconnect switch induced transients and trapped charge in gas-insulated substations", *IEEE Transactions on Power Apparatus and Systems*, vol. PAS - 101, No. 10, 1982.

[6] S.A. Boggs, N. Fujimoto, M. Collod, E. Thuries, "The modelling of statistical operating parameters and the computation of operation-induced surge waveform for GIS disconnectors", *CIGRE Paper 13 - 15*, Paris 1984.

[7] N. Fujimoto, H.A. Stuckless, S.A. Boggs, "Calculation of disconnector induced overvoltages in gas-insulated substations", *Gaseous Diel. IV*, Pergamon Press, 1986.

[8] N. Fujimoto, F.Y. Chu, S.M. Harvey, G.L. Ford, S.A. Boggs, V.H. Tahiliani, M.Collod, "Development in improved reliability for gas-insulated substations". *CIGRE Paper 23-11*, Paris 1988.

[9] A. Edlinger, G. Mauthe, F. Pinekamp, D. Schlicht, W. Schmidt, "Disconnector switching of charging currents in metal-enclosed SF_6-gas insulated switchgear at EHV", *CIGRE Paper 13-14*, Paris 1984.

[10] H. Murase, I. Oshima, H. Aoyagi, I. Miwa, "Measurement of transient voltages induced by disconnect switch operation", *IEEE Transactions on Power Apparatus and Systems*, vol. PAS - 104, No. 1, 1985.

[11] J. Lalot, A. Sabot, J. Kieffer, S.W. Rowe, "Preventing earth faults during switching of disconnectors in gas insulated voltage transformer", *IEEE Transactions on Power Delivery*, vol. PWRD-1, Jan. 1986.

[12] R. Witzman, "Fast transient in gas insulated substations (GIS) - modelling of different GIS component", *V Int. Symp. on HV Eng.*, Braunschweig, 23-28 Aug. 1987.

[13] K.C. Lee, H.W. Dommel, "Addition of modal analysis to the UBC Line Constants Program", *This is a research report to B.C. Hydro and Power Authority published by the Electrical Department of the University of British Columbia*, Jan. 1980.

[14] J.R. Carson, "Wave propagation in overhead wires with ground return", *Bell.Syst.Tech.J.*, vol. 5, 1926.

[15] F. Pollaczek, "On the field produced by infinitely long wire carrying alternating current", *Elektrische Nachrichtentechnik (in German)*, vol. 4, pp. 18-30, 1927 (French translation in RGE), vol 29, 1931.

[16] R. Mahmutćehajić, "Proračunavanje parametara kabela: I dio, Opća formulacija", *XI simpozij o energetskim kabelima*, Svetozarevo, Oct. 1988.

[17] R. Mahmutćehajić, "Proračunavanje parametara kabela: II dio, Frekventna ovisnost parametara". *XI simpozij o energetskim kabelima*, Svetozarevo, Oct. 1988.

[18] CIGRE WG 13.04 of SC 13, "Requirement for switching tests of metal enclosed switchgear", *Electra*, No. 110.

Zijad Haznadar was born in Banja Luka, (Bosna, Yugoslavia) on October 29, 1935. He received the B.Sc. and Ph.D. degrees from Zagreb University, in 1959 and 1964, respectively. He has been with the Faculty of Electrical Engineering of Zagreb University, Yugoslavia, since 1960, and he is currently professor at the same faculty. He has published about 250 professional papers on different topics from power system engineering. His teaching and research responsibilities involve theory of calculation of electromagnetic fields, power system analysis and CAD/CAM. Dr. Haznadar is a member of CIGRE and an adviser of Yugoslav Government Institution for science and technology.

Salih Čaršimamović was born in Sarajevo (Bosna, Yugoslavia) on January 23, 1946. He received a B.S. degree in electrical engineering from the Sarajevo University, M.S. degree and Ph.D. degree from Zagreb University, in 1970, 1980 and 1990, respectively. From 1970 to 1973 he was with the Power Industry "Energoinvest" (Sarajevo, Yu). From 1973 to 1977 he was with the Safety Institute of Sarajevo University working on the protection against electrical shock hazard. In 1977 he joined the Electrical Power Institute "Energoinvest" working in the Power Laboratory and he is now a fellow researcher in power system transients. He has published about 30 professional papers on different topics from power system engineering.

Rusmir Mahmutćehajić was born in Stolac (Bosna, Yugoslavia), on June 29, 1948. He received B.S. degree in electrical engineering from the Sarajevo University, M.S. degree and Ph.D. degree from Zagreb University, in 1973, 1975 and 1980, respectively. From 1973 to 1979 he was with Safety Institute of Sarajevo University working as researcher in power system analysis and as institute manager. From 1979 to 1985 he was with Ergonomics Institute of Sarajevo University working as manager. In 1985 he joined the Faculty of Electrical Engineering of Osijek University (Yugoslavia) working as a professor of electrical engineering. From 1989 he held the position of dean of the faculty. He has published 80 professional papers on different topics in power system analysis.

IEEE Transactions on Power Delivery, Vol. 8, No. 1, January 1993.

DEVELOPMENT AND VALIDATION OF DETAILED CONTROLS MODELS OF THE NELSON RIVER BIPOLE 1 HVDC SYSTEM

P. Kuffel, Non-member K.L. Kent, Member G.B. Mazur, Non-member M.A. Weekes, Member

Manitoba Hydro

Winnipeg, Manitoba, Canada

ABSTRACT

With the Nelson River Bipole 1 mercury arc valve group replacement project and planning for the expansion of the Nelson River HVDC system with a third bipole underway, it was decided to pursue a program to develop and validate detailed models of the existing HVDC transmission facilities and their associated ac systems for use in system studies. The first phase of the program concentrated on the development of detailed controls models associated with the Bipole 1 transmission facility. Based on previous experience at Manitoba Hydro with the Electromagnetic Transient DC simulation program (EMTDC), it was decided that model development and validation would use this program.

This paper presents the reasons behind the development of detailed models, the methods used in developing models related to Bipole 1, results of validation tests, difficulties encountered during the process, and the overall benefits resulting from the project. An example of applying the models to investigate a low frequency oscillation which has occurred on the dc system in the past is also presented.

INTRODUCTION

A greater dependence on digital simulation as a study tool for the increasingly complex Manitoba Hydro HVDC system [1], has resulted in the need to develop increasingly complex controls models. Previous efforts at model development [2] produced models which were generally representative of the overall control functions but still required some degree of adjustment to produce results in good agreement with known system responses. These models proved to be suitable for studies which investigated the system response under known conditions, the results of which could be compared to actual system traces.

Experience gained during the Bipole 2 line fault recovery optimization study [3], where site performance was compared to model performance using generic controllers, suggested the need for more detailed control models. With participation in the testing of the Cigre HVDC Benchmark Model [4] it became clear that the degree of detail in control modeling played a crucial role in the ability of a single general model to provide accurate results for a range of studies. These combined experiences demonstrated that if enough detail was incorporated into the controls models, good agreement with actual system performance could be attained, for different study conditions, without the need to fine tune controls models for individual studies.

92 WM 289-9 PWRD A paper recommended and approved by the IEEE Transmission and Distribution Committee of the IEEE Power Engineering Society for presentation at the IEEE/PES 1992 Winter Meeting, New York, New York, January 26 - 30, 1992. Manuscript submitted August 1, 1991; made available for printing January 13, 1992.

REASONS FOR DETAILED MODEL DEVELOPMENT

With the present Bipole 1 mercury arc valve group replacement and the planning for the expansion of the HVDC transmission facilities to include a third bipole underway, the need to have a model capable of predicting the existing HVDC system response accurately for various disturbances and for a range of studies, became imperative. Based on this foreseen need it was decided to undertake the task of developing and validating detailed models of the existing HVDC and associated ac transmission facilities, the first phase of which dealt with the Bipole 1 system controls. The ultimate goal of the project is to develop a single validated model of sufficient detail for use in a variety of studies with a high degree of confidence. The single model will also ensure a high level of consistency in study results throughout the entire corporation.

PROCEDURE FOR MODEL DEVELOPMENT

During the early stages of the project, the following objectives were identified:

- the general model should contain as much detail as possible, simplifications to the models for specific studies could then be made in the future once the effects of the simplifications were examined, based on comparison to the most accurate model available,
- the overall structure of the models should resemble the real control circuits as much as possible,
- signal levels within the model control paths should be equivalent to corresponding signals in the actual controls,
- means for correlating designated monitoring points in the actual controls to the model should be provided,
- validation of the model should be carried out at each stage of development of the individual control paths.

Based on these objectives, the procedure adopted for the development and validation of the overall Bipole 1 controls models was as follows:

- perform detailed circuit analysis at the printed circuit board level based on drawings and inspection,
- develop functional block diagrams of the individual control paths,
- develop model code for the individual control paths,
- develop "as coded" block diagrams for the individual control paths,
- bench test individual printed circuit boards or groups of boards,
- validate individual control path models by duplicating tests carried out on individual printed circuit boards or groups of boards,
- perform overall system tests,
- validate overall model by duplicating system tests with the model.

As an example of model development for a specific path, the model for the α (firing angle) advance upon detection of a commutation failure circuit within the valve group controls shown in Figure 1 is presented. The transfer function for the circuit as derived from detailed circuit board analysis is shown in the Functional Block Diagram in Figure 2. The Functional Block Diagram of the circuit provides a means of understanding the particular circuit and was the starting point for all model development. Based on the Functional Block Diagram, code for the model was developed and the As Coded Block Diagram shown in Figure 3 was constructed. The As Coded Block Diagram is a direct representation of the model code, and is meant to provide a simple reference for understanding the model. Together, the two block diagrams provide the means to validate a particular model and simplify its use by providing a direct path between points in the model code and the corresponding points on the actual controls. This direct path was also possible since signal levels and polarities within the model were set equal to the corresponding signals in the actual controls being modeled.

The first stage of validation for a given control path was at the individual printed circuit board level. Figure 4 shows digitized responses obtained from bench tests of the printed circuit board containing the path in question along with results of identical tests performed on the corresponding model. A comparison of the results shows the model accurately represents the actual circuit. By carrying out validation tests at various stages of the model development the number of possible problem areas within the overall model was greatly reduced.

Once the individual paths within the controls were modeled and validated wherever possible, validation of the remaining paths and the overall control model was carried out. Overall model validation was accomplished by

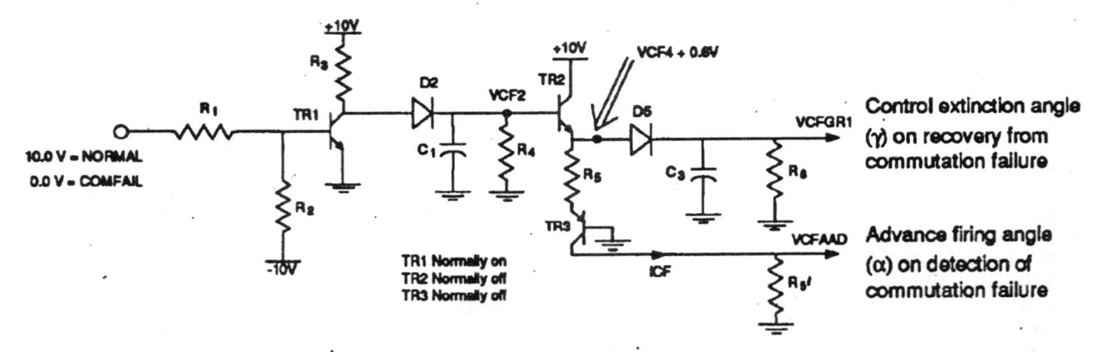

Figure 1 α Advance on Detection of Commutation Failure Circuit

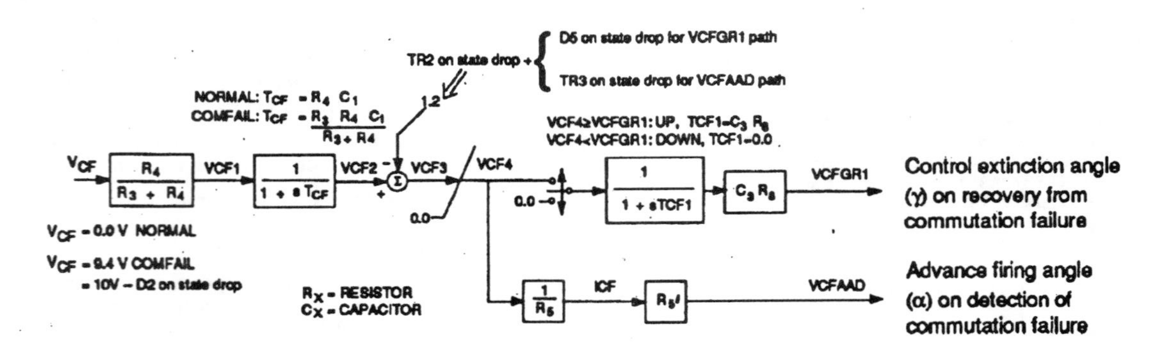

Figure 2 Functional Block Diagram of α Advance on Detection of Commutation Failure

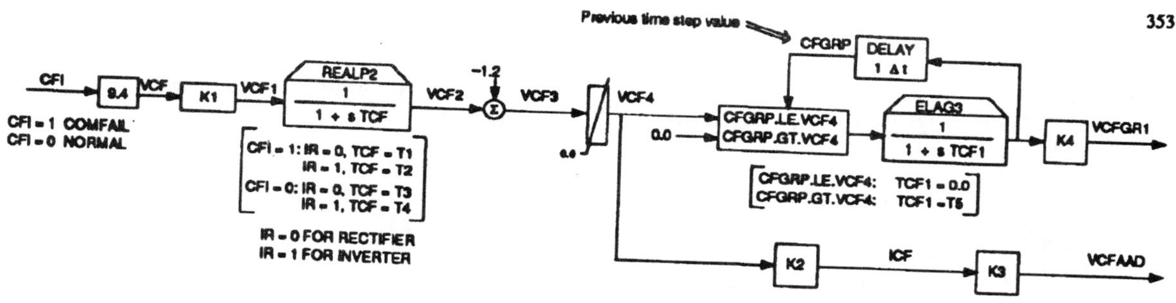

Figure 3 As Coded Block Diagram of α Advance on Detection of Commutation Failure

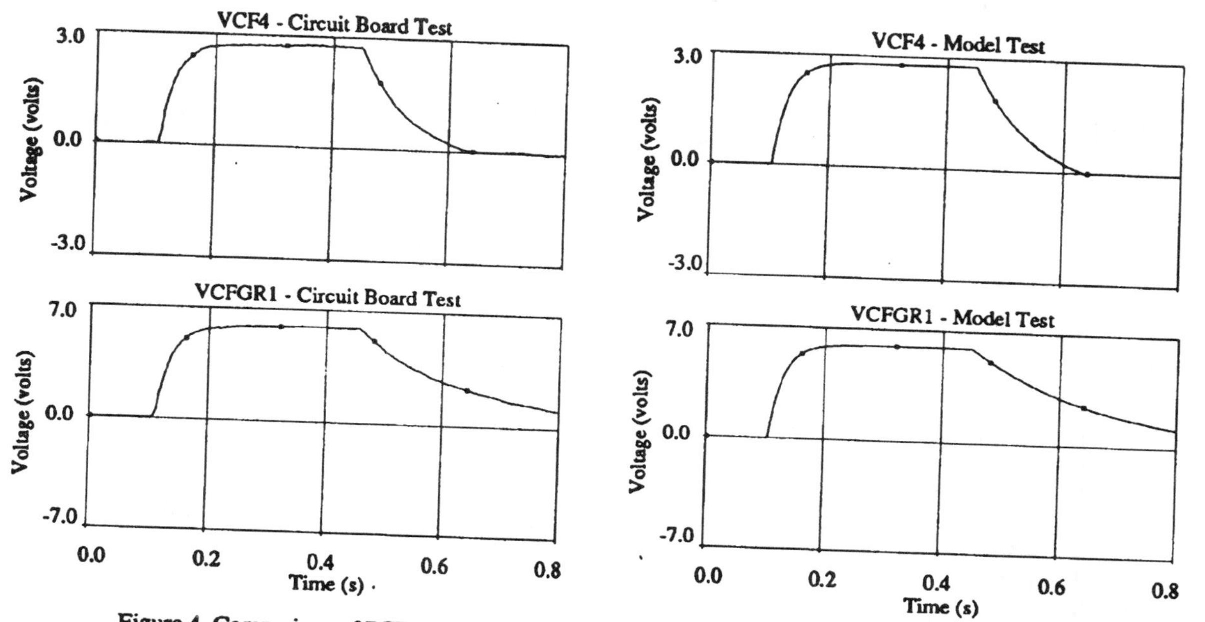

Figure 4 Comparison of PCB and Model Test Results of α Advance on Detection of Commutation Failure Circuit for Commutation Failure Indication from 100ms to 450ms.

performing a number of tests on the operating system under known conditions and duplicating those same tests on the model. The tests performed were selected since they minimized the interaction of Bipole 1 and Bipole 2 through the ac system, thus allowing the response of the Bipole 1 controls to be isolated and verified. This was necessary since no validated models of Bipole 2 and the ac system were available. Tests performed on the system included:

- Initiation of a commutation failure by suppressing the firing pulse to a valve.
- Application of a step change in current order.
- Valve group block.
- Valve group deblock.

Model performance was evaluated based on the comparison of model and system responses to identical disturbances with no fine tuning of the model for the different disturbances. Tests were conducted on the operating system with Bipole 1 configured +450kV, -150kV with a power order of 450MW, and Bipole 2 configured +250kV, -250kV with a power order of 450MW.

In order to provide the basis for a good comparison of results it was necessary to perform the model tests on a system configuration which resembled the conditions under which the tests were conducted on the operating system. This was complicated due to the lack of validated models for the ac systems and Bipole 2 and necessitated the use of simplified representations of Bipole 2 and the ac system. Therefore it was decided that the ac systems at both converters should be modeled as R-L-L equivalents [4] with the short circuit ratio (SCR) at the rectifier set to 5 at an angle of -85°, and the SCR at the inverter set to 10 at an angle of -80°. The simplified ac equivalents chosen for the model tests were deemed to be acceptable representations of the short circuit ratios existing when the system tests were performed, however it was realized that ac sys-

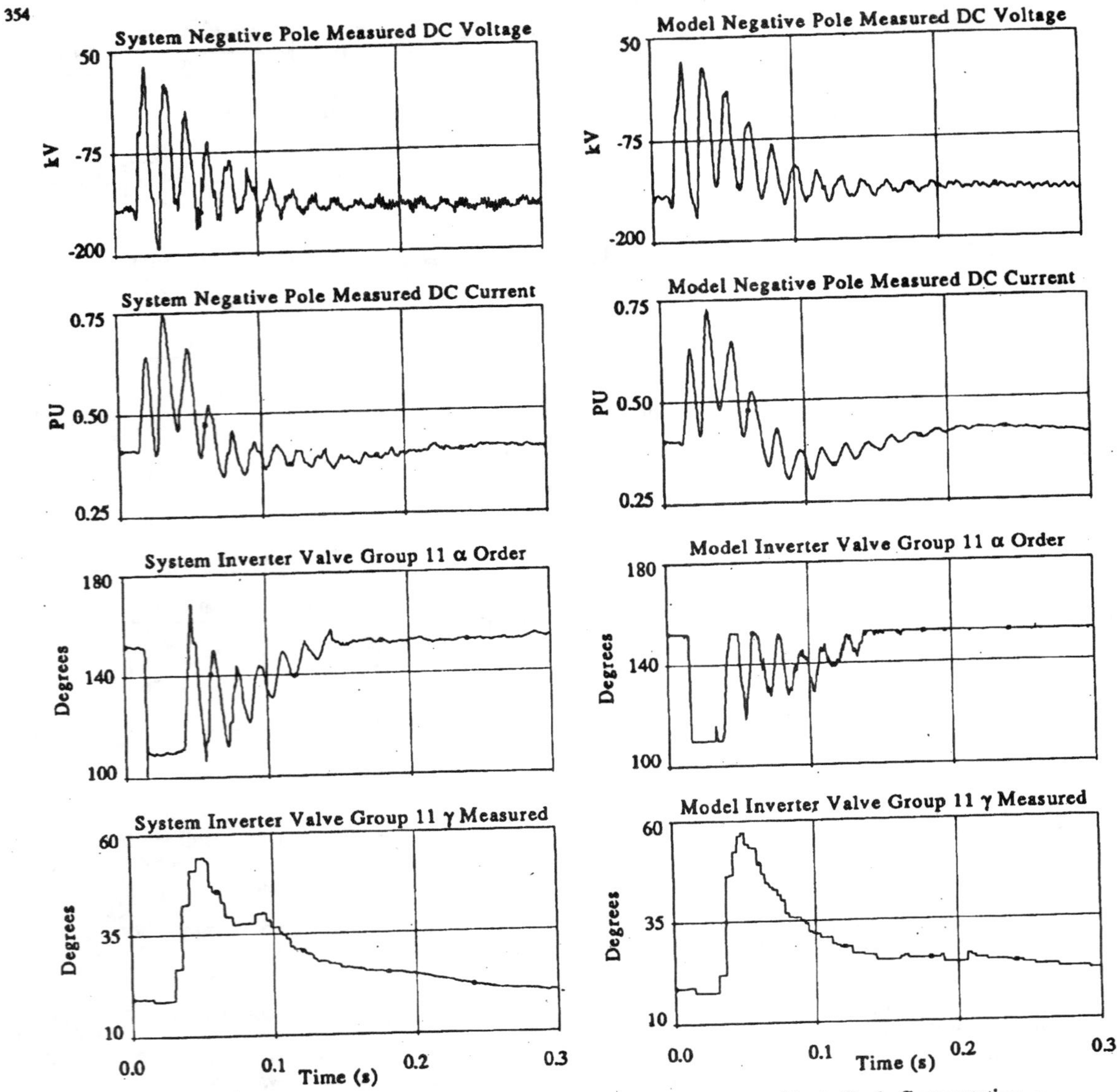

Figure 5 Comparison of System and Model Test Results for a Single Cycle Commutation Failure Initiated by the Suppression of a Valve Firing Pulse

tem dynamics may not be properly represented. Bipole 2 was modeled by constant dc voltage sources at both stations. This simplification was justified on the basis that no inter-bipole interactions were observed while performing the system tests.

The curves in Figure 5 represent the digitized responses of the actual system and the model for the case of a commutation failure in one inverter valve for one cycle. The commutation failure was initiated in the inverter negative pole valve group by suppressing the firing pulse to valve 5 for one cycle. The results in Figure 5 show good agreement between system and model responses. The model responses indicate a slightly lower initial 60 Hz ringing and a slightly larger underlying low frequency swing which was due partially to deficiencies in the transmission line model (which are presently being corrected) and partially to the simplified ac system representations. A small difference in system and model response for the inverter negative pole valve group measured γ was observed over the time span 80ms to 120ms. This difference was attributed to simplifications in the ac system model representations. The clamping of the valve group α order, which is more evident in the model response, was the result of applying the limit to the controller in the model as a hard limit. In the actual system the limit is applied by introduc-

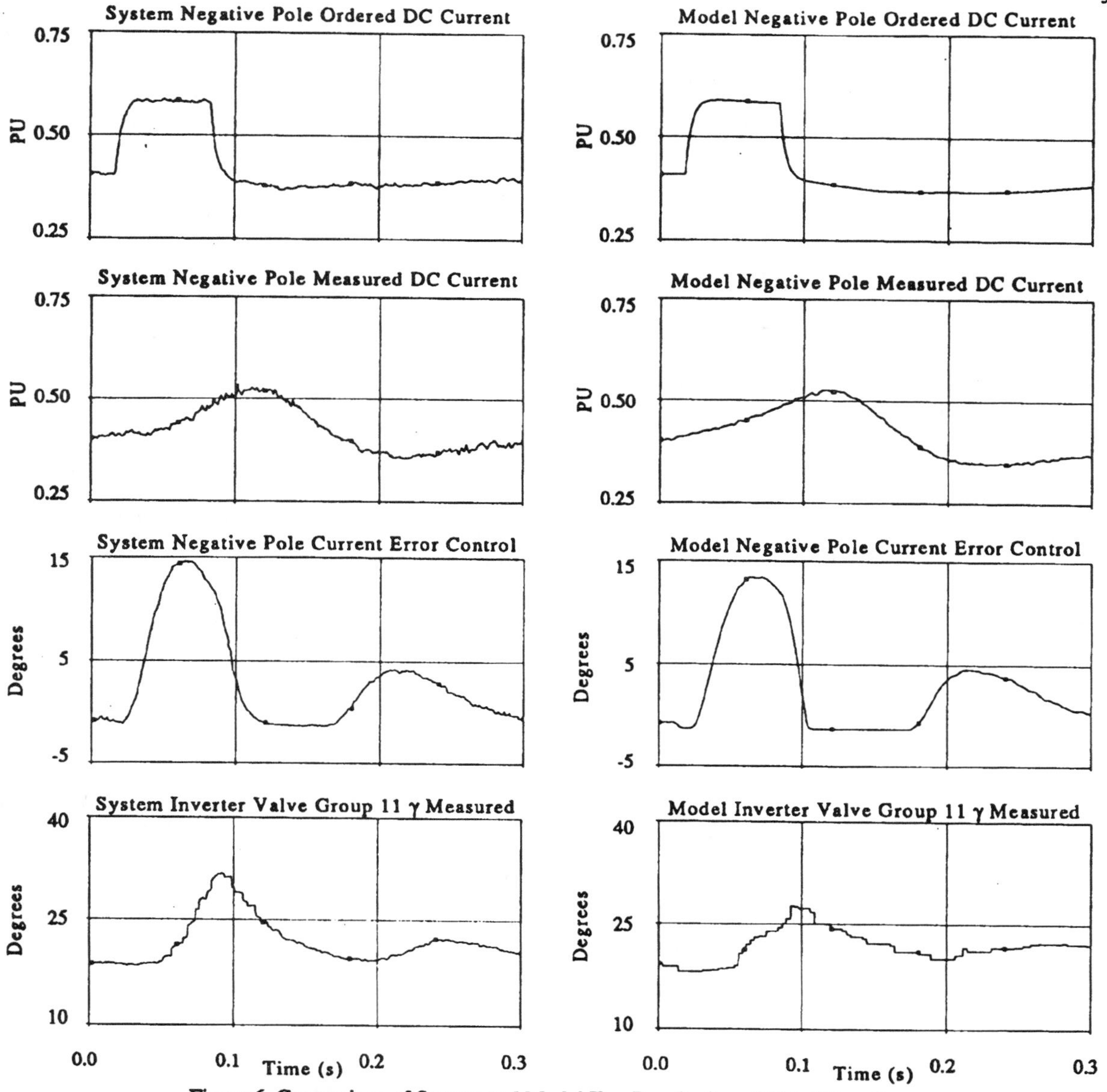

Figure 6 Comparison of System and Model Test Results for a 333A,100ms Step Increase to the Negative Pole Current Order

ing an input to the controller once its output has exceeded the limit value, resulting in a soft limit. The use of a hard limit in the model was deemed acceptable based on the results of tests performed on the actual controller.

The results of system and model responses to a step change in current order are presented in Figure 6. A step increase of 333A was introduced into the negative pole current order at the inverter for 100ms within the master controls. The results of Figure 6 show close correlation between system and model responses. Again a slight discrepancy can be seen between the system and model inverter negative pole valve group γ measured which is believed to be the result of the simplified ac system representations used in the model.

DIFFICULTIES ENCOUNTERED

Much of the equipment in Bipole 1 which was modeled is over twenty years old, and consequently, it was often necessary to examine in-service circuit boards during equipment outages to ensure that the model was an accurate representation of the actual controls in operation. The lack of available test data for such equipment as measurement transductors also caused problems during the validation of the overall model.

Other obstacles were encountered in trying to model complex, highly non-linear analogue circuits digitally. In some cases approximations were required which had to be thoroughly investigated to ensure that they did not reduce the accuracy of the overall model. Some of the approximations made which seemed acceptable during the individual path validation stage were found to be unacceptable during the overall system validation stage, and had to be reevaluated. In some situations it was necessary to employ state space modeling due to the mode switching in some circuits, for example diode limits when op-amps saturate. This added significant time to the overall model validation process.

Due to the high degree of detail incorporated into the controls models, particular attention was also required to accurately represent the measurement of control variables. For instance, the measurement of the valve extinction angle is initiated based on a measurement of the rate of change of current. The discrete nature of the digital model introduces errors into the derivative calculation. The introduction of artificial smoothing to stabilize the derivative function resulted in unacceptable behavior of the overall model at times, requiring reevaluation. This again added considerable time to the model validation process.

Small numerical errors were encountered which were dependent on the magnitude of the solution time step, because of the discrete nature of digital simulation. Based on past experience it was decided that a solution time step of 50μs would provide sufficient accuracy for most studies while not resulting in excessively long simulation run times. In instances where the simulation time step size was small in comparison to the circuit time constants, the error resulting from the use of previous time step values in feedback loops was found to be negligible. However, in other instances unacceptable numerical jitter was found to result from the introduction of one time step delays. In order to maintain solution integrity while retaining the 50μs time step it was necessary to incorporate linear interpolation techniques at these key points within the controls models. Based on sensitivity studies it was determined that the parameters most requiring interpolation were the voltage controlled oscillator which determines the instant of valve firing, and extinction angle measurement circuits.

PROJECT RESULTS

The main products resulting from the project include:

- development of a single general model which can be used with a high degree of confidence in a broad range of studies,
- development of a detailed set of complete, up to date block diagrams.

Other benefits realized include:

- establishment of a procedure for future model development,
- increased knowledge of existing system controls,
- further experience with digital simulation which has led to improvements to EMTDC.

MODEL PERFORMANCE

A complete representation of Bipole 1 consisting of three valve groups per pole, per converter station, along with all associated controls, all ac and dc filters, distributed parameters transmission line, and simple ac equivalents has been constructed. Simulation run times in the order of 20 minutes real time per second of simulation are typical for the complete model using a SUN SPARCstation 2. Model performance in the investigation of past system disturbances has been excellent. The availability of a complete model which can be quickly and accurately reconfigured to represent the system for specific conditions has proven very valuable in reducing study set up and execution times. The complete model has also increased the consistency of study results.

EXAMPLE OF MODEL USE

The spontaneous development of low frequency oscillations in the 8 Hz range has been experienced on the HVDC system in the past. In some instances these oscillations have been related to low system strength, high converter bus angle damping control gain and high dc power [5]. In other cases, the oscillations were believed to be due to the intermittent loss of the rate of change of pole dc current (dI/dt) signal to one rectifier pole controller, as shown in Figure 7. The influence of the angle damping controls on the oscillations has been investigated in the past, however, due to the lack of adequate Bipole 1 dc control models, it was not possible to accurately determine the role of the dc controls in the oscillations. A study was undertaken using the newly developed and validated controls models in order to investigate the influence of the dc controls on the 8 Hz oscillations.

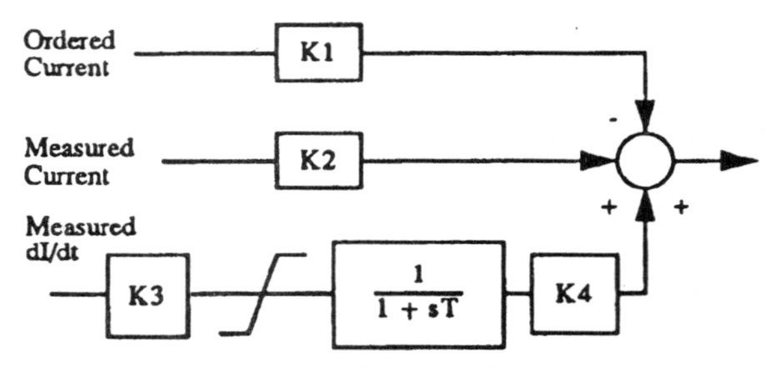

Figure 7 Portion of Rectifier Pole Controls Illustrating Configuration of dI/dt Signal

The main objective of the study was to determine the effects of the loss of one rectifier pole controller's dI/dt signal. Figure 8 illustrates the measured dc current in the negative pole in steady state and with the negative pole controller's dI/dt signal removed at 100ms. The results in Figure 8 clearly show the immediate development of sustained system oscillations in the 8 Hz range, a result which had been predicted based on field experience, but had eluded previous simulations. Further studies are now under way using the digital model to investigate possible control solutions to the 8 Hz oscillations.

FUTURE DIRECTION

Based on the success of the development of the Bipole 1 controls models an effort is well underway to accomplish the same task for Bipole 2. In addition, efforts are still ongoing to complete the outstanding controls and protection functions associated with Bipole 1. To date, approxi-

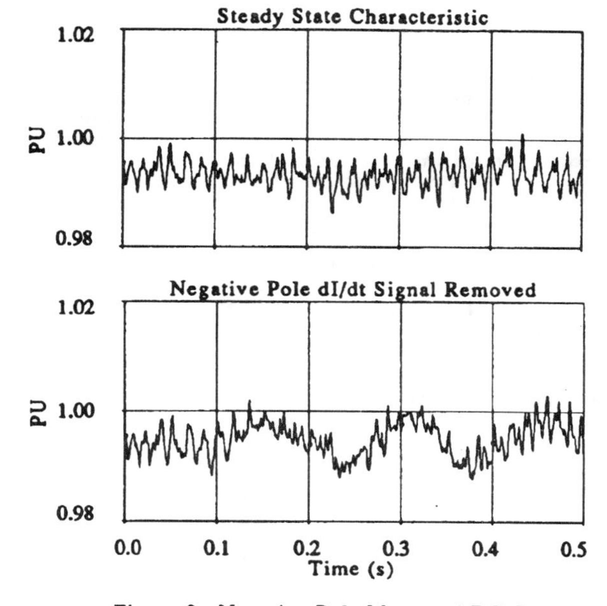

Figure 8 Negative Pole Measured DC Current

mately two person-years has been spent on the development and validation process, and it is felt that an equivalent amount of time is still required to complete the task for the existing HVDC transmission facilities and their associated ac systems. As the model development proceeds to encompass more of the overall system, further system testing aimed at providing a means to validate the interaction of the bipoles and ac system will be pursued. The ultimate goal is to develop, validate, and maintain a complete HVDC and associated ac system model including future additions to the system. To this end, the Bipole 1 valve group replacement specification required the manufacturer to provide models of the replacement controls. This was intended to simplify the modification of the overall Bipole 1 model to reflect the equipment changes, and it is a practice that will be recommended for all future projects.

The controls models developed will also be used to produce detailed general models for stability studies using a similar approach to that described in reference [6].

CONCLUSIONS

A detailed model of the Nelson River Bipole 1 HVDC system controls has been developed and validated. A process for detailed model development and validation was established which resulted in a model that provides a high degree of confidence and consistency in study results, for different study conditions, without the need to fine tune the control models for individual studies.

REFERENCES

[1] C.V. Thio and J.B. Davies:"New Synchronous Compensators for the Nelson River HVDC System - Planning Requirements and Specification" IEEE Trans. PAS, Vol. 6, No. 2, April 1991, pp. 922-928

[2] D.A. Woodford: "Validation of Digital Simulation of DC Links" IEEE Trans. PAS, Vol. 104, No. 9, Sept. 1985, pp. 2588-2595

[3] M.A. Weekes and D.P. Brandt: "Use of EMTDC Modeling for Enhancement of Nelson River Bipole 2 DC Line Fault Recoveries" Proceedings of the Second HVDC System Operating Conference, Winnipeg, Manitoba, Canada, Sept. 18-21, 1989

[4] M. Szechtman et al: "First Benchmark Model for HVDC Control Studies" Electra, No. 135, April 1991

[5] J.B. Davies and D.G. Chapman: "Recent AC Control Enhancements of the Nelson River HVDC Links" Proceedings of the International Conference on DC Power Transmission, Montreal, Quebec, Canada, June 4-8, 1984

[6] R.A. Hedin, K.B. Stump and B.A. Vossler: "Field Test Comparison, Control, and Modeling of the Sidney Converter Station" IEEE Trans. PAS, Vol. 6, No. 2, May 1991, pp. 536-541

BIOGRAPHIES

P. Kuffel was born in Manchester, England in 1964, and graduated with a B.Sc. and M.Sc. in Electrical Engineering from the University of Manitoba, Canada, in 1985 and 1987 respectively. He joined Manitoba Hydro in 1986 and is presently an HVDC Controls Engineer in the Transmission Planning Division. His interests include power system control and simulation.

K.L. Kent was born in Winnipeg, Manitoba in 1963, and graduated with a B.Sc. in Electrical Engineering from the University of Manitoba in 1986. Since 1986 he has been working at Manitoba Hydro, and is presently an HVDC Planning Engineer in the Transmission Planning Division. He is currently working towards an M.Eng. degree at the University of Manitoba.

G.B. Mazur was born in St. Boniface, Manitoba, in 1949, and was raised in the rural community of Brokenhead, Manitoba. He received his B.Sc. in Electrical Engineering in 1971, and M.Sc. in Electrical Engineering in 1986, both from the University of Manitoba.

From 1971 to 1975 he was employed by Bristol Aerospace where he was a control and instrumentation systems engineer. From 1975 to 1980 he worked for Manitoba Hydro as a DC Control and Protection Systems Design Engineer. In 1980 he joined Teshmont Consultants and worked on many dc projects as a DC Control Systems Specialist until 1988. He returned to Manitoba Hydro and is presently working in the HVDC Design Department as an HVDC Specialist.

He has been a Registered Professional Engineer in the Province of Manitoba since 1974.

M.A. Weekes was born in Bridgetown, Barbados in 1960. He received the B.Sc. and M.Sc. degrees in Electrical Engineering from the University of Manitoba, in 1983 and 1986 respectively. He is presently working for Manitoba Hydro as an HVDC Controls Engineer in the Engineering and Construction Division. His interests include HVDC, power system control, and simulation.

Discussion

M. H. Baker (Stafford, UK): In addition to the Bipole 1 and Bipole 2 controllers discussed, there will be a new controller available in the near future for the Bipole 1 replacement programme, where three out of six mercury avc valves are to be replaced by thyristor valves. Do you plan to develop a similarly detailed model of this new equipment?

Are there any conclusions or lesson that you have drawn from this minute examination of the controls that would lead you to make any recommendations for the future?

Manuscript received February 3, 1992.

D.A. Woodford (Manitoba HVDC Research Centre). The ability to exactly model HVDC controls in an electromagnetic simulation using EMTDC has been effectively demonstrated by the authors. Presumably detailed simulation of SVC controls could also be modelled in a similar fashion. The benefits to the utility transmission engineers in having an accurate simulation model is immense since the mystery of the controls can be stripped away and an understanding of their workings revealed. When the manufacturer's warranty expires, the utility may be dependent on the supplier for continued engineering services involving controls. However, with a detailed simulation model and an understanding of the workings and performance of the controls, the utility engineer can be in the driving seat directing all developments.

Would the authors comment on the benefit in having the supplier of HVDC or SVC systems provide either as part of the main contract, a working simulation model of controls on EMTP or EMTDC or, whether the utility engineer should build the detailed model in-house with data contracted from the supplier? Would modern digital controls be easier to model than the twenty year old analog controls reported in this paper? What is an estimate of the manpower expended in developing the simulation model for the Nelson River Bipole One controls?

Manuscript received February 13, 1992.

GERHARD JUETTE

Siemens Energy & Automation, Inc.
Alpharetta, GA 30202

The authors should be commended for an excellent paper and presentation.

In the context at hand, it seems important to point out that also with line commutated HVDC converters, simultaneous and independent reactive and real power modulation are, in fact, used. As an example, the Virginia Smith Converter Station at Sidney, Nebraska can vary the reactive power consumption by over 50 MVAr and hold the real power at any level between minimum and rated output (200 MW) in either direction. This capability is used for ac bus voltage control in combination with switched reactor and capacitor banks. On the other hand, real power modulation ("remedial action") for power system swing damping is provided, using a remote generator speed signal, while the voltage control function via reactive power remains active. Both the active and reactive power control loops are very fast from any practical system stability point of view.

Dynamic and reactive power modulation will, of course, impact both ac sides (rectifier and inverter). For steady-state purposes, the reactive power and ac voltage control on both sides are effectively decoupled via the transformer tap changers.

Manuscript received February 13, 1992.

P. KUFFEL, K.L. KENT, G.B. MAZUR, M.A. WEEKES

The authors would like to thank the discussers for their interest in our paper and for their valuable comments.

In response to Mr. Baker's comments, the work presented here reflects the first stage of an extensive model development and validation program undertaken by Manitoba Hydro. The ultimate goal is to produce similarly detailed models of the entire Manitoba Hydro HVDC and associated ac systems. To this end, we are currently developing and validating models of Bipole 2 controls and the new controllers accompanying the Bipole 1 replacement valves. Based on the experience gained during the initial phase of work presented in this paper, it was found that in order to ensure accurate models it was necessary to rigorously follow the established model development procedure and avoid simplifications unless their impact is fully investigated. Thorough documentation in the form of detailed functional block diagrams and descriptions are essential in order to ensure consistent model development in a variety of simulation packages and to simplify their use. Studies using the models described within the paper are presently ongoing to determine possible solutions to problems recently experienced on the Bipole 1 system.

In response to Mr. Woodford's comments, one of the major benefits of undertaking the model development program was the increased knowledge gained of in-service equipment. Therefore having a supplier provide a working simulation model is not as beneficial as having the models developed inhouse. As stated above, thorough documentation is critical in order to ensure accurate model development, and therefore suppliers should provide detailed functional block diagrams and descriptions. This would give the utility engineer a basis for model development and ensure consistent models regardless of the simulation tool used.

Modeling of modern digital controls will in general be simpler than modeling analog controls because of the problems associated with modelling complex analog circuits digitally. Approximately two person-years was spent on the development and validation of the Bipole 1 controls discussed within the paper. A substantial portion of this time was spent on the gathering and verification of schematics and component data and the development of the functional block diagrams.

Manuscript received April 8, 1992.

IEEE Transactions on Power Delivery, Vol. 8, No. 1, January 1993.

CONTROLS MODELLING AND VERIFICATION FOR THE PACIFIC INTERTIE HVDC 4-TERMINAL SCHEME

A. Hammad, Fellow, IEEE
R. Minghetti, Non Member
J. Hasler, Non Member
P. Eicher, Non Member
ABB Power Systems
Baden, Switzerland

R. Bunch, Member, IEEE
D. Goldsworthy, Member, IEEE
Bonneville Power Administration
Portland, OR

Abstract: A detailed digital model for the actual control system of the Pacific Intertie HVDC scheme is presented. The scheme is operated as multi-terminal bipole HVDC with four terminals in parallel. Each pole comprises two separately located converter stations with independent converter controls at each end of the transmission line. The control model includes bipole, pole, station and converter control systems. Special control techniques for providing safe and stable operation of the parallel converters are described. The techniques also result in fast recovery of the HVDC transmission scheme following severe ac and dc system disturbances. Verification of the completeness and accuracy of the model are made using field tests made on the actual HVDC scheme.

Keywords: HVDC, Parallel Converters, Controls, Modelling, EMTP, Field Tests.

INTRODUCTION

Introducing multiple HVDC systems (whether in the form of multi-terminal or several dc converters or links) in power systems has increased the complexity of analyzing the different interaction phenomena with the interconnected ac networks. Both manufacturer system developers and utility planners need the proper tools for their investigations. Today, many digital simulation and analytical tools exist as demonstrated in Refs. [1-7]. These techniques are used for analyzing different phenomena such as:

- stresses on system components
- overall ac/dc system transient performance
- control stability
- electromechanical stability
- sub and super-synchronous torsional interactions and shaft stresses
- voltage stability
- harmonic and sub-harmonic stabilities, etc.

92 WM 292-3 PWRD A paper recommended and approved by the IEEE Transmission and Distribution Committee of the IEEE Power Engineering Society for presentation at the IEEE/PES 1992 Winter Meeting, New York, New York, January 26 - 30, 1992. Manuscript submitted August 30, 1991; made available for printing December 18, 1991.

The correct and accurate modelling of system components, particularly HVDC with its controls, is essential in all these phenomena. Therefore, validation of such models is crucial to maintain and to improve the credibility of the digital tools.

This paper presents an elaborate control model for the 4-terminal HVDC transmission scheme of the Pacific Intertie in mono-pole and bi-pole configurations. All principal control and protection functions and actual system main hardware components are faithfully represented. The model is primarily used for transient simulations using the well known EMTP (electro-magnetic transient program). It can, however, be used with other digital simulation tools for ac/dc systems. All possible modes of operation and configurations of the HVDC scheme as well as all possible faults within and outside the HVDC converter stations can be correctly simulated. Numerous field tests made during the commissioning phase of the HVDC scheme have been used to verify the completeness and accuracy of the digital model.

PACIFIC INTERTIE HVDC SCHEME

The Pacific HVDC Intertie links the Pacific Northwest (at Celilo) in Oregon with Southern California (at Sylmar) near Los Angles - a distance of 1360 km. The original HVDC bipolar overhead transmission dates back to 1970 when it was commissioned as ± 400 kV, 1800 A scheme. The terminals, based on the valve technology of the time, have three 133 kV, 6-pulse mercury arc converter groups per pole. These are denoted GR1,3,5 on Pole 3 and GR2,4,6 on Pole 4 as shown in Fig. 1. Both terminals are connected to the 230 kV ac bus at Big Eddy and Sylmar respectively. In 1979, the current rating of the stations was raised to 2000 A. In 1985, the terminals were upgraded to ± 500 kV by the addition of a 6-pulse 100 kV solid state converter group in series to each pole (GR7 & GR8 in Fig. 1). This scheme was comprehensively modeled by Goldsworthy & Vithayathil [8]. The latest expansion of the scheme was commissioned in 1989 with the addition of modern 12-pulse thyristor converter terminals in parallel with the old terminals [9]. The new bipole converters are rated ± 500 kV and 1100 A and are shown as C1 & C2 in Fig. 1. The two HVDC converter stations at each pole (C1&C3 at P3 and

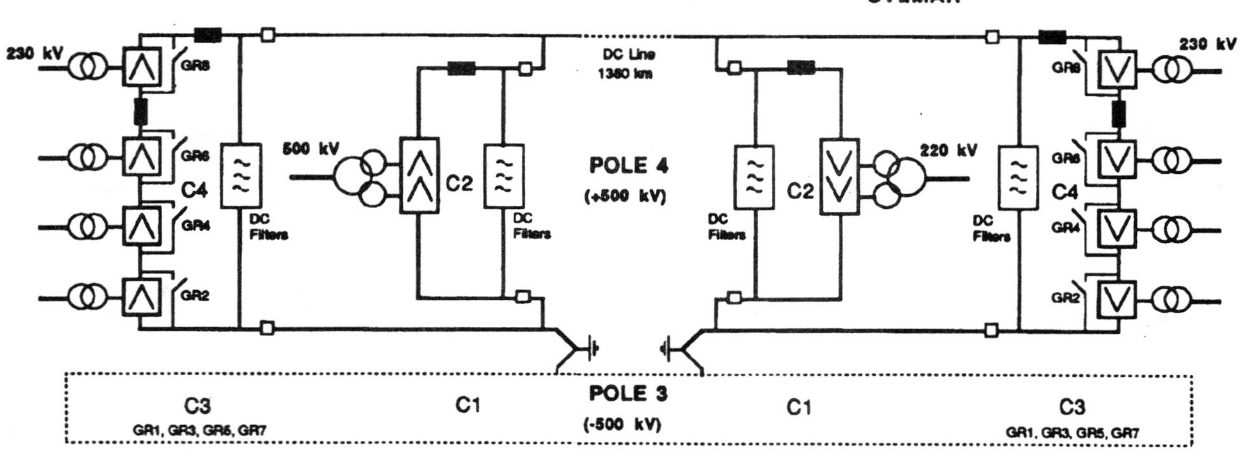

Fig. 1 Pacific Intertie 4-Terminal HVDC Scheme Schematic Diagram

C2& C4 at P4 respectively) are operating in parallel with independent controls. On the ac side, however, the new Celilo terminal is connected to the Big Eddy 500 kV bus and the new Sylmar terminal is connected to the SCE 220 kV bus. At present the nominal capacity of the scheme is 3100 MW with an overload capacity of 550 A provided by the new terminals.

The HVDC scheme has a large number of operating modes with various combinations of converter groups [9]. Such combinations can be grouped into five operating modes:

- Bipolar balanced operation with stand-alone or with parallel converters
- Bipolar unbalanced operation with stand-alone or with parallel converters
- Monopolar ground-return with stand-alone or with parallel converters
- Monopolar metallic-return with stand-alone or with parallel converters
- Reduced dc voltage operation.

Furthermore, due to the versatile and reliable new digital controls, the HVDC scheme can adapt quickly to all system disturbances by changing its configuration or operating voltage and currents in order to ensure continuous power transmission [10]. As an example, in the event that a mercury arc valve group is blocked or out of service, the expansion converter in parallel to that group will operate continuously at reduced dc voltage of 367 kV utilizing high firing angle control and converter transformer tap changer.

MODEL CONTROL FUNCTIONS

Fig. 2 shows an overall structure of the controller model for the entire HVDC Pacific Intertie scheme. The HVDC controller consists of the following basic functional levels:

1. Bipole controls
2. Pole controls
3. Station pole controls
4. Adaptation control to interface the new controls to the old pre-expansion converter controls (ECA)
5. Converter controls (new and old)

BPS Controls

The bipole, pole and station pole controls (BPS) calculate and allocate current orders based on the power order. They coordinate the current between the two poles, between the stations and between the parallel converters.

The appropriate dc voltage reference, based upon maximum rated voltage of the end stations and the configuration of the groups and converters in operation, is established by these controls.

The BPS controls also contain functions for fast power change and reversal, converter and electrode line overload limits, current order mismatch protection, voltage and firing angle dependent current limits, pole and inverter current margin compensations to properly control and allocate current margins, pole voltage control and pole fault recovery sequences

With reference to Fig. 2, the given bipole power order is divided by the calculated bipole rectifier

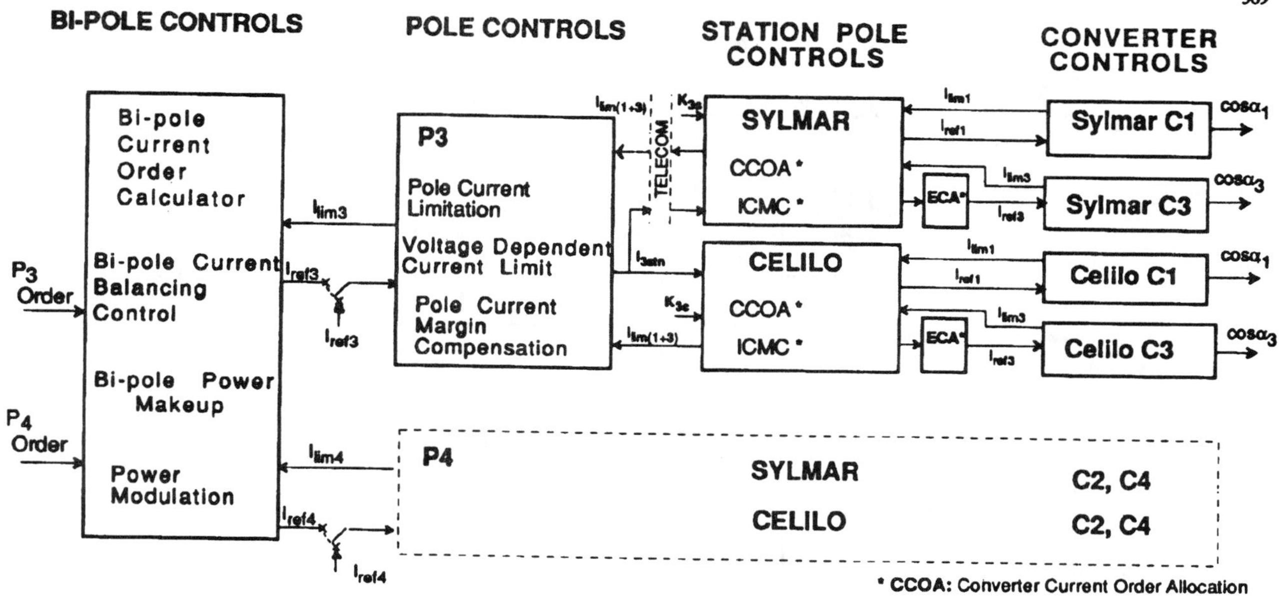

Fig. 2 Controller Model Overall Structure for The 4-Terminal HVDC Scheme

voltage to obtain the pole current order. A supplementary control loop is provided to correct any unbalance between the current orders in the two poles to minimize the ground current in balanced operation mode. Power makeup function maintains the total bipole power whenever one pole is subject to a disturbance and incapable of producing its power share. The difference between the desired current and the encountered current limit is automatically transferred to the healthy pole.

BPS controls allow as well operation in constant power or constant current modes. The current order can be selected from manual control if desired. The current order is compensated by current margin in case of inverter changing to current control due to any rectifier limitations.

Under persistent low dc voltage or detection of 60 Hz component on the dc side (commutation failure) the VDCL is activated to temporarily reduce the current reference. This is necessary to limit thermal overloading of the thyristor valves. The final pole current order Istn is exchanged between Celilo and Sylmar stations in a fast telecom channel.

A current split factor, Ks for Sylmar and Kc for Celilo, is used to distribute the dc pole current between the two parallel converters at each station respectively. Depending upon the converter configuration, the CCOA function (in Fig. 2) enforces the maximum and minimum current limits for each converter. The Inverter Current Margin Compen-sation ICMC has a relatively slow action (0.5 s) and is active only when the station is in inverter mode. This function ensures a stable operation with the desired current sharing between the two parallel converters by moving the operating point of the expansion converter to the adequate location on the Vd/Id characteristics.

Converter Controls

Each converter control (CC) receives the current reference value Iref from the BPS controls and generates the delay angle order for its converter. In addition, it performs the block/deblock sequences for its respective converter. Other functions include filtering of measured quantities (such as Id, Vd and γ), current, voltage and gamma controls as well as mixed current/voltage and mixed current/gamma controls as shown in Fig. 3. These controls are locally independent from each other and have the capability for local activation without connection to the BPS controls.

During steady state operation with 4-terminals and power transfer say from north to south, the Celilo converters C1, C2, C3 & C4 (rectifiers) and Sylmar expansion converters C1 & C2 (inverters) are basically in current control mode. In this case the dc voltage is determined by the Sylmar converters C3 & C4 (inverters).

The steady state characteristics for C1 & C2 in inverter operation are defined by:

- γ (min) = 17°
- I margin = 150 A.

The steady state characteristics for converters C3 & C4 in inverter operation are defined by:

- γ (min) = 16° for mercury arc groups
- γ (min) = 18° for thyristor group
- I margin = 180 A.

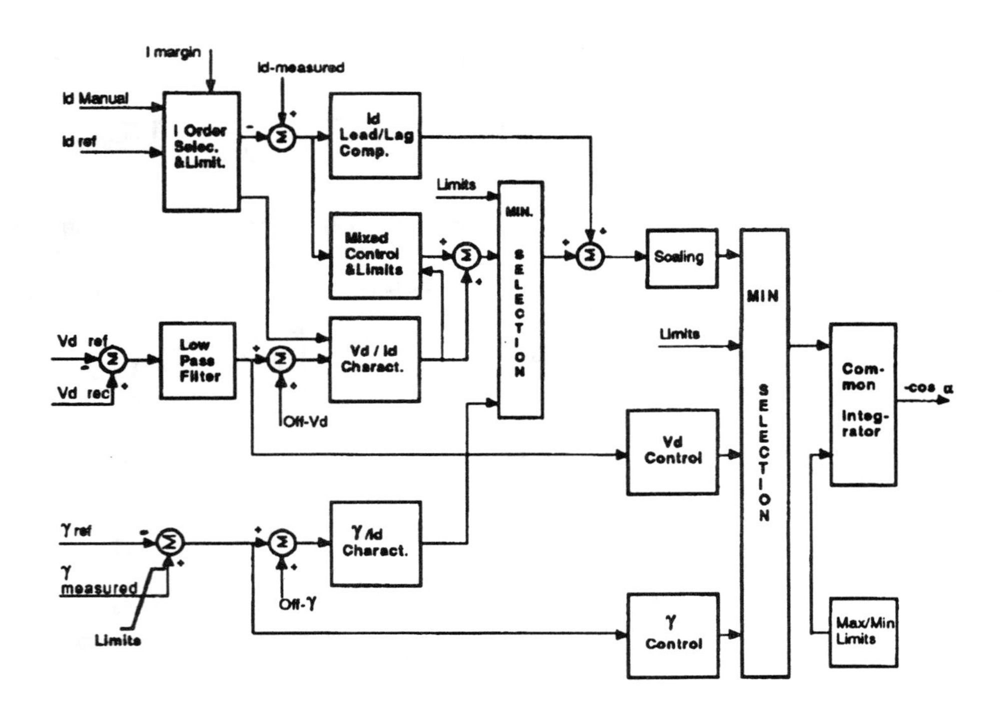

Fig. 3 Converter Control Model for Expansion Terminals C1 & C2

During transient operation all CC converter control loops play a dominant role in providing the scheme with a stable and fast response. By means of the mixed controls, a combined current margin and voltage limiting technique is realized using a predefined Vd/Id slope and dc voltage-offset for rectifier and γ/Id characteristics and gamma-offset for inverter terminals. Fig. 4 depicts such steady state characteristics at certain transformer tap positions.

The primary objective of the mixed controls function of the expansion CC is to minimize changes in the operation or overloading of the pre-expansion converters. In addition, this function greatly enhances the recovery performance of the HVDC scheme from severe transients. Without the mixed controls, if the dc voltage has been depressed by some external event and then allowed to recover, the driving terms for the controller to recover the dc voltage and current are based upon the current errors only. Such errors would be fractions of the current margins. With fixed control parameters, this would result in a very slow recovery unless the current margins are very high.

One method for making the current margins high without degrading the performance is to temporarily allow the set point current in the inverter to be as low as Imin and in the rectifier to be as high as Imax at low voltages. This technique is realized by modifying the Vd/Id characteristics of the expansion converters as shown in Fig. 5.

At low voltages, the effective current margin is in the order of 1000 A, which provides a strong drive to the current controllers. This is very effective in transient operation. However, if the voltage is depressed for too long, such action would be like an opposite action of a VDCL on the rectifier and would result in a very low current on the inverter of the expansion converters. This may drive the currents in C3 & C4 inverters well beyond their maximum

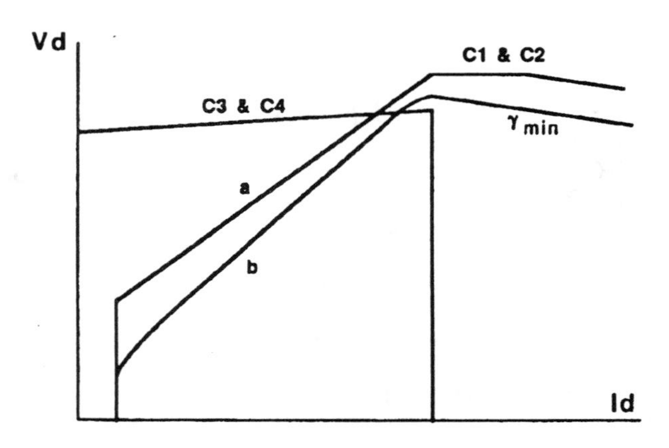

Fig. 4 Converters Steady State Characteristics
(a) Voltage/Current Characteristics
(b) Gamma/Current Characteristics

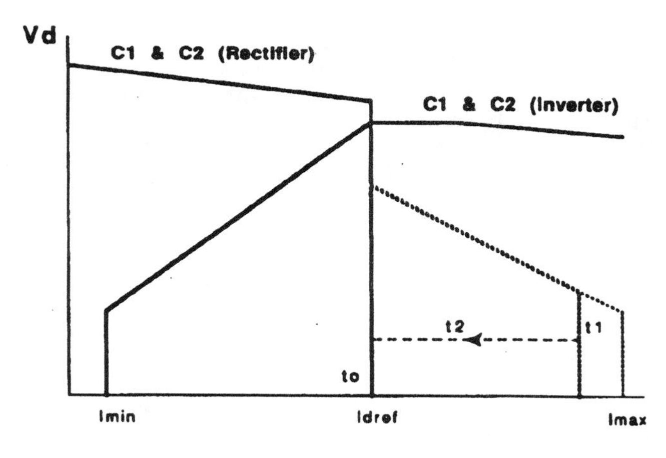

Fig. 5 Expansion Converters Characteristics During Disturbances

allowed limits. Hence, an additional integral current error is formed by means of the mixed control and added to the Vd/Id characteristics that slowly drives the current back to their reference values. This way a smooth and stable performance is always ensured even in case of losing telecommunication between the end stations.

Fig. 5 shows an example of the expansion rectifier current set points when the dc voltage is depressed on a long term basis at the inverter end, e.g. due to temporary blocking of mercury valve groups. At pre-disturbance time (to) the current set point is Idref. When the dc voltage drops at (t1) the Vd/Id characteristics controller modifies the current set point as shown in Fig. 5. Depending on the severity of the voltage drop the set point can reach Imax. Finally the set point is returned to Idref after some delay (t2) due to the integrator action of the mixed controller.

The simulation case presented in Fig. 6 illustrates the transient effects of the converter controls. A remote 3-phase ac fault is simulated near Sylmar station while operating as inverter with 1400 MW total power. The fault is cleared in 3 cycles. Note how the expected high dc currents at fault commencement are diverted mainly to the expansion converters C1& C2 in order to avoid overloading C3& C4. Also note the remarkable fast recovery of dc power (50 ms) after fault clearing.

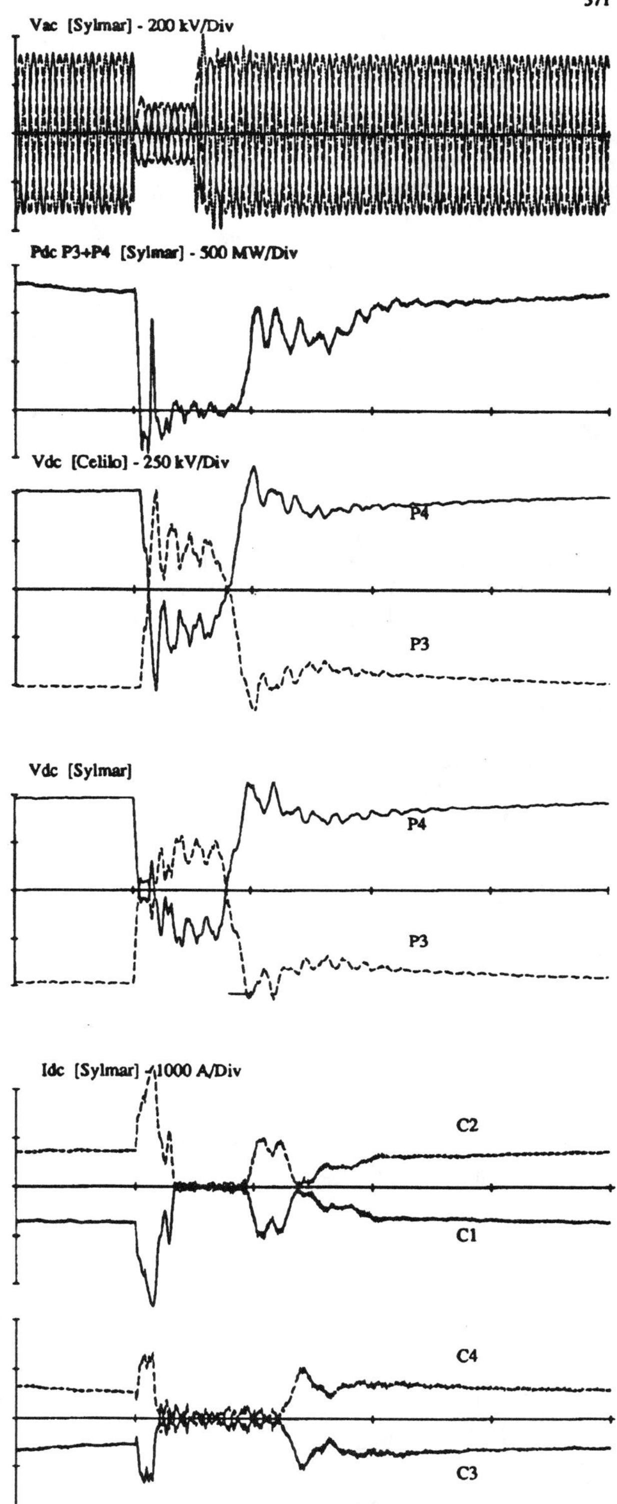

Fig. 6 Simulated Test for 3-ph Fault Near Sylmar

Adaptation Controls

Because of the need for coordination between pre and post-expansion converters, some of the functions which were originally performed by the old control logic are taken over by the new BPS controls. The current order signals are provided to the analogue control amplifiers of C3&C4 from the

BPS. The control amplifier, as explained in [8], including feedback, current limits and some special functions related to mercury arc valves remained unmodified. The old controller is integrated in the new controls by means of the control adaptation unit (ECA) which modifies the control amplifier output voltage and consequently the valve firing angle. In the new control system, the ECA provides the interface to BPS and contains additional functions necessary for parallel operation [11].

The new control system is a multi-processor (digital) system. The BPS, CC and ECA controls require high speed and multiple signal processing which are provided by a Programmable High Speed Controller PHSC described in Ref. [12].

Gate Firing and Protection

The old grid timing system delay angle determinator for C3& C4 was modeled in Ref. [8]. The new gate control for C1& C2 is a PLL equidistant firing and is modeled using the principles outlined in Ref. [7].

The model for dc line protection follows that described in [8]. It provides the retard and restart signals to all new and old controllers. Voltage rate of change and level are used to determine the occurrence of a line fault. Fast detection of line fault (5 ms) is possible. With telecommunications in operation, line protections at both terminals are active with the inverter sending a force retard signal to the rectifier when a fault is detected. Protections at parallel converters are active with the first to detect a fault initiating the retard for the entire pole. The same protection is also active without telecom but with an additional slower backup (50 ms) which operates at the rectifier to detect faults near the inverter terminals.

SIMULATION VERIFICATION CASES

Cases representing various HVDC scheme operating modes under all possible configurations have been simulated by the model and verified against commissioning field tests. These include the following operating modes in stand-alone and in parallel converter configurations:

- bipolar balanced and unbalanced operation
- monopolar earth return and metallic return
- reduced dc voltage operation

The good agreement between simulated and real life cases ensured the accuracy and completeness of the digital model despite the relatively simplified representation used for the ac networks. Figs. 7 through 10 depict two of such comparisons.

Figs. 7 & 8 show the respective simulation and field test cases for a dc line arcing fault on pole 3 near Celilo. The HVDC scheme is in bipole operation with only converters C1& C2 in operation. After detecting

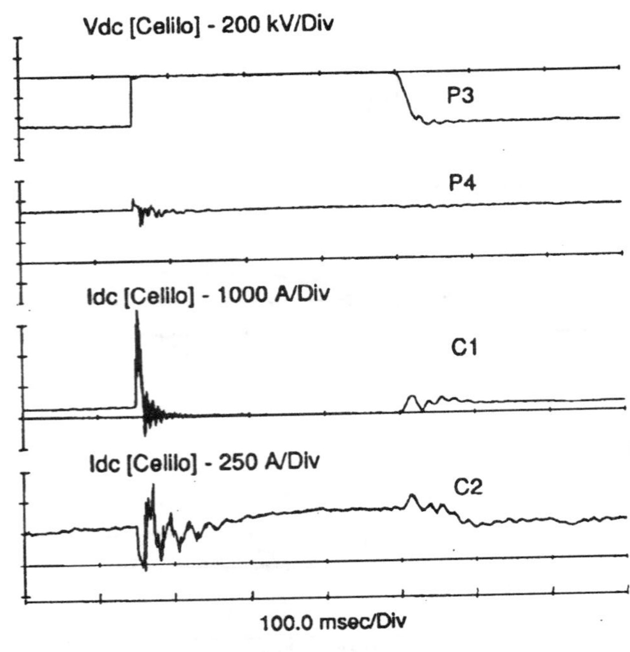

Fig. 7 Simulated Test for dc Line Fault at Celilo

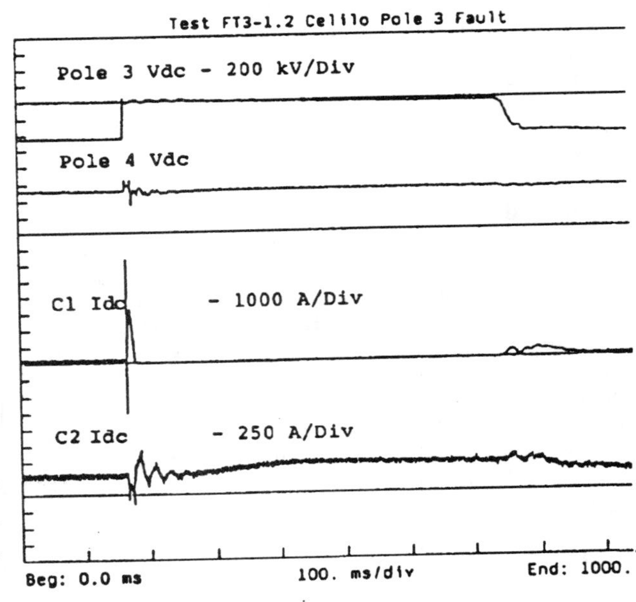

Fig. 8 Field Test for dc Line Fault at Celilo

line fault, the dc line protection activates the force retard signal in 5 ms. During the fault period (deionization time) the current on the faulted pole is reduced to zero. The current on the healthy pole is simultaneously increased in order to maintain the prefault power. After fault clearing, the dc voltage and current on the affected pole recover within 20 ms and 50 ms respectively. At the same time the original current of the healthy pole slides back to its prefault value.

Figs. 9 & 10 show respectively the simulation and field tests for a SLG fault near Big Eddy 500 kV bus when the dc power flow is from Celilo to Sylmar. All converters C1, C2, C3 & C4 are in operation with full dc voltage. During the fault period (3.5 cycles) the faulted phase (A) voltage drops to 20%. After fault clearing, the prefault power is restored in less than 50 ms with the new converters showing a faster recovery than the old ones.

CONCLUSIONS

A comprehensive digital model representing the controls of the Pacific Intertie HVDC bipole scheme with 4 terminals is developed. It includes sufficient details on actual system main components and control functions.

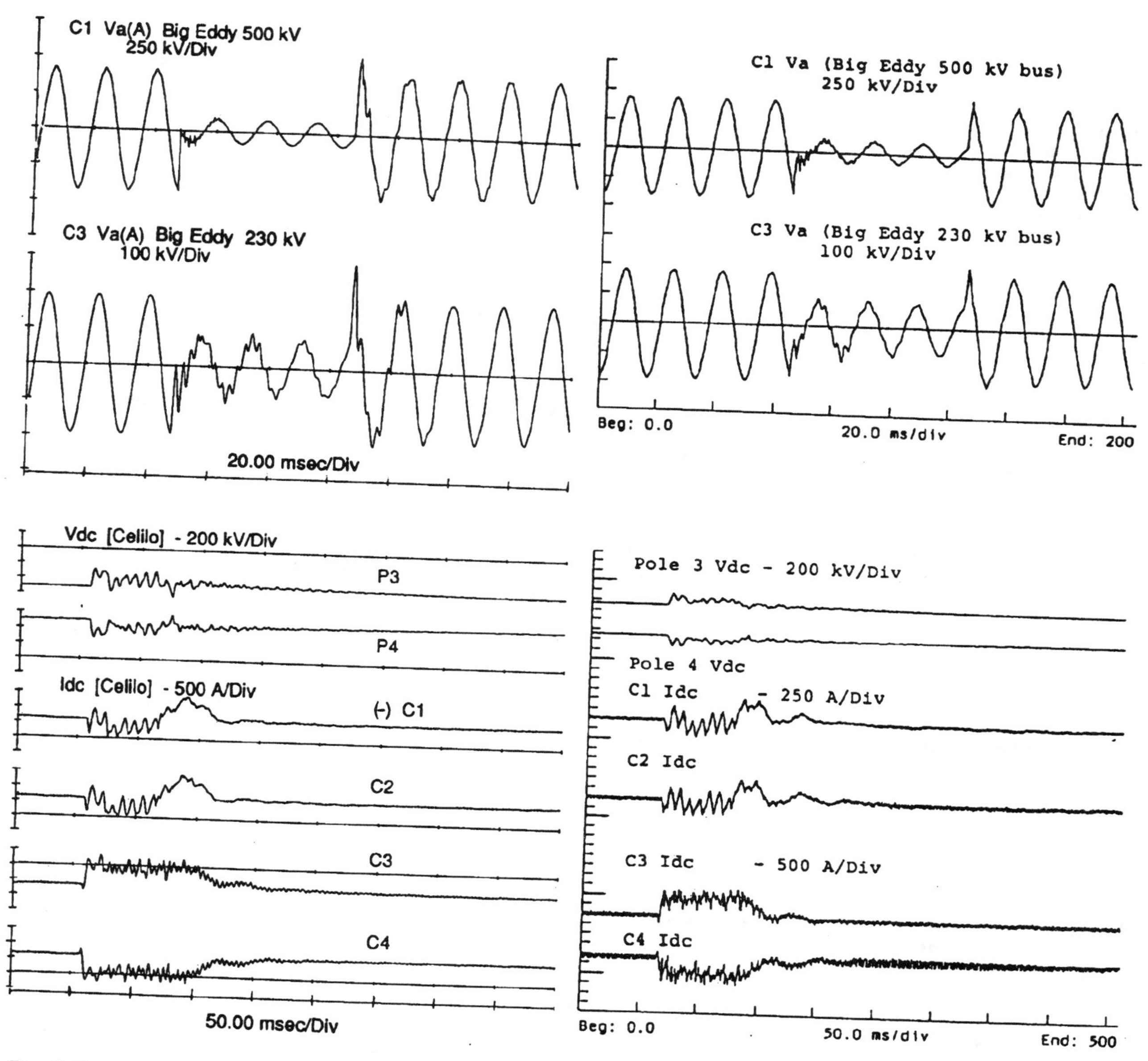

Fig. 9 Simulated Test for SLG F. at Big Eddy 500 kV bus

Fig. 10 Field Test for SLG fault at Big Eddy 500 kV bus

Advanced techniques utilizing mixed current, current/voltage and current/gamma controls can be used to ensure a stable operation of the parallel converters with or without telecommunication between end stations. These techniques enhance the ac/dc system dynamic performance and produce a fast HVDC system recovery after severe faults.

The model has a modular structure and is suitable for transient simulations using EMTP and other ac/dc study tools. Accuracy of the model has been verified against field staged fault tests. The exact similarity between simulated and actual system performance proves that today's available tools can be a good basis for future development.

ACKNOWLEDGEMENT

The authors wish to acknowledge all the excellent and useful contributions made to the control scheme and its model development by Prof. R. Baker of Washington State University, USA, M. Drummond and L. Pilotto of CEPEL, Brazil, L. Bui of IREQ and D. Brandt of MH, Canada.

REFERENCES

[1] EMTP User Manual: April 1990.

[2] M. Heffernan, et. al., "Computation of AC-DC System Disturbances", IEEE Trans., Vol. PAS-100, Nov. 1981, pp 4341-4363.

[3] J. Reeve, R. Adapa," A New Approach to Dynamic Analysis of AC Networks Incorporating Detailed Modelling of DC Systems", IEEE Trans. PWRD, Vol. 3, Oct. 1988, pp 2005-2019.

[4] D. Woodford, et. al," Digital Simulation of DC Links and AC Machines", IEEE Trans., Vol. PAS-102, June 1983, pp 1616-1623.

[5] K. Kruger, R. Lasseter," HVDC Simulation Using Netomac", MONTECH, Sept. 1986, pp 47-50.

[6] K. Padiyar, et. al, "Study of HVDC Controls Through Efficient Dynamic Digital Simulation of Converters", IEEE T-PWRD, Oct. '89, pp 2171-2178.

[7] S. Lee, S. Bhattacharya, T. Lejonberg, A. Hammad, S. Lefebvre," Detailed Modeling of Static VAr Compensators Using Electromagnetic Transients Program", IEEE T&D Conf., Dallas, Sept. 1991.

[8] D. Goldsworthy, J. Vithayathil," EMTP Model of an HVDC Transmission System", MONTECH, Montreal, Sept. 1986, pp 39-46.

[9] R. Bunch, et al,"Completion and Initial Operation of the Expanded 4-Converter Celilo-Sylmar Pacific Northwest Southwest Intertie", CIGRE 1990 14-202

[10] L. Blahous, et al, "Extension of the Pacific Intertie by Two Parallel Bipolar Stations: Concept and Factory Simulation Tests", CIGRE 1990 14-205.

[11] A. Hammad, et al, "Modelling of the Pacific Intertie 4-Terminal HVDC Scheme in EMTP", IEE 5th International Conference on AC & DC Power Transmission, London, Sept. 1991, pp 362-367.

[12] H. Stemmler ,G. Gueth," A New High Speed Controller with Simple Program Language for the Control of HVDC and SVC", IEEE Int. Conf. on DC Power Trans., Montreal, June 1984, pp 109-115.

BIOGRAPHIES

Adel Hammad received his B.Sc.(hons.) and M.Sc. from Cairo University and the Ph.D. from University of Manitoba, Canada, all in Electrical Engineering, in 1972, '74 and '78 respectively. From 1975 to 1978 he worked in the system planning div. of Manitoba Hydro, Winnipeg, Canada. In 1978 he was the chief electrical engineer of UNIES Engg. Consultants in Winnipeg. Since 1979 he is with Asea Brown Boveri where he is now manager of HVDC and SVC system applications in ABB Power Systems, Switzerland.

Dr. Hammad is a fellow of IEEE and a member of its Power System Engineering Committee. He serves on the IEEE System Dynamic Performance and the DC Transmission subcommittees and is active on several IEEE and CIGRE working groups. Dr. Hammad is a registered Prof. Engineer in the Province of Manitoba.

Roberto Minghetti received his Bachelor degree in Electrical Engineering from the University of Genova, Italy in 1980 and his M.Sc. from the University of Illinois, Urbana-Champaign, USA in 1983.

From 1982 to '83 Mr. Minghetti worked with GE Co. in Urbana and Schenectady on research for the US DOE. From 1983 to '85 he was with Italimpianti, Genova as a power system engineer. Since 1986 he is with Asea Brown Boveri where he is now a senior engineer in the HVDC and SVC system applications dept. of ABB Power Systems in Baden, Switzerland.

Jean-Philippe Hasler received the B.Sc. from the Engineering School of St.-Imier in 1981 and the M.Sc. in Electrical Engineering from the Federal Polytechnical Institute, Lausanne, Switzerland in 1986.

In 1986 Mr. Hasler joined Asea Brown Boveri where he worked in the drives division for the development of control and protection systems for SVC and HVDC systems. Since 1987 he is a senior control engineer in ABB Power Systems, Baden, Switzerland.

Pierre-Alain Eicher received the B.Sc. from the Engineering School of St.-Imier in 1981 and the M.Sc. in Electrical Engineering from the Federal Polytechnical Institute, Lausanne, Switzerland in 1986.

Before joining Asea Brown Boveri in 1986, Mr. Eicher spent one year with LONGINES watches in its time keeping dept. In ABB he worked in the drives div. for the development of control and protection systems for SVC and HVDC systems. Since 1987 he is a senior control engineer in ABB Power Systems, Baden, Switzerland.

Daniel Goldsworthy received his B.S. degree in Civil Engineering in 1978 and M.S. in Electrical Engineering in 1980 from Montana State University, Bozeman, Montana, USA.

Since 1980 Mr. Goldsworthy has been with the Bonneville Power Administartion in Portland, OR. His work has centered around transient simulation of ac and dc systems using the EMTP. He is a member of IEEE and is a registered Professional Engineer in the state of Oregon.

Richard Bunch received the BS degree in Electrical Engineering from Walla Walla College in 1963.

From 1963 to 1971 Mr. Bunch was employed by the Bonneville Power Administration as a power system communications and control engineer. From 1971 to 1981 he was assistant superintendent and control, protection and maintenance engineer of the Celilo HVDC terminal. Since 1981 he is a principal engineer for HVDC operations and maintenance at the Bonneville Power Administration Dittmer control center.

D.A. Woodford (Manitoba HVDC Research Centre). The authors have prepared an excellent paper demonstrating the feasibility of precision modelling of HVDC controls using in this case, EMTP as the simulation platform. A trend is being established where exact simulation models of complex power system controls are being demanded by the user utility of HVDC and SVC systems. In order to assess the effort involved, would the authors please provide an estimate of the manpower expended in developing the HVDC controls on EMTP? Furthermore, how long is the simulation run time and on what computer?

Manuscript received February 21, 1992.

A. Hammad, R. Minghetti, J. Hasler, P. Eicher(Asea BROWN BOVERI, Baden, Switzerland): The authors thank Mr. D. Woodford for his valuable comments and discussion.

Certainly with the rapid (almost daily) advances made in the computer hardware and software fields, engineers are encouraged to venture into domains that were almost prohibitive some years ago. With advanced digital techniques that exist today for simulating power systems(see e.g. Ref. 7, A, B & C) it is possible to develop exact models of complex controls such as those used for HVDC and SVC systems. Development of such complex models is a continuous process that always depend on the previous (pre-built) sub-models. One can hardly imagine that a complex model is built today starting from scratch.

The model described in our paper was developed almost 3 years ago at the time when only the original EMTP (Version M39) was available. Modelling of controls was based on the original effort of Ref. [8] using the well known TACS features. Most of the manpower expended was practically struggling to fit the "true digital" controller into TACS which are basically digitized "analog" control models. That manpower effort was about 5 man-months including documentation. With todays tools a similar task would probably need one-tenth of such man power requirements.

It should, however, be emphasized that verification of the model adequacy is a major and significant effort that cannot be overlooked.

The digital simulation cases reported in the paper were run on a VAX Station 3100. A typical case with the entire bipole scheme modeled with all old 16 converter groups and new 4 parallel converters had the following software requirements:

- number of model network nodes : 1053
- number of model network branches : 1802
- working space for TACS : 60181 floating points
- integration time step : 40 us
- CPU time for data input : 4910 s
- CPU time to reach steady state (300 ms) : 4040 s
- CPU time for fault case (500 ms) : 7220 s

Once again we thank Mr. Woodford for his interest in our paper.

REFERENCES

[A] L. Bui, S. Casori, G. Morin, J. Reeve, "EMTP TACS-FORTRAN Interface Development for Digital Controls Modelling", paper no. 91 SM 417-6 PWRS presented at IEEE/PES 1991 Summer Meeting, San Diego, July 1991.

[B] D. Falcao, E. Kaszkurewicz, H. Almeida, "Application of Parallel Processing Techniques to the Simulation of Power System Electromagnetic Transients", paper no. 92 WM 287-3-PWRS, presented at IEEE/PES 1992 Winter Meeting, New York, Jan. 1992.

[C] Manitoba HVDC Research Centre, "Digital TNA", Centre Journal, Vol. 5, No. 1, Jan. 1992.

Manuscript received April 13, 1992.

MODELING OF THE HYDRO-QUEBEC - NEW ENGLAND HVDC SYSTEM AND DIGITAL CONTROLS WITH EMTP

G. Morin, Member
L. X. Bui, Senior Member
S. Casoria, Member
Hydro-Québec
Montréal, Canada

J. Reeve
Fellow
University of Waterloo
Waterloo, Canada

Abstract

The commissioning of the Québec - New England multiterminal dc transmission system underscores the increasing need to develop numerical models for simulation purposes. The modern microprocessor-based HVDC converter controls used in this scheme make the modeling a real challenge since Electro-Magnetic Transient Program (EMTP) usually employed for simulation purposes and its Transient Analysis of Control Systems (TACS) was designed to simulate analog, not digital controls.

This paper gives details about the modeling of digital controls with a modified EMTP version and presents some comparisons with HVDC simulator results which show good agreement between the two power system analysis tools. A case of validation with field test result is also included in the paper.

Keywords : HVDC, digital control, EMTP.

INTRODUCTION

The Radisson-Sandy Pond HVDC system, which will come into operation in 1992 is the world's first large-scale multiterminal dc system. Originally designed to be operated with five terminals in service, it was reduced to a four-terminal system following NEPOOL's decision to withdraw Comerford station from the scheme.

Hydro-Québec has gained considerable experience in the domain of HVDC transmission with its Châteauguay 1000 MW back-to-back, Madawaska 350 MW back-to-back and Des Cantons 690 MW schemes over the last ten years. However, the unprecedent nature of the Radisson-Sandy Pond scheme makes it essential to gain a deep understanding of the power system behavior. Nevertheless, this can not be achieved without developing a numerical model for study purposes and validating the results on the HVDC simulator prior to commissionning tests in the field. The simulator is a very powerful tool which has been used extensively in recent years. Its main disadvantage is that it requires a tremendous amount of on-line equipment and is not necessarily available for every potential user at the required time. In fact, most engineers use digital programs to study various transient and dynamic performances of the power system. Software such as EMTP and stability programs are often employed to make preliminary studies and understand how the system behaves.

92 WM 274-1 PWRD A paper recommended and approved by the IEEE Transmission and Distribution Committee of the IEEE Power Engineering Society for presentation at the IEEE/PES 1992 Winter Meeting, New York, New York, January 26 - 30, 1992. Manuscript submitted August 6, 1991; made available for printing December 23, 1991.

Simultaneous field tests and simulator studies provide a unique opportunity to develop and validate a model for numerical studies. There are two important aspects to numerical modeling. First, it allows a deeper insight into the various control functions and the overall behavior of the control system. Second, engineers can use the model whenever they need it without having to wait for the simulator to become available.

The Québec-New England multiterminal operation was discussed in recent papers [1-3, 8] which dealt mainly with design aspect, control philosophy and results from simulator tests. So far, however, no control model has been made available for numerical transient studies. Aspects of analog control and application to HVDC converters using TACS have been published in the past [4-5]. On the other hand, the modern design trend using digital controls gained impetus with the availability of fast and relatively inexpensive microprocessors such as in the case of Québec-New England HVDC scheme. At the same time, the adaptation of analysis tools to this new technology is a real challenge [5-7]. In fact, since it is mostly digital, this control system is quite difficult to model in EMTP with TACS. The need to solve this particular problem when modeling digital controls with EMTP has been investigated and a new solution incorporating a TACS-FORTRAN interface to model digital controls has been developed [7]. This made it possible to model the Québec-New England HVDC system and digital controls with EMTP. The results obtained are presented in this paper.

CONTROL FUNCTIONS MODELED

DIGITAL CONTROLS

In the Québec-New England scheme, the digital controls (Fig. 1) and protection system are divided into subsystems, contained on microprocessor-based boards operating in parallel. The software is written using mainly a high-level language, PL/M. Exceptionally, when speed is at a premium, low-level assembly language and Digital Signal Processor (DSP) circuits are used. A few functions, representing very fast processes in the converter firing control, still use hard-wired logic and must be modeled as analog control functions.

The software runs under a custom-made real-time operating system (RTOS) which manages the different tasks constituting a function or process which in its simplest form, is a program comprising a task for input handling, a calculation task, and another one for output handling. The many jobs within a function are executed efficiently using a CPU in a multi-tasking mode. Since tasks often run for brief periods, then idle for relatively long periods, other tasks could use the CPU.

The allocation of CPU time between the tasks is scheduled by the RTOS. Among the several possible methods of doing this [7] the so-called "mosaic scheduler with more than one interrupt" was selected. In the basic mosaic scheduler, a periodic clock with a specific cycle time causes a timer interrupt which is inputted into a special task at the end of the process. The only raison d'être for that task is to wait for the interrupt before the cycle restarts. In the mosaic scheduler with more than one interrupt, hardware interrupts may start some priority tasks. The tasks are sorted in levels having different priorities with levels

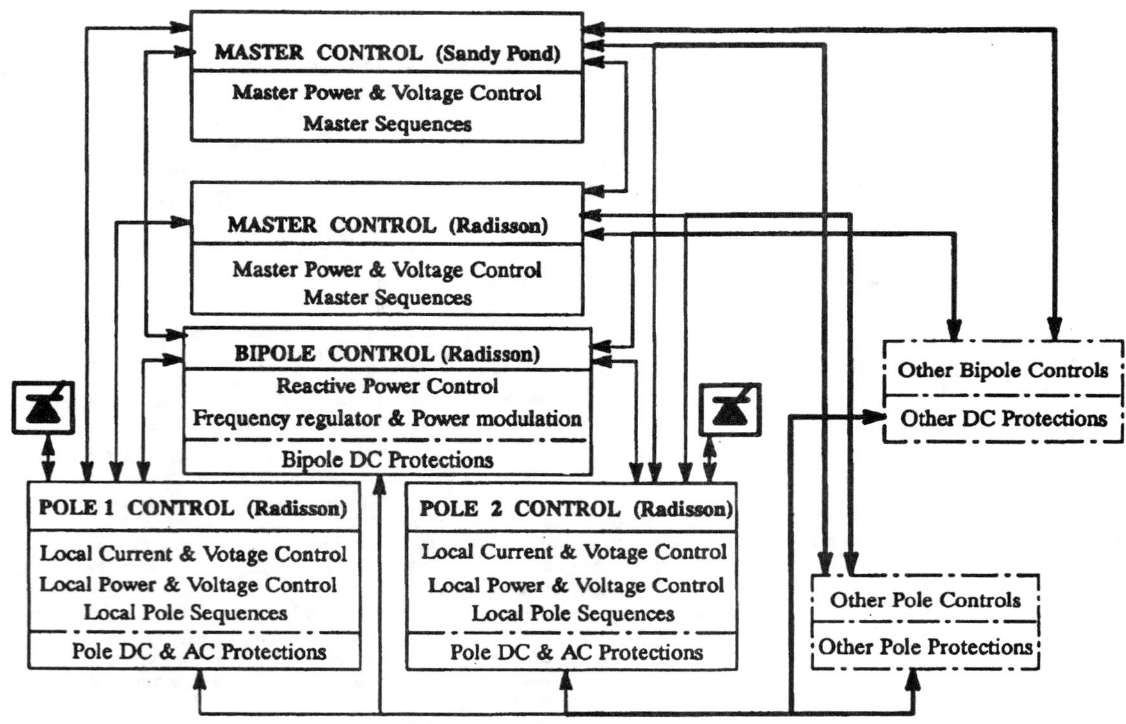

Figure 1 : Control and protection sub-systems structure.

that are hardware-interrupted having a higher priority than those interrupted by the timer.

The operating system also provides frequently used basic functions such as filters, flip-flops and limits.

CONTROL SUBSYSTEMS

The control subsystems are structured as shown in Fig. 1. Only functions relevant to the dynamic operation of the system are modeled. The outputs of non-simulated functions, considered constant in term of the simulation time (e.g. dynamic overload limit, tap changer control), are entered as fixed parameters.

The master control coordinates the control of the terminals (e.g. balancing power and current orders, ramping orders, switching, etc.) via the master telecommunications system. In future, telecommunication delays will be modeled. The Master Power and Voltage Control (MPVC) computes the current order to each converter and the voltage reference for the voltage setting terminal (the one that determines the dc voltage on the line). The Master Sequences (MSEQ) coordinates the switching of each pole of the multiterminal system. The orders received at the local station then initiate a local sequence if system conditions permit. Two master controls are provided in case the two systems (Québec and New England) ever operate independently. Normally, when power is exchanged between Québec and New England, only one master control is active; the other is therefore continuously updated and remains in standby mode.

The bipole control consists of functions that act on both poles, such as the Reactive Power Control (RPC) and the Frequency Regulator and Power Modulation (FRPM) (at Radisson only). The RPC is used to control the ac voltage or the reactive-power exchange in a station. The FRPM maintains the ac system frequency at Radisson when the latter operates in islanded mode, but can also maintain stability on the Hydro-Québec grid following a severe disturbance by modulating the dc power at Radisson when the latter is in synchronized with Hydro-Québec's ac system.

The pole control comprises subsystems acting locally at the pole level which are the same on each pole. These subsystems are the Local Current and Voltage Control (LCVC), the Local Power and Voltage Control (LPVC), the Local Pole SEQuences (LPSEQ), and the protection.

The LCVC consists mainly of the Converter Firing Control (CFC), the current (or voltage) regulator, the Voltage Dependent Current Order Limiter (VDCOL), and the Decentralized Dynamic Voltage Controlled Recovery (DDVCR) task. The latter, essential for a MTDC system, determines the converter current orders after the occurrence of a fault if the station is not under master supervision (i.e. during a telecommunication failure or in local control mode).

The LPVC includes the Tap Changer Control (TCC), the Dynamic Overload Calculation (DOC) and functions to define the current and voltage orders (Current Order Memory (COM) and Voltage Reference Memory (VRM)) in case the station is in local control mode. The DOC is not modeled. In fact, the overload capability, which is important for dc modulation at Radisson, has a characteristic curve such that the current order limit decreases slowly (over 15 min) to the steady-state value. Hence, the limit is justifiably considered to be constant during the transient simulation. For the same reason, and also because the transformer model in EMTP has a fixed tap during the simulation, the TCC is not modeled.

The LPSEQ consists of functions for the control and switching operations, block/deblock of the converter, etc., which usually require an interlocking scheme or are performed in a determined sequence. The pole protection includes DC Line-to-ground fault Protection (DCLP), and Short Circuit and Commutation Failure Protection (SCCFP). Both are simulated at the present stage. The DCLP detects ground faults on the dc line, extinguishes fault current and restores the power transmission while the SCCFP detects faults in the converter due to phase to phase-to-phase short circuits or commutation failures. In the latter case, action is taken to prevent further failures. A dedicated telecommunication system links the protection signals of one station with all the others.

EMTP MODEL

A new FORTRAN interface implemented in EMTP [7] made it possible to create a control structure similar to that used in real controls thereby extending TACS capabilities to the modeling of digital controls. Since the RTOS task scheduling is simulated, the tasks can be executed exploiting fully the property of "execution only when needed". The analog parts of the control, which have to be executed at every time step, are represented in TACS. Furthermore, useful basic functions included in the RTOS are modeled and used as part of a toolbox in the form of a FORTRAN library. Lastly, since the PLM control programming language is somewhat similar to FORTRAN, it was relatively easy to implement the software subsystems.

The modeling work completed so far covers the essential parts of the control system and protection: the complete LCVC, which includes the converter firing control, the DCLP and SCCFP protection. Some functions less relevant to the simulation (e.g. the functions in the LPVC defining the current and voltage orders in local control mode) are neglected without affecting the validity of the results. Thus, the simulation is restricted to the operation of controls under master control supervision (i.e. master control mode). Approximately, there are 40 digital software functions (i.e subroutines) modeled in FORTRAN including basic ones such as integrator, filter and delay units, some of them using z-transform control system theory. In order to give an idea on how large is the digital control system modeled, the compiled library of FORTRAN subroutines for simulation of 2 terminals operation takes 1.66 megabytes of disk memory. Of course, this is not including the high speed functions such as dc line voltage slope detector, overlap calculator, user's input data etc. which are modeled in TACS. Overall, the control modeled in EMTP corresponds to a total of 50 individual microprocessor based control functions.

Figure 2 shows an example of an EMTP data file. The $INCLUDE feature allows blocks to be added as needed to represent the different control functions associated with a specific converter station and pole. The modular structure allows pole control functions to be added or removed very easily. Moreover, these blocks facilitates the exchange of parameter values between TACS and FORTRAN.

```
          BEGIN NEW DATA CASE
C              DC CONTROLS
$INCLUDE  TACS . USERIN,
C     RADISSON CONVERTER POLE 1
C     LOCAL CURRENT AND VOLTAGE CONTROL
$INCLUDE  TACS . LCVC . R1,
C            PROTECTIONS
$INCLUDE  TACS . SCCFP . R1,
$INCLUDE  TACS . DCLP . R1,
C     SANDY POND CONVERTER POLE 1
C     LOCAL CURRENT AND VOLTAGE CONTROL
$INCLUDE  TACS . LCVC . S1,
C            PROTECTIONS
$INCLUDE  TACS . SCCFP . S1,
$INCLUDE  TACS . DCLP . S1,
C            AC/DC  SYSTEM DATA
$INCLUDE  ,  SYSTEM . DATA,
```

Fig. 2 EMTP structure of data file

The use of FORTRAN code implies that the user, in certain circumstances, may be obliged to recompile the source code to modify the control system. To minimize this, a user's input data file is created so that the user can change parameter values from run to run. For example, in this multiterminal bipolar scheme, if a two terminal monopolar system is required to be modeled, the simplest way is to transfer a variable relating to the number of poles required from TACS to FORTRAN. This variable, "npcode", is then decoded in the FORTRAN subroutine to call in only the required poles. The first two digits of the variable indicate that Radisson converter pole 1 is in service whereas pole 2 is out of service, the next six digits that Nicolet, Des Cantons and Comerford stations are out of service, and the last two digits that Sandy Pond pole 1 is in and pole 2 out of service. This provides a full description of all possible multiterminal configurations. A brief example corresponding to a TACS.userIN data file is shown in Fig. 3.

```
C                 .........RADISSON POLE 1
C                 . .........NICOLET POLE 1
C                 . . ......DES CANTONS POLE 1
C                 . . . ...COMERFORD POLE 1
C                 . . . .  SANDY POND POLE 1
C                                .
88NPCODE  $       1000000010
C                  ......RADISSON POLE 2
```

Fig. 3: TACS UserIn data file

FORTRAN subroutine tacsfo, as seen in Fig. 4, reads "npcode" and executes, as often as needed, the control and protection functions associated with each pole declared to be in service. All variables in the FORTRAN subroutines are vectors having a dimension of 10, corresponding to the total number of poles in the system; each number from 1 to 10 of the index represents the value associated with a particular converter and pole.

```
      SUBROUTINE TACSFO
      NPCODE = USERIN ( ' NPCODE ' )
      DO 10 NP = 1, 10
      IF ( MOD(NPCODE/10**(10 - NP ) , 2 ) ) THEN
         CALL LCVC
         CALL PROTECTIONS
      END IF
   10 CONTINUE
      RETURN
      END
```

Figure 4: FORTRAN subroutine calling different control and Protection functions.

AC AND DC NETWORK MODEL

The network data for both the real-time simulator and the EMTP simulation are stored in a common database in a computer. An interface program can be used to convert the data into an EMTP input file. This in-house software facilitates comparison between EMTP and simulator study results, since errors in network data used for EMTP and simulator studies are eliminated. Because of the complexity of the networks and the large number of elements involved, this method helps the user to avoid mistakes in data manipulation and accelerates data input file creation.

For the results presented in this paper, the ac system representation on the rectifier side has been simplified into a 2 resistor and one inductor circuit. The short circuit level on this side is 7902 MVA. But the ac system on the inverter side is represented in much more details (using multiport network) so that the possibility of having commutation failures is reproduced accurately so that comparison can be done with the simulator results. The equivalent short circuit level on this side is approximately 11000 MVA. Space limitations prohibit inclusion of the multiport network diagram. But it is available on request to the authors. The Power Line Carrier (PLC) circuit is modeled by an

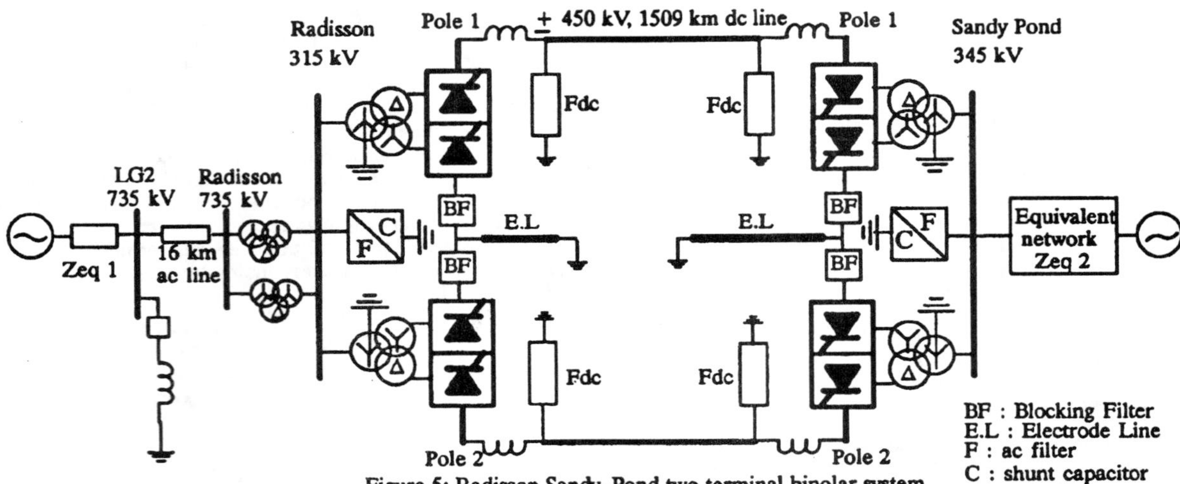

Figure 5: Radisson Sandy–Pond two terminal bipolar system

equivalent impedance while the ac filters have been represented in detail. The dc system is also modeled in detail and includes distributed parameters lines, dc filters, blocking filters, etc. A two terminal bipolar scheme (Fig. 5) is used to compare EMTP results to simulator and field test results. Some of the network parameters have been published [1–2; 8], others are given in the Appendix.

INITIALIZATION

In a real system, converter start-up takes many seconds to reach rated current. The converter current ramp-up time to 10% of rated current is approximately 10 s. As far as the EMTP model is concerned, the time to reach steady state at rated current should be minimized in order to save CPU time. Consequently, the converter start-up procedure has been modified to speed up initialization to approximately 200 ms.

SIMULATION RESULTS

The results presented in this paper cover some basic control performance tests aimed at validating the system dynamic behavior. Since the paper length is limited only results from some typical tests are presented. Basically, two types of comparisons have been made:

COMPARISON WITH SIMULATOR RESULTS

Since simulator results are easier to obtain, most of our comparisons were done with the simulator tests. Mainly, the EMTP results are validated against identical tests on the dedicated real-time simulator of IREQ with the following tests:

- Three phase fault at Radisson 315 kV bus (rectifier side)
- Single phase fault at Sandy Pond 345 kV bus (inverter side)
- Monopolar dc line-to-ground fault at Sandy Pond (inverter side)

The comparison is done by superposition of EMTP results (full line traces) on the simulator results (broken line traces). These tests are shown in figure 6, 7 and 8 respectively. It is shown from these traces that excellent agreement between results of the two tools are obtained. Some differences were found, for instance in the trace of the current order (variable Iorder Radisson, pole 1). After closer examinations, it was found that the measurement of this variable from simulator has a small time delay which clarified the difference.

COMPARISON WITH FIELD TEST RESULTS

After proving the accuracy of numerical simulation (EMTP) from comparison with physical simulator test results, the digital model has been used to reproduce a case of instability obtained from a test done on site. The test consists of switching on a shunt reactor of 165 MVAR at the LG2 735 kV bus as shown in figure 5. At the switching moment, short time imbalance in the three phase voltages (due to the non-simultaneous closing of the three poles of the breaker) creates a dc offset in the neutral current of the nearby power transformers. This, added to the inrush current into the inductor, results in an appreciable component of 2nd harmonics in the ac feeding voltage to the converters. If this component is sustained long enough, the system could develop into instability as shown by traces of current and voltages (Ud, Id) obtained from field test in figure 9. The same test repeated on EMTP with the digital model reproduces surprisingly the same phenomenon despite the fact that no data has been given either for the shunt reactor or for the closing times of the three poles of the breaker. Typical values of 1.4 p.u knee level and a 40% saturation slope have been assumed for the shunt reactor. The breaker closing of the three poles has been done simutaneously on the EMTP simulation. It is to be noted that on the field test, the dc system is blocked at about 1300 ms to avoid eventual damage on the equipment (but not the EMTP simulation).

COMPUTER RUNNING TIME

With the detailed system simulated in figure 5 and a very detailed model of the dc digital control system, the running time to produce a simulation of one second using a time step of 23 µs (0.5 electrical degree based on 60 Hz system) with a fairly large amount of ouput variables takes approximately 14 hours of CPU time on a SPARC station 1. No attempt of "snap shot" (option START AGAIN in EMTP) has been tried to save the running time (this could conservatively cut off 1/3 of the running time). It is to be noted that the digital control modeled is extremely large: the executable file of the EMTP modified version once optimally compiled takes 4.9 megabytes in which at least 1 megabytes of disk memory is attributable to the digital control model. New trend of computer with faster work station such as SPARC station 2 or even super computer could cut down the running time by a factor of 2 or more. Even in the present state, the model provides a excellent tool to study transient phenomena of the system with accuracy and provides a complementary analysis tool for the simulator in many circumstances.

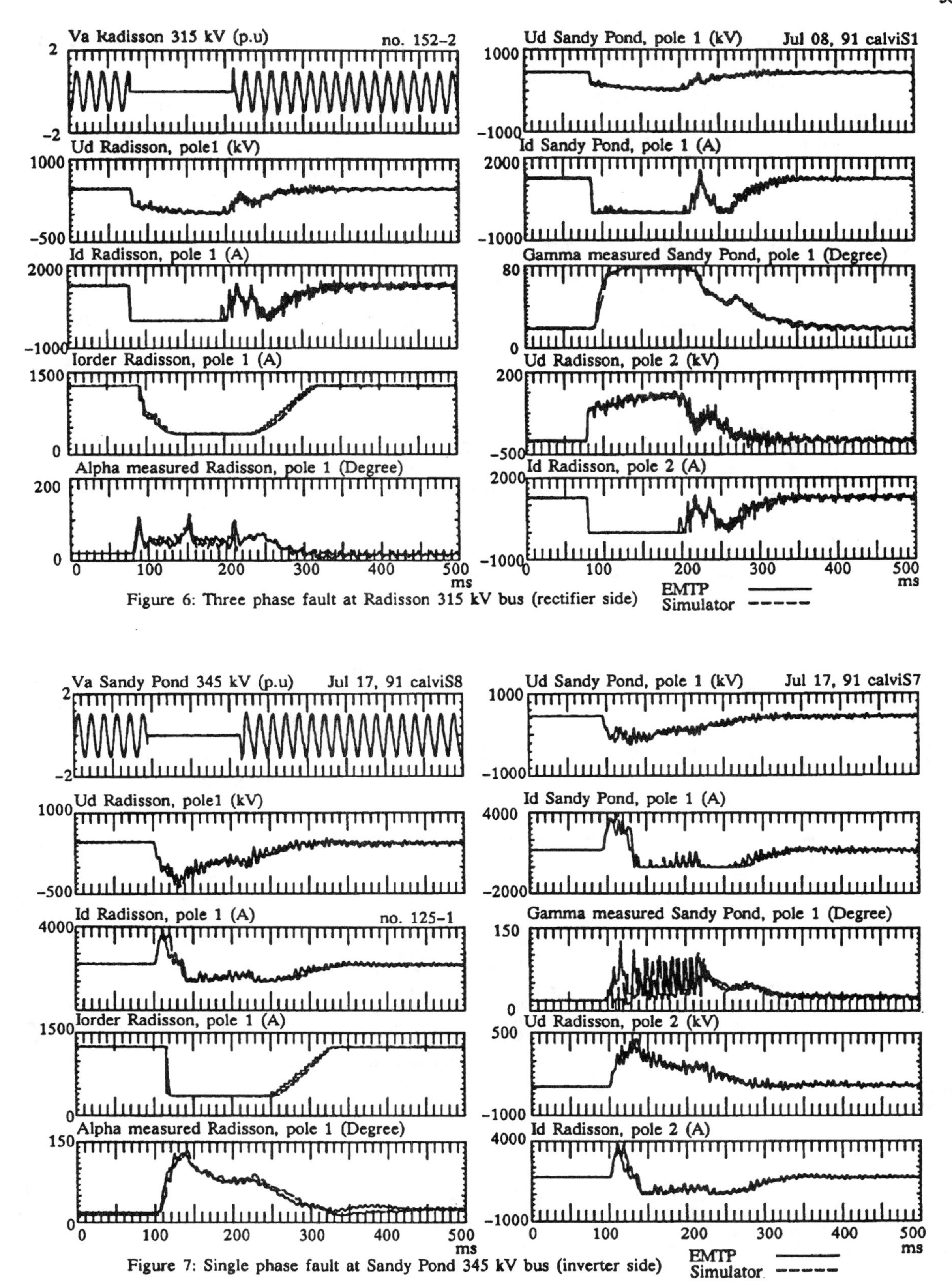

Figure 6: Three phase fault at Radisson 315 kV bus (rectifier side)

Figure 7: Single phase fault at Sandy Pond 345 kV bus (inverter side)

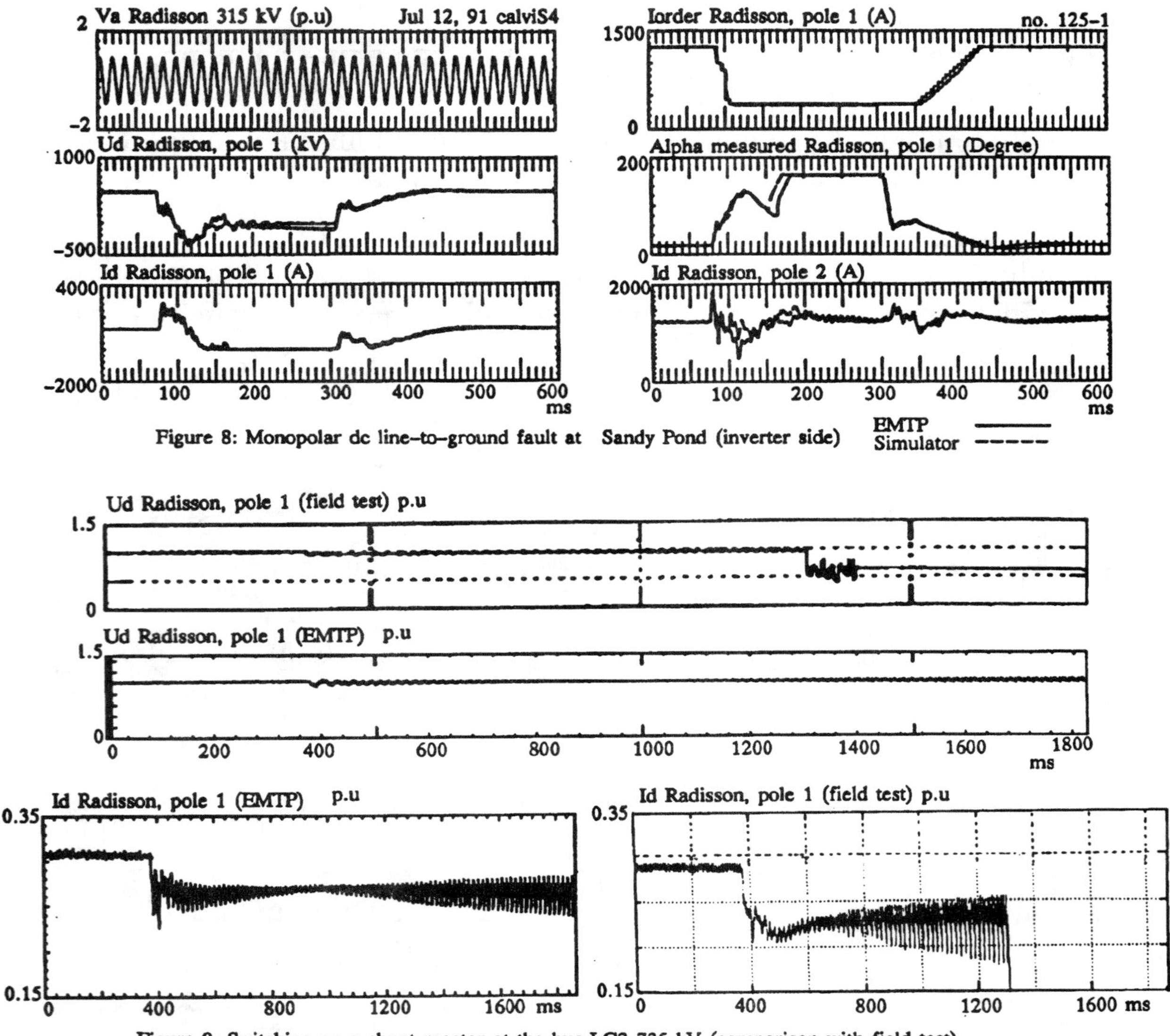

Figure 8: Monopolar dc line-to-ground fault at Sandy Pond (inverter side)

Figure 9: Switching on a shunt reactor at the bus LG2 735 kV (comparison with field test)

COMPILATION OF FORTRAN CODE

EMTP source code and FORTRAN subroutines source code are catalogued in completely separated libraries so that the EMTP source code is not to be recompiled every time a change in FORTRAN subroutines is required. Moreover, the FORTRAN subroutines could be catalogued in sub-libraries and only the ones with modifications are recompiled. Usually, such recompilation is very quick taking often less than two minutes of CPU time. Linking of all libraries to produce an executable EMTP version requires approximately 2 minutes. The very large case studied takes about 5 minutes in total to produce a new EMTP executable file thanks to the structure of library and catalogue in FORTRAN which is transparent to the users.

FUTURE WORK

The modeling work completed so far covers the principal features of the control system. Consequently, all the functions forming the LCVC have been represented without any simplification and all basic protection functions are modeled. Further modeling work is still needed to fully accommodate multiterminal operation with and without telecommunications. The second stage of modeling will be different in that only essential functions will be represented. It is clear, nevertheless, that any supplementary control loop can be implemented to cover a specific application. Future work will focus on modeling the remaining parts of pole control, bipole control and master control for the multiterminal operation. Another important aspect of future work will be model validation with complex ac systems or with low-frequency resonances close to the 2nd harmonic. Also, synchronous machine dynamics will be modeled to cover the islanded mode of operation of the multiterminal scheme. The need to accelerate initialization of the ac/dc system simulation becomes more crucial in the multiterminal scheme.

CONCLUSIONS

Since this is the first time in Hydro-Quebec, a commercial HVDC digital control system has been modeled using EMTP, and in anticipation of application to a real system, it was important to validate the model. Comparisons have been done showing good agreement of EMTP, simulator and field test results during the commissioning period confirming the accuracy of the digital control model and provides encouragement to the use of EMTP for power system transient analysis.

ACKNOWLEDGMENTS

The authors wish to express their thanks to P. C. S. Krishnayya, D. McCallum and G. Sybille at IREQ for their help, encouragement and simulator test results during this work and Lesley Kelly-Régnier for editing of the text. Special thanks are also due to D. Soulier, G. Croteau, G. Moreau and O. Bourgault from Hydro-Québec for providing traces of field test results and opportunity to validate the EMTP simulation.

REFERENCES

[1] M. Hegi et al, "Control of the Québec-New England multiterminal HVDC system," CIGRÉ, 1988 session, paper No. 14-04.

[2] M.P. Bahrman, G.A. Sweezy, "Quebec - New England Phase II HVDC transmission system, Part I : Steady-state multiterminal control strategy," CIGRE Symposium on AC/DC Transmission Interaction and Comparisons, Boston, Sept. 1987.

[3] M.P .Bahrman, D.E. Martin, G. A. Sweezy, "Québec-New England Phase II HVDC transmission system, Part II: " multiterminal control strategy for dynamic performance," CIGRÉ Symposium on AC/DC Transmission Interaction and Comparisons, Boston, Sept. 1987.

[4] L. Dubé, H.W. Dommel, "Simulation of Control System in an ElectroMagnetic Transient Program with TACS," IEEE Transactions on Power Industry and Computer Applications Proceedings, 1977, Publication No. 77 CH 1131-2-PWR, pp. 266-271

[5] D. Goldsworthy, J. J. Vithayathil, "EMTP Model of an HVDC Transmission System," Proceedings of the IEEE Montech '86 Conference on HVDC Power Transmission, September 29 - October 1, 1986, pp. 39-46.

[6] J.Reeve, S. P. Chen, "Versatile Interactive Digital Simulator based on EMTP for AC/DC Power System Transient Studies," IEEE Transactions on Power Apparatus and Systems, Vol. 103, No. 12, pp. 3625-3633, December 1984.

[7] L. X. Bui, G. Morin, S. Casoria, J. Reeve "EMTP TACS-FORTRAN Interface Development for Digital Controls Modeling," Paper 91 SM 417-6 PWRS, IEEE Summer Meeting, San Diego, July 1991.

[8] M. Granger, M. Tessier "Description du Système Multiterminal Radisson-Nicolet-Des Cantons et des Principaux Choix d'Arrangement des Postes Convertisseurs," Paper 91presented in IEEE MONTECH conference proceedings, pp. 8-12, Montreal, Sept. 1991.

BIOGRAPHIES

L.X.Bui (S.M, 85) - graduated with a B.A.Sc. from Ecole Polytechnique de Montréal, in 1975 and received a M.Eng. degree from McGill University, Montréal in 1978. From 1975-1978, he was with the Planning Department of Hydro-Quebec. Since 1979, he has been at IREQ. His current interests are in HVDC transmission and SVC systems. He is currently Chairman of the IEEE Power Enginnering and Industry Applications, Montreal Section.

G. Morin (M, 85) - received his B.Sc.A in electrical engineering from University of Sherbrooke in 1978 and his M.Sc.A from Ecole Polytechnique in 1983. He is with Hydro-Québec since 1978 working in the Power System Operation Department. His main interests are harmonics, dc systems and power system transients. He is author of publications ralated to harmonic overvoltages during power system restoration. He is member of IEEE Power System Restoration Task Force, IEEE Switching Surge Group, IEEE Working Group 15.15.02 and Chairman of the Frequency Dependent Equivalents Task Force.

S. Casoria (M, 78) - received the B.Ing. and M.Sc.A degrees in Electrical Engineering from Ecole Polytechnique de Montréal in 1977 and 1981 respectively. Since 1981, he is with Institut de Recherche d'Hydro-Québec (IREQ). From 1988 to 1989, he was a trainee at Asea Brown Boveri (ABB) plant in Ludvika, Sweden. During this period, he developed software subsystems for the Québec-New England HVDC multiterminal transmission system. His area of professional interest include HVDC transmission and control systems.

J. Reeve (F, 81) - received the B.Sc., M.Sc., Ph.D and D.Sc. degrees from the University of Manchester, England. After employment with the English Electric Company, Stafford, involved in the development of protective relays, and 6 years as a faculty member at the University of Manchester Institute of Science and Technology, he has been with the University of Waterloo, Ontario, Canada since 1967, currently a Professor in the Department of Electrical and Computer Engineering. His research interests for 30 years have been centered mainly on aspects of dc power transmission. He is a Past Chairman of the IEEE dc Transmission Subcomittee and is currently a member of several IEEE and CIGRE Working Groups and task forces concerned with dc transmission. Dr. Reeve is President of John Reeve Consultants Ltd. He recently completed a one year assignment at IREQ concerned with the simulation of the Hydro Québec-New England multiterminal dc system.

APPENDIX

1- The equivalent network at Radisson (short-circuit level = 7902 MVA) is:

2- The dc line impedance is:
Z0 = 0.33530000 + j 1.0274000 Ohms / km
B0 = 3.6805999 micro-mhos
Z1 = 0.01130000 + j 0.2984 Ohms / km
B1 = 5.4254 micro-mhos
Length = 1509.5 km

3- The 735-315 kV transformer characteristic is:
2 units of 1650 MVA each, Yg-Yg-Delta connection
Knee level = 1.2 p.u
Leakage inductance = 12.3 %
Saturation slope = 40.0%C Elements de type Tr

4- The Radisson converter transformer characteristic is:
2 units of 1212 MVA each, Yg-Y-Delta connection
Knee level = 1.2 p.u
Leakage inductance = 24.0 %
Saturation slope = 37.0%C Elements de type Tr

5- The Sandy Pond converter transformer characteristic is:
2 units of 1060 MVA each, Yg-Y-Delta connection
Knee level = 1.2 p.u
Leakage inductance = 22.0 %
Saturation slope = 37.0%

6- The short circuit level at Sandy Pond is 11000 MVA.

7- The PLC circuit is represented by a 4 mH inductance

8- The 60 hz blocking filter is:

X = 150. mH
C = 46.9 micro-farad

Discussion

J. Rittiger (Siemens AG, Germany): The authors have presented a very timely subject of comparison between simulator results and digital calculations with excellent correspondence. We also model digital HVDC-control equipment in great detail.

We would like to ask you some additional questions concerning your solution.

(i) What in particular were difficulties that it was necessary to extend the TACS-part of EMTP? (Modelling of sample and hold circuits; realisation of z-transform, etc.)

(ii) Is it possible to use the EMTP control representation also for stability studies without changes?

(iii) How complicated is the DC and the AC network (number of nodes and branches), because the calculation times seems to be long? We use for one of our projects an AC/DC network with 1400 branches, electrical machines and a precise model of digital controls. The calculation time for 1 sec real time on a workstation is approximately 80 minutes with a time step of 100 μsec at 50 Hz. This greater time step is possible because NETOMAC ensures that the continuity conditions are always satisfied using a variable time step.

(iv) How many simulator cases were investigated with the digital model? Was the correspondence for all cases as good as for the cases presented in the paper?

Manuscript received February 18, 1992.

G. Morin, L. X. Bui, S. Casoria, J. Reeve: The authors would like to thank Mr. J. Rittiger for his interest in our paper and for the valuable comments.

For the first question, in modeling digital control with EMTP prior to our development of the TACS-FORTRAN Interface, the principal difficulty is the incapability of TACS to accommodate FORTRAN statement other than a limited number of algebraic and logical FORTRAN expressions and some simple branching devices in TACS. The reason is that TACS has been designed to model mainly analog control type. For more details, please refer to our paper titled "EMTP TACS-FORTRAN Interface Development for Digital Controls Modeling" and presented at the 1991 Summer Meeting in San Diego, USA.

The answer to the second question, referring to the possibility to use EMTP control representation in stability studies without modifications, depends on what control the question is referred to. For HVDC control, this is quasi impossible since this type of control needs three phase commutation voltages for the generation of firing pulses to the converters (Y and Δ) and for some protections such as short circuit and commutation failure. Generally, EMTP HVDC control must be adapted to be used in stability simulation which usually models only single phase network assuming a perfect balance operation. Other type of control such as generator excitor control may be used without modification in stability studies.

For the third question, the simulated network is composed of a 400 nodes, 1000 branches network and a complete, detailed structure of the HVDC multi-terminal control system. Although in the paper's application, only a two terminal, bipolar operation network was used. The simulated HVDC control system includes also protection such as commutation failure, overvoltage protection. In fact, the simulated control in this paper is not only a detailed representation, but also a replica of essential parts of the real digital control system, capable of reproducing the control/network harmonic interaction phenomena, since the microprocessor-based codings of the real digital control system is imported directly into EMTP through the TACS-FORTRAN interface. In this way, there is no question about the validity of the HVDC control representation. This results in an enormous 123800 variables TACS array only for the HVDC control purposes. Another important point which increases the total CPU time is the number of output. Since the development was under testing, over 500 output variables have been requested during the simulation in order to achieve a complete validation of the control system behavior. Moreover, the integration step in our simulation is 23 μs being less than one fourth of your integration step. All these reasons add up to a larger CPU time in our simulation. In our opinion, quadrupling the integration time step and reducing the output variables to a reasonable value will bring down the CPU time close to the number given in the discussion.

Finally, for the last question, there was a large amount of cases investigated in the simulator study. Due to the limitation of pages allowed for the paper, only some important cases were reported. Generally, the excellent agreement between EMTP and the simulator results are well observed in other cases since the HVDC control representation in EMTP is a replica of the real, on site digital control. In some cases, because simulator components are lossy, EMTP results correspond rather better to field tests results than those from the simulator.

Manuscript received February 18, 1992.

IEEE Transactions on Power Delivery, Vol. 5, No. 3, July 1990

A STATIC COMPENSATOR MODEL FOR USE WITH ELECTROMAGNETIC TRANSIENTS SIMULATION PROGRAMS

A.M.Gole, Member
Dept. of Electrical Engineering
University of Manitoba
Winnipeg, Manitoba.
CANADA. R3T 2N2

V.K.Sood, Senior Member
Hydro-Quebec (IREQ)
1800 Montee Ste. Julie
Varennes, Quebec.
CANADA. J0L 2P0

Abstract – A static var compensator (SVC) model based on state variable techniques is presented. This model is capable of being interfaced to a parent (or host) electromagnetic transients program, and a stable method of interfacing to the EMTDC program, in particular, is described. The model is primarily that of a thyristor controlled reactor (TCR) and a thyristor switched capacitor (TSC). Capacitor switchings within the TSC have been handled in a novel way to simplify storage and computation time requirements. During thyristor switching, the child SVC model is capable of using a smaller timestep than the one used by the parent electromagnetic transients program; after the switching, the SVC model is capable of reverting back to a (larger) timestep compatible with the one used by the parent program. Other features that are considered include the modeling of a phase-locked-loop based valve firing system. The paper ends with the discussion of an application of this model in the simulation of a SVC controlling the ac voltage at the inverter bus of a back-to-back HVdc tie.

Keywords : Static compensator model, Transient simulation, Variable timestep, HVdc transmission.

INTRODUCTION

Some of the popular electromagnetic transients simulation programs [1,2] utilize the modeling algorithm proposed by Dommel [1], in which an inductor, capacitor, transmission line or other device is represented as a parallel combination of a resistor and a current source; the values of these depend on the past history of current and voltage in the device. This representation is used to solve for node voltages at any time t, based on known values from the previous simulation instant ($t-\Delta t$), Δt being the simulation timestep. This approach has the great advantage of simplicity, because an entire network can be quickly reduced to one containing resistors and current sources, and from which the network admittance matrix Y is readily constructed. There is no need for the user to actually write down the differential equations associated with the network to study its transient behaviour. There are two major disadvantages of this approach, however.

Firstly, it requires a relatively small timestep to avoid spikes (numerical) whenever switching due to convertor/compensator models within the network takes place. This requires a premium in terms of CPU time for the overall simulation. It is difficult to change the timestep dynamically during the simulation run, because that would mean re-calculation of all resistor values and current sources and a re-inversion of the network matrix. (This may be possible if these electromagnetic transients programs are re-written, and this feature introduced, but that would entail a great deal of effort.) One approach to overcome this limitation is that used by the 'NETOMAC' simulation program [5]. In this approach the timestep is not changed but the history terms in the trapezoidal algorithm are interpolated and modified so that the simulation timestep can be synchronized with the switching instant.

Secondly, since the Dommel algorithm is not state variable based, it leads to a poorer conditioning of the network admittance matrix Y due to its unequal treatment of inductors and capacitors within the network. For example, the equivalent resistance associated with an inductor L is $2L/\Delta t$, and for a capacitor C it is $\Delta t/2C$. Thus, elements in the network admittance matrix for L and C are affected in opposite ways when Δt is reduced leading to a poorer conditioning of the Y matrix. Although this algorithm usually works, at times this drawback may cause the system to show numerical instablity. A state variable formulation, on the other hand, always integrates the differential equations of the system and solves the equations for capacitor voltages and inductor currents, not for node voltages as in the Dommel method. Thus all the variables have the same dependance on the timestep, and the resulting matrices are better conditioned and less likely to give numerical instability. Another advantage of the state variable approach is that the dependence of the system matrices on the value of the timestep is directly proportional, and the timestep can be readily changed during a run. The major disadvantage, however, of the state variable approach is that it is difficult to write code for the automatic generation of the state equations given the network connectivity information; the Dommel algorithm, on the other hand, handles this with ease.

For the above mentioned reasons it was decided to use a compromise approach in modeling the static compensator and its associated external network by developing a stand-alone (child) state variable model for the static compensator, but retaining the powerful network modeling capabilities of the parent transients program for modeling the external network. The compensator model program interfaces as a Norton current source with the transient simulation program. The state variable based static compensator module results in a robust algorithm for calculating the currents and voltages internal to the static compensator, and also allows for features such as the introduction of a variable local timestep required for the switching elements to be modeled adequately. This approach is similar to that proposed by other authors [11] who have interfaced a dc convertor model in an electromagnetics transients program (EMTP) to a Transient Stability Program where the two programs run with different (though not variable) timestep; interfacing between programs was achieved by using Norton and Thevenin equivalents.

The authors' program differs from the NETOMAC approach and merely selects a smaller timestep which is a submultiple of the parent program's timestep. Also, unlike NETOMAC, the authors' approach uses a state variable based formulation which has the advantages mentioned earlier. In addition the authors' approach can be used to develop models which can be interfaced to electromagnetic transients simulation programs (such as EMTP or EMTDC) that do not have the time mesh shifting capability of NETOMAC.

Many authors [3,4] have successfully used state variable based modeling for HVdc convertors, but in their approaches even the associated external systems have been modeled in state variable form. These external systems cannot therefore be easily changed to a different topology (without rewriting the equations) as is possible with the Dommel method.

The following sections of this paper deal with details of modeling the SVC, the method of making a stable interface between the child model and the parent EMTDC (although the same could be done with EMTP, or any other programs based on the algorithm proposed by Dommel) and present some sample results. The paper concentrates on the modeling of the SVC itself because the bulk of the controls are simulated with the control system building blocks (EMTDC CSMF functions or EMTP TACS functions) of the parent program. Some modeling details of the phase-locked-loop (PLL) firing system used here, are also discussed because this is modeled internally in the SVC module. The paper concludes with the simulation of a SVC regulating the ac voltage bus at the inverter end of a back-to-back HVdc tie.

90 WM 078-6 PWRD A paper recommended and approved by the IEEE Transmission and Distribution Committee of the IEEE Power Engineering Society for presentation at the IEEE/PES 1990 Winter Meeting, Atlanta, Georgia, February 4 - 8, 1990. Manuscript submitted August 18, 1989; made available for printing November 17, 1989.

MODELING AND TESTING

The Basic Model

Figure 1 shows a schematic diagram of the SVC. The SVC transformer is modeled by nine coupled windings on the same core with three windings representing the primary winding, and three each representing the wye and delta secondaries. The special case of 3 single phase transformers can also be represented with the proper selection (of zeroes in the appropriate locations) of the inductance matrix.

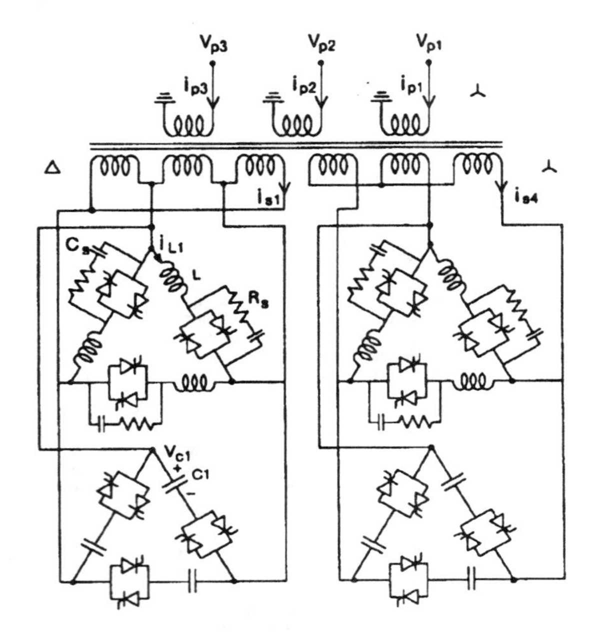

Figure1 : SVC Circuit Diagram.

The TCR elements are connected in delta, with the thyristor switches modeled as changing resistances. The snubber circuits are modeled as R-C elements in parallel with the thyristors.

The TSC branches are modeled as capacitors. Regardless of the number of TSC branches in operation at a given time, all of these are represented together as an equivalent single capacitor per phase. The value of this equivalent capacitor and its initial voltage are adjusted when the TSC switching logic indicates the turning on or off of a capacitor bank. The advantage of this approach is that only one state variable per phase (6 phases in all) is required. With a TSC having many stages, each extra stage would increase the number of state variables by 6 (3 for the wye, and 3 for the delta), and the number of state variables would rapidly become very large. This approach is exact only if the TSC branch is comprized of a pure capacitance. It was decided not to include a series inductance (at this stage) because it would require each arm to be modeled separately. Since the aim was to obtain a model for relatively long duration studies (typically 100 -2000 ms), and since capacitors are switched only a few times in a simulation run, this was a small price to pay in making the program fast.

Saturation of the transformer is represented by flux-dependent current sources in parallel with the transformer windings. The flux is calculated from the integral of the voltage across the winding. A flux magnetizing -current relationship is then consulted to determine the extra magnetizing current that must be injected for that particular flux level. At present, there is no representation of the hysteresis loop, but there is a provision for the representation of some core losses via a selectable shunt resistor across each winding. The details of the equations used are presented in Appendix. These equations are obtained by graph theory techniques [6]. The state variables (variables to be integrated) are the primary, secondary-delta and secondary-wye currents ($i_p, i_{s\Delta}, i_{sy}$); the TSC capacitor voltages ($v_{c\Delta}, v_{cy}$); the TCR inductor currents ($i_{l\Delta}, i_{ly}$); and the snubber capacitor voltages ($v_{sc\Delta}, v_{scy}$) Each of these except $v_{c\Delta}$, v_{cy} and i_{ly} have three state variable components, corresponding to each individual phase. Since the capacitors are connected in delta, the three voltages add up to zero, so that only two capacitor voltages can be chosen as independent state variables. Likewise, in the wye connected transformer secondary winding, the three currents sum to zero, thus allowing for only two independant state variables. One could have introduced a fictitious (large) resistance to ground at the neutral point to make the formulation more symmetrical, with all three currents as state variables, but such solutions often lead to numerical problems. Instead, the algorithm is made more robust by eliminating variables when capacitor loops or inductor cutsets are present. Thus, the SVC model has 21 state variables in all.

The matrices in the equation are functions of the states of the thyristors, and the appropriate entries are re-calculated whenever the firing logic requires the turning on or off of a thyristor. Also, the capacitor voltages and the capacitance values are initialized whenever the TSC operates.

Method of Integration

The trapezoidal rule has been extensively used in electromagnetic transient programs [1]. However, implementing it here would have entailed a matrix inversion every time a switching occurred, or whenever the timestep had to be changed. For this reason, an Adam's second order closed formula [7] was used, the numerical stability of which is identical to that of the trapezoidal rule. Equations 1A, 2A – 5A in the Appendix can be written as

$$\dot{\mathbf{X}} = \mathbf{A}.\mathbf{X}(t) + \mathbf{B}.\mathbf{u}(t) \tag{1a}$$

where $\mathbf{X}$ is the state variable vector, and $\mathbf{u}$ the vector of inputs (primary voltages). To apply the chosen method of integration, the increment in $\mathbf{X}$ is defined as

$$\Delta\mathbf{X} = [\mathbf{A}.\mathbf{X}(t-\Delta t) + \mathbf{B}.\mathbf{u}(t)]\Delta t \tag{1b}$$

Then the state vector estimate is updated as

$$\mathbf{X}_{e1}(t) = \mathbf{X}(t-\Delta t) + \Delta\mathbf{X} \tag{1c}$$

Equation 1b is re-evaluated with $\mathbf{X}(t-\Delta t)$ replaced by $\mathbf{X}_{e1}(t)$, to find a new update $\mathbf{X}_{e2}(t)$, which is used to re-evaluate $\Delta\mathbf{X}$ again, until there is negligible change in the updated $\mathbf{X}_{en}(t)$. In practice, for the range of timestep values used (25 – 50 μs), 3 iterations are found to be sufficient. Notice also, that no matrix inversions are involved. An application of the trapezoidal rule would have involved only one evaluation intead of 3, but would have required a matrix inversion, every time the value of matrix A changed with a switching or change of timestep.

Variable Timestep for handling thyristor switching

Since the simulation algorithm works with discrete timesteps, the zero crossing of a current often falls in-between two timesteps as in Figure 2a. When the turnoff of a thyristor in series with a current carrying inductor (such as in a TCR) is being simulated, there is the distinct possibility of a spurious voltage spike appearing in the thyristor and inductor voltages. Figures 3a(i) and (ii) show simulated TCR current and voltage waveforms respectively for a simple SVC system (Figure 6), with timestep Δt = 50 μs.

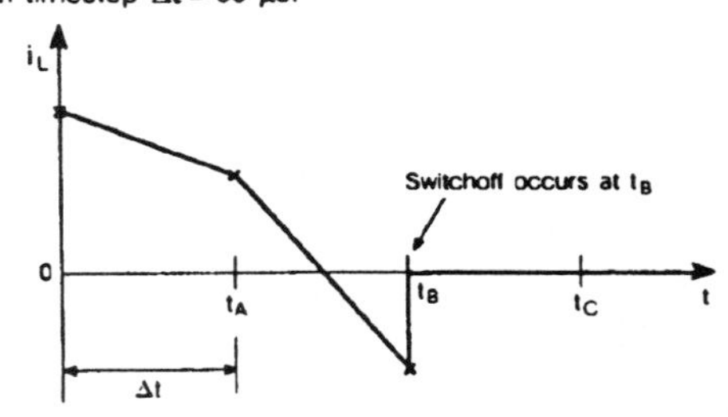

Figure 2a : Thyristor switchoff with Fixed Timestep.

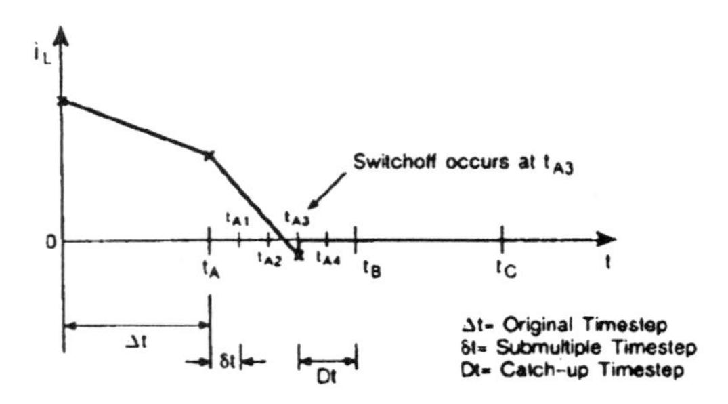

Figure 2b : Thyristor switchoff with Variable Timestep.

Many solutions to this problem are possible, and one of them includes increasing the snubber capacitance but at the expense of excessive snubber circuit losses. Another way is to reduce the timestep, but this entails a heavy computer CPU time penalty, particularly because the smaller timestep may not be necessary while simulating the transient in between switching instants.

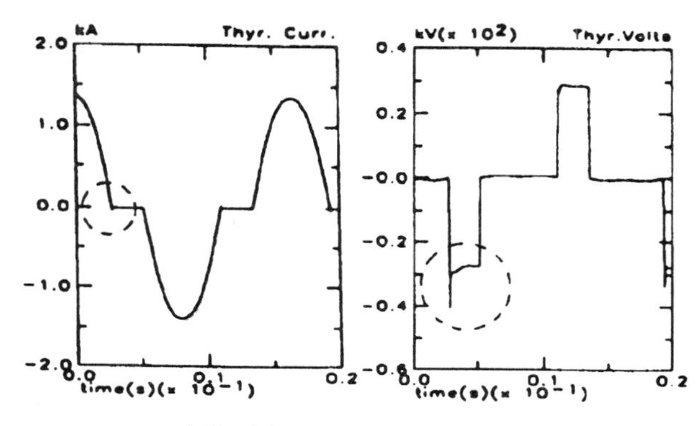

a) Fixed timestep with Δt = 50 μs.

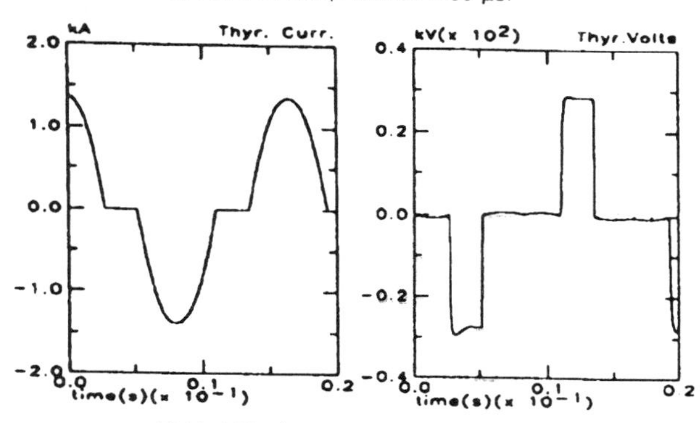

b) Variable timestep with Δt = 50 μs and δt = 10 μs.

Figure 3 : TCR waveforms for fixed and variable timesteps.

The authors' approach to this problem is to use a smaller timestep where required (Figure 2). Consider that the program has been operating with timestep Δt (Figure 2a). At point t_A, the current in an inductor i_L is positive, but at the next simulation instant t_B, it is negative. If a thyristor in series with the inductor attempted to switch off at either t_A or t_B, a voltage spike would result because of the chopping of a large inductive current. When a change of sign in the current value i_L is detected, the program does not update the state variables, but reverts back to point t_A, and resumes simulation with a submultiple timestep $\delta t = \Delta t/n$ (points $t_{A1}, t_{A2}, \ldots, t_{An}$ etc.). The switching instant is now more finely straddled depending on the number of submultiple timesteps; a value of n between 2 to 5 is recommended. The thyristor switches off at time (say) t_{A3}, which results in a smaller spurious voltage spike because the current magnitude being chopped, is smaller. The program then takes one more 'catchup' step Dt to come back in synchronism with the original sampling instant t_B. The values of Δt and δt are under the user's control. Figure 3b shows results from a re-simulation of the test but with Δt = 50 μs and δt =10 μs; the voltage spike is no longer evident.

In the proposed method, the smaller timestep is used only where required. Since switchings within the child program are being more precisely simulated by this smaller timestep, it is more economical than changing the timestep for the entire program (parent plus child). In the case of EMTDC or EMTP (standard versions), it is not possible to use variable timestep anyway. As an assessment of the benefits and overhead of the use of the variable timestep strategy, the simulation time for a 100 ms run increased from 40.90 s to 43.26 s or roughly 5.5%, which is much smaller than the 500% increase in time which would have resulted if the timestep had been set to 10 μs throughout the run.

Modeling the Firing System

The external control system controlling the compensator generates two signals: i) a firing angle order for the TCR, and ii) a capacitor switch on/switch off order for the TSC. The exact instant of valve firing for the TCR is determined when a reference angle derived from a phase locked loop (PLL) equals the ordered angle. A built-in dq0-transformation based PLL, referred to as of the 'Transvektor' type [8] is provided as the default PLL. The user may replace it with his own model. To do this, the user must build his PLL model by selecting building blocks from the parent program control system functions, and provide12 ramps, in the correct time sequence, for firing the SVC thyristors.

Figure 4 shows a block diagram of this type of PLL. The three phase synchronising voltages derived from the commutating bus are Va, Vb and Vc. Using a 3-phase to 2-phase transformation, the direct and quadrature axes voltages, Valpha and Vbeta respectively, are derived according to the following equations:

$$\text{Valpha} = (2/3)\text{Va} - (1/3)\text{Vb} - (1/3)\text{Vc} \tag{2}$$

$$\text{Vbeta} = (1/\sqrt{3})(\text{Vb} - \text{Vc}) \tag{3}$$

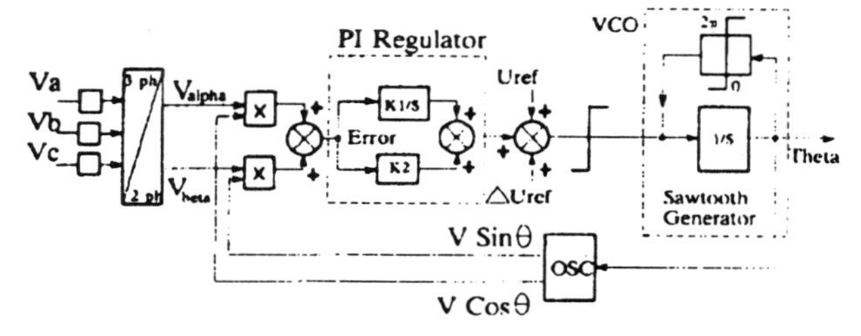

Figure 4 : Default built-in PLL.

An Error signal is generated according to the following equation:

$$\text{Error} = \text{Valpha}.\text{Vsin}\Theta - \text{Vbeta}.\text{Vcos}\Theta \tag{4}$$

where Θ is the phase output of the VCO. The Error signal is acted upon by a PI controller with proportional gain K1 and integral gain K2. This is followed by a VCO to derive a control signal Theta Θ for a Sine-Cosine Oscillator; the nominal frequency of the VCO is controlled by a reference voltage Uref; dynamic modulation of this reference voltage can be accomplished by the input ΔUref. The outputs of the Sine-Cosine Oscillator, VsinΘ and VcosΘ, are fed back to derive the Error, as indicated in equation (4). The output of the VCO, which is limited between 0 and 180 degrees, generates the timing Sawtooth waveform Θ (derived as in eq.5), which is utilised to derive the firing pulses for the valves of the compensator/convertor.

$$\Theta = [(\text{K1}/s + \text{K2})(\text{Error}) + \text{Uref} + \Delta\text{Uref}]/s \tag{5}$$

Twelve ramps for the 12 thyristor firings are generated from this basic ramp, and each is compared with the firing angle order. Thus it is even possible to modulate the firing angle order to provide controlled individ-

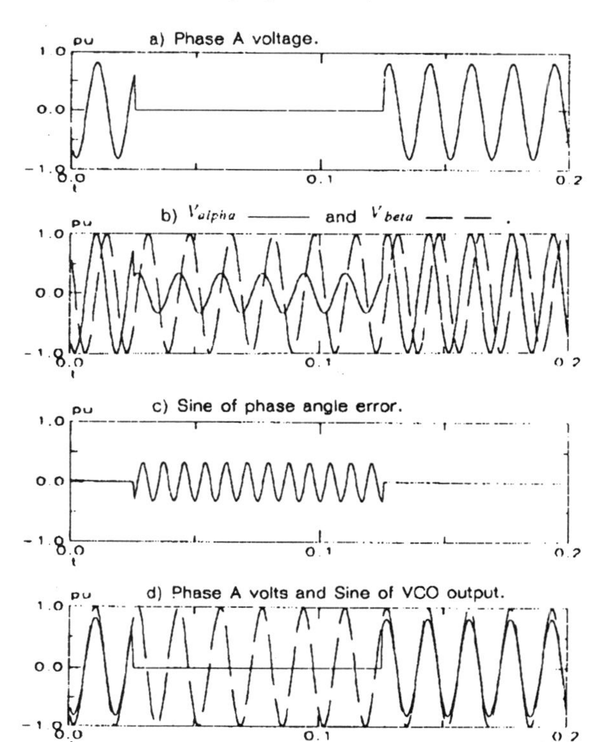

Figure 5 : PLL performance under single line to ground fault.

ual phase firing such as may be required for, say damping a given harmonic in a system [9]. The PLL also shows improved immunity to system harmonics and rapidly regains a lock on phase following faults causing loss of synchronising voltage [13].

Figure 5, is a typical simulation of this PLL where a single phase to ground fault on Phase a reduced Va to zero. The following results are presented in Figure 5: (a) Va Single Phase AC voltage input signal (b) Valpha and Vbeta signals, (c) Error signal as defined by eq.(4), (d) Va(fund) and VsinΘ signals. The fault resulted in a reduced magnitude for Valpha to one third of its original value, and no change in Vbeta. A second harmonic component is observed in the Error signal. These results are consistent with theoretical predictions. The synchronization of the fundamental component of Va, Va(fund), and VsinΘ signals is rapid and achieved in less than one cycle after the fault is removed (Figure 5d). Two-phase and three phase faults gave similar results.

Modeling the Switching of Capacitors

As discussed in the introduction, a novel method of modeling the thyristor switched capacitors has been used. The order of the model would quickly increase if each of the capacitor stages were modeled separately (6 new state variables per extra branch). Instead the switching on/off of a capacitor bank is carried out by re-adjusting the initial charge on the capacitors, and changing their capacitance values. This approach is exact only if there are no series inductances which are often present in practical TSCs. This simplification was deemed a worthwhile tradeoff between modeling detail and simulation speed since the primary objective of this model was to study power system transients.

Capacitors are switched on when the voltage difference between the system and the capacitor to be switched is a minimum. Capacitors are switched off only at a current zero.

Figure 7 shows (a) the 3-phase currents in the primary of the SVC transformer, (b) 3-phase secondary voltages of SVC transformer, (c) 3-phase capacitor currents in the delta TSC for the switching on/off a capacitor bank, (d) the reactive power supplied by the TSC together with the capacitor switching order (switch on at 20 ms and switch off at 60 ms), and (e) the capacitor votage in the simple test system of Figure 6. Note that the TCR pulses have been blocked so that only the transients associated with the capacitor switching are present. From Figure 7e, it is evident that a capacitor switch on occurs at a voltage zero (at 20 ms); also a capacitor switch off occurs at a current zero (Figure 7c) which corresponds to a maximum capacitor voltage (at 60 ms). The voltages on switched off capacitors are caused to decay at a user specified rate (barely evident in Figure 7e). The detailed equations for capacitor switching are presented in the Appendix.

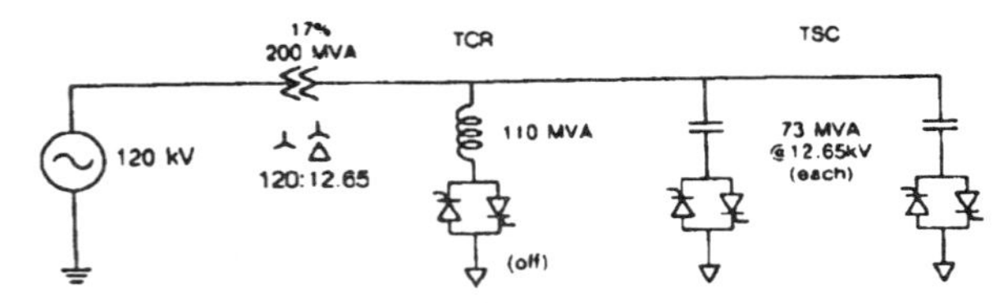

Figure 6 : Test system for capacitor switching.

Interfacing to the parent Simulation Program

The child SVC model is a separate stand alone subroutine written in FORTRAN. Consequently, features such as the variable timestep etc., can readily be introduced without impacting on the parent program. The only drawback is that the information between the SVC model and the parent program is exchanged at the end of a parent program timestep, and the calculations within that timestep by the SVC model and the parent program each have to work with one timestep old information about the other. Care is necessary to avoid numerical instability when interfacing the SVC model to the parent program. This instability may arise because the model interfaces to the parent program as a current source whose value is not affected by the results of the calculations in the parent program until one timestep later. Therefore for the duration of the present timestep, this current source appears like an incremental open circuit. Such 'open circuit' terminations have a destabilizing influence on the simulation particularly if they occur at the end of inductive branches [2]. To get around this problem, a technique used before with success in interfacing machines and HVdc valve groups to transient simulation programs [2,10] is employed. Figure 8 shows the SVC current for some phase $I_s(t)$ calculated by the model for injection into EMTDC. A fictitous resistance to ground R_c is introduced in the main program,so that there is no longer an open termination at this node and the numerical instability problem can be avoided.

The value for R_c is chosen to approximate the very short term behaviour of the SVC (for example $R_c = 2l''/\Delta t$ where l'' is the zero sequence inductance of the SVC transformer). The exact value chosen is not too important because an extra current i_c is injected to compensate for any errors introduced by R_c. This compensation current is estimated from the most recent (and therefore stale by one timestep) voltage information available to the SVC model. If $i_c = V(t-\Delta t)/R_c$ is added to $I_s(t)$ and injected into the parent program, a current $V(t)/R_c$ bleeds into R_c and the current entering the rest of the system is $I_s(t) + [V(t-\Delta t) - V(t)]/R_c$. Note that the second term of this current vanishes in the limit $\Delta t \to 0$, and so the current has a value very nearly equal to $I_s(t)$, as was intended.

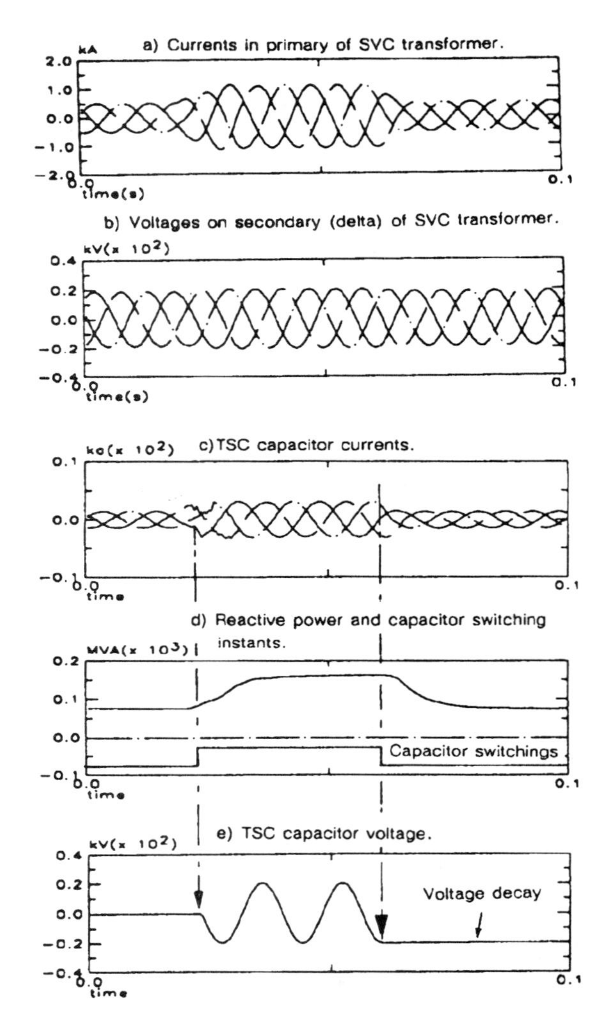

Figure 7 : Capacitor switchings in system of Figure 6.

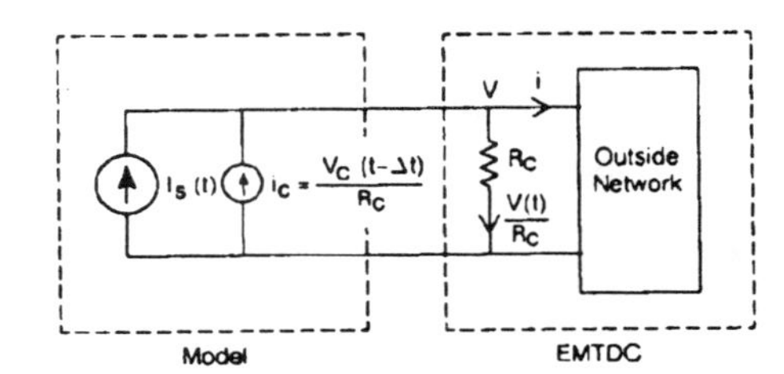

Figure 8 : Interfacing to the EMTDC program.

In addition to this termination the phase voltages applied to the SVC model can also be estimated for the next timestep. For example if Va, Vb, and Vc are the most recent instantaneous values (with the common zero sequence removed), then the SVC model uses Va' instead of Va where

$$V_a' = V_a + [w.\Delta t(V_c - V_b)][1 - (w.\Delta t)^2]/1.732 \quad (6)$$

where w is the system frequency and Δt the timestep. This small correction (which vanishes as $\Delta t \rightarrow 0$) is a good estimate of the expected value of Va one timestep later, and further reduces errors associated with the one timestep delay.

USE OF THE MODEL IN A SIMULATION STUDY

Figure 9 shows a 500 MW pole of a back-to-back HVdc tie between sending and receiving end systems of short circuit ratios (SCR) 4.5 and 2.5 respectively. (The tie rating is based on one pole of the Chateauguay link in Quebec). The rectifier side ac bus is rated at 315 kV and has 250 MVAR of harmonic filters, and the inverter ac bus is rated at 120 kV and also has 250 MVAR of filtering. In addition, a SVC rated at +167 (with 2 stages of switched capacitance) MVAR/-100 MVAR is present at the inverter ac bus to regulate the voltage.

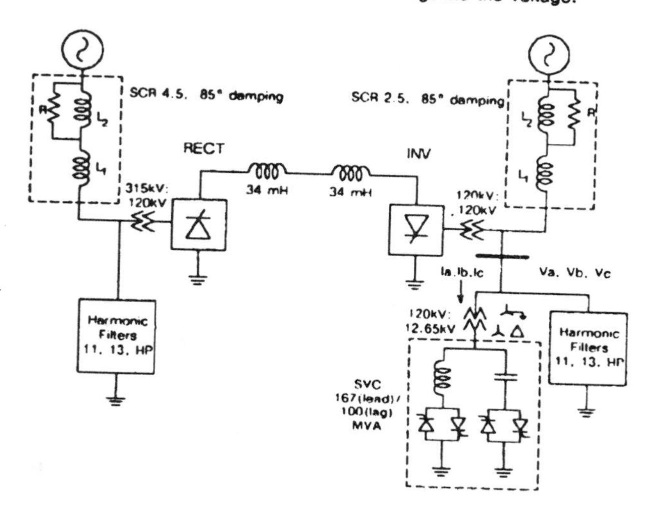

Figure 9 : Back-to-back dc tie with SVC.

The block diagram of the SVC controls is shown in Figure 10a [12], and a steady state control characteristic typical of SVC model is shown in Figure 10 b. The SVC controls are identical to the compensator used at Chateauguay. It consists of a three phase bus voltage measurement block and a three phase current measurement block. These measurements are used to indicate the reactive power Q_{SVC} of the compensator. The magnitude of the voltage measurement V_L is used to derive a droop. This is followed by a block that adds the droop, proportional to the reactive current of the SVC, to the magnitude of the measured voltage. This signal is then filtered. The error between this filtered signal and the reference bus voltage V_{ref} is passed through a PI controller that results in a reactive power order (BSVS) to the SVC. This order is split by the 'allocator' into a capacitor on/off signal for the TSC and a reactive power demand (BTCR) from the TCR. Hysteresis between capacitor stages is built into the model. A relationship representing the non-linear dependence between BTCR and the required firing angle α is consulted and the required α-order passed on to the PLL-based firing system in the SVC model. All these control blocks are modeled with the control system modeling facilities of the parent program (CSMF in EMTDC). In addition there are the usual set of controls (based on Chateauguay tie) for the dc system which are similarly modeled but not described here. The HVdc convertors and the SVC are modeled as 12 pulse valve groups.

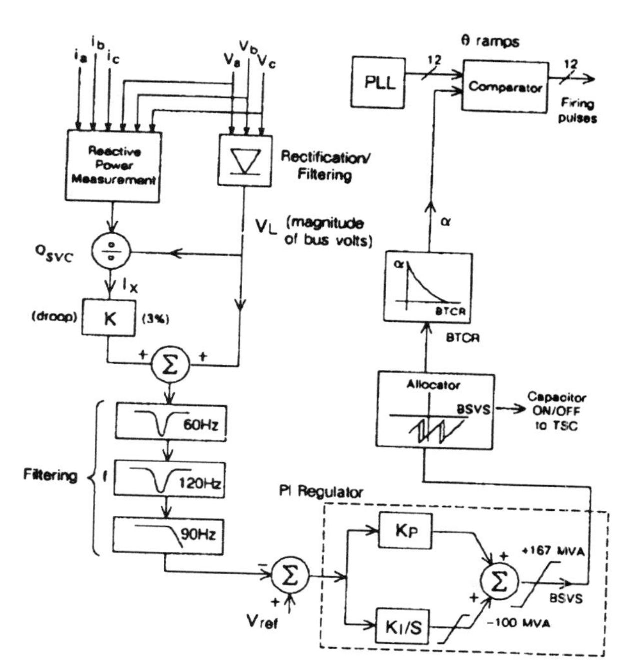

Figure 10a : SVC Controls.

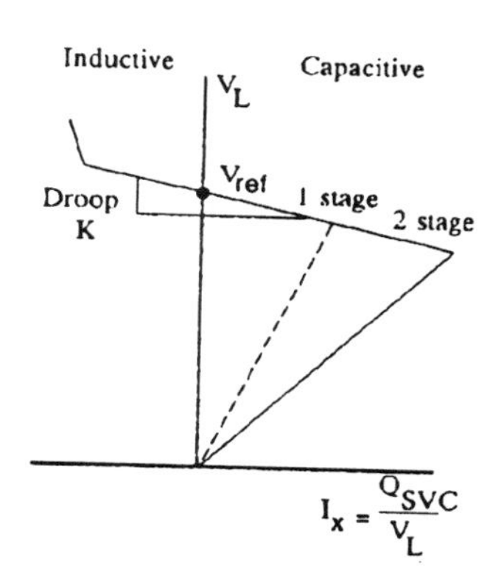

Figure10b : SVC Regulation Characteristic.

Figures 11 & 12 show the system's response to a 8% voltage reduction of the equivalent source on the inverter side; this perturbation was selected to show essential features of the SVC behaviour. This perturbation is meant to simulate a 12 cycle remote three phase ac fault. During the fault period (0.05 s to 0.25 s in the simulation), the SVC switches on one more capacitor bank (Figure 11g) in order to regulate the inverter ac voltage (Figures 11a and b). On removal of the undervoltage at 0.25s there is a subsequent transient overvoltage which is again regulated by the TSC switching off a capacitor bank and the TCR generating the appropriate inductive current (Figures 11 f,e). Also shown in Figure 11 are total demanded reactive power from the SVC (Figure 11c) and the TCR firing angle α (Figure 11d). Figures 11e & f show the TCR and total SVC currents. The step change in TCR firing angle α when a capacitor is switched is a result of the action taken by the allocator (Figures 11d & g). Figures 12a, b and c show the response of the dc system variables (dc current, dc voltage, rectifier firing angle and inverter extinction angle). The ramping down of the ordered dc current during the undervoltage is a consequence of certain dc control functions which are not discussed here.

The computer CPU time for the above example on a SUN 3/60 workstation with 8 MB memory is approximately 45 minutes for 1 second of real time using a 25 μs timestep. The simulation time for the SVC with a simple external system (similar to one shown in Figure 6) was 7 minutes 13 seconds for 1 second of real time using a 50 μs timestep.

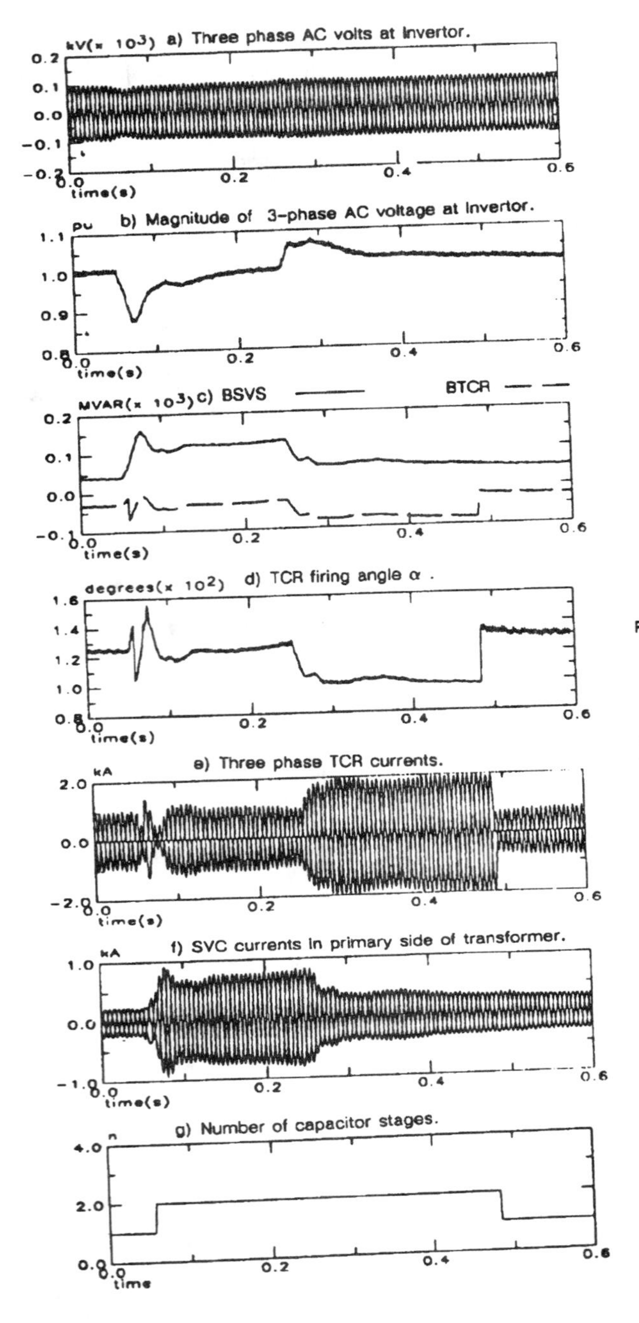

Figure 11 : SVC waveforms for simulation study.

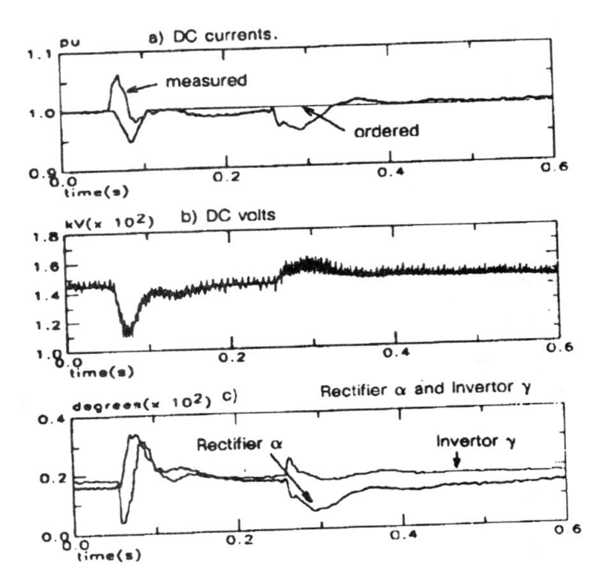

Figure 12 : DC system waveforms for simulation study.

CONCLUDING REMARKS

The state variable technique is a powerful tool for modeling power electronic circuits, even though accomodating topological changes in these circuits means rewriting the state equations. The Dommel algorithm (used in popular electromagnetic transients programs) has the ability to accomodate such changes with ease, but often suffers from the numerical problems associated with poorly conditioned matrices. In addition, many of these modern programs do not allow the variable timestep feaure which economises computer CPU usage when simulating power electronic circuits. A technique has been presented in this paper which uses the advantages of both these techniques.

A detailed state variable based model of a SVC has been presented and modeling details of the firing control circuit and the novel treatment of capacitor switching are explained.

With proper care, models such as the one developed here can be interfaced to commercially available transient simulation programs and provide the benefits of a variable timestep and stable numerical performance. From the simulation example presented in the last section, it is evident that the model can be used succesfully in very detailed simulation studies.

Further work: The techniques presented here can also be used for detailed representation of other switching models such as an HVdc convertor.

ACKNOWLEDGEMENTS

This program was developed at IREQ where A.Gole was on sabbatical leave from the University of Manitoba. The authors are grateful to IREQ for providing the facilities and to the University of Manitoba for approving Dr.Gole's sabbatical leave. The financial assistance of NSERC Canada is also appreciated. The authors benefited greatly from numerous discussions with other IREQ research staff, and in particular they express special thanks to Alpha Oumar Barry and Lewis Vaughan.

REFERENCES

[1] Dommel,H.W.,"Digital Computer Solution of Electromagnetic Transients in Single and Multiphase Networks", IEEE Trans. PAS, Vol. PAS-88,No.4, April 1969, pp 388-399.

[2] D.A.Woodford, A.M.Gole and R.W.Menzies; "Digital Simulation of Dc Links and AC Machines", IEEE Trans. PAS, Vol. PAS-102, No.6, June 1983, pp 1616-1623.

[3] M.D.Heffernan, K.S.Turner, J.Arrilaga and C.P.Arnold; "Computation of AC-DC System Disturbances Part I", IEEE Trans. PAS, Vol. PAS-100, No.11, Nov.1981, pp 4341-4363.

[4] K.R.Padiyar, Sadchidanand, A.G.Kothari, S.Bhattacharya and A.Srivastava; "Study of HVDC Controls through efficient Dynamic Digital Simulation of Converters", IEEE PES Winter Meeting, New York, Jan.29-Feb.5,1989, Paper No. 89 WM 113-2 PWRD.

[5] K.H.Kruger, R.H.Lasseter,"HVDC Simulation Using NETOMAC", IEEE MONTECH '86 Conference, Montreal, 29 Sept.- 1 Oct. 1986, Proceedings pp 47-50.

[6] N.Balbanian and T.A.Bickert; "Linear Network Theory; Analysis, Properties, Design and Synthesis", Matrix Publishers,1981. ISBN 0-916460-10, pp 104-147.

[7] Robert W.Hornbeck; "Numerical Methods", c Quantum Publishers, Prentice Hall, 1975. pp 198-199.

[8] K.Bayer, H.Waldmann and M.Weibelzahl; "Field Oriented Closed Loop Control of a Synchronous Machine Using the New Transvektor Control System", Siemens Review, XXXIX,No.6, 1972, pp 220-223.
[9] G.B.Mazur and R.W.Menzies; "Advances in the Determination of Control Parameters for Static Compensators", IEEE PES Winter Meeting, New York, Jan.29-Feb.5, 1989, Paper No. 89 WM 052-2 PWRD.
[10] A.M.Gole, R.W.Menzies, D.A.Woodford and H.Turanli; "Improved Interfacing of Electrical Machine Models in Electromagnetic Transients Programs", IEEE Trans. PAS, Vol PAS-103, No.9, Sept.1984, pp 2446-2451.
[11] J.Reeve and R.Adapa, "A new approach to Dynamic Analysis of AC Networks incorporating Detailed Modeling of DC systems - Parts I and II," IEEE Trans. on Power Delivery, Vol.3, No.4, Oct. 1988.
[12] Static Compensators for Reactive Power Control, Editor R.M.Mathur, Canadian Electrical Association Publication, 1984. Ch. 5.
[13] A.Gole, V.K.Sood and L.Mootoosamy, " Validation and Analysis of Grid Control System using dqz Transformation for Static Compensator Systems", Canadian Conference on Electrical and Computer Engineering, Sept. 17-20, 1989, Montreal, Canada.

APPENDIX

Legend

$v_p = [v_{p1}, v_{p2}, v_{p3}]^T$ – primary voltages,

$i_p = [i_{p1}, i_{p2}, i_{p3}]^T$ – primary currents,

$v_{s\Delta} = [v_{s\Delta1}, v_{s\Delta2}, v_{s\Delta3}]^T$ – delta secondary voltages,

$v_{sy} = [v_{sy1}, v_{sy2}, v_{sy3}]^T$ – wye secondary voltages,

$v_{C\Delta} = [v_{C1}, v_{C2}]^T$ – delta secondary capacitor volts, (only two voltages linearly independant)

$v_{Cy} = [v_{C4}, v_{C5}]^T$ – wye secondary capacitor volts, (only two voltages linearly independant)

$v_{Cs\Delta} = [v_{Cs1}, v_{Cs2}, v_{Cs3}]^T$ – Snubber capacitor voltages (delta),

$v_{Csy} = [v_{Cs4}, v_{Cs5}, v_{Cs6}]^T$ – Snubber capacitor voltages (wye),

$i_{l\Delta} = [i_{l1}, i_{l2}, i_{l3}]^T$ – TCR currents (delta winding),

$i_{ly} = [i_{l4}, i_{l5}, i_{l6}]^T$ – TCR currents (wye winding),

$i_{s\Delta} = [i_{s1}, i_{s2}, i_{s3}]^T$ – secondary currents (delta),

$i'_{sy} = [i_{s4}, i_{s5}, i_{s6}]^T$ – secondary currents (wye), – (only two currents linearly independent)

$i_{sy} = [i_{s4}, i_{s5}]^T$ – secondary currents (wye),

R – Loss resistor,

$C_1, C_2, C_3, C_4, C_5, C_6$ – TSC capacitors,

$r_{t1}, r_{t2}, r_{t3}, r_{t4}, r_{t5}, r_{t6}$ – Thyristor equivalent resistors,

L – TCR inductor,

C_s – Snubber Capacitor,

R_s – Snubber Resistor,

$L_{11}, \ldots\ldots, L_{99}$ – Transformer inductance matrix elements.

Summary of state variable equations

Using standard graph theory techniques [6] one obtains for the SVC equivalent circuit (Figure A1) the following five state equations:

$$\frac{d}{dt}\begin{bmatrix} \vec{i}_p \\ \vec{i}_{s\Delta} \\ \vec{i}_{sy} \end{bmatrix} = \mathcal{L}^{-1}[K_{PSC}]\begin{bmatrix} \vec{v}_{c\Delta} \\ \vec{v}_{cy} \end{bmatrix} + \mathcal{L}^{-1}\begin{bmatrix} v_p \\ 0 \\ 0 \end{bmatrix} \tag{1A}$$

where

$$\left.\begin{aligned} \mathcal{L}_{ij} &= L_{ij}, & i &= 1,6 \quad j = 1,6 \\ \mathcal{L}_{ij} &= L_{ij} - L_{i9}, & i &= 1,6 \quad j = 7,8 \\ \mathcal{L}_{ij} &= L_{ij} - L_{9j}, & i &= 7,8 \quad j = 1,6 \\ \mathcal{L}_{ij} &= L_{ij} - L_{9j} - L_{97} - L_{99}, & i &= 7,8 \quad j = 7,8 \end{aligned}\right\} \tag{1B}$$

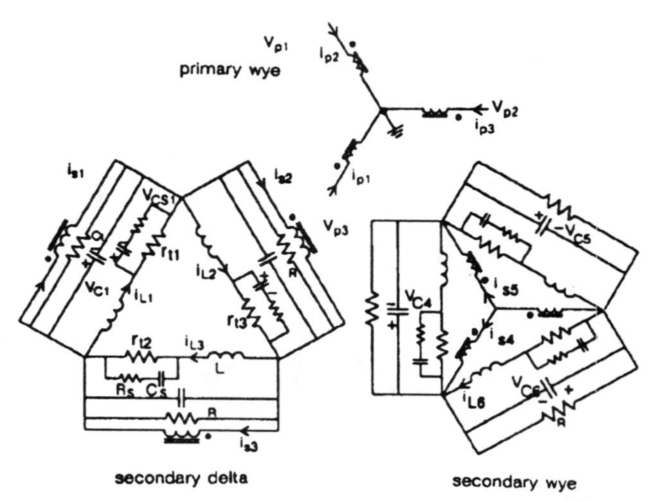

Figure A1: Equivalent circuit of SVC .

$$\vec{K}_{PSC} = \begin{bmatrix} 1 & 0 & 0 & 0 \\ 0 & 1 & 0 & 0 \\ -1 & -1 & 0 & 0 \\ 0 & 0 & -1 & -1 \\ 0 & 0 & 0 & -1 \end{bmatrix} \tag{1C}$$

$$\frac{d}{dt}\begin{bmatrix} \vec{v}_{cs\Delta} \\ \vec{v}_{csy} \end{bmatrix} = [K_{VCSL}]\begin{bmatrix} \vec{i}_{l\Delta} \\ \vec{i}_{ly} \end{bmatrix} + [K_{VCSC}]\begin{bmatrix} \vec{v}_{cs\Delta} \\ \vec{v}_{csy} \end{bmatrix} \tag{2A}$$

where

$$[K_{VCSL}]_{ij} = \delta_{ij}/[C_s(1 + R_s/r_{ti})] \tag{2B}$$

$$\delta_{ij} = \begin{cases} 0, & i \neq j \\ 1, & i = j \end{cases}$$

$$[K_{VCSC}]_{ij} = -\delta_{ij}/[C_s(R_s + r_{ti})] \tag{2C}$$

$$\frac{d}{dt}\begin{bmatrix} \vec{i}_{l\Delta} \\ \vec{i}_{ly} \end{bmatrix} = [K_{LC}]\begin{bmatrix} v_{c\Delta} \\ v_{cy} \end{bmatrix} + [K_{LL}]\begin{bmatrix} \vec{i}_{l\Delta} \\ \vec{i}_{ly} \end{bmatrix} + [K_{LCS}]\begin{bmatrix} \vec{v}_{cs\Delta} \\ \vec{v}_{csy} \end{bmatrix} \tag{3A}$$

where

$$K_{LC} = \begin{bmatrix} 1 & 0 & 0 & 0 \\ 0 & 1 & 0 & 0 \\ -1 & -1 & 0 & 0 \\ 0 & 0 & 1 & 0 \\ 0 & 0 & 0 & 1 \\ 0 & 0 & -1 & -1 \end{bmatrix}[1/L] \tag{3B}$$

$$[K_{LL}]_{ij} = \delta_{ij}(-R_s)/[L(1 + R_s/r_{ti})] \tag{3C}$$

$$[K_{LCS}]_{ij} = -\delta_{ij}/[L(1 + R_s/r_{ti})] \tag{3D}$$

$$\frac{d}{dt}[\vec{v}_{c\Delta}] = [K_{CS\Delta}]\vec{i}_{s\Delta} + [K_{CS\Delta}]\vec{i}_{l\Delta} + [K_{CC\Delta}]\vec{v}_{c\Delta} \tag{4A}$$

where

$$[K_{CS\Delta}] = \frac{1}{\sum\limits_{\substack{j \neq i \\ 1,2,3}} C_i C_j}\begin{bmatrix} (-C_2 - C_3) & C_3 & C_2 \\ C_3 & (-C_1 - C_3) & C_1 \end{bmatrix} \tag{4B}$$

$$[K_{CC\Delta}] = \frac{-1}{R\sum\limits_{\substack{j \neq i \\ 1,2,3}} C_i C_j}\begin{bmatrix} (C_2 + C_3) & (C_2 - C_3) \\ (C_1 - C_3) & (2C_1 + C_3) \end{bmatrix} \tag{4C}$$

$$\frac{d}{dt}[\vec{v}_{cy}] = [K_{CSY}]\vec{i}_{sY} + [K_{CLY}]\vec{i}_{lY} + [K_{CCY}]\vec{v}_{cy} \quad (5A)$$

where

$$[K_{CSY}] = \frac{1}{\sum_{\substack{j \neq i \\ 4,5,6}} C_i C_j} \begin{bmatrix} C_5 & -C_6 \\ C_4 & (C_4 + C_6) \end{bmatrix} \quad (5B)$$

$$[K_{CLY}] = \frac{1}{\sum_{\substack{j \neq i \\ 4,5,6}} C_i C_j} \begin{bmatrix} (-C_5 - C_6) & C_6 & C_5 \\ C_6 & (-C_4 - C_6) & C_4 \end{bmatrix} \quad (5C)$$

$$[K_{CCY}] = \frac{-1}{R \sum_{\substack{j \neq i \\ 4,5,6}} C_i C_j} \begin{bmatrix} (2C_5 + C_6) & (C_5 - C_6) \\ (C_4 - C_6) & (2C_4 + C_6) \end{bmatrix} \quad (5D)$$

Note that if

$$\vec{X} = \left[\vec{i}_p,\ \vec{i}_{s\Delta},\ \vec{i}_{sy},\ \vec{v}_{cs\Delta},\ \vec{v}_{csy},\ \vec{i}_{l\Delta},\ \vec{i}_{ly},\ \vec{v}_{c\Delta},\ \vec{v}_{cy}\right]^T \quad (6)$$

is chosen as a state variable, equations 1A – 5A have the form:
$\dot{\vec{X}} = A\vec{x} + B\vec{u}$ with $\vec{u} = \vec{v}_p$ which is the standard state variable form.

Accounting for capacitor switching

Consider Figure A2 in which the capacitor C_{2A} of initial voltage v_{2A0} is switched in parallel to capacitor C2. Initial voltages on capacitors C1, C2 and C3 are v_{C10}, v_{C20} and v_{C30} respectively. The switching causes a charge Δq to trasfer from each capacitor to the next, as in the figure. Assuming q_1, q_2 and q_3 to be the post-switching charges, we have:

$$\Delta q = q_1 - q_{10} = q_2 - q'_{20} = q_3 - q_{30} \quad (7)$$

where $q'_{20} = q_{20} + C_{2a} * v_{2A0}$
There are thus two independent equations:

$$q_1 - q_{10} = q_2 - q'_{20} \quad (8a)$$

$$q_2 - q'_{20} = q_3 - q_{30} \quad (8b)$$

in the three unknowns q_1, q_2 and q_3. Kirchoffs voltage law provides the third equation:

$$v_1 + v_2 + v_3 = 0 \quad (9)$$

i.e. $$q_1/C_1 + q_2/C'_2 + q_3/C_3 = 0 \quad (10)$$

where $C'_2 = C_2 + C_{2A}$

On solving for q_1, q_2 and q_3 and adjusting the after switching voltages on C1, C2 and C3, the values on the capacitors are also changed. C2 now becomes $C'_2 = C_2 + C_{2A}$. Removing a capacitor bank of value C_{2A} is equivalent to adding a capacitance of value $-C_{2A}$.

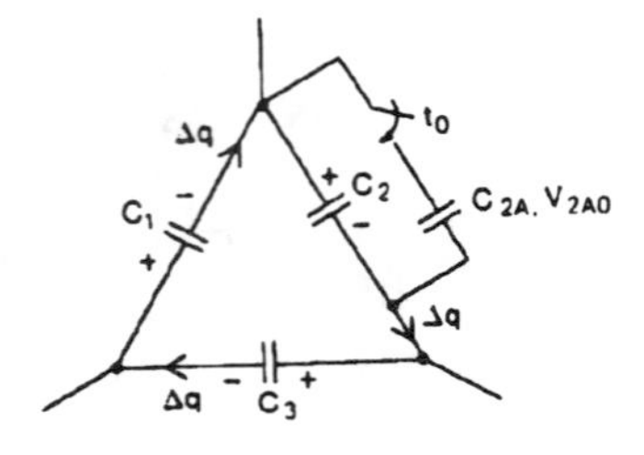

Figure A2 : Switching on a Capacitor C_{2A} in phase 2.

Data for the system used in simulation study

AC network equivalents
Sending end: 315 kV, SCR 4.5 at 85 degrees damping angle.
L1 = 22.2 mH, L2 = 95.4 mH, R = 13.1 ohms.
Receiving end: 120 kV, SCR 2.5 at 85 degrees damping angle.
L1 = 10.49 mH, L2 = 21 mH, R = 10.8 ohms.

Convertor Transformers
Sending end: 315 kV/120 kV, 610 MVA, XL = 18 %.
Receiving end: 120 kV/120 kV, 610 MVA, XL = 18 %.
Saturation (both) : knee 1.2 pu, slope 0.4 pu.

AC Filters
Sending end:
11 th harmonic: R = 1.13 ohms, L = 27.4 mH, C = 2.12 uF.
13 th harmonic: R = 0.25 ohms, L = 19.5 mH, C = 2.12 uF.
C = 2.4 uF.

Receiving end:
11 th harmonic: R = 0.163 ohms, L = 4 mH, C = 14.6 uF.
13 th harmonic: R = 0.14 ohms, L = 2.85 mH, C = 14.6 uF.
C = 16.58 uF.

SVC
Transformer: 120 kV/12 65 kV, 200 MVA.
X LD = X LY = 17 %, X LDY =2.1 %,
knee level = 1.2 pu, Saturation slope 0.4 pu.

TCR: L = 22 mH (100 MVA)
TSC: C= 202 uF/stage (83 MVA).

DC system
Rating 140.6 kV, 506 MW, 3.6 kA.
Smoothing Reactor 34 mH on each side.

BIOGRAPHIES

A.M.Gole (S'77,M'82) obtained the B.Tech.(E.E.) degree from the Indian Institute of Technology Bombay in 1978 and the Ph.D.degree in Electrical Enginnering from the University of Manitoba in 1982.

Dr. Gole is currently an Associate Professor in the Dept. of Electrical Engineering at the University of Manitoba. From Sept. 1988 to Aug.1989 he was on sabbatical leave at IREQ in Varennes, Quebec. His research interests include HVdc transmission, power system transients simulation and the use of optimization techniques.

Dr. Gole is a Registered Professional Engineer in the Province of Manitoba.

V.K.Sood (SM 1985) obtained the B.Sc. (Ist Class Honours) degree in Electrical Engineering from University College, Nairobi (Kenya) in 1967, the M.Sc. degree from University of Strathclyde, Glasgow (Scotland) in 1969 and the Ph.D. degree from University of Bradford (England) in 1977.

From 1969-76, he worked at the Railway Technical Centre, Derby (UK) carrying out research in DC Choppers for rapid transit applications. Since 1976, he has been at the Hydro-Quebec research institute (IREQ) in Varennes, Quebec working with the HVdc analog simulator. His present research interests include HVdc transmission, Forced Commutation and analog/numerical simulation techniques.

Dr.Sood is Senior Member of the I.E.E.E. , Member of the I.E.E. and a Registered Professional Engineer in the Province of Quebec. He is also an Adjunct Associate Professor in the Department of Electrical and Computer Engineering at Concordia University.

Discussion

D. Povh and H. Tyll (Siemens AG, Dept. EV NP, P. O. Box 3240, D-8520 Erlangen, Germany):

Authors are to be commented for a very interesting paper on the modelling of SVC in the electromagnetic transient simulation program presenting new ideas to improve the EMTP and EMTDC programs. We understand that the main issue of the work is to overcome the difficulties at the calculations of SVCs and HVDC in these programs.

As mentioned in the paper the program NETOMAC has in contrary to other programs the ability to interpolate the time step to switch at current zeros and to adjust the firing instant of thyristors exactly according to the instant given by the firing pulse control. The method used in NETOMAC is superior to other methods and enable calculation of SVC and HVDC without any risks of numerical instabilities. This has been also very impressive demonstrated in the reference 5 of the paper.

The automatic reduction of time step when approaching current zero or when high frequency oscillations occur in the circuit was first used in our program ADIEU /1/,which was predecessor of NETOMAC and was used in our company for years. Our experiences with time step reduction are, however, similar to the authors'. It is mostly more convenient to calculate the transients continuously with the reduced step. In NETOMAC this possibililty is also incorporated.

An example of use of NETOMAC to calculate SVC performance is shown in Fig. 1 which is taken out of reference /2/. The SVC simulated consists of one TCR and two TSC branches of different ratings together with ac filters on the secondary side of the main transformer. The control of the SVC was represented very detailed. Fig. 1 shows the comparison of NETOMAC calculations and simulator (TNA) readings using original control cubicles. The case presented shows the linear change of the SVC admittance and shows an excellent agreement between measurements and calculations.

To the authors' suggestion to simulate the SVC we have following questions:

(i) How is the SVC control represented? Is the numerical solution in the child model for the circuit and for the control in the same step?

(ii) If more SVCs are connected to the network and possibly two SVCs on the same ac bus, are any additional measures provided to avoid instability or numerical oscillations between the SVCs?

(iii) The transients in the TSC capacitor currents of Fig. 7c show rather strong damping. Our experience is based on field measurements that these harmonic currents oscillate longer and have impact on the capacitor voltage. Does this damping result from the representation in Fig. 8? How is the leakage reactance of

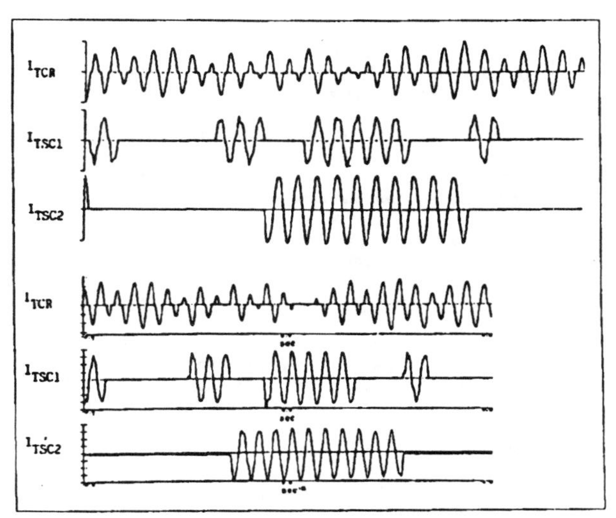

Fig. 1 : Response of SVC following admittance characteristic predetermined by the control NETOMAC calculations (a) and measurements on simulator (b)

the transformer represented which should result in transient TSC currents of approx. harmonic order of 4.

(IV) It was understood that the child model can be used only for calculation of SVC performance in the network because on the secondary side of the transformer an equivalent circuit is used for capacitors. Does it mean that the important studies on stresses and performance at the internal faults can not be done? Are the corresponding values of TSC currents and secondary side voltages also calculated?

/1/ D. Povh
Berechnung von Ausgleichsvorgängen in elektrischen Netzwerken insbesondere zur Ermittlung der Spannungsbeanspruchungen der Ventile einer Hochspannungs - Gleichstrom - Übertragungs - Anlage bei Erdkurzschluß
Dissertation der TH Darmstadt, 1971

/2/ D. Poch; H. Tyll
Static Var Compensators for High-Voltage Systems
II. Symposium of Specialists in Electric Operational and Expansion Planning
(II. SEPOPE), Sao Paulo, August 1989

Manuscript received February 28, 1990.

Subroto Bhattacharya and Sachchidanand (ABB Advanced Systems Technology, Pittsburgh and University of Western Ontario, London, Canada): The authors are to be complimented for a well-written paper on a SVC model using the state variable approach. We would appreciate the authors' response to the following questions and comments:

1. The state variable approach is suitable for handling event-driven systems. A switching event occurs when a thyristor current goes zero. To find the instant of current zero, the proposed algorithm restarts its simulation from the previous time value with a timestep ($\delta t = \Delta t/n$) which is an integer sub-multiple of the original timestep (Δt). With this procedure the current zero instant is approximated within a submultiple time interval (Figure 2b) and is not determined precisely. Also, we feel that this part of the algorithm is time consuming. The

references [4, A, B] find the time instant (t_c) at which the current through a thyristor goes to zero between the two integration timesteps (t_A and t_B) by using linear interpolation of thyristor current. All other state-variables are interpolated to determine their values at t_C. This procedure does not result in any "numerical" abnormalities even when the original timestep (Δt) is kept around 2 electrical degrees (or approximately 100 μs for a 60 Hz system). Once t_C is obtained the program can perform the "catchup" step. Have the authors tried some such approach and if so, what has been the experience?

2. The child program is interfaced with any parent program by using Norton and Thevenin equivalents. Does the child program receive a Thevenin's equivalent (i.e., a 3 phase voltage source and an impedance matrix) from the parent program on every timestep? Does the parent program wait for the child program to compute the current sources?
3. It is mentioned in the text that the parent and the child programs use one timestep old information. Is there an inherent delay between the child and the parent program? In case of the EMTP any such interconnection between the parent (EMTP) and the child (non-linear or linear network elements) programs can be achieved as described in Sections 12.1.2.1 to 12.1.2.3 of [C] to avoid any delay if it may so exist.
4. It is possible to run the child program with a larger timestep than the parent program?
5. What factors prompted the authors to choose a timestep of 25 μs for the case shown in Figure 9? Did a larger timestep than 25 μs cause numerical instability?

As a final remark, the paper reinforces the use of state-variable approach for representing power semiconductor circuits using graph theory. Earlier, this approach has been used to model HVDC converters [4, A] and industrial converters [B]. The authors have made a comment in the paper about [4] with reference to which we would like to mention that the state-variable model of the HVDC converter proposed in [4] does not warrant representation of the associated external system (ac system or dc network) in the state-variable form. Hence, it is not a limitation of the modeling approach. Instead, it shows that a flexibility exists to model each subsystem (ac system, converter system and dc network) in any desired manner and to any desired degree of detail. Norton-Thevenin interface between power semiconductor circuits and an external system has been presented earlier [4, A].

References

[A] K. R. Padiyar and Sachchidanand, "Digital simulation of multi-terminal HVDC system using a novel converter model", IEEE Transaction on PAS, Vol. 102, June 1983, pp. 1624–1632.

[B] S. P. Yeotikar, S. R. Doradla and Sachchidanand, "Digital Simulation of a three phase AC-DC PWM converter-Motor system using a new state space converter model", Proceedings of IEEE—Industry Applications meeting, 1986, pp. 672–679.

[C] H. W. Dommel, "ElectroMagnetic Transients Program Reference Manual (EMTP Theory Book)", prepared for Bonneville Power Administration, Portland, OR, USA August 1986.

A.M.GOLE and V.K.SOOD, (University of Manitoba, Winnipeg, Manitoba and Hydro-Quebec, Varennes, Quebec): The authors thank the Discussers for their kind compliments, discussion and interest.

In response to the discussion by S.Bhattacharya and Sachchidanand, we have the following comments:

1. The variable timestep feature described in the paper is only one alternative to get around the numerical instability problem due to switching of components. Other techniques exist, as noted by the discussers, but it should be pointed out that all methods whether interpolative or not, are never-the-less approximations as to the point of zero-crossing. The great advantage of the variable timestep feature described here is that it is simple and yet effective; moreover, it is totally contained within the child model where, incidentally, it is most needed. Furthermore, it permits the child program to be interfaced to a standard parent program such as EMTP or EMTDC without any modification of the code in the parent program. We feel the variable timestep technique described here is a computationally efficient means to alleviating the numerical instability problem, although not completely overcoming it. The time overhead of the algorithm is estimated as below:

Using a typical value of 10 μS for the submultiple timestep and a 50 μS main timestep, an average number of 2.5 extra steps per cycle are required. Thus, for a 12 pulse converter 30 extra timesteps per cycle are required. At 60 Hz, one cycle has 333 timesteps; hence an average overhead of 30/333 i.e. 9% is required. We feel that this is quite acceptable.

2. Only the parent program receives a Norton Equivalent since the interfacing technique requires it only in one direction.
3. The problem of one timestep delay is inherent to the procedure of interfacing. One way to avoid this delay is to incorporate the child model within the parent, in which case the flexibility due to the child-parent relationship is lost. Another possibility is to iterate the parent-child solutions within one timestep, which is time-consuming.
4. In the present version, it is not possible to run the child program with a larger timestep than the parent timestep. This is, however, not a limitation of the method, and should a need exist then this could be implemented.
5. The choice of timestep of 25 μS for case shown in Figure 9 was determined by reasons of simulation faithfulness during commutation failures. Timesteps of 50 μS were used and no numerical instability was apparent.

As a final remark, we agree with the discussers that the state variable technique is a powerful tool for developing robust models of power electronic elements such as the one described in the paper.

We are particularly pleased to acknowledge the discussion from Povh and Tyll to share their experience with NETOMAC and state variable modeling. It confirms that the straightforward application of Dommel Algorithm based programmes to simulate power electronic circuits has certain difficulties. The program NETOMAC has been one succesful approach to this problem ; here we have presented an alternative approach that has worked very well. It is encouraging to see also that their experience with state variable modelling is similar to ours. In response to their specific questions:

1. Only the innermost control loop – the phase locked loop based firing system – has been modeled inside the SVC child program and works in the same timestep. The other control system components (upto the generation of the α order in Figure 10a) have been modeled with the control system building blocks (CSMF) of the EMTDC parent program, where a one timestep delay is not critical.
2. The simulation also works well with more than one SVC on the same bus. Indeed, the case simulated in the paper is with one SVC and two HVDC converters on the same bus. We attribute the stability of the solution directly to the interfacing technique used.
3. The SVC transformer has been modeled with the entire 9x9 matrix representing the coupling between all 9 windings; and so the leakage impedance is modeled accurately. We do not attribute the extra damping (if any) to the interfacing technique used because the interface has been carefully designed to offset only the negative damping introduced by the numerical procedure. In our case (for the timestep used) the compensation resistor carried less than 0.3 % of the main current and hence should not contribute to the damping. If there is more damping we attribute it to the data used by us such as the quality factor of the SVC capacitor and the values of the snubber circuit elements, and not to the model itself. We can demonstrate less damping by changing this data.
4. All currents and voltages on both the primary and secondary sides of the SVC transformer are calculated and are available for observation. In addition, TCR and TSC thyristor voltages and currents are also similarly available as outputs. Although the model has been designed primarily for system side studies, certain faults that do not affect the circuit topology of the SVC can be studied. These include l-l short circuits and valve misfires. However, faults such as a secondary side l-g short circuit cannot be simulated because it would mean a change in circuit topology and a change in the state equations describing the circuit. The program could be modified to allow for such simulations but for the requirements of our simulations, we were not interested in doing so. At present the circuit topology must conform to that in Figure 1a of the appendix.

Manuscript received March 29, 1990.

DETAILED MODELING OF AN ACTUAL STATIC VAR COMPENSATOR FOR ELECTROMAGNETIC TRANSIENT STUDIES

A.N. VASCONCELOS A.J.P. RAMOS J.S. MONTEIRO M.V.B.C. LIMA H.D. SILVA L.R. LINS
Senior Member

CHESF, Recife, Brazil

ABSTRACT

This paper presents a detailed model of an actual Static VAR Compensator for digital simulation of electromagnetic transients. It also demonstrates that this model is adequate to reproduce SVC transient behaviour as verified in TNA studies carried out with a replica of the SVC control system. The SVC control system is described, with emphasis on some special blocking schemes needed to meet particular requirements of the power system transient performance. Complete system and SVC data are also included.

KEYWORDS - Static VAR Compensator (SVC), Thyristor Controlled Reactor (TCR), Electromagnetic Transient Program (EMTP), Transient Network Analyser (TNA).

INTRODUCTION

Time domain simulation is the most important tool there is for power system analysis. Powerful, efficient programs now available, as the EMTP [1] enable simulation to the required level of detail of practically all transients of a power system. With such programs, sophisticated, detailed models of, for example, SVC control systems can be developed to investigate complex transients of a power system. Transient Network Analysers have also been used in the analysis of system transients. These may use physical models of power system components, as reactors, capacitors etc, or electronic models. They are not commonly available within utility companies, where digital programs are the usual tool for simulating power systems.

The simulation facilities and modeling potential that such digital programs provide have assuredly met the requirements of power systems made increasingly complex by equipment of new, sophisticated technology. Various types of SVCs, HVDC transmission and their control systems, for instance, cause new, complex phenomena that substantially modify system transients. Understanding the interrelated physical phenomena, and sorting the effects which should be taken into account from those which can be disregarded are prime requirements for modeling system equipment. The definition of the level of detail required by a specific phenomenon in the system demands familiarity with the equipment and with the characteristics of the system.

For synchronous machines, it has been possible to set up general guidelines for the level of detail required by models for various types of investigation [2]. Owing to differences of design concept and type between SVCs, however, guidelines for them have proved difficult to set up. Furthermore, interaction between the SVC and inherent system characteristics affects the modeling requirements [3]. False assumptions can lead to the suppression in the simulation of relevant, real effects or to the appearance of non-existing effects.

This paper presents the experience of modeling an actual SVC whose detail requirements are remarkably strict. It is made up of a thyristor-controlled reactor and a fixed capacitor. It operates in a regional subsystem that is part of the Northern-NorthEastern interconnected power system of Brazil, in the Fortaleza substation. Its Effective Short Circuit Ratio - ESCR is 1.67. As its rating is so high compared with the short-circuit power level of the connected bus, the SVC control system can be said to practically dictate the behaviour of system transients. Short-circuits and subsequent load rejections demand an effective, suitable SVC response, which is determined by the strategy and parameters of the SVC control system.

Studies to define, for example, the parameters of the control system of an SVC demand an appropriately detailed modeling of its control system, of its synchronising, and of its firing system.

In the present work, each section of the SVC structure is described in detail. Some special features of the SVC control system, as the Undervoltage Blocking Scheme (UBS) and the BOD Blocking Scheme (BBS) are commented on. Detailed modeling of the synchronising and firing systems for EMTP/TACS (Transient Analysis of Control Systems) is presented. Results of simulations with the EMTP compared with TNA simulations are also presented.

91 WM 211-3 PWRS A paper recommended and approved by the IEEE Power System Engineering Committee of the IEEE Power Engineering Society for presentation at the IEEE/PES 1991 Winter Meeting, New York, New York, February 3-7, 1991. Manuscript submitted June 12, 1990; made available for printing January 3, 1991.

2 - DESCRIPTION OF FORTALEZA SVC

2.1 - Power Circuit

Fortaleza SVC has two sections, each one with one capacitor bank and one TCR connected to the low voltage side of a 230/26/26kV step down transformer, as shown in Figure 1.

Power components of the SVC are represented with the conventional models available in the EMTP. Complete data of the SVC and transmission system studied, shown in Figure 4, are presented in the appendix.

2.2 - Control System

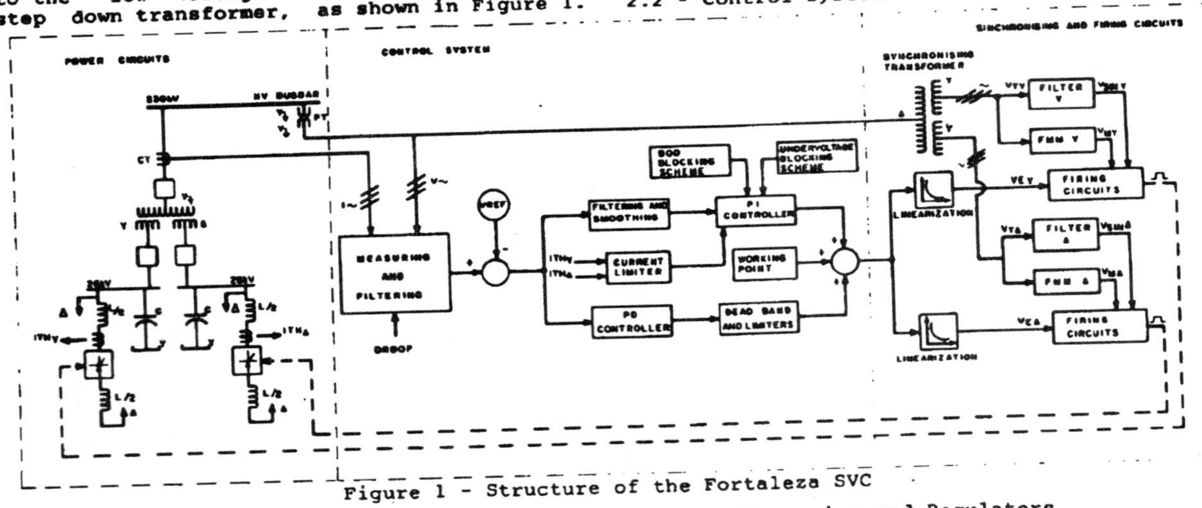

Figure 1 - Structure of the Fortaleza SVC

The SVC is dimensioned to be capable of drawing -140Mvar (inductive) to +200Mvar (capacitive) at high voltage bus, when operating with both sections in a twelve pulse mode or from -70Mvar to +100Mvar when operating with only one section, for any value of HV busbar voltage within the range 230kV ± 5%.

2.2.1 - Measuring and Regulators

Three phase voltages and currents measured by 230kV potential transformers (PTs) and current transformers (CTs) are used to evaluate w(t) = v(t) + ki(t), where k is the regulating droop, adjustable between 0 and 10%, as shown in Figure 2. These signals are

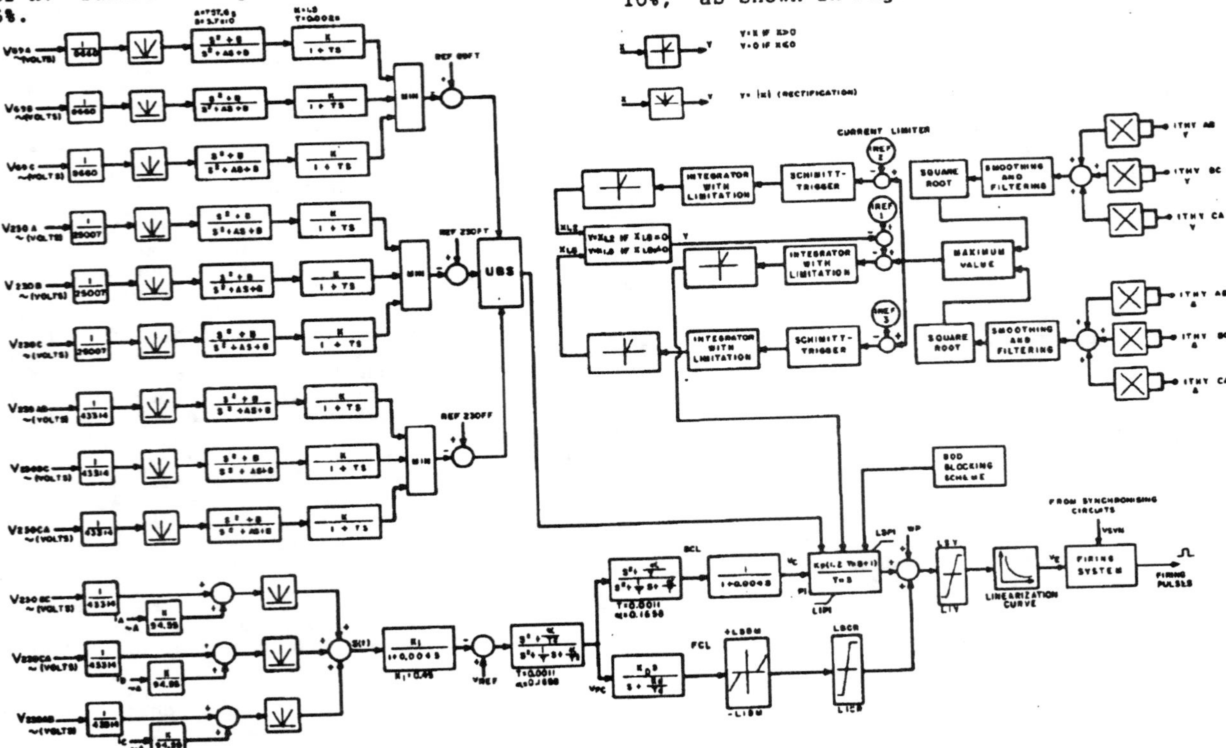

Figure 2 - Block diagram of the SVC control system

rectified, smoothed and summed to obtain U. U is then compared with the reference value to give the error signal E. The control action is determined by two control loops, known as Basic Control Loop (BCL) and Fast Control Loop (FCL), whose structures are shown in Figure 2. The BCL, which operates continuously, uses a PI regulator that is the main regulator. The FCL operates only for large disturbances due to the use of Dead-Band blocks. The magnitude of the FCL control action is also adequately limited.

for each SVC section Y and Δ. The output of the linearization curves are voltage signals proportional to the firing angle "α". The linearity between UR and the corresponding Y is tested on site during commissioning.

2.2.2 - Undervoltage Blocking Scheme

During faults in the transmission system, for which the system voltages go down, the SVC voltage regulator would attempt to control the busbar voltage. This would take the SVC to the

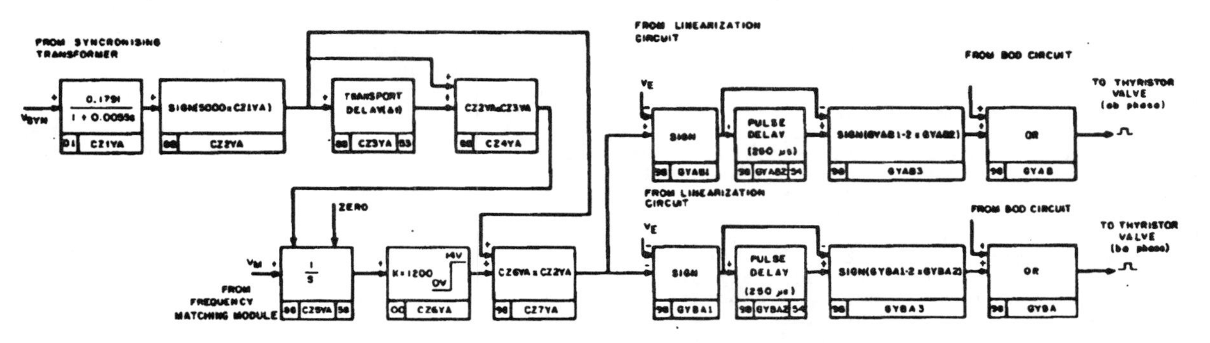

Figure 3 - Modeling of firing system in term of TACS/EMTP language

The total control UR is made up of two control loops plus a DC value (WP), so that 0 Volt input of the linearization curves gives zero Mvar at 230kV busbar, corresponding to a firing angle equal to 119°. UR is a voltage signal proportional to the SVC admittance (Y) for fundamental frequency; here, 60Hz.

The linearization curves needed to compensate the non-linear relationship between admittance and firing angle are tuned to produce the desired effect of admittance as seen from the 230kV busbar, as shown in Figure 2. Two exactly equal curves are provided, one

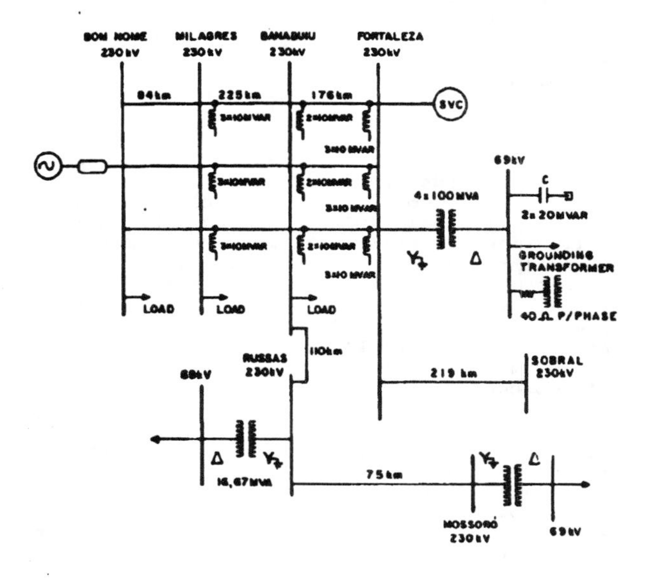

Figure 4 - Simplified one-line diagram of the subsystem as used in simulations.

maximum capacitive operating point with disastrous results after fault clearing, especially when load rejection occurs. To avoid this, an Undervoltage Blocking Scheme (UBS) is provided, to force the SVC to zero Mvar operating point, whenever any of the 230kV phase-phase, 230kV phase-ground or 69kV phase-ground voltages drops below set values.

The UBS works by discharging the integrator of the PI regulator, blocking its output at zero until 10ms after the blocking conditions are removed. With the PI regulator blocked, the SVC operates around 0 Mvar. Such characteristics of the SVC control system can easily be modeled with TACS/EMTP special block facilities [1].

The UBS plays an essential role in SVC transient performance. Voltage control equipment that handles 340 Mvar in a system busbar with a power short-circuit equal to only 600Mvar must be appropriately intelligent and reliable.

2.2.3 - Current Limiter

In order to avoid thyristor valve current overloads, an additional current limiter control loop (CL) is provided.

The rms values of thyristor valve currents are measured in both sections of the SVC, for each phase. The highest value of the twelve measured currents is selected, filtered and smoothed, as shown in Figure 2. Comparing this signal with reference values enables the identification of an overload condition that requires CL control.

Under thyristor current overload, the CL reduces the inductive limit of the total regulator output.

2.2.4 - BOD Blocking Scheme

TNA simulations have shown that under some operating conditions, faults on the 69kV busbar with total load rejection may give rise to overvoltages on the thyristor valves high enough to cause BOD firing. In these cases, the transient thyristor valve currents are high enough to activate the current limiter (CL) control loop. The BOD Blocking Scheme (BBS) coordinates BOD firing with CL actuation. It operates as follows:

- When protective firing occurs, the BBS forces the PI regulator output to an adjustable value for 10ms. After 10ms, the PI is released to operate normally.

The TNA studies demonstrated that this BBS action prevents oscillations between CL and BOD. In simulations where BOD firing is likely to occur, the CL and BBS must therefore be adequately modeled.

2.3 - Synchronising and Firing Circuits

The synchronising circuit gives the firing system voltage signals that are in phase with the 26kV busbar voltages applied to the thyristor valves. These voltages are obtained from the 230kV busbar by means of a 230kV/115V potential transformer and by means of a special "synchronising transformer" (Fig. 1). The synchronising voltages are filtered by a low-pass, first-order filter that introduces a lag of 60°. The six voltage signals so obtained are free of harmonic distortion; a requisite condition for a precise firing.

The firing system is based on the zero-crossing technique, for which it uses the TCA 780 standard IC. The synchronising voltages and firing angle voltage signals are used by an electronic circuit called TRIGGER SET to emit the firing pulses. According to the TRIGGER SET characteristic, -2.3 V and 4.2V inputs correspond to 92.5° and 170° firing angle respectively. The firing pulses are treated by many other electronic circuits before they reach the thyristor gates. Such circuits, however, which have no effect on the functionality of the process, need not be modeled. Inaccuracies of firing can be guaranteed to be under 2°. In the present application, the firing angle is limited to the range of 92.5° to 170°.

In steady-state operation, the firing system uses a Phase-Locked-Loop (PLL), that has a firing error under 0.2°. When the PLL and zero-crossing system pulses present a shift equal to or greater than 2°, the firing system switches to the zero-crossing system. This can be caused by system transients, when abrupt phase shifts in the power system voltages are produced by short-circuits, line removal, load rejection etc. For digital simulation, the representation of the zero-crossing system process is sufficient. The synchronising and firing system modeling in term of TACS language is shown in Figure 3 for phase AB of section Y. A special circuit called Frequency Matching Module - FMM gives a signal to compensate eventual frequency deviation. In electromagnetic transient simulations the FMM was not considered, because frequency is assumed to be constant.

3 - SVC MODELING WITH THE EMTP

3.1 - General Comments

Fortaleza SVC, owing to its specific control schemes needed to meet important performance requirements, has a relatively complex control system.

Defining the level of detail required in modeling the SVC for electromagnetic transient studies demands a familiarity with the system, from the electronic equipment to the power system transient phenomena. Such work usually calls for people with different experience.

With a digital program as EMTP, practically all known controls can be adequately modeled. Yet obviously we can only model controls when they are fully known, and when we are sure of how they will operate under all actual system conditions. In our present case, the power supply for the SVC electronic circuits is obtained from normal auxiliary service. Accordingly, HV disturbances caused by a system fault will affect voltage supply. This means that the control system performance must be verified under fault conditions.

A replica of the Fortaleza SVC control system, purchased from the manufacturer of the SVC, was installed in the TNA of CEPEL, the Brazilian Electric Power Research Center in Rio de Janeiro. The simulation of severe contingencies was carried out with the SVC properly interfaced with the system in the TNA.

This SVC model was developed from the electronic circuits and with a judicious evaluation of their performance under actual condition.

It seems to us that experience with many simulations using a very detailed model, as here described, can lead to justifiable simplifications in later studies. This is important for the system under study, as two further SVCs are planned to come into operation close to Fortaleza SVC. Digital studies with such detailed representation of three SVCs would be computationally prohibitive.

3.2 - Initialization

In our case, a network resonance close to the second harmonic makes the start-transient excessively slow. Unbalanced, distorted 26kV busbar voltages in the initial instants result in wrong firing which, in turn, contributes to the generation of harmonics. This self-sustaining cycle leads to a long time transient. To accelerate the network transient, fictitious ideal voltage sources, whose voltage magnitude and angle are exactly equal to the corresponding steady-state value obtained from loadflow results, are connected to the 26kV busbar for 30ms.

The SVC control system would attempt to control terminal voltage during the network initialization transient. This would make this transient longer, especially if a high regulator gain is used. To avoid this, the PI regulator input is switched off and the SVC operates as in manual. The operating point corresponds to the PI initial condition.

The integration time step around 20 microseconds seems to be sufficient for most simulations.

The representation of the snubber circuits improves physical modeling. At the same time, it avoids numerical oscillations associated with the switching of inductive currents. As suggested by [4] the time step should be, at least 0.5Tsn, where Tsn is the snubber circuit time constant. Here Tsn = 216 microseconds.

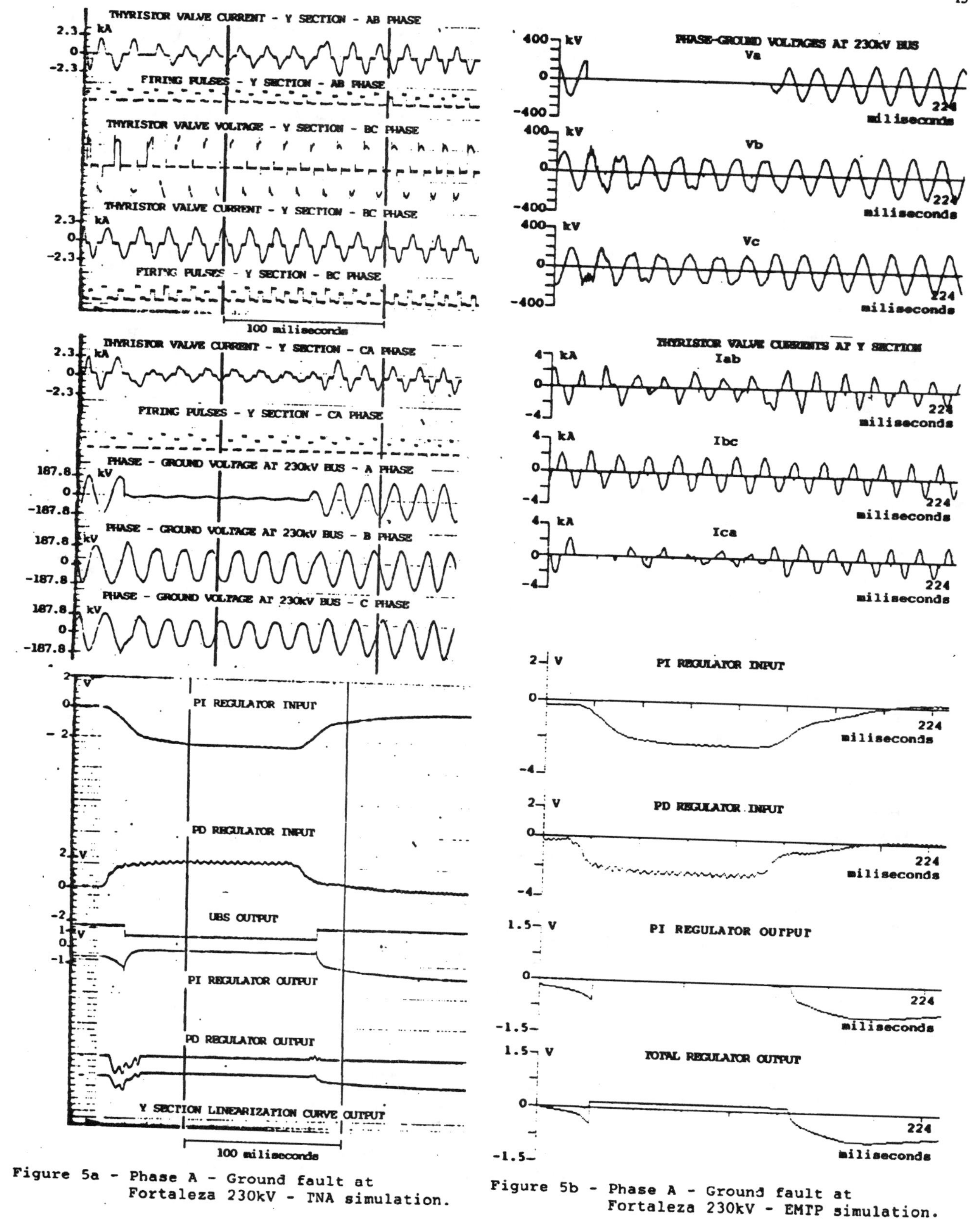

Figure 5a - Phase A - Ground fault at Fortaleza 230kV - TNA simulation.

Figure 5b - Phase A - Ground fault at Fortaleza 230kV - EMTP simulation.

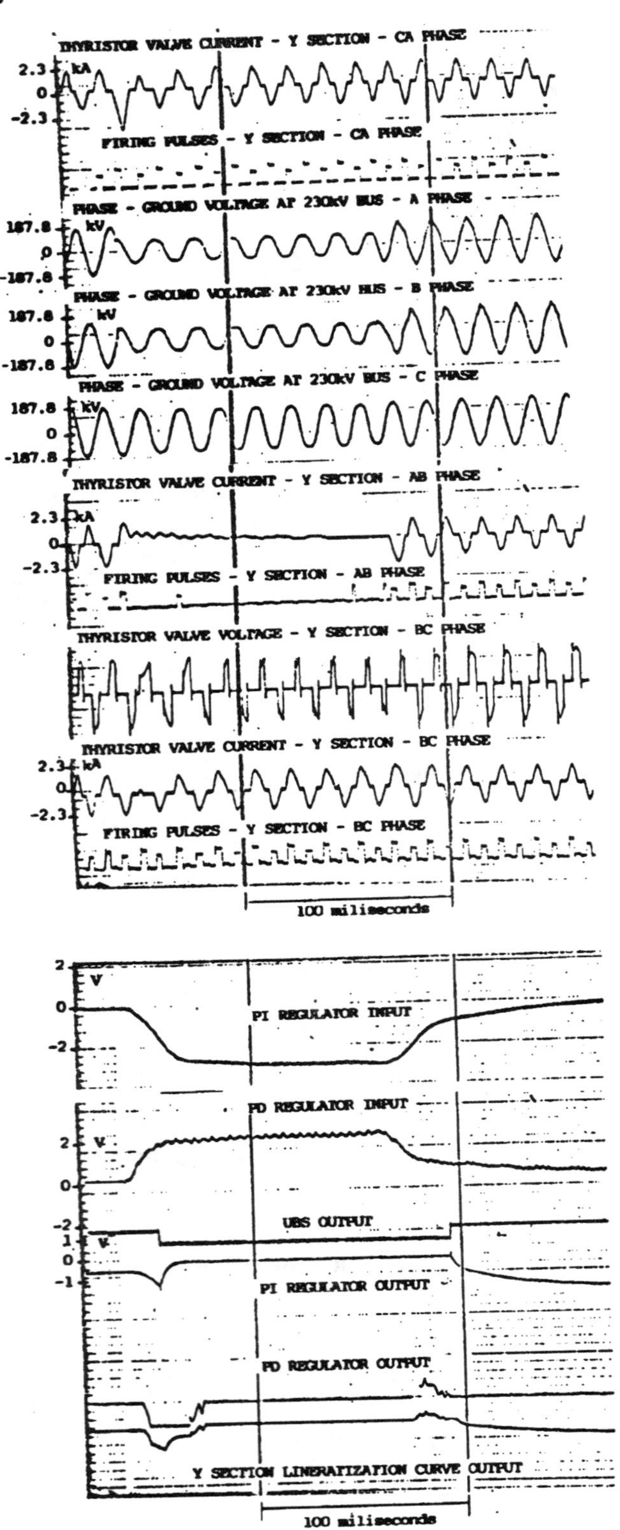

Figure 6a - Phase A-phase B fault at Fortaleza 230kV. TNA simulation.

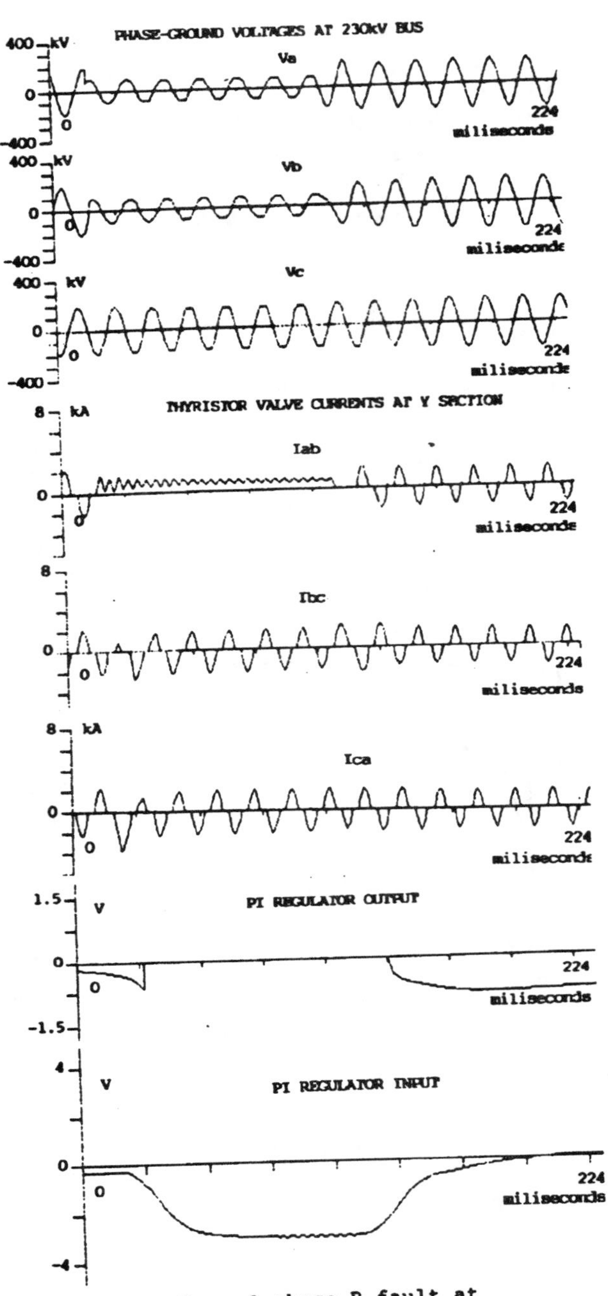

Figure 6b - Phase A-phase B fault at Fortaleza 230kV. EMTP simulation.

4. SIMULATION RESULTS WITH EMTP

Simulation results from EMTP were compared with those from the TNA study. Some differences between EMTP and TNA models merit prior consideration.

a - Transmission system

Real quality factors of reactors and transformers that can be modeled with EMTP are excessively low in TNA, which exaggerates damping.

While we can be sure of the same instant of fault initiation with both EMTP and TNA , they give different instants of fault clearing, owing to their different damping of fault current.

b - Control system and firing circuits.

Thyristor valves in TNA have resistance values disproportionately great, where as EMTP can model the real values. Owing to this, DC components of the TCR transient currents decay faster in TNA simulations.

In EMTP, the internal electronic transients need not be represented. For example, PI blocking is performed by jumping the capacitor of the corresponding integrator. This gives rise to an RC circuit transient. EMTP blocks PI abruptly.

Figure 5 shows some selected variables of the SVC control, 230kV busbar voltages and TCR current for a single-phase short-circuit, as obtained from EMTP and TNA simulations. The fault clearing is obtained by removing one of the Fortaleza-Banabuiu transmission line. Figure 6 shows the result of a phase A - phase B short-circuit.

The results of many simulations have proved the SVC model developed for EMTP reliable enough to use in detailed studies, as SVC parameter tuning, overvoltage investigation etc.

5 - CONCLUSIQNS

At the expansion planning stage, a general model of SVCs may be adequate for long-term studies. Short-term studies, however, demand detailed modeling of SVCs. This is particularly so where the SVCs have a preponderant influence on system transient behaviour. In such cases, the control, synchronising and firing systems must be modeled in detail adequate to reproduce their transient behaviour under real system conditions.

6 - REFERENCES

1 - Rule Book, "Electromagnetic Transient Program (EMTP) Bonneville Power Administration, Portland, 1984.

2 - F. P. de Mello, "Power System Dynamics - Overview "Symposium and Adequacy and Phylosophy of Modeling: Dynamic System Performance. Presented at the IEEE Power Engineering Society, 1975 Winter Meeting.

3 - A. J. P. Ramos and H. Tyll, "Dynamic Performance of a Radial Weak Power System with Multiple Static VAR Compensators", IEEE Trans on Power System vol. 4, pp. 1316-1325, November 1989.

4 - CIGRÉ Working Group 38-01, Task Force nº 2 on SVC, "Modeling of Static VAR Compensator in Power System Studies". CIGRÉ Book "Static VAR Compensators", pp. 29-40, September 1985.

7 - APPENDIX - System Data

. Transmission Lines - per circuit, 230kV
..r1 = 0.09 ohm/km, x1 = 0.5 ohm/km,
..y1 = 3.2E-6 S/km (positive sequence).
..r0 = 0.5 ohm/km, x0 = 1.5 ohm/km,
..y0 = 2.0E-6 S/km. (zero sequence)
. Transformers
.. Fortaleza: 100MVA, 230/69kV, X = 14% tap = 0.95 (four equal units)
.. Russas: 16,67MVA, 230/69kV, X=38%, tap=0.95
.. Mossoró: 39MVA, 230/69kV, X = 38%, tap=0.88
. Loads (constant impedance type)
Bom Nome 230kV: 22.6MW, -2.9Mvar; Milagres 230kV: 61.7MW, 42.3Mvar; Banabuiu 230kV: 5.6MW, 2.8Mvar; Fortaleza 69kV: 190MW, 46.5Mvar; Russas 69kV: 11.7MW, 0.4Mvar; Mossoró 69kV: 21.7MW, 7.4Mvar.
. SVC data
.. Transformer: 200MVA, 230/26/2kV, (H-L1-L1) XH = 1.6%, XL1 = XL2 = 26%
.. Fixed capacitors: 373 microFarad per phase,
.. Surge capacitor: 0.1 microFarad, per phase,
.. TCR: 2x11.4 mH per phase branch, Q = 247.
.. Snubber circuits: 864 ohm, 0.25 microFarad.
. SVC Control system (see block diagram Fig.2)
.. K (regulating droop) = 0.0
.. PI regulator (BCL): Kp = 0.32V/V; Tn = 11ms; LSPI = 2.82V; LIPI = -2.77V.
.. PD regulator (FCL): KD = 0.74V/V; TD = 19ms; LSCR = 0.50V; LICR = -0.90V; LSBM = 3.0V; LIBM = -3.0V; Working Point, WP = 0.0V
.. Current Limiter (CL); IREF1 = 2.20pu; IREF2 = 1.50pu; IREF3 = 1.00pu.
.. Undervoltage Blocking Scheme (UBS)
REF230FF = 70%; REF230FT = 50%; REF 69FT = 50%

ALEXANDRE NAVARRO DE VASCONCELOS was born in Recife, Brazil, in 1950. He received the B.S degree in Electrical Engineering from the Federal University of Pernambuco, Brazil in 1975. In 1976 he was at Federal Engineering School of Itajubá for a pos-graduation program. He joined CHESF in 1975. His area of interest is SVC design and performance.

ÁLVARO JOSÉ PESSOA RAMOS was born in Recife, Brazil, in 1951. He received the B.S degree in Electrical Engineering from the Federal University of Pernambuco, Brazil in 1973, and the MSc from the Federal Engineering School of Itajubá, Brazil in 1975. In 1974 he joined CHESF. His area of interest is power system dynamics. He is a member of the IEEE.

MANFREDO VELOSO BORGES CORREIA LIMA was born in Recife, Pernambuco, Brazil,in 1957. In 1979 he was gratuated in Electrical Engineering from the Federal University of Pernambuco. In 1978, he joined CHESF, where he is engaged on SVC design and performance.

JOSIVAN SAMPAIO MONTEIRO was born in Campina Grande, Paraiba, Brazil in 1952. He received the B.S degree in Federal University of Paraiba, Brazil in 1977. In 1980 he was at Federal Engineering School of Itajubá for a pos-graduation program. He joined CHESF in 1978. His area of interest is power system transients.

HUMBERTO DÓRIA SILVA was born in Maceió, Brazil in 1951. He received the B.S degree in Electrical Engineering in 1974 and the MSc in Computer Sciences in 1980, both from Federal University of Pernambuco, Brazil. In 1975 he joined CHESF. This area of interest is computer methods and system modeling for power system studies.

LUCIANO RODRIGUES LINS was born in Garanhuns, Brazil in 1952. He received the B.S degree in Electrical Engineering in 1977. In 1982 he was at Federal Engineering School of Itajubá for a pos-graduation program. He joined CHESF in 1978. His area of interest is power system dynamics and SVC performance.

Discussion

J. Schwartzenberg and R. Fischl, (Drexel University, Philadelphia, PA): We wish to congratulate the authors for a very interesting paper on SVC modeling and would appreciate their response to the following comments and questions:

1. Have the authors encountered any problems with numerical stability of their EMTP model, when performing transient studies, arising from errors in the prediction of the instant when the thyristor is turned off as discussed in References [1] and [2]?
2. Have the authors obtained any field data to verify the accuracy of their EMTP or TNA SVC models?
3. It would have been useful if the authors plotted the errors between the EMTP and TNA SVC models. Of particular interest are the values of the maximum absolute errors and at what points in time they occur. Specifically, if the largest errors occur during turn-on or turn-off periods, then one can evaluate the accuracy of the EMTP's prediction of the SVC response.
4. When examining the responses of the TNA and EMTP models in Figures 5a & 5b, it seems that the PD regulator input signals are opposite in polarity.

References

[1] A. M. Gole, V. K. Sood, "A Static Compensator Model For Use With Electromagnetic Transients Simulation Programs," IEEE 1990 PES Winter Meeting, Atlanta, GA, February 4-8, 1990, Paper No. 90 WM 078-6 PWRD.

[2] K. H. Kruger, R. H. Lasseter, "HVDC Simulation Using NETOMAC", IEEE MONTECH '86 Conference, Montreal, Sept. 29-Oct. 1, 1986, Proceedings pp. 47-50.

A. N. Vasconcelos, A. J. P. Ramos, J. S. Monteiro, M. V. B. C. Lima, H. D. Silva and L. R. Lins (CHESF, Recife, Brazil): The authors thank the discussers for their interest on the paper. We address the questions in the same order, as follows:

1 - We didn't encounter any numerical oscillation when the existing snubber circuits are modeled. In our case, there are power system transients with overvoltages sufficient high to operate the thyristor protective firing (BOD - Breakover Diode). Since the BOD operates based on thyristor valve instantaneous voltages, the thyristor recovery voltage following a turn-off must be properly reproduced for correct operation of the BOD. Thus, our problem is not to eliminate numerical oscillation but to reproduce adequately the turn-off transients. The method used by NETOMAC [2] to avoid numerical oscillation is a convenient way to reduce computer costs when the thyristor recovery voltage is not of interest.

We intend to develop an improved thyristor valve model so that the thyristor recovery voltage, as registered in field, be reproduced adequately, as shown in Figure 1. This model must consider the thyristor dynamic characteristics (recovery charge Qrr) [1]. It requires a time step much smaller than that required to simulate system transients. In this case the approach proposed by Gole [1] will be very advantageous. Unfortunately it is not an available facility of the EMTP. The use of time step smaller than 20 microseconds suppresses or attenuates excessively the voltage peak so that we never obtain the voltage transient as verified in field (overshoot aproximately 20%). Time steps between 20 and 30 microseconds give rise to recovery voltage transients with overshoot in the range of 0 to 20%, depending on how close time increment is with respect to the point of current zero crossing. Figure 2 shows the thyristor recovery voltage obtained with the EMTP model.

2 - We have field data obtained from a single-phase short circuit applied at the end of a 230kV transmission line circuit. This field test showed a significant influence of the dynamic characteristic of the load. The active and reactive load don't restore immediately after fault removal, but slowly according to its dynamic characteristic. As the TNA and EMTP simulation make use of impedance type load, the agreement between simulation and field test with regard to regulator performance was poor for such large disturbance.

3 - The simulation carried out in the TNA intended to verify SVC control system performance and parameter tunning. The several variables concerned with power system and SVC control system were registered with conventional paper oscillographs. No care was taken to obtain precise data to compare with results from the EMTP model, that was developed some years later.

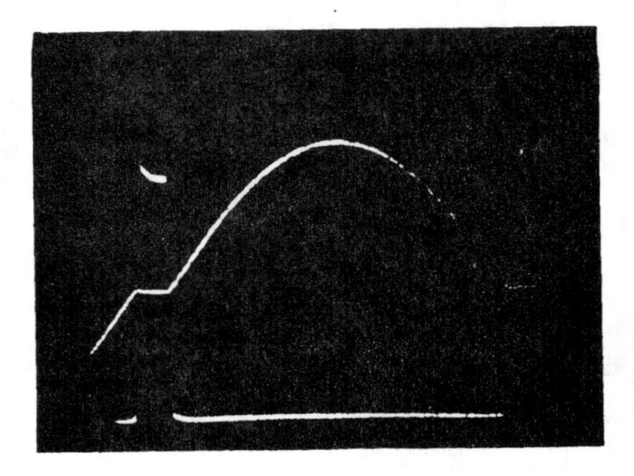

Figure 1 - Field Test - Thyristor valve voltage and current for firing angle 100°.

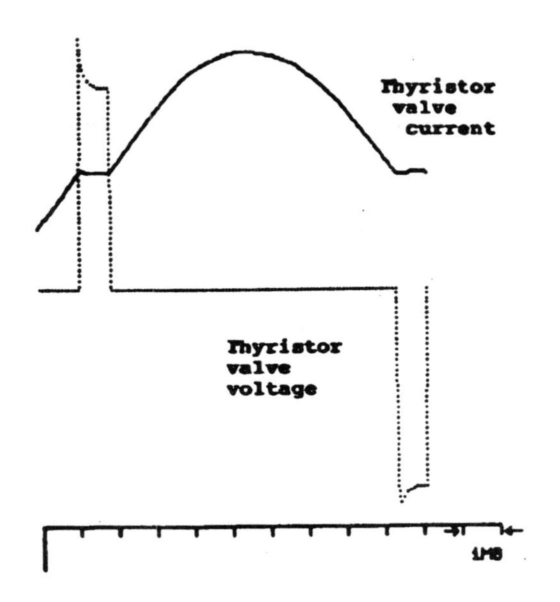

Figure 2 - EMTP Model - Thyristor valve voltage and current for firing angle 100°. Time step 30 microsecond.

4 - The PD input signal in Figure 5a, 5b and 6a, 6b are, in fact, with opposite polarity. The oscillograph channels for such signals were connected with opposite polarity for more convenient use of available space an the oscillograph paper.

REFERENCE

[I] Grafham, D. R. and Hey, J. C.: SCR Manual, Fifth Edition General Electric Co., Syracure, N. Y. 1972.

Manuscript received August 29, 1991.

A STATIC COMPENSATOR MODEL FOR THE EMTP

S. Lefebvre (Member)
Hydro-Québec
Vice-Présidence Recherche
Varennes, Québec, Canada

L. Gérin-Lajoie
Hydro-Québec
Vice-Présidence Planification du Réseau
Montréal, Québec, Canada

Abstract- A static VAR compensator model based on nodal analysis is presented. The model is integrated in the ElectroMagnetic Transients Program (EMTP) with minimal interface error and by taking into account initialization. The model is basically a generic compensator using Thyristor Controlled Reactors. The models are modular to represent adequately compensators of different designs while being detailed enough for predicting possible harmonic interactions between the AC system and the SVC. Typical applications on the Hydro-Québec system are described.

INTRODUCTION

The use of Static VAR Compensators, or SVC, in many locations are key elements for maintaining the stability of the Hydro-Québec transmission system. They improve system properties such as steady-state and dynamic stability, voltage profile, reactive power flow and fault situation (overvoltages, load rejections and other) due to rapid response to voltage deviation. Their response is however limited by considerations of stability within the SVC voltage control loop, for example interaction with the many resonances of the transmission network. These issues have been covered recently in [1,2].

Four of the eleven SVC in the Hydro-Québec network can be assumed to consist of Fixed Capacitor (FC) banks in parallel with Thyristor Controlled Reactors (TCR) using reverse parallel connected thyristors. But, in the present version of the EPRI/DCG EMTP, which is widely used for detailed transient studies, built-in models for SVC are not available.

There have been various concepts in the literature regarding SVC models [3,4,5]. However none of the authors have considered providing a general purpose built-in self-initializing EMTP model which would be directly applicable to the simultaneous simulation of an arbitrary number of SVC located at various locations in the network.

This paper concerns the integration of such a module in the EMTP. A TCR-FC model, which is based on nodal analysis is presented. The model is integrated in the ElectroMagnetic Transients Program (EMTP) with minimal interface error and by taking into account initialization. The model is basically a generic compensator using Thyristor Controlled Reactors. It may run for several seconds or even minutes without drift. It is fully implemented in the EMTP through a user-defined data module concept [6] which offers true modeling flexibility. The manner in which the model is implemented and initialized is transparent to the user.

91 SM 461-4 PWRS A paper recommended and approved by the IEEE Power System Engineering Committee of the IEEE Power Engineering Society for presentation at the IEEE/PES 1991 Summer Meeting, San Diego, California, July 28 - August 1, 1991. Manuscript submitted January 28, 1991; made available for printing June 18, 1991.

A major feature of the SVC module is that initialization is smooth with the system initially near a state possibly predetermined by a snap-shot of a transient stability run. Different initial MVAr output production is available.

The paper organization is as follows. The first part describes the major components of a typical SVC which is used at four of the existing SVC locations in the Hydro-Québec network. Then the model implementation and initialization is described. Typical simulation results are presented.

DESCRIPTION OF SVC

In their simplest form, the elements of a TCR-FC are shown in Figure 1. They include a reactor in series with a bi-directional thyristor-valve pair. The reactor is split into two units, with one unit on either side of a valve, in order to limit valve fault currents. The valves conduct on alternate half-cycles of the supply frequency depending on their firing angle. Full conduction, at a conduction angle σ of 180 degrees, is obtained with a firing angle of 90 degrees. The zero crossing of valve voltage defines the zero value of firing angle. In this case, the current is the same as that obtained if the valves were short circuited. Partial conduction is achieved with firing angles between 90 degrees and 180 degrees. The firing angle is controlled to maintain the ac voltage at the set point V_{ref}. Firing angles between 0 degree and 90 degrees are not permitted as they produce asymmetrical currents with DC components which saturate the transformer core. If α is the firing angle and X the total reactance of the reactor, the TCR appears as a controllable susceptance $B(\alpha)$:

$$B(\alpha) = (2(\pi - \alpha) + \sin(2.\ \alpha)) / (\pi . X) \quad (1)$$

$$\sigma = 2(\pi - \alpha) \quad (2)$$

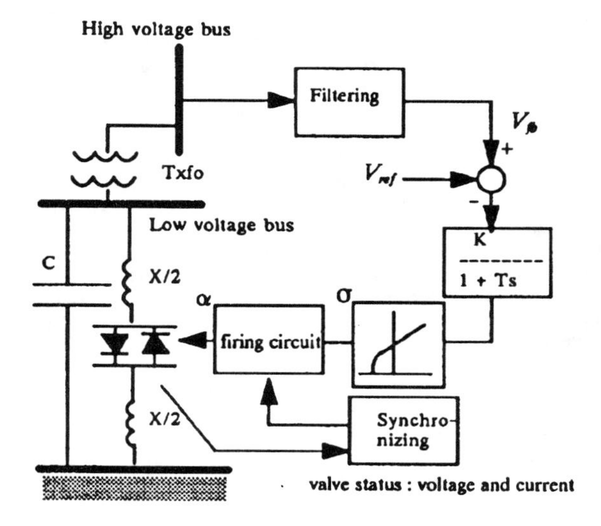

Figure 1: Basic elements of TCR-FC

The voltage control action provided by the SVC can be interpreted in terms of the shunt susceptance seen by the network to which it is connected. Control signals derived from system voltage at the SVC location and the SVC reactive power drawn define a voltage-control function of specifiable slope.

The TCR generates harmonic currents for partial conduction. With balanced operation, these are of odd-order. In a balanced three-phase system, where the three single-phase TCR elements are connected in delta to form a 6-pulse TCR-FC, as shown in Figure 2, only harmonics of the order 6n+1 and 6n-1 exist. Zero sequence current harmonics of the order 3,9, etc. circulate in the closed delta and are absent from the line currents. By increasing the pulse order, harmonics are lowered. A basic form of 12-pulse SVC can be derived from two separate TCR bridges of the kind in Figure 2, and with transformer connections, Txfo in Figure 1, which introduce a phase shift between the supplies to the different bridges. This is normally achieved with a Yg-Y and Yg-Δ arrangement. The 12-pulse SVC produce harmonics of the order 12n+1 and 12n-1.

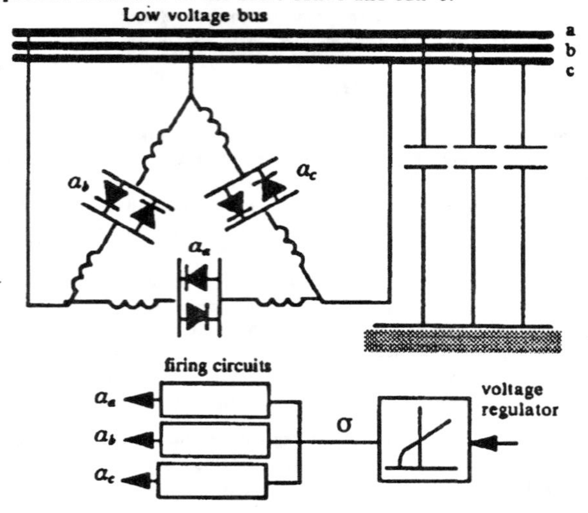

Figure 2: 6-pulse TCR-FC

The dynamic response of the TCR alone is fast, less than 5-10 ms, but the delays in measurement and control circuits as well as the system impedance may impose settings that give slower response times for control loop stability reasons of typically around 60-250 ms. Speed of response is mainly controlled by the adjustable time constant T of the voltage regulator in Figure 1.

Elements to be modeled include the valves, the synchronization circuit and the firing circuit. Overvoltage protection (fast over-ride), low current characteristics as well as duty cycle limits are not generic but they do not generally need to be modeled for low-frequency interaction studies. A closed-loop voltage regulator is required, as well as filter and measurement devices. The voltage regulator output is the required conduction angle which is passed on as a control signal to the firing circuit which generates firing pulses synchronized to achieve the requested conduction angle.

The three SVC phases are equipped with their own firing and synchronization circuits, but the voltage regulator is common to the three phases. The three line-to-neutral bus voltages are rectified in a 12-pulse bridge, not shown in Figure 1, and its output is filtered to remove harmonics before feeding the voltage regulator.

Models using 6-pulse and 12-pulse representations are useful. Simulation with the 6-pulse model is much faster and may be sufficient for 12-pulse SVC's, depending of the nature of the study (e.g. SVC control loop stability with a 6-pulse model, and harmonic studies with the 12-pulse model):

IMPLEMENTATION STRATEGY

The EMTP is a nodal analysis program based on the fixed time-step trapezoidal integration method. Discretized network equations are given by $Y_n V_n = I_n$, where Y_n is the nodal admittance matrix, V_n is the node to ground voltage vector and I_n represents node current injections including current sources for the integration history terms.

Valves can be directly modeled in the EMTP as switch-components and the SVC is easily incorporated into the Y_n matrix. The switches close after the anode voltage becomes larger than the cathode voltage, as soon as a firing signal is received at the appropriate firing or conducting angle. The switches open as soon as the current goes through zero from a positive value. The time-varying topology created by the switches operation requires that Y_n be rebuilt and re-triangularized. Irrespective of the type of solution method, the simulation equations have to be re-adjusted to take into account switch operation. When each SVC is modeled as a separate module, re-triangularization or re-formulation is confined to this device. When the SVC's are not distinct from the system equations, re-triangularization could be confined to nodes connected to switches, but for a large number of switches this does not lead to an efficient implementation.

Most transients programs, such as the EMTP, use the trapezoidal rule of integration. Its advantages, such as simplicity, absolute numerical stability and reasonable accuracy, usually outweigh its disadvantages. However, it also may cause sustained but bounded numerical oscillations at switch openings due to incorrect initialization at the discontinuity.

A general power converter simulation module based on hybrid analysis is described in [7]. The SVC module can be designed in the same manner, with the robustness, numerical oscillations suppression and initialization advantages described in [7]. Hybrid analysis enables easy automatic generation of the device equations which yield acceptable conditioning. To provide a module without interface error, a key requirement, the EMTP needs to be substantially modified and adapted in terms of EMTP/module and TACS/module interface, which is a different endeavor. This is not suitable to production code development in the short term since it would require considerable work.

State variable component modeling of an SVC has also been advocated as an alternative to nodal or hydrid analysis [3]. The reasons for such a choice, even though it is difficult to automatically generate proper trees to write device equations and even if it results in time step interface errors in the authors implementation, were to avoid numerical oscillations created by valve switching or poor conditioning that could result from nodal analysis of L-C circuits. Practically, the EMTP is quite robust for the simulation of L-C circuits due to its implementation of the solution techniques. Regarding the instabilities associated to valve operation, which cannot be eliminated by simply reducing the time step, it has been shown [8] that the backward Euler method and re-initialization (Critical Damping Adjustment) will cure them at modest cost for the EMTP. Practically, voltage spikes on the valves are of less concern than interface errors.

Production code development prefers to rely on an approach where minimal changes are required to the existing EMTP and good performance is expected. Valves are thus modeled as switch-components. The SVC are separate modules, but the SVC are integrated in the Y_n matrix in order to reduce the interface error. A straightforward and flexible strategy for implementing such models into the EMTP is to represent a device with the basic components of EMTP and TACS [9] but in such a way as to form an SVC module which can be called upon request for SVC's of different ratings located at any number of buses.

The interface error is between EMTP and TACS where one time step delay exists, however all non-control equipment is solved simultaneously with the network itself. The structure of the simulation is shown in Figure 3.

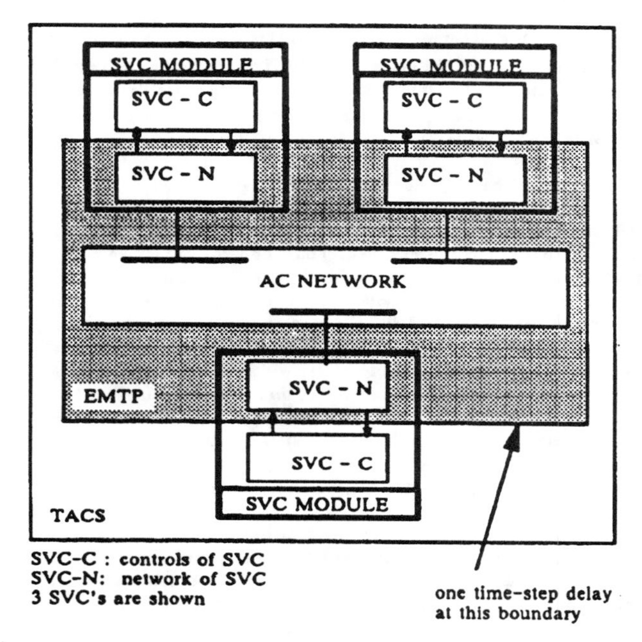

Figure 3: Integration of SVC's in the EMTP

EMTP MODELING

The concept of nested EMTP Data Modules (EDM) [6] is key to the modular SVC's implementation as EDM gives a facility to substitute variables in a module. A module may be built without compilation and linking using either models from basic EMTP data, or models from TACS. It does not require hierarchical data-type ordering since data can appear in any order.

A module is simply an ASCII file describing EMTP data with additional pointer information to allow alphanumeric substitution. An EDM is easily accessed through an include specification, therefore SVC's of different ratings can be connected at many arbitrary system buses by proper design of the SVC module. All equations associated with the EDM are automatically discretized either into the Y_n matrix of EMTP, or that of TACS.

SVC controls must represent the actual equipment, as the details of saturation, losses, voltage drops, firing controls, voltage sensing, regulator processing, synchronizing and valve firing controls can make a notable difference on dynamic performance. The SVC model must be able to handle accurately system switching events, fault initiation and fault clearing, asymmetric operation under unbalanced system conditions, and load rejection. It must precisely predict interaction of the SVC controls with the power system and it may run for several seconds or minutes.

TACS is used to model the basic TCR-FC controls, while the non-control equipment (SVC network) is modeled by EMTP devices. An SVC module is assembled from basic modules consisting of transformers, thyristor valves (switches and dampers), controlled reactor, shunt capacitance banks, harmonic filters, and initialization statements. The model includes 3rd harmonic ac filters in the specific Hydro-Québec SVC's simulated. The SVC-EDM is designed in such a way that a single include statement, followed by the EDM parameters, suffices to load an SVC. All other EDM needed by the SVC-EDM are included in it.

Transformers

Transformer EDM consist of interconnected single-phase saturable transformers and linear leakage reactances all simulated by standard and validated EMTP devices. For Hydro-Québec applications, the SVC 12-pulse transformers are Yg-Y and Yg-Δ 735/22 kV. They use a leakage of 15% on the high-side with little on the low-side, while the saturation knee is around 1.2 pu on the low-side with a slope of 40%.

Thyristor valves

Thyristor valves are modeled as grid-controlled switches. The backward Euler method is not yet fully implemented in the production version of the EMTP. Therefore it is required to use a damping circuit in parallel with the valves to remedy to some numerical oscillations. This is simply a R-C circuit whose parameters are selected to damp numerical oscillations. It is also useful to place resistances in series (to partly account for the losses) and in parallel with the TCR in each phase to further damp the oscillations. This is not to be confused with the real snubber circuit of the valve. Since there are two switch changes per cycle, there is sufficient time in between to damp the oscillations with damping resistances which are large enough so as not to distort the circuit behavior and the losses.

The valves are connected in delta as in Figure 2. The required grid signal, which is a 100μsec pulse train, is computed by the control-EDM which consists of TACS statements.

Control system

The control system is also a module in itself. It consists of several modules entirely described by TACS statements. It includes measurement circuits, the voltage regulator shown in Figure 1, the linearization circuit between the voltage regulator output and the conduction angle order σ, the synchronizing circuits and the firing pulse generator.

It is possible in the control-EDM to change the main settings of the input filters, as well as the slope and the time constant of the regulator SVC. If the control structure needs modification, it is relatively simple to create a new sub-module in the EDM.

Firing and synchronizing. The firing and synchronization units for several Hydro-Québec SVC's are shown in Figure 4. Their modeling follows the principles delineated in [5] as it turns out to be a good representation. Other devices can be accommodated by replacing this module. The input σ is provided by the regulator shown in Figure 1.

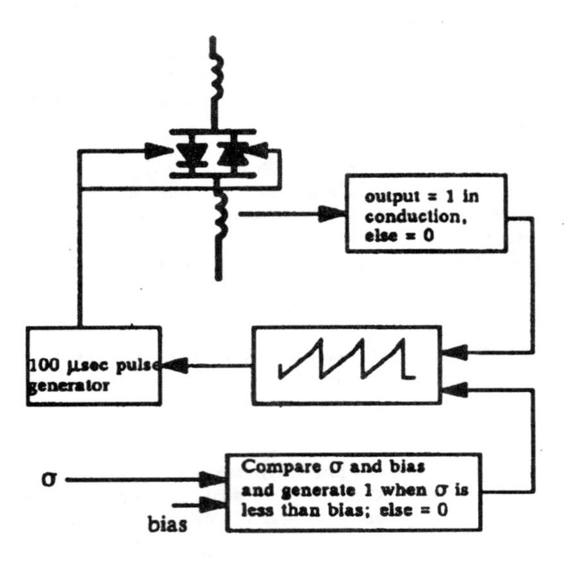

Figure 4: Firing and synchronizing

Filtering of voltages. Figure 5 provides further detail on the filtering applied to the measured voltages illustrated in the overall model of Figure 1. Details may vary somewhat but general characteristics are indicated. The dc side 60-120-360Hz notch filters are quite standard. The combination of notch filters (80Hz and 96 Hz) on the AC side of the voltage transducer and 57Hz low-pass filtering on the dc side were added to the Hydro-Québec SVC's when they were first installed to suit the system needs [10]. The additional user-defined ac side high-pass filters are useful to eliminate interaction with 5-20Hz oscillation modes resulting from the series compensation of the high voltage lines [1].

Detailed filtering characteristics are described in [1,10]. The module can be user-modified to take into account different filtering strategies. The output signal of the module is scaled such that 735kV on the high voltage bus translates into a signal V_ϕ of 1pu (5V). This scale factor adjustment is important, due to the closed-loop SVC regulation action, to reproduce system operating conditions at the exact value of MVAr and conduction angle.

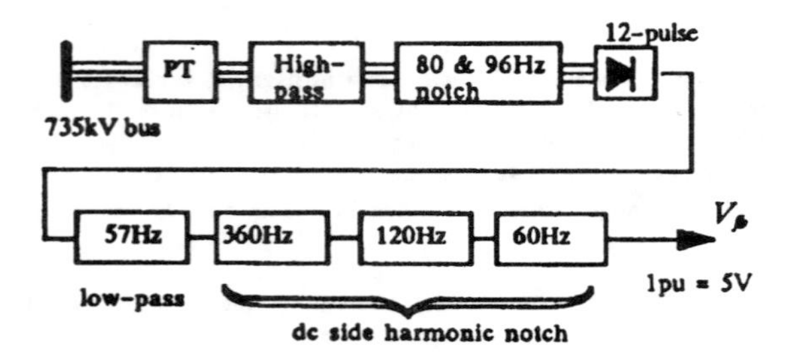

Figure 5: Filtering of measured voltages

INITIAL CONDITIONS

Initialization is a key feature of the model and requires use of an auxiliary program to compute required SVC's parameters and initial conditions.

It is important to carry over initialization in such a way as to approach as much as possible the specified steady-state condition at time zero. Failure to do so may result in unacceptable cpu time since the system simulated is large, abnormal modes may occur, and small signal interaction studies can be difficult if there exist poorly damped modes of oscillation.

Combined initialization of SVC's and the ac network is complex, not only by some EMTP and TACS limitations, but also by the nature of the nonlinear cyclic time-varying circuit that constitutes SVC's. Although it is possible to have an estimate of the valves initial conduction pattern since the SVC topology is known, it is necessary to initially model a current injection that corresponds to a given MVAr injection corresponding to the desired steady-state value of the conduction angle.

Initialization is also complicated by the fact that a floating SVC uses a conduction angle σ of 101 degrees (for the Hydro-Québec choice of X and C in Figure 1), thereby making even the MVAr sensitive to small errors in σ. Harmonic generation is large at such an angle, further complicating the issues.

Phasor solution representation of TCR

All switches grid-controlled by TACS are initially open unless their initial state is differently specified. Thus, only the capacitive part of the SVC's would be connected during the EMTP phasor solution: this results in incorrect MVAr generation at time zero. If switches are assumed close, then the full TCR reactance is used in the phasor solution, which again is incorrect. To properly represent the cyclic action of the switches in steady-state, which modifies the effective value of TCR reactance, it would be required to replace the TCR by a set of admittances at different frequencies and to proceed to multi-frequency initialization. The admittances are function of X and the value of firing angle α, as described by equation (1) for the fundamental frequency. This is an approximation.

Instead of modifying the EMTP itself, it was chosen to use a more approximate but simpler TCR model for the phasor solution. All valves are assumed initially close, and the TCR is replaced by an inductance L1 whose value corresponds to the TCR fundamental susceptance at the desired conduction or firing angle.

This is only approximate since harmonics are neglected. The method is thus quite simple and avoid undesirable dc bias in the TCR currents. In the time step loop of the EMTP, or after a few time steps, a reactor L2 is connected in parallel with L1 such that the correct value of TCR inductance is obtained, and the switches are allowed to operate under the control of the firing devices of Figure 4, which itself needs some time to initialize.

The EMTP-TACS initialization deficiencies are therefore alleviated in a simple manner, compatible with the coming releases of the EPRI/DCG EMTP.

For a network not showing poorly damped modes of oscillation, initialization only requires a few cycles. In comparison, starting from arbitrary initial conditions would often require a few seconds to reach an equivalent steady-state.

EMTP/TACS limitations

There is a time step delay between TACS and EMTP. In the current EPRI/DCG EMTP version, TACS devices using an EMTP variable as input are assumed to have zero initial conditions, unless these are provided separately. It is necessary to resort to such initialization by providing at least approximate instantaneous ac voltages to the filtering circuits of Figure 5.

Unnecessary TACS constraints are being removed and multi-frequency initialization is to come in a future EMTP release.

Auxiliary program

An auxiliary program eliminates the burden of computing the required initial values, including those of L1, L2. This program, associated with the SVC module, is designed such that on user-input of the high voltage magnitude V_{ac} and angle δ at the SVC bus, the initial MVAr generation Q_0, the pulse-number, the desired SVC regulation slope S, and the SVC rating, the program provides all parameters and constants needed.

Explicit use of the SVC topological and control information is made in order to assist initialization by EMTP/TACS. The auxiliary program algorithm is based on the SVC steady-state operating equations. It is briefly described by:

a) Given V_{ac} and Q_0, find the net SVC current on the low voltage side of the transformer.

b) Given Q_0 and the computed losses, compute the net reactive power Q_{22} on the low voltage side (22kV) of the transformer.

c) Given Q_{22}, find the net susceptance B_{net} on the low voltage side of the transformer.

d) Given B_{net} and the total SVC shunt capacitance, deduct the TCR reactance $B_{inductive}$.

e) Given $B_{inductive}$ and the nominal TCR reactance X, use equations (1) and (2) to obtain the conduction angle σ from eqn(1-2).

f) Compute L1 to correspond to $B_{inductive}$.

g) Compute L2 such that the parallel combination of L1 and L2 yields the nominal TCR reactance.

h) Given the slope S, V_{ac} and Q_0, find the reference voltage V_{ref} of Figure 1.

i) Estimate all other useful initial conditions, including initial TCR applied voltages.

The auxiliary program uses some specific SVC information and must be tailored to the devices used, for example 735kV/22kV transformers.

Example of initialization

The TCR currents at initialization in one 6-pulse bridge are shown in Figure 6 for a difficult but realistic start-up condition with the series compensated Hydro-Québec system. The sequence of events is as follows.

For the phasor solution, the valves are short-circuited and L1 replaces the TCR. This initializes the network at the correct voltage and SVC MVAr generation. At the first time step, the valves are allowed to operate normally, however there is still an error in the synchronization signals which results a strong imbalance between the phases in the delta of TCR. The time at which the firing pulses are put in service can be optimized. Later, at one cycle, the TCR reactance assumes its normal value and the conduction angle is adjusted to steady-state.

Although not perfect, initialization is quite rapid. Sufficient time must be given to the system for recovering from the initial stress imposed.

Within two cycles, the SVC is operating a large value of conduction angle and the current imbalance is much reduced. The series compensated system is more difficult to initialize due to less damped system response.

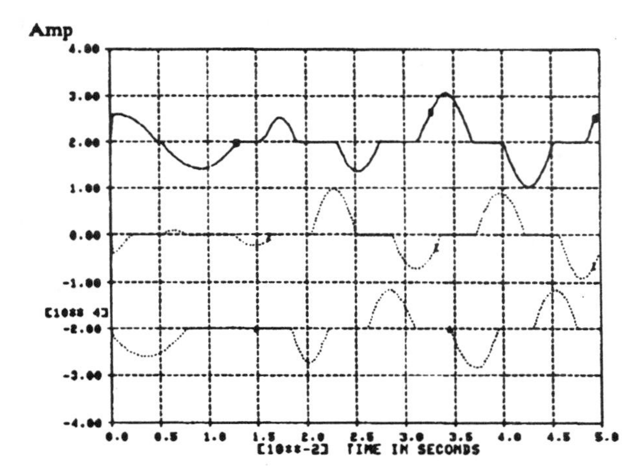

(the curves are offseted along the y-axis)

Figure 6: TCR currents at initialization

CASE STUDIES

The developments reported in this paper lead to analysis and evaluation facilities for a very wide range of transient operating modes in systems which include TCR-FC's. For example, the model allows TCR harmonics to be evaluated closely for any given operating state, including those of phase-voltage unbalance. The evaluation is made directly from wave-forms in the time-domain and Fourier transform analysis.

Performance of the model is illustrated on the Hydro-Québec network. System data to model the Hydro-Québec high voltage network consist of an EMTP data file of roughly 450 three-phase branches with approximately 30 000MW of winter peak load. The data include all of the parameters of the transmission network needed for transient analysis.

Of special importance are the load models which match field measurements, since these provide a major portion of the inherent damping of the shunt network resonant modes[2]. Damping is also influenced by low-frequency resistance of the major transmission lines, of the shunt reactors and of the major generators. Resistive loads are only important when in an area with little or no motor load.

SVC's can be represented as 6-pulse or 12-pulse models representing 1, 2 or 4 basic units110MVAr inductive and 330MVAr capacitive. Due to the choice of the TCR value, in the floating mode, the conduction angle σ is expected to be near 101 degrees. This angle is limited between 25 degrees (fully capacitive mode) and 156 degrees (inductive mode).

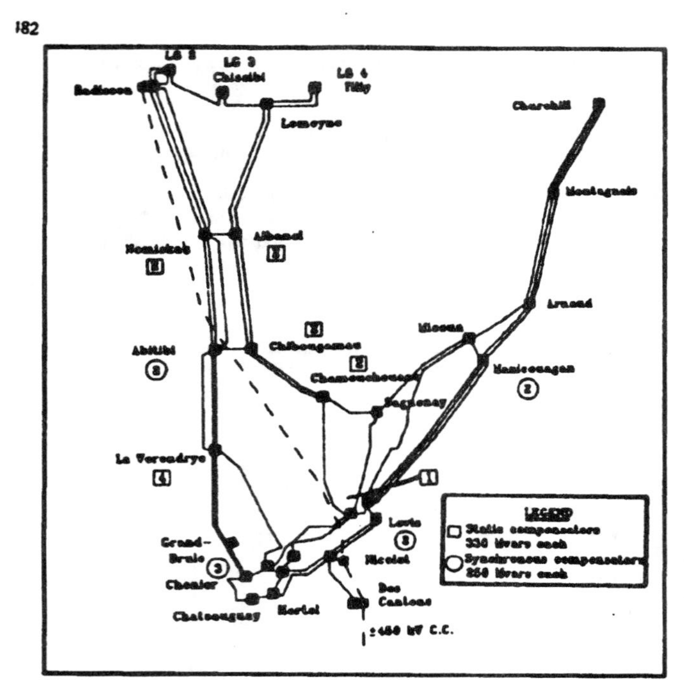

Figure 7: Hydro-Québec high voltage network

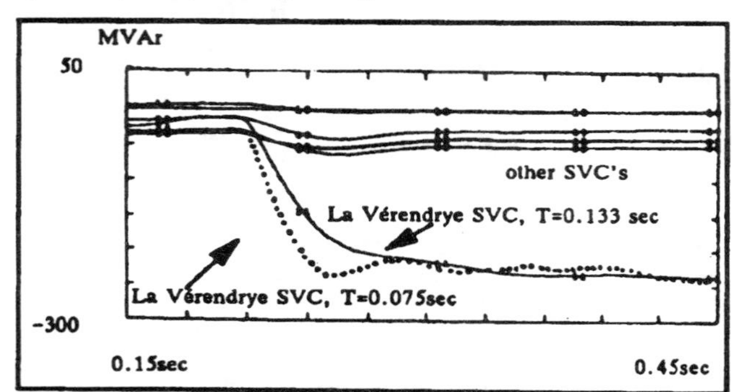

Figure 8: Connection of 330MVAr reactor

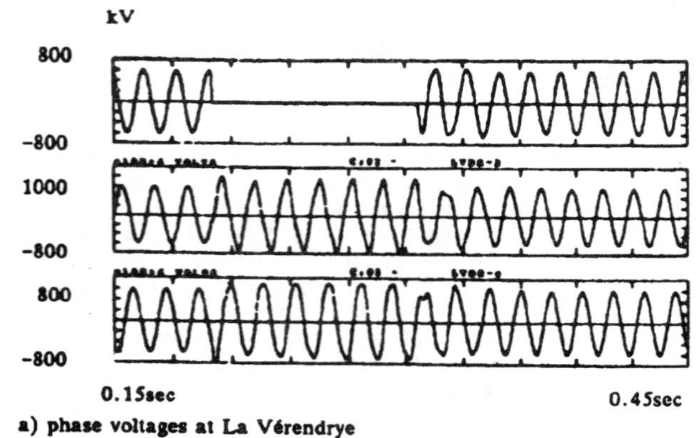

a) phase voltages at La Vérendrye

Figure 9: Single-phase fault at La Vérendrye

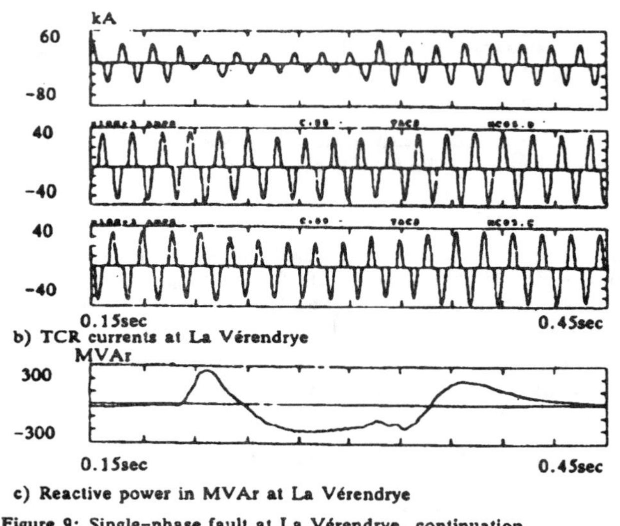

b) TCR currents at La Vérendrye

c) Reactive power in MVAr at La Vérendrye

Figure 9: Single-phase fault at La Vérendrye, continuation

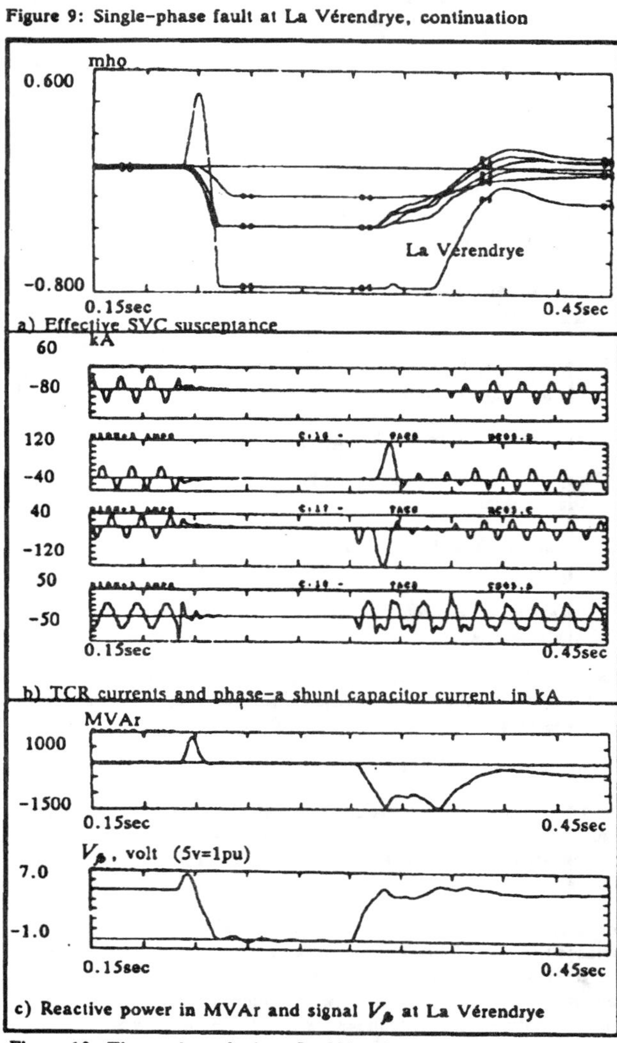

a) Effective SVC susceptance

b) TCR currents and phase-a shunt capacitor current, in kA

c) Reactive power in MVAr and signal V_ϕ at La Vérendrye

Figure 10: Three-phase fault at La Vérendrye

In the case under investigation, there are six 6-pulse SVC, representing 13 units, distributed throughout the James Bay transmission corridors, see Figure 7. All major transmission lines are shown in the figure. Three type of events are described: connection of a 330MVAr reactor, single-phase and three-phase faults. These events occur at La Vérendrye, where an SVC representing 4 units is modeled. All SVC's are adjusted with a regulation slope of 3%. No series compensation is used here in order to focus on SVC performance with simpler system behavior.

Case 1: Connection of a 330MVAr shunt reactor

Figure 8 shows the SVC's MVAr production for this event when the time constant T (Figure 1) of the voltage regulator is specified as 0.133sec. The MVAr sharing among the SVC's is function of their location and characteristics. Here 80% of the MVAr are provided by the La Vérendrye SVC. The figure also illustrates a more oscillatory and faster behavior with T of 0.075sec instead of 0.133sec.

Case 2: 6-cycle single-phase fault

Figures 9a and 9b show the behavior for a fault applied on phase a. During the fault, voltages on phases b and c increase. This initially reflects as larger La Vérendrye TCR currents than in pre-fault in two of the branches. After a few cycles the inductive currents start to diminish though. Figure 9c confirms that the MVAr production is inductive for about 2.5 cycles, before becoming capacitive for the rest of fault duration.

Case 3: 6-cycle three-phase fault

The SVC's respond quickly to a three-phase fault as they all go to their limit as shown in Figure 10a which illustrates the apparent SVC's susceptance. The initial susceptance overshoot is due to a corresponding overshoot in ac voltage signal provided by the measurement circuits. The SVC's then return slowly to their initial floating state when the fault is cleared. Figure 10b shows the three TCR currents, as well as the shunt capacitance current of phase-a. When the fault is eliminated there is a sudden burst of current in two of the TCR currents. As expected, the harmonic content of the capacitance current is high since here SVC's are modeled as 6-pulse.

Figure 10c indicates the MVAr production and the measured voltage V_ϕ (5V = 1pu). When the fault is cleared, the V_ϕ signal reaches 0.97pu only after 3cycles. This is the threshold of the capacitance mode (regulation slope is 3%). Hence when the fault is cleared, there is a delay before the effective susceptance is modified.

Comparison with field tests

A 6-cycle single-phase fault is used to compare field test results (only 2 SVC's) with simulation. System conditions and operating points are not strictly the same, but they allow a qualitative comparison. All SVC's are initially floating. The SVC's are 12-pulse units, but they are simulated as 6-pulse. Note also that the field test uses a time constant for the voltage regulator in Figure 1 of T = 0.100 sec, whereas the simulation uses T=0.133 sec.

Results are given in Figure 11 for a fault located at one of the SVC high voltage bus, Figure 11a for the field test and Figure 11b for the simulation.

The measured voltage V_ϕ (5V = 1pu), in both the field test and the simulation, shows an initial 10% overshoot even though the phase-a voltage has collapsed to zero. There is another overshoot when the fault is cleared, this is followed by a first order voltage decay to steady-state.

Other comparisons, both from field tests and analytical studies, have been successly performed.

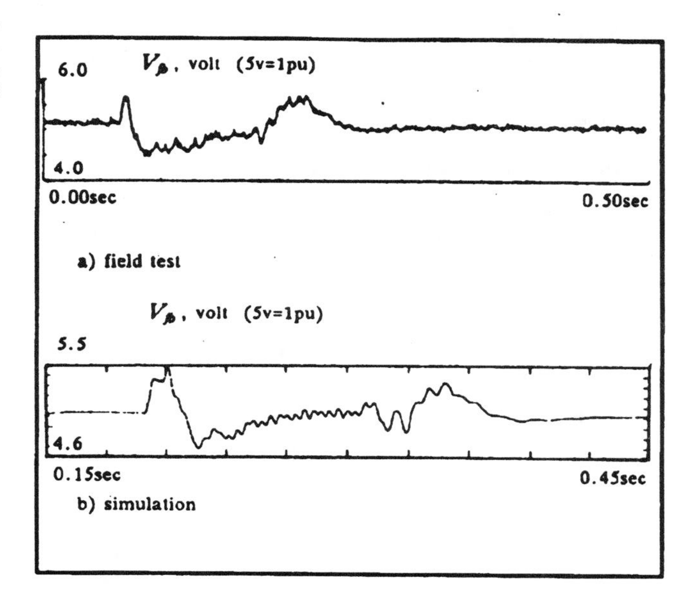

Figure 11: Single-phase fault,
a) field test,
b) simulation

CONCLUSIONS

The developments of the paper extend further the scope of detailed electromagnetic transient analysis and simulation facilities since the SVC model is integrated in the current EPRI/DCG EMTP.

The work of this paper has been carried out partly as Hydro-Québec collaborative contribution to the modeling in the EPRI/DCG EMTP program. The model will be available in a future program release.

The modeling methods make provision for representing any particular SVC as may be required without the need of user intervention inside the EMTP itself and without the need to re-compile and link the program. Initialization of the SVC has been described and the ideas are applicable to a range of models.

Model behavior has been checked against separate field results from commissioning tests, and against analytical studies.

REFERENCES

[1] Gérin-Lajoie L., Scott G., Breault S., Larsen E.V., Baker D.H., Imece A.L., "Hydro-Québec multiple SVC application control stability study", IEEE/PES 1990 Winter Meeting, Atlanta, Paper 90 WM 079-4 PWRD.

[2] Larsen E.V., Baker D.H., Imece A.L., Gérin-Lajoie L., Scott G., "Basic aspects of applying SVC's to series-compensated ac transmission lines", IEEE/PES 1990 Winter Meeting, Atlanta, Paper 90 WM 080-2 PWRD.

[3] Gole A.M., Sood, V.K., "A static compensator model for use with electromagnetic transients simulation programs", IEEE/PES 1990 Winter Meeting, Atlanta, Paper 90 WM 078-6 PWRD.

[4] Nguyen T.T., "TCR and SR compensators in composite electromagnetic transient simulation", IEE Proceedings- series C, Vol 136, No 3, pp. 195-205, 1989.

[5] Lasseter R.H., Lee S.Y., "Digital simulation of static Var system transients", IEEE Transactions on Power Apparatus and Systems, Vol PAS-101, No 10, pp. 4171-4177, 1982.

[6] Meyer W.S., "EMTP data modularization and sorting by class: a foundation upon which EMTP data bases can be built", EMTP Newsletter, Vol 4, No 2, November 1983.

[7] Mahseredjian J. Lefebvre S., Mukhedkar D., "Power converter simulation module connected to the EMTP", IEEE/PES 1990 Summer Meeting, Minneapolis, Paper 90 SM 454-9 PWRS.

[8] Marti J.R., Lin J., "Supression of numerical oscillations in the EMTP", IEEE Transactions on Power Delivery, Vol 4, No 2, pp. 739-747, 1989.

[9] EMTP rule book, EPRI report EL-4541s-CCMP, 2 volumes.

[10] Czech P., Hung S.Y.M., Huynh N.H., Scott G, "TNA study of static compensator performance on the 1982-1983 James Bay System", International Symposium on Controlled Reactive Compensation, sponsored by Hydro-Québec and Electric Power Research Institute, September 1979, pp. 323-347.

Serge Lefebvre (M'76) received the BScA and MScA degrees in electrical engineering from École Polytechnique de Montréal in 1976 and 1977 respectively, and a Ph.D. from Purdue University (Indiana) in 1980. He is working at the Research Department of Hydro-Québec since 1981 while being an associate professor at École Polytechnique de Montréal. His research interest are in power system analysis techniques, computer applications, and dc systems. Dr. Lefebvre is Chairman of the IEEE working group "Dynamic performance and modeling of dc systems".

Luc Gérin-Lajoie was born in Montréal, Québec in 1958. He received his BScA degree in electrical engineering from École Polytechnique de Montréal in 1982. From 1982 to 1985, he worked for Hydro-Québec in system planning of the high voltage network. After one year as an associate for a firm involved in design and manufacturing of electronic control systems, he joined the control and protection department of Hydro-Québec in the system planning group. His responsibilities include analytical studies relating to the performance of SVC's on the bulk transmission system, as well as control, stability, short circuit, load flow studies and protection specifications for transmission system.

Discussion

A.M.GOLE and V.K.SOOD, (University of Manitoba, Winnipeg, Manitoba and Hydro-Quebec, Varennes, Quebec) : We have read the paper presented by the authors with interest. We complement the authors on their effort in modelling an SVC model that can be initialized. A number of issues have been raised by the authors, and we would like to clarify some of these :

1. The concept of the self-contained, state variable, modular (child) models of any convertor (SVC or HVDC) interfaced to a standard general purpose EMT (parent) program proposed in [3] is ideally suited for an arbitrary number of SVC placed at various locations in the network, although in the example in [3] we presented only one compensator at an HVDC inverter bus. We are at a loss to understand the claim made by the authors in their introduction.

2. The reason for the stand-alone model described in our paper [3] is not due to the lack of chatter elimination methods in the host program EMTDC which utilizes a method [A] very similar to the 'Backward Euler' method [8] available in some versions of the EMTP. The paramount reason for the selection of our model was that it allows for the switching of the SVC valves to occur as close as possible in time to the ideal switching instants. In several EMTP type programs there can be firing and turnoff errors up to one timestep in duration which are a potential source of spurious non-characteristic harmonics. This aspect was particularly important for the Chateauguay System model described in our paper because of the harmonic resonances at the inverter bus. Had we used a model that was embedded in the main system Y matrix such as the one the author's have used, we would have had to operate with very small timesteps for our particular system. Instead our approach uses a smaller timestep in the neighborhood of switching instants and moves to a larger timestep otherwise. One other method of implementing this feature is to use interpolation of history terms as is done in the NETOMAC program [B], but we chose the variable timestep approach instead in which the external module could be interfaced to any standard host program such as EMTDC or EMTP. The overhead of using a variable timestep was marginal as shown in [3] and below :

Using a typical value of 10 μS for the sub-multiple timestep and a 50 μS main timestep, an average number of 2.5 extra steps per cycle are required. Thus, for a 12 pulse converter 30 extra timesteps per cycle are required. At 60 Hz, one cycle has 333 timesteps; hence an average overhead of 30/333 i.e. 9% is required.

3. Algorithms for generating proper trees for a state variable based formulation are available [C] but were not used in our program because for the specific SVC topology the equations were readily written by hand. We have presented in [3] an extremely simple interfacing technique that eliminates numerical errors due to one-timestep interfacing delays and results achieved demonstrate this. Our experience is therefore contradictory to what is mentioned in the 5th paragraph under the heading 'Implementation Study' which suggests that such problems exist with our model.

4. We must point out that the authors themselves have used a fictitious damper circuit across the SVC valves and resistors in parallel to the TCR to damp out numerical oscillations. As discussed in [3], such problems can arise due to the inherent non-canonical nature of the Trapezoidal algorithm used in EMTP type programs. The state variable method on the other hand is more robust in such circumstances.

We believe however that numerical errors caused either by component switching or interfacing timestep delays should be of concern to all simulations. The importance one attaches to these errors should be judged by the results obtained for the simulation at hand

References

[A] Irwin, G.D. & Woodford, D.A., "EMTDC - High Performance Electromagnetic Digital Simulation" Paper accepted for publication in Electrosoft, 1991.

[B] Kulicke, B., "Simulation program NETOMAC: Difference Conductance Method for Continuous and Discontinuous Systems", Siemens Research & Development Reports, Vol. 10, No. 5, pp. 299-302, (1981).

[C] Cheung, R.W.Y., Jin, H., Wu, B. & Lavers, J.D., "A Generalized Computer-Aided Formulation for the Dynamic and Steady-State Analysis of Induction Machine Inverter Drive Systems", IEEE Trans. on Energy Conversion, Vol 5, NO. 2, June 1990, pp 337-343.

Manuscript received August 27, 1991.

S. Lefebvre and L. Gérin-Lajoie: We thank our colleagues for their continued interest in our work.

The model development originates from the need to have readily available detailed static compensator models in the DCG/EPRI release of the ElectroMagnetic Transients Program (EMTP) which is a basic study tool for the Hydro-Québec engineers. The model presented in the paper is the result of intensive R&D efforts. Modeling the Static Var Capacitor (SVC) itself was the simplest task. Initialization was a difficult key feature to implement. It is needed by the type of studies and the specific use intended for the model. The model was then validated to match commissioning studies.

In addition to the SVC presented in the paper, the next release of the DCG/EPRI EMTP, will also contain a TCR-TSC SVC, which has been recently described in [A]. This will complete the set of SVC models, previously not existent in EMTP.

Here are some specific replies on the questions raised in the discussion.

Question 1 : Generic or detailed SVC models, other than those described above, are simply not available in the EMTP. Furthermore, the SVC model proposed in [3] cannot be considered a generic model. Firstly, its equations must be re-calculated and re-sequenced for an even slightly different circuit topology, whereas our approach handles topological changes. Secondly, there is no discussion in [3] regarding the self-initialisation of the SVC module. Thirdly, a sequential interface is used in [3], with no discussion of network impact.

Questions 2 and 3 : The Critical Damping Adjustment procedure is presently being implemented in the production version of the DCG/EPRI EMTP. The accurate detection of the change of state in thyristor valves is important for improving numerical stability. The use of the reduced time-step as originally applied in the SACSOTR [B] program and discussed in [C] (pp. 229-231), is also applied in [3] but within the sequential interface of the SVC module. When the time-step changes in the module, the diakoptic equations of the parent network are affected, but the method in [3] does not account for this. An heuristic procedure may work for some networks but not for others. The SVC module in [3] is considered as a pseudo-nonlinear element in the EMTP, the limitations and difficulties of this type of modeling are discussed in [D] (pp. 8-1 and 8-2). Numerical errors can of course be judged from the simulation results obtained, but the numerical simulation tool should not be used in such a trial and error procedure. In any case, there must be specific criteria for the selection of the intermediate time-steps and network modifications such as the resistance R_c, to account for limitations of the interface methodology.

Question 4 : We have used fictitious dampers to alleviate numerical oscillations since, as described previously, the CDA procedure of EMTP is now being implemented in the production version of DCG/EPRI EMTP. The reason why the numerical oscillations can be eliminated when the state variable method is used and solved by the trapezoidal algorithm, is the reformulation of the state variable equations at each network topology change. With no reformulation and the replacement of valves by binary resistors, extraneous eigenvalues are introduced and numerical oscillations will occur [E]. It can be demonstrated that when the state variable equations are assembled for nodal variables and not branch variables as usually (also in [3]), then these equations are equivalent to those obtained from the nodal analysis formulation based on the associate discrete circuit model.

References

[A] S. Lee, S. Bhattacharya, T. Lejonberg, A. Hammad, S. Lefebvre, "Detailed Modeling of Static Var Compensators Using the Electromagnetic Transients Program (EMTP), IEEE T&D Conference, September 1991.

[B] B. Hébert, SACSOTR Simulation Program, Université du Québec à Trois-Rivières, April 1983.

[C] V. Rajagopalan, Computer aided analysis of power electronic systems, Marcel Dekker Inc., 1987.

[D] EMTP Revised Rule Book, Version 2.0, Vol 1, EPRI-6421-L

[E] J.G. Kassakian, "Simulating Power Electronics Systems- A New Approach," Proc. IEEE, Vol 67, No 10, Oct 1979.

Manuscript received November 12, 1991.

IEEE Transactions on Power Delivery, Vol. 10, No. 3, July 1995

An EMTP Study of SSR Mitigation Using the Thyristor Controlled Series Capacitor

W. Zhu, R. Spee, R.R. Mohler, G.C. Alexander
Department of Electrical and Computer Engineering
Oregon State University
Corvallis, OR97331

W.A. Mittelstadt
Bonneville Power Administration
Portland, OR97208

D. Maratukulam
EPRI
Palo Alto, CA94103

Abstract This paper presents an EMTP (Electro-Magnetic Transient Program) simulation study of the SSR mitigation effect of Thyristor Controlled Series Compensation (TCSC) operated in the vernier mode, based on a simplified model of the North-Western American Power System (NWAPS). The study shows that TCSC vernier operation provides significant mitigation of SSR in some cases. An analysis of the equivalent TCSC impedance with respect to different frequencies is used to supplement these studies.

1. Introduction

It is known that series capacitor compensation benefits power systems in more than one way, such as enhancing transient stability limits, increasing power transfer capability, etc.[1]. It is also known that fixed series compensation may cause Subsynchronous Resonance (SSR) in power systems, which can lead to severe problems, such as damage to the machine shaft[1]. However, it has been noted that the newly developed Thyristor Controlled Series Compensation (TCSC) operated in vernier mode benefits the mitigation of SSR[2,3]. This is among several TCSC benefits to power systems, which include enhancing transient stability limits to higher values than using fixed series compensation, making load flows more flexible, and controlling loop flows[2, 3]. A detailed simulation study of the SSR mitigation effect of TCSC vernier operation, compared with fixed series compensation case is necessary for each application. Also, it is desirable to base the simulation on realistic power system models and to use standard power system simulation programs, such as EMTP. Moreover, analytic techniques to integrate and explain the underlying characteristics are needed. Assuming slow dynamics of SSR, one approximate method of analyzing the mitigation effect is to view TCSC as an electrical element in the power system and study the frequency domain characteristics of the equivalent TCSC impedance for a typical level of excitation. In this paper, we present our studies as mentioned. The studies focus on the SSR mitigation effect of TCSC vernier operation compared with fixed series compensation. This phenomenon has been observed repeatedly in our simulation studies, and is supported by previous studies[2]. The study of the modal damping provided by TCSC vernier operation is currently underway. A field test of the TCSC device installed[3] is currently being planned by Bonneville Power Administration (BPA) with a view to assessing TCSC performance.

94 SM 477-0 PWRD A paper recommended and approved by the IEEE Transmission and Distribution Committee of the IEEE Power Engineering Society for presentation at the IEEE/PES 1994 Summer Meeting, San Francisco, CA, July 24 - 28, 1994. Manuscript submitted December 30, 1993; made available for printing April 20, 1994.

2. Model Considerations

The simulation studies discussed are based on a model representing the 500kV-level network of the North-Western American Power System (NWAPS) and using the standard EMTP.

A diagram of the power system model used is shown in Fig.1 . The TCSC is installed between SLATT and BUCKLEY. It has been noted that when the generator at BOARDMAN, shown in Fig.1, is radialized to BUCKLEY and beyond, the generator is most vulnerable to the excitation of SSR at one or more of its shaft modes through the effect of the SLATT series compensation and the fixed series compensation beyond GRIZZLY in the NWAPS. Thus, in the model used, the BOARDMAN generator is modelled in detail by a system of differential equations while the remaining 21 generators are modelled by voltage sources. The BOARDMAN generator is radialized to BUCKLEY and beyond by operating the breakers at SLATT and GRIZZLY. Both transient and subtransient electrical sub-system dynamics of the BOARDMAN generator are represented. Five masses, namely, mass #1 through mass #5, are mounted on the shaft of the generator. The mass #1 through mass #5 represents high-pressure turbine, low-pressure-1 turbine, lower-pressure-2 turbine, generator, and exciter, respectively. The mechanical damping values used represent nominal values. The BOARDMAN generator field tests are needed to establish actual damping constants. In addition to a 2.2 Hz local *swing mode*, the BOARDMAN generator has the following 4 shaft modes:

mode 1 - 12.5 Hz; *mode 2* - 25.0 Hz;
mode 3 - 29.0 Hz; *mode 4* - 50.0 Hz

The transmission lines are described by distributed parameters, while the fixed series compensation installations in the system are modelled accordingly. The load flow of the system is regulated such that it represents the heavy load condition of the NWAPS. The real and imaginary power output of the BOARDMAN generator is 540MW and -62MVar, respectively.

A simplified TCSC model used in this simulation study is simulated using the "MODELS" feature of EMTP. An earlier version of the model is described in[4, 5]. As shown in Fig.2, the TCSC model used in this study is represented by an 8 ohms internal capacitor equivalently (The SLATT TCSC consists of six segments of 1.33 ohms each), in parallel with a path including thyristor switches and commutation inductor. The TCSC also has a bypass breaker, as shown in Fig.2. The TCSC can be operated at three basic modes. In the

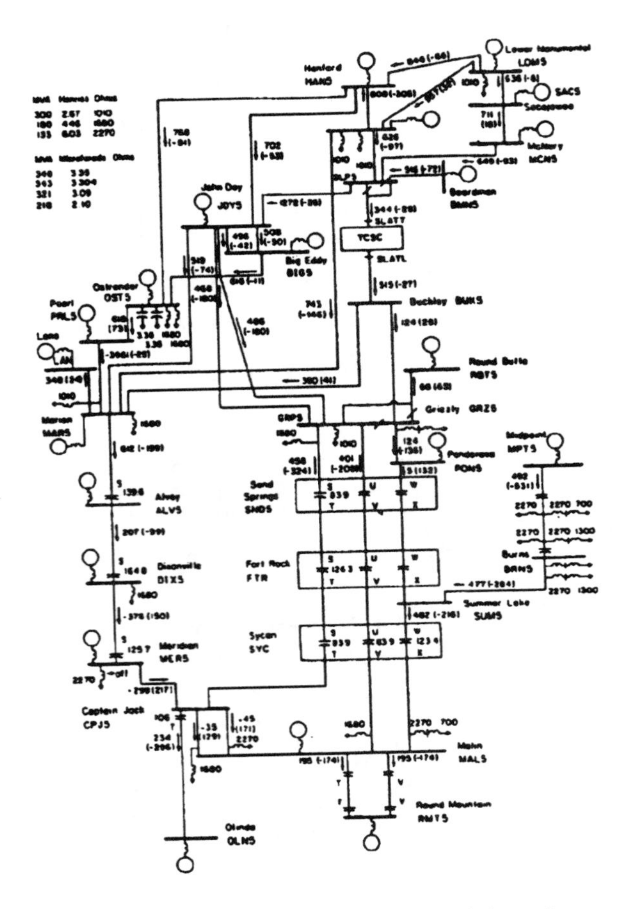

Fig.1 NWAPS model used in the SSR simulation study

bypassed mode, the thyristor path is conducting continuously. In the blocked mode, the thyristor path is blocked continuously, which is equivalent to the fixed series compensation at the capacitor reactance, i.e., 8 ohms in this study. Finally, in the vernier mode, the thyristor path is partially conducting to achieve a specified ohms order. When the TCSC is operated in the vernier mode, thyristor firing results in a loop current flowing through the inductor in the opposite direction of the internal capacitor current[6, 7], as shown in Fig.2. This loop current results in an increase of the equivalent TCSC impedance over the internal capacitor reactance with respect to synchronous frequency. The term "ohms order", or "Xorder" is used to describe the series compensation capability of the TCSC[6, 8]. In this paper it is defined as the ratio of the 60 Hz equivalent TCSC impedance to the internal capacitor reactance, i.e., 8 ohms in this study, under steady-state operating conditions. The larger the Xorder, the higher the series compensation level with respect to synchronous frequency in the steady state. Ohms order is limited by a number of practical considerations[6]. In this study, the values of Xorder is limited between 1.0 p.u. and 3.0 p.u., with a base value equal to the internal capacitor reactance, i.e., 8 ohms.

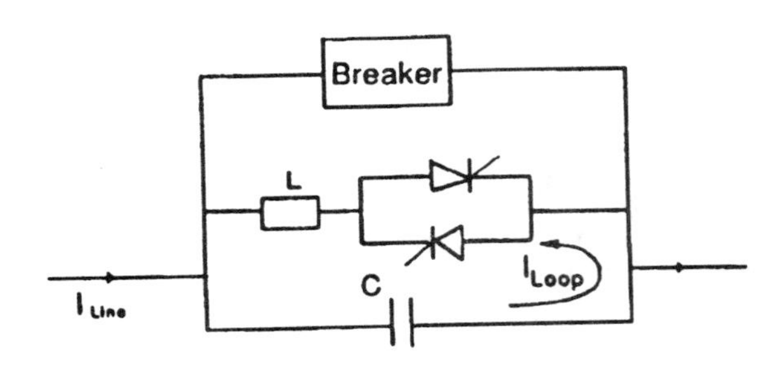

Fig.2 TCSC model

3. Simulation Results

In this section, the results of the simulation cases for two different disturbances with different Xorders of the TCSC vernier operation are presented. Comparison is made of the post-disturbance responses with the TCSC in vernier mode to the response with the fixed series compensation.

Case I

A disturbance is applied by switching in the series compensation initially. The operation scenario is as follows:

In the case of the TCSC,
At t = 0-, TCSC is bypassed;
At t = 0+, TCSC is switched into vernier mode,

and in the case of the fixed series compensation,
At t = 0-, TCSC is bypassed;
At t = 0+, TCSC is blocked.

Figs.3 through 6 show the speed responses of the BOARDMAN generator shaft elements after the disturbance. Fig.3 shows the responses of the system with 8 ohms fixed compensation after the disturbance. It can be seen that the SSR at shaft mode 4 of 50.0 Hz is dominantly excited after the disturbance, as shown in the mass #4 and #5 speed responses. Also, oscillation at shaft mode 1 of 12.5 Hz can be observed in the mass #1 and mass #2 speed responses, and oscillation at shaft mode 2 of 25.0 Hz can be observed in the mass #3 speed response. Figs.4 through 6 shows the speed responses of the system with the TCSC at Xorder = 1.5 p.u., Xorder = 2.0 p.u., and Xorder = 3.0 p.u., respectively, after the disturbance. It is shown that with the TCSC vernier operation at the studied Xorder values, the amplitude of the dominant shaft mode 4 oscillation becomes significantly smaller compared with the case of the fixed compensation. This results from the TCSC vernier operation changing the network characteristics affecting the SSR. Figs.4 through 6 also show that the shaft mode 1 and shaft mode 2 oscillations are better mitigated with increasing Xorder, as seen by comparing the mass #1 through mass #4 speed responses in the figures.

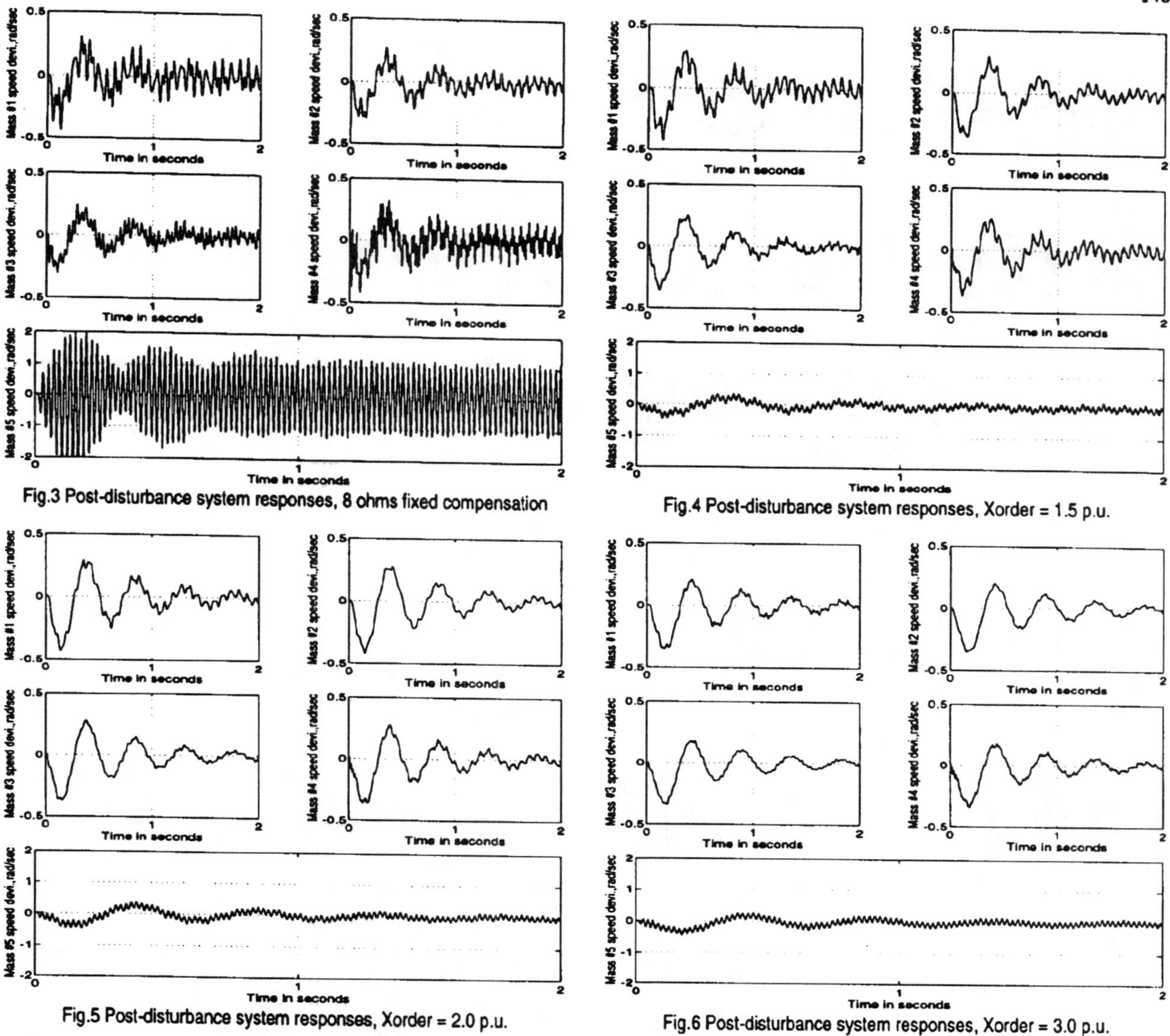

Fig.3 Post-disturbance system responses, 8 ohms fixed compensation

Fig.4 Post-disturbance system responses, Xorder = 1.5 p.u.

Fig.5 Post-disturbance system responses, Xorder = 2.0 p.u.

Fig.6 Post-disturbance system responses, Xorder = 3.0 p.u.

Figs.7 through 9 show the electrical responses of the system after the disturbance. Fig.7 shows the blocked TCSC (which is equivalent to 8 ohms fixed compensation) voltage of the system, after the disturbance. The electrical oscillation due to the interaction between the electrical and mechanical subsystems after the disturbance can be seen clearly from the response during the first 0.5 second simulation time. Fig.8(a), Fig.8(b), and Fig.8(c) shows the TCSC voltage of the system with the Xorder of 1.5 p.u., 2.0 p.u., and 3.0 p.u., respectively. The increase of the TCSC voltage reflects the increase of Xorder. It is shown that the electrical oscillation is well mitigated in the cases of the TCSC vernier operation, in contrast to the case of the fixed series compensation. Fig.9 shows the thyristor currents when TCSC is in vernier mode at the studied Xorder values.

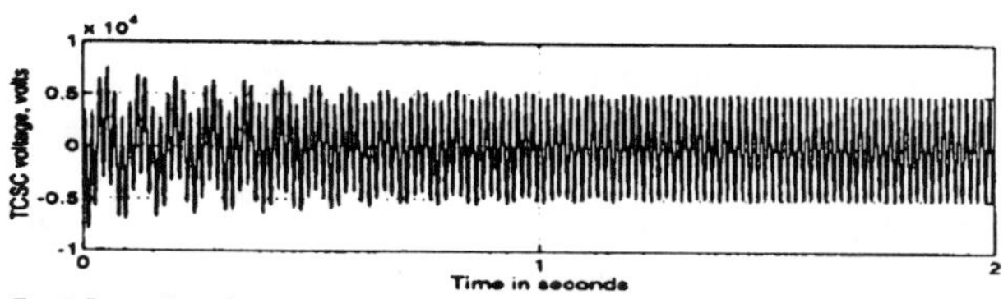

Fig.7 Post-disturbance response of the system, 8 ohms fixed compensation

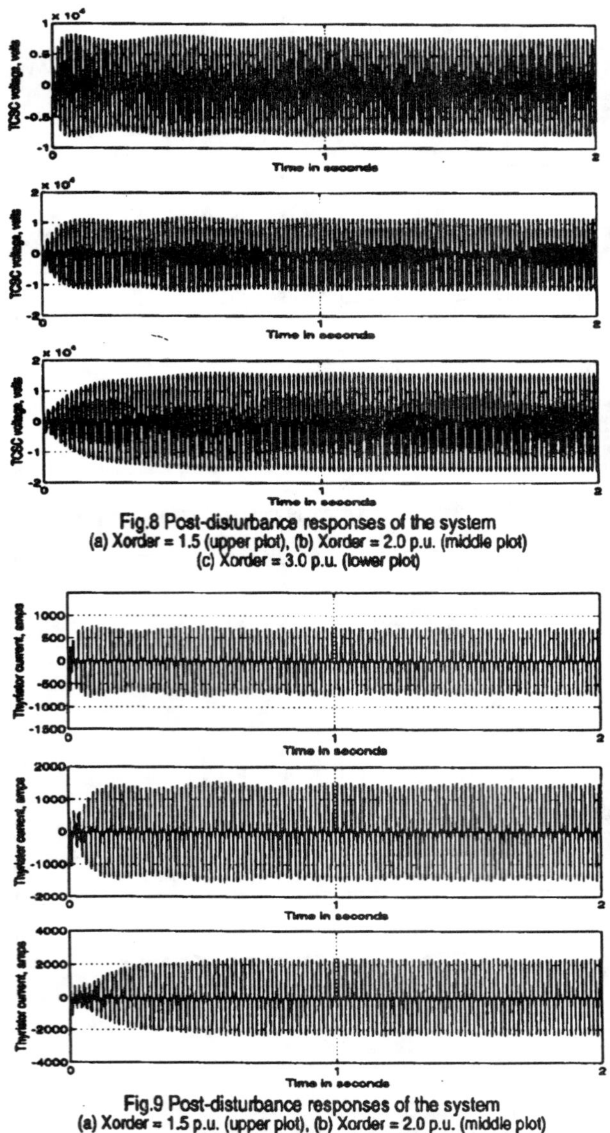

Fig.8 Post-disturbance responses of the system
(a) Xorder = 1.5 (upper plot), (b) Xorder = 2.0 p.u. (middle plot)
(c) Xorder = 3.0 p.u. (lower plot)

Fig.9 Post-disturbance responses of the system
(a) Xorder = 1.5 p.u. (upper plot), (b) Xorder = 2.0 p.u. (middle plot)
(c) Xorder = 3.0 p.u. (lower plot)

Case II

In this simulation case, the *Case I* disturbance is enlarged by switching out a small shunt impedance branch at SLATT initially. The operation scenario is as follows:

At t = 0-, TCSC is bypassed & the shunt branch at SLATT is in the system;

At t = 0+, TCSC is either blocked (in the case of fixed compensation), or switched into vernier mode; & the shunt branch is switched out of the system

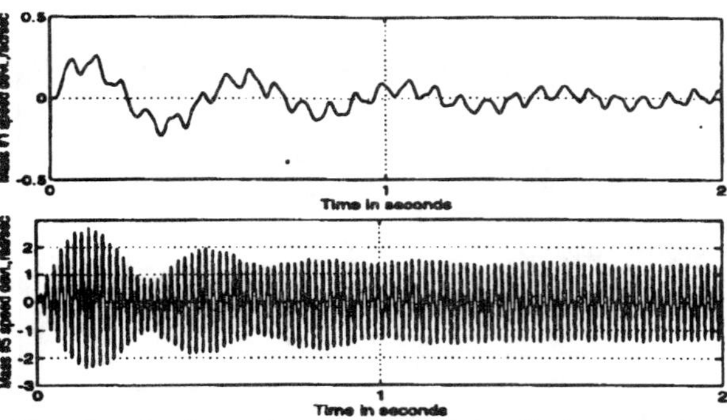

Fig.10 Post-disturbance system responses, 8 ohms fixed compensation

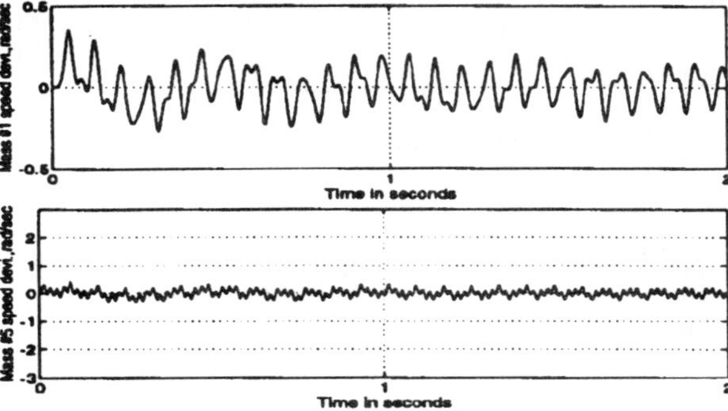

Fig.11 Post-disturbance system responses, Xorder = 1.5 p.u.

In this case, shaft mode 4 is dominantly excited, as in the *Case I*, which can be seen from the mass #5 speed response in Fig.10. Oscillation at shaft mode 1 can also be observed, as shown in the mass #1 speed response in Fig.10. Fig.11 shows the system responses with a TCSC Xorder of 1.5 p.u.. It can be seen that compared with the fixed series compensation, the TCSC vernier operating provides significant mitigation of the shaft mode 4 oscillation. However, although the magnitudes of the shaft mode 1 oscillation are small in both the case of the fixed compensation and the case of the TCSC vernier operation, the oscillation is notably more active in the latter case, as seen by comparing the mass #1 speed responses shown in Figs.10 and 11. An investigation of the cause of this phenomenon is currently underway. But it may be that this mode is excited more by initiating the disturbance with the TCSC vernier operation.

4. Frequency Domain Study of the Equivalent Impedance of TCSC

An analysis of the equivalent TCSC impedance can offer an explanation for the effectiveness of the TCSC at different oscillation frequencies. In the following, we present the frequency domain characteristics of the equivalent TCSC impedance using a 1.33 ohms internal capacitor reactance (one segment at SLATT).

A network including the TCSC in series with the network impedance and the voltage sources at both synchronous frequency and subsynchronous frequency is simulated in EMTP. The relationship between the voltage across the TCSC and the current through the TCSC at synchronous frequency and at different subsynchronous frequencies has been studied to determine the equivalent impedance of the TCSC with respect to different frequencies. The network for the simulation is shown in Fig.12 . The magnitudes of the subsynchronous voltage sources are chosen to be much smaller than the magnitude of the synchronous voltage source.

As an example, voltage sources at frequencies of 60 Hz and 10 Hz are considered with a thyristor conducting time of 70 electrical degree within half cycle. The simulated TCSC voltage is shown in Fig.13 . The square root of the Power Density Spectrum (PDS) of the TCSC voltage is shown in Fig.14. It can be seen that the two main frequency components of the response are at 60 Hz and 10 Hz. Similarly, the TCSC current, that is, line current, has the two main frequency components at 60 Hz and 10 Hz. The filtered 60 Hz and 10 Hz components are shown in Fig.15, where a Chebychev second-order filter was used. This analysis can be done conveniently using design packages, such as MATLAB[9]. From the magnitude and phase angle relationship between the voltage and current at a given frequency, we can obtain the equivalent TCSC impedance at these frequencies for the excitation level used. In this case, the equivalent TCSC impedances at 60 Hz and 10 Hz are $1.65\angle{-90.0^\circ}$ ohms and $3.40\angle{-8.5^\circ}$ ohms, respectively. The Xorder is 1.2 (=1.65/1.33) p.u., with a base value equal to the internal capacitor reactance, i.e., 1.33 ohms, in this case.

Based on the simulation studies and the analysis procedure illustrated, we obtain values of the equivalent TCSC impedance for different frequencies and Xorder values. Fig.16 shows the resulting real and imaginary parts of the equivalent TCSC impedance. It can be seen that for the studied values of Xorder, the equivalent TCSC impedance has negative(capacitive) imaginary part and positive (resistive) real part, with respect to subsynchronous and synchronous frequencies. Thus, if we view the TCSC as an electrical element, it can be represented as a resistor in series

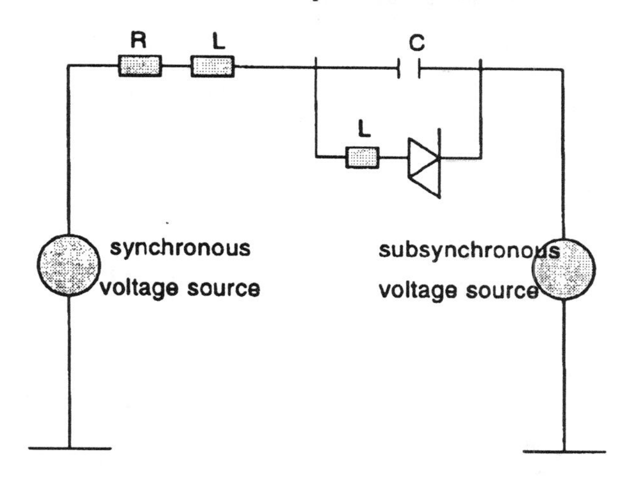

Fig.12 Network for the simulation study of the TCSC equivalent impedance

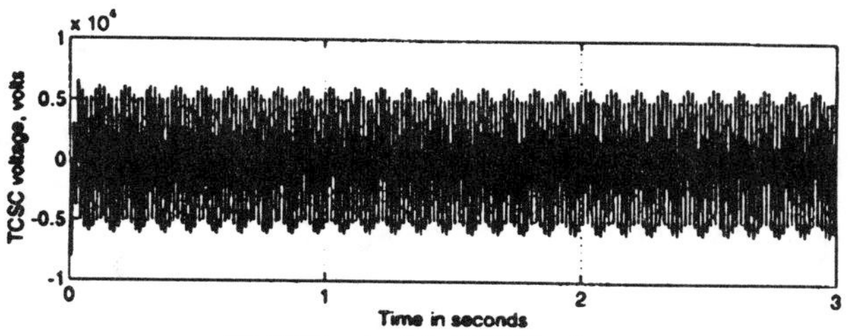

Fig.13 The simulated TCSC voltage

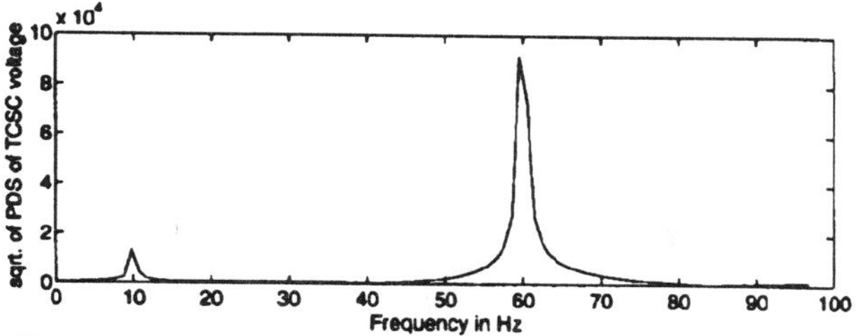

Fig.14 The square root(sqrt.) of the Power Density Spectrum (PDS) of the simulated TCSC voltage

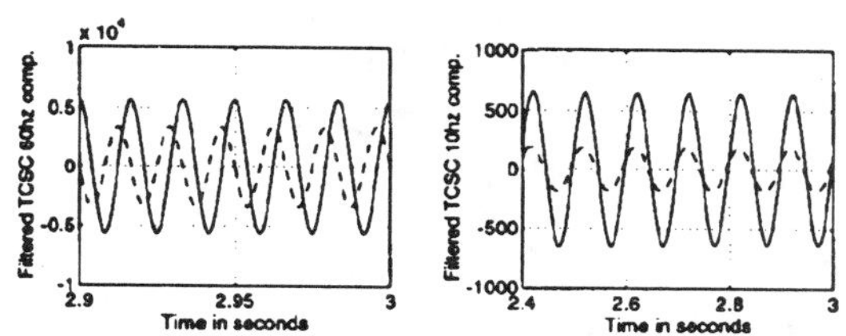

Fig.15 Filtered frequency components of the system responses
__ voltage in volts, --- current in amps

with a capacitor as follows:

$$Z_e(\omega) = R_e(\omega) + jX_e(\omega),$$

where the subscript "e" stands for "equivalent", and $R_e \geq 0$, $X_e < 0$, the frequency domain characteristics of which are described in Fig.16. Because the TCSC operation involves nonlinearity, these characteristics may also depend on excitation level.

Fig.16(a) shows that $R_e(\omega)$ tends towards zero at synchronous frequency for any studied Xorder. Fig.16(b) shows that the equivalent TCSC reactance at synchronous frequency is enlarged over the internal capacitor reactance, which reflects the increased series compensation capability. Thus, it can be reasoned that with respect to synchronous frequency, the TCSC in vernier mode behaves as a lossy capacitor in an average sense.

Fig.16(a) shows that $R_e(\omega)$ is nonzero and positive with respect to subsynchronous frequencies for any studied Xorder. It can be reasoned that the TCSC provides resistive damping to SSR. Fig.16(b) shows that for any studied Xorder, the frequency domain characteristic of $X_e(\omega)$

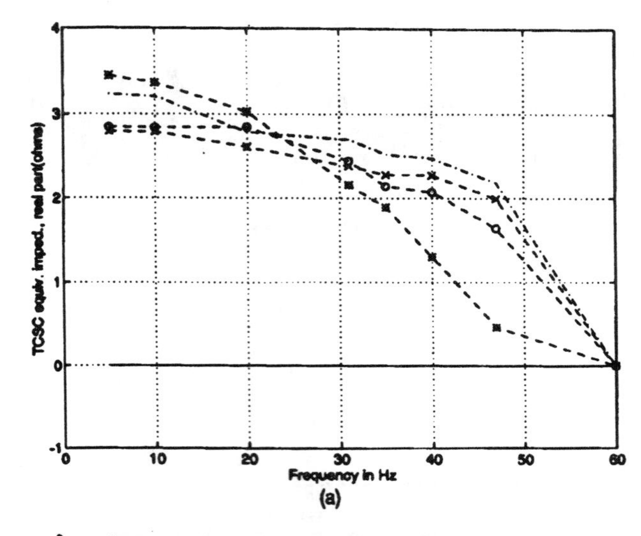

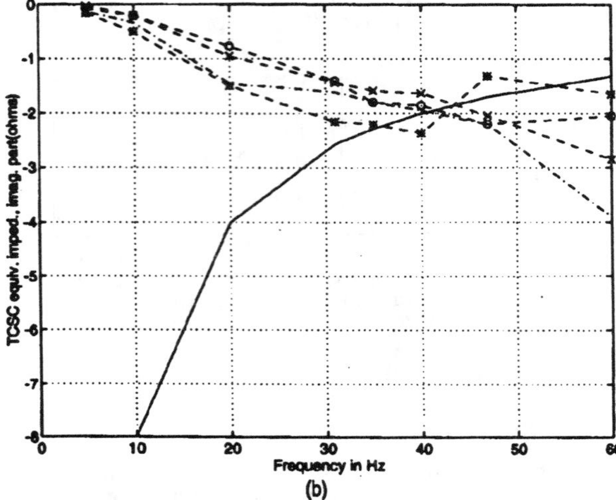

Fig.16 Frequency domain characteristics of the real part & imaginary part of the TCSC equivalent impedance

——	TCSC internal capacitor
-·*·-	Xorder = 1.2
--o--	Xorder = 1.5
--x--	Xorder = 2.1
·-·-·-·	Xorder = 2.9

deviates from that of the TCSC internal capacitor reactance with respect to subsynchronous frequencies, as well as with respect to synchronous frequency. Thus, it can be reasoned that the TCSC operated in vernier mode avoids a resonant condition by changing the capacitive reactance at SSR frequencies and by introducing equivalent resistive damping.

Fig.16(a) shows that the value of $R_e(\omega)$ is somewhat larger at lower network frequencies for a given Xorder. Fig.16(b) shows that the deviation of $X_e(\omega)$ from the internal capacitor reactance is somewhat more significant at lower network frequencies for a given Xorder. Since the electrical frequencies of SSR are the complementary frequencies of the shaft modes[1, 10], it can be reasoned that the SSR mitigation effect of the TCSC vernier operation may be more significant with respect to higher frequency shaft modes, e.g., shaft mode 4 of 50.0 Hz in the studied system. We note further that for network frequencies between 28 and 60 Hz, the value of $R_e(\omega)$ increases as Xorder is increased, as shown in Fig.16(a). This may indicate that higher Xorder introduces a larger effective resistance at relatively low shaft mode frequencies (high network frequencies), e.g., at shaft mode 1 of 12.5 Hz.

5. Conclusions

A simulation study of the SSR mitigation effect of TCSC vernier operation has been performed using a model of the NWAPS and the standard EMTP simulation program. The frequency domain characteristics of the equivalent TCSC impedance has been used to give an explanation of the issues involved. The study serves as an exploration of the SSR mitigation effect provided by TCSC vernier operation in a realistic power system model. It is shown that the TCSC vernier operation provides significant mitigation of 50 Hz shaft mode that was dominantly excited with the fixed compensation.

The equivalent TCSC impedance study shows that the TCSC in vernier mode no longer behaves as only a capacitor in an average sense with respect to subsynchronous frequencies for the studied Xorders. The TCSC vernier operation may benefit the mitigation of SSR in two aspects. One is that the TCSC vernier operating provides equivalent resistive damping to subsynchronous oscillations, and the other is that the equivalent reactance of the TCSC in vernier mode deviates from the internal capacitor reactance with respect to subsynchronous frequencies, as well as at synchronous frequency. These effects lead to a change of the system characteristics and suppression of SSR.

The frequency domain analysis also shows that the SSR mitigation effect of the TCSC vernier operation may change with the dominant SSR frequencies. Both the simulation study *Case I* and the equivalent TCSC impedance study suggest that with respect to SSR at shaft modes of relatively low frequencies, e.g., shaft mode 1 of 12.5 Hz in the studied system, the SSR mitigation effect of the TCSC vernier operation increases with increasing Xorder. Small Xorders can provide good mitigation of SSR at shaft modes of relatively high frequencies, e.g., shaft mode 4 of 50.0 Hz in the studied system.

As in any dynamical system studies, the results of the simulation studies presented can be influenced by a number of factors, such as, system configurations, system parameters, TCSC operation conditions, etc. Thus, the conclusions presented are related to the specified problems studied. Additional study is needed to account for the larger mode 1 amplitude with TCSC control (Fig.11), and to verify the relationships suggested. Also, the study of the modal damping provided by TCSC vernier operation is currently underway.

6. References

[1] IEEE Committee Report, "Reader's Guide to SSR," IEEE Transactions on Power Systems, Vol.7, No.2, 150-157, February 1992

[2] E.V. Larsen, C.E.J. Bowler, B.L. Damsky, S.L. Nilsson, "Benefits of Thyristor-Controlled Series Compensation," CIGRE Paper 14/37/38-04, Paris, 1992

[3] J. Urbanek, R.J. Piwko, E.V. Larsen, B.L. Damsky, B.C. Furumasu, W.A. Mittelstadt, J.D. Eden, "Thyristor Controlled Series Compensation Prototype Installation at the Slatt 500 kV Substation," IEEE PES Paper 92-SM-467-1PWRD, Seattle, July 1992

[4] R. Spee, W.Zhu, "Flexible AC Transmission Systems - Simulation and Control", Proc. Africon, Swaziland, 1992

[5] Advanced Concepts Studies Related to FACTS: Operation and Nonlinear Control - Phase II, OSU ECE Technical Report FACTS 9201, February, 1992, EPRI Contract No. RP 4000-06, BPA Contract No. DE B179-90BP08423

[6] E.V. Larsen, K. Clark, S.A. Miske, Jr., J.Urbanek, "Characteristics and Rating Considerations of Thyristor Controlled Series Compensation," IEEE PES Paper 93-SM-433-3PWRD, Vancouver, B.C., Canada, July, 1993

[7] Scott G. Helbing, G.G. Karady, "Investigation of an Advanced Form of Series Compensation," IEEE PES Paper 93-SM-431-7PWRD, Vancouver, B.C., Canada, July, 1993

[8] S. Nyati C.A. Wegner, R.W. Delmerico, R.J. Piwko, D.H. Baker, A. Edris, "Effectiveness of Thyristor Controlled Series Capacitor in Enhancing Power System Dynamics: an Analog Simulator Study," IEEE PES Paper 93-SM-432-5PWRD, Vancouver, B.C., Canada, July, 1993

[9] MATLAB, The Mathwork, Inc., 24 Prime Park Way, Natick, MA 01760

[10] IEEE Power System Engineering Committee Report, "Terms, Definitions and Symbols for Subsynchronous Oscillations," IEEE Transactions on PAS, Vol. PAS-104, No.6, 1326-1334, June 1985

Acknowledgement *This research was supported by NSF Grant ECS9301168, and EPRI contract RP3573-05. The first author gratefully acknowledge the support by BPA during summer '93.*

We would like to thank the BPA engineer Jerry Northtrom for his help in discussing the modelling of the NWAPS. We would also like to thank Dr. Scott Meyer, Dr. Tsu-huei Liu in BPA for their help in discussing EMTP. Also, we would like to thank Mr. J. Vithayathil for his helpful discussion about the work. We would like to thank Mr. E.V. Larsen for his helpful discussion and valuable comments on the work.

Wenchun Zhu received her BS degree in electrical engineering in 1988, and the MS degree in electrical engineering in 1990, both from Tsinghua University, Beijing, P.R. China. She is currently a Ph.D candidate at Oregon State University, Corvallis. Her research interests include power system dynamics, subsynchronous resonance, simulation, power electronics, and nonlinear control.

Rene Spee(S'84,M'88,SM'92) was born in Stuttgart, West Germany. He attended the University of Stuttgart and Oregon State University, where he received the M.S. and Ph.D. degrees in electrical engineering in 1984 and 1988, respectively.

In 1988 he joined the Department of Electrical and Computer Engineering at Oregon State University, where he is currently an Associate Professor. His areas of interest include ac adjustable-speed drives, power electronic systems, and power system applications.

Dr. Spee is a member of the Industrial Drives and Electric Machines Committee of the IEEE Industry Applications Society.

Ronald R. Mohler (M'59, SM'79,F'80) received the BS degree from Pennsylvania State University, University Park, PA, in 1956, the MS degree from the University of Southern California, Los Angeles, CA, in 1958, and the PhD degree from the University of Michigan, Ann Arbor, MI, in 1965. He is professor of Electrical and Computer Engineering at Oregon State University since 1971 and has over ten years of industrial type experience at the Los Alamos National Laboratory, the Hughes Aircraft Corp., and Rockwell International (Textile Machine Works). He has been an industrial consultant since 1965. Since 1971, he has been Professor of Electrical and Computer Engineering at Oregon State University where he was Department Head from 1971 to 1978. Formerly he was Professor of Electrical, Aerospace, Mechanical, and Nuclear Engineering as well as Chairperson of Information and Computing Science at the University of Oklahoma, and was Associate Professor of Electrical Engineering and Computer Science at the University of New Mexico. He has held visiting positions at the University of Rome, University of London, Australian National University, University of California at Los Angeles, Naval Postgraduate School, and the International Institute for Applied Systems Analysis. His interests include nonlinear control, power systems, biomedical engineering, aerospace applications, and signal processing. He is the author of four books and editor of four others on these topics.

G. C. Alexander Gerald C. Alexander is an Associate Professor of Electrical Engineering in the Department of Electrical and Computer Engineering at Oregon State University, Corvallis, Oregon. He received the B.S. in Electrical Engineering from Oregon State University; the ScM from the Massachusetts Institute of Technology; and the Doctor of Philosophy from the University of California, Berkeley. His current interests are in traveling wave phenomena, power system dynamics, fields in special machines and power electronics applied to high power applications.

William A. Mittelstadt is currently Principal Transmission Planning Engineer at the Bonneville Power Administration (BPA) in Portland, OR. He is responsible for coordinating transmission planning efforts of the Northwest Reinforcement Project associated with the California-Oregon Transmission Project, a new 500 kV circuit from California to Oregon. He was also responsible for coordinating transmission planning efforts for the Celilo-Sylmar HVDC Terminal Expansion Project which went into service in early 1989. Mr. Mittelstadt is industry advisor for the Flexible AC Transmission System effort sponsored by the Electric Power Research Institute. He is past Secretary of CIGRE Study Committee 38, Power System Analysis and Techniques.

Mr. Mittelstadt received his Bachelor's degree from Oregon State University in 1966 and his Master's degree from the same institution in 1968. Since starting work after graduation he has held supervisory and technical positions at BPA including Manager of Research and Development. He is a member of Eta Kappa Nu, Sigma Xi, and Tau Beta Pi and is a Fellow of IEEE.

D.J. Maratukulam is currently Manager, Power System Design, Power Systems Planning and Operations Program, Electric Power Research Institute, Palo Alto, CA. Mr. Maratukulam received the B.Tech degree from the Indian Institute of Technology, Madras, India. He also received M.S. (Material Science) and M.Eng. (Electrical Engineering) degrees from the University of Washington, Seattle (1972) and the University of British Columbia, Vancouver, Canada (1974), respectively. Before joining EPRI in 1987, he worked at Systems Control Inc., Palo Alto, CA, B.C. Hydro, Vancouver, Canada, and Bharat Heavy Electricals, Trichy, India. His areas of interest at EPRI include flexible AC transmission systems (FACTS), voltage stability, power system losses, power system protection, and parallel processing.

400 MW SMES Power Conditioning System Development and Simulation

Ibrahim D. Hassan, *Senior Member*, Richard M. Bucci, *Senior Member*, and Khin T. Swe, *Member, IEEE*

***Abstract*—A conceptual design for a 22 MWh superconducting magnetic energy storage (SMES) system engineering test model (ETM) has recently been developed. The objectives of the SMES-ETM are to demonstrate the feasibility of using a SMES system to perform load-leveling and system stabilization for commercial utilities and to supply 400 MW power pulses for ground-based defense systems. This paper presents the performance requirements and configuration of the proposed 22 MWh SMES-ETM and its power conditioning system. The power conditioning system consists of a dc–dc chopper linked to a GTO-based voltage source converter interfacing the superconducting energy storage coil to the ac power system. The SMES system operation in the charging and discharging modes is described and the results of digital simulations demonstrating the feasibility of the proposed power conditioning system and exploring its overall behavior under normal and fault conditions are presented.**

I. Introduction

SUPERCONDUCTING magnetic energy storage (SMES) systems have been the subject of active investigation for the past 30 years. Since the early 1970's active programs of research were begun by the University of Wisconsin and by The Department of Energy at Los Alamos National Laboratory [1]. In the early 1980's the Electric Power Research Institute (EPRI) sponsored a study to investigate the technical and economic feasibility of a SMES system for utility use. In 1983, the Department of Energy in conjunction with the Bonneville Power Administration tested a 30 MJ SMES unit on the Western U.S. Power System to dampen power oscillations on the Pacific Intertie [2].

In the mid-1980's the Department of Defense became interested in SMES technology as a potential source of pulsed power to be supplied to ground-based defense systems [3]. As a result a program was undertaken to design, fabricate and test a 22 MWh–400 MW SMES Engineering Test Model (ETM). The ETM has the following two objectives:

1) To demonstrate the feasibility of using a SMES system to perform load-leveling and system stabilization for commercial electrical utilities.

2) To assess the feasibility of using the same SMES system as a pulsed power source to power ground-based defense systems.The development of full-scale 1000 MWh and 5500 MWh SMES systems with the capability of discharging at up to 1000 MW is planned to follow the demonstration of the successful performance of the ETM.

This paper provides a description of the overall SMES-ETM configuration and performance requirements. A brief overview of the superconducting magnetic energy storage coil, conductor and leads will be provided. This will be followed by a detailed description of the power conditioning system (PCS) selected to interface the superconducting coil to the ac power system. The performance requirements, configuration, principles of operation, and control technique are described. The results of digital simulations of the SMES-ETM to demonstrate the selected power conditioning system performance and to explore the overall system behavior under normal and transient conditions are also presented. The Electromagnetic Transients Program (EMTP) was used for the simulations.

A. SMES-ETM Power Conditioning System Performance Specifications

The following performance specifications were established for the PCS to meet the two objectives of the SMES-ETM:

1) The utility load-leveling function requires a daily charge/discharge cycle with minimum losses and the PCS shall have optimal efficiency.

2) The utility system stabilization function requires the SMES to supply or absorb adjustable levels of real and reactive power and the PCS shall be capable of four quadrant operation with fast and independent control of real and reactive power.

3) The strategic nature of the ground-based defense system requires that the SMES be capable of discharging with the utility system unavailable and the PCS shall be self-commutating.

4) Since demonstrating the capability of the ETM is a prerequisite for developing a full-scale SMES, the PCS and its components shall be modular and/or scalable.

5) During normal operating conditions, the PCS shall interface the coil with the utility transmission system and shall isolate the ground-based defense distribution system. Operation of the PCS during normal operation shall be automatically controlled from the host utility system control and dispatch center.

6) The PCS shall automatically short the coil whenever a failure prevents normal operation in order to preserve the stored energy and provide an uninterrupted path for the coil current.

7) The PCS design shall minimize the harmonic voltage levels impressed on the superconducting coil to minimize coil losses.

Manuscript received June 17, 1991; revised November 12, 1992. The Superconducting Magnetic Energy Storage Engineering Test Model Development Program was supported by the Defense Nuclear Agency under Contract DNA-001-88-C-0027. This work was performed in cooperation with the Westinghouse Science and Technology Center.

The authors are with Ebasco Services Inc., New York, NY 10048-0752.

IEEE Log Number 9209301.

8) The SMES coil due to its size and structure has a complex pattern of distributed capacitances and resonance frequencies. The PCS shall not generate harmonics that may excite such resonances or resonances that may exist between the coil and the converter or converter transformer stray capacitances.

The PCS in this context includes the following components necessary to interface the coil with the ac power system: converters, converter transformers, ac and dc harmonic filters, electromechanical switches and circuit breakers, and regulators and controllers.

B. Overview of the SMES-ETM

The SMES-ETM consists of a superconducting magnetic energy storage coil connected to a power conditioning system (PCS) by vapor-cooled leads. The PCS interfaces the coil to the utility transmission system and the ground-based defense distribution system. The basic electrical design parameters of the SMES-ETM are as follows:

Total stored energy	=	22 MWh
Usable energy	=	20 MWh
Residual energy (9%)	=	2 MWh
Maximum power capability	=	400 MW
Utility power demand	=	50 MW
Coil maximum operating current	=	50 000 A
Maximum coil voltage (at minimum residual energy)		
at Maximum Power (400 MW)	=	13 330 V
at Utility Power Demand (50 MW)	=	1670 V

C. Superconducting Coil

The SMES-ETM coil is a four-layer solenoid with 104 turns per layer. It is wound in a modular configuration with two modules (upper and lower). The inductance of the four-layer 50 kA coil is 61.2 H. Each module is a separate parallel current path with the coupling between the two modules being 0.75. From an operational standpoint, the coil is equivalent to a two-layer 100 kA, 15.3 H inductor.

The coil is immersed in a superfluid helium bath and is contained in a helium vessel that is surrounded by, and supported from, a vacuum vessel. Between the helium vessel and the vacuum vessel is a nitrogen shroud surrounding the helium vessel. The vacuum vessel is set into a trench partially below ground and is structurally supported by the trench wall and base. A cross section of the coil, vessels, and support structures is shown in Fig. 1.

The coil design parameters are as follows:

Aspect ratio (height/radius)	=	0.033
Radius	=	67.0 m
Height, top to bottom conductor	=	4.087 m
Nominal operating current	=	50 000 A
Nominal operating temperature	=	1.80 K
Maximum operating temperature	=	1.90 K
Maximum field including self field	=	4.8 T

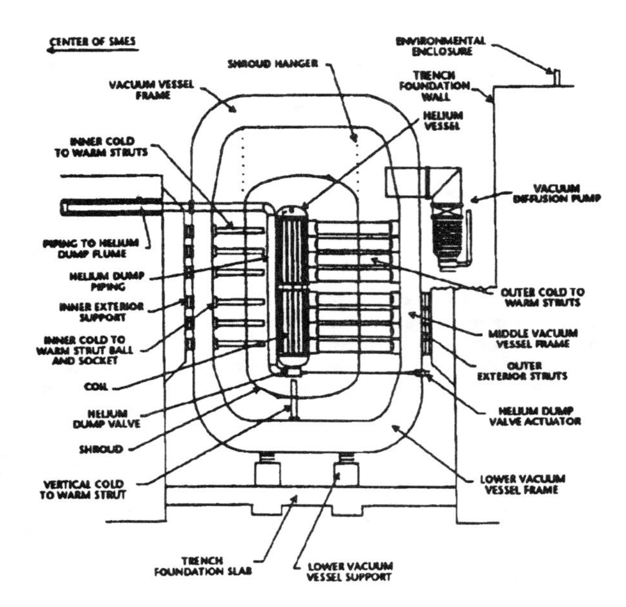

Fig. 1. Cross section of the coil, vessels, and support structure.

Conductor self field	=	1.2 T
Radius with field density (B) = 100 G	=	154 m
Radius with field density (B) = 10 G	=	311 m

Conductor

The conductor is constructed from eight superconducting strands soldered into slots in a high purity aluminum stabilizer. The superconducting strand diameter is 0.110 . The high-purity aluminum stabilizer is a 1 in. diameter extrusion with eight slots for the strands. The strands are composed of NbTi filaments embedded in copper. The nominal strand current is 6250 A at a maximum magnetic field of 4.8 T. The conductor maximum-allowed hot-spot temperature is 350 K.

D. Coil Bus and Leads

Vapor cooled leads and cryogenic buses provide an interface between the supercooled coil and the power conditioning equipment operating in a room temperature environment. The power conditioning equipment is located in a building outside the 100 G magnetic field line to allow unrestricted human access. The vapor cooled leads operate at a temperature between 4.2 and 300 K. The bus stubs provide the electrical connection between the 4.2 K temperature region and the 1.8 K coil temperature region.

II. Power Conditioning System

A. Overall Configuration

The SMES-ETM coil is made of two groups of two series-connected layers, each rated for 50 000 A to store half of the total SMES-ETM energy. The two coil groups are closely coupled magnetically and their currents must be closely matched. Therefore, two separate but identical power converter groups are used with each rated for half the maximum total

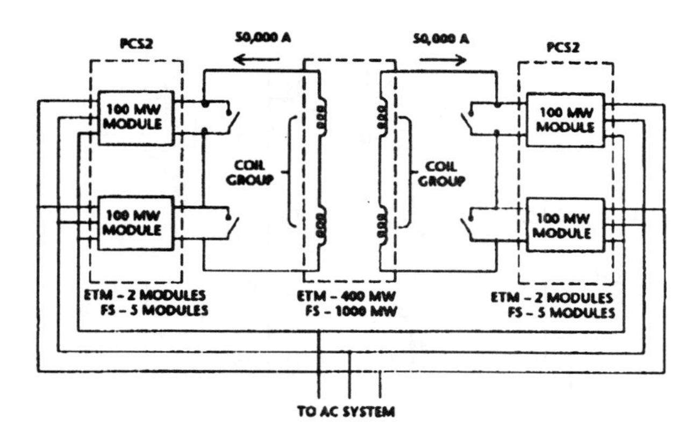

Fig. 2. Overall SMES-ETM configuration.

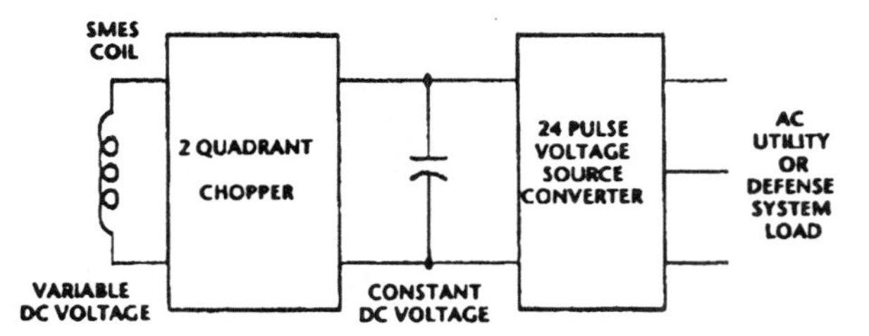

Fig. 3. 100 MW module converter configuration.

required power. The two converter groups are controlled to balance the currents in the two coil groups.

Generally, the power requirements of the ground-based defense system and the utility system vary widely (1000 and 200 MW, respectively, for the full-scale SMES). Since operating efficiency is a major requirement during load-leveling operation, it is undesirable to use one converter rated for maximum power to be continuously operated at a small percentage of its rating. It is also undesirable to use two separate converters since the need for supplying power pulses to the ground-based defense system is infrequent. A separate converter for the ground-based defense system cannot be continuously tested. Therefore, to meet the requirements for high operating efficiency, testability and scalability, a modular design was selected.

The modular design makes use of the fact that the coil maximum operating current is the same in the two modes of operation (i.e., utility mode and pulse mode). Moreover, the coil maximum operating current is the same for the ETM and the full-scale SMES. Therefore, power modulation can be achieved by increasing or decreasing the coils terminal voltage and system modularity can best be achieved by the use of series connected modules. As more power is required, more modules are connected in series. Also, to allow testing at full rating, a basic module rating equal to one half of the utility requirement (in the full-scale SMES) is selected. This implies a module rating of 100 MW with two such modules connected in series for the ETM as shown in Fig. 2. For the full-scale SMES five such modules connected in series are used. By using the individual modules on a cyclic basis during the load-leveling charge/discharge cycle, all of the modules can be tested at full power on a regular basis.

B. Converter Type

Since the coil is a stiff current source, the use of a current source converter may appear as the natural choice. Such current source converters have been widely used in high-power HVDC and other industrial uses. However, the coil current varies from 50 kA at full charge to 15 kA at minimum residual energy. Moreover, the voltage across the coil varies up to 13.33 kV at maximum power and minimum residual energy. These wide variations in the coil current and voltage made the use of the conventional current source converter impractical. A conventional current source converter must be rated for the maximum current of 50 kA and the maximum voltage of 13.33 kV resulting in a total capacity of 667 MVA for discharging a maximum power of only 200 MW. In addition, its use imposes high levels of reactive power demand on the utility system or requires high-capacity reactive power compensation means. Therefore, the conventional current source converter was not considered for the SMES.

To replace the current source converter, detailed analyses were made to evaluate hybrid current source converters (CSC) and voltage source converters (VSC) [4]. The use of either of these converter types allows independent control of real and reactive power and would result in a significant reduction in the converter MVA capacity relative to the conventional current source converter. A hybrid CSC consists of two CSC's with their ac terminals connected in parallel. One CSC is line commutated and consumes lagging reactive current while the other is self-commutated and can be controlled to generate leading reactive current. When the two converters are operated at complementary firing angles, the reactive current components cancel out and only the real power current component flows in the converter transformer to the utility system. One byproduct of this scheme is the harmonic distortion of the total ac current flowing in the converter transformer. Due to the harmonic currents, the hybrid CSC requires a transformer rated 140% of a transformer used with a VSC of the same kVA capacity. Moreover, the hybrid CSC requires 5376 GTO's while the VSC scheme requires 768 GTO's for the converter plus 1856 GTO's for a dc–dc chopper. This results in the total cost of the switching devices in the hybrid CSC to be 173% of the cost of the switching devices and power diodes required for the VSC and the chopper. Also, loss evaluations concluded that a voltage source converter has an efficiency comparable with the hybrid current source inverter. On these bases it was concluded that the voltage source converter is more cost effective. Moreover, a voltage source converter has a better self-commutating capability and injects considerably lower harmonic currents in the utility system than a comparable current source converter due to the effect of the interface reactance.

C. Converter Configuration

The basic concept of the 100 MW module converter configuration chosen is shown in Fig. 3. The converter consists of a quasi-24 pulse voltage source converter (VSC) interfacing with the ac power system and a dc-dc chopper interfacing with

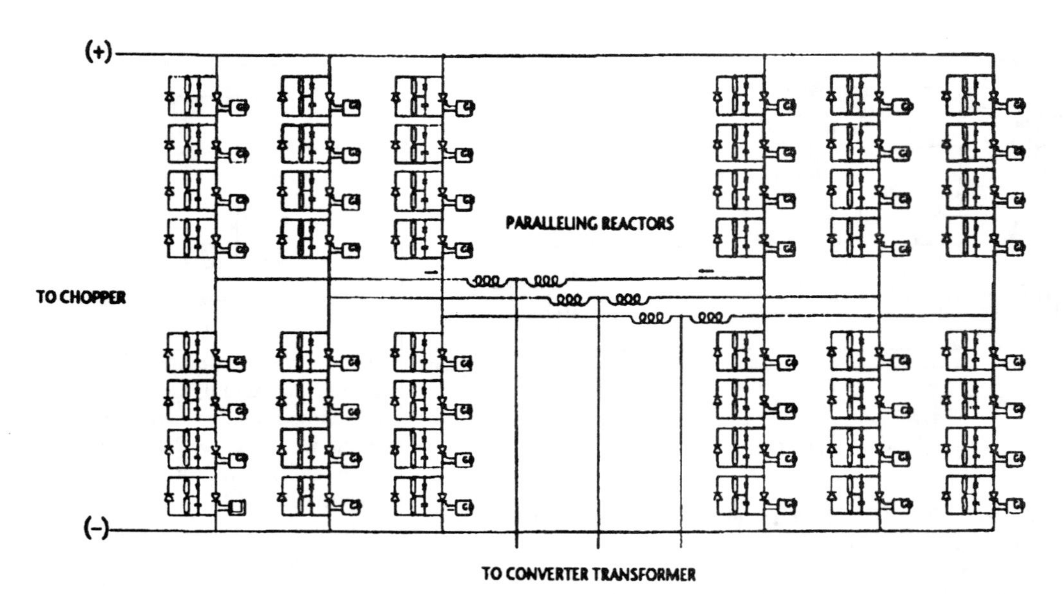

Fig. 4. 25 MW 6–pulse VSC module.

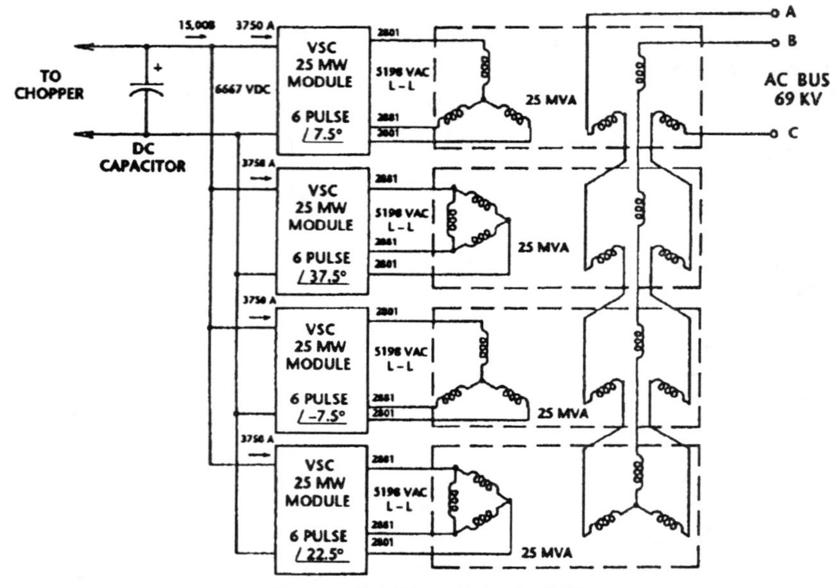

Fig. 5. 100 MW quasi-24–pulse VSC.

the SMES coil. The VSC and the dc-dc chopper are linked by a dc link capacitor. The dc link capacitor behaves as a stiff but controllable dc voltage source providing the characteristics required by the VSC and the dc–dc chopper. This configuration also significantly decouples the SMES coil from the utility system and reduces the coil exposure to disturbances on the utility system.

D. Voltage Source Converter (VSC)

The VSC is made up of four 6 pulse, 25 MW modules with the configuration of each 6 pulse bridge as shown in Fig. 4. The four 25 MW modules are connected as shown in Fig. 5. Each bridge leg is made up of a string of diodes in reverse parallel with GTO's to block voltage in one direction and conduct current in both directions to allow the flow of reactive power. The ac terminals of each 6 pulse module are connected to a three-phase 25 MVA transformer. The primaries of the transformers are series connected and tied to the ac power system. The transformer series-connected primaries ensure current sharing in the parallel-connected bridges. The four 6 pulse converter bridges and transformers constitute two 12 pulse converters. By controlling the firing

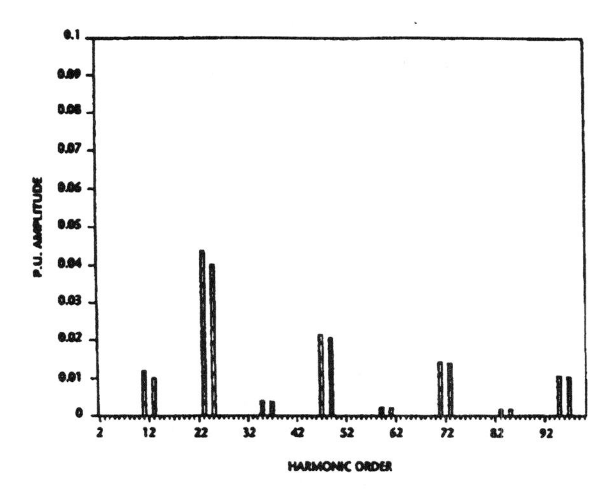

Fig. 6. Harmonic spectrum of quasi-24–pulse system.

angles of the GTO's, the two 12 pulse bridges are phased 15° apart and produce a quasi-24 pulse ac voltage waveform. This configuration employs combinations of standard wye- and delta-connected transformers to eliminate the need for costly special transformers required to produce a true 24 pulse waveform.

The transformer connections chosen produce an output waveform that has evenly spaced steps with slight amplitude differences from a true 24 pulse waveform. This waveform has low levels of residual 12 pulse harmonics. The magnitude of the 11th and 13th harmonics are approximately 1% and can easily be filtered if necessary. The harmonic spectrum of the selected quasi-24 pulse system is illustrated in Fig. 6.

The fundamental component of the quasi-24 pulse voltage waveform $(V)_{cnv}$ is directly proportional to the dc voltage across the dc link capacitor $(V)_{dc}$, or

$$V_{cnv} = 0.780\, V_{dc} \tag{1}$$

The converter transformers rated secondary voltage is selected at 5200 V rms to produce 6666 V dc at the VSC output.

The VSC is capable of operation in all four quadrants and can independently control real and reactive power flow [5]. This can be illustrated using the converter single-phase equivalent circuit shown in Fig. 7.

The three-phase real power (P) and reactive power (Q) can be expressed as:

$$P = (V_{ac})\,(V_{cnv}\sin a)/(X) \tag{2a}$$
$$= (V_{ac})\,(0.780V_{dc}\sin a)/(X) \tag{2b}$$
$$Q = (V_{ac})\,(V_{ac} - V_{cnv}\cos a)/(X) \tag{3a}$$
$$= (V_{ac})\,(V_{ac} - 0.780V_{dc}\cos a)/(X) \tag{3b}$$

where

V_{ac} = ac system line voltage (rms)

α = converter firing angle

X = converter transformer leakage reactance.

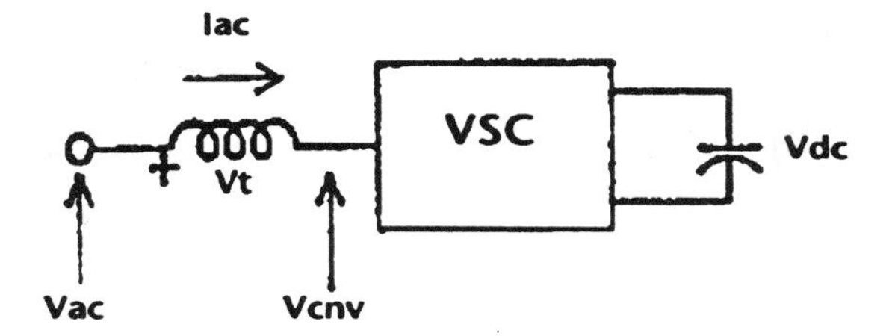

Fig. 7. VSC single phase equivalent circuit.

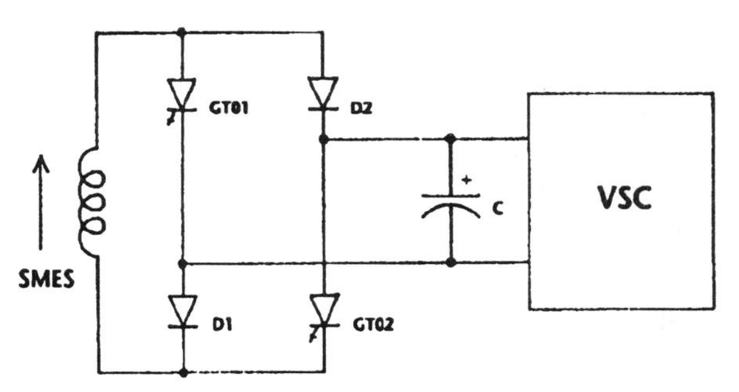

Fig. 8. Two quadrant chopper and commutation of crowbar switch.

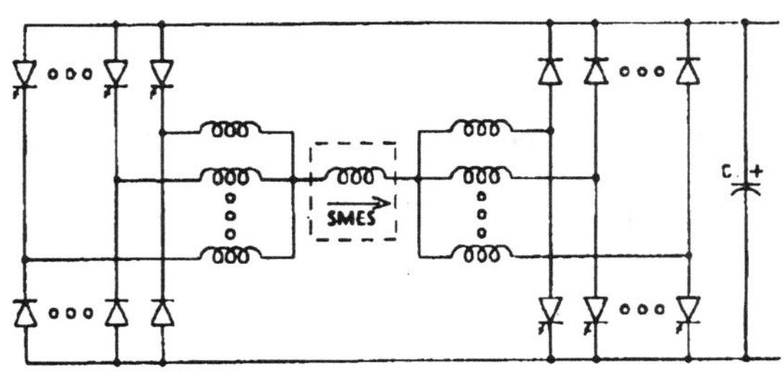

Fig. 9. Two quadrant multiphase chopper.

Therefore, the real power flow can be controlled by adjusting the phase angle between the converter voltage and the ac system voltage through adjusting the converter GTO firing angle. The reactive power can be controlled by controlling the relative magnitudes of the dc link capacitor and the ac system voltage.

E. DC–DC Chopper

Controlling the SMES coil rate of charge/discharge requires varying the coil voltage magnitude and polarity according to the coil state of charge (i.e., the coil current). The dc–dc chopper is provided to convert the essentially constant VSC dc output voltage to the adjustable voltage required across the coil terminals.

The dc–dc chopper is a GTO based 58-phase two-quadrant chopper. The configuration of each phase is shown in Fig. 8 and the multiphase configuration is shown in Fig. 9. The inductors at the output of each chopper phase allow current sharing and gradual buildup of output current in the SMES coil-shorting switch described in the next section. From Fig. 8, when the two GTO's are fired simultaneously, the diodes are reverse biased, and the coil can be charged. When the two

GTO's are turned off, the diodes become forward biased and the coil discharges. The coil voltage is regulated by controlling the GTO conduction time over the switching cycle.

The voltage and current relationships within the chopper can be expressed as follows: where

$$V_{smes} = [1 - 2(1/A)]V_{dc} \tag{4}$$

$$I_{dc} = [1 - 2(1/A)]I_{smes} \tag{5}$$

V_{smes} = average voltage across the SMES coil
I_{smes} = coil current
V_{dc} = dc link capacitor voltage
I_{dc} = average VSC dc current
1/A = duty cycle (GTO conduction time/period of one switching cycle)

At a duty cycle of 0.5, the SMES coil average voltage and the VSC average dc current are both zero and no power is transferred. At a duty cycle larger than 0.5 the coil is charged while at a duty cycle less than 0.5 the coil is discharged. Therefore, control of the rate of charge/discharge is accomplished by controlling the duty cycle or the timing of the GTO gating signals.

The multiphase chopper design avoids the use of a large number of parallel paths to handle the maximum coil current of 50 kA. The multiple stages are controlled identically and operate in a phase shifted mode. The multiphased chopper design also eliminates the low-order ripples in the chopper output voltage and input current. The lowest ripple frequency is the chopper switching frequency times the number of chopper phases. In addition, the chopper GTO's form a fast-acting momentary duty solid-state coil shorting switch that augments a continuous duty electromechanical shorting switch.

F. Shorting Switch (Crowbar)

During periods when the SMES coil is charged but there is no need for the energy, the coil should be shorted to avoid power dissipation and draining the stored energy. In addition, to protect the coil from excessive voltages and allow reconfiguration of the PCS, a crowbar switch is connected across the coil. One crowbar switch is required for each of the coil groups. For the ETM, this switch must be rated for 50 kA continuously and must withstand up to 13.33 kV.

Solid-state switches have the advantage of fast operation in this application. However, they cause excessive power dissipation and require many parallel paths to conduct 50 kA. For this reason mechanical switches are selected. The mechanical switch is made up of six, three-pole 15 kV class vacuum circuit breakers connected in parallel. Each single pole breaker is rated for a continuous current of 3000 A and has a 48 kA rms 3 s current carrying capability. The close and latch rating is 77 kA rms. With such high short-time ratings, there is no concern about timing mismatch among the individual breakers.

To ensure current sharing among the breakers, a resistor is connected in series with each breaker pole. The resistor is rated to develop a 92.6 mV voltage drop. This crowbar switch configuration is shown in Fig. 10.

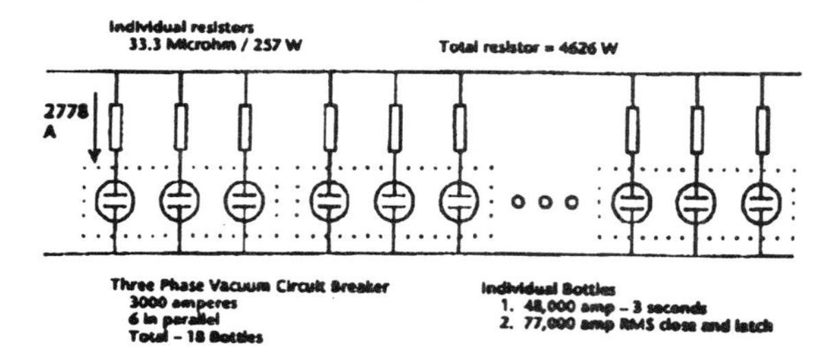

Fig. 10. Crowbar switch configuration.

The crowbar switch can be opened only when no current is flowing through it. Therefore, to open the crowbar the dc–dc copper is used to build-up a current through the switch opposing the SMES coil current as shown in Fig. 8. When the two currents are equal and the net current through the switch is zero, it can be opened.

G. Filters

AC filter banks consisting of tuned and wide-band filters are provided at the converter transformer ac windings to limit the utility system voltage distortion. The filters also serve to filter out harmonic voltages at the PCS output and to supply the ground-based defense system with a harmonic free voltage.

A capacitor is connected across each of the SMES coil groups to absorb voltage ripples generated at the dc–dc chopper output. Such ripples if not filtered may produce excessive losses within the coil.

The complete PCS for the SMES-ETM including the utility tie, the power distribution system, and test loads is shown in Fig. 11.

H. Control System

The PCS controller is required to ensure the following:

1) That the PCS negotiates real and reactive power from the utility system according to commanded values and within the maximum coil voltage and current limits

2) Dynamic balance between the currents in the two closely coupled SMES coil groups

3) PCS operation in a stand-alone or black-start mode when the utility system is not available and the SMES system is the only source supplying the ground-based defense system. In this case the controller is required to regulate the output voltage and frequency.

Vector concepts are used in formulating the control structure for the converter. Basically, the proposed control concept is to control the output real and reactive power by means of the voltage source converter firing angle and the dc link capacitor voltage. The real power exchanged with the SMES coil is controlled through the chopper duty cycle and the dc link capacitor voltage and is regulated by controlling the differential between ac-side real power and the SMES coil. The SMES coil currents are dynamically balanced by means of a differential real power command delivered to the two parallel converter groups in response to observed differences between the two SMES coil currents. In the black start mode

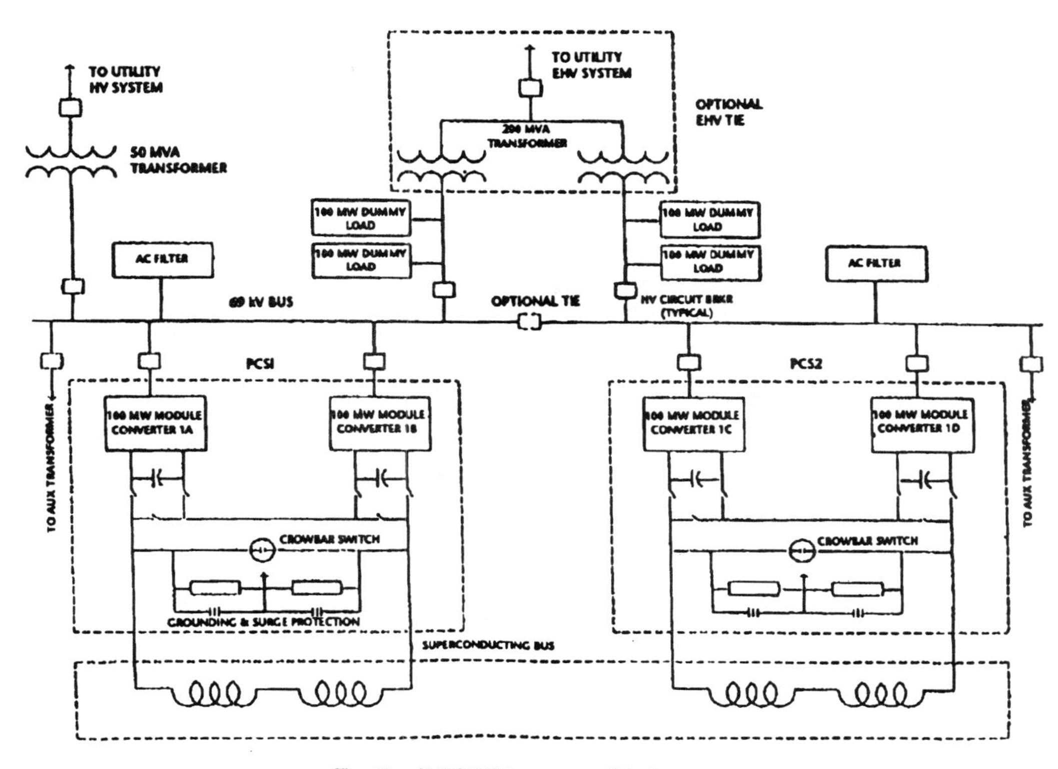

Fig. 11. SMES-ETM power conditioning system.

the controls are re-configured to regulate the output voltage and frequency and to force load sharing between the two converter groups.

The control strategy is based on constantly adjusting the dc–dc chopper duty cycle and the VSC firing angles to maintain the required dc-link capacitor voltage and meet the real and reactive power demand at all levels of the SMES coil current.

Modularity is also considered in configuring the PCS control system. The controls for the SMES-ETM are divided into one controller for each 100 MW module and one main controller to integrate the complete PCS.

III. SMES-ETM Operating Modes

The SMES-ETM has the following four operating modes:

- Startup from the shorted SMES coil configuration
- SMES coil charge
- SMES coil discharge while connected to the utility system
- Black Start (utility system unavailable)

Startup from the Shorted SMES Coil Configuration: In this mode the coil is initially charged, the crowbar switch is closed, and both the dc–dc chopper and the VSC are off. Startup from this mode is as follows:

1) With the GTO's blocked, the VSC is operated as a rectifier to charge the dc link capacitor through the diodes.

2) After the capacitor is charged, the dc–dc chopper is operated with a duty cycle larger than 0.5 to gradually charge the chopper inductors and circulate current through the crowbar switch opposite to the coil current.

3) The crowbar switch is opened when its net current is zero.

4) Following the crowbar switch opening, the chopper duty cycle is reduced to 0.5 and the VSC GTO's are fired at zero angle such that no energy is exchanged with the SMES coil and the SMES is available for charge or discharge operation.

SMES Coil Charge: In this mode, the crowbar switch is open, the dc-dc chopper and the VSC are operated. The chopper duty cycle is larger than 0.5 to match the required charging rate and maintain the capacitor voltage required by the reactive power demand. The VSC firing angle is set leading the ac system voltage and to match the required charging rate.

SMES Coil Discharge While Connected to The Utility System: In this mode, the crowbar switch is open, the dc–dc chopper and the VSC are operated. The chopper duty cycle is set less than 0.5 to match the required discharging rate and maintain the capacitor voltage required by the reactive power demand. The VSC firing angle is set lagging the ac system voltage and to match the required discharging rate.

Black Start: In this mode, the coil is initially charged, the crowbar switch is closed, the dc–dc chopper and the VSC are off, and the utility system is not available. Startup from this mode is as follows:

1) The dc-link capacitor is charged with minimal energy from an auxiliary source such as a small diesel-generator or a storage battery.

2) After the capacitor is charged, the dc–dc chopper is operated with a duty cycle larger than 0.5 causing a resonant

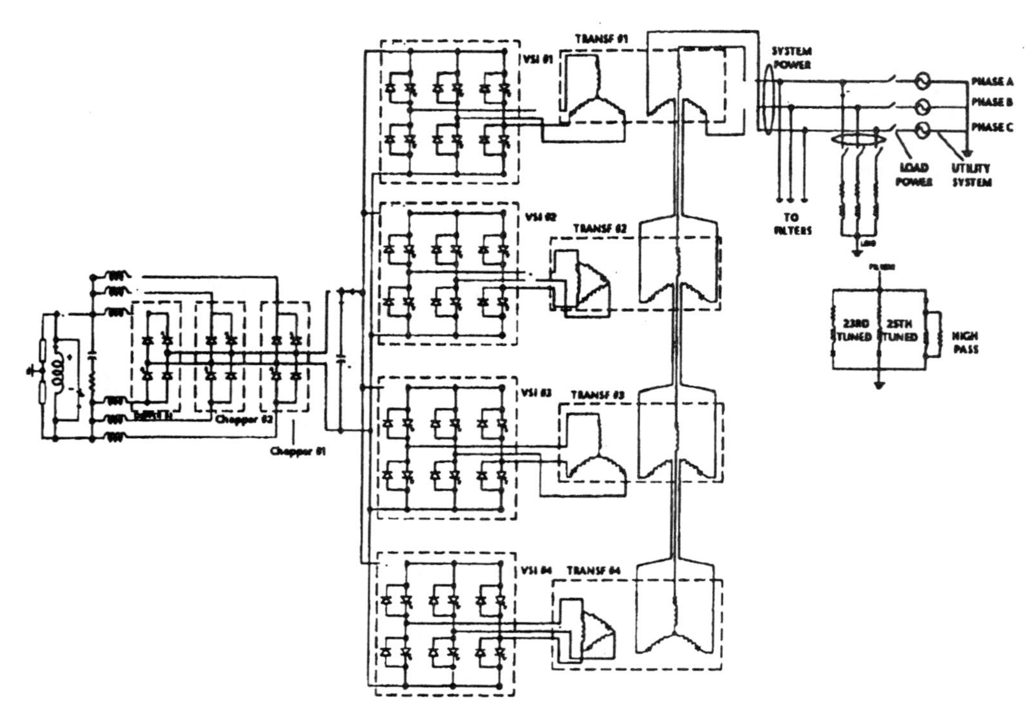

Fig. 12. Simulation model of SMES-ETM.

transfer of energy from the capacitor to the chopper inductors and the crowbar switch. As soon as the crowbar switch net current is zero, it is opened in a single-pulse commutation mode.

3) After crowbar switch opening, the chopper is operated at a duty cycle less than 0.5 to recharge the dc-link capacitor to the appropriate level.

4) After the dc-link capacitor is charged, the VSC GTO's are fired and ac power is supplied to the load.

IV. Simulations

A. SMES-ETM Model

Modeling the SMES-ETM requires modeling two magnetically coupled coils with each coil interfaced to the ac power system through two 100-MW power conditioning modules. However, except for the interaction between the two magnetically coupled coils, the behavior of the SMES system can be predicted with reasonable accuracy by modeling one coil. Therefore, at the conceptual design stage the SMES system is modeled as one 100 MW PCS module interfacing one SMES coil with the ac power system as shown in Fig. 12. Details of this model are as follows:

1) The SMES coil is represented as one lumped inductor grounded through a center tap high-resistance grounding system.

2) The ac power system is represented by a Thevinens equivalent with a short-circuit ratio (SCR) of 18.

3) The load is represented by a variable equivalent impedance.

4) The ac filter is represented by two tuned filters (tuned to the 23rd and 25th harmonics) and one damped high-pass filter.

5) Firing signals for the VSC GTO's are generated by a signal generator. The signal generator modeled consists of a synchronization scheme and an equidistant firing scheme to synchronize the firing signals to the fundamental component of the ac system voltage [6].

6) The dc–dc chopper is modeled as a three-phase chopper with 300 Hz switching frequency. The firing signals of the chopper GTO's are generated by an equidistant firing scheme.

7) A capacitor is added across the SMES coil to absorb the ripples generated at the chopper output. A small resistor is added in series with the capacitor to dampen voltage oscillations associated with changes in the chopper duty cycles.

8) The proposed feedback control scheme is not modeled at this conceptual design stage. An open-loop control strategy based on FORTRAN expressions is used to regulate the system operating conditions between discrete points by correlating the required power level to the coil current, the chopper duty cycle and the VSC GTO firing angle.

B. Startup from the Shorted SMES Coil Configuration

The simulated sequence of events for startup from the shorted SMES coil configuration with the coil current at 15 kA is as follows:

1) The dc capacitor has been precharged to 6.3 kV.

2) The VSC GTO's are blocked and the capacitor is further charged through the VSC diodes to the normal operating voltage.

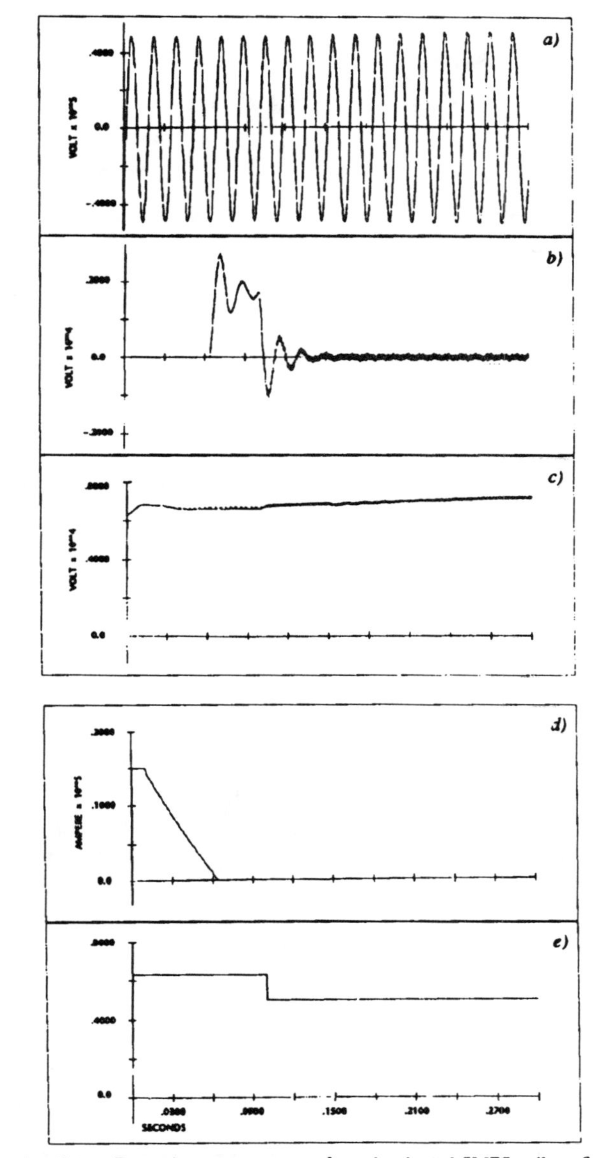

Fig. 13.(a) Dynamic response-startup from the shorted SMES coil configuration (coil current=15 kA) a)Phase A-C line voltage. b) DC voltage across the SMES coil. c) DC voltage across the dc-link capacitor. d) Crowbar switch current. e) Chopper duty cycle.

3) At 10 ms the choppers start switching with a duty cycle > 0.5 to charge the chopper inductors and circulate current through the crowbar switch opposite to the coil current.

4) The crowbar switch opens when the net current is zero.

5) Following the crowbar switch opening, the chopper duty cycle is reduced to 0.5 at 100 ms so that no energy is exchanged with the SMES coil.

6) At 150 ms the VSC GTO's are switched with zero firing angle to complete startup and put the SMES in standby mode.

The simulation results of this sequence are shown in Fig. 13. Fig. 13(b) shows the equal current sharing among the three

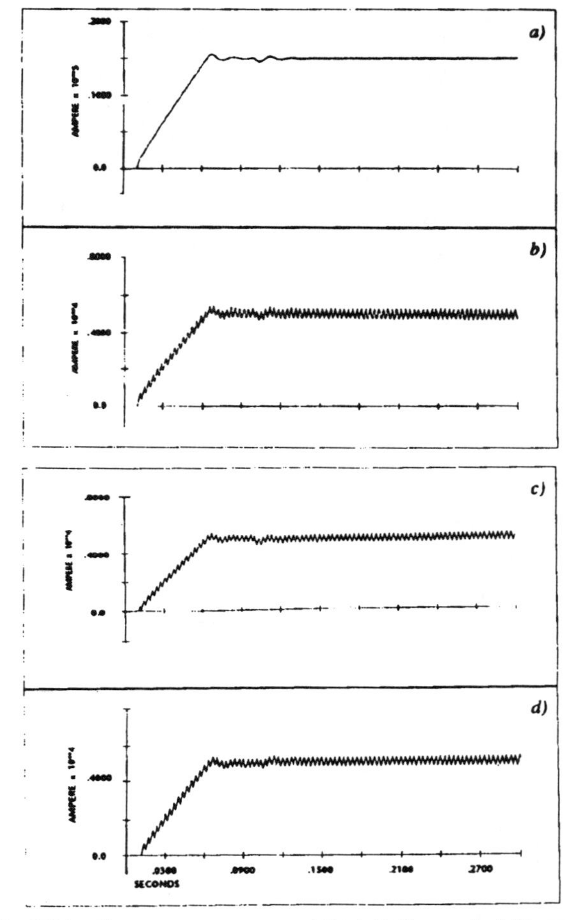

Fig. 13(b). Choppers output currents. a) Total. b) Chopper 1. c) Chopper 2. d) Chopper 3.

chopper phases and the effect of the multiphase chopper on reducing the ripples in the total chopper output.

C. SMES Coil Discharge

The simulated sequence of events during the discharge mode of operation is as follows:

1) The SMES system is initially in the standby mode with no exchange with the utility power system. In this mode the chopper duty cycle is 0.5 and the VSC firing angle is zero. The SMES coil initial current is 49 kA.

2) At 550 ms, a load of 25 MW is switched and the chopper duty cycle and VSC firing angle are changed accordingly.

3) At 750 ms, the load is increased to 50 MW and the chopper duty cycle and VSC firing angle are changed accordingly.

4) At 950 ms, the load is further increased to 100 MW and the chopper duty cycle and VSC firing angle changed accordingly.

The simulation results of this sequence are shown in Fig. 14.

D. Loss of Utility System during SMES Coil Discharging

The loss of the utility power system during discharging at 100 MW was simulated to demonstrate the stand alone

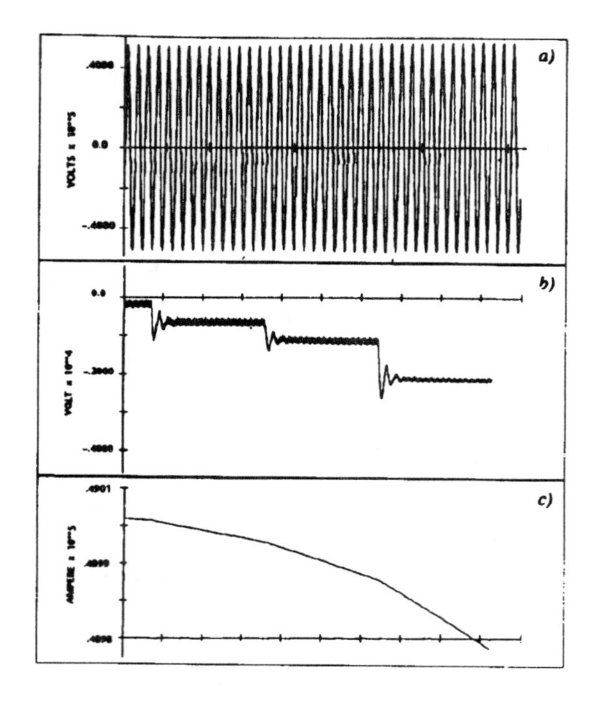

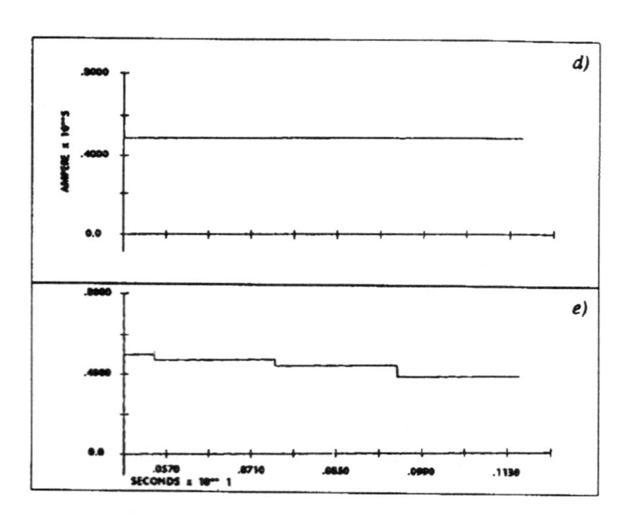

Fig. 14(a). Dynamic response during SMES coil discharge. a) Phase A-C line voltage. b) DC voltage across the SMES coil. c) Coil current. d) Total chopper output current. e) Chopper duty cycle.

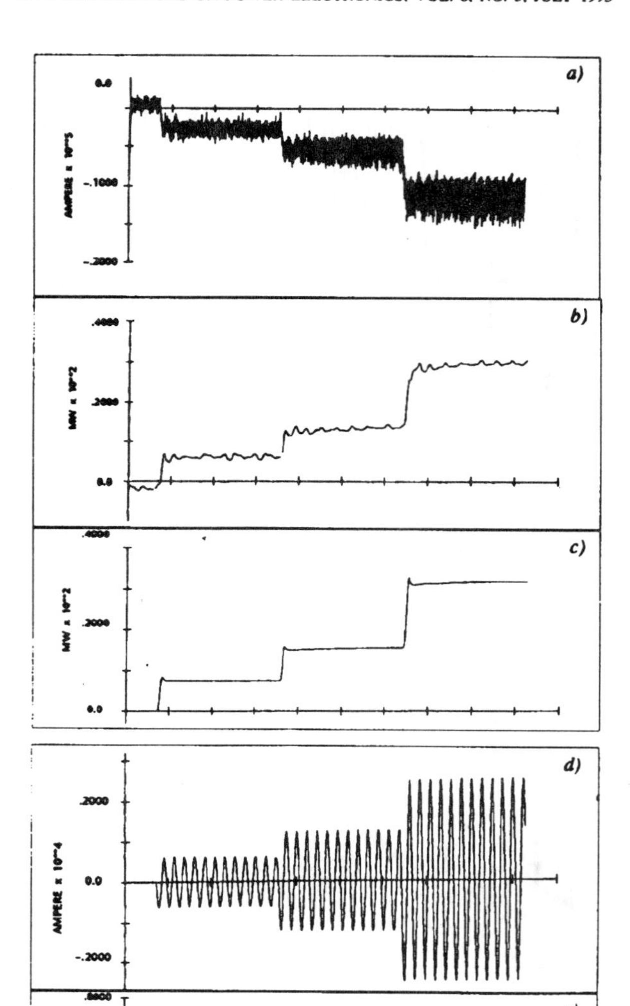

Fig. 14(b). Dynamic response during SMES coil discharge. a) Total VSC output current. b) SMES output power. c) Load power. d) Load current. e) Chopper duty cycle

capability of the PCS. The simulation results of this event are shown in Fig. 15.

The loss of the utility system is simulated at 1.20 s. The simulation results illustrate the capability of the SMES system to supply the load independent of the utility system. The slight drop in the SMES output power is attributed to the absence of the feedback control system to regulate the dc link capacitor voltage.

E. Dynamic Response to Faults

Selected fault conditions were simulated. These faults are initiated at 1.2 s during the discharge mode at a power level of 100 MW with the SMES coil current at 20 kA. The dynamic response of the SMES coil voltage during the simulated faults is shown in Fig. 16.

The fault simulations do not include simulation of control or protective devices actions. Rather, the faults were simulated to

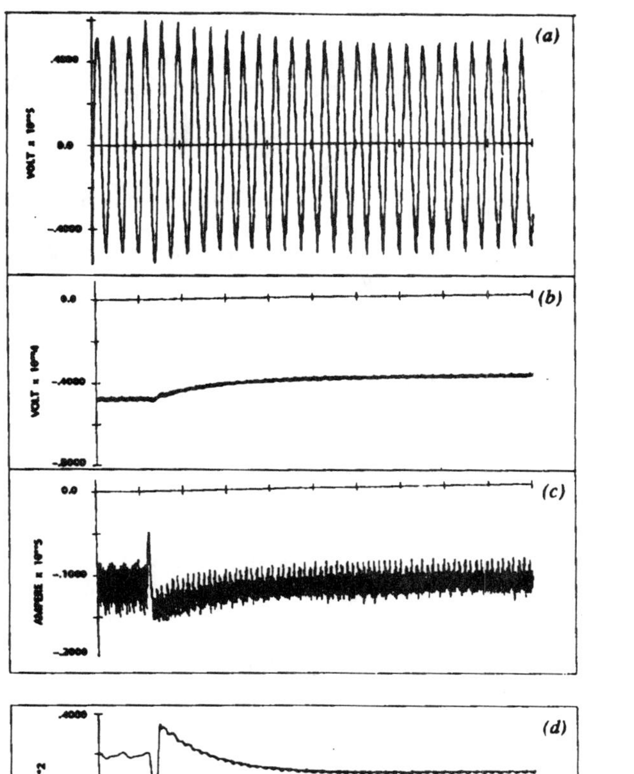

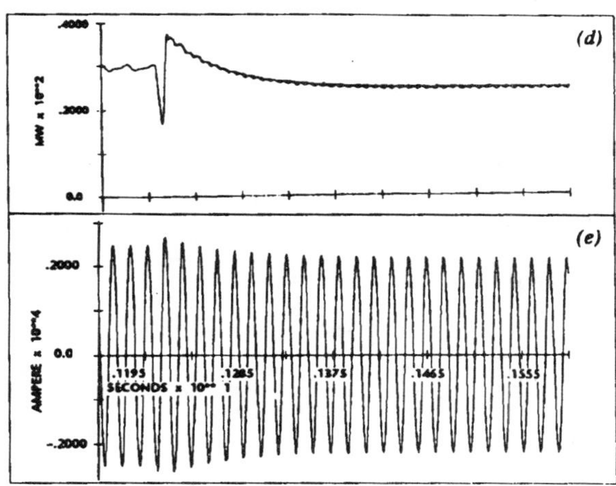

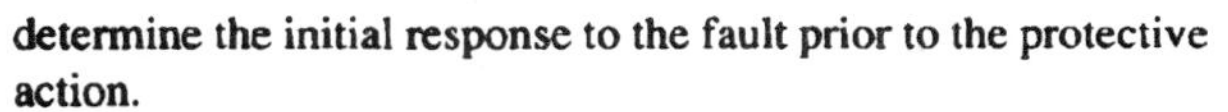

Fig. 15. Dynamic response to loss of utility during discharging. (a) Phase A–C line voltage. (b) DC voltage across the SMES coil. (c) Total VSC current. (d) Phase output power. (e) Load current.

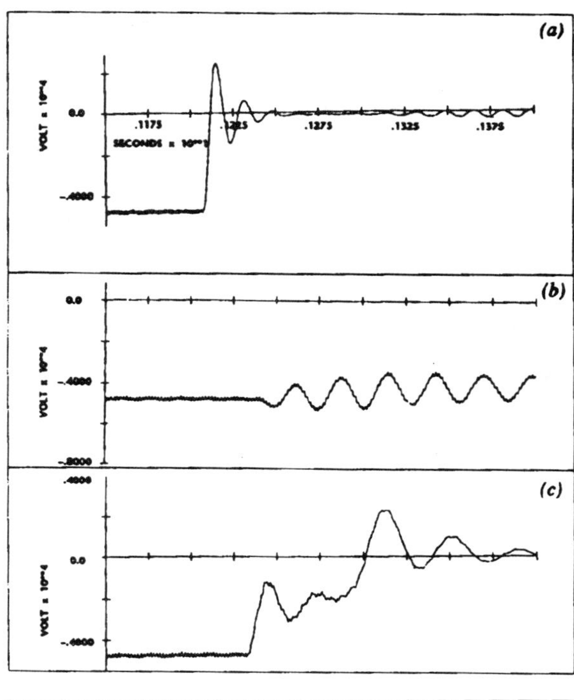

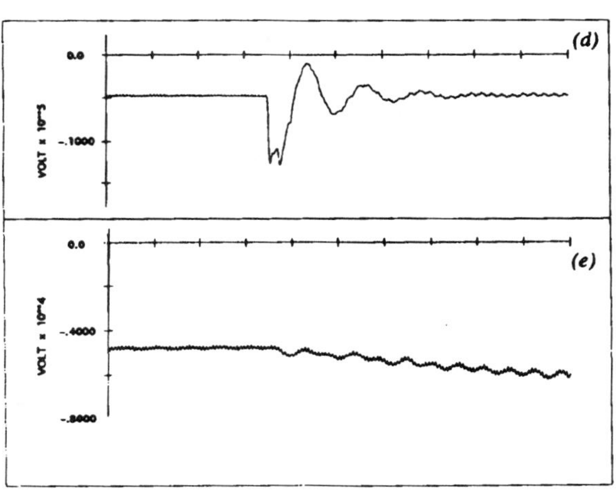

Fig. 16. Coil voltage dynamic response to faults during discharge. (a) One VSC bridge leg shorted. (b) One VSC bridge leg open. (c) One chopper GTO leg shorted. (d) One chopper GTO leg open. (e) Loss of one utility system phase.

determine the initial response to the fault prior to the protective action.

Also simulated is a single line to ground fault initiated at the VSC output during charging at a level of 25 MW with the coil current at 15 kA. The voltage across the SMES coil during the simulated single line to ground fault condition is shown in Fig. 17.

The fault simulations revealed that under some fault conditions the SMES coil would be subjected to voltage transient with a rather steep rate of rise. Such voltage transients would cause high-voltage gradients along the coil windings if not accounted for. The fault simulations also reveal that unbalances within the ac system and unbalanced faults would impress high levels of ac voltage across the coil. Such ac voltages can substantially increase the coil conductor ac losses.

F. Results Summary

The simulation results reveal the following:

1) EMTP can be used to adequately simulate the SMES system and the proposed PCS. The model developed can be easily expanded to include two mutually coupled coils, additional PCS modules, the proposed control system, and other details.

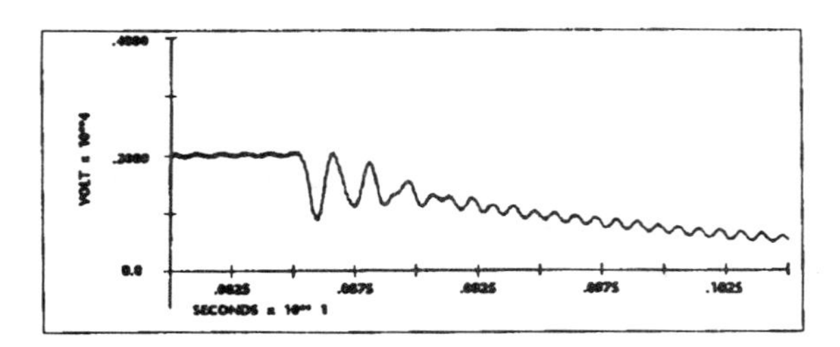

Fig. 17. Coil voltage dynamic response to a single line-to-ground fault during discharging.

2) The simulations demonstrate the feasibility of interfacing the SMES coil to the utility and the ground-based defense system through the proposed PCS.

3) The simulations shown in Fig. 13 and demonstrate the feasibility of the proposed crowbar shorting switch. The time required to open the crowbar switch is in the order of two to four cycles depending on the initial coil current.

4) The overall simulation results demonstrate equal current sharing among the chopper phases and the VSC bridges during normal and transient conditions. The incorporation of the proposed feedback control scheme will further improve the current sharing.

5) No transient overvoltages are experienced across the SMES coil during normal conditions. Incorporation of damping techniques and the feedback control system is required to dampen the voltage over-shoots across the SMES coil associated with sudden changes in the chopper duty cycle.

6) Surge suppressors are required to protect the SMES coil from transient overvoltages experienced during some fault conditions.

7) The fault condition simulations provide data that aid in understanding the system response to faults and will aid in developing the required protective systems.

V. Conclusion

This paper presents the conceptual design of a 22 MWh SMES-ETM. The SMES-ETM power conditioning system is rated 400 MW and is of modular design. Each module is rated 100 MW and consists of a 58-phase dc–dc chopper linked to a quasi-24-pulse voltage source converter. The voltage source converter produces a relatively constant dc output voltage at all levels of power transfer. The dc–dc chopper converts this relatively constant dc voltage to regulate the voltage across the coil to satisfy the required power transfer. This configuration overcomes the problems associated with the wide variations in the coil current and terminal voltage.

The GTO-based voltage source converter is capable of operating in a stand-alone mode and of four quadrant operation with independent control of real and reactive power. The multiphase dc–dc chopper avoids subjecting the SMES coil to low-order voltage ripples. The chopper GTO's can serve as a fast-acting solid-state shorting switch to protect the coil.

The dynamic response of the SMES-ETM during startup, charge, discharge, and fault conditions is included. Digital simulation results demonstrate the practicality of the power conditioning system and aid in understanding the overall SMES-ETM behavior.

It is planned to expand the model in the SMES-ETM detailed design stage to include the two mutually coupled coil groups, the feedback control system, protective devices, a detailed SMES coil model and utility system dynamics. The expanded model will be used to:

1) Verify the SMES-ETM operation, controllability and stability
2) Determine the interactions between the utility transmission system and the SMES coil
3) Investigate potential resonances
4) Determine transient voltage distributions along the SMES coil.

Acknowledgment

The authors would like to express their appreciation for the support and advice received from Dr. W. Scott Meyer and Dr. Tsu-huei Liu of the Bonneville Power Administration on the use of the Alternative Transients Program (ATP) version of EMTP. The concept of a chopper linked to a voltage source converter was proposed by the late P. Wood of Westinghouse Science and Technology Center. K. Mattern, also of the Westinghouse Science and Technology Center, is credited for much of the detailed engineering of this concept.

References

[1] R. W. Boom and H. A. Peterson, "Superconductive energy storage for power systems," *IEEE Trans. Magn.*, vol. MAG-8, pp. 701–703, 1972.
[2] J. D. Rogers *et al.*, "30-MJ superconducting magnetic energy storage system for electric Utility transmission stabilization," *Proc. IEEE*, vol. 71, pp. 1099–1107, 1983.
[3] S. M. Schoenung, W. V. Hassenzahl, and P. G. Filios, "U.S. program to develop superconducting magnetic energy Storage," in *Proc. 23rd Int. Society Energy Conversion Engin. Conf.*, vol. 2, 1988, pp. 537–540.
[4] R. H. Lasseter and S. G. Jalali, "Power conditioning systems for superconductive magnetic energy storage," presented at the IEEE/PES 1991 Winter Meeting, NY, Feb. 3–7, 1991.
[5] W. McMurray, "Feasibility of gate-turn off thyristors in high-voltage direct-current transmission systems," EPRI EL- 5332, Project 2443–5 Final Report, Aug. 1987.
[6] R. H. Lasseter, *Electromagnetic Transients Program (EMTP) Workbook IV (TACS)*, EPRI EL-4651, Project 2149–6, Final Report, Apr. 1988.
[7] H. W. Dommel, *Electromagnetic Transients Program Reference Manual (EMTP Theory Book)* Bonneville Power Administration, Portland, OR, Aug. 1986.

Ibrahim D. Hassan (SM'89) was born in Egypt in 1943. He received the B.Sc. degree from Cairo University and the M.S. and Postgraduate Degree of Engineer from the New York Polytechnic Institute, Brooklyn, in 1965, 1971, and 1980, respectively, all in electrical engineering.

Since 1977 he has been with Ebasco Services Incorporated where he is presently a Senior Consulting Engineer in the Nuclear and Advanced Technology Group. His work there deals with system analysis, simulation, and application of power converters and rotating machinery. He has taught theory of electrical machinery as a visiting associate professor at Pratt Institute, NY.

Mr. Hassan is a member of Sigma Xi and is a registered professional engineer in New York state.

Richard M. Bucci (SM'88) was born in Brooklyn, NY, in 1949. He received the B.S.E.E. degree from Pratt Institute, NY, and the M.SE.E. (power systems) degree from the New York Polytechnic Institute, Brooklyn.

He has been with Ebasco Services Incorporated since 1974, working in the areas of power generation, transmission, and distribution systems. He is presently Manager, Systems and Equipment Engineering in the Electric Power Systems Division, where he is responsible for systems planning and equipment studies.

Mr. Bucci is a member of the Systems Planning and Dynamic Performance Subcommittees of the IEEE PES Power System Engineering Committee and the Operations Testing and Surveillance Subcommittee of the Nuclear Power Engineering Committee. He is a registered professional engineer in New York state.

Khin T. Swe (M'80) received the B.S. and M.S. degrees from the New York Polytechnic Institute, Brooklyn, in 1980 and 1984, respectively, both in electrical engineering.

Since 1980 she has been with Ebasco Services Incorporated. She is presently a Senior Engineer in the Nuclear and Advanced Technology Group. Her work there includes development of software for system analysis and simulation of power systems.

Ms. Swe is a member of Tau Beta Pi and Eta Kappa Nu.

IMPACT OF UTILITY SWITCHED CAPACITORS ON CUSTOMER SYSTEMS -

MAGNIFICATION AT LOW VOLTAGE CAPACITORS

M. F. McGranaghan R. M. Zavadil
Members, IEEE
Electrotek Concepts, Inc
Knoxville, Tennessee

G. Hensley T. Singh
Members, IEEE
Pacific Gas and Electric Company
San Ramon, California

M. Samotyj
Senior Member, IEEE
Electric Power Research Institute
Palo Alto, California

Abstract - This paper analyzes the potential for magnified transient voltages at customer buses during capacitor switching on the primary distribution system. The various factors affecting this phenomena are analyzed in detail through extensive sensitivity analysis simulations. These factors include the switched capacitor size, short circuit capacity at the switched capacitor, customer step down transformer size, low voltage power factor correction, and customer load characteristics. The impacts of these transients on customer equipment are described and possible solutions to control the transients are presented.

Keywords: power quality, capacitor switching, magnification, arrester duties

INTRODUCTION

Transient overvoltages are always a concern when capacitor switching is involved. Each time a capacitor is energized, a transient oscillation occurs between the capacitor and system inductance. The result, illustrated in Figure 1, is a transient overvoltage which can be as high as 2.0 pu at the capacitor location. The magnitude is usually less than this due to the damping provided by system loads and losses.

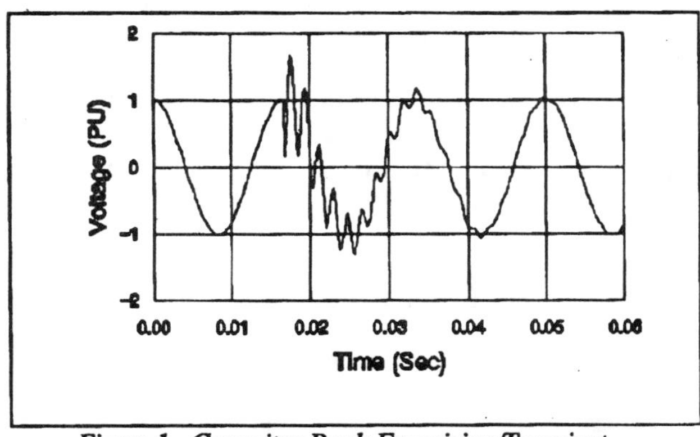

Figure 1 - Capacitor Bank Energizing Transient

The transient overvoltages caused by capacitor energizing have not been of concern to the power supplier because they are below the level at which surge protective devices will operate (1.8 pu or above). There are a couple of trends, however, which are forcing a reexamination of this relatively common power system phenomena. Customer loads are becoming increasingly sensitive due to a move to power electronics equipment for increased energy efficiency and flexibility. Utility customers are also adding power factor correction capacitors to avoid rate penalties and further reduce energy costs. The combination of these trends is now resulting in increased customer power quality problems due to capacitor switching events on the utility system.

Transient overvoltages from utility capacitor switching can be magnified in the customer facility if the customer has low voltage capacitor banks for power factor correction. Magnification occurs when the transient oscillation initiated by the capacitor energizing can excite a series L-C circuit formed by the customer step down transformer and low voltage capacitors. In this case, a higher magnitude transient oscillation can occur at the low voltage bus.

The potential for magnified transients during capacitor switching was analyzed in the classic paper by

0-7803-0219-2/91/0009-0908$01.00 ©1991IEEE

Schultz, Johnson, and Schultz [1]. The concern has been recognized in subsequent publications describing capacitor switching concerns [2,3] but has not been described in detail as it relates to switching capacitor banks on distribution systems and the resulting transients that can occur within customer facilities.

This paper describes the magnification phenomena and provides a detailed analysis of the potential concern for transient overvoltages within customer facilities. All of the important parameters affecting the magnification phenomena are characterized and possible solutions to the problem are presented. This concern has become particularly important as utilities are instituting higher power factor penalties and encouraging customers to add their own power factor correction. The simulations and case studies performed to characterize this phenomena were carried out under EPRI contracts RP2935-13 and RP2935-99.

THE BASIC PHENOMENA

The circuit used to analyze the magnification problem is illustrated in Figure 2. Magnification occurs because the natural frequency of the transient oscillation during capacitor energizing on the primary excites a series LC circuit formed by the customer step down transformer and the low voltage capacitors. The circuit in Figure 2 was developed to analyze the different parameters that can affect this magnification. In order to perform the analysis, a model was developed for the Electromagnetic Transients Program (EMTP). The base conditions for the analysis were as follows:

- System Source Strength at the Substation = 200 MVA
- Switched Capacitor Bank Size = 3 MVAr
- Total Feeder Load = 3 MW
- Customer Transformer Size = 1500 kVA (6% Impedance)
- Customer Power Factor Correction = 300 kVAr
- Customer Resistive Load = 300 kW

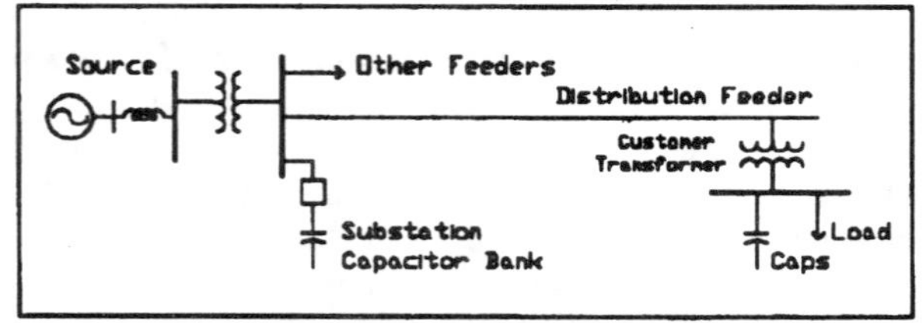

Figure 2 - One Line Diagram for Example System

Distribution feeder length was not considered in the parametric study. Because of the relatively low natural frequencies associated with switching utility capacitor banks, the distribution feeder can be adequately represented as a series resistance and inductance. Given that, the impedance can be combined with the customer stepdown transformer impedance. Long distribution feeders would have an impact on the transient magnification by increasing the equivalent inductance of the secondary LC circuit. However, to limit the number of variables in the parametric investigation, the customer location is assumed to be electrically very near the utility capacitor location (i.e. no distribution line length was represented).

The magnification phenomena is illustrated by the waveforms in Figure 3. Figure 3a gives the transient at the switched capacitor location and Figure 3b shows the magnified transient caused by the 300 kVAr capacitor bank at the customer's 480 Volt bus.

The waveforms in Figure 3 illustrate that the magnification can be quite significant. It is not uncommon for magnified transients at the low voltage capacitors to be in the range of 2.0-3.0 pu. These transients have significant energy associated with them and are likely to cause failure of protective devices, electronic equipment (SCR's, etc.), capacitors, or other devices. Adjustable-speed drives (ASDs) are particularly susceptible to these transients because of the relatively low peak inverse voltage (PIV) ratings (sometimes only 1200 Volts) of the semiconductor switches and the low energy ratings (less than 100 Joules) of the MOV's typically used to protect the power electronics [4].

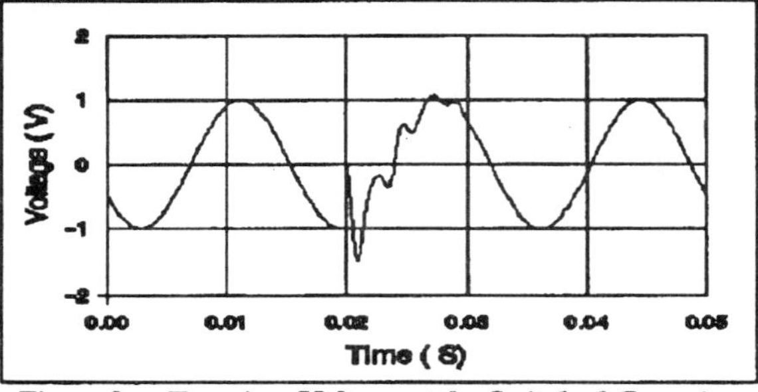

Figure 3a - Transient Voltage at the Switched Capacitor

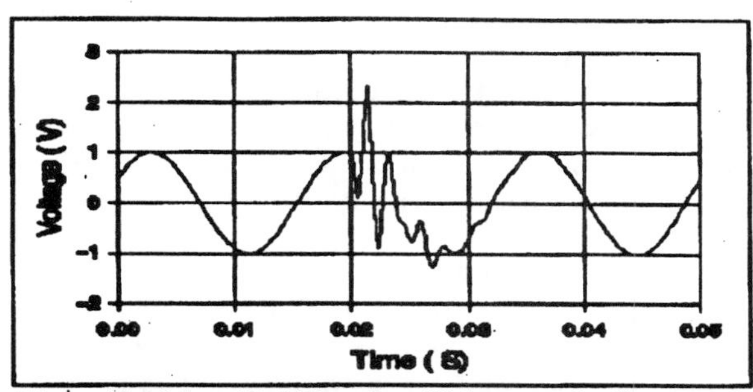

Figure 3b - Magnified Transient Voltage at the Customer's 300 kVAr Capacitor Bank

The characteristics of the magnified transient experienced by the customer depend on a number of factors. Previous analyses have concluded that the magnification is most severe when the following conditions exist [2]:

- The capacitor switched on the higher voltage system is much larger (kVAr) than the capacitor at the low voltage bus.
- The frequency of oscillation which occurs when the high voltage capacitor is energized is close to the resonant frequency formed by the step-down transformer in series with the low voltage capacitor.
- There is little damping provided by loads on the low voltage system, as is usually the case for industrial plants (motors do not provide significant damping of these transients).

However, these basic characteristics do not provide enough information to evaluate a specific set of conditions and determine whether or not there is a possible problem. The following section illustrates the effect of the different parameters which affect the severity of the magnified transient and provides curves which can be used to evaluate specific conditions.

PARAMETRIC ANALYSIS

Factors affecting the magnification include the source strength at the switched capacitor, the switched capacitor size, the customer step down transformer size, the customer power factor correction capacitor size, and the system damping. The most important of these parameters are analyzed here.

Since the magnification phenomena is related to two different LC circuits (the LC circuit for the switched capacitor transient and the LC circuit formed by the low voltage capacitor and the inductance between the two capacitor banks), it is important to consider the interrelationships between the different parameters. The level of power factor correction at the customer's low voltage bus is probably the single most important parameter affecting the potential for magnified transients. Figures 4 and 5 present the results of numerous sensitivity analysis simulations to illustrate this effect in combination with the effect of the other parameters influencing the two LC circuits.

Figure 4 gives the peak transient magnitude at the customer bus as a function of the low voltage capacitor size for a range of different switched capacitor sizes. These curves illustrate that low voltage capacitor size resulting in the worst magnification depends on the switched capacitor size. As the switched capacitor size increases, the potential for magnified transients occurs over a wider range of low voltage capacitor sizes (broader response).

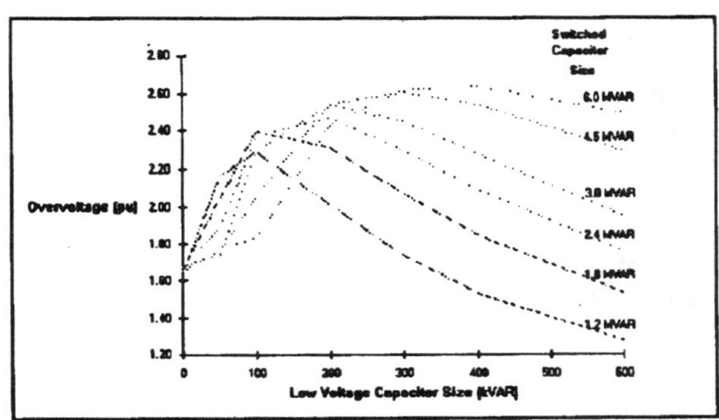

Figure 4 - Transient Voltage Magnitude at the Low Voltage Bus as a Function of Low Voltage Power Factor Correction Level

Figure 5 gives the peak transient magnitude at the customer bus as a function of the low voltage capacitor size for three different customer step down transformer sizes. The transformer inductance dominates the inductance between the two capacitors and, therefore, determines the resonant frequency for the second LC circuit. The curves illustrate that, as the step down transformer size is increased, the worst case magnification occurs for higher values of low voltage capacitors. However, the most important observation from these curves is that the magnified transients can occur over a wide range of low voltage capacitor sizes.

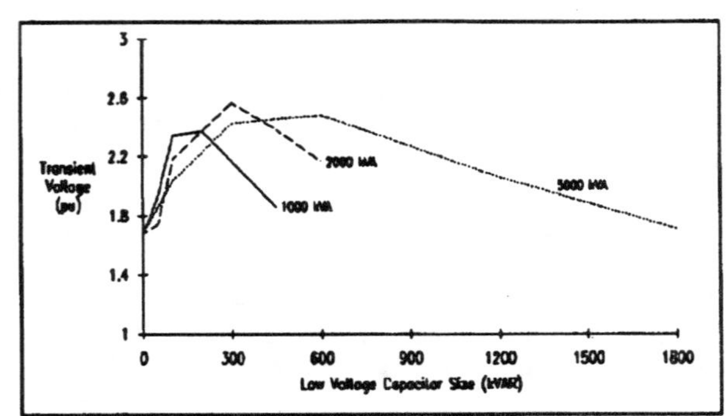

Figure 5 - Transient Voltage Magnitude at the Low Voltage Bus as a Function of the Customer Step Down Transformer Size

The magnified transient is also affected by the level of damping in the system. Resistive load on the distribution system provides damping which reduces the magnitude of the energizing transient on the distribution primary. However, the magnification can still result in severe transients at the customer bus.

The damping provided by the customer load is most important for reducing the magnitude of the magnified transient. Unfortunately, many industrial customer loads are dominated by motors which provide only minimal damping at these transient frequencies due to their inductive characteristics. Figure 6 illustrates the effect of both resistive and motor load on the magnified transient magnitude for the base case system conditions. Note that the magnified transient is still greater than 2.0 pu even with 1000 kW of resistive load on the 1500 kVA transformer.

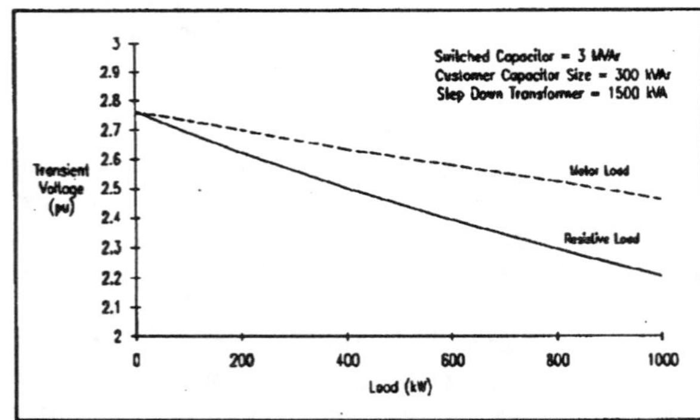

Figure 6 - Transient Voltage Magnitude at the Low Voltage Bus as a Function of Customer Load Characteristics

EFFECT OF LOW VOLTAGE ARRESTERS

Low voltage metal oxide varistors (MOV's) are frequently used to protect electronic equipment in industrial environments. Adjustable speed drives and dc drives with thyristor rectifiers will typically come equipped with MOV's for protection of the SCR's. The problem is that the energy levels for these magnified transients greatly exceed the capabilities of most of the small MOV's used in electronic equipment.

Figure 7 illustrates the waveforms for an MOV arrester operation during the magnified transient for the base case conditions. Figure 7a illustrates the effect of the arrester on the 480 Volt bus voltage (line-to-ground connected arresters result in the phase-to-phase voltage being clipped at about 2.0 pu). Figure 7b shows the current surge in the arrester to achieve this clipping. Figure 7c is a plot of the arrester energy duty as a function of time. The energy of approximately 1100 Joules would definitely cause the failure of small MOV's used in electronic equipment.

Figure 8 gives the arrester duty as a function of the low voltage capacitor size for a number of different switched capacitor sizes. The potential duty on the low voltage arrester increases substantially as the size of the switched capacitor is increased. This must be taken into account when selecting an arrester rating for protection of the 480 Volt bus.

Arrester duties resulting from the magnification phenomena, as indicated in Figure 8, can be quite severe. High-energy MOV arresters with energy capability sufficient for these duties are commercially available, and can be a low-cost solution to the magnification phenomena in many instances. Normal arrester protective levels, however, may still be above the protective level necessary for adequate protection. This is especially true for some power electronics equipment, where device PIV levels may be below normal arrester protective levels. For instance, the common 1200 v rating of semiconductor switches used in some adjustable-speed drives translates to a 1.77 pu transient on a line-to line basis. Coordinating arrester protective levels with equipment requirements may be difficult under these circumstances.

Some equipment may be affected by transient voltage levels far below that which may be achieved by arrester application [4]. Nuisance tripping of adjustable-speed drives has been shown to occur for transient overvoltages of only 1.2 pu. Here, the equipment will be sensitive to most utility capacitor switching events even without magnification of the overvoltages at the load bus.

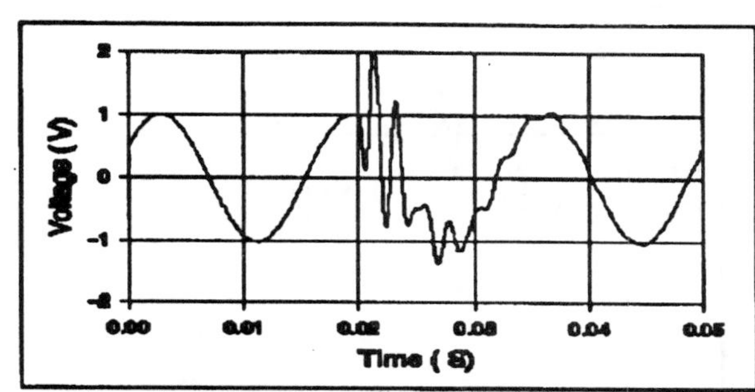

Figure 7a - 480 Volt Bus Voltage with MOV Arrester

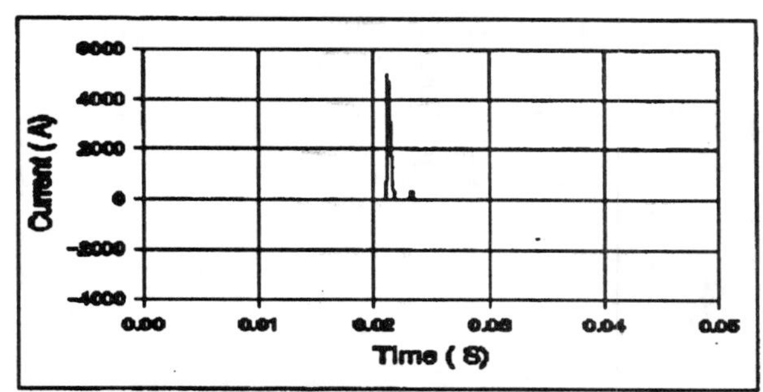

Figure 7b - Arrester Current (Phase A)

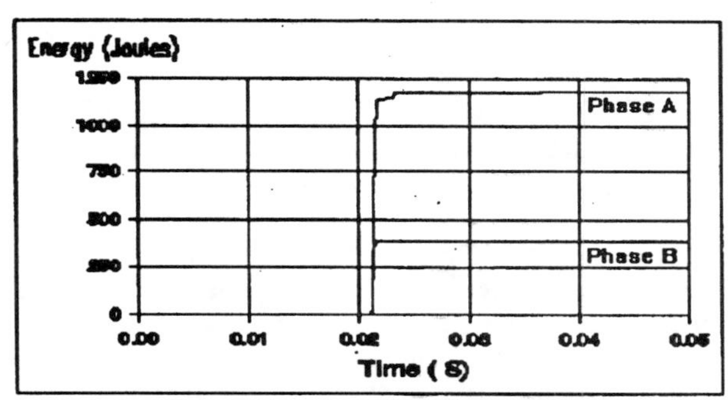

Figure 7c - Arrester Energy vs. Time

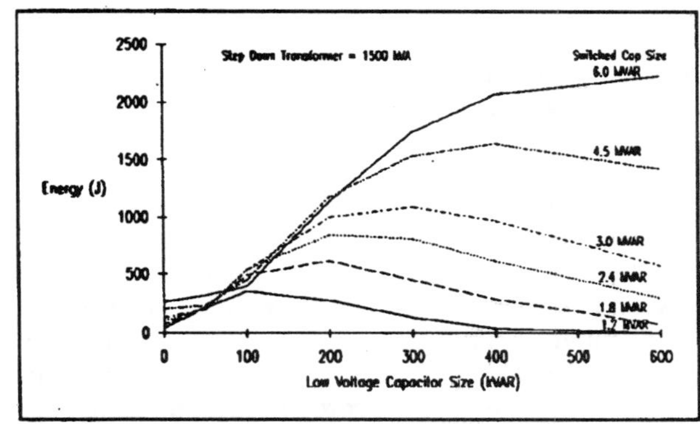

Figure 8 - Low Voltage Arrester Duty as a Function of the Low Voltage Capacitor Size

CONTROLLING THE TRANSIENT AT THE SWITCHED CAPACITOR

The ideal solution to the transient voltage magnification problem would be to minimize the capacitor switching transient that excites the LC circuit resulting in magnification. There are basically two methods to accomplish this:

1. The capacitor switch closing instants can be controlled so that each phase of the capacitor bank is energized when the voltage across the switch is zero. This could be accomplished easily with solid state switches. However, the only conventional switching device that is fast enough to permit implementation of a synchronous closing control is a vacuum switch or vacuum breaker [5]. Figure 9 illustrates the effect of the closing instant on both the primary distribution system transient and on the magnified transient for the base case conditions. Note that the magnified transient is reduced just by separating the closing instants of the three phases because it is the phase-to-phase transient that gets magnified.

2. The capacitor switching device can include closing resistors optimized to reduce the capacitor energizing transient. Figure 10 illustrates the effect of resistor size on the energizing transient and on the magnified transient at the customer bus for the base case conditions. Unfortunately, optimum size closing resistors are not commonly available for distribution switching devices making this a difficult solution to implement.

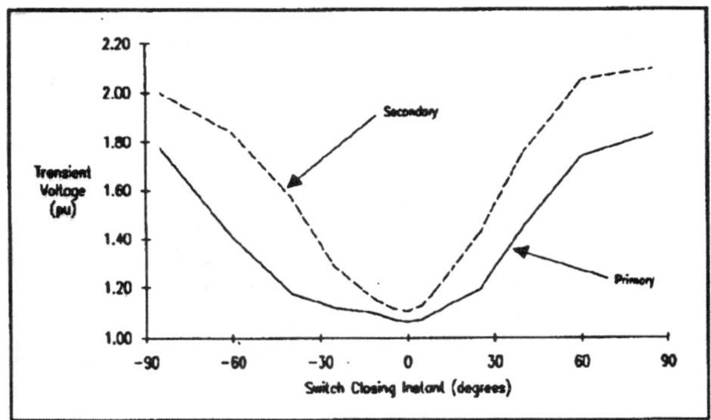

Figure 9 - Capacitor Switching Transient Voltages as a Function of the Switch Closing Instants (time of closure on each phase relative to the voltage zero crossing)

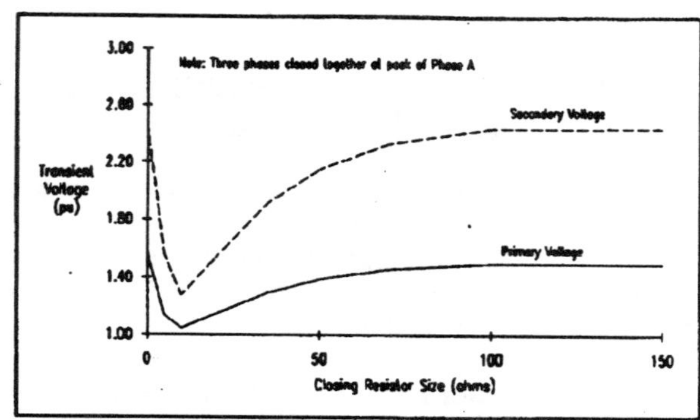

Figure 10 - Capacitor Switching Transient Voltages as a Function of Resistor Size Used in the Capacitor Switching Device

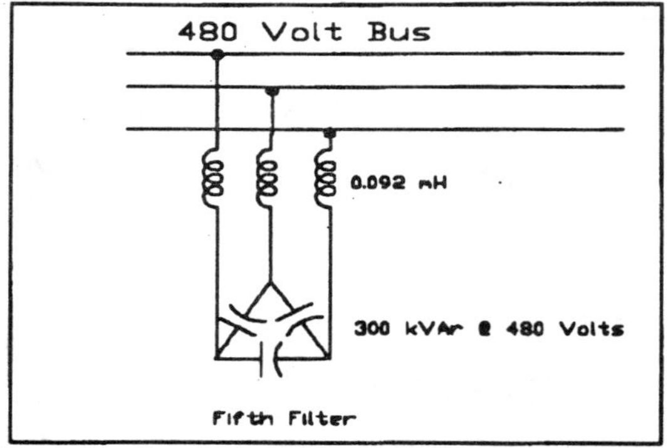

Figure 11 - Low Voltage Harmonic Filter Configuration (300 kVAr Filter)

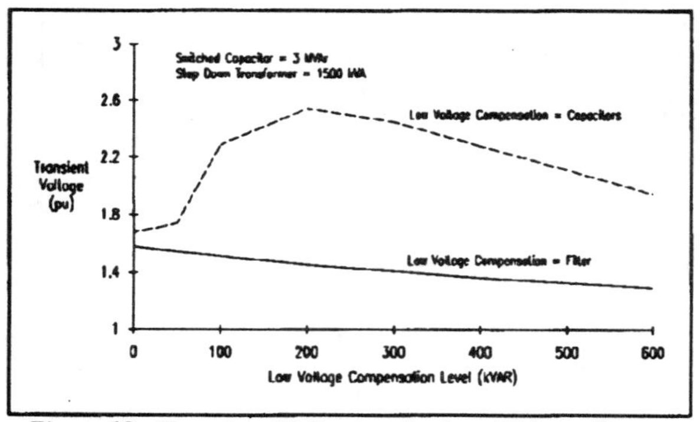

Figure 12 - Transient Voltage at the Low Voltage Bus vs. Compensation Level Illustrating the Effect of Harmonic Filters (Switched Capacitor Size = 3 MVAr)

EFFECT OF HARMONIC FILTERS

Based on the discussions thus far, it appears that the customer has very few options for controlling magnified transients if he needs to add power factor correction and the utility has switched capacitor banks. Low voltage arresters can be applied but they must have high energy ratings and they still may not coordinate with the withstand levels of equipment in his plant. It would be better to eliminate the magnification completely.

In order to eliminate the magnification, the series LC circuit that is being excited from the distribution primary must be detuned. In the base case, this series LC circuit consists of the step down transformer and the low voltage capacitors. The resonance for this series combination is usually in the range 400-800 Hz, which is in the same range as the oscillation created by the primary capacitor energizing.

One way to detune the series LC circuit is to add a significant inductance in series with the power factor correction capacitors. For instance, if the power factor correction is applied as harmonic filters tuned to the 4.7th harmonic (a common practice for control of harmonic distortion levels), the resonance for the total series combination looking from the primary is reduced to less than 285 Hz. This lower frequency resonance is not as easily excited by the primary capacitor switching. The filter has the additional benefit of moving the maximum transient voltage from the 480 Volt bus to the capacitor side of the tuning reactors. Figure 11 illustrates the most typical configuration for low voltage harmonic filters. Figure 12 shows the effect of applying the power factor correction as harmonic filters instead of just capacitors. It can be seen that magnified transient voltages are not a problem, regardless of the kVAr level.

It is important to note that all of the low voltage power factor correction must be applied as harmonic filters if this solution is to be effective. If harmonic filters are applied in combination with shunt capacitors, a new resonance is introduced at a higher frequency which can easily be excited by the primary capacitor switching. The transient waveform in Figure 13 illustrates this problem for power factor correction consisting of a 200 kVAr filter and an additional 100 KVAr capacitor. The magnified transient is again greater than 2.0 pu.

It should be noted that while the transient overvoltage at the load bus will be limited by the harmonic filter, the transient voltages across the filter inductor and capacitor will be higher. However, the transient duty on the capacitor bank in a harmonic filter should be no higher than if it were connected directly to the bus and magnification occurs. Surge withstand capability of the filter components needs to be considered in the design stage for harmonic filters, even when they will not be exposed to utility capacitor switching. Capacitors with 600 v ratings are many times used in filter applications due to increased duties from harmonic currents, which provides some additional margin for overvoltages due to utility capacitor switching.

If harmonic filtering is not required to limit voltage distortion, as in situations where most of the plant load is linear and and it has been determined that harmonic resonance problems will not occur, high-energy MOV arresters may be the preferred solution to the magnification problem.

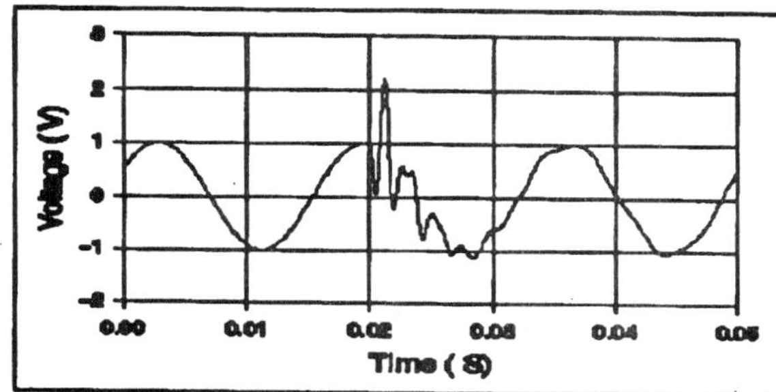

Figure 13 - Transient Voltage at the 480 Volt Bus ith Power Factor Correction Consisting of a 200 KVAr Filter and a 100 KVAr Capacitor Bank (Switched Capacitor Size = 3 MVAr)

CONCLUSIONS

1. The addition of power factor correction capacitors at low voltage buses can result in magnified transient voltages during capacitor switching on the primary distribution system. Magnified transients in the range 2-3 pu are possible over a broad range of low voltage capacitor sizes. Problems with these transients depend on the sensitivity of the customer equipment. Some types of power electronic devices can be particularly susceptible to these transients.

2. Low voltage arresters (MOV's) used for protection of customer equipment can be exposed to severe energy duties as a result of the magnification phenomena. The arrester duties can be in the range 2000-3000 Joules as compared to a capability of less than 100 Joules that is typical for many MOV's used to protect power electronic devices. High energy MOV arresters are available but their protective characteristics must still be evaluated with respect to the withstand characteristics of the equipment being protected.

3. It is possible to control the capacitor energizing transient on the primary distribution system to avoid severe transients in customer plants. This can be done with a synchronous closing control on vacuum switching devices or with appropriately sized closing resistors. In many existing installations, either of these solutions could require complete replacement of the switchgear.

4. The best solution to the magnification problem is to apply low voltage power factor correction as harmonic filters rather than as just capacitors. The filter configuration detunes the circuit, preventing magnified transients. The filters are also effective in limiting harmonic distortion levels due to nonlinear loads, such as adjustable speed drives.

REFERENCES

[1] A. J. Schultz, I.B. Johnson, and N. R. Schultz, "Magnification of Switching Surges," *IEEE Transactions on Power Apparatus and Systems*, Vol 77, February 1959, pp 1418-1425.

[2] S. S. Mikhail and M. F. McGranaghan, "Evaluation of Switching Concerns Associated with 345 kV Shunt Capacitor Applications," *IEEE Transactions PAS*, Vol. 106, No. 4, pp 221-230, April, 1986.

[3] M. F. McGranaghan, W. E. Reid, S. W. Law, and D. W. Gresham, "Overvoltage Protection of Shunt Capacitor Banks Using MOV Arresters," *IEEE Transactions PAS*, Vol. 104, No. 8, pp 2326-2336, August, 1984.

[4] G. Hensley, T. Singh, M. Samotyj, M. McGranaghan, and T. Grebe, "Impact of Utility Switched Capacitors on Customer Systems, Part II - Adjustable Speed Drive Concerns," Submitted for the 1991 IEEE-PES Winter Power Meeting.

[5] R. W. Alexander, "Synchronous Closing Control for Shunt Capacitors," *IEEE Transactions PAS*, Vol. 104, No. 9, pp 2619-2626, September, 1985.

IEEE Transactions on Power Delivery, Vol. 9, No. 4, October 1994

ARC-FURNACE MODEL FOR THE STUDY OF FLICKER COMPENSATION IN ELECTRICAL NETWORKS

G. C. Montanari[1], M. Loggini[1], A. Cavallini[1], L. Pitti[1], D. Zaninelli[2]
Senior Member Member Non Member Non Member Member

[1]Istituto di Elettrotecnica Industriale, Università di Bologna
[2] Dipartimento di Elettrotecnica, Politecnico di Milano
Italy

Abstract - This paper presents an arc-furnace model consisting of non-linear, time varying resistance where two different time-variation laws of arc length are considered. One consists of a periodic, sinusoidal law, the other of a band-limited white-noise law. The arc-furnace model is implemented by EMTP, referring to actual electric-plant configurations. Simulations are reported where the values of flicker sensation and short-term flicker severity, P_{ST}, are determined according to UIE specifications. It results that the model based on sinusoidal time-variation law can be useful for worst-case approximations, while the model using white-noise law is able to fit flicker measurements made in electric plants supplying arc furnaces. The models are used to investigate the effect on flicker compensation of the insertion of series inductors at the supply side of the furnace transformer. It is shown that considerable reduction of P_{ST} is obtained at the point of common coupling by series inductor installation at constant furnace active power.

Keywords: Arc Furnace, EMTP simulation, Flicker.

INTRODUCTION

Arc furnaces used for steel production are a main cause of voltage fluctuations in electrical networks, which may give rise to the flicker effect. Voltage fluctuations, due to ramdom arc-length variations during scrap melting, have typical frequencies in the range 0.5-25Hz.

Flicker consists of luminosity variations of lamps which may affect the human visual system, depending on their frequency and intensity. For example, voltage-amplitude variations of about 0.3% at frequency 10 Hz are sufficent to get over the mean human perceptivity threshold [1-3].

Since voltage fluctuations are not limited to electric plants supplying arc furnaces, but may affect the HV network to a large extent, several MV and LV customers can be disturbed by flicker, so that electricity-supply companies must take care of this problem. Generally, the solutions able to reduce flicker are [4]

i) to decrease furnace power,

ii) to increase short-circuit power at the point of common coupling (PCC),

iii) to install apparatus for flicker compensation.

As is obvious, solution i) is not economically valid, except for short periods, while ii) depends on network management.

Solution iii) is more easily available for the arc-furnace customer and preferable for the electricity supplier. Often, static-var-systems (SVS) are installed in plants feeding arc furnaces in order to compensate both voltage fluctuation and voltage distortion. However, this solution can be quite expensive, depending on the plant size, and, moreover, fluorescent-light flicker have been addressed to SVS-network interactions [4].

On-field experiences, as well as computer simulations, have proven the effectiveness of inserting series inductors at the supply side of the furnace transformer [5, 6], but in depth investigations on this topic are still lacking.

The choice of the most convenient solution for flicker compensation requires availability of accurate arc-furnace models. This would allow simulating the electric plant supplying the arc furnace and studying the effects of compensation systems on arc and flicker behavior, as well as on voltage and current distortion factor, power factor, furnace active and non-active power. At present, electric arc is usually modelled by voltage generators which provide fundamental and harmonic voltages whose amplitudes are time-modulated to describe arc-length variations and, hence, network-voltage fluctuations. The modulation law can be sinusoidal at frequencies typical of flicker [6, 7]. However, it has been shown in [8] that this solution is not fully satisfactory, being a linear representation of non-linear phenomena, which is unable to take into account the effect on electric-arc behavior of changes of electric-plant configuration (due to, e.g., insertion of series inductors or shunt filters). Therefore, an improved model was proposed in [8], where the electric arc is described by a non-linear, time-varying resistance.

In this paper, the effect on flicker compensation of the installation of series inductors at the supply side of the furnace transformer is investigated, using EMTP simulation [9]. The electric arc is described by a non-linear resistance and two different time-variation laws (based on sinusoidal and white-noise functions) are considered. Comparisons of the proposed models with measurements made on north Italian plants are reported.

NON-LINEAR, TIME-VARYING ARC MODELS

A typical circuit diagram of an electric plant supplying an arc furnace is shown in Fig. 1. The furnace is connected to bus 1, the PCC, by means of a HV/MV tranformer (T1) and is fed by a MV/LV tranformer (T2). The furnace-side of this transformer usually has adjustable voltage in order to vary the furnace power. X_{LSC} is the short-circuit reactance at the PCC, X_P the series reactance inserted for flicker compensation (both X_{LSC} and X_P are varied in the simulations; the reference value of the short-circuit power at PCC is 3500 MVA). X_C and R_C are reactance and resistance of the connection line between

94 WM 086-9 PWRD A paper recommended and approved by the IEEE Transmission and Distribution Committee of the IEEE Power Engineering Society for presentation at the IEEE/PES 1994 Winter Meeting, New York, New York, January 30 - February 3, 1994. Manuscript submitted June 24, 1993; made available for printing December 6, 1993.

furnace electrodes and MV/LV transformer, which generally make a significant contribution to the total impedance seen by the arc furnace.

The values attributed to the plant parameters for the EMTP simulations presented in this paper are typical of arc-furnace plants installed in northern Italy. The HV/MV and MV/LV transformers (220/21 kV and 21/0.9÷0.6 kV) have rated power 95 MVA and 60 MVA, respectively, per cent values of short-circuit voltage and losses 12.5% and 0.5% (HV/MV), 10% and 0.5% (MV/LV). The values of the lead reactance and resistance are $X_C=3\ 10^{-3}$ Ohm and $R_C=3\ 10^{-4}$ Ohm [7, 10].

As mentioned above, simulation of the electric arc is realized by means of a non-linear model. The arc voltage-current characteristic,

$$V_a = V_a(I_a) \tag{1}$$

can be described by the following relationship

$$V_a = V_{at} + \frac{C}{D + I_a} \tag{2}$$

where V_a, I_a are arc voltage and current, V_{at} is the threshold value to which voltage tends when current increases, C and D are constants whose values (C_a, D_a and C_b, D_b) determine the difference between the increasing and decreasing-current parts of the v-i characteristic (eq.(2) is written for $I_a>0$, but can be easily arranged for $I_a<0$ as well). Figure 2 shows the arc voltage-current characteristic obtained by the model of eq.(2), implemented in the TACS (Transient Analysis Control System) section of EMTP [9], with reference to an arc length giving $V_{at}=200$ V, $C_a\approx190000$ W, $C_b\approx39000$ W, $D_a=D_b=5000$ A [8].

In the event that the arc length, l, would not change with time ($l=l_0$), the arc voltage-current characteristic would be time-invariant. Then, the furnace would not give rise to flicker at the PCC, but only to voltage and current harmonics due to intrinsic non-linearity of the arc characteristic. Simulating, by EMTP, the electric plant feeding the arc furnace as well as the electric arc (according to eq.(2)), for different values of arc length, the characteristic curves of the arc furnace can be plotted for both single-phase and three-phase plant configurations. The characteristic curves generally consist of the graphs of adimensional quantities S/S_{SC}, P/S_{SC}, P_{arc}/S_{SC}, Q/S_{SC}, PF, plotted as function of I/I_{SC}. S, P and Q are apparent, active and non-active power, respectively [11], P_{arc} is arc active power, PF is the power factor, S_{SC} and I_{SC} are the apparent power and current at bus 2 of Fig. 1 when the furnace operates in short-circuit conditions, I is the rated load current.

In this paper, single-phase investigations are performed, since the main purpose is to propose arc-furnace models suitable for the study of the effect on flicker compensation of series inductors. However, the range of values of I/I_{SC} where the EMTP simulations are realized assure continuous conduction both in single-phase and three-phase configurations. This would allow referring single-phase results to three-phase evaluations, even if quantitative estimates of voltage fluctuation are affected by the fact that the P/S_{SC} and Q/S_{SC} curves are slightly different for single and three-phase configurations (contrarily to adimensional apparent power) [8, 12]. Three-phase estimates of voltage fluctuation, based on single-phase models, are also affected by the implicit assumption of balanced load which often is not the actual operating condition.

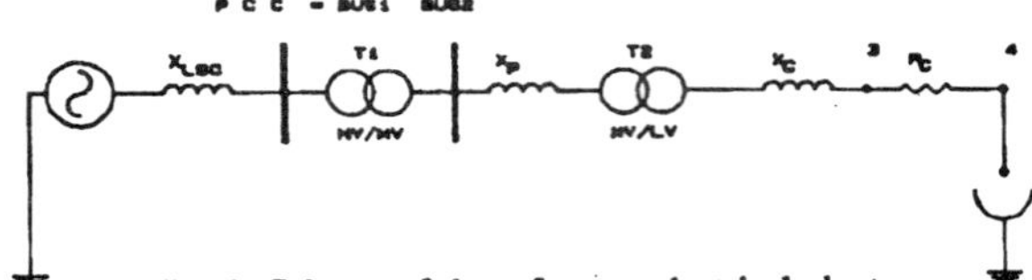

Fig. 1. Scheme of the reference electrical plant.

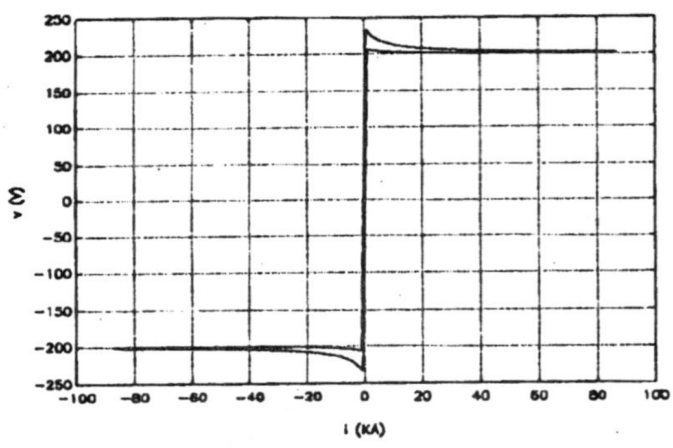

Fig. 2. Dynamic voltage-current characteristic of the furnace electric arc obtained by the proposed modellization implemented by EMTP.

Figure 3 shows the characteristic curves of the electric arc fed by the circuit of Fig. 1, with the values of plant parameters previously reported (MV/LV transformer ratio 21/0.6 kV).

In order to bring the stationary arc-model to give rise to voltage fluctuations, cause of flicker, the voltage-current characteristic must undergo time variations which correspond to a time dependence of the arc length.

This can be realized rewriting eq.(1) as [8]

$$V_a = k\ V_{a0}(I_a) \tag{3}$$

where V_{a0} is the arc voltage corresponding to the reference length, l_0. Parameter k is the ratio of the threshold arc voltage corresponding to a length l, $V_{at}(l)$, to that relevant to the reference length, $V_{at}(l_0)$ [8, 12]. Since the relationship between threshold voltage and arc length can be explained as

$$V_{at} = A + B\ l \tag{4}$$

where l is the arc length (in centimeters), A is a constant taking into account the sum of anode and cathode voltage drops ($A\approx40$ V), B represents the voltage drop per unit arc length ($B\approx10$ V/cm), then k is given by

$$k = \frac{V_{at}(l)}{V_{at}(l_0)} = \frac{A + B\ l}{A + B\ l_0} \tag{5}$$

In the TACS section of EMTP, k has been set as an input variable which allows realizing simulations with arc length either constant or time variable, once appropriate time-variation laws have been singled out.

TIME-VARIATION LAWS FOR FLICKER SIMULATION

The fast variations of the current absorbed by an arc furnace during melting time are connected with arc-length variations which are mainly caused by metal-scrap adjustments, electrodynamic forces and arc-electrode variable displacement.

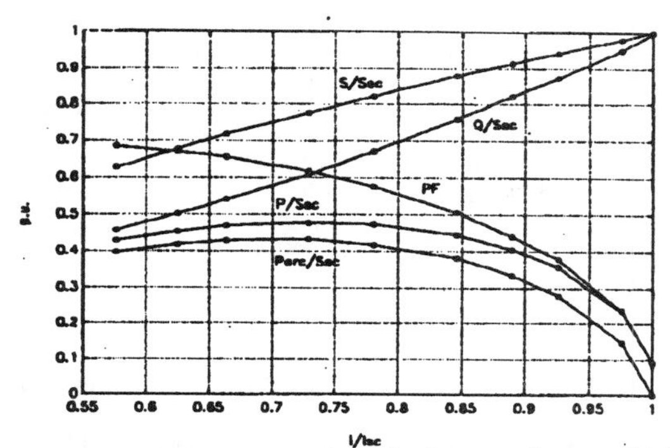

Fig. 3. Characteristic curves of single-phase arc furnace obtained by EMTP simulation, with non-linear arc resistance. The voltage ratio of the furnace transformer is 21/0.6 kV.

The complex nature of these phenomena does not favor a physical approach to the study of arc-length variation. Therefore, flicker investigations have been performed on the basis of deterministic, [6, 8], or stochastic, [13], assumptions for the time-variation law.

The deterministic approach considers that the arc length is subjected to sinusoidal time variations, with frequency suitably chosen in the range of those typical of flicker (i.e. 0.5-25 Hz). This assumption does not clearly represent a normal working condition for furnaces, but has the advantage of being easily manageable by computer simulation and require short-time simulations due to the periodicity of the law.

The stochastic approach is supported by the observation that the arc-length time-variation can be considered a random phenomenon. In fact, extensive measurements of active and reactive power, voltage and current in plants feeding arc furnaces have shown that voltage fluctuations, as well as reactive power variations, at the PCC and the furnace bus, behave as a band-limited white noise, with time-varying amplitude [13]. Therefore, random-variation laws should be attributed to arc-length. Simulation times are longer than in the previous case, due to the the fact that the signal is not periodic. On the other hand, standards introduce weighted stochastic indices to evaluate flicker severity, such as P_{ST} and P_{LT}, which are based on 10 minutes, or more, of recording time [3].

The above-proposed non-linear model (eqns. (2)-(5)) is able to fit both deterministic and stochastic approaches providing that appropriate laws are attributed to arc length, $l = l(t)$.

Sinusoidal law for time variation

In order to approach periodic flicker behavior, simulations can be made attributing to arc length a sinusoidal law with frequency close to the most sensitive for flicker perceptivity. For example, the frequency of 10 Hz can be chosen, which lies in the center of the sensitivity range, close to the minimum of the flicker perceptivity threshold curve for sinusoidal voltage fluctuations [3].

With reference to eqns. (2)-(5), the arc-length time-variation law can be expressed as

$$l(t) = l_0 - (Dl/2)\,(1+\sin wt) \qquad (6)$$

where Dl is the maximum variation of arc length. The time variation of the arc voltage-current characteristic thus becomes, from (3), (5) and (6)

$$V_a(I_a) = k(t)\,V_{a0}(I_a) \qquad (7)$$

with

$$k(t) = \frac{A+Bl(t)}{A+Bl_0} = 1 - \frac{(B\,Dl/2)\,(1+\sin wt)}{A + B\,l_0} \qquad (8)$$

The whole procedure for ATP modellization of the non-linear arc resistance with time-varying sinusoidal law can be implemented in the TACS section of EMTP, as reported in [8]. Figure 4 shows the voltage and current waveforms at the arc-furnace bus, obtained by EMTP simulation on the basis of the proposed model, with sinusoidal time-variation law at frequency 10 Hz. The arc length varies in a wide range, corresponding to values of arc threshold voltage $40V \le V_{at} \le 240V$ and continuous conduction. The furnace transformer has secondary voltage of 600 V. Under these conditions, the relative voltage variation at the PCC, DV/V, is 1.35%.

Actually, the UIE flickermeter has an output which is representative of flicker sensation S(t). This quantity is obtained referring the voltage fluctuations to the threshold-perceptivity curve by means of a weighting filter having minimum attenuation at the frequency of 8.8 Hz which is about that of maximum eye sensitivity to light emitted by incandescent lamps [3]. One unit of output corresponds to the visual perceptivity threshold of flicker occurrence.

The computer simulations based on the proposed model, with deterministic sinusoidal time-variation load, should conveniently provide flicker estimates which are directly comparable with the values given by the UIE flickermeter. For this purpose, the flickermeter can be implemented by EMTP, as shown in [8], so that the 8.8Hz equivalent voltage variation, DV_{eq}/V, can be obtained. With reference to the simulation providing Fig. 4, the value 1.29% of DV_{eq}/V is thus derived (the small difference with respect to the value of DV/V above reported is due to the choice of the frequency of the sinusoidal law, which is very close to 8.8 Hz).

Figures 5A and 5B show the harmonic analysis of the voltage at bus 2 (Fig. 1), obtained by simulations with time-varying (frequency of the sinusoidal law 10 Hz) and constant arc length. As can be seen comparing the figures, the non-linearity of the model causes the presence of characteristic and non-characteristic harmonics (Fig. 5B) [14, 15], while interharmonics (i.e. non-multiple harmonics [16]) are generated when the sinusoidal time-variation law is considered (fig. 5A). These side-bands of the multiple harmonics are always expected in the presence of voltage modulation due to arc-length time variation. The interharmonics responsible for the flicker effect are mainly those included in the sidebands of the

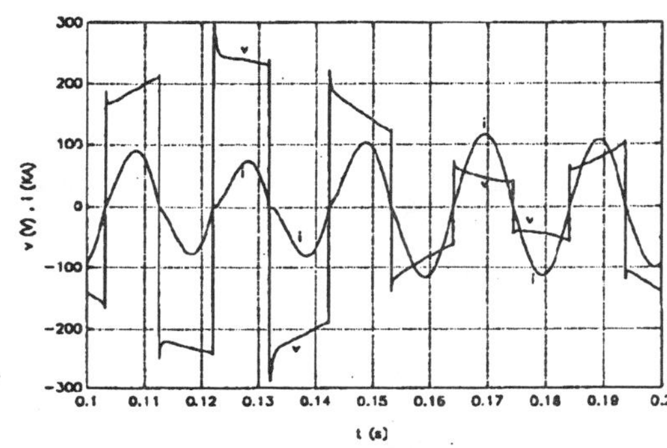

Fig. 4. Voltage and current waveforms at the arc-furnace bus (point 4 of fig. 1) obtained by arc-furnace simulation made by non-linear, time varying resistance.

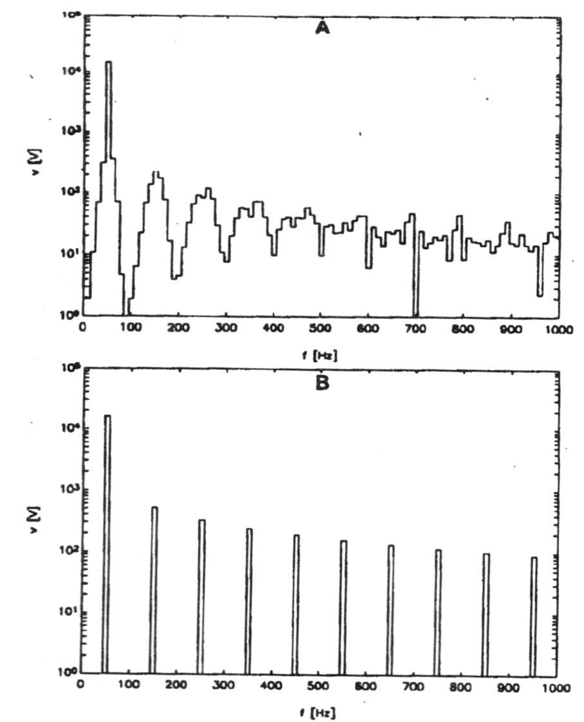

Fig. 5. Harmonic analysis of the voltage at bus 2 of Fig. 1, for sinusoidal time-varying arc-length (A) and constant arc-length (B).

fundamental frequency, i.e. 50 Hz (on the other hand, the UIE flickermeter, [3], prevailingly takes into account fundamental-voltage fluctuations).

White-noise time variation

With reference to the proposed model, the time dependence of the arc-length can be expressed as

$$l(t) = l_0 - r(t) \tag{9}$$

where l_0 is the maximum arc length compatible with continuous conduction and r(t) is the law of arc-length variation with respect to the reference condition l_0. Signal r(t) is a white noise with band in the frequency range where voltage fluctuations produce flicker.

The time variation of the arc voltage-current characteristic has again expression (7), where k(t) becomes, for the white-noise time-variation law of eq. (9),

$$k(t) = \frac{A+Bl(t)}{A+Bl_0} = 1 - \frac{B\, r(t)}{A + B\, l_0} \tag{10}$$

Even in this case, the procedure for EMTP simulation of variable arc length can be implemented in the TACS section resorting to three blocks, that is, a random-number generator, a pass-band filter with lower and upper cut frequencies 4 Hz and 14 Hz (according to [13]), and the third block where eq.(9) provides k(t) as output signal.

With respect to the case of periodic time-variation laws (e.g. that sinusoidal previously accomplished), considerably longer simulation times are required for the random arc-length time law here considered. Moreover, the flicker sensation, S(t), should be processed in order to obtain the comprehensive quantity also available at the UIE flickermeter output, the so-called short-term flicker severity, P_{ST} [3]. It consists of a weighted sum of percentiles of the cumulative probability distribution of S(t), having the purpose to provide objective information on the flicker-severity level independently of the type of flicker, its time law and evolution. P_{ST} is thus defined as

$$P_{ST} = (0.0314\, S_{99.9\%} + 0.0525\, S_{99\%} + 0.0657\, S_{97\%} + 0.28\, S_{90\%} + 0.08\, S_{50\%})^{1/2} \tag{11}$$

where the percentiles 50%, 90%, 97%, 99% and 99.9% of S(t) are considered.

P_{ST} estimates are based, according to UIE recommendations, on 10-minute observations (another quantity is also proposed in [3], that is, long-term flicker severity, P_{LT}, referred to two-hour observations), but the results here reported are relevant to one-minute simulations for the sake of computing-time saving.

Figures 6A and 6B show the time behavior of non-active power (i.e. $Q=(S^2-P^2)^{0.5}$) and the corresponding flicker sensation, S(t), at the PCC obtained by computer simulation based on the white-noise time variation of arc length. The plant parameters are the same as the previous simulations, pertinent to sinusoidal time law, with arc-threshold voltage, V_{at}, varying in the range 40-240 V, where continuous conduction, as well as wide arc-length variations, are allowed. The value of P_{ST} calculated for one-minute simulation is 1.6 (according to [3], P_{ST} values exceeding 1 indicate flicker disturbance).

The results of these simulations proved to reproduce quite-well real cases for both flicker and voltage distortion-factor evaluations (considering also the approximations due to single-phase simulation). Indeed, measurements made on a plant with characteristics very close to those used for the simulations provided values of P_{ST} approaching 1.6 in the starting melting period. In simulations longer than one minute, the arc-length variation range could be changed with time, such as to describe the voltage-fluctuation decrease that generally occurs increasing the quantity of melted metal.

Model discussion

A direct comparison of the two kinds of arc-length time-variation laws presented above can be performed once the results obtained by the sinusoidal law, that is, DV_{eq}/V, are converted into P_{ST}. Indeed, P_{ST} calculation is meaningless for a deterministic signal and, moreover, shorter times than 1 or 10 minutes are needed to evaluate flicker effect for periodic signals. However, for the sake of comparison it can be observed that the probability distribution of a deterministic, sinusoidal signal is stepwise ([3]), hence from (11)

$$P_{ST} \approx (0.5096\, S_{MAX})^{1/2} \tag{12}$$

where S_{MAX} is the maximum value of flicker sensation. Under these premises, the value of P_{ST} derived from the simulations with sinusoidal time-variation law pertinent to Figures 4, 5 (providing $DV_{eq}/V=1.29\%$) is 3.7.

Therefore, the value of P_{ST} obtained by the white-noise time-variation law is significantly lower (1.6) than that derived by the sinusoidal law (3.7), for the same plant and furnace working conditions.

It can be argued that the sinusoidal law provides limit conditions for furnace operation, that is, a worst-case approximation, which enables determination of maximum flicker sensation caused at the PCC by a furnace of known

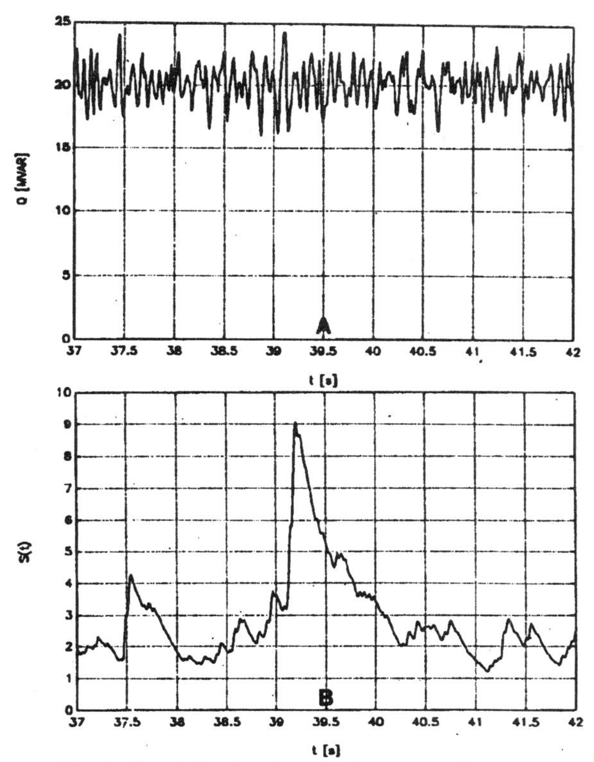

Fig. 6. Time behavior of non-active power (A) and the corresponding per-unit flicker sensation (B) at the PCC obtained by computer simulation based on the white-noise time variation of arc length.

characteristics (similar results are expected by modulation with rectangular law). This approximation seems more effective than the criterion used in [12, 13], where DV/V is evaluated referring to the limit cases of not-feeded furnace and short-circuit conditions. Moreover, the proposed model allows to realize investigations on the effects of flicker-compensation systems in worst-case conditions, and provides the resulting flicker evaluations in terms of S(t) or P_{ST}, according to UIE instrumentation.

When the white-noise time-variation law is assumed for the arc length, real working conditions can be approached and a sort of average P_{ST} estimated for the studied plant. This analysis can be useful to evaluate flicker-compensation strategies in the design stage of electric plants supplying arc furnaces or in existing plants where measurements show the need to resort to flicker-compensation systems.

FLICKER COMPENSATION BY SERIES INDUCTORS

The calculations reported up to this point are relevant to secondary voltage of the MV/LV transformer of 600V, absence of the series inductor (i.e. $X_P=0$ in Fig. 1) and short-circuit ratio (SCR) at PCC equal to 58 (the short-circuit ratio is defined as the ratio of the short-circuit power to the mean apparent power required by the load).

In order to envisage the effects of the installation of series inductors in the plant feeding the furnace, the presence of a series reactance, X_P, at the supply-side of the MV/LV furnace transformer can be considered, as shown by Fig. 1 (the equivalent reactor resistance is neglected). However, insertion of series inductors gives rise to decrease of furnace power so that actions are required to avoid significant reductions of furnace productivity. Mainly, the transformer turns ratio of the MV/LV transformer can be changed, taking profit of the adjustable secondary voltage (varying from 600V to 900V, step 60V, in our simulations). This causes arc-length variations, too.

Two design criteria are here compared in order to look in detail at the behavior of series inductors. One consists of keeping constant the power absorbed by the system series reactance-transformer-furnace with short-circuited electrodes (so that the SCR at PCC does not significantly change with inductor insertion). The other, which better conforms with the above requisites of furnace efficency, is to keep constant the mean active power absorbed by the furnace. In both cases, simulations with several values of series reactance have been realized, changing the furnace-side voltage of the MV/LV transformer in accordance to the design criteria. As a consequence of the assumed arc-furnace model, an increase of furnace-side voltage, due to insertion of series inductor, causes arc lengthening. According to on-field observations, [17], the maximum arc-length variations, Dl, have been assumed independent of arc length in the range of values used for the simulations (i.e. corresponding to continuous conduction). Hence, longer arcs provide smaller variation of relative length Dl/l. Both sinusoidal and white-noise laws for arc-length time variation have been considered in the simulations (the former with frequency 10Hz).

Tables 1 and 2 report the values of the series reactance, X_P, inserted in the plant of the characteristics above described, together with the secondary voltage of the furnace transformer, the equivalent voltage variation, DV_{eq}/V, and the corresponding P_{ST}, for the two design criteria. The ratios of the series reactance to the total reactance, X_t, as seen upstream the arc-furnace electrodes, are also given. The sinusoidal law of arc-length time variation is assumed. In Figure 7 the graph of voltage variation vs ratio X_P/X_t, relevant to the data of Table 2, is drawn.

The simulation results show that by both design criteria the insertion of series inductors at the supply side of the furnace transformer can significantly reduce voltage fluctuations, and, therefore, flicker, at PCC. Taking advantage of the maximum voltage adjustment allowable at the secondary side of the furnace transformer, the voltage variation at PCC decreases to 80% and 60% with respect to the values determined in the absence of series reactance. Hence, significant compensation possibilities are provided by the design criterion of constant mean power absorbed by the furnace (Table 2), but also the other criterion, where the short-circuit power at bus 2 of Fig. 1 is practically kept constant, provides non-negligible flicker compensation. In the former case, however, higher series-reactance values are required for the same transformer voltage.

The dependence of the series-reactance compensation effect on the short-circuit ratio is depicted in Figure 8, referred to the criterion of constant furnace power and sinusoidal modulation. The surface shows that increasing SCR (SCR=58 corresponds to short-circuit power, S_{SC}, of 3500 MVA), the P_{ST} decreases for any value of series reactance, as expected. In fact, a simple relationship, derived under the approximate assumption that varying the short-circuit power, the furnace current does not considerably change (being X_{LSC} very small with respect to the total reactance of the line feeding the furnace) points out the inverse relationship between voltage variation (or P_{ST}) and short-circuit power:

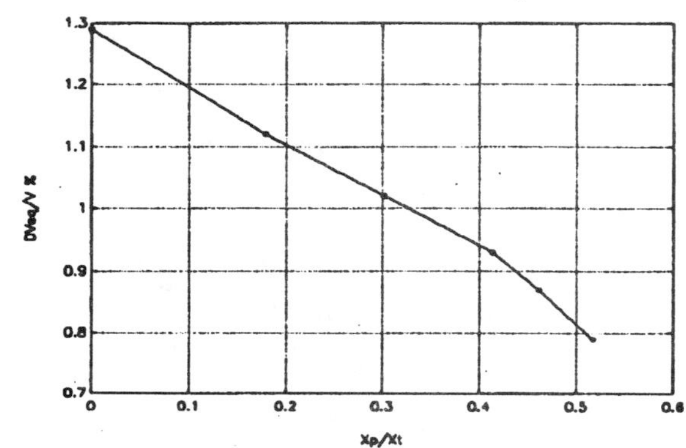

Fig. 7. Per cent voltage variations, $DV_{eq}/V\%$, as a function of the ratio between series reactance and total reactance, X_p/X_t, for a design based on the criterion of constant power absorbed by the system reactance-transformer-furnace with short-circuited electrodes. Sinusoidal time-variation law.

$$(DV_{eq}/V)_2 = (DV_{eq}/V)_1 \ (S_{SC2}/S_{SC1}) \tag{13}$$

where $(DV_{eq}/V)_2$ and $(DV_{eq}/V)_1$ correspond to S_{SC2} and S_{SC1}, respectively.

Similar results have been obtained for white-noise time-variation law of the arc length, even if the values of P_{ST} are smaller than those reported in Tables 1, 2 (valid for sinusoidal law). For example, in the case of constant mean active power absorbed by the furnace, P_{ST} varies from 1.6, in the absence of series inductor, to 1.28 for $X_p/X_t=3.29$.

Comparisons of these results with on-field measurements are not easily performable, referring to literature data. Electrical plants feeding arc furnaces generally show similar configurations, but various parameter values and, moreover, flicker measurements are not always reported in comparable conditions as measurement time and unit, melting process, or, otherwise, incomplete information on measurement conditions are given.

However, in [17] some results of P_{ST} measurements are reported pertinent to plants with parameters similar to those used for our simulations. Indeed, the flicker severity is given by the maximum values of P_{ST}, measured for several times in a plant at different operating conditions, exceeded with probability 10% (that is, the 90% cumulative probability of the distribution of the peak P_{ST} values). Figures 9 and 10 report the behavior of the maximum P_{ST} associated to exceeding probability 10% as function of the values of series reactance and short-circuit power, respectively. Both Figures show qualitative good agreement with the results of our simulations, summarized in Figs. 7, 8. As regards quantitative evaluations, under the above premises on comparison problems, it can be seen that Fig. 9 shows P_{ST} variations with X_p very close to those reported in Table 2 and Fig. 8. On the other hand, the distribution of the maximum values of P_{ST} could be reasonably compared with the simulation results relevant to sinusoidal time law, which provides, as previously mentioned, worst-case evaluations. Likewise, Fig. 10 does not remarkably differ from the intersections of the surface of Fig. 8 for $X_p/X_t=0$ and $X_p/X_t=0.52$

Therefore, it can be concluded that the insertion of series inductors has significant effect on flicker compensation, and that the models here proposed can provide approximate, but meaningful, evaluations of this effect.

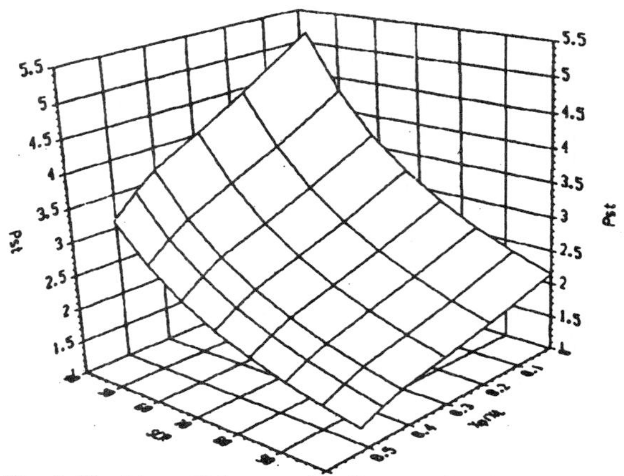

Fig. 8. Short-term flicker severity, P_{ST}, as a function of the ratio series reactance to total reactance, X_p/X_t, and the short-circuit ratio, SCR, for a design based on the criterion of constant mean power absorbed by the furnace. Sinusoidal time-variation law.

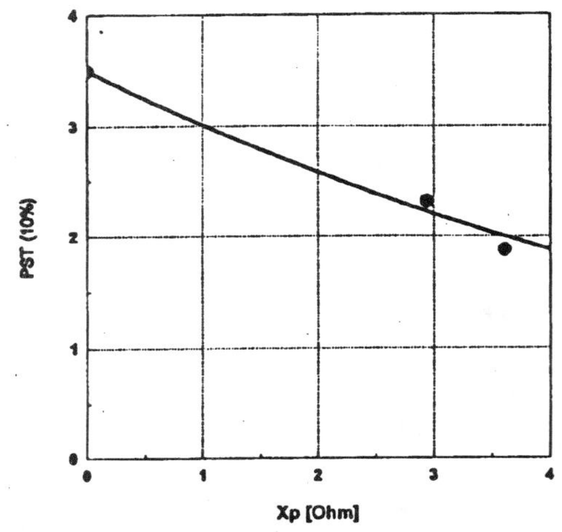

Fig. 9. Behavior of maximum P_{ST} exceeded with probability 10% as a function of the values of series reactance. After [17].

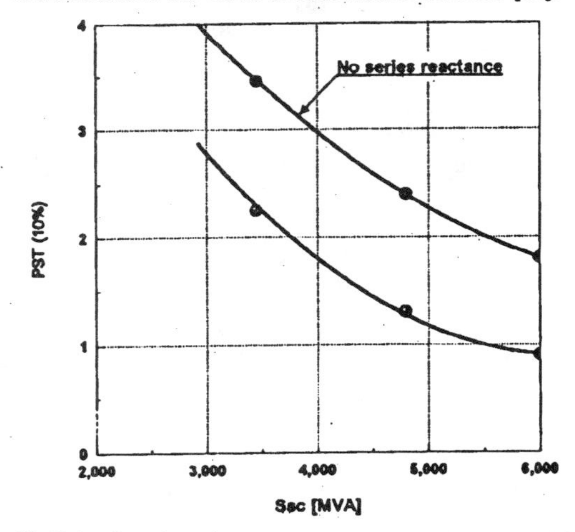

Fig. 10. Behavior of maximum P_{ST} exceeded with probability 10% as a function of the short-sircuit power at the PCC, with or without series reactance. After [17].

Table 1. Series-reactance values (X_P), ratios between series reactance and total feeding-line reactance (X_P/X_t), secondary voltages of the furnace transformer (V_2), voltage variation (DV_{eq}/V) and short-term flicker severity at the PCC (P_{ST}) for design realized at constant power absorbed by the system reactance-transformer-furnace with short-circuited electrodes. Sinusoidal time-variation law of arc length.

X_P (Ohm)	X_P/X_t	V_2 (kV)	DV_{eq}/V %	P_{ST}
0	0	0.60	1.29	3.68
0.63	0.12	0.66	1.21	3.45
1.12	0.22	0.72	1.15	3.28
1.50	0.29	0.78	1.10	3.14
1.79	0.35	0.84	1.06	3.03
2.04	0.40	0.90	1.02	2.91

Table 2. Series-reactance values (X_P), ratios between series reactance and total feeding-line reactance (X_P/X_t), secondary voltages of the furnace transformer (V_2), voltage variation (DV_{eq}/V) and short-term flicker severity at the PCC (P_{ST}) for design realized at constant mean power absorbed by the furnace. Sinusoidal time-variation law of arc length.

X_P (Ohm)	X_P/X_t	V_2 (kV)	DV_{eq}/V %	P_{ST}
0	0	0.60	1.29	3.68
0.98	0.18	0.66	1.12	3.20
1.73	0.30	0.72	1.02	2.91
2.37	0.41	0.78	0.93	2.66
2.86	0.46	0.84	0.87	2.48
3.29	0.52	0.90	0.79	2.26

SERIES INDUCTORS AND VOLTAGE-CURRENT DISTORTION

Arc furnaces are well-known harmonic voltage and current sources, owing to their non-linear characteristic. Multiple harmonics are injected in the feeding plant, due to waveform distortion with respect to the sinusoidal frame, besides non-multiple harmonics, or interharmonics, [16], which are caused by arc-length time variation [5, 7, 10]. Figure 5A, relevant to sinusoidal modulation, and Figure 11, pertinent to white-noise time-variation law, show that voltage and current in plants supplying arc furnaces have almost continuous frequency spectrum (the harmonic analysis of Fig. 1 is relevant to the entire simulation of one minute).

However, calculation of distortion factor should conveniently separate the effect of interharmonics generation, which is computed by flicker measurements, by the waveform deviation from sinusoidal shape, which is well taken into account by the usual expression for total harmonic distortion (THD) calculation, recommended by IEEE 519 [14], that is

$$THD = \frac{(\sum_{h=2}^{N} A_f^2)^{1/2}}{A_1} 100\% \qquad (14)$$

where A_f is the amplitude of multiple harmonic voltages or currents, A_1 is the amplitude at the fundamental frequency (50 or 60Hz), N is normally lower than 50.

It was recognized that insertion of protection reactances in electrical plants supplying static power converters generally causes reduction of voltage distortion at the supply bus [18, 19]. In the case of arc furnaces, the opposite occurs for both design criteria (as shown by Tables 3, 4). Insertion of series reactances enhances arc length (due to feeding voltage adjustement), so that the current THD (THD_I) increases. Consequently, the voltage THD (THD_V) at bus 2, and then at the PCC, increases as the series-reactance value rises. However, only a slight increase is detected for the constant furnace-power criterion, which is likely the sought condition, for evident economical advantages (Table 3). It is interesting to observe that current THD does not depend on design criterion, but only on the arc-length variation range, determined by the value of furnace voltage (the same values of THD_I are, in fact, obtained for V_2=900V in Tables 3 and 4).

On the other hand, insertion of capacitors and/or shunt filters can compensate for non-active power due to reactive and distortion powers [7], while they do not provide noticeable contribution to flicker reduction. Use of filters is promoted to avoid dangerous and uncontrolled resonances that might occur when capacitor banks are used, and, in addition, to contribute to distortion compensation [20]. However, even the use of filters should be carefully regarded due to the almost-continuous harmonic spectra of voltage and current.

CONCLUSIONS

The description of flicker behavior in electrical plants supplying arc furnaces by the models proposed in this paper seems quite satisfactory. The non-linear arc model provides voltage and current waveforms as well as arc characteristics which are similar to those observed in actual plants

By means of suitable time-variation laws attributed to arc length, quite accurate evaluations have been made on the effects of the insertion of series inductors at the supply side of the furnace transformer, showing that the short-term flicker severity, that is, the voltage variations cause of flicker, can be significantly reduced. Clearly, a limit value for inductor size is conditioned by arc stability, that is, the range of continuous conduction.

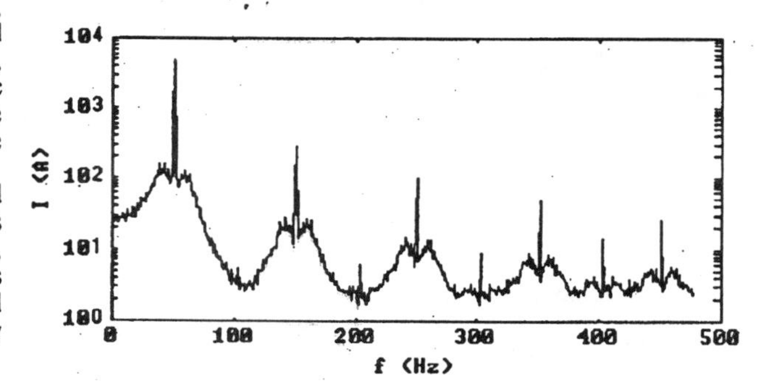

Fig. 11. Harmonic analysis of the arc-furnace current at the supply-side of the furnace transformer, for white-noise law of arc-length time variation. X_P=3.29 Ohm, transformer secondary voltage 900 V (bandwith for the measurement data 1 Hz).

Therefore, the use of series inductors, associate with capacitors and/or filters for non-active power compensation, seems to be, in certain plant conditions and after technical-economical evaluations, an alternative solution to installation of static var systems.

Improvements of modelling accuracy would be achieved working on three-phase simulations, where imbalances of the arc-furnace operations as well as appropriate furnace characteristic curves can be taken into account and proper filter-effectiveness investigation realized.

Table 3. Series-reactance values (X_p), ratios between series reactance and total feeding-line reactance (X_p/X_t), secondary voltages of the furnace transformer (V_2), per cent current and voltage total harmonic distortion at bus 2 of Fig. 1 (THD_I% and THD_V%, respectively) for design realized at constant power absorbed by the system reactance-transformer-furnace with short-circuited electrodes. Upper limit of summation for THD calculation N=20. Sinusoidal time-variation law of arc length.

X_p (Ohm)	X_p/X_t	V_2 (kV)	THD_I %	THD_V %
0	0	0.60	3.99	1.61
0.63	0.12	0.66	4.34	1.74
1.12	0.22	0.72	4.72	1.85
1.50	0.29	0.78	5.07	1.98
1.79	0.35	0.84	5.41	2.10
2.04	0.40	0.90	5.72	2.21

Table 4. Series-reactance values (X_p), ratios between series reactance and total feeding-line reactance (X_p/X_t), secondary voltages of the furnace transformer (V_2), per cent current and voltage total harmonic distortion at bus 2 of Fig. 1 (THD_I% and THD_V%, respectively) for design realized at constant mean power absorbed by the furnace. Upper limit of summation for THD calculation N=20. Sinusoidal time-variation law of arc length.

X_p (Ohm)	X_p/X_t	V_2 (kV)	THD_I %	THD_V %
0	0	0.60	3.99	1.61
0.98	0.18	0.66	4.34	1.62
1.73	0.30	0.72	4.71	1.64
2.37	0.41	0.78	5.05	1.68
2.86	0.46	0.84	5.39	1.72
3.29	0.52	0.90	5.72	1.74

REFERENCES

[1] J.J. Koenderink, A.J. Van Doorn, "Visibility of unpredictably flickering lights", *Journal of the Optical Soc. of America*, Vol. 64, n. 11, November 1974.

[2] J.J. Koenderink, A.J. Van Doorn, "Detectability of power fluctuation of temporal visual noise", *Vision Res.*, Vol. 18, pp. 191-195, Pergamon Press, 1978.

[3] UIE Disturbances WG, *Flicker measurements and evaluation*, 1992.

[4] B. Bhargava, "Arc furnace masurements and control", *IEEE Trans. on Power Del.*, Vol. 8, n. 1, pp. 400-409, January 1993.

[5] L. Bisiach, L. Campestrini, C. Malaguti, "Technical and operational experiences for mitigating interferences from high-capacity arc furnaces", *Int. Conf. on Large High-Voltage El. Sys.*, CIGRE', Paris, France, September 1992.

[6] M. Loggini, G.C. Montanari, L. Pitti, E. Tironi, D. Zaninelli, "The effect of series inductors for flicker reduction in electric power systems supplying arc furnaces", *IEEE/IAS Ann. Meeting*, Toronto, Canada, October 1993.

[7] W.S. Vilcheck, D.A. Gonzalez, "Measurements and simulation-combined for state-of-the-art harmonic analysis" *IEEE/IAS Ann. Meeting*, pp. 1530-1534, Pittsburgh, USA, October 1988.

[8] A. Cavallini, G.C. Montanari, L. Pitti, D. Zaninelli, "ATP simulation for arc-furnace flicker investigation", to be published in *ETEP*, 1993.

[9] *ATP Rule Book*, Leuven EMTP Center, July 1987.

[10] S.R. Mendis, D.A., Gonzalez, "Harmonic and transient overvoltage analyses in arc furnace power systems", *IEEE Trans. on Ind. Appl.*, Vol. 28, n.2, pp. 336-342, April 1992.

[11] R. Sasdelli, G.C. Montanari, "The compensable power: its definition for electrical systems in nonsinusoidal conditions", *IEEE Trans. on Instr. and Meas.*, 1993.

[12] L. Di Stasi, *Electric furnaces* (in Italian), Patron ed., Padova, Italy, 1976.

[13] G. Manchur, C.C. Erven, "Development of a model for predicting flicker from electric arc furnaces", *IEEE Trans. on Power Del.*, Vol. 7, n.1, pp. 416-426, January 1992.

[14] IEEE Publ. 519, *IEEE recommended practices and requirements for harmonic control in electric power systems*, 1991.

[15] IEC TC 33 (Secretariat) 148, *Guide for a.c. harmonic filters for industrial applications*, October 1992.

[16] A.E. Emanuel, J.A. Orr, D. Cyganski, "Review of harmonics fundamentals and proposals for a standard terminology", *3rd ICHPS*, pp. 1-7, Nashville, USA, September 1988.

[17] L. Campestrini, L. Lagostena, G. Sani, A. Bellon, R. Manara, E. Nazarri, "Flicker control in high power arc furnaces and cumulative flicker analysis in HV networks", *11th Int. Conf. on Electricity Distribution*, Liege, Belgium, April 1991.

[18] G.C. Montanari, M. Loggini, "Voltage-distortion compensation in electrical plants supplying static power converters", *IEEE Trans. on Ind. Appl.*, Vol. 23, n. 1, pp. 181-188, February 1987.

[19] G.C. Montanari, M. Loggini, "Filters and protection reactance for distortion compensation in low-voltage plants", *IEEE IAS Annual Meeting*, pp. 1488-1494, Pittsburgh, USA, October 1988.

[20] D.A. Gonzalez, J.C. McCall, "Design of filters to reduce harmonic distortion in industrial power systems", *IEEE Trans. on Ind. Appl.*, Vol. 23, n. 3, pp. 504-511, June 1987.

Biographies

Gian Carlo Montanari (M'86, SM'91) was born in Bologna, Italy, on November 8, 1955. He received his Doctor's Degree in Electrical Engineering in 1979 from the University of Bologna. In 1983, he joined the University of Bologna as researcher and has become professor of Electrical technology in 1986. He has worked since 1979 in the field of aging and endurance of solid insulating materials and systems.
He is also engaged in the fields of harmonic compensation in electrical power systems, power electronics and statistics. He is an IEEE Senior member and member of IEC 15B and IEC TC 33.

Andrea Cavallini was born in Mirandola, Italy, on December 21, 1963. He received his Doctor's Degree in Electrical Engineering in 1990 from the University of Bologna. At present, he is Ph.D. student at the Institute of Industrial Electrotechnic of the University of Bologna/Italy. His interest fields are power systems harmonics, reliability of electrical systems and power electronics.

Mauro Loggini (M'86) was born in Grosseto, Italy, on August 29, 1938. He received his Doctor's Degree in Electronic Engineering from the University of Bologna. In 1970, he joined the same University as an assistant professor of Electrical technology. At present, he is professor of Industrial electrical applications and works in the field of harmonic compensation in electrical power systems and power electronics. He is an IEEE member and member of IEC TC 33.

Luca Pitti was born in Arezzo, Italy, on October 15, 1963. He received his Doctor's Degree in Electrical Engineering in 1992 from the University of Bologna. He is presently cooperating with the University of Bologna, and is private consultant in electrical plant design, working in Arezzo, via Nazario Sauro 32. His interest fields are power systems harmonics, reliability of electrical systems and power electronics.

Dario Zaninelli (M'88) was born in Romano di Lombardia on April 3, 1959. He received the Ph.D. degree in Electrical Engineering at the Politecnico di Milano in 1989 and became researcher at the Electrical Engineering Department of the Politecnico di Milano on 1990. His areas of research include power system harmonics and power system analysis. He is an IEEE member.

Discussion

S. Bhattacharya and **W. Wong** (ABB Transmission Technology Institute): We would like to congratulate the authors on their paper. The paper presents an arc furnace model. The authors use this model to study the effects on voltage flicker by adding a series inductor at the supply side of a furnace transformer. The paper presents two different time-variation laws that are used by the model. It also mentions the modulation with rectangular law and it provides similar results. Could the authors provide more information on the modulation with rectangular law and contrast its effect on the flicker sensation with respect to the white-noise time variation?

In the text, the authors comment that the simulation results are difficult to compare with the on-field measurements. However in the conclusion section, the authors mention that the waveforms and the arc characteristics obtained from the model are similar to those observed in actual plants. Would the authors elaborate on the techniques (that overcome the comparison difficulties) they have used to compare the simulation results with the on-field measured waveforms? Would the authors also share a few on-field measured waveforms?

Inserting a series inductor between the PCC point and MV/LV transformer reduces the voltage flicker. However, the voltage variation at the arc furnace increases. The authors suggest that the secondary voltage of the MV/LV transformer can be adjusted to compensate for the voltage variation. We foresee that without an on-load tap-changer the voltage regulation will be poor. For instance, if the secondary tap is set for the full-load operation, it will create overvoltage during the light-load condition. Do the authors assume that an on-load tap-changer is employed to adjust the secondary voltage? If so, would they comment on the frequency of operation of the tap-changer and the resulting wear-tear?

To obtain an optimum performance and economy, an arc furnace requires a stable and steady voltage supply. Higher steady voltage provides shorter meltdown times, reduces energy cost and extends electrode life. Many installations effectively use static var compensator (SVC) to reduce flicker, to improve power factor and arc furnace efficiencies. The authors' comment on the effect of the series inductor concept on the power factor compensation and the arc furnace efficiencies will be appreciated.

Manuscript received February 24, 1994.

G. C. Montanari, M. Loggini, A. Cavallini, L. Pitti, and **D. Zaninelli**: We thank the discussers for their congratulations and the stimulating questions. Before answering to each point raised by the discussers, we would like to point out that the proposed arc furnace model is a step forward in the investigation of flicker and distortion compensation in electrical plants supplying arc furnaces. The authors are well aware that the model does not allow a perfect simulation of furnace behavior during the whole melting cycle, but the comparison of previous models and experimental results lead to consider the proposed model a satisfactory compromise between the need to approach the problem of power quality.in plants supplying arc furnaces and the difficulty to know the actual furnace working conditions at any time. For this reason, the model resorts to eqns. (6)–(9) where the evident simplification of steady power absorbed by the furnace (beyond the voltage fluctuations responsible by flicker) is made. Moreover, simulations are made resorting to the maximum range allowable by the condition of continuous conduction for arc length. However, the implemented model is flexible and, compatible with TACS size, more complex laws for furnace power and voltage variations can be considered, depending on the amount of data available for the studied plant.

But let us answer in detail each question of the discussers.

1- Both sinusoidal and rectangular laws for arc-length modulation are deterministic laws, and should be regarded as worst-case approximations for flicker estimation. The difference in P_{ST} values obtained by the two laws mainly depends on the behavior of demodulation and weighting filters of flickermeter [3]. In fact, on the basis of [3] the following, approximate expressions can be deducted, which relate the equivalent voltage variation, DV_{eq}/V to the flicker sensation S(t):

$$S(t) = \left(100(DV_{eq}/V)/0.25\right)^2$$

and

$$S(t) = \left(100(DV_{eq}/V)/0.20\right)^2$$

for sinusoidal and rectangular laws, respectively. Hence, the perceptivity limit ($S = 1$) is given by $DV_{eq}/V \approx 0.3$ and 0.2 for sinusoidal and rectangular laws, respectively. Moreover, eq. (12) holds for both laws, since in both cases the probability distribution of the signal S(t) is stepwise. In conclusion, the rectangular modulation provides slightly higher values of S(t) and P_{ST} than the sinusoidal one, so that both laws can be considered worst cases with respect to the white-noise time variation law, which should more closely simulate the actual arc-length variations.

2- As mentioned in the paper, quantitative comparisons of the proposed model with on-field measurements cannot be easily made, due to the single-phase simulation and the difficulties to know in any time the exact furnace working conditions. This led to the use of a stochastic model based on white-noise modulation. By this way, values of P_{ST} close to the average measured during the starting melting periods of an arc furnace were obtained. Clearly, accuracy can increase resorting to three-phase simulations (which is the last achievement of the research) and to more accurate reproduction of the actual working conditions of the arc furnace. As regards the voltage and current waveforms, in steady conditions the model can well reproduce the typical waveforms of an electric arc, often displayed in literature. Considering a plant which supplies an arc furnace, there are so many different operating conditions, involving randow laws, that an accurate reproduction of actual waveforms can be seldom obtained by simulation based on white-noise law. However, the proposed model provides voltage and current waveforms which resemble those observed in plants during periods of arc operation, as it is also confirmed by the values of THD. As an example, Figs. C1 and C2 report the voltage and current waveforms measured and simulated at the bus 2 of the plant, when the series reactance ($X_p/X_t \approx 0.3$) is inserted. Simulation considers white-noise law with time variation close to that detected by the measure. Amounts of plots of active and non-active power could also be provided (see Fig. 6 of the paper), but their apparent fitting the simulations does not give anymore contribution to prove the model validity.

3- Plants where the solution of series inductors for flicker reduction is employed are not uncommon in North-Italy. On the basis of on-field measurements, the observations of the discusser seem appropriate. The capability of secondary-voltage regulation of the MV/LV transformer limits the size of the series inductor, since insertion of the series inductor forces to increase the secondary voltage (if the furnace active power must be kept constant), but reduces consequently, the tap-regulation range. Both simulation and experimental data show that increasing the arc voltage (i.e., lengthening the arc), arc stability decrease, but this can be properly taken into account by electrode control. Hence, the frequency of secondary-voltage adjustments should not significantly vary with respect to operation without series inductor. Indeed, it must be clarified that the series inductor is kept steadily inserted in the plant, so that the increase of the rate of commutation of tap changer is related only to the operating conditions of the furnace. In the paper, this topic was not dealt with in depth, since the purpose was to show how the series inductor works in the plant.

4- Insertion of series inductor affects power factor, PF, due to the contrasting effects of increased line inductance (that lowers PF) and changed average working point of the furnace (the same furnace power is obtained by lower I/I_{sc} values and, hence, higher power factor, as shown by Fig. 3 of the paper). On the whole, power factor variations due to series-inductor insertion are relatively small, so that connection of filter or capacitor banks is needed. Three-phase simulations would be appropriate to investigate the influence of filters on THD, PF and P_{ST}. The last development of the authors research, leading to three-phase plant simulation, show that filters reduce voltage THD (in the absence of significant resonance amplification, i.e., for well-designed filters), compensate for PF, but do not improve P_{ST}. For example, P_{ST} varies from 0.85 to 0.88 after filter insertion (tuned to 3rd harmonic), for a plant with $x_p/x_t = 0.4$, while voltage THD varies from 1.78 to 0.83. In general, it can be argued that the insertion of filters and series reactance can constitute an interesting solution, alternative and cheaper than SVC, in plants where problems of P_{ST} and THD are not too dramatic, that is, the standard limits for these quantities are exceeded for a limited amount.

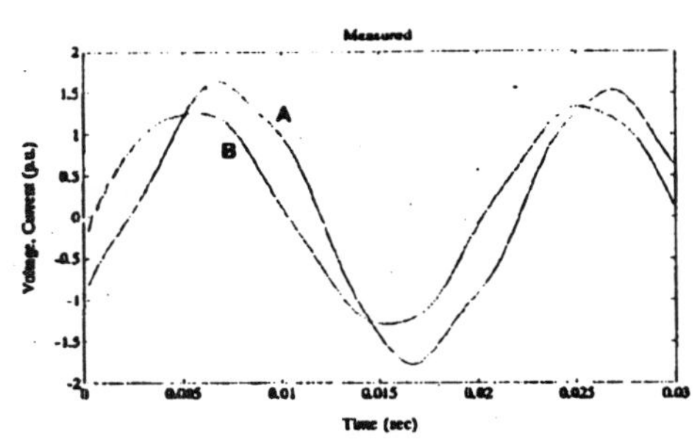

Fig. C1. Voltage and current waveforms measured at the PCC of the plant (B = voltage, A = current).

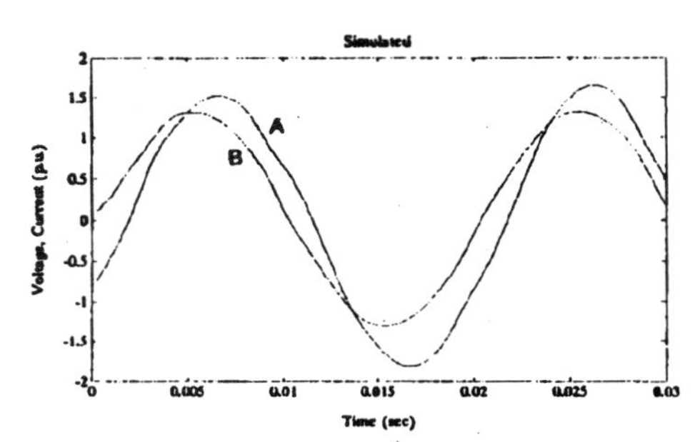

Fig. C2. Voltage and current waveforms obtained by simulation with white-noise law at the PCC of the plant (B = voltage, A = current).

Manuscript received April 11, 1994.

Evaluation of Harmonic Impacts from Compact Fluorescent Lights On Distribution Systems

Rory Dwyer Afroz K. Khan
Mark McGranaghan Le Tang
Electrotek Concepts, Inc

Robert K. McCluskey Roger Sung
Thomas Houy
Southern California Edison

ABSTRACT

Compact fluorescent lamps (CFLs) have the potential to increase the overall harmonic distortion levels on distribution systems. Measurements of the harmonic characteristics of different types of CFLs are presented and the possible impacts on a distribution system are analyzed. The analysis uses a combination of EMTP simulations and conventional harmonic analysis simulations to predict the distribution system distortion levels as a function of system characteristics, the CFL characteristics, and the CFL penetration level.

1.0 INTRODUCTION

Compact fluorescent lights (CFLs) provide significant energy savings over incandescent lighting. As a result, CFLs are being promoted as part of energy conservation programs at many electric utilities. A concern associated with compact fluorescent applications is that some types of CFLs exhibit high levels of harmonic current distortion. A recent study by Emmanuel [1] evaluated the impacts of high distortion CFLs on a typical distribution system. The results indicated that relatively low CFL penetration levels could cause the feeder voltage distortion to exceed 5%.

Previous studies have used relatively simple system models and have not included the effects of harmonic cancellation within a household or between households. More recent analytical work by Grady [2,3] has shown the harmonic cancellation that can exist with this type of load due to different impedances and due to the effect of the load on the voltage distortion.

95 WM 105-7 PWRS A paper recommended and approved by the IEEE Power System Engineering Committee of the IEEE Power Engineering Society for presentation at the 1995 IEEE/PES Winter Meeting, January 29, to February 2, 1995, New York, NY. Manuscript submitted August 1, 1994; made available for printing January 5 1995.

The work presented in this paper uses the Electro-Magnetic Transients Program (EMTP) to help verify the cancellation effects within a household and for CFLs in multiple households. The EMTP model is also used to develop a representation for this type of load in linear models for harmonic simulations. More conventional harmonic simulations are then used with a detailed three-phase distribution system representation to look at the cumulative effect of CFLs throughout the system on the overall voltage distortion levels.

Measurements were performed to identify typical harmonic characteristics for different types of CFLs. These characteristics were then used as a starting point to derive harmonic injection characteristics for households and groups of households, including the effects of cancellation. The distribution system voltage distortion levels resulting from the increased harmonic injection are derived for different distribution system characteristics.

2.0 COMPACT FLUORESCENT CHARACTERISTICS

Measurements were performed for a wide variety of different CFLs. The harmonic characteristics of these CFLs were divided into three general categories. Typical waveforms and harmonic spectrums representing each of these categories are provided below.

2.1 High distortion electronic ballast CFLs

This category represents the CFLs with the highest distortion levels. The characteristics of these CFLs are similar to the characteristics presented in previous publications [1]. The harmonic current distortion for CFLs in this category can be on the order of 140%, as shown by the example waveform and harmonic spectrum in Figure 1.

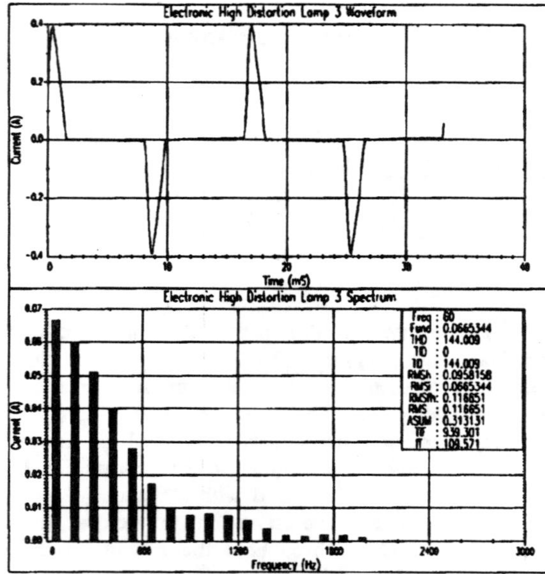

Figure 1. High distortion electronic ballast

2.2 Low distortion electronic ballast CFLs

Some CFLs with electronic ballasts include harmonic control as part of the design. The harmonic control can be passive (e.g. input filter or choke) or active. The CFLs with the lowest distortion levels use active filtering technologies. An example of a more common waveform is given in Figure 2, with a current distortion level of about 30%.

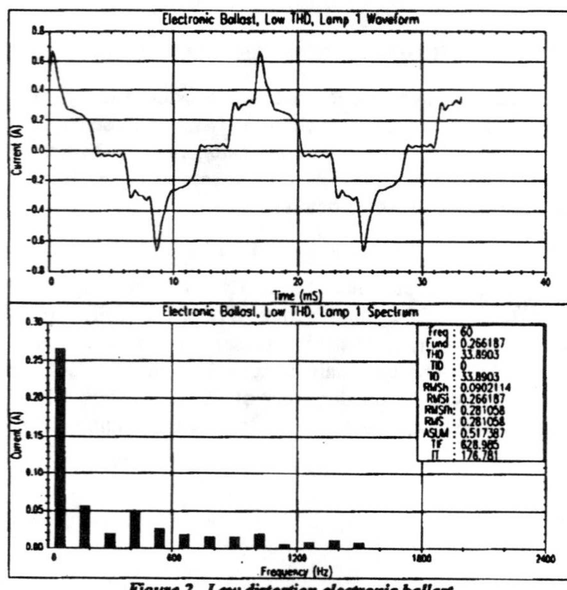

Figure 2. Low distortion electronic ballast

2.3 Magnetic ballast CFLs

The magnetic ballast CFLs tested typically have distortion levels below 20%. The harmonics are caused by the nonlinearity of the lamp arc itself in series with the ballast. The waveform for these CFLs is very similar to the waveform for larger fluorescent lights with magnetic ballasts. Figure 3 below is an example.

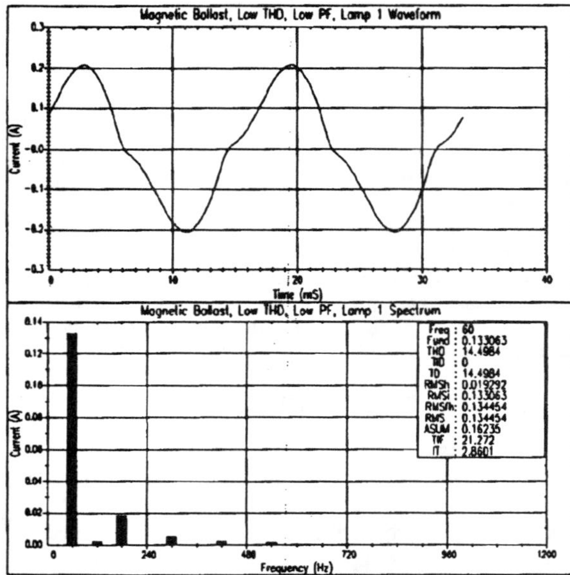

Figure 3. Magnetic ballast CFL without power factor correction

3.0 ANALYSIS OF HARMONIC CANCELLATION AT THE DISTRIBUTION TRANSFORMER SECONDARY

An EMTP model was developed based on the simplified circuit given in Figure 4 to evaluate the effect of harmonic cancellation within and between households. Four residences are represented on a single distribution transformer and six CFLs are included in each residence. Other loads within the residence included a 100 W electronic load (TV), 150 Watts of incandescent lighting, and 100 Watts of refrigerator motor load.

With the impedances used for the house wiring and the small currents associated with the CFLs, there was virtually no cancellation of lower order harmonic currents between the six CFLs within a single residence. The assumed waveform for the TV load was also similar to the CFL waveforms and effectively added at the lower order harmonics. The THD in the current is reduced from 56% to 43% due to the impedances between the different houses. The harmonic reduction is greatest at frequencies above the seventh harmonic.

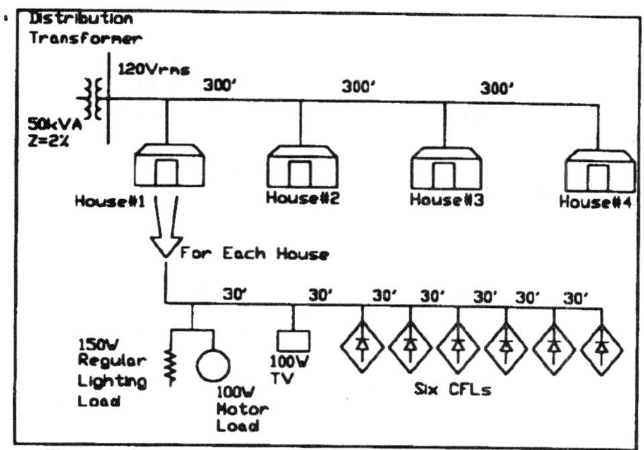

Figure 4. Simulated system for evaluation of harmonic cancellation

The EMTP results are combined with expected cancellation due to different impedances as calculated by Grady [2] to build an equivalent harmonic spectrum that could be injected throughout the feeder for the CFL loads. The harmonic spectrum representing the high distortion electronic ballasts, including cancellation effects, is given in Table 1 below.

Table 1. Assumed harmonic spectrum for high distortion, electronic ballast CFL loads, including cancellation.

Harmonic	Magnitude	Angle
1	100.0%	-11.5
3	82.0%	-38
5	53.4%	-62.5
7	31.6%	-85.1
9	16.6%	-103.3
11	8.2%	-111
13	1.5%	-99.1
15	1.0%	-88.4
17	0.6%	-95.5
19	0.5%	-109.5
21	0.3%	-120.5
23	0.1%	-107.5
25	0.1%	-67.4

4.0 DESCRIPTION OF DISTRIBUTION SYSTEM STUDIED

A simplified one line representation of the system studied is provided in Figure 5. This system consists of six parallel feeders serviced by a 44.8 MVA substation transformer. There is a 6 Mvar substation capacitor bank that can be either in or out of service and there are numerous feeder capacitors throughout the system.

The feeders on this system consist of both residential and commercial load. The commercial load is typically supplied through delta-wye transformers. The residential load can be supplied either line-to-line (overhead customers) or line-to-ground (most underground customers) on the primary.

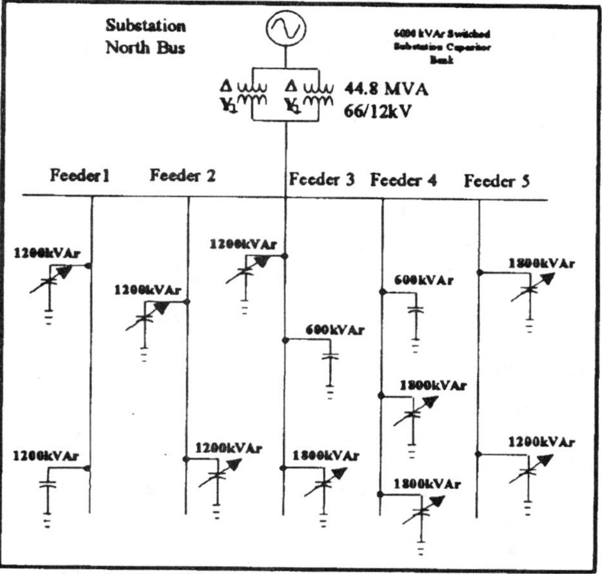

Figure 5. Single line diagram for distribution system studied

5.0 FREQUENCY RESPONSE CHARACTERISTICS

The distribution system frequency response characteristics determine the impact of the harmonic currents on the system voltage distortion. The frequency response is determined by the system strength, the capacitors in service, and the amount of load (damping). The load representation is particularly important, especially as the portion of the load made up of power electronics increases. The different factors affecting the frequency response are described briefly here.

5.1 Base Case Characteristics

The base case condition corresponds to a detailed representation of only one of the feeders (Feeder 1) shown on the one line diagram. Both feeder capacitors are in service for this simulation. This base case condition was selected because previous simulations of CFL impacts [1] did not include representation for parallel feeders off the same bus. Two cases with different loading were investigated, heavy loading and light loading. Heavy loading circumstances were based upon 79% of the maximum feeder load as residential loads and 21% of the maximum feeder load as commercial loads. Light loading excluded all commercial loads and utilized 50% of maximum loading for residential loads.

Figure 6 gives the frequency response looking from the end of the feeder for both load levels. The additional damping at heavy load can be seen. It is also clear that system resonances at low order harmonics are possible for typical capacitor configurations.

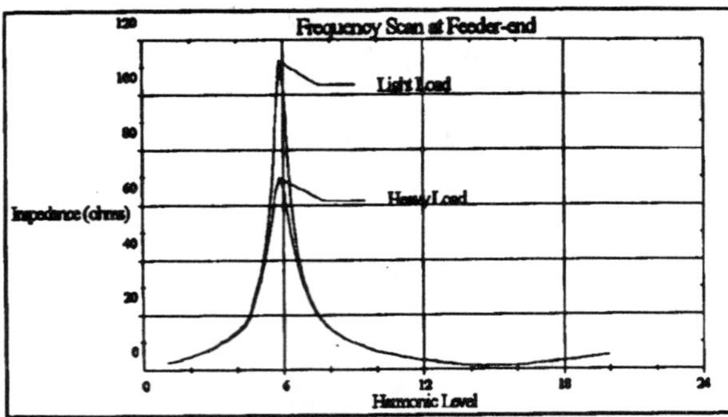

Figure 6. Base case frequency response from end of Feeder 1

5.2 Effect of Parallel Feeders

The frequency response characteristics with all five feeders on the distribution system was investigated at heavy and light loading. The response from the same location at the end of Feeder 1 is shown in Figure 7. The parallel feeders with their capacitors results in a lower frequency resonance that represents all of the system capacitors in parallel with the source equivalent at the substation. The major resonance as seen from the end of the feeder is still caused by the feeder capacitors. It is clear that it is important to represent all of the feeders connected to the substation bus when evaluating the overall system response.

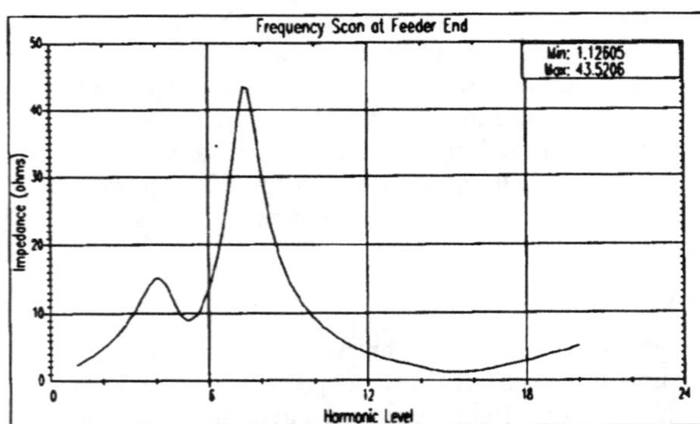

Figure 7. Frequency response from end of Feeder 1 with all feeders modeled.

5.3 Effect of Capacitor Banks

Figure 8 illustrates the frequency response characteristics with the substation capacitor bank in service along with some of the banks on the parallel feeders (scans from the substation and the end of feeder 1 are included). The substation capacitor bank moves the primary system resonant peak to the 5th harmonic. This case was selected to represent worst case realistic circuit conditions since it resulted in a system resonance at the fifth harmonic. Again, the effect of damping at heavy load conditions is apparent. This condition is used as the basis for evaluating the effect of CFL penetration levels.

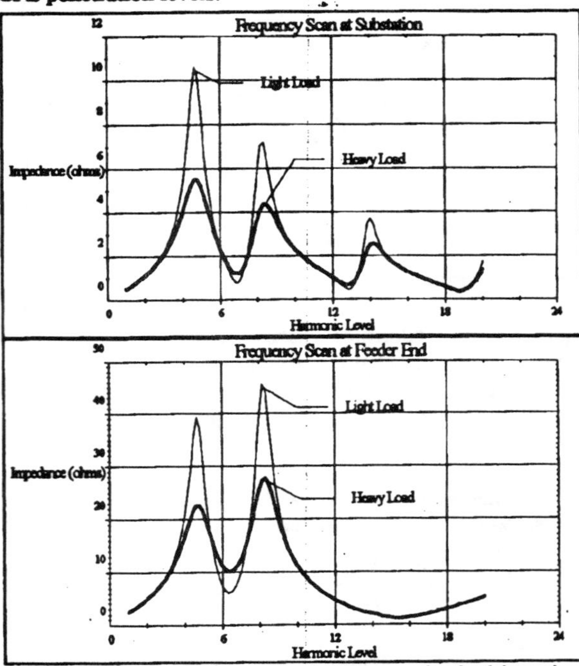

Figure 8. Frequency response with substation capacitor bank in service

6.0 DEVELOPING AN ELECTRONIC LOAD REPRESENTATION FOR SYSTEM RESPONSE CALCULATIONS

One important effect that was not considered in previous frequency scan cases is the representation of the load, especially when the load includes a significant amount of power electronics equipment (power supplies). The load representation used in the previous cases consisted of resistive and reactive components calculated from the loading and power factor. This is definitely not an accurate representation for power electronics type of load.

6.1 EMTP Simulations

EMTP simulations were performed to develop approximate models that can be used in the frequency domain simulations to evaluate the impact of these loads on the overall system frequency response characteristics. The circuit shown previously in Figure 4 was used as a representative case. The method used to develop an equivalent load representation for this circuit was as follows:

1. Run an EMTP case for the steady state conditions represented by the circuit given in Figure 4. Record the steady-state voltage and current waveforms at the distribution transformer primary.
2. Repeat the case and record waveforms during energizing of a capacitor bank on the transformer primary.
3. Subtract the voltage and current waveforms recorded in the first case from corresponding voltage and current waveforms recorded in the second case to obtain Δv(t) and Δi(t).
4. Perform FFT on Δv(t) and Δi(t) to get V_h and I_h correspondingly.
5. Derive the load impedance Z_h from
   $Z_h=V_h/I_h$ at each harmonic frequency.

Once the load impedance vs. frequency characteristic is obtained the parameters for an equivalent model can be approximated and utilized in the frequency domain simulations. This is only an approximation since the actual impedance vs. frequency for a load of this type will actually be different in different parts of the 60 Hz cycle.

The load impedance characteristics as seen from the high voltage side of the distribution transformer is illustrated in Figure 9. The impedance characteristic actually exhibits a series resonance (low impedance) close to the fifth harmonic. This is caused by the capacitors in the electronic load power supplies. In fact, the fifth harmonic impedance of the load is less than one third of the impedance at the fundamental frequency. After this series resonance, the characteristic is dominated by the inductance and resistance of the load. The combination of many loads with this characteristic will have a very dramatic impact on the overall distribution system frequency response characteristics.

6.2 Equivalent Circuit

The load impedance vs. frequency characteristics can be approximated using a passive circuit with a structure as shown in Figure 10. This circuit is used as an equivalent for the load connected to the secondary of a distribution transformer (Figure 4).

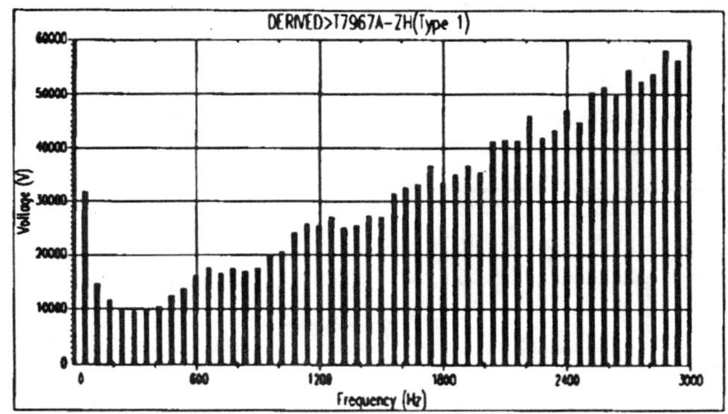

Figure 9. Z_h- Derived load impedance vs. Frequency with electronic loads

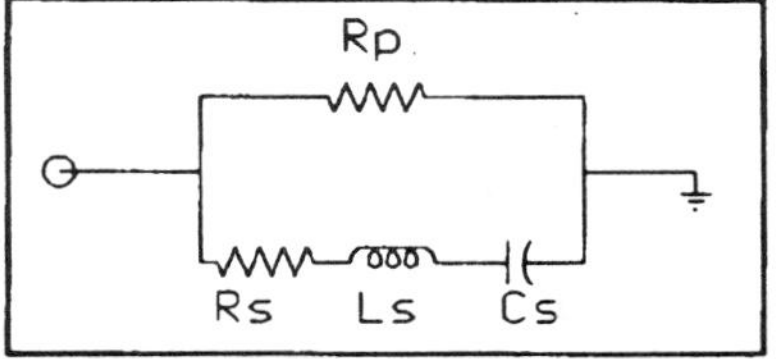

Figure 10. Equivalent circuit for load impedance vs. frequency representation for a single house

In order to approximate the load impedance characteristics, a trial and error method was used to determine parameters Rp, Rs, Ls and Cs. Cs is based on the estimated dc capacitance of electronic power supplies. Rp is based on the equivalent dc resistance at the output of these power supplies. Rs and Ls result in the damped series resonance at 300 Hz. The parameters selected to represent the equivalent at the secondary of a 50 kVA transformer are as follows:

Rs = 1.02 Ω   Rp = 156 Ω   Ls = 1.96 uH   Cs = 267 mF

Figure 11 illustrates the approximated load impedance magnitude with these parameters compared with the original Zh magnitude.

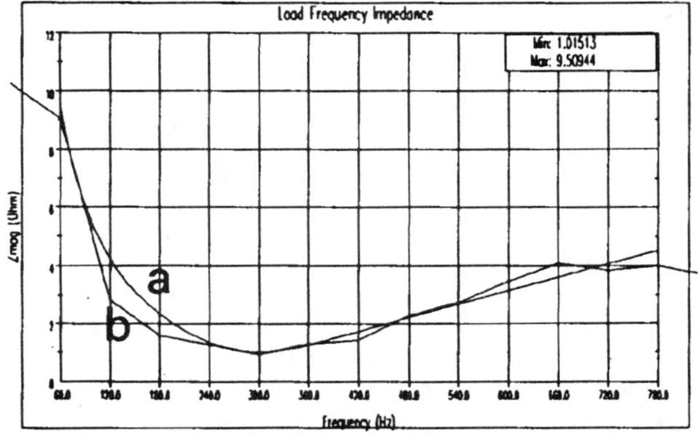

Figure 11. Comparison of equivalent circuit frequency response (curve-a) with simulated frequency response(curve-b).

6.3 Effect on System Frequency Response

The load representation was updated to include the effect of electronic load using the equivalent circuit derived above. Figure 12 illustrates the effect on the system impedance vs. frequency characteristics from the middle of Feeder 1.

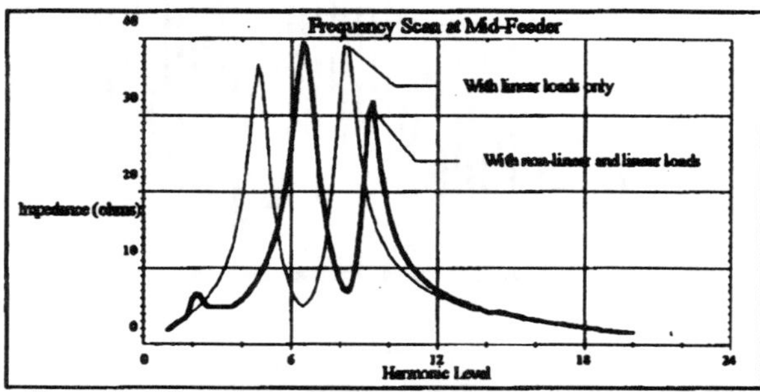

Figure 12. Effect of load representation on system frequency response looking from the middle of Feeder 1.

It is obvious that the electronic load can have a very dramatic effect on the overall system frequency response characteristic. With the load representation that includes the electronic load, the system resonance near the fifth harmonic is eliminated by the low impedance of the loads and the first significant parallel resonance of the system is closer to the seventh harmonic. This effect will also be important in representing some commercial buildings and industrial facilities that have a high percentage of electronic load.

7.0 EVALUATION OF EXPECTED HARMONIC DISTORTION LEVELS DUE TO CFL PENETRATION

7.1 Base Case Characteristics

The first step in evaluating the impact of the CFL penetration is to establish a base case condition representing distortion levels due to existing load characteristics. The worst case system configuration that results in a resonance near the fifth harmonic (Figure 8) is used to evaluate these characteristics. The existing characteristics are patterned after the load representation of Figure 4 without the CFLs. The expected harmonic distortion levels with this load assumption are shown in Figure 13. For this evaluation, the residential loads are connected line-to-neutral. This will result in worst case distortion levels. The effect of other primary transformer connections is discussed later.

Figure 13 shows the expected voltage distortion levels at three locations along Feeder 1 as a range. The minimum values in the range are for heavy load conditions where resistive load provides damping of the resonance and the maximum levels in the range are for light load conditions when the harmonic sources are still on but there is less damping. The expected distortion levels are within IEEE 519 guidelines [4] of 5% distortion despite the fifth harmonic resonance.

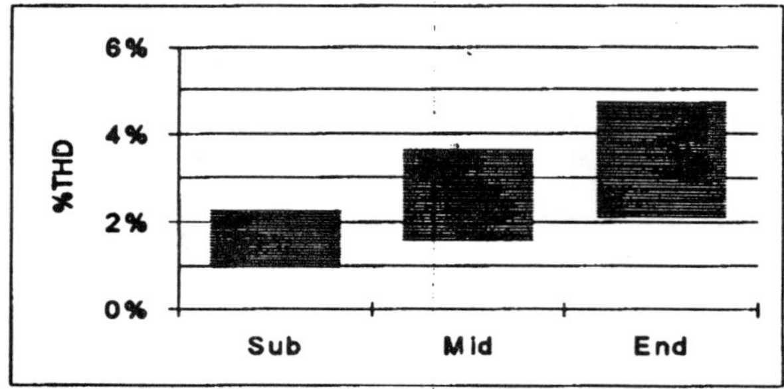

Figure 13. Voltage distortion along Feeder 1 for existing load conditions

7.2 Voltage Distortion vs. CFL Penetration Level

Distortion levels for average CFL penetrations of 50W, 100W, and 150W per household are shown in Figure 14. Simulated distortion levels can be very high - up to 16% for the case of maximum CFL penetration, unfavorable capacitor configuration, and minimum feeder loading. This is a very conservative estimate, for it assumes that all CFLs in the system use the high THD electronic ballasts described earlier. It is important to note that these results are in good agreement with previous study results [1].

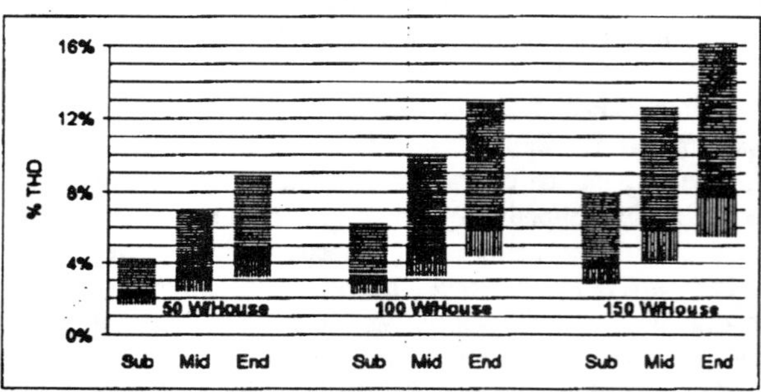

Figure 14 - Effect of CFL penetration on voltage distortion along Feeder 1

Effect of System Frequency Response

The worst case conditions in Figure 14 are indicated by the ranges with horizontal lines. This is for a capacitor configuration that results in a resonance near the fifth. The vertical lines represent a capacitor configuration (all caps in service) that results in a more "favorable" system response at corresponding lower distortion levels.

Effect of the Electronic Load Model

Including the representation of the electronic load derived previously also significantly improves the voltage distortion levels on the feeder by moving the system resonance away from the fifth. A comparison of simulation results using the conventional load representation and results with the equivalent for the electronic load is provided in Figure 15. The electronic load representation reduces the expected distortion levels but they still can be high because the more detailed load model causes a resonance at the ninth harmonic (see Figure 12) that increases ninth harmonic voltage distortion on the system.

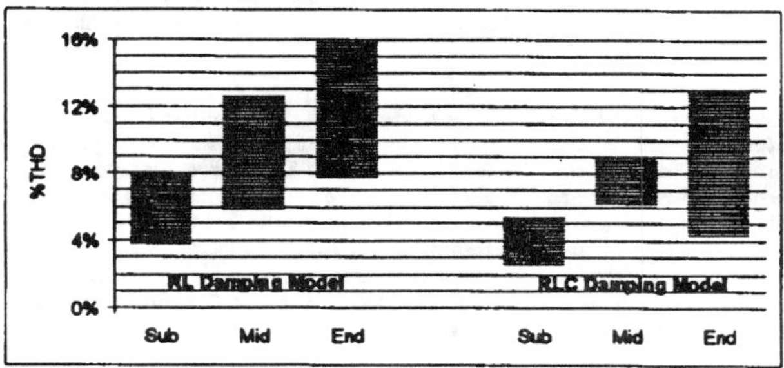

Figure 15. Effect of load representation on Feeder 1 voltage distortion for the 150W/house penetration level.

Effect of CFL Type

The previous results were all based on the worst CFL harmonic injection characteristics - the high distortion electronic ballast in Figure 1. As described in Section 2, other types of CFLs have lower harmonic distortion levels. Figure 16 below illustrates the expected distortion levels on the feeder at the high penetration level of 150W per household for all three ballast types. Harmonic voltage distortion should only be a problem if there is significant penetration of the high distortion electronic type of ballasts. Voltage distortion levels due to CFL penetration should not exceed 5% for the other types investigated.

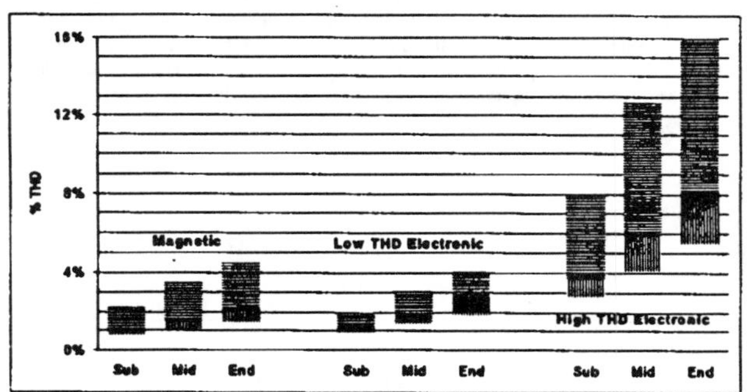

Figure 16. Effect of CFL type on expected feeder distortion levels.

Effect of Transformer Connections

The preceding cases have assumed that residential service transformers are connected line-neutral throughout the system. However, most underground supplies and some of the overhead supplies use line-to-line connections for the service transformers on these circuits. The line-to-line connection can significantly reduce distortion levels by providing cancellation of balanced third harmonics. In this case, the voltage distortion is dominated by the fifth harmonic system resonance so there is little benefit from the cancellation of triplen harmonics.

6.0 CONCLUSIONS

The conclusions from this investigation are in basic agreement with the conclusions of previous studies. High penetration levels of CFLs that use high distortion electronic ballasts can result in unacceptable voltage distortion levels on the distribution systems. This conclusion is true even when the effects of cancellation between the various loads is considered. CFLs with lower current distortion levels (30% or less) should not cause voltage distortion problems on the feeder, even at the high penetration level of 150W per household.

The distortion levels that can be expected are very dependent on system conditions. Capacitor configurations will cause system resonances that can magnify specific harmonic components. The worst case distortion levels simulated are for system conditions that result in a fifth harmonic system resonance. The system response is also significantly impacted by the representation of electronic loads. The capacitance in the power supplies of these loads can change the system response.

Figures 17a and 17b summarize the expected voltage distortion level as a function of CFL penetration for a case with a problem resonance and a case without a problem resonance. The expected distortion from the substation to the end of the feeder is shown for each CFL penetration level. These curves are for the worst case electronic ballast characteristics.

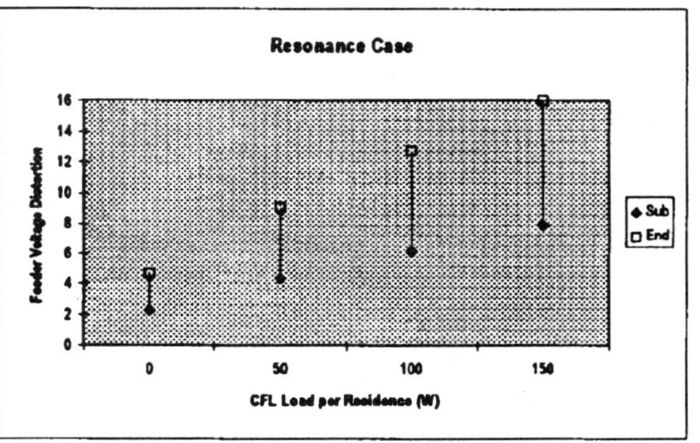

Figure 17a. Summary of expected voltage distortion levels along the feeder as a function of CFL penetration level per household - with a problem resonance

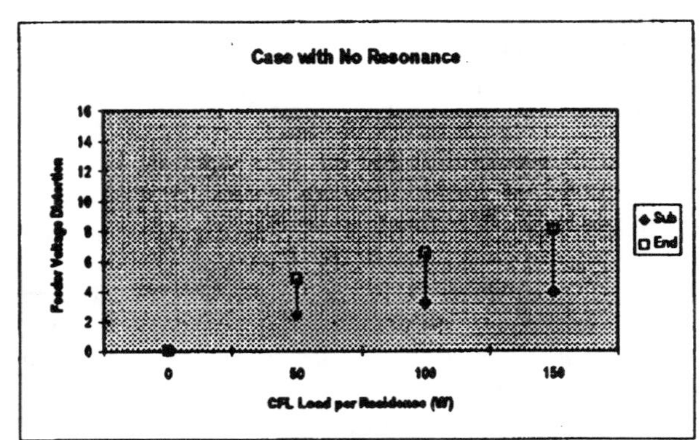

Figure 17b. Summary of expected voltage distortion levels along the feeder as a function of CFL penetration level per household - without a problem resonance

REFERENCES

[1] A. E. Emmanuel, T. J. Gentile, D. J. Pileggi, E. M. Gulachenski, C. E. Root, "The Effect of Modern Compact Fluorescent Lights on Voltage Distortion", Presented at the *IEEE/PES 1992 Summer Power Meeting*, Seattle, WA, July, 1992.

[2] A. Mansoor, W. M. Grady, A. H. Chowdhury, and M. J. Samotyj, "An Investigation of Harmonics Attenuation and Diversity among Distributed Single-Phase Power Electronic Load", *Proceedings of the 1994 Power Engineering Society Transmission and Distribution Conference*, Chicago, IL, May, 1994, pp. 110-116.

[3] A. Mansoor, W.M. Grady, R.S. Thallam, M.T. Doyle, S. Krein, and M.J. Samotyj, "Effect of Supply Voltage Harmonics on the Input Current of Single-Phase Diode Bridge Rectifier Loads," Presented at the *1994 IEEE/PES Summer Power Meeting*, San Francisco, July, 1994.

[4] IEEE Standard 519-1992. *Recommended Practice for Harmonic Control and Reactive Compensation in Power Systems.*

BIOGRAPHIES

Rory Dwyer received a BS and MEng in Electric Power Engineering from Rensselaer Polytechnic Institute. Presently Rory is Senior Power Systems Engineer at Electrotek where his primary responsibility is power quality measurement and analysis. Over the past four years, he has investigated a wide variety of power quality problems in utility distribution systems, commercial buildings, industrial facilities, mines, shipboard and other systems. These problems have included overvoltages, voltage sags, lightning transients, and high harmonic levels.

Afroz K. Khan received a BSEE from the University of Alaska Fairbanks and her MEng in Electric Power Engineering from Rensselaer Polytechnic Institute. She performs case studies that involve the investigation of equipment problems due to capacitor switching transients and voltage sags. She has also conducted detailed studies investigating the harmonic impact of advanced lighting techniques on power systems. Afroz is presently involved in the renewable energy research effort including the modeling of wind and solar energy devices.

Mark McGranaghan received his BSEE and MSEE from the University of Toledo and his MBA at the University of Pittsburgh. As Manager of Power Systems Engineering at Electrotek, Mark is responsible for a wide range of studies, seminars, and products involving the analysis of power quality concerns. He has worked with electric utilities and end users throughout the country performing case studies to characterize power quality problems and solutions as part of an extensive Electric Power Research Institute (EPRI) project. He has also been involved in the EPRI Distribution Power Quality Monitoring Project which is establishing the baseline power quality characteristics of U.S. distribution systems through a multi-year monitoring effort.

Le Tang received his BSEE from Xiang Jiantong University in China and his Meng and PhD in Electric Power Engineering from Rensselaer Polytechnic Institute. Le is presently Senior Power Systems Engineer at Electrotek. He conducts harmonic and transient analyses of power systems engineering phenomena, including capacitor switching transients, lightning phenomena, arrestor energy duties, ferroresonance, circuit interruption, feeder inrush characteristics, cable energization and fault clearing. He is an expert in the modification and application of the Electro-magnetic Transients Program (EMTP) relating to power system analysis.

Robert K. McCluskey received his BSEE from Iowa State University. He currently manages the Power Electronics Team in Transmission and Distribution Systems Research, and is the current project manager for power quality research at SCE. His professional background of twenty-three years in the electric utility industry includes work experience in the planning, siting, design, construction and start-up of major steam electric generating plants, energy service facilities, and pollution control apparatus, as well as research and development associated with a wide variety of emerging advanced technologies.

Roger Sung received his BS in Chemistry from Iowa State University and an MS in both Physical Organic Chemistry and Civil/Sanitary Engineering. He attained his PhD in Chemical Engineering from UC Davis. Dr. Sung is a Senior Research Engineer and a Program Manager in the Research Development and Demonstration Department of SCE. Prior to this he was the Program Manager for Power Quality, responsible for the development of a comprehensive Research Power Quality program. Dr. Sung has over twenty-five years of broad professional engineering experience including working for utilities, government agencies, municipalities, and consulting firms.

Thomas Houy received a BSEE and MBA from California State Polytechnic University. He is currently conducting research into the mitigation of power quality problems impacting customers as well as utilities. These areas include voltage sags/surges, switching transients and harmonics. His main expertise lies in the optical detection of high voltage and currents utilizing Pockel and Kerr effect materials.

Discussion

L. Pierrat, Senior Member, IEEE, (Electricité de France, Division Technique Générale, 37 Rue Diderot, 38040 Grenoble, France): The effect of electronic loads (CFL) in harmonic disturbances is analyzed and an equivalent model is proposed to these type of loads. However, a representative load model is hard to determine [A].

The response of proposed model seemed a good compromise in a given range of frequencies (especially at first serie resonance, i.e. 300Hz).

Considering the comparison presented in Fig. 11, could the authors give more precision about the background to choose the topology in Fig. 10 ? How this topology can be used in order to take into account the first parallel resonance (appears about at 700 Hz)?

In our opinion, to overcome this problem another model based on second Foster topology with two branches RLC in parallel where no capacitor in first branch can be used [B].

[A] CIGRE Working Group 36-05, "Harmonics, characteristics parameters, methods of study, estimates of existing values in the network," Electro Review, No. 77, 1981, pp. 35-54.

[B] Pierrat,L., Budi, Wang, Y.T., "Frequency domaine modelling of aggregated loads for harmonic studies," 1st European Conf. on Power Syst. Trans., EPST'93, Lisbone, June 17-18, 1993, pp. 245-249.

Manuscript received March 3, 1995.

A. K. Khan and L. Tang, Members, IEEE, (Electrotek Concepts, Inc.): The authors would like to thank Mr. L. Pierrat for his discussion of the paper and agree with him that the topology in Fig. 10 cannot be used to take into account the first parallel resonance at 700 Hz. As stated by L. Pierrat, another model of Foster topology would be necessary to accomplish this. The topology in Fig. 10 was utilized to illustrate an important result of this analysis. The impedance vs. frequency curve developed from a capacitor switching transient exhibits a series resonance close to the fifth harmonic. In fact, the fifth harmonic impedance of the load is less than one third of the impedance at the fundamental frequency. This load characteristic partially explains some observations from field measurements in which the 5th and 7th harmonic contents are not as high as expected based on the simple summation of multiple nonlinear loads. The low impedance around 300 Hz makes the load act like a series filter tuned near the 5th harmonic. This filtering effect could significantly reduce harmonic voltage distortion levels on to feeder circuits. As a result of this, we were specifically interested in primarily developing a test base case model based upon this phenomena.

Manuscript received May 2, 1995.

IEEE Transactions on Power Delivery, Vol. 11, No. 3, July 1996

Voltage Notching Interaction Caused by Large Adjustable Speed Drives on Distribution Systems with Low Short Circuit Capacities

L. Tang, Member, IEEE
M. McGranaghan, Member, IEEE
Electrotek Concepts, Inc.
Knoxville, Tennessee

R. Ferraro, Member, IEEE
FOA
Knoxville, Tennessee

S. Morganson
Alberta Power
Edmonton, Alberta

B. Hunt, Member, IEEE
Moonlake Electric
Roosevelt, Utah

Abstract - This paper describes voltage notching associated with large adjustable speed drives. The notching is a normal characteristic of a phase-controlled rectifier but this paper illustrates problems that can occur on systems with low short circuit levels where the voltage notching can excite the natural frequency of the distribution system and cause significant distortion in the supply voltage. The notching characteristics and the interaction with the distribution system frequency response characteristics are described, along with possible solutions, using two actual examples where problems were encountered.

Keywords: ASD, Harmonics, Voltage Notching

INTRODUCTION

Adjustable speed ac and dc drives are used with very large motors (e.g. 1,000-20,000 hp) for a variety of reasons. The drives can result in significantly improved efficiency when the driven load is variable. For large motor applications on weak systems, the drives may be required for motor starting to avoid high inrush currents.

Adjustable speed ac drives (ASDs) 1000 hp and larger typically use phase-controlled rectifiers (SCRs) and a large dc link inductor to supply a relatively constant dc current to the inverter. This is known as a current source inverter (CSI) configuration. The input rectifier may be configured as a six pulse, twelve pulse, or even higher pulse number rectifier, depending on harmonic control requirements. For dc drive applications, phase-controlled rectifiers are used to supply the dc current directly to the dc motor.

95 SM 388-9 PWRD A paper recommended and approved by the IEEE Transmission and Distribution Committee of the IEEE Power Engineering Society for presentation at the 1995 IEEE/PES Summer Meeting, July 23-27, 1995, Portland, OR. Manuscript submitted January 3, 1995; made available for printing May 11, 1995.

The voltage notching discussed in this paper is caused by the commutating action of the controlled rectifier. Whenever the current is commutated from one phase to another, there is a momentary phase-to-phase short circuit through the rectifier switching devices (SCRs, in this case). For a six pulse converter, this happens six times each cycle. The voltage notch is defined by its duration and its depth. The duration (commutation period) is determined by the source inductance to the drive and the current magnitude. The depth of the notch is reduced by inductance between the observation point and the drive (e.g. isolation transformer or choke inductance). An example waveform illustrating simple notches resulting from a drive operation is shown in Figure 1.

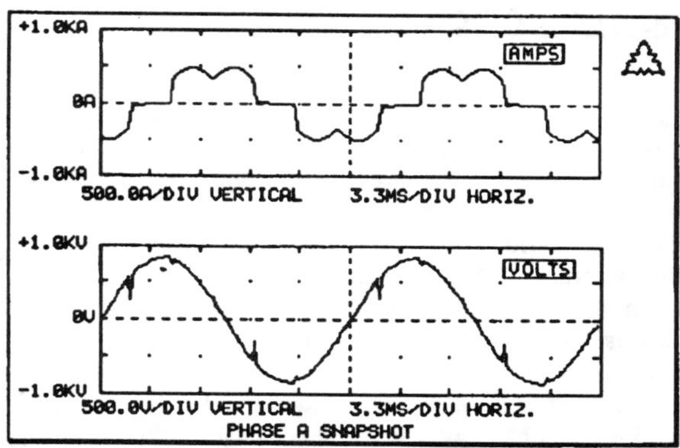

Figure 1. Example of voltage notches caused by converter commutation.

PROBLEM DESCRIPTION

On most systems, problems with voltage notching can be minimized by applying sufficient isolation reactance at the drive. This limits the notch magnitude on the source side of the isolation reactance. However, on some systems, the notches that appear at the system level can still be significant. If there is not much resistive load on a system like this, the notches can excite the natural frequency of the distribution system (determined by the capacitance of lines, cables, and capacitor banks in parallel with the system source inductance) and cause significant distortion in the voltage waveform.

Numerous papers have described the voltage notching phenomena in industrial facilities and sizing isolation reactance to limit the notching effect on other loads [1-5]. However, there has been little literature describing the potential for voltage notching to excite natural frequencies of the distribution system. The high frequency oscillations that result can cause problems with communication interference and sensitive customer loads. This paper describes the concern and possible solutions using two examples where problems were encountered. The methodology for evaluating these problems and the solutions implemented should be valuable to anyone else encountering this problem.

It is important to note that the notching problem described should only exist with large adjustable speed drives with current-source inverter configurations or with dc drives. With other types of ASDs that use voltage source inverters (e.g. pulse width modulation), the rectifier does not have a constant dc current that needs to be commutated from one switching device to another. It is this current being commutated that that essentially looks like an injection of a disturbing current into the distribution system.

EXAMPLE 1: LARGE INDUCTION MOTOR DRIVE

The first example system is illustrated by the one line diagram in Figure 2. The 25 kV distribution system is supplied through a 10 MVA transformer from the 144 kV transmission system. The customer causing the notching problems has a 6000 hp induction motor supplied through an adjustable speed drive. This drive is at a 4.16 kV bus supplied through a 7.5 MVA transformer. Harmonic filters (5th, 7th, 11th) are included to control the lower order characteristic harmonics of the six pulse drive.

Another customer on a parallel feeder supplied from the same 25 kV bus has motor loads at both 4.16 kV (800 hp motor) and 480 volts. The 800 hp motor includes surge capacitors for transient protection. The customer also has power factor correction capacitors at the 480 volt bus. These lower voltage surge capacitors and power factor correction capacitors have the potential to magnify the oscillations which occur on the distribution system.

Operation of the 6000 hp motor and drive resulted in significant oscillations on the 25 kV supply system. These oscillations caused clocks to run fast at the customer with the 6000 hp motor (clocks were fed separately from the 25 kV system) and failure of surge capacitors on the 800 hp motor at the customer located on the parallel feeder.

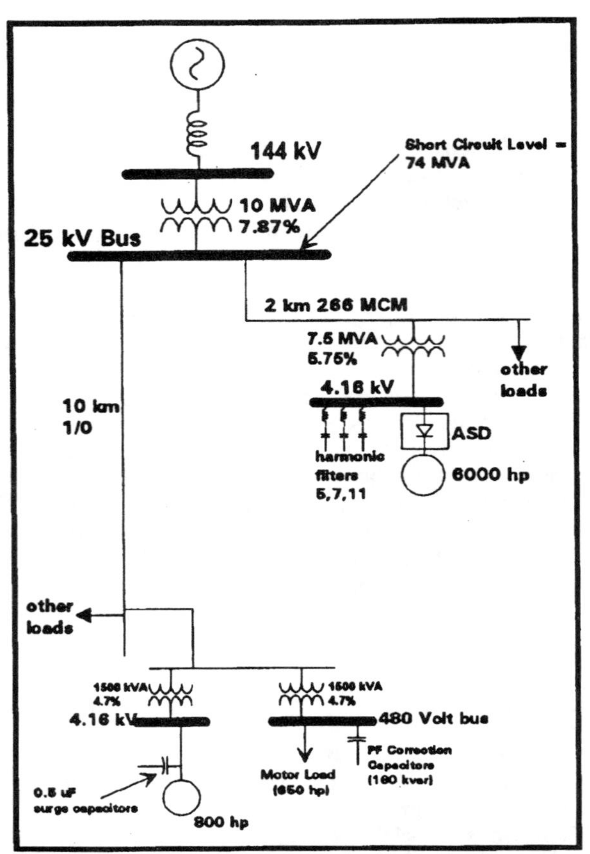

Figure 2. One line diagram for the first example system.

Figure 3 illustrates the measured waveforms on the 25 kV supply system. The oscillations have a primary frequency component near the 60th harmonic. In this case, the natural frequency is the result of the line capacitance from approximately 12 km of overhead line in parallel with the system source inductance. Note that the oscillations are excited six times per cycle corresponding to the six-pulse operation of the drive.

System Frequency Response

A model was developed using the electro-magnetic transients program (EMTP) to evaluate the magnification at the surge capacitor location and to evaluate possible solutions to the problem. First of all, the steady state frequency response of the system was simulated to illustrate the natural frequency that can excite the oscillations illustrated above. Figure 4 shows the voltage on the 25 kV system as a function of a 1 amp source at the 4.16 kV bus where the drive is located. The system resonance just above the 60th harmonic is apparent in the figure. Note also the lower order series and parallel resonances caused by the harmonic filters.

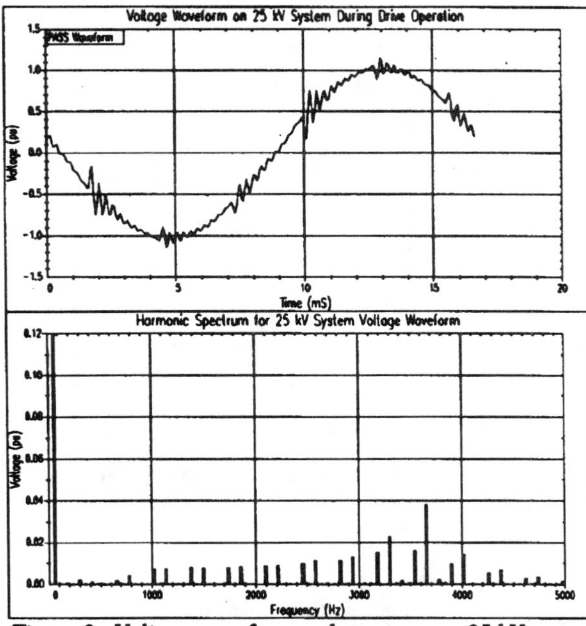

Figure 3. Voltage waveform and spectrum on 25 kV system during drive operation.

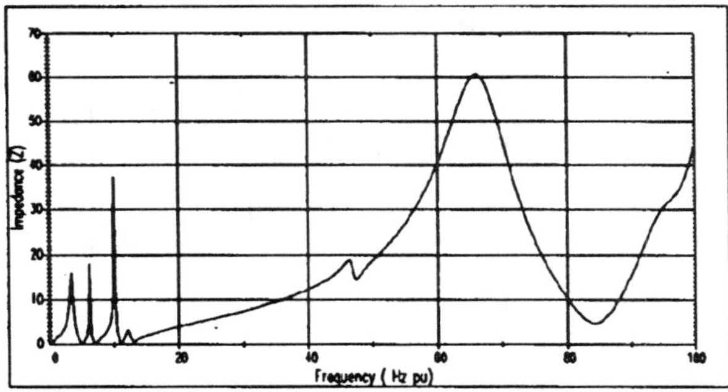

Figure 4. Voltage vs. frequency at 25 kV bus due to 1 amp source at 4.16 kV drive location

Effect of Customer Low Voltage Capacitors

Next, the actual adjustable speed drive and motor load were represented to reproduce the notching oscillations observed in the measurements. The worst notching problems are associated with a firing angle at about 70% load. The simulated waveform for the 25 kV bus voltage is shown in Figure 5 below. The oscillations at each commutation point are in good agreement with the measurement results.

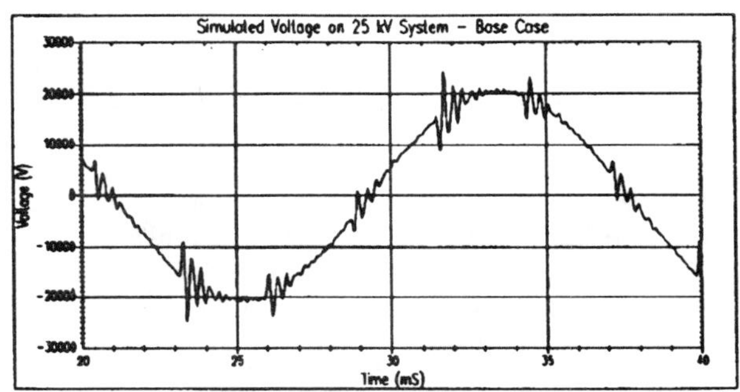

Figure 5. Simulated 25 kV system voltage with drive operating.

The model was then used to evaluate the voltage waveforms at the customer bus located on the parallel feeder. Figure 6 illustrates the voltage waveform at the 4.16 kV bus where the 800 hp motor surge capacitors cause magnification of the oscillations. The potential for problems at this location is quite evident. The surge capacitor failures typically occured during startup of the drive when the firing angles went through this worst case condition. Figure 7 shows the waveform at the 480 volt bus where the power factor correction capacitors damp out the high frequency oscillation. The power factor correction capacitors are much larger than the surge capacitors and result in a much lower resonant frequency. No problems were encountered with loads on the 480 volt bus.

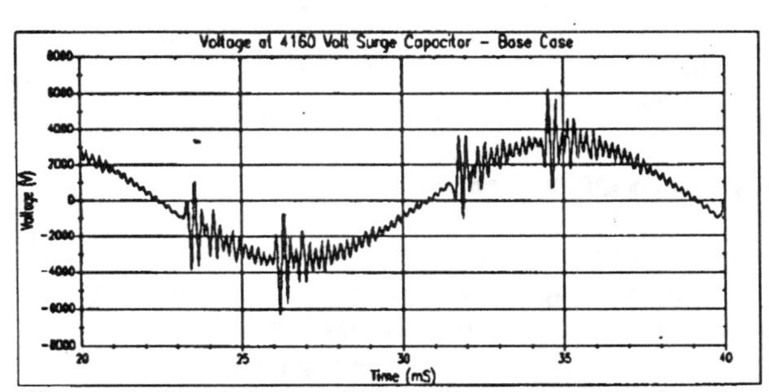

Figure 6. Simulated waveform at surge capacitor location (4.16 kV bus of customer on parallel feeder).

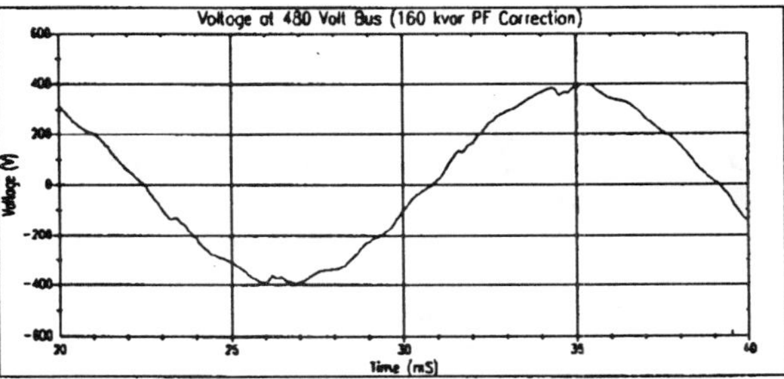

Figure 7. Simulated waveform at power factor correction capacitors (480 volt bus of customer on parallel feeder).

Possible Solutions

A number of possible solutions to the high frequency oscillation problem were evaluated. These included:

1. A larger choke inductance at the input to the adjustable speed drive. This approach could be effective if the choke impedance was included as part of the drive's initial design. As a retrofit, it is impractical because the size required would cause voltage regulation problems at the drive.

2. Larger surge capacitors at the 800 hp motor of the parallel customer. This approach is feasible to protect the individual motor and surge capacitors of concern. However, large surge capacitors would be required and they do not eliminate the oscillations on the 25 kV system that are the source of the problem. Note that the surge capacitors were removed as a temporary solution for the parallel customer.

3. Modification of the filtering at the adjustable speed drive to include a high pass filter instead of just tuned branches at the 5th, 7th, and 11th. This approach proved to be ineffective because the high pass filter cannot provide sufficient damping at the higher frequency resonance.

4. Addition of a capacitor bank on the 25 kV system. This approach has the advantage of being the least expensive and the most practical to implement. Simulations show that this solution can be effective.

Effect of a 25 kV Capacitor Bank

Adding a capacitor bank to the 25 kV system changes the system frequency response to prevent the high frequency oscillation shown on the previous waveforms. However, the capacitor bank creates a new system parallel resonance at a lower frequency that could result in magnification of the lower order harmonic components created by the adjustable speed drive.

The first capacitor bank size tried was 1200 kvar. This created a resonance that magnified the thirteenth harmonic component on the system if the power factor correction capacitors at the parallel customer were out of service. Figure 8 gives the measured voltage on the 25 kV system for this condition. Note that the notching oscillation problem is solved (no high frequency components) but the thirteenth harmonic component in the voltage is approaching 5%.

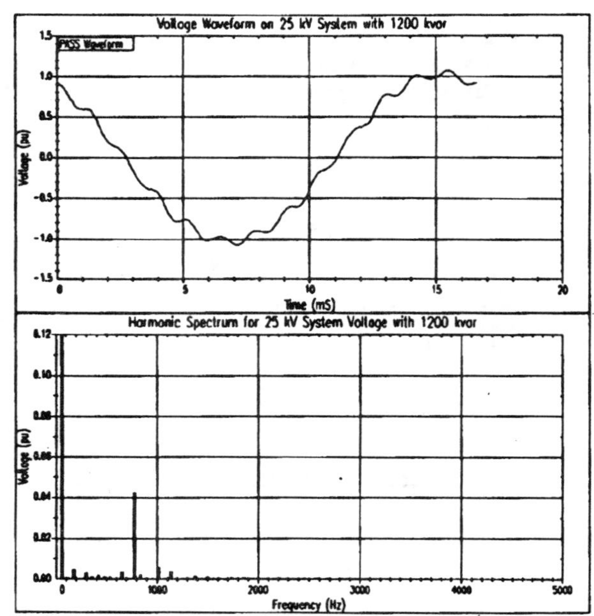

Figure 8. Voltage waveform and spectrum on 25 kV system with a 1200 kvar capacitor bank in service and the customer power factor correction capacitors off.

After examination of the system frequency response with the drive filters and the existing load power factor correction capacitors, the capacitor size was increased to 2400 kvar to solve the thirteenth harmonic resonance problem. Figure 9 illustrates the frequency response at the 25 kV bus for a 1 amp source located at the 4.16 kV drive location.

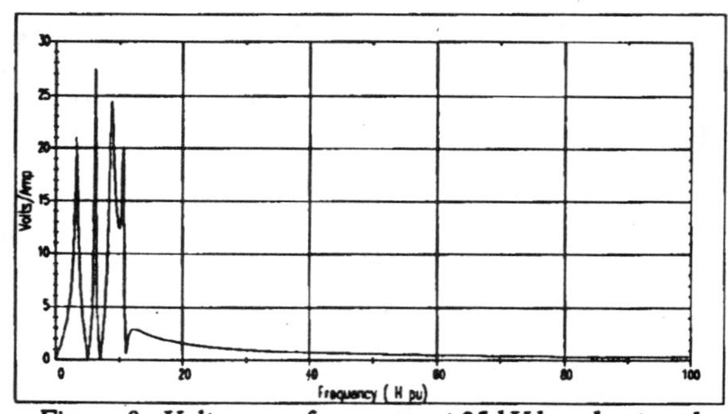

Figure 9. Voltage vs. frequency at 25 kV bus due to a 1 amp source at the 4.16 kV drive location with 2400 kvar capacitor in service at the 25 kV bus.

With a 2400 kvar capacitor bank, all of the system resonances that could cause magnification are located below the eleventh harmonic and are at frequencies that are not characteristic harmonics of the drive. Figure 10 gives the measured voltage waveform and harmonic spectrum with the 2400 kvar capacitor operational. The voltage distortion is less than 2% with the 2400 kvar capacitor in service.

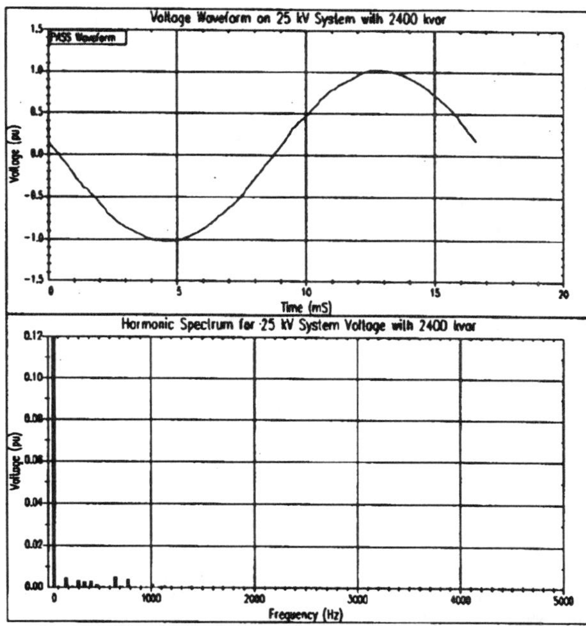

Figure 10. Voltage waveform and spectrum on 25 kV system with a 2400 kvar capacitor bank in service.

EXAMPLE 2: LARGE DC DRIVE APPLICATION

The second example involves large dc drives used to pump a slurry in an industrial process. There are three sets of tandem 1000 hp dc drives used in the process. Initially, this customer was the only customer on the distribution system. Then, the distribution system was extended to supply a new customer with electronic controls for large gas turbine-driven compressors. When this new customer came on line, numerous problems were encountered, including motor heating and failures of control circuits. A one line diagram illustrating the system configuration is provided in Figure 11.

Measurements on the system indicated that harmonic distortion levels were unacceptably high on the 34.5 kV system due to the operation of the dc drives. As a first step to try and solve the problem, the customer replaced the existing transformers supplying the dc drives with new transformers that had phase shifts designed to achieve 18 pulse operation for the overall facility (0 degrees, plus 20 degrees, and minus 20 degrees for the three transformers). This was effective in reducing the lower order harmonic components (when all three drives were operating) but it did not reduce the high frequency oscillation that was occuring as a result of a system resonance near the 35th harmonic. Figure 12 gives a measured waveform and spectrum from the 34.5 kV system after the phase shifting transformers were put in service. Note that the distortion is spread over a range of harmonic components because the oscillation is not an exact multiple of 60 Hz. The corresponding voltage from the 480 volt service entrance at the new customer is given in Figure 13.

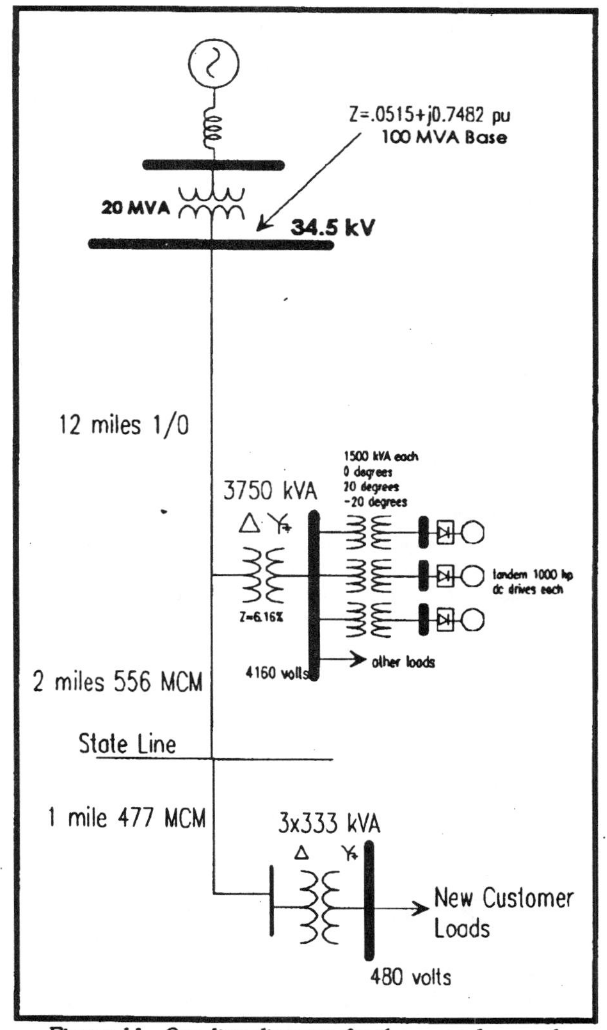

Figure 11. One line diagram for the second example system.

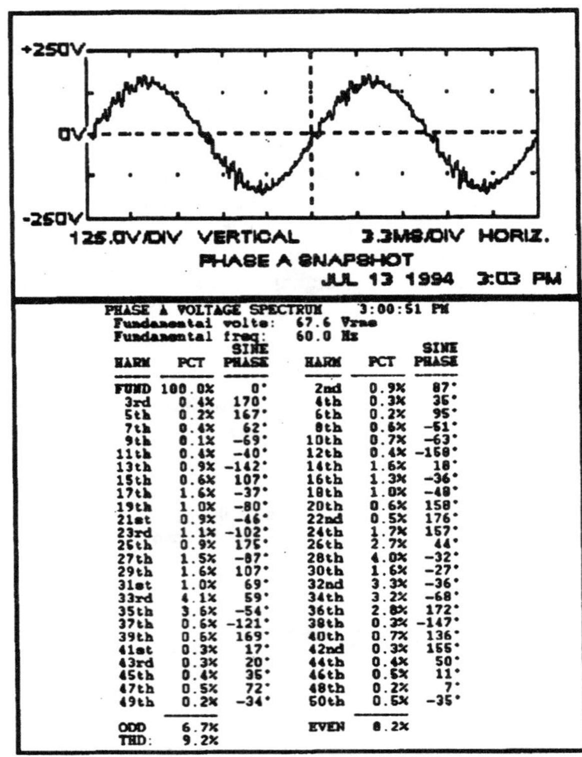

PHASE A VOLTAGE SPECTRUM 3:00:51 PM
Fundamental volts: 67.6 Vrms
Fundamental freq: 60.0 Hz

HARM	PCT	SINE PHASE	HARM	PCT	SINE PHASE
FUND	100.0%	0°	2nd	0.9%	87°
3rd	0.4%	170°	4th	0.3%	35°
5th	0.2%	167°	6th	0.2%	95°
7th	0.4%	62°	8th	0.6%	-51°
9th	0.1%	-69°	10th	0.7%	-63°
11th	0.4%	-40°	12th	0.4%	-158°
13th	0.9%	-142°	14th	1.6%	18°
15th	0.6%	107°	16th	1.3%	-36°
17th	1.6%	-37°	18th	1.0%	-48°
19th	1.0%	-80°	20th	0.6%	158°
21st	0.9%	-46°	22nd	0.5%	176°
23rd	1.1%	-102°	24th	1.7%	157°
25th	0.9%	175°	26th	2.7%	44°
27th	1.5%	-87°	28th	4.0%	-32°
29th	1.6%	107°	30th	1.6%	-27°
31st	1.0%	69°	32nd	3.3%	-36°
33rd	4.1%	59°	34th	3.2%	-68°
35th	3.6%	-54°	36th	2.8%	172°
37th	0.6%	-121°	38th	0.3%	-147°
39th	0.6%	169°	40th	0.7%	136°
41st	0.3%	17°	42nd	0.3%	155°
43rd	0.3%	20°	44th	0.4%	50°
45th	0.4%	35°	46th	0.5%	11°
47th	0.5%	72°	48th	0.2%	7°
49th	0.2%	-34°	50th	0.5%	-35°
ODD	6.7%		EVEN	8.2%	
THD:	9.2%				

Figure 12. Measured voltage on 34.5 kV system after installation of phase shifting transformers

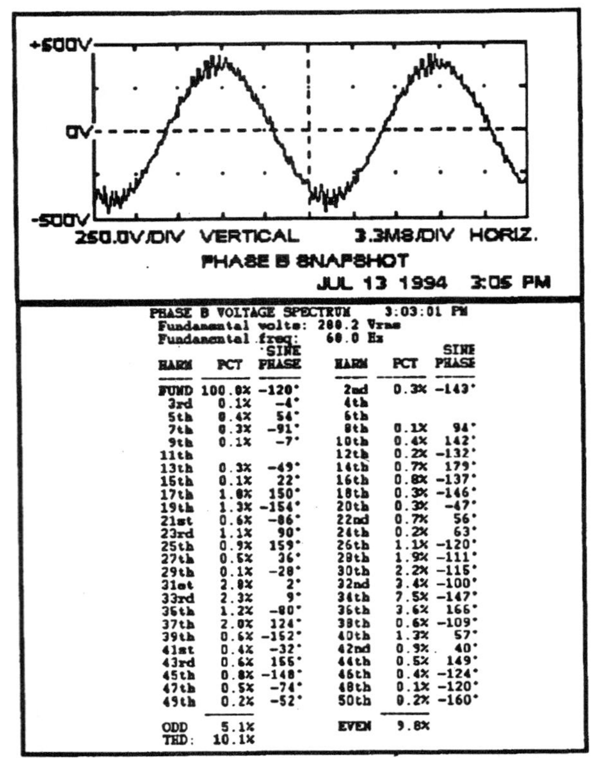

PHASE B VOLTAGE SPECTRUM 3:03:01 PM
Fundamental volts: 288.2 Vrms
Fundamental freq: 60.0 Hz

HARM	PCT	SINE PHASE	HARM	PCT	SINE PHASE
FUND	100.0%	-120°	2nd	0.3%	-143°
3rd	0.1%	-4°	4th		
5th	0.4%	54°	6th		
7th	0.3%	-91°	8th	0.1%	94°
9th	0.1%	-7°	10th	0.4%	142°
11th			12th	0.2%	-132°
13th	0.3%	-49°	14th	0.7%	179°
15th	0.1%	22°	16th	0.8%	-137°
17th	1.8%	150°	18th	0.3%	-146°
19th	1.3%	-154°	20th	0.3%	-47°
21st	0.6%	-86°	22nd	0.7%	56°
23rd	1.1%	90°	24th	0.2%	63°
25th	0.9%	159°	26th	1.1%	-120°
27th	0.5%	36°	28th	1.9%	-111°
29th	0.1%	-28°	30th	2.2%	-115°
31st	2.8%	2°	32nd	3.4%	-100°
33rd	2.3%	9°	34th	7.5%	-147°
35th	1.2%	-80°	36th	3.6%	166°
37th	2.0%	124°	38th	0.6%	-109°
39th	0.6%	-152°	40th	1.3%	57°
41st	0.4%	-32°	42nd	0.9%	40°
43rd	0.6%	155°	44th	0.5%	149°
45th	0.8%	-148°	46th	0.4%	-124°
47th	0.5%	-74°	48th	0.1%	-120°
49th	0.2%	-52°	50th	0.2%	-160°
ODD	5.1%		EVEN	9.8%	
THD:	10.1%				

Figure 13. Voltage from 480 volt service entrance at new customer after installation of phase shifting transformers

Effect of 34.5 kV Capacitor Bank

Again, the best solution to eliminate the high frequency oscillations that are excited by notching from the dc drives is to change the frequency response of the distribution system. This is most easily done by adding a shunt capacitor bank at the 34.5 kV level. Figure 14 gives the base case frequency response characteristic showing a resonance near the 35th harmonic, along with the response for three different capacitor sizes added to the 34.5 kV system near the customer with the dc drives.

At this facility, the dc drives are not always operated together. During lighter load pumping requirements, there may only be one or two of the tandem dc drives in service. During these conditions, cancellation of the lower order harmonic current components is not achieved and resonances at the 5th, 7th, 11th, or 13th harmonics could be excited. For this reason, it is important that the capacitor addition at the 34.5 kV level not cause a resonance near one of these characteristic harmonics.

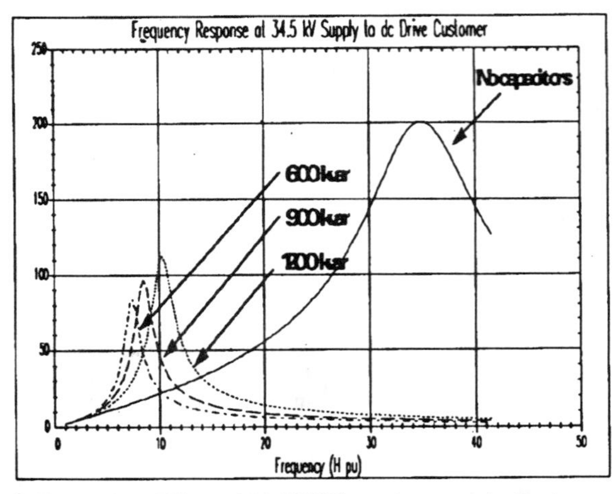

Figure 14. Effect of 34.5 kV Capacitors on the System Frequency Response Characteristics

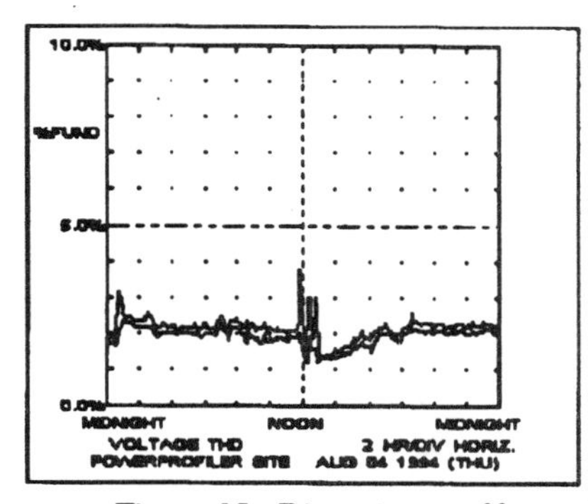

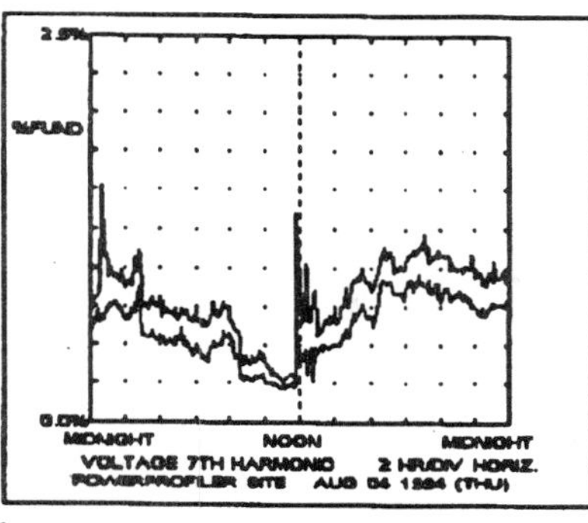

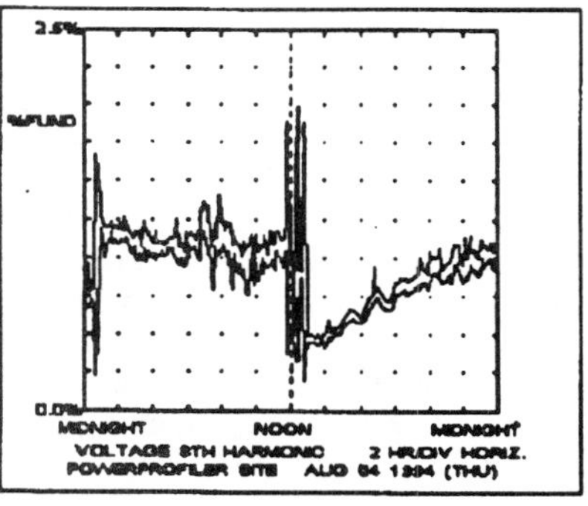

Figure 15. Distortion profiles at the new customer service entrance with 900 kvar feeder capacitor in service

A capacitor size of 900 kvar was selected because it results in a resonance between the 7th and 11th harmonics (see Figure 14). This capacitor is most effective when the drives are operating in the full three pump mode and there is cancellation of the lower order harmonics. When in the two pump mode, generation of lower order harmonics increases the distortion on the system but the total harmonic distortion is still less than 5%. Plots of typical profiles of the total harmonic distortion, 7th harmonic component, and 8th harmonic component over a 24 hour period are shown in Figure 15. Note that the total harmonic distortion is normally less than 4%.

With this configuration, the new customer has been able to operate without problems. However, a better long term solution would involve the addition of 5th, 7th, and 11th harmonic filters at the 4.16 kV bus supplying the dc drives. These filters would prevent resonance problems at the low order harmonics during two pump operation, even if system changes in the future (e.g. another capacitor bank added to the distribution system or at a customer location) change the 34.5 kV system frequency response characteristics.

SUMMARY AND CONCLUSIONS

- The paper illustrates the characteristics of high frequency oscillations that can result from the commutation notches of large adjustable speed ac or dc drives operating on systems with relatively low short circuit levels.

- The system oscillations can be magnified at customer locations where low voltage capacitors exist. Small capacitors, such as motor surge capacitors can be a particular problem. Even capacitors in the power supplies for electronic loads or smaller adjustable speed drive rectifiers can be affected by these oscillations.

- The problem can be solved by careful selection of a capacitor size for the primary distribution system. The capacitor should not introduce a new resonance at one of the characteristic harmonics of the adjustable speed drive. The interaction of the proposed capacitor with other system capacitors and harmonic filters must be evaluated.

REFERENCES

1. IEEE Std. 519-1992, *IEEE Recommended Practices and Requirements for Harmonic Control in Electrical Power Systems,* IEEE, New York, 1993.

2. J.C. Read, "The Calculation of Rectifier and Converter Performance Characteristics," *Journal of the IEE,* vol. 92, pt. II, 1945, pp 495-509.

3. E.F. Christensen, et. al., "Analysis of Rectifier Circuits," *AIEE Transactions,* vol. 63, 1944, pp. 1048-1058.

4. R.A. Adams, et. al., "Power Quality Issues Within Modern Industrial Facilities, " IAS Textile Film and Fiber Meeting, Atlanta, GA., 1990.

5. D.A. Jarc and R.G. Schieman, "Powerline Considerations for Variable Frequency Drives, " *IEEE Transactions on IAS,* vol. 1A-21, no. 5, 1985.

Le Tang is a Senior Consulting Engineer with Electrotek Concepts, Inc. He received his BS degree from Xian Jiaotong University in Electrical Engineering, 1982 and his ME and Ph.D. from Rensselaer Polytechnic Institute in Electric Power Engineering in 1985 and 1988 respectively. His areas of interest include power system transient and harmonic analyses, power electronics and machine simulation, renewable energy development and applications.

<u>Mark McGranaghan</u> is General Manager of Power Systems Engineering at Electrotek Concepts, Inc. He received his BSEE and MSEE degrees from the University of Toledo in 1977 and 1978, respectively. Mark is responsible for a wide range of studies, seminars, and products involving the analysis of power system transients, harmonics, and power quality concerns. He is Chairman of IEEE P519A which is developing an application guide for IEEE 519-1992.

<u>Ralph Ferraro</u> established the private consulting firm of Ferraro, Oliver & Associates, Inc. (FOA) in 1990. He has a BSEE from Newark College of Engineering. His interests include power electronics applications, adjustable speed drives, and power quality considerations.

<u>Bruce L. Hunt</u> is presently employed as an electrical engineer for Moon Lake Electric in Roosevelt, Utah, where he works extensively in providing technical assistance in the application of customer-owned equipment. He also analyzes the integration of various market-driven requirements with the efficient use of resources. Mr. Hunt has a degree in Physics, is a registered Professional Engineer, and is a member of IEEE.

<u>Sharon Morganson</u> received a BSEE degree from the University of Calgary in 1983. She works for Alberta Power Company where she is responsible for investigation of power quality concerns and development of power quality standards for the utility/customer interface. She has drafted an internal standard for implementing harmonic limits at the customer interface.

DISCUSSION

VITALY FAYBISOVICH, FPS Consulting,Los Angeles, CA.:

This paper describes the high frequency oscillations in the distribution systems with low short circuit levels. Those oscillations are initiated by the voltage notching associated with large adjustable speed drives. The authors provide this phenomenon analysis along with possible solutions.

The authors should to be complimented on a fine contribution to this important topic of the power quality problem. The authors' comments to the following questions would be greatly appreciated:

1) The described high frequency oscillations have frequency range up to 3,600 Hz. Can the authors provide more details about high frequency models for overhead lines, thansformers, motors and ASD that were implemented in EMTP study?

2)What kind of measuring equipment was used for such oscillations' registration and what frequency range of this equipment?

3)For Example 1 the authors provide results of simulation for critical buses (4.16 kV buses) before installation of the additional capacitors at 25 kV buses. Are results of simulation or field tests for critical buses after capacitors' installation available?

4)What ITHDs have 6000 hp ASD from Example 1 and DC drives from Example 2 and how this equipment meets IEEE Std 519-1992 requirements for those particular low short circuit capacities conditions?

Manuscript received August 18, 1995.

L. Pierrat (Electricite De France, Division Technique Generale, Grenoble Cedex, France) and Y. Baghzouz (University of Nevada, Las Vegas, NV): The authors presented two practical cases where voltage notching from line commutation of adjustable speed drives cause voltage quality problems to other customers along the distribution feeder. This valuable article is one of the first to report the high-frequency voltage oscillations that can result from commutation notches. It was made clear that such events generally occur in weak ac systems with low short-circuit capacities.

The authors' comments on the following points will be greatly appreciated:

1) The high-frequency "ringing" in the voltage waveforms is solved by capacitor placement on the high-voltage side of the distribution networks. Capacitor sizes are selected so that all system resonances do not occur at frequencies that are characteristic harmonics from notches. In other words, the system resonance frequencies are shifted from high values (60-70 pu in example 1 and 30-40 pu in example 2) to low values (below 11 pu in both examples). Since the dominant harmonics of most nonlinear loads are of low order, doesn't this shift cause new problems if other loads excite the network at these frequencies?

2) Capacitor or passive filter placement are probably still the most economical and practical solution to harmonic resonance problems. At the mean time, these devices are often the cause of harmonic amplification [A]. Over the past decade, there has been significant technological advances in active power filters (APF) which appear to be the ideal solution to most harmonic problems. Have the authors considered the installation of an APF at the harmonic source as a possible alternative solution? Is the cost and unavailability of these devices still a problem?

3) It is noted that the measured voltages in example 2 (figures 12 and 13) are richer in even harmonics than in odd harmonics. What are the origins of these unusually large even harmonics? Is it possibly the unsymmetrical commutation due to voltage imbalance? Is the amplification of even harmonics near the network resonance frequency (26-36 pu) caused uniquely by the converter commutation? What do the authors mean by "the voltage oscillation is not an exact multiple of 60 Hz"?

REFERENCES

[A] IEEE Task Force (V.E. Wagner, Chair), "Effect of Harmonics on Equipment," *IEEE Trans. on Power Delivery*, Vol. 8, No. 2, 1993, pp. 672-80.

Manuscript received August 21, 1995.

M. McGranaghan. I will address the comments of Mr. Faybisovich first.

1. The waveforms on the distribution system caused by voltage notching include frequency components up to 3600 Hz (60th harmonic). These frequency components are caused by the natural frequency of the distribution system formed by the line capacitance and the system source inductance. As indicated in the first question, these higher frequency components result in some additional requirements for modeling the distribution system components in EMTP. Most importantly, the line models used for the distribution lines must accurately represent the distributed nature of the line capacitance and the losses should be based on the dominant frequencies in the transient. These requirements can be met with pi models for the line with parameters calculated for the primary frequencies in the transient. There are no special modeling requirements for the transformers (line capacitance is significantly greater than the transformer stray capacitance. Accurate modeling of the ASD for these cases is more difficult - snubber circuits and exact firing angles can significantly influence the transients. Further discussion of this topic would require much more

detail. There is currently a task force in IEEE developing guidelines for power electronics modeling in transient studies.

2. Monitoring equipment used to characterize the distribution system transients must also be capable of resolving the higher frequency components. The equipment used for the system with frequency components around the 60th harmonic sampled the voltage waveform with 256 samples/cycle (BMI PQNode). The sampling is synchronized to the fundamental frequency for maximum accuracy at harmonic components.

3. The paper provides examples of waveforms before and after the installation of capacitors on the 25 kV system for Example 1. Before installation of capacitors, the system resonance results in high frequency oscillations on the system and at customer buses (Figures 3-6). After installation of a 1200 kvar capacitor bank, the high frequency oscillations are gone but the 13th harmonic component in the voltage is excessive (Figure 8). Finally, after installation of a 2400 kvar bank, the system voltage is very clean - both the high frequency oscillations and the 13th harmonic resonance are gone.

4. In Example 1, compliance with IEEE 519 was part of the facility design. The customers meet IEEE 519 requirements through the installation of 5th, 7th, and 11th harmonic filters. The higher frequency oscillations are excited by relatively low levels of harmonic current injection. Example 2 is similar except they were not required to meet IEEE 519 as part of the original facility design. However, lower order harmonic components were significantly reduced by implementing an 18 pulse configuration for the three dc drive buses.

Mr. Pierrat and Dr. Baghzouz also raised a number of good questions. These are discussed here.

1. The question is whether or not other problems could be caused by the low order resonance that is created by the addition of capacitors to solve the high frequency oscillation problem. This is a very important concern. In order to avoid problems at low order harmonics, the capacitor size should be selected to avoid a resonance at one of the characteristic harmonics of nonlinear loads on the system (e.g. 5th or 7th). Hence the selection of a capacitor size that results in a 9th harmonic resonance in Example 2. In Example 1, the exact size was not as critical because the nonlinear loads included harmonic filters at the 5th, 7th, and 11th components. If other important nonlinear loads were added to the system, the same concern would exist.

2. The cost and availability of active filters is definitely a problem. In these cases, the problem had to be solved as quickly as possible because there were operational problems with customers on the systems. Active filters may be an alternative in the future. However, most active filter configurations currently being considered involve a parallel connection with an isolation reactor. These configurations would not be capable of solving the high frequency problems described in this paper.

3. The question relates to the even harmonic components in the voltage waveform spectrums for Example 2. There are two causes for this. The first relates to the asymmetries in firing of the converters used in the dc drives. Slight asymmetries are greatly exaggerated at these high frequencies (26th to 36th harmonic). The other cause is the nature of the FFT itself. The FFT is representing the waveform sampled with frequency components that are exact multiples of the fundamental (because the sample has exactly 256 samples per cycle). If there is an oscillation in the waveform that is not exact multiple of the fundamental, it will be represented by components that are multiples of the fundamental. These will include both even and odd components. This occurs in the case of voltage notching because the oscillation that occurs during each notch is at the natural frequency of the distribution system (not necessarily of multiple of the fundamental).

The authors thank the discussers for their valuable input.

Manuscript received November 28, 1995.